Reliability, Safety and Hazard

ADVANCES IN RISK-INFORMED TECHNOLOGY

Reliability, Safety and Hazard

ADVANCES IN RISK-INFORMED TECHNOLOGY

Chief Editors
HS Kushwaha
VK Raina
AK Verma

Editors
PV Varde
A Srividya
VVS Sanyasi Rao
Ashok Chauhan

Narosa Publishing House
New Delhi Chennai Mumbai Kolkata

Chief Editors

HS Kushwaha
Director
Health, Safety and Environmental Group
Bhabha Atomic Research Centre
Mumbai, India

VK Raina
Associate Director
Reactor Group
Bhabaha Atomic Research Centre
Mumbai, India

AK Verma
Indian Institute of Technology Bombay
Mumbai, India

Editors

PV Varde
Reactor Group
Bhabha Atomic Research Centre
Mumbai, India

A Srividya
Indian Institute of Technology Bombay
Mumbai, India

VVS Sanyasi Rao
Health, Safety and Environmental Group
Bhabha Atomic Research Centre
Mumbai, India

Ashok Chauhan
Nuclear Power Corporation of India Limited
Mumbai, India

NAROSA PUBLISHING HOUSE PVT. LTD.

22 Daryaganj, Delhi Medical Association Road, New Delhi 110 002
35-36 Greams Road, Thousand Lights, Chennai 600 006
306 Shiv Centre, D.B.C. Sector 17, K.U. Bazar P.O., Navi Mumbai 400 703
2F-2G Shivam Chambers, 53 Syed Amir Ali Avenue, Kolkata 700 019

www.narosa.com

Printed from the camera-ready copy provided by the Editors

ISBN 81-7319-729-6

Published by N.K. Mehra for Narosa Publishing House Pvt. Ltd., 22 Daryaganj, Delhi Medical Association Road, New Delhi 110 002, and printed at Rajkamal Electric Press, New Delhi 110 033, India

FOREWORD

It gives us great pleasure to release the book compiled from the papers presented in the International Conference on Reliability, Safety and Hazard, 2005 (ICRESH05) organized in Mumbai during December 1 – 3, 2005. This book has the unique distinction of presenting a collection of research papers from all important areas of reliability and safety engineering applications, especially from the strategic and commercial sectors. This book also includes many papers which reflect the latest developments in the field of reliability, probabilistic safety assessment, deterministic safety analysis, hazard analysis, software quality and reliability, risk management in nuclear industry and fuzzy-reliability engineering applications. The editors have done an excellent job of ensuring good job formatting, trimming and fine tuning of the papers to ensure high readability thus enhancing the reference value of the book. The papers, have gone through the reviewers' desk which has ensured their quality and relevance to the industry, R&D and academia. The contribution of papers from experts from leading research and industrial organizations from India and abroad has provided a healthy mix of insights, both theoretical and applied in nature. We, compliment the authors of the papers for their contributions in making this conference a success.

Chief Editors
HS Kushwaha
VK Raina
AK Verma

PREFACE

Reliability and Safety are the core issues to be addressed during the design and operation of engineering systems like nuclear power plants, refining industries, software systems, security systems, aircraft systems, transportation systems, launch & delivery systems and systems of strategic importance. In this age of globalization, great competitive advantage can be derived through guaranteed performance of products and services on a sustained basis. In fact, a look at the trends of development, the world over, shows that progress in those countries has been phenomenal who took the lead in investing resources in research and development in the area of quality, reliability and safety.

Nuclear industry has pioneered the efforts towards addressing the issues related to reliability and safety from very early stages of its development. However, the two major nuclear accidents, Three Mile Island in 1979 and Chernobyl in 1986, provided new insights into the accident phenomenology & management and added thrust and focus to the frontier research areas of safety, reliability and hazard. Though the established principles of defence-in-depth to ensure the safety of the plants remained the governing consideration, it was recognized that Probabilistic Safety Assessment methodology provides a systematic and effective framework for safety assessment. PSA provided added safety insights into system design and operation and thus complementing / supplementing the conventional deterministic methods.

The state of the art in PSA is now considered mature enough to address several real-time complex issues related to design, operation and regulatory aspects of power and process industries. The emergence and wider acceptability of risk-informed approach, where apart from the conventional defence-in-depth philosophy; the insight obtained from PSA is also utilized for taking decisions is a testimony to successful application of PSA methods. Risk-informed approach offers an effective framework in support of decision making for real-time scenarios where conventional deterministic methods alone are not adequate to handle the complex issues. The issues include man-machine interaction, assessment of imprecise and uncertain data and information, impact of common cause failures, assessment of the sensitivity of component and human performance on the overall safety of the plant, etc.

This book containing papers presented during the proceedings of the International Conference on Reliability, Safety and Hazard 2005 (ICRESH05) held during December 1 – 3, 2005 in Mumbai, is addressed exactly to this highly relevant theme of advances in risk-informed / risk-based technology. The objective has been to discuss the research and development work being performed in research organizations, academic institutes & operating organizations to draw the road map for future work to be performed towards achieving the goal of integrated asset management for engineering systems. Integrated asset management approach addresses aspects related decision making by considering the performance indicators which not only include safety but the economic aspects of system operations like cost of production, revenue generation and profit as well.

The chapters are broadly classified into three main subjects, reliability, safety and hazard. There are quite a few papers that relate to the reliability and safety issues of mission critical and safety critical systems like nuclear power plants, software systems and systems of societal applications. The book deals with the important topics like application of reliability & safety models, fuzzy logic, artificial neural networks, fault tolerant approach, integrated asset management, residual life assessment or ageing studies which are potential candidates to support development of risk-informed methods. Going by the wide variety of application domains covered and relevant safety and reliability issues addressed, this book is bound to find appreciation amongst the power and process industries. Reliability and safety aspects of societal systems and domestic products have also been discussed. This book is addressed to professionals working in the area of power plant safety, process industry, graduate students and researchers in Universities and Laboratories specially in strategic sectors of Nuclear, Defence, and Aerospace industry.

Editors

PV Varde
A Srividya
VVS Sanyasi Rao
Ashok Chauhan

ACKNOWLEDGEMENTS

The editors wish to express their appreciation and sincere gratitude to Shri H.S. Kushwaha, Director, Health Safety and Environment Group, Bhabha Atomic Research Centre (BARC); Shri A.K. Tikku, Director, Reactor Group, BARC and Shri V.K. Raina, Associate Director, Reactor Group, BARC for their constant encouragement and guidance during the conduct of the ICRESH 05 Conference from the proceedings of which this book has evolved. The Programme Technical Committee deserves special appreciation for the excellent work of thoroughly reviewing of all the papers to ensure high technical quality. We thank Dr. A.K. Ghosh, Head, Reactor Safety Division of BARC who guided the whole review team with his rich technical knowledge and insight of the subject matter. We also thank Dr. (Ms) Gopika Vinod, S/Shri Durga Rao K., Santosh, M Hariprasad of Reactor Safety Division, BARC and Shri Alok Mishra of NPCIL and Shri N.S. Joshi of RRSD, BARC for providing the technical support. Special thank to Prof. P.K. Kapur, Dr. Amit Gupta and their Colleagues at Delhi University for taking care of coordination of printing jobs for bringing out this book in time. Publication of this book within a short span of time would not have been possible without the active support of Narosa Publishing, House, New Delhi.

CONTENTS

Reliability, Safety and Hazard
Editors: P.V. Varde et al.

MODELING AND ANALYZING FAILURE LOGIC OF PRODUCTS AND SYSTEMS

Seppo Virtanen, Per-Erik Hagmark and Jussi-Pekka Penttinen

Tampere University of Technology, Machine Design and Operation Laboratory
Korkeakoulunkatu 6, FI-33101 Tampere, Finland

ABSTRACT

Since 1996 eleven Finnish companies have participated in a research project whose objective was to develop a computer supported probabilistic based method for the development of the equipment's and systems' reliability and safety. The ideal design approach is one of those problem areas delaying production and weakens equipments' and systems' reliability, availability and safety, are discovered and solved during the design stage. The participating companies are both manufacturers and users in metal, energy, process and electronics industries. Their products and systems have to correspond to high safety and reliability demands. The research project was carried out by Tampere University of Technology and it was completed in February 2005. The research work was divided in three parts: modeling and analysis of failure logic of product and system, specification and allocation of reliability and availability requirements, simulation and calculation of reliability performance and maintenance costs. This paper presents a computer-supported method for modeling and analyzing failure logic of products and systems. The developed method is one of the main results from a nine-year research project. Artekus Oy (www.artekus.fi) is responsible for commercializing, marketing and providing technical support of the developed computer software.

Keywords: Cause, Consequence, Event, Logic, Modeling, Analysis, Simulation

1. INTRODUCTION

Modeling and analysis of failure logic form a foundation for investigation of the reliability, safety and risks related to a design entity. The general term "entity" or "design entity" can stand for function, system, equipment, mechanism, or any kind of part.

In the developed method, failure logic is divided into two types of trees; a "cause tree" and "consequence tree". The cause tree consists of such (well-defined) causes and interconnected causalities that can lead to the occurrence of a TOP-event. Thus, a cause tree structure forms a basis for a failure logic model of the design entity in question. The consequence tree again describes the possible chains of consequences initiated from a TOP-event. A consequence may further cause other consequences, either exclusively or independently. Finally, a combination of cause trees and a consequence tree, illustrated in Figure 1, will be called a "cause-consequence tree". A cause-consequence tree may for example contain several separate chains of events that lead to the same consequence. (Note the chains to consequences 1 and 2 in Figure 1.)

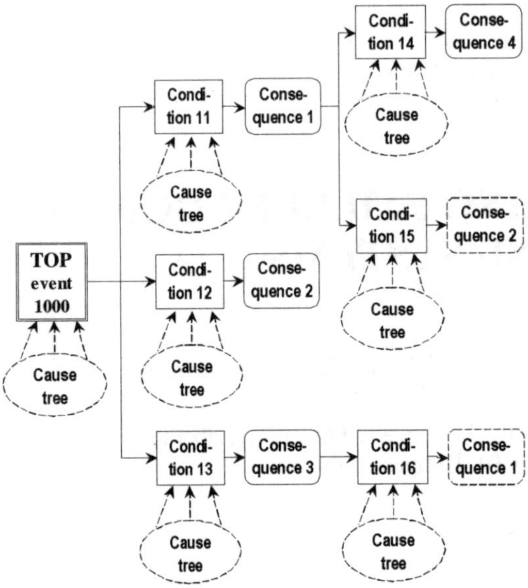

Fig. 1: Cause-consequence Structure

The cause tree model is used to define the occurrence of the TOP-event, from which the consequences to be studied originate. Conditional relations between consequences may also be modeled precisely by using cause trees. The developed method can further describe relations and shared causes between cause and consequence structures. The consequence tree does not offer any additional logical structure, but it makes it possible to model such consequences, which have conditional relations to the cause tree structures. It is also possible to model and analyze several TOP-events simultaneously.

For the analysis of causes and consequences of failures, the root cause probabilities and the gate probabilities are first estimated, and then the modeled failure logic is analyzed through stochastic simulation. The developed method is simple enough to be applicable also for the analysis of very large models. Notwithstanding, it is still capable to produce exact and useful results.

The structure of the paper is as follows: In section 2, the developed cause tree model is introduced, and in section 3 the developed method for modeling and analyzing a consequence tree is presented. In section 4 there is a short conclusion.

2. THE CAUSE TREE MODEL

2.1 The Gate Model

An event alone or together with other events can cause a new event. In other words, the *state* of an input event is either $x = 1$ (occurring, true), or $x = 0$ (not occurring, false), and the equivalent state of the *gate* (or *gate event*) is determined with a partly logical and partly stochastic mechanism. The gate mechanism is characterized by a data column

$$(G,\ a,\ b,\ p,\ C_1,\ C_2,\ \ldots,\ C_n)^T \qquad (1)$$

- G ID-number of the gate (event), positive integer
- a, b Logic parameters, $0 \leq a \leq b \leq n$, integers
- p Conditional probability, $0 < p \leq 1$
- $|C_i|$ ID-number of input (event), i = 1, 2,..n
- $C_i<0$ Input (i) is first negated (NOT operator).

The state of a gate (gate event) G is a random variable depending on the states of the input events:

$$x_G = \Phi\left(a \leq \sum_{i=1}^{n} \left| x_{|C_i|} - \Phi(C_i < 0) \right| \leq b \right) \cdot \Phi(U \leq p) \qquad (2)$$

where U is a random variate from the uniform distribution on the unit interval, and the truth function Φ ("statement") equals 1 if "statement" is true, and otherwise 0. Briefly: The logic of the gate is true, if at least a, and at most b of the inputs are true. If so, the gate event is true with probability p.

Example. The gate $(8,\ 1,\ 2,\ 0.9,\ 1,\ -3,\ 4)^T$ in Figure 2 is an example of a non-monotonic (-3) and stochastic (0.9) gate. The variety generation formula (2) takes now the form:

$$x_8 = \Phi(1 \leq x_1 + (1 - x_3) + x_4 \leq 2) \cdot \Phi(U \leq 0.9)$$

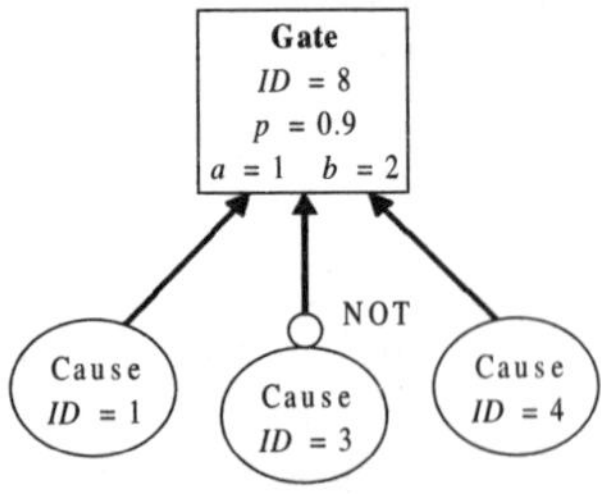

Fig. 2: A gate

By varying the parameters *a, b, p*, and by using the NOT operator (C_i<0), it is possible to model logically complicated and stochastic gates. For example, "inhibit" structure [1] is embedded in our gate model by definition. The probability of the conditional event is just the model parameter $p < 1$ (e.g. gate 8 in fig.2). Further, by choosing $a = b$ in our model we have a natural generalization of the XOR (exclusive-OR) gate. The normal logical XOR gate [2] corresponds of course to the stronger choice $a = b = 1$, $p = 1$. The following table lists the simplest gate types.

Type of gate	*a*	*b*	*p*
OR	1	*n*	1
AND	*n*	*n*	1
Voting, k/n ($0 \le k \le n$)	*k*	*n*	1
Generalized Inhibit	*a*	*b*	<1
Generalized XOR ($1 \le k < n$)	*k*	*k*	1

2.2 Cause Tree Structure

A cause tree is a network of events, where the causally directed connections between the events are assumed not to alter from time to time. The mechanism of the occurrence of a gate event was defined in section 2.1.1. If an event is not an input to any gate, it is called a *TOP-event*. If an event has no inputs, it is called a *root cause* (or basic event). If wanted, a root cause can also be interpreted as a gate, whose logic is always true: a=0, b=∞.

Example. Different sequences of two causes can be modeled to have different consequences, Figure 3. Observe the Priority AND structures.

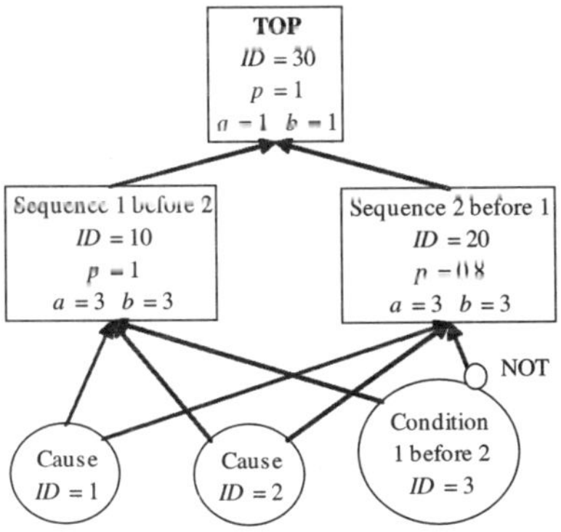

Fig. 3: A cause tree with Priority-AND gates

A cause tree can be fully expressed with a structure matrix as follows. Each column of the matrix is the data column of a gate (1). Because of the number of gates' inputs various, the shorter ones are filled with zeros to obtain equal lengths for all columns of the matrix. For example, a *structure matrix* of the cause tree in Figure 3 is

$$\begin{bmatrix} G & 10 & 20 & 30 \\ a & 3 & 3 & 1 \\ b & 3 & 3 & 1 \\ p & 1 & 0.8 & 1 \\ C_1 & 1 & 1 & 10 \\ C_2 & 2 & 2 & 20 \\ C_3 & 3 & -3 & 0 \end{bmatrix} \qquad (3)$$

The order of columns is sufficiently determined by the following principle: If a gate is an input to another gate, then its column is located to the left. The simulation of the tree proceeds in the matrix from left to right, so the inputs of a gate are always generated before the gate itself. In other words, the simulation order will not contradict the chronological order.

2.3 Construction of Cause Tree Graphic

During the research project, we have developed the Event Logic Modeling and Analysis Software (ELMAS) tool. After the identification of the events related to the TOP-event, experts examine the generated event list one by one and indicate the event's cause and consequence connections with the other events. Based on the expert's decisions, ELMAS draws the logic diagram on the screen. The same cause can occur in many places in the logic. On the computer screen the expert can drag and drop the events in to the right position based on his/her best understanding of the logic. If the event is moved so that it leads to a loop in the tree, ELMAS gives a warning and rejects the choice. After the causes and consequences of events are determined, the types of gates are defined (Figure 4).

The same cause tree model (2.2) is also used in software for allocation of reliability and availability requirements (RAMalloc), and simulation of reliability and maintenance costs (RAMoptim).

2.4 Simulation of the Failure Logic

If the logic in the modeled cause tree is simple enough, it can be directly studied by using analytical means, for example minimal cut sets. When the causes and their interconnected causalities are complex and the logic is not monotonic, stochastic simulation is highly preferred. The simulation data leads to a variety of useful results.

Before the simulation, the probabilities of the occurrence of the root causes are estimated. For the estimation of these probabilities, several different types of expert judgment methods are integrated into ELMAS. Each of these methods defines the probability within a short period of time, dt, or at a random moment.

At the beginning of a simulation process, the random occurrence of the root causes is simulated. After this, the occurrences of the gates are generated in a "chronological" order based on the cause tree matrix. One simulation round defines the complete state of the entire cause tree within a period of length dt.

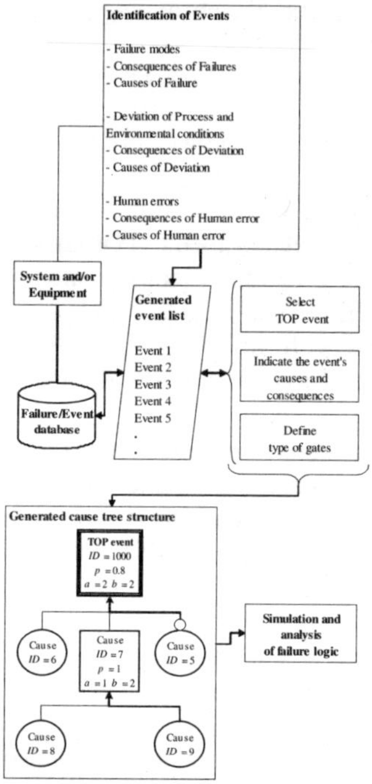

Fig. 4: Principle of modeling the cause tree structure related to the selected TOP-event with ELMAS

After a sufficient number of simulation rounds, the statistical information about the behavior of the studied object is representative. The developed software assists in defining how many simulation rounds are required. During the simulation, it is possible to see on the screen (a) how many of all the possible combinations have already occurred and (b) an estimation of the maximum probability of the combinations that have not yet occurred.

2.5 Results of the Failure Logic Analysis

The raw simulation data can be refined to useful results. The simplest is the (estimated) probability of an event:

$$P_A = \frac{n_A}{n} \tag{4}$$

P_A the probability of event A

n_A the number of occurrences of A

n the total number of simulated rounds.

It is also possible to estimate conditional probabilities. The condition can be the occurrence of an event or a combination of events:

$$P_{A|X} = \frac{n_{AX}}{n_X} \tag{5}$$

$P_{A|X}$ conditional probability of A under condition X

n_{AX} the number of times A occurred under condition X

n_X the number of occurrences of condition X.

Finally, conditional probabilities can also be calculated for combinations of the states of specific events:

$$P_{C|X} = \frac{n_{CX}}{n_X} \tag{6}$$

$P_{C|X}$ conditional probability of combination C under condition X

n_{CX} the number of times C occurred under condition X

n_X the number of occurrences of condition X.

2.6 Importance Measures

An importance measure describes correlative relations between two events. The events to be studied may be chosen freely from the model, except that the causality order is required (otherwise, the results would not be meaningful). We consider several importance measures [3]. They all attach slightly different "importance values" to the events, and they may even lead to a different order of importance. Some of them can be found in Figure 5, [4]. All of these importance measures are also integrated into ELMAS.

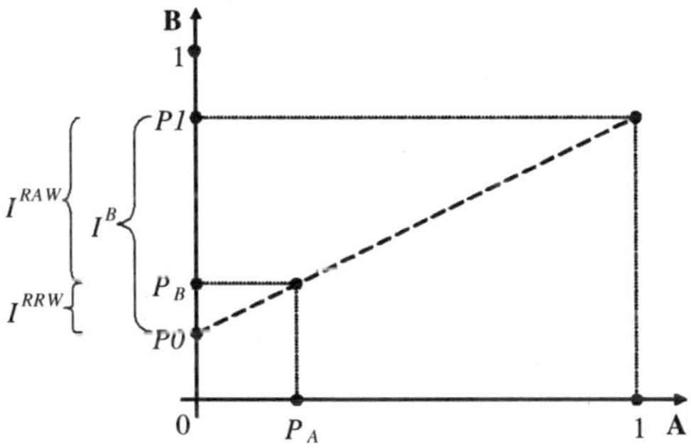

Fig. 5: Importance measures

The *importance* of event A (cause) is determined from the event B (consequence) point of view. The current probabilities, P_A and P_B, are assessed from simulation data as described in section 2.5. If P_A will increase to 1, which means that A is always true, P_B will increase to (say) $P1$. Correspondingly, if P_A will reduce to 0, which means B will never occur, P_B will reduce to $P0$.

The probabilities $P0$ and $P1$ are conditional for event B, when event A is occurred or not occurred. Their difference is called the *Birmbaum's importance measure*

$$I^{B} = P1 - P0 \tag{7}$$

This figure expresses the scope of how much P_A can affect P_B in general. The *Risk Reduction Worth* describes how much the probability of event B reduces at most, i.e., if the probability of A can be reduced to zero:

$$I^{RRW} = P_B - P0 = I^{B} \cdot P_A \tag{8}$$

Similarly, the *Risk Achievement Worth* describes how much the probability of B will increase at most, i.e., if the probability of A happens to increase to one (9):

$$I^{RAW} = P1 - P_B = I^{B} \cdot (1 - P_A) \tag{9}$$

The *Criticality importance* again describes the probability that A is the main cause to the occurrence of B (10).

$$I^{CR} = 1 - \frac{P0}{P_B} = \frac{I^{B} \cdot P_A}{P_B} = \frac{I^{RRW}}{P_B} \tag{10}$$

2.7 Risk Analysis

All events can be given commensurate losses, which represent what will happen if the event occurs. When the losses have been assessed for the selected causes, all needed information for risk analysis is at hand:

$$R_A = P_A \cdot C_A^{\alpha} \tag{11}$$

R_A the risk value of event A
P_A probability of event A
C_A the extent of loss that event A causes.

The parameter α in equation (11) is used to weaken or strengthen the importance of large damages [5]. If $\alpha > 1$, the risks of large damages are amplified. It is also possible to compare the risks of different causes or different combinations of causes by using the conditional probabilities.

3. CAUSE-CONSEQUENCE TREE

The definition of the occurrence of selected TOP-event and the extent of all possible consequences of course forms the basis for performing example a complete risk analysis for the design entity. The *consequence tree* is an additional structure to be used for modeling the consequences of the TOP-event, which itself is modeled with a cause tree. The cause-consequence structure (Figure 1), where conditions between consequences can be modeled with cause trees, makes it possible to create very general models for the propagation of consequences.

The main contribution is here the possibility to define various conditional levels. These levels do not need to be independent. Interconnections can be taken into account. Some causes in the cause tree model can perhaps be defined only after some consequences have occurred. This may also lead to a situation where the occurrence of some consequence determines the occurrences of some root cause in a cause tree.

3.1 Modeling Consequences of Events

The consequence tree consists of events, which have two states like the events in the cause tree model. When a consequence is true (occurring), it may cause other consequences. The propagation of consequences, starting with a TOP-event, proceeds in a way similar to the cause tree. A consequence "gate" can ramify in a pre-defined set of consequences (outputs). Our model offers two branching types, independent or exclusive.

Independent branching means that several alternatives can occur at the same time and with their own probability. This type is typical for the first level just after TOP. For example, if the TOP-event is a certain type of failure, several types of damage can follow the same failure (human, environmental, property, business, etc).

Exclusive branching means that at most one alternative follows. This type is typical for the second level. For example, human damage can be classified as "fatal", "severe" or "minor", and business damage can be classified as enormous", "big", "average" or "minor".

3.2 Modeling Conditionality Between Consequence Events

The binary state of a condition between consequences is modeled with a cause tree. This way it is of course possible to describe very complicated conditions. It is also possible to model more conditional levels than only one level before failure and one after. The first conditional level is usually a calendar time level. If the studied object should always operate, this level is trivial. Otherwise it is possible to precisely define the planned non-operational times, for example, maintenance time, by using cause tree structure.

The next level can be e.g. the normal operation level of the studied entity. At this level, all events related to the normal operation are defined. Some of these events are used as conditions in the next levels. The occurrence of these conditions can cause some consequences, which are pre-conditions to the next levels. These pre-conditions may be e.g. different types of failures of the object operation or just situations, where operation of the object is started.

With the developed method, it is thereby possible to model different levels of operation of the studied entity before getting into the failure levels. These operation levels can be, for example, different running modes of the entity. The failures can also be divided into different levels. For example, some failure may cause system overheating, overheating may cause fire, and fire may spread to different parts of the system. The current running mode of the system may affect how the fire will spread. In other words, there could also be connections between an operating level and a failure level.

3.3 Consequence Matrix

A consequence tree can be described precisely with the corresponding matrix. This consequence matrix is created similarly as the cause tree matrix. Every column of the matrix describes the consequences of a certain consequence event, and the columns are arranged in a chronological order, the simulation order. Every consequence has its own identity code (ID), and the corresponding column contains information about the next level consequences:

$$(ID, excl, (cond_1 \mid conseq_1), \ldots, (cond_n \mid conseq_n))^T \quad (12)$$

The field *excl* in the column (12) defines whether the subsequent consequences are exclusive or independent (section 2.1). The field ($condi \mid conseq_i$) contains two values. The first one is the ID of the condition of the next level consequence, that is, the ID of some cause from the cause tree structure. The latter value is the ID of the possible next level consequence.

An example of a consequence tree is shown in Figure 6. The corresponding consequence matrix and the cause structure matrix are given by (13) and (14), respectively. The first consequence ID=100 has the second level consequences ID=200 and ID=300, which are exclusive.

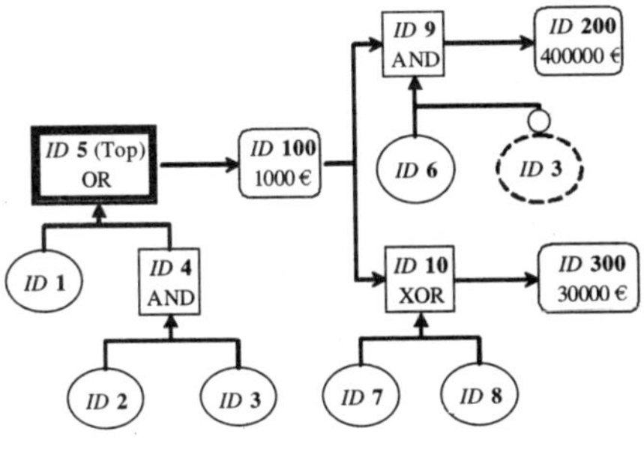

Fig. 6: A cause-consequence tree

$$\begin{bmatrix} ID & 0 & 100 \\ exlusive & FALSE & TRUE \\ cond_1 \mid conseq_1 & 5 \mid 100 & 9 \mid 200 \\ cond_2 \mid conseq_2 & & 10 \mid 300 \end{bmatrix} \quad (13)$$

$$\begin{bmatrix} G & 4 & 5 & 9 & 10 \\ a & 2 & 1 & 2 & 1 \\ b & 2 & 2 & 2 & 1 \\ p & 1 & 1 & 1 & 1 \\ C_1 & 2 & 1 & 6 & 7 \\ C_2 & 3 & 4 & -3 & 8 \end{bmatrix} \quad (14)$$

3.4 Simulation and Analysis of Cause-consequence Tree

Exactly the same simulation method that is used in the simulation of cause trees is applied to simulate the state of conditions between consequences. The simulation results are also identical with the results of the cause tree simulation. The only difference is that a level of the consequence structure is simulated only if its pre-condition is true.

By the calculation of a consequence level is meant the handling of a column in the consequence matrix. During a simulation round all conditional levels are treated in chronological (i.e., matrix column) order. At least one level is always handled (matrix column ID=0), because the first consequence of the tree is always true. The next level is handled if the consequence of that level has occurred.

In example Figure 6, the causes under gate 5 are at the first level. The causes under gates 9 and 10 are at the second level, which is handled only if consequence 100 is true. Note that root cause 3 belongs to the first level and it is used in the second level, too, as an input to gate 9. There it promotes the occurrence of consequence 200 if it is in the "not occurring" state. After all levels have been handled, one simulation round is finished.

The example (Figure 6) was analyzed by using the following root cause probabilities:

ID	1	2	3	6	7	8
Prob.	0.0001	0.02	0.03	0.005	0.002	0.04

The simulation of consequence structure consists of only simple testing. Therefore, it does not have remarkable effect on the calculation time. The results from 100000000 simulation rounds were the following:

ID	Loss	Probability	Risk
100	1000	7.03E-4 ± 6.0E-6	~0.703
200	400000	4.6E-7 ± 1.4E-7	~0.884
300	30000	2.95E-5 ± 1.1E-6	~0.184

4. CONCLUSION

The cause-consequence tree method presented above makes it possible to explain and describe precisely the relations between causes and consequences of failures. The causes can be ranked from the probability and/or risk point of view. Results from the analysis help designers to identify both the most probable causes and chains of causes leading to the TOP-event, and the most significant consequences and chains of consequences. After ranking the causes, a more detailed root cause analysis can be performed by applying the event-cause-consequence method FMEA, which is integrated into ELMAS.

The applicability of the developed methods and software has been tested in the companies, which have been involved in the research project. Most of the participating companies have started to apply the proposed method and software for modeling and analysis of failure logic for their products and systems. The application of the method forces experts to identify all potential component hardware failures, human errors, possible disturbances and deviations in the process, and environmental conditions related to the selected TOP-event. Based on experience, and with the help of the methods, it is possible to find out those problem areas of the design stage, which can delay product development and/or reduce safety and reliability.

5. REFERENCES

1. *Fault Tree Handbook. U.S. Nuclear Regulatory Commission, Nureg-0492. (Co-author David Haasl) ISBN/ISSN 1051-H-02. 1981, p 209.*
2. *E. J. Henley, H. Kumamoto, Probabilistic Risk Assessment and Management for Engineers and Scientists, Second Edition, IEEE Press Piscataway, NJ. 1996, p 597.*
3. *M.Cheok, G. Parry, R. Serry, "Use of Importance Measures in Risk-Informed Regulatory Applications". Reliability Engineering and System Safety, 1998.*
4. *J. P. Penttinen. Analysis of failure logic using simulation, Master' Thesis, Tampere University of Technology. 2005, p 94.*
5. *J. Norman, McCormick, Reliability and Risk Analysis, Methods and Nuclear Power Applications. Academic Press. Inc. 1981, p 446.*

Reliability, Safety and Hazard
Editors: P.V. Varde et al.

PRODUCT RELIABILITY: BRIDGING THE GAP BETWEEN THEORY AND PRACTICE

D.N.P. Murthy

Division of Mechanical Engineering, The University of Queensland, Brisbane, Australia
Norwegian University of Science and Technology, Trondheim, Norway

ABSTRACT

The paper looks at the gap between reliability theory and practice and discusses some new topics for research to bridge the gap.

1. INTRODUCTION

All products and systems are unreliable in the sense that they degrade with time and fail when they are unable to perform their intended function. When a failure occurs, no matter how benign, its impact is felt. Failures occur in an uncertain manner and are influenced by factors such as design, manufacture or construction, maintenance and operation.

Prior to World War II, the notion of reliability was largely intuitive, subjective and qualitative. A more quantitative (or mathematical) and scientific approach to reliability grew out of the demands of modern technology and particularly out of the experiences in the Second world war with complex military systems (Barlow and Proschan (1965)). Since the appearance of this classic book, the theory of reliability has grown at a very rapid rate, as can be seen by the large number of books and journal articles that have appeared on the subject.

However the gap between theory and practice is still very large. We first look at some of the reasons for this and then discuss new research needed to bridge the gap by looking at reliability from a product life perspective.

2. PRODUCT RELIABILITY – A HOLISTIC VIEW

Reliability of a product (system) conveys the concept of dependability, successful operation or performance and the absence of failures. Unreliability (or lack of reliability) conveys the opposite. The reliability of a product (system) is the probability that the product (system) will perform its intended function for a specified time period when operating under normal (or stated) environmental conditions. Reliability theory deals with the interdisciplinary use of probability, statistics and stochastic modelling, combined with engineering insights into the design and the scientific understanding of the failure mechanisms, to study the various aspects of reliability. As such, it encompasses issues such as (i) reliability modelling, (ii) reliability analysis and optimization, (iii) reliability engineering, (iv) reliability science, (v) reliability technology and, (vi) reliability management.

A proper framework for the study of reliability needs a holistic approach where the different aspects are integrated in an effective manner. Figure 1 shows the holistic view for standard products (consumer durable, industrial and commercial) where the interaction between the technical variables (design, manufacturing etc which determine product reliability) and the commercial variables (sales, costs, customer satisfaction etc which are influenced by product reliability) is seen clearly.

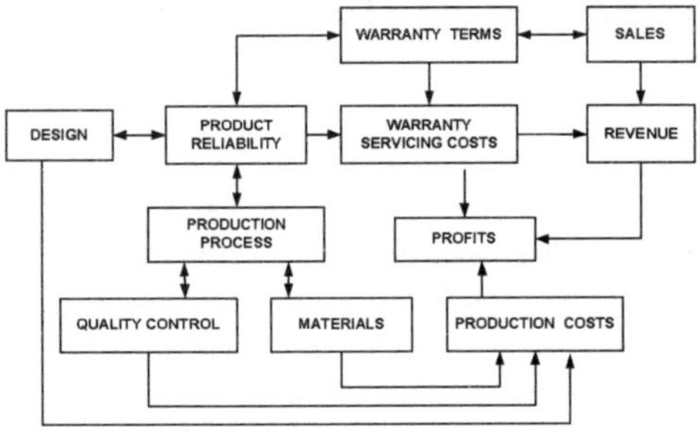

Fig. 1: Holistic view for study of reliability

Reliability decisions need to be done in the product life cycle context. In the simplest characterisation, the life cycle is comprised of three stages with several phases in each. The three stages are — (i) Pre-production, (ii) Production and, (iii) Post-production. The reliability of a product over the life cycle is shown in Figure 2 (from Blischke and Murthy (2000)). A feasibility study is carried out using the specified target value for product reliability. During the design stage, product reliability is assessed in terms of part and component reliabilities. Product reliability increases as the design is improved. However, this improvement has an upper limit. If the target value is below this limit, then the design using available parts and components achieves the desired target value. If not, then a development program to improve the reliability through test-fix-test cycles is necessary. Here the prototype is tested until a failure occurs and the causes of the failure are analysed. Based on this, design and/or manufacturing changes are introduced to overcome the identified failure causes. This process is continued until the reliability target is achieved.

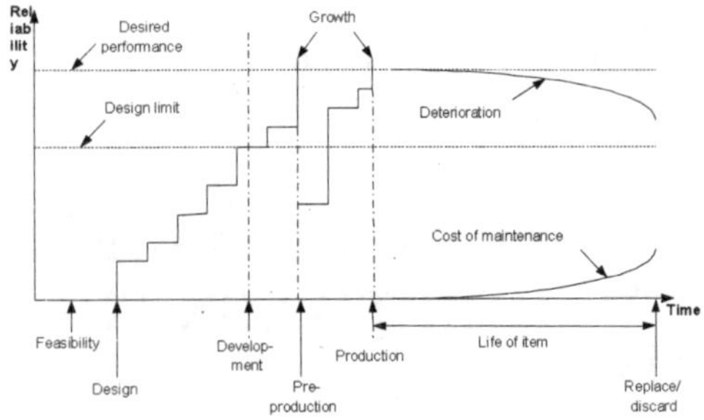

Fig. 2: Reliability: Product life perspective

The reliability of items produced during the pre-production run is usually below that for the final prototype. This is caused by variations resulting from the manufacturing process. Through proper process and quality control, these variations are identified and reduced or eliminated and the reliability of items produced is increased until it reaches the target value. Once this is achieved, full-scale production commences and the items are released for sale.

The reliability of an item in use deteriorates with age. This deterioration is affected by several factors, including environment, operating conditions and maintenance. The rate of deterioration can be controlled through preventive maintenance as shown in Figure 2.

3. BRIDGING THE GAP BETWEEN THEORY AND PRACTICE

The gap between theory and practice is large. Some reasons for this are the following.

- Very few undergraduate programs deal with reliability and as such most engineers have very poor understanding of reliability in the context of design and operation.
- Very few universities offer postgraduate programs in reliability and most of these do not offer a holistic view of reliability.
- Most researchers tend to be narrow and discipline specific and most models are very simple and not very realistic.
- The real world problems are often complex and, require an interdisciplinary approach where one needs to integrate technical and commercial issues

- Data for model building is limited and messy so that standard approaches are of limited use

In the remainder of the paper we identify some research topics that need to be addressed to bridge the gap between theory and practice.

4. PRE-PRODUCTION STAGE

4.1 Evaluating Reliability of Current Products

Most new products (e.g., mobile phone) are an improvement over an existing product and involve several modules to carry out different tasks (sending voice, text, picture etc in the case of mobile phone). Each module involves several components. The usage pattern across consumer population varies considerably. Some component failures are not detected and others not reported. This implies that the failures reported do not give a true picture of the reliability of a component. How does one determine which component needs improvement? How does one go about getting new data to obtain better estimates of reliability? What are the costs and the benefits?

4.2 Assessing Reliability during Development

One needs to use accelerated testing to assess the reliability of a component (or module) during the development stage. Most models assume a deterministic scaling relationship to link failure times at two different stress levels. This is often not true. Also, there are new modes of failure at higher stress levels and items that fail under this mode provide very little information. How does one decide on the test plan (stress levels, number of items to be tested etc) taking into account these issues?

4.3 Reliability Design

The literature deals mainly with "stress-strength" models at component level. Design requires dealing with different notions of reliability performances (see, Figure 3) and a hierarchy of specifications (a set of statements that form the basis for building the product to achieve the desired performance) as shown in Figure 4 (see, Murthy et al (2007) for more details). How to derive the different specifications to ensure the desired reliability performance?

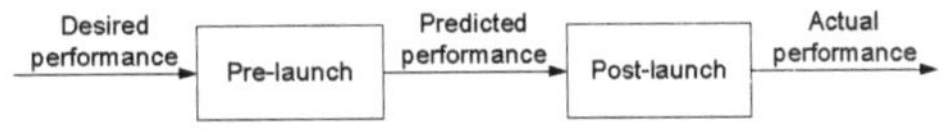

Fig. 3: Performance and specification in design

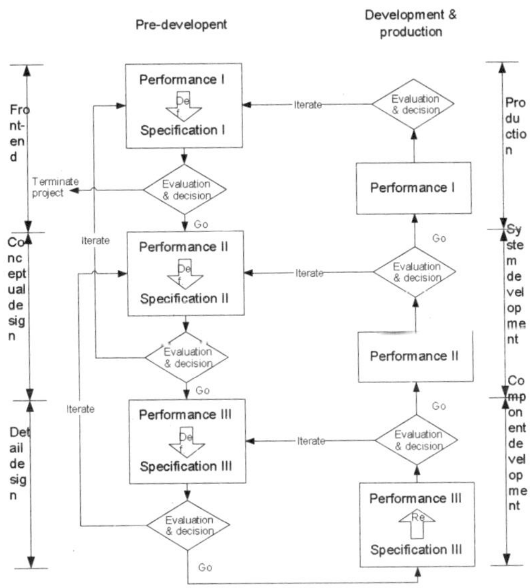

Fig. 4: Hierarchy of performance and specification in design

4.4 Assessing Field Reliability

During the development stage, items are tested in a continuous mode. In real life items are often used in an intermittent mode. Also the stress levels vary from the nominal value used in the design and during the development testing. How does this affect the field reliability of the product? This has tremendous implications for warranty costs and customer satisfaction.

4.5 Out-sourcing of Design

There is growing trend towards outsourcing of design activity. In this case we have a Principal (OEM) and an agent (external designer) and their interests are different. This raises several issues indicated in Figure 5. New research into these issues is needed – especially in the formulation of contracts.

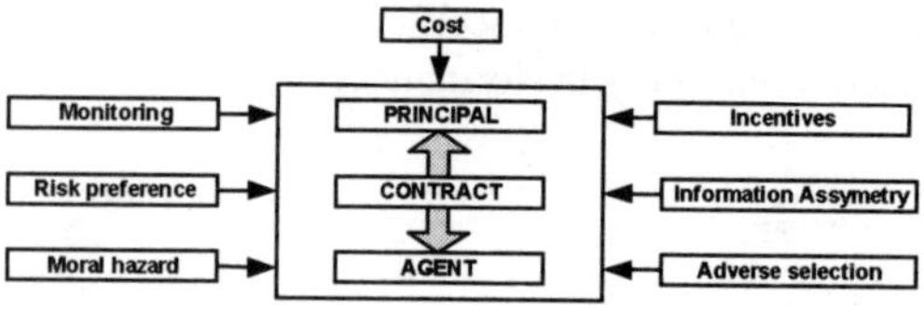

Fig. 5: Issues in principal-agent problem

5. PRODUCTION STAGE

Often components are bought from external suppliers. Again we have a Principal-Agent problem. How does one formulate the contract that ensures desired reliability?

6. POST-PRODUCTION STAGE

6.1 Warranty Logistics

The cost of servicing warranty varies from 2 – 10% of sale price. The servicing cost depends on the warranty logistics and the key elements affecting the costs are shown in Figure 6 (see, Murthy et al (2004) for more details).
Some of areas that need further study are the following:

1) Material flow issues
2) External service agents: Again we have a Principal-Agent problem.

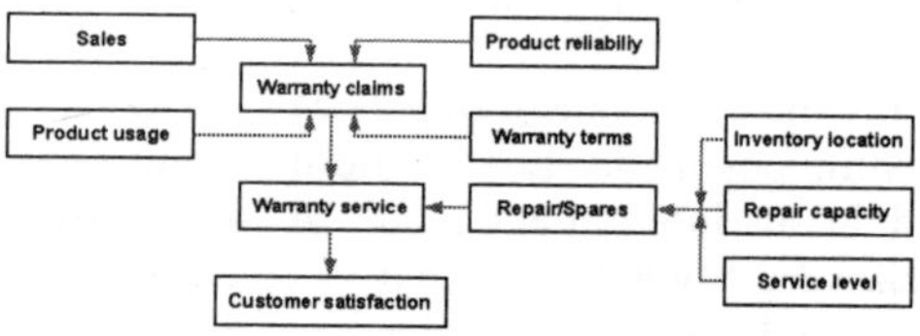

Fig. 6: Warranty servicing logistics

6.2 Maintenance of Unreliable Systems

The degradation of systems depends on the operating load that is dictated by commercial considerations. The maintenance of complex systems needs to take this into account. Murthy et al (2002) proposed the strategic maintenance management and it involves linking the technical and commercial issues as indicated in Figure 6. The modelling of some of the elements is still an open area for research.

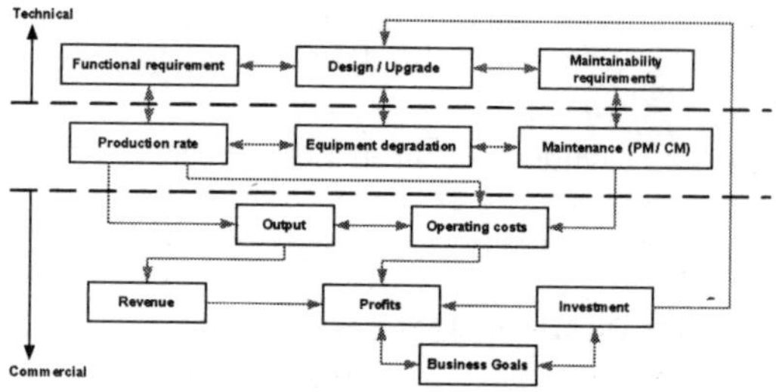

Fig. 7: Strategic maintenance

6.3 Out-sourcing of Maintenance

There is a growing trend toward outsourcing of maintenance. The key elements for proper decision-making are shown in Figure 7 and there is need for building new and appropriate models to determine optimal strategies for both the owner and the service agent.

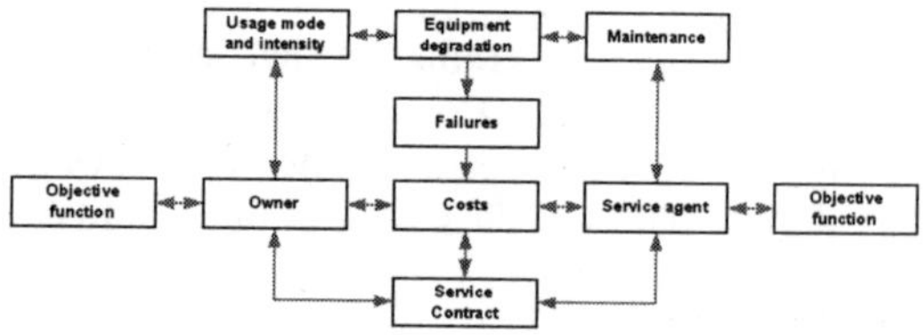

Fig. 8: Out-sourcing of maintenance

6.4 Reliability Data Collection and Analysis

Industry collects lot of reliability related data (e.g., warranty data) but the analysis is done poorly. This is an area that needs more attention to bridge the gap between theory and practice.

7. CONCLUSIONS

In this paper we have defined some topics that need investigation if one is interested in bridging the gap between theory and practice in product reliability

8. REFERENCES

1. Barlow, R.E. and Proschan, F. (1965), *Mathematical Theory of Reliability*, John Wiley & Sons, Inc.,New York.
2. Blischke, W.R. and Murthy, D.N.P. (2000), *Reliability*, Wiley, New York.

3. Murthy, D.N.P., Atrens, A and Eccleston, J.A. (2002), Strategic maintenance management, *Journal of Quality in Maintenance Engineering*, **8**, 287-3.
4. Murthy, D.N.P., Solem, O. and Roren, T (2004) Product Warranty logistics: Issues and challenges, *European. Journal of Operational Research*, **156**, 110-126.
5. Murthy, D.N.P., Osteras, T. And Rausand, M. (2007), *Reliability Performance and Specification in New Product Development*, under preparation for publication by Springer Verlag, London.

Reliability, Safety and Hazard
Editors: P.V. Varde et al.

COST ANALYSIS OF A TWO-STAGE PRODUCTION SYSTEM WITH TWO PARALLEL MACHINES IN THE FIRST STAGE AND SUBJECT TO INSPECTION

M.N. Gopalan and J.M. Subramanya

SDM Institute of Management Development, Mysore

ABSTRACT

The paper deals with the expected number analysis of a two-stage production system with two production machines in parallel in the first-stage and subject to inspection. The time taken to process a job on a machine and the time taken for inspection of the finished job are random governed by probability distributions which are quite general. Semi-finished jobs processed by the machines in the first-stage are transferred to the machine in the second stage for further processing. Machine in the second-stage processes two type of products based on the semi-finished products transferred from the machines in the first-stage. Inspection is carried out of the items processed by machine in the second stage and only good items are accepted, while others are scrapped. The analysis is carried out by employing the state-space method and the regeneration-point technique. Several measures of system performance, transient in nature, are evaluated. A cost-analysis is also carried out by using some of the measures of system performance obtained.

Keywords: Production system, Expected number analysis, State-space method, Semi-regenerative phenomena, Transient behaviour of the system, Cost analysis

1. INTRODUCTION

Gopalan, Usha A. Kumar and Subramanya [1] have recently carried out the stochastic analysis of a two-stage production system with two parallel machines in the first-stage and subject to inspection and have obtained expressions for system characteristics namely,

1. Expected duration machine j (j=1, 2, 3) is busy with processing in [0, t].
2. Expected duration machine j is blocked in [0, t].
3. Expected duration machine 3 is free in [0, t].
4. Expected duration machines 1 and 2 are blocked together in [0, t].
5. Expected duration inspection station is busy in [0, t].

and have carried out a cost-analysis of the system based on some of the system characteristics obtained above.

The aim of the present paper is to carry out the expected number analysis of the system discussed in [1] and carry out the cost-analysis under the assumption that the machine in the second-stage produces two different types of products based on the semi-finished products transferred to it by the machines in the first-stage by employing the state-space method and the regeneration-point technique. Fig. 1 shows the schematic representation of the system under consideration.

2. ASSUMPTIONS

1. At instant t=0, the system starts operating.

2. There is a buffer of infinite size in front of stage 1 of the system. As a result, the machines in the first-stage do not starve. However, they may get blocked.
3. The transfer of semi-finished items from machines 1 and 2 to machine 3 in the second-stage is instantaneous.
4. Machines are reliable. They do not fail while in operation or otherwise.
5. The size of machine 3 is sufficient to process jobs supplied by the machines in the first-stage.
6. The times taken to process jobs on the three machines are random, independent and governed by probability distributions, which are general in nature.
7. Jobs processed by the machine in the second stage are inspected and only good items are accepted, while others are rejected.
8. The time taken to inspect a job processed by the machine in the second stage is random, governed by a probability distribution which is quite general.

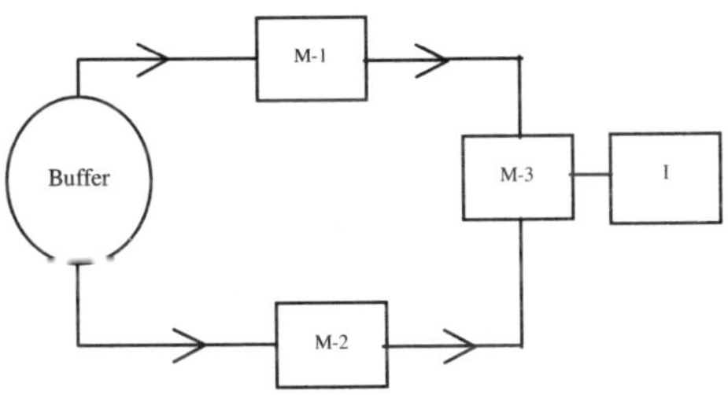

Fig. 1: Schematic Representation of the System

3. NOTATIONS

pdf / cdf / sf: Probability density / Cumulative distribution / Survivor function

f_j (.): pdf of time taken to process a job on machine j =1, 2

g_j (.): pdf of time taken by machine 3 to process a semi-finished job processed on machine j =1, 2

h_j (.): pdf of time taken by machine 3 to inspect a semi-finished job produced by machine j in the first stage, j = 1, 2

F_j (.) / G_j(.) / H_j (.): cdf corresponding to f_j(.) / g_j(.) / h_j (.)

F_j (.) / G_j (.) / H_j (.): sf corresponding to f_j(.) / g_j(.) / h_j(.)

X(t) State variable characterising the state of the system at instant t. At t=0, the system starts operating.

© : Convolution ; f(t) © g(t) = $\int_o$ f(u) g(t-u)du

4. STATE-SPACE OF THE SYSTEM

The system can be found in any of the states listed in the Table below at any instant.

It can be observed that the transitions from state 1 to states 4, 5, from state 4 to states 5, 10, from state 5 to states 4, 11, from state 10 to state 5 and from state 11 to state 4 are semi-regenerative. The epochs of entry to the semi-regenerative states are indicated by a dot. The transitions are listed below.

1-2- •4 1-3- •5
4-6- •10 4-8-3- •5
4-8-10 •5 5-7- •11
5-9-2- •4 5-9-11- •4
10- •5 11- •4

dot (•) indicates the epoch of transition to the state is semi-regenerative.

Fig. 2 represents the one-step transitions between the states 1 to 11. The system can be found in any of the states listed in the Table 1 at any instant.

5. THE ANALYSIS

As mentioned in the Introduction, the system will be analysed by using the state-space method and the regeneration-point technique to obtain expressions for the following measures of system performance:

1. Expected number of times machine j (j=1, 2, 3) is busy with processing in [0, t].
2. Expected number of times machine j is blocked in [0, t].
3. Expected number of times machine 3 is free in [0, t].
4. Expected number of times machines 1 and 2 are blocked together in [0, t].
5. Expected number of times inspection station is busy in [0, t].

A cost analysis will be carried out by using some of the system characteristics mentioned above.

5.1 Expected number of times machine j (j=1, 2, 3) is busy with processing in [0, t].

Let $M_1^{P(j)}(t)$ be the expected number of times machine j (j=1,2,3) is busy with processing a job at instant t, given that the system started operating at instant t=0.
We have

$$M_1^{P(j)}(t) = a_{1,4}^{P(j)}(t) © M_4^{P(j)}(t) + a_{1,5}^{P(j)}(t) © M_5^{P(j)}(t) + A_{1,4}^{P(j)}(t) + A_{1,5}^{P(j)}(t) + a_{1,0}^{P(j)}(t)$$

$$M_4^{P(j)}(t) = a_{4,5}^{P(j)}(t) © M_5^{P(j)}(t) + a_{4,10}^{P(j)}(t) © M_{10}^{P(j)}(t) + A_{4,5}^{P(j)}(t) + A_{4,10}^{P(j)}(t) + a_{4,0}^{P(j)}(t)$$

$$M_5^{P(j)}(t) = a_{5,4}^{P(j)}(t) © M_4^{P(j)}(t) + a_{5,11}^{P(j)}(t) © M_{11}^{P(j)}(t) + A_{5,4}^{P(j)}(t) + A_{5,11}^{P(j)}(t) + a_{5,0}^{P(j)}(t)$$

$$M_{10}^{P(j)}(t) = a_{10,5}^{P(j)}(t) © M_5^{P(j)}(t) + A_{10,5}^{P(j)}(t) + a_{10,0}^{P(j)}(t)$$

$$M_{11}^{P(j)}(t) = a_{11,4}^{P(j)}(t) © M_4^{P(j)}(t) + A_{11,4}^{P(j)}(t) + a_{11,0}^{P(j)}(t)$$

$$M_1^{P(3)}(t) = {}_1M_1^{P(3)}(t) + {}_2M_1^{P(3)}(t)$$

5.2 Expected number of times machine j is blocked in [0,t].

Let $M_1^{B(j)}(t)$ be the expected number of times machine j is blocked at instant t, given that the system started operating at instant t=0.
We have

$$M_1^{B(j)}(t) = a_{1,4}^{B(j)}(t) © M_4^{B(j)}(t) + a_{1,5}^{B(j)}(t) © M_5^{B(j)}(t) + A_{1,4}^{B(j)}(t) + A_{1,5}^{B(j)}(t) + a_{1,0}^{B(j)}(t)$$

$$M_4^{B(j)}(t) = a_{4,5}^{B(j)}(t) © M_5^{B(j)}(t) + a_{4,10}^{B(j)}(t) © M_{10}^{B(j)}(t) + A_{4,5}^{B(j)}(t) + A_{4,10}^{B(j)}(t) + a_{4,0}^{B(j)}(t)$$

$$M_5^{B(j)}(t) = a_{5,4}^{B(j)}(t) © M_4^{B(j)}(t) + a_{5,11}^{B(j)}(t) © M_{11}^{B(j)}(t) + A_{5,4}^{B(j)}(t) + A_{5,11}^{B(j)}(t) + a_{5,0}^{B(j)}(t)$$

$$M_{10}^{B(j)}(t) = a_{10,5}^{B(j)}(t) © M_5^{B(j)}(t) + A_{10,5}^{B(j)}(t) + a_{10,0}^{B(j)}(t)$$

$$M_{11}^{B(j)}(t) = a_{11,4}^{B(j)}(t) © M_4^{B(j)}(t) + A_{11,4}^{B(j)}(t) + a_{11,0}^{B(j)}(t)$$

5.3 Expected number of times machine 3 is free in [0, t].

Let $M_1^{F(3)}(t)$ be the expected number of times machine 3 is free at instant t, given that the system started operating at instant t=0.
We have

$$M_1^{F(3)}(t) = a_{1,4}^{F(3)}(t) © M_4^{F(3)}(t) + a_{1,5}^{F(3)}(t) © M_5^{F(3)}(t) + A_{1,4}^{F(3)}(t) + A_{1,5}^{F(3)}(t) + a_{1,0}^{F(3)}(t)$$

$$M_4^{F(3)}(t) = a_{4,5}^{F(3)}(t) © M_5^{F(3)}(t) + a_{4,10}^{F(3)}(t) © M_{10}^{F(3)}(t) + A_{4,5}^{F(3)}(t) + A_{4,10}^{F(3)}(t) + a_{4,0}^{F(3)}(t)$$

$$M_5^{F(3)}(t) = a_{5,4}^{F(3)}(t) © M_4^{F(3)}(t) + a_{5,11}^{F(3)}(t) © M_{11}^{F(3)}(t) + A_{5,4}^{F(3)}(t) + A_{5,11}^{F(3)}(t) + a_{5,0}^{F(3)}(t)$$

$$M_{10}^{F(3)}(t) = a_{10,5}^{F(3)}(t) © M_5^{F(3)}(t) + A_{10,5}^{F(3)}(t) + a_{10,0}^{F(3)}(t)$$

$$M_{11}^{F(3)}(t) = a_{11,4}^{F(3)}(t) © M_4^{F(3)}(t) + A_{11,4}^{F(3)}(t) + a_{11,0}^{F(3)}(t)$$

5.4 Expected number of times machines 1 and 2 are blocked together in [0, t].

Let $M_1^{B}(t)$ be the expected number of times machines 1 and 2 are blocked together at instant t, given that the system started operating at instant t=0.
We have

$$M_1^{B}(t) = a_{1,4}^{B}(t) © M_4^{B}(t) + a_{1,5}^{B}(t) © M_5^{B}(t) + A_{1,4}^{B}(t) + A_{1,5}^{B}(t) + a_{1,0}^{B}(t)$$

$$M_4^{B}(t) = a_{4,5}^{B}(t) © M_5^{B}(t) + a_{4,10}^{B}(t) © M_{10}^{B}(t) + A_{4,5}^{B}(t) + A_{4,10}^{B}(t) + a_{4,0}^{B}(t)$$

$$M_5^{B}(t) = a_{5,4}^{B}(t) © M_4^{B}(t) + a_{5,11}^{B}(t) © M_{11}^{B}(t) + A_{5,4}^{B}(t) + A_{5,11}^{B}(t) + a_{5,0}^{B}(t)$$

$$M_{10}^{B}(t) = a_{10,5}^{B}(t) © M_5^{B}(t) + A_{10,5}^{B}(t) + a_{10,0}^{B}(t)$$

$$M_{11}^{B}(t) = a_{11,4}^{B}(t) © M_4^{B}(t) + A_{11,4}^{B}(t) + a_{11,0}^{B}(t)$$

5.5 Expected number of times inspection station is busy in [0, t].

Let $M_1^{I}(t)$ be the expected number of times inspection station is busy at instant t, given that the system started operating at instant t=0.
We have

$$M_1^{I}(t) = a_{1,4}^{I}(t) © M_4^{I}(t) + a_{1,5}^{I}(t) © M_5^{I}(t) + A_{1,4}^{I}(t) + A_{1,5}^{I}(t) + a_{1,0}^{I}(t)$$

$$A_4^{I}(t) = a_{4,5}^{I}(t) © M_5^{I}(t) + a_{4,10}^{I}(t) © M_{10}^{I}(t) + A_{4,5}^{I}(t) + A_{4,10}^{I}(t) + a_{4,0}^{I}(t)$$

$$M_5^{I}(t) = a_{5,4}^{I}(t) © M_4^{I}(t) + a_{5,11}^{I}(t) © M_{11}^{I}(t) + A_{5,4}^{I}(t) + A_{5,11}^{I}(t) + a_{5,0}^{I}(t)$$

$$M_{10}^{I}(t) = a_{10,5}^{I}(t) © M_5^{I}(t) + A_{10,5}^{I}(t) + a_{10,0}^{I}(t)$$

$$M_{11}^{I}(t) = a_{11,4}^{I}(t) © M_4^{I}(t) + A_{11,4}^{I}(t) + a_{11,0}^{I}(t)$$

$$M_1^{I}(t) = {}_1M_1^{I}(t) + {}_2M_1^{I}(t)$$

The various notations used in the analysis carried out above are explained in the Appendix.

The integral equations established in the analysis are of convolution type and can be solved by using numerical techniques suggested by J.G. Jones [4]. However, if the distributions involved in the analysis are of Erlang type, one can employ Laplace transform technique to obtain explicit expressions for the system characteristics obtained in the analysis

6. COST ANALYSIS

Let C(t) be the expected expenditure spent in operating the system during [0,t].
We have

$C(t) = \Sigma\, c_j^P\, M_1^{P(j)}(t) + \Sigma\, {}_jc_3^P\, {}_jM^{P(3)}(t) + \Sigma\, {}_jc_3^I\, {}_jM_1^I(t)$

where

c_j^P = expected cost incurred per unit in operating machine j, j = 1, 2

${}_jc_3^P$= expected cost incurred per unit in operating machine 3 using semi-finished product processed on machine j, j = 1, 2

${}_jc_3^I$ =expected cost incurred per unit in inspecting a product processed on machine 3 using semi-finished product processed on machine j, j = 1, 2

${}_jM_1^{P(3)}$ (t) Expected number of units produced in [0,t] of a product processed on machine 3 using a semi-finished product processed on machine j, j = 1,2.

${}_jM_1^I$ (t) Expected number of inspections carried out in [0,t] of a product processed on machine 3 using a semi-finished product processed on machine j, j = 1,2.

7. REFERENCES

1. Birolini A., (1985), *"On the use of stochastic processes in modelling reliability problems"*, Lecture Notes in Economics and Mathematical Systems, Springer-Verlag, Berlin,.
2. Cinlar E., (1975), *"Introduction to stochastic processes"*, Prentice-Hall Inc., New Jersey, USA.
3. Gopalan M.N., Usha A. K., & Subramanya J.M., (Communicated), *"Stochastic analysis of a two-stage production system with two parallel machines in the first-stage and subject to inspection"*.
4. Jones J.G., (1961), *"On the numerical solution of convolution integral equation and system of such equations"*, Mathematics of Computation **15**, 131-142.

Table 1 System States

X(t)	1	2	3	4	5	6	7	8	9	10	11
M-1	O_1	B_1	O_1	O_1	B_1	B_2	B_1	O_1	B_1	B_2	B_1
M-2	O_2	O_2	B_1	B_1	O_2	B_1	B_2	B_1	O_2	B_1	B_2
M-3	F	F	F	O_{31}	O_{32}	O_{31}	O_{32}	I_{31}	I_{32}	I_{31}	I_{32}

Legend : O = Operating, F = Free, B – Blocked, I – Inspection

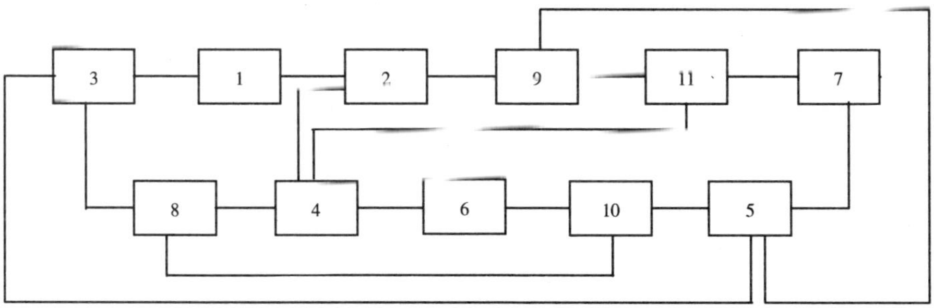

Fig. 2: One-Step Transition Diagram Indicating the Transitions between The States 1 To 11

APPENDIX – I (NOTATION)

$a_{1,4}^{P(j)}(t) = f_2(t) F_1(t)$ $\quad A_{1,4}^{P(j)}(t) = (\delta_{1j} + \delta_{2j}) \int a_{1,4}^{P(j)}(u)\, du$

$a_{1,5}^{P(j)}(t) = f_1(t) F_2(t)$ $\quad A_{1,5}^{P(j)}(t) = (\delta_{1j} + \delta_{2j}) \int a_{1,5}^{P(j)}(u)\, du$

$a_{4,5}^{P(j)}(t) = f_1(t) \int g_1(u) H_1(t-u)\, du + \int g_1(u) h_1(t-u)\, du \int_u f_1(v)\, dv$

$A_{4,5}^{P(j)}(t) = (\delta_{1j} + {}_1\delta_{3j}) \int a_{4,5}^{P(j)}(u)\, du$

$a_{5,4}^{P(j)}(t) = f_2(t) \int_o g_2(u) H_2(t-u)\, du + \int g_2(u) h_2(t-u)\, du \int_u f_2(v)\, dv$

$A_{5,4}^{P(j)}(t) = (\delta_{2j} + {}_2\delta_{3j}) \int a_{5,4}^{P(j)}(u)\, du$

$a_{4,10}^{P(j)}(t) = h_1(t) F_1(t)$ $\quad A_{4,10}^{P(j)}(t) = (\delta_{1j} + {}_1\delta_{3j}) \int a_{4,10}^{P(j)}(u)\, du$

$a_{5,11}^{P(j)}(t) = h_2(t) F_2(t)$ $\quad A_{5,11}^{P(j)}(t) = (\delta_{2j} + {}_2\delta_{3j}) \int a_{5,11}^{P(j)}(u)\, du$

$a_{10,5}^{P(j)}(t) = h_1(t)$ $\quad a_{11,4}^{P(j)}(t) = h_2(t)$

$A_{10,5}^{P(i)}(t) = H_1(t)$ $\quad A_{11,4}^{P(i)}(t) = H_2(t)$

$a_{1,0}^{P(j)}(t) = \delta_{1j} F_2(t) F_1(t) + \delta_{2j} F_2(t) F_1(t)$

$a_{4,0}^{P(j)}(t) = {}_1\delta_{3j} G_1(t) F_1(t) + \delta_{1j} G_1(t) F_1(t)$

$a_{5,0}^{P(j)}(t) = {}_2\delta_{3j} G_2(t) G_2(t) + \delta_{2j} G_2(t) G_2(t)$

$a_{1,0}^{B(j)}(t) = \delta_{1j} F_1(t) F_2(t) + \delta_{2j} F_1(t) F_2(t)$

$a_{4,0}^{B(j)}(t) = \delta_{1j} \{G_1(t) F_1(t) + \int g_1(u) H_1(t-u) \int_u f_1(v)\, dv\, du\}$

$a_{5,0}^{B(j)}(t) = \delta_{2j} \{G_2(t) F_2(t) + \int_0 g_2(u) H_2(t-u) \int_u f_2(v)\, dv\, du\}$

$a_{1,0}^{F(3)}(t) = 1 - F_1(t) F_2(t)$ $\quad a_{4,0}^{F(3)}(t) = F_1(t) \int_o g_1(u) H_1(t-u)\, du$

$a_{5,0}^{F(3)}(t) = F_2(t) \int_o g_2(u) H_2(t-u)\, du$

$a_{4,0}^{B}(t) = G_1(t) F_1(t) + \int g_1(u) H_1(t-u) \int_u f_1(v)\, dv\, du$

$a_{5,0}^{B}(t) = G_2(t) F_2(t) + \int g_2(u) H_2(t-u) \int_u f_2(v)\, dv\, du$

$a_{4,0}^{I}(t) = {}_1\delta_{3j} \int f_1(u) \int g_1(v) H_1(t-v)\, dv\, du$

$a_{5,0}^{I}(t) = {}_2\delta_{3j} \int f_2(u) \int g_2(v) H_2(t-v)\, dv\, du$ $\quad a_{10,0}^{I}(t) = {}_1\delta_{3j} H_1(t)$

$a_{11,0}^{I}(t) = {}_2\delta_{3j} H_2(t)$ (Note : $a_{ij}^{P(j)}(t) = a_{ij}^{B(j)}(t) = a_{ij}^{F(3)}(t) = a_{ij}^{I}(t)$; $j \neq 0$; All the remaining terms appearing in sections 5.1 to 5.5 are zeros.)

Reliability, Safety and Hazard
Editors: P.V. Varde et al.

SYSTEMS INTEGRATION: THE HIDDEN WEAK LINKS & THEIR IMPACT ON SAFETY AND RELIABILITY

Pritesh Patel

P.O. Dahej, Ta: Vagra, Dist:Bharuch, Gujarat 392 130

ABSTRACT

Import aspect like safety, reliability and availability are already considered in the design stage of critical control system and safety shutdown system itself Standards for the same is also available And also there are standards available to end user for proper selection and to achieve the desired level of safety, reliability and availability of selected system. This paper discusses about one more"hidden weak link", which is often overlooked or ignored resulting degradation of system in a totality and huge looses.

Keywords: Redundancy, System integration

1. SYSTEM INTERCONNECTION

Important aspect like safety, reliability and availability are already considered at the design stage of critical control system like DCS and safety shutdown system like PLC. For designers of DCS, PLC and all other critical control systems, standards are available to achieve the above mentioned important aspect. Also there are standards available to end user for proper selection of critical system and to achieve the desired level of safety, reliability and availability of selected system. Generally great care is taken while selections of important system like DCS and PLC but they are often ignored while making interconnection with their associated accessories. The important aspect like online status availability and online replacement are not considered while making the interconnection between main system and its associated accessories using Bus Bar, MCB, fuses diodes, TB and interconnection cable. Also interfacing of peripherals like printer, HMI with PLC/DCS is not taken seriously.

The cost of peripherals or associated accessories is very less compared to main system. But the failure of them will have same impact as that of main system. For example, if grounding aspect is not considered while interfacing Printer and HMI with PLC/DCS the grounding fault in Printer/HMI will propagate in main system and may damage them. This paper discusses about such issues, which are often overlooked or ignored. The ignorance of them may result in degradation of redundant system in a totality and huge looses. The more example of such **"weak-link"** in power supply distribution scheme, control room wiring, and interfacing of subsystem with main system are discussed here. Also **"weak-link"** in field instrumentation, like wrong interconnection in Field SOV using tubing is discussed here.

(1) Weak-Link in Power Supply Distribution

Electrical power is basic requirement for all control room instruments. The power supply distribution scheme is very important aspect for all control room instruments like DCS/PLC systems, Barriers, Repeaters, Isolators, 110Vac to 24Vdc Power supply units etc. In case of power supply failure, the instruments being fed (like DCS, PLC, Barriers, 110Vac to 24Vdc power supply etc.) also stop functioning. To tolerate such failure and allow the normal functioning of instrument, fault tolerant scheme is required. Usually dual redundancy scheme is used to tolerate such single failure. But often one or more **weak link** is found in redundant-scheme due to which the sole purpose of redundancy and fault tolerance will be lost. The weak-link may be at input side, at out put side or anywhere in the loop. The source of weak-link may be Main-UPS, Feeder, MCB, PSU (110Vac to 24Vdc) or even the simple termination. The few such examples are discussed here.

(a) Weak-Link at Input Side of PSU (110Vac to 24Vdc)

A few examples are shown here in which redundancy at input side is not properly considered. So even after using the redundant 110Vac to 24Vdc power supply unit, the sole purpose of redundancy is defeated.

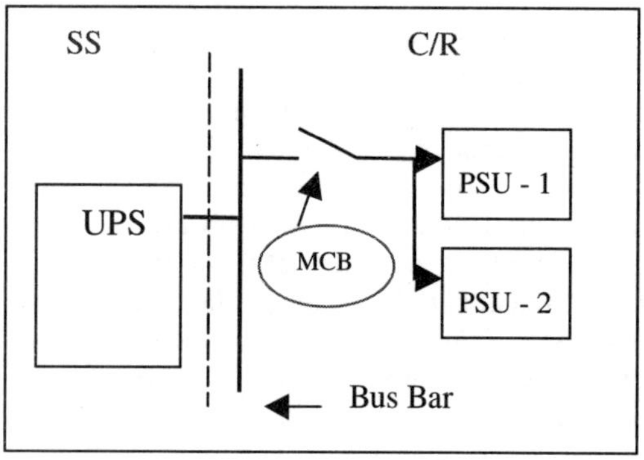

Fig. 1:

In figure: 1 two 110Vac to 24Vdc power supply (PSU-1 and PSU-2) are used as redundant power supply to feed the barriers. PSU-1 and PSU-2 are feed form same MCB. The MCB is common between them. If the MCB fails or malfunction, the both redundant PSU -1 and PSU- 2 will fail leading to failure of power to all barriers. Thus here Common MCB is a **Weak-Link.**

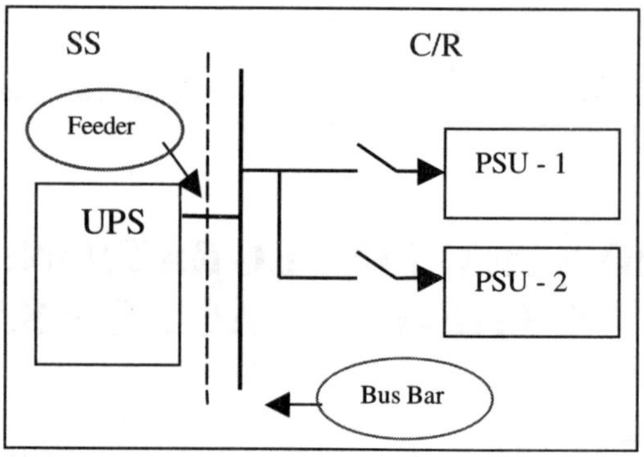

Fig. 2:

In figure: 2 the Weak-Link is modified and separate MCB is used to feed each PSU. Here failure or malfunction of one MCB will only affect the respective PSU but does not affect the other PSU. Barriers will continue to receive power supply with one MCB failure. But here the Bus Bar/Feeder is common between redundant PSU-1 and PSU-2. So the problem in Bus Bar or Feeder will lead to the failure of both PSU-1 and PSU-2.

In following Figure 3 both weak link MCB and Bus Bar/Feeder are modified.

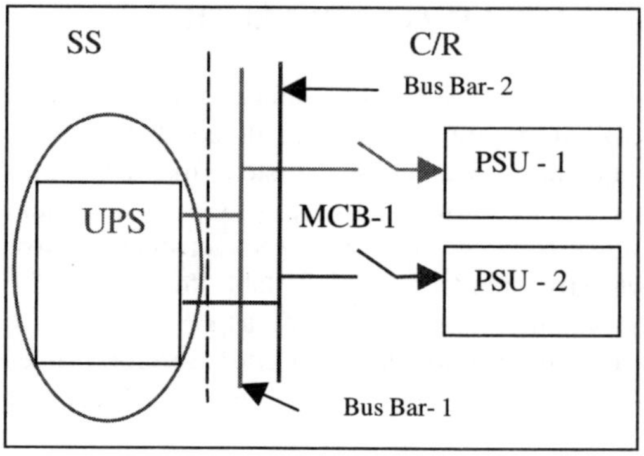

Fig. 3:

In figure: 3 separate MCB from separate Bus Bar is used to feed each PSU. Here failure or malfunction of one MCB or Bus Bar will not affect the other PSU. But here the UPS is common between both redundant PSU. The problem in UPS will lead to the failure of both PSU-1 and PSU-2.Thus here the Weak link s UPS.

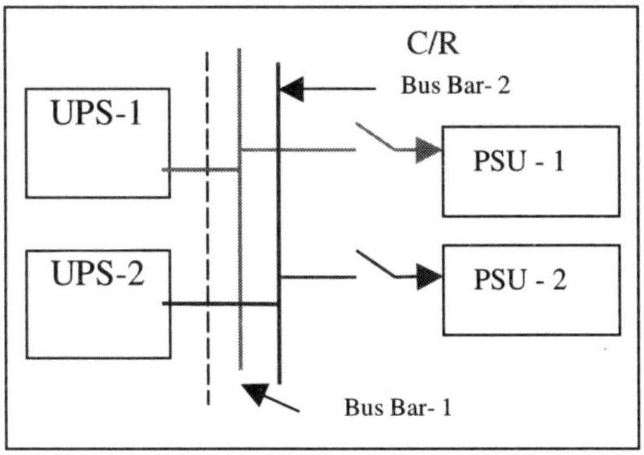

Fig. 4:

In figure 4 all weak-links are like MCB Bus Bar, Feeders and UPS are modified and the complete redundant system is shown in which the separate UPS, Feeder, Bus bar and MCBs are used to feed each PSU. There is no common weak link between PSU-1 and PSU-2. Failure of any of UPS, Feeder, Bus bar or MCB of one PSU will not affect the other PSU. So barriers will continue to receive power supply in any failure.
As shown in figure 4 input sides of PSU-1 and PSU-2 is totally redundant but if output side is not properly designed then any failure at output side will lead to a failure and defeat the sole purpose of redundancy.

(b) Weak-Link at Output Side of PSU (110Vac to 24Vdc)

A few examples are shown here in which redundancy at output side is not properly considered. So even after using the redundant power supply unit and complete redundancy at input side, the sole purpose of redundancy is defeated.

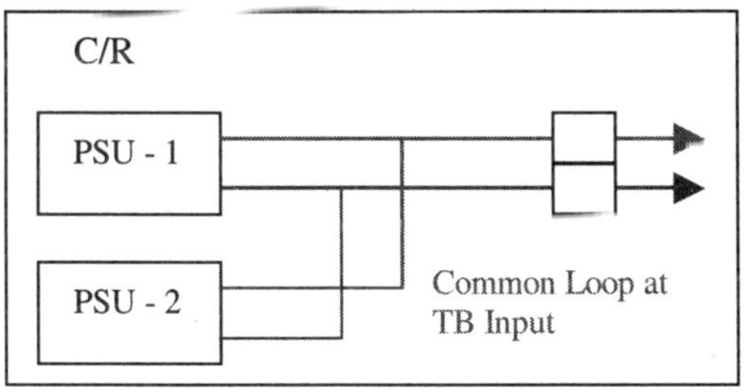

Fig. 5:

In figure 5 two 110Vac to 24Vdc power supply (PSU-1 and PSU-2) are used as redundant power supply to feed the barriers. Hear the outputs of both PSU-1 and PSU-2 are looped without using Diode. The momentarily surge or output failure of one PSU will sink output of other PSU and all barriers being fed will loose the power. Thus here the common looping is weak link.
In figure 6 this weak link is removed by providing diode at output side. The diode at output of each power supply will protect them against surge or short circuit at output of other PSU by not allowing reverse flow of current.

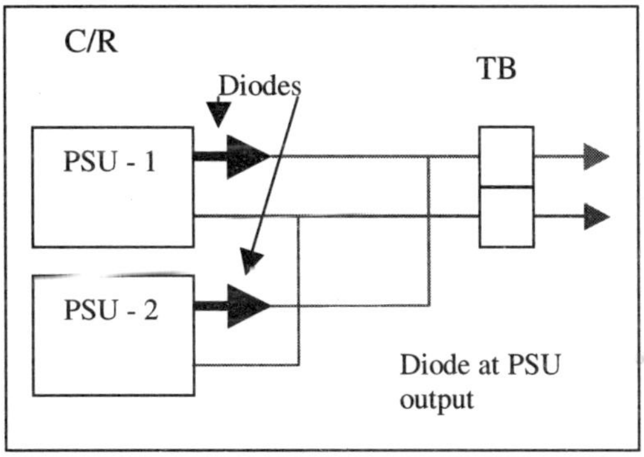

Fig. 6:

As shown in Figure 6 Common looping at TB after diode will survive with surge or short circuit at one PSU output and system continues to run. But here also one weak link is still remaining and that is **"Non availability of Online Replacement"**. Since on line replacement is not possible, in case of failure of one unit we have to run with only one PSU in line and wait for shutdown to replace to faulty PSU. Since chances of shutdown are very rare system has to run without redundancy for long time after one failure. And it is very worst because you know that you are running without redundancy!!

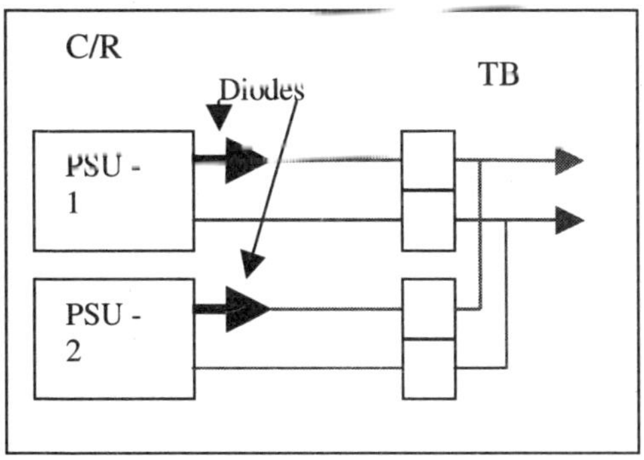

Fig. 7:

In figure 7 inputs at TB are separately connected and looped at output this simple arrangement makes online replacement of faulty PSU. We need not to wait for shutdown to replace the faulty PSU.

But in figure 7 also, there is one thing missing and that is availability of on line status of PSU healthiness. Let's take an example of LCP (Local Control Panel). Two 110Vac to 24Vdc redundant PSU (PSU-1 and PSU-2) are installed in LCP to feed barriers. Since the system is fully redundant is continue to run normally even after failure of any one component in loop like MCB, Bus Bar Feeder and PSU. Generally LCP are not opened frequently for maintenance so the first failure of will not be noticed. And if there is a second failure the total system will fail. Here even though on line replacement of faulty component was possible, it could not taken up because of non availability of on line status. This draw back is rectified in figure 8.

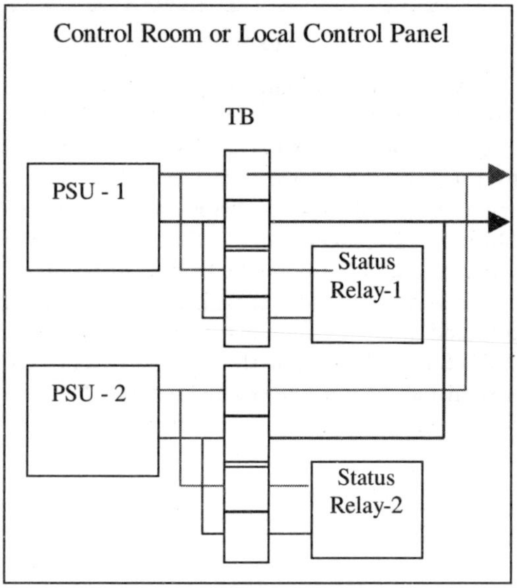

Fig. 8:

The figure 8 shows the complete redundant system with on line status and on line replacement facility. The relay provides the online status while careful wiring done at output of PSU makes the online replacement possible.

The combination of figure 4 and 8 should be used to achieve the complete redundancy and fault tolerance system. It can tolerate any single fault right from UPS to PSU and also online replacement makes it easy to replace the fault component. It as is shown in figure 9

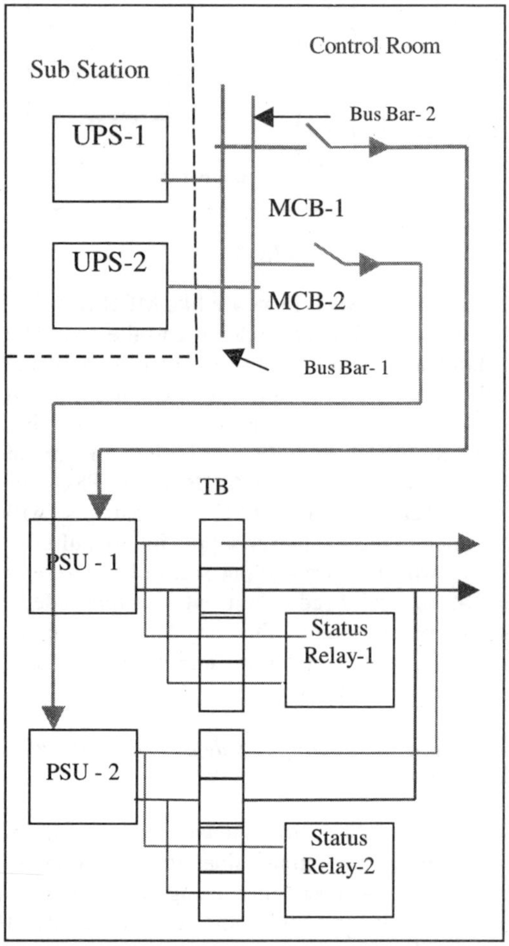

Fig. 9:

Figure 9 shows complete redundant power supply system. Here in case of failure of any component related to PSU-1 loop is available through Status Relay-1 and of failure of any component related to PSU-2 loop is available through Status Relay-2. Also careful interconnection makes the online replacement of any faulty component.

(2) Weak-Link in PLC output side

Normally in plant ESD system, Dual redundancy or TMR scheme is used. They are standard redundant scheme. But one or more weak-link is found in them. A few classic examples of weak link in Dual Redundant system and TMR system are discussed here.

(a) Weak-Link in PLC CPU

There are two processors in dual redundant system, One Processor is called Primary Processor and the other Processor is called Secondary Processor. Both processor scan input and perform ladder logic calculation but only Primary Processor will update the output. In case of failure of Primary Processor, Secondary will take control and start controlling output.

Usually the Processor Status is not available as a hard-wired alarm. Since on the failure of Primary Processor, Secondary will take over smoothly there want be any upset in the process being monitored. And so the failure of Primary Processor remains unnoticed. In case of Secondary Processor failure there want be any change and it also remain unnoticed. If hard-wire alarm is provided the failure of processor can be noticed and corrective action can be taken.

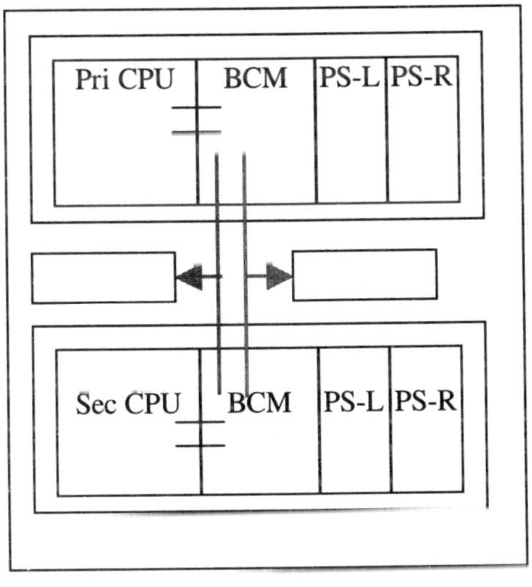

Fig. 10:

Here in figure 10, the example of Allen Bradely is shown in which hard wire alarm for Primary Processor failure and Secondary Processor failure is provided. For Primary Processor NO contact of BCM connected with Secondary Processor card is used and for Primary Processor NC contact of BCM card connected with Secondary Processor.

(b) Weak-Link in PLC Digital Output

TMR system maintains Tipple modular redundancy form input module to Processor to output module. The TMR system can tolerate and compensate any single fault in input module, processor or at output module.

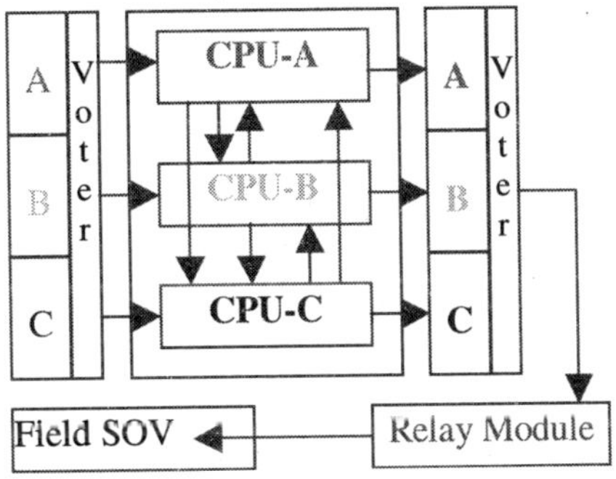

Fig. 11:

In figure 11 an example of TMR system is shown. Here the Tipple Modular Redundancy is maintained from DI to CPU to DO. Since the contact rating of DI module is not sufficient to derive the SOV, Relay module with higher contact is used between DO and SOV. The Relay module used is not TMR. Here the system can tolerate fault in DI module, CPU and DO module but cannot tolerate failure in Relay module and hence the sole purpose of redundancy is not served. Here the Relay module is weak –link. The total system cannot tolerate and compensate the single relay failure and single relay failure will initiate the false tripping. The modified system is shown in figure 12 in which DO with higher contact is used which can directly drive the field SOV.

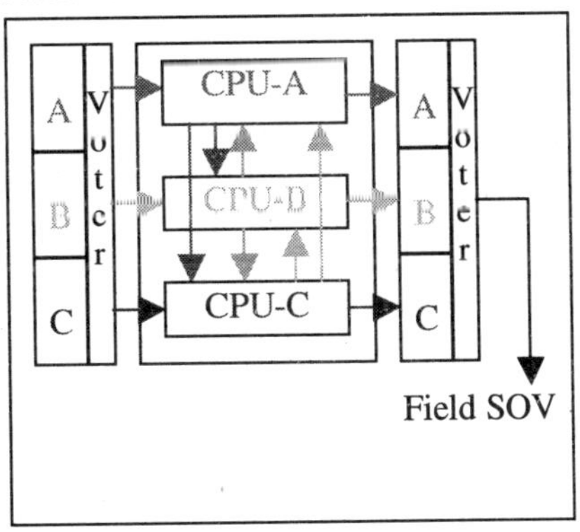

Fig. 12:

(3) Weak-Link in DCS output side

In control system like DCS generally single output of DCS is used to drive field I/P converter. If the AO (Analog Output) of DCS then input to I/P become zero leading to upset the process being controlled. The problem can be solved by using redundant AO. But in case of redundant AO also the single barrier become weak link and its failure will also leads to the same situation as that of DCS AO fails.

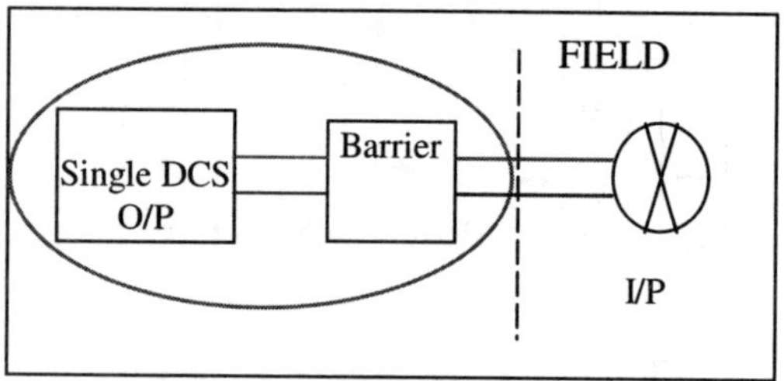

Fig. 13

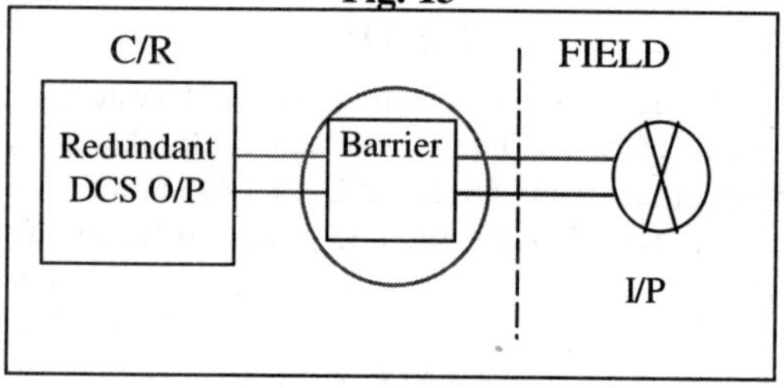

Fig. 14:

In figure 15/16 a full proof scheme is shown which can tolerate the DCS AO failure as well as barrier failure. Here a trip amplifier is connected in series with barriers. It monitors the barrier output. If barrier out put become zero, the trip amplifier activate and De energized the relay and I/P is controlled by second barrier.

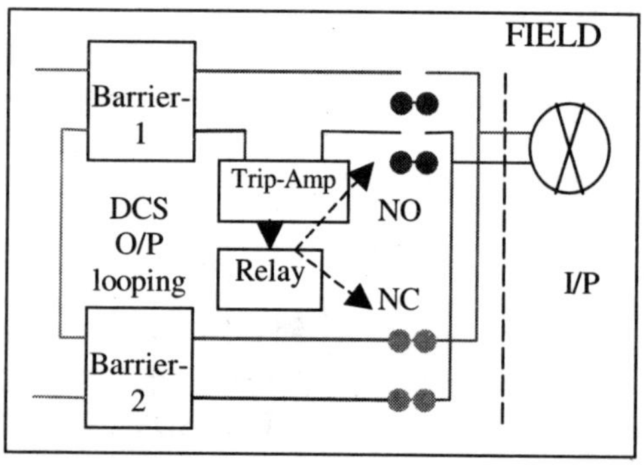

Fig. 15:

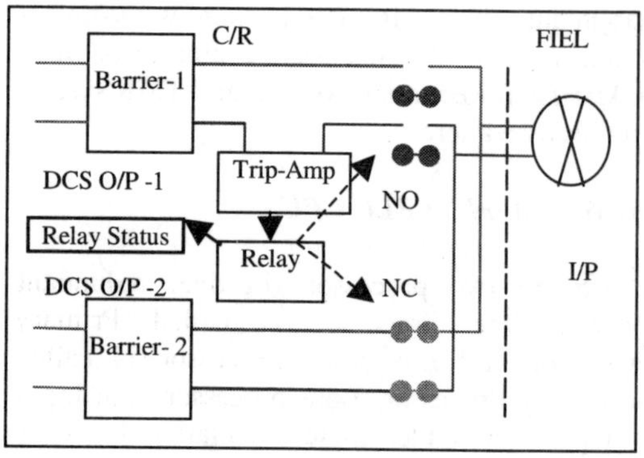

Fig. 16:

(6) Rules to avoid Weak-Link

We have discussed many cases of weak-link but there are two simple rules to be followed to avoid the failure in case of redundancy. They are mentioned below:

(a) Rule-1: Availability of on-line status

For any dual redundancy or TMR system on line status of each component should be made available. The online status availability makes sure that the failure does not remain unnoticed.

(b) Rule-2: On Line Replacement of Faulty Component

Ensure proper interconnection wiring among various component so that on line replacement of each faulty component will be possible without affecting the system-performance.
To ensure the desired level safety, reliability and availability of redundant system in totality, apply the above rules and verify that the system complies these two rules.

2. CONCLUSION

Since there is no standard, often we find one or more weak link while making interconnection between the system and its peripherals. They can be easily identified at any stage of project (like Design, FAT and SAT) by testing by above mentioned two rules and it can be removed easily by minor changes.

3. REFERENCES

1. Wiring Diagram PSA - UOPIA - 1488-B-162
2. Logic diagram and functional schematic EIL - Job no – 3271
3. Logic diagram Stone and Webster - Job No - 02964.10
4. Control Panel diagram Woodward - Item No - 5413-347
5. Anti-surge control system diagram - CCC - 7716

Reliability, Safety and Hazard
Editors: P.V. Varde et al.

RELIABILITY TEST OPTIMISATION USING DESIGN OF EXPERIMENTS

D. Damodaran[1], T. Anbupalam[2] and R. Muthukumar[1]

[1] Centre for Reliability, VSI Estate, Ministry of Communications & Information Technology Chennai 600 041

[2] Department of Statistics, University of Madras, Chennai 600 005

ABSTRACT

Reliability Testing is carried out at all phases of Product development right from the prototype stage to full-scale production. One phase, which is very vital for failure reduction, is the production process. During this phase Reliability Test specifications can be framed based on a number of experimentations undertaken with various levels of different stress parameters. "Efficiency Test" is run to check for the desired failure yield using Taguchi's Orthogonal Arrays. While checking for the efficiency of the level chosen, deliberate defect seeding is done and the defective products tested along with good units. At the end of the testing, the defective units are detected by normal functional test. The project "point of sale machine" was taken as a case study for conducting the reliability testing with three parameters viz. temperature cycling, voltage frequency and random vibration each with three levels. Based on the L9 Orthogonal Array, nine experiments were conducted out of which experiment no.7 (i.e. experiment with parameter 1 of 3rd level, 2 of 1st level and 3 of 3rd level) worked out to be the earliest best stimulator of all the 5 deliberately seeded defects. Therefore this methodology is useful in arriving at the optimized Reliability test specification, which can be implemented in production process to detect the manufacturing defects.

Keywords: Reliability Testing, Prototype, Efficiency Test, Weak Population, Defect Seeding, Temperature Cycling, Point of Sale Machine (POS), Random Vibration, Voltage Frequency, Design of Experiments (DoE), Stress parameter, Orthogonal array

1. INTRODUCTION

Reliability testing is an important activity carried out at all phases of product development right from the prototype stage to full-scale production. By this testing the reliability issues are systematically addressed and the defects minimized. One phase, which is very vital for failure reduction, is the production process. Production process is highly failure prone due to human and machine handling of devices and materials. The production industry consists of various processes like, inspection, electrical verification, lead cutting, tinning automatic component insertion, fluxing, soldering, assembly, potting, coating etc. These processes pass on unnecessary stresses on to the devices, which remain unrelieved in the normal course. If specifications can be framed by identifying the right failure stimulating stresses, which are product unique, and the stress levels based on a number of experimentations. To test the efficiency of the Reliability test specification designed, "Efficiency Test" is run to check for the desired failure yield. The efficiency test is done by experimentation by selecting the stress parameters and their different levels. Therefore it is felt that the utility of design of experiments using orthogonal arrays helps in arriving at a proper test specification more quickly. In this paper the application of design of experiments for reliability testing is discussed and a case study of applying the methodology to a point of sale machine and the derived results are provided.

2. OPTIMISATION OF RELIABILITY TEST

Optimization of reliability test is carried out in the industries to detect process defects attributed to production process. These defects are created out of the weak population of the components along with the impending manufacturing anomalies of the assemblies. As the assemblies are in the advanced stage of production, the testing process shall not degrade the life of the products. At the same time the defective units have to be tested out for improving the field performance. As the situation is vital, there is a requirement to exactly optimize the test specifications to ensure application of correct stresses and their respective stress levels. In order to arrive at the exact stresses repeated experiments are conducted in the optimization process. As lot of time may be required to conduct all the experiments it is appropriate to restrict the number of experiments. To conduct minimum number of experiments it is decided to use design of experiments [5] using orthogonal arrays. While checking for the efficiency of the stress levels chosen, deliberate defect seeding is done and the defective products tested along with good units. At the end of the testing the defects are detected by normal functional testing. This ensures the efficiency of the testing and also since the optimum number of cycles only is undetected these weakened parts can fail in the field. Designing an optimized Reliability test can control the production of such defects normally called the infant mortality [3]. The reliability test applied lot of time is saved during the production life cycle.

3. CASE STUDY

A Product

In order to check for efficiency of the test specifications, a case study was taken up and the technique outlined above was applied. A point of Sale (POS) machine was taken up for the study. The POS machine is used in shopping complex and other multi business entities to enable quick billing and delivery of items. The machine has a disk memory for data storage and retrieval, a colour monitor, RAM, thermal printer assembly, Data Entry Key Board (DEKB) and Task Centric Key Board (TCKB). The machine is normally mains operated, but some times driven by generator supply which feeds varying frequencies. The product has a wide application of usage even in rural areas in grocery shops and medical dispensaries. A key skin is also provided for both the keyboards. All these aspects are taken into account while framing the specifications for the experiments.

B Parameters and their levels

The stress parameters chosen for the product reliability testing are temperature cycling, random vibration and frequency margining [2].

As the product functions in the harsh shop environment, during power cuts, they are fed with generator power with varying input supply frequency in the order of 40 to 60 Hz, frequency margining was also considered as one of the combination of stresses. Combination of stress parameters always detects the infant mortality defects quickly. Thermal cycling and random vibration can detect about 60 to 70% [4] of the defectives in general whereas in our POS case a change in input supply frequency was also selected considering the nature of the products application. The available stress levels arrived at, based on the operational, functional and environmental specifications of the point of sale machines (POS) are provided in the following table.

Table 1.1 Stress Parameters and Levels

Stress Parameters	Levels
1. Temperature Cycling	L1 = -20 °C to + 70° C; L2 = -30 °C to + 80° C L3 = -40 °C to + 100° C
2.Voltage Frequency	L1 = 40 Hz L2 = 50 Hz L3 = 60 Hz
3. Random Vibration	L1 = 14 Grms L2 = 20 Grms L3 = 24 Grms

C Defects seeding and detection

After choosing the parameters and their stress levels suitable enough to detect the defects, some defects are identified and seeded in some of the units. These units are now combined with good units and the testing was conducted.

Table 1.2 Defects seeded in the samples

Defect Id	Name of the Defect
A	Poor epoxy coating in thermal print head.
B	A defective box capacitor.
C	A poor sealed crystal oscillator,
D	Improperly designed power supply cable holder
E	Dry solder in resistor network

The details of the defects seeded in the units for detection is provided in the table above.

From the above inputs the optimum stress levels is to be designed using orthogonal array. There are three parameters and three levels and hence there shall be 27 combinations [6] of these parameters and levels are to be experimented. Instead, by resorting to Taguchi's orthogonal arrays, the L9 array [1] is chosen which reduced the number of experiments to only nine. Each column of the L9 array is assigned to a stress parameter and there are only three stress parameters. Hence the fourth column of the array needs to be ignored. The nine experiments are conducted and checked for detection of all the defects seeded in the point of sale machines within a specified time of exposure.

Table 1.3 L9 Orthogonal Array

Exp.No.	Col 1	Col.2	Col.3	Col.4
1	1	1	1	1
2	1	2	2	2
3	1	3	3	3
4	2	1	2	1
5	2	2	3	1
6	2	3	1	2
7	3	1	3	2
8	3	2	1	3
9	3	3	2	1

Based on the above L9 orthogonal array, experimentations were conducted by selecting three levels for the three stress parameters in their combinations as provided in the array. The three levels of temperature cycling were chosen based on the input obtained from the specification limits provided for the product. The specification limit for the POS machine was - 10 °C to + 60 °C. From the specification limit to the upper limit on the level was arrived at based on a step-by-step principle of advancing the stress level by 15 to 20 %. For random vibration the product does not have a specification limit. It has a sine vibration limit. But random vibration proved to be an effective tool in detecting the defects in the electronics industry [4]. Therefore going by past experience, the levels of random vibration are chosen from a low value of 14 Grms and a high value of 24 Grms. The vibration testing frequency profile was from the typical product spectrum of 10 to 1.8 KHz with provision for energy decay at the rate of 3 dB per octave from 1.8 KHz to 2 KHz. For supply frequency, the stress levels are chosen based on actual measurement with a generator used in commercial shops ranging from 40 Hz to 60 Hz.

Table 1.4 Practical Experimentations Conducted

ExpNo.	Temperature Cycling	Voltage Frequency	Random Vibration
1	-20 °C to + 70° C	40 Hz	14 Grms
2	-20 °C to + 70° C	50 Hz	20 Grms
3	-20 °C to + 70° C	60 Hz	24 Grms
4	-30 °C to + 80° C	40 Hz	20 Grms
5	-30 °C to + 80° C	50 Hz	24 Grms
6	-30 °C to + 80° C	60 Hz	14 Grms
7	**-40 °C to + 100° C**	**40 Hz**	**24 Grms**
8	-40 °C to + 100° C	50 Hz	14 Grms
9	40 °C to + 100° C	60 Hz	20 Grms

The nine trials were conducted separately on defect-seeded samples along with good units repeatedly in order to find out the best combination that will bring out all the seeded defects in one go with minimum time of testing. Out of the nine experiments, experiment number seven (7) worked out to be the earliest best stimulator of all the five defects. The trial results are provided below in the table along with the defects detected in each trial. The defect identification is done in the second column as given in table 2.

Table 1.5 Results of the experimental trials

Trials Number	Defects Detected
1	A only
2.	A and C
3.	A and C
4.	A, B and C
5.	A, B, C and E
6.	A, B, C and E
7.	**A, B, C, D and E**
8.	A, B and C
9.	A, B and C

Therefore the experiment number 7 in which all the five defects were detected is considered to be the efficient combination of all the three stress parameters in their given levels. The specification worked out to be, temperature cycling in the range -40 °C to + 100° C; voltage frequency maintained at 40 Hz and random vibration conducted at 24 Grms.

4. CONCLUSION

The minimum experiments conducted on the point of sale machines based on the Taguchi's orthogonal array for this optimization process yielded the desired results of detecting the five types of failures deliberately injected to prove the efficiency of the reliability test. Instead of conducting a number of experiments to arrive at the right test specifications it is appropriate to use this methodology to complete the testing exercise within a short duration and at the same time bring out the process defects.

5. REFERENCES

1. [1] Dr.Genichi Taguchi, 1986. Introduction to Quality Engineering, Asian Productivity Organisation.
2. Wong, K. 1989. A new environmental stress screening theory for electronics. Proceedings of the Institute of Environmental Sciences and Technology (IEST): 45 – 56.
3. O'Connor, P.D.T., 1996. Practical Reliability Engineering, John Wiley and Sons, New York.
4. Silverman, M. 1997. Summary of HALT and HASS results at an accelerated reliability test center. Nepecon West Conference, 823 – 841.
5. Montgomary, D.C., 2001, Design and Analysis of Experiments, John Wiley & Sons, New York.
6. D.Raghava Rao, Construction & Combinatorial Problems in DOE, Wiley & Sons Inc.

Reliability, Safety and Hazard
Editors: P.V. Varde et al.

STUDIES ON SIMULATION BASED SYSTEM RELIABILITY EVALUATION

K. Durga Rao[a], H.S. Kushwaha[a], A.K. Verma[b] and A. Srividya[b]

[a] Health, Safety and Environment Group, Bhabha Atomic Research Centre, Mumbai 400 085
[b] Indian Institute of Technology Bombay, Mumbai 400 076

ABSTRACT

Reliability indices provide a measure of performance evaluation of engineering designs/systems. Analytical techniques such as Reliability Block Diagram, Fault Tree Analysis and Markov Models are widely used in evaluation of reliability indices for engineering systems with the assumption of exponential time to failure probability density function. Analytical methods are effective only for non-repairable simple systems. While it can give time dependent models for non-repairable systems, it can't give the same for repairable complex scenarios. The output of the analytical methods to repairable systems is usually limited to expected values only and also these are based on simplified mathematical models with the questionable assumptions in most of the practical cases. Stochastic Monte-Carlo Simulation (SMC) can be used to overcome these difficulties for system reliability evaluation of any kind of complex scenario. This paper explains the concepts of simulation and demonstrates the application to two case studies.

Keywords: Stochastic simulation, Fault tree analysis, Markov model

1. INTRODUCTION

The reliability indices (Ex: Availability, MTBF, Failure frequency, etc.) of a system can be evaluated using one of two basic approaches, direct analytical techniques or stochastic simulation [1]. Reliability Block Diagram (RBD), Fault Tree Analysis (FTA), and Markov models fall under analytical techniques. Analytical techniques represent the system by a mathematical model, which is often simplified, and evaluate the reliability indices from this model using direct mathematical solutions.

Simulation technique, on the other hand, estimates the reliability indices by simulating the actual process and random behaviour of the system. Simulation is a very valuable method which is widely used in the solution of real engineering problems [2]. This method treats the problem as a series of real experiments conducted in a simulated time. It estimates the probability and other indices by counting the number of times an event occurs in simulated time. The disadvantage with the analytical approach is that the model used in the analysis is usually a simplification of the system; sometimes to an extent it becomes totally unrealistic. In addition, the output of the analytical methods is usually limited to expected values only. Analyses that involve repairable systems with multiple additional events and/or other maintainability information are very difficult (if not impossible) to solve analytically. In these cases, analysis through simulation becomes necessary.

The Simulation approach overcomes the disadvantages of the former method by incorporating and simulating any system characteristic that can be recognised. It can provide a wide range of output parameters including all moments and complete probability density functions. It can handle very complex scenarios like in constant transition rate multi state systems and time dependent reliability problems. However, the solution time is usually large and there is uncertainty from one simulation to another. But the recent studies show the demerits of simulation can be easily overcome with few modifications in the simulation [3]. It is to be noted that the experimentation required is different for different types of problems and it is not possible to precisely define a general procedure that is applicable in all circumstances. However, the simulation technique provides considerable flexibility in solving any type of complex problem.

In this work, stochastic simulation approach is applied to two reliability problems, viz., Typical Class IV supply system (four component series system) and emergency injection pumps branch(two component parallel system) in any Nuclear Power Plant. All the components are considered to be repairable in both the cases of application. Steady state unavailability is considered as the required reliability indices. In each iteration (random experiment), from simulated random behaviour of the system, uptime and down time are calculated based on system success logic. After sufficient number of iterations, unavailability is estimated from the results. By varying the mission time and repeating the simulation in the similar manner, exact probability density functions can be estimated for the desired system characteristic. The same problems are compared with the RBD and Markov model approaches.

2. PERFORMANCE EVALUATIONS WITH SIMULATION

In simulation, random failure/repair times from each components failure/repair distribution are generated. These failure/repair times are then combined in accordance with the way the components are reliability-wise arranged within the system. The overall results are analyzed in order to determine the behaviour of the entire system. Sound understanding of the system behaviour is the prerequisite for system success/failure logic. It is assumed that the reliability values for the components have been determined using standard (or accelerated) life data analysis techniques, so that the reliability function for each component is known. With this component-level reliability information available, simulation can then be performed to determine the reliability of the entire system. The random failure/repair times of components is obtained using uniform random numbers and converting these into required density function as per the component PDF. The simulation procedure is explained below with the help of the two above mentioned examples.

Typical Class IV supply system consists of supply from the Grid, circuit breaker, transformer and bus. The success criterion is the availability of power supply at the bus, which demands the successful functioning of all the components. So, it is simple four component series system. The reliability block diagram (Functional diagram) is shown in figure 1. Typical emergency core cooling system of Nuclear Power Plant consists of a two unit injection pump active redundant system. One pump operation is sufficient for the successful operation of the system. The failure of the system occurs when both the pumps fail. The reliability block diagram (Functional diagram) is shown in figure 2. In addition to success/failure logic, failure and repair Probability Density Functions (PDF) of components are the input to the simulation. The PDF of failure and repair for the components are taken from IAEA TECDOC 478 [4] and Modarres [5] (see Table 1). It is assumed that all the component failure/repair PDFs are following exponential distribution. However, the procedure is same even when component PDFs are non-exponential except that the random variants will be different as per the PDF. The simulation procedure is as follows [1]:

Step 1: Generate a random number
Step 2: Convert this number into a value of operating time using a conversion method on the appropriate times to failure distribution (exponential in the present case)
Step 3: Generate a new random number
Step 4: Convert this number into a value of repair time using conversion method on the appropriate times to repair distribution
Step 5: Repeat step 1-4 for a desired period of operating life.
Step 6: Repeat steps 1-5 for each component
Step 7: Compare the sequences for each component and deduce system failure times, frequencies and other parameters.

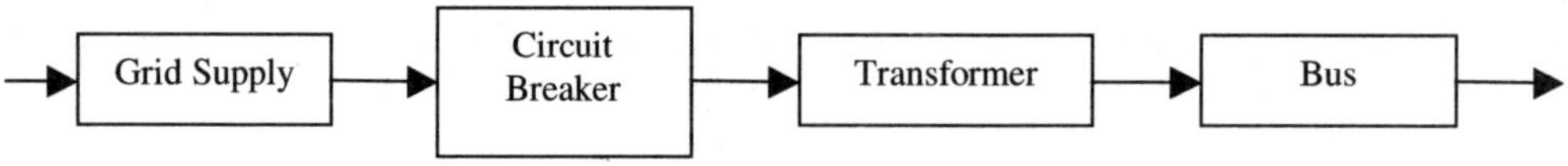

Fig. 1: Functional Diagram of typical Class IV supply system (Problem 1)

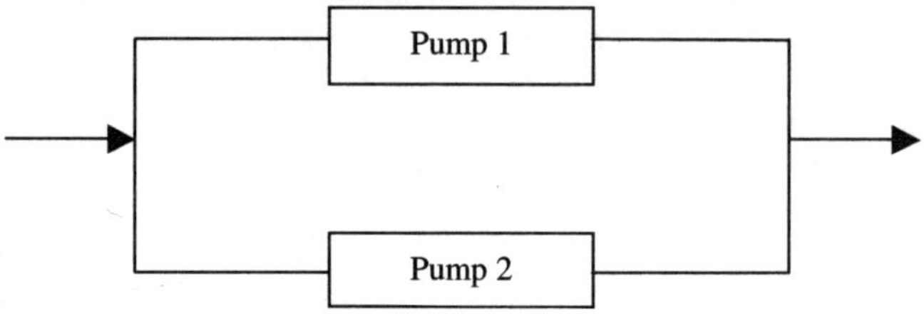

Fig. 2: Functional block diagram of two unit pump active redundant system (Problem 2)

Table 1 Failure and Repair data of components[4]

S. No.	Component	Failure rate (/hr)	Repair rate (/hr)
1.	Grid Supply	2e-4	2.59
2.	Circuit Breaker	1e-6	0.166
3.	Transformer	2e-6	0.0925926
4.	Bus	1e-7	0.0925926
5.	Pump (1 & 2)	3.7e-5	0.0925926

Table 2 Summary of Results

S. No	Problem 1		Problem 2	
	Analytical	Simulation	Analytical	Simulation
Average Unavailability	1.059e-4	1.059e-4	1.59e-7	1.61e-7
Avg. No. of Failures	2030.78	2031.031	2.955	2.997
Failure Frequency(/hr)	2.031e-4	2.031e-4	2.95e-8	2.99e-8
Mean time between failure (hrs)	4924.21	4923.51	33.84e+6	33.36e+6
Mean time to repair (hrs)	0.5214	0.5214	5.39	5.37

Step 8: Repeat steps 1-7 for the desired number of simulations

Overview of up and down states for Class IV supply system is shown in Fig. 3. System failure time is sum of the component failure times if they are mutually exclusive. If there is any simultaneous failure of two or more components, failure time of component having largest value is taken for system failure time.

Overview of up and down states for emergency injection pumps branch having two pumps in parallel is shown in Fig. 4. System failure time is the time when there is simultaneous failure of two pumps.

3. PERFORMANCE EVALUATION WITH ANALYTICAL APPROACH

In the analytical (or algebraic analysis) approach, the system's PDF/other reliability indices are obtained analytically from each component's failure distribution using probability theory. In other words, the analytical approach involves the determination of a mathematical expression that describes the reliability of the system in terms the reliabilities of its components.

When considering only the failure characteristics of the components, the analytical approach is generally used. However, when both

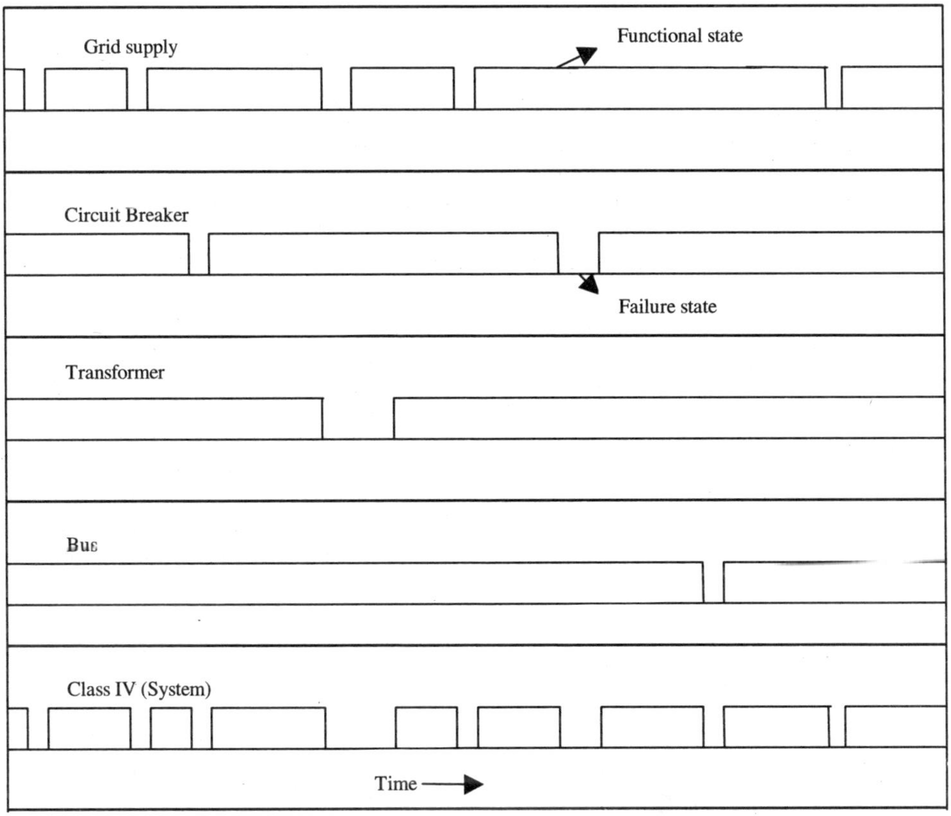

Fig. 3: Overview of up and down states for Class IV supply system having four components in series

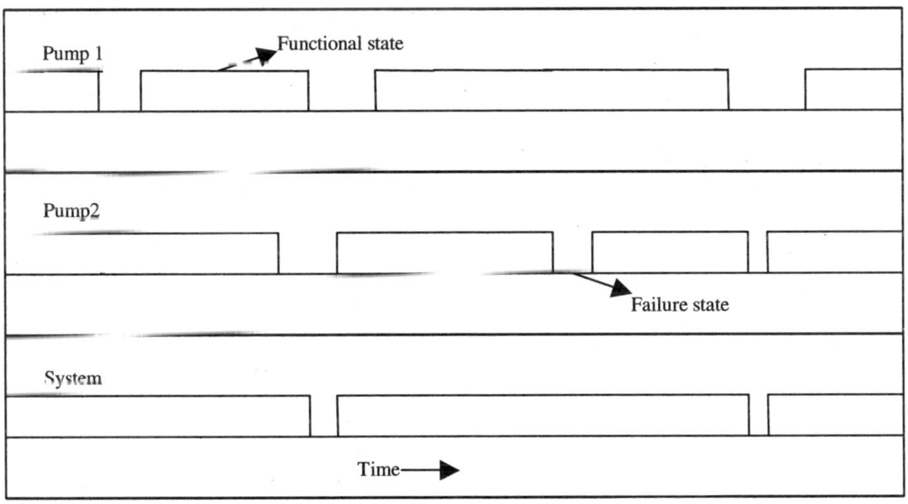

Fig. 4: Overview of up and down states for emergency injection pumps branch having two pumps in parallel

the failure and maintenance characteristics need to be considered, the simulation method must be utilized to take into account the additional events. This is due to the fact that the uncertainties that arise due to simplification by the analytical mathematical models can be eliminated with simulation. Nevertheless, the experimentation required in simulation is different for different types of problems and it is not possible to precisely define a general procedure that is applicable in all circumstances. Due to this reason and simplicity that are offered by Fault tree approach and Markov Models; they are widely used in the performance evaluation of reactor systems in Level-1 Probabilistic Safety Assessment (PSA) of Nuclear Power Plant. However, the uncertainty in modelling is always there with these analytical approaches. The models are only applicable with exponential failure/repair probability density functions. They are invalid for components having non-exponential failure/repair PDFs [2].

Considering components to be independent the availability expression for Class IV system (A) is given by the following expression (1):

$$A = A_1 A_2 A_3 A_4$$
$$A_i = \frac{\mu_i}{\mu_i + \lambda_i} \tag{1}$$
$$A = \frac{\mu_1 \mu_2 \mu_3 \mu_4}{(\mu_1 + \lambda_1)(\mu_2 + \lambda_2)(\mu_3 + \lambda_3)(\mu_4 + \lambda_4)}$$

A_i, λ_i, and μ_i are the availability, failure rate and repair rate of ith component (i=1,2,3 and 4).

Considering components in the second case of application, two-unit active redundant parallel pump system, to be independent, the unavailability (Q) is given by the following expression (2):

$$Q = Q_1 Q_2$$
$$Q_i = \frac{\lambda_i}{\mu_i + \lambda_i} \tag{2}$$
$$Q = \frac{\lambda_1 \lambda_2}{(\mu_1 + \lambda_1)(\mu_2 + \lambda_2)}$$

Q_i, λ_i, and μ_i are the unavailability, failure rate and repair rate of ith component (i=1 and 2).

4. RESULTS AND DISCUSSION

As RBD approach is limited to non-repairable systems and the considered problems (1&2) are repairable systems, it is not considered in the modelling. Fault tree approach and Markov model can be used to solve both the problems analytically. The later method has advantage of giving time dependent expressions. Nevertheless, Markov model can not be used when there is large number of components in the system and PDFs are non-exponential. Although, fault tree approach can be used for complex systems, but incorporation of dynamics in the event occurrences (both failure and maintenance) and flexibility with different types of PDFs is very difficult if not impossible. Fault tree approach can give only point estimated only. But, Monte-Carlo simulation approach is having flexibility in solving any kind of complex reliability problem[6-8]. It can solve problems of dynamic in terms of sequence occurrences, time dependent, having any kind of component PDF and it can give the required system attribute. However, the solution time is usually large and there is uncertainty from one simulation to another. But the recent studies show the demerits of simulation can be easily overcome with few modifications in the simulation. It is to be noted that the experimentation required is different for different types of problems and it is not possible to precisely define a general procedure that is applicable in all circumstances. However, the simulation technique provides considerable flexibility in solving any type of complex problem.

The modelling of both the problems in the analytical approach has been done using ISOGRAPH software. In-House computer code has been developed for carrying out time dependent simulation for repairable systems. Table 2 gives the comparison of both the approaches, analytical and simulation, for the two problems. In addition to the parameters such as average unavailability, expected number of

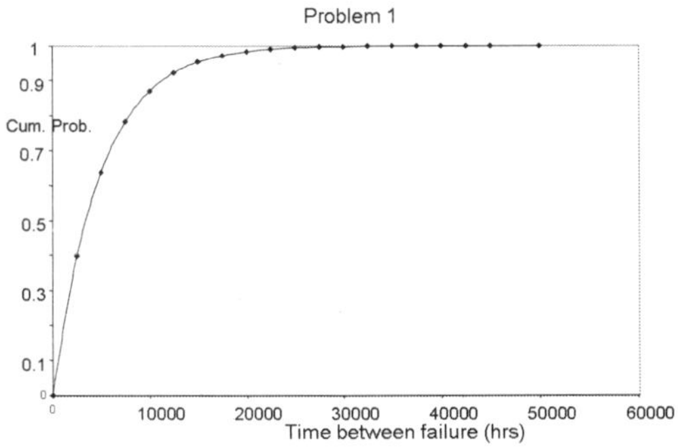

Fig. 5: CDF of Problem 1

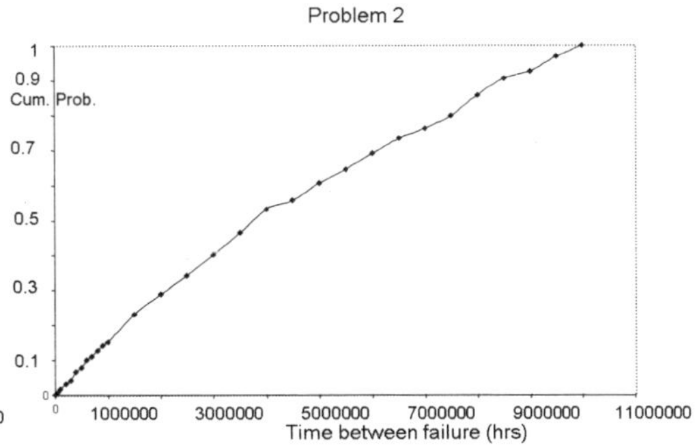

Fig. 6: CDF of Problem 2

failures, failure frequency, Mean time between failures and mean time to repair, simulation can give Cumulative Density Function (CDF) of random variable, time between failures for the system under consideration. The CDF of problem 1 and problem 2 are shown in Fig. 5 and Fig. 6 respectively. Mission times of 10^7 hrs and 10^8 hrs are considered for problem 1 and problem 2 respectively. Simulation results are from 1000 iterations in both the cases.

The incredible development in the computer technology for data processing at unprecedented speed levels are further emphasizing the use of simulation approaches to solve reliability problems. Use of simulation approach eliminates many of the assumptions that are inevitable with analytical approaches. In order to simplify the complex reliability problems, analytical approaches make lot of assumption to make it to a simple mathematical model. On the contrary, Monte-Carlo simulation based reliability approach, due to its inherent capability in simulating the actual process and random behaviour of the system, can eliminate the uncertainty in system reliability modelling. One shall not forget the Einstein's quotation in this regard, "A theory should be as simple as possible, but no simpler."

5. REFERENCES

1. Roy B. & Ronald N. A., (1992), *Reliability Evaluation of Engineering Systems*, Plenum Press, New York.
2. *ISOGRAPH Software*, Isograph Inc., USA.
3. *System Analysis Reference*, ReliaSoft Publishing, 2003.
4. *IAEA TECDOC 478*, (1988), *Component Reliability Data for Use in Probabilistic Safety Assessment*, IAEA Agency, Vienna.
5. Modarres M., (1993), *What every Engineer should know about Reliability and Risk Analysis*, Marcel Dekker Inc., New York.
6. Adolfo C. M., Antonio S. H., & Benoit, I., (2005), *Monte Carlo-based assessment of system availability*, Reliability Engineering and System Safety **88,** 273–289.
7. Enrico Z., Luca P., & Valerie Z., (2005), *A combination of Monte Carlo simulation and cellular automata for computing the availability of complex network systems*, Reliability Engineering and System Safety (Article in press), 1–10.
8. Marseguerra M., Zio. E., Devooght J., & Labeau P. E., (1998), *A concept paper on dynamic reliability via Monte-Carlo Simulation*, Mathematics and Computers in Simulation **47**, 371-382.
9. Misra K. B., (1993), *Reliability Analysis and Prediction*, Elsevier Science Publishing, Amsterdam.
10. Yanez J., Ortuno T., Vitariona B., (1997), *A simulation approach to reliability analysis of weapon systems*, European Journal of Operational Research **100,** 216-224.

Reliability, Safety and Hazard
Editors: P.V. Varde et al.

RELIABILITY ANALYSIS OF DIVERSE SAFETY LOGIC SYSTEMS OF FAST BREEDER REACTOR

Bh. Ravi Kumar[a], L. Srivani[b], P.R. Apte [a], S. Ilango Sambasivan[b] and P. Swaminathan[b]

[a] Indian Institute of Technology Bombay, Mumbai 400 076
[b]Electronics and Instrumentation Group,
Indira Gandhi Centre for Atomic Research, Kalpakkam

ABSTRACT

Safety Logic for Fast Breeder Reactor (FBR) is designed to initiate safety action against Design Basis Events. Based on the outputs of various processing circuits, Safety logic system drives the control rods of the shutdown system. So, Safety Logic system is classified as safety critical system. Therefore, reliability analysis has to be performed. This paper discuses the Reliability analysis of Diverse Safety logic systems of FBRs. For this literature survey on safety critical systems, system reliability approach and standards to be followed like IEC-61508 are discussed in detail. For Programmable Logic device based systems, Hardware Description Languages (HDL) are used. So this paper also discusses the Verification and Validation for HDLs. Finally a case study for the Reliability analysis of Safety logic is discussed.

Keywords: Safety Logic, FBR, CPLD, FPGA, HDL, HVL, IEC, Diverse, PLD, V & V, Safety Critical, Safety Related, SDS

1. INTRODUCTION

Fast Breeder Reactor is the second stage of Indian Atomic Energy Programme, which is aimed at preparing the country for utilization of the extensive Thorium reserves and providing means to meet the large demands of electrical energy. Since safety requirement of FBRs is very high, two independent diverse shutdown systems are preferred for ensuring the safe shutdown of the reactor even when one system is not available. The purpose of reactor shutdown systems is to promptly terminate the Fission chain reaction and thereby ensure safety during Design Basis Events. This categorizes the shutdown system a safety critical system. Safety Logic system drives the control rods of the shutdown system, based on the outputs of various processing circuits. Since Safety Logic systems are related to shutdown systems, they are also categorized as Safety Critical systems. So, the Reliability analysis for Safety logic systems becomes inevitable. This paper discusses the reliability analysis of safety logic systems and the IEC-61508 standard for functional safety. Finally this paper includes a case study on reliability analysis of safety logic systems of Prototype Fast Breeder Reactor.

2. SAFETY CRITICAL SYSTEMS

The safety critical systems are provided to assure safe shutdown of the reactor, during undesired events Even if the safety critical systems fail, they should fail in fail-safe manner. The Safety Critical systems, for example Safety logic systems, require a very high reliability.

Generally the target values will be specified in terms of Probability of failure on demand and the spurious failure rate. The probability of failure on demand is defined as "the average system safety function unreliability due to random hardware failures". The availability of the system will decrease with the increase in the spurious failure rate. For example, AERB has specified that the shutdown system should have less than 10E-3 failures per demand and 0.1 spurious failures per year.

Many standards [1] are introduced to evaluate the functional safety of the systems. IEC 61508 is one of the popular standard among them.

3. IEC STANDARD

IEC 61508 [2] is a seven-part standard that defines specific direction and guidelines on the functional safety of electrical/ electronic/ programmable electronic (E/E/PES) safety-related systems. Accordingly to the standard, Safety function is defined in terms of both functionality and Safety integrity level (which is the probability that the specified action will be carried out in order to achieve the required risk reduction). IEC 61508 uses "Safety Life cycle" to formulate its own requirements.

Even after following a systematic way to develop a system, there may be many functional deficiencies in the system. These have to be identified before the system is kept in operational state.

Qualitative analysis gives limited insight into the system. Quantification of the achieved safety [3] is necessary for compliance with IEC 61508. This is to ensure that the effect of random failures, common cause failures of hardware/software and the failure of communication equipment is less than the target failure measure.

The standard gives only guidelines on how to quantify the loss of safety. It allows us to use different quantitative techniques. It also provides guidelines on the quantification by giving examples in part-6.

4. RELIABILITY ANALYSIS

There are many techniques to perform safety and Reliability analysis like Failure modes and effects analysis (FMEA), Preliminary Hazard analysis, Fault tree analysis (FTA), Markov modeling, Reliability block diagrams (RBD), Event trees etc., Different analysis techniques cover different aspects [5] of the system behavior. And in turn, this may lead to different calculation results. A comparison of this can be found in [4] paper.

For evaluating the reliability in compliance to IEC 61508 standard, three steps [1] have to be followed:

The three steps are:

1. FMEA has to be performed on the system, which will give an idea about the fraction of safe and dangerous failures in the system. MIL-1629 can be used as guidelines to perform this analysis.
2. Failure rate of the system is calculated using the reliability prediction. MIL-217-F (Notice-2) standard can be used for this purpose.
3. The resultant failure rate has to be divided into safe and dangerous failures and the average probability of loss of safety (PFD) has to be calculated. This can be calculated either by using the formulas given in IEC 61508 part2 or using any markov model developed for the system or using the fault tree technique [6]. Tieling Zhang et al [8] discussed in detail about PFD and its associated terms.

FMEA is a bottom up approach for analyzing system design and performance. To begin a FMEA, the lowest levels of the system are outlined. This can be the individual components (referred to as a *piece part FMEA*) or the lowest level assemblies in the system (referred to as a *functional FMEA*). For each lowest level, a list of potential failure modes is generated. Effects of each potential failure mode are then determined. A FMECA (criticality analysis) can use failure rate calculations that were performed during the Reliability Prediction portion of an analysis to determine probability of occurrence.

4.1 Failure Rate Fractions

The total failures in the system can be divided into Safe (see ref [2], [7]), and dangerous failures.

$$\lambda = \lambda_S + \lambda_D \quad (1)$$

Among the unsafe failures, few may be detected by the online diagnostic techniques built inside the system. They are called *"Dangerous failures Detected"*. Finally few may be left undetected. These are called *"Dangerous failures Undetected"*.

It was given in IEC 61508 standard that "for complex systems, where a detailed analysis of each failure mode is not possible, a division of failures into 50% safe and 50% dangerous is generally acceptable".

$$\lambda_S = \lambda_D = \frac{\lambda}{2} \quad (2)$$

Fault coverage will have a big impact on the system safety and reliability. Accurate determination of the coverage factor is very much essential. In reference [9] the concept of fault coverage factor is explained in detail and FMEA is proposed to arrive at a coverage factor value. An extension to FMEA could be used to estimate the diagnostic coverage factor. But, it won't be possible to find out all the faults in a complex programmable electronic system using FMEA. A detailed Verification and validation must be performed on the system. This is explained in the next section.

$\lambda_D \rightarrow Dangerous\_failures$

$\lambda_{DU} \rightarrow Dangerous\_failures\_undetected$

i.e. failures undetected by online diagnostics

$\lambda_{DD} \rightarrow Dangerous\_failures\_detected$

i.e. failures detected by online diagnostics.

DC is diagnostic coverage factor.

$$\lambda_{DU} = \lambda_D \times (1 - DC), \quad (3)$$

$$\lambda_D = \lambda_{DU} + \lambda_{DD} \quad (4)$$

$$\lambda_{DD} = \lambda_D \times DC \quad (5)$$

$$Coverage\_factor = \frac{Failures\_detected}{Total\_no\_of\_failures} \quad (6)$$

Common mode failure (among redundant systems) is another factor that will have significant impact on the achieved safety. Beta factor model is suggested by IEC standard to model the common mode failures. Latter, MBF model is proposed by Per Hokstad and Kjell Corneliussen [7]. This value has to be carefully chosen.

The next section discusses the verification and validation techniques for the design of re-configurable hardware.

4.2 Verification and Validation of HDLs

Programmable Logic Device (PLD) is an integrated circuit used to program digital circuits. Hardware Description Languages (HDLs) are meant for formal description of electronic circuits. HDLs can describe the circuit's operation and verify its operation by means of simulation. They are used to program the PLDs. In contrast to a software programming language, an HDL's syntax and semantics include explicit notations for expressing time and concurrency which are the primary attributes of hardware. HDLs are needed to both model hardware and to define a test bench to verify the model. As design sizes increases, the amount of verification required also increases dramatically.

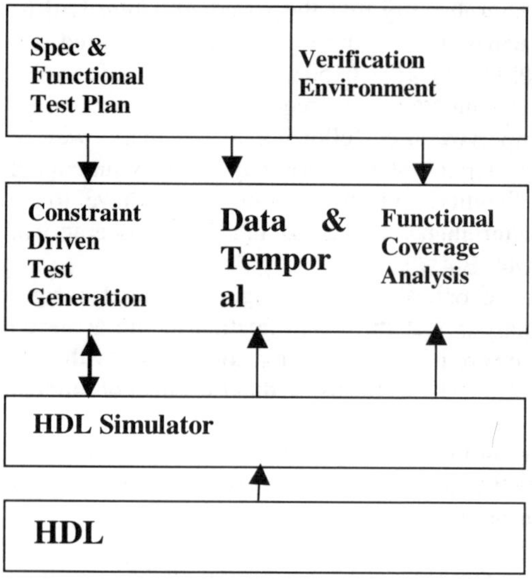

Fig. 1: Verification and validation of HDL

The first step towards improving the efficiency of functional verification for large designs is to raise the level of abstraction of the verification environment to the specification level as shown in Figure-1. The common specifications driving the verification process are the design specification, the interface specification and the functional test plan. The verification team can develop and implement a comprehensive verification strategy only when the rules defined in these specifications can be captured in an executable form. For this verification tools are available. So this completes the Hardware reliability analysis.

5. CASE STUDY

Prototype Fast Breeder Reactor (PFBR), a 500MWe reactor under construction at Kalpakkam is provided with two independent diverse shutdown systems SDS1 and SDS2. Safety logic with Fine Impulse Test (SLFIT) system and Pulse Coded Safety Logic (PCSL) system are the respective safety logic systems of the SDSs. SDS1 uses conventional solid-state logic SLFIT facility where as Safety logic in SDS2 employs PCSL (which is self-diagnostic in nature). Diversity in safety logic systems is maintained by implementing the concepts like Self-diagnostics and fault tolerance. Programmable electronic systems are increasingly finding their way in latest systems. The safety logic systems are developed using various types of Programmable Logic device (PLD) s and many discrete components.

The Safety logic systems monitor the important parameters of the reactor and generate trip signal in case of any undesired event. The reliability of each of the safety logics shall be better than 0.1 spurious failures per year and 1.84 E-4 /de (as per AERB guidelines). The scope of this project is to perform Reliability analysis on both the systems to check whether they are meeting the required reliability target or not.

First, Reliability prediction has been performed using US-MIL (military) handbook 217f-2 on SLFIT and PCSL systems. The failure rates of the SLFIT and PCSL systems are well within the target limits.

Critical components in the system are identified and functional FMEA is performed on the SLFIT and PCSL systems. The effects of the failure modes of each component on the system have been documented. No potential uncovered faults have been discovered using FMEA. Using the IEC 61508 standard, the average system unreliability of SLFIT and PCSL systems are found to be less than 10^{-5}/de respectively. It was also found out that, there will occur 0.02 spurious failures in SLFIT and PCSL systems per year. The results are meeting the reliability targets. Finally, the influence of coverage factor on the availability of the SLFIT system has been studied using a Markov model.

5.1 Markov Model of SLFIT

Markov models are a flexible system modeling that enables an analyst to consider scenarios that are not traditionally supported by other classes of analyses. Markov models are used to consider systems where future performance is determined solely by its current state. They are particularly useful where a simulated answer to a complex problem is not acceptable. A typical analysis could begin by determining, the states of operation under consideration, operating condition of each state (good/failed/capacity), connection between different states and transition rates for each connection.

5.2 Assumptions

Failure rate of Safety logic (λ_S)= $5\text{x}10^{-6}$/hr
Failure rate of testing system (λ_T) =10^{-6}/hr
Coverage factor (DC) = 99%
Test rate (r_T)= 20/sec
50% safe 50% dangerous failures
Proof test interval ($1/r_{PTI}$)= 8 months
Mean Time To Repair ($1/r_R$) = 4 hrs

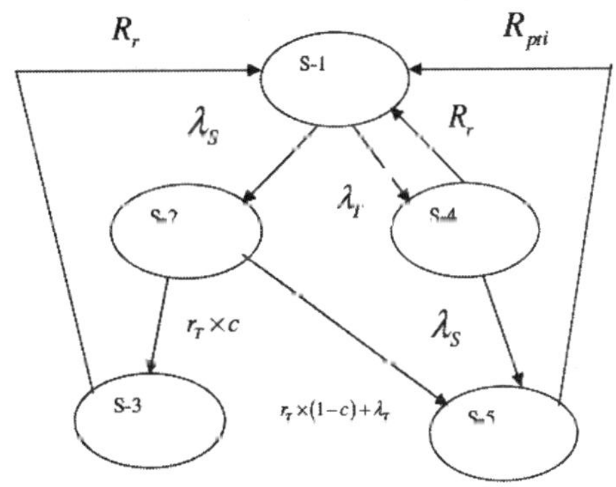

Fig. 2: Markov Model of SLFIT

The system is modeled using a markov modelas shown in Figure- 2

State 1: Safety Logic, Test system both functioning
State 2: Safety Logic failed, Test system functioning
State 3: Safety Logic failed. Test system detects the fault
State 4: Safety Logic Functioning, Test system failed
State 5: Safety Logic Failed, Failure not detected.

State 1 and State 4 are safe states where system is available. State 5 is dangerous state. In State 3 the system is in safe state, but it is unavailable. State 2 is an intermediate state.
According to the formulas given in part-6 of the standard, for a 1oo1 system, the Probability of failure on demand is given by

$$Pfd = 1 - e^{-\lambda_D \times t_{CE}} \approx \lambda_D \times t_{CE} \quad .(7)$$

Dangerous failure=2.5 E-6/hr
t_{CE} is system down time given by the formula

$$t_{CE} = \frac{\lambda_{DU}}{\lambda_D}\left(\frac{T_1}{2} + MTTR\right) + \frac{\lambda_{DD}}{\lambda_D} \times MTTR \quad (8)$$

=32.80 hrs
Pfd = 8.2 E -5 /de
Spurious failure rate is 0.0216/yr
The availability of the system (state 1) with respect to time at various coverage factor values is plotted in Figure-3.

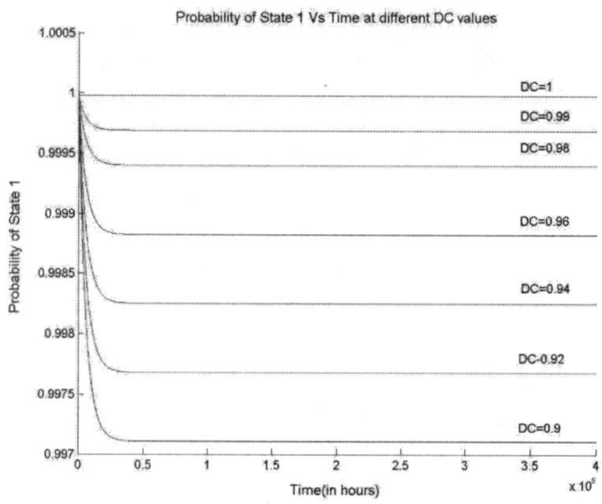

Fig. 3: Probability of state 1 Vs Time at Different DC values

6. CONCLUSION

Safety logic systems of SDS are categorized as Safety critical systems. Hence reliability analysis for these systems is a must. The literature survey on safety critical systems, system reliability approach and standards to be followed like IEC-61508 are discussed. In this paper the reliability approach for the safety logic systems is discussed with a case study.

7. REFERENCES

1. J. P. Rooney, "IEC61508: An Opportunity for Reliability", *Reliability and Maintainability Symposium*, pp 272 – 277, 22-25 Jan. 2001.
2. IEC61508. Functional safety of electric/ electronic/ programmable electronic safety-related systems, Parts. 1–7; Oct 1998– May 2000.
3. D. M. Karydas and A. C. Brombacher, "Reliability certification of programmable electronic systems", *Reliability Engineering and System Safety*, pp 103-107, Issue 2, Nov 1999.
4. J. L. Rouvroye and A. C. Brombacher, "New quantitative safety standards: different techniques, different results?", *Reliability Engineering and System Safety*, pp 103-107, issue 2, Nov 1999.
5. J. L. Rouvroye, E. G. van den Bliek, "Comparing safety analysis techniques", *Reliability Engineering & System Safety*, pp 289-294, vol 75, issue 3, Mar 2002
6. Dallas. S. Green, Arthur M. Dowell, "How to design, verify and validate emergency shutdown systems", *ISA Transactions,* pp261-177, 1995.
7. Per Hokstad, Kjell Corneliussen, " Loss of safety assessment", *Reliability engineering and system safety,* 1999. [8] T.Zhang et al, "Availability of systems with self diagnostic components applying markov model to IEC 61508-6", *Reliability Engineering & System Safety,* vol 80, pp 133-141, issue 2, May 2003.

Reliability, Safety and Hazard
Editors: P.V. Varde et al.

FMEA – THE NUCLEUS TO RELIABILITY SOLUTIONS

Anil Kumar and R. Radhika

Centre for Reliability, Government of India, Chennai

ABSTRACT

Failure Modes, Effects Analysis (FMEA) is one of the most powerful and versatile tools for reliability analysis. It forms "The Nucleus to Reliability Solutions", which can be applied to all types of systems and processes in various engineering disciplines. This technique can be applied throughout the life cycle of the product. It can be performed during design, production or maintenance phase of a product / process. FMEA finds out the root cause of the problem and helps in providing solutions. It identifies preventive or corrective measures to mitigate / minimize the effect of failure. The solutions are directed towards applying various Quality and Reliability tools -- Fault tolerance, Thermal design, Design of Experiments, Quality Function Deployment, Fault Tree Analysis, Worst Case Analysis, Accelerated Life Testing, Statistical Process Control, Reliability Centered Maintenance, Spares etc. In this paper a clear picture of how FMEA forms the nucleus to reliability solutions is brought out. Eight case studies highlighting how FMEA solutions direct towards various Q&R techniques are illustrated.

Keywords: FMEA, Reliability task, Failure mode, Design reliability, Quality, Product & process improvement, Maintenance, Spares, Safety, Accelerated testing, Solutions

1. INTRODUCTION

In this paper we are bringing out the potential of FMEA in a different perspective rather than the conventional approach. FMEA investigates the failure till the root cause is found and then identifies means or ways to eliminate it. This can be applied to a product design or a manufacturing process and can be extended to maintenance. In other words, FMEA provides engineering solutions through out the life cycle of the product.
Today, industry needs to be competitive in terms of time to market, reduced warranty failures, increased warranty period, quick fault diagnosis, reduced maintenance cost, high customer satisfaction etc. FMEA as a structured technique provides solutions to the above. The most important aspect of FMEA is the timeframe for implementation of the suggestions and recommendations of the analysis. Usually, it is meant to be "before-the-event" action not an "after-the-fact" exercise.
Apart from engineering disciplines, FMEA can be used in many walks of life - Banking, Hospitals, Training, Building, Insurance, Finance and Human etc. Many industries have successfully applied FMEA for systems and facility development. The healthcare industry has recently begun using FMEA as a process improvement tool. St.Joseph's Community hospital in West Bend, WI, USA has used FMEA to create a replacement facility aimed at reducing errors and promoting patient safety and satisfaction through design.

2. WHAT IS FMEA?

FMEA is a procedure used to identify potential failure modes and resulting effects in an effort to assess and implement measures to contain or eliminate the root causes of failure. It provides a methodology for determining the relative risk associated with each failure mode so that resources can be allocated to eliminate or reduce the likelihood of the critical failure modes. It is a dynamic tool. It is iterative in nature and leads to continuous product and process improvement. There are three components that help define the priority of failures – Severity (S), Occurrence (O), Detection (D). The priority of the problems is articulated via Risk Priority Number (RPN). This number is a product of S*O*D ranking. The value by itself should be used only to rank order concerns of the system, design, product, process and service. Figure 1. is a flow chart depicting FMEA procedure.

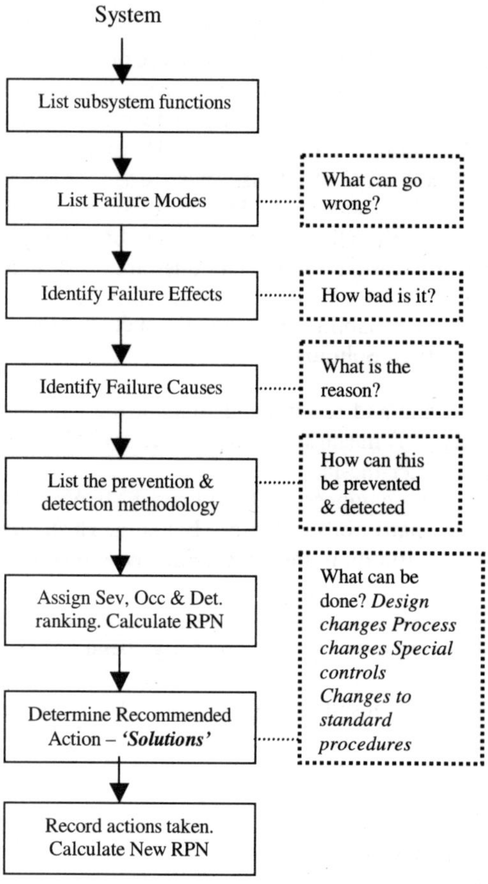

Fig. 1: FMEA Procedure

3. FMEA THE NUCLEUS

Whenever an FMEA is performed, it is presented finally in a prescribed format. If we notice in this format, we find that for each failure mode / cause the column on 'Recommended Actions' provides the solutions to minimize or eliminate the failure effect. If we closely examine this column we find that the solutions range from the application of various Quality & Reliability tasks to Maintenance tasks. It is really interesting to note that by performing an FMEA on the product, whatever the stage of the life cycle of the product we are in – be it Design, Production or Maintenance, the solutions are there right in front of you. Figure 2. depicts various Quality, Reliability and Maintenance tasks that are documented as recommended actions in the FMEA worksheet. Failure Mode and Effects Analysis constitutes the nucleus --- in which ***Failure Modes*** can be visualized as ***Neutrons*** and ***Failure Effects*** as ***Protons***. Further, the ***Solutions*** directed by FMEA forms the **Electrons**. Hence, it is quite evident that FMEA forms 'The Nucleus to Reliability Solutions'. It can be said that FMEA is in the center or the core and it directs to or points at various tasks to be followed for eventual solutions.

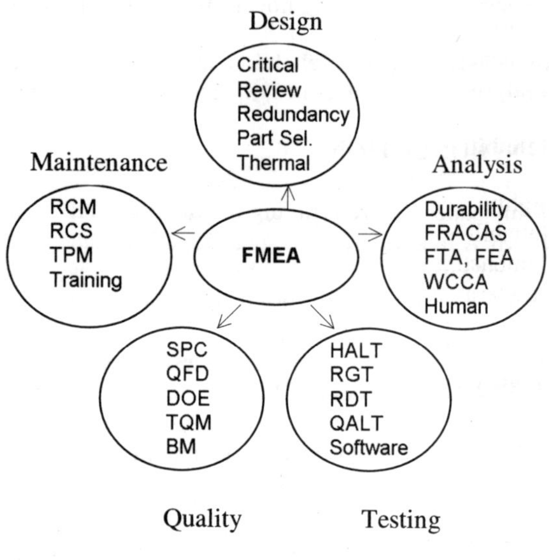

Fig. 2: The Nucleus and Solutions

4. SOLUTIONS THAT FMEA DIRECTS TO

FMEA provides a road map for selecting appropriate reliability tasks for product and process improvement. The objective of reliability tasks is to design and manufacture a reliable product in a cost effective manner. Reliability tasks should be integrated into the design of the product and manufacturing process. Early program and design decisions can have a major impact in defining overall product life cycle cost. Reliability tasks should be performed as early as possible in the product development life cycle and iterated as necessary to effectively impact the product design and process.

Design Reliability Tasks

Critical Item, Derating, Design Review, Redundancy, Parts Selection and Thermal Design.

Reliability Analysis Tasks

Durability Analysis, Failure Reporting, Analysis & Corrective Action System (FRACAS), Fault Tree Analysis (FTA), Finite Element Analysis (FEA), Worst Case Circuit Analysis (WCCA) and Human Reliability.

Reliability Testing Tasks

Highly Accelerated Life Testing (HALT), Reliability Growth Testing (RGT), Reliability Demonstration Testing (RDT), Quantitative Accelerated Life Testing (QALT) and Software Reliability.

Quality Tasks

Statistical Process Control (SPC), Quality Function Deployment, Design of Experiments, Total Quality Management and Benchmarking.

Maintenance Tasks

Reliability Centered Maintenance (RCM), Total Productive Maintenance (TPM) and Reliability Centered Spares (RCS).

5. CASE STUDIES

The case studies given below illustrate how FMEA forms the nucleus to solutions that lead towards applying various Q&R techniques. The examples have been drawn from various fields viz. product, process, human, software, service, building, hospitals, etc. in order to demonstrate the wide applicability of this tool to enhance reliability solutions in design and process. FMEA worksheets for the case studies are shown in Tables 1.1and 1.2.

Case Study 1

An extract of the FMEA analysis on printer driver mechanism and turn around drive of a 24-pin dot matrix printer are discussed here. The printer driver drives the print mechanism consisting of chain, gear, spool, sprocket, motor etc. It is seen that chain breakage was quite frequent which resulted in loss of print. This was because the print head misses limit switch at the end. Another reason for chain breakage was stretching of the chains due to normal wear. Recommendations were made to replace limit switches with heavy-duty type (*Parts Selection*), to install tension detector, cycle counter and monitor cycles (perform *RCM*). It was also recommended to provide adequate number of *Spares* for chains.

The turn around drive turns the fixture around and back to original position. Here, the important part is the sprocket. Sprocket wear-out results in loosening of sprocket causing misalignment. This is because the setscrew loosens due to shock and vibrations. It was suggested to provide *Redundancy* i.e. provide two screws; conduct *HALT* on turn around drive for shock and vibration stresses.

Case Study 2

In semiconductor wafer processing it is not uncommon for the cause of a potential failure mode to be an inherent by-product of the process itself. During deposition of silicon dioxide film, for instance, the quartz reaction chamber itself becomes a receiving surface for the deposited layer. After a certain period of time so much silicon dioxide will be deposited on the chamber walls that it will flake off and contaminate the silicon wafers. The flaking of quartz ware leads to particulate contamination reducing the strength of the oxide layer. The flaking problem was acknowledged and increased cleaning reduced the occurrence. The solution lies with *TPM* -- Increase frequency of quartz cleaning.

Case Study 3

In automobile industry corrosion is quite evident on the body (chassis) of the vehicle. This reduces the life of the vehicle (chassis) and gives poor appearance, which seriously damages the reputation of the manufacturer. This could be due to various reasons -- non-uniform paint coverage over the surface, insufficient paint thickness, inappropriate paint composition etc. Probably one of the remedies is to perform *HALT* – conduct laboratory accelerated corrosion testing.

Another reason for corrosion is related to the process of manual application of paint by means of spraying. Insufficient paint coverage leads to poor appearance due to rust formation. This can be due to clogging of spray head because of high viscosity, low temperature and low pressure. It is recommended to conduct *DOE – Design of Experiments* on viscosity vs temperature vs pressure.

Case Study 4

A traveling crane is a system comprising of two distinct sections: mechanical section consisting of components such as ropes, gears, pulleys, hook etc and an electrical section. The crane hook being the load suspension device is in itself a highly critical component in the whole system. Hook breakage leads to free-fall of the load – a catastrophic failure - resulting in death of human. The failure could be largely attributed to the poor material strength or that load exceeds material strength. A *Redesign* is a must in order to incorporate safety measures. Perform *FTA* with load free fall as the top event and incorporate design changes. Conduct *Durability* testing for the hook material. *Redundant* hook may be thought of.

Case Study 5

A cooling fan system provides cool air to the surrounding. It consists of various parts such as motor, bearing, coils, shroud, blades etc. During FMEA each part is analyzed to determine its criticality to the operation of the fan. One such critical part is the motor. Motor provides mechanical power to fan. Motor burn out or bearing failure will stop the fan resulting in loss of cooling air. This is due to overheating or lack of air circulation. The solution is to provide vent holes in the motor case (*Thermal design)* and conduct *Durability* testing for motor.

Case Study 6

This example illustrates how FMEA can be applied for 'Service' sectors – Training. The training center imparts training to customers to enable them to succeed through integrated application of knowledge-based solutions. Inadequate number of workshops or delays or problems due to lack of direction in training center will impact revenue. Implementing *QFD* to understand customer's voice, and Quality Management system – *TQM, ISO 9000* etc., could sort out the above problems.

Case Study 7

Human Errors happen everyday at work and while most are harmless, some can result in fatalities, injury or plant damage. Human Errors can occur in operation, maintenance or emergency tasks. The example given is an extract of the analysis performed for crane operation by a human operator. The task is to move a container by crane to a new location. This task is critical because the only route available for the load is over a production area. If the load drops during transit, it could break vessels and pipe work releasing high flammable gas. The recommendations to avoid *human error* is focused on relieving the time pressure problem and also improving communications between those involved in the task and making changes to the procedure for checking lifting equipment.

Case Study 8

Software FMEA determines the software failure modes that are likely to cause failure events. This example presents a bulk food transportation process, which is controlled by embedded software. The software program instructions perform many tasks like: Bucket reads sensor position and comes to a stop; Activates drum to drop bulk product; Reads scale position and stops at scale; Since the item handled is food, utmost care should be taken to prevent contamination. If

Table 1.1 FMEA Worksheet

Sl.No	Product / Process	Function	Failure Mode	Failure Effect	S	Failure Cause	O	D	RPN	Recommended Actions *'Solutions'*
Case Study: 1 1.1	Printer Driver	Drives the printer mechanism	Chain breaks	Loss of Print	6	Misses Limit switch Chain worn -out	8	7	336	a. Replace limit switch with heavy duty type – ***'Parts Selection'*** b. Install tension detector and cycle counter, Monitor cycles – ***'RCM'*** c. Provide Spares for chain and supporting block -- ***'RCS'***
1.2	Turn around drive	Turns the fixture around and back to original position	Sprocket worn out	Sprocket becomes loose causing misalignment	5	Set screw loosens due to shock and vibrations	7	6	210	a. Routine inspection – ***'RCM'*** b. Provide two screws ***'Redundancy'*** c. Spares – chain & sprocket - ***'RCS'***
Case Study: 2 2.1	Deposition of oxide layer	To deposit defect free oxide layer of desired thickness	Particulate contamina-tion	Reduces strength of the oxide layer	6	Flaking of quartz ware	7	3	126	a. Increase frequency of quartz cleaning. – ***TPM'***
Case Study: 3 3.1	Manual paint spray	Manual application of paint	Insufficient paint coverage	Poor appearance due to rust	7	Spray heads clogged due to high viscosity, low temp and pressure	5	5	175	a. Conduct Design of Experiments on viscosity vs temp. vs pressure. –***'DOE'***
3.2	Two wheeler automobile chassis	Covers the body and provides protection from severe environment	Corrosion	Deteriorated life of the chassis due to corrosion. Poor appearance	7	Insufficient paint thickness over the chassis	6	7	294	a. Conduct laboratory accelerated corrosion testing – ***'HALT'*** b. Conduct Design of Experiments on paint thickness – ***'DOE'***
Case Study: 4 4.1	Crane Hook	Suspend load	Hook breakage	Fall of load (Catastrophic – may lead to death)	10	Poor material strength Load exceeds material strength	3	8	240	a. Redesign - incorporating safety measures – ***'FTA'*** b. List Hook as ***'Critical Item'***. c. Conduct ***'Durability testing'***
Case Study: 5 5.1	Cooling fan motor	Provide mechanical power to fans	Motor burn out, bearing or brush failure	Loss of cooling	5	Overheating, lack of air circulation	2	5	50	a. Vent holes in motor case –***'Thermal Design'*** b. Conduct ***'Durability Test'***

Table 1.2 FMEA Worksheet

Sl. No	Product / Process	Function	Failure Mode	Failure Effect	S	Failure Cause	O	D	RPN	Recommended Actions *'Solutions'*
Case Study: 6	Training Center	Impart training	Inadequate number of workshops	Impacts revenue	6	Lack of directions on topics	5	5	150	a. Market survey – ***'QFD'***
6.1			Delays & problems in training center	Impacts revenue	6	No project manager assigned	7	5	210	a. Implement Quality Management System – ***'TQM"***
6.2	Training Center building	Place and facilities for training	Non-compliance with city code	City corporation shuts down training center on audit	8	No. of restroom in building was out of compliance.	4	5	160	a. ***Redesign*** to comply with city code regulations
6.3	Training Center building	Place and facilities for training	Injury to human	Lawsuit??	10	Tripping hazards, Corner / edges are sharp	3	5	150	a. Perform FTA for human injury and incorporate safety measures – ***'FTA'***
Case Study: 7 7.1	Moving container by crane	Attach lifting equipment	Use incorrectly rated sling	Load drops because sling is too light for the load	10	Supervisor did not check that lifting equipment is appropriate	2	4	80	a. Equipments are colour coded and the codes are well known. -- ***Training*** b. Test lift (raising load 0.5m initially to check that the load hangs level and that lifting equipment is sound). -- ***Testing*** c. Appoint two persons – one for checking and the second for approving – '***Reduces Human Error'***
7.2	Moving container by crane	Traverse container	Move too rapidly	Loose control of load and crash into equipment Load shifts within the container	9	Crane is set at high speed by key system which supervisor retains	6	5	270	a. Operator competence needs to be improved. – ***'Training'*** b. Relieve the time pressure problem and improve communications between those involved in the task. – ***'Human Reliability'***
Case Study: 8 8.1	Bulk Food Transporta-tion Process	Software reads bucket position	Wrong bit in the memory position	Buckets move empty and pile on each other	7	Error handling for wrong bit not incorporated	4	5	140	a. Incorporate error detection code – ***'Software Reliability'***
8.2	Bulk Food Transporta-tion Process	Activates drum to drop bulk product	Tilt mechanism switch may fail after the drum is tilted	Product over spills on production floor	10	Improper switch No indication of failure.	5	7	350	a. Design additional switch-***'Redundancy'*** b. Software to flag failure of any switch – ***'Software Testing'***

the software malfunctions the process gets disrupted and there is even chances of over spilling of food on production floor. Incorporate error detection code – *Software Testing.*

6. CONCLUSION

We have seen through the case studies, FMEA links to task oriented solutions relating to design, production, quality, maintenance, human and software. The recommended tasks of the analysis MUST be implemented with full vigor and sincerity to derive maximum benefit of FMEA. This will ensure robust design and reliable product, improved and controlled manufacturing process, optimized & cost effective maintenance. **"FMEA Is A Wonderful Coordinating Tool."**

7. REFERENCES

1. Mark Hatty and Norm Owens, Potential Failure Modes and Effects Analysis – A Business Perspective Motorola Inc., Quality Engineering, 7(1), p169-p186, 1994-95.
2. R.Subburaj, Total Quality Management, Tata McGraw-Hill Publishing Company Limited, New Delhi, (2005), ISBN 0-07-059090-7.
3. Potential Failure Mode and Effects Analysis – Reference Manual, Third Edition, July 2001, (SAE J-1739), QS-9000.
4. John G.Reiling and Barbara L.Knutzen with Mike Stoecklein, FMEA –The Cure for Medical Errors, Quality Progress, p67-p71, August 2003.
5. Rai Chowdhary, FMEA – Saves Business from Shutdown, Quality and Productivity Research Conference, Austin, Texas, May 22-25, 2001.
6. D.H.Stamatis, Failure Mode and Effect Analysis: FMEA from Theory to Execution, Productivity Press (India) Pvt Ltd, Madras, first Indian edition 1997, ISBN81-85985-30-8.
7. Blueprints for Product Reliability, RBPR-1 & 3, Reliability Analysis Centre, NY, USA. 1996.
8. Human Factors, briefing notes no-12, The Institute of petroleum, London.

Reliability, Safety and Hazard
Editors: P.V. Varde et al.

RELIABILITY ANALYSIS OF WATER DISTRIBUTION NETWORK –A CASE STUDY

T. Nagaraj and B.V. Mudgal

Centre for Water Resources, Anna University, Chennai 600 025

ABSTRACT

The reliability of a Water Distribution System (WDS) is important as it not only gives indications about the quality of service and management practices adopted, but also means to improve it. The existing WDS network at Anna University is identified to be a branched pipe network with one supply node and multiple demand nodes. The system is divided into sub-systems for each demand node and supply node. Network analysis is done using Hazen-Williams formula to determine the frictional losses. Pipes are the main components considered for the reliability analysis of the sub-systems. Poisson distribution is assumed to compute the probability of failure for each pipe and the reliability is computed using fault-tree-analysis.

Keywords: Water distribution system, Fault-tree-analysis

1. INTRODUCTION

A water distribution system is designed to provide adequate quantity of water at demand nodes. The capacity of the system is its resistance and the actual flow in the network is the load on the system. The carrying capacity of the pipe network mainly depends upon the condition of the pipes and its surface roughness. The flow in the system is a variable depending upon the hourly, daily, weekly and seasonal demand. The reliability of the system can be estimated from the interaction of the load and resistance. The reliability of water distribution system is composed of (a) hydraulic reliability and (b) mechanical reliability. Hydraulic reliability refers to consistently providing the specified quantity of water at required pressures and at the time and location desired, while mechanical reliability relates to the operability of the equipment (Mays and Tung, 1993).

The evaluation of reliability of a water distribution network is complex as the total parameters are large. Some of these parameters are; quality and quantity of water at the source, failure rates of pumps, supply outages, flow capacity of transmission mains, which in turn varies with the surface roughness of the pipes, pipe breaks, valve failures, pipe breaks, variation in hourly, daily, weekly and seasonal demands as well as demand growth over the years.

The objectives of the present study are:

(a) To determine the demand at each node based on the population and hence to determine the pressure-head required at each of the nodes by pipe network analysis.

(b) To determine the reliability of each pipe (component reliability) and hence evaluate the system reliability using fault-tree-analysis.

2. WATER DISTRIBUTION NETWORK ANALYSIS

The existing water distribution system of Anna University Main Campus at Chennai is taken up for the present study. This system is identified to have one supply node (pumping station) and multiple demand nodes with no interconnections between the various junction nodes. Thus the network is of branched type, with pipe sizes ranging from 25 mm to 175 mm diameter. The population of the campus is estimated to be nearly five thousand including students, teaching and non-teaching staff. The topography of the campus is generally plain with marginal undulations and the ground level varying within elevation of 5.5 m to 10 m above mean sea level. The campus is spread over an area of 100 hectares with about 50% built-up area.

The discharge and the pressure required at each of the demand nodes is known based on the population and the minimum head required. The water is distributed intermittently with selected demand nodes supplied by rotation. The analysis is carried out starting from a demand node and progressing in the backward direction to the supply node (backward analysis). Neglecting the minor losses, the major loss i.e., the head loss due to friction is calculated using the Hazen-Williams formula

$$h_f = 10.654\left(\frac{Q}{C_{HW}}\right)^{1.85}\left(\frac{L}{D^{4.87}}\right) \qquad (1)$$

Where, h_f is the head loss due to friction (m), Q is the discharge (m^3/s), C_{HW} is the Hazen –Williams coefficient, L the length and D the diameter of the pipe (m). The computed head loss for the pipe network is shown in Table.1. The schematic water distribution network is shown in Figure.1

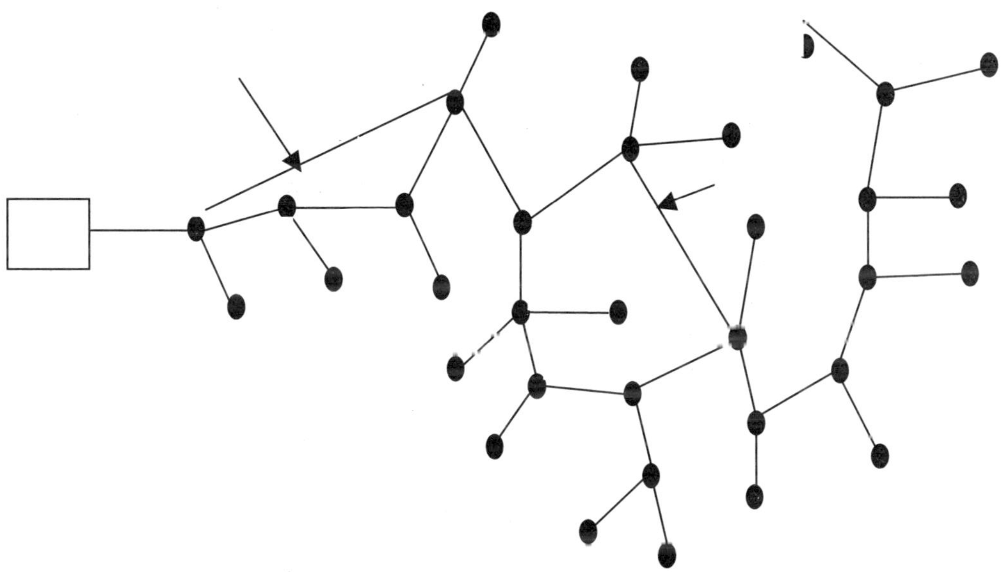

Fig. 1: Schematic sketch of the existing water distribution network at Anna University

3. RELIABILITY ANALYSIS

Component Reliability

The reliability $\alpha(t)$ of a component is defined as the probability that the component experiences no failure during the time interval zero to time t, given that it is new or repaired at time zero

$$\alpha(t) = \int_t^{\infty} f(t)\,dt \qquad (2)$$

Where f(t) is the probability density function of the time-to-failure of the component (Mays and Tung, 1992)

In the present analysis, only the failure of the pipes are considered in evaluating the reliability. The probability of failure of individual pipes is calculated using the Poison probability distribution (Goulter and Coals, 1986; Su et.al., 1987).

Although pipe breaks depend upon several factors like, diameter and thickness of pipes, material, age, surrounding soil type, water pressure etc., the pipe diameter is considered to be the most important. Goulter and Coals [1] suggest the following expression

$$P_i = 1 - \exp(-\beta_i) \qquad (3)$$

Where P_i is the failure probability of pipe i, and β_i is the expected number of failures for pipe i. β_i is given by

$$\beta_i = r_i\, l_i \qquad (4)$$

Where r_i and l_i are the expected number of failures per year per unit length of pipe i, and length of the pipe i respectively.

The reliability of pipe i is given by

$$\alpha_i = \exp(-\beta_i) \qquad (5)$$

The results of the pipe component reliabilities are given in Table.1.

Water Distribution System Reliability

As the water distribution network is of branched type with one supply node and multiple demand nodes, the entire system is decomposed into multiple sub-systems with one supply node and one demand node. The reliability of each of the sub-systems is evaluated using the fault-tree-analysis. Since, the pipes in a branched network are in series we get mainly the OR nodes in the fault tree. Thus the failure of any one of the k pipes in the sub-system would result in the failure of the sub-system. The probability of union of k events is given by

$$P\left(\bigcup_{i=1}^{k} A_i\right) = \sum_{i=1}^{k} P(A_i) - \sum_{i}^{k}\sum_{\neq\, j}^{k} P(A_i, A_j) + \sum_{i}^{k}\sum_{\neq\, j}^{k}\sum_{\neq\, l}^{k} A_i, A_j, A_l - \cdots + (-1)^{k+1} P(A_1, A_2, \ldots, A_k) \qquad (6)$$

Assuming the events to be independent i.e., the failure of one pipe is independent of the other, the probability of failure for each of the sub-systems is evaluated from the above equation using a computer code. The results are shown in Table.2

It is seen that the sub-system containing node-21 has the least reliability of 0.353. This is taken to be the reliability of the entire system. The reliability of the system is proposed to improve by converting part of the branched network to looped network. Hence, sub-links are proposed between nodes 13-25, and 28-34. It is found that the reliability of the system increases to 0.756 from 0.353.

Table 1 Water Distribution System Network analysis and Pipe Component Reliability (α_i) computations

Pipe No	D mm	L m	Q m^3/s	C_{HW}	h_f m	P_i	α_i
P1	80	78.4	0.0055	110	1.987	0.107	0.893
P2	50	3.2	0.0053	110	6.203	0.005	0.995
P3	80	27.5	0.0107	110	2.413	0.039	0.961
P4	80	131.5	0.0064	110	4.423	0.174	0.826
P5	80	30.6	0,0096	110	2.197	0.043	0.957
P6	50	40.3	0.0032	110	3.778	0.059	0.941
P7	80	27.3	0.0107	110	2.395	0.039	0.961
P8	40	65.9	0.0045	140	21.585	0.097	0.903
P9	80	5.9	0.0107	110	0.518	0.009	0.991
P10	40	27.7	0.0056	110	21.508	0.042	0.958
P11	80	39.4	0.0107	110	3.457	0.056	0.944
P12	80	27.9	0.0045	110	0.493	0.040	0.960
P13	80	110.2	0.0107	110	9.669	0.148	0.852
P14	80	23.5	0.0032	110	0.219	0.034	0.966
P15	80	38.2	0.0046	110	0.71	0.054	0.946
P16	80	15.9	0.0078	110	0.78	0.023	0.977
P17	125	137.5	0.0107	100	1.638	0.169	0.831
P18	80	89.1	0.0064	110	2.96	0.121	0.879
P19	125	68.5	0.0107	100	0.816	0.088	0.912
P20	80	11.5	0.0057	110	0.314	0.017	0.983
P21	80	267.5	0.0059	110	7.698	0.322	0.678
P22	125	157.2	0.0116	100	2.161	0.191	0.809
P23	50	63.9	0.0043	110	10.214	0.091	0.909
P24	50	209.2	0.0041	110	22.235	0.190	0.810
P25	80	128.3	0.0084	110	4.591	0.167	0.833
P26	175	153.4	0.0116	100	0.41	0.149	0.851
P27	80	45.6	0.0051	110	1.002	0.064	0.936
P28	175	114.5	0.0116	100	0.306	0.113	0.887
P29	80	192.6	0.0056	110	5.097	0.244	0.756
P30	175	89.2	0.0116	100	0.238	0.089	0.911
P31	62	60.4	0.0064	110	6.943	0.087	0.913
P32	175	60.5	0.0116	100	0.162	0.062	0.938
P33	80	61.2	0.0054	110	1.479	0.085	0.915
P34	175	199	0.0116	100	0.604	0.189	0.811

4. CONCLUSIONS

From the present study, it is concluded that the reliability of a branched water distribution network can be taken to be the least reliability of

Table 2 Reliability computations at demand nodes using fault tree analysis

S. No.	Demand Node No	Probability failure of Demand node	Reliability of Demand node
1	1	0.540	0.460
2	2	0.571	0.429
3	4	0.536	0.464
4	6	0.574	0.426
5	8	0.565	0.435
6	10	0.550	0.450
7	12	0.530	0.470
8	14	0.618	0.382
9	15	0.610	0.390
10	18	0.630	0.370
11	20	0.624	0.376
12	21	0.647	0.353
13	23	0.551	0.449
14	24	0.463	0.537
15	27	0.488	0.512
16	29	0.552	0.448
17	31	0.391	0.609
18	33	0.338	0.662

the sub-systems and can be evaluated by fault-tree-analysis. The reliability of the entire system can be improved by converting a branched network into a looped network considering the demand node which has the least reliability.

The present work can be improved by considering various other components like pumps, valves etc.

5. REFERENCES

1. Goulter, I.C. and Coals, A.V. (1986). "Quantitative approaches to reliability assessment in pipe networks." J. Transp. Engrg. ASCE, **112**(3), pp. 287-301.
2. Mays, L.W. and Tung, Y-K. *Hydrosystems Engineering and Management.* McGraw Hill, NY, 1992.
3. Su, Y.C., Mays, L.W., Duan, N. and Lansey, K.E. (1987). "Reliability-based optimization model for water distribution systems." J. Hydraul.Engrg., ASCE, **114**(12), pp. 1539-1556.

Reliability, Safety and Hazard
Editors: P.V. Varde et al.

PROBABILITY OF EXCESSIVE LEAK RATE THROUGH MULTIPLE DEFECTS IN DEGRADED NUCLEAR STEAM GENERATOR TUBES

Leon Cizelj

"Jožef Stefan" Institute, Ljubljana, Slovenia

ABSTRACT

A probabilistic approach aimed at prediction of the probability of excessive leakage through multiple flaws in tubes is proposed in the paper. The excessive leakage is assumed to occur when the sum of all individual leak rates exceeds the predefined acceptable value. Some estimators used in the field to obtain conservative estimates of the 95th percentile (e.g., upper limit) of the total leak rate are given. The conservatism of those estimators is quantified through failure probability in the two numerical examples analyzing the behaviour of the model for both small and large flaw sizes.

1. INTRODUCTION

Through-wall flaws in ductile pressurized components (e.g., pressure vessels, piping etc.) may at appropriate conditions lead to detectable leaks long before the structural integrity of the component is challenged (e.g., Roos 1999). Through-wall leaks should in most cases be treated as a reliable call for a corrective action (repair or replacement). However, in some specific cases continued operation with controlled leaks within the safety, legal, economical or other constraints might be acceptable.

The Alloy 600 mill annealed (MA) steam generator tubes in pressurized water reactors may serve as a good example of allowable continued operation with limited leakage. Steam generator tubes represent a major fraction of the reactor coolant pressure boundary surface area and are susceptible to corrosion damage, which may have detrimental consequences to the plant safety and reliability (Shah and MacDonald 1993). Two potential failure modes of the tubes have received special attention: (i) single or multiple tube rupture and (ii) excessive primary-to-secondary leakage without tube rupture (Banic et al. 1996).

A sufficient safety margin against tube rupture used to be the basis for a variety of maintenance strategies, which were developed to maintain a suitable level of plant safety and reliability (see (Banic et al. 1996) and (Shah and MacDonald 1993) for more details). This topic will not be pursued further here. It will merely be noted that some of the maintenance strategies justified sufficient margin also for the tubes with through-wall flaws. Consequently, several through-wall flaws may remain in operation and potentially contribute to the total primary-to-secondary leak rate. Cases with up to 2000 flaws within a single steam generator (typically, in the order of 5000 tubes of about 20 mm in diameter and 1 mm wall thickness) have been reported to operate successfully (Cuvelliez and Roussel 1997). Quantification of the expected leak rate and demonstration of the available safety margin against the associated radiological consequences was required in such cases (Nuclear Regulatory Commission 1995).

The quantification of the total leak rate through a single component (for example steam generator) essentially requires summation of leak rates through individual through-wall flaws. The information about the number and other relevant characteristics of the individual flaws is in practical situations obtained from non-destructive examinations with significant data scatter inherent to the method. In addition, the empirical correlations and the uncertainties attached to the leak rate model are also significant sources of variability in the calculated leak rate even through well-defined individual defects. Our goal is therefore to estimate a sum of a random number of random variables.

Appropriate probabilistic method has already been proposed by the authors (Cizelj et al, 1998), aiming at predicting the probability of excessive leaks in SG tubes, i.e. the probability that the sum of individual leak rates exceeds the allowable leak rate (given operating conditions and the distribution of defects in the steam generator). It was based on the analogy with probabilistic models estimating the tube burst probability of tubes (Cizelj, Dvoršek, and Mavko 1996), (Cizelj, Mavko, and Vencelj 1996). In this paper, the probabilistic approach is revisited and generalized towards other potential applications. Some estimators of the 95th percentile of the total leak rate are also formulated and their properties discussed.

Two realistic examples differing in the size and number of defects are provided to illustrate the performance of the proposed probabilistic approach and estimators of the 95th percentile. Input data represents the outside diameter stress corrosion cracks in the 3/4 inch Alloy 600 MA tubes of already replaced Belgian Doel-4 and Slovenian Krško steam generators are addressed.

2. PROBABILITY OF EXCESSIVE LEAKAGE

Main assumptions

Throughout this section we will assume the existence of a two-state model, which for each defect characterized by V_i estimates conditional individual leak rate Q_i:

$$Q_i = \begin{cases} Q(V_i, \Delta p_i, \Delta T_i, \ldots) & if \quad P_i > 0 \\ 0 & if \quad P_i = 0 \end{cases}, \qquad (1)$$

and probability P_i that this particular defect will leak ($Q_i > 0$):

$$P_i = P(V_i, \Delta p_i, \Delta T_i, \ldots) \qquad (2)$$

The essential parameters in eqs. (1) and (2) include, but may not be limited to the defect characteristics denoted by V_i (e.g., defect size, roughness, complexity of the leak path etc.) and variables describing the fluid conditions such as pressure difference Δp_i and temperature difference ΔT_i. Their respective stochastic properties are described with the probability density functions $f_i(V_i)$, $f_i(\Delta p_i)$ and $f_i(\Delta T_i)$. With eqs. (1) and (2) it is now possible to derive the probability density function of the conditional individual leak rate $f_i(Q_i)$.

It is beyond the scope of this paper to discuss further details on the leak rate modeling.

The total leak rate Q_T through n defects is then obtained as sum of individual contributions from a random number of l leakers and $n-l$ leak tight defects:

$$Q_T = \sum_{i=1}^{n} Q_i = \sum_{i=1}^{l} Q_i\Big|_{P_i > 0} + \sum_{j=1}^{n-l} Q_j\Big|_{P_j = 0} \qquad (3)$$
$$\text{where } Q_j\Big|_{P_j = 0} = 0$$

The discrimination between leaking and leak tight defects is made according to the rule in eq. (2). Some additional useful and reasonable assumptions were used:

- all individual leak rates Q_i are members of the same population. This assumption requires for example that the same degradation mechanism is responsible for the development of all defects considered.
- individual leak rates are statistically independent. This is fulfilled in fluid systems where any change in the energy stored in the system caused by the total leak rate is negligible as compared to the same system without leaks. The pressure, temperature and mass flow rates in such system shall therefore not depend on the total leak rate.

Probabilistic model

It is reasonable to expect that continued operation with controlled leak rates be limited by an acceptable leak rate Q_{MAX}. Its value may be governed by safety, legal, economical or other constraints and may be either constant or a stochastic variable.

It is then useful to define that the event "failure" takes place whenever "the total leak rate Q_T exceeds the allowable leak rate Q_{MAX}". Recall that the total leak rate Q_T is simply the sum of all individual leak rates Q_i (eq (3)). The failure probability P_f is then given by:

$$P_f = P(Q_T \geq Q_{\text{MAX}}) = P\left(\sum_{i=1}^{n} Q_i \geq Q_{\text{MAX}}\right) \quad (4)$$

The individual leak rate through a single defect has already been described as a random variable, denoted by Q_i (eq.(1)) with probability density function $f_i(Q_i)$.

Using the well-known estimators from statistical reliability theory (Madsen, Krenk, and Lind 1986), the failure probability P_f can be calculated by the integration of the probability content of the failure space $(Q_T > Q_{\text{MAX}})$:

$$P_f = \int_{g(Q_1,Q_2,\ldots,Q_n)\leq 0} \prod_{i=1}^{n} \left(f(Q_i)dQ_i\right) \quad (5)$$

$g(Q_1, Q_2, \ldots)$ represents the failure function, which is by definition positive in the safe space and negative in the failure space ($Q_T > Q_{MAX}$). Therefore, it is given here as:

$$g(Q_1, Q_2, \ldots, Q_n) = Q_{\text{MAX}} - \sum_{i=1}^{n} Q_i \quad (6)$$

Numerical solution

For the purpose of this paper, eq (5) was solved using direct Monte Carlo simulation. Only the leakers (recall eq (2)) need to be considered in the simulation. A reasonably reliable estimate of failure probability in the order of 10^{-2} required about 1000 simulations.

3. MODELING OF THE LEAK RATES

The example pursued in this paper is application of repair criteria dedicated to defects recognized as Outside Diameter Stress Corrosion Cracking - ODSCC - at the tube support plate intersections (Nuclear Regulatory Commission 1993), which were detected in the Alloy 600 MA steam generator tubes in nuclear power plants with pressurized water reactors worldwide (Shah and MacDonald 1993).

The state-of-the-art modeling of leak rates through individual ODSCC defects is based on regression analysis of experimentally determined leak rates as a function of flaw size (Nuclear Regulatory Commission 1993) (Electric Power Research Institute 1993) at selected thermodynamical conditions representing loading during a hypothetical design basis accident (e.g., Δp_i and ΔT_i are assumed constant and identical for all defects!). This is, of course, connected with relatively large uncertainties.

This section briefly summarizes the main characteristics of the defects followed by the discussion on the defect size, probability of leak given the defect size and conditional leak rates through individual defects.

ODSCC defects in SG tubes

The ODSCC defects at tube support plates are usually seen as complex networks of cracks. The network is initiated at the outside surface of the tube in the chemically aggressive crevice between the tube and tube support plate and then slowly branches through the wall thickness. The prediction of leak rates through such networks is rather uncertain, even for defects with well-defined morphologies. More detailed discussion of sources of uncertainty is given elsewhere (Electric Power Research Institute 1993).

The extent of information, which can be obtained from the state-of-the-art inspection methods, may also represent a serious constraint. For ODSCC in particular, the eddy current technique is believed to have reasonably good detection capabilities (Harris 1996). On the other hand, the information about crack morphology obtained from the inspections is currently limited to length and orientation of single cracks. The networks of cracks, which are typical for ODSCC, cannot yet be accurately described.

Thus, a relatively simple and robust model which relies on experimental analysis of a sample of tubes including those pulled from operating steam generators was developed and proposed by EPRI (Electric Power Research Institute 1993) that is further detailed below.

Size of an ODSCC defect

The field inspections of ODSCC defects are performed by bobbin coil probes (eddy current technique). The result of the inspection, which is assumed to indicate the defect size, is the amplitude of the signal (measured in Volts) obtained from the bobbin coil. As a very general reference, the signal amplitude indicates the volume of the lost material in the sense that its value depends on the crack length, crack depth and crack opening.

The models described below are based on experimentally determined leak rate data from the steam generator tubes. Some of the tubes analysed were pulled from operating steam generators, other defects were grown in the laboratory conditions. The defect sizes were measured with appropriate bobbin coil method prior to the measurement of leak rates. It is important to note that a well-defined calibration estimator should be used together with the bobbin coil eddy current technique in order to comply with the model outlined in this Section.

In operation, the defect size is generally a time dependent variable. In this analysis, two points in time are of special concern and essentially define the value of defect size: (1) beginning (BOC) and (2) end (EOC) of the cycle between two consecutive inspections. Thus, the prediction of the EOC defects sizes includes stochastic combination of BOC defect sizes and defect growth.

At present, the growth of the crack network is assumed to be described by the voltage increase between two successive inspections of the ODSCC defects. For the purpose of the present analysis, it may be predicted from the statistical analysis of results of consecutive inspections (Cizelj et al. 1995).

Probability of leak through a defect

The probability of leaking $P_i = P(V_i)$ is determined given the size of the defect V_i:

$$P(V_i) = \frac{1}{1-\exp\left[-\left(\eta_0 + \eta_1 \log(V_i)\right) + z\sigma_\eta\right]} \qquad (7)$$

The parameters η_0 and η_1 were obtained by regression analysis of measured values (Electric Power Research Institute 1993). The uncertainty of the regression is accounted for in parameters z and σ_η. The measure of uncertainty $z\sigma_\eta$ is a random variable, which describes the scatter around the regression curve (logit function).

The physical interpretation of eq (7) is rather simple. It namely splits the population of all defects in two parts: (1) those that exhibit measurable leaks and (2) those, which remain leak-tight. As an example, consider an infinite number of defects of size V_i. Then, a fraction of $P(V_i)$ is expected to leak while the rest $1-P(V_i)$ will not leak. It is however beyond the scope of this paper to enter the physical reasons for leak tightness of a significant number of defects.

Leak rate through an individual defect

For the subpopulation of leakers, the individual leak rates Q_i can be obtained given the defect sizes V_i:

$$\log(Q_i) = b_0 + b_1 \log(V_i) + \varepsilon \qquad (8)$$

The parameters of the regression model (b_0, b_1 and ε) were obtained from measured values and are given elsewhere (Electric Power Research Institute 1993). The uncertainty of the regression model is quantified by b_0, b_1 and ε. Correlated random variables b_0 and b_1 denote the intercept and slope of the regression line. ε denotes the scatter around the regression line.

4. ESTIMATORS OF THE TOTAL LEAK RATE (95TH PERCENTILE)

For the field use, EPRI (Electric Power Research Institute 1993) proposed a relatively simple estimator aiming at a deterministic estimate of the 95th percentile (e.g., upper limit) of the total leak rate with the property:

$$P_f = P\left(\sum_{i=1}^{n} Q_i \geq Q_T\right) \leq 5\% \qquad (9)$$

The EPRI estimator and two additional estimators proposed by (Cuvelliez and Roussel 1997) are described below. Further discussion of their properties is given in the section of numerical examples.

EPRI 95%/95% Estimator

The whole sample of defects detected and sized by the non-destructive examination is divided in classes of size 0.1 V. The number of defects in class k [k-0.1 V, k V] is denoted by m_k. A representative individual leak rate for the particular defect size is then defined as:

$$Q_k\big|_{95\%} \approx \left[P(V_k)Q(V_k)\right]_{961|1000} \quad (10)$$

The subscript 961|1000 means that 1000 values of $\left[P(V_k)Q(V_k)\right]$ are generated using Monte Carlo simulation (see eqs (7) and (8)) and finally sorted in ascending order. The 961st value of this ordered set is chosen as a representative value for the entire class of defects. This is assumed to yield an estimate of the 95% limit with 95% confidence (Electric Power Research Institute 1993).

Eq (10) essentially averages the leak rates over both leakers and non-leakers. This approach is reasonable given a large number of defects of approximately the same size or samples with $P(V_k) \approx 1$. On the other hand, use of eq (10) may lead to an underestimation of Q_T when applied to small samples or even individual leaks ($m_k = 1$) with $P(V_k) << 1$. This is indeed the case in some practical applications as illustrated in the numerical examples.

The total leak rate through the steam generator Q_T is given by summing contributions from all n_{class} classes:

$$Q_T\big|_{95\%} \approx \sum_{j=1}^{n_{class}} m_j \left[P(V_j)Q(V_j)\right]_{961|1000} \quad (11)$$

The computational effort involved in evaluation of eq (11) is simply given by 1000 n_{class} evaluations of eqs (7) and (8) followed by 1000 n_{class} sorts to obtain the 961st value.

AVN Estimators

The AVN estimators (Cuvelliez and Roussel 1997) are aiming at extending the statistical analysis further than the determination of the 95th percentiles of individual leak rates (EPRI, eq (11)). In order to do that, the total leak rate is considered as a random variable and the upper limit of the total leak rate is determined from its probability distribution.

AVN estimator #1

Assume that the defects are divided in small classes as in the EPRI estimator (eq. (11)). Then, the 95% limit of the total leak rate is defined from the set of calculated total leak rates rather than from a set of class limits:

$$Q_T\big|_{95\%} \approx \left[\sum_{k=1}^{n_{class}} P(V_k)Q(V_k)m_k\right]_{961|1000} \quad (12)$$

The estimate of 95% limit of the total leak rate is the major difference as compared to the EPRI estimator. The practical consequence of this difference is that eq. (12) would always yield lower or at least equal estimates of Q_T than eq. (11), which can be explained by the Chebishev inequality. In other words, the AVN#1 estimator uses less conservative way to estimate the 95% limit of the Q_T than the EPRI estimator.

The computational effort is reduced with respect to EPRI estimator (eq (11)). Although there are 1000 n_{class} individual leak rate calculations, there is only one sorting of results (EPRI requires n_{class} sorts).

The AVN estimator #2

This estimator excludes grouping of defects in voltage classes ($m_j = 1$). Each individual leak rate is considered as a random variable and the total leak rate is then the sum of these individual leak rates over the whole set of n defects:

$$Q_T\big|_{95\%} \approx \left[\sum_{i=1}^{n} P(V_j)Q(V_j)\right]_{961|1000} \quad (13)$$

The AVN estimator #2 (eq (13)) is essentially the limit of the AVN estimator #1 (eq (12)) given $n_{class} \to n$. A consequence of interest in this analysis is again consistently lower estimate of the Q_T obtained by AVN#2 than by both AVN#1 and EPRI estimators. Using individual data points rather using the upper limits of data classes as representative defect sizes causes this particular reduction of conservatism.

The computational effort of the AVN #2 estimator is larger than by EPRI and AVN#1 estimators and is proportional to 1000 n .

5. NUMERICAL EXAMPLES

Two numerical examples were chosen to illustrate the performance of the proposed probabilistic approach and estimators of the 95th percentile of the total leak rate. They are based on the inspection data obtained from Slovenian

Krško and Belgian Doel-4 steam generator tubes (3/4 inch tubes made of Alloy-600 Mill Annealed).

Krško

Data

The distribution of defect sizes at the BOC is given in Fig. 1. In the calculations, empirical and fitted lognormal distributions of defect sizes at BOC were used. The differences between failure probabilities obtained with different defect size distributions were comparable to the statistical noise of the Monte Carlo simulation. Nevertheless, the empirical distribution consistently leads to higher values of the probability of exceeding the allowable leak rate and was therefore selected as representative input data model for subsequent analysis. The total number of defects detected was 492.

Allowances for defect growth (with 52,6 % of defects exhibiting nonnegative growth) and measurement errors were also provided and yielded the defect size at EOC. The Krško power plant specific values were used (detailed description is given in (Cizelj et al. 1995)).

The statistical stability of the total leak rate estimate given by (eq (11)) was analyzed using Monte Carlo simulation (Table 1 and was considered reasonable. A big scatter is however observed in maxima of sums. This scatter is mainly caused by the large influence of the uncertain regression parameters in the leak rate model (eqs (7) and (8)). The absolute maximum total leak rate (lower right cell of Table 1) always exceeded the 95% upper limit predicted by the EPRI estimator (eq (11)) by a few orders of magnitude.

Based on results in Table 1, the value of 25 l/h was chosen to be representative for the EPRI estimator (eq (11)) for comparison with probabilistic calculations. Representative values for both AVN estimators were chosen in an analogous way to be: 20 l/h for AVN #1 (eq (12)) and 15 l/h for AVN #2 (eq (13)).

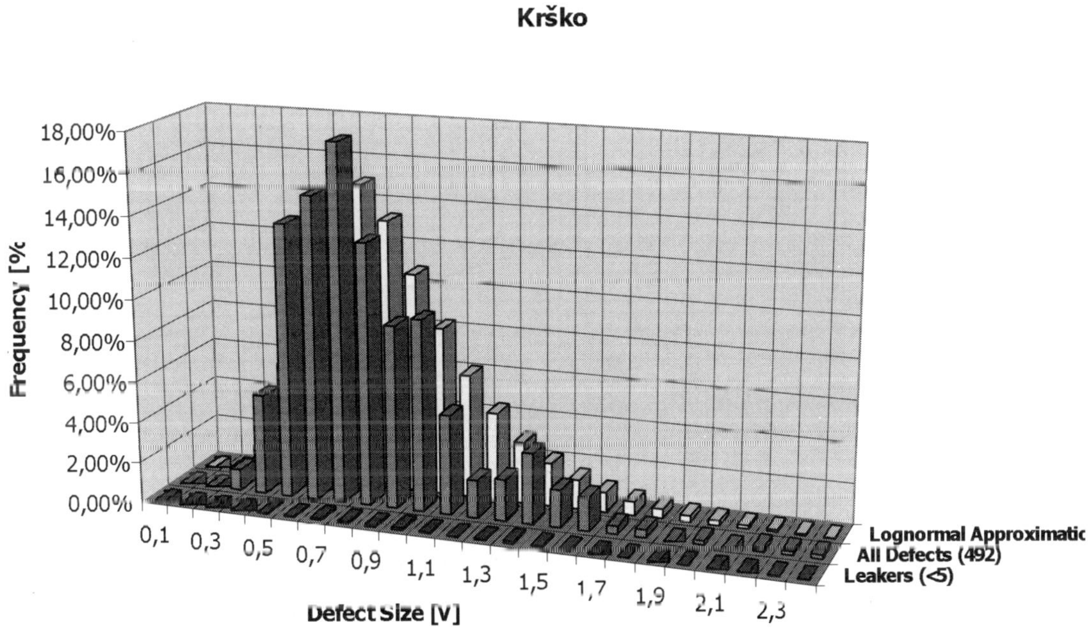

Fig. 1: Distribution of BOC defect sizes-Krško plant

Results

Fig. 2 depicts probability of excessive leakage as a function of allowable total leak rate. Three different curves are given to illustrate the effect of defect progression: BOC (no defect progression), EOC with 52,6% of defects exhibiting growth (which is consistent with field observations in Krško) and EOC with 100% of defects exhibiting growth. The Krško specific growth rate is used in the analysis (Cizelj et al. 1995) and is shown to contribute less than one order of magnitude.

The statistical standard error of Monte Carlo simulations (represented by error bars in Fig. 2) is in the order of 0.1% and has negligible influence on the quality of results.

Simulations revealed that only about 5 (1%) out of 492 potential leakers would really leak given assumed model (recall eqs (7) and (8)) and postulated hypothetical accidental conditions - steam line break. A distribution of leakers as obtained from the Monte Carlo simulations (eq (7)) is depicted inFig. 1. Although this is not clearly seen, larger defects tend to leak more frequently which is in accordance with eq (7).

The results of the three estimators are visualized by vertical lines at appropriate values of total leak rates (Fig. 2). Thus, for EPRI estimator, the failure probability is defined by the intersection of curves denoted by EOC, 52,6% growth and EPRI (25 l/h). Its value is about 3.5%. The AVN estimators #1 and #2 tend to 5 and 11%, respectively.

The main conclusion is that the EPRI estimator yields about the 95th percentile of the total leak rate value. Thus, its should be treated as realistic and not conservative, as anticipated in (Electric Power Research Institute 1993). Both AVN estimators underestimate the 95th percentiles.

The failure probability is decreasing relatively slowly with increasing allowable total leak rate. This again indicates considerable scatter in the leak rate model.

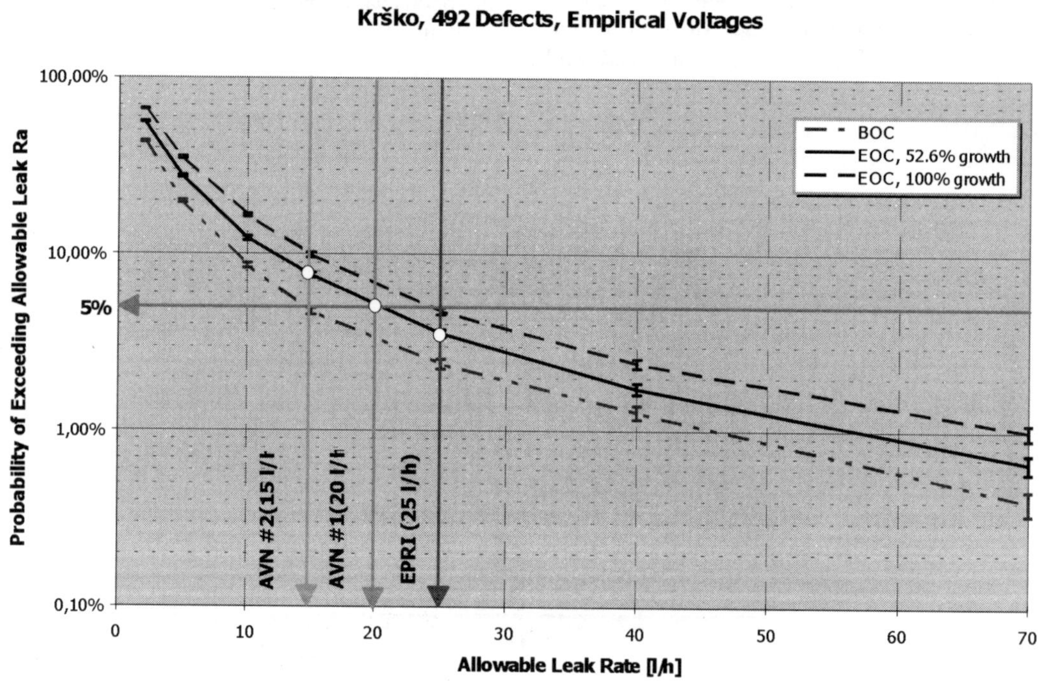

Fig. 2: Probability of Exceeding Allowable Leak Rates - Krško case

Table 1 Selected statistical parameters of total leak rates obtained by EPRI estimator (eq (11)) - Krško steam generator

Total leak rate based on eq (11)	Total leak rates from 100 numerical experiments [l/h]				
	Mean	Median	Std.Dev.	Minimum	Maximum
min (1|1000)	$<10^{-4}$	$<10^{-4}$	$<10^{-4}$	$<10^{-4}$	$<10^{-4}$
median (501|1000)	0.35	0.35	0.01	0.32	0.37
95% (961|1000)	23.83	23.90	1.17	21.44	26.60
max (1000|1000)	1,298.24	1,148.14	652.24	507.68	4,465.07

Doel-4

Data

The distribution of defect sizes from Belgian Doel-4 plant is given in Fig. 3. The number of defects detected is 1960 (significantly more than in Krško). Also, the defect sizes are rather large and poses a maximum of 18.9 V. It should be however noted that Belgian inspection standards differ from EPRI standards. Thus, appropriate correlation between "Belgian" and "EPRI" bobbin coil signal amplitudes was implemented to generate input data for this paper. This may be an additional source of uncertainties, which was however not further investigated. The presence of defects with large voltage is due to the use of a specific repair criterion allowing defects of about 10V to remain in service. The distribution of defects shown in Fig. 3 is the raw distribution as given by bobbin coil inspection. Also, a fit of lognormal distribution to the empirical distribution is depicted in Fig. 3. No defect progression was considered in the calculations.

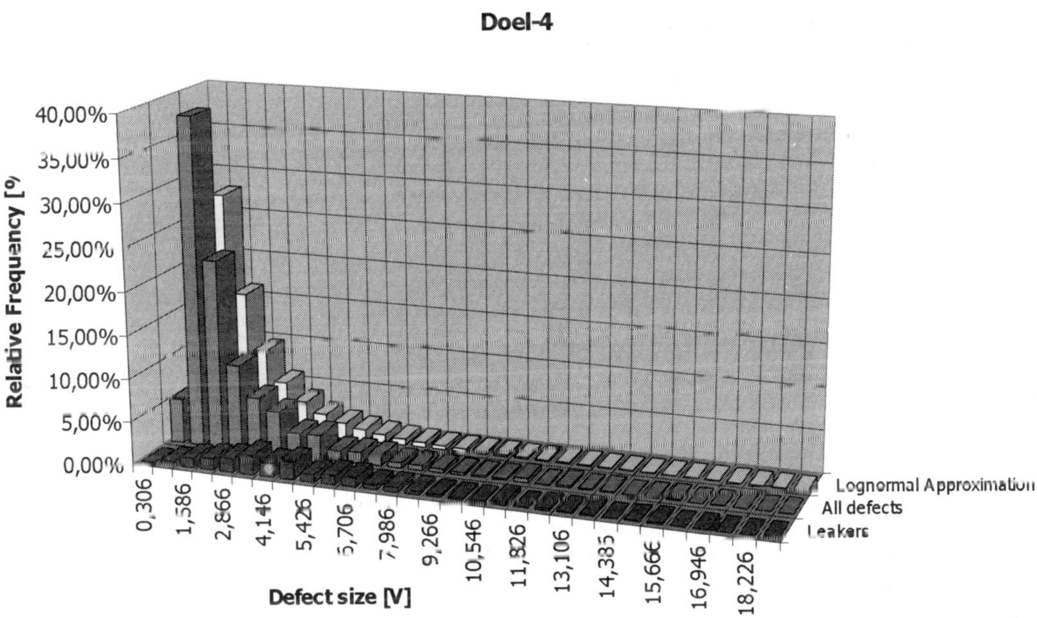

Fig. 3: Distribution of BOC defect sizes - Doel-4 plant

Table 2: Selected statistical parameters of total leak rates obtained by EPRI estimator (eq. (11) - Doel-4 steam generator

Total leak rate based on eq (5):	Total leak rates from 100 numerical experiments [1000 l/h]				
	Mean	Median	Std.Dev.	Minimum	Maximum
min (1|1000)	3.1 10^{-3}	3.1 10^{-3}	0.3 10^{-3}	2.4 10^{-3}	3.9 10^{-3}
median (501|1000)	**2**.128	**2**.125	**0**.027	**2**.074	**2**.206
95% (961|1000)	**82**.044	**82**.082	**1**.938	**77**.141	**87**.184
max (1000|1000)	**3,434**.731	**3,039**.900	**1,959**.141	**2,142**.990	**19,449**.900

In addition, Fig. 3 gives the sample distribution of defects, which exhibit leak (eq (7)). The simulations revealed that no more than 330 (17%) would leak in this particular case. It is clearly shown in Fig. 3 that defects with larger voltages tend to leak more frequently, which is in accordance with eq (7).

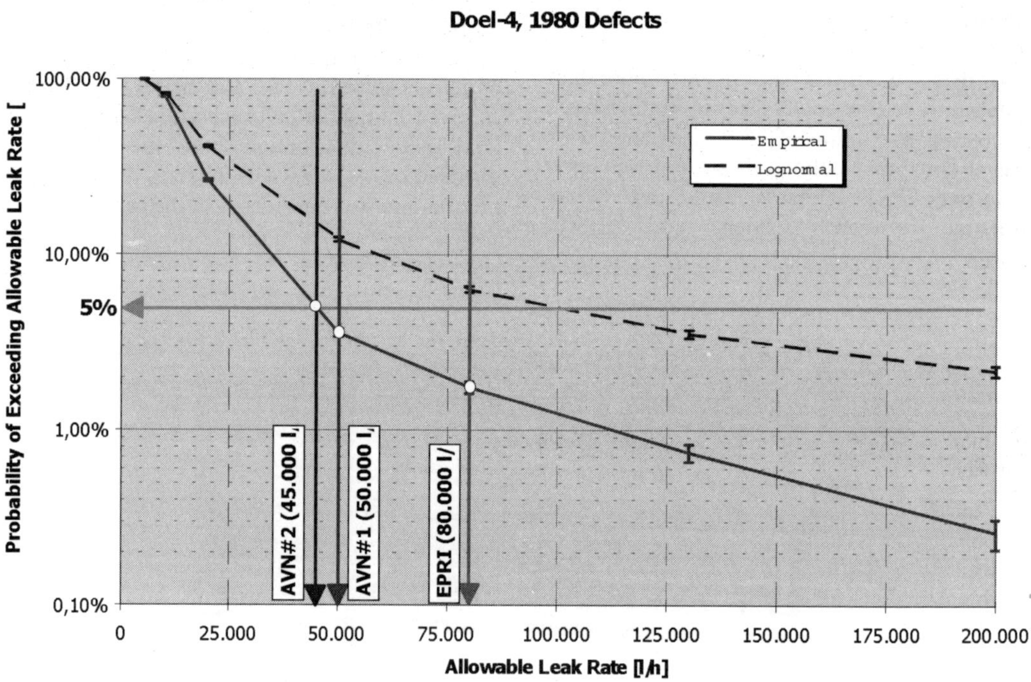

Fig. 4: Probability of Exceeding Allowable Leak Rates – Doel-4 case

Table 3was generated to illustrate the numerical stability of the EPRI estimator (eq. (11)) for the Doel-4 data. Again, a rather stable estimate of about 80,000 l/h was obtained in 100 numerical experiments. It is however indicative that it is possible to obtain leak rates up to about 20 10^6 l/h.

A value of 80,000 l/h was chosen as representative value of EPRI estimator for comparison with probabilistic calculations. Similar considerations lead to the representative values of 50,000 and 45,000 l/h for AVN#1 and AVN#2 estimators, respectively.

Results

The probability of excessive leakage is depicted in Fig. 4 as a function of allowable total leak rate. Results obtained by two different input distributions of defect sizes are shown. The histogram of all defects shown in Fig. 3 served as the empirical distribution. On the other hand, the lognormal distribution was fitted to the raw inspection data. The rather large difference between both curves is caused by two facts:

- lognormal distribution explicitly allows for rare events. In other words, it allows for occurrences of voltage values, which exceed the measured maximum of 18.9 V.
- the probability of leakage (eq (7)) is approaching 1 as the defect size is approaching 20 V. Thus, the uncertainties in the leak rate model are not dominant in this region which increases the sensitivity of the input data as compared to the Krško case.

Again, the results of all three estimators are comparatively depicted in Fig. 4. All of them rely on empirical distributions of the defect sizes, which suggests the comparison with probabilities of excessive leakage obtained from empirical distribution. The results of two estimators appear to be slightly conservative (EPRI 1.8%, AVN#1 3.5%) while AVN#2 tends to the expected 5%. On the other hand, if the lognormal distribution is used, EPRI estimator seems to be realistic (leading to failure probability of 6%) while the AVN estimators lead to failure probabilities exceeding 10% (AVN#1 11%, AVN#2 16%).

Estimators vs. Probabilistic Approach

The main differences between the probabilistic approach and estimators are discussed here:

- Probabilistic approach explicitly splits the population of defects on leakers and non-leakers (eq. (7)), while estimators average the leak rates over both leakers and non-leakers (eq. (10)). The error of eq. (10) is rather serious underestimate of total leak rates (at 95% probability level) for small defects where the probability of leak (eq. (7)) is relatively small (e.g., Krško case). However, it levels out for relatively large defects with probabilities of leak close to 1 (Doel-4 case).
- The differences between estimators are different estimates of the total leak rate at 95% probability level (EPRI vs AVN #1 and AVN #2) and classification of defect sizes (EPRI and AVN #1 vs AVN #2). All conservativisms accumulated in EPRI estimator seem to be appropriate to compensate the underestimates caused by averaging of leaks over both leakers and non-leakers (eq. (7)) for small defect sizes (Krško Case). On the other hand, the least conservative AVN #2 estimator seems to perform well for relatively large defects with probabilities of leak close to 1 (Doel-4 case).

Based on above considerations, the use of the probabilistic approach is recommended because it is not sensitive on the defect size and the inherent assumptions of the estimator used.

Sensitivity of The Results Obtained

The qualitative discussion on the sensitivity is mainly based on the significant difference in the number and size of defects analyzed in both numerical examples. In addition, some interesting observations were made during the analysis of misplaced simulations.

For rather low defect sizes (Krško case), the failure probability is most sensitive to the uncertainties in the leak rate model (eqs (7) and (8)). Those uncertainties are strong enough to generate more extremely high leak rates than large defects would do. Consequently, the probabilities of exceeding the allowable leak rate are not particularly sensitive to the different representations of defect size distributions. This leads to the conclusion that the probability of failure is mostly dominated by uncertain and correlated parameters of regression line (eq. (8)) which is followed by the probability of leak (eq (7)). Beneficial consequences of further experimental work aiming at reducing the uncertainties of the regression curves are obvious.

Table 3: Sensitivity to the number of defects

	Krško		Doel-4	
Fraction of defects considered	# of defects	Probability of exceeding 25 l/h	# of defects	Probability of exceeding 80.000 l/h
50% of defects	246	1.48%	980	0.44%
100% of defects	492	3.56%	1980	1.75%
200% of defects	984	9.01%	3920	4.77%

In the case of large defect sizes (Doel-4 case), the influence of the probability of leak is vanishing as its value approaches 1. The uncertainty in the representation of input data becomes as important as the uncertainty of the regression model.

In both cases, there is a noticeable influence of allowable total leak rate (see Fig. 2 and Fig. 4). The effect of the number of defects on the other hand is shown in Table 4 interesting to note that a factor of two change in the number of defects triggers a factor of about two and half in the corresponding failure probabilities. This may be attributed to rather small number of active and dominant leakers (from 1% in case of Krško to 17% in case of Doel-4).

Another important observation in case of Krško is that the maximum individual leak rate tends to dominate the entire sum of leak rates within the steam generator (about 5 leak rates together). Similar effect can also be observed in the Doel-4 case especially when relatively large total leak rates are observed. This leads to the implementation of extreme value statistics, which is planned in future.

6. CONCLUSIONS

A probabilistic approach aiming at prediction of the probability that a total leak rate through degraded steam generator tubes exceeds the allowable value is proposed. The approach is compatible with the current probabilistic fracture mechanics practice.

Existing estimators, which were developed to determine conservative predictions of total leak rates, outlined and verified against the probabilistic approach. The probabilistic approach gives a reliable measure of their conservativity.

In particular, two numerical examples based on data from Slovenian Krško and Belgian Doel-4 power plants were studied. It was shown that the existing estimators are conservative for large defect sizes (>10V), but not for small defect sizes (<3 V).

It is recommended to introduce probabilistic methods in the field use. This is because probabilistic methods give better insight in the problem, which includes a large number of uncertainties. On the other hand, the computational effort involved is comparable to the estimators.

The probability of excessive leakage is most sensitive to the probability of leakage at lower defect sizes (Krško case) and global uncertainties inherent to the applied regression models at any defect size.

The main topics suggested for the future work are implementation of extreme value statistics in order to reduce the computational effort and further experimental work aimed at reducing the uncertainties of the regression models.

7. REFERENCES

1. Banic, M. J et al, 1996. *Assessment and Management of Ageing of Major Nuclear Power Plant Components Important to Safety: Steam Generator.* IAEA Safety Series. Vienna, Austria: International Atomic Energy Agency.
2. Cizelj, Leon, et al, 1995. Safety Analysis of Krško NPP Steam Generator Tubes Damaged by Outside Diameter Stress Corrosion Cracking at Tube Support Plates, IJS-DP-7175, Rev. 0. Institut "Jožef Stefan", Ljubljana, Slovenia, Proprietary.
3. Cizelj, Leon, Tomaž Dvoršek, and Borut Mavko. 1996. Efficiency of Defect Specific Maintenance Approaches in SG Tubes. Proceedings of 15th EPRI Steam Generator NDE Workshop, 9.1-9.10Palo Alto, Ca., USA: Electric Power Research Institute.

4. Cizelj, Leon, Borut Mavko, and Peter Vencelj. 1996. Reliability of Steam Generator Tubes With Axial Cracks. Journal of Pressure Vessel Technology 118: 441-46.
5. Cizelj, L., I.Hauer, G. Roussel, Ch. Cuvelliez. 1998. *Probabilistic Assessment of Excessive Leakage through Steam Generator Tubes Degraded by Secondary Side Corrosion,* Nuclear Engineering and Design 185 (1998) 347-359
6. Cuvelliez, Charles, and Guy Roussel. 1997. Assessment of the Leak Tightness Integrity of the Steam Generator Tubes Affected by ODSCC at the Tube Support Plates. Proceedings of the CNRA/CSNI Workshop on Steam Generator Tube Integrity in Nuclear Power Plants, Rockville, Md., USA: Nuclear Regulatory Commission.
7. Electric Power Research Institute. 1993. Outside Diameter Stress Corrosion Cracking (ODSCC) of Steam Generator Tubing at Tube Support Plates-A Database for Alternate Repair Limits, Vol. 2: 3/4 Inch Diameter Tubing, EPRI TR-100407, Rev. 1. Electric Power Research Institute, Palo Alto, Ca., USA.
8. Harris, Douglas D. 1996. Capabilities of Eddy Current Data Analysts to Detect and Characterize Defects in Steam Generator Tubes. Proceedings of 15th EPRI Steam Generator NDE Workshop, 8.1-8.8Electric Power Research Institute.
9. Madsen, Henrik O., Sven Krenk, and N. C. Lind. 1986. Methods of Structural Safety.Prentice Hall.
10. Nuclear Regulatory Commission. 1993. Voltage-Based Interim Plugging Criteria for Steam Generator Tubes, NUREG-1477. U.S. Nuclear Regulatory Commission, Rockville, Md., USA.
11. Nuclear Regulatory Commission. 1995. Voltage-Based Repair Criteria for Westinghouse Steam Generator Tubes Affected by Outside Diameter Stress Corrosion Cracking, Nuclear Regulatory Commission. NRC Generic Letter 95-05. U.S. Nuclear Regulatory Commission, Rockville, Md., USA.
12. Roos, E. 1999, Integrity Analysis of Piping Systems Using Leak-Before-Break Concept, Transactions of SMIRT-15, Aug. 15-20, 1999, Seoul, Korea.
13. Shah, Vikram N., and Phillip E. MacDonald. 1993. Aging and Life Extension of Major Light Water Reactor Components.Elsevier.

Reliability, Safety and Hazard
Editors: P.V. Varde et al.

STATE PROBABILITY CALCULATION FOR A CRITICAL COMPONENT DUE TO CREEP AND FATIGUE – OFFLINE AND ONLINE

S. Guin, M.K. Samal and B.K. Dutta

Reactor Safety Division, Bhabha Atomic Research Centre, Mumbai 400 085

ABSTRACT

Creep and fatigue are two of the important degradation mechanisms in critical power plant components (e.g., elbows in a thermal power plant). Therefore it needs to be addressed by a realistic reliability model for reducing the frequency of in-service inspection (ISI) activities in such components. Present paper considers a suitable Markov Model which demands for the state transition rates, associated to the component due to crack growth (e.g. creep and fatigue) and plastic instabilities to determine the different state probabilities. First Order Reliability Method (FORM) has been used to determine the state transition rates. These rates are fed to a set of differential equations to get the state probabilities. With the data of initial crack size, crack growth after certain time and their statistical variations, the state probabilities are calculated in an offline manner and their variations with plant operational time have been presented. Also, a strategy of online calculation of the state probabilities has also been suggested with actual online results. After the state probabilities are known, risk calculation corresponding to each state can be used to adopt a proper risk-informed in-service-inspection (RI-ISI) strategy.

Keywords: Creep and fatigue, Markov model, FORM, RI-ISI

1. INTRODUCTION

The electricity supply industry in recent times is facing a challenging competition scenario to serve people. More efficient technologies are being adopted to enhance the plant performance under demand load and to extend the plant life to economize the total project plans. The maintenance schedule also plays an important role in determining the running cost of a power plant. To reduce the maintenance cost, one should reduce in-service inspections as much as possible, but keeping safety and integrity of the critical components in view. To prioritize the components under maintenance plan, it is necessary to evaluate the risks associated with each component. This risk calculation in turn demands for the failure probability associated with a component under operating conditions. In this article, performance of a critical component in a thermal power plant has been analyzed from probabilistic point of view. Fatigue and creep crack growth data being available from prior analysis of plant data, the current probabilistic analysis calculates the different state transition rates and the state probabilities (e.g., leak and rupture).

2. COMPONENT DESCRIPTION

An ageing in-service Hot Reheat (HRH) pipe bend before Intermediate Pressure (IP) Stop/Control Valve in thermal power plant, Dadri, was identified for crack growth calculation under creep and fatigue (Figure1). The HRH piping including low pressure bypass line incorporating various supports such as directional restraints, constant weight hangers and spring hangers, was modeled using straight and bend elements (Figure2). The importance of hot reheat piping in an operating plant emerges from the concern of its critical function as a transporter of high-energy steam to the IP turbine. This piping system, one of the static critical power plant components has a finite service life. The piping supplies high-energy superheated steam (design temperature and pressure: 540°C at 46.3 Kg/cm^2) to its intermediate pressure turbine [1]. It is one of the important static critical power plant components. This component (made of low alloy ferritic steel – ASME SA 335 P22) was found to have an apparent bulge at the intrados during routine in-service inspection and therefore was considered for fitness for purpose of service assessment and subsequent reliability analysis.

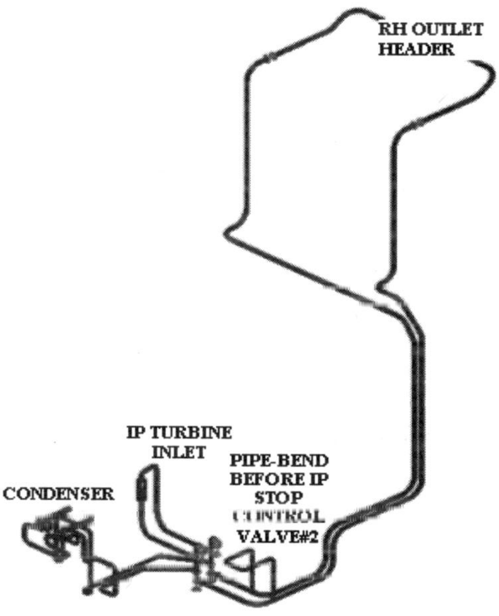

Fig. 1: HRH piping model

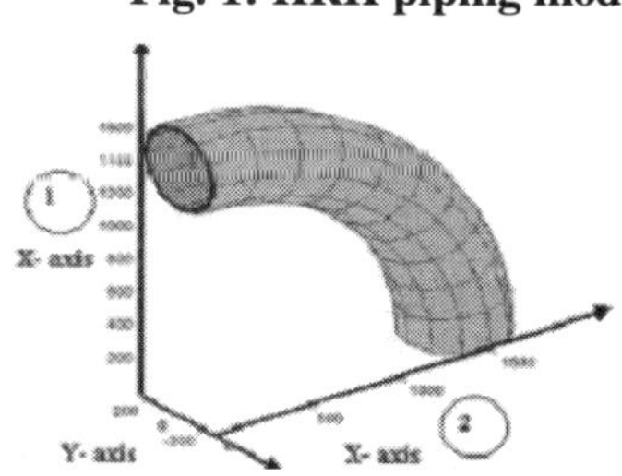

Fig. 2: Finite Element model of the pipe bend

3. MARKOV MODEL STRATEGY TO FIND THE STATE PROBABILITIES

Description of the Model

Markov modelling for a pipe segment under single or multiple degradation mechanism(s) suggests how the degradation process(s) are responsible to put the segment in discrete mutually exclusive states during plant operational life, keeping chances of flaw detection and leak repair in view. These states, i.e., success (S), flaw (F), leak (L) and rupture (R), are schematically shown in Figure3, along with the criterion for each state [4]. Now, creep and fatigue are two degradation mechanisms, which reduce the effective pipe thickness due to crack growth and hence a pipe segment is expected to undergo the different states during the course of plant life. To be conservative rupture has been assumed to take place either from initial undetected flaw or flaw/leak state. Here, the state of 80% reduction in thickness due to crack growth is taken equivalent to a leak.

Estimation of Flaw and Leak Repair Rates

These two rates are themselves dependent on several probabilities as follows:
Flaw repair rate is:

$$\omega = P_{FI}P_{FD} \,/\, (T_{FI}+T_{FR}) \quad (1)$$

where P_{FI} = probability that a piping segment will be inspected for flaw per inspection (dependent on what fraction of the segment is inspected), P_{FD} = probability that an existing flaw under the inspected segment will be detected (dependent on the reliability of inspection technique used), T_{FI} = mean time between two inspections (taken according to inspection codes, e.g., ASME Section XI), T_{FR} = mean time to repair a flaw.
Similarly, leak repair rate is given by:

$$\mu = P_{LI}P_{LD} \,/\, (T_{LI}+T_{LR}) \quad (2)$$

where P_{LI} = probability that a piping segment will be inspected for leak per inspection, P_{LD} = probability that an existing leak will be detected (dependent on the leak detection system), T_{LI} = mean time between two inspections (taken according to inspection codes), T_{LR} = mean time to repair a leak.

It is assumed here that a flaw or leak, once detected, will be repaired.

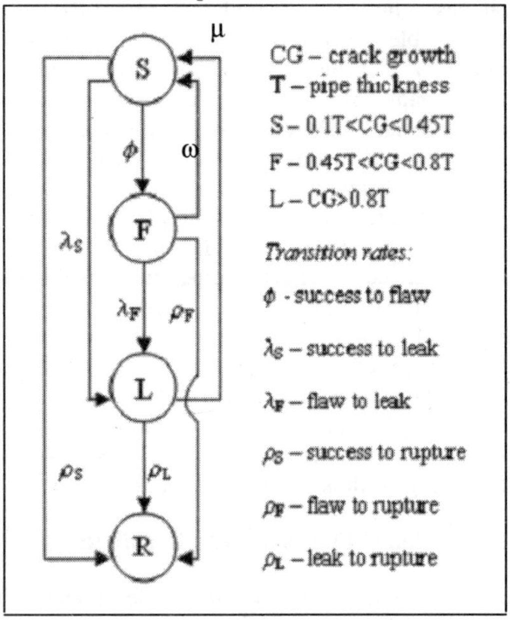

Fig. 3: Markov model for piping states

Limit State Functions (LSF) for Transition between States

These functions will be used to calculate the probability of occurrence of a particular state from another state. As shown in Figure3, there are six such possibilities (except the repair paths) which will be discussed now.

Success to Flaw (ϕ)

Assuming the maximum possible depth of an undetected flaw could be 10% of the thickness, the LSF corresponding to this transition is clearly given by:

$$LSF_1 = 0.45\times t - (d_1 - CGR\times T) \quad (3)$$

Here t = pipe thickness, d_1 = undetected flaw depth = 0.1t, CGR = crack growth rate, T = plant operation time

Success to Leak (λ_S)

Assuming the same initial flaw size, the LSF corresponding to this transition is:

$$LSF_2 = 0.8\times t - (d_1 - CGR\times T) \quad (4)$$

t = pipe thickness, d_1 = undetected flaw depth = 0.1t, CGR = crack growth rate, T = plant operation time

Flaw to Leak (λ_F)

An existing undetected flaw, which has already crossed a depth of 45% of the thickness, now can go for creating a hypothetical leak, which corresponds to a depth 0.8t. Thus, the second LSF is:

$$LSF_3 = 0.8\times t - (d_2 - CGR\times T) \quad (5)$$

where $d_2 = 0.45t$

Success, Flaw or Leak to Rupture (ρ_S, ρ_F, ρ_L)

Since rupture is inherently associated to material strength, instead of putting the thickness function directly as the LSF, we use a function based on the structural instability caused by the loading. Thus,

$$LSF_{4\&5\&6} = \sigma_y - \sigma_{ref} \quad (6)$$

where σ_y is the material yield stress and σ_{ref} is the reference stress under particular loading and component geometry, calculated as per RCC-MR A16 code[5]. When this σ_{ref} becomes equal to the yield stress, rupture is assumed to take place. After calculating the probabilities of failure corresponding to each LSF by First Order Reliability Method, they are divided by respective time periods to get the state transition rates.

Determination of the State Probabilities

Once the state transition rates are known, we can set four differential equations, one for each state, as follows:

$$\left\{\begin{aligned} \frac{dS}{dt} &= -(\varphi+\lambda_S+\rho_S)S + \omega F + \mu L \\ \frac{dF}{dt} &= \varphi S - (\lambda_F + \rho_F + \omega)F \\ \frac{dL}{dt} &= \lambda_S S + \lambda_F F - (\rho_L + \mu)L \\ \frac{dR}{dt} &= \rho_S S + \rho_F F + \rho_L L \end{aligned}\right. \quad (7)$$

Since these pipe states are mutually exclusive and a complete set, the following condition is satisfied for any time t:

$S\{t\} + F\{t\} + L\{t\} + R\{t\} = 1$

In vector form these equations (7) can be rewritten as:

$$\frac{d\mathbf{X}}{dt} = \mathbf{AX}$$

, where

$$\mathbf{X}(t) = \begin{bmatrix} S(t) \\ F(t) \\ L(t) \\ R(t) \end{bmatrix}$$

and

$$\mathbf{A} = \begin{bmatrix} -(\varphi+\lambda_S+\rho_S) & \omega & \mu & 0 \\ \varphi & -(\lambda_F+\rho_F+\omega) & 0 & 0 \\ \lambda_S & \lambda_F & -(\rho_L+\mu) & 0 \\ \rho_S & \rho_F & \rho_L & 0 \end{bmatrix}$$

The solution of this linear matrix differential equation is given by:

$$\mathbf{X}\{t\} = C_0\mathbf{E_0}e^{z_0 t} + C_1\mathbf{E_1}e^{z_1 t} + C_2\mathbf{E_2}e^{z_2 t} + C_3\mathbf{E_3}e^{z_3 t} \tag{8}$$

where the eigenvalues are given by: $\mathbf{z} = \{z_0, z_1, z_2, z_3\}$, and the eigenvectors are given by $\{\mathbf{E_0}, \mathbf{E_1}, \mathbf{E_2}, \mathbf{E_3}\}$. The constants C_j, j = 0,1,2,3, are constants of integration and can be determined by the following boundary conditions (assuming that the pipe segment is free of detectable flaws at t=0):

$$\mathbf{X}(t=0) = \{1\ 0\ 0\ 0\}^1$$

The eigenvalues are determined by solving the fourth order polynomial equation given by [2]:

$$\det[\mathbf{A} - \mathbf{zI}] = 0$$

4. CASE STUDY – OFFLINE CALCULATION

As already mentioned, the hot reheat pipe bend in a 210 MWe coal fired power plant unit is considered for analysis. Table-1 shows the real-life offline data used for the analysis.

Assuming all the statistical parameters are uncorrelated and are following normal distribution, the state transition rates have been determined using FORM through the in-house code MARKOV. Once the transition rates are obtained, the state probabilities can be determined using the solution of the equation (8). A study has been performed to get the variation in these state probabilities during plant operational time, assuming a plant life of 40 years. Figure4 shows how the state probabilities for flaw, leak and rupture vary with plant operational time, in logarithmic scale. At any time, the success probability can be obtained by subtracting the total of these probabilities from 1.

Table 1. Statistical and other input Parameters

PARAMETER	VALUE
Mean and standard deviation (% of mean) of pipe thickness (mm)	30;10
Initial crack depth (% of thickness)	10
Mean and standard deviation (% of mean) of crack growth (mm)	5.6×10^{-03};10
Crack growth time (operational Hrs)	4024
Mean and standard deviation (% of mean) of pipe inner radius (mm)	224;5
Mean pipe and standard deviation (% of mean) of bend radius (mm)	1500;5
Mean and standard deviation (% of mean) of fluid pressure (MPa)	3.5;10
Mean and standard deviation (% of mean) of fluid temperature (^{0}C)	540;5
Mean and standard deviation (% of mean) of yield stress (MPa)	200;10
Flaw inspection time (yrs)	10
Leak inspection time (yrs)	1
Flaw inspection chance at crack location	0.25
Flaw detection probability	0.9
Leak inspection chance	0.9
Leak detection probability	0.9
Flaw or leak repair time (Hrs)	200

5. ONLINE STATE PROBABILITIES

In case of offline calculation, the mean values of the associated parameters don't change over time. In reality, they do change with time and sometimes only offline analysis may produce non-conservative results. To cater for this problem, one can think of the whole calculation procedure, discussed till now, to be put on a real-time based monitoring system, which takes care of the changes in parametric values with plant operational time [3]. The analysis procedure in online strategy will be slightly different from offline procedure in a way described by the flowchart in Figure6. Here, after every small time interval, the present operational data will be considered to calculate the probabilities and the new crack length will be modified to feed to the calculation for the next loop.

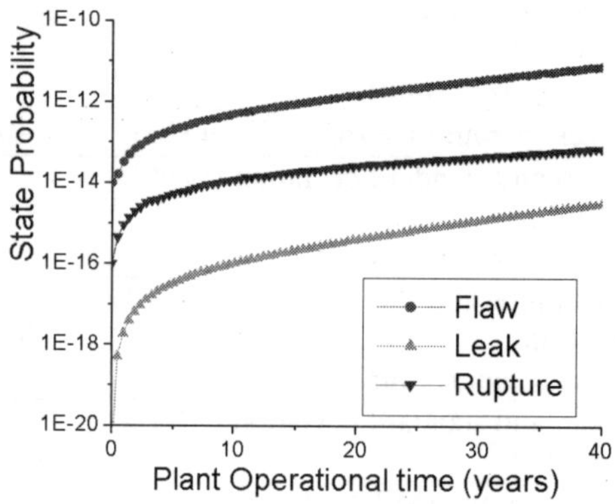

Fig. 4: Offline State probabilities over operational time

The component under analysis has been attached with a real-time data acquisition system to get the readings of fluid temperature and pressure after regular time intervals. Thus we can get the online data for our study of state probabilities. We have considered actual plant data for two years in this calculation. Following the procedure given in flowchart (Figure5), we carried out the probabilistic analysis in the online environment. Figure6 shows the online variation of state probabilities (except for success which can be determined if other three probabilities are known) over a plant operational time period of two years.

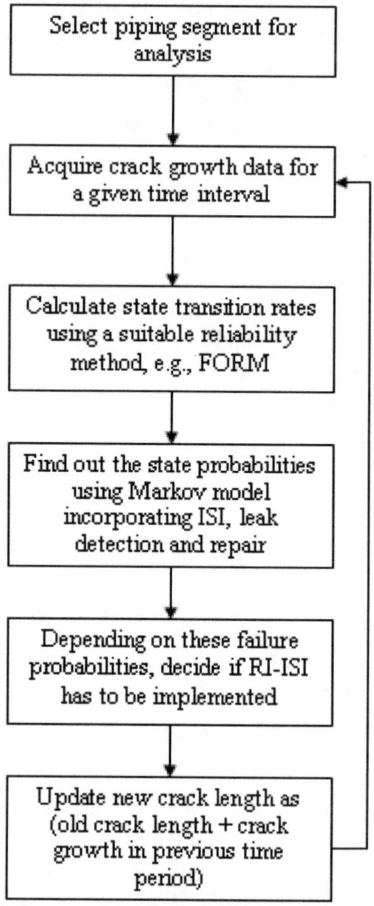

Fig. 5: Online monitoring of state probabilities

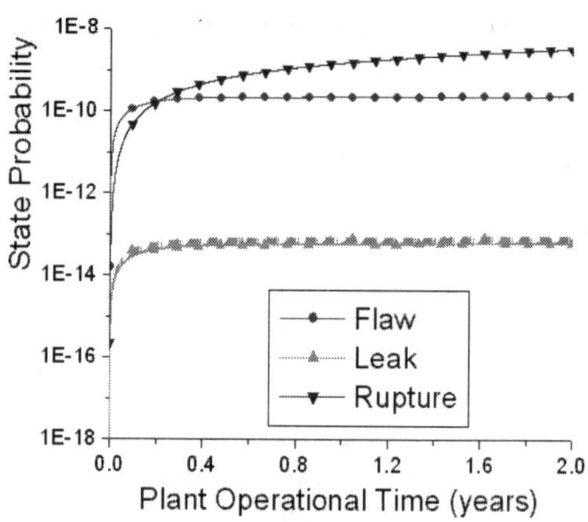

Fig. 6: Online state probabilities over two years

6. DISCUSSION

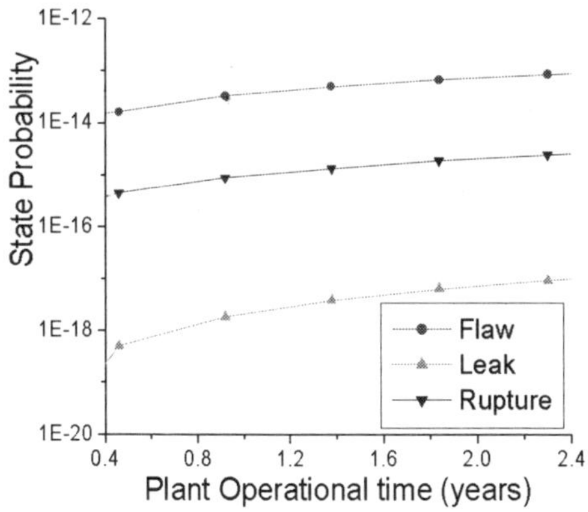

Fig. 7: Offline state probabilities over two years

Figure7 shows the offline variation of state probabilities for the same period. A comparative study reveals that online state probabilities are much higher (except success) compared to the online ones. This is because the online analysis takes care of the actual variation in pressure and temperature (shown in Figure8) which sometimes may go high compared to the offline mean value. In these particular cases, the probability of the instability causing rupture becomes high and effectively the state probabilities rise. Once these probabilities rise, with time they go on rising. The more is this fluctuation of pressure and temperature from the mean value, the more will be the rise in state probabilities. The comparison also urges to use the online analysis data to monitor the structural health of the critical components instead of using mere offline mean parameters for analysis.

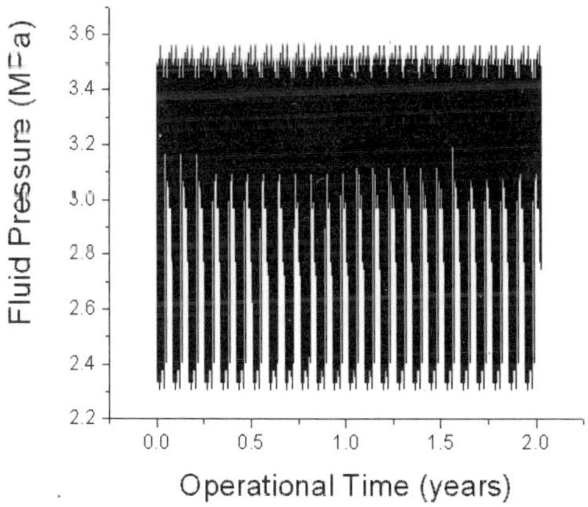

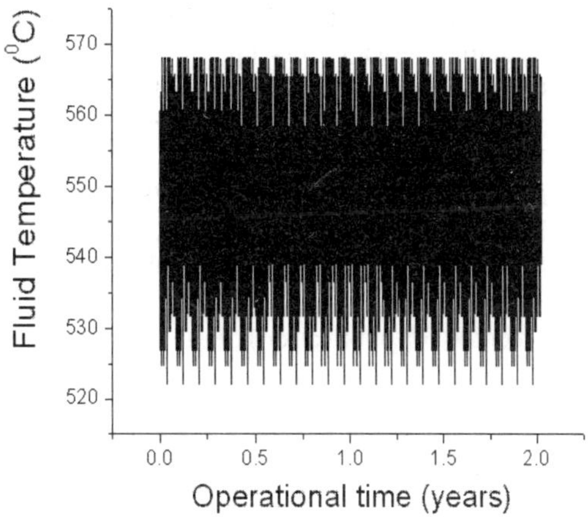

Fig. 8: Variation of fluid pressure and temperature with plant operational time

7. CONCLUSIONS

From the figures 4 and 6, it is seen that the state probabilities corresponding to flaw, leak and rupture increase with time. Since the sum of all the state probabilities at any point of time is one, the probability of success goes down.

A beauty of this model is that it takes care of the possibilities to go back to success stage from flaw or leak, with some adherent probability. Corresponding to each state (S, F, L or R), if the consequences are known, the risk associated to a particular component and a particular state can be determined. For a number of critical components in a plant, this kind of similar studies can be worked out. After the risks are quantified for each component and state, the components can be arranged in order of decreasing risk. This helps in prioritizing the components for risk informed in service-inspection (RI-ISI) strategy. Thus, unnecessary inspections may be avoided while putting safety also as a high priority, thereby minimizing the inspection time and cost.

8. REFERENCES

1. Dutta B.K., Samal M.K.,Guin S.; On-Line Health Monitoring of Power Plant Components; ASME Mechanical Engineering Congress and Exposition, 2005 (paper accepted for publication)
2. Fleming K. N.; Markov models for evaluating risk-informed in-service inspection strategies for nuclear power plant

piping systems; Reliability Engineering & System Safety 83 (2004) p. 27-45.

3. Guin S., Samal M. K., Dutta B. K.; Piping State Probabilities in Primary Heat Transport (PHT) System of Pressurized Heavy Water Reactor (PHWR) due to Erosion Corrosion – Offline and Online; International Conference on Reliability and Safety Engineering INCRESE 2005 (paper accepted for publication)
4. Vinod G., Bidhar S. K., Kushwaha H. S., Verma A. K., Srividya A.; A comprehensive framework for evaluation of piping reliability due to erosion–corrosion for risk-informed inservice inspection; Reliability Engineering & system safety 82 (2003) p. 187-193.
5. RCC-MR Appendix A16, AFCEN, Paris, 2002.

Reliability, Safety and Hazard
Editors: P.V. Varde et al.

IN SERVICE/FIELD PRODUCT RELIABILITY TRACKING/ANALYSIS

S. Visalakshi, V.S. Prakash, Sandeep Baliga, Rajesh Kumar S., Ramji Sethu and Nitesh Lall

Honeywell Technology Solutions Lab, 151/1, Doraisanipalya, Bannerghatta Road
Bangalore 560 076

ABSTRACT

This paper highlights the methodology of "In service/field Product Reliability tracking/analysis". It provides an insight to the approach followed to evaluate a system/product life with the usage of available field data, Reliability mathematics, tools & techniques. The field data structure is an important requirement in field data analysis. Data quality issues are of concern in any field data analysis. The paper gives in detail the requirements of data structure, the process of using Weibull analysis to model the life of the Components based on in-service field data. This paper also covers the forecasting methods used to predict the present and future risk. Identifying batch problems, method to estimate component life with no failures, interpreting the Weibull data is also addressed. An overview of the Reliability modeling approach followed to evaluate the System life with respect to the life model of the sub components is provided. Weibull Analysis is used to model the Components Life based on the in-service field data. In this approach, the lives of the top nonconformance components are considered and the life of each component is varied to study the sensitivity of the system life. Reliability growth models like Crow AMSSA used to evaluate the Product trends have been addressed. Different topics of field data analysis mentioned above will be presented as case studies. The goal is to address field data analysis, data interpretation and use of reliability methods to propose design improvements that are tangible, producing significant system improvements. This also helps to improve decision-making based on Reliability/ Cost benefit tradeoff. Field data analysis is an aid to address customer warranty claims, spares management and Proactive Reliability tracking.

Keywords: Reliability, In-service/field data, NC – Nonconformances, R&O – Repair & overhaul, Modelling, Weibull, Forecast, Risks, Crow AMSAA and Cost

1. INTRODUCTION

Today the customers expect that suppliers understand their product reliability and trends but also expect support of their products and spares management in light of this knowledge. The effect of unreliability creates costs for the company and our customers through poor performance, repair costs, warranty etc. It also affects the business prospects to secure or win proposals.

Field data analysis contributes primarily in tracking the performance of the fielded components. It also supports critical decision making for proposal, secure/win proposals and after market purposes. If problems are detected early customer satisfaction can be managed and company risk due to warrantees, maintenance service agreements and reliability guarantees can be minimized.

2. RELIABILITY TRACKING OF IN SERVICE PRODUCTS

For reliability studies of in service products, field data is very much important to track and record the product performance. This data basically refers to the findings during the products return to Repair and Overhaul (R/O) shops for failure or service. This data must be captured thoroughly for further investigations and to support reliability analyses. The kind of information recorded also plays an important role for the reliability analyses. The product removal may be an Unscheduled; scheduled with fault or No Fault found. A sample structure of the field data information to be collected is presented in the Fig 1.

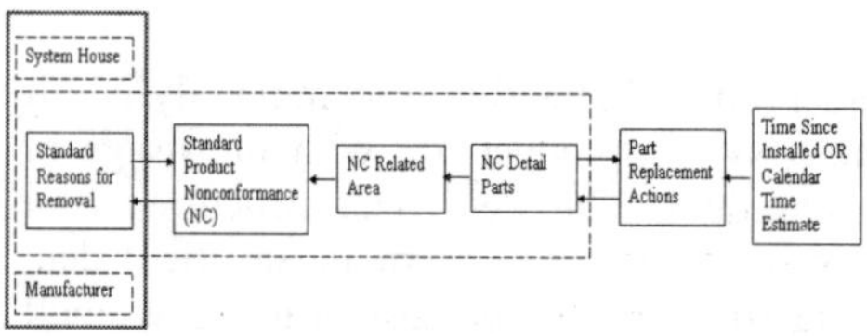

Fig. 1: Structure of the R/O Data

Data Quality Concerns

This is one of the persistent problems with the field data faced by many of the reliability engineers for Reliability-based field data analyses. The data capture process is affected by the volume of product returns, numerous R/O shops, partial product repairs, third party product repairs, man power constraints, different methods/tools & global presence of R/O shops. Imperfect findings data entered by technicians rather than engineers, no proper standard format of the non-conformances or findings descriptions, improper evaluation of fault /no fault affect data quality and Reliability analyses. It is very much required that the technicians at the R&O shops be trained well on the diagnosing the reasons for return of components. They should be educated on the importance of the information they are collecting and the impact of this on the performance of the system and the customer satisfaction.

A standard format for the non-conformances description should be adopted. This will help in preventing the technician from entering the findings description in different ways for the same removal cause or reason. This will also help reliability engineers in identifying the actual problem with the fielded components and mitigating the issues.

3 WEIBULL ANALYSIS AND MODELLING

Weibull Distribution is the most generic distribution and widely used for life data analyses. This distribution can take the characteristics of any other distributions depending on the shape parameter [1].

Weibull Analysis is basically used to characterise failure modes of any system based on the life data. The two parameter Weibull is widely used to characterise the failure modes. The two parameters are η eta, the characteristic life and β beta, the slope of the Weibull line. The cumulative density function is given by the formula [1]:

$$F(t) = 1 - e^{-(t/\eta)^{\beta}} \qquad (1)$$

To arrive at the Weibull parameters, the life data of the fielded components are very much essential. This information can be readily made available in the field data (time since installed) or can be estimated from the installed date and the failed date using utilisation rate for a particular product. The utilisation rate depends on various factors (in Aerospace – Fleet, operators, Engine hours etc). Along with the failure time for the products it is also required to consider the life data of those samples of population which have never failed and continue to perform in the field. These components are referred as suspended. The suspended data can be obtained from shipped /installed dates for product which never failed/returned. This failure time and suspension time data for the entire population will be fed in, to arrive at the eta and the beta values, which statistically represents the characteristics of the product under consideration. The two methods: RR (rank Regression) or MLE (maximum likelihood) is used for the weibull analysis which depends on the sample size.

Weibull Analysis is performed to:

1) To evaluate System/product life.
2) The field Reliability trends.

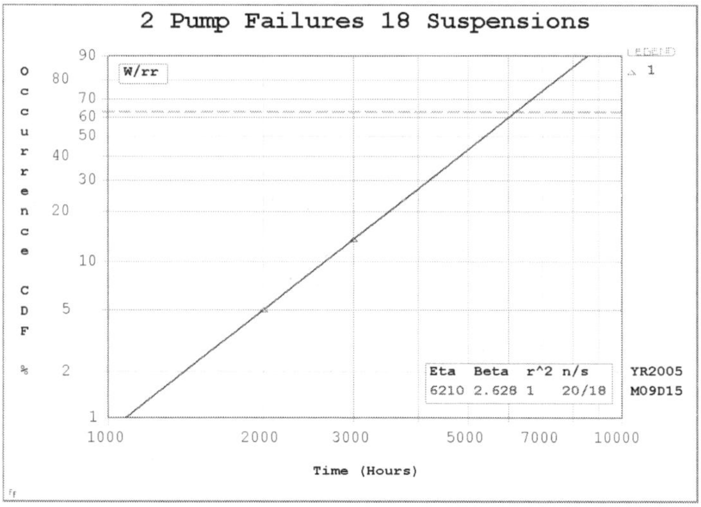

Fig. 2: Weibull Analysis for Pump Failures

3) How a New design/service change affected the product performance
4) Soft time (Overhaul/scheduled) removal recommendations.
5) Warranty period.
6) Inventory & Spares management.
7) To assess dominating failure modes.
8) To support Product weibull model.

Weibull Interpretation: The slope of the Weibull plot, beta (β), determines which member of the Weibull failure distributions best fits or describes the data.

Table 1 Beta Vs Failure Distributions

Beta (β)	Failure Distribution
β = 1.0	Exponential
β = 0.5	Log Normal
β = 3.44	Normal

Along with this it also indicates which class of failure is present.

Table 2 Beta Vs Failure Types

Beta (β)	Failure Type
β < 1.0	Infant Mortality
β = 1.0	Random
β > 1.0	Early Wear Out
β > 4.0	Old Age Wear Out

Risk forecasting can be done based on the Weibull analysis. A risk analysis or failure forecast predicts the expected number of incidents that may occur in a specific period of time. This analysis can be used to calculate the:

- Expected Failures
- Future Failures

Expected Failures Now [1]: If there is a population of N items and each has accumulated t_i hours or cycles, the expected number of failures from this population is the probability of failure by time t_i summed over the number of units, N, including both failures and suspensions. The expected number of failures is:

$$= \sum_{i=1}^{N} \left[1 - e^{(t/\eta)^{\beta}} \right] \quad (2)$$

Case Study: 2 Pump Failures for a fleet of 20 pumps.

Table 3 Expected Failure Calculation

N	t	F(t)	F(t) * N
5S	1000	0.0082	0.041
4S + 1F	2000	0.0496	0.248
4S + 1F	3000	0.1374	0.687
5S	4000	0.2700	1.350

Expected Failures Now = 2.326.

The expected failure approximates the observed 2 failures. If the expected-failures-now is much larger than the observed number of failures, the Weibull may not apply to the entire population. This is a "batch" problem, i.e., the entire fleet does not have the "illness". This is the main reason for estimating the expected failures.

Failure Forecast: All future failures will occur on today's survivors. Failure forecasts can be made for any interval. Considering the same pump example, the expected number of failures over the 12 months can be predicted.

If pump i has accumulated t_i hours to date without failure, and will accumulate u additional hours in a future period, the failure forecast for the 18 pumps at risk is given by the equation:

$$= \sum \frac{F(t_i + u) - F(t_i)}{1 - F(t_i)} \tag{3}$$

Table 4 summarises the one year failure forecast calculation for the 18 pumps at risk with a one year usage of 1000 hours.

Table 4 12 Month Pump Failure Forecast

N	t	t+u	a = F(t+u)	b = F(t)	PF = (N*(a-b))/(1-b)
5	1000	2000	0.0496	0.0082	0.2087
4	2000	3000	0.1374	0.0496	0.3695
4	3000	4000	0.2700	0.1374	0.6149
5	4000	5000	0.4321	0.2700	1.1100

12 Month failure forecast = 2.303.

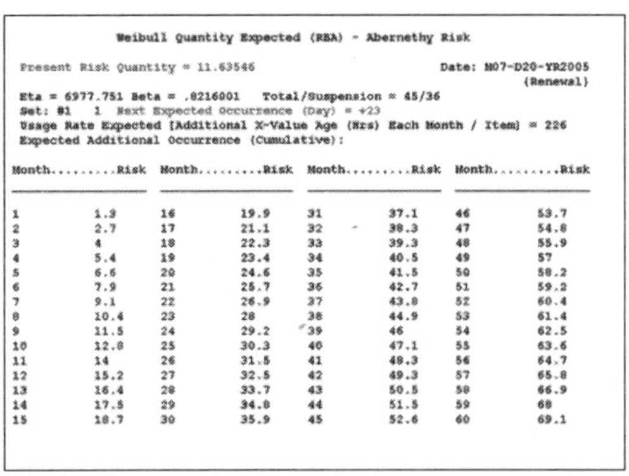
Weibull Quantity Expected (RBA) - Abernethy Risk

Present Risk Quantity = 11.63546 Date: M07-D20-YR2005 (Renewal)

Eta = 6977.751 Beta = .8216001 Total/Suspension = 45/36

Set: #1 1 Next Expected Occurrence (Day) = +23

Usage Rate Expected [Additional X-Value Age (Hrs) Each Month / Item] = 226

Expected Additional Occurrence (Cumulative):

Month	Risk	Month	Risk	Month	Risk	Month	Risk
1	1.3	16	19.9	31	37.1	46	53.7
2	2.7	17	21.1	32	38.3	47	54.8
3	4	18	22.3	33	39.3	48	55.9
4	5.4	19	23.4	34	40.5	49	57
5	6.6	20	24.6	35	41.5	50	58.2
6	7.9	21	25.7	36	42.7	51	59.2
7	9.1	22	26.9	37	43.8	52	60.4
8	10.4	23	28	38	44.9	53	61.4
9	11.5	24	29.2	39	46	54	62.5
10	12.8	25	30.3	40	47.1	55	63.6
11	14	26	31.5	41	48.3	56	64.7
12	15.2	27	32.5	42	49.3	57	65.8
13	16.4	28	33.7	43	50.5	58	66.9
14	17.5	29	34.8	44	51.5	59	68
15	18.7	30	35.9	45	52.6	60	69.1

Fig. 3: Risk Forecasting – Winsmith Weibull

Risk forecasting is used to check or validate the weibull analysis, to see if the Present Risk forecasted is close to the in field failures, is there a batch problem, Spares & Inventory management.

Weibayes

To track reliability trends of new products /design changes, there may be few or no failure available. This makes difficult to plot a weibull to understand the product trend. We can make use of the slope value of the similar product family, which is in field to evaluate the performance. This method is called the weibayes technique. As and when the returns of the new product are gathered the weibayes plot is updated to study if the new design is better than the old.

Reliability/Weibull Modeling & Cost Benefit Analysis [2]

Reliability modelling approach followed to evaluate the System life with respect to the life model of the sub components. Field data analysis support Modeling & cost benefit analysis. This Modeling approach using in-field data is a significant improvement over traditional "take-away Predictions. Helps to identify the design changes in the product that will support management decision-making with respect to product Reliability improvement

The problem context was to identify a process when a product Reliability fell below expectations and had repeated improvements over time.

1) Supporting Analysis

- With Performance Engineering, identified "Standard" reason for removal for the product - meaningful to System house and product manufacturer
- Performed Fault Tree Analysis to identify possible failure causes supporting Reason-for-Removal
- Performance analysis Physics-of-Failure simulation done to determine likelihood of each failure-cause leading to end

event/primary failure and grouping these causes into functional groups

2) **Sample Selection and Scoring**
   - Sample size was selected from shop findings based on significant operators using the latest configuration product
   - Shop findings were scored for "Reason"
   - Shop findings removals were scored based on performance analysis for primary failures.

3) **Weibull Analysis:** Analysis was performed for all primary failures / functional groups. The Weibull analysis included suspensions based on parts replacement actions

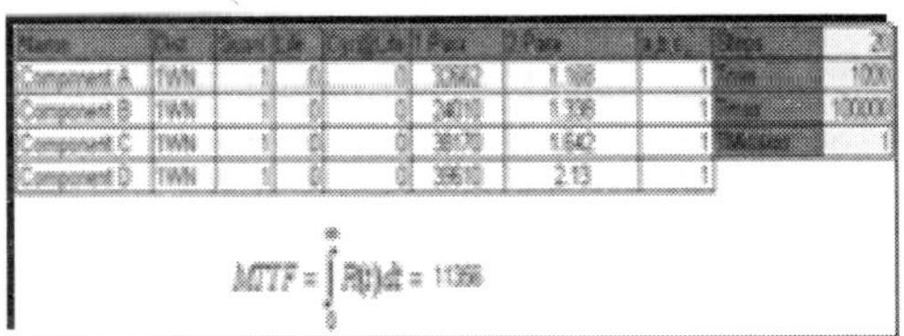

Fig. 4: Reliability Modeling Tool

4) **"Product Enhancement Analysis"**

- Component design alternatives were assessed.
- Evaluated design changes for introduction of new failure modes

5) **Cost Benefit Analysis**

- Expected component life improvements were assessed based on Physics-of-Failure
- Impact of System Reliability MTTF improvement was assessed based on series model
- Costs were associated to each design alternative.
- Impact of cost to benefit in system reliability was analysed.

4. CROW-AMSAA-DUANE RELIABILITY GROWTH MODEL:

The Crow-AMSAA (CA) model is recommended to track the progress of reliability growth in a development process to determine when the configuration is good enough to put into production. With the Weibull distribution the individual times-to-failure are used, but the CA uses cumulative time. Weibull is one failure mode at a time but CA handles mixtures of failure modes. CA models a process where the overall reliability may be changing with cumulative time.

The CA model follows the equation:

$$n\,(t) = \lambda t^{\beta} \tag{4}$$

The log of cumulative failure events n (t) versus log cumulative time (t) is a linear plot.

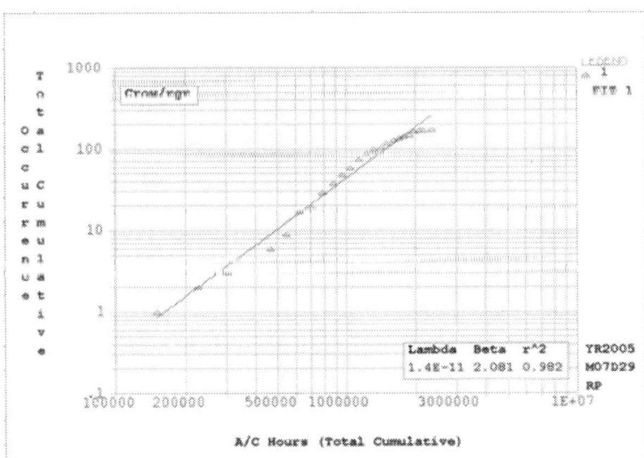

Fig. 5: Crow AMSAA (CA) Analysis

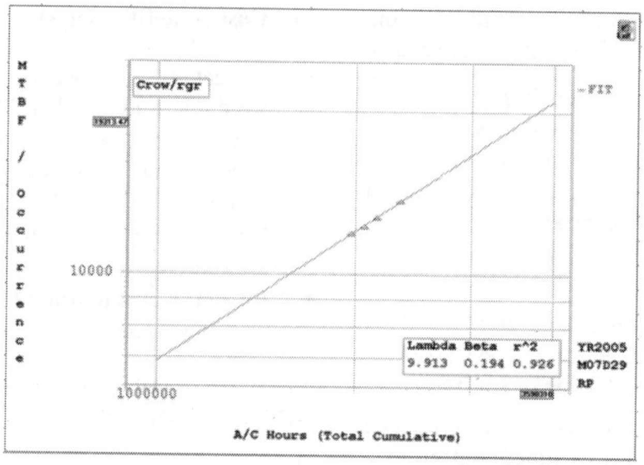

Fig. 6: CA Analysis Latest trend forecast

This analysis can be used for:

- Growth Trends
- Failure forecasting based on the latest trends.
- Spares and inventory management.
- To identify design improvements, seasonal variations in the reliability trend

5. REFERENCES

1. "The New Weibull Handbook" Fourth Edition – Dr. Robert Abernethy.
2. "Reliability Modeling Impact on Design"- Paper presented at "Applied Reliability Symposium 2005", San Diego, CA. – Visalakshi Subramaniam, Prakash VS, Sandeep Baliga, Ramji Sethu, Honeywell Technology Solutions Lab, Bangalore.

Reliability, Safety and Hazard
Editors: P.V. Varde et al.

RELIABILITY ASSESSEMENT OF SHUT-OFF ROD DRIVE MECHANISM FOR TAPP-3&4 AND 'CRITICAL FACILITY' THROUGH LIFE CYCLE TESTING

Manjit Singh[1], *D.N. Badodkar*[1], *N.K. Singh*[1], *N.S. Dalal*[1], *M.K. Mishra*[1], *G. Veda Vyas*[1], *C.B. Kothari*[1], *V.V.S.S. Rao*[2] *and R.K. Saraf*[2]

[1]Division of Remote Handling & Robotics (DRHR), BARC, Mumbai 400 085
[2]Reactor Safety Division (RSD), Bhabha Atomic Research Centre, Mumbai 400 085

ABSTRACT

Shut-off rod drive mechanism forms a safety critical system of a nuclear reactor. It is the space constraints for the given reactor layout, which makes design of shut-off rod drive mechanism (SRDM) a custom built design. Design of SRDM adopts fail-safe, replaceability and the simplicity criterion ensuring very high reliability of its operation.

Shut-off rod drive mechanism for TAPP-3&4 and 'Critical Facility' have been recently designed and developed at Division of Remote Handling & Robotics (DRHR), BARC. These are designed with a number of advanced features and these are significantly different than those used in Dhruva and 220 MWe PHWRs. Design of SRDM is qualified through prototyping and life cycle testing on a full-scale test set-up. This paper gives details of qualification and life cycle test data for prototype SRDM for TAPP-3&4 and 'Critical Facility' and reliability assessment.

Keywords: Primary Shutdown System (PSS), Shutdown System (SDS), Shut-off Rod Drive Mechanisms (SRDM), Reactivity Mechanism

1. INTRODUCTION

Control Mechanism Section (CMS) of DRHR has been working on design and development of reactivity mechanisms for research as well as power reactors. Recently, state-of-the-art new generation cable winch type electromechanical drive mechanisms have been designed for shut-off rods of TAPP-3&4 (India's first 540 MWe PHWRs), incorporating a number of advanced features. These mechanisms are significantly different from the mechanisms used in Dhruva and 220 MWe PHWRs. After successful completion of this development work, shut-off rod drive mechanisms having similar features have also been designed and tested for 'Critical Facility'. This is a low power research reactor, used for conducting lattice physics experiments for validating physics design parameters for future PHWRs and advanced heavy water reactor (AHWR).

Shut off rod drive mechanisms (SRDM) constitute safety critical system of the reactor. Its purpose is to shutdown the reactor under any emergency condition as well as under normal operating condition and to hold the reactor in shutdown state for prolonged period of time.

Design of SRDM adopts fail-safe, replaceability and simplicity criterion ensuring very high reliability of its operation.

These mechanisms are custom built to suit the given reactor layout. Being safety critical system, design of shut-off rod drive mechanism is qualified for reactor use through prototyping and its life cycle testing on a full-scale test station.

2. FUNCTIONAL REQUIREMENTS

Shut-off rod drive mechanism is designed to meet passive fast insertion of the absorber rod into the reactor core in a given short time limit. The drive mechanism is required to raise and hold the rod at the given park up position indefinitely. Design adopts 'fail-safe' non-reverse scram characteristics i.e. any failure in the drive mechanism will cause fast insertion of rod into the reactor core and the rod can not move out of the core on its own. At the reactor start up, rod withdrawal also meets limited reactivity addition capability i.e., rod withdrawal speed is limited and is decided by reactor physics considerations.

The drive mechanism is built with simple construction having modularity and interchangeability for ease of maintenance. Its design ensures high reliability giving long service life. Design includes provision for rod continuous position indication, end-limit indications and trips/alarms to check its safe operation and healthiness.

3. BRIEF DESCRIPTION OF THE SHUT-OFF ROD DRIVE MECHANISM ASSEMBLY

Shut-off rod drive mechanism is an electromechanical type of drive with rope and sheave arrangement. The absorber rod is attached to the drive mechanism using stainless steel wire rope having spherical shaped end terminals crimped at both ends using special tool. Cadmium is used as absorber element, which is sandwiched in stainless steel tubes in case of TAPP-3&4 and in Aluminium tubes in case of 'Critical Facility'.

The drive mechanism assembly consists of an electric motor, worm & worm gear, electromagnetic clutch (EM), gear reduction stages, mechanism housing with rope sheave arrangement, potentiometer for rod position as well as dashpot shaft position, hydraulic dashpot and an arrangement for direct and indirect rod end limit position indication.

SRDM for TAPP-3&4 and 'Critical Facility' are designed with a number of advanced features which include: Modular construction giving ease of maintenance; 90% free-fall giving consistency in rod drop performance, Hydraulic dashpot-two vane construction giving balanced design, feature requiring less force while re-setting, spiral spring for re-setting the dashpot during rod withdrawal; Use of rotary mechanical shaft seals; Provision for lubrication of rotating parts, etc.

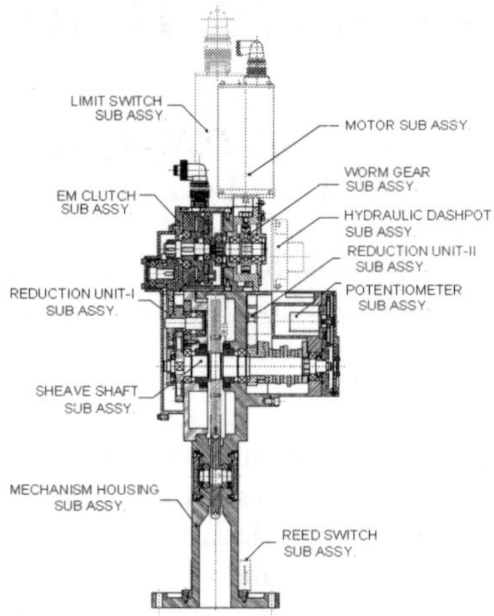

Fig. 1: General Assembly of SRDM for TAPP-3&4

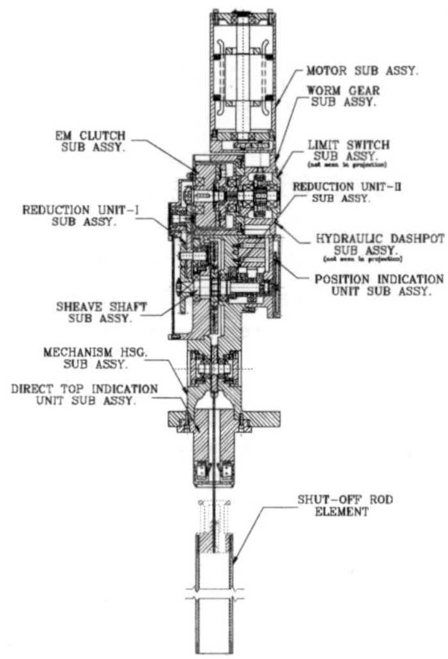

Fig. 2: General assembly of SRDM for 'Critical Facility'

During reactor start up, the rod is withdrawn at fixed speed such that maximum reactivity addition rate is limited from reactor safety consideration. After reaching the top end limit (park up position), motor power supply is cut-off and the rod is held in position through energized clutch and the irreversible feature of worm gear unit (built by design). On actuation of reactor trip, power supply to the clutch is cut-off causing absorber rod to fall freely into the core under gravity. This feature makes it 'fail-safe'. A hydraulic dashpot is used to bring the rod velocity smoothly to zero at the end of rod travel. Set of pick-up rings are used, which lapse one over the other, such that dashpot comes into action at about 90% rod travel. As 90% travel is under free fall condition, this feature results in consistent rod drop performance

For the purpose of reactor safety, non-availability of two most effective rods is considered in case of TAPP-3&4 while non-availability of single most effective rod is considered in case of 'Critical Facility'.

Table 1 gives relevant technical specifications for SRDM for TAPP-3&4 and 'Critical Facility'.

4. QUALIFICATION AND LIFE CYCLE TESTING

Being safety critical system, design of shut-off rod drive mechanism is qualified through prototyping and life cycle testing on the full-scale test set-up. Prototypes of SRDM for TAPP-3&4 and 'Critical Facility' have also undergone life cycle testing before the design was qualified for reactor use.

During cycle testing, rod actuation delay and 90% drop time were recorded and drop was monitored for abnormal noise, if any. Rod drop profile (sheave shaft rpm Vs time) & Dashpot oil pressure Vs time were recorded intermittently. Details of qualification and life cycle testing are given below:

Table 1 Technical Specifications

Parameter	SRDM for TAPP-3&4	SRDM for 'CF'
No. of rods	28	6
Total rod worth	72 mk	83 mk (for reference core)
Absorber element	Cd in SS tubes	Cd in Al tubes
Active length of the absorber	5500 mm	1500 mm
Total rod travel	6600 mm	2400 mm
Rod weight	50 Kgs.	10 Kgs
Time to raise the rod for full travel	150 ± 10 sec.	132 ± 5 sec
Rod drop time for 90% travel (without actuation delay)	1.8 ± 0.2 sec	1 ± 0.2 sec
Ambient temp.	65°C (max)	Normal ambient temp.
Radiation field (at mechanism location)	~ 10 R/hr	~ 10 R/hr

Testing of SRDM for TAPP-3&4

Prototype has been tested for more than 5000 drops with normal ambient at mechanism location and tank water (simulating calandria). Subsequently, same mechanism was further tested for more than 1000 drops with 65°C at mechanism location (using custom built oven) and 80°C of tank water. Rod drop performance for 90% drop time was found to be consistent with normal as well elevated temperature of 65°C at mechanism location. Also, effect of tank water temperature from normal ambient to 80C was found to have negligible effect on the rod drop performance. No maintenance was required during this testing. After completion of this testing, mechanism was dis-assembled and no significant wear of components was observed.

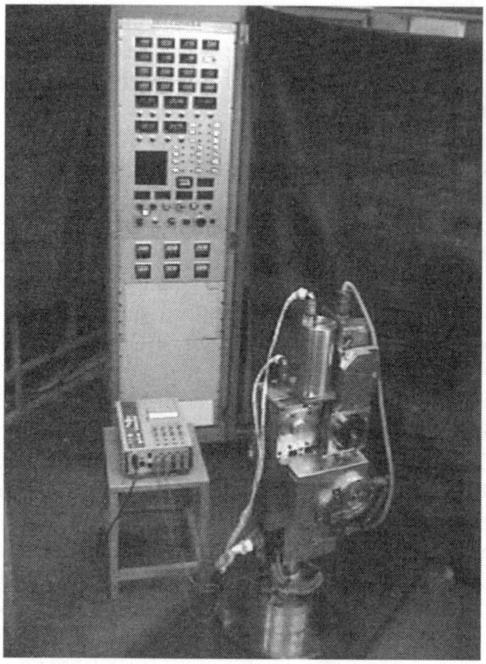

Fig. 3: Prototype SRDM for TAPP-3&4 with full-scale test facility at BARC

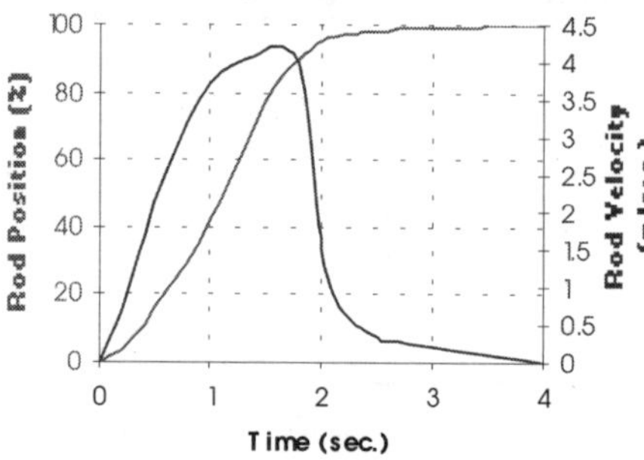

Fig. 4: Typical Rod Drop Profile for Shut-off Rod of TAPP3&4

TAPP-3&4 being power reactor, shut-off rods are held at park up position for long duration of time. Hence, in order to check their healthiness and availability on demand, SRDMs are provided with on-line test facility. In this, the clutch supply is cut-off for a short time such that rod leaves the park up position but does not enter the core region. Clutch supply is restored and rod is again taken up to its park up position. Prototype has been qualified for more than 3000 cycles (three successive demands per cycle).

Testing of SRDM for 'Critical Facility'

Prototype has been tested for more than 7000 cycles with normal ambient at the mechanism location as above. After testing, mechanism assembly was inspected and no significant wear of components was observed.

From rod drop data as above: (Ref. Fig. 7)

(90% drop time includes delay time of test console, which is in the range of 100 to 150 msec)

Minimum 90% drop time	:	1.010 sec.
Maximum 90% drop time	:	1.198 sec.
Mean of 90% drop time	:	1.107 sec.
Standard deviation	:	0.015 sec.

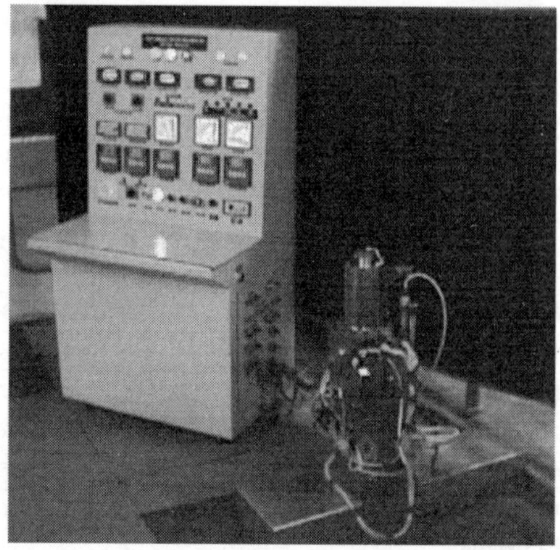

Fig. 5: Prototype SRDM for 'Critical Facility' with full-scale test facility at BARC.

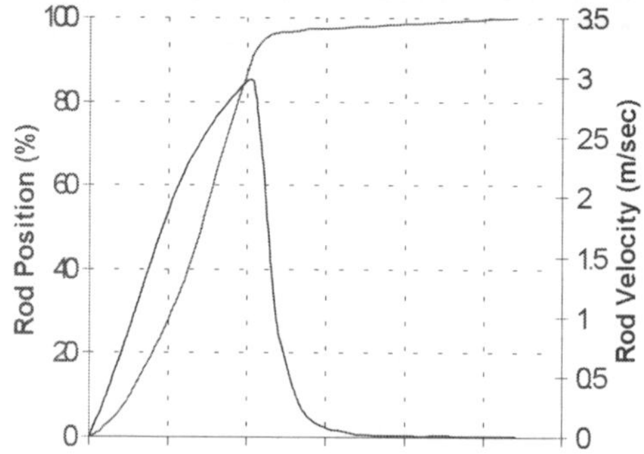

Fig. 6: Typical Rod Drop Profile for Shut-off Rod of 'Critical Facility'

5. DETAILS OF FAILURE MODES OF OPERATION

The shut-off rod drive mechanism assemblies are functionally tested at reactor site before installation, in an out-of-pile test station. After installation, assemblies are checked as per the check list (pre-requisites for operation) before these are put into operation.

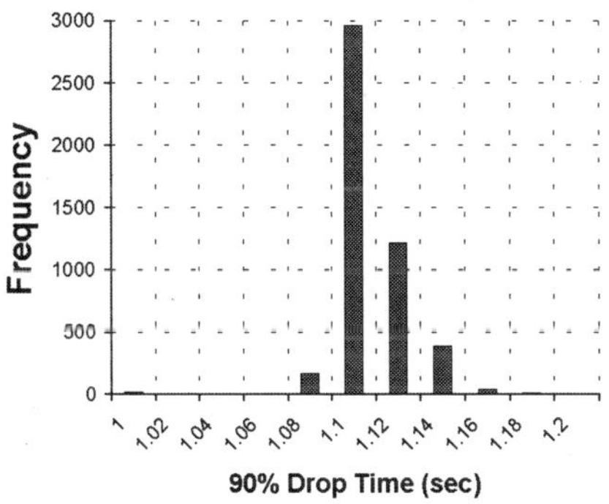

Fig. 7: A typical histogram for 4800 drops for shut-off rod of 'Critical Facility'

System operation under upset condition (Level-B) is considered as rope snap. Under this condition, absorber will fall into the guide tube. Guide tube is designed to withstand the impact load arising out of this event. If rope snaps with absorber under park up position, this event will be indicated by actuation of top limit switch and also reduction in reactor power. If rope snap occurs during rod withdrawal, then top limit switch will not actuate for 100% position of potentiometer. However, under this condition, motor gets cut-off through the mechanical stopper provided in the mechanism.

System operation under operating basis earthquake (OBE) is considered for Clutch failure, Dashpot failure, Bearing/Gear jamming and the Motor failure. During reactor operation, if clutch slippage takes place, then rod will leave the park up position and actuation of top limit switch will generate trip signal under 'All rod not out' condition in case of 'Critical Facility', however, with one or two rods leaving the park up position, reactor continues to operate in TAPP-3&4. Assembly is inspected after reactor shutdown. In case supply to clutch is cut-off and clutch fails to disengage it will result in rod failure to drop. The clutch is designed with release spring for its quick release under de-energised condition. However, under clutch failure to drop, shutdown system is built with a capability to make reactor subcritical with two most effective rods failing to drop in case of TAPP-3&4 and one most effective rod failing to drop in case of 'Critical Facility'.

Dashpot is used to bring the rod velocity smoothly to zero at the end of travel to limit the loads seen by mechanism components. In this design, dashpot comes into action only after 90% rod travel. Failure of the dashpot will not affect the rod drop performance, however, absence of dashpot may cause rope snap and damage of the mechanism components. Under this condition, after shutdown, mechanism assembly can be taken up for maintenance.

Failure of bearing or gear jamming is similar to clutch jamming event. Failure of the drive motor does not affect rod drop capability. However, under this condition, rod can not be withdrawn (for reactor start up) till the motor is repaired or power supply to the motor is restored.

6. RELIABILITY ASSESSMENT FROM EXPERIMENTAL DATA

The variation in drop time (with mechanism tested at normal ambient and specified design temperature) is within acceptable range given in the technical specification and there is no event of failure observed during the testing, the number of test demands is 6000 for TAPP-3&4 and 7000 for 'Critical Facility'. As mentioned above no significant deterioration of the components of the mechanism that can affect the performance was observed.

Upper limit of demand failure probability is estimated using F-Distribution for zero failures in the given demands as below:

$$P_{UC} = \frac{F_{0.95}(2,2D)}{D + F_{0.95}(2,2D)}$$

Where,

D – No. of demands

$F_{0.95}$ – 95% confidence limit for the variable from F-distribution table

Based on these test demands and zero observed failures the probability of failure on demand of the rod turns out to be less than 5 x 10^{-4} for TAPP-3&4 and 4.3 x 10^{-4} for 'Critical Facility' with 95% confidence.

7. CONCLUSION

Testing of prototype shut-off rod drive mechanisms for TAPP-3&4 and 'Critical Facility' on full-scale test set-up at DRHR, BARC has been successfully done for large number of drops. Testing shows high reliability for these mechanisms. Mechanisms are designed with fail-safe feature and built with higher safety margins to achieve longer service life and high reliability of its operation.

Reliability assessment from the experimental data (life cycle testing) shows rod demand failure probability to be of the order of $5x10^{-4}$ year per year with 95% confidence.

Cycle testing was carried out simulating reactor operating condition i.e. elevated temperature at mechanism location and tank water (simulating calandria) temperature to 80°C in case of TAPP-3&4, except radiation level. It is expected that presence of radiation field under actual operating conditions will not affect rod drop performance. Also, materials used in the mechanism are selected to have very high radiation resistance of the order of 10^7 - 10^8 rad, giving life of more than 100 years. However, in case of material degradation, if any, these are replaceable during annual shutdown.

8. REFERENCES

1. Design Basis Report on Shut-off Rod Drive Mechanism, No. TAPP-3&4/DBR/31450/01, Rev. 2 (NPCIL).
2. Manjit Singh and G. Govindarajan, 'Evolution of shutdown mechanism for PHWRs', *Workshop on Reactor Shutdown System*, March 1997, IGCAR, Kalpakkam.
3. D. N. Badodkar, et. al. , 'Design & Development of Drive Mechanisms for Adjuster Rods, Control Rods and Shut-off Rods of TAPP-3&4', *BARC Newsletter*, No. 209, June 2001, PP 10-16.
4. Design Basis Report on Shut-off Rod Drive Mechanism for 'Critical Facility', Report No. DRHR/CMS/CF/2004/01, Rev. 1, Issue June 2004. (BARC)
5. Summary of Design Qualification and Life Cycle Tests of Prototype Shut-off Rod, Control Rod and Adjuster Rod drive Mechanism for TAPP-3&4 at DRHR, BARC', Issue: Dec. 2004. (BARC)
6. N. K. Singh et.al., 'Design and Development of Shut-off Rod Drive Mechanism for Critical Facility', *BARC Newsletter*, No. 247, Aug. 2004, PP 1-10.
7. A. K. Babar, 'Probabilistic Safety Assessment and Equipment Qualification, *National Conference on Nuclear Reactor Technology*, BARC, Mumbai, Nov. 2002.
8. Gopika Vinod, R. K. Saraf, A. K. Babar, V. V. S. S. Rao, Rajiv Tharani, 'RELDATA-A Tool for Reliability Database Management', Report No.: BARC/2000/E/041.

Reliability, Safety and Hazard
Editors: P.V. Varde et al.

ELIMINATING PROCESS DEFECTS THROUGH HIGHLY ACCELERATED STRESS SCREENING (HASS)

R. Muthukumar and D. Damodaran

Centre for Reliability, Department of Information Technology, Dr.V.S.I. Estate, Thiruvanmiyur, Chennai 600 041

ABSTRACT

Highly Accelerated Stress Screening (HASS) eliminates automated process defects called infant mortality. Temperature cycling and random vibration were chosen as stress screens for a product Automatic Voltage Regulators used in alternators and generators and the experiment conducted. Temperature cycling specifications are: -40° C to + 90° C, dwell time of 15 minutes, rate of change of temperature of at least 20° C/min., No. Of cycles = 10. The Random Vibration specifications are: 12 Grms, frequency range 10 Hz to 2KHz in 2 axes with 15 minutes duration in each axes. Identified defects were deliberately injected in some of the units and screening conducted along with good units. Defective units were detected after the screening. The optimized temperature cycling screen was repeated over 270 times. During this time the wear out failure mechanism in the metallisation of LM 324D, which is a SO – 14 package, electrolyte totally drying up in electrolytic capacitor and resistive element erosion in a film resistor, were noticed. This proved that only a thermal cycle of 10 cycles in the specifications arrived in above, no way affects the fatigue life of the product. Thus this HASS screen proved to be an effective methodology to eliminate process defects from the automated production process and also does not eat away the fatigue life of the units.

Keywords: HASS, POS, EOF, Thermal Cycling, Random Vibration, Grms, Automatic Voltage Regulator, Optimization, Limit determination, Dwell time

1. INTRODUCTION

The prototype hardware design is earmarked for automated full-scale production in huge volumes after the hardware attains maturity. The full-scale production process involving heavily automated machines and fabricating equipments produces infant mortality along with good population [7], which need to be eliminated before the products shipment. A process called Highly Accelerated Stress Screening (HASS) can do the elimination of these defects. The product Automatic Voltage Regulator (AVR) with a bill of materials of about 252 components, used in regulation of supply to the starter unit of alternators and generators was chosen for experimentation. This product is produced in huge volumes with automated production process. Infant mortality failures are reported from the field for this product. The process of screening these defects involves choosing the right stress screen among the available screens like thermal cycling, random vibration, power cycling, voltage margining, frequency margining etc. suitable enough to stimulate failure mechanisms of different modes. Optimizing the screens for maximum efficiency evidenced by detection of the seeded defects and applying the screen to the production samples, eliminates the defective units. **Care shall be taken not to consume the fatigue life of the product**. Repeated cycling till wear out mechanisms are activated ensures non-consumption of fatigue life. Thus HASS screen was found to be an effective methodology to screen out infant mortality and process defects in the production samples and proved to be an effective tool in Reliability improvement of the products.

2. HIGHLY ACCELERATED STRESS SCREENING (HASS)

Highly Accelerated Stress Screening (HASS) is an accelerated aging process [5] sufficiently optimized to eliminate infant mortality population from a population of units manufactured by automated process and built with infant mortality. Design of HASS specification limits for various stresses viz. temperature cycling, random vibration, power cycling, voltage margining, frequency margining etc., involves the limit determination exercise. Limit determination [1] is the study of operating limits and destructive limits of the units under screening for each of the stresses chosen for screening. The limits are single sided and sometimes double sided also. The various limits determined are specification limits, operating limits and destructive limits. The optimum level of the HASS will be above the operating limits and well below the destructive limits. The experimentation starts with step stress approach from the operating limit until an optimum level with good percentage of defect fall out is reached. Then Proof of Screen (POS) [5] is run to verify that the screen is efficient and also it does not eat away much of the fatigue life of the product. Firstly, efficiency of the screen is checked by deliberately injecting defective products along with good samples and when subjected to the optimized screen simple functional testing identifies the bad units. Secondly the life consumption is checked by repeating the optimized screen for over 250 times to check for end of life failures (EOF) cropping up and at that time it is better to stop the screening and the required number of cycles to initiate wear out mechanisms is noted. A portion of the total number of cycles is only adopted as a regular screening cycle and adopted on 100% basis on the production samples. Thus we know how much life is consumed during the screening for infant mortality. This will ensure that enough fatigue life is still left over in the product for good endurance in the customer hands. Thus HASS screening methodology is an effective control tool to eliminate infant mortality defects.

3. CASE STUDY

A. Product

For experimentation, an Automatic Voltage Regulator (AVR) used to protect the Generators, Alternators and voltage or frequency sensitive loads, was taken for the experimentation. The AVRs have a bill of materials of 212 electronic components and materials. The unit has dimensions of 10 x 9.3 x 3.5 cms and weighs approximately 200 gms. The component list includes resistors, capacitors, driver and switching transistors, regulating diodes and comparators. A unique fixture was designed to accommodate six units with separate input supply and output load exciter coils so that it is easy to handle them during testing in chambers and vibration tables. The units were numbered from G1 to G8 and functional testing was carried out individually for all the units after every exposure. Experience guided us to select temperature cycling and random vibration [6] as suitable stress screens for the product which can stimulate out infant mortality failures. For these two screens the limit determination exercise was performed to arrive at the operational and destructive limitations.

B. Limit Determination

Limit determination [1] was carried out to arrive at the operating limits and the destructive limits of the units, for the stimulating stresses viz. temperature and random vibration. The experimentation was started from the specification limit as provided for the product which is from – 10° C to + 70° C. As the specification is two sided the limit determination has to proceed in both the directions of +ve and –ve sides. It is a step-by-step process of attaining the limitations of the product. The fixing up of the steps goes by a thumb rule of 15 to 20% more than the previous stress level. At each level of stress functional testing was carried out and the health of the units monitored. For Vibration stress the specification limits provided for the product is 5g in each perpendicular planes. As for stimulating purpose Random Vibration is the best-chosen stress, the limit determination was arrived at for random vibration alone. Also the specification is one sided only of having no negative component. Eventually we started from low values of Grms (unit of random vibration). Thus the limits arrived at are given in the table below.

Table 1.1 Operating Limits (OL) and Destructive Limits (DL) determined for the AVRs

Limitations	Level
High Temp OL	+ 90° C
Low Temp OL	- 15° C
Random Vibration OL	10 Grms
High Temp DL	> 110° C
Low Temp DL	< - 55° C
Random Vibration DL	14 Grms

C. Optimization Process

The optimization process is arriving at the best-suited stress screen given the type of screen equipment, its capabilities, the fixtures etc. Conventional chambers and Vibration machines did not work well for the stress-screening program. The parameters optimized for the chambers for carrying out thermal cycling are a) High temperature limit b) Low temperature limit c) Rate of change of temperature d) Dwell time at each temperature limits e) the air velocity around the samples in the workspace and f) number of cycles. High temperature and Low temperature limits are arrived at based on the limit determination exercise. Rate of change is determined in such a way that lengthy stabilization times and inadequate product temperature changes are avoided. For this, mounting mobile sensors on carefully chosen thermally lagging devices did a thermal survey exercise. This was found to be at least 20° C/min. and the dwell time was determined to be 15 minutes in each of the temperature limits. The air velocity within the workspace in the direction of air flow made laminar by honey comb structure is maintained at 1.8 meters/sec to maintain the rate of change of temperature and the heat transfer rate commensurate to the thermal mass of the AVRs. The number of cycles was arrived at as 10, as beyond 10 cycles the screen efficiency at that level of stress appeared to be stabilizing. The temperature cycling profile looks as below with very less dwell time and rapid rate of change of temperature.

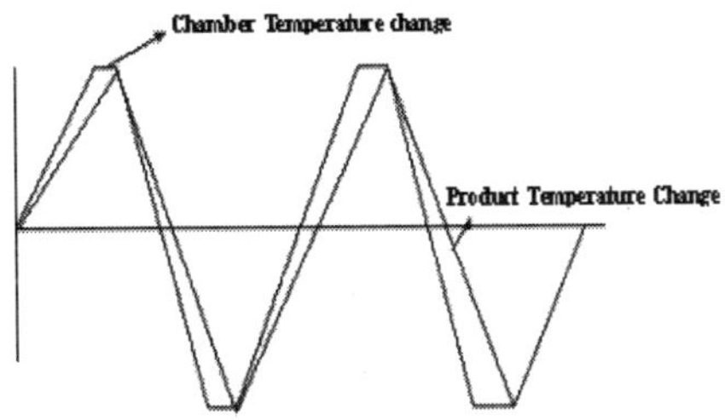

Fig. 1: Thermal cycling profile.

Fast changes of stresses in the above fashion results in high efficiency of screening in terms of defect fallout [6].

The parameters optimized for the random vibration testing are a) the frequency range, b) Grms (Acceleration Spectral Density) level, c) the duration of exposure and d) the axes of vibration. The frequency range was selected based on the study made on the possible natural frequencies of the components used in the units. This was found to be in the range of 10 Hz to 2 KHz. The optimization was done for the Grms level by starting from 10 Grms, which is the operating limit. It was found that 12 Grms level is enough to bring out the soldering defects and the heat sink design defects. The duration of exposure was determined as 15 minutes as this was found to be more yielding and beyond this time of exposure was not advantageous. Two axes of vibration one in the vertical and the other in the horizontal direction (X and Y) was found to be adequate enough to precipitate the flaws. Fixtures were designed to hold the printed circuit boards in a stable condition, as random vibration exposure is high energy content. Care was taken to prevent the shaker from going into resonance. The response from the units was obtained by mounting an accelerometer, as feedback and continuously monitored during the screening.

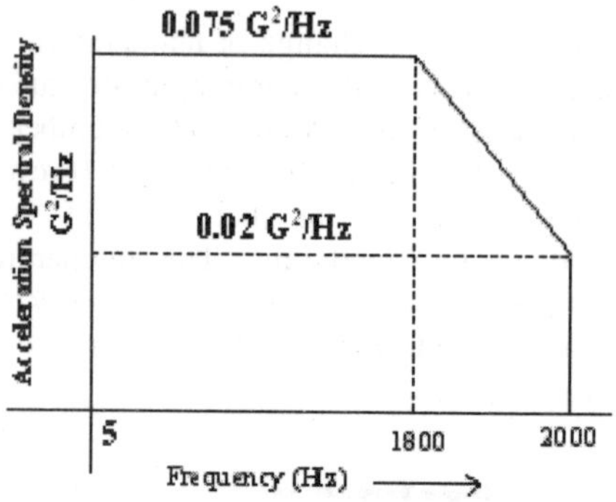

Fig. 2: Random Vibration profile for 12 Grms

Table 1.2 Thermal Cycling screen specifications

Factors	Specification Range
Temperature Range	-40°C to +90°C
Dwell Time	15 minutes
Rate of Change of temperature	Atleast 20°C/min.
Air Velocity within the Chamber	1.8 meters/sec
Number of cycles	(20) 10 + 10
Conditioning	Energized for 10 cycles and Un-energised for 10 cycles

The Random Vibration screen profile for 12 Grms Acceleration Spectral Density (ASD) in the frequency range of 1 to 1.8 KHz is given above. Above 1.8 KHz, there is a 3dB/Octave fall in the ASD level, indicating less fallout in this frequency band. The critical frequencies at which maximum amplitude can be obtained for components fall in this band.

Table 1.3 Random Vibration screen specifications

Factors	Specification Range
Vibration Level	12 Grms
Frequency Range	10 Hz to 2 KHz
Rate of Change	3 dB/Octave rise and fall
Number of Axes	2 (Horizontal and Vertical)
Duration	15 minutes in each axis
Conditioning	Un-energised

D. The Recommended HASS Screen

The HASS screen specification arrived at for the AVRs is given in the table below.

The thermal cycling and Random Vibration specification has been arrived at based on experimentation by adopting the step-by-step process of incrementally advancing the stress from the operating limit. The failure criteria for the units for continuous monitoring of the health of the units were fixed as fluctuation in load current proportionate to the reduction in AC voltage. The first step was from -20°C to +70°C and followed with random vibration at 6 Grms. This progressed up to -40 °C to +90°C, (90V to 270V) and 12 Grms while all the defective units were detected. The units were subjected to Thermal Cycling as stimulator and Random Vibration as detector of the failure mechanisms alternately [4].

The step-by-step procedure adopted for arriving at the suitable screen specification is given above in the table for units 1 and 2. The Step Stress process for units 3 to 8 is given below in the table 1.4.

Table 1.4 Step Stress Process

	Temperature cycling / Voltage variation (Duration: 20 minutes in each of the temperature extreme) and Random Vibration (Frequency 1 Hz to 2 KHz)		
Stress Levels	-20 °C to +70°C / 6 Grms	-30 °C to +80°C / 10 Grms	-40 °C to +90°C (90V to 270V) / 12 Grms
Unit 1	Okay	Okay	Okay
Unit 2	Okay	Fluctuation in load current from –0.049A to –0.062A and proportionally there was reduction in AC voltage	Fluctuation in load current from –0.049A to –0.062A and proportionally there was reduction in AC voltage
	Temperature cycling / Voltage variation (Duration: 20 minutes in each of the temperature extreme) and Random Vibration (Frequency 1 Hz to 2 KHz)		
Stress Levels	-20 °C to +70°C / 6 Grms	-30 °C to +80°C / 10 Grms	-40 °C to +90°C (90V to 270V) / 12 Grms

Unit 3	Okay	Fluctuation in load current from –0.050A to –0.062A and proportionally there was reduction in AC voltage	Fluctuation in load current from –0.050A to –0.062A and proportionally there was reduction in AC voltage
Unit 4	Okay	Fluctuation in load current from –0.054A to –0.062A and proportionally there was reduction in AC voltage	Fluctuation in load current from –0.049A to –0.062A and proportionally there was reduction in AC voltage
Unit 5	Okay	Okay	Okay
Unit 6	Okay	Okay	Okay
Unit 7	Okay	Okay	Fluctuation in load current from –0.049A to –0.061A and proportionally there was reduction in AC voltage
Unit 8	Okay	Okay	Okay

From the two tables above it can be seen that three steps are involved before we arrive at the optimum specification for stress screen. Units 2, 3, 4 and 7, which were seeded with the defects, as outlined above have been detected to fail after the application of the stress screens. Unit 7 failed only when the stress level have been taken to the extent of -40 °C to +90°C and followed with 12 Grms random vibration. Further during the thermal cycling under energized condition, the voltage was varied from 90V to 270V. This was optimum enough to bring out the failure in Unit 7. The units which are loaded with the defects such as wide tolerance resistors, low-grade filter capacitors with low range of voltage ratings in the field current conditioning & controlling section of the AVR, resulted in fluctuation in load current and proportionately there was reduction in AC output voltage. This demonstrated that the HASS screen chosen was adequate enough to bring out these process related failures. Therefore the screen can be applied as a full-scale production screen.

E. Proof of Screen (POS)

Proof of Screen (POS) [5] is an exercise to verify the effectiveness of the HASS screen as designed above in precipitating the defects out of the production units. Further the optimized thermal cycling screen shall be repeated for over 250 times in order to initiate End Of Life (EOF) failures in the devices of the units under screen. This is to prove that enough fatigue life is still left in the AVRs for customer use even after 10 cycles of screen operation. In order to prove the efficiency of the optimized screen, FOUR units were identified for deliberately injecting the known failures and subject them to HASS screen along with the good ones. The screen shall bring out the injected failures without missing them [3]. The defects seeded were:

a) SCR Firing section: - A low-grade box capacitor is used which generates pulses with a peak voltage of 4 V, peak current 12mA and duration of 1mS. This capacitor is degraded faster because of the automated assembly process by machines, which pass on stresses on to these devices due to their fragile quality. These stresses if remains unrelieved, they show up in the field, which we earnestly propose to eliminate during the screening program. b) Controller Section: - A low-grade electrolytic capacitor, box capacitor and some low temperature version resistor were placed, which are degraded due to the automatic processing and machine handling. c) Field current conditioning & controlling Section: - Wide tolerance resistors, low-grade filter capacitors with low range of voltage ratings, which upon degradation show up during the screening exercise. The unit failure criteria were fixed as Fluctuation in load current proportionate to the reduction in AC voltage due to any one of the above defects in the units. The proof of screen was conducted on the 8 units supplied by the customer and the defects as seeded above in the four units were detected easily by performing the full functional test using the test bench provided by the manufacturer at the end of the stress screen application. For executing this screen on 100% basis automatic screening equipments have to be put in place to ensure screening as a routine procedure in the shop floor. Due to test equipment shortcomings the screen specifications shall not be down graded, which will be a fruitless exercise. This HASS exercise will take care of the change in vendors of components, process changes, technology changes etc., which are envisaged by the manufacturer in the future.

Fig. 3: Metal migration seen in the metallisation of IC LM324 D

The second part of POS is that the minimum fatigue life deterioration in the AVRs is to be substantiated. This can be done by conducting the optimized thermal cycling screen repeatedly for over 250 times. Thus the cycle was repeated 252 times and the EOF failures viz. metal migration [2] in IC LM324 D, electrolyte totally drying up in electrolytic capacitor C5 and film erosion in resistor R4, are reported at this time. This is quite a longer period for the 10 number of cycles optimized and proposed as the effective screening procedure for the AVRs.

Fig. 4: Dry rolled element indicating long duration of operation in capacitor C5.

With the appearance of these failure mechanisms the experimentation was stopped. This substantiates that 10 recommended cycles is only 4% of the total fatigue life for the end of life mechanisms to get initiated, which is a reasonable sacrifice for the betterment of the product.

The first point of termination of the thermal cycling occurred while the fluctuation in the output load current happened owing to the metal migration defect due to wear out of the aluminium metallisation. This was substantiated by a detailed component physical failure analysis activity. Further when the field current and conditioning section was examined, an electrolytic capacitor was misbehaving. This was found to be due to total drying up of the electrolyte as can be seen in the photograph, as otherwise normally the rolled element is wet.

4. CONCLUSION

The Automatic Voltage Regulators were HASS screened for possible process defects, which can pose Reliability problems in the field. The proof of screen was run and demonstrated that the screen specification adopted were effective in bringing out the process problems. The suggested screen as given above may be implemented on a 100% basis in the production line of automatic voltage regulators. This proved the effectiveness of the HASS screen in eliminating infant mortality failures caused due to process defects.

5. REFERENCES

1. Silverman, M. 1997. Summary of HALT and HASS results at an accelerated reliability test center. Nepecon West Conference, 866 – 892.
2. Pecht, M., Lall, P., and Whelan, S.J. 1990. Temperature dependence on microelectronic device failures. Quality and Reliability Engineering International, 6:275 – 284.
3. Hobbs, G., and McLean, H. 1994. Is uniformity and repeatability essential to vibration and temperature screening? Sound and Vibration (April): 22 – 23.
4. Caruso, H. 1983. Significant subtleties in stress screening. Proceedings of the Institute of Environmental Sciences (IEST), 154 – 158.
5. Harry W. McLean. 2000. HALT, HASS & HASA Explained: 31 – 53.

6. Wong, K. 1989. A new environmental stress screening theory for electronics. Proceedings of the Institute of Environmental Sciences (IEST): 43 – 65.

7. O' Connor, P.D.T., 1996. Practical Reliability Engineering, John Wiley and Sons, New York.

Reliability, Safety and Hazard
Editors: P.V. Varde et al.

RELIABILITY EVALUATION STUDIES OF AERIAL GAMMA SPECTROMETRY SYSTEM

Probal Chaudhury, M.S. Kale, G. Krishnamachari, N. Raman, V.V.S. Sanyasi Rao[1], Pradeepkumar K.S. and D.N. Sharma

Radiation Safety Systems Division, [1]Reactor Safety Division
Bhabha Atomic Research Centre, Mumbai 400 085

ABSTRACT

The Aerial Gamma Spectrometry System (AGSS) is a field instrument developed in Bhabha Atomic Research Centre for quick qualitative and quantitative assessment of radioactive contamination over large area on ground in case of a major nuclear / radiological emergency. This necessitates the system to be operated for aerial surveys above complex topography and harsh environmental conditions. The AGSS has been designed taking into consideration all operational aspects with allowance for variations in parameters such as current, voltage, temperature, humidity etc. The measures taken at the design stage of the AGSS are intended to ensure the reliability of the system during the survey operations under various environmental conditions. The AGSS has been built around an Industrial Computer and consists of NaI(Tl) Detector, Signal Processing Module and Data Storage and Interface Module, which are PC add-on cards and a Global Positioning System (GPS). The Industrial Computer, GPS and the detector are commercially available and they have undergone reliability assessment by the manufacturer. The reliability assessment studies of the Signal Processing and the Data storage and Interface modules have been done in-house. This paper describes the details of the reliability study carried out on AGSS components. The results of reliability evaluation show that the AGSS will provide reliable service if operated well within the limits of operating conditions during aerial / field surveys. The chances of functional failure of AGSS are found to be less than two times in thousand operations of eight hours duration each.

Keywords / Phrases: Aerial Gamma Spectrometry System, Reliability Evaluation, Aerial Survey, Radiological Emergency, Airborne Rotary Winged Environment

1. INTRODUCTION

The Aerial Gamma Spectrometry System (AGSS) [1] has been developed in Bhabha Atomic Research Centre for quick qualitative and quantitative assessment of radioactive contamination over large area on ground resulting from large-scale radioactive release. AGSS is a field instrument to be operated for aerial surveys above complex topography and harsh environmental conditions. Environmental factors such as vibration, thermal cycling and thermal shock accelerate failures in electronic components and assemblies in an airborne environment [2]. The AGSS has been designed taking into consideration all operational aspects with allowance for variations in parameters such as current, voltage, temperature, humidity etc. The measures taken at the design stage are intended to ensure the reliability of the system during the surveys under various conditions. This paper describes the details of the reliability study carried out on various AGSS components that are developed in-house.

2. SYSTEM DESCRIPTION

The AGSS has been configured around an Industrial Computer and consists of NaI(Tl) Scintillation Detector, Signal Processing Module (SPM) for spectral data acquisition and Data Storage and Interface Module (DSIM), which are PC add-on cards and a Global Positioning System (GPS). The system also consists of software [3] for the system control, data acquisition and data processing operations developed in-house. The block diagram of the system is shown in the Figure 1.

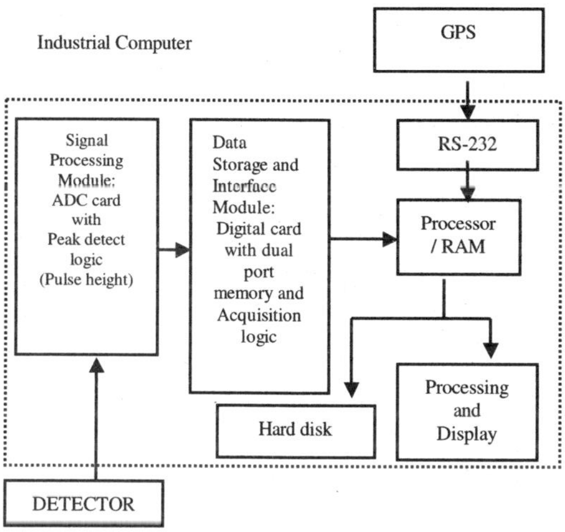

Fig. 1: Block diagram of Aerial Gamma Spectrometry System.

The signal-processing module receives signal form the detector, which is suitably amplified and shaped before Analog to Digital conversion. The amplified signal is digitized using nuclear ADC and is processed by the data storage and interface module and the spectral data is stored in the dual port memory. At the end of acquisition time spectrum is transferred to the computer memory and is tagged with the positional information received from the GPS receiver through RS-232 interface. The data is processed and displayed on the screen and is stored in the hard disk for detailed offline analysis and generation of contours [4].

3. RELIABILITY EVALUATION

The system is a series system from reliability point of view i.e. proper individual functioning of all the subsystems is needed for the successful performance of AGSS. The system reliability R_{SYS} is expressed as

$$R_{SYS} = R_{SPM} \times R_{DSIM} \times R_{PC} \times R_{DET} \times R_{GPS}$$

where, R_{SPM} is Signal Processing Module Reliability, R_{DSIM} is Data Storage and Interface Module Reliability, R_{PC} is Industrial Computer Reliability, R_{DET} is Detector Reliability and R_{GPS} is GPS Reliability. As the system is a series system common cause failures are not considered.

For the evaluation of reliability, the Signal Processing Module and the Data Storage and Interface Module, which are developed in-house, are only considered. The Industrial Computer, GPS receiver and the detector are commercially available and they have undergone reliability assessment by the respective manufacturers. The power supply for the system is used from the standard 28 V DC available in the aircraft for aerial surveys and is not considered here for reliability evaluation. The reliability analysis of the system software is being evaluated separately and is not covered in this paper.

The reliabilities are evaluated for a given mission time. The various subsystems of the two modules consist of mainly electronic components. The failure rates of the electronic components over the operational range are constant. Hence the reliability of the subsystems can be expressed as

$$R_{SMP} = \exp(-\lambda_{SPM} \times T),$$
$$R_{DSIM} = \exp(-\lambda_{DSIM} \times T),$$
$$R_{PC} = \exp(-\lambda_{PC} \times T),$$
$$R_{DET} = \exp(-\lambda_{DET} \times T) \text{ and}$$
$$R_{GPS} = \exp(-\lambda_{GPS} \times T)$$

where 'T' is the mission time and λ_X is the failure rate of the subsystem X. The mission time is usually around 2 to 3 hours. However, conservatively a mission time of eight hours is assumed in this evaluation. The system reliability expression turns out to be

$$R_{SYS} = \exp(-\lambda_{SYS} \times T)$$

where $\lambda_{SYS} = \lambda_{SMP} + \lambda_{DSIM} + \lambda_{PC} + \lambda_{DET} + \lambda_{GPS}$. The various types of components used in the two PC add-on cards are resistors, potentiometers, ceramic capacitors, electrolytic capacitors, mica capacitors, other types of capacitors such as paper, metallized paper, metallized plastic etc.,

TTL ICs, Linear ICs, Voltage Regulator ICs, Memory ICs, CMOS ICs, IC Sockets, Diodes, Transistors, Fuses, Crystals, and Connectors. All the components of the subsystem are required to function for a successful and trouble free operation of the subsystem. Hence the subsystem failure rate is expressed as the sum of the failure rates of the components it consists of using procedures specified in reference [5].

Thus, $\lambda_{subsys} = \sum_{i=1}^{n} \lambda_{comp.i}$

where, n is the no of components in subsystem and $\lambda_{comp.i}$ is the failure rate of the ith component. Only some modes of component failure lead to system failure. However, the failure rate corresponding to all modes of component failure is considered in evaluating subsystem failure rate. The operating voltage and power levels of these components are well below the rated or designed values. Hence the chances of either open or short circuit across resistors and/or across capacitors are minimized. However, the contribution due to these failures is taken into consideration in evaluating the failure rates of the subsystems.

Airborne rotary winged environment has been taken into consideration in modifying the base failure rate as the system is intended for use in a helicopter. The other factors considered are quality factor corresponding to the quality of manufacturer of commercial grade components, ambient temperature of 50°C, ratio of operating voltage to rated voltage and ratio of operating power to rated power of the components. The component failure rates are evaluated at their respective operating conditions. Fixed metal film resistors with higher power ratings than the actual dissipation have least temperature coefficient and are more stable. Hence these have been extensively used in the circuits. Proper soldering of components, anchoring of the add-on cards and their interconnections have been assumed for this analysis. This assures the overall reliability of the system. The component wise failure rate arrived at are as shown in Table 1 and Table 2 for the Signal Processing Module and the Data Storage and interface Module respectively. These were derived from data specified in MIL-HDBK-217F [5].

Table 1 Failure rate and Reliability of Signal Processing Module

Component Name	Failure Rate* of Individual Component A	No. of Components B	Failure Rate* A × B
Metal Film Resistors	0.27	150	40.5
Ceramic Capacitors (10 nF)	0.35	79	27.65
Electrolytic Capacitors (100 μF)	1.41	4	5.64
Silver Mica Capacitors (33 μF)	0.0019	18	0.0342
TTL IC's (14-PIN)	0.078	15	1.17
TTL IC's (16-PIN)	0.091	6	0.546
TTL IC's (20-PIN)	0.13	3	0.39
Linear IC's (8-PIN)	0.056	25	1.4
Linear IC's (14-PIN)	0.091	2	0.182
IC Socket's (8-PIN)	0.024	25	0.6
IC Socket's (14-PIN)	0.032	17	0.544
IC Socket's (16-PIN)	0.036	6	0.216
IC Socket's (20–PIN)	0.042	3	0.13
Crystal (10 MHz)	25.58	1	25.6
ISA Bus Connector	0.42	1	0.42
		Total	105.02
Failure rate (λ_{SPM}) = 105.02 x 10^{-6} /h Reliability (R_{SPM}) = exp ($-\lambda_{SPM} \times$ T), T: Mission Time = 8 h = 0.9992			

* per million hours

The Failure Rate contribution of the Industrial Computer (λ_{PC}) is 1 / MTBF. The Mean Time Between Failure (MTBF) of the Industrial Computer is 100,000 hours as specified by the manufacturer. Hence, $\lambda_{PC} = 10 \times 10^{-6}$ / h and the reliability for Industrial Computer (R_{PC}) is 0.99992 for a mission time of 8 hours.

NaI(Tl) scintillation detectors are manufactured according to the highest standards of design and production. These are intended for use in a normal laboratory environment and will operate reliably between -5^0 C and 45^0C. The detectors used for aerial applications along with the AGSS are well insulated from both mechanical and thermal shocks and so the failure rate of the detectors is extremely small, in comparison with all other electronics. However, a value of 10^{-6} /h for detector failure rate (λ_{DET}) has been assumed. Hence, the reliability of the detector (R_{DET}) is 0.999992.

Table 2 Failure rate and Reliability of Data Storage and Interface Module

Component	Failure Rate* of Individual Component A	No. of Components B	Failure Rate* A × B
Metal Film Resistors	0.27	33	8.91
Ceramic Capacitors (10nF)	0.35	38	13.3
Electrolytic Capacitors (100 μF)	1.41	4	5.64
TTL IC's (14-PIN)	0.078	10	0.78
TTL IC's (16-PIN)	0.091	13	1.18
TTL IC's (20-PIN)	0.13	11	1.43
Memory IC's (48 PIN)	0.30	4	1.2
IC Socket's (14 - PIN)	0.032	10	0.32
IC Socket's (16 – PIN)	0.035	13	0.46
IC Socket's (20 – PIN)	0.042	11	0.46
IC Sockets (48 – PIN)	0.096	4	0.38
Crystal (1 MHz)	14.97	1	14.97
ISA Bus Connector	0.42	1	0.42
		Total	49.45
Failure rate (λ_{DSIM}) = 49.45 x 10^{-6} /h Reliability (R_{DSIM}) = exp ($-\lambda_{DSIM}$ x T), T: Mission Time = 8 h = 0.9996			

* per million hours

The GPS is a navigational aid frequently used for marine purposes and is highly reliable. Since this equipment meets the marine equipment standards, the failure rate of the GPS (λ_{GPS}) is considered to be 10^{-6} /h on a conservative basis and the reliability of GPS (R_{GPS}) is 0. 999992.

The AGSS reliability, R_{AGSS}

$= R_{SPM} \times R_{DISM} \times R_{PC} \times R_{DET} \times R_{GPS}$

$= 0.9992 \times 0.9996 \times 0.99992 \times 0.999992 \times 0.999992$

$= 0.9987$

Hence, the chances of failure of AGSS to meet the performance requirements are found to be thirteen times in ten thousand operations of eight hours duration each i.e. less than two times in thousand operations.

4. CONCLUSIONS

Taking into consideration the use of proper components and the above reliability analysis, it could be concluded that the system works reliably when operated well within the specified ratings. Several successful aerial and land-based surveys (~50) have been conducted and this system has been found to work reliably. This type of reliability analysis gives the confidence about the design of the system for use in field conditions usually encountered during surveys.

5. ACKNOWLEDGEMENTS

We are grateful to Shri H S Kushwaha, Director, Health, Safety and Environment Group, BARC for his constant encouragement and guidance. The help rendered by Shri N. Padmanabhan and Shri Amit Jain during the course of this work is gratefully acknowledged.

6. REFERENCES

1. A Compact Industrial PC based Aerial Gamma Spectrometry System, G. Krishnamachari, M.S. Kale, Probal Chaudhury, Amit Jain, M.M.K. Suri, D.N. Sharma and M.C. Abani, Proceedings of Symposium on Intelligent Nuclear Instrumentation (INIT-2001), February 6-9, 2001, Mumbai. pp 58 63.
2. Electronic Failure Analysis Handbook, Perry L. Martin, McGraw-Hill
3. An integrated software package for control and data analysis of Aerial Gamma Spectrometry System (AGSS), Probal Chaudhury, M.S. Kale, N. Raman, M.M.K. Suri, G. Krishnamachari, K.S. Pradeepkumar and D.N. Sharma, Bulletin of Radiation Protection and Environment, Vol. 20, No. 3, July-Sept 1997, pp 111-118.
4. Contouring of irregularly spaced data, Rajvir Singh, Probal Chaudhury and D.N. Sharma, Fourteenth National Symposium on Radiation Physics (NSRP-14) Guru Nanak Dev University, Amritsar, Nov. 1-3, 2001, pp 426 – 430.
5. Military Handbook of Reliability Prediction of Electronic Components, MIL-HDBK-217F, U.S. Department of Defense, 1991.

Reliability, Safety and Hazard
Editors: P.V. Varde et al.

ELECTRONIC SYSTEM RELIABILITY PREDICTION AND ANALYSIS OF EFFECT OF REDUNDANCY AND REPAIR ON SYSTEM RELIABILITY

Manoj Kumar[1], *V. Ponnar*[1] *and A.K. Verma*[2]

[1]Control Instrumentation Division, Bhabha Atomic Research Centre, Mumbai 400 085
[2]Reliability Engineering, IIT Bombay, Mumbai 400 076

ABSTRACT

Domain of electronic systems is increasing day by day. To use them in safety critical applications, such as safety function of nuclear reactor, high reliability of these systems need to be ensured. As reliability measurement is costly and time consuming, models are the only solution. Various methods and models exist for reliability prediction, applicable at different stages. In this paper, a unified approach by integrating different methods to estimate system reliability from its basic components is presented. Techniques to increase the reliability of the system such as redundancy and repair are discussed. A method to determine number of spares, required for a given replaceable module for the mission time is presented. An example is taken to show the effect of redundancy and repair on the system reliability. Paper also discussed the effect of depletion of spare inventory and proposed method to incorporate this in system reliability estimation.

Keywords: MIL-HDBK-217F, Reliability Block Diagram, Continuous Time Markov Chain, M out-of N, Renewal process, spare prediction

1. INTRODUCTION

Electronic system reliability estimation is quite involved task. Measuring the system's reliability is difficult because of cost, short time to design, and rapid innovation in the field. So, solution is reliability models. The steps are to start from components failure date and aggregate them to get the system's reliability. Standards such as MIL-STD-217F [1] and Bellcore [2] have failure rate database of a large number of electronic components and illustrate method to derive component failure rate for the use environment.

Electronic system can be divided into sub-systems, which further can be divided into modules/PCB composed of basic electronic component. So the reliability estimation can be performed by bottom-up approach.

To increase reliability and achieve fault-tolerance, redundancy is used at various levels (Hardware, Software and, Information). Fault-tolerance is the capability of a system to operate even after occurrence of faults. For achieving fault-tolerance, redundancy is necessary and sufficient. While addition of redundancy does not always increase reliability, means, redund-ancy is necessary but not sufficient. Consider the following cases, (i) Reliability of an entity is ≤ 0.5, then configuring a MooN (M out-of N) system ($M > 2$, $N > 1$) will not improve the system reliability; (ii) replicating entities in parallel (or

MooN), will not appreciably increase the system reliability after 4 or 5th replication.

Repair means replacing the faulty units from the system with healthy one. It requires fault-detection, location, replacement and restoration [4,5]. So repair requires system to have fault-tolerance and hence redundancy. Repair improves the reliability of a system, as system is bring-back to the same state, it was at some time back.

The purpose of this paper is to provide a set of methods to estimate reliability of electronic system from its component failure data, and to discuss techniques to increase reliability. The main contribution is a CTMC (Continuous Time Markov Chain) based method to estimate the number of spare required for mission time, and method to incorporate finite probability of inventory depletion in reliability estimation. The paper is organized as follows: in Section 2 methods and models used at various stages and bottom-up approach of electronic system reliability estimation are discussed. In Section 3, an example is taken and examined the effect of redundancy and repair. In section 4, CTMC based method for spare prediction is explained, and its effect on system reliability is discussed. Concluding remarks follow in section 5.

2. BACKGROUND

2.1 Reliability Analysis Methods and Models

2.1.1 MIL-HDBK-217F

Military Standard Handbook MIL-HDBK-217F establishes and maintains consistent and uniform methods for estimating the inherent reliability of military electronic equipment and systems [1]. It provides necessary information and data for reliability prediction of electronic components, modules and systems. This handbook considers Part count and Part stress reliability prediction methods. These methods vary in degree of information needed to apply them. The Part Stress Analysis method requires a greater amount of detailed information and is applicable during the later design phase when actual hardware and circuits are designed. The Part Count Method requires less information, generally part quantities, quality level and the application environment. This method is applicable during the early design phase and proposal formulation. In general, the Part Count Method will usually result in a more conservative estimate of the system reliability than the Part Stress Method.

2.1.2 Bellcore

Bellcore's [2] (Telcordia) TR-332, provides reliability prediction models for electronic components in commercial applications. This standard considers three methods, (i) Part Count, (ii) Part count and updating with result with laboratory data, and, (iii) Part Stress Method.

2.1.3 RBD (Reliability Block Diagram)

RBD is a graphical tool consisting of blocks and individual blocks may represent single component, module, sub-system, and/or logical blocks etc. The blocks are connected in a manner to depict the reliability-wise relationship among the blocks. For example, pair of shoes is reliability-wise in series, while a pair of eyes is in parallel. For the system to be successful in its operation at least one path must exist between input block and output block. The RBD can be analyzed using analytical or simulation methods. Analytical methods include, series-parallel method, cut-set/path-set method, MooN (M out-of N) and Bayes Method. Simulation method includes Monte-Carlo simulation. For complex system with large number blocks and/or complex interaction among blocks simulation is used.

2.1.4 Continuous Time Markov Chain (CTMC)

Markov chains are widely used for modeling and analysis problems of stochastic nature. Markov stands for memory-less property of the stochastic processes. Means, the next state of the process is independent of the history of the process, it only depends upon its current state. The processes of this nature are termed as Markov process. Eqn (1) describe a Markov Process [3,0].

$$\Pr[X(n)=j \mid X(1)=i_1, X(2)=i_2, \ldots\ldots\ldots, X(n-1)=i_{n-1}] = \Pr[X(n)=j \mid X(n-1)=i_{n-1}] \quad (1)$$

where

X(n): Random variable at nth trial

$i_1, i_2 \ldots, j$: outcome of random variable

A Markov process is characterized by its states and transitions. Time to transition and states may be discrete or continuous, independent of each other. So a Markov process can further be divided into four domains. A Markov process with discrete states and continuous-time to transition is termed as "*Continuous Time Markov Chain*".

CTMC is well suited for reliability analysis of electronic equipments, as their failure process is characterized as Markov. In reliability analysis states of the CTMC depicts the state of the system and transition depicts the failure rate. CTMCs are capable of taking redundancy and repair activity into account.

CTMCs are analyzed by solving the set of differential equation, along with initial and boundary condition.

2.1.5 Fault Tree Analysis (FTA)

The FTA is an event-oriented analysis in contrast to structure-oriented models discussed above [0]. Structure-oriented methods can only model hardware failures, while event-oriented methods can model hardware failures as well as failures on account of software failure, human errors, operation and maintenance errors and environment influences on the system.

FTA identifies relationships between an undesired system event and the subsystem failure events that may contribute to its occurrence. Fault tree development employs a top-down approach, descending from the system level to more detailed levels of subsystems and component levels. It is well suited to evaluate the reliability considerations at each stage of the system design.

Fault tree provides both qualitative or quantitative system reliability. Environmental and other external influences can easily be considered in FTA. It provides visual and graphical aid to the analyst.

2.2 Reliability Estimation of Electronic systems

The flow of reliability estimation is shown in Fig.1. For electronic component reliability, any one from the two well-known methods, MIL-HDBK-217F [1] or Bellcore can be used.

For predicting failure rates of various components, failure models are used. Both the standards contain failure models for a broad variety of components/parts used in electronic systems. Failure rate of a component is calculated as product of base failure rate and application based multiplying factors [1,6]. Thus the base failure rate of an electronic component is modified to 'use environment' failure rate by multiplying it with application based factors.

Components' failure rate data at specified environment condition is used for estimating the modules failure rate/MTBF/MTTF. Reliability block diagram or fault tree can be used at this stage.

A module (or PCB) is an aggregation of electronic components. The failure scenario of a module is defined, based on the functions it is required to perform. The failure rate of the module can be estimated by following methods, (i) Sum of failure rates, i.e. add the failure rates of all the components, including PCB and connectors etc. (ii) RBD or FTA. RBD/FTA requires detailed information about the module i.e. various functions performed, redundancy and failure criteria etc.

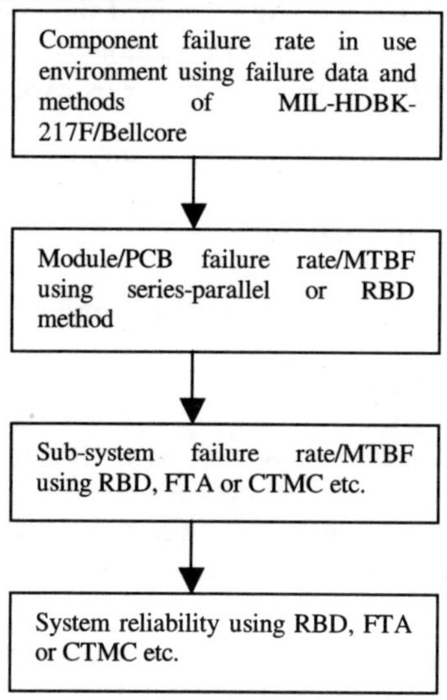

Fig. 1: Bottom-up approach of reliability estimation

Subsystems consist of various modules in a defined configuration to perform certain functions. The modules failure rates can be used to estimate subsystems reliability. RBD, FTA, or CTMC can be used for subsystem reliability estimation.

A system consists of various subsystems connected together in a defined configuration to perform the intended functions. The subsystems' reliability is used to estimate the system reliability.

Reliability of a basic component is a non-increasing function of time. Reliability will decrease as time passes and eventually result in component failure.

A system without repair and restoration of sub-system, or modules, also exhibits non-increasing reliability function w.r.t. time. A typical system with redundancy will have reliability function as shown in eqn (2).

$$R(t)=\sum_{i=1}^{n} a_i \left(\prod_{j=1}^{m} R_j(t) \right) \tag{2}$$

where :

a_i : constant ($a \in I$)

$R(t)$: system reliability at time t

$R_j(t)$: j^{th} module reliability at time t

So, the system's reliability will decrease as the time passes even if it is having redundancy at the beginning. Fig. 2 compares the reliability of system in various configuration (Simplex, 1oo2, 2oo3 and 3oo5) w.r.t. time. Therefore we infer, the reliability can be increased by two means. First, put very high redundancy at design time. Second, keep replacing the redundant faulty unit with working one when it fails. The first approach is very costly, while second requires optimum redundancy with fault detection and monitoring the status of the system and maintenance staff. Second seems to be cost effective, but it requires monitoring and repair throughout systems' life.

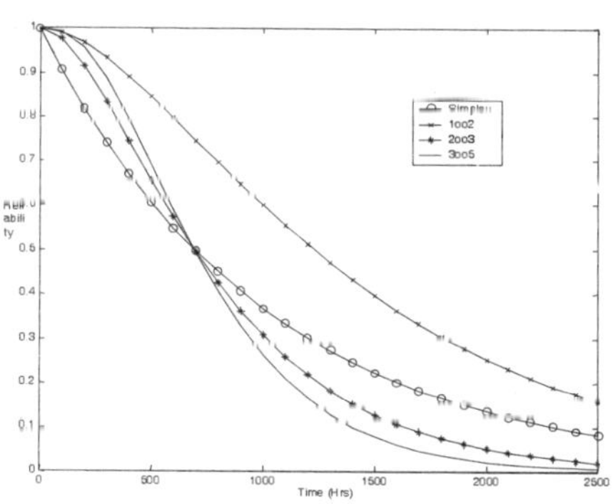

Fig. 2: Comparison of reliability of various configurations with time

When a failure is detected, repair activity renews the system, by replacing the faulty entity with a working one. This way system's reliability does not decrease rapidly. The next section takes an example to show the effect of repair on reliability.

3. Examples

3.1 Effect of Redundancy

We take an example of an electronic control system, used for "protection against temperature rise" in a research nuclear reactor. The control system takes neutron flux signal from Ion Chambers (ICs), process them to determine power (Neutronic), and manipulate the adjustor rods to control the power. This is a safety related function and has a reliability requirement of 0.99 for a period 3 months (2000 Hrs). The system environment is classified as Ground, Benign, components are of mil-grade and system's ambient temperature is 45°C.

Modules failure rate is estimated using MIL-HDBK-217F. The modules of the system along with their failure rates and quantity in system are given in Table 1. The reliability block diagram (RBD) of the system without redundancy is shown in Fig. 3a. RBD of the system with redundancy is shown in Fig. 3b. Some new modules for selection and switching are introduced in the system with redundancy. Both the RBDs are analyzed using analytical techniques for a mission time of 2000 Hrs and results are summarized in Table 2.

Table 1 Modules along with their failure rates

S. N.	Identifier	Module Description	Failure Rate		In system
			λ_O (10^{-6})	λ_S (10^{-8})	
1	IC	Ion Chamber	4.08	4.08	3
2	HVPS	HV Power Supply	18.80	25.80	3
3	LLA	Linear & Log Amp.	7.50	14.20	3
4	PL	Power Limiter	8.40	25.34	3
5	SCC	Switching control Circuit	5.04	22.86	15
6	PS7	Power Supply 7	8.80	18.90	6
7	PS6	Power Supply 6	8.80	18.90	10
8	CC	Cascade Control	19.20	50.50	5
9	PD	Pulse Distributor	7.98	82.92	5
10	DA	Drive Arrest	5.90	6.00	5
11	PS8	Power Supply 8	8.80	18.90	10
12	ARL	Adjustor Rod Logic	2.90	5.00	4
13	EMR	EM Relay	3.90	6.00	2
14	PS2	Power Supply 2	8.80	18.90	4
15	ARM	Adjustor Rod Manipulator	0.25	0.25	4

3.2 Effect of repair

We observe from the results at Table 2, that reliability target of the system is not met by introduction of redundancy. Now we will introduce repair that means as soon as a fault is detected, faulty unit is replaced with a spare working one.

Literature says time to carry out repair follows exponential distribution, for electronic equipment with an effective built-in test capability along with a rapid remove and replace concept.

The system in example also, has built-in test capability to detect failure up to module level, and modules are available in the spare inventory. So, here repair time is assumed to follow exponential distribution with mean time to repair (MTTR), 15 mins. The repair rate μ, is reciprocal of MTTR.

For analysis one repair station is assumed. To analyze a system with repair CTMCs are best suited. Following algorithm is followed for estimating reliability with repair:

1. Identify the MooN blocks with $M < N$. Parallel blocks are considered 1oo2.
2. Solve them using CTMC shown in Fig.4, for 2000 Hrs. Reliability of the block, R_B, is given by:

$$R_B(t) = \sum_{i=M}^{i=N} P_i(t) \qquad (3)$$

where :

$P_i(t)$: Probability of being in state i at time t

3. Put these results in the RBD, with this RBD reduce to a series of blocks, and solution gives the reliability with repair.

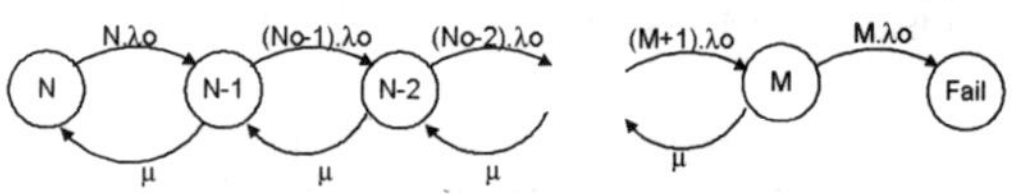

Fig. 4: CTMC for calculating reliability of MooN block with repair

The reliability of the system with repair is shown in Table 2. There is quite a good improvement in the reliability of the system by introduction of the repair activity. Here it is assumed that spares are available throughout the life.

In this fast changing world, technologies and products are short lived, so sufficient spare inventory must be there for the lifetime of the system. Also in case of offshore systems, sufficient spares must be on board for the mission time to achieve high reliability with repair. In next section we discuss methods to predict number of spares.

Table 2: Reliability estimates for the mission time

Without Redundancy	With Redundancy	With Repair
0.817	0.947	0.998

4. SPARES

4.1 Spares Prediction

The number of spares required for the mission time to fulfill the system requirement with a given probability is modeled as the number of failures expected in the system and spare inventory during the mission time. As repair of the system is a renewal process, renewal methods are used to determine the number of failure in the system during the mission time. Since new spares are not coming to spare inventory for the mission time, so failure in spare inventory or use of spares will deplete the inventory. It can be modeled as a CTMC.

Renewal process can be non-homogeneous Poison process (NHPP) or homogenous Poison process depending upon the rate of occurrence of failure [8]. If the rate of occurrence of failure shows a trend, then the failure process is NHPP, on the other hand if there is no trend then failure process is HPP. For electronic equipment failure process shows no trend and hence it is assumed a HPP. With this assumption, number of failure in system can be Poison-counting process. So probability of number of failures by time t, is given by eqn.(4).

$$P(N(t) \le j) = \sum_{i=0}^{i=j} \frac{(N_o \lambda t)^i e^{-N_o \lambda t}}{i!} \qquad (4)$$

where :

N(t) : number of failures by time t

N_o : number of modules in operation

λ : failure rate of the replaceble module

j : number of failures

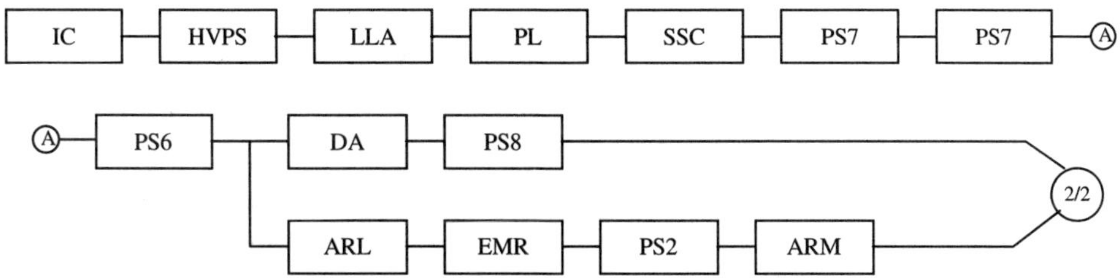

Fig. 3a: RBD of the system without redundancy

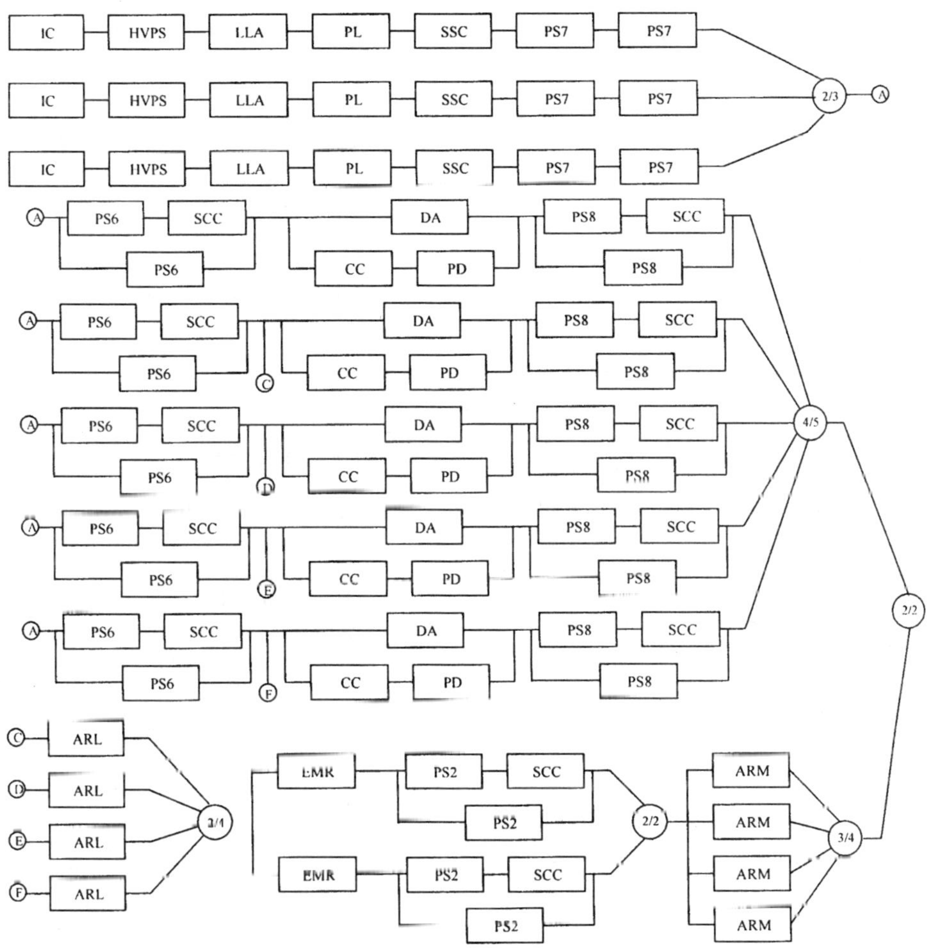

Fig. 3b: RBD of the system with redundancy

Let N_S be the number of spares for module, then,

$$N_s = n + m \tag{5}$$

where :

n : no. of failures in operating modules

m : no. of failures in spare inventory

Let N_T be the total number of modules, then

$$N_T = N_o + N_s \tag{6}$$

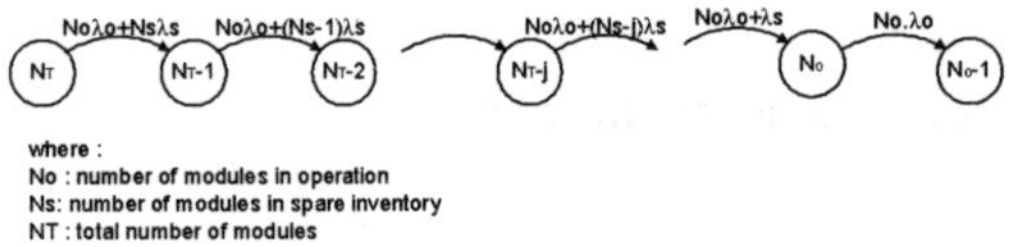

Fig. 5: CTMC for calculating number of spares

The algorithm to predict number of spares of a module for a mission time t, with a probability of spare availability p, is as follows:

1. Find the number of failure in operating modules, n, using eqn. (4) for probability p
2. Set $N_S = n$
3. Solve the CTMC shown in Fig.5 for state probabilities for the mission time.
4. Sum the state probabilities N_o to N_T.
5. If sum is greater than or equal to p then N_S is the number of spares, otherwise increment N_S by 1go to step (3).

In Table 3 module wise spare requirement with a probability of spare availability, 0.9 is given.

Table 3 Spare requirement

S. No.	Module Identifier	No of Spares		S. No	Module Identifier	No of Spares
1.	IC	0		9.	PD	1
2.	HVPS	1		10.	DA	1
3.	LLA	1		11.	PS8	1
4.	PL	1		12.	ARL	0
5.	SCC	1		13.	EMR	0
6.	PS7	1		14.	PS2	1
7.	PS6	1		15.	ARM	0
8.	CC	1				

4.2 Effect of Spares

If the spares get depleted then system cannot be repaired and restored back, and system's will fail when all redundancy have failed. To limit the initial cost, space and weight requirements, availability of spares is limited to certain probability less than1 for the mission time. To take in to account this factor of limited availability of spares on system reliability, repair rate is modified to μ′,

$$\mu' = p.\mu \tag{7}$$

where :

p : probability of spare availability

μ : failure rate of module

5. Conclusion

In this paper, we have discussed methods available for estimating electronic system reliability starting form basic components. We discussed these methods using an example also. For repair time we have assumed exponential distribution, which may give erroneous result, as repair requires certain time, but it greatly simplifies the calculations. We have also described method to estimate the number of spares required for a given mission time and spare availability probability.

6. ACKNOWLEDGEMENT

The authors are thankful to Shri S. Bhattacharya (Head, CnID, BARC) and U. Mahapatra and P.P. Marathe for their encouragement and support.

7. REFERENCES

1. MIL-HDBK-217F, Military Handbook: Reliability Prediction of Electronic Equipment, *Department of Defense*, Washington DC, Dec.1991.
2. www.telcordia.com/services/genericreq/digest/invitation/sr332.html
3. Kishor S. Trivedi. *Probability & Statistics with Reliability, Queueing, and Computer Science Applications,* PHI, New Delhi, 2001.
4. B. W. Johnson, *Design and Analysis of Fault-Tolerant Digital Systems,* Addison Wesley Publishing Company, 1989.
5. P. K. Lala, *Fault Tolerant and Fault Testable Hardware Design,* PHI, 1985.
6. S.K. Khobare et al, *Reliability analysis of Micro computer modules and computer based control systems important to safety of nuclear power plants,* Reliability Engineering & System Safety, pp 253-258, Vol 59 (2), Feb. 1998.

7. K.B. Mishra, *Reliability Analysis And Prediction*, Elsevier, 1992.
8. J.L. Hurtado et al, *Generalized Renewal Process: Models, Parameters Estimation and Application to Maintenance Problems,* Performability Engineering, pp 37-50, Vol. 1(1), July 2003.

Reliability, Safety and Hazard
Editors: P.V. Varde et al.

IMPROVING RELIABILITY IN WIRELESS SENSOR NETWORKS

Bhupendra Kumar Meena[1] *and Vallieswaran V.*[2]

[1]Reliability Engineering Technology, Indian Institute of Technology Bombay, Mumbai 400 076
[2]K.R. School of Information, Indian Institute of Technology Bombay, Mumbai 400 076

ABSTRACT

This paper focuses on Optimal routing tries to maximize the duration over which the sensing task can be performed, but requires future Knowledge. Two different scenarios with blue noise based selection of nodes for routing are simulated. One with all nodes having equal functionality and other with a node acting of coordinator summarises data and replies queries. The algorithm for selection of coordinator and query replying is presented. The number of messages for queries (in networks) and network life time of different sizes is monitored using a java simulation. This paper concludes that sensors networks life time depends on data sent per query and number of nodes in network.

Keywords: Optimal routing, Wireless sensor networks

1. INTRODUCTION

Wireless sensor nodes can be deployed on a battlefield and organize themselves in a large-scale ad-hoc network. Traditional routing protocols do not take into account that a node contains only a limited energy supply.

Optimal routing tries to maximize the duration over which the sensing task can be performed, but requires future Knowledge. This paper considers that applications of wireless sensor networks where a spatially band-limited physical phenomenon (e.g., temperature, pressure, low-frequency vibrations) can be sufficiently monitored by a subset of the nodes (randomly) deployed in the environment. In other words, the total number of sensors deployed is such that if the sensors were placed uniformly within the area, the Nyquist criteria[1] would be met in terms of spatial frequency. While such a random distribution does not provide ideal uniform coverage, it does guarantee within a reasonable likelihood that the area is covered densely enough to meet, if not exceed, the Nyquist criteria requirements within most sub regions.

This paper proposes a method that determines which sensors within the more densely covered sub regions should be selected to acquire data from the environment and which nodes should remain inactive in order to conserve energy. The proposed method is especially suitable for applications where it is desirable to trade spatial resolution for sensor network longevity.

Our proposed method chooses sensor subsets such that the sensor positions can be mapped into the blue-noise binary patterns that are used in many image processing applications [1]. The method guarantees that the subsets would be chosen such that each subset provides a near optimal signal-to-noise ratio for the given sensor distribution and desired number of active sensors [2]. Meanwhile, the method also guarantees that the sensor nodes with the minimum residual energy are the primary candidates for deselect ion (i.e., they are the first to be turned off). And then routing should be performed through high energy nodes.

This work investigates how a blue noise sampling techniques [4][5][6] being used to trace out highly residual energy nodes.

1.1 Steps are as Follows

1. Let us first consider an application where *N* sensor nodes are randomly deployed inside the area to be observed such that in all sub sets(regions) of the area [6], it is likely that the density of the sensor nodes is more than necessary to meet the requirements of the Nyquist criteria [1] of sampling [7].
2. An image grid pattern[3][7] within the observed area that is dense enough so that each sensor's location can be precisely mapped into a single grid point and each point on the grid is associated with no more than one sensor is assumed.
3. If a sensor is associated with a grid point, the grid point is assigned a value of 1; otherwise, it is assigned a value of 0. The resulting binary pattern should have white noise spectral characteristics because of the way in which the sensor nodes are deployed.
4. Then a low pass filter relaxation algorithm is being applied [1] [8] to remove all unnecessary nodes (here low pass filter have coefficient that is being adjusted so that nodes falls within range).
5. The proposed algorithm determines which point in the filtered grid has the largest spatial low- frequency content (the point with the highest value after the filter has been applied) and removes the most nearby sensor from the set of active sensors. Likewise, the high residual energy nodes in all the regions are being determined.

Here, in the Figure 1 all the selected nodes (high residual energy nodes (N)) are marked 1 and low residual energy or non selected nodes (n) are represented as 0 in the image grid.

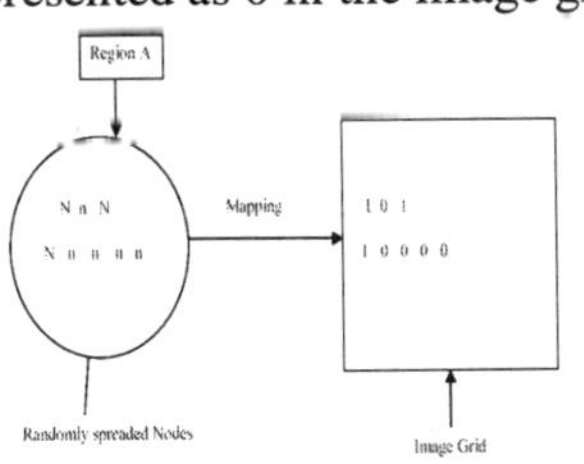

Fig. 1: Image grid representation

2. PROPOSED RELATED WORK:-

Let's suppose following is a wireless sensor network having three regions as shown:

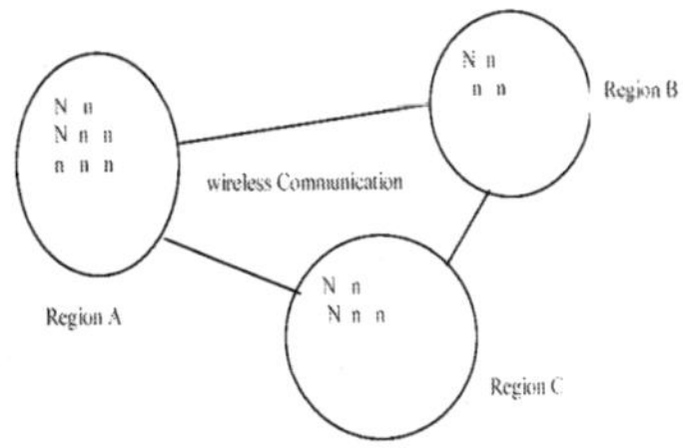

Fig. 2: A wireless sensor network with three regions as A, B, C

and each region contains their high residual energy nodes(marked .N.) and low residual energy nodes(marked .n.). [8][10]

*Below is the image grid representation of the above Wireless sensor network. [7]

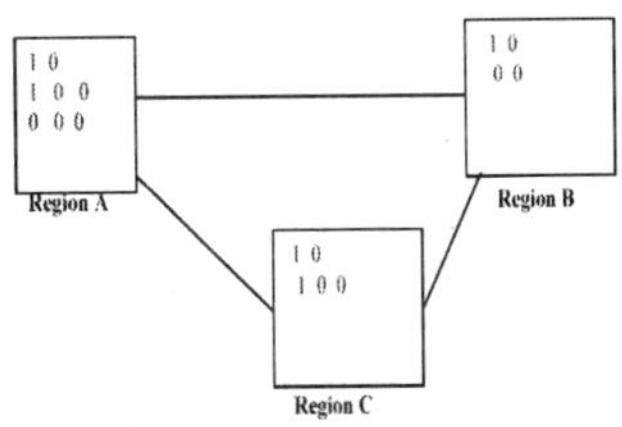

Fig. 3: Image grid representation of the above WSN

Firstly, the high residual energy nodes (as marked .1. by image grid) of each region have to be selected.

3. PROPOSED ENERGY EFFICIENT ALGORITHM

Now, after getting the high residual energy nodes[8] (as marked 1 by image grid) of each region,[1][3] proposed protocol works like as follows:

Step 1

Initially each selected nodes of regions will transmit an empty packet to all the other nodes of the same region.

Step 2

Then after getting reply from others, each node will have to maintain a routing table containing information about its nearest neighbour.

Step 3
Step 2 must be followed for all the regions in similar fashion. Then,

Step 4
Now, all the selected nodes of each region got an idea about their nearest neighbour. Then, all the transmission in any region will be proceeds in an energy efficient way with covering shortest paths and through higher residual energy nodes. So, it will increase the life time of the operation of a wireless sensor network by a great amount.

This algorithm considers only selection of nodes according to residual energy of the node. This dose not considers routing energy cost. The selection of this nodes have to communicated this also consumes energy largely. Every message send from one node to other reduces the energy stored in the entire system.

Message power = transmitter power * Time (1)

The Power spent on transmitting one message in transmitter power and time for transmitting the message. If the number of messages in the system increases the life of the sensor network decreases.

To reduce the number of message and selection complexity the following improved algorithm is used.

Step 1
Initially each selected nodes of regions will transmit an empty packet to all the other nodes of the same region.

Step 2
Then every node in the region replies with number of number of empty packets received. The node with maximum number is the coordinator of the region

Step 3
Then after getting reply from others, each node will have to maintain a routing table to the node with coordinator of the region & information about its nearest neighbour.

Step 4
Step 2 and 3 must be followed for all the regions in similar fashion. Then,

Step 5
Now, all the selected nodes of each region got an idea about their nearest neighbour and leader.

Step 6
Then, all the transmission in any region will be proceeds in an energy efficient way with covering shortest paths and through higher residual energy nodes to coordinator of the region.

Step 7
The region coordinator does the processing of the query and send reply to the other region with result in a single message instead of all sensors messages.

Operations are like maximum/ minimum/ average sensed values of the region this can be done by simple Digital circuit added to the sensor.

4. SIMULATION RESULTS

From the paper [3] results, increase in longevity of network by choosing high energy nodes for routing using blue noise sampling technique.

Sensors are simulated as objects in java and blue nose based sensor selector as global object where it stores the energy level of every node and dictates selection based on power formula given in paper [3] for 100x100 grid.

The following assumption are made

- Introduced one query to get data from the sensors in each region in 100x100 grids.
- One sensor will last for 1000 hrs without any packet transfer.
- Power for 1min will be consumed by 100 packet transfer from the node.
- Number of sensors data will be used to get accurate result for a query is also considered.

The number of messages per query (on average) and life time of sensor network with uniformly distributed sensors in the image grid is calculated.

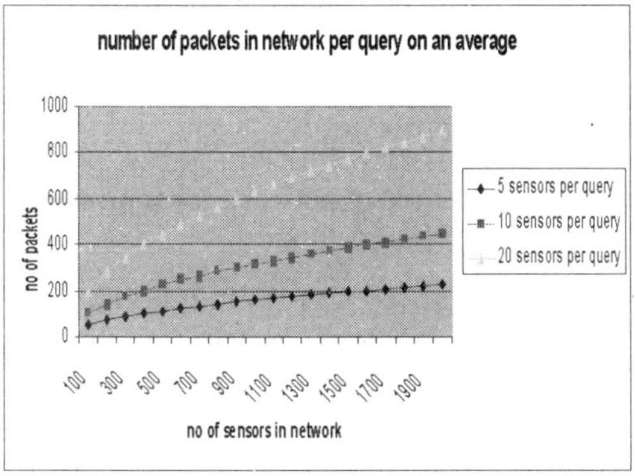

Fig. 4: Number of packets in network per query on an average

Here x-axis represents the number nodes; y-axis represents the number of packets per query and the three lines shows for different number of nodes data used in each query.

In Figure 5. x-axis represents the number nodes, y-axis represents the network life time in number of hours and the three lines shows for different number of nodes data used in each query.

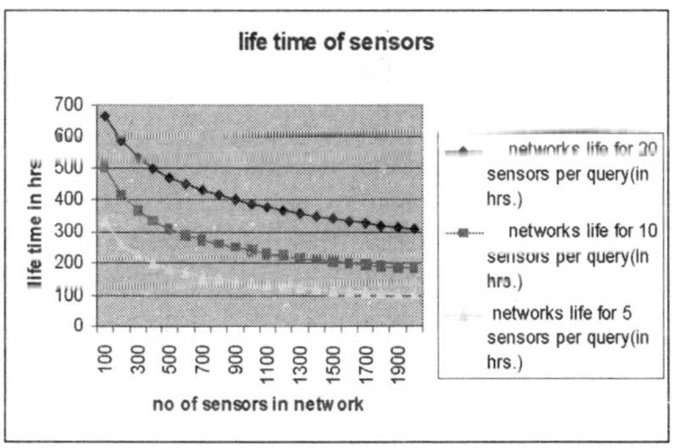

Fig. 5: Life time of sensors

The life time of network is calculated till the majority of sensors are down. In our simulator every sensor has assumed life time of 1000 hrs. When ever a packet is transmitted 600ms will be reduced as one minute goes down also life time reduces. The simulation considers one second as one minute and it initiates a query for each region. Once in 2 minutes blue noise based sensor selector will be called. The result of life time the network lasted is shown Figure 5.

The steps for selecting coordinator among the high energy nodes selected by blue noise function are implemented. And improved the sensors with little calculation operations such as average, maximum and minimum of arrived data. The coordinator aggregates data and send single result for the query. Here, the number of sensors required for consistency is present in single region is considered.

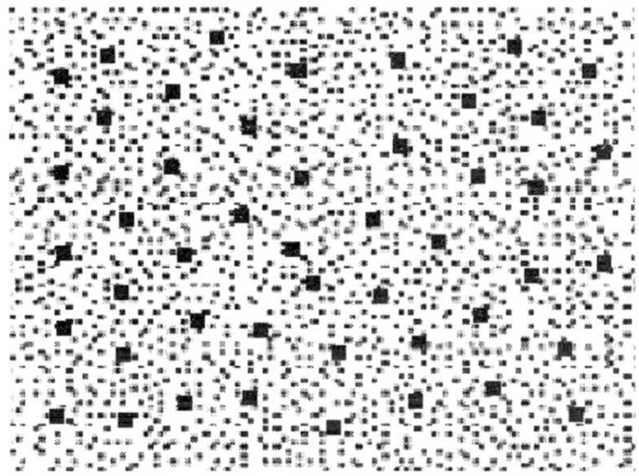

Fig. 6: A typical plot from blue noise selector and coordinator selecting techniques.

The result of the coordinator selection is shown in the diagram figure 6. The lager dots represent the coordinator and all other nodes are sensors. Usually the region formed resembles a circular or hexagonal around the coordinator. The figure 6 is generated by Grid plot using the positions stored as 1's 0's and 2's for coordinators.

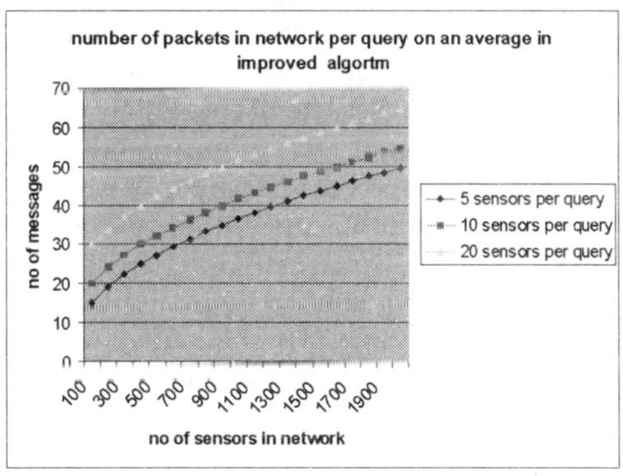

Fig. 7: Number of packets in network per query on an average

In Figure 7, x-axis represents the number nodes, y axis represents the number of packets per query and the three lines shows for different number of nodes data used in each query.

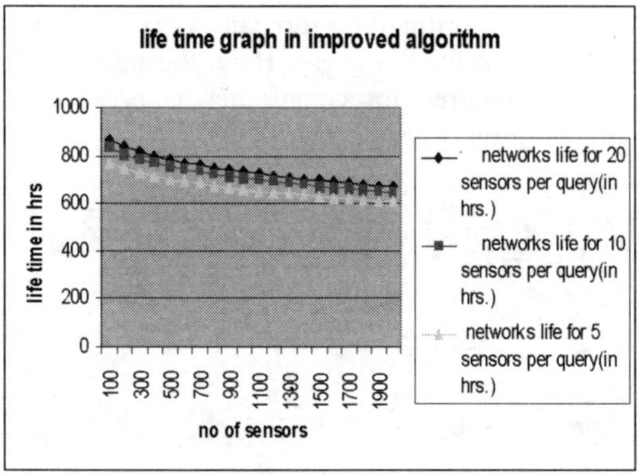

Fig. 8: Life time of sensors

The number of messages per query(on average) and life time of sensor network is calculated with uniform distribution of sensors in the grid as in the previous one but the number of messages drastically reduced due to elimination of broadcasting of data.

In figure 8, x axis represents the number nodes, y axis represents the network life time in number of hours and the three lines shows for different number of nodes data used in each query.

Table 4.1 shows the life time of sensor network when five sensors data used per query the data in the two algorithms.

The table 4.1 shows improvement life time with respect algorithms due to reduction in broadcast of data. But the factor (how much message transfer affect the life of network) has two components one is data sent per query and other is number of nodes used up in processing of query.

Comparing the graphs, the algorithm shows that decrease in network life time is in proportional to data sent per query. In the improved algorithm, the decrease in network life time is comparatively constant, but it affects initial value.

In both the cases, with respect to increase in number of nodes there is decrease in life time. The decrease is in continuously increasing fashion (but not linear). This gives us an idea that, with increase in number of nodes to the decrease in life time, may be in promotional to root of the number of nodes.

Table 4.1 Lifetime of network for improved algorithm

No. of sensors	Networks life for 5 sensors per query(in hrs.) for algorithm in paper [3]	Networks life for 5 sensors per query(in hrs.) for our improved algorithm
100	666.6	869.56
200	585.8	839.6
300	535.9	817.66
400	500	800
500	472.1	785.5
600	449.6	772.7
700	430.6	763
800	414.2	750.1
900	400	740.7
1000	387.4	732
1100	376.2	724.1
1200	366	716.1
1300	356.8	709.2
1400	348	702.2
1500	340.5	695.9
1600	333.3	689.6
1700	326.7	683.8
1800	320.4	678.4
1900	314.5	673.4
2000	309	668

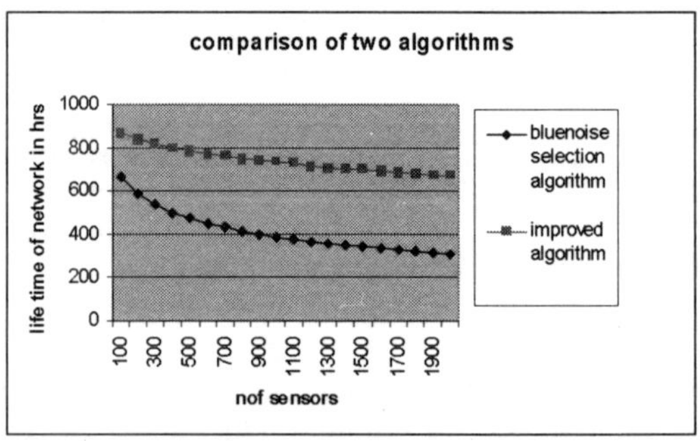

Fig. 9: Number of packets in network per query (on an average)

5. CONCLUSIONS

1. Our proposed algorithm gives the better approach of using the blue noise sampling technique.
2. By using our improved algorithm the life time of Wireless sensor networks gets increased up to 32 percent in compared with the old blue noise selection method (as shown in figure 9) .
3. As the number of selected nodes decreases then the reliability and operability of the WSN goes down.
4. The life time is inversely proportional to number of sensors selected for each query and some root function of number of sensors in network.

6. FUTURE WORK

Our plan is to extends this blue noise sampling techniques[3] to include various other parameters of wireless sensor networks such as pressure, temperature and other environmental and WSN factors so that the selection of the nodes and their power Consumption calculations can be made more accurate. Currently this paper considers only packets transfer but, in future, we would like to include interference, collision and retransmission occurring in message transmissions.

REFERENCES

1. Mark Perillo, Zeljko Ignjatovic and Wendi Heinzelman, An energy conservation method for wireless sensor networks employing a blue noise sampling technique. IPSN, 2004
2. S. Hiller, O. Deussen, and A. Keller, .Tiled Blue Noise Samples,.in Proceedings of Vision, Modeling, and Visualization, 2001, p.265.271.
3. Zeljko Ignjatovic, Mark Perillo, Wendi Heinzelman, Mark Bocko, An energy conservation method for wireless sensor network using blue noise spatial sampling technique.ACM, 2004
4. T. Mitsa and K. J. Parker, .Digital Halftoning Technique Using a Blue-Noise Mask,.in Journal of the Optical Society of America,vol. A:9, Nov. 1992, pp. 1920.1929.
5. R. A. Ulichney, .Dithering with Blue Noise,. in Proceedings of the IEEE, vol. 76:1, Jan. 1988, pp. 56.79.
6. J. chou and D. Pterovic, A Distributed and adaptive signal processing approach to reducing energy consumption in Sensor networks, IEEE (INFOCOM), 2003.
7. H.G. Feichtinger and K.H. Grochenig, Theory and practice of irregular sampling, CRC press 1994.
8. P. Ishwar, A. Kumar and K. Ramachandran, Distributed sampling for Dense Sensor Networks,(IPSN), 2003.
9. S. S. Pradhan, J. Kusuma, and K. Ramachandran, Distributed Compression in a Dense Microsensor Network. IEEE Signal Processing Magazine, 19(2):51–60, March 2002.
10. R. Stasinski and J. Konrad. Improved POCS-based image reconstruction from irregularly-spaced samples. In Proceedings of the XI European Signal Processing Conference, 2002.

Reliability, Safety and Hazard
Editors: P.V. Varde et al.

SAFETY AND RELIABILITY ASPECTS OF SEAWATER REVERSE OSMOSIS DESALINATION PLANTS

S. Prabhakar, Saly T. Panicker and P.K. Tewari

Desalination Division, Bhabha Atomic Research Centre, Mumbai 400 085

ABSTRACT

Being a societal support system, desalination plants have to stand the test of operational safety and reliability of performance. The system design has to take into account all the safety related features in terms of mechanical integrity under high pressure operation and corrosion point of view. Provision of standby equipment, proper instrumentation & controls, periodical maintenance and following all the safe operational procedures ensure the reliability and safety. In this paper, the safety and reliability aspects of seawater reverse osmosis desalination plants are discussed in detail, including the case study of a 1800 m^3/day seawater RO plant, integrated with a nuclear desalination system.

1. INTRODUCTION

Increasing demand for fresh water has resulted in the installation of a number of desalination plants, particularly in coastal areas. These desalination systems are mostly based on, membrane (RO) and thermal (MSF & MED) processes. As continuous supply of water is essential for industry or for human consumption, the reliability of a desalination plant is very important. For ensuring the maximum availability of water and to reduce the cost of production, reliability of a plant has to be high. When the desalination plant is integrated to a nuclear power plant, safety measures are to be more stringent, in order to achieve smooth operation and public acceptance.

2. REVERSE OSMOSIS PLANT

Reverse osmosis (RO) is the process of separating water from a solution, through a semi-permeable membrane, using pressure. As the membrane is sensitive to suspended & colloidal matters and scaling constituents, the water to be treated by RO has to be seasoned physically and chemically. Thus an RO plant essentially constitutes a pretreatment system, high pressure pump and membrane modules. The 1800 m^3/day seawater RO (SWRO) plant which is chosen for the case study, is a part of the Nuclear Desalination Demonstration Project (NDDP) and is complement to another desalination plant based on thermal process. NDDP is integrated with the 220 MW(e) PHWR, Kalpakkam.

Safety Factors

In the normal case, no hazard is expected to happen in the RO plant, as the stream handled in bulk is only seawater. Being a high pressure system, integrity of the components is important. Another possibility of hazard is in the event of any leakage of concentrated acid from the storage tank and is taken care by providing shielded acid storage area. The following criteria in operation and monitoring of the plant are normally adopted in an RO plant.

- Safety issues are identified to comply with the site requirement.
- Operational prerequisites are established and the plant is operated in accordance with manufacturers' procedures.

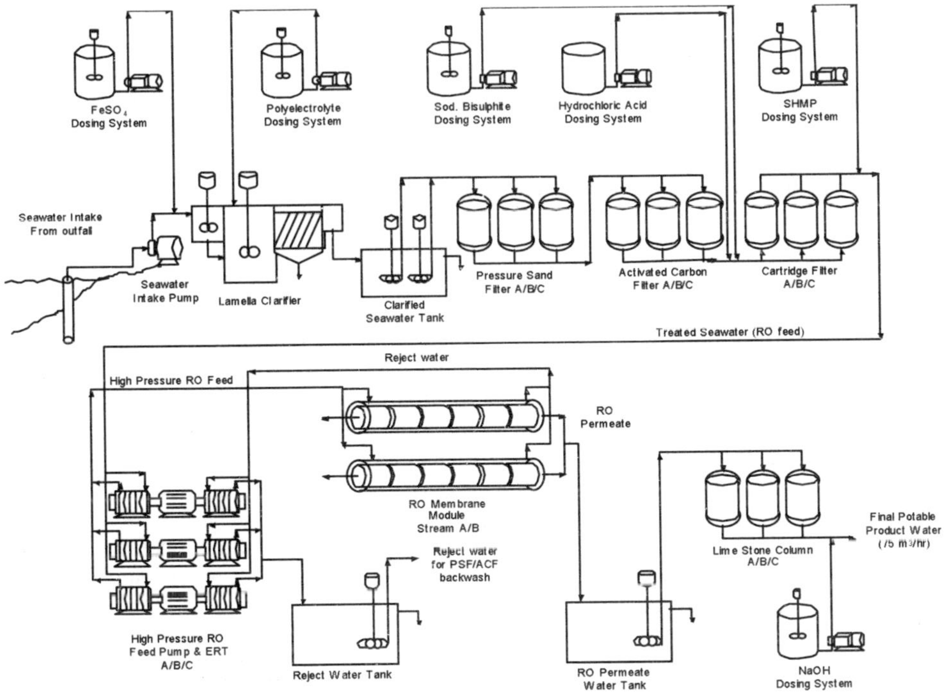

Fig. 1: 1800 m^3/day Seawater Reverse Osmosis Plant

- All the components are tested as per the standards, including stage-wise inspections.
- Plant is continuously monitored to detect deviations from normal operating conditions.
- Corrective actions are taken to rectify abnormalities.
- Plant integrity and personnel safety are maintained as per the standard procedures.

Case Study of the SWRO Plant, Kalpakkam

Figure1. shows the flow sheet of the plant. The safety factors incorporated in the design and operation of the SWRO plant, Kalpakkam are discussed below.

Electrical System

The basic aim of electrical system design is to provide maximum availability and reliability of power during operation. For loss of power due to system break down or grid supply failure, provision for alternate source of power is made through diesel generators and AC – DC converters with battery back up. None of the areas in which electrical equipment are installed is having flammable liquids, gases or dusts. Liberal safety margin in design of systems including weatherproof and corrosion resistant enclosures of equipment are provided. Since saline water is handled in the plant, equipment and panels are designed to withstand corrosive environment. Wall mounted plugs and receptacles provided for testing of power supply are above 1 meter high from ground level and contacts are shrouded. Power supply to the loads is metered to check their performance and keep watch on the rating limits.

The loads of pump motors, workshop machinery, lighting, ventilation and air conditioning are fed from 6.6 KV and 415 V buses. For operational reliability during outage/fault/maintenance of one transformer, a second transformer is provided to have 2x 100% flexibility. Suitable motor controlled centers (MCCs) are there with facility for remote ON-OFF and lamp indication at control room.

Protective networks are designed with the objective of providing primary protection and thereafter supply a form of back protection, operating in the event of failure of primary protection. All MCCs are metal enclosed, single front, sheet steel cubicles.

Power and control cables are laid on cable trays in trench or buried underground.

Instrumentation and Control

On-line instruments, local control panels as well as transmitters to control room are provided to measure operating parameters such as, flow rates, pressures, differential pressures, conductance, pH, oxidation – reduction potential (ORP) and turbidity of all streams and levels of water tanks and chemical dosing tanks. Operation of the plant is mainly from the control room. Pumps and compressors are started from the local control panels. Green colour indicates running condition and flashing of lamp and horn indicate tripping of any machinery. Interlocks are provided in such a way that any machinery tripped due to power failure will not restart automatically on resumption of power, thus ensuring safety. Machines can be stopped from the control room in emergency. All the operating parameters are continuously indicated and recorded in the control room. Critical parameters are controlled by controllers through auto or manual mode. The control valves will automatically go to 'fail safe' mode in the event of power or instrument failure.

Advanced automation system: The conventional automation system in an RO plant could perform the following actions only.

- Turn the plant on and off in a specific order.
- Emergency shut down
- Control of product conductivity, low suction and high discharge pressures of the high pressure pump and low and high levels of tanks
- Flushing and cleaning of the elements

Sometimes, some of the on-line meters can fail to function, which may go unnoticed immediately. Better safety is ensured if additional facilities are provided to sense the critical parameters. Such facility should be based on standard, reliable, user-friendly, proven and modern technologies. Thus, in addition to the conventional control system, SCADA [1] with controllers of direct interface facilities also are incorporated. SCADA accomplishes, on-line data acquisition, event sequence recording and computation. The main advantage of this system is its flexibility. The system can be remotely operated from longer distances.

Pump Operation

Low level switches are provided with all the pumps to prevent dry running when the level in the tanks go down the set minimum values. Interlock is provided between the high pressure pump and the clarified water pump so as to ensure that the high pressure pump will start only when the clarified water pump is in operation, so that water is always available at the suction line. Pressure switches are provided to sound pre- and trip alarms to switch off the high pressure pump beyond the maximum operating pressure. Whenever a pump trips, provision is so given that all the pumps in series will stop. The module-flushing pump is connected to a DG set. Whenever a power failure occurs, this pump automatically starts and sends permeate water to the membrane elements to wash off the seawater held in them.

Pressure Vessels of RO Elements

The FRP pressure vessels holding the RO elements are fabricated and tested as per ASME Boiler and Pressure Vessel Code – Section X.

Valve Operation

Two critical quality control points are (i) the inlet of high pressure pump and (ii) permeate line of RO modules. Unless the required quality of treated water coming from the pretreatment section is as per the requirement in terms of any of the parameters such as pH, turbidity or ORP, the water is sent back to the clarified water tank through an auto-operating solenoid valve. The same way, if the conductance of the permeate water goes up beyond the set value, the air-operated butterfly valve in the permeate line to the permeate water tank gets closed and the valve in the drain line opens and the water is redirected to the feed tank.

Handling of Chemicals

The only hazardous chemical used in the plant is 33% HCl. Acid proof brick wall (1000 mm high) is provided all around the acid storage area. If any leakage occurs, the acid will be neutralized with caustic soda, diluted with water and discharged to the drain. Personnel handling acid use, rubber gloves, gumboots and goggles for protection. The other chemicals used are $FeCl_3$, poly electrolyte, sodium bisulfite, organic or inorganic anti-scalants, limestone and sodium hydroxide, which are non-toxic.

Product Water Quality

The product water quality is maintained as per the WHO standards, in terms of the total dissolved solids as well as the individual constituents. The RO permeate which is devoid of calcium content and at acidic pH is further passed through a limestone column to maintain the pH between 7 and 8 and also to have a calcium level of not less than 20 ppm. LSI of the final product is monitored to assess the level of corrosiveness or scaling to take care of the distribution pipes.

Reject Disposal

The reject is around 1.7 times more concentrated, compared to the incoming raw seawater. Though it contains the chemicals added during the pretreatment, they are eco-friendly. To take care of the higher levels of concentration, this stream is mixed with the coolant return water of the thermal desalination plant, which is the raw seawater.

Reliability Factors

Nature of seawater

Before designing an RO plant, thorough knowledge of the seawater going to be used is essential, in terms of the physical and chemical characteristics. Variation of these characteristics with different seasons could be in the range of up to 30 %. RO performance and the life of the elements are very sensitive to the quality of the feed seawater. Availability of an RO plant depends to a great extent on the performance of its pretreatment system, which in turn depends on the physical nature of the water in terms of suspended, colloidal and biological loads. Normally a plant is designed based on the most probable characteristics, with some flexibility to take care of the expected variation. Thus, long term analytical studies of the water are required before starting the design.

Seawater Intake Point

The seawater intake system is an area where problems could be high. The intake system should be designed, installed and maintained in such a way that, it can take care of all the conditions in the sea, including, the highly corrosive water and the environment and also the carry over of solids. The intake pump and pipe lines should be properly supported to withstand hydrostatic and dynamic pressures and water hammer.

Corrosion Problems

The materials used for intake pumps and pipes should be corrosion resistant, as the salinity is too high and also sulphate reducing bacteria could be present. Precautions against leakage of seawater should be taken to avoid corrosion. Seawater RO plants being at coastal areas, the environment itself is so corrosive that protection of the external surfaces of equipment has to be done, by way of regular painting etc. The most frequently failing components are pumps and valves in an RO plant, mainly due to corrosion. Considering the corrosiveness of seawater, utmost attention has to be given in design, selection of material of construction, operation, maintenance etc. Though costly, corrosion resistant materials should be used for pumps, especially high pressure pumps. As far as possible, standby pumps shall be provided. Failures requiring pump dismantling should be avoided as it would affect the availability of the plant.

To enhance the plant reliability of the SWRO plant at Kalpakkam, the following factors were considered.

- RO process adopted for desalination and the equipment used are simple, well tested and commercially proven.
- The physical and chemical characteristics of the seawater were analysed throughout the year and arrived at the nominal values based on which the final plant design was done.
- Three options of intake seawater, such as, surface water, reactor outfall water and water from beach wells were initially

considered. The out fall water is the coolant return from the process water heat exchanger loop of the power reactor. Long term studies including, bathymetry of deep sea, assessment of water yield and stability of beach wells etc. were conducted to decide the best option. The surface water contained a lot of suspended load, jelly fish etc. Moreover, to pump water from the sea, elaborate pumping facility, including a jetty had to be built, which was more involving, time taking and expensive. The physical quality of the beach well water was much better and the salinity also was less, compared to the seawater. But, the water yield was not sustaining. In the case of the outfall water, structural facilities were already existing, which required minimum modifications to incorporate the additional pumping system. Thus, it was found to be simple, reliable and more economical. In addition, this water was devoid of any big particles or water animals, as it had once passed through the screens and meshes of the pretreatment system of the nuclear reactor.

- The design and construction of the plant have included strict technical and performance specifications and also measures for quality control.
- Materials those are prone to corrosion like unlined mild steel was totally avoided, as the corrosion products will cause problems.
- As far as possible, the plant is in continuous operation, to avoid corrosion and growth of microorganisms during shut down.
- RO modules and the pipe lines are flushed with RO permeate before long shut downs. To take care of the RO elements even during emergency shutdowns due to power failure, suck back tanks at higher elevations are provided to store the permeate, which can flush the elements by gravity, in addition to generator - operated flushing facility .
- The staff operating and maintaining the plant are fully trained. New personnel go through a training programme.
- Clearly written operation manuals are available in the plant.
- Preventive maintenance programmes are planned in advance.
- Stock of consumables, spare parts etc. is maintained.
- Based on the operational experience, modifications are included in the operational schemes and even in design.
- Manual as well as auto modes of operation and control are provided.
- In addition to the electronic data acquisition facility, manual data logging system also is followed in the plant.
- Causes of abnormal functioning of any of the components or the total plant are identified by analyzing the technical and operational information in a logical and sequential manner.
- If required, advice from external agencies is sought for the smooth operation of the plant.

3. CONCLUSION

RO plants in general are highly reliable and safe. The three years of operational experience of the SWRO plant at Kalpakkam proves that if properly designed, constructed, operated and maintained as per the standard codes and manufacturers' guidelines, smooth running of any plant is ensured.

4. REFERENCES

1. S.A. Avlonitis, M. Pappas, K. Moutesidis, D. Avlonitis, K. Kouroumbas, N. Vlachakis, PC based SCADA system snd additional safety measures for small desalination plants, Desalination 2004,165, 165-176.
2. Ibrahim I. Kutbi, Zeinab A. Sabri, Abdo A. Hussein, Selection for Desalination Processes for Dual-Purpose Nuclear Plants, Desalination 1986, 58, 113-134.

Reliability, Safety and Hazard
Editors: P.V. Varde et al.

SAFETY AND RELIABILITY IN FUEL HANDLING AND STORAGE SYSTEM OF DHRUVA

Kunal Chakrabarty, J. Tiwari and H.G. Gujar

Reactor Operation Division, Bhabha Atomic Research Center, Mumbai 400 085

ABSTRACT

The fuel handling and storage system of Dhruva is provided to ensure safe handling of fresh fuel component; storage, inspection, assembly and out-of-pile testing of the components; handling the fresh and irradiated assembly in reactor core and storage areas; various under water handling operations with irradiated fuel assemblies and finally shipment of irradiated fuel assemblies. The most significant safety problems that arise during handling and storage of fuel are those concerned with inadvertent criticality, high levels of radiation exposure and unacceptable release of radioactivity. Accordingly, certain general design features related to these problems are identified and practiced at every stage. Based on the operating experience gathered over around 20 years, additional safety up gradations was incorporated to enhance reliability and availability of fuelling machine. This paper briefly describes about the operating experiences of fuel handling during the last 20 years and various measures taken towards enhancing safety during fuel handling operations.

1. INTRODUCTION

The role of research reactors for the development of nuclear programme of any country is well established. Apart from providing large volume of neutron source for carrying out a variety of experiments, the research reactor forms the basic training ground for grooming up scientists and engineers for the nuclear power programme. Research reactors are utilized to produce radioisotopes and offer irradiation facilities for testing various nuclear materials such as fuel and structural materials.

Dhruva one of the high flux research reactors, was built in 1985 and it offers a maximum neutron flux level of 1.8 X 10^{14} n/cm^2/sec. It uses natural metallic uranium clad with aluminium as fuel and heavy water as coolant, moderator and reflector. The main reactor vessel is made up of stainless steel. 146 positions arranged in a square lattice, are provided to accommodate fuel assemblies, experimental and irradiation assemblies and shut off rods. The reactor vessel is housed vertically inside water filled stainless steel lined heavy density concrete vault. Cadmium absorbers are used as shut off rod material. The heavy water moderator level in reactor vessel is varied to achieve the desired reactor power. Heavy water coolant flowing from bottom to top through each assembly removes fission heat. The heat is transferred to de-mineralized water, which is called secondary coolant and subsequently to the tertiary seawater coolant in the respective heat exchangers. Each fuel assembly is in the form of cluster of seven pins to maximize the heat transfer area. The reactor and the associated primary coolant (heavy water) system piping and equipments are located inside the reactor building containment. The high-density concrete containment structure of the reactor building helps in containing the radiation and radioactivity within the reactor building in case of unlikely event of an accident.

The fuel handling and storage system of Dhruva is provided to ensure safe handling of fresh fuel component; storage, inspection, assembly and out-of-pile testing of the fuel assembly; handling the fresh and irradiated assemblies in reactor core and storage areas; various under water handling operations with irradiated fuel assemblies and finally shipment of irradiated fuel sub assemblies. Apart from handling of fuel assemblies, other in-core assemblies such as shut-off-rod, isotope production assemblies and other special irradiation assemblies are also handled by fuel handling system.

2. BRIEF DESCRIPTION OF FUELLING OPERATION

One fuelling machine (FM) is used for handling heavy water cooled assemblies like fuel assemblies and isotope tray rods and another machine is used for other assemblies. A separate machine for heavy water cooled assemblies help in maintaining the isotopic purity of heavy water in the system. Fuelling machine is provided with two barrels for handling an irradiated fuel assembly in one barrel and fresh assembly in other barrel. The entire fuelling operation consists of picking up fresh fuel from a dry storage bay, transportation of the fresh fuel to the reactor pile, picking up irradiated fuel from the reactor pile into the fuelling machine under continuous cooling, installation of fresh fuel into the reactor under continuous cooling, installation of fresh fuel in to the reactor, transfer of the irradiated fuel to spent fuel storage bay (SFSB) ensuring proper cooling during transit period, dumping of heavy water coolant to fuelling machine storage tank and lowering of irradiated fuel in light water bay.

The major steps involved in fuelling operations are

1. Receiving the components of fuel assemblies in three parts
2. Assembling the fuel components, carrying out qualification tests and making it ready for pile.
3. Transferring the fresh assembly to dry storage bay (DSB).
4. Picking up the fresh assembly from DSB into FM.
5. Aligning FM over particular fuel channel and lowering the shielding skirt.
6. Coupling the magazine to the channel by lowering the snout and carry out seal test.
7. Opening the main valve to make free passage from magazine to snout.
8. Lowering and coupling of extractor to the fuel assembly
9. Unlocking the fuel assembly from the channel.
10. Raising the extractor with the fuel assembly.
11. Indexing the magazine to interchange the position of spent fuel and fresh fuel.
12. Lowering the fresh fuel fully in to fuel channel.
13. Locking the fuel assembly to the channel.
14. Decoupling and lifting of extractor fully into the magazine
15. Closing the main valve to isolate magazine and snout.
16. Recovering out interspace heavy water and carry out seal test.
17. Raising the snout and skirt to decouple the machine from fuel channel.
18. Taking FM to storage block for disposal of irradiated fuel to underwater storage

The refueling operation has become highly complex in view of the facts that the fuel assembly with or without guide tube being handled is around 10 m long. The machine has to accommodate the entire assembly and provide uninterrupted cooling with adequate shielding arrangement for beta-gamma as well as neutrons. The overall weight of the machine including the shielding flask, carriage and trolley structures is around 300 tonne and require alignment accuracy with respect to channel being refueled to be within $\pm$ 0.25mm. The controls / operations are of semi automatic nature with human interventions for each sequential operations. The operator has to have knowledge of plant systems other than the machine itself particularly, power supply distribution and coolant and cover gas system etc., since prompt actions are required in case of any transients, unusual occurrences. The checklist for fuel handling operation is quite exhaustive and some of operations are of repetitive nature making it vulnerable to operator errors.

Further, during fuelling operation, coolant system boundary is extended to fuelling machine magazine and connected systems. Special checks and precautions are required to ensure that the integrity of coolant and cover gas system boundary is not breached. Since this machine handles irradiated fuel assembly, operator should be well conversant with various cooling modes and their implications. After dumping the coolant from the barrel, the irradiated fuel should he discharged to the light water bay, within prescribed time limit. Also, since hardware interlock control circuit permissive is required for any activity to be performed, the hardware control circuit should be highly reliable. For many crucial operations, the operator has to perform manual checks to rule out the possibility of deviations due to control instrument/mechanical component malfunctioning.

2.1 General design and safety principles

The most significant safety problems that arise during handling and storage of fuel in general are those concerned with inadvertent criticality, high levels of radiation exposure and unacceptable release of radioactivity. Accordingly certain general design features related to these problems are identified and practiced at every stage. These salient features are listed below:

1. The layout and arrangement of fuel storage facilities are designed such that inadvertent criticality is prevented (even under accident conditions) mainly by the use of geometrically safe configurations.
2. Appropriate combination of physical protection/ safety control logic and administrative controls are provided to prevent radioactive components from being lifted and placed into unshielded positions.
3. Adequate provisions are made for inspecting fresh and irradiated fuel.
4. Fuel handling equipments are designed in such a way as to prevent the dropping of fuel assemblies or the imposition of unacceptable handling stress on the fuel.
5. Care is taken during the design to limit the reactivity addition or removal rate within specified limits.
6. The fuel assemblies, storage racks and channel positions in reactor have a smooth profile for self-guiding of the assembly during raising or lowering. Sharp corners are avoided.
7. Factor of safety for design of handling equipments are taken conservatively.
8. Proper operating procedures are implemented to prevent the loss of fuel or isotope capsules.
9. Storage facilities are designed in such a way that inadvertent draining of water from the pools is not possible and fuel sub-assemblies always remain immersed in water even in case of breakage in pool recirculation line.
10. Handling and storage facilities are provided with adequate heat removal provisions.
11. Instrumentation to monitor the release of radioactivity in and around the working areas and the exposure to the operating personnel during handling or storage should be provided in all the relevant handling equipments and areas.
12. Ventilation arrangement to ensure adequate number of air changes and maintaining appropriate pressure differentials to ensure air flow from low active to high active zones.
13. Operating procedures are simple, to the extent possible to minimize risk of error.

2.2 Handling and storage of fresh fuel

Well-designed storage facilities are available at reactor sites to store fuel components and fuel assemblies. These facilities are provided with proper ventilation to avoid entry of dust into these areas. Suitable racks to store the fuel sub-assemblies and other sealing and shielding components in vertical mode are provided. These areas are provided with adequate security measures to avoid any damage or pilferation of the strategic materials. Use of natural uranium fuels rule out requirements of shielding during handling of fresh fuels. Furthermore, it completely eliminates any inadvertent criticality. The procedures involved in handling fresh fuel include the following:

1. Fuel identification and verification against documents
2. Visual inspection after removal from transport container and recording the observations on fuel history cards
3. Checking of dimensional tolerances, which might get affected during transport, and handling
4. The sub assemblies are made in parts for their easy transportation from the fabricators place to the assembly station of the plant. Suitable jigs, fixtures and cranes are provided in assembly station for handling of the fuel in safe manner.
5. A tilter is always used whenever a fuel assembly is made vertical from horizontal position or vice versa to avoid bending.
6. The assembled fuel is flow tested under simulated reactor hydraulic conditions. Subsequently the fuel assembly is dried to avoid degradation of heavy water coolant of reactor before making it ready for pile.
7. Each fuel assembly is identified with suitable numbers clearly marked on it.
8. Fuel assembly handling activities are carried out only by trained personnel and supervised by authorized personnel.
9. The movements of fuel assemblies are tracked at each stage through a system known as "Transfer Slip" and documented.

2.3 Handling and storage of irradiated fuel and other core components

In addition to the criteria for fresh fuel assemblies, following points are considered with regard to handling and storage of irradiated fuel and other core components.

Fuelling machine is designed taking into consideration of cooling and shielding requirements and the reactivity effects.

A back-up system is provided for uninterrupted cooling of the assembly being handled in the fuelling machine. Handling operations are kept to a minimum and most of the operations are automated to the extent possible to improve reliability for prompt action and to reduce the strain on the operating staff. All auto drives are backed up by manual drive mechanism for emergency handling and power failure conditions. Important parameters such as coolant flow through the assembly, coolant temperature, the location of the fuel assembly inside the machine and the radiation field outside the shielding are continuously monitored during the fuelling operation. Initial raising and final lowering of the assemblies is done on slow speed to prevent any damage to the assembly being handled or to the core components.

Since unintentional movement of a fuelling machine coupled to a fuel channel could lead to breach of primary coolant boundary suitable mechanical and electrical interlocks to prevent the same is provided. The coolant used in fuelling machine is of the same quality as the coolant used for the assembly in pile. Filters and ion exchangers in the coolant system of the fuelling machine are housed in suitable shielded containers to minimize exposure of the operating personnel.

Before commencing any refuelling operation all the associated systems of fuelling machine are tested in a test channel to ensure effectiveness of the all logic and controls. The deficiency noticed if any is corrected and the machine is retested before commencing the operations.

All the handling in pile is done only as per approved agenda and procedures. Radiation survey with necessary monitoring equipments is carried out at all times during fuel handling.

The facility for storing and handling of irradiated fuel are generally filled with de-mineralised water, which removes the heat due to radioactive decay from the irradiated fuels and also provides the adequate shielding against radiation. As natural uranium fuels are stored in light water filled bay, inadvertent criticality has been totally ruled out. Salient features of storage facility are listed below

1. The storage pool facility is well ventilated and has enough capacity to accommodate all the assemblies in core, in case the same has to be unloaded, in addition to the storage capacity required for the irradiated fuels removed during normal operation.
2. A separate storage pool is available adjacent to reactor core for accommodating one full charge of reactor to cater to emergency unloading

of core. The pitch of this storage pool is chosen in such a way, that the system cannot become critical even if the light water of the pool is replaced with heavy water.

3. Shielding requirements for the above-mentioned storage facilities are calculated assuming that all the storage positions are occupied by spent fuel, which has undergone maximum design burn up and minimum decay time.
4. The water pools are lined with SS plates to prevent water seepage. In order to get good visibility and minimize the corrosion of fuel assemblies' components and the liner metal, the storage pool water is kept under continuous purification. Provisions are available to detect the liner leak. The pool water temperature is also maintained within safe limits.
5. Under water equipments are installed with simple mechanisms so that they can be removed without difficulty.
6. Suitable isolating gates are provided in spent fuel storage bay so that the compartments housing the handling equipments can be isolated and drained off water to facilitate major maintenance activities leaving the compartment with active fuel element full of water.
7. It is possible to fill the bays with alternate, source of water in case of emergency like major breakage of bay drain line or non availability of normal make up water.
8. The pool water level is always maintained within a specified band. In case the level drops below the specified limit the same is annunciated in control room alarm panel.
9. Radiation monitors are provided in all the critical operating areas and in case the radiation level exceeds the set value, the alarm is annunciated in control room.
10. Spent fuels are transferred to shipping cask under water. Shipping cask lids are opened or closed under water. Suitable facilities are available for decontaminating the shipping cask. The crane facilities provided for handling the shipping cask is serviced and checked at periodic intervals. Before transporting an irradiated fuel in a cask it is ensured that the assemblies have seen sufficient cooling period.
11. The tools are designed in such a way that their handling is easier, safer and will have positive locking fuel during handling. Only approved and tested tools are used for fuel handling so that the fuel is not raised too close to the water level and the submergence of assemblies is maintained to ensure adequate shielding.
12. Health Surveyor is always present at site during handling operations.

2.4 Isotope Handling

Handling and storage facilities for fuel assemblies mentioned so far are common to most of the isotope production assemblies also. In addition facilities like hot cell, master-slave manipulafor etc. are provided to handle radioactive isotopes. Facilities are available to irradiate the target material for a period varying from a few second to several years. Most of the safety criteria for handling and storage of fuel assemblies are applicable for handling isotope production assemblies also. In addition following safety criteria is also considered for isotope production.

1. Target material is filled in a capsule and sealed properly; the integrity of the sealing is tested and certified before loading the sample into the reactor.
2. Assessment of target material is done with respect to reactivity effect, nuclear heating, induced activity, material compatibility and shielding requirements before loading the same into the reactor.
3. Dimensional tolerances of sample capsule are checked. The capsule is also passed through simulated routes outside the core to ensure the smooth movement of the sample.
4. All operations on irradiated capsules are remotely carried out to avoid radiation exposure and spread of contamination

5. Hot cells are well ventilated and maintained at lower pressure compared to the adjacent areas.
6. Master-slave manipulator, isotope, assembly handling buggy and the hot cell doors are mechanically/electrically interlocked to maintain correct sequence of isotope handling. This is to avoid damage to handling equipments and isotope assemblies and to prevent personnel exposure to radiation.
7. Inventory of all the isotope capsules is done whenever the isotope assembly is lowered into the hot cell. This ensures that no capsule is dislodged from its position during irradiation/handling.

3. RELIABILITY UP GRADATION BASED ON OPERATING EXPERIENCE TO ENHANCE SAFETY

3.1 Improvement of extractor drive interlock circuit

In the last 20 years, FM has handled around 3500 fuel assemblies safely. On a few occasions, after dumping the heavy water coolant from barrels to storage tank of machine, extractor with irradiated fuel assembly could not be lowered into water filled bay immediately by hydraulic drives of machines as per the normal procedure, which ensures sequential completion of required actions. The irradiated assembly was immediately doused with emergency light water coolant and was lowered manually into the water filled bay. On all the occasions, integrity of fuel could be ensured. Probable reasons for this could be non-fulfilment of certain interlock(s) or malfunctioning of hydraulic drive circuit or mechanical friction or jamming. One supervisory lamp indication was introduced to indicate fulfilment of all interlocks of extractor lower circuit in advance. With this modification, if a similar type of failure happens, the nature of the fault can be established early and suitable corrective actions can be taken quickly without restoring to light water dousing.

3.2 Reliability of machine alignment over channel

Alignment of FM over channel has to be reliable and repetitive to ensure safe handling of fuel assembly. Drift in alignment, if any, during refuelling operation, may lead to unsafe situation and hence has to stopped. Over the years suitable modification for example providing guide rollers for skirt to limit machine movement on skirt lowering to bare minimum, aligning the machine on fuel channels with hydraulic lines kept loaded; etc have been carried out.

3.3 Improvement in Coolant system of FM

The assembly being handled in fuelling machine is cooled by heavy water supplied by the coolant pumps mounted on the machine itself. The coolant circuit consists of two pumps (one standby), heat exchangers, storage tank and purification-circuit. Both the pumps were provided with class-III electrical power supply initially.

In case of non-availability of heavy water coolant in machine due to any reason particularly when machine is in transit with irradiated fuel, heavy water can be valved in from the tap off point provided in main coolant system through quick connecting coupling and flexible hoses. Two booster pumps (one standby) with class-II electrical power supplies provided in the main coolant system ensures adequate pressure at the above tap off points.

Provision to supply light water cooling to an irradiated assembly in fuelling machine is also made. This facility caters to an emergency-cooling requirement while discharging an irradiated fuel assembly into the storage pools after draining heavy water from the barrels to storage tanks of the machine.

To enhance the reliability of the cooling system during extended class-III power outage, one of the coolant pumps of the machine was provided with class-II supply. This further avoided the need of immediate light water dousing there by increasing the availability of the machine.

Earlier the emergency cooling to the machine through booster pump tap off line (heavy water) and light water lines was being connected to same lines in the machine. However this required some valve manipulation in the machine depending on the mode of injection. Light water injection circuit was segregated completely there by improving the reliability of the emergency injection. Further redundancy of the system was enhanced by providing one extra tap off point for light water injection.

3.4 Improvement in access control

During initial commissioning trials with dummy fuel, it was realized that possibility of fuel assembly remaining unlocked from a channel couldn't be ruled out. In Dhruva upward force exerted on the fuel by the coolant flow when main coolant pumps are in operation is more than the weight of the fuel and an unlocked fuel can get lifted up. In view of this apprehension, provisions were made to deny access to fuel channels when main coolant pumps are in operation in the form of physical barriers and interlocks. In addition special locking gadget was provided on top of each fuel assembly whose design does not permit the assembly to be unlocked while coolant pumps are in operation.

3.5 Reliable handling of tray rod

Regular refuelling operations are carried out only with main coolant pumps in off conditions. The isotope tray rods were handled every week to deliver radioisotopes. This necessitated shutting down of reactor every week. Weight of tray rod assemblies was increased by lead filled shielding section to counter the up thrust due to coolant flow. With this provision the tray rods cannot get ejected from the channel even if they remain unlocked in pile with main coolant pumps operating. After incorporating this passive safety feature tray rods are handled every week with reactor in operating condition with main coolant pumps on.

Since 1990, around 1300 on-power isotope tray rod assemblies were handled safely by the FM. Every week, radioisotopes used in various field of medicine, agriculture and industry are being supplied to various agencies.

4. PROCEDURES AND RECORDS

All the fuel handling and isotope handling activities are to be carried out as per approved procedures. All related in-core operations are done only after the same is evaluated from reactivity, cooling and shielding considerations and approved by the plant management. All the handling is done only by trained personnel and supervised by authorized supervisors.

Records pertaining to the fuel assemblies detail, testing results, in-core performance data, details of maintenance activities carried out on various handling equipments meticulously maintained for future reference. Preventive maintenance on handling equipments is done at regular intervals to minimize breakdown of the equipments.

A full-fledged software simulator of the fuelling machine interlocks and fuelling steps was developed in house and is being successfully used for manpower training. The software was developed using visual basic 6.0 and is compatible with all Microsoft operating systems. The software provides a visual feel of the fuelling machine/operation and is equipped with suggestions for emergency actions in case of unforeseen transient during fuelling. It is proposed to link the software with fuelling machine in future. This will reduce the possibility of human error during actual fuelling. Software will take inputs from the machine and keep a track of all the actions carried out by the operator. In case operator deviates from the standard procedure and misses some steps, an audio-visual alarm will be generated and suggested correct step will be displayed in front of the operator.

5. CONCLUSION

Fuel and isotope handling equipments are designed and operated essentially with both fuel safety and personnel safety in mind. Good operating procedures and maintenance of records for movement of the assemblies are very essential. Routine testing and regular preventive maintenance of the handling equipments are essential for assuring reliable and safe operation. Strict adherence to safety norms and standard procedures are necessary to minimize radiation exposures and avoid spread of contamination during handling. Enhancement of safety is an on going process. From over 20 years of experience in the field of handling and storage of irradiated fuel from Dhruva we have incorporated several changes towards enhancing safety and improving system performance. These include design modification, procedural changes, and provision of physical barrier.

6. REFERENCES

1. Safety criteria for handling fuel and radio isotopes; S.Duraisamy (Training course on research reactor safety principles)

2 Design manual; Fuelling Machine-A (R-5 project; Fuel handling & storage group)

Reliability, Safety and Hazard
Editors: P.V. Varde et al.

OVERVIEW OF HUMAN PERFORMANCE IMPROVEMENT INITIATIVES IN NUCLEAR POWER PLANTS (NPPs)

Ashok Kumar Sharma

Nuclear Power Corporation of India Limited, 10-N-28, Vikram Sarabhai Bhavan
Anushaktinagar, Mumbai 400 094

ABSTRACT

Nuclear Power Plants (NPPs) are very complex systems. Diverse, multiple and redundant technological systems are used for effective control and safety of the NPPs. The increased numbers of such systems require increased operator attention. Additionally, the control stations (man-machine interfaces) are to be kept manageable in size. This sometimes reduces the scope for truly ergonomic design. These limitations, coupled with the shortcomings of human nature, led to unintended human performance problems and errors resulting into poor plant performance worldwide. Some organisational weaknesses, managerial decisions and latent errors also aided and abetted human errors. In view of this, a need was felt for development of performance culture at all levels in NPP organisations. Towards this end, ready-to-use performance improvement tools were developed and used for individual performers, supervisors and managers in the NPPs. This paper describes the experiences of the global nuclear electricity generating industry towards human performance improvement and error reduction.

1. NUCLEAR ELECTRICITY GENERATION

Nuclear energy is a major source of safe, clean, economical, and reliable energy worldwide, providing 16 percent of the world's electricity from 441 Nuclear Power Plants (NPPs). Many countries rely heavily on nuclear energy for their electricity needs. France obtains 78 percent of its electricity from nuclear energy (IAEA, 2005). India has 15 operating reactors, three of which figure in the 50 best performing NPPs worldwide (USNRC, 2003). These reactors provide about 2.8% electricity supply to the country. As environmental awareness grows, nuclear power is poised to become an ever more compelling energy option (USDoE/NPI, 2004).

In the recent years, a renewed interest in the nuclear power is witnessed globally. USA plans to add 50,000 megawatts (MWe) of new nuclear generating capacity to the US grid by 2020 (NEI, 2001). India too is headed for an eight fold increase resulting in the installed nuclear electricity capacity of 20,000 MWe by the year 2020 AD. Globally there are 24 reactors under construction at present; eight of these are in India (IAEA, 2005).

Complex Socio-Technical System

In addition to the usual hazards present in the large industrial units using high pressure and temperature processes, NPPs use radioactive processes for generation of heat energy. Therefore, they are needed to be remotely controlled. Many of the safety systems in NPPs are controlled automatically. There are teams of operators called 'crews' that operate and control the NPPs from the Main Control Rooms in round-the-clock shifts.

Maintenance on the plant equipment and systems is done by a specialised group of people using high-tech monitoring, inspection and diagnostic techniques in view of restrictions in accessing the equipment due to radiation. All important tasks are performed according to approved task procedures that require performer's self-verification or independent verification for quality assurance. As such, about 600 personnel are typically needed in twin unit NPPs in different functions like operation, maintenance, engineering support, training, quality assurance, reactor physics, station chemistry, health physics, industrial safety etc. These personnel have varied education, training and job-qualification or job-license as applicable. These considerations make the NPPs very complex socio-technical systems.

The Challenges Encountered

Some twenty years back, the performance indicators of the NPPs were not as impressive as they are today. The unit capability factors were in the range of 60-70 percent, automatic scrams (reactor shutdowns) were about 2 per 7000 hours critical (approximately one year), and industrial safety record too – although good in absolute terms – was four times less impressive than what it is today. Studies conducted on NPP events and causal factors indicated that there were many undesirable events (not accidents) even during the mid nineties that had origin in the inadequacies of human performance as well as unintentional human errors (WANO, 2001).

The Paradigm Shift

In the late seventies, the nuclear electricity generating industry began to notice human performance problems, including human errors, against the backdrop of numerous hardware or equipment problems. As a result, the human performance problems got overshadowed by equipment problems. Any event taking place due to human error was mistakenly assigned a technological or equipment malfunction tag. This used to happen because the implications of human element involved in plant performance were neither understood nor measured and evaluated. Logically, the industry focussed only on improved understanding of nuclear technology while operating and maintaining NPPs. In this process, the industry kept on improving its hardware or equipment performance significantly. In the meanwhile, the supporting industries also started providing equipment and services of much better quality. The sharp focus maintained by the NPPs on technological understanding gave rise to elaborate and comprehensive technical training to performers at all levels (Bhattacharya, 1998). So, the overall performance indicators of the industry became pretty good in the mid nineties.

When the technological understanding and equipment performance became better, the human performance problems and errors got highlighted as significant performance problems. Therefore, a pioneering remedial work on human performance improvement was taken up by the Institute of Nuclear Power Operators (INPO) in USA. INPO is jointly funded by the NPPs of North America. This work was adopted and further developed by the nuclear electricity generating industry globally.

2. HUMAN PERFORMANCE AND ERRORS

Human performance is defined as a series of behaviours an individual executes to accomplish task objectives (results) according to some standard. The key elements of human performance are behaviours and results. If appropriate behaviours are executed by the performer but the results are not achieved, then the human performance is considered less than adequate. It is possible to have appropriate behaviours in an inappropriate task environment (e.g. error prone jobsite, abnormal job condition, faulty job procedure and defective tools etc.) leading to non-achievement of results. On the other hand, if the appropriate behaviours are not executed by the performer, then certainly the results will not be achieved. Human performance, as such, will be less than adequate and will lead to poor overall business results in both the cases. Human error is defined as an action that unintentionally departs from an expected behaviour according to some standard. For example, sometimes when we want to switch-on the fan in our living room, we mistakenly switch-on the lights. Such unintended human errors do take place in all industrial facilities, including NPPs. Elaborate error tolerant designs, performance support systems, as well as verification culture are needed to reduce the human errors.

Model of an Event

Human performance and error were needed to be understood in the organisational context by the industry. Accordingly, the model of an event (Reason, 1990) shown below was adopted for understanding the mechanism that leads to occurrence of an event in NPPs. This model has five major elements – worker action, weak defences, error conditions, hidden organisational weaknesses and the event itself. Many events are caused by worker action. The worker action could be in error either due to error conditions prevailing at the jobsite or due to unintended active error by the worker. Even when the worker action is inappropriate, the multiple layers of defences (defences-in-depth) built by the NPPs should avoid any event. But the events take place when defences are weak too. Both weak defences and error conditions are caused by hidden organisational weaknesses. In the context of this model it is important to clarify meaning of a few terms. **Event** is an undesirable consequence to the plant generally in terms of reduced safety margins. Event may not necessarily cause harm to the plant, personnel, public or environment. Mere reduction in the margin of safety, which is usually maintained by the NPPs, would also be called an event. **Worker action** is an action by an individual (worker, supervisor or manager), either correct or in error, that may result in a plant event.

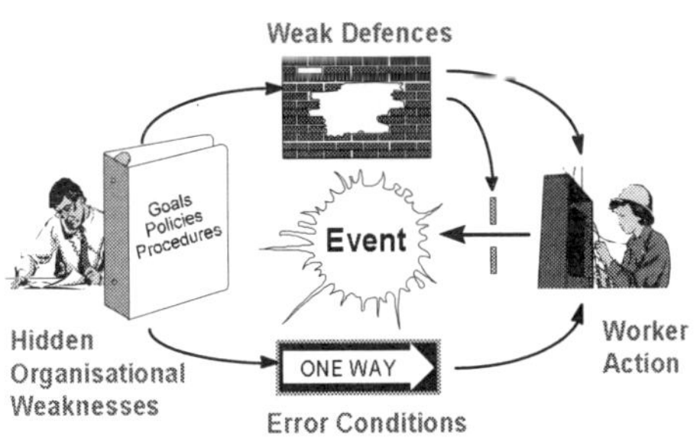

Active errors are those errors that have immediate, observable, undesirable outcomes in the plant. For example, starting pump # 1A in place of pump # 2A by mistake leading to overheating of reactor. The industry was concerned with errors, not violations (deliberate deviations from approved procedures). **Latent errors** are those errors that result into undetected organisation related weaknesses or equipment flaws that lie dormant. For example, writing a wrong step in a task procedure will remain dormant till the procedure is used by an individual to perform the task inappropriately. **Defences** (physical or administrative) are the measures that protect against various hazards (radiation, chemical, heat etc), mitigate the consequences of the hazards, or encourage consistent behaviours. Administrative permissions, independent verifications for quality assurance, equipment interlocks, locked out valves and doors are some examples of defences. **Weak defences** are those which, under the right circumstances, may fail to protect equipment or people against hazards or fail to prevent the occurrence of active errors. Whenever an event occurs, either there are flaws in the existing defences, or appropriate defences are not in place. In every case, the flaws exist before the event occurs. Therefore, it is possible to prevent events by taking a more careful look at defences before beginning a task. **Defence-in-depth** implies use of multiple barriers to protect. However, multiple defences can add complexity to the job and make performance more difficult. Yet, while dealing with radiological hazards, it is essential that multiple barriers are put in place. **Error conditions** are the prior conditions at the jobsite that increase the probability for error when performing a specific action (error-likely situations). Error conditions exist in the work place before the error occurs. If discovered before the error, the conditions can be changed to reduce the possibility of error.

Hidden organisational weaknesses are undetected weaknesses in the management control processes (strategy, policies, work control, training, and resource allocation) or values (shared beliefs, attitudes, norms, and assumptions). These weaknesses can create workplace conditions that provoke error and weaken the defences.

Before understanding and adopting this model of an event, popular misconception persisted in the NPPs that people must lack proper motivation when they make a mistake or act without clear judgment. However, the analysis of industry events causal factors (WANO, 2001) revealed that events occurred more often due to error-likely tasks and error-likely work environments than from error-prone individuals lacking appropriate motivation. As a corollary, it followed that the decisions and activities of the managers and supervisors determine what is done, when it is done, and how well it is done. This either contributes to the health of the NPPs or further weakens their resistance to errors and events. Therefore, managers and supervisors are also expected to perform their duties with the same caution for error-likely work environments as the workers do at the jobsite.

The Error Iceberg Model

In Nuclear Power Corporation of India Limited an error iceberg model was developed for understanding and prevention of human errors (Sharma, 2002). The figure below depicts the error iceberg model.

The Error Iceberg Model
(Sharma, 2002)

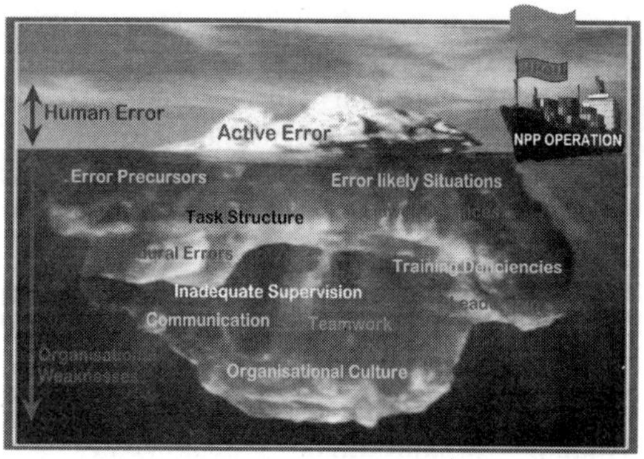

The error iceberg model indicated that the industry selectively saw only the active errors that led to immediate undesirable consequences. A lot of other factors that contributed to errors were not visible like the submerged part of an iceberg. These factors include error precursors, error-likely situations, flawed defences, inappropriate task structures, procedural errors, training deficiencies, inadequate supervision, leadership, communication, teamwork and organisational culture. Collectively these factors were called organisational weaknesses (Sharma, Dayal, & Rajendren, 2001).

In yet another study of 37 incidents involving human performance issues, US Nuclear Regulatory Commission (USNRC, 2001) also concluded that "latent errors (i.e., errors committed prior to the event whose effects are not discovered until an event occurs) were present four times more often than were active errors (i.e. those occurring during event response)."

These studies sensitised the industry towards the significance and contribution of the latent organisational weaknesses. Consequently, the industry prominently positioned this issue in its strategy for human performance improvement and error reduction.

Need for Organisation-wide Intervention

It was established by these models and studies that human performance can be improved by improving work behaviours of all the individuals. This is due to the fact that the jobsite conditions including the defences are created, maintained and improved by workers, supervisors and managers together. Similarly, organisational processes and values are mainly created by supervisors and managers with reciprocal social, cultural and collective bargaining inputs from workers. Therefore, human performance of all individuals is needed for meaningful and sustained improvement in plant results. Accordingly, the industry initiated a broad-based human performance movement involving lowest to highest level of individuals in the organisational hierarchy. The term *leader* was also redefined by the industry to make it broader in its coverage. **Leader** is an individual who takes personal responsibility for his or her performance and the plant's performance and attempts to influence the processes and values of the organisation (NANT, 1999). However, the individuals at different levels had different developmental backgrounds, intellectual capabilities, work allocations and performance modes. Therefore, different sets of ready-to-use behavioural tools were needed for execution of leadership behaviours by individual performers, supervisors / managers of individuals and functional managers. In response to this need, the industry adopted, developed and provided specific tools that are described after individual performance modes.

Individual Performance Modes

There are three basic performance modes of individual performance – skill-based, rule-based and knowledge based performance. Individuals perform in one of these three modes depending on the familiarity with the task and consequent attention paid by them to the task (NANT, 1999). **Skill-based performance** refers to highly practiced actions that are successfully performed as a matter of routine. These actions are usually performed from memory without much conscious thought in a familiar situation. Inattention could lead to error in this performance mode. A representative probability of such errors could be about one in ten thousand performances. **Rule-based performance** refers to behaviour based on selecting predefined procedural steps (or rules). Misinterpretation of the procedure could lead to error in this performance mode. A representative probability of such errors could be one in one thousand performances. **Knowledge-based performance** refers to behaviour in response to an unfamiliar situation. There are no skills or procedures (rules) available to guide the performer. The performer must rely on understanding of the system, scientific principles and theory related to the system. A representative probability of such errors could be one in two performances.

Tools for Individual Performers

In the year 1999, a study of 200 human error incidents in nuclear power plants by Central Research Institute of Electrical Power Industry (CRIEPI) of Japan, it was concluded that "inappropriate tightening", "mis-writing and binding" as well as working on "wrong unit [one of the few identical units within a multi-unit station for which the work is not planned], train [incorrect series of active components out of many] and component" were some of the frequently observed error types executed by the individuals. The most frequent causal factors identified were internal factors, including "individual psychological / physiological factors as well as experience and competence." These findings were typical of the nuclear electricity generating industry worldwide.

As discussed before, those individuals who are responsible for their own performance and performance of the plant are considered leaders. In their role as individual performers, their work behaviours should be appropriate and error-free so that the types of errors discussed above are not repeated. Towards this end a few specific behavioural tools were adopted by the industry and provided to the individual performers. Some of these are discussed below.

STAR is a self-checking tool. STAR is an acronym for **S**top, **T**hink, **A**ct and **R**eview. This tool is used for its speed breaking and thought initiating mechanism to avoid the errors of inattention in familiar situations. **Peer-check** refers to an informal on-line checking of task performance by a colleague on request of the performer. By this tool, the performer gets to know from the colleague (peer) when (s)he is about to make a mistake. **Verification** and double verification are formal checking of the task performance either during the performance or after the performance by independent observers. **Three-way communication** is a behavioural tool to reduce chances of error in all the performance modes due to faulty communications. While using this tool the sender sends a message, the receiver repeats the message back to the sender by paraphrasing general content and repeating the task instructions verbatim using standard phonetic alpha-numerals. This process repeats till the sender gets satisfied by the feedback and communicates permission to the receiver to start performing the task. **Verbalisation of actions** is a behavioural tool for rule-based performance by a team of performers in which a performer declares aloud his intention to perform procedural steps in advance. The co-workers working in the vicinity thereby get a chance to stop the performer from executing those procedural steps that may be inappropriate from their perspectives. **Timeout** is a behavioural tool for knowledge-based performance. While using this tool, a team of performers stops working and takes time out to discuss the problem and alternative solutions to improve shared understanding by pooling the knowledge available with individuals of the team. This tool is also referred as **Stop When Uncertain** for seeking expert inputs during team as well as individual task performance in knowledge mode. **Questioning** is yet another behavioural tool whereby the performers fully understand the task, its context, specific assumptions, and expected results before starting to perform it. No task is performed before fully understanding it. **Advocacy** is a behavioural tool whereby the individuals in a team performance openly (and candidly when needed) advocate their point of view without fear so that erroneous assumptions and overlooked perspectives are brought to the attention of all and "groupthink" (the tendency of members of a committee, profession, etc. to conform to those opinions or feelings prevailing in their group) can be avoided while working in teams in knowledge mode.

Tools for Supervisors

The CRIEPI (1999) study also indicated that "recent increasing [causal] factors are human-machine interface and communication. The former includes ambiguous demarcation [identification / labelling] and malfunctioning indicators, the latter specifies communication between different hierarchy / organisations." These findings too were very commonly experienced by the nuclear electricity generating industry worldwide. Supervisors are responsible for planning, organising, coaching, verifying and controlling tasks of people in their teams. Along with their own performance, they are responsible for the performance of their team members. Additionally, in line with the sense of ownership generated by the current human performance paradigm, they are responsible for the plant performance and should take necessary actions to modify jobsite conditions to ensure excellence in performance. For such role holders, the following behavioural tools were adopted by the industry.

Anticipating and communicating error-likely situations refers to a behavioural tool whereby the supervisors review the jobsite and the task steps in advance with a view to anticipate what may go wrong while performing the task or to find what error traps exist at the jobsite. These findings are communicated to the performers. Additional defences are also installed when the error-likely situations become too challenging to allow successful task performance. **Pre-job briefing** as a behavioural tool refers to a structured communication to the performers by their supervisor (or team leader) before they perform a task that is non-routine and safety-critical. Pre-job briefing is also used for routine work when change is involved at the jobsite. Pre-job briefing generally covers the critical elements of the task, performance modes and error conditions, defences, potential consequences and response strategy in abnormal conditions. **Post-job debriefing** as a behavioural tool refers to a structured discussion by the performers with their supervisor after a task is performed. The purpose of this tool is to learn from the experience so that in future the task can be performed in a better way to achieve task standards. It is particularly used when a task is not performed as per standards and rework is involved. It is also used when a critical and non-routine task is performed well. **Matching task demands and performer capabilities** refers to a planning strategy whereby supervisors list task demands and select a worker who has capabilities to meet them. Mental, physical and team competencies are considered in this exercise. They include knowledge, skills, experience, fitness and attitude, need for supervision, worker initiative for the task, formal task qualification etc.

The purpose of reinforcement is to increase desired behaviours and decrease / eliminate undesirable behaviours. **Reinforcing desired individual and team behaviours** at the jobsite by feedback and positive reinforcement is considered appropriate. Negative reinforcement and extinction are also used selectively. With reference to human performance improvement, a non-punitive policy is widely publicised and practiced by the industry. Therefore, punishment is never used as a part of the reinforcement strategy. Desired results in terms of quantity, quality, cost and timeliness are also paired with desired behaviours for instrumental conditioning.

Tools for Managers

Functional managers representing operation, maintenance, engineering support, training, quality assurance functions along with other top executives have a prominent role in developing, shaping, maintaining and interpreting organisational processes and values. Additionally, all the individual performers, supervisors and socio-political formations in the organisation reciprocally influence these processes and values. Individuals (irrespective of formal role descriptions) practicing leader behaviours especially influence these processes and values. For this the following behavioural tools were adopted by the industry for functional managers.

Provide role model refers to a series of behaviours that strengthen the organisational communications by demonstrating performance to exacting standards, communicating expectations clearly, providing feedback, welcoming input, and supporting the front line workers. In essence, this tool guides the leaders to demonstrate excellence in their own behaviours so that others can understand and follow their example. Closely related tool is **Clarify vision, mission, objectives and imperatives** to the followers through collaborative envisioning exercises. These tools help in providing direction as well as synergising the collective human potential at the NPPs. **Conservative decision making for safety culture** refers to a behavioural tool that emphasises safety as an overriding priority. It guides the safety philosophy of the leaders such that they decide to reduce or stop production to ensure safety. **Defence-in-depth approach** as a behavioural tool guides the leaders to establish multiple barriers to guard against human errors. These barriers need to be evaluated, renovated and reinforced from time to time. All important tasks are performed according to reviewed and approved task procedures so that all administrative barriers and physical barriers are followed. **Identifying and eliminating root causes** of recurring problems refers to a system of diagnosing the problems with specialised techniques that reveal root causes as well as causal factors. Appropriate corrective actions, to deal with the root causes and causal factors, eliminate recurrence of the problems. **Train and qualify performers as per task standards** is a tool to ensure that the performers, when required to perform specific tasks for which they are qualified, will perform the tasks to predefined task standards. Elaborate systems of training, job-qualification and job-licence (Sharma, 1997) have been developed by the industry using Systems Approach to Training (SAT). These systems use a wide range of training settings such as classroom training, workshop training, simulator training, on-the-job training guided by the task performance based checklists, walkthrough in the plant as well as personal interview and coaching (Bhattacharya, 1998). Managers, supervisors, and front-line workers in the industry are trained in human performance fundamentals to recognise error-likely situations by cascade training for multiplier effect (Sharma, 2001). **Use and share operating experience** is a tool that guides the managers to use, and promote the use of, operating experience of the global nuclear industry. The global industry associations have codified, structured and computerised the operating experience so that the managers can get ***just-in-time*** inputs. **Self-assessment** and **Benchmarking** are closely related tools that have been effectively used by the managers in the industry. Self-assessment is a structured, objective and visible process or set of processes whereby individuals, groups and managers within a plant evaluate the effectiveness of their own operational processes against predetermined targets, goals and other performance expectations. Benchmarking through peer reviews is also an assessment process with the only difference that the assessment is conducted by a team of experts from the other NPPs.

Other significant tools used by the NPPs are – human performance organisation within NPPs, human performance meetings, human performance indicators and their trending, human performance clocks prominently displaying the time elapsed since the last human performance event, human performance group incentives, human performance observation cards for supervisors and managers, job site report cards for managers, human performance posters, slogans, competitions and awards.

3. IMPROVING TRENDS IN NPPs

The entire nuclear electricity generating industry has been enormously benefited by the use of the human performance improvement tools discussed above. According to recent data from the industry, nuclear plant performance and safety has improved consistently over the last fifteen years. These data also show major strides in plant capacity, reliability, and economic performance. A few representative indicators (worldwide averages) of capacity utilisation, reliability and safety (WANO, 2004) of NPPs are discussed next.

Unit capability factor: Unit capability factor indicates energy actually supplied as the percentage of maximum energy generation that a plant is capable of supplying to the electrical grid, limited only by factors within control of plant management. A high unit capability factor indicates effective plant programmes and practices to minimise unplanned energy losses and to optimise planned outages. As seen in the chart below, there is a steady rise in unit capability factors of NPPs over the last fifteen years.

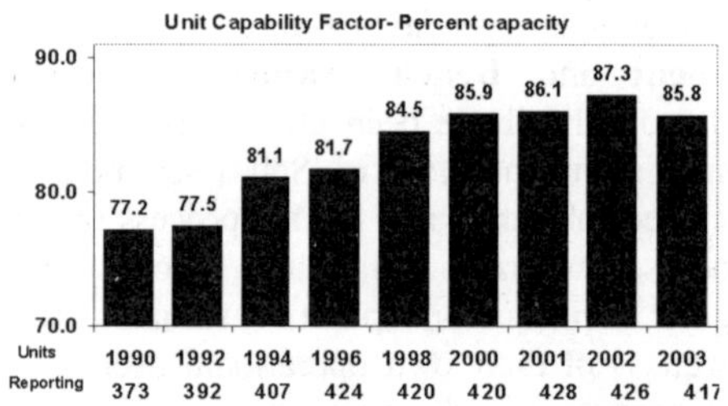

It would be appropriate to emphasise here that the data given in the chart are world averages. There were fifty NPPs that had operated between 95 to 106 % capacity factors during the FY 2003. Thirteen of these fifty NPPs operated above 100% (USNRC, 2003) capacity factors.

Unplanned automatic scrams per 7,000 hours critical: The unplanned automatic scrams per 7,000 hours critical indicator tracks the mean scram (automatic plant shutdown) rate for approximately one year (7,000 hours) of operation. Unplanned automatic scrams result in interruption of electricity generation and subject the plant systems to the thermal and hydraulic transients. So, the unplanned automatic scrams are undesirable occurrences. As seen in the chart below, there has been a steady decrease in this indicator of the NPPs over the last fifteen years. This indicates improved reliability of electricity supply and reduced challenges to the plant systems.

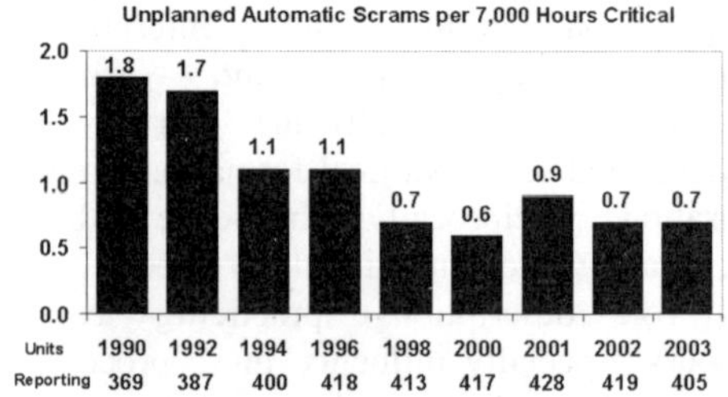

Industrial safety accident rate: The industrial safety accident rate tracks the number of accidents that result in lost work time, restricted work, or fatalities per 200,000 work-hours. As the chart below indicates, the nuclear industry continues to provide one of the safest industrial working environments.

Over the last fifteen years, the industrial safety accident rate has reduced four times indicating towards the enormous success achieved by the industry in developing leadership for human performance improvement and error reduction. The US Bureau of Labour statistics show (NEI, 2004) "that it is safer to work at a nuclear power plant than in the manufacturing sector and even in the real estate and finance industries."

4. CONCLUSION

In addition to usual industrial challenges, the nuclear electricity generating industry faces challenges of managing complex systems handling radioactive processes and materials in an environment of high public scrutiny. Additionally, operational events in any NPP have global impact on the entire nuclear electricity generating industry. It became obvious to the industry in the mid nineties that human performance improvement and error prevention through leadership development holds the key to further improvement in plant performance. Therefore, the industry funded collaborative studies towards this end, while working in heavily regulated business environments and national sensitivities.

These studies were successful in modelling the mechanisms that are instrumental in human performance problems including errors. These models explained roles of individual performers, supervisors and mangers in improving human performance and reducing human errors. Consequently, practical behavioural tools for human performance improvement were devised and implemented by the industry. The industry associations have helped in quickly spreading the use of these tools to over 400 NPPs globally. World Association of Nuclear Operators (WANO), in particular, has been very effective in this through repeated training programmes and peer reviews. This has resulted in exemplary performance improvement in the nuclear industry that is worth emulation.

Lessons learnt in the nuclear electricity generating industry, with regard to leadership development for human performance improvement and error reduction may be relevant to the thermal power, petrochemical, fertiliser and pesticide plants as well as other hazardous chemical processing industries.

5. REFERENCES

1. Bhattacharya, A. S. (1998). Nuclear Power Operations Training and Human Resources Development. *INS NEWS Vol. 11 No. 1,2&3 (Jan – Sep. 1998)*
2. Central Research Institute of Electrical Power Industry [CRIEPI]. (1999). *Analysis of human error incidents in domestic nuclear power plants: Principal research results.* Retrieved July 19, 2004, from http://criepi.denken.or.jp /en/ e_publication/a1999/99seika4.pdf
3. IAEA. (2005). *Nuclear share in electricity generation in 2004.* Retrieved September 19, 2005, from http://iaea.org/ cgi-bin/ db.page.pl/ pris.nucshare.htm
4. National Academy of Nuclear Training [NANT]. (1999). *Human performance fundamentals professional development seminar participant handout*
5. Nuclear Energy Institute [NEI]. (2001). Vision 2020: Powering Tomorrow With Clean Nuclear Energy Retrieved July 19, 2004, from http://www.nei.org
6. NEI. (2004). NEI - Plant safety. Retrieved October 10, 2004, from http://www.nei.org/ index.asp?catnum=3&catid=768
7. Reason, J. (1990). *Human Error.* Cambridge, UK: Cambridge University Press
8. Sharma, A. K. (1997). *Manual of instructional system design of Nuclear Power Corporation of India Limited, Mumbai*
9. Sharma, A. K., Dayal, S., & Rajendren, T. K. (2001). Overview of human performance improvement system in NPCIL , *Proceedings of WANO Tokyo centre workshop on measures adopted to improve human performance in NPPs Mumbai, India. Jan 10-11, 2001*
10. Sharma, A. K. (2002). Effective HRD Practices in Nuclear Power Corporation of India Limited. *Proceedings of IAEA regional workshop on upgrading training capabilities to response to emerging needs, Daejeon, Korea, 2 ~ 6 September 2002*
11. U. S. Department of Energy / Nuclear Power Industry [USDoE/NPI]. (2004). *Strategic plan for light water reactor research and development. First edition.* Retrieved July 19, 2004, from www.nei.org
12. U.S. Nuclear Regulatory Commission [USNRC]. (2001). *Review of findings for human error contribution to risk in operating events (NUREG August 2001 INEEL /EXT-01-01166)* Retrieved July 19, 2004, from www.nrc.gov
13. U.S. Nuclear Regulatory Commission [USNRC]. (2003). *Information digest- 2003 edition (NUREG -1350, Vol. 15)* Retrieved July 19, 2004, from www.nrc.gov
14. World Association of Nuclear Operators [WANO]. (2001). Student handouts, WANO OE central team training: Improving human performance. Narora NPP, India. February 2001
15. WANO. (2002). Principles for excellence in human performance (WANO GL 2002-02)
16. WANO. (2004). *WANO 2003 performance indicators.* Retrieved July 19, 2004, from www.wano.org.uk

Reliability, Safety and Hazard
Editors: P.V. Varde et al.

A HUMAN ERROR MODEL IN ACCELERATOR SAFETY ANALYSIS

Lekha M. Chowdhury and P.K. Sarkar

Health Physics Unit (VECC), ARSS, RSSD, BARC, 1/AF, Bidhannagar, Kolkata 700 064

ABSTRACT

The high power and high energy particle accelerators that are now being installed, in good numbers inherently increase the risk due to human error for such hazardous situations in these accelerators like high beam loss, target rupture, inadvertent trapping or access of personnel in high radiation zones etc.[1]. This paper describes the different aspects of human performance that influence risk levels in a particle accelerator. Human tasks and the way they can affect safety is discussed here along with the factors influencing human actions. In the present analysis we propose a human error evaluation model based on expert elicitations of stress-strengths of individuals or working groups.

Keywords: Human error, Expert elicitation, Accelerator safety, PSA

1. INTRODUCTION

Recently there are many high power accelerators being installed for various purposes. For operational reasons, some of them need be located in populated areas. Such accelerators are associated with inherent risks when operated under faulty conditions because of human error. In this paper, we intend to study the interaction between the human operator and the different systems of the accelerator to evaluate the possible causes of human error that can lead to hazardous situations.

There are numerous variations and designs of accelerators existing. However, in general an accelerator consists of an ion source, the main accelerator and the target [1]. There are magnet systems, with high voltage and high current power supplies, cooling arrangements, vacuum systems, ventilation systems and safety interlock systems. The accelerator beam is transported to the target using the beam transport system. The material, design and dimensions of the target vary depending on the end use and user requirements, experiments planned, the quality and quantity of beam to be extracted. A qualitative human reliability analysis for particle accelerators are discussed alongwith a proposed model based on expert opinion of the operators handling the tasks.

2. HUMAN TASK ANALYSIS

The first step in any Human Error analysis is Human Task Analysis. There are three stages of operation in the accelerator as follows [2].

a) Preoperation tasks:

Before starting of operation, testing of various systems at different periodicities is made alongwith a checklist to ensure compliance. Difference in work culture and different confidence levels may result in variable approach to the same job by different people.

b) During operation the main tasks are

1) Start up beam tuning - For a typical three sector cyclotron there are 13 trim coils and valley coils to be tuned. Several optimal conditions are possible as beam tuning depends on multiple variables.

2)Switching to required beam line – Here an unintentional error is possible.

3)Maintaining a steady beam on target - There can be beam fluctuations due to lack of stability in power supply. Beam current output may reduce and the quick and easy solution adopted to circumvent the loss is to increase input current, which directly increases the output current. This will lead to more beam loss which is not desirable. Any system degradation or failure will also have an impact on the beam.

All the above tasks would very much depend on the operator. It is recommended that proper system diagnostic tools be used to detect the cause of any beam loss. Though the procedure is standard time required and the accuracy with which the job is done will depend on individual experience, skill and knowledge. When an operator follows a written set of procedures then it becomes a rule-based task and this can lead to faster beam tuning, lower human error and beam losses.

i) Shutdown tasks

Breakdown maintenance is the main task here.

3. HUMAN FACTORS AND ERROR ANALYSIS

Our analysis will deal with the main tasks during operation. A listing of possible errors by operators while carrying out the plant operation task include [2]:

1. ***Decision errors:*** This occurs due to inappropriate knowledge, poor choice or procedural error. During preoperational tasks and shutdown improper operation planning and improper remedial action is possible.
2. ***Skill based errors:*** These occur without significant conscious thought and include failures of attention, memory and/or technique.
3. ***Perpetual errors:*** These occur as a response to illusions. During prestart up, beam tuning and maintaining steady beam erroneous response may occur under variable situations due to degraded sensory input and misreading of the beam tuning parameters.

If we take an actual initiating event in an accelerator, beam loss is one of the most important events that depend very much on the operator. As can be seen from the fault tree in figure 1., The major causes of Beam losses are Human error or Equipment error. If we see the causes of Human error as identified in the fault tree there are three main stages where there can be an error. Firstly the operator has to set the parameters for beam tuning. This he has to do in stages for different stages. After the beam is tuned it is extracted and switched on to the beam line where it is required.

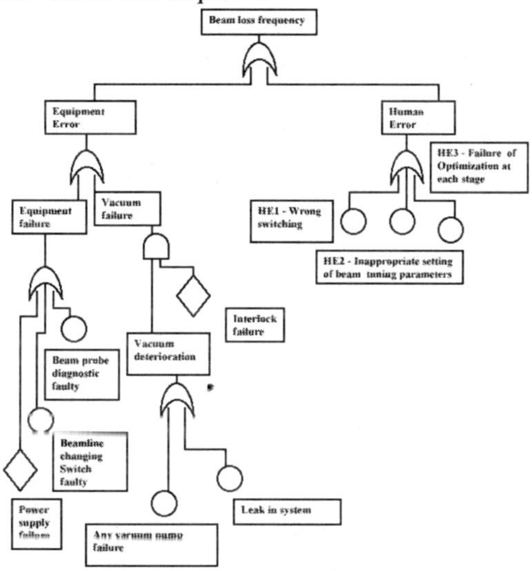

Fig. 1: Beam loss fault tree

4. HUMAN ERROR MODEL

Our model for evaluating Human error possibility is based on expert elicitation of operator capability. This is because the inherent strength of the operator can be estimated by experts depending on his past performance, skill, knowledge and response to a situation. As an example, each expert (Ex1-Ex3) is asked to grade all available operators, numbered p1 to p5 in our case, on a scale of 0-1, their possibility of committing an error. The gradings are shown in Table1. P_{AV} indicates average possibility of error for each person.

Table 1 Error possibility on a scale of 0-1

Ex	P1	P2	P3	P4	P5
Ex 1	0.3	0.6	0.5	0.2	0.1
Ex 2	0.4	0.5	0.4	0.1	0.2
Ex 3	0.4	0.5	0.3	0.2	0.1
P_{AV}	0.37	0.53	0.4	0.17	0.17

As we can see from Table 1, maximum chances of error is by operator p2 (0.53) and minimum by operator p4 (0.17) and p5(0.17). In the model we presume that the possibility of error by the operator is also modified by the stress under which they are operating. This stress depends on the event for which the operator is taking action. Experts evaluate the stress levels for events V1 to V4 on a scale of 0-1 as shown in Table 2.

Table 2 Event stress on a scale of 0-1

Ex	V1	V2	V3	V4
Ex 1	0.5	0.4	0.2	0.3
Ex 2	0.6	0.3	0.1	0.3
Ex 3	0.7	0.5	0.2	0.2
S_{AV}	0.6	0.5	0.17	0.27

In figure 2 the possibility of error is plotted along x axis while the y axis shows the membership value (0-1) for the function. For each person we assign to his average possibility of error a membership value of one. The minimum membership value of zero is given to the possibility of error values outside the range of (P_{AV}-3σ) to (P_{AV}+3σ), where σ is the standard deviation of the data from experts. A linear membership function is assumed between the maximum and minimum values. We get a triangle showing the membership function $\mu_P(x)$ of the possibility of error for a person.

On the same figure, we plot the stress exerted by a particular event. For the membership function $\mu_V(x)$ of stress of an event, we assign a value of one for possibilities (1-S_{AV}) and above. A linear function is assumed between zero and (1-S_{AV}). This curve gives us the error forced on an operator by the stress caused by a particular event. The intersection of the inherent weakness of the operator and the stress forcing to cause an error will estimate the actual possibility of error.

Thus, the possibility of error for any operator for a given event can be estimated from the membership function $\mu_f(x)$, which is the intersection between the two membership functions $\mu_P(x)$ and $\mu_V(x)$ which can be represented mathematically as :

$$\mu_f(x) = \mu_{P\wedge V}(x) = \min[\mu_P(x), \mu_V(x)]$$

Possibility of error x_{AV}

$$= \int \mu_f(x)dx \Big/ \int \mu_P(x)dx \text{ for } \{x_{AV} \in X\}$$

where the integration is done from P_{AV} -3σ to P_{AV} +3σ . In the present case, the membership function $\mu_\phi(x)$ is having three straight line segments, two of which are given by the weakness function (P1 in Fig.2) and the other by the stress function(V1 or V3 in Fig.2).

In the fault tree (Fig.1), we can see that human error can occur in three situations, namely, HE1, HE2 and HE3. In our proposed model, V1, V2, V3 refers to the three basic events leading to HE1, HE2 and HE3. Evaluating the possibility of error as explained above we obtain 5 values for 5 operators for each error HE1, HE2 and HE3. The value to be used in the fault tree will be the average of the five estimations.

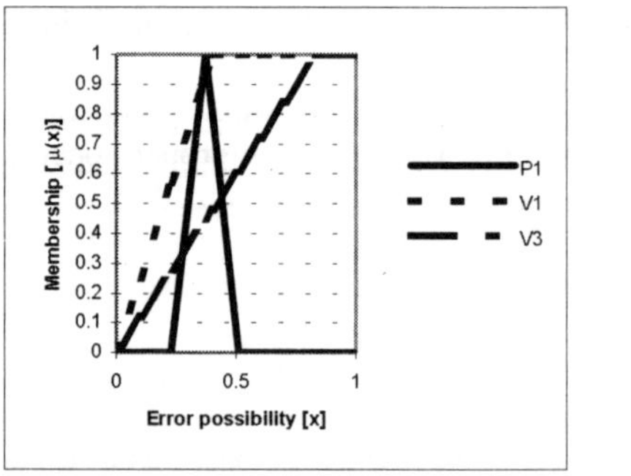

Fig. 2: Plot of error possibility versus membership values for person p1 and events V1 and V3

5. RESULTS

We take the above example to give some numerical results. The mean value $P1_{AV}$ based on 3 experts is 0.37 for which we have assigned a membership value of 1. The standard deviation σ_{p1} = 0.14. Membership values for p1 are 0 below 0.23, (P_{AV}-3σ) and above 0.51, (P_{AV}+3σ). We select two events V1 and V3 for which (1-S_{AV}) values are 0.4 and 0.83 respectively, above which the membership values are 1. Using our model we get intersected trapezoids for events V1 [(0.23,0), (0.353,0.8825), (0.378,0.945), (0.51,0)] and for V3 [(0.23,0), (0.277,0.333), (0.4364,0.5258), (0.51,0)]. The area for trapezoid of V1 is 0.545 and V2 is 0.0944 respectively. The area of the triangle for p1 is 0.14. Therefore the possibility of error for p1 is (0.545/0.14=) 0.984 in case of V1 and (0.0944/0.14=) 0.674 for V2. This possibility of error is the combined effect of event stress and intrinsic weakness of the person. It may be noted that the values of stress and weakness used in this example are deliberately made very high to demonstrate the model.

6. CONCLUSIONS

In the present analysis, we propose a human error evaluation model based on expert elicitations of weakness of individuals (operators) and stress induced by events. This is a preliminary proposition and has not been tested for all possible situations. A complimentary model based on the strength of the individual and S_{AV} will yield the same results. The membership functions assumed are simple. A gaussian distribution for the weakness/strength and Boltzmann or Weibull distribution for the stress functions would have been more appropriate. In case of conflicts between opinion of experts and extreme value opinion of experts, there must be a way of resolving the difference in opinion and arriving at an agreeable value. Any human intervention can also reduce or eliminate the impact of any incident in a plant by acting proactively. Further detailed study is required in this area to strike a balance between computerized automation and human intervention levels and limits.

In addition to all estimations regarding operator error it must be noted that: " However, many accidents have their roots high within the organization. It is the decision made by those at the top that often influence the safety culture contributing significantly to the preconditions for unsafe acts. In order to understand the causal genesis of an accident, we must "peel the proverbial onion back, layer by layer"[3].

7. REFERENCES

1. Chowdhury L.M. and Sarkar P.K, "Probabilistic Safety Analysis in particle accelerators", International Conference on Topical Issues in Nuclear Installation Safety, Beijing, China, 18-22 October, 2004.
2. Chowdhury L.M. and Sarkar P.K , "Effect of human reliability on safety in a particle accelerator", Indian Particle Accelerator Conference - 2000, VECC, Kolkata, India, 01-05 March,2005.
3. Reason.J, *Human Error*. New York: Cambridge University Press, 1990

Reliability, Safety and Hazard
Editors: P.V. Varde et al.

APPLICATION OF EXPERT ELICITATION TECHNIQUES IN HUMAN RELIABILITY ASSESSMENT

V.V.S. Sanyasi Rao, R.K. Saraf, A.K. Ghosh and H.S. Kushwaha

Health Safety and Evnironment Group, Bhabha Atomic Research Centre, Mumbai 400 085

ABSTRACT

Expert elicitation techniques are being used, in the area of technological forecasting, in estimating data needed for analysis when it is either difficult to arrive at the data by experimental means or when it is quite involved to plan and conduct the experiment. In this study, expert elicitation techniques are applied to the evaluation of the frequencies of the various accident sequences that can result from the initiating event (IE) "High Pressure Process Water (HPPW) system failure" in typical Indian Pressururised Heavy Water Reactor (IPHWR) of the older generation. The Operating Procedure under Emergency Conditions (OPEC) for this IE involves human actions according to a pre- defined procedure. The Human Error Probabilities for all these human actions are obtained using expert elicitation techniques. These techniques aim at eliciting the opinion of the experts in the area of interest with regard to the issue in question. The uncertainty is analysed by employing the measure of dissonance and the most probable range of human error probabilities are arrived at by maximizing this measure. These values are combined using the same procedures mentioned above to yield a distribution representing the uncertainty associated with the predictions.

Keywords: Human reliability, Expert elicitation, High pressure process water system

1. INTRODUCTION

Expert elicitation techniques are being used, in the area of technological forecasting, in estimating data needed for analysis when it is either difficult to arrive at the data by experimental means or when it is quite involved to plan and conduct the experiment. In this study, expert elicitation techniques were applied to the evaluation of the frequencies of the various accident sequences that can result from the initiating event (IE) "High Pressure Process Water (HPPW) system failure" in typical Indian Pressururised Heavy Water Reactor (IPHWR) of the older generation. Since there are ncertainties associated with the opinions of the experts, for each of the event under consideration a low, medium and high value are obtained. These values aggregated using appropriate methods and a distribution representing the uncertainty is obtained.

2. DESCRIPTION OF HIGH PRESSURE PROCESS WATER SYSTEM

High pressure process water system is a support system designed to remove heat from various components like moderator pumps, primary heat transport system pumps, diesel generator coolers, auxiliary boiler feed pumps etc. There are five

HPPW pumps (three running and two on auto standby). The loads on this system are distributed in various locations/buildings at the site. An emergency storage tank is provided for this system to take care of minor leakages. In case of major leaks from the system, fire water is injected into the HPPW headers to provide the necessary cooling water supply. Details of this system are available in [1].

3. DESCRIPTION OF THE OPERATING PROCEDURE UNDER EMERGENCY CONDITION (OPECS) FOR HIGH PRESSURE PROCESS WATER SYSTEM FAILURE

HPPW system failure occurs due to
i. all the three running pumps fail and the standby pumps fail to start on 'auto'.
ii. Failure of HPPW pump discharge header
The salient steps of the OPEC dealing with these situations are outlined below.

3.1 Cues for Diagnosis and Steps in Diagnosis

***(a)* HPPW system pressure 'low' annunciation in control room.**

The operator needs to confirm the low HPPW pressure from the following.
- Motor current of process water pumps
- Voltage of class III buses
- HPPW pump discharge valve position
- 'Open' indication of the outlet valve of the emergency storage tank
- Standby pumps started on 'auto' and are running indication
- Discharge valves of standby pumps are open indication

***(b)* Process water emergency storage tank level 'low' and 'very low' annunciations.**

These annunciations occur only when the HPPW system pressure cannot be restored even after the following operator actions.
- Starting of standby HPPW pumps if these have failed to start on 'auto'
- Opening the discharge valves of the pumps if these valves have failed to open on 'auto'.

Even after carrying out the above actions, if the low HPPW system pressure annunciation persists, the operator has to ensure that the fire water injection valve to HPPW headers is open (auto/manual). The persistence of low HPPW system pressure, in spite of the above actions, is an indication of heavy leakage /pipe rupture in the system. Under such conditions, *leak identification and isolation are necessary for selecting the strategy for ensuring secondary heat removal.* The reactor has to be shutdown (or the reactor will be shutdown on uncleared set back).

The leak identification in different areas is possible by the following indications.

(i) floor beetle alarm in reactor building, if it comes, indicates HPPW leak in reactor building (RB)

(ii) if the leak is not in RB, checking for leak in turbine building (TB) and isolation of the leak

(iii) if the leak is neither in RB or TB, checking for the leak in pump house and isolation of the leak

(iv) if the leak is in none of the above three locations, checking for and isolation of the leak in the open areas (This is revealed by flooding and leaks through manhole covers).

Depending on the leak location and possibility of its isolation, 'valving in' of shutdown coolers or fire water injection to steam generators (SGs) has to be resorted to by the operator in manual mode for ensuring decay heat removal.

As can be noticed from the above, the human interaction in HPPW system low pressure incident is a dynamic task requiring a higher level of man-machine interaction like decision making, keeping track of the situation and controlling several functions

4. HUMAN RELIABILITY ANALYSIS

The human actions involved in executing this OPEC are assessed using the Operator Action Tree (OAT) as shown in Figure 1. Out of the seventeen sequences shown in the figure, only seven sequences are likely to affect core cooling. It is intended to analyse the frequency of these sequences using the expert elicitation techniques. Expert elicitation techniques find a place in situations where the data are scarce.

4.1 Human actions involved in executing the OPEC [2]

The various human actions involved in executing this OPEC are as follows:

a. Diagnosis of the occurrence of incident
b. Manual restoration of HPPW system pressure. The following actions need to be performed.
   - check whether the standby pump is on
   - if not, start it manually
   - check whether the discharge valve is open
   - if not, open manually the discharge valve
   - check HPPW header pressure

The actions in this set are judged to be completely dependent on each other. The stress is moderately high. The actions are considered to be critical procedural actions.

c. Fire water injection to HPPW headers. The following actions need to be performed.
   i. check whether Motor Operated Valve (MOV) for injection of Fire Fighting Water (FFW) to HPPW header is open
   ii. if not, open the valve
   iii. check HPPW header pressure

These actions are similar to those in (b) above.

d. Leak detection, identification and isolation of leak location. The actions needed are as follows.
   i. Check for beetle alarm in RB. If alarm exists, isolate the leak. The time needed is assessed to be 10 minutes taking into consideration an additional five minutes.
   ii. If there are no beetle alarms, the operator is instructed to check for leak in TB area and isolate it if it exists. The time for this action is about 15 minutes (10 minutes for checking and 5 minutes for isolation).
   iii. If the leak is neither in RB nor in TB, check for leak in pump house area. If it is found, isolate the leak. The assessed time for this action is about 15 minutes (10 minutes for checking and 5 minutes for isolation).
   iv. If the leak does not exist in any of the above areas, check for flooding in open areas (observed as leaks through manhole covers) and isolate the leak. The assessed time for this action is about 15 minutes.

Thus the estimated range of times of leak detection and identification is 10-45 minutes. It may be noted that the total time available for this event is around 60 minutes (injection of alternate supply for the essential loads of HPPW system has to be initiated before ~60 minutes). Leak isolation may be considered to be a critical procedural action under moderately high stress.

e. 'Valving in' of shutdown cooling system

The actions involved in valving in of shutdown cooling system are

i. closing of bleed condenser isolation valves (two numbers)
ii. opening of shutdown cooling line warm-up valves (two numbers)
iii. closing of shutdown cooling line warm-up valves (two numbers)
iv. starting of one shutdown cooling pump

The time available for carrying these actions depends on whether the system has to be valved in during hot PHT conditions or after the PHTS has been cooled to around 150° C. For this initiating event, it can be assumed that SDC is going to be valved in only when the PHTS is cooled to around 150° C. The operator is well trained and these actions are performed routinely whenever the reactor is shutdown for maintaining any equipment in the primary heat transport system.

f. 'Valving in' of fire fighting water to steam generators

Fire fighting water can be injected to steam generators in case boiler feed pumps and auxiliary boiler feed pumps are not available. These pumps will not be available due lack of cooling to the pump motors as a result of failure of HPPW system. For injection of fire fighting water to the steam generators, the following actions are needed.

i. opening fire water injection valves to steam generators (two numbers)

Even though the operator is well aware of the situation, he is not trained in this operation. Time stress may not be high in this case.

Taking these human actions into consideration, the frequencies of various accident sequences shown in figure 1 need to be assessed

5. EXPERT ELICITATION TECHNIQUES

An expert opinion can be defined as the formal judgment of an expert on a matter, also called an issue. Also, an opinion could mean a judgment or belief that is based on uncertain information or knowledge. An opinion is a subjective assessment, evaluation, impression, or estimation of the quality or quantity of something of interest that seems true, valid or probable to an expert's own mind. The expert should be most knowledgeable about issue.

The primary reason for using expert-opinion elicitation is to deal with uncertainty in selected technical issues related to a system of interest. Issues with significant uncertainty, issues that are controversial and/or contentious, issues that are complex, and/or issues that can have a significant effect on risk are most suited for expert-opinion elicitation. The value of the expert-opinion elicitation comes from its initial intended uses as a heuristic tool, not a scientific tool, for exploring vague and unknown issues that are otherwise inaccessible.

Even though efforts towards consensus are desirable, as is done in case of Delphi method, in this analysis aggregation of the expert opinions is carried out using the methods discussed by Bilal Ayyub [3]. The measure of dissonance (a state of contradiction between claims, beliefs or interests) and the measure of confusion (characterized by the multitude of subsets supported by evidence, in this case expert opinions, are considered to represent the uncertainty associated with the predictions.

The measure of dissonance, D(m), is defined as:

$$D(m) = \sum m(A_i)\log_2(Pl(A_i))$$

where the summation is carried out over the range of subsets supported by the evidence (number of experts). Here $Pl(A_i)$ stands for the plausibility measure which represents the total evidence supporting the claim the assignment belongs to the set A_i (as defined by the expert). $m(A_i)$ is the value of basic assignment/belief which is interpreted as the degree of evidence that the element(here the true value belongs to the set A_i. Measure of confusion can be defined

HPPW failure	Diagnosis of HPPW header pressure	Auto/Manual restoration of HPPW pressure	FFW injection to HHPW header	Restoration of HPPW pressure	Leak detection	Leak isolation	SDC	FFW to SGs

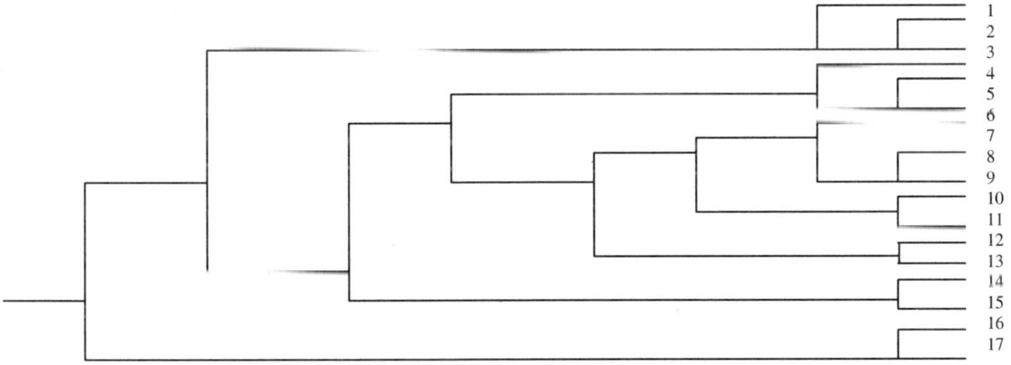

Fig. 1: Event Tree for HPPW System Failure

Table 1 Parameter for expert elicitation

S.No.	Question	Expert -1	Expert -2	Expert -3	Expert -4	Expert -5
1.	In your opinion, what are the chances that the operator will not be able to diagnose the situation? (He has to infer the IE from the cues given in section 3.1)	H: 0.01 M: 0.005 L:0.001	H: 10^{-4} M: 10^{-6} L: 10^{-8}	H: 10^{-2} M: 10^{-3} L: 10^{-4}	H: 10^{-3} M: $5*10^{-4}$ L: 10^{-4}	H: 0.002 M: 0.001 L: 0.000
2.	What are the chances of auto/manual restoration of HPPW system? (For this, the operator has to follow the steps mentioned in section 4.1.b).	H:0.05 M:0.01 L:0.005	H:0.999999 M:0.9999 L: 0.999	H: 0.999 M: 0.5 L: 0.25	H: 0.9999 M: .9995 L: 0.999	H: 0.99 M: 0.95 L: 0.91
3.	What are the chances of fire water injection to HPPW headers? (For this, the operator has to follow the steps mentioned in section 4.1.c)	H:0.1 M:0.05 L:0.01	H: 0.999999 M:0.9999 L: 0.999	H: 0.999 M: 0.5 L: 0.25	H: 0.9999 M: .9995 L: 0.999	H: 0.99 M: 0.95 L: 0.91
4.	What are the chances of restoration of HPPW system pressure?	H:0.05 M:0.01 L:0.005	H: 0.999 M: 0.99 L: 0.9	H: 0.999 M: 0.5 L: 0.25	H: 0.9999 M: .9995 L: 0.999	H: 0.99 M: 0.95 L: 0.91
5.	What are the chances of leak detection? (For leak detection, the operator has to follow the steps mentioned in section 4.1.d)	H:0.005 M: 0.002 L:0.001	H: 0.9999 M: 0.999 L: 0.99	H: 0.999 M: 0.5 L: 0.25	H: 0.9999 M: .9995 L: 0.999	H: 0.99 M: 0.95 L: 0.91
6.	What are the chances of leak isolation? (This involves closure of proper valves so that leak can be arrested.)	H:0.05 M:0.01 L:0.005	H: 0.999 M: 0.99 L: 0.9	H: 0.9 M: 0.5 L: 0.2	H: 0.9999 M: .9995 L: 0.999	H: 0.99 M: 0.95 L: 0.91
7.	What are the chances of valving in shutdown cooling system? (For this, the operator has to follow the steps mentioned in section 4.1.e)	H:0.1 M:0.09 L:0.05	H: ~1.0 M:0.99999999 L:0.999999	H: 0.95 M: 0.55 L: 0.25	H: 0.9999 M:0.9995 L: 0.999	H: 0.99 M: 0.95 L: 0.91
8.	What are the chances of injecting fire water system to steam generators? (For this the operator has to follow the steps mentioned in section 4.1.f)	H:0.1 M:0.09 L:0.05	H:0.999999 M:0.99999 L: 0..9999	H: 0.95 M: 0.55 L: 0.25	H: 0.9999 M:0.9995 L: 0.999	H: 0.95 M: 0.90 L: 0.85

in a similar manner. However, belief function instead of plausibility function is used.

6. RESULTS AND DISCUSSION

For the problem under consideration, the issues are classified as shown in table-1. For each of the issues high, middle and low values are sought from the experts. These values are expected to indicate a reasonable value and the uncertainty associated with this value in the opinion of the experts. The experts chosen are plant operators, who have a good familiarity with the system under consideration, technical support personnel and system designers with adequate operational experience and PSA practitioners and regulators. This wide of expertise considered in selecting the experts is indicated in a fairly good agreement obtained in the estimates of the middle values of the estimates. It can be noted from the table that the agreement regarding uncertainty range is not consistent among the experts.

The three values given by the experts are treated as representing a triangular representation of the membership function. For obtaining the measure of dissonance, the function D(m) is integrated over the range of values provided by each expert and over all the experts. This represents entropy like measure. This measure is evaluated for different values of lower and upper ranges of integration. Initially the lower limit is taken as lowest of the lower values defined by all the experts and the upper limit the highest of high values defined. Later, keeping the upper limit constant, the lower limit is varied. These evaluations are repeated for different for different values of the upper limit. The maximum entropy values defined the range of the parameters obtained for all the issues considered in this study. A computer program was written in C language to carry out these evaluations. Rounding of the entropy to the first decimal place, the minimum range obtained for all the systems is shown in table-2. Using these values of basic assignments (which are treated as probability), the uncertainty in the conditional probabilities of the various sequences shown in fig. 1 is evaluated. These are shown in Table 3.

Table 2 Ranges of the reliability parameters for all the systems

System	Range
1	4.0e-4 – 9.6e-3
2	2.65e-1 -9.85e-1
3	2.5e-1 – 1.0
4	2.65e-1 – 9.85e-1
5	2.65e-1 – 9.85e-1
6	2.32e-1 – 9.84e-1
7	2.65e-1 – 9.85e-1
8	2.5e-1 – 9.85e-1

From table-3 it can be observed that except for the sequences 14 and 15, the range seems to be reasonable. The large error band in case of sequences 14 and 15 primarily arises due to the high value of success probability (i.e.1.0) for the system 3 whose failure probability appears in these sequences.

Table 3 Range of Conditional Probabilities of Accident sequence

AS	Conditional probability Range	AS	Conditional probability Range
2	4.8e-2 – 1.1e-1	10	1.5e-3 – 4.5e-3
3	1.3e-2 - 1.5e-2	11	1.7e-4 – 2.0e-2
4	1.3e-2 – 1.1e-1	12	1.5e-3 – 2.5e-2
5	8.9e-2 – 1.3e-2	13	1.7e-4 – 7.4e-2
6	1.5e-2 – 2.7e-2	14	0.0 – 0.14
7	2.2e-3 – 1.3e-2	15	0.0 – 0.41
8	1.45e-3 – 1.51e-3	16	1.0e-4 – 8.9e-3
9	1.1e-3 – 4.5e-3	17	4.0e-4 – 9.6e-3

AS: Accident Sequence

7. CONCLUSIONS

Expert elicitation techniques have been applied to the evaluation of the frequencies to accident sequences associated with the initiating event "High Pressure Process Water System Failure" The solicited opinions are combined using the measure of dissonance and the range is obtained by maximizing this measure. This gives assurance regarding the application of the methods of expert elicitation to engineering problems. However, these methods should be applied when the problem is complex and is not amenable to easy solutions.

8. ACKNOWLEDGEMENTS

The authors express their appreciation to Ms. Rajee Guptan, S/Shri Sayee Lakshman,, G. Gawande, and of NPCIL and R. Venkatraman and U.K Paul of Atomic Energy Regulatory Board for participating in the expert elicitation process and offering their valuable comments.

9. REFERENCES

1. Madras Atomic Power Station Safety Report, Vol.1, 1984
2. Operating Procedures for Emergency conditions, MAPS
3. Elicitation of Expert Opinions for Uncertainty and Risks, Bilal M. Ayyub, CRC Press, 2001

Reliability, Safety and Hazard
Editors: P.V. Varde et al.

PROBABILISTIC SAFETY ASSESSMENT BASED EXPERT SYSTEMS IN SUPPORT OF DYNAMIC RISK ASSESSMENT

P.V. Varde, U.L. Sharma, S.K. Marik, V.K. Raina and A.C. Tikku

Reactor Group, Bhabha Atomic Research Centre, Mumbai 400 085

ABSTRACT

Probabilistic Safety Assessment (PSA) studies are being performed, world over as part of integrated risk assessment for Nuclear Power Plants and in many cases PSA insight is utilized in support of decision making. Though the modern plants are built with inherent safety provisions, particularly to reduce the supervisory requirements during initial period into the accident, it is always desired to develop an efficient user friendly real-time operator advisory system for handling of plant transients / emergencies which would be of immense benefit for the enhancement of operational safety of the plant. This paper discusses an integrated approach for the development of operator support system. In this approach, PSA methodology and the insight obtained from PSA has been utilized for development of knowledge based or rule based experts system. While Artificial Neural Network (ANN) approach has been employed for transient identification, rule-base expert system shell environment was used for the development of diagnostic module in this system. Attempt has been made to demonstrate that this approach offers an efficient framework for addressing requirements related to handling of real-time / dynamic scenario.

Keywords: Probabilistic Safety Assessment, Artificial Neural Network, Operational Safety, Transient Identification, Knowledge Based Systems, Diagnostics

1. INTRODUCTION

Probabilistic Safety Assessment (PSA) methods have gradually evolved over the years and have now matured enough to address many real-time issues. The insight obtained from a plant specific PSA study are increasingly being accepted and used for taking regulatory and operational decisions as part of risk-informed approach world over. While on one hand the results of Level 1 PSA studies are used for planning of plant emergencies the results of Level 2 and 3 PSA studies show potential applications for various on-site and off-site plant emergency planning and preparedness.

The study of many accidents reveals that the human error are one of the major contributory factors for a large number of nuclear or non-nuclear accidents world over. This calls for incorporation of an efficient operator aid system which can improve operator's performance during plant emergencies. Nuclear plants are complex in nature and to identify the thrust areas for improving operator's performance requires an integrated framework. PSA results in conjunction with deterministic studies has great potential to provide a superior framework for the identification of emergency scenario which may have potential safety implications. The Expert System approach which includes ANN and rule based expert systems; provide a convenient environment to develop an operator advisory system for handling various plant emergencies by facilitating incorporation of knowledge of various experts and hard copy emergency procedures. Apart from this the plant symptoms and the knowledge gathered from PSA and deterministic studies and plant emergency procedures are also utilized for creating operator support systems which cater to plant emergencies.

2. ROLE OF PSA FOR DEVELOPMENT OF OPERATOR SUPPORT SYSTEM

In this study, PSA framework has been utilized for development of an operator support system. While on one hand PSA methodology provides an elegant mechanism for knowledge representation scheme in development of experts systems, the insight obtained from PSA, on the other hand, has been effectively utilized as an input for identifying the vital issues, human actions and components important to safety.

The knowledge representation scheme follows both model based approach and fault tree and event tree models in the diagnostic rule based module of the rule-based expert system – DIRES (Diagnostic Rule Based Expert System). The knowledge base of the expert system can be broadly divided in to two parts, static and dynamic knowledge base. The knowledge in the rule-based expert systems is represented in the form of IF-AND-THEN rules. The fault tree and event tree approach provide an elegant mechanism to represent this knowledge in the diagnostic module of the rule-based expert system. For instance, IF condition A AND B are true THEN the 'Consequence' is true. This is analogous to fault tree deductive logic for analysing an undesired event. Accordingly, fault tree knowledge can be readily captured in the form of rules in the rule-based expert system for performance of diagnosis.

While on one hand the fault tree methodology as mentioned above provides a framework for knowledge representation in diagnostic module, the event tree approach on the other hand provides an efficient framework for event progression in a dynamic manner. This representation allows use of PSA knowledge on accident sequences to track the plant condition. The on-line estimation of nodal conditional probability of event tree makes it possible to have prediction of the evolving scenario.

The frequency or probability of component failures and safety actions have been used as an input for prioritizing the rule as part of conflict resolution strategy. This also includes the prioritization of Emergency Operating Procedures (EOP) for a particular accident sequence.

The database developed on Common Cause Failure (CCF) has special importance in prioritization of the thrust areas from the point of view of safety. Similarly, safety significant human action analysis provides a vital input in working out the time window available for finishing an emergency task, which is an important consideration while developing an expert system.

3. OPERATOR SUPPORT SYSTEM FOR DYNAMIC RISK ASSESSMENT

Although the modern nuclear plants are designed with the advanced features like passive safety, inherent safety, which includes the core physics and thermal hydraulic design, which imparts additional level of safety to the plant. The supervisory role of plant operator during normal operation of the plant in general and emergency condition in particular requires an operator support system which help to reduce the cognitive load on the operator particularly during the on-set of a deviation or accident condition. Hence, provision of an efficient operator-plant interface system is expected to be effective in arresting deviations to propagate into accidents. This operator interface should be able to cater to the operator support system requirements related to transient conditions and handling a dynamic accident scenario. This system should offer transient identification capability during plant upsets and also inherit the diagnostic capability to get to the root cause of the disturbance. These features will effectively enhance operator action reliability, by supporting the diagnostics related to plant emergencies. The integrated approach developed uses Artificial Neural Network (ANN) for transient identification while knowledge based or rule based framework has been employed for performing diagnostics related to system failure and assessment of plant condition [1].

3.1 Artificial Neural Network (ANN) Approach for Plant Transient Identification

The application of artificial neural network (ANN) or more commonly referred as the Neural Network (NN) to the operation of nuclear power plants has become an important subject for the researcher world wide. This is due to the advantages offered by this approach for the problems, which, earlier seemed intricate in nature and could have been difficult to solve using other mathematical methods or heuristic logics. It is evident from the available literature that neural networks are increasingly being applied to nuclear power plant operations [2]. Various applications of ANN in nuclear power plant include, monitoring, measurement of operational parameters, signal validation, diagnosis of normal conditions, modelling plant thermodynamics, fuel loading pattern optimization, reactivity surveillance, classification and prediction of critical heat flux, etc.

Based on the experience in the area of development of various applications of ANN, it could be analysed that ANN is an efficient tool to solve the complex problems which pose challenges that include a) the problem to be analysed is non-linear in nature, b) the input data arrives in a dynamic manner, i.e. the input data set shows variation in time and spatial domain, c) the input data to be processed in a parallel manner for the purpose of identification of pattern, d) available time is limited for identification of the a signature from among the input data set, etc. Keeping in view the characteristics of the problem of identification of reactor state in a dynamic manner, ANN was employed for reactor state identification.

3.1.1 ANN Model

The earlier work carried out on transient identification [3] and the subsequent survey of literature on similar problems have amply demonstrated that the application of feed forward multilayer ANN, as given in Fig. 1, comprising of one input layer, one hidden layer and one output layer forms effective architecture for reactor status monitoring. The input nodes which receive the signals from the plant are connected through the hidden nodes to the output nodes, which finally gives the ANN output. The input layer consists of a set of variables (x_1, x_2, x_3, ...x_n) and the output layer consists of set of activation values (o_1, o_2, o_3, ...o_n). If the target output is to predict only one variable as a function of the variables x_i's, then the output is a single variable o. The relationship between these associations can be represented as follows:

$$o = C_o + \sum_{i=1}^{n} C_i \phi_i (x_1, x_2, x_3, \ldots, x_n) \qquad (1)$$

where, Φ are nonlinear polynomial functions. Instead of modeling an explicit functional form the ANN relates o_i with x_i using a network of connection weights between pairs of nodes of adjacent layers. The node in the ANN sums the product of the products of the input and connection weights from the nodes of the previous layer and then limits it by a nonlinear threshold function. The weighed sum of the inputs for the jth node in a layer k is given by

$$net_j = \sum_i w_{ij}^k x_i^{(k-1)} + b_j^{(k)} \qquad (2)$$

where $w_{ij}^{(k)}$ is the connection weight between the ith node in the k-1 layer and j^{th} node in the k

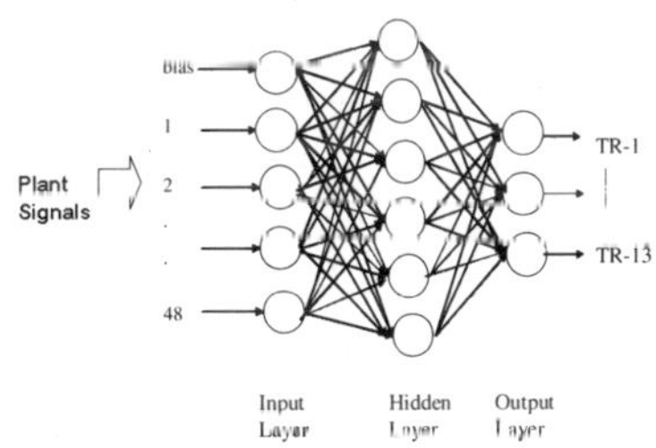

Fig. 1: ANN model

layer, $x_i^{(k-1)}$ is the output from the ith node in the *k-1* layer, and $b_j^{(k)}$ is the bias associated with the jth node of layer k and it produces the effect similar to adjusting the threshold function of the processing node.

The underlining principle used for the reactor status monitoring is that each reactor state can be associated using a 'unique pattern' of the plant symptoms. These symptoms include the reactor pre-trip, trip and analogue instrument readings of various plant parameters available in the control room of the plant. Whenever a plant transient occurs, the affected input signals coming to the control room of the plant change their states and form a pattern which can be used as an indicator of the stabilized state of the plant.

Accordingly, the signals required for identifying various reactor states were identified. The plant parameters listed in Table 1 were used to generate the input pattern. One pattern composed of total 48 plant inputs and one bias node (Total 49 inputs) which include reactor pre-trips, trips and analog readings as shown in Table 2. It may be noted that the digital reactor protection systems in APR (Advanced Pressurised Water Reactor) has been designed using four redundant channel and follows 2/4 coincidence logic for enhanced safety, reliability and availability [4]. However, this study being demonstration exercise only inputs from Channel A have been used.

Table 1 List of parameters as input for ANN

Parameter (Unit)	Nominal Full Power	Trip Setting
Variable Overpower (%)	100	125
Hi Log-rate (%)	0	6
Hi LPD (kw/ft)	< 14	20
Lo DNBR	1.79	> 1.2
Hi Pressurize Pr. (psia)	2250	2425
Lo Pressurize Pr. (psia)	2250	1750
Lo SG Water level (%)	82	45
Lo SG Pr. (psia)	1070	870
Hi Cont. Pr. (psia)	0 - 5	14*
Hi SG Water level (%)	82	94
Lo Reactor Coolant flow (%)	80	*60

*Legends: Hi: High, DNBR: Departure from Nucleate Boiling Ratio; LPD: Local Power Density; SG: Steam Generator; Pr.: Pressure. * Assumed value*

Table 2 Input vector / pattern parameters

No.	Code	Symptom description
1	PT-1	SG 1 low pressure
2	PT-2	SG 2 low pressure
3	PT-3	Containment pressure high
4	PT-4	SG 1 Low level
5	PT-5	SG 2 Low level
6	PT-6	SG 1 differential pressure low
7	PT-7	SG 2 differential pressure low
8	PT-8	SG 1 level high
9	PT-9	SG 2 level high
10	PT-10	Pressurizer pressure low
11	PT-11	Pressurizer pressure high
12	PT-12	Log rate high
13	PT-13	DNBR low
14	PT-14	Local Power Density high
15	PT-15	Variable over power
16	PT-16	Containment pressure v. high
17	T-17	SG 1 low pressure
18	T-18	SG 2 low pressure
19	T-19	Containment pressure high
20	T-20	SG 1 Low level
21	T-21	SG 2 Low level
22	T-22	SG 1 differential pressure low
23	T-23	SG 2 differential pressure low
24	T-24	SG 1 level high
25	T-25	SG 2 level high
26	T-26	Pressurizer pressure low
27	T-27	Pressurizer pressure high
28	T-28	Log rate high
29	T-29	DNBR low
30	T-30	Local Power Density high
31	T-31	Variable over power
32	T-32	Containment pressure v. high
33	R-33	SG 1 low pressure
34	R-34	SG 2 low pressure
35	R-35	Containment pressure high
36	R-36	SG 1 Low level
37	R-37	SG 2 Low level
38	R-38	SG 1 differential pressure low
39	R-39	SG 2 differential pressure low
40	R-40	SG 1 level high
41	R-41	SG 2 level high
42	R-42	Pressurizer pressure low
43	R-43	Pressurizer pressure high
44	R-44	Log rate high
45	R-45	DNBR low
46	R-46	Local Power Density high
47	R-47	Variable over power
48	R-48	Containment pressure v. high

Only. Table 2 shows the plant parameters which forms input vectors for ANN training. Once the processing of the input data was over, the data were supplied to the ANN. for ANN training to identify 13 reactor states The list of the 13 reactor states for which the ANN was trained has been shown in Table 3.

Considering the above, the final matrix was prepared for identification of 13 reactor conditions. As shown in Fig. 1, a multilayer feed forward network architecture was selected with three layers; one each for input layer, hidden layer and output layer. Based on the requirements of reactor status monitoring the ANN configuration with 49 nodes in the input

Legends: PT: Pre-trip, T: Trip, R: Instrument Readings

layer, ~ 80 nodes in the hidden layer and 13 nodes in the output layer was used to start the parametric study.

Weighed sum of the inputs available from equation (2) is processed using an activation threshold function. There are many functions available for this purpose. However, depending on the nature of problem these functions are applied in the ANN. The sigmoid function is widely used as an activation function. For this problem the sigmoid function can be represented as follows:

$$x_j^{(k)} = \frac{1}{1+\exp(-net_j^{(k)})} \tag{3}$$

Apart from Sigmoid function two more activation functions were used for the problem under study. These functions are 'Tansigmoid' and a linear function called 'Pureline'. The use of these functions in the design of this network has been discussed in the succeeding section of this paper.

3.1.2 Plant Transients and ANN Architecture

Table 2 gives the list of the signals available in the plant control room of APR. Individual input patterns for corresponding 13 plant condition was characterised by assigning applicable value to each parameter. In an individual pattern the For and Pre-Trips, values 0 and 1 was taken for signal to be OFF and ON respectively. For analog parameters normalized values were used in assigning the value of instrument reading for pattern formulation.

The list of the 13 reactor states for which the ANN was trained has been shown in Table 3. Though the design of plant is in advanced stage, there were some areas where the required information was not enough to finalize the ANN modeling requirements.

Governing considerations for formulating the input patterns for ANN training were as follows:

a) While formulating the patterns it was assumed that all the trip parameters will have corresponding pre-trips and for each trip parameter there will be analog instrumentation available in control room of the reactor.
b) Pre-trip setting was assumed to be 10% lower than the corresponding trip setting.
c) During the normalization process it was observed that some parameters instead of having fixed nominal value vary between a range of values. For such parameter the mean value of the indicated range was considered as the value of the node for normal operation.

Table 3 Nomenclature of the reactor states

Transient No.	Transient Description
TR-1	Reactor Operation Normal
TR-2	Feedwater Temp. decrease
TR-3	Feedwater flow increase
TR-4	ISOGADV
TR-5	Turbine trip
TR-6	Loss of condenser vacuum
TR-7	Loss of feed water flow
TR-8	Loss of RC flow / 1 RCP seizer
TR-9	RCP shaft break
TR-10	Uncontrolled CEA withdrawal at low power
TR-11	Inadvertent de-boration
TR-12	CEA ejection
TR-13	LOCA

Legend: ISOGADV: Inadvertent operation of steam generator atmospheric dump valve, CEA: Control Element Assembly, LOCA Loss of coolant accident.

d) Only Channel A parameters and trips are considered for this study. It was assumed that the pre-processor module will handle 2-out-of-4 logic and normalization of the trips / variables. This has helped to reduce the size of the ANN. One could expect ANN to be more robust with pre-processor module installed.

Considering the above the final matrix table was prepared for 13 reactor conditions. For each transient the patterns / vector formulated for ANN training. For identifying these transients a 13×13 matrix was employed.

3.1.3. Network Training

The Back-propagation network (BPN) algorithm was considered appropriate for implementation of a training scheme for the network. Initial assignment of the weights was random in the range +0.1 to -0.1. The BPN algorithm attempts to minimize the overall mean square error (MSE) between the desired and the actual output for all the output nodes over all the input patterns by iteratively adjusting the weights. The training was carried out in batch mode in using the Neural Network Toolbox in MATLAB environment.

Table 4 Final parameter of ANN

Parameter description	Value
Number of layer	3
Number of nodes in input layer	49
Number of node in hidden layer	100
Number of nodes in output layer	13
Training rate	0.2
Number of epoch required for training	6248
CPU Time required for training (min)	12
Error target	1e-5
Activation function - input layer	Sigmoid
Activation function – hidden layer	tansig
Activation function – output layer	pureline

The objective here was to select the network parameters, the training rate, momentum coefficient and number of nodes in the hidden layer such that it allows the convergence of the network to achieve a target MSE value of 1E-5. This required a parametric study to be performed using the above said parameters and the network does not get stuck in the local minima. Apart from this the other and rather secondary objective was to arrive at a network configuration such that the convergence of MSE is achieved in optimum number of epochs. As part of parametric study, repeated trials were made with the activation functions, viz, 'sigmoid', 'tansigmoid' and the linear function 'pureline', in the three layers of the network. These functions are available in the MATLAB library. It may be noted that successful convergence could be achieved only after arriving with the configuration as shown in Table 4, i.e. sigmoid, tansig and pureline transfer function in the input, hidden and output layer, respectively. The convergence of MSE during the training of the network could be achieved in 6248 iterations or epochs as shown in Fig. 2.

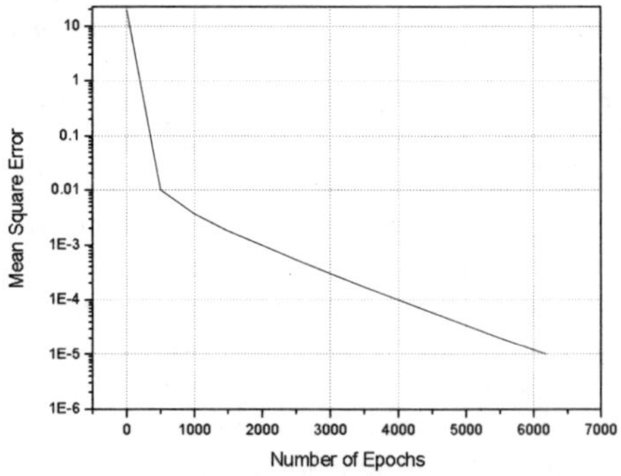

Fig. 2: Convergence of RMS Error during Training

3.1.4 Testing Of Network by Simulating Near Real-Time / Dynamic Scenarios

The tests were carried out using a simulator specially developed for simulating plant transient scenario in temporal and spatial domain. The network was trained using patterns comprising of stabilized values varying between 0 and 1 at the end of the transient at each node in the input layer. Keeping in view the requirements related to dynamic scenario / real-time processing of data the simulation test pattern was fired in a manner such that it exhibits the characteristics of a scenario evolving over a period of time. In actual scenario during a transient the signals keep coming for a finite duration and the stabilized pattern appears (for which the network was trained) only at the end of the transient. To bridge this gap the signals in a pattern were fired in time domain as it happens during an actual transient. Refer Fig. 3 for better understanding of the process of simulation. Please note that only part of pattern (only 15 nodes out of 49) has been shown due to space limitation. On right side part of the pattern used for training has been shown. The plant signals registers in time domain, for instance the 12th signal from the pattern registers first at the onset of the transient again the signal 2 registers during second epoch in time and at the end of the transient all signals as given in the trained patterns registers. The variation of mean square error with time has been shown in Fig. 4 for this transient. As could be seen the transient could be identified with an error value of ~ 10^{-5}. Another aspects of the real-time / dynamic scenario is that the network may encounter learned pattern or an unlearned scenario for which it was not trained. The capability of ANN was checked by performing the recall tests on the trained ANN as follows:

3.1.4.1 Recall Tests with Learned patterns

The first requirements which come from the plant is that the network should be capable of identifying all the pre-defined scenarios for which the Emergency Operating Procedure (EOP) exists. To cater to these requirements the recall tests were performed on the trained network using the same pattern for which ANN was trained. As it was expected, ANN was able to identify all the transients successfully. The RMS error computed for each of the 13 pattern was very low.**3.1.4.2 Recall Tests for Unlearned Patterns**

These patterns were generated based on the real-time operational experience and information derived from plant EOPs. The objective was to check the network behaviour for situation for which the network had not encountered earlier Accordingly, four simulation tests were performed as follows:

1. Sensor failed data
2. Noisy data
3. Presence of single trip before the transient sets in
4. Calibration Error

A general observation was that ANN could judiciously differentiate between actual transient even sometimes using derived signals when the direct or main signal was absent or there was a deviation from the original signature for which the ANN was trained. The confidence of ANN

Plant sensor signals / Signal sampling epochs →	1	2	3	4	5	6	7	8	9	10	11	12	13	14	15	16	17	18	Pattern used for training
1	0	0	0	0	0	0	0	0	0	0	0	0	0	0	0	0	0	0	0
2	0	1	0	0	0	0	0	0	0	0	0	0	0	0	0	0	0	0	1
3	0	0	0	0	0	0	0	0	0	0	0	0	0	0	0	0	0	0	0
4	0	0	0	0	0	0	0	0	0	0	0	0	0	0	0	0	0	0	0
5	0	0	0	0	0	0	.7	0	0	0	0	0	0	0	0	0	0	0	.8
6	0	0	0	0	0	0	0	0	0	0	0	0	0	0	0	0	0	0	0
7	0	0	0	0	0	0	0	0	0	0	0	0	0	0	0	0	0	0	0
8	0	0	0	0	0	0	0	0	0	0	0	0	0	0	0	0	0	0	0
9	0	0	0	0	0	0	0	0	0	0	0	1	0	0	0	0	0	0	1
10	0	0	0	0	0	0	0	0	0	0	0	0	0	0	0	0	1	0	1
11	0	0	0	0	0	0	0	0	0	0	0	0	0	0	0	0	0	0	0
12	1	0	0	0	0	0	0	0	0	0	0	0	0	0	0	0	0	0	1
13	0	0	0	0	0	0	0	0	0	0	0	0	0	0	0	0	0	0	0
14	0	0	0	0	0	0	0	0	0	0	0	0	0	0	0	0	0	0	0
15	0	0	0	0	0	0	0	0	0	0	0	0	0	0	0	0	0	0	0

T= 0.0 (Transient start) Time ⟶ (Transient end) T=t

Fig. 3: Depiction of real-time scenario for transient identification

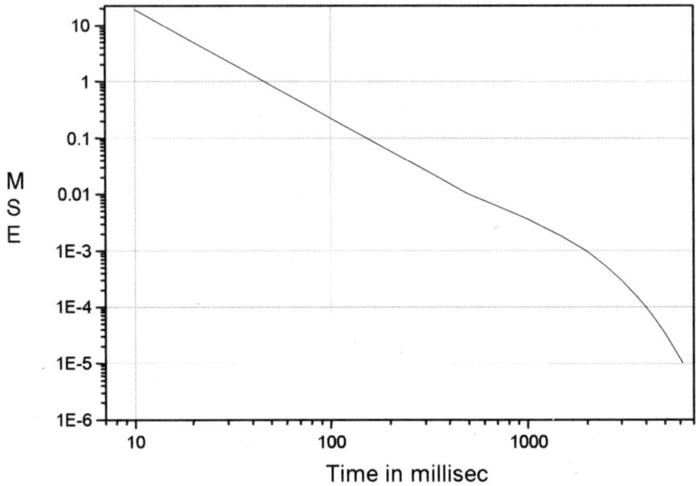

Fig 4: Convergence of error with time during transient simulation

output was reflected in the error estimates generated during the recall tests.

3.2 Rule Based System For Performing Diagnostics

PREDICATE <frame-1><slot-1><type><Value>
PREDICATE <frame-2><slot-2><type><Value>
;
PREDICATE <frame-n><slot-n><type><Value>
N
action-1
;
Action-n
Implementation of fault tree knowledge in the expert system is achieved as follows:
Rule 1
IF
Component A = Fail OR
Component B = Fail OR
Component C = Fail
THEN
Sub-System H = Fail
Rule 2
IF
Component D = Fail AND
Component E = Fail
THEN
Sub-System I = Fail
Rule 3
IF
Sub-System H = Fail OR
Sub-System I = Fail
THEN
System 'X' = Fail

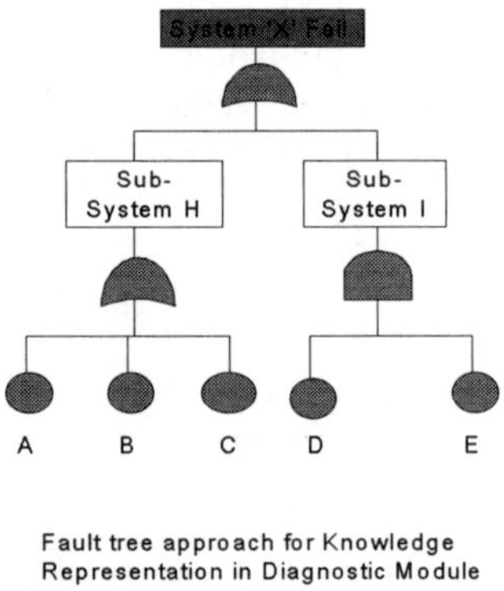

Fig. 5. Approach for deriving production rules system fault tree

Once the transient has been identified the control passes to the diagnostic module of the operator support system. The experience gained on development of expert systems suggests that the rule based expert system offers an effective framework for creating diagnostic module for an operator support systems. A Diagnostic Rule Based Expert System (DIRES) has been developed. The salient features of this systems are, a) Knowledge Base comprised of 'frames' and production rules, b) Inference mechanism is forward chaining, c) the system poses no limits on number of rules, d) there is explanation generation module which tracks the decision process that goes on while operator performs the diagnostics, and e) search strategy has been implemented by reducing the state space.

In DIRES the syntax of the rule in DIRES is as follows:

Fig. 5 demonstrate the process of deriving the rules from a system fault tree. Similar approach is followed in the development of event progression module also. In this module event tree framework is employed for knowledge representation of various events and the objective is to index the application procedure for a given accident sequence. To check the efficacy of this module in support of plant diagnostics many case studies were performed. The objective was to check the performance of this system for the control room applications. The performance of the system was found to be satisfactory.

4. CONCLUSIONS

An integrated approach has been presented for the development of operator support system for a nuclear power plant application. The objective has been to address the issues related to the real-time / dynamic scenario modelling.

A general observation was that the PSA tools like fault tree and event tree offer an elegant mechanism for the representation of knowledge in the experts system. Apart from this the insight gained from a PSA study that includes data on CCF, human action probabilities, importance factors were utilized for development and fine tuning the knowledge base of the operator support systems.

ANN approach for modeling the plant transient and its identification using a three layer neural network has been studied. The recall tests performed demonstrate the feasibility of using this approach as part of operator support system for nuclear power plant. The testing carried out also brings out the limitation of this methodology.

5. REFERENCES

1. Varde, P.V., "An Integrated Approach for Research Reactor Operations and Fault Diagnosis Through Connectionist Framework and PSA Based Knowledge Based System". Ph.D. Thesis, Indian Institute of Technology, Bombay, 1996 .
2. Uhrig, R.E., "Potential Application of Neural Networks to the Operation of Nuclear Power Plants, Nuclear Safety, Vol. 32 (68), 1992.
3. Ragheb, M. and Campas, T. "Nuclear Power Plant Transients Pattern Recognition using Neural Network Methodology", Trans. On Americal Nuclear Society, 62 (129), 1990.
4. Varde, P.V., Seop, Hur, Lee D.Y., Moon B.S. and Han, J.B., "Proposed Neural Network Approach for Monitoring Plant Status in Korean Next Generation Reactors", International Journal of Fuzzy Logic and Intelligent Systems, Vol 3 No. 1, June 2003.

Reliability, Safety and Hazard
Editors: P.V. Varde et al.

CORRELATION BETWEEN CODING STANDARDS COMPLIANCE AND SOFTWARE QUALITY

Wojciech Basalaj and Richard Corden

Programming Research, 9/11 Queens Road, Hersham, Surrey KT12 5LU, UK

ABSTRACT

Software Quality has different meaning to different people. The ISO 9126 standard was developed to introduce clarity and establish a framework for quality to be measured. This paper aims to explore how Internal Quality characteristics of a software system (source code) can be measured effectively. Instead of relying on traditional software metrics, which have been shown to be a poor predictor of underlying software quality, we advocate measuring compliance to a coding standard. This approach will be illustrated using a real-world C++ project, where improved compliance to the High Integrity C++ Coding Standard can be correlated with objective improvement between two versions of the product.

Keywords: Software Quality Modelling, Coding Standards

1. ISO 9126 QUALITY MODEL

The ISO 9126-1 standard 5 has been introduced to formalise the notion of Quality of a Software System. 3 distinct aspects are considered:

- Internal Quality measured for a non-executable form of the Software System, e.g. its source code.
- External Quality, which pertains to the run-time behaviour of the system, as experienced during dynamic test.
- Quality in use, which addresses the degree to which user goals and requirements are fulfilled.

Internal and External Quality can be further categorised into 6 separate characteristics:

- Functionality
- Reliability
- Usability
- Efficiency
- Maintainability
- Portability

Each of these 6 characteristics can be further subdivided, and there are 27 sub-characteristics in total.

Quality in Use has been divided into 4 characteristics:

- Effectiveness
- Productivity
- Safety
- Satisfaction

ISO 9126-1 advocates measuring each of these characteristics, but does not specify how. Examples of suitable metrics are given in Technical Reports: 9126-2 6, 9126-3 7, 9126-4 8. The standard stipulates that with suitable choices of metrics Internal Quality should predict, or in other words correlate with External Quality, which in turn should predict Quality in Use.

In this study we will be focusing on the Effectiveness Quality in Use characteristic. We will attempt to demonstrate that this characteristic can indeed be predicted by measuring Internal Quality of a software system, see Section 4. We will also be examining empirical evidence of a correlation between Internal and External Quality measures, see Section 3.

Prior to conducting such a study we needed to settle on suitable metrics for Internal Quality.

ISO 9126-3 7 is a Technical Report that proposes such metrics. The vast majority of them are of the following form: percentage of items (functions, variables, etc.) meeting a specific requirement. There are a number of problems with such a definition of metrics. Their calculation cannot be easily automated, and their value needs to be determined by comparing implementation and design documents with specification. These metrics indicate how much work on the project has been completed, rather than underlying quality of the implementation. Such metrics represent good project management practise for green-field projects, and cannot be applied when part of the system is re-engineered. Lastly, quality or lack thereof is not seen as attribute of source code, as none of the proposed metrics are based on direct measurements on source code. This is against the guidance of ISO 9126-1 5 page 15.

Prior to ISO 9126 there has been a vast amount of research devoted to software metrics 3. These traditional metrics, such as Cyclomatic Complexity or Estimated Static Path Count, are concerned with the structure of a function, vocabulary of a source file, etc. Therefore, they may yield the same values for drastically different versions or stages of a Software Product, e.g. Cyclomatic Complexity for pseudo code stage may be the same as for the final implementation. Moreover, there is no well-defined and substantiated mapping for these metrics to ISO 9126 characteristics.

2. CODING STANDARDS

Nowadays, increasingly more emphasis is given to following best practise, and defining and enforcing coding standards, especially for high cost of failure software projects. Compliance to a coding standard is often treated as a pass/fail test. However, a different approach is possible, where the level of compliance is measured, either as the absolute number of violations for a particular source file, module or component, or normalised by the size of the entity, e.g. number of lines of code. This would allow correlating compliance with measurements of other aspects of the product, e.g. run time behaviour or user experience.

The most popular coding standard in the public domain for the C language is MISRA-C 1213. It constitutes a subset of the C language that restricts usage of poorly defined or unsafe constructs. Less emphasis is given to presentational aspects: naming conventions and layout. No such definitive coding standard exists for the C++ language, the most complete is probably High Integrity C++ 14. Other C++ guidelines tend to focus on specific programming aspects 4111617.

The rules of these coding standards represent common pitfalls with developing in the corresponding programming language, and have been derived either from experience or on theoretical grounds, by examining the language specification 910. Therefore, counting the number of violations of such rules in a Software Product appears well founded, and intuitively corresponds to a measure of its Internal Quality.

3. QUALITATIVE RESULTS

We wanted to verify the proposal for measuring Internal Quality of a software product with real-world examples. We have engaged with some of our key customers, to find out what tangible benefits enforcement of a coding standard has given them. They were able to offer broad qualitative statements, and these are documented in Section 3.1 and 3.2. However, they could not provide, in time for publication of this paper, any numerical data that would allow us to compare for example faults found in the field and compliance to a coding standard of specific software modules. To this end we have decided to use our own software product for such a comparison, details of which are provided in Section 4.

3.1 Company A

They have been using MISRA-C:1998 12 ever since historical process data have been collected. Some extra rules are enforced to do with naming conventions and limiting undefined behaviour. Typically, approximately 90% of the rules of this combined coding standard are adhered to for a project.

For a number of specific projects porting from one platform to another was required, and this was achieved with hardly any re-coding. This result was attributed to restricting undefined and implementation defined behaviour in their coding standard.

In the development process unit testing occurs on a parallel track to coding, review and bench testing. By examining process data it was found that all the faults found in unit testing were also identified in the development track during code review (of which coding standard compliance is a part) or bench testing stages. Therefore, unit testing, despite being part of industry best practise, did not yield any new issues, apart from fulfilling its secondary role of verifying the specification. Subsequently, for some projects unit testing has been limited or dropped altogether in preference to proceeding straight to integration/system test.

3.2 Company B

The AUTOSAR2 subset of MISRA-C:1998 12 is used, as well as other proprietary coding standards, depending on the project, and this is mandated contractually.

The software projects are large, typically around 500KLOC. By defining a software platform, and making it conform to stricter rules on limiting implementation defined behaviour, they were able to migrate from one compiler and micro controller combination to another in a matter of weeks. This result is similar to that of Company A, see section 3.1.

Reuse is very common across projects, and coding standard rules on layout and naming conventions were found to be helpful in this regard.

4. QUANTITATIVE RESULTS

We have decided to base the numeric comparison between Internal Quality and Quality in Use on our own Software Product: QAC++, a static analyser for C++. Between the current 2.1 and the previous 2.0 releases we have modified a number of analysis checks to improve their signal to noise ratio, that is to minimise the occurrence of false positives, while at the same time maximising the occurrence of genuine warnings. This equates to improving Quality in Use: Effectiveness characteristic, and we decided to measure it with the following formula:

$$S\text{-}N \qquad (1)$$

where S is the number of occurrences of genuine warnings for a particular check, and N is the number of false positives.

At the outset we concluded that it would be inappropriate to measure such interpretation of Quality in Use on anything but ordinal scale, so the exact choice of a formula was unimportant, as long as it achieved the desired ranking of measurements.

As we were primarily interested in verifying enforcement of a coding standard as a viable measure for Internal Quality of a Software Product, we chose density of warnings to capture this. Density has the advantage that it will not be affected by the addition of a large number of new lines of code to a source file/module being measured, as long as the number of warnings per line of code is similar in the existing and new portions of the source file. This was an important property when considering that improvement in effectiveness of analysis checks often leads to an increase in complexity of their implementation, to handle various edge cases.

Whilst Quality in Use: Effectiveness is measured for a check on arbitrary source code, Internal Quality (enforcement of a coding standard) is measured for the source files implementing this analysis check in QAC++. The coding standard that we have chosen and apply in QAC++ development is High Integrity C++ (HIC++) 14.

To establish the level of improvement in effectiveness of QAC++ we used a sizeable body of C++ source code: The ACE Toolkit 1. It is a C++ library facilitating cross platform, OO network programming. It totals some 1300 source files and 200KLOC. When analysing this library with QAC++, 16 checks that have been modified between version 2.0 and 2.1 are triggered, i.e. have at least one instance being reported. Therefore, Quality in Use improvement for a particular check c can be calculated on ordinal scale with the following formula:

$$(S_{c,2.1}\text{-}N_{c,2.1})\text{-}\ (S_{c,2.0}\text{-}N_{c,2.0}) \qquad (2)$$

The improvement in HIC++ enforcement can be calculated for a given check c as follows:

$$Density_{c,2.0}\text{-}\ Density_{c,2.1} \qquad (3)$$

as we anticipate improvement in coding standards compliance between successive releases of QAC++.

For each of the 16 checks that have been modified, and thus can have different values for Eqn. 2 and 3, we have gathered their respective values. Calculation of Eqn. 2 is manual, and requires every single occurrence of a warning to be evaluated for whether it is signal or noise. QAC++ analysis checks can be used to automate enforcement of a coding standard, and for HIC++ 65% of the rules can be enforced. Therefore, we ran QAC++ over its own source code to calculate Eqn. 3. Figure 1 is a representation of the measurements on ordinal scale, by only considering their relative rank.

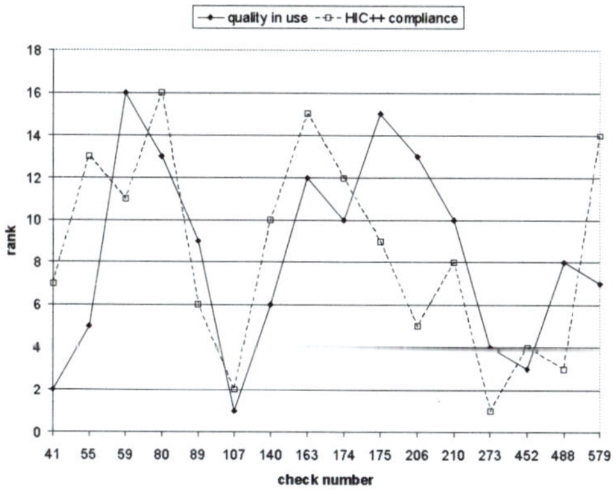

Fig 1: Correlation between Quality in Use and Internal Quality.

As can be seen in Figure 1, Effectiveness and Internal Quality appear to be correlated; however, we have decided to use a more rigorous statistical technique to confirm this. We needed a non-parametric test, i.e. operating on ranks of the observations, and one that would not make any assumptions about the distribution of their underlying variables (Effectiveness, and coding standards compliance). Kendall's τ (tau) 15 has these properties, and has the advantage that its interpretation is intuitive: the calculated value is the difference between probability that variables are correlated versus that they are not. The value of τ for the 16 measurements displayed in Figure 1 is 0.353. Considering the number of measurements this gives significance level p=2.8% for rejecting the H_0 (null hypothesis) that the variables are not correlated. Therefore, the visual interpretation of Figure 1 is confirmed as statistically significant.

This result can be interpreted as follows: the improvement in effectiveness of an analysis check in QAC++ is largely proportional to improvement in compliance to HIC++ of the underlying implementation of this check. Therefore in this instance, enforcement of a coding standard has been shown to represent the quality of the software product closely.

5. SUMMARY

In this paper we have proposed using coding standards compliance as a measure of Internal Quality of a Software System. The validity of this metric has been confirmed on a real-world software project, as it was found to correlate with a metric for Quality in Use: Effectiveness characteristic. Also, compliance to a coding standard has been found by 2 separate organisations to positively impact External Quality: Portability characteristic of their software.

More investigation is needed into the extent each of the ISO 9126 Internal Quality characteristics are affected by coding standards compliance. For example Functionality is unlikely to be affected by following generic rules typically found in a coding standard, and instead is determined by fulfilling project specific requirements.

6. REFERENCES

1 The ADAPTIVE Communication Environment (ACE) Toolkit, www.cs.wustl.edu/~schmidt/ACE.html

2 Automotive Open System Architecture, www.autosar.org

3 N.E. Fenton, S.L. Pfleeger. *Software Metrics: A Rigorous Approach.* 2nd edition. PWS, Boston, 1998

4 M. Henricson, E. Nyquist, N. Erik. *Industrial Strength C++: Rules and Regulations.* Prentice Hall, 1997

5 ISO/IEC 9126-1:2001. *Software engineering – Product quality – Part 1: Quality model.*

6 ISO/IEC TR 9126-2:2003. *Software engineering – Product quality – Part 2: External metrics.*

7 ISO/IEC TR 9126-3:2003. *Software engineering – Product quality – Part 3: Internal Metrics.*

8 ISO/IEC TR 9126-4:2004. *Software engineering – Product quality – Part 4: Quality in use metrics.*

9 ISO/IEC 9899:1990. *Programming languages – C.*

10 ISO/IEC 14882:2003. *Programming languages – C++.*

11 S. Meyers. *Effective C++: 55 Specific Ways to Improve Your Programs and Designs*. 2nd edition. Addison Wesley, Boston, 2005
12 MIRA, *MISRA-C:1998 - Guidelines for the Use of the C Language in Vehicle Based Software*. www.misra-c.com, 1998
13 MIRA, *MISRA-C:2004 - Guidelines for the use of the C language in critical systems*. www.misra-c.com, 2004
14 Programming Research. *High Integrity C++ Coding Standard Manual*. www.codingstandard.com, 2004
15 S. Siegel. *Nonparametric Statistics for the Behavioral Sciences*. McGraw-Hill Book Company, Berkshire, 1956.
16 H. Sutter, A. Alexandrescu. *C++ Coding Standards: 101 Rules, Guidelines, and Best Practices*. Addison Wesley, Boston, 2004
17 H. Sutter. *Exceptional C++*. Addison Wesley, 1999

Reliability, Safety and Hazard
Editors: P.V. Varde et al.

FLEXIBLE SOFTWARE RELIABILITY GROWTH MODELS

P.K. Kapur

Department of Operational Research, University of Delhi, Delhi

ABSTRACT

Numerous Software Reliability Growth Models (SRGMs), which relate the number of failures (faults identified/removed) and execution time, have been discussed in the literature. These models are used to predict fault content and reliability of the software. It has been observed that the relationship between the testing time and the corresponding number of faults removed is either Exponential or S-Shaped or the mix of the two There is another important class of SRGMs, known as flexible SRGMs, which can depict, depending upon parameter values, both exponential and S-Shaped growth curve In this paper, we start with the discussion on existing three flexible SRGMs due to Obha [1984], Bittanti et. al. [1988] and Kapur and Garg [1992] and later discuss their discrete equivalents. We have also included two other discrete equivalents of Kapur and Garg model existing in the literature due to Innoue [2004] and Satoh [2000]. However they are subsequently shown to be equivalent to a mathematically and computationally much simper discrete equivalent of Kapur & Garg [1992]. The derivation of continuous SRGMs from the equivalent discrete model has been also shown. The interesting part of the paper starts with the introduction of a new concept of Power Logistic learning function which proves to be very flexible in the sense that it represents various curve types – Exponential, Rayleigh, Weibull or simple Logistic. This flexible nature of power logistic function gives the flexible SRGM higher degree of accuracy and wider applicability. It makes the model adaptable to any type of software dataset say Exponential, S-Shaped or Logistic type. In the next section we take further the concept of generalized SRGMs by defining the software fault detection rate as the convex combination of power logistic and the Weibull functions. From the numerical illustrations, we see that the newly defined class of generalized SRGMs yield considerably improved results with better predictability because of lower MSE and higher R^2.

1. INTRODUCTION

Due to the incredible growth in information technology and science, computers and computer-based systems have invaded every sphere of human activity e.g. in Nuclear research, Telecommunications, Space research Programmes, Aviation, transportation, Banking Education and Health-care etc. With the world of information technology changing at a staggering rate & proliferation of Internet, more and more systems are being automated, thus an increase in the dependence of mankind on computers.

Though this technology revolution has made our lives better, concern for safety and security has never been greater. There are already numerous instances where failure of computer controlled systems have led to colossal loss of human lives and money. With increased complexity of products design, shortened development cycles and higher customer satisfaction of quality and highly destructive consequences of software failures a major responsibility lies in the areas of Software Debugging, Testing and Verification Moreover, the competition in the software market is intense and because of the nature of applications involved, purchasers look for quality in terms of reliability of the software. The nature and complexity of software requirements have drastically changed in the last few decades and users all over the world have become much more demanding in terms of cost, schedule and quality.

Computer based systems typically consist of hardware and software. Quality hardware can now be produced at a reasonable cost but the same cannot be said about software. Software development consists of a sequence of activities where perfection is yet to be achieved. Hence there is every possibility that fault can be introduced and can remain in a software. These faults can lead to failures with catastrophic results. Hence a lot of emphasis is put on avoiding introduction of faults during software development and to remove dormant faults before the product is released for use. The only way to verify and validate the software is by testing. The software testing involves running the software and checking for unexpected behavior of the software output. During the testing phase test cases that simulate the user environment are run on the software and any departure from specifications or requirements is called a failure and immediately an effort is made to remove the cause of that failure (i.e. fault). The successful test can be considered to be the one, which reveals the presence of the latent faults. Therefore, the software should be thoroughly tested to expose as many software faults as possible. Testing phase is an extremely important component of Software Development Life Cycle (SDLC) where around 50 % of the developmental resources are consumed Therefore, testing the software through executing all its statements and paths (even at least once) is not practically possible in a large-scale software system. A compromise testing approach is used which involves dividing the software into blocks and executing each block at least once. The potentially troublesome combinations of blocks are executed many times. However, testing techniques can be categorized into two main techniques:

1. Top-down testing involves starting the test at the sub-system level. Then, the modules, which comprise the sub-system, are tested. The procedure is recursively repeated until the test reaches the lowest level software component (function, object).
2. Down-Top testing involves reversing the previous process.

Each testing technique has its merits and demerits. Choosing the testing approach mainly depends on the software development technique. The software testing is a destructive process as it aims at forcing the software to behave abnormally under some conditions. For this reason, the software programmers subconsciously avoid bringing their product into this stage. Therefore, it is preferred that an independent team tests the software. This testing method is called Independent Validation and Verification (IVV), which mean that the testing team is functionally independent from the development team and they test the software from the user's point of view. This method is also called black box method. The process of locating the faults and designing the procedures to remove them is called the debugging process. The process of fault removal (repair) involves rewriting the code if the fault is due to coding and design error or changing the requirements (which require doing major repair). The chronology of failure occurrence and fault removals can be utilized to provide an estimate of the software reliability and the level of fault content. In this light, there is a need to develop a tool that can utilize this information to help the software engineers and managers in monitoring the process of the testing. The Software Reliability Model (SRM) is the tool, which can be used to evaluate the software quantitatively, develop test status, schedule status and monitor the changes in reliability performance.

Numerous Software Reliability Growth Models (SRGMs), which relate the number of failures (faults identified/removed) and execution time, have been discussed in the literature [7, 10, 13]. These models are used to predict fault content and reliability of the software. It has been observed that the relationship between the testing time and the corresponding number of faults removed is either Exponential [4] or S-Shaped [17] or the mix of the two There is another important class of

SRGMs, known as flexible SRGMs, which can depict, depending upon parameter values, both exponential and S-Shaped growth curves [7]. In these models the role of the learning process during the testing phase is taken into account, which comes from for the experience gained with the progress of software testing.

In this paper, we start in section 2 with the discussion on existing three flexible SRGMs due to Obha [12], Bittanti et. al. [2] and Kapur and Garg [6] and later discuss their discrete equivalents in section 3. We have also included two other discrete equivalents of Kapur and Garg [6] model existing in the literature due to Innoue [19] and Satoh [15]. However they are subsequently shown to be equivalent to a mathematically and computationally much simper discrete equivalent of Kapur & Garg [6]. The derivation of continuous SRGMs from the equivalent discrete model has been also shown. The interesting part of the paper starts in the section 4 with the introduction of a new concept of Power Logistic learning function which proves to be very flexible in the sense that it represents various curve types – Exponential, Rayleigh, Weibull or simple Logistic. In the same section we take further the concept of generalized SRGMs by defining the software fault detection rate as the convex combination of power logistic and the Weibull functions. From the numerical illustrations, we see that the newly defined class of generalized SRGMs yield considerably improved results with better predictability because of lower MSE and higher R^2.

2. FLEXIBLE SRGMS

2.1. Inflection S-shaped SRGM [Ohba, 12]

This model has been developed under the assumption that more errors we detect, the more undetected errors become detectable.

$$\frac{d}{dt}m(t) = b(t)\big(a - m(t)\big)$$

Where $b(t) = b\varphi(t)$

And $\varphi(t) = r + (1-r)\dfrac{m(t)}{a}$

The solution of the above differential equation with the initial condition $m(t = 0) = 0$ is,

$$m(t) = a\left(\frac{1 - e^{-bt}}{1 + \frac{1-r}{r}e^{-bt}}\right) \tag{1}$$

Where r, $(0 \le r \le 1)$, is proportion of independent errors and is inflection parameter. The model becomes exponential when $r = 1$ and takes the logistic form when $r = 0$. It implies that initially r is almost zero but as most of faults are detected and removed it approaches unity. Here $\varphi(t)$ reflects the gradual increase in testing effort and the skill gained by the testers as the testing progresses.

2.2 Flexible SRGM [Bittanti et al. , 2]

The model is based on the following differential equation:

$$\frac{d}{dt}m(t) = k(m)\big(a - m(t)\big)$$

Where $k(m) = k_i + (k_f - k_i)\dfrac{m(t)}{a}$

Here k_i and k_f are initial and final values of Fault Exposure Coefficient. If ki = k_f , then it reduces to Exponential model. If $k_f >> k_i$; the failure growth curve takes S-shape. If k_f is very small as compared to k_i that it is almost equal to zero, the failure growth curve become flat at the end.

The solution of equation with initial condition $m(t = 0) = 0$ is :

$$m(t) = a\left(\frac{1 - e^{-k_f t}}{1 + \frac{k_f - k_i}{k_i}e^{-k_f t}}\right) \tag{2}$$

For different values of k_f and k_i , it describes different growth curves.

2.3. SRGM for an Error Removal Phenomenon [Kapur and Garg, 6]

This model is based upon the assumption that the detection of errors also results in detection of some of the remaining errors without these errors causing any failure.

The differential equation for this model is given by

$$m(t) = p(a - m(t)) + q\left[\frac{m(t)}{a}\right](a - m(t))$$

The solution is,

$$m(t) = a\left(\frac{1 - e^{-(p+q)t}}{1 + \frac{q}{p} e^{-(p+q)t}}\right) \quad (3)$$

where

p : Failure occurrence rate.

q : Fault removal rate of additional removed faults.

This model was developed to account for some additional fault being detected without them causing failure. Here we can observe that if $q = 0$; then on each failure, only the error causing the failure is removed and it correspond to Exponential model. The failure growth curve defined by above model is S-shaped whose nature depends on q/p.

2.4. One Stage equivalent model for (1), (2) and (3)

All the three models i.e Ohba [12] Bittanti [2], Kapur and Garg [6] can be written in General Form (one stage process)

$$\frac{d}{dt} m(t) = b(t)\,(a - m(t))$$

With $b(t) = \dfrac{b}{1 + \beta e^{-bt}}$

Upon solving we get:

$$m(t) = a\left(\frac{1 - e^{-bt}}{1 + \beta e^{-bt}}\right) \quad (4)$$

By giving different forms to b and β, the above three models can be derived.

Here, in the alternate formulation, it may be observed that the fault detection rate for the above specified models is a logistic function and $b(t) \to b$ as $t \to \infty$. Here β is the learning factor and it represents the skills and experience gained by the testers during testing. If $\beta = 0$, then $b(t) = b$, i.e. constant.

3. DISCRETE SRGMS

NHPP based SRGMs described above are continuous time models which use the execution time (i.e., CPU time) or calendar time. The other group contains models, which use the test cases as a unit of fault removal period. Such models are called discrete time models, since the unit of software fault removal period is countable. A test case can be a single computer test run executed in an hour, day, week or even month. Therefore, it includes the computer test run and length of time spent to visually inspect the software source code. A large number of models have been developed in the first group while fewer are there in the second group due to the difficulties in terms of mathematical complexity involved.

Nevertheless, the utility of discrete SRGMs cannot be under estimated. As the software failure data sets are discrete, these models many times provide better fit than their continuous time counterparts. Therefore, in spite of difficulties in terms of mathematical complexity involved, discrete models are proposed regularly [7, 18].

Now we develop a flexible discrete SRGMs on the lines of above mentioned continuous time SRGMs using Probability Generating Function (P.G.F). Two other forms of the model are also given and it can be shown that the three forms are equivalent. It is further shown, how continuous time SRGM can be derived from the discrete model. As most of the software failure data sets are discrete, therefore, it is suggested that discrete SRGM should be used as compared to the corresponding continuous time SRGMs.

3.1 Discrete Time Flexible SRGM for Error Removal [6]

Most of the software reliability growth models assume that the fault removal phenomenon also describes the failure phenomenon. In reality this may not always be true. The test team can remove some faults in the software, without these faults causing any failure; although this may involve some additional effort. Fault, which is removed consequent to a failure, is known as a leading fault. While removing the leading faults, some other faults are removed which may have caused failure in future. These faults are known as dependent faults.

The difference equation under above assumptions for the SRGM is given by:

$$\frac{m(n+1)-m(n)}{\delta}$$
$$= p[a-m(n)]+\frac{q}{a}m(n+1)[a-m(n)]$$

where δ is constant time interval.

The solution can be obtained as follows (using PGF)

$$(1-\delta q)m(n+1)$$
$$-(1-\delta p)m(n)+a\delta p-\delta\frac{q}{a}m(n+1)m(n)$$

Now by multiplying both sides z^n and sum over n from 0 to ∞, we get

$$(1-\delta q)\sum_{n=0}^{\infty}z^n m(n+1)=(1-\delta p)\sum_{n=0}^{\infty}z^n m(n)$$
$$+a\delta p\sum_{n=0}^{\infty}z^n-\delta\frac{q}{a}\sum_{n=0}^{\infty}z^n m(n+1)m(n)$$

Or

$$\frac{1}{z}(1-\delta q)\sum_{n=0}^{\infty}z^{n+1}m(n+1)$$
$$=(1-\delta p)\sum_{n=0}^{\infty}z^n m(n)+a\delta p\sum_{n=0}^{\infty}z^n$$
$$-\delta\frac{q}{a}\sum_{n=0}^{\infty}z^n m(n+1)m(n)$$

Denote $\sum_{n=0}^{\infty}z^n m(n)=P(z)$ and solving we get:

$$\frac{(1-\delta q)}{z}P(z)=(1-\delta p)P(z)$$
$$+a\delta p\sum_{n=0}^{\infty}z^n-\delta\frac{q}{a}\sum_{n=0}^{\infty}z^n m(n+1)m(n)$$

$$\frac{(1-\delta q)-z(1-\delta p)}{z}P(z)$$
$$=a\delta p(1+z+z^2+z^3+........)$$
$$-\delta\frac{q}{a}(zm(1)m(2)+z^2m(2)m(3)+........)$$

$$[(1-\delta q)-(1-\delta p)z][zm(1)+z^2m(2)+...]$$
$$=a\delta p(z+z^2+z^3+........)$$
$$-\frac{\delta q}{a}(z^2m(1)m(2)+z^3m(2)m(3)+...)$$

Comparing the Coefficients of z^n and at $t=0$,

$$m(0)=0$$

$$m(1)=\frac{a\delta p}{(1-\delta q)}$$

$$m(2)(1-\delta q)-(1-\delta p)m(1)$$
$$=a\delta p-\frac{\delta q}{a}m(1)m(2)$$

Substituting the value of $m(1)$ we get

$$m(2)=a\left[\frac{1-\{1-\delta(p+q)^2\}}{1+\frac{q}{p}\{1-\delta(p+q)^2\}}\right]$$

By induction, we can write,

$$m(n)=a\left[\frac{1-(1-\delta(p+q))^n}{1+\frac{q}{p}(1-\delta(p+q))^n}\right] \tag{5}$$

Here as $n\to\infty, m(n)\to a$

Above discrete SRGM is equivalent to continuous time SRGM of Kapur & Garg [6] when $\delta\to 0$.

Let us define $t = n\delta$

As $\delta \to 0$

$$a\left[\frac{1-(1-\delta(p+q))^n}{1+\frac{q}{p}(1-\delta(p+q))^n}\right]$$

$$\to a\left[\frac{1-e^{-(p+q)t}}{1+\frac{q}{p}e^{-(p+q)t}}\right]$$

3.2. Other Discrete SRGMs

Various other discrete SRGM have been reported in the Literature corresponding to continuous model. One of them is due to Yamada & Inoue [19]. It is based on discrete Riccati Equation.

It is given by

$$\frac{m(n+1)-m(n)}{\delta}$$

$$= pa + (q-p)\left[\frac{m(n)+m(n+1)}{2}\right]$$

$$+\frac{q}{a}m(n).m(n+1)$$

The exact solution of above difference equation is given by:

$$m(n) = a\left[\frac{1-\left[\frac{\left(1-\frac{1}{2}\delta(p+q)\right)}{\left(1+\frac{1}{2}\delta(p+q)\right)}\right]^n}{1+\frac{q}{p}\left[\frac{\left(1-\frac{1}{2}\delta(p+q)\right)}{\left(1+\frac{1}{2}\delta(p+q)\right)}\right]^n}\right] \quad (6)$$

Similarly, using Satoh [15] approach another discrete equivalent of continuous model (3) can be obtained.

$$\frac{m(n+1)-m(n-1)}{2\delta}$$

$$= p\left[a - \frac{m(n+1)+m(n-1)}{2}\right]$$

$$+\frac{q}{a}\left[a\left(\frac{m(n+1)+m(n-1)}{2}\right) - m(n+1).m(n-1)\right]$$

The exact solution as given by Satoh is

$$m(n) = a\left[\frac{1-\left(\frac{1-\delta(p+q)}{1+\delta(p+q)}\right)^{n/2}}{1+\frac{q}{p}\left(\frac{1-\delta(p+q)}{1+\delta(p+q)}\right)^{n/2}}\right] \quad (7)$$

Again, it can be shown that as $\delta \to 0$, m(n) given by (6) and (7) converge to corresponding continuous SRGM given by (3).

3.3. Equivalence of Discrete SRGMs Given by Equation (5), (6) and (7)

Consider (7)

$$m(n) = a\left[\frac{1-\left(\frac{1-\delta(p+q)}{1+\delta(p+q)}\right)^{n/2}}{1+\frac{q}{p}\left(\frac{1-\delta(p+q)}{1+\delta(p+q)}\right)^{n/2}}\right]$$

Here $0 < \delta\,(p+q) < 1$

Using $(1-x)^n \approx 1 - nx$, we get

$$m(n) = a\left[\frac{1-\left(\frac{1-\frac{\delta}{2}(p+q)}{1+\frac{\delta}{2}(p+q)}\right)^n}{1+\frac{q}{p}\left(\frac{1-\frac{\delta}{2}(p+q)}{1+\frac{\delta}{2}(p+q)}\right)^n}\right]$$

which is same as equation (6).

In the above equation we can use the following expansion

$$\left(\frac{1-\frac{\delta}{2}(p+q)}{1+\frac{\delta}{2}(p+q)}\right)^n$$

$$=\left[1-\frac{\delta}{2}(p+q)\right]^n\left[1+\frac{\delta}{2}(p+q)\right]^{-n}$$

$$=\left[\left(1-\frac{\delta}{2}(p+q)\right)\left(1+\frac{\delta}{2}(p+q)\right)^{-1}\right]^n$$

$$\approx\left[\left(1-\frac{\delta}{2}(p+q)\right)\left(1-\frac{\delta}{2}(p+q)\right)\right]^n$$

$$\approx\left[1-\delta(p+q)\right]^n \qquad (8)$$

As $0<\delta(p+q)<1$,

$$(1+\frac{\delta}{2}(p+q))^{-1}\approx 1-\frac{\delta}{2}(p+q)$$

Using (8), equation (6) can be written as:

$$m(n)=a\left[\frac{1-(1-\delta(p+q))^n}{1+\frac{q}{p}(1-\delta(p+q))^n}\right]$$

It is same as equation (5).

Discrete equivalent to (3) given by (5) has been observed to be easier and faster to compute as compared with (6) and (7).

4. A NEW CLASS OF GENERALIZED KAPUR & GARG MODEL

4.1. Generalized Model-1

Now we present an interesting variation of continuous time SRGM due to Kapur & Garg [6] that adds extra flexibility and wider scope to the already flexible model.

Consider the case when the failure growth phenomenon also depends upon the testing time in addition to number of faults remaining in the software as well as the number of faults already identified. [6]. Based on these assumptions the differential equation for fault identification / removal can be written as:

$$\frac{dm}{dt}=t^k\left(p+q\frac{m}{a}\right)(a-m) \qquad (9)$$

Here $k\geq 0$. It can be easily seen that if $k=0$, then it is same as the differential equation for SRGM due to Kapur & Garg [6].

Solving it with the initial condition $m(0)=0$ we get :

$$m(t)=a\left(\frac{1-e^{-(p+q)\frac{t^{k+1}}{k+1}}}{1+\frac{q}{p}e^{-(p+q)\frac{t^{k+1}}{k+1}}}\right) \qquad (10)$$

The solution obtained above is capable of capturing different type of variations shown by the curve of $m(t)$ depending upon different values of k. It can be easily shown that if $k=0$ & $q=0$ then the model simply takes Exponential form and if $k=0$ & $q\neq 0$ it reduces to Kapur and Garg [6]

4.2. Alternate Formulation for Generalized Model-1

4.2.1. Alternate Formulation –1

The number of failures during testing is dependent upon the number of instructions executed. Here number of instructions executed is a function of testing time.

Using these assumptions, the failure removal phenomenon can be expressed as:

$$\frac{dm}{dt}=\frac{dm}{de}\frac{de}{dt}$$

Let us define

$$\frac{dm}{de}=\left(p+q\frac{m}{a}\right)(a-m)$$

Let the number of instructions executed be a power function of testing time

$$\frac{de}{dt} = t^k$$

By combining above two factors, we can write:

$$\frac{dm}{dt} = t^k \left(p + q\frac{m}{a} \right)(a - m)$$

which is same as (9)

4.2.2. Alternate Formulation –2

Now we describe another alternate formulation for SRGM given by (10), which will allow us to define a new type of logistic function later on.

Let us consider:

$$\frac{dm}{dt} = \frac{bt^k}{1 + \beta e^{-b\frac{t^{k+1}}{k+1}}}(a - m) \qquad (11)$$

Then solving it with the initial condition $m(0) = 0$, we get the same result as obtained in (9).

Here $b(t) = \dfrac{bt^k}{1 + \beta e^{-b\frac{t^{k+1}}{k+1}}}$ can be termed as power logistic rate which reduces to ordinary logistic rate if $k = 0$.

4.3. Generalized Model-2

The basic levels of testing are –White Box testing and Black Box testing. White box testing is performed to find faults, which lie within the internal structure of the software. It requires the tester to have detailed knowledge of the internal structure. The main objective of white-box testing is to ensure that test cases execute every path throughout the code. The importance of white-box testing is expressed in terms of test or code coverage metrics, which measure the fraction of code executed by test cases. Black box testing is carried out to judge how well software meets the user's requirements.

Testing coverage is defined as the extent or the degree to which software is executed by the test cases. During testing, if few numbers of bugs are encountered, then it necessarily does not indicate the high quality of the coding but on the other hand it implies the poor designing of test cases. To ensure better testing, the testing coverage analysis is used to assess the quality of test cases. Testing Coverage is actually a structural testing technique in which the software performance is judged with respect to specification of the source code. The basic testing coverage measures are:

[1] Statement Coverage: It is defined as the proportion of lines executed in the program. If we assume that the faults are uniformly distributed throughout the code, then percentage of executable statements covered shows the percentage of faults discovered.

[2] Decision / Condition Coverage: This measure indicates whether Boolean expressions tested in control structures evaluated to both true and false.

[3] Path Coverage: This measure shows the percentage of all possible paths existing in the code exercised by the test cases.

[4] Function Coverage: This measure indicates the proportion of functions/ procedures influenced by the testing

Testing coverage provides an important criterion for the optimal release policy based on available testing resources and the importance of risk-free and safe operations of the software on implementation. Therefore, safety-critical systems have a high coverage objective.

Now, we present a software reliability growth model, which incorporates the concept of testing coverage in the model building. We have used the concept of convex combination of two different types of functions to represent the testing coverage function. This approach helps us to capture various type of coverage function simultaneously depending upon the various possible values of the parameters. The goodness of the fit has been tried on two real software failure datasets. The results obtained are fairly accurate and shows considerable improvement & better fit.

Let $c(t)$ defines the proportion of the software executed by the test cases. So, $1 - c(t)$ defines the proportion of the code which is yet to be covered by the test cases. Then, the first order derivative of $c(t)$, denoted by $c'(t)$, represent the testing coverage rate. The ratio of two i.e.

$\frac{c'(t)}{1-c(t)}$ will be taken as measure of the fault detection rate

Let

$$c(t) = 1 - \exp(-k\lambda\, t^{u} - (1-k)(e^{\gamma t^{b}} - 1)) \quad (12)$$

Here $0 \le k \le 1$ and $\lambda, u, b, \gamma > 0$

Where *u* and *b* are the shape parameters

λ, γ are the scale parameters.

Using (12)

$$\frac{c'(t)}{1-c(t)} = k\lambda\, u t^{u-1} + (1-k) b t^{b-1} \gamma e^{\gamma t^{b}} \quad (13)$$

Consider the model having the fault detection rates as defined above in (13) respectively:

$$\frac{d}{dt} m(t) = \frac{c'(t)}{1-c(t)} (\text{a} - \text{m}(\text{t})) \quad (14)$$

When we observe the model, it can be seen that failure intensity depends not only on the number of remaining faults in the software but also on ratio of rate at which the remaining faults are covered & the present proportion of uncovered faults.

Using fault detection rate as defined in (13), we get:

$$\frac{d}{dt} m(t) = (k\lambda\, u t^{u-1} + (1-k) b t^{b-1} \gamma e^{\gamma t^{b}})(\text{a} - \text{m}(\text{t})) \quad (15)$$

Using initial condition $m(0) = 0$, we get:

$$m(t) = a\left[1 - \exp(-k\lambda\, t^{u} - (1-k)(e^{\gamma t^{b}} - 1))\right] \quad (16)$$

where $m(t) \to a$ as $t \to \infty$.

Here if k=1 , then it is same as Weibull model due to Yamada.[20]

5. MODEL VALIDATION

To check the accuracy of the generalized SRGM-1 given by (10) with power logistic and generalized SRGM-2 given by (16) with convex combination of two different types of function with respect to Flexible SRGM of type (4), parameter estimation on real software failure datasets has been done.

Data Set-1 (DS-1):

The data was obtained from Musa et. al.. [9]. The software was a real time command and control system, which was tested for 92 days (21 weeks). The delivered object instructions were 21,700 involving 9 programmers. 136 faults were removed during testing.

The estimation results are given in the Table 1.

Data Set – II (DS-II):

The data are cited from Brooks and Motley [3]. The fault data set is for a radar system of size 124 KLOC (kilo lines of code) tested for 35 months in which 1301 faults were identified.

The estimation results are given in the Table-2.

Table 1

Data Set	Parameter Estmation	Models Under Comparison		
		Flexible Model	Generalized SRGM-1	Generalized SRGM-2
DS-1	a	168	140	142
	b	33242	00403	.0105
	β	193.55	8.073	-
	k	-	1.7103	.000025
	λ	-	-	.0595
	u	-	-	4.81
	γ	-	-	.162
Comparison Criterion	R^2	.99310	.99738	.99745
	MSE	16.354	6.21	6.037

Table 2

Data Set	Parameter Estmation	Models Under Comparison		
		Flexible Model	Generalized SRGM-1	Generalized SRGM-2
DS-2	a	1331	1322	1305
	b	.20059	.1234	.423
	β	20.16	11.15	-
	k	-	.1678	.0029
	λ	-	-	.177
	u	-	-	2.594
	γ	-	-	.0262
Comparison Criterion	R^2	.99904	.99911	.99909
	MSE	204	190.35	193.86

It has been observed that the generalized SRGM-1 and 2 given by (10) and (16) has always provided results better than or same as Flexible SRGM of type (4). This is due to the flexibility rendered to the model with the presence of 'k' factor. The increased accuracy achieved shows the capability of the model to capture different types of failure datasets e.g. Exponential, S-shaped Rayleigh or Weibull types etc.

5.1 Goodness of Fit Curves for the Proposed Generalized SRGMs

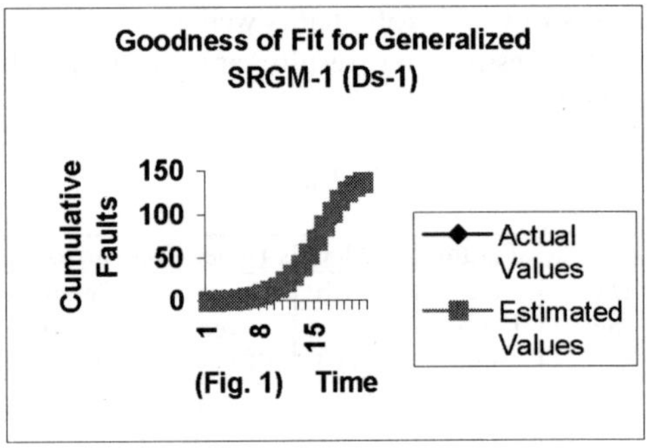

(Fig. 1)

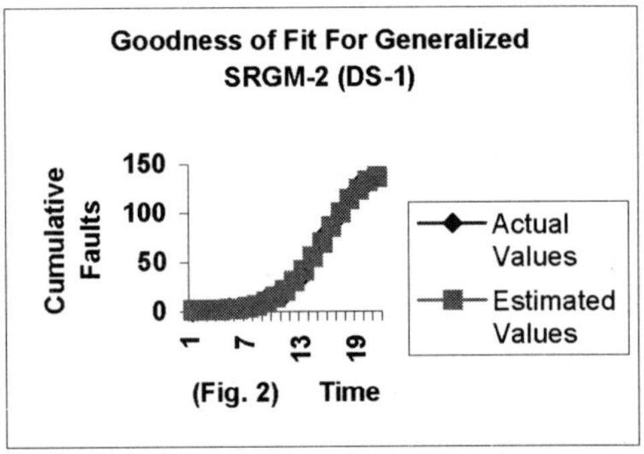

(Fig. 2)

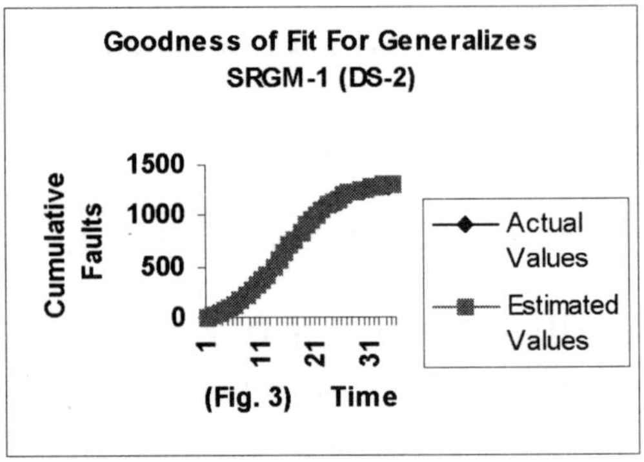

(Fig. 3)

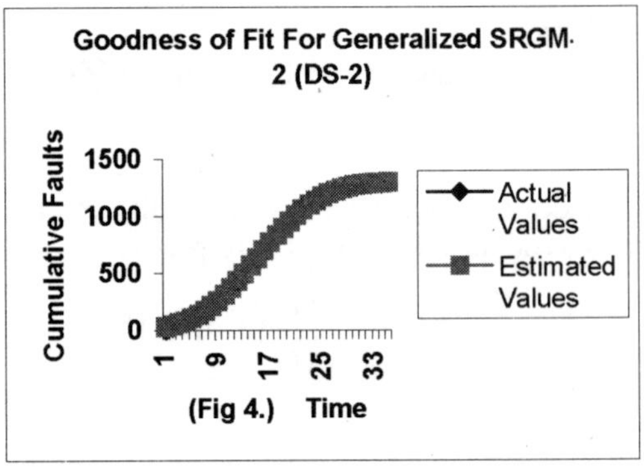

(Fig 4.)

6. CONCLUSION

Software Reliability Engineering is a fast emerging as a very important field of study in the area of Information Technology. The mathematical models play a significant role in its growth as these models provide quantitative tools to assess the reliability of the software by software developers. In this talk we have tried to assess an important class of SRGMS, known as 'Flexible SRGMs'. Later on we have added new dimensions to the flexible modelling by introducing the concept of 'k' factor and convex combination, which subsequently, proves to be more flexible and yield better predictability and higher degree of accuracy.

The usability of these newly introduced concept is not limited to these SRGMs only. It can be extended to improve the results of any SRGM –whether Exponential, S-shaped or any other type e.g. multi stage models or SRGMs related with Distributed Environment.

7. REFERENCES

1. Bass FM. "A new product growth model for consumer durables" *Management Science* **1969**; 15(5): 215-224.
2. Bittanti S, Bolzern P, Pedrotti E, Scattolini R. "A flexible modelling approach for software reliability growth" *Software Reliability Modelling and Identification(Ed.) G. Goos and J. Harmanis*, Springer Verlag, Berlin, **1988**, 101-140.
3. Brooks WD, Motley RW. "Analysis of discrete software reliability models - *Technical Report (RADC-TR-80-84)*" **1980**; New York: Rome Air Development Center.
4. Goel AL, Okumoto K. "Time dependent error detection rate model for software reliability and other performance measures" *IEEE Transactions on Reliability* **1979**; R-28(3): 206-211.
5. Jelinski Z, Moranda PB. "Software reliability research" In: *Freiberger W, (Ed.) Statistical Computer Performance Evaluation* **1972**; New York: Academic Press: 465-497.
6. Kapur PK, Garg RB. "A software reliability growth model for an error removal phenomenon" *Software Engineering Journal* **1992**; 7: 291-294.
7. Kapur PK, Garg RB, Kumar S. "Contributions to hardware and software reliability" **1999**; Singapore: World Scientific Publishing Co. Ltd.
8. Kapur PK, Shatnawi O and Singh O "Discrete imperfect software reliability growth models under imperfect debugging environment". In: Rajaram NJ and Verma AK *(Editors) Proceedings of the International Conference on Multimedia and design*; Arena Multimedia & IIT: Mumbai, **(2002)**. Vol (II): pp. 114-29.
9. Musa JD. "Validity of execution time theory of software reliability" *IEEE Transactions on Reliability* **1979**; R-28: 181-191.
10. Musa JD, Software Reliability Engineering **1999**, McGraw-Hill.
11. Musa JD, Iannino A, Okumoto K. "Software reliability: Measurement, Prediction, Applications" **1987**; New York: Mc Graw Hill.
12. Ohba M. "Software reliability analysis models" IBM Journal of Research and Development **1984**; 28: 428-443.
13. Pham H. "Software Reliability" **2000**; Springer-Verlag Singapore Pte. Ltd.
14. R. Hirota, "Nonlinear partial difference equation. V. Nonlinear equations reducible to linear equations" *Journal of the Physical Society of Japan,* 46, **1979**; 312-319.
15. Satoh D, "A discrete bass model and its parameter estimation", Journal of Operations Research Society of Japan, 44(1):1-18.
16. Xie M. "Software reliability modelling" **1991**; World Scientific.
17. Yamada S, Ohba M., Osaki S. "S-shaped software reliability growth modelling for software error detection" *IEEE Trans. On Reliability*, **1983**; R-32(5): 475-484.
18. Yamada S and Osaki S. "Discrete Software reliability growth models" *Applied Stochastic models and data analysis* **1985**; 1: pp. 65-77.
19. Yamada S, Inoue S and Yamamoto T "Software Reliability Growth Modeling Based on Testing-coverage" To appear in *International Journal of Quality, Reliability and Safety Engineering,* **2004**.
20. Yamada, S., J. Hishitani and S. Osaki , "Software Reliability Growth Model with Weibull testing effort: A model and application", *IEEE Trans. on Reliability,* **1993**; R-42, 100-105.

Reliability, Safety and Hazard
Editors: P.V. Varde et al.

FLEXIBLE TESTING-DOMAIN DEPENDENT SOFTWARE RELIABILITY GROWTH MODELS

P.K. Kapur, Sunil Khatri[1], Amit Gupta* and V.B. Singh[2]*

*Department of Operational Research, University of Delhi, Delhi 110 007
[1] Mother Teresa Institute of Management, G.G.S.I.P. University, Delhi 110 092
[2] Delhi College of Arts & Commerce, University of Delhi, Delhi 110 023

ABSTRACT

In the testing phase of software development, there is a set of paths and functions which are influenced by executed test cases is called a testing-domain. The growth of isolated testing-domain in the software system is closely related to the number of executed test-cases and their quality. The number of detectable faults is determined by the growth of isolated testing-domain. In this paper, we propose flexible software reliability growth models (SRGMs) which describe the time-dependent behavior of the isolated testing-domain in the software system. The models are formulated by Non-Homogenous Poisson Process (NHPP). We propose general flexible models which describe the growth of isolated testing-domain. The models are applied to fault data observed in the real life projects and the comparisons of the goodness-of-fit with the conventional SRGMs are shown. The results show improved goodness.

Keywords: Software Reliability Growth Model, Non Homogeneous Poisson Process (NHPP), Testing-Domain, Logistic Function

1. INTRODUCTION

Developing software system is generally a quite complex and time-consuming process. Moreover, the nature and complexity of software requirements have drastically changed in the last few decades and users all over the world have become much more demanding in terms of cost, schedule and quality[9,12].

Software Engineering Methodologies constitute the framework that guides software developers in optimally developing the software systems. These frameworks define the different phases of software development (such as planning, requirements analysis, design testing and maintenance). The choice of which methodology to use in a specific development process is closely related to the size, complexity, reliability and maintainability of the software, and to the environment it is supposed to function in [9,12].

The ability and quality of a software depends heavily on it's reliability. The reliability of any software relies on testing phase. While execution, if a system fails, it implies that there are errors (faults) inherent in the system. This reduces the reliability of the software system. It is difficult to produce an error-free software system due to human prone to errors. Therefore, the testing phase focuses on detecting and correcting the faults in the software. Software error detection phenomenon can be regarded as a software reliability growth process. As a mathematical model for the reliability growth process, a software reliability growth model (SRGM) has been studied by many researchers. The software reliability growth model describes the relationship between the cumulative number of detected software errors and the time span of software testing.

In the testing phase of software development, there is a set of paths and functions which are influenced by executed test cases is called a testing-domain. The growth of isolated testing-domain in the software system is closely related to the number of executed test-cases and their quality. The number of detectable faults is determined by the growth of isolated testing-domain.

In this paper, we discuss flexible SRGMs based on two kinds of testing-domain. In Section 2, we propose two kinds of the testing-domain models represented by the basic testing domain and testing-domain with skill factor. In Section 3, the SRGMs are formulated based on an NHPP by using the two testing-domain models discussed in Section 2. Finally, in Section 4, these models are applied to the fault data observed in the actual software development projects, Results show improved goodness-of-fit as compared with the existing models and the comparisons of goodness-of-fit with the existing models are performed.

2. SOFTWARE TESTING- DOMAINS

2.1 Basic Testing-Domain[1,2,5]

During the testing phase, the software developer executes many test-cases in order to verify functions. These test-cases influence a set of functions, which is called a testing-domain. The isolated testing-domain expands with the increasing number of test-cases and as a result number of detectable faults is also increased. The testing-domain growth rate is closely related to the quality and quantity of the executed test-cases

In this paper, we assume the following .

1) Correction of faults detected does not introduce any new fault.
2) Latents faults in the software are uniformly distributed.
3) For the isolated testing-domain, number of errors detected in the testing time interval $(t, t + \Delta t]$ is proportional to the number of errors remaining in the software system outside of the isolated testing domain.

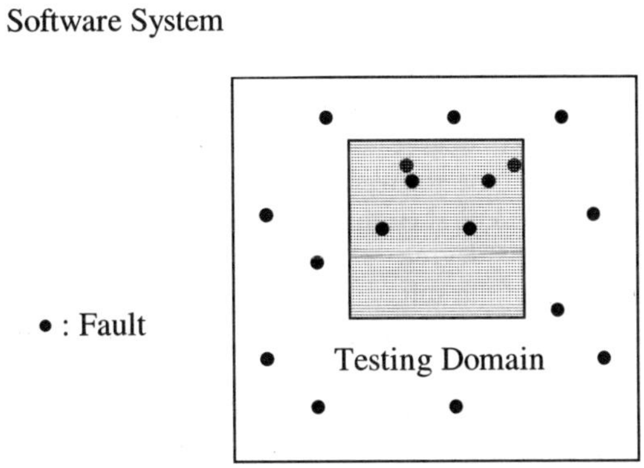

Fig. 1: Testing Domain

From the above assumptions, we get the following differential equation :

$$\frac{du_a(t)}{dt} = \lim_{\Delta t \to 0} \frac{u_a(t+\Delta t) - u_a(t)}{\Delta t} = v[a - u_a(t)], \qquad (a>0, v>0) \qquad \ldots (1)$$

where a is the faults latent in the software, v is the testing-domain growth rate and $u_a(t)$ is the total number of detectable faults in the isolated testing-domain at testing time t. Under the initial condition $u_a(0) = 0$, equation (1) yields the following :

$$u_a(t) = a\left(1 - e^{-vt}\right), \qquad \ldots (2)$$

where $(1 - e^{-vt})$ is the testing-domain growth ratio to the final testing-domain to be covered. $u_a(t)$ is called the basic testing-domain function.

Equation(2) assumes that error distribution is uniform. In other words, we can say that testing-domain growth rate is uniform. But, in practice, error distribution is not uniform. As a result, the equation (2) becomes :

$$u_b(t) = a\left(1 - pe^{-vt}\right), (1 \geq p > 0) \qquad \ldots (3)$$

where p is the uniformity factor in error distribution. When $p = 1$, that is, when error distribution is uniform, equation (3) reduces to equation (2).

2.2 Testing-Domain with Skill Factor[1,2,5]

Testing-skill of the designer affects the testing-domain growth rate and ultimately the number of detectable faults in the system. In general, there are two cases of test-skills :

Low Skill : Personnel being inexperienced have lower degree of knowledge of internal structure of the software.

High Skill : Highly experienced personnel have high degree of knowledge of internal structure of the software.

If the test-case designers have low skill, then the number of modules covered by the test-cases is limited and the testing-domain will not spread. On the other hand, if the test-case designers have high level of skill, many modules will be covered by test-cases and the testing-domain will grow to the complete software system.

We get the following differential equations

$$\frac{ds(t)}{dt} = v[a - s(t)], \quad (a > 0, v > 0) \quad \ldots (4)$$

$$\frac{du(t)}{dt} = v[s(t) - u(t)], \quad \ldots (5)$$

where a is the faults latent in the system, v is the testing-domain growth rate, $s(t)$ is the number of faults detectable at time t and $u(t)$ is the number of faults existing in the isolated testing-domain at time t. Solving equations (4) and (5) simultaneously with initial condition $u(0) = 0$, we get :

$$u_c(t) = a\{1 - (1 + vt)\exp(-vt)\}, \quad \ldots (6)$$

Equation (6) shows that testing-domain is not existing at the starting point of testing phase since skill of test-case designer is very low. It also shows that testing-domain growth curve is S-shaped.

Let us now consider the situation where the skill of test-case designer is very high. Persons involved in testing are expert because of which fault detection rate is very high from beginning. And the testing-domain can spread over to the entire software system very quickly. Under the initial condition $u(0) = a(1 - p)$, where p is the skill factor :

$$u_d(t) = a\{1 - p(1 + vt)\exp[-vt]\}, \quad (1 \geq p \geq 0) \quad \ldots (7)$$

Skill factor $p = 0$ indicates that test-case designers are expert and experienced leading to high potential of detecting the faults in initial stages of testing. On the other hand, when $p = 1$, equation (7) becomes equation (6) which signifies that designers have low level of skill.

3. RELIABILTY GROWTH MODELS

A software system is subject to software failures caused by the errors remaining in the system. We propose flexible software reliability growth models based on the testing-domain discussed above, which describe software error detection phenomena during software testing. The models have following fundamental assumptions :

1. A software failure is caused by an error.
2. The error causing the failure can be immediately removed.
3. Correcting detected errors does not introduce any new error.
4. Faults detected exist in the isolated testing-domain.
5. No. of faults detected in the small time interval (t , $t + \Delta t$) is proportional to the number of faults remaining in the testing-domain at testing time t.

Let $\{N(t), t > 0\}$ denote a counting process representing the cumulative number of errors detected up to testing time t ($t \geq 0$). Then, a software reliability growth model for such an error detection process can be described by a Non Homogenous Poission Process (NHPP) as

$$\Pr\{N(t) = n\} = \frac{\{m(t)\}^n}{n!}.\exp[-m(t)], \quad (n = 0,1,2,\ldots), \quad \ldots (8)$$

where $m(t)$ is a mean value function representing the expected cumulative number of errors detected upto testing time t, i.e., $m(t)$ is the expected value of $N(t)$ [11,13,14]. It is assumed that $m(t)$ has a boundary condition $m(0)=0$. From assumption 5, since the number of errors detected in the small interval (t , $t + \Delta t$) is proportional to the detectable errors in the isolated testing-domain at time t, we have the following differential equation:

$$\frac{dm(t)}{dt}=\lim_{\Delta t\to 0}\frac{m(t+\Delta t)-m(t)}{\Delta t}=b(t)[u(t)-m(t)]\ ,\quad (b>0) \qquad (9)$$

where $b(t)=\dfrac{b}{1+\beta e^{-bt}}$.

The logistic function is taken instead of constant b as proposed by Yamada, Ohtera and Ohba(1992) which makes the model more flexible.

With b as constant, equation (9) with respect to four kinds of $u(t)$ i.e., $u_a(t)$ in (2), $u_b(t)$ in (3), $u_c(t)$ in (6) and $u_d(t)$ in (7), respectively has following SRGMs as was given by Yamada, Ohtera and Ohba[1,2,5]:

SRGM - 1

$$m_a(t)=a\left[1-\frac{1}{(v-b)}\left(ve^{-bt}-be^{-vt}\right)\right]\ ,\quad (v\neq b) \qquad \ldots (10)$$

SRGM – 2

$$m_b(t)=a\left[1-\frac{1}{(v-b)}\left((v-b+bp)e^{-bt}-bpe^{-vt}\right)\right],\ (v\neq b, 1\geq p>0) \qquad \ldots (11)$$

SRGM – 3

$$m_c(t)=a\left[1-\left(\frac{v}{v-b}\right)^2 e^{-bt}+\frac{b}{v-b}\left(vt+\frac{2v-b}{v-b}\right)e^{-vt}\right],(v\neq b) \qquad \ldots (12)$$

SRGM – 4

$$m_d(t)=a\left[1+\frac{bp}{(v-b)}\left(vt+\frac{2v-b}{v-b}e^{-vt}\right)-\left(1+bp\frac{2v-b}{(v-b)^2}\right)e^{-bt}\right]$$

$$(v\neq b, 1\geq p>0) \qquad \ldots (13)$$

Solving equation (9) with respect to four kinds of $u(t)$ i.e., $u_a(t)$ in (2), $u_b(t)$ in (3), $u_c(t)$ in (6) and $u_d(t)$ in (7), respectively we propose the following flexible testing-domain dependent SRGMs [3,4,6,7,8] :

SRGM – 5 (Proposed for SRGM - 1)

$$m_a(t)=\frac{a}{1+\beta e^{-bt}}\left[1-\frac{1}{(v-b)}\left(ve^{-bt}-be^{-vt}\right)\right]\ ,\ (v\neq b) \qquad \ldots (14)$$

SRGM - 6 (Proposed for SRGM - 2)

$$m_b(t)=\frac{a}{1+\beta e^{-bt}}\left[1-\frac{1}{(v-b)}\left((v-b+bp)e^{-bt}-bpe^{-vt}\right)\right],$$

$$(v\neq b, 1\geq p>0) \qquad \ldots (15)$$

SRGM - 7 (Proposed for SRGM - 3)

$$m_c(t)=\frac{a}{1+\beta e^{-bt}}\left[1-\left(\frac{v}{v-b}\right)^2 e^{-bt}+\frac{b}{v-b}\left(vt+\frac{2v-b}{v-b}\right)e^{-vt}\right],$$

$$(v\neq b) \qquad \ldots (16)$$

SRGM - 8 (Proposed for SRGM - 4)

$$m_d(t)=\frac{a}{1+\beta e^{-bt}}\left[1+\frac{bp}{(v-b)}\left(vt+\frac{2v-b}{v-b}e^{-vt}\right)-\left(1+bp\frac{2v-b}{(v-b)^2}\right)e^{-bt}\right],$$

$$(v\neq b, 1\geq p>0) \qquad \ldots (17)$$

Clearly, in the case of $v = b$ for (14), we have the mean value function of a delayed S-shaped software reliability growth model.

The reliability growth parameters a, b, v, p and β in the mean value functions $m_a(t)$, $m_b(t)$, $m_c(t)$ and $m_d(t)$ can be estimated by a method of maximum-likelihood. Suppose that the data on the cumulative number of detected errors y_k in a given time interval $(0,t_k]$ (k=1, 2, . . . , n; $0 < t_1 < t_2 < \ldots < t_n$) are observed. Then, based on an NHPP model described by (8), the joint probability mass function i.e., the likelihood function, for the observed data, is given by

$$L\equiv\{N(t_1)=y_1, N(t_2)=y_2,\ldots,N(t_n)=y_n\}$$

$$=\prod_{k=1}^{n}\frac{\{m(t_k)-m(t_k-1)\}^{y_k-y_{k-1}}}{(y_k-y_{k-1})!}.\exp(-m(t_n)) \qquad \ldots (18)$$

where $t_0 \equiv 0$ and $y_0 = 0$. Therefore, substituting $m_a(t)$ in (14), $m_b(t)$ in (15), $m_c(t)$ in (16) and $m_d(t)$ in (17) for $m(t)$ in (18), the reliability growth parameters a, b, v, p and β can be estimated by maximizing the likelihood function in (18).

4. GOODNESS-OF-FIT CRITERIA

4.1 Estimation of Parameters

In this section, we analyze actual fault data by using the proposed SRGMs in this paper. We apply these models to two sets of software fault-detection data : DS1 and DS2. DS1 is in the form (t_k, y_k) (k=1,2, ... , 35; t_k (months)), which is cited by Brooks and Motley [16]. DS2 is in the form (t_k, y_k) (k=1,2, ... , 19;t_k (weeks)) which was cited by Ohba [15].

4.2 Comparisons of Models

1. The Mean Square Fitting Error (MSE) :

The model under comparison is used to simulate the fault data, the difference between the expected values, $\hat{m}(t_i)$ and the observed data x_i is measured by MSE as follows.

$$MSE = \sum_{i=1}^{k} \frac{(\hat{m}(t_i) - x_i)^2}{k} \quad \ldots \qquad (19)$$

where k is the number of observations. The lower MSE indicates less fitting error, thus better goodness of fit.

2. The Akaike Information Criterion (AIC) :

It is defined as :

AIC = -2(The value of the maximum log likelihood function) + 2(The number of the parameters used in the model)... (20)

This index takes into account both the statistical goodness of fit and the number of parameters that are estimated. Lower values of AIC indicate the preferred model, i.e. the one with the minimum number of parameters that still provides a good fit to the data.

3. Coefficient of Multiple Determination (R^2) :

This goodness of fit measure can be used to investigate whether a significant trend exists in the observed failure intensity. We define this coefficient as the ratio of the sum of squares resulting from the trend model to that from constant model subtracted from 1, that is R^2=1 – residual SS/corrected SS. R^2 measures the percentage of the total variation about the mean accounted for the fitted curve. It ranges in value from 0 to 1. Small values indicate that the model does not fit the data well. The larger R^2, the better the model explains the variation in the data.

5. MODEL VALIDATION

To check the validity of the proposed model to describe the software reliability growth, it has been tested on two Data Sets.

Data Set 1

This data is cited from Brooks [16]. The software was tested for 35 months during which 1301 faults were removed. Based upon the estimated parameters, parameters of SRGMs were estimated. The estimation results are given in Table 1. The goodness of fit results is given in Table 3. The curves for goodness of fit are depicted in Fig 2 –Fig 9.

Data Set 2

This data is cited from Ohba [15]. The fault data set is for a system tested for 19 weeks in which 328 faults were removed. Based upon the estimated parameters, parameters of SRGMs were estimated. The estimation results are given in Table 2. The goodness of fit results is given in Table 4. The curves for goodness of fit are depicted in Fig 10 –Fig 17.

The performance of an SRGM is judged by its ability to fit the past software failure / fault removal data (goodness of fit) and to predict satisfactorily the future behavior of the software failure occurrence / fault detection process from present and past data (predictive validity)

It can be observed from the estimation results that the proposed models gives the best estimates in terms of goodness of fit for both the data sets. The plot of actual values and the estimated values (of testing effort and fault removal) for the two data sets are depicted in figures 2 to 17.

6. CONCLUSION

In this paper, we have discussed two kinds of testing-domains. We have proposed flexible testing-domain dependent SRGMs based on an NHPP in comparison to the testing-domain SRGMs proposed by Yamada, Ohtera and Ohba. Models were applied on the actual fault data observed during testing phase and compare in terms of goodness-of-fit to the existing models.

We are planning to develop the discrete testing-domain dependent models in the future.

Table 1 For DS-1

Models	A	b	v	p	β
SRGM-1	1689.3	0.0899	0.8985	-----	----
SRGM-2	1500.0	0.0883	0.1102	0.8950	----
SRGM-3	1448.8	0.1662	0.1663	-----	----
SRGM-4	1457.1	0.1656	0.1630	0.9866	----
SRGM-5 (proposed)	1425.4	0.1715	0.1096	----	4.0055
SRGM-6 (proposed)	1325.1	0.2082	0.0935	0.0001	22.841
SRGM-7 (proposed)	1335.6	0.1950	2.1235	-----	17.880
SRGM-8 (proposed)	1342.0	0.1938	0.2214	0.3045	21.229

Table 2 For DS-2

Models	a	B	v	p	β
SRGM-1	588.4	0.0481	1.4973	-----	----
SRGM-2	431.2	0.1386	0.1386	0.7313	----
SRGM-3	605.9	0.0458	3.4746	-----	----
SRGM-4	387.6	0.3640	0.2109	0.8067	----
SRGM-5 (proposed)	383.7	0.1668	3.0755	-----	2.274
SRGM-6 (proposed)	423.2	0.1761	0.1186	0.7103	0.353
SRGM-7 (proposed)	393.9	0.1542	4.3905	-----	1.926
SRGM-8 (proposed)	388.3	0.3694	0.20904	0.8051	0.0430

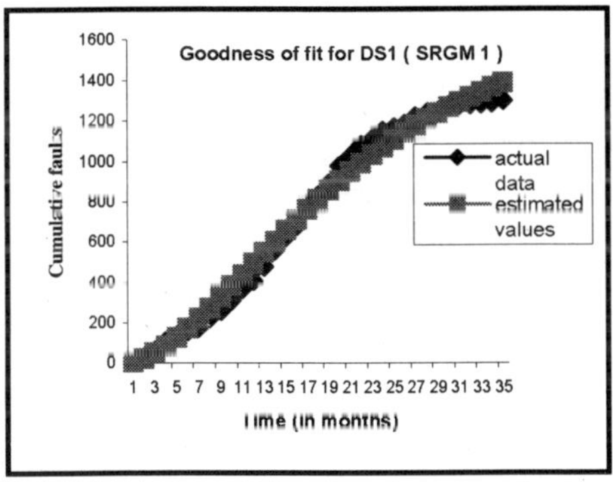

Fig. 2: Goodness-of-fit for SRGM1

Table 3 For DS-1

Models	R^2	MSE	AIC
SRGM – 1	0.9872	2713.22	506.91
SRGM – 5 (Proposed)	0.9959	862.43	418.34
SRGM – 2	0.9734	206.87	512.16
SRGM – 6 (Proposed)	0.9989	203.39	343.56
SRGM – 3	0.9842	1228.19	475.58
SRGM – 7 (Proposed)	0.9986	284.76	339.99
SRGM – 4	0.9943	284.86	451.24
SRGM – 8 (Proposed)	0.9987	284.69	405.47

Table 4 For DS-2

Models	R^2	MSE	AIC
SRGM – 1	0.98822	121.58	219.99
SRGM – 5 (Proposed)	0.99103	121.49	208.79
SRGM – 2	0.99055	97.55	215.99
SRGM – 6 (Proposed)	0.99074	96.33	217.01
SRGM – 3	0.98815	121.67	220.03
SRGM – 7 (Proposed)	0.99066	96.41	210.46
SRGM – 4	0.99146	99.86	213.20
SRGM – 8 Proposed)	0.99244	99.69	215.08

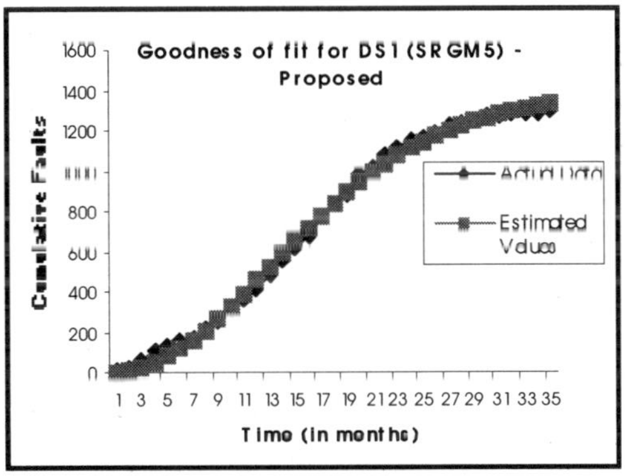

Fig. 3: Goodness-of-fit for SRGM5

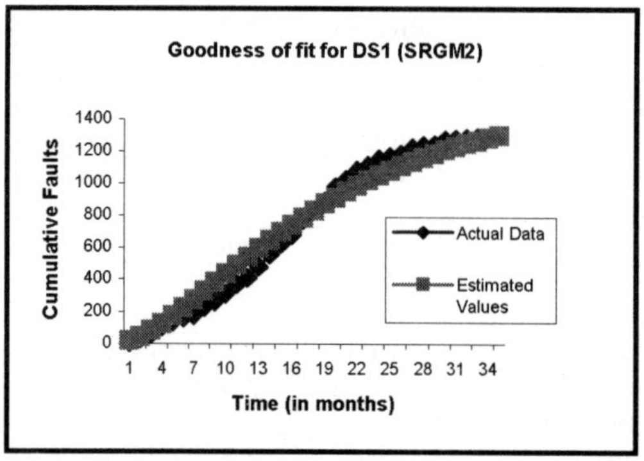

Fig. 4: Goodness-of-fit for SRGM2

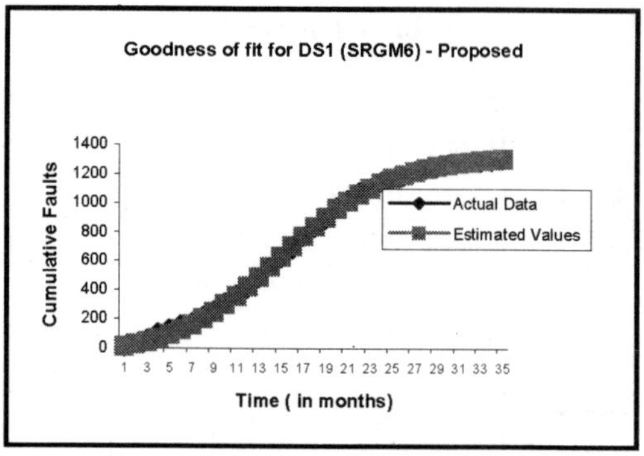

Fig. 5: Goodness-of-fit for SRGM6

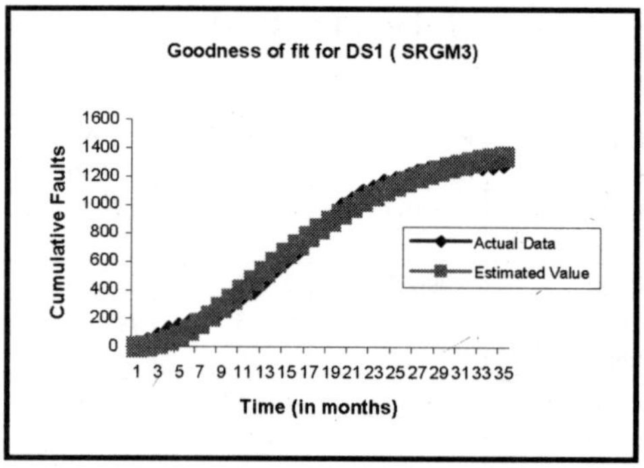

Fig. 6: Goodness-of-fit for SRGM3

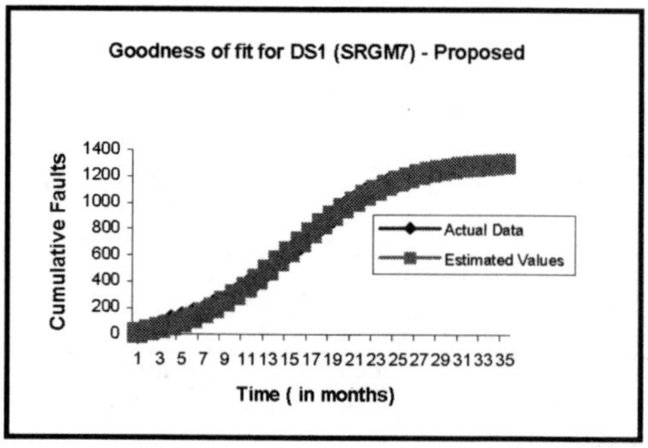

Fig. 7: Goodness-of-fit for SRGM7

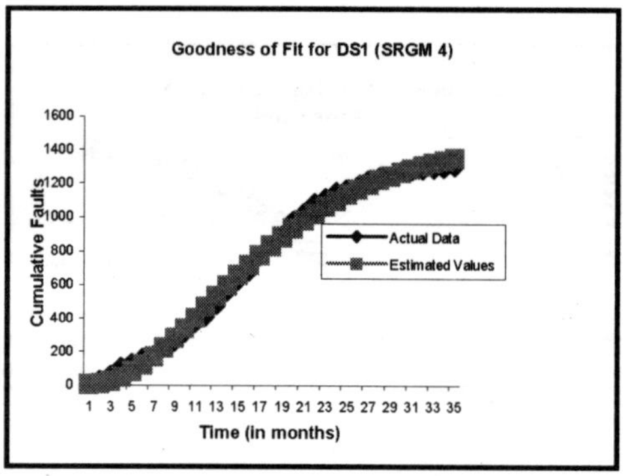

Fig. 8: Goodness-of-fit for SRGM4

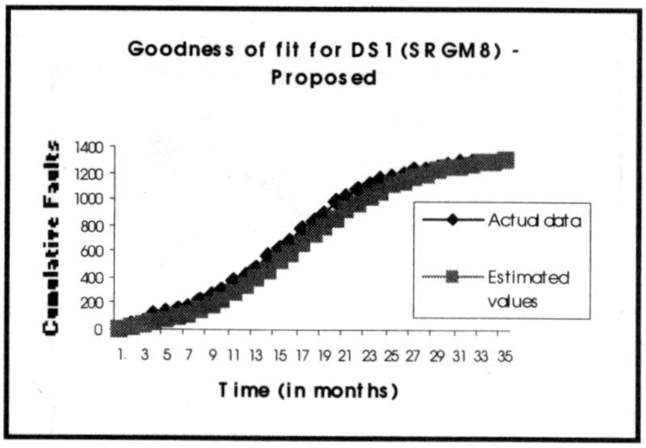

Fig. 9: Goodness-of-fit for SRGM8

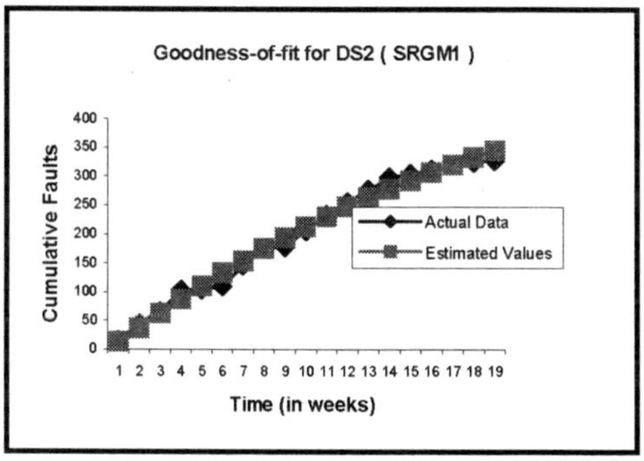

Fig. 10: Goodness-of-fit for SRGM1

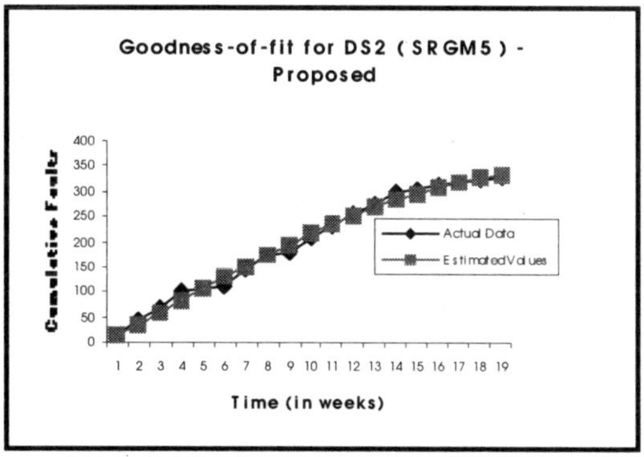

Fig.11: Goodness-of-fit for SRGM5

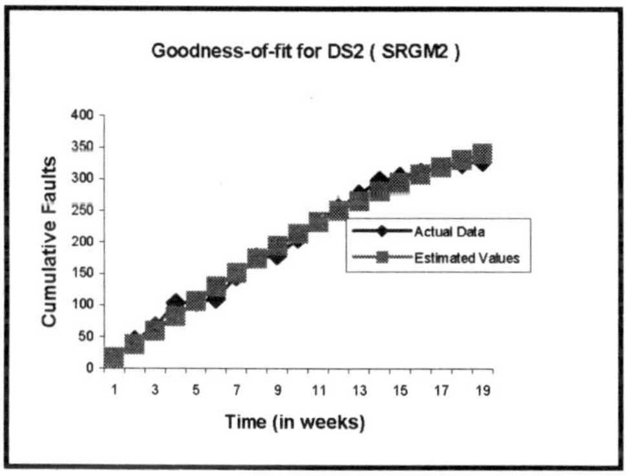

Fig. 12: Goodness-of-fit for SRGM2

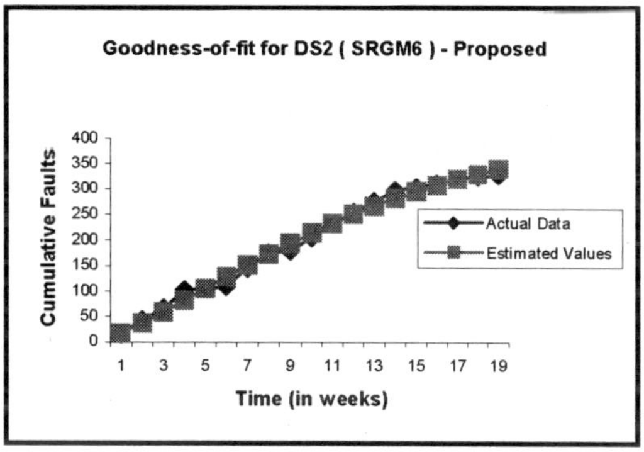

Fig.13: Goodness-of-fit for SRGM6

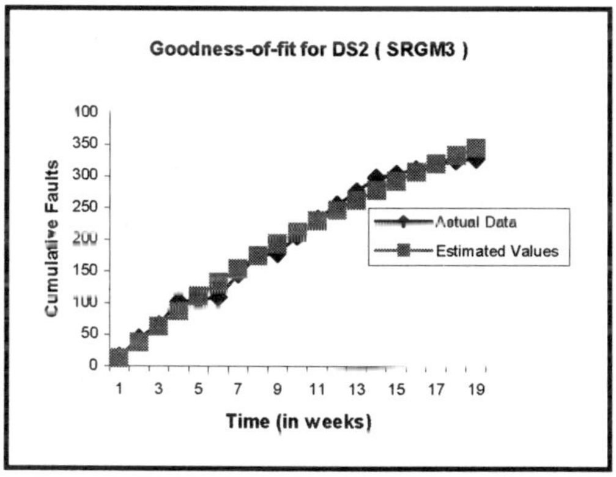

Fig. 14: Goodness-of-fit for SRGM3

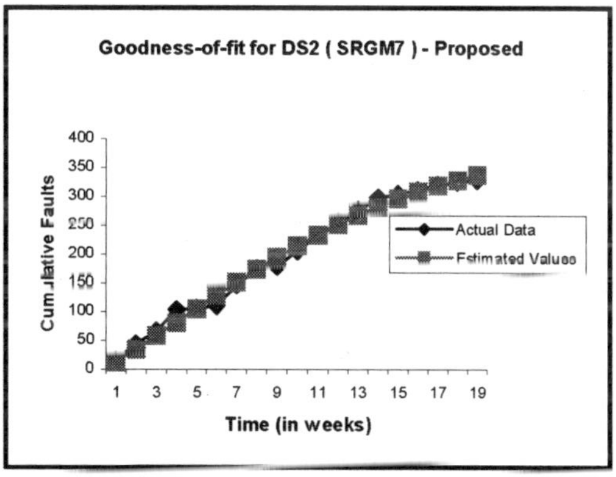

Fig.15: Goodness-of-fit for SRGM7

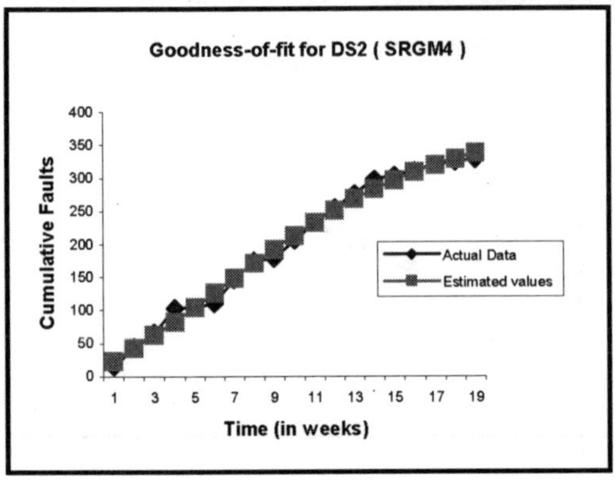

Fig. 16: Goodness-of-fit for SRGM4

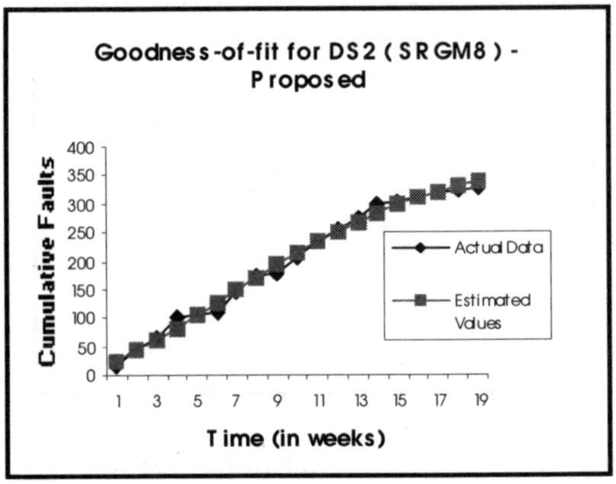

Fig.17: Goodness-of-fit for SRGM8

7. REFERENCES

1. S.Yamda, H. Ohtera and M.Ohba , "*Testing-Domain Dependent Software Reliability Models*", (Pergamon Press 1992).
2. S.Yamda and T.Fujiwara, "Testing - Domain Dependent Software Reliability Growth Models and their Comparison of Goodness–of–fit" (World Scientific Publishing Company, 2001).
3. P. K. Kapur and A. K. Verma, "Quality, Reliability and Information Technology, Trends and Future Directions", (Narosa Publications, 2005).
4. P. K. Kapur,"Recent Developments in Quality Reliability and Information Technology", (IMH, 2004).
5. T. Fujiwara and S. Yamada, "Software Reliabilty Growth Modeliing based on Testing-Skill Characteristics : Model and Application", (Electronics and Communications in Japan, Part 3, Vol. 84, No. 6, 2001).
6. P. K. Kapur and R. B. Garg, "A Software Reliability growth Model for an Error Removal Phenomenon", (Software Engineering Journal 1992; Vol. 7,pp. 291-294).
7. P. K. Kapur, R. B. Garg and S. Kumar, "Contributions to Hardware and Software Reliability", (1999, Singapore:World Scientific Publishing Co. Ltd.).
8. P. K. Kapur, D. N. Goswami and A. Bardhan, "A Flexible Delayed S-shaped Software Reliability Growth Model", (Communicated to the Journal of Computer Society of India, 2003).
9. Roger S. Pressman, "Software Engineering-A Practitioneer's Approach", (Fourth Edition, McGraw Hill).
10. S.Yamda, "Software Reliability Models Fundamentals and Application" (JUSE Press, Tokyo , 1994).
11. A. L. Goel and K. Okumoto, "A time-dependent error detection-rate model for a large scale software system", (Proc. 3rd USA-Japan Computer Conference, San Francisco, CA 1978).
12. SL Pfleeger , "Software Engineering-Theory and Practice", (New Jersey: Prentice Hall,1998).
13. H. Ohtera, S. Yamada and H. Narihisa, "An optimal release problem based on a testing-effort dependent software reliability model", (Trans. IEICE of Japan, E71(11), 1988).
14. S. Yamada and S. Osaki, "Software reliability growth modellling : Models and applications", (IEEE Trans. Software Engineering, SE-11(12), 1985).
15. M. Ohba, "Software Reliability analysis models", (IBM J. Research Development 28(4), 1984).
16. W. D. Brooks and R. W. Motley, "Analysis of discrete software reliability models", (IBM Corporation, Final Technical Report,RADC-TR-80-84, Rome Air Development Center, New York, 1980).

Reliability, Safety and Hazard
Editors: P.V. Varde et al.

FLEXIBLE TESTING EFFORT DEPENDENT SOFTWARE RELIABILITY GROWTH MODELLING WITH TESTING COVERAGE FOR DISTRIBUTED SYSTEMS

P.K. Kapur, D.N. Goswami[1] and Archana Kumar

Department of Operational Research, University of Delhi, Delhi 110 007
[1]S.O.S in Computer Science, Jiwaji University, Gwalior 474 011

ABSTRACT

The paper analyzes and models the software development process and presents the field experience for distributed systems. Testing coverage is the measure to evaluate the quality of testing and tested software products. Software testing is a critical element of software quality assurance and represents the ultimate review of specification, design and coding. This paper proposes a software reliability model for distributed systems that incorporates testing effort with testing coverage information. The model is based on Non Homogenous Poisson Process (NHPP) and can be used to estimate and predict the reliability of the software products. The goodness-of-fit of the proposed model is compared with other existing NHPP models. The new proposed model provides the significant improved goodness-of-fit results.

Keywords: Software Reliability, Non Homogenous Poisson Process (NHPP), Testing-Coverage, Distributed Development Environment, Fault Severity

1. INTRODUCTION

Software reliability engineering is a fast growing field. More than 60% of critical applications are dependent on software. The complexity of business software application is also increasing. Customers need products with high performance that can be sustained over time. Due to high cost of fixing failures, safety concerns, and legal liabilities organizations need to produce software that is reliable. There are several methodologies to develop software but questions that need to be addressed are how many times will the software fail and when, how to estimate testing effort, when to stop testing and when to release the software. Also, for a software product we need to predict / estimate the maintenance effort; for example, how long must the warranty period must be, once the software is released, how many defects can be expected at what severity levels, how many engineers are required to support the product, for how long etc. Software reliability engineering addresses all these issues, from design to testing to maintenance phases. Software Reliability is defined as the probability of failure free software operation in a defined environment for a specified period of time [11]. Software Reliability Engineering (SRE) is a standard, proven best practice that makes testing more reliable, faster, and cheaper. A failure is the departure of software behavior from user requirements. Since software does not deprecate like hardware, the reliability of software stays constant over time if no changes are made to the code or to the environmental conditions including the user behavior.

Testing is the process of executing a program with the intent of finding an error [14].

Testing process focuses on the logical internals of the software, assuring that all statements have been tested, and on the functional externals that is conducting tests to uncover errors and ensure that defined input will produce actual results that agree with required results. Although it sometimes seems that way, the purpose of testing is not to use up all the remaining budget or schedule resources at the end of a development effort. The goal of testing is to ensure that the software performs as intended, and to improve software quality, reliability and maintainability. Testing Coverage is the measure to evaluate the quality of testing and tested software products. According to IEEE definition Test Coverage is the degree or extent to which a given component meet in order to pass a given test [12]. It is the process of finding areas of a program not exercised by a set of test cases, Creating additional test cases to increase coverage, and to determine the quantitative measure of test coverage, which is an indirect measure of quality. Testing coverage analysis assures the quality of the set of tests, not the quality of the actual product.

Many Software Reliability Growth Models (SRGMs) [4] have been proven to be successful in estimating the software reliability and the number of errors remaining in the software. It has been observed that the relationship between the testing time and the corresponding number of faults removed is either Exponential or S-Shaped or the mix of the two [2,5,8,9,13]. The actual removal of a fault is done after a failure is reported and corresponding fault is isolated. In the first phase, the failure identification team isolates a failure. In the second phase another team primarily consisting of programmers removes the fault causing that failure. Kapur et al [8,9] proposed an SRGM with three types of fault. For each type, the Fault Removal Rate per remaining faults is assumed to be time independent. The first type is modeled by an Exponential model of Goel and Okumoto [5]. The second type is modeled by Delayed S-shaped model of Yamada et al. [19]. The third type is modeled by three stages Erlang model proposed by Khoshogoftaar [16]. The total removal phenomenon is again modeled by the superposition of the three SRGMs. Later they extended their model to cater for more types of faults [8].

In this paper we proposes a software reliability model for distributed systems [21] that incorporates testing effort with testing coverage information. The model is based on Non Homogenous Poisson Process (NHPP) and can be used to estimate and predict the reliability of the software products. The goodness-of-fit of the proposed model is compared with other existing NHPP models. The new proposed model provides the significant improved goodness-of-fit results. Section 2 derives the proposed models in distributed environment which are more flexible. Sections 3, 4 and 5 give the method used for parameter estimation and criteria used for validation and evaluation of the proposed model. We conclude this paper in section 6.

2. SOFTWARE RELIABILITY GROWTH MODELLING

Yamada et al. [21] have proposed an SRGM based upon NHPP for a distributed software system consisting of n reused and m newly developed software components assuming that cumulative number of detected faults is an exponential curve when a software system consists of several used software components; while on the other hand, the cumulative number of faults is described by an S-shaped growth curve when the newly developed software components are used. Under this assumption the SRGM based on NHPP for a distributed development environment with testing effort can be formulated as

$$H(t)=a\left[\sum_{i=1}^{n} p_i\left(1-e^{-b_i W(t)}\right)+\sum_{j=1}^{m} p_{n+j}\left(1-\left(1+b_{n+j}W(t)\right)e^{-b_{n+j}.W(t)}\right)\right] \quad (1)$$

Where

- a Expected number of initial inherent faults
- b_i Software failure rate per inherent fault for the ith software module
- p_i Weight parameter which expresses the proportion of the total testing load for the software components and $\left(\sum p_i = 1, p_i > 0\right)$

2.1 Framework for Modelling

The NHPP models are based on the assumption that 'Software failures occur at random times during testing caused by faults lying dormant in the software'. NHPP is used to describe the failure phenomenon during the testing phase. The counting process $\{N(t), t \geq 0\}$ of an NHPP process is given as follows.

$$\Pr.\ \{N(t)=k\}=\frac{\{m(t)\}^k}{k!}e^{-m(t)},\ k = 0,1,2, \quad (2)$$

and $$m(t)=\int_0^t \lambda(x)dx$$

The intensity function $\lambda(x)$ (or the mean value function $m(t)$) is the basic building block of all the NHPP models existing in the software reliability engineering literature. In models proposed by Yamada et al. [20] and Trachtenberg [15], the effect of intensity of testing effort on the failure phenomenon has been studied. Faults if present in the software are exposed when the software is run. A general framework for model development has been proposed.

2.2 Model Notations:

m, $m(t)$ Expected number of faults identified in the time interval (0,*t*] during testing phase

e, $e(t)$ Expected number of instructions executed on the software in the time interval (0,t]

W, $W(t)$ Cumulative testing effort in the time interval (0,t]

$$\frac{d}{dt}W(t) = w(t)$$

a Constant, representing the total number of faults lying dormant in the Software at the beginning of testing.

a_i Initial fault content of type *i* reused component

a_j Initial fault content of type *j* newly developed component with hard faults.

a_k Initial fault content of type *k* newly developed component with complex faults.

b Constant; failure rate

b_i Proportionality constant failure rate/ fault isolation rate rate per fault of ith reused component

b_j Proportionality constant failure rate / fault isolation rate per fault of jth newly developed component

b_k Proportionality constant failure rate / fault isolation rate per fault of kth newly developed component

β Constant

d_i Constants

$i = 1, 2$

2.3 Model Assumptions

1. The Non-homogeneous Poisson Process (NHPP) can describe software failure phenomenon.
2. The number of failures during testing is dependent upon the number of faults remaining in the software at that time. It is also dependent upon the rate of testing coverage.
3. Testing coverage increases due to testing effort.
4. As soon as a failure occurs the fault causing that failure is immediately identified. Identified faults are removed perfectly and no additional faults are introduced during the process.
5. The number of instructions executed is a function of testing effort / number of users.
6. Testing effort / number of users is a function of time.

Using the above assumptions the failure phenomenon can be described with respect to time as follows:

$$\frac{dm}{dt}=\frac{dm}{de}\frac{de}{dW}\frac{dW}{dt} \quad (3)$$

We discuss below each component on the right hand side of the above expression individually.

Component-1

The rate at which failures occur depends upon the number of faults remaining in the software [5,8]. As the coverage of the software is increased more faults are removed. The rate at which additional faults are identified is directly dependent upon the rate at which software is covered through additional test cases being run. It is also dependent upon the size of the uncovered portion of the software. Based on these facts the differential equation for fault identification / removal can be written as:

$$\frac{dm}{de} = d_1 \frac{c'}{1-c}(a-m) \quad (4)$$

c : percentage of the software that has been covered through testing.

$1-c$: percentage of the software that has not covered that has not been through testing.

c' : the coverage rate.

Component-2

The second component of expression (3) relates the number of instructions executed with the testing effort or the number of users of the software. For the sake of simplicity we assume it to be constant.

$$\frac{de}{dW} = d_2 \quad (5)$$

Substituting (3) and (4) in (2) we get

$$\frac{dm}{dt} = d_1 \frac{c'}{1-c}(a-m)d_2 \frac{dW}{dt} \quad (6)$$

The testing coverage function is non-negative and non-decreasing function. For logistic fault removal rate, let

$$c = \frac{1-(1+bW)e^{-bW}}{1+\beta e^{-bW}} \quad (7)$$

Therefore, $\frac{c'}{1-c} = \frac{b(bW+\beta-\beta e^{-bW})}{(1+\beta e^{-bW})(1+\beta+bW)}$ (8)

Substituting the value of $\frac{c'}{1-c}$ in (6) and assuming $d = d_1.d_2$ we get

$$\frac{dm/dt}{dW/dt} = \frac{d.b(bW+\beta-\beta e^{-bW})(a-m)}{(1+\beta e^{-bW})(1+\beta+bW)} \quad (9)$$

Solving equation (9) with initial conditions $m(0)=0$ and $W(0)=0$ we get

$$m(W(t)) = \frac{a((1+\beta e^{-bW(t)})^d - (1+\beta+bW(t))^d e^{-bdW(t)})}{(1+\beta e^{-bW(t)})^d} \quad (10)$$

Let this be Flexible Delayed S-Shaped Model with Testing Coverage.

2.4 Modelling Testing Effort

As the time increases, testing effort also increases. The testing effort rate is proportional to the testing resources available.

$$\frac{dW(t)}{dt} = v(t)[\alpha - W(t)] \quad (11)$$

Where $c(t)$ is the time dependent rate at which testing resources are consumed, with respect to remaining available resources. Solving equation (11) under the initial condition $W(0) = 0$

$$W(t) = \alpha\left[1 - \exp\left\{\int_0^t v(x)dx\right\}\right] \quad (12)$$

When $v(t) = v$, a constant

$$W(t) = \alpha(1-e^{-vt}) \quad (13)$$

If $v(t) = v.t$, (11) gives Rayleigh type curve

$$W(t) = \alpha(1-e^{-vt^2/2}) \quad (14)$$

Huang et al. [6] developed an SRGM, based upon NHPP with a logistic testing effort function. A logistic curve model is described as $\frac{dW(t)}{dt} = \frac{l}{\alpha}W(t)(\alpha - W(t))$ .The cumulative testing effort consumed in the interval $(0,t]$ has the following form

$$W(t) = \frac{\alpha}{1+ve^{-lt}} \quad (15)$$

Where p, r and l are constants. SRGM with logistic testing effort function provides better result on some failure data sets.

Yamada et al. [19] described the time dependent behavior of testing effort expenditure by a Weibull curve while proposing a SRGM.

$$W(t) = \alpha(1-e^{-vt^l}) \quad (16)$$

Exponential and Rayleigh curves become special cases of the Weibull curve for l = 1 and 1 = 2 respectively.

2.5 Modelling Total Fault Removal Phenomenon

The proposed model for distribution environment is the superposition of the Flexible

Delayed S-Shaped Model with Testing Coverage (10), Flexible Delayed S-Shaped Model [7] and Generalised Erlang Model with logistic function [10]. Thus, the mean value function of superposed NHPP is:

$$m(w(t)) = \sum_{i=1}^{p} m_{ir}(t) + \sum_{j=p+1}^{p+q} m_{jr}(t) + \sum_{k=p+q+1}^{p+q+s} m_{kr}(t) \tag{17}$$

$$m(u(t)) = \sum_{i=1}^{p} \left(\frac{a_i\left(\left(1+\beta e^{-b_i W(t)}\right)^{d_i} - \left(1+\beta+b_i W(t)\right)^{d_i} e^{-b_i d_i W(t)}\right)}{\left(1+\beta e^{-b_i W(t)}\right)^{d_i}} \right) + \sum_{j=p+1}^{p+q} \left(\frac{a_j\left(1-\left(1+b_j.W(t)\right)e^{-b_j.W(t)}\right)}{1+\beta e^{-b_j t}} \right) + \sum_{k=p+q+1}^{p+q+s} \left(\frac{a_k\left(1-\left(1+b_k.W(t)+\frac{b_k^2 W^2(t)}{2}\right)e^{-b_K.W(t)}\right)}{1+\beta e^{-b_k t}} \right) \tag{18}$$

Case 1:

Further, We consider the simple case where software system consists of one reused (i.e. p=1) and two newly developed components (one of which contains errors represented by a two stage error removal phenomenon while other contains errors represented by a three stage error removal phenomenon i.e. p=1, q = 1 and s = 1 respectively). Accordingly equation (19) can be rewritten as

$$m(w(t)) = \sum_{i=1}^{1} \left(\frac{a_i\left(\left(1+\beta e^{-b_i W(t)}\right)^{d_i} - \left(1+\beta+b_i W(t)\right)^{d_i} e^{-b_i d_i W(t)}\right)}{\left(1+\beta e^{-b_i W(t)}\right)^{d_i}} \right) + \sum_{j=2}^{2} \left(\frac{a_j\left(1-\left(1+b_j.W(t)\right)e^{-b_j.W(t)}\right)}{1+\beta e^{-b_j t}} \right) + \sum_{k=3}^{3} \left(\frac{a_k\left(1-\left(1+b_k.W(t)+\frac{b_k^2 W^2(t)}{2}\right)e^{-b_K.W(t)}\right)}{1+\beta e^{-b_k t}} \right) \tag{19}$$

Where $a_1 + a_2 + a_3 = a$. Call this expression as 1-1-1.

3. PARAMETER ESTIMATION

The success of mathematical modeling approach to reliability evaluation depends heavily upon quality of failure data collected. Hence efforts should be made to make the data collection more explicit and scientific. Usually data is collected in one of the following two ways. In the first case the times between successive failures are recorded. Though this type of data collection is more desirable, it may not be simple. The other easier and commonly collected data type is known as the grouped data. Here testing intervals are specified and number of failures experienced during each such interval is noted. For both these data types method of least squares and method of maximum likelihood have been suggested and widely used for estimation of parameters of SRGMs.

4. COMPARISON CRITERIA FOR SRGMS

The performance of SRGMs are judged by their ability to fit the past software fault data [11].

4.1 Goodness of Fit criteria

a. ***The Mean Square Fitting Error (MSE) [8]:*** The model under comparison is used to simulate the fault data, the difference between the expected values, $\hat{m}(t_i)$ and the observed data x_i is measured by MSE as follows.

$$MSE = \sum_{i=1}^{k} \frac{(\hat{m}(t_i) - x_i)^2}{k} \tag{20}$$

Where k is the number of observations. The lower MSE indicates less fitting error, thus better goodness of fit.

b. The Akaike Information Criterion (AIC) [1,8]:

It is defined as AIC = -2(The value of the maximum log likelihood function) + 2(The number of the parameters used in the model).

This index takes into account both the statistical goodness of fit and the number of parameters that are estimated. Lower values of AIC indicate that model is good.

c. Coefficient of Multiple Determination (R^2)[8]:

We define this coefficient as the ratio of the sum of squares resulting from the trend model to that from constant model subtracted from 1, that is R^2=1 –residual SS/corrected SS. R^2 measures the percentage of the total variation about the mean accounted for the fitted curve. It ranges in value from 0 to 1. Small values indicate that the model does not fit the data well. The larger R^2, the better the model explains the variation in the data.

5. MODEL VALIDATION

To check the validity of the proposed model and to find out its software reliability growth, it has been tested on two Data Sets. SRGMs obtained through Exponential, Rayleigh, Logistic and Weibull testing effort function has been validated on the datasets given below.

Data Set 1

This data is cited from Wood [17]. The software was tested for 20 weeks during which 10000 computer hours were used and 100 faults were removed. The estimation results for Exponential, Rayleigh, Logistic and Weibull function are given in table 1.Based upon the estimated parameters, parameters of SRGM (10,19) were estimated. The proposed model in (10) was compared with Flexible Delayed S-Shaped Model [7]. The Parameter Estimation and the goodness of fit results are given in table 2. The Parameter Estimation result and the goodness of fit results for the distributed combination (1-1-1) are given in table 4.

Data Set 2

This data is cited from Brooks and Motley (1980) [3]. The fault data set is for a radar system of size 124 KLOC (Kilo Lines of Code) tested for 35 weeks in which 1301 faults were removed. The estimation results for Exponential, Rayleigh, Logistic and Weibull function are given in table 1. Based upon the estimated parameters, parameters of SRGM (10,19) were estimated. The proposed model in (10) was compared with Flexible Delayed S-Shaped Model [7]. The Parameter Estimation and the goodness of fit results are given in table 3. The Parameter Estimation result and the goodness of fit results for the distributed combination (1-1-1) are given in table 5.

Different distributed combinations for the proposed model were tried. The Proposed model in combination with Flexible Delayed S-Shaped Model [7] and Generalised Erlang Model with logistic function [10] gives better result as compared to other functions. It can be observed from the estimation results that the proposed model (1-1-1) given by (19) gives the best estimates in terms of goodness of fit. The proposed model is highly S-Shaped; therefore it fails to give good results for exponential function. For other Functions like Rayleigh, Logistic and Weibull the results are fairly encouraging. The logistic function gives the best results for the proposed model for both the datasets. In case of distribution environment, the Logistic function gives the best result for the combination 1-1-1 for the dataset DS-I and the Weibull function gives the best result for the combination 1-1-1 for the dataset DS-II.

Table 1

Datasets	Testing Effort Function	Parameter Estimation		
		α	v	l
DS-I	Exponential function	35386	.0172	-
	Rayleigh function	10000	.0175	-
	Logistic function	9974	13	.288
	Weibull function	11711	.0235	1.46
DS-II	Exponential function	1020677	.000047	-
	Rayleigh function	2873	.00173	-
	Logistic function	2067	38.64	.161
	Weibull function	2670	.000773	2.069

Table 2 for DS-I

Models under Comparisons	Parameter Estimation and Comparison Criteria				
		Testing Effort Function			
	Parameters	Exponential	Rayleigh	Logistic	Weibull
Flexible Delayed S-Shaped SRGM	a	107	98	101	103
	b	.0004	.0005	.0005	.0004
	β	.0001	.0001	.0001	.0001
	R^2	.97114	.87961	.96728	.93614
	MSE	23.46	97.86	26.59	51.91
	AIC	112.46	240.60	93.02	157.34
Proposed Model	a	104	112	132	103
	b	.6666	.0212	.9970	.6946
	β	.00001	.00001	.1	.00001
	d	.00035	.01035	.00015	.00035
	R^2	.94058	.94031	.99754	.94933
	MSE	48.30	48.52	1.99	41.91
	AIC	97.35	153.38	80.57	107.04

Table 3 for DS-II

Models under Comparisons	Parameter Estimation and Comparison Criteria				
		Testing Effort Function			
	Parameters	Exponential	Rayleigh	Logistic	Weibull
Flexible Delayed S-Shaped SRGM	a	1361	1303	1282	1298
	b	.0036	.0032	.0035	.0035
	β	7.38	.0001	.00001	.00001
	R^2	.99696	.98867	.99519	.98726
	MSE	647.43	2412.5	1023.5	2713.5
	AIC	395.27	723.37	452.77	770.25
Proposed Model	a	1660	1363	1334	1377
	b	.1411	.0120	.0124	.0168
	β	.00001	.00001	.1	.00001
	d	.00035	.01025	.00017	.00035
	R^2	.98929	.99117	.99810	.99114
	MSE	2281.1	1880.8	404.4	403.1
	AIC	693.48	698.41	403.16	729.95

Table 4 for DS-I

Models under Comparisons	Parameter Estimation and Comparison Criteria				
		Testing Effort Function			
	Parameters	Exponential	Rayleigh	Logistic	Weibull
Proposed Model With combination (1-1-1)	a	132	182	132	184
	b_1	.44	.44	.44	.44
	b_2	.55	.55	.83	.55
	b_3	.994	.492	.996	.523
	β	1.037	.1	.1	1.034
	d	.00015	.00045	.00015	.00045
	R^2	.98238	.99730	.99755	.99296
	MSE	14.32	2.19	1.99	5.72
	AIC	103.29	92.66	88.59	96.68

Table 5 for DS-II

Models under Comparisons	Parameter Estimation and Comparison Criteria				
		Testing Effort Function			
	Parameters	Exponential	Rayleigh	Logistic	Weibull
Proposed Model with combination (1-1-1)	a	1557	1317	1372	1318
	b_1	.0023	.0087	.4400	.0089
	b_2	.9300	.00399	.8300	.00390
	b_3	.0079	.3898	.0172	.4350
	β	1.037	1.037	1.027	1.055
	d	.1055	.1051	.1045	.2065
	R^2	.99094	.99972	.99914	.99973
	MSE	1929	59.34	183.60	58.37
	AIC	470.94	315.38	415.81	316.01

6. CONCLUSION

In this paper, we have compared Software Reliability Model that incorporates testing effort with testing coverage information with Flexible Delayed S-Shaped Software Reliability Growth Model [7]. The model is based on Non Homogenous Poisson Process (NHPP). The distributed combinations of the proposed model with Flexible Delayed S-Shaped Model [7] and Generalised Erlang's Three Stage Model [10] with logistic function has been developed and it is found out that the total fault removal phenomena of the proposed 1-1-1 model is better when compared with other Software Reliability Growth Model based on Non Homogenous Poisson Process (NHPP) for testing phase. The new proposed model provides the significant improved goodness-of-fit results. The results of the proposed model are fairly encouraging when compared with other model developed under similar environment.

7. REFERENCES

1. Akaike H., A New Look at Statistical Model Identification-IEEE Transactions on Automatic Control 1974; 19:716-723.

2. Bittanti S., P. Bolzern, E. Pedrotti and R. Scattolini; (1998), A Flexible Modeling Approach for Software Reliability Growth, in G. Goos and J. Harmanis, eds.: Software Reliability Modelling and Identification, Springer Verlag, Berlin, 101-140.
3. Brooks WD and Motley RW; Analysis of Discrete Software Reliability Models-Technical Report (RADC-TR-80-84)(1980) New York: Room Air Development Center.
4. Downs T and Scott A.; Evaluating The Performance of Software Reliability Models, IEEE Transactions on Reliability (1992) 41(4): 532-538.
5. Goel AL and Okumoto K.; Time Dependent Error Detection Rate Model for Software Reliability and Other Performance Measure IEEE Transactions on Reliability (1979); R-28 (3): 206-211.
6. Huang, C-Y.; Kuo, S-Y and Chen J.Y.; (November 1997) Analysis of a Software Reliability Growth Model with Logistic Testing Effort Function, Proc. 8th International Symposium on Software Reliability Engineering; pp 378-388.
7. Kapur P.K., Goswami D.N., Bardhan A.K.;(2003), A Flexible Delayed S-Shaped Software Reliability Growth Model, Communicated to the Journal of Computer Society of India.
8. Kapur P.K., R.B. Garg and S. Kumar; (1999), Contributions to Hardware and Software Reliability, World Scientific, Singapore.
9. Kapur PK and R.B. Garg; (1992), A Software Reliability Growth Model for an Error Removal Phenomenon, Software Engineering Journal 7, 291-294.
10. Kapur P.K. Younes S and Agarwala S.; (1995), Generalised Erlang Model with n types of faults, CSI Bulletin.
11. Musa, J.D., Iannini, A.and Okumoto, K.;(1987) Software Reliability: Measurement, Prediction, Applications, McGraw Hill.
12. Nina S Godbole; Software Quality Assurance- Principles and Practice Reliability; Narosa
13. Ohba, M.; (1984), Software Reliability Analysis Models, IBM Journal of research and Development 28, 428-443.
14. Pressman Roger S., Software Engineering-A Practitioner's Approach- Fourth Edition, McGraw Hill.
15. Trachtenberg, M., (1990) A General Theory of Software Reliability Modeling, IEEE Trans. on Reliability; Vol 39(1), pp.92-96.
16. Woodcock TG and Khoshogoftaar TM; (1991), Software Reliability Model Selection: A case study, Proceedings of the International Symposium on Software Reliability Engineering, pp.183-191.
17. Wood, A., (1996) Predicting Software Reliability, IEEE Computers; Vol. 11, pp. 69-77.
18. Yamada S, Hishitani J. and Osaki S.; (1993), Software Reliability Growth Models with a Weibull Test Effort, A Model and application; IEEE Trans on Reliability; Vol R-42: pp. 19-23.
19. Yamada S, Ohba M, and Osaki S.; (1984), S-Shaped Software Reliability Growth Models and their Applications, IEEE Transactions on Reliability, 1984; R-33: 169-175.
20. Yamada, S., Ohtera, H. and Narihisa H., (1986) Software Reliability Growth Models with Testing-Effort, IEEE Trans. n Reliability; Vol R-35, pp. 19-23.
21. Yamada S, Tamura Y and Kimura M ;(2003), A Software Reliability Growth Model for a Distributed Development Environment-Electronics & Communication in Japan Part 3, Vol. 83, pp 1446-53.

Reliability, Safety and Hazard
Editors: P.V. Varde et al.

SOFTWARE RELIABILITY GROWTH MODEL WITH NUMBER OF INSTRUCTIONS EXECUTED DEPENDENT ON TESTING EFFORT

P.K. Kapur, Ompal Singh and Anu Gupta

Department of Operational Research, University of Delhi, Delhi 110 007

ABSTRACT

With the advances in science and information technology, software is used to run complex systems. To ensure the safe and risk free working of systems, software are tested and debugged. The latent faults must be found and removed through additional testing effort, if the safety of the system is not to be compromised. In the practical software development, the number of remaining faults judges the quality and safety of the software. To ensure that number of remaining faults lie within the pre-specified safety value, efforts are assigned for software testing. In this paper, we propose a new methodology for software reliability modelling which relates the fault removal rate to the number of instructions executed and testing effort employed. From real life studies, it has been observed that more and more faults are detected as testing grows and as additional (testing) effort are used. It is also assumed that number of instructions executed is a power function of testing effort. The applicability of this model is demonstrated through real software test datasets. The results show that the proposed SRGM with testing effort has quite good accurate prediction capabilities and gives results closer to real values. This SRGM can be quite useful for systems where safety is the prime concern.

1. INTRODUCTION

With the boom in Information technology sector, the computer software are getting their place everywhere ranging from critical infrastructures, such as space research programmes, transportation, defense, and telecommunication systems, to our daily life services such as education, health-care, banking, entertainment, home appliances etc. Given the potentially highly risky and costly impact of software failures for many of these applications, it is extremely important to ensure the accuracy and safety of the software and has sound methods to judge and measure the software reliability.

With the world of information technology changing at a staggering rate, the design and complexity of the software systems have increased manifold. Software reliability is the most important criterion for the selection of software products by the users. It is defined as the probability that the system will operate without any failure for a given time in a given environment. Various models have been developed to measure, estimate and predict the reliability

Software testing plays an important role in a software product life cycle. It is estimated that on the average around 50 percent of the software development effort is spent on testing. The objective of testing is to discover the residual design errors before delivery to the customer. The failure data during the testing process are taken down in order to estimate the software reliability.

Software Testing is the process of executing a program or system with the intent of finding errors. Regardless of the limitations, testing is an integral part in software development. It is broadly deployed in every phase in the software development cycle. Typically, more than 50% percent of the development time is spent in testing. Testing is basically performed to improve quality and verification and validation of the software.

There are various types of testing methods and testing techniques available, serving multiple purposes in different life cycle phases. Classified by purpose, software testing can be divided into: correctness testing, performance testing, reliability testing and security testing. Classified by life-cycle phase, software testing can be classified into the following categories: requirements phase testing, design phase testing, program phase testing, evaluating test results, installation phase testing, acceptance testing and maintenance testing. By scope, software testing can be categorized as follows: unit testing, component testing, integration testing, and system testing.

Correctness is the minimum requirement of software, the essential purpose of testing. Under Correctness testing, the tester may or may not know the inside details of the software module under test, e.g. control flow, data flow, etc. Therefore, either a white-box testing or black-box testing is considered.

Software reliability refers to the probability of failure-free operation of a system. Guided by the operational profile, software testing (usually black-box testing) is used to obtain failure data, and an estimation model is then used to analyze the data to estimate the present reliability and predict future reliability. Therefore, based on the estimation, the developers can decide whether to release the software & when, and the users can decide whether to adopt & use the software.

It has been observed that there is a very close relationship between reliability of software and the amount of testing it has been subjected to. The resources, which are involved during testing of any software basically, include manpower used for fault detection and subsequent fault removal and CPU time spent in executing software under test. Greater the amount of testing effort faster is the testing process. Many SRGMs have been proposed but few explicitly study the impact of software testing on its reliability. In this paper we present the relationship between software reliability growth modelling and the number of instructions executed dependent on power function of testing effort The results show that the proposed SRGM with testing effort has quite good accurate prediction capabilities and gives results closer to real values. This SRGM can be quite useful for systems where safety is the prime concern.

2. NOTATIONS

m, $m(t)$: Expected number of faults identified in the time interval $(0, t]$

e, $e(t)$: Expected number of instructions executed on the software in the time interval $(0, t]$

W, $W(t)$: Cumulative testing effort in the time interval $(0, t]$

a: Constant, representing the number of faults lying dormant in the software at the beginning of testing.

α, β, δ, γ, l, r, k: Constants

k_i, i = 1,2,3: Constants

3. FRAMEWORK FOR MODELLING

The SRGM presented in this paper is based upon Non Homogenous Poisson Process (NHPP). The NHPP models are based on the assumption that the software sy^tem is subject to failures at random times caused by manifestation of remaining faults in the system.

The proposed model is based upon the following basic assumptions.

1. Software failure phenomenon can be described by the Non-homogeneous Poisson Process (NHPP).
2. The number of failures during testing is dependent upon the number of faults remaining in the software as well as on the number of faults already identified. at that time
3. As soon as a failure occurs the fault causing that failure is immediately identified. Identified faults are removed perfectly and no additional faults are introduced during the process.
4. The number of instructions executed is a power function of testing effort
5. Testing effort is a function of time.

Using the above assumptions the failure phenomenon can be described with respect to time as follows:

$$\frac{dm}{dt} = \frac{dm}{de}\frac{de}{dW}\frac{dW}{dt} \quad (1)$$

We discuss below each component on the right hand side of the above expression individually.

Component-1

During testing instructions are executed on the software and the output is matched with the expected results. If there is any departure from the specifications then a failure is said to have occurred. Effort is made to identify and subsequently remove the cause of the failure. The rate at which failures occur depends upon the number of faults remaining in the software as well as the number of faults already identified [4]. Based on these assumptions the differential equation for fault identification / removal can be written as:

$$\frac{dm}{de} = (k_1 + k_2 \frac{m}{a})(a - m) \quad (2)$$

Here k_1 is the rate at which residual faults cause failure. It is a constant as each one of these faults has an equal probability of causing failure. k_2 is the rate at which additional faults are identified without them causing any failure which is again a constant.

Component-2

The second component of expression (1) relates the number of instructions executed with the testing effort employed for the software. In the SRGMs proposed earlier in the literature () this component has often been taken as a constant. In the SRGM proposed here we define it as a power function of testing effort:

$$\frac{de}{dW} = k_3 W^K \quad (3)$$

Here $k \geq 0$. It can be easily seen that if k = 0, then $\frac{de}{dW} = k_3$ (i.e. a constant)

This case is more general and flexible in nature in the sense that it represents different type of relationships between e and W depending on different values taken by k
Substituting (2) and (3) in (1) we get:

$$\frac{dm}{dt} = \left(k_1 + k_2 \frac{m}{a}\right)(a - m) k_3 W^k \frac{dW}{dt} \quad (4)$$

It is a first order differential equation. Solving it with the initial condition $m(0)$ = 0 and $W(0) = 0$, we get:

$$m(t) = a\left(\frac{1 - e^{-\delta \frac{W(t)^{k+1}}{k+1}}}{1 + \gamma e^{-\delta \frac{W(t)^{k+1}}{k+1}}}\right) \quad (5)$$

Here $\delta = k_3(k_1 + k_2)$ and $\gamma = (k_2 / k_1)$.

4. SPECIAL FEATURES OF THE PROPOSED SRGM

The solution obtained above has a very interesting feature of capturing the different type of variations shown by the curve of $m(t)$ depending upon different values of k. It can be easily shown that if $k=0$ & γ = 0 then the model simply takes Exponential form and if k = 0 & $\gamma \neq 0$ it reduces to Ohba type of form. If k=1 & γ = 0, it takes Rayleigh form while for any higher value of k with γ = 0 it takes Weibull form.
Another interesting aspect of the model is its alternate formulation.
If we define

$$\frac{dm}{dt} = \frac{\delta W^k}{1 + \gamma e^{-\delta \frac{W(t)^{k+1}}{k+1}}}(a - m) \quad (6)$$

Then solving it with the initial condition $m(t=0)=0$ and $W(t=0)=0$, we get the same result as obtained in (5)

Now we describe the mathematical models for testing effort of software.

5. MODELING TESTING EFFORT

The most frequently used function to explain the testing effort are-

- Exponential
- Rayleigh
- Logistic
- Weibull

The first two can be derived from the assumption that, "the testing effort rate is proportional to the testing resources available".

$$\frac{dW(t)}{dt} = c(t)[\alpha - W(t)] \quad (7)$$

Where $c(t)$ is the time dependent rate at which testing resources are consumed, with respect to remaining available resources. Solving equation (5.1.8) under the initial condition $W(0)=0$, we get

Case 1:

When $c(t)=c$, a constant

$$W(t) = \alpha\left(1 - e^{-ct}\right) \quad (8)$$

Case 2:

If $c(t)=c.t$, we get Rayleigh type curve

$$W(t) = \alpha\left(1 - e^{-ct^2/2}\right) \quad (9)$$

Huang et al. [3] developed an SRGM, based upon NHPP with a logistic testing effort function.

If we define

$$\frac{dW(t)}{dt} = l\ \frac{W(t)}{\alpha}[\alpha - W(t)]$$

On solving, the cumulative testing effort consumed in the interval $(0,t)$ is given by

$$W(t) = \frac{\alpha}{1+re^{-lt}}$$ where

$$W(0) = \frac{\alpha}{1+r}$$

Where α, r and l are constants.

Yamada et al. [10] described the time dependent behavior of testing effort expenditure by a Weibull curve while proposing a SRGM.

$$W(t) = \alpha\left(1 - e^{-\beta t^k}\right) \quad (10)$$

Exponential and Rayleigh curves become special cases of the Weibull curve for $k = 1$ and $k = 2$ respectively. To study the testing effort process, one of the above functions can be selected.

6. PARAMETER ESTIMATION:

For parameter estimation we have used the technique of least square, which consists of minimizing the sum of squares of the deviations between the actual and estimated values.

The Comparison Criteria used in this paper are:

6.1. The Mean Square Fitting Error (MSE)

Using the estimated values of the parameters, the estimated values for failure data are calculated. The difference between the estimated values, $m(t_i)$ and the observed values y_i is measured by MSE as follows

$$MSE = \frac{\sum_{i=1}^{k}(m(t_i) - y_i)^2}{k}$$

Where k is the number of observations. Lower value of MSE indicates less fitting error, thus better goodness of fit.

6.2. Coefficient of Multiple Determination (R^2)

This goodness of fit measure can be used to investigate whether a significant trend exists in the observed failure intensity. We define this coefficient as the ratio of the sum of squares resulting from the trend model to that from constant model subtracted from 1

$$R^2 = 1 - \frac{\text{residual SS}}{\text{residual SS}}$$

R^2 measures the percentage of the total variation about the mean accounted for the fitted curve. It ranges in value from 0 to 1. Small values indicate that the model does not fit the data well. The larger R^2, the better the model explains the variation in the data.

7. MODEL VALIDATION

To check the accuracy of the proposed SRGM with power logistic with respect to different type of testing effort function, parameter estimation on three real software failure datasets obtained during testing phase have been done. The comparison has been performed between exponential (Yamada), exponential with logistic (Kapur et. al.) and the proposed model

Data Set-1 (DS-1):

The data was obtained from Musa et. al.. [6]. The software was a real time command and control system, which was tested for 92 days (21 weeks). The delivered object instructions was 21,700 involving 9 programmers. 136 faults were removed during testing. The estimation results are given in Table-1 and Table-2.

Table 1

Testing Effort Funtion	Parameter Estimation		
	α	v	l
Exponential	35395	0.000020681	-
Rayleigh	22128	0.000004397	-
Logistic	45.15	0.24610954	148
Weibull	26.78	0.0000000103	6.39

Data Set – II (DS-II):

The data are cited from Brooks and Motley [1]. The fault data set is for a radar system of size 124 KLOC (kilo lines of code) tested for 35 months in which 1301 faults were identified. The estimation results are given in Table-3 and Table-4.

Table 2

Models under Comparison	Parameter Estimation	Type of Testing effort function			
Yamada Model		Exponential	Logistic	Rayleigh	Weibull
	a	5756	387.65	2825.6	150.56
	b	.00118	.01957	.00235	0.0858
	R^2	.73924	.97951	.95475	.99129
	MSE	618.16	48.56	107.28	20.62
Ohba Model	a	151.01	139.21	146.19	150.56
	b	0.5693	0.2344	0.2793	0.0858
	β	583.37	7.697	21.07	0
	R^2	.99716	.99724	.99717	.99129
	MSE	6.73	6.54	6.71	20.62
Proposed SRGM	a	143.52	140.37	142.14	150.56
	δ	0.093	0.0478	0.2771	0.0858
	γ	41.02	0.685	22.55	0
	k	0.782	0.5186	0.033	0
	R^2	.99750	.99762	.99748	.99129
	MSE	5.92	5.65	5.98	20.62

Table 3:

Testing Effort Function	Parameter Estimation		
	α	v	l
Exponential	1020677	.00005	-
Rayleigh	2873	.00173	-
Logistic	2067	38.64	161
Weibull	2670	.00077	2.07

Table 4

Models under Comparison	Parameter Estimation	Type of Testing effort function			
		Exponential	Logistic	Rayleigh	Weibull
Yamada Model	a	10708	1552	1560	1547
	b	.00008	.00109	.00107	.001095
	R^2	.95810	.99111	.99396	.99470
	MSE	8924.1	1892.3	1287.3	1129.2
Ohba Model	a	1331	1308	1348	1356
	b	.0042	.0028	.0024	.0022
	β	20.19	2.13	1.58	1.33
	R^2	.99904	.99816	.99776	.99755
	MSE	204	393	478	522
Proposed SRGM	a	1323	1312	1348	1356
	δ	.0012	.00102	.0024	.0022
	γ	10.639	.889	1.58	1.33
	k	.191	.1364	0	0
	R^2	.99909	.99859	.99776	.99755
	MSE	194.47	300.23	478	522

Data Set – III (DS-III):

This data is cited from Misra [9]. The software was tested for 38 weeks during which 2456.4 computer hours were used and 231 faults were removed. Based upon the estimated parameters, parameters of SRGM () were estimated. The estimation results are given in Table-5 and Table-6.

Table 5

Test effort Function	Parameter Estimation		
	α	v	l
Exponential	20237	0.003109	-
Rayleigh	2241	0.004035	-
Logistic	2836	10.49313	0.099
Weibull	52610000000	0.000000001	1.009

Table 6

Models under Comparison	Parameter Estimation	Type of Testing effort function			
		Exponential	Logistic	Rayleigh	Weibull
Yamada Model	a	560.5	447.2	305	458
	b	.00022	.00029	.0005	.00031
	R^2	.99464	.98865	.94508	.99442
	MSE	19.90	42.14	203.89	20.72
Ohba Model	a	240	299	305	458
	b	.0014	.00078	.0005	.00031
	β	2.17	.9082	0	0
	R^2	.98634	.98899	.94508	.99442
	MSE	50.71	40.87	203.89	20.72
Proposed SRGM	a	284	306	305	
	δ	0.0009	.00051	.0005	.00031
	γ	1.203	.5001	0	0
	k	.00181	.0384	0	0
	R^2	.99163	.99064	.94508	.99442
	MSE	31.06	31.74	203.89	20.72

8. CONCLUSION

In this paper, a new concept of Power Logistic function has been introduced which proves to be very flexible in the sense that it represents various curve types – Exponential, Rayleigh , Weibull or simple logistic function. This flexible nature of power logistic function gives the proposed SRGM higher degree of accuracy and wider applicability. It makes the model adaptable to any type of software dataset say Exponential, S-Shaped or Logistic type. Introduction of power factor puts the model above the many other well-known SRGMs related to Testing effort. The higher level of accuracy and better predictability for the reliability of the software being tested makes the model quite valuable for the situations where safe and risk free operation of the system are the high priority.

The usability of this newly introduced concept is not limited to the SRGM with testing effort. It can be extended to improve the results of any SRGM –whether Exponential, S-shaped or any other type e.g. multi stage models or SRGMs related with Distributed Environment.

9. REFERENCES

1. Brooks, W. D. and R.W. Motley, "Analysis of discrete software reliability models-Technical Report (RADC-TR-80-84)", *Rome Air Development Center*, **1980**; New York
2. Goel AL, Okumoto K. "Time dependent error detection rate model for software reliability and other performance measures" *IEEE Transactions on Reliability* **1979**; R-28(3): 206-211.
3. Huang, C-Y, S-Y. Kuo and J.Y. Chen, "Analysis of a software reliability growth model with logistic testing effort function", in Proceeding of 8th International Symposium on software reliability engineering, **1997**, 378-388.
4. Kapur PK and R.B. Garg, "A software reliability growth model for an error removal phenomenon", *Software Engineering Journal*, **1992**, 7, 291-294.
5. Kapur P.K., R.B. Garg and S. Kumar, "Contributions to Hardware and Software Reliability", World Scientific, Singapore.**1999**.
6. Musa, J.D., A. Iannino and K. Okumoto , "Software reliability: Measurement, Prediction, Applications",**1987** , Mc Graw Hill, New York.
7. Ohba M. "Software Reliability analysis Models" IBM Journal of Research and Development **1984**: 28:428-443.
8. Trachtenberg M. "A general theory of software reliability modelling" IEEE Trans. On Reliability. **1990**: 39(1), 92-96.
9. Yamada S, Ohba M., Osaki S. "S-shaped software reliability growth modelling
10. for software error detection" *IEEE Trans. On Reliability*, **1983**; R-32(5): 475-484

11. Yamada, S., J. Hishitani and S. Osaki , "Software Reliability Growth Model with Weibull testing effort: A model and application", *IEEE Trans. on Reliability*, **1993**; R-42, 100-105.

12. Yamada H, Ohera H, and Narihisa H. "Software reliability growth models with testing effort" *IEEE Trans. On Reliability* **1986**; R-35(1): 19-23.

Reliability, Safety and Hazard
Editors: P.V. Varde et al.

FLEXIBLE SOFTWARE RELIABILITY GROWTH MODEL DEFINING ERRORS OF DIFFERENT SEVERITY

P.K. Kapur, Archana Kumar, Rubina Mittal and Anu Gupta

Department of Operational Research, University of Delhi, Delhi 110 007

ABSTRACT

Several Software Reliability Growth Models have been developed in the literature. Some exhibit exponential reliability growth whereas others are S-Shaped. They are broadly classified in these two categories. Some of the existing SRGMs define errors of different severity though implicitly which strengthens the fact that an SRGM should explicitly defines errors of different severity. Most of the models developed can be termed as Black Box Models, which are used without having any information about the nature of the software being tested. Whereas if one has to develop what is called White Box Model, one needs to know about software technology which has been used to develop the software. Thus it is imperative to clearly understand the software development environment and accordingly there is need to develop a model, which can explicitly explain the software technology that has been used to develop the software. Thus it becomes all the more important that the SRGMs should explicitly take into account errors of different severity. Such an approach can capture the variability in the growth curves depending on the environment it is being used. In this paper we have developed a very general flexible SRGM accounting for errors of different severity. It should prove especially useful for safety critical systems. The model has been validated on several software datasets and it is shown that that the proposed model fairs comparatively better than the existing ones.

1. INTRODUCTION

The role and functionality of software in modern computer-based systems is growing at a tremendous rate. At the same time, pressures are mounting on software developers to deliver software of better quality, and to actively monitor the field performance of their software. Current experience indicates that software failures are increasing in proportion to system failures as organizations create more complex systems, while the information about these failures is frequently less than complete, uniform or precise.

IEEE 610.12-1990[5] defines reliability as "The ability of a system or component to perform its required functions under stated conditions for a specified period of time." IEEE 982.1-1988[6] defines Software Reliability Management as "The process of optimizing the reliability of software through a program that emphasizes software error prevention, fault detection and removal, and the use of measurements to maximize reliability in light of project constraints such as resources, schedule and performance." Using these definitions, software reliability is comprised of three activities:

- Error prevention
- Fault detection and removal
- Measurements to maximize reliability, specifically measures that support the first two activities

Software failures may be due to errors, ambiguities, oversights or misinterpretation of the specification that the software is supposed to satisfy, carelessness or incompetence in writing code, inadequate testing, incorrect or unexpected usage of the software or other unforeseen problems. A system may fail either because it does not comply with the specification, or because the specification did not adequately describe its function. A failure occurs when an error reaches the service interface and alters the service. A fault is the adjudged or hypothesized cause of an error. A fault is active when it produces an error; otherwise it is dormant. Fault is the defect in the code that can be the cause of one or more failures. Failure is the inability of a system or system component to perform a required function with specified limits. A failure is a departure of system behavior in execution from user needs. Failure intensity is the number of failures per unit time.

Testing is basically of two types. Black Box Testing and White Box Testing [14]. Black box testing does not focus on the internal details of the program but uses external requirements. Black-box tests serve to detect deviations of a test object from its specification. The selection of test cases is based on the specification of the test object without knowledge of its inner structure. White box testing is a test case design technique that makes use of the knowledge of the internal program logic. For a limited number of program paths, which usually suffices in practice, the test permits correct manipulation of the data structures and examination of the input/output behavior of test objects.

In this paper we proposes a software reliability model defining errors of different severity with power logistic function. The model is based on Non Homogenous Poisson Process (NHPP) and can be used to estimate and predict the reliability of the software products. The goodness-of-fit of the proposed model is compared with Flexible Software Reliability Growth Model [8]. The Flexible Software Reliability Growth Model is combination of exponential model [4], Flexible delayed S-shaped model [7] and Generalised Erlang model with logistic function [10]. The new proposed model provides the significant improved goodness-of-fit results. Section 2 presents the proposed models. Sections 3,4 and 5 give the method used for parameter estimation and criteria used for validation and evaluation of the proposed model. We conclude this paper in section 6.

2. SOFTWARE RELIABILITY GROWTH MODELLING

2.1 Model Development

Many Software Reliability Growth Models (SRGMs) [3] have been proven to be successful in estimating the software reliability and the number of errors remaining in the software. It has been observed that the relationship between the testing time and the corresponding number of faults removed is either Exponential or S-Shaped or the mix of the two [2,4,8,9,13]. It has been observed that the software contains different types of faults and each fault requires different strategies and different amount of testing effort to remove it. Ohba [13] proposed Hyper-exponential SRGM. He suggested that the fault removal process for each module separately and that the total fault removal phenomenon is the addition of the fault removal process of all the modules. Yamada et al. [18] proposed a modified exponential SRGM assuming the software contains two types of faults. The model is based on the observation that in the early stages of the software phase, the testing team removes a large number of simple faults (faults that are easy to remove) while the hard faults are removed in the later stages of the testing phase. Accordingly, they assumed the fault removal process to be the superposition of two NHPP, the first NHPP models the removal of the simple fault while the second models the removal of the hard faults. The time delay between the failure observation and the subsequent fault removal is assumed to represent the severity of the fault. The more severe the fault, the more the time delay. The fault is classified as simple if the time delay between failure observation, fault isolation and fault removal is negligible. If there is time delay, it is classified as a hard fault. If the removal of a fault after its isolation involves an even greater time delay, it is classified as a complex fault. In the models of Ohba [13] and Yamada et al. [18], the assumption of constant FRR per remaining fault still holds for each fault type of faults. The total FRR per remaining fault is, however, a function of time. Kapur et al [8,9] proposed an SRGM with three types of fault. For each type, the FRR per remaining faults is assumed to be time independent. The first type is modeled by an Exponential model of Goel and Okumoto [4]. The second type is modeled by Delayed S-shaped model of Yamada et al. [17]. The third type is modeled by three stages Erlang model proposed by Khoshogoftaar [15]. The total removal phenomenon is again modeled by the superposition of the three SRGMs[11]. Later they extended their model to cater for more types of faults [8]. The proposed model incorporates a power logistic function to capture the learning phenomenon during the removal phase of hard and complex fault type.

2.2 Model Assumptions And Notations Used

2.2.1 Model Assumptions

The proposed model has the following explicit assumptions.

1. Failure observation / fault removal phenomenon is modeled by NHPP.
2. Software is subject to failures during execution caused by faults remaining in the software.
3. The faults existing in the software are of three types: simple, hard and complex. They are distinguished by the amount of testing – effort needed to remove them and are modeled by one-stage two-stage and three-stage processes respectively.
4. Each time a failure is observed, an immediate (delayed) effort takes place to decide the cause of the failure in order to remove it. The time delay between the failure observation and its subsequent removal is assumed to represent the severity of faults. The more severe the fault, more the time delay.
5. The fault removal process i.e., the debugging process is perfect.
6. Fault removal rate of the simple fault is power function, whereas for hard and complex faults it is a power logistic learning function, as it is expected that the learning process will grow with time.
7. The expected number of faults removes in $(t, t+\Delta t)$ is proportional to the number of faults remaining to be removed.

2.2.2 Model Notations

m, $m(t)$	Expected number of faults identified in the time interval $(0,t]$ during testing phase
e, $e(t)$	Expected number of instructions executed on the software in the time interval (0,t]
a	Total fault content
a_i	Initial fault content of type i with type-I faults(simple faults)
a_j	Initial fault content of type j with type-II faults(hard faults)
a_k	Initial fault content of type k with type-III faults(complex faults)
b_i	Proportionality constant represents failure rate / fault isolation /fault removal rate per fault for fault type i
b_j	Proportionality constant represents failure rate / fault isolation /fault removal rate per fault for fault tvpe j
b_k	Proportionality constant represents failure rate / fault isolation /fault removal rate per fault for fault type k
$b_j(t)$	Power Logistic learning process function, i.e, fault removal rate per fault for hard faults.
$b_k(t)$	Power Logistic learning process function, i.e, fault removal rate per fault for complex faults.
$m_{ir}(t)$	Mean number of faults removed from Type I component by time t
$m_{jf}(t)$	Mean number of failures caused by Type II component by time t
$m_{jr}(t)$	Mean number of faults removed from Type II component by time t
$m_{kf}(t)$	Mean number of failures caused by Type III component by time t
$m_{ku}(t)$	Mean number of faults isolated from Type III component by time t
$m_{kr}(t)$	Mean number of faults removed from Type III component by time t
η	A constant parameter in the logistic learning function
d	constant
p_i	constant

$i = 1, 2$

2.3 Model Formulation

2.3.1 Modelling the Type-I, Type-II and Type-III Faults

Modelling the Type-I Faults(Simple Faults)

It is assumed that faults in the Type-I components are simple faults, which can be removed instantly as soon as they are observed. Hence fault removal of Type-I Components are modeled as one-stage processes:

$$\frac{d}{dt} m_{ir}(t) = b_i(t)(a - m_{ir}(t)) \quad (1)$$

Where $b_i(t) = b_i t^d$

In this paper we have take fault removal rate to be function of testing time; at d=0 ; $b_i(t)=b_i$. For other values of d, we try to capture different types of curves shown as failure growth. Solving the differential equation (1) under the boundary condition $m_{ir}(t = 0) = 0$, we get:

$$m_{ir}(t) = a_i\left(1 - e^{-b_i \frac{t^{d+1}}{d+1}}\right) \quad (2)$$

Modeling the Type-II Faults(Hard Faults)

The Type-II faults consume more testing effort when compared with faults of Type-I components. The removal process for such faults is modeled as a two-stage process:

$$\frac{d}{dt} m_{jf}(t) = b_j t^d (a_j - m_{jf}(t)) \quad (3)$$

$$\frac{d}{dt} m_{jr}(t) = b_j(t)(m_{jf}(t) - m_{jr}(t)) \quad (4)$$

Where

$$b_j(t) = \frac{b_j t^d}{1 + \eta e^{-b_j \frac{t^{d+1}}{d+1}}} \quad (5)$$

$b_j(t)$ is as defined above have been termed as power logistic function. Thus $b_j(t)$ is very flexible in the sense that for d=1, $b_j(t)$ is simple logistic rate. If $\eta = 0$ (i.e no learning) then $b_j(t)$ is same as fault rate for simple faults. The first stage of the two-stage process is given by the equation (3). This stage describes the failure observation process. The second stage of the two-stage process given by equation (4) describes the delayed fault removal process. During this stage the fault removal rate is assumed to be time dependent. The reason for this assumption is to incorporate the effect of learning on the removal personnel. Power Logistic learning function is used again to represent the knowledge gained by the removal team.

Solving, the above differential equations under the boundary condition, $t = 0$, $m_{jf}(t = 0) = 0$ and $m_{jr}(t = 0) = 0$, we get:

$$m_{jr}(t) = a_j \frac{\left(1 - \left(1 + b_j \cdot \frac{t^{d+1}}{d+1}\right) e^{-b_j \cdot \frac{t^{d+1}}{d+1}}\right)}{1 + \eta e^{-b_j \frac{t^{d+1}}{d+1}}} \quad (6)$$

Modelling the Type-III Faults(Complex Faults)

There can be components still having Type-II faults or Type-III faults. These faults can require more effort for removal after isolation. Hence they need to be modeled with greater time lag between failure observation and removal. The third stage added below to the model serves the purpose.

$$\frac{d}{dt} m_{kf}(t) = b_k t^d (a_k - m_{kf}(t)) \quad (7)$$

$$\frac{d}{dt} m_{ku}(t) = b_k t^d (m_{kf}(t) - m_{ku}(t)) \quad (8)$$

$$\frac{d}{dt} m_{kr}(t) = b_k(t)(m_{ku}(t) - m_{kr}(t)) \quad (9)$$

Where

$$b_k(t) = \frac{b_k t^d}{1 + \eta e^{-b_k \frac{t^{d+1}}{d+1}}} \quad (10)$$

The first stage of the three-stage process is given by the equation (7). This stage describes the failure observation process. The second stage given by equation (8) describes the fault isolation process. The third stage given by equation (9) describes the fault removal process. During this stage the fault removal rate is assumed to be time dependent. Power Logistic learning function is used again to represent the knowledge gained by the removal team.

Solving, the above differential equations under the boundary condition, t = 0, $m_{kf}(t = 0) = 0$, $m_{ku}(t = 0) = 0$ and $m_{kr}(t = 0) = 0$, we get:

$$m_{kr}(t) = \frac{a_k\left(1-\left(1+b_k\frac{t^{d+1}}{d+1}+\frac{b_k^2\left(\frac{t^{d+1}}{d+1}\right)^2}{2}\right)e^{-b_k\frac{t^{d+1}}{d+1}}\right)}{1+\eta e^{-b_k\frac{t^{d+1}}{d+1}}} \quad (11)$$

2.3.2 *Modeling total fault removal phenomenon*

The proposed model is the superposition of the NHPP of Type-I, Type-II and Type-III components. (2), (6) and (11) are mean value functions of respective NHPPs [8,10]. Thus, the mean value function of superposed NHPP is:

$$m(t) = m_{ir}(t) + m_{jr}(t) + m_{kr}(t) \quad (12)$$

$$m(t) = a_i\left(1-e^{-b_i\frac{t^{d+1}}{d+1}}\right) + a_j\frac{\left(1-\left(1+b_j.\frac{t^{d+1}}{d+1}\right)e^{-b_j\frac{t^{d+1}}{d+1}}\right)}{1+\eta e^{-b_j\frac{t^{d+1}}{d+1}}} + \frac{a_k\left(1-\left(1+b_k\frac{t^{d+1}}{d+1}+\frac{b_k^2\left(\frac{t^{d+1}}{d+1}\right)^2}{2}\right)e^{-b_k\frac{t^{d+1}}{d+1}}\right)}{1+\eta e^{-b_k\frac{t^{d+1}}{d+1}}} \quad (13)$$

Where $a_i + a_j + a_k = a$

2.4 Alternate Formulation of Proposed Srgm

The failures depend upon number of instructions executed and the number of instructions executed is a power function of time.

$$\frac{d}{dt}m_{ir}(t) = \frac{dm}{de}\frac{de}{dt} \quad (14)$$

Let $\frac{dm}{de} = p_1(a-m)$ and $\frac{de}{dt} = p_2t^d$

So $\frac{d}{dt}m_{ir}(t) = p_1p_2t^d(a-m)$

Let $b = p_1p_2$

$$\frac{d}{dt}m_{ir}(t) = bt^d(a-m_{ir}(t)) \quad (15)$$

Solving the differential equation (16) under the boundary condition $m_{ir}(t = 0) = 0$, we get:

$$m_{ir}(t) = a_i\left(1-e^{-b_i\frac{t^{d+1}}{d+1}}\right) \quad (16)$$

For Two Stage:

$$\frac{d}{dt}m_{jf}(t) = b_jt^d(a_j - m_{jf}(t)) \quad (17)$$

$$\frac{d}{dt}m_{jr}(t) = b_j(t)(m_{jf}(t) - m_{jr}(t)) \quad (18)$$

Where

$$b_j(t) = \frac{b_jt^d}{1+\eta e^{-b_j\frac{t^{d+1}}{d+1}}} \quad (19)$$

Solving, the above differential equations under the boundary condition, t = 0, $m_{jf}(t = 0) = 0$ and $m_{jr}(t = 0) = 0$, we get:

$$m_{jr}(t) = a_j\frac{\left(1-\left(1+b_j.\frac{t^{d+1}}{d+1}\right)e^{-b_j\frac{t^{d+1}}{d+1}}\right)}{1+\eta e^{-b_j\frac{t^{d+1}}{d+1}}} \quad (20)$$

If $d = 0$ in (19) we get the ordinary logistic equation for $b_j(t)$. On Solving it reduces to Flexible Delayed S-shaped Model [7]. If $d = 0$ and $\eta = 0$ in (19) we get $b_j(t) = b$. On Solving it reduces to Exponential Model [4]. The general nature of d allows us to have greater accuracy over other widely accepted SRGMs.

Similarly we can derive the result for the three-stage model.

3. PARAMETER ESTIMATION

The success of mathematical modeling approach to reliability evaluation depends heavily upon quality of failure data collected. The parameters of the SRGMs are estimated based upon these data. Hence efforts should be made to make the data collection more explicit and scientific. Usually data is collected in one of the following two ways. In the first case the times between successive failures are recorded. Though this type of data collection is more desirable, it may not be simple. The other easier and commonly collected data type is known as the grouped data. Here testing intervals are specified and number of failures experienced during each such interval is noted. For both these data types method of least squares and method of maximum likelihood have been suggested and widely used for estimation of parameters of SRGMs.

4. COMPARISON CRITERIA FOR SRGMS

The performance of SRGMs are judged by their ability to fit the past software fault data (goodness of fit) and to predict satisfactorily the future behavior of the software fault removal process (predictive validity)[12].

4.1 Goodness of Fit Criteria

a. The Mean Square Fitting Error (MSE) [8]:

The model under comparison is used to simulate the fault data, the difference between the expected values, $\hat{m}(t_i)$ and the observed data x_i is measured by MSE as follows.

$$MSE=\sum_{i=1}^{k}\frac{(\hat{m}(t_i)-x_i)^2}{k} \qquad (21)$$

Where k is the number of observations. The lower MSE indicates less fitting error, thus better goodness of fit.

b. The Akaike Information Criterion (AIC) [1,8]:

It is defined as AIC = -2(The value of the maximum log likelihood function)+2(The number of the parameters used in the model). This index takes into account both the statistical goodness of fit and the number of parameters that are estimated. Lower values of AIC indicate the preferred model.

c. Coefficient of Multiple Determination (R^2)[8]:

We define this coefficient as the ratio of the sum of squares resulting from the trend model to that from constant model subtracted from 1, that is R^2=1 –residual SS/corrected SS. R^2 measures the percentage of the total variation about the mean accounted for the fitted curve. It ranges in value from 0 to 1. The larger R^2, the better the model explains the variation in the data.

4.2 Predictive Validity Criterion [12]

The number of faults removed by time t_k can be predicted by the SRGM and compared to the reported fault removal, i.e. y_k. The difference between the predicted value $\hat{m}(t_k)$ and the reported value measures the fault in prediction. The ratio $[(\hat{m}(t_k)-y_k)/y_k]$ is called the Relative Prediction Error (RPE). If the RPE is negative (positive) the SRGM is said to underestimate (overestimate) the fault removal process. Portions of the failure data are sequentially chosen to calculate the RPE. Values close to zero for RPE indicate more accurate prediction, thus more confidence in the model. Value is acceptable if it is within $\pm$ 10.

5. MODEL VALIDATION

To check the validity of the proposed model and to find out its software reliability growth, it has been tested on two Data Sets.

Data Set 1

This data is cited from Wood [16]. The software was tested for 20 weeks during which 100 faults were removed. Based upon the estimated parameters, parameters of SRGM (13) were estimated. The Parameter Estimation result and the goodness of fit results for the proposed model are given in table 1. The RPE (Relative Prediction Error) results are given in table 3.

Data Set 2

This data is cited from M.Ohba [13]. The software was tested for 19 weeks during which 328 faults were removed. Based upon the estimated parameters, parameters of SRGM (13) were estimated. The Parameter Estimation result and the goodness of fit results for the proposed model are given in table 2. The RPE (Relative Prediction Error) results are given in table 4.

DS-I and DS-II are truncated into different proportions and used to estimate the parameters of the proposed model. For each truncation, one value of relative (percent) is obtained. It is observed that the predictive validity of the model varies from one truncation to another. The relative error of the proposed model overestimates in DS-I the fault removal process, except when the testing progress ratio is about 60%, 70% and 80% it underestimates the process as seen in table 3. The relative error of the proposed model overestimates in DS-II the fault removal process, except when the testing progress ratio is about 50% and 60% it underestimates the process as seen in table 4.

Table 1 for DS-I

Parameter Estimation and Comparison Criteria	Proposed Model	Flexible SRGM for Distributed Systems
a	136	102
b_1	.1	.1
b_2	.109	.84
b_3	.639	.40
β	3.36	18.23
d	.5511	-
R^2	.99924	.99874
MSE	.62110	1.021
AIC	82.62	83.63

Table 2 for DS-II

Parameter Estimation and Comparison Criteria	Proposed Model	Flexible SRGM for Distributed Systems
a	423	354
b_1	.18	.32
b_2	.06	.33
b_3	.23	.29580
β	7.41	11.31
d	.8384	-
R^2	.99755	.99460
MSE	25.28	55.70
AIC	176.56	209.83

It can be observed from the estimation results that the proposed model given by (13) gives the best estimates in terms of goodness of fit and predictive validity criteria for both the data sets.

Table 3 for DS-I

RPE Calculations in percentage	Proposed Model	Flexible SRGM for Distributed Systems
100%	0.403	.738
90%	0.534	2.001
80%	-0.165	3.021
70%	-2.149	.0172
60%	0.617	-1.408
50%	0.791	.868

Table 4 for DS-II

RPE Calculations in percentage	Proposed Model	Flexible SRGM for Distributed Systems
100%	0.123	1.67
90%	1.909	2.978
80%	-0.316	6.310
70%	5.500	9.353
60%	-1.912	11.288
50%	-3.271	-3.276

6. CONCLUSION

In this paper, we have presented a Software Reliability Growth Model (SRGM) defining errors of different severity. The time delay between the failure observation and the subsequent fault removal for the proposed model represent the severity of the fault. The proposed model is compared with Flexible Software Reliability Growth Model [8]. The model takes into account different types of faults (Type I, Type II and Type III). The power factor in the proposed model provides flexibility to the model and increases the reliability for the proposed model. The proposed model has been validated, evaluated and compared with the well-documented NHPP model by applying them to an actual software development project. The results of the proposed model are fairly encouraging. We feel that the proposed model depicts better result.

7. REFERENCES

1. Akaike H., A New Look at Statistical Model Identification-IEEE Transactions on Automatic Control 1974; 19:716-723.

2. Bittanti S., P. Bolzern, E. Pedrotti and R. Scattolini; (1998), A Flexible Modeling Approach For Software Reliability Growth, in G. Goos and J. Harmanis, eds.: Software Reliability Modelling and Identification, Springer Verlag, Berlin, 101-140.
3. Downs T and Scott A.; Evaluating The Performance Of Software Reliability Models, IEEE transactions on reliability (1992) 41(4): 532-538.
4. Goel AL and Okumoto K.; Time Dependent Error Detection Rate Model For Software Reliability And Other Performance Measure-IEEE Transactions on Reliability (1979); R-28 (3): 206-211.
5. IEEE Standard 610.12-1990 Glossary of Software Engineering Terminology.
6. IEEE Standard 982.2-1988 Guide for the Use of Standard Dictionary of Measures to Produce Reliable Software.
7. Kapur P.K., Goswami D.N., Bardhan A.K.;(2003), A Flexible Delayed S-Shaped Software Reliability Growth Model, Communicated to the Journal of Computer Society of India.
8. Kapur P.K., R.B. Garg and S. Kumar; (1999), Contributions to Hardware and Software Reliability, World Scientific, Singapore.
9. Kapur PK and R.B. Garg; (1992), A Software Reliability Growth Model For An Error Removal Phenomenon, Software Engineering Journal 7, 291-294.
10. Kapur P.K. Younes S and Agarwala S. ; (1995), Generalised Erlang Model with n types of faults, CSI Bulletin.
11. Kapur P.K. Younes S and Grover P.S. ; (1995), Software Reliability Growth Model with Errors of Different Severity. Coputer Science and Informatics (India) 25(3):51-65
12. Musa, J.D., Iannini, A.and Okumoto, K.;(1987) Software Reliability: Measurement, Prediction, Applications, McGraw Hill.
13. Ohba, M.; (1984), Software Reliability Analysis Models, IBM Journal of research and Development 28, 428-443.
14. Pressman Roger S., Software Engineering-A Practitioner's Approach- Fourth Edition, McGraw Hill.
15. Woodcock TG and Khoshogoftaar TM; (1991), Software Reliability Model Selection: A case study, Proceedings Of The International Symposium On Software Reliability Engineering, pp.183-191.
16. Wood, A., (1996) Predicting Software Reliability, IEEE Computers; Vol. 11, pp. 69-77.
17. Yamada S, Ohba M, and Osaki S.; (1984), S-Shaped Software Reliability Growth Models And Their Applications, IEEE Transactions on Reliability, 1984; R-33: 169-175.
18. Yamada S, Osaki S. and Narihisa H.;(1985), Software Reliability Growth Models with Two Types of Errors-Recherche Operationnelle /Operations Research (RAIRO) 19:87-104.

Reliability, Safety and Hazard
Editors: P.V. Varde et al.

MODELLING SOFTWARE RELIABILITY GROWTH IN OPERATIONAL PHASE (AN APPLICATION OF INNOVATION AND DIFFUSION MODEL IN MARKETING)

P.K. Kapur, Rubina Mittal, Anu Gupta and Mashaalla Basirzadeh

Department of Operational Research, University of Delhi, Delhi 110 007

ABSTRACT

This paper deals with the fault removal and reliability growth phenomena for the software during its operational use. No software is bug free when it is released in the market for customer use. The residual fault content present in software at the time of release causes failure during the operational use. Whenever a fault is encountered, an error report is generated and it is sent to the software developer for the bug fixing. In this paper we assume that fault removal depends on the number of instructions executed which in turn grows as the power function of usage. To describe the usage function, the Innovation and Diffusion model from marketing has been used. The usage function so defined takes care of S-shaped user- growth curve. Initially when the software is released, users adopt the product on the basis of external influence only (i.e. innovators) but as the software gets diffused into the user population, the user growth rate is higher (i.e. Innovators and Imitators) which later tails off because of various factors like- change in user requirements, availability of similar software in the market etc. In the proposed paper we suggest a Software reliability growth model, which combines Innovation-Diffusion model in Marketing with Software Reliability so as to represent usage of the software in operational phase. Applicability and effectiveness of the model have been carried out for different software failure datasets in operational phase. Mean Squared Error (MSE) has been used as the measure of 'Goodness of fit'. It has been observed that the results are fairly accurate and near to observed values.

1. INTRODUCTION

Basically SRGMs help us in estimating the number of initial faults and understanding the effects of faults on software operation. They also provide the estimate for the reliability of the software, the quality and help in predicting the residual fault content in the software at the time of release. Research efforts done in last three decades in the area of software Reliability engineering mainly concentrated on the development of Software reliability Growth Models for the testing phase [4,7,10,11] and very little seems to be done in the area of modelling the software failure phenomenon observed in its operational phase. It is generally assumed that operational phase is similar to the testing phase, which may not be the case in practice. In this paper we develop a software reliability growth model, which describes reliability growth during operational phase by taking into account its number of instructions executed and usage. The rate for number of instructions executed is assumed to be power function of its usage. To incorporate the usage in SRGM, we have combined Innovation and Diffusion model from Marketing [1]. The usage function so defined can also take care of increasing, linearly decreasing trends as a function of time, which implies slow start but gain in growth rate; a big beginning and tailing off in the usage growth.

In commercial software development organizations, increased complexity of products, shortened development cycles and higher customer satisfaction of quality have placed a major responsibility in the areas of Software Debugging, Testing and Verification. At the time of release, no software is bug free. Hence there is a need for providing support to customers during its operational use. Selling software is not a one-off deal. It involves cultivating long-term relationship with the purchasers. So Organizations spend significant resources tracking customer satisfaction and managing customer reported defects and change requests. The after sales debugging support helps in building long term relationship with purchasers, results in improved quality and capturing and retaining market share. As manual systems are increasingly being automated a failure due to software can lead to loss of money, goodwill and even human lives. The competition in the commercial software market is intense and because of the nature of applications involved purchasers look for quality in terms of reliability of the software.

During the testing phase test cases that simulate the user environment are run on the software and any departure from specifications or requirements is called a failure and immediately an effort is made to remove the cause of that failure (a fault in the software). Testing goes on till the management is satisfied with the reliability of the software. But software cannot be tested exhaustively within a limited time period. This is the reason, why we often hear about failures of software in operation and sometimes even for safety critical systems. These failures are caused by the faults that remain even after testing. Hence it is important to study how these failures occur in the user phase. Software Reliability Growth Models (SRGMs) have been widely used to estimate the reliability of software during testing [5,7,8]. Many authors have even tried to extend SRGMs to represent the failure phenomenon during the operational phase, typically used in release time problem of software [10]. But this approach is not correct when usage of software is different from during testing, which is actually the case for most of the commercial software. The intensity with which failures would manifest during the operational use is dependent upon the number of times the software is used and not much has been done in the literature for the situation [10]. An attempt has been made in this paper to model the reliability growth linking it to the number of users in the operational phase.

Though commercial software products are there in the market for the last two decades identifying the target customers with certainty is impossible. Hence a product, which may be similar in many respect to another one when launched in the market, behaves as a new product or an innovation. Bass model for innovation diffusion [1] in marketing has satisfactorily been used for this dynamic market of software products. This model explicitly categorizes the customers into innovators and imitators. Innovators have independent decision making abilities whereas imitators make the purchase decisions after getting first hand opinion from a user. Here it is assumed that purchasers or users whose number with respect to time can be modeled as an innovation diffusion phenomenon are those who can report a failure caused by the software to the developer. Such a model can correctly describe the users growth in terms of

a) a slow start but gain in growth rate
b) a constant addition of users
c) a big beginning and tail off in the usage rate as pointed out by Kenny [10].

Once the number of users of the software is known, the rate at which instructions in the software are executed can be estimated. The intensity with which failures would be reported depends upon this usage.

Applicability and effectiveness of the model have been carried out for different software failure datasets in operational phase. Mean Squared Error (MSE) has been used as the measure of 'Goodness of fit'. It has been observed that the results are fairly accurate and near to observed values.

2. NOTATIONS

m, $m(t)$: Expected number of faults identified in the time interval $(0,t]$ during operational phase

e, $e(t)$	:	Expected number of instructions executed on the software in the time interval $(0,t]$
W, $W(t)$	:	Cumulative usage in the time interval $(0,t]$
A	:	Constant, representing the number of faults lying dormant in the software at the time of release
α, β, δ, γ, k	:	Constants
k_i, $i = 1,2,3$	:	Constants
$\bar{W}$	:	Constant, representing the saturation point for the user growth of software

3. FRAMEWORK FOR MODELLING

The SRGM presented in this paper is based upon Non Homogenous Poisson Process (NHPP). The NHPP models are based on the assumption that the software system is subject to failures at random times caused by manifestation of remaining faults in the system.

In models proposed by Kenny [10] and Kapur et. al. [9], the effect of user growth on the failure phenomenon has been studied. Faults if present in the software are exposed when the customers use the software. Hence the rate at which failures would occur depends upon its usage (number of users in the operational phase [9,10]. Hence SRGMs should incorporate the effect of usage. But this may give rise to more complication and confusion as a number of functions exist in the literature that describes the user growth with time. In this paper an attempt has been made to relate the Innovation Diffusion model of Marketing to the Failure phenomenon. The proposed model is based upon the following basic assumptions.

1. Software failure phenomenon can be described by the Non-homogeneous Poisson Process (NHPP).
2. The number of failures during operation is dependent upon the number of faults remaining in the software at that time as well as on the number of faults already identified at that time
3. As soon as a failure occurs the fault causing that failure is immediately identified. Identified faults are removed perfectly and no additional faults are introduced during the process.
4. The number of instructions executed is a power function of number of users.
5. Number of users is a function of time.

Using the above assumptions the failure phenomenon can be described with respect to time as follows:

$$\frac{dm}{dt} = \frac{dm}{de}\frac{de}{dW}\frac{dW}{dt} \tag{1}$$

We discuss below each component on the right hand side of the above expression individually.

Component-1

During operations instructions are executed on the software and the output is matched with the expected results. If there is any discrepancy a failure is said to have occurred. An Error report is generated and is sent to the software developer for failure rectification. The rate at which failures occur depends upon the number of faults remaining in the software as well as on the number of the faults already identified [6,7,9]. Based on these assumptions the differential equation for fault identification / removal can be written as:

$$\frac{dm}{de} = (k_1 + k_2 \frac{m}{a})(a - m) \tag{2}$$

Here k_1 is the rate at which residual faults cause failure. It is a constant as each one of these faults has an equal probability of causing failure. k_2 is the rate at which additional faults are identified without them causing any failure.

Component-2

The second component of expression (1) relates the number of instructions executed with the number of users of the software. In the SRGMs proposed earlier in the literature [6,9] this component has often been taken as a constant. In the SRGM proposed here we define it as a power function of usage:

$$\frac{de}{dW} = k_3 W^K \tag{3}$$

Here $k \geq 0$. It can be easily seen that if $k = 0$, then $\frac{de}{dW} = k_3$ (i.e. a constant)

This case is more general and flexible in nature in the sense that it represents different type of relationships between e and W depending on different values taken by k

Substituting (2) and (3) in (1) we have:

$$\frac{dm}{dt} = \left(k_1 + k_2\frac{m}{a}\right)(a-m)k_3W^k\frac{dW}{dt} \qquad (4)$$

It is a first order differential equation. Solving it with the initial condition $m(0) = 0$ and $W(0) = 0$, we get:

$$m(t) = a\left(\frac{1-e^{-\delta\frac{W(t)^{k+1}}{k+1}}}{1+\gamma e^{-\delta\frac{W(t)^{k+1}}{k+1}}}\right) \qquad (5)$$

Here $\delta = k_3(k_1+k_2)$ and $\gamma = (k_2/k_1)$

4. MODELLING THE USER GROWTH

Kenny [10] used the power function to describe the growth in the user population of software.

$$W(t) = \frac{t^{(k+1)}}{(k+1)} \qquad (6)$$

W(*t*) here is the number of users of the software in the operational phase at time *t*. The function can correctly describe the users growth in terms of

a) a slow start but gain in growth rate
b) a constant addition of users
c) a big beginning and tail off in the usage rate.

The growth in number of users with respect to time can also be described by the Bass model of innovation diffusion [1,6,9]. For applying the Bass model it is assumed that there exists a finite population of prospective users who with time increasingly become actual users of the software (no distinction is made between users and purchasers here as Bass model has been successfully applied to describe the growth in number of both of them). In each period there will be both innovators and imitators using the software product. The innovators are not influenced in their timing of purchase by the number of persons who have already bought but they may be influenced by the steady flow of nonpersonal promotion. As the process continues the relative number of innovators will diminish monotonically with time. Imitators are however influenced by the number of previous buyers and increase relative to the number of innovators as the process continues.

The combined rate of first purchasing of innovators and imitators are given by the term $\left[\alpha + \beta\frac{W(t)}{\overline{W}}\right]$ and increases through time because *W*(*t*) increases through time. In fact the rate of first purchasing is shown as a linear function of the cumulative number of previous first purchasers. But the number of remaining non-adopters, given by $\left(\overline{W} - W(t)\right)$, decreases through time. The shape of the resulting sales curve of new adopters will depend upon relative rate of these two tendencies. If software product is successful, the coefficient of imitation is likely to exceed the coefficient of innovation i.e. $\alpha < \beta$. On the other hand, if $\alpha > \beta$, the sales curve will fall continuously.

The following mathematical model known as Bass model [1] in the marketing literature describes the above situation.

$$\frac{dW(t)}{dt} = \left(\alpha + \beta\frac{W(t)}{\overline{W}}\right)\left(\overline{W} - W(t)\right) \qquad (7)$$

The solution of (7) for *W*(*t* = 0) is

$$W(t) = \overline{W}\,\frac{1-\exp(-(\alpha+\beta)t)}{1+(\beta/\alpha)\exp(-(\alpha+\beta)t)} \qquad (8)$$

5. ESTIMATION METHOD

For parameter estimation we have used the technique of least square, which consists of minimizing the sum of squares of the deviations between the actual and estimated values. The Comparison Criteria used in this paper are:

5.1 The Mean Square Fitting Error (MSE)

Using the estimated values of the parameters, the estimated values for failure data are calculated. The difference between the estimated values,

$m(t_i)$ and the observed values y_i is measured by MSE as follows

$$MSE = \frac{\sum_{i=1}^{k} \left(m(t_i) - y_i \right)^2}{k} \qquad (9)$$

Where k is the number of observations. Lower value of MSE indicates less fitting error, thus better goodness of fit.

5.2 Coefficient of Multiple Determination (R^2)

This goodness of fit measure can be used to investigate whether a significant trend exists in the observed failure intensity. We define this coefficient as the ratio of the sum of squares resulting from the trend model to that from constant model subtracted from 1

$$R^2 = 1 - \frac{\text{residual SS}}{\text{residual SS}} \qquad (10)$$

R^2 measures the percentage of the total variation about the mean accounted for the fitted curve. It ranges in value from 0 to 1. Small values indicate that the model does not fit the data well. The larger R^2, the better the model explains the variation in the data.

6. NUMERICAL ILLUSTRATIONS

To validate the models two real software failure data sets during operations have been chosen. The datasets used here are based on failure reports of softwares in operational use.

Data set-1

This failure data set [11] is for an Operating system in its operational phase. The software consists of hundreds of thousands of delivered object code of instructions. 112 faults were reported during the observation period of around five months

Data set-2

This failure data set [11] is for a time-sharing Operating system in its operational phase. The software consists of hundreds of thousands of delivered object code of instructions. 192 faults were reported during the observation period

Kenny user function as well as Bass model (1) has been chosen to estimate the Usage growth. Estimation results for both data sets are summarized in Table-1 and Table-2. From the estimated results, we see that there is a considerable improvement in the results in terms of SSE and R^2 provided by the proposed SRGM. This is due to the flexibility rendered to the model with the presence of 'k' factor.

Table 1 For Kenny Usage Function

Data Set	1		2	
Models under comparison	Kapur et. al.	Proposed SRGM	Kapur et. al.	Proposed SRGM
a	115.26	114.58	458.97	352.02
b	.00578	.00555	.00017	.00269
z	1.281	1.2404	1.24	6.402
r	.33078	.157	.386	.0129
k	-	.1654	-	.03833
SSE	1166.71	1160.77	2894.34	2291.56
R^2	.99003	.99008	.99509	.99611

Table 2 For Bass Usage Function

Data Set	1		2	
Models under comparison	Kapur et. al.	Proposed SRGM	Kapur et. al.	Proposed SRGM
a	112.487	113.83	327.65	334.65
b	.008	.00609	.084	.336
z	6.035	1.834	7.58	7.28
k	-	.275	-	.00101
$\overline{W}$	8779.31	8831.45	9199.24	9861.72
$(\alpha+\beta)$	.00098	.00027	.000077	.00111
β / α	1.19	.874	14.11	96.4
SSE	1340.95	1210.64	2294.19	2223.5
R^2	.98855	.98966	.99617	.99623

7. CONCLUSION

In the paper proposed here, we have introduced the concept of Power Logistic function in the operational phase of the Software usage. The presence of 'k' factor adds greater flexibility and higher accuracy to the model, which is quite useful in capturing the various types of failure curves observed during the user phase. From the numerical illustrations, we see that the model provides with considerably improved results with better predictability because of lower MSE and higher R^2.

8. REFERENCES

1. Bass FM. "A new product growth model for consumer durables" *Management Science* **1969**; 15(5): 215-224.
2. Bittanti S, Bolzern P, Pedrotti E, Scattolini R. "A flexible modelling approach for software reliability growth" *Software Reliability Modelling and Identification (Ed.) G. Goos and J. Harmanis*, Springer Verlag, Berlin, **1988**, 101-140.
3. Brooks WD, Motley RW. "Analysis of discrete software reliability models - *Technical Report (RADC-TR-80-84)*" **1980**; New York: Rome Air Development Center.
4. Goel AL, Okumoto K. "Time dependent error detection rate model for software reliability and other performance measures" *IEEE Transactions on Reliability* **1979**; R-28 (3): 206-211.
5. Jelinski Z, Moranda PB. "Software reliability research" In: *Freiberger W, (Ed.) Statistical Computer Performance Evaluation* **1972**; New York: Academic Press: 465-497.
6. Kapur PK, Bardhan AK, Butani NL. "Modelling failure phenomenon of a commercial software in operational use" In *A.K. Verma (Chief Editor) Proceedings of the International Conference on Quality, Reliability and Control (Communication and Information Systems)* **2001**; IETE and IIT: Mumbai, R-59.
7. Kapur PK, Garg RB. "A software reliability growth model for an error removal phenomenon" *Software Engineering Journal* **1992**; 7: 291-294.
8. Kapur PK, Garg RB, Kumar S. "Contributions to hardware and software reliability" **1999**; Singapore: World Scientific Publishing Co. Ltd.
9. Kapur PK, Jha PC, Goswami DN et. el. "General framework for modeling software reliability growth during testing and operational use" Recent Developments in Quality Reliability and Information Technology; **2004** ; 187-198.
10. Kenny GQ. "Estimating defects in a commercial software during operational use" IEEE Trans. On Reliability, **1993**; 42(1); 107-115.
11. Musa JD. "Software reliability data" **1980**; Data and Analysis Center for software, USA and www.dacs.dtic.mil.
12. Musa JD, Iannino A, Okumoto K. "Software reliability: Measurement, Prediction, Applications" **1987**; New York: Mc Graw Hill.
13. Ohba M. "Software reliability analysis models" IBM Journal of Research and Development **1984**; 28: 428-443.
14. Pham H. "Software Reliability" **2000**; Springer-Verlag Singapore Pte. Ltd.
15. Xie M. "Software reliability modelling" **1991**; World Scientific.

Reliability, Safety and Hazard
Editors: P.V. Varde et al.

DATA COLLECTION CHALLENGES TO DETERMINE SOFTWARE RELIABILITY: PRACTITIONER'S PERSPECTIVE

Mahesh Kuruba, A. Srividya and A.K. Verma

Reliability Engineering Group, Electrical Engineering Department
Indian Institute of Technology Bombay, Mumbai 400 076

ABSTRACT

Researchers have been developing various reliability models for practitioners to determine the software reliability. However, the usage of reliability models in the industry is very limited. The reasons have been attributed to lack of knowledge about models and their applicability. There is yet another dimension that poses problem while determining reliability is 'Data'. Data collection is a major challenge for practitioners and makes it difficult to use the reliability models for determining the reliability of software systems In this paper various challenges involved in collecting data to determine software reliability from the practitioner's perspective are discussed. Also, this paper highlights the factors, which are missing in the reliability models, and suggests new avenues for research in software reliability.

Keywords: COTS product reliability, Data collection challenges, Software reliability, Software reliability data

1. INTRODUCTION

Various software reliability models have been proposed and applied in the literature [1-21]. A survey conducted by American society for quality control reported that only 4% of the practitioners, developers and users use the reliability models [22]. Low usage of reliability models is attributed to lack of knowledge about selecting and applying the model. It has been observed that the suitability of reliability model varies based on project and organization [23]. However, there is a need to look at another dimension in the usage of software reliability models or determining software reliability, which is related to data collection. There are various challenges in obtaining the reliable data to compute software reliability in the industry. In this paper, some of these challenges are highlighted from practitioner's perspective based on the industry scenario.

In this paper the data requirements for reliability models are discussed by type of the data required. Also, some practical problems pertaining to the availability of data in the global software development scenario is discussed. Shortcomings in the current reliability models are also discussed.

2. DATA REQUIREMENTS

The data required for computing software reliability using various models in the literature has been listed as follows [1-21]:

1. Software failures/faults: The number of software failures/faults during a certain period of time, which can be either elapsed time or calendar time where the software is

in operation or testing, is termed as Failure-count.

2. Time: The time in which the software is in operation or test. The time may be execution time or elapsed time or calendar time. *Execution time* is the CPU time that is actually spent by the computer in executing the software. *Calendar time* is the time people normally experience in terms of years, months, weeks, days etc.,. *Clock time* is the elapsed time from start to end of computer execution in running the software. Software failure time is the time at which the software has failed to perform its operation. Software failure time gives the duration between two failures, which is termed as time-between-failures.
3. Operational profile: Operational profile of software [23] is defined as the set of operations that the software can execute along with the probability with which they occur. Operational profile challenges are not discussed in this paper, as the data required can be obtained from server logs. However, if the server software is new then the server log itself may not be reliable.
4. Competency: The design and development of the software is still human-intensive and human dependent. Since the software development is human dependent, competency of the team plays vital role in the software reliability. Relatively competent people inject less number of defects thus increasing the software reliability. Utilization of competent people for identifying and fixing the defects is essential. Competency of the people can be determined using Competency Index (CI) metric [24] as represented in eq 1.

$$CI_{\text{Team member}} = \sum_{i=1}^{k}\left[\left(\frac{C_L}{4}\right)_i * (C_W)_i\right] \quad (1)$$

Where

C_L is competency level in a competency, which is rated on a scale of 0-4

C_W is the weightage of the particular competency, which is computed using AHP

k is the number of competencies in a particular role

The competency levels in a particular competency are classified as:
Level 0: Absence of knowledge
Level 1: Exposure – Has an exposure to the knowledge area but not practiced it
Level 2: Experience – Has an exposure to the knowledge area and has practiced it
Level 3: Expertise – Has an exposure to the knowledge area, has practiced over a period of time, and is a perfectionist.
Level 4: Excellence – Is a researcher and master in the particular knowledge area

The current reliability models may be refined incorporating CI metric.

3. DATA COLLECTION CHALLENGES

Based on the data requirements identified in the previous section, this section of the paper discusses the challenges involved in collecting such data.

3.1 Software failures/faults:

Project teams identify defects in requirements, design and code. They classify the identified defects based on their severity and type. Commonly used defect severities such as high, medium and low or severity 1, severity 2, severity 3 and severity 4 etc. and defect types such as functional, performance, standards etc.,. The teams do not distinguish defect as an error or fault or failure. High severity defect may not always be a failure. So, it becomes difficult for the practitioners while determining reliability to segregate the failures/faults.

The defects found by the development team in reviewing or testing the software can be captured. However, team members may not report all the defect data, as it might impact their compensation or they may not have time to log all the defects.

Customers or end-users also find defects in the software. Customers adopt several mechanisms to communicate these defects like e-mail, telephone, problem report systems etc. In absence of problem reporting systems it becomes difficult for managers to consolidate all the defects and determine reliability.

3.2 Time

Data collection strategy of time related data varies based on the type of software (like real time, client-server, web based, embedded, batch etc.,). For instance, collection of data for execution time in a client-server environment needs to be collected at the server as well as the client CPU time, which may be difficult to collect in the absence of automated data collection system. Collection of data for calendar time primarily depends upon the effort spent by the software users, which is difficult to collect, as the users may not be using the software continuously. The effort spent by the development team in reviewing or testing the software can be captured. However, it is difficult to capture the effort of the team members spent on a task, as there are cognitive factors that are involved and other interrupting activities (mails, teleconferences, meetings etc.,) affect the work of the team members. Team members may work over time and may still not report the extra effort.

Capturing the software failure time will be possible with server logs and problem reporting systems. Capturing the software failure that is perceived by the user may be difficult. Users may not report failures due to oversight or lack of time or knowledge.

Using effort and defect related data, which is reported by the team member, for the compensation might impact reliability of data. The data reliability will reduce, as people tend to report incorrect data. It depends upon the organization culture and support from the management to report the reliable data. Capturing the accurate data is another challenging task, for which the management need to understand and support the team members for providing the reliable data. Also, type of software needs to be considered while choosing the appropriate reliability model.

3.3 Competency of the team

Competency of the team plays a significant role in injecting defects in to the design or code of the software. It has been observed in the industry that experienced designers or programmers inject relatively less number of defects. Also, from fig 1, it can be observed that as the competency in the organisation increases the defect density has decreased. However, reliability models do not consider competency of the team while determining software reliability.

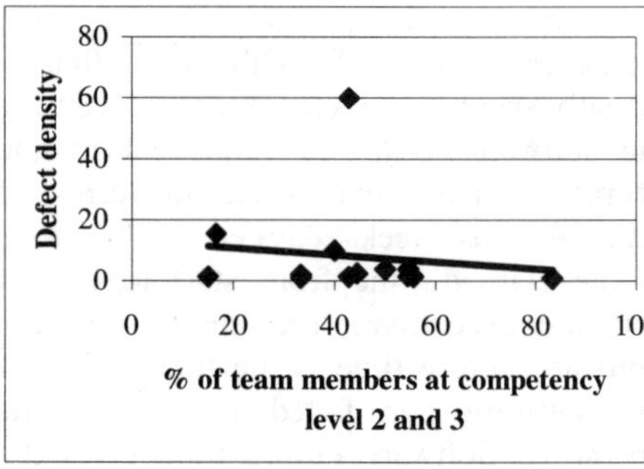

Fig. 1: Impact of competency on defect density

4. OTHER CHALLENGES

Some challenges that affect data collection has been discussed in this section.

4.1 Review Defects Data

The two basic techniques to uncover defects are review and testing. Critical defects pertaining to design and code are uncovered during the review. These identified defects can lead to software failure.

There are several methods of review varying from inspection to walkthrough. Inspection is the rigorous approach of uncovering defects in contrary walkthrough is an informal approach. Thus review effectiveness varies from inspection to walkthrough. Higher the review effectiveness lower number of defects will seep out. So, higher the review effectiveness higher is the software reliability. So, there is a need to consider the type of review and its effectiveness while determining software reliability.

Defects are also identified in the testing phase. Historically, several reliability models have been considering faults and failures identified during the test and operation phases of software. Considering the review defects will vary reliability prediction. Hence, there is a need for considering review defect data, review method and its effectiveness in reliability determination.

Practitioners categorize the review defects in to several types and severities. Attributing review defects that leads to software failure is difficult.

4.2 Multiple Vendor Scenarios

Organizations receive IT related services from multiple global vendors. The vendor who develops software may be different from the vendor who maintains the software. Multiple vendors may be involved in one software development. The software failure data of these sub-systems of software will be available with the vendor itself. Also, these sub-systems may have dependency on the sub-system developed by a different vendor. A fault caused in one of the sub-system may cause a failure of the software. Also, the failure caused in a system may be because of the other sub-system that was developed by other vendors. So, it becomes difficult to collate the failure data from multiple vendors. Usually, the customers are concerned about the software and its related artifacts, failure information may be of low concern to them. So, the failure data of the software may not be available to determine software reliability. In another scenario where the vendor maintains the software will not have any failure data during development of the software or software components to assess reliability.

Organisations purchase COTS products and build applications based on them. The COTS vendors will not share the failure data for the credibility reasons. Since the reliability of the application developed depends upon the COTS product also, it becomes difficult to determine the software reliability. This scenario is applicable for the reuse of product components as well. Organisations may not maintain the component related failure data to determine the reliability of the product that is developed using reusable components.

4.3 Verification of Predicted Reliability

Predicted number of failures using reliability models needs to be verified against the actual failures of the software in operation. As discussed in the multiple-vendor scenario, capturing failure data and coupled with mechanisms obtaining failure data is a challenging task. So, there is a need to devise the failure data collection mechanisms.

5. RESEARCH AREAS

As discussed in the earlier sections of the paper,

- Competency of the development team impacts software reliability. There are no reliability models that consider competency of the team. So, researchers may refine the existing models incorporating competency.
- Review defects are not being considered in several reliability models. Researchers might refine the models incorporating the review defects as well.
- Researchers need to devise mechanisms to obtain the failure data from multiple vendors and various phases of software development, to predict and validate the predicted software reliability.

6. CONCLUSION

In this paper various data required to determine software reliability has been listed and the challenges in collecting such data has been discussed. Based on the challenges discussed, new avenues for research in software reliability are discussed. This sets the path for the future research in the software reliability domain thus making reliability models more amenable to implement in practice.

7. REFERENCES

1. Moranda P.L, Jelinski Z, Final report on Software reliability study, McDonnell Douglas Astronautics Company, MADC report number 63921, 1972.
2. Goel A.L, Okumoto K, "Time dependent error-detection rate model for software and other performance measures", IEEE Trans on Reliability, Vol 28, No 3, Aug 1979, pp:206-211.
3. Schneidewind N.F, "Analysis of error process in computer software" Sigplan note, Vol 10, no 6, 1975, pp: 337-346.
4. Musa J.D, " Progress in software reliability measurement", IEEE Transactions on Software engineering, Vol SE-1, No 3, 1975, pp: 312-327.

5. Musa J.D, "A theory of software reliability and its applications", Second software lifecycle management workshop, published by IEEE, New york, 1978, pp: 153-155.
6. Musa J.D, "Validity of execution – Time theory of software reliability", IEEE Trans on Reliability, Vol R-28, no 3, 1979, pp: 181-191.
7. Musa J.D, "Software reliability measures applied to systems engineering", Proceedings of the National computer conference, 1979, pp: 941-946.
8. Musa J.D, "The measurement and management of software reliability", Proceedings of the IEEE, Vol 68, No 9, 1980, pp: 1131-1143.
9. Musa J.D, Iannino A, Okutomo K, Software reliability – measurement, prediction, application, McGraw Hill, New York, 1987.
10. Ohba M, "Software reliability analysis models", IBM Journal of research and development, vol 21, no 4, July 1984, pp: 428-443.
11. Musa J.D, Okumoto K, "Software reliability models: concepts, classification, comparisons and practice", Electronic systems effectiveness and life cycle costing, J.K.Skwirzynski (ed) NATO ASI series, F3, Springer-Verlag, Heidelberg, pp: 395-424.
12. Yamada S, Ohba M, Osaka S, "S-Shaped reliability growth modeling for software error detection", IEEE Transactions on Reliability, vol R-32, no.5, December 1983, pp: 475-478.
13. Duane J.T, "Learning curve approach to reliability monitoring", IEEE Transactions on Aerospace, Vol 2, 1964, pp: 563-566.
14. Moranda P.B, "Predictions of software reliability during debugging", Proceedings of the annual reliability and maintainability symposium, Washington D.C, 1975, pp: 327-332.
15. Moranda P.B, "Event-Altered rate models for general reliability analysis", IEEE Transactions on reliability, vol R-28, no 5, 1979, pp: 376-381.
16. Musa J.D, Okumoto K, "A logarithmic poisson execution time model for software reliability measurement", Proceedings 7th International conference on software engineering, Orlando, Florida, pp: 230-238.
17. Littlewood B, Verrall J, "A Bayesian reliability growth model for computer software", Journal of the royal statistical society, series C, vol 22, no 3, 1973, pp: 332-346.
18. Littlewood B, "Validation of a software model", software lifecycle management workshop, Atlanta, Georgia, 1978, published by the International Business services, Inc, pp: 146-152.
19. Littlewood B, "The Littlewood-Verrall model for software reliability compared with some rivals", Journal of systems and software, vol 1, no 3, 1980, pp: 251-258.
20. Gaffney J.E, Davis C.F, "An approach to estimating software errors and availability", SPC-TR-88-007, version 1, March 1988, Proceedings of the Eleventh Minnowbrook workshop on software reliability, July 1988.
21. Gaffney J.E, Pietrolewicz J, "An automated model for Software Early Error Prediction (SWEEP)", Proceedings of the 13th Minnowbrook workshop on software reliability, July 1990.
22. Hoang Pham, "Software reliability and testing", IEEE Computer society press, 1999.
23. Michael R.Lyu, "Handbook of software reliability engineering", IEEE Computer society press, 1995.
24. Mahesh Kuruba, Dr.Gargi Keeni, Chandrasekhar Bandaru, "Competency tracking system for project managers", European SEPG 2004, London, June 2004.

Reliability, Safety and Hazard
Editors: P.V. Varde et al.

SOFTWARE METRICS TAXONOMY FOR METRICS GUIDED SOFTWARE DEVELOPMENT

Piyush Mehta, A. Srividya and A.K.Verma

Reliability Engineering Group, Electrical Engineering Department
Indian Institute of Technology Bombay, Mumbai 400 076

ABSTRACT

Measurement plays an important role in different phases of the software development. In the current work, state of the art literature review is done to identify the software metrics proposed by the researchers. Contextual information and the attributes from these metrics are extracted. The software metrics are classified according to the subclasses like life cycle phases, product attributes, process attributes, project type, technology type and other segmentation factors. The software metrics classification information is used to develop the Software Metrics Taxonomy. Proposed Software Metrics Taxonomy is represented in the abstract form of logical diagram. The paper concludes with the benefits of the proposed software metrics taxonomy in software development.

Keywords: Software Measurement, Software Metrics Taxonomy (SMT), Software Project Management, Software Quality

1. INTRODUCTION

Importance of the software component in the large scale systems has increased many folds in the last couple of decade. Measurement plays an important role in different phases of the software development. Software measurement is used to assess, predict & control the various attributes of the software components. Software measurement gauges the developmental activities early in the development cycle. Analysis of the measurement triggers the corrective action and change of course of action required to meet the established deadline and committed quality within agreed budget. Measurement of the specification product in upstream development cycle is critical for project management and quality control. A very large number of software metrics are suggested by the various researchers in different contexts. These proposed metrics serves different stakeholders of the project needing different view of the software development. It becomes difficult for the project managers to understand the various metrics suites and select the appropriate metrics suitable for their project. A classification of the software metrics is required which can facilitate the project managers to select the appropriate metrics.

In the current work, state of the art literature review is done to identify the metrics proposed by the researchers. Contextual information and the attributes from these metrics are extracted. The software metrics are classified according to the subclasses like life cycle phases, product attributes, process attributes, project type, technology type and other segmentation factors.

The software metrics classification information is used to develop the Software Metrics Taxonomy (SMT). Proposed SMT is represented in the abstract form of logical diagram. The paper concludes with the benefits and usage of the proposed software metrics taxonomy in software development.

2. LITERATURE REVIEW

Several metrics suites have been proposed by researchers in couple of decades. The metrics proposed are suitable for different software development environment. It is difficult to attribute the origin of the metrics, as the metrics have been used widespread in many industries.

The 14 well known studies have been chosen and the metrics discussed in the studies have been summarized in table 1.

Table 1 Summary of metrics and models

Authors	Productivity	Quality	Effort	Cycletime	Size	Cost	Change	Cusomter
Barnard [1]	X							
Burr [2]	X	X		X	X			X
Daskalantonakis [3]	X	X	X	X		X		
Davidson [4]	X	X	X	X		X		X
Florac [5]	X	X	X	X	X	X	X	
Garrison [6]			X	X		X		
Grady [7-8]	X	X	X	X	X			
Hayes [9]	X	X	X	X	X			
Herbsleb [10]	X	X		X		X		
Jones [11]	X	X	X	X	X			
Kan [12]	X	X			X		X	X
McGibbson [13]			X	X	X	X		
Rico [14-15]	X	X	X	X				
Rosenberg [16]	X		X		X		X	

The metrics listed by the above authors were classified into the generic head like productivity, quality effort, cycle time, size, cost, change and customer. These generic heads are applicable to all the projects and cover the most of the metrics published by various authors.

3. SOFTWARE MEASUREMENT TERMINOLOGY

In order to establish the understanding of the various measures proposed by the practitioners, a sound understanding of the software measurement terminology is required. Following paragraph defines the software terminology and becomes the foundation for proposed SMT.

Entity: Entity is any object or an event of interest for measurement. The objects are tangible output and can be termed as product, while event can represent any process, task, activity. E.g. software module, software programme, review and testing.

Attribute: Attribute is a feature or property of an entity.

As above, entities are classified into product or process. This classification extends as product or process attribute.

Product Attribute: Feature or property of an intermediate/final product. eg. Feature or property of software requirements specification, relational data model, specification products, code and user manual

Process Attribute: Feature or property of the process. The process is followed in turn to create the product. The process can also be classified into management process, engineering process and support processes.

Basic classification of product and process attributes can be spawned into direct and derived attributes.

Direct/Internal Attribute: Attributes that are measured on its own. eg. size of the code, document etc. These are simple attributes.

Derived/External Attribute: Derived attributes are based on the combination of two or more attributes. The combination can be between processes, products or process & product. These are complex and are used in many instances.

Attributes are measured by gauging the behaviour of the entity with respect to its environment. External attributes are difficult to measure. These are more important from the customer or manager's viewpoint.

For instance, *Defect density* is the derived attribute of the product. It is based on the number of defects and size of the product. *Effort variance* is the derived attribute of process. It is based on the planned and actual effort. *Productivity* is the derived product and process attribute, it combines the product size and effort.

4. SOFTWARE METRICS TAXONOMY

A number of software metrics are reported by the researchers. Practitioners/new project managers are often confused by various software metrics proposed. They are not able to understand the importance/benefits of these metrics and hence are not able to choose the appropriate metric set for the project.

The classification scheme of the software metrics is needed in order to understand what practices are prevalent in the industry. After classifying the metrics in the various categories and creating the precise definition, the data collection can be done. This promotes the understanding of the software measures and will aid in the data analysis and interpretation stage.

The metrics reported by the researchers are used to find the generic classification attributes as shown in table 2. These generic classifiers form the basic building blocks for the software metrics. Generic classifiers provide the clear understanding of the underlying elements of the proposed software metrics and their interrelationship.

Table 2 Generic Classification of the software attributes

Type of Measure	Product	Process
Base Measures	Size	Effort
	Complexity	Duration
	Defects	Review
		Competency
Derived	Productivity	
	Cost	
Segmentation Factors	Technology Type 3GL 4GL Product Type	Process Type Software Engineering Process Management Process Support Process Phase wise classification
	Project Type	Project Type
	Project Management	Project Management
	Tools (Debugging, IDE)	Tools (CASE, Process Workbench)

There are number of measures which can be derived from generic classifier attribute. Combination of these attributes unfolds the possibility of generating many measures each quantifying a particular view of the product, process or combined.

Software Metrics Taxonomy (SMT) is generated by using the generic classifier attribute and representing their relationship in the form of logical diagram. This logical diagram is the abstraction of the metrics classification as shown in figure 1 (appended in the end). Proposed SMT is generic and can be used by any IT organisation. It clearly depicts the relationship between the various attributes of software development. Proposed SMT provides flexibility to add the new classifying attributes depending on the organizational need. The same taxonomy can be extended to form the metrics collection, consolidation and analysis tool in the organization.

Table 3 Instantiation of size and product type attributes

Sr. No.	Product Type	Size Attribute
1.	Software Requirements Specification	Number of pages
2.	Software Requirements Specification	Number of unique requirements
3.	Software Requirements Specification	Number of use cases
4.	High Level Design	Number of pages
5.	High Level Design (3 GL)	Number of Bubbles
6.	High Level Design (4 GL)	Number of Classes
7.	High Level Design	Function Point Estimates
8.	High Level Design (4 GL)	Number of Use Cases
9	High Level Design	Number of Feature Points
10.	Low Level Design	Number of Pages
11.	Low Level Design	Number of Bubbles (Fan-in, Fan out)
12.	Low Level Design	Number of Classes
13.	Low Level Design	Number of FP
14.	Low Level Design	Number of Use Cases
15.	Low Level Design	Number of Feature Point
16.	Code (3GL)	Lines of Code
17.	Code (3GL)	Function Points
18	Code (4GL)	Function Points
19.	Code	Feature Point
20.	Code	Use Case Point
21.	Code	Components
22.	User Manual	Number of Pages
23.	Other Documents	Number of Pages

Each composition in the SMT can be extended to create a view on the software metrics. This view can be termed as the instantiation of the composition. For example

Size (Product Base Measure Dtl in Figure 1) can be combined with Product Type (Product Type Sec Fct Dtl in Figure 1) to generate the instantiation shown in table 3.

In the similar way all the compositions in the SMT can be instantiated. These instantiations can be used to report the key performance indicators.

5. BENEFITS OF PROPOSED SOFTWARE METRICS TAXONOMY:

The benefits of the proposed Software Metrics Taxonomy are as follows:

(1) The SMT is generic enough to be used among software development industries promoting the consistency.

(2) The SMT is flexible enough to include future classifier attribute or organization specific customisations.

(3) The metrics are classified as base and derived measures along with the segmentation. Organisations intending to incorporate new metrics can categorise them under base or derived measure. This will ease the implementation of software metrics in the organization keeping the integrated view of measurable entities.

(4) Quite often, practitioners collect various metrics which consume enormous effort. Classification of metrics as base and derived measures will help practitioners to understand what base measure data need to be collected to arrive at the metrics. This reduces the effort and confusion among the practitioners.

(5) The SMT provides consolidated view of the software attributes and their interrelationship, thus facilitating the quick understanding as oppose to going through whole metrics documentation.

(6) The consolidated view in SMT provides an opportunity for developing the tool to collect, consolidate and analyze the software metrics at organisational level.

(7) Hypothesis established by data collection and analysis can be visually validated by observing the relationships in the SMT.

6. CONCLUSION

Literature review of 14 well known software metrics study is used to extract the generic classification attributes and segmentation factors. Software Metrics Taxonomy (SMT) is developed using these generic classification attributes and their relationship. Logical model is used to represent the SMT. SMT can be instantiated into various software metrics view. As an example, one such composition of the SMT is instantiated into the software metrics view for size and product type attribute. The paper concludes with the key benefits of the proposed software metrics taxonomy.

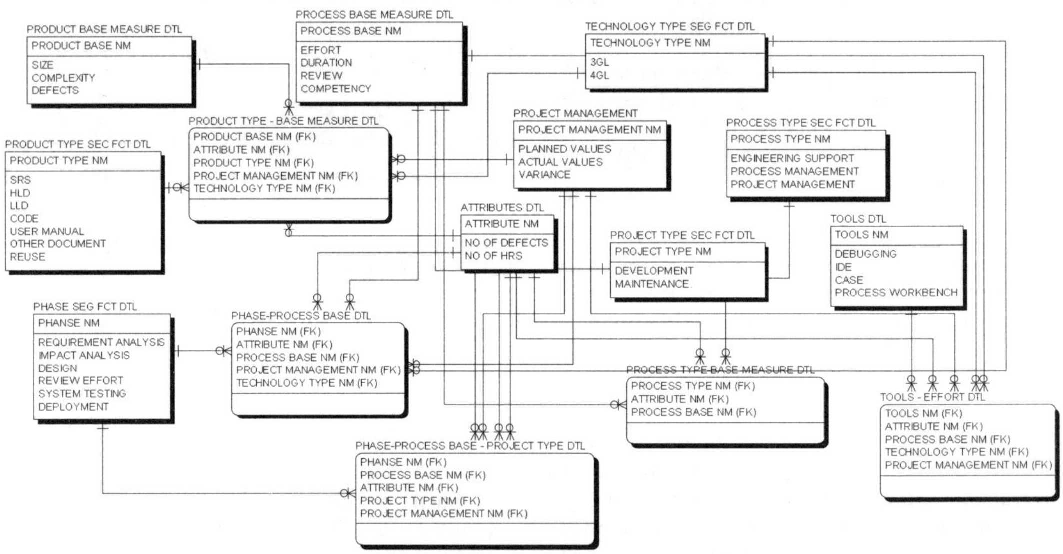

Fig. 1: Software Metrics Taxonomy (SMT)

7. REFERENCES

1. Barnard J and Price A, "Managing code inspection information", IEEE Software, 1994, Vol 11, No 2, pp: 59-69.
2. Burr A and Owen M, "Statistical methods for Software quality: Using metrics for software process improvement", 1996 Boston, MA: International Thomson publishing.
3. Daskalantonakis M.K, "A Practical view of software measurement and implementation experiences within Motorola", IEEE Transactions on Software engineering, Vol 18, No 11, Nov 1992, pp: 998-1010.
4. Davidson W.H, "Beyond reengineering: The three phases of business transformation", IBM systems journal, Vol 32, No 1, pp: 65-79.
5. Florac W.A, Carleton A.D, "Measuring the software process: Statistical process control for software process improvement", 1999, MA: Addison Wesley.
6. Garrison R.H and Noreen E.W, Systems Design: Activity based costing and quality management, Managerial accounting, pp:178-237.Boston, MA:McGraw Hill.
7. Grady R.B, Successful software process improvement, Saddle River, 1997, NH: Prentice Hall.
8. Grady R.B and Caswell D.L, Software metrics: Establishing a company wide program, Englewood cliffs, 1986, NJ:Prentice Hall.
9. Hayes W, Over J W, "The Personal Software Process: An empirical study of the impact of PSP on individual engineers", CMU/SEI-97-TR-001, 1997, Pittsburg, PA: Software Engineering Institute.
10. Herbsleb J, Carleton A, Rozum J, Siegel J, Zubrow D, "Benefits of CMM-based software process improvement: Initial results", CMU/SEI-97-TR-013, 1994, Pittsburg, PA: Software engineering institute.
11. Jones C, Activity based costs: Polishing the Software process, Software development, 1997, pp:47-54.
12. Kan S.H, Metrics and models in software quality engineering, 1995, Addison Wesley.
13. McGibbson T, "A Business case for software process improvement", Contract number: F30602 -92-C-0158), Rome, NY: Air force research laboratory – Information Directorate, (AFRL/IF), Data and Analysis Center for Software (DACS) 1996.
14. 14.Rico D F, "Software process improvement: Impacting the bottom line by using powerful solutions", 1998, http://davidfrico.com/spipaper.html
15. Rico D F, "Software process improvement: Impacting the bottom line by using powerful solutions", 1996, http://davidfrico.com/spipaper.html
16. Rosenberg L H, Sheppard S, Butler S A, "Software process assessment", 3rd International symposium on space mission operations and ground data systems, USA, 1994.

Reliability, Safety and Hazard
Editors: P.V. Varde et al.

APPLICATION OF NEURAL NETWORK FOR MODELING SOFTWARE DEFECT PREDICTION

Piyush Mehta, Mahesh Kuruba, A. Srividya and A.K. Verma

Reliability Engineering Group, Electrical Engineering Department
Indian Institute of Technology Bombay, Mumbai 400 076

ABSTRACT

Early software defect prediction is beneficial for project managers and organizations. Several defect prediction models are proposed in the literature. In this paper an attempt has been made to model the test defect prediction using neural networks. The model has been trained for 80 software projects of an organization and the prediction is done for 20 projects. Multiple regression model variants are built on same 80 projects data. The regression models are used for predicting the 20 projects defect prediction. The error values of the prediction models are compared. It has been observed that accuracy of the linear regression model is better than neural network model and warrants for further fine tuning of the neural network model.

Keywords: Artificial Neural Network, Defect Prediction, Software Quality

1. INTRODUCTION

Defect detection techniques are important corner stone in software development to improve software quality. Defect detection techniques can be broadly classified into reviews and testing. In the current software development trend reviews and testing constitute of 40% of the software development effort. Defect prediction model provides the quantitative support for the effective execution of defect detection techniques. Defect Prediction Models are helpful in planning, monitoring and control of defect detection techniques. Important advantages of the defect prediction model are as under:

- Prediction of defects leads to the detection of the defect in the upstream development life cycle.
- Alarms about the deviations from normal conditions
- Preventive measures can be adopted
- Comparison of the predicted and actual defects provides early assurance about the quality being built into the product through various development phases.

The purpose of this paper is to discuss various traditional defect prediction models, understand the limitations of these models. An attempt has been made to model the defect prediction based on the application of Artificial Neural Network (ANN). Different ANN models are developed. Comparison is done between the traditional and neural network models. The paper concludes with the comparison of multiple linear regression model and neural network model and sets the direction of future research in this area.

2. BACKGROUND

Several models have been presented in the literature. These models can be broadly classified as Prediction based on Size and complexity [1-8], Prediction based on testing process [9-11], defect seeding, statistical process control (SPC) based estimates, Rayleigh model and Closed Loop Defect Removal Model [12]. Fenton and Neil [13] provide a detailed critique of software defect prediction models. The essential problem is the oversimplification that is generally associated with the use of regression models.

Defect prediction based on the size and complexity takes into account the size of software in Line of Code (LOC) and/or code complexity. Defect seeding model is not a practical for every project. The accuracy of the Statistical Process Control (SPC) based models depends upon estimation of size, a parameter used in the defect density calculation. However the SPC model is based on the assumption that detection techniques are in control and do not consider the uncovered defects. Closed loop defect removal model is useful for in-process quality management and defect prediction.

The conventional methods of defect prediction need exhaustive effort to refine the models continuously. Neural networks address the limitations of the conventional methods. ANN is the promising technique of the machine learning paradigms and has been successfully applied in other areas of engineering and science. ANN has also been applied to software engineering to a limited extent. Hence, ANN has been used to predict software defects in this paper.

3. ARTIFICIAL NEURAL NETWORK MODEL

3.1 Artificial Neural Network Modelling

Artificial Neural Network model are parallel systems inspired by the architecture of biological neural networks, comprising simple interconnected units called neurons. Neural network model developed by McCulloch and Pitts [14] is shown in fig 1, each input is multiplied by corresponding weight W, and sums the weighted inputs. If this sum is greater than a predetermined threshold value, the output is one; otherwise, it is zero.

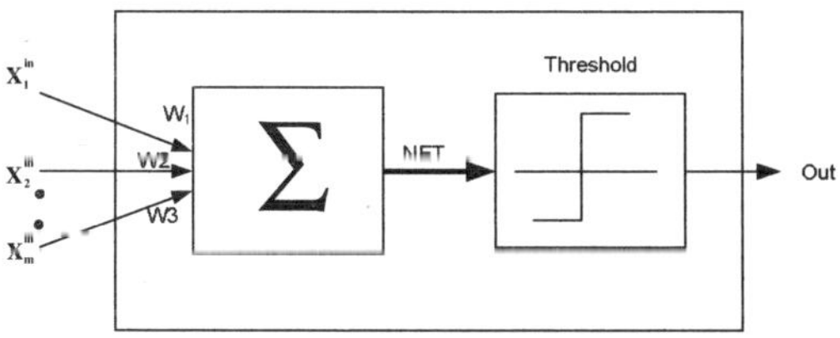

Fig. 1: McCulloch and Pitts Neural Network Model

As represented, the neuron receives signals from several input connections each having different weights thereby defining the coupling strength among various input connections. These weights are established by a learning process, in which weights are modified according to given patterns and specified learning rule. An input function (summation in this case) is employed to consolidate all the inputs into a single value. This single value is fed into the activation function. The result from activation function is transmitted to connecting network structure via an output function. A variety of activation functions like Linear, Sigmoid, Tanh, Modified Sigmoid, Parabola etc are available and can be chosen based on the type of application. Based on the type of connections made between neurons, different neural network structure can be conceived. Structure of Feed forward neural network model is shown in fig 2. The hidden and output layer neurons are each connected to all of the units in the preceding layer. It is possible to define networks that are partially-connected to only some units in the preceding layer; however fully-connected networks are found to be better in most applications.

3.2 Multilayer Perceptrons with backpropagation training algorithm

The multilayer perceptron is a type of network model in which the neurons are arranged in layers, whereby the neurons of a layer are connected with the neurons of the following layer. Connections exist only from the input layer in the direction of the output layer and is called feedforward network. First layer which is subjected to the input values is called the input layer. The layer producing the final result is the output layer. There can be one or many hidden layers between the input and output layers. The input is given to neural network and is propagated to next layer by means of chosen activation function.

There are two phases when using multilayer perceptron neural network model to any problem. In the training phase the historical data of input and output parameters is given to neural network model. As the input and output data of the training phase is given to neural network model, it is called as supervised training. The network is initialized with the random weights on the neural nodes. Based on the differences in the calculated/predicted output and the actual output, the weights at the different nodes in various layers are adjusted. Error signals are calculated at the output layer and propagated back to the hidden layers till the input layer is arrived. Backpropagation algorithm developed by Rumelhart and McClelland [15] implements the same concept in training the neural network mode. The error of the network is defined as the square distance between the stipulated output pattern and the actual output pattern generated on the basis of current network weights [16]. This results in an error function which is dependent on the weighting factors through the activation function. The training is continued till the predefined criteria like error threshold etc. are achieved. Once the training is complete, in the prediction phase the neural network model is subjected to the test data. The error function is calculated for the test data which determines the suitability of the neural network model for the identified problem. If the error is high, the neural network model need to be further trained or the network configurations need to be changed to address the problem at hand. The modified training and prediction phase are shown in fig 2 and 3 respectively [17].

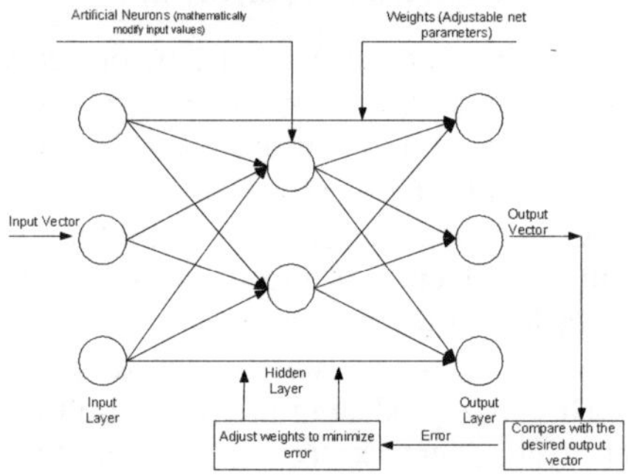

Fig. 2: Neural Network Training Phase

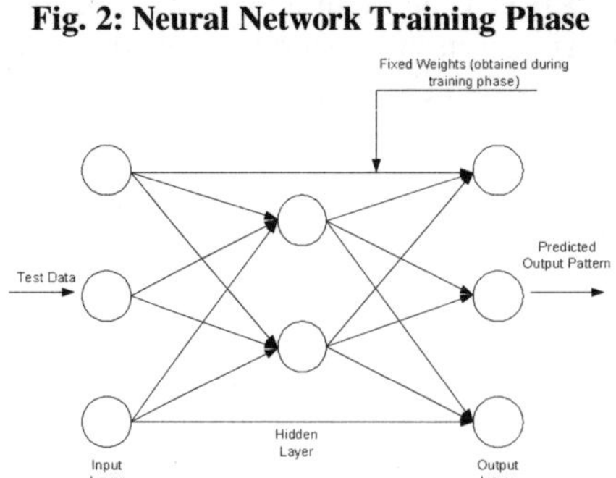

Fig. 3: Neural Network Prediction Phase

3.3 Model Errors Types

The four model error types are as follows:

Mean Absolute Error (MAE):

$$\mathbf{MAE} = \frac{1}{n}\sum_{i=1}^{n} \left| actualvalue_i - forecastedvalue_i \right| \tag{1}$$

Where the N is the number of the forecasted values

Mean Squared Error (MSE):

$$\mathbf{MSE} = \frac{1}{n}\sum_{i=1}^{n} (actualvalue_i - forecastedvalue_i)^2 \tag{2}$$

Root Mean Square Error (RMSE):

$$\mathbf{RMSE} = \sqrt{\frac{1}{n}\sum_{i=1}^{n} (actualvalue_i - forecastedvalue_i)^2} \tag{3}$$

Mean Absolute Percentage Error (MAPE):

$$\mathbf{MAPE} = \frac{1}{n}\sum_{i=1}^{n} \left| \frac{actualvalue_i - forecastedvalue_i}{forecastedvalue_i} \right| \tag{4}$$

4. TEST DEFECT PREDICTION MODEL

Test defect prediction model is used to predict the number of test defects.

4.1 Benefits of Test Defects Prediction:

The following are the benefits of test defect prediction:

1. Project managers can plan (resource, time and duration) for test phase of the software development project
2. It guides an project managers in making decision about continuation or abandonment of testing
3. Project managers can make decision about the delivery of product to the customer
4. Project managers can predict the defect density, service level agreement performance etc.

Proposed Test Defect Prediction Model:
Test Defects = F (Size, Requirements Defects, Design Defects, CWT Defects)
CWT = Code Walkthrough
The criteria variable is test defects, while all the remaining variables are the predictors. The data set is divided into two sets, first one containing the 80 projects list. The model is fitted with the 80 projects list. The model is applied to the remaining 20 projects. Mean Absolute Error,

Mean Square Error, Root Mean Square Error and Mean Absolute Percentage Errors are calculated.

4.2 Multiple Linear Regression Model for Test Defect Prediction

The multiple regression model is built on the project data and the model summary is shown in table 1. The predictor variables explain the variance of 97.8% in the dependent variable.

Table 1 Model summary of multiple linear regression model

Model Summary[b]

Model	R	R Square	Adjusted R Square	Std. Error of the Estimate
1	.989[a]	.978	.977	15.45744

a. Predictors: (Constant), CWT_DEF, REQ_DEF, SIZE_FP, DES_DEF

b. Dependent Variable: TEST_DEF

Table 2 ANOVA of Multiple linear regression model

ANOVA[b]

Model		Sum of Squares	df	Mean Square	F	Sig.
1	Regression	810330.8	4	202582.705	847.866	.000[a]
	Residual	17919.930	75	238.932		
	Total	828250.8	79			

a. Predictors: (Constant), CWT_DEF, REQ_DEF, SIZE_FP, DES_DEF

b. Dependent Variable: TEST_DEF

As shown in table 2 the prediction model is statistically significant as significance <0.05

Coefficients of Regression:

Based on table 3, the regression equation for the multiple linear regression model is:

Predicted number of Test Defects = -2.062 + 0.228 FP -0.157 REQ_DEF + 0.603 DES_DEF + 0.653 CWT_DEF (5)

The regression equation in the standard score form is:

Z predicted number of Test Defects = 0.069 FP - 0.046 REQ_DEF + 0.426 DES_DEF + 0.547 CWT_DEF (6)

Table 3 Coefficients of Multiple linear regression model

Coefficients[a]

Model		Unstandardized Coefficients B	Unstandardized Coefficients Std. Error	Standardized Coefficients Beta	t	Sig.
1	(Constant)	-2.062	4.169		-.495	.622
	SIZE_FP	.228	.174	.069	1.310	.194
	REQ_DEF	-.157	.161	-.046	-.975	.333
	DES_DEF	.603	.106	.426	5.661	.000
	CWT_DEF	.653	.113	.547	5.778	.000

a. Dependent Variable: TEST_DEF

4.3 Test Defects Prediction Model - Multiple Regression with Backward Elimination Method

Backward elimination is variable selection procedure in which all variables are entered into the equation and then sequentially removed. The variable with the smallest partial correlation with the dependent variable is considered first for removal. If it meets the criterion for elimination, it is removed. After the first variable is removed, the variable remaining in the equation with the smallest partial correlation is considered next. The procedure stops when there are no variables in the equation that satisfy the removal criteria. The variables entered and removed are shown in table 4. Criteria chosen for the model was F>0.10 for removal. Summary of the model is shown in table 5.

Table 4 Variables entered and removed in the model

Variables Entered/Removed[b]

Model	Variables Entered	Variables Removed	Method
1	CWT_DEF, REQ_DEF, SIZE_FP, DES_DEF		Enter
2		REQ_DEF	Backward (criterion: Probability of F-to-remove >= .100)
3		SIZE_FP	Backward (criterion: Probability of F-to-remove >= .100).

a. All requested variables entered.

b. Dependent Variable: TEST_DEF

Table 5 Model summary of multiple regression with Backward Elimination method

Model Summary[d]

Model	R	R Square	Adjusted R Square	Std. Error of the Estimate	Change Statistics				
					R Square Change	F Change	df1	df2	Sig. F Change
1	.989[a]	.978	.977	15.45744	.978	847.866	4	75	.000
2	.989[b]	.978	.977	15.45245	.000	.951	1	75	.333
3	.989[c]	.977	.977	15.57930	-.001	2.269	1	76	.136

a. Predictors: (Constant), CWT_DEF, REQ_DEF, SIZE_FP, DES_DEF
b. Predictors: (Constant), CWT_DEF, SIZE_FP, DES_DEF
c. Predictors: (Constant), CWT_DEF, DES_DEF
d. Dependent Variable: TEST_DEF

As seen from the model summaries, it can be concluded that removal of REQ_DEF do not have impact on the R Square. Similarly the Size_FP is also having minimal impact.

Table 6 Coefficients of Multiple regression with Backward Elimination method

Coefficients[a]

Model		Unstandardized Coefficients		Standardized Coefficients	t	Sig.
		B	Std. Error	Beta		
1	(Constant)	-2.062	4.169		-.495	.622
	SIZE_FP	.228	.174	.069	1.310	.194
	REQ_DEF	-.157	.161	-.046	-.975	.333
	DES_DEF	.603	.106	.426	5.661	.000
	CWT_DEF	.653	.113	.547	5.778	.000
2	(Constant)	-2.162	4.166		-.519	.605
	SIZE_FP	.258	.171	.078	1.506	.136
	DES_DEF	.584	.105	.413	5.578	.000
	CWT_DEF	.607	.103	.509	5.910	.000
3	(Constant)	.450	3.819		.118	.906
	DES_DEF	.594	.105	.420	5.638	.000
	CWT_DEF	.686	.089	.575	7.721	.000

a. Dependent Variable: TEST_DEF

Based on table 6, the regression equation for the model is:

TEST_DEF = 0.450 + 0.594 DES_DEF + 0.686 CWT_DEF (7)

The regression equation for the model in the standardized score form is:

TEST_DEF = 0.420 DES_DEF + 0.575 CWT_DEF

4.4 Test Defects Prediction Model - Multiple Regression with Forward Selection Method

Forward Selection is a stepwise variable selection procedure in which variables are sequentially entered into the model. The first variable considered for entry into the equation is the one with the largest positive or negative correlation with the dependent variable. This variable is entered into the equation only if it satisfies the criterion for entry. If the first variable is entered, the independent variable not in the equation that has the largest partial correlation is considered next. The procedure stops when there are no variables that meet the entry criterion. The variables entered and removed are shown in table 7. Criteria chosen for the model was F<0.05 for entry. Summary of the model is shown in table 8.

Table 7 Variables entered and removed

Variables Entered/Removed[a]

Model	Variables Entered	Variables Removed	Method
1	CWT_DEF	.	Forward (Criterion: Probability-of-F-to-enter <= .050)
2	DES_DEF	.	Forward (Criterion: Probability-of-F-to-enter <= .050)

a. Dependent Variable: TEST_DEF

Table 8 Model summary of multiple regression with forward selection method

Model Summary[c]

Model	R	R Square	Adjusted R Square	Std. Error of the Estimate	Change Statistics				
					R Square Change	F Change	df1	df2	Sig. F Change
1	.984[a]	.968	.968	18.39855	.968	2368.778	1	78	.000
2	.989[b]	.977	.977	15.57930	.009	31.784	1	77	.000

a. Predictors: (Constant), CWT_DEF
b. Predictors: (Constant), CWT_DEF, DES_DEF
c. Dependent Variable: TEST_DEF

Table 9 Coefficients of multiple regression with forward selection method

Coefficients[a]

Model		Unstandardized Coefficients		Standardized Coefficients	t	Sig.
		B	Std. Error	Beta		
1	(Constant)	.066	4.509		.015	.988
	CWT_DEF	1.174	.024	.984	48.670	.000
2	(Constant)	.450	3.819		.118	.906
	CWT_DEF	.686	.089	.575	7.721	.000
	DES_DEF	.594	.105	.420	5.638	.000

a. Dependent Variable: TEST_DEF

Based on table 9, the regression equation for the model is

TEST_DEF = 0.450 + 0.594 DES_DEF + 0.686 CWT_DEF (9)

The regression equation for the model in the standardized score form is

TEST_DEF = 0.420 DES_DEF + 0.575 CWT_DEF (10)

The forward selection and backward removal model has resulted into the same regression model for test defect prediction. This may vary based on the data set.

4.5 Artificial Neural Network Models for Test Defects Prediction

Two ANN models were generated for the above variables with the same 80 projects data. The configuration selected for the ANN model is chosen as follows:

Table 10 Neural Network parameters for test defects prediction

Sr. No.	Neural Network Parameter	Neural Network Model I	Neural Network Model II
1	Number of layers	3	3
2	Number of Neurons on Input Layer (linear transfer function)	4	2
3	Hidden Layer (tanh transfer function)	5	5
4	Output Layer (linear transfer function)	1	1
5	Learning Method	Backpropagation	Backpropagation
6	Learning Strategy	Single Step Delta	Single Step Delta
7	Presentation Order	Sequential	Sequential
8	Learning Rate	Hidden Layer: 0.1 and Output Layer 0.1	Hidden Layer: 0.1 and Output Layer 0.1
9	Weight initialization	Upper bound: 0.1, Lower bound: 0.1	Upper bound: 0.1, Lower bound: 0.1
10	Stop Conditions	Best training set based on RMS training error & number of epochs divisible by 1000	Best training set based on RMS training error & number of epochs divisible by 1000

The project data was divided into three data sets namely training, testing and recall. The training dataset is of 60 projects. Test data is of 20 datasets and recall data set is also of 20 datasets.

Table 11 Neural network V1.0 training state for test defect prediction

Sr. No.	Training Parameter	Neural Network Model I	Neural Network Model II
1	Epoch	168	110
2	Max Training Error	46.56	37.79
3	RMS Training Error	9.78	10.31
4	Max Test Error	32.11	28.31
5	RMS Test Error	10.84	11.48

Table 12 Neural network V1.1 training state for test defect prediction

Sr. No.	Training Parameter	Neural Network Model I	Neural Network Model II
1	Epoch	976	197
2	Max Training Error	44.72	37.47
3	RMS Training Error	9.83	10.43
4	Max Test Error	29.36	26.85
5	RMS Test Error	10.26	11.91

Using eq 1, 2, 3 and 4 the errors are compared for MAE, MSE, RMSE and MAPE and shown in table 13.

Table 13 Error comparison of various defect prediction models

Method Name	MAE	MSE	RMSE	MAPE
Stepwise Linear Regression	2.79	10.41	3.22	0.28
Backward Elimination	2.85	10.62	3.26	0.29
Neural Network Model I version 1.0	6.88	85.01	9.22	1.97
Neural Network Model II version 1.0	3.27	16.83	4.10	0.31
NN_Model I Version 1.1	5.99	78.27	8.84	0.47

5. DATA ANALYSIS FOR TEST AND FIELD DEFECT PREDICTION

From the above data analysis it is evident that multiple linear regression is able to fit the defect data and is having minimum errors as compared to neural network models. This may be due to stable data used for the defect prediction, which may not be the case in every analysis. Further fine tuning of neural network models is required to minimize the errors. Application of the traditional statistical methods and NN model can be applied to different organisational data sets to compare the defect prediction accuracy. This forms basis for future research in this area.

6. REFERENCES

1 Stephen Kan, Metrics and models in Software Quality engineering, Addison Wesley.

2 Sunita Chulani, Constructive Quality Modeling for Defcet Density Prediction: COQUALMO .

3 F. Akiyama, "An Example of Software System Debugging," Information Processing, vol. 71, pp. 353-379, 1971.

4 A.E. Ferdinand, "A Theory of System Complexity," Int'l J. General Systems, vol. 1, pp. 19-33, 1974. M.H. Halstead, Elements of Software Science. Elsevier, North-Holland, 1975.

5 L.M. Ottenstein, "Quantitative Estimates of Debugging Requirements," IEEE Trans. Software Eng., vol. 5, no. 5, pp. 504-514, 1979.

6 M. Lipow, "Number of Faults per Line of Code," IEEE Trans. Soft ware Eng., vol. 8, no. 4, pp. 437-439, 1982.

7 J.R. Gaffney, "Estimating the Number of Faults in Code," IEEE Trans. Software Eng., vol. 10, no. 4, 1984.

8 T. Compton, and C. Withrow, "Prediction and Control of Ada Software Defects," J. Systems and Software, vol. 12, pp. 199-207, 1990.

9 M.A. Cusumano, Japan's Software Factories. Oxford Univ. Press, 1991.

10 K. Koga, "Software Reliability Design Method in Hitachi," Proc. Third European Conf. Software Quality, Madrid, 1992.

11 K. Yasuda, "Software Quality Assurance Activities in Japan," Japanese Perspectives in Software Eng., pp. 187-205, Addison-Wesley, 1989.

12 Achamma Jose, Anju N.K, S.K.Pillai, "Closed-loop defect removal model using statistical process control", Vol 3, No 1, pp:39-47.

13 N. Fenton and M. Neil "A Critique of Software Defect Prediction Research", IEEE Trans. Software Eng., 25, No.5, 1999.

14 McCulloch W. W., and Pitts W. (1943). "A logical calculus of the ideas imminent in nervous activity", Bulletin of the Mathematical Biophysics, 5, pp: 115–133.

15 McClelland James L, Rumelhart, David E (1989), "Explorations in parallel distributed processing : a handbook of models, programs, and exercises" , Cambridge : MIT Press.

16 Rumelhart D., Hinton G., and Williams R (1986)., Learning internal representations by error propagation, in Parallel Distributed Processing, vol.I, MIT Press, Cambridge, MA, 318- 362.

17 Khoshgoftaar Taghi, Pandya A., More H.(1992), "A neural network approach for predicting software development faults", Proceedings of Third International Symposium on Software Reliability Engineering, pp. 83-88.

Reliability, Safety and Hazard
Editors: P.V. Varde et al.

CONCEPT OF SOFTWARE RELIABILITY IN MOBILE PHONE SOFTWARE SYSTEMS

R. Anil

L&T InfoTech Ltd, Mhapae, Navi Mumbai 400 701

ABSTRACT

Nowadays, we are using many systems, which contain one or another form of software. If a system is analysed in the reliability perspective, one can easily identify that the weakest link is software. In the telecommunication domain this is particularly true. In order to overcome this problem, extensive reliability based testing must be performed for the software systems. Generally, the testing of software from the user perspective is seldom the first priority than from the perspective of the developer.

In this paper, the usage of the software reliability concept for the mobile phone software testing from the user perspective is discussed with the help of a case study.

Keywords: Software reliability, Mobile phone, Operational profile

1. INTRODUCTION

Nowadays, most of the systems which we are using contain one or another form of software. These systems have become an integral part of our life. This increasing dependency reduces the tolerance to failures and thus there is a need for more reliable systems. If a system is analysed in the reliability perspective, one can easily identify that the weakest link is software. In the era of embedded systems, this is a major concern.

In the telecommunication domain this is predominantly true. The hardware parts of the telecommunication systems like mobile handsets, mobile antenna towers and the basic server configurations are proven for their high reliability. Due to the increase in use of software to manage the telecommunication systems more effectively and efficiently, the software reliability has to be increased for achieving high overall reliability. Similarly, one can find extensive reliability based testing in case of the mobile phone hardware. But in case of its software, this trend is not visible. In order to overcome this problem, extensive reliability based testing must be performed for the software parts just like the hardware parts.

The testing of software in the present scenario is not based on the usage of the software. Rather the testing is from the perspective of the developer. However, the software is developed based on the user requirements. While testing the software, the main focus is to see whether the software written is meeting the requirements or not. The testing of software from the user perspective is seldom the first priority.

If the testing of the mobile phone software is from the user perspective, one can avoid lot of potential defects, which may be leaked to the customer, and thus increase the customer satisfaction.

The reliability based testing, in brief, is done by determining the probability of usage of the functionalities (or operational profile of the functionalities) of the software. The testing time will be determined based on the probability. i.e., more testing will be conducted on that functionality which will be used more frequently. One point is to be noted here is that the reliability based testing will start only after the unit testing of the functionalities. This method is more appropriate at the time of validation testing.

The testing will be conducted in a random manner. i.e., the sequence of testing of different functionalities will be in random, which is closer to the usage by the user.

In order to find out the sequence of different functionalities while in operation, an nXn matrix of the different functions can be used to find out the valid sequence of functions. For example, the sequence of receiving a call while making a call is valid in case of mobile phone, where as the reverse is not valid

2. RELIABILITY BASED MODELS

In order to make the testing environment as close to reality as possible it is important to have good models of how the system is used. According Lyu[1] the most referred usage specifications are the domain-based model, the algorithmic model, operational profiles, Markov model, grammar model and the state hierarchy model. Since the operational profile based testing is considered more relevant in this context, the decision is to go for the operational profiles. Hence, the last three are not covered in this paper.

Domain-based model

The domain-based model is the most basic one. It has a probability attached to every function of the system that the user has access to. The functions are executed randomly according to the probability, and mostly unchanged probabilities are used with this model. The main drawback of the domain-based model is too simple to use in more complex software systems.

Algorithmic model

The algorithmic model is an improvement of the domain-based model. The improvement is that this model uses the input history to calculate the probabilities of the functions. To make this possible the input history must be divided into a set of classes. These classes have their own probabilities for each function, see figure 1 with m classes and n different functions, the usage profile is described in an mXn matrix, where the values are the probabilities for the function with a given input history.

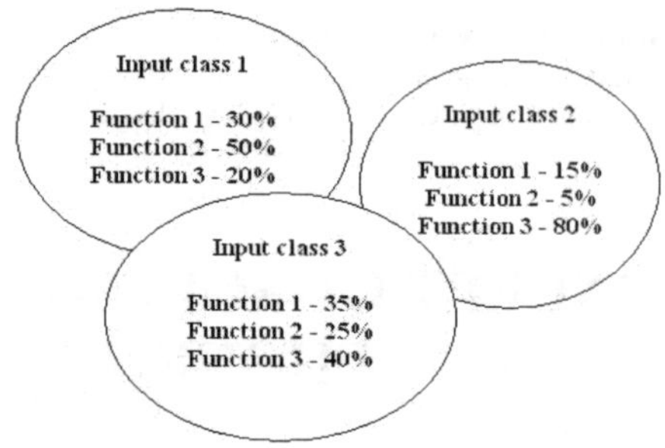

Fig. 1: E.g. input classes in the algorithmic model

Operational profile model

The model of operational profiles is a further refinement of the two previous models. The solution to the problem with selecting classes is "built into" the model. It is divided into five levels, where each one is more detailed than the parent level, see figure 2. On every level there exist alternatives that together have a probability of 100%. The upper most level is called "customer profile". A customer is a person or institute who obtain the system. The next level is the "user profile" which consists of, for example, age groups or different business segments within the chosen customer. The users are the ones that actually employ the system. If the customers are the same as the users, only one of these levels is needed. The next level is the "system-mode profile", which describes the different states of the system. Below the "system-mode profile" is the "functional profile", where all functions available to the user are gathered. On the last level, the "operational profile" contains the operations that implement the functions.

In figure 2, an example is shown to illustrate the model. The advantages here are that it is fairly simple to achieve a realistic model if sufficient usage information is available and that it can handle large systems. The downside however is, that it does not support the detailed behaviour of the user.

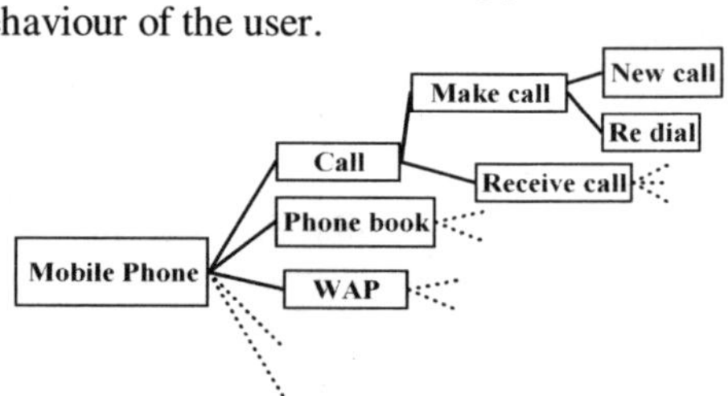

Fig. 2: Operational profile model

3. DEVELOPMENT AND STRUCTURE OF OPERATIONAL PROFILES

The formation of an operational profile is made by looking at the usage from customer down to operation and quantifying how often the elements of each step will be used. But one cannot say that the usage is the only thing to take into account. In such case, infrequently used, high-risk operations could be left untested, e.g. emergency landing system of an aircraft. One solution is to consider both usage and criticality. During the development of the operational profile, several sub-profiles will be created. In this context, a profile is defined as a set of alternatives, of which only one can occur at a time, and their occurrence probability. An operational profile can involve as many as five levels, as shown in figure 3, although not every application will require all levels. The steps are to create the customer profile, the user profile, the system-mode profile, the functional profile and the operational profile.

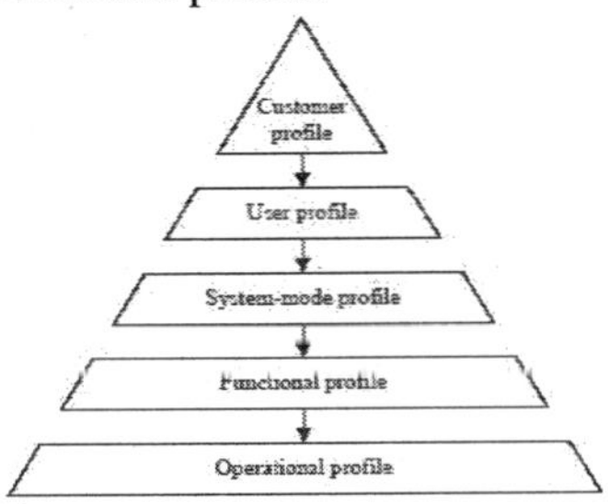

Fig. 3: System Profiles

Customer profile

The first step is to find the customer profile. A customer can be a person, group or institution that acquires the system, and the customer profile is a set of customers or groups and their associated occurrence probabilities. Information about potential customers can be obtained from market research or similar techniques. The probability of each customer group is depending on the proposition of the usage. In the mobile phone scenario, the customers are the general people distributed around the world.

User profile

The second step is to find the user profile. A user can be a person, group or institution, but the user employs the system. The user profile is a set of users or user groups and their associated occurrence probabilities. Similar user groups among different customer groups should be combined for convenience, and the occurrence probability is the proportion of use the user group represents within its customer group. In our case, we have considered the customers itself as users.

System-mode profile

As a third step, the system-mode profile needs to be created. A system-mode is a number of functions or operations that are grouped to simplify the analysis of the execution behaviour. The system-mode profile is a set of system-modes and their associated occurrence probabilities. A system-mode could be based on a user group, a significant environment condition, an operational architecture structure, criticality, or user experience. In our case, the mobile phone users are considered as a set of users with almost similar knowledge about the system. Hence we have not divided them into different system mode profiles.

Functional profile

To create the functional profile, the system has to be broken down into tasks, and an occurrence probability should be assigned to each task or function. Since the functional profile is based on the features and functions built into the software, the highest efficiency is usually achieved if it is developed during the requirements phase, and preferably becoming a part of the requirements document.

Two tasks should be defined as different functions if they can be developed with different priorities and resource allocations, and especially if they differ considerably in frequency of use and criticality.

Operational profile

At the bottom level of figure 3 is the operational profile. Unlike the more user oriented functional profile, the operational profile is based on the operations that implement the functions. An operation should be considered as a task performed by the system, and is therefore more specific than a function. Each function may in other words involve more than one operation, which results in generally greater number of operations than the functions [5]

Just like the other profiles, the operational profile is a set of operations and their associated occurrence probabilities.

4. CONSIDERATIONS WHILE CREATING OPERATIONAL PROFILES

Below, given are the five considerations [2, 3] that are the most important when creating operational profiles.

1. Operational profiles should not be too complex since the cost increases at the same rate as the number of profiles and operations. The approach is to set aside operations, which are rarely used and less important to the system. If there are faults in these rare operations, it will not be critical to the system. It is of greater importance that more often used and more vital operations are tested. If there are operations, which are hardly ever used although essential, they must of course be tested.

2. A good way to resolve the occurrence rates is to do a survey among expected users with, for example, a questionnaire. According to [2, 4], "in sampling users, the same guidelines and statistical theory used for polling and market surveys apply; a sample of 30 or even fewer users may suffice to generate an operational profile with acceptable accuracy." The best though, is to use "Delphi techniques" [2]. This means individual estimations by experts, which are reviewed by an expert group. If none of the above is possible for some reason, the occurrence rates must be estimated by the developers or whoever knows the customers best. In our case, we have used Delphi techniques to find out the probabilities

3. There are some risks regarding the statistics collection of the usage, which can result in imbalanced values This can of course result in inverted imbalanced values and may need to be edited manually to make sure these functions get reasonable values.

4. When using operational profiles, it should not be handled with test cases using, for example, random sequences of inputs to force an error. It should rather be dealt with using test cases similar to other operations tested, but with the difference that they contain a few erroneous inputs. These test cases should then be added as ordinary operations but with much lower probabilities. The size of the probabilities can be acquired by measuring the frequency of human typing errors.

5. In the software that is being developed, the number of possible combinations can be huge and the testing therefore unrealizable. The way to handle this is to examine which combinations are most common. Usually only a few combinations out of a large portion of the combinations are used in the market. This means that the least common combinations can be discarded, alternatively be briefly tested.

5. CASE STUDY

L&T InfoTech is developing software for different mobile phone manufactures. Normal mobile phone software typically consists of many modules like call, phonebook, games, messages etc. Each module contains different sub modules or scenarios, which will be used by the users.

When a normal software application is being tested, normally the test cases are prepared based on the requirements. As mentioned earlier, the main drawback here is that the usages of different requirements are not considered. When the requirements are given to the developers who develop and test the software, the thought process will be mainly from the developer's angle. By this, the complicated or advanced sub modules as per developers' angle will be tested more rigorously. It may also be possible that the user rarely uses these complicated or advanced sub-modules. This creates a possibility that in spite of the software is tested vigorously, the user may find out defects in the very early stage of the usage and may occur very frequently.
Hence it was decided that during the integration testing, the software reliability concepts would be used.

Steps involved

1. Finding all the possible scenarios.
The mobile phone software contains many sub modules and scenarios. Using the requirements captured from the customer, al the possible scenarios can be captured. It is possible that

these scenarios can be triggered at the same time. For example it is possible to receive a call while typing a message. These types of scenarios have to be captured for the foolproof testing. Part of the nXn matrix created for collecting all these scenarios is shown below This matrix is prepared by placing sub modules column wise as well as row wise. Then the interception point defines one complex scenario.

Table 1 nXn matrix to find out all scenarios

Modules	Sub Module (Parameters)	Make Voice Call Using Key Pad	Make Voice Call Using Call Log	Make Voice Call Using Phone Book	Receive Voice Call
Call	Make Voice Call Using Key Pad	T1	T2	T3	T4
Call	Make Voice Call Using Call Log	T10	T11	T12	T13
Call	Make Voice Call Using Phone Book	T19	T20	T21	T22
Call	Receive Voice Call	T28	T29	T30	T31
Call	Reject Incomming Call	F	F	F	F
Call	Volume Level	F	F	F	T38

If the generated scenario is valid then it is numbered as per some naming convention. E.g. all valid scenarios are marked here as T1, T2, T3 etc.

It is found that individual scenarios are valid but their combination may not be always valid. For e.g. Trying to run following two scenarios, "Reject Incoming Call" & ""Receive Voice Call". In this case, its interception happens to be invalid. Invalid scenarios are marked as "F".

*2. **Assigning the probability***

The main purpose of this testing is to conduct the testing from users' angle. To do this, the probability of usage for each modules and sub modules are to be captured. An internal survey has been conducted and the probabilities were decided.

Table 2 Probabilities of modules

	Total Time	180 min	
Sr.No.	Module Name	Probability	Time (min)
1	Call	0.5	180 x 0.5 = 90
2	Phone book	0.2	180 x 0.2 = 36
3	File browser	0.1	180 x 0.1 = 18
4	Media player	0.1	180 x 0.1 = 18
5	Organiser & tools	0.05	180 x 0.05 = 9
6	Phone profile	0.05	180 x 0.05 = 9

Before testing, based on the past experience in the similar software testing, the team decided that the testing time would be 180 minutes for each iteration.(Table 2) The time for each sub-module will be derived form this time considering the individual probabilities.

After finding the probabilities for each module, the probabilities of each module were split again to find out the testing time for each sub module. (Table 3)

Table 3 Probabilities of sub modules

Sr No.	Module Name (min)	Sub Module	Sub Probability	Sub Cumulative	Testing Time (min)
1	Call (90)	Receive Voice Call	0.3	0.3	27
		Make Voice Call Using Key Pad	0.2	0.5	18
		Make Voice Call Using Call Log	0.2	0.7	18
		Make Voice Call Using Phone Book	0.2	0.9	18
		Volume Level	0.1	1	9
2	Media Player (18)	Play Music File	1	1	18
3	Phone Book (36)	Adding Contacts	0.5	0.5	18
		Assign Ringtone	0.2	0.7	7.2
		Editing Contacts	0.2	0.9	3.6
		Deleting Contact	0.1	1	7.2
4	File Browser (18)	Create Folder	0.5	0.5	9

Now, each sub module is split into the valid scenarios. There exist single scenarios as well as combination of scenarios as mentioned above. These valid scenarios are clubbed into the respective sub modules and the probabilities of the respective scenarios are calculated. An example of the scenarios under the sub module receive voice call is given in table 4.

Table 4 Scenarios under software sub modules

Module	Sub module	Scenario
CALL	Receive Voice Call	
0.5	0.3	Receive a Incoming Voice Call, and Receive another Incoming Voice Call
0.5	0.3	Receive a Incoming Voice Call And Reject another Incoming Voice Call
0.5	0.3	Receive a Incoming Voice Call And Make a Outgoing Voice Call Through Phone Book
0.5	0.3	Receive a Incoming Voice Call And Make a Outgoing Voice Call Through Call Log
0.5	0.3	Receive a Incoming Voice Call And Make a Outgoing Voice Call By Dialling Key Pad
0.5	0.3	Receive a Incoming Voice Call
0.5	0.3	Reject a Incoming Voice Call

After determining the operational profiles, the cumulative probabilities were determined. With the above data, the total system cumulative probabilities were determined and the testing was conducted by generating random numbers. The random numbers are also generated to find out which sub module has to be tested. Another set of random numbers is also generated to find out which scenario under the selected sub module has to be tested.

6. CONCLUSION

The benefits of software reliability based testing are more precise satisfaction of customer needs, better effort and schedule control, increase in productivity, more systematic management of project etc.

The piloting of this method on the initial release of the actual software was a huge success. Many undetected defects during the normal testing were detected in this test. It was observed that the defects, which were leaked to the customer, were reduced drastically.

7. FUTURE WORK

With the acceptance of this method by the developers and customers, the final release of the software, which contains much more modules and sub modules than the software under test, will be tested with this methodology. The determination of reliability of the software at the time of delivery is also planned. The results of these will be used to develop a model for the same type of software is also under anvil, which will predict the reliability at the beginning of the software development itself.

8. REFERENCES

1. M.R Lyu (Ed), "Handbook of software reliability Engineering", IEEE computer society press, 1996
2. Musa JD, Iannino A, Okumoto K., "Software reliability: measurement, prediction, application". New York: McGraw-Hill, 1987.
3. J D Musa, "Introduction to software reliability engineering and testing" 8th International Symposium on Software Reliability Engineering (Case Studies), 1997
4. J D Musa, http://members.aol.com/JohnDMusa/index.html
5. B. Yang, M. Xie, "A study of operational and testing reliability in software reliability analysis", Reliability Engineering and System Safety, 2000 323-329

Reliability, Safety and Hazard
Editors: P.V. Varde et al.

A NEW APPROACH TO SOFTWARE RELIABILITY PREDICTION BASED ON FAULT POTENTIAL

Vinod M. Kumre[1], *R.B. Misra*[2] *and Praveen V. Sasatte*[2]

[1]Entity Solutions Limited, Pratap Nagar Industrial Area, Udaipur
[2]Reliability Engineering Centre, I.I.T. Kharagpur

ABSTRACT

In conventional software reliability analysis, an equal importance is given to all faults and more importance to the 'time to failure' instead of potential of fault to cause number of failures in field. Hence with conventional approach it is not possible to find true reliability improvement due to a fault removal. In this paper a new approach to software reliability prediction is proposed, wherein more importance is given to the potential of fault instead of 'time to failure'. The faults are examined for their potential of influencing the reliability of software when removed and when not removed from the software. The fault potential considered is the number of times a fault can cause software failure if not removed and the magnitude of reliability improved after removing them.

Keywords: Software reliability, Fault Potential

1. INTRODUCTION

The potential of faults to cause number of failures in the field during the operational period of software is of a great concern because of its magnitude of influencing the reliability of software. With the conventional software reliability analysis methods it is not possible to estimate the true reliability of software which is small in size and/or does produce small number of failures. As per the existing reliability growth models like Musa Basic, Musa Okumoto, Jelinski-Moranda, etc. [1, 2] it requires around 40-50 failures for application of these models to have a reasonably accurate reliability estimate. These models do not take care of the potential of a fault for getting a failure in the field based on the testing data. The reliability estimated with these models during the testing phase can not be correlated with the field reliability. This paper primarily aims to address the existing software reliability problems and suggests a new software reliability prediction method based on the fault potential. To elaborate the proposed concept, let us consider, in software testing we found two faults. In between these two faults, first fault can cause system to fail daily in field if we don't remove it whereas second fault can cause system to fail once in six month in field if we don't remove it. In the above example the first fault has great influence on reliability of software and if we remove it then software reliability will improve by large magnitude and if we don't remove it then we will experience a low reliability from software. However the second fault is having low influence on reliability of the software. Its removal or no removal can't affect the reliability of software by large magnitude. The following sections of this paper describe the proposed methodology in detail.

2. PROPOSED METHODOLOGY

In the proposed software reliability analysis, more importance is given to the potential of fault instead of 'time to failure'. This can be explained as follows.

In a software development process let us consider two copies of the same software are obtained for the testing. First copy of software is used for testing with debugging and five 'faults' has been found. (Assume that software is fault free after removing these five faults). Second copy of software is tested without removing the faults and purpose of this testing to count the number of repeated failures caused by each of the faults. The five faults and their probability to cause failure in field is descried in Table 1, this probability is obtained using second type of testing described above.

Table 1 Probability of fault to cause failure in field

Fault No.	Occurrence Probability
Fault-1	One failure in two month
Fault-2	One failure in three month
Fault-3	One failure in Four month
Fault-4	One failure in six month
Fault-5	One failure in one year

If all five faults in software are not removed then user will observe total sixteen failures in one year operational period. If the fault-1 is removed from the software then reliability of software will be improved by larger magnitude (user will experience only 10 failures in operational life of one year) on other hand if fault-5 is removed from the software then reliability of software will not be improved as much observed in earlier case (user will experience 15 failures in operational life of one year). Hence the fault-1 has more influence on software reliability than the fault-5.

3. RELIABILITY ESTIMATION

T_o = Operational life of software = the time span in which each operation of software is executed at least once.

N_0 = Total number of faults in software at the beginning of testing

Nr = Total number of faults removed in testing

N = $N_0 - N_r$ = Total number of residual faults in software. Here we have assumed that (N0 > Nr)

n_0 = Total number of failures that will be experienced by user if software is used without testing and debugging

n_r = Total number of failures that will not be experienced by user due to removal of Nr faults

n = $n_0 - n_r$ = Total number of failures that will be experienced by user after Nr faults are removed in testing Here we have assumed that $(n_0 > n_r)$

$$MTTF_o = \frac{T_o}{n_0} = \text{MTTF of software if software is used without testing and debugging}$$

$$MTTF_r = \frac{T_o}{n} = \frac{T_o}{n_0 - n_r} = \text{MTTF of software after } N_r \text{ faults are removed in testing}$$

Our objective is to estimate the $MTTF_r$. To estimate $MTTF_r$, n_r is estimated from number faults removed in testing and their occurrence probability and n_0 can be estimated as described in following section.

Estimation of Reliability Parameters

Estimation of N_0

The N_0 is estimated from any of the fault estimation model. The 3 sigma standard deviation upper bound of the expected number of faults is taken as a value of N_0.

Estimation of n_r

To estimate n_r following steps are followed;

***Step 1*:** Map the identified testing faults (N_r) to the operations in operational profile in such a way that the 'Fault j' is observed in i'th operation.

***Step 2*:** The yearly occurrence rate (O_i) of the operation is identified for each operation

***Step 3*:** The probability of each fault to result in a failure (P_j) in single use of operation is identified.

***Step 4*:** The expected number of failures (n_r) that will not be experienced after removal of N_r faults, is calculated based on the yearly occurrence of the operation i and the probability of the fault j to result into a failure in single use of operation i.

Table 2 presents details related with the estimation of n_r

Table 2 Example of estimation of n_r after removal of N_r faults in testing.

Operation	O_i (Hrs)	Faults mapping	P_j	n_r
Op 1	100	Fault 1	0.1	10
		Fault 2	0.2	20
		Fault 3	0.05	5
Op 2	75	Fault 1	0.2	15
		Fault 2	0.1	8
		Fault 3	0.05	4
Op 3	40	Fault 1	0.25	10
		Fault 2	0.3	12
Op 4	15	-	-	-
Op 5	12	Fault 1	0.45	5
		Fault 2	0.3	4
Op 6	12	Fault 1	0.2	2
		Fault 2	0.04	0
		Fault 3	0.08	1
Op 7	8	Fault 1	0.2	2
Op 8	4	Fault 1	0.5	2
Op 9	2	0	0	0
Op 10	1	0	0	0
k =10	**269**	**Faults =15**	-	**100**

Where, i= 1 to k; j = 1 to m; k=10
k – Total number of operations
m_i = Total number of faults in i'th operation

The average value of P_j can be established from past projects.
From Table 2, n_r = 100

Estimation of n_0

It can be assumed on the basis of experience that

$n_0 = f(N_0, k, m_i, O_i, P_j)$ (1)

Where,
k = Total number of operations
m_i = Total number of faults in i'th operation
O_i = yearly occurrence count of i'th operation
P_j = Probability of j'th fault to result in a failure of single use of i'th operation
To estimate n_0, following steps are followed;

Step 1: List the number of operations and their yearly occurrence count.
Step 2: Calculate occurrence probability of each operation
Step 3: Establish the average probability of fault to result into failure from past projects
Step 4: Distribute the estimated number of faults (N_0) as per the operation probability
Step 5: To estimate n_0 operation wise, multiply the occurrence count, allocated faults and the average probability of a fault to result into failure.
Step 6: Total number of estimated failures (n_0) will be the summation of the individual operation wise failures.
Table 3 presents details for estimation of n_0

Table 3 Example of estimation of n_0 after considering N_0 = 20 and average probability of fault to result into failure be 0.1 (average value of P_j from past projects)

Operation	O_i (Hrs)	P_{OCC}	Faults mapping m_j =(P_{OCC} * N_0)	n_0 (O_i* m_j * 0.1)
Op 1	100	0.3717	7.4349	74.3494
Op 2	75	0.2788	5.5762	41.8216
Op 3	40	0.1487	2.9740	11.8959
Op 4	15	0.0558	1.1152	1.6729
Op 5	12	0.0446	0.8922	1.0706
Op 6	12	0.0446	0.8922	1.0706
Op 7	8	0.0297	0.5948	0.4758
Op 8	4	0.0149	0.2974	0.1190
Op 9	2	0.0074	0.1487	0.0297
Op 10	1	0.0037	0.0743	0.0074
k = 10	**269**	**1**	**20**	**132.5**

Where, i= 1 to k; j = 1 to m; k=10.
P_{OCC} = Probability of occurrence of a operation
From Table 3, the empirical estimation of the total number of failures n_0 = 132

Estimation of $MTTF_r$

For the given example, T_0 =269 hrs meaning that the software is going to be used for 269 hrs in one year

Average use of software per day –
269/365 = 0.736986 hrs/day

$$MTTF_r = \frac{T_o}{n} = \frac{T_o}{n_0 - n_r}$$

= 269/ (132-100)
= 8.4 hrs. in 269 hrs of use

$MTTF_r$ = 8.4 / 0.736986
= 11.4 days

The proposed methodology could be verified using the actual projects data.

4. CONCLUSION

The proposed methodology for predicting the software reliability based on fault potential gives the actual reliability that the software may experience in the field. It predicts the field reliability based on the operational profile and the faults observed in testing.

5. REFERENCES

1. Lyu M. R., (1995), *Handbook of Software Reliability Engineering,* New York: McGraw-Hill, pp. 71-117.
2. Musa J. (1998), *Software Reliability Engineering,* New York: McGraw-Hill, pp. 259-340.

Reliability, Safety and Hazard
Editors: P.V. Varde et al.

A METRIC FRAMEWORK FOR CMMI$^{(SM)}$ IMPLEMENTATION

N.J. Rajaram and Anjali Mogre

Syntel, Unit 112, SDF IV, SEEPZ, Mumbai 400 096

ABSTRACT

Capability Maturity Model Integration – CMMI$^{(SM)}$ – Level 4 process area – Quantitative Project Management demands the process capability and product quality to be measured quantitatively. However the model do not specify the list of metrics that should be used. It is left to the organizations to decide on a set of metrics it would like to measure, monitor and improve.

This paper suggests a framework for a comprehensive set of metrics for some process areas in the model. Organizations can select suitable metrics as per their own goals and requirements. The paper also provides techniques to select the metrics and guidelines for successful metrication program.

1. INTRODUCTION

Software measurement is a challenging but essential component of a healthy and highly capable software engineering culture. Effective project management mandates measurement of capability of the processes followed to develop the product. Software measures should help in ensuring and predicting the quality of output. They facilitate in driving deliveries as per schedule and in budget.

Software measurement provides better control on the project, raises alarms at distinct points in the various phases of software development life cycle to take corrective actions. They also aid in helping to predict product quality, generating correct effort and cost estimates and enabling proactive preventive measures.

Software metrication programs not only improve the software quality but also reduce the overall software costs. The following graph depicts the relationship between software quality and software cost for Health and Pathological SDLC. If one follows healthy SDLC model – regular reviews, controls and a proper software metrication program at every phase - the cost may be higher in initial stages but the overall cost is significantly less. In Pathological SDLC - where the reviews and controls are not regular but done only if the symptoms are seen and no systematic software metrication program – the initial software development cost might be lower but the overall development cost in much higher.

Fig. 1: Relationship between Software Quality and Cost [1]

2. CMMI – Capability Maturity Model Integration$^{(SM)}$

CMMI- Capability Maturity Model Integration $^{(SM)}$ developed by Carnegie Melons Software Engineering institute (SEI), provides a structured view of process improvement across organization. The CMMI$^{(SM)}$ model integrates systems and software disciplines into one process improvement framework. The model defines 5 maturity levels; namely; Initial, Managed, Defined, Quantitatively Managed and, Optimizing.

3. SOFTWARE METRICS

A metric quantifies a characteristic of a process or product. Metrics define what is to be measured. Software metrics can be evolved to measure the capability of the processes used to create the software product and the quality of the product delivered or the service provided. The current measures of performance need to be used to improve the future work estimates; otherwise the future estimates will just be speculations. The project status cannot be tracked meaningfully unless the actual effort and time spent on each task are compared to planed effort and time respectively. It is not possible to decide whether the product is stable enough to ship unless one tracks the rate at which the development team is finding and fixing defects. One cannot quantify how well the new development processes are performing without the measures of the current performances and baselines to compare against. Metrics help in controlling of software projects and learning more about the way organization works.

Software metrics are typically used for

- Size, Effort and Cost estimation
- Productivity measurements
- Reliability measurements
- Quality modeling
- Management of tasks and projects

4. METRICS CLASSIFICATION

In the approach for metrication, we try to identify and classify metrics into three different categories. The metrics categories are operational, managerial, and business oriented metrics.

Operational Metrics:

Operational metrics are those metrics that the projects are expected to track frequently during project execution. These metrics help project manager to take appropriate decisions to keep the project in control.

Managerial Metrics:

Managerial metrics are tracked at regular intervals and are used to help in senior management decisions. If operational metrics are not tracked and controlled on time and if they exceed the threshold values then they start reflecting in the managerial metrics.

Business Metrics:

Business metrics are related to Cost, schedules and customer related measures. If managerial metrics are not controlled at the stipulated time, then they start impacting business metrics; it may result in loss of revenue, loss of business and reduced customer satisfaction.

A business metrics may have a number of managerial metrics. Each managerial metrics will have a set of operational metrics, which the project leads will monitor at the regular intervals.

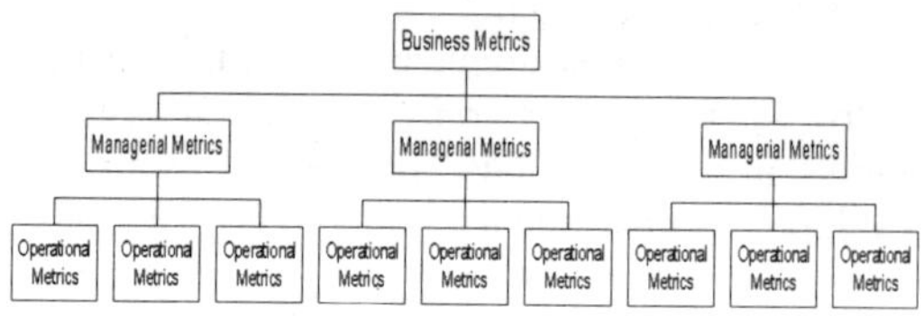

Fig. 2: Relationship between Business, Managerial and Operational Metrics

Considering the case of Requirement Development Process Area (SP1.1.2), the Business Metrics, Managerial metrics and Operational Metrics can be derived as furnished in the table given below.

Table 1.1 Metrics for Requirement Development Process Area SP1.2

SP 1.1-2	Business Metrics	Managerial Metrics	Operational Metrics
Requirement Development (RD) Elicit stakeholder needs, expectations, constraints, and interfaces for all phases of the product life cycle.	Effort / Cost on RD	1. Size of Req 2. Count of Req 3. # of requests / unit time 4. Time spent by customer	1. # of Stakeholder needs 2. # of constraints 3. # of interfaces 4. Time spent on requirement elicitation 5. # of methods used for requirement elicitation

5. SELECTION OF METRICS

Methodology

The methodology suggested here is based on the <u>GQM</u> paradigm.

Goals → Questions → Metrics

Fig. 3: Metrics selection Methodology – GQM

To select suitable metrics for an organization, first have a look and organizational goals [2]. From each organizational goal derive a set of questions that needs to be answered to achieve the particular goals. The answers of these questions will provide you a set of metrics that the organization should focus upon.

6. METRICS FOR CMMI FRAMEWORK

Taking the case of two engineering related process areas; namely Requirement Management (RM) and Requirement Development (RD); we derive the following operational metrics and establish relationship with the managerial and business metrics.

The measurement goal of the metric is defined based on the Process Area and the Specific / Generic Practice. It is seen that this goal relates to the operational metric(s), which converges to a set of managerial metric(s), which in turn leads to a set of Business metrics.

7. IMPLEMENTING THE SOFTWARE METRICATION PROGRAM

Collecting the data for measurement, analyzing and interpreting results takes considerable amount of time and effort. Organizations need to carefully decide the set of metrics that should be tracked, monitored and improved. The above table facilitates this part of the activity. Tools may assist in the collection of operational metrics and deriving the Managerial and Business metrics.

The metrication program should raise alarms at the completion of work product and force to take suitable corrective or preventive action resulting in well-controlled software process and good quality product. The following guidelines will help in the implementation of a good metrication program.

8. GUIDELINES FOR SUCCESSFUL METRICATION PROGRAM [2]

<u>Start Small</u>. First select a basic set of initial metrics. Once the metrication program gets momentum introduce new metrics to better manage your business.

<u>Explain Why</u>. Clearly explain the project teams the benefit of collecting and monitoring metrics. Use a case study to demonstrate.

<u>Share the Data</u>. Share the result with the team. Share summaries and trends with the team at regular intervals and tell them what the data is telling.

<u>Define Procedures</u>. Define clearly the operational definitions of data as well as formulas that are used for metrication. Build necessary validations in the data acquisition tool.

<u>Understand Trends</u>. Trends in the data over time are more significant than individual data points. Trends may have multiple interpretations; carefully interpret the trend based on the definitions of data.

Table 1.2 Metrics for Requirement Management Process Area

Requirements Management					
	Specific Practice	**Measurement Goal**	**Operational Measures**	**Managerial Measures**	**Business Measures**
SG1	**Requirements are managed and inconsistencies with the project plans and work products are identified**				
SP 1.1-1	**Develop an understanding with the requirements providers on the meaning of the requirements**	Extent of understanding of requirements.	# of Requirements # of Review comments Time spent of requirement review Review Efficiency Score of Requirement Checklist % Time spent on requirement review Ratio of Planned Requirement effort / Planned coding effort	Effort spent on developing requirements Effort spent on Req review % of effort spent on Requirement wrt to estimated efforts Ratio of Actual requirement effort / Estimated requirement efforts	Investment if Requirement (Mgmt. + Dev.) Customer complaints related to Req.
SP 1.2-2	**Obtain commitment to the requirements from the project participants.**	Extent of commitment to requirements	% of approved requirements to total Req.	% of Req. "not approved"	Effort in Gold plating
SP 1.3-1	**Manage changes to the requirements as they evolve during the project.**	Extent of changes to Req.	# of requirements that are changed # of new requirements	% of New Req. % of Req. changed % effort increased due to Req. change	Project Req. creep per unit time Effect of project cost
SP 1.4-2	**Maintain bi-directional traceability among the requirements and the project plans and work products.**	Reduced rework due to maintenance traceability	Effort spent on maintaining traceability	# of work products developed per Req. Correlation of effort spent on requirement and rework effort	Revenue Realized per work product
SP 1.5-1	**Identify inconsistencies between the project plans and work products and the requirements**	Consistency of requirements	Req. vs. work product vs. plan inconsistency	# Requirements not fulfilled Additional work required to resolve inconsistencies	Work Realized

Table 1.3 Metrics for Requirement Development Process Area

Requirements Development					
	Specific Practice	**Measurement Goal**	**Operational Measures**	**Managerial Measures**	**Business Measures**
SG1	**Stakeholder needs, expectations, constraints, and interfaces are collected and translated into customer requirements.**				
SP 1.1-2	**Elicit stakeholder needs, expectations, constraints, and interfaces for all phases of the product life cycle.**	Stakeholders involvement in RD	# of Stakeholder needs # of constraints # of interfaces Time spent on requirement elicitation # of methods (interviews, Prototype, surveys, use cases, product domain analysis) used for requirement elicitation	Size of Req. Count of Req. # of requests / unit time Time spent by customer	Effort / Cost on RD
SP 1.2-1	**Transform stakeholder needs, expectations, constraints, and interfaces into customer requirements.**	Rigor of RD process	# of needs identified # of Req. defined # Product component defined	# Requirements per customer need	Cost per product component
SG2	**Customer requirements are refined and elaborated to develop product and product-component requirements**				
SP 2.1-1	**Establish and maintain product and product-component requirements, which are based on the customer requirements.**	Rigor of RD process	# of product and product component requirements # of external interfaces	# Req. per product component Cost performance tradeoffs of requirements and of lifecycle phases considering business objectives (No. of requirements cancelled, postponed)	Cost per product component
SP 2.2-1	**Allocate the requirements for each product component.**	Distribution of requirements	# of requirements per product component	Effort per product component	Cost of fulfilling Requirement
SP 2.3-1	**Identify interface requirements.**	Complexity of requirements	# of external Interface requirements	# of Req. per interface	Cost of fulfilling Interface Req.
SG3	**The requirements are analyzed and validated, and a definition of required functionality is developed.**				
SP 3.1-1	**Establish and maintain operational concepts and associated scenarios.**	Rigor of Requirement validation process	# of prototypes / simulations to be developed Estimated Effort and time required for generating prototypes	# scenarios fulfilling Req.	Cost of prototype / Proof Of Concept
SP 3.2-1	**Establish and maintain a definition of required functionality.**	Rigour of Requirement validation process	# of functional requirements # of Non-functional requirements # of special requirements	% Functional Req changed	Cost of building application

			# of User Interface related requirements # of functional requirements that directly relate to product operation		
SP3.3-1	**Analyze requirements to ensure that they are necessary and sufficient.**	Rigour of Requirement validation process	Distribution of different types of Req (such as 'must', 'nice to have', technical, interface, business etc.) Requirement checklist score	Requirement completeness measure on a 10 point scale	Risk of Incomplete Req
SP3.4-3	**Analyze requirements to balance stakeholder needs and constraints.**	Rigour of Requirement validation process	# Constraints	# risk along with impact and probability associated with requirements	Ratio of risks associated with requirements / total risks Predicted impact of requirement risks on cost and schedule
SP3.5-2	**Validate requirements to ensure the resulting product will perform as intended in the user's environment using multiple techniques as appropriate.**	Rigour of Requirement validation process	Effort spent on Requirement Validation # of techniques to demonstrate requirements functionality (e.g., prototypes, simulations, analyses, scenarios, and story boards)	Ratio of requirement validation efforts to total requirement efforts	Cost of Prototype / Proof Of Concept

9. CONCLUSION

Good Practices alone need not improve quality of the software product and services. They have to be accompanied by a good set of pragmatic metrics. The paper offered a framework for metrics definition and their classification for the purpose they serve. This paper also provides a methodology and guidelines for the selection and implementation of metrication program in a matured software institution. Here, a case has been discussed taking into account the specific practices for Requirement Management and Requirement Development, in the Engineering area of the CMMI framework. The framework can be extended to other process areas as well. The model clearly establishes the interrelationships between operational, managerial and business metrics.

10. REFERENCES

1. Capers Jones – Chief Scientist Emeritus (2002), "Software Quality in 2002 – A survey of the state of the art", P 26.
2. Karl E. Wiegers (1999), "A Software Metrics Primer Software Development" Software Development, July 1999, P 5.
3. Stephen H Kan (2002), "Software Quality metrics Overview", Addison Wesley Professional., P 2

11. BIBLIOGRAPHY

Norman E Fenton, Shari Lawrence Pfleeger (2002), *"Software Metrics: A rigorous and Practical Approach"* Thompson Asia Ovt. Led., Singapore,

Reliability, Safety and Hazard
Editors: P.V. Varde et al.

VVER REACTOR PRESSURE VESSELS DOSIMETRY FOR NPP LIFETIME MANAGEMENT

B. Petrov, K. Ilieva, S. Belousov and T. Apostolov

Institute for Nuclear Research and Nuclear Energy (INRNE), Bulgarian Academy of Sciences
72 Tzarigradsko shosse Blvd., 1784 Sofia, Bulgaria

ABSTRACT

Assessment of the current state and lifetime of the Kozloduy NPP VVER-type reactors has been carried out by monitoring the irradiation exposure of reactor vessels and metal surveillance specimens. The neutron fluence determination is based on three-dimensional neutron transport calculation by the TORT code using the problem-oriented neutron cross-section library BGL, and by the MCNP code using the continuous energy dependence library DLC200. The fluence data are validated by measurements of the induced activity of threshold foil detectors. Reactor core loading patterns are determined to achieve minimal reactor vessel neutron exposure.

Keywords: Reactor Pressure Vessel, Lifetime, Neutron Fluence Determination, Loading Schemes

1. INTRODUCTION

The Kozloduy NPP in Bulgaria is the largest power plant on the Balkan Peninsula. It is located at the Bulgarian northern border with Romania, on the Danube river bank. About half of the country's electricity is being produced by Kozloduy NPP. It consists of four reactor units of type VVER-440/V-230 (Units 1 through 4) and two ones of later-generation type VVER-1000/V-320 (Units 5 and 6).

The reactor pressure vessel (RPV) integrity of VVER-type reactors is a topical problem because of the need for justification of their safety and service lifetime.

The steel embrittlement of a reactor pressure vessel caused by the neutron irradiation during the operation of NPP could be the reason for a break of the vessel integrity when accident shutdown of the unit is performed. The assessment of the fast neutron fluence on the RPV is therefore required for evaluation of the vessel steel degradation and for revision of the safety margins limits. These limits determine the lifetime of the reactor relevant to the irradiation embrittlement.

Following the requirements for safety operation a Reactor Pressure Vessels Surveillance Program is being carried out at Kozloduy NPP. According to this program assessment of the current state and lifetime of RPV has been carried out by monitoring the irradiation conditions of RPV and metal surveillance specimens (SS) from the same vessel metal irradiated during the reactor operation. The RPV neutron fluence evaluation methodology has been developed and applied on Units 1 through 6 of Kozloduy NPP since 1989. Reactor core loading patterns are determined so that minimal RPV neutron exposure is achieved.

2. EVALUATION OF THE METAL EMBRITTLEMENT UNDER NEUTRON IRRADIATION

The Russian standard (Gosatomenergonadzor, 1989) for VVER reactors uses the following relation for determination of the temperature of neutron embrittlement:

$$T_k = T_{k0}{}^B + \Delta T_F, \qquad (1)$$

where T_{k0}^{B}, ^{0}C, is the initial temperature of neutron embrittlement after annealing, and ΔT_F, ^{0}C, is the shift of the temperature of neutron embrittlement. The nondestructive method for evaluation of the metal embrittlement under neutron irradiation is based on the relation between ΔT_F and the fluence F, cm^{-2}, of neutrons with energy above 0.5 MeV:

$$\Delta T_F = A_F (F/F_0)^{1/3}, \qquad (2)$$

where A_F, ^{0}C, is a chemical coefficient of neutron embrittlement, and $F_0=10^{18}$, cm^{-2}.

3. NEUTRON FLUENCE CALCULATIONAL METHODOLOGY

Calculational methodology has been developed for the evaluation of the RPV neutron fluence of VVER-1000 and VVER-440 (Ilieva et al., 1994; Ilieva et al., 1994; Ilieva et al., 1996; Belousov and Ilieva, 1997; Ilieva et al., 1996; Bucholz et al., 1996; Belousov et al., 1998; Apostolov et al., 1998; Belousov et al., 1998; Ilieva et al., 1999).

The distribution of the neutron flux and its responses on the RPV and adjacent zones of VVER are obtained as a solution of the kinetic equation of neutron transport. Because of the complexity of neutron production and transport from the core to the RPV the task is divided into two parts. The first part describes the generation of fission neutrons and their distribution in the reactor core, and the second part describes the neutron transport with fixed sources. These sources are determined by the fission neutrons generated in the core, and are appropriate for neutron fluence and induced activity calculation (Ilieva et al., 1996).

The determination of the power distribution and the fuel burnup distribution in the core taking into account the operational history has been carried out using the three-dimensional codes PYTHIA (Thomas, 1996) for VVER-440, and TRAPEZ (Schulz, 1997) for VVER-1000. These codes use libraries of effective diffusion cross sections generated by the spectral code NESSEL (Schulz, 1994).

The neutron fluence determination methodology of INRNE is based on three-dimensional neutron transport calculation by the discrete-ordinates code TORT (Rhoades and Childs, 1987) with multigroup energy dependence approximation using the neutron cross-section library BGL (Bucholz et al., 1996), and without approximation of the energy dependence by the Monte Carlo code MCNP (Briesmeister, 2000) using the neutron cross-section library DLC200 (Briesmeister, 2000). The problem-oriented BGL library was especially created for VVER-1000 and VVER-440 types of reactors, by collapsing the fine-group library VITAMIN-B6 (199 neutron and 42 gamma groups) to 67-group structure (47 neutron and 20 gamma groups). The libraries consider the features (detailed one-dimensional geometry and material compositions) of the respective reactor and contain upscattering data for the five thermal-energy groups. The order of scattering of the Legendre expansion is P5. Activation data from the IRDF-90 (International Atomic Energy Agency) file are used for these calculations.

The fixed source needed for the discrete-ordinates calculation is prepared by the code package DOSRC (Belousov et al., 2001) developed especially for this purpose. Besides the source preparation, the package provides calculational geometry model visualization that makes it very useful for calculational geometry modeling and for Quality Assurance applications.

Input data sensitivity analyses for choice of spatial coordinates' meshes, order of velocity angle quadratures S_n, polynomial neutron scattering anisotropy approximation P_n, convergence constant value, and neutron constants libraries have been carried out to reduce the calculational method errors (Ilieva et al., 1994; Ilieva et al., 1994; Belousov and Ilieva, 1997; Ilieva et al., 1996; Belousov et al., 1998). The neutron fluence is calculated at the surveyed places of VVER-1000 RPV (seam welds 3 and 4, and base metal, on 8 deg, which is the direction of maximum neutron exposure) and VVER-440 RPV (seam weld 4 and base metal, on 30 deg for standard core loading, and on 13 deg for loading with dummy cassettes in the periphery).

The neutron fluence evaluation methodology is being developed under research projects of IAEA, EC, the Bulgarian National Funds, and in close collaboration with working teams from Russia, the Czech Republic, Germany, Spain, Belgium, Hungary, and Ukraine, particularly within the WGRD-VVER group activity. In the framework of international collaboration, RPV benchmarks for the VVER-440 and VVER-1000 vessels have been created on the basis of the critical assembly LR0 at NRI Rez in the Czech Republic and have been used to validate and improve the methodology for RPV neutron fluence evaluation.

4. VALIDATION OF THE NEUTRON FLUENCE EVALUATION

The reliable calculation of the neutron fluence is limited by the data (nuclear and structural) used in the neutron transport calculations. That is why it is always advisable to compare calculational with measured values. Since the neutron flux is not directly measurable, flux-related values, such as radioactivity of an isotope generated by the neutron interaction, are measured and compared. The verification and adjustment of the fluence on surveillance specimens is performed by measured activity of detectors from the surveillance containers and Mn-54 activity of the specimens themselves.

4.1. Neutron fluence on RPV

The Kozloduy NPP RPV fluence data are validated by measurements of the induced activity of threshold foil detectors placed in the air cavity behind the RPV.

Following the RPV Surveillance Program verification/validation of the calculated neutron fluence on the vessel for each operated cycle is being carried out by the use of the experimental data of induced activity of ex-vessel iron, niobium, and copper detectors. The detectors have been disposed in the air cavity behind the vessels on all units of Kozloduy NPP. A rack device supports a band with foil detectors comprising a 60-deg sector, and a vertical one set along the core height (Ilieva et al., 1996). The calculated ex-vessel detector activities of ^{54}Mn, ^{60}Co, and ^{93m}Nb are compared with the measured ones. The difference between calculated and measured values is a basis for estimation of the fluence evaluation uncertainty.

The verification/validation of the calculated neutron fluence on the VVER-440/230 pressure vessels is very important, in particular because there are no surveillance assemblies in VVER-440/230. Additional activities have been carried out in order to assess experimentally the RPV metal embrittlement and the RPV lifetime. Templates from the inner wall of the Unit 2 vessel (1992), and scraps (1995) and templates (1996) from the Unit 1 vessel have been taken out. Good consistency (within 9%) (Apostolov et al., 1998) between calculated and measured ^{54}Mn activity of scraps has been found.

4.2. Neutron fluence on surveillance specimens

There are six sets of surveillance assemblies in the VVER-1000/320 type of reactors. Each set is placed within 60-deg sector of symmetry of the reactor core, and contains five cylindrical assemblies located at different azimuth positions on the baffle, about 30 cm over the reactor core upper edge. Each assembly contains six cylindrical containers with SS and neutron iron, niobium, and copper foil detectors. The containers are arranged at two axial levels, symmetrically around the axial axis of the assembly (Belousov et al., 1998).

The strong neutron flux radial and axial gradients through each surveillance assembly do not allow to obtain a set of specimens with close neutron irradiation needed for metallurgical analysis. The measured ^{54}Mn activity of the specimens was applied to determine the experimental neutron fluence values. The neutron fluence in the middle of the specimens varies around the axis of each assembly within 15% for the upper level and 35% for the lower level (Ilieva et al., 1996; Belousov et al., 1998). The results obtained for the neutron fluence of metal surveillance specimens have been used for assessment of the current embrittlement and lifetime prognosis of the Kozloduy NPP Units 5 and 6.

5. RPV LIFETIME PROGNOSIS

The RPV residual lifetime relevant to the neutron irradiation is limited by the maximum neutron fluence F_{max} and depends on the fluence of the operation time. The residual lifetime (RLT) in cycles is determined as

$$RLT = (F_{max} - F)/F_{avr}, \qquad (3)$$

where F_{avr} is the prognostic fluence of standard cycle of 290 FPD. The maximum neutron fluence is determined by the maximum critical temperature T_{ka} which is a characteristic of the vessel:

$F_{max}=(T_{ka}/A_F)^3$. (4)

The results obtained for the residual lifetime relevant to the RPV neutron embrittlement of Kozloduy NPP Units 1 through 6 are presented in Table 1.

6. CONTROL APPLICATION AND MANAGEMENT DECISIONS

On the basis of the RPV neutron fluence results obtained taking into account the local power distribution history, the operation of the 16th cycle of the Unit 2 reactor, and the 17th and 18th cycle of the Unit 1 reactor were substantiated. The RPV fluence results were used in licensing procedure of Units 3 and 4 by the Bulgarian Nuclear Regulatory Agency.

The calculational and experimental data obtained for metal specimens and neutron detectors of the Units 5 and 6 surveillance sets have been used for modification and improvement of the Surveillance Program of Kozloduy NPP Units 5 and 6.

Table 1 Residual Lifetime Relevant to the RPV Neutron Embrittlement

Unit / Reactor Type	Start Year	Design Lifetime, Cycles	Operated Cycles till 2005	Residual Lifetime, Cycles
1 / VVER-440	1974	30	23	2
2 / VVER-440	1975	30	24	8
3 / VVER-440	1980	30	20	6*
4 / VVER-440	1982	30	19	> 45
5 / VVER-1000	1987	40	11	29
6 / VVER-1000	1991	40	9	31

*after annealing more cycles could be operated

Reactor core loading patterns have been determined in a way that leads to achieving minimal RPV neutron exposure. Such an approach is a prerequisite for effective management and planning of the RPV lifetime. Diverse low-leakage loading schemes with burned-up assemblies in the periphery of the reactor core are applicable. Low-fluence loading (LFL) is the scheme when burned-up assemblies are put only in direction of maximum exposure. In this case the reduction of the neutron fluence is about 20 to 30% toward the standard core loading (Table 2). Low-leakage-and-fluence loading (LLFL) is the scheme with 36 burned-up assemblies. It reduces the neutron fluence with about 30 to 40% toward the standard loading. Such loading schemes have been applied at Kozloduy NPP Unit 4 with VVER-440 and Units 5 and 6 with VVER-1000.

Even greater reduction of the neutron exposure (up to 70% toward the standard loading) has been achieved at Units 1 through 3 with VVER-440. Loading schemes with 36 dummy (steel) cassettes put in the reactor core periphery instead of fuel assemblies have been applied. In this case the direction of maximal azimuthal neutron exposure is shifted from 30 deg to 13 deg and 47 deg (in a 60-deg sector of symmetry). The combination of dummy cassettes and burned-up assemblies in the periphery of the reactor core additionally diminishes the neutron fluence.

Table 2 Neutron Fluence Reduction

Loading Scheme	Description	Reduction Effect
LLL (low leakage)	IN-IN-OUT* & max OUT-IN-IN	Insignificant
LFL (low fluence)	OUT-IN-IN** & max IN-IN-OUT	~ 20-30%
LLFL (LL & LF)	IN-IN-OUT	~ 30-40%
Dummy	Dummy cassettes in the periphery of the reactor core	Shift max ~ 70%
Dummy & LL	Dummy & LL	≥ 80%

* IN-IN-OUT - the assembly is in the periphery during its third or fourth cycle in the core

** OUT-IN-IN - the assembly is in the periphery during its first cycle in the core

7. CONCLUSION

Calculational methodology for the evaluation of the RPV neutron fluence of VVER-1000 and

VVER-440, based on advanced methods, codes, neutron constants libraries, and experimental verification by induced activity measurements, has been developed.

The neutron fluence validated data have been used for assessment of the vessel metal degradation and the lifetime relevant to the RPV neutron embrittlement of Kozloduy NPP Units 1 through 6.

For effective management and extension of the RPV lifetime, appropriate core loading patterns must be applied so that the RPV neutron exposure be minimized.

8. REFERENCES

1. Apostolov, T., Ilieva, K., Belousov, S., Petrova, T., Penev, I., Popova, I., Antonov, S. (1998), *Experimental Assessment by Scraps Activity of Neutron Fluence onto the Reactor Vessel of Unit 1 at Kozloduy NPP*, Nuclear Engineering and Design 186, 361-368.
2. Belousov, S., Ilieva, K. (1997), *ASYNT - Adjoint Synthesis Method for Neutron Irradiation Assessment of VVER/ Pressurized Water Reactor Pressure Vessel*, Nuclear Science and Engineering 126 2, 239-244.
3. Belousov, S., Ilieva, K., Antonov, S. (1998), *Influence of Iron Data from Different Neutron Cross Section Libraries on VVER PV Fluence Calculation*, Reactor Dosimetry, Hamid A. Abderrahim, Pierre D'hondt, Bohumil Osmera, Eds., World Scientific Publishing Co. Pte. Ltd., Singapore, 473-479.
4. Belousov, S., Ilieva, K., Popova, I. (1998), *Calculation of Neutron Flux/ Responses at the Surveillances and Pressure Vessel of VVER-1000/320 Type of Reactor*, Proc. of RP&S Topical Meeting, Nashville, USA, ANS, 1 440-447.
5. Belousov, S., Kirilova, D., Ilieva, K. (2001), *Development of the Processing Software Package for RPV Neutron Fluence Determination Methodology*, International Meeting "Nuclear Power in Eastern Europe: Safety, European Integration, Free Electricity Market," Bulgarian Nuclear Society, Varna, Bulgaria, 17-20 June 2001, CD-ROM, No 62.
6. Briesmeister, I. F., Ed. (2000), *MCNP - A General Monte Carlo N-Particle Transport Code, Version 4C*, LA-13709-M, RSICC ORNL, Code Package, CCC0700.
7. Bucholz, J., Antonov, S., Belousov, S. (1996), *BGL440 and BGL1000 Broad Group Neutron/ Photon Cross Section Libraries Derived from ENDF/B-VI Nuclear Data*, IAEA, INDC(BUL)-15.
8. Gosatomenergonadzor (1989), *Regulations and Calculation Norms for Equipment and Tubing Strength of the Nuclear Energy Facilities* (PNAE G-7-002-86, PNAE G-7-008-89), Energoatomizdat.
9. Ilieva, K., Antonov, S., Belousov, S. (1996), *Calculation Modeling of Detector Activity in the VVER/ PWR Reactor Pressure Vessel Surveillance*, Nuclear Science and Engineering 122 1, 131-135.
10. Ilieva, K., Apostolov, T., Belousov, S., Antonov, S. (1994), *Neutron Fluence Determination for VVER-440 Reactor Pressure Vessel Ageing Surveillance*, Reactor Dosimetry, ASTM STP 1228, Harry Farrar IV, E. Parvin Lippincott, John G. Williams, and David W. Vehar, Eds., American Society for Testing and Materials, Philadelphia, 38-44.
11. Ilieva, K., Apostolov, T., Belousov, S., Penev, I., Popova, I. (1996), *Validation of Neutron Fluence Calculation Methodology for VVER-440 Vessel*, Nuclear Science and Engineering 125, 243-247.
12. Ilieva, K., Belousov, S., Antonov, S., Zaritsky, S., Brodkin, E. (1994), *Test Calculation of the Neutron Flux on VVER-1000 Reactor Pressure Vessel*, Reactor Dosimetry, ASTM STP 1228, Harry Farrar IV, E. Parvin Lippincott, John G. Williams, and David W. Vehar, Eds., American Society for Testing and Materials, Philadelphia, 343-347.
13. Ilieva, K., Belousov, S., Apostolov, T. (1999), *Testing of Multigroup Neutron Cross Section Libraries on Neutron Transmission through VVER-440 Vessel*, Nuclear Science and Engineering 131, 282-285.
14. International Atomic Energy Agency, *IRDF90 v.2, the International Reactor Dosimetry File*, IAEA-NDS-141.

15. Rhoades, W. A., Childs, R. L. (1987), *TORT Three-Dimensional Discrete Ordinate Neutron/ Photon Transport Code with Space-Dependent Mesh and Quadrature*, ORNL-6268.
16. Schulz, G. (1994), *NESSEL Version 6.0*, Berlin, K.A.B.
17. Schulz, G. (1997), *Code Manual TRAPEZ Version 6.07a*, Berlin, K.A.B.
18. Thomas, S. (1996), Programm PYTHIA zur Berechnung des Makroabbrands in Druckwasserreaktoren vom Typ WWER, Berlin, K.A.B.

Reliability, Safety and Hazard
Editors: P.V. Varde et al.

MINIATURE SPECIMEN TESTING – A TOOL FOR ESTIMATING THE RESIDUAL LIFE OF SAFETY CRITICAL COMPONENTS AND STRUCTURES

Raghu V Prakash

Machine Design Section, Department of Mechanical Engineering
Indian Institute of Technology Madras, Chennai 600 036

ABSTRACT

Extending the operating life of a component, structure or a system of components with a high level of reliability has been one of the important issues discussed by technologists and plant operators worldwide. Mechanical properties of structural materials used in plant components provide the basis for residual life assessment. Base material properties loose their significance when there is a suspicion of embrittlement and toughness degradation due to prolonged exposure of the components at high temperatures or under corrosive environments. In such cases, a small amount of material is scooped out of working components and miniature specimens are fabricated out of them. It is pertinent to examine the results obtained through miniature specimens and compare the same with data obtained from standard specimens. This paper discusses some of the miniature specimen test methods used for evaluating the tensile properties, impact energy, fracture toughness properties and fatigue properties along with some results.

Keywords: Mechanical property evaluation, Miniature specimen, Tension, Impact energy, Fracture toughness, Fatigue properties

1. INTRODUCTION

Small specimen test method has gained increased significance in the recent years due to the desire to extend the useful life of currently operating power-plants which are at the final stages of their designed life. Downtime associated with shutting down and rebuilding an equivalent power-plant makes life-extension as an alternative option. As life extension involves safety of operation coupled with a high degree of reliability, a scientific approach is required for estimation of residual life of critical components.

Conventionally, fatigue damage is estimated using Palmgren-Miner's rule, while creep damage is assessed through the examination of material microstructure. In cases where there is a combined creep-fatigue interaction, empirical correlation is arrived (for e.g. ASME Code N-47) to estimate the residual life. Figure 1 [Ref. 1] presents schematic view of life estimation.

In life estimation calculations, common practice is to consider the base-material property (in virgin state) as input data. However, it is well known that the mechanical properties of materials change as a function of usage, due to interaction with external environment with mechanical loading of components, leading to embrittlement of material. Mechanical property for embrittled materials is not usually available.

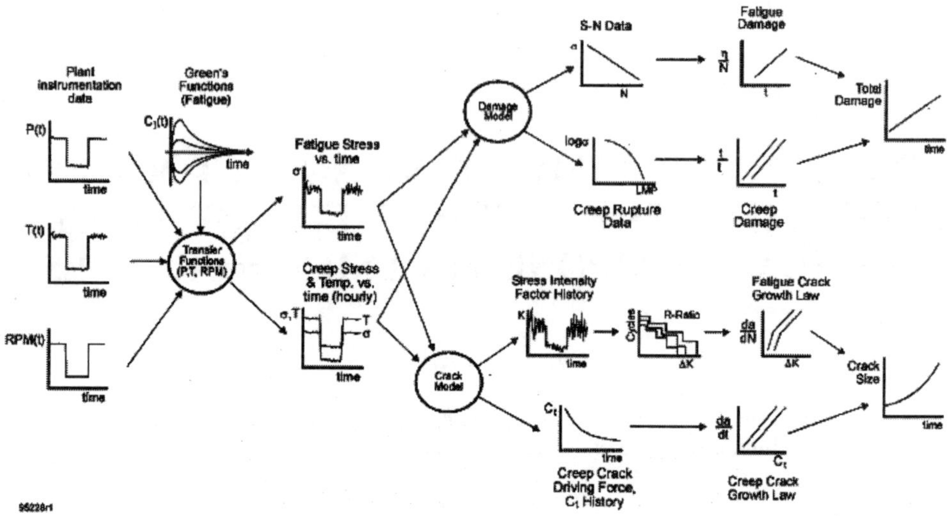

Fig. 1: Schematic of life predication under combined fatigue-creep damage [1]

In order to evaluate the mechanical properties of embrittled materials, small pieces of material are cut from an existing component of a plant and mechanical properties evaluated. Miniature specimens prepared from this are used to evaluate the tension, impact, fracture toughness property of material. Thus, miniature specimen testing permits characterization of mechanical behavior of in-service components with minimum material use. Scope of miniature specimen testing ranges from failure analysis, life-extension for safety critical components and structures, pressure vessels used in power-plants, irradiated/corrosion embrittled components [2].

Experimental methods to evaluate the mechanical properties of materials using standard and small specimen test technique can be broadly classified as those based on:

a) mechanical property to be estimated (tensile strength, impact strength, fracture toughness, crack growth) and
b) Specimen fabrication method (miniaturization Vs reconstitution specimen).

2.1 Miniature Tensile Specimens

Several researchers have studied mechanical property evaluation using small tensile specimens and compared the results with that of standard specimen. Figure 2a [Ref. 3] presents a typical miniature tensile specimen used for material property characterization. Sharpe et al [4] have used a micro-tensile specimen having a gage section width of 0.2 mm and an overall specimen length of 3.1 mm to evaluate the tensile properties (Fig. 2b). Tensile data obtained on multiple specimens suggest an acceptable consistency in data between repeat tests (Fig. 3). However, the absolute values are found to be different from the ASTM E-8 standard specimen data – the yield strength and ultimate tensile strength was found to be lower than standard specimen data by approx. 5% of reference value. It is possible to establish a correlation between the standard specimen data and miniature specimen data.

Table 1 presents the tensile test results (yield strength and UTS) for all micro-tensile specimens tested. Mean and standard deviation data are also included in the Table 1. It may noted that the coefficient of variation (CV) which is the ratio of standard deviation to mean value appears to be high in an absolute sense for micro-tensile samples. However, as the CV ratio is similar to /within the limits specified in ASTM A 20 standard - i.e. 5% for yield strength and 3% for UTS for steel plate, the data is acceptable.

Based on multiple studies on miniature specimen tension tests, it is observed that minimum specimen dimension of at least five times the grain size is required for correlation of tensile data of small specimens with standard specimen data [5].

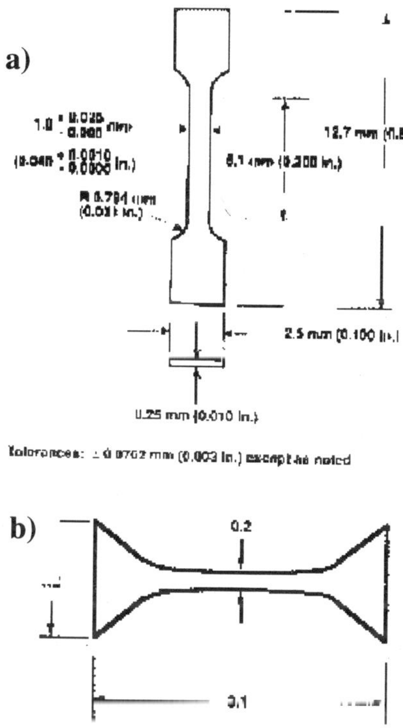

Fig. 2: Typical a) mini-tensile [3] and b) micro-tensile test specimen [4].

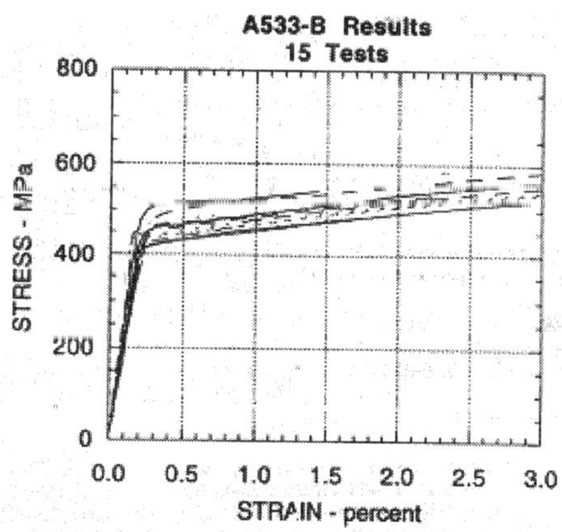

Fig. 3: Tensile stress-strain data evaluated through micro-tensile specimens [4].

2.2 Alternate Methods To Evaluate Tensile Property

2.2.1 Shear Punch Test

Shear punch test method is increasingly been used as a miniature specimen test method to evaluate the yield strength, UTS and percentage elongation of materials. A small disk (of the size of a TEM specimen stage) is used. Test involves blanking of flat specimens that are scooped from components having a thickness in the range of 0.25 – 0.5 mm with a 1 mm diameter flat cylindrical punch at a slow, constant speed. The load Vs punch displacement trace is used for the identification of yield, ultimate tensile strength and elongation at fracture.

Table 1: Tensile test data obtained from micro-tensile specimens and comparison with standard specimen data [4].

Microspecimen Tests									Macrospecimen Tests		
Test No.	RD Yield-MPa	RD Ult-MPa	Test No.	S Yield-MPa	S Ult-MPa	Test No.	T Yield-MPa	T Ult-MPa	Test No.	T Yield-MPa	T Ult-MPa
A15	463	596	B11	451		B12	438	634	T2	458	643
A16	428		B13	463	649	B14	467	657	T4	441	648
A17	432		B21	424		B22	439		T6	454	643
A18	515	590	B23	437	590	B24	417		T7	450	645
A19	458		B31	406	611	B32	425	626	T9	454	648
A21	457	624	C11	464		C12	497				
A22	459		C13	438		C14	457				
A23	470		C21	438	596	C22	425	587			
A24	437	614	C23	444	580	C24	419	578			
A25	495	645	C31	413	556	C32	406	556			
A26	501	664									
A27	441	599									
A28	464	632									
A31	441	[illegible]									
A32	450	616									
A33	430										
A34	476	644									
A35	458										
A36	432	595									
A38	432	573									
A41	419										
A42	419	574									
A43	458	619									
A44	450	585									
A45	457										
A46	432	612									
A47	432	589									
AVG MPa	453	609		438	597		432	606		451	645
SD MPa	24	25		19	31		28	39		6.6	2.5
CV %	5.3	4.1		4.3	5.2		6.5	6.4		1.5	0.4

Figure 4 presents the schematic of shear punch test setup and typical load-displacement trace obtained during testing. Experimental results [6] suggest that the predicted yield strength to be lower than standard specimen data, while the UTS values are marginally higher. This could be due to the difficulties in identification of non-linearity in load-displacement trace.

Identification of yield point stress has been refined by Kasiviswanathan et al [7] by the use of acoustic emission technique during testing. Yield point has been identified as the point where there is a peak in AE activity.

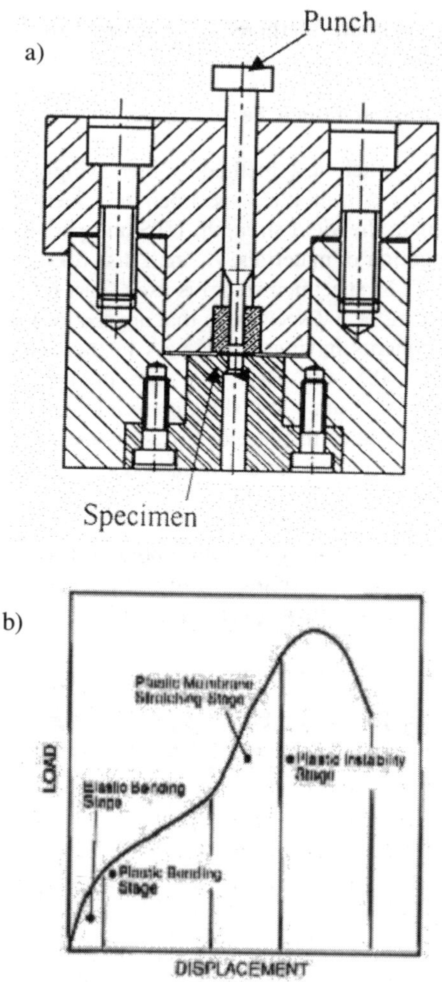

Fig. 4: a) schematic of shear punch test setup [7], b) – typical load-displacement trace obtained during shear punch test [6]

Table 2 presents the comparison of yield stress as identified by the non-linearity method (1st Criteria) and AE method. It may be noted that the results obtained through non-linearity method are subject to wide scatter. Figure 5 presents the comparison of yield stress obtained by testing standard tensile specimen Vs values obtained through shear punch test and data analyzed by AE technique. This figure suggests that the data obtained through a careful experimentation is in good comparison with standard tensile specimen data. This implies that scatter in data is more due to data analysis procedure than the physical phenomenon.

Table 2: Comparison of yield strength obtained through Shear Punch Test using non-linearity method and AE method [7]

MATERIAL	YS USING 1st CRITERIA (Mpa)		YS USING PEAK AE ACTIVITY (MPa)	
	Mean	Standard Deviation	Mean	Standard Deviation
9Cr-1Mo Steel	428	101.1	567	23.8
2.25Cr-1Mo Steel	446	48.2	490	30.0
AISI 316 SS	249	57.0	272	24.0
AISI 304 SS	235	56.74	195	18.6
10% CW 304 SS	443	54.8	510	31.2
20% CW 304 SS	610	76.8	677	29.4
40% CW 304 SS	688	80.2	895	27.3
50% CW 304 SS	776	45.2	975	56.4

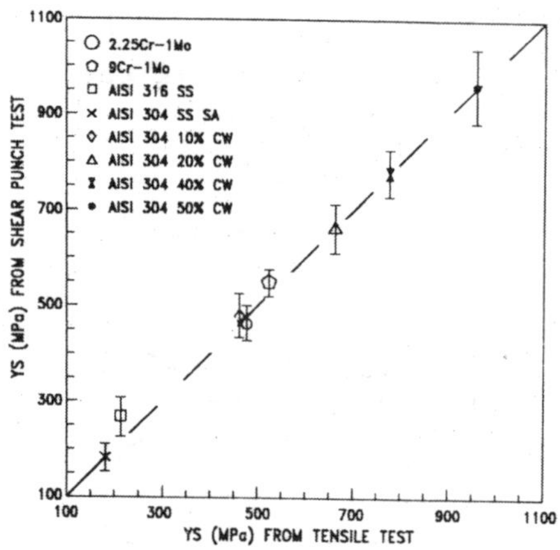

Fig. 5: Comparison of yield strength data obtained by shear punch test method (AE Method of data deduction) with standard tensile test [7]

2.2.2 Ball Indentation Test

Automated Ball Indentation (ABI) test method has been promoted as a non-destructive and non-invasive test method that can be used to characterize the tension, hardness, fracture toughness properties of metals [8]. In this method, a small tungsten-carbide spherical ball attached to rigid fixture is used to make repeated indents at the same location of a test-specimen. The depth of indent is often controlled through servo-loop control and a series of load-unload cycles are applied on the specimen. Depth of indent/penetration Vs maximum compressive load applied is used to correlate the tensile properties of material.

2.3 Impact Energy/Fracture Toughness Evaluation

2.3.1 Charpy Impact Test

Embrittlement of materials and subsequent loss of strength can be evaluated through impact tests. Specimen miniaturization often results in a decrease in ductile-brittle transition temperature (DBTT) and upper shelf energy (USE). It is important to establish a normalization of the impact properties obtained using sub-sized specimens to those of standard specimens.

Quarter size (2.5 mm x 2.5 mm), one-third size and half-size Charpy impact samples compared to conventional 10 mm x 10 mm have been evaluated and observed to provide consistent test results [9] that correlate with standard specimen data.

2.4 Fracture Toughness Properties

Miniature compact tension specimens are used to evaluate the fracture toughness properties. Miniature disk compact tension [C(T)] specimens with an overall diameter of 10 mm as shown in Fig 6(a) [Ref. 10] have been used. Special displacement gage is used to estimate crack size during testing.

It may be noted that in many cases, obtaining a stock of material from an existing component to fabricate a miniature compact tension specimen may not be feasible. To minimize material requirement, reconstituted specimens have been tried to build a standard test specimen [11] with small volume of test piece material that needs property characterization. Usually a small piece of embrittled specimen is placed in the critical section of standard test specimen and fused/welded so that the resultant specimen looks like a standard test specimen. The critical loading section (for example, crack path in impact, C(T) sample) is embedded with material that is environmentally degraded / embrittled. Care has to be exercised in this reconstitution method to ensure that failure zone is not shifted from base material to fuse-welded region. Figure 6(b) presents a schematic of reconstituted specimens.

a)

b)

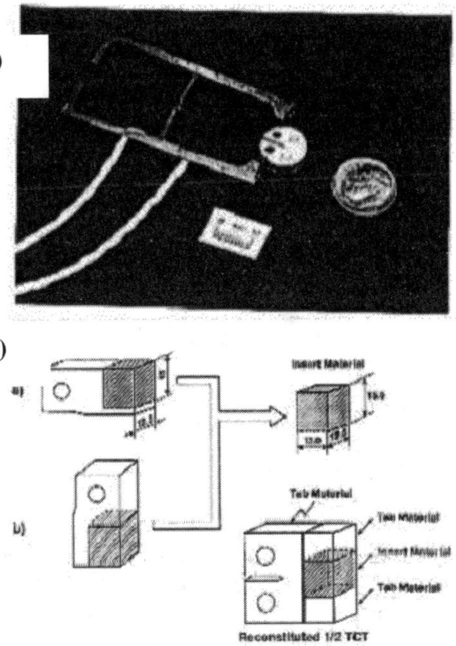

Fig 6:a) Miniature disk C(T) specimen [10] and b) reconstituted C(T) specimen [11]

2.4.1 Novel Specimen Geometries

Some researchers have tried other novel geometries such as cylindrical rods of small diameter [12, 13] to study the crack propagation rates and fracture toughness values. This method appears quite reasonable as it permits simulation of presence of a fatigue crack as well as constraint / free-edge simulation. However, care has to be taken while evaluating the stress intensity factor through finite element technique as the stress intensity factor is dependent on the crack shape, aspect ratio [14].

2.5 Fatigue Property Evaluation

2.5.1 Failure Life Data

Fatigue is an important consideration to be taken into account during residual life assessment. Failure life data for embrittled material has been evaluated using miniature low cycle fatigue specimens and their results compared with those obtained using standard specimens. Figure 7 presents the schematic of miniature low cycle fatigue specimen and a comparison of results obtained through miniature specimens.

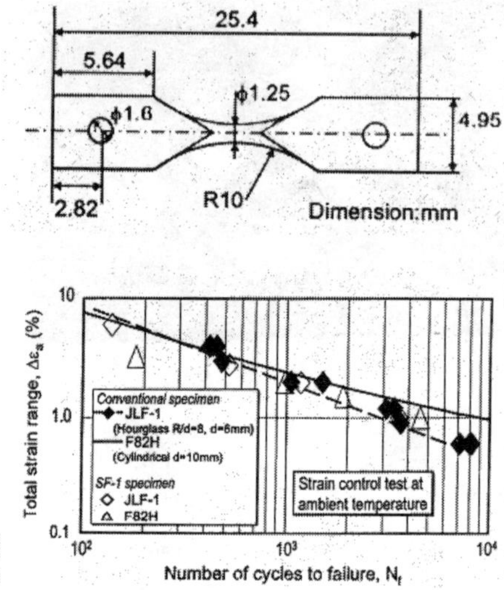

Fig 7: Small Low Cycle Fatigue Specimen and comparison of experimental results with standard LCF specimen data [15]

2.5.2 Fatigue Crack Growth

Murase et al [16] have evaluated the fatigue crack growth characteristics using miniature single edge notch specimen as shown in Fig. 8 (a) to study the effect of irradiation on crack growth. The overall specimen length is 43 mm, with a width of 10 mm and a thickness of 0.15 mm.

Li et al [17] have used a 3 point bend specimen for fatigue crack growth studies (Fig. 8b). The crack length is monitored by a CCD camera. Crack growth rate of miniature bend specimen was observed to be lower than the standard specimen in a few materials (304 stainless steel, Inconel 718), but the slope in the Paris regime is observed to be similar for standard and miniature specimen.

3. SUMMARY

Experimental results obtained through small specimen testing suggest the existence of a correlation between standard specimen data and miniature specimen data. However, the scatter in data appears to be much high compared to the data obtained through standard specimens. This could due experimentation errors, data analysis procedures, the influence of microstructure, grain size, and constraint effects on mechanical properties. Further work is required in establishing

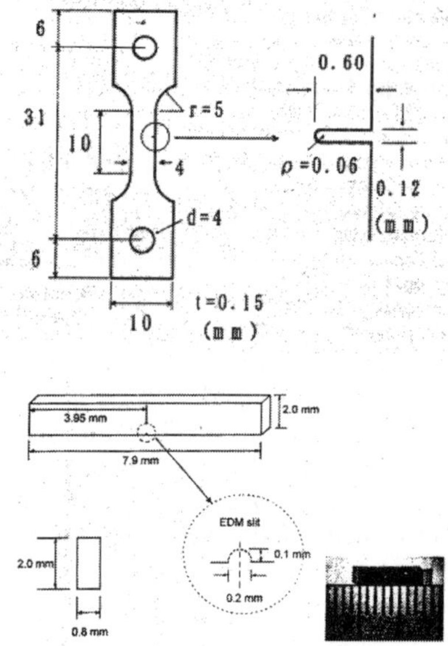

Fig. 8: a) Single edge notch tension specimen [16] and b) bend specimen [17] used for fatigue crack propagation studies using miniature specimens.

a consistent data to enable its use with high levels of confidence in residual life estimation. Automation of miniature specimen testing could ensure better quality and consistent results and is yet another area that needs attention.

4. REFERENCES

1. Rosario, D.A and Tilley, R. M., EPRI Intl. Conf. on Adv. Plant Life Assessment, March 2002, Orlando, FL.
2. Manahan, M.P., Williams, J., and Mariakanitz, R.P., *in* ASTM STP 1204, ASTM, Pa, 1993, pp 62-76.
3. Hamilton, M.L., Boltter, M.A., and Edwards, D.J., *in* ASTM STP 1204, ASTM, Pa, 1993, pp 368-385.
4. Sharpe, W.N. Jr., Danley, D. and LaVan, D.A. *in* ASTM STP 1329, ASTM, Pa, 1998, pp 497-512.
5. Byun, T.S., Kino, J.H., Cla, S.H., and Hong, J. H., *in* ASTM STP 1329, ASTM, Pa, 1998, pp 575-586.
6. Foulds, J.R., Wu, M., Srivastav, S., Jewett, C.W., *in* ASTM STP 1329, ASTM, Pa, 1998, pp .

7. Kasiviswanathan, K.V., Hotta, S. K., Mukhopadhyay, C.M. and Baldev Raj *in* ASTM STP 1329, ASTM, Pa, pp 523-538.
8. Haggag, F.M., Server, W.L., Lucas, G.E., Odette, G.R., Sheckherd, J.W., *in* J. Testing and Evaluation, Vol 1, No: 1, pp 62-69, 1990.
9. Rosinski, S.T., and Kumar, A.S., Cannon, S., and Hamilton, M.L., *in* ASTM STP 1204, ASTM, Pa, 1993, pp 405-416.
10. Alexander, D.J., *in* ASTM STP 1204, ASTM, Pa, 1993, pp 130-142.
11. Tomimatsu, M., Kawaguchi, S, and Iida, M., *in* ASTM STP 1329, ASTM, Pa, 1998, pp 470-483.
12. Scibetta,M., and Chaouadi, R., *in* ASTM STP 1329, ASTM, Pa, 1998, pp 363-379.
13. Giovanola, J. H., Klopp, R.W., Crocker, J. E., Alexander, D.J., Corwin, W.R., and Nanstad, R.K., *in* ASTM STP 1329, ASTM, Pa, 1998, pp 328-351.
14. Carpinteri, A., Brighenti, R., and Spagnoli, A., *in* Fatigue and Fracture of Engg. Mater. Struct., Vol 21, No: 18, pp 1027-1035.
15. Hirose, T., Tanigawa, H., Ando, M., Suzuki, T., Kohyama, A., Katoh, Y., and Narui, M., *in* ASTM STP 1418, ASTM, Pa, 2002, pp 181-194.
16. Murase, Y., Nagakawa, J., and Yamamoto, N., *in* ASTM STP 1418, ASTM, Pa, pp 211-220.
17. Li, M and Stubbins, J. F., *in* ASTM STP 1418, ASTM, Pa, pp 321-335

Reliability, Safety and Hazard
Editors: P.V. Varde et al.

ESTIMATION OF LIFE-SPAN OF INSTRUMENTATION & CONTROL CABLES USED IN NUCLEAR POWER PLANTS USING THE TENSILE TESTING METHODOLOGY

P.K. Ramteke, A.K. Ahirwar, N.B. Shrestha, V.V.S. Sanyasi Rao
R.K. Saraf and A.K. Ghosh

Reactor Safety Division, Bhabha Atomic Research Centre, Mumbai 400 085

ABSTRACT

Nuclear Power Plants (NPPs) contain myriads of electrical cables of all sizes, voltage ratings and lengths delivering electrical power to much vital, as well as non-vital equipment. These cables are insulated with some form of polymeric insulation. These insulating materials will gradually age from exposure to heat, radiation, moisture and chemicals. Hence, on-site testing of the conditions of these cables is necessary to ensure their continued reliability in service and to predict their remaining life. In addition these cables must operate reliably in the harsh steam environments produced by postulated design-basis events (e.g. LOCA). Conservatively, this event is assumed to occur at the end of the planned service life of the NPPs. Evidence must be provided prior to their installation that the affected cables can meet the specified functional requirements. Usually this evidence is provided by the results of the qualification tests, simulating operational aging, the LOCA and the post accident phase. Research efforts are continuing, through out the world, for developing effective testing methods which can assess the present condition of a cable's electrical property and can also predict the remaining life and LOCA survivability with some assurance. It has been found that the insulation resistance of cables are not much affected by ageing. Experience has shown that the failure of nuclear cables is primarily due to the hardening and embrittlement of the insulation resulting in the formation of micro cracks, a loss of dielectric strength and high leakage currents. The elongation at break measurements carried on various cables subjected to accelerated thermal ageing were used to arrive at Arrhenius parameters based on an acceptable 65% elongation at break criteria. Life estimation have been carried at operating temperatures using these parameters.

1. INTRODUCTION

Safety related power, control and instrumentation cables are critical to the safe operation of nuclear power plants and require environmental qualification (EQ) to provide assurance that common mode failures do not occur during service or during a postulated design basis accident (DBA). The qualification of the cables for use in NPPs has generally been based on IEEE-323 (1974) and IEEE-383 (1974). These standards aim to demonstrate the ability to survive a desired service life combined with design basis event (i.e. LOCA). For this purpose, accelerated tests are carried out at high temperature and dose rates to reduce the test time during simulation followed by a LOCA test. Such accelerated tests may not adequately simulate the behaviour of cables in long-term operational conditions due to approximations (such as high dose rates, synergistic effects of combined environments like temperature, radiation and humidity and hot spots) used in the design of the test.

2. BACKGROUND

2.1 Tensile Property (Elongation-at-Break and Tensile Strength) Measurements: Basic Tool for Correlation

Experience has shown that the failure of nuclear cables is primarily due to the hardening and embrittlement of the insulation resulting in the formation of micro cracks, a loss of dielectric strength and high leakage currents. Micro cracks can occur in the embrittled insulation due to thermal expansion and contraction of the insulation, or the manipulation of the cable during maintenance. Consequently the end of life criteria for insulation can be either a percentage decrease of elongation from the as received value or an absolute elongation at break.

Tensile strength and elongation tests are performed in accordance with ASTM-D2633-82, using a Universal Testing Machine (UTM) equipped with pneumatic grips and having an extensometer clamped to the sample. Special tensile specimens (dumbbells for larger cables or cylinders for smaller cables) of the insulation or jacket materials without the copper conductor are needed for these tests. Because of the difficulties in removing these conductors from an aged cable, samples of the polymeric material are prepared before exposing them to thermal or radiation condition. These tests are destructive, and therefore, many samples are required if tests are conducted regularly. Furthermore, for statistical purposes 3-5 samples often are tested at each time to obtain an average value.

3. CABLE IDENTIFICATION FOR RESEARCH PROGRAMME

3.1 Selection of Cable Samples and Retrieval of Data

For life span estimation the following selection criteria are employed while chosing the cables:

3.1.1 Generic types of Instrumentation & Control (I&C) cables that represent the largest volume of power plant cables with emphasis on Environmental Qualification (EQ) cables.

3.1.2 Cables known to have a significant continuous electrical loading or frequent inrush transients.

3.1.3 Similar cables in different locations with different environments.

3.2. Details of cable samples chosen for the ageing study

Three types of cables, EPR, HTXLPE & LTXLPE, were selected for the comprehensive ageing studies. These cables are the majority of the Instrumentation & Control systems. The Dumbbell specimens have been prepared from the cable insulating materials for carrying out tensile test. Tensile test have been carried out using UTM.

4. AGEING MODELS

Ageing models are useful to

- develop accelerated ageing procedures
- Establish a qualified life

The Arrhenius model addresses thermal ageing, and the equal dose/equal damage model addresses radiation ageing. Here only Arrhenius Model is considered.

Thermal Ageing Models

The Arrhenius model is used for addressing time-temperature ageing effects. Other models have also been used. For example, a simple model, the 10°C rule, states that chemical reaction rates double and the material life decreases by one-half for every 10°C increase in temperature. However, Arrhenius model is preferred over the 10°C rule, since the later is simply a rough approximation of the former. This equation is used in chemical kinetics to predict the rate process. Its application has been extended to reliability on the assumption that failure rate is proportional to reaction rate. In reliability studies this equation takes into account not only the chemical reactions but also physical and physico-chemical reactions. This model is represented by the following equation.

$$t = Ae^{\phi/kT} \qquad (1)$$

where ϕ = activation energy (eV)

(1 eV = 23.06 Kcal/mole= (1.6 x 10^{-12} ergs
k = Boltzmann's constant = 8.617 x 10^{-5} eV/°Kelvin

= 1.38 x 10^{-16} ergs/°k

t = Days, [Life at temperature T (in °K)]

A = Constant (Days)

To find out the values of A & ϕ, accelaretd thermal ageing tests have to be conducted at least at two temperature levels. Accelerated tests reduces the test durations. The accelerated temperatures for these tests should be fixed such that no new failure mechanisms, which are not present under natural ageing, are introduced. Thermal ageing is always present. Percentage elongations-at-break are derived from measurements by tensile tests on cable insulation materials using UTM. The trend by utilities is to specify an end of life as 65% absolute elongation. This value will provide a sufficient margin for the insulation to function without decreasing electrical performance and pass a design basis event like LOCA. The ϕ can be calculated using the following equation derived from equation 1, using two test temperatures.

$$\frac{t_1}{t_2} = \exp\left[\frac{\phi}{k}\left(\frac{1}{T_1} - \frac{1}{T_2}\right)\right] \qquad (2)$$

t_1 & t_2 = Thermal ageing time at temperature T_1 & T_2 (in °K) respectively to reach end of life criterion (e.g. 65% elongation at break)

5. TEST RESULTS

Tensile testing has been carried out on different types of cable insulation materials namely EPR, HTXLPE & LTXLPE. To establish the ageing characteristics of the cable insulation materials, all the three types of cables specimens were thermally aged at 120°C and 130°C separately. Tensile test on insulation specimens was carried out periodically till the elongation-at-break is observed to be 65% or below, which is the end life criterion. The test results are shown in Table-1.

An ageing curve was drawn for % elongation versus ageing test duration for EPR cable as shown in Figure 1. Similar ageing curves can be drawn for other cable materials. The criteria to establish the end of life was chosen as 65% elongation. From the ageing curves test duration in days at 120°C and 130°C was obtained and these values were used to draw the Arrhenius Plot to extrapolate the life at normal operating temperature i.e. 60°C. Arrhenius Plot for EPR cable is shown in Fig. 2. The constants A & ϕ. were evaluated from this Arrhenius plot.

Table 1 Thermal ageing test results for various types of cable insulation materials

Test Temp., °C	EPR Cable Insulation		LTXLPE Cable Insulation		HTXLPE Cable Insulation	
	Test duration, Days	E-at-B, %	Test duration, Days,	E-at-B, %	Test duration, Days,	E-at-B, %
120°C	Initial	985	Initial	912	Initial	925
	10	801	10	775	10	766
	20	671	20	625	20	655
	30	552	30	525	30	560
	40	453	40	435	40	476
	50	371	50	360	50	404
	60	297	60	290	60	341
	70	239	70	249	70	287
	80	191	80	191	80	240
	90	159	90	175	90	200
	100	105	100	123	100	164
	110	76	110	102	110	133
	120	67	120	80	120	107
	130	37	130	61	130	84
	140	21	140	46	140	65
	150	12			150	47
130°C	Initial	985	Initial	912	Initial	925
	10	672	10	630	10	629
	20	455	20	420	20	445
	30	285	30	269	30	307
	40	189	40	185	35	261
	50	111	50	97	40	201
	60	73	55	89	45	162
	70	37	60	77	50	127
	80	17	65	65	55	111
	-	-	70	49	60	89
	-	-	75	35	65	69
	-		-	-	70	43
	-		-	-	75	31
	-		-	-	80	21

Eqn. 1&2 can also be used to evaluate the constants A & ϕ. Using these, the life was estimated at normal operating temperature i.e. 60°C.

For EPR Cable Insulation Material:

$$\frac{t_1}{t_2} = \exp\left[\frac{\phi}{k}\left(\frac{1}{T_1} - \frac{1}{T_2}\right)\right]$$

Where t_1 and t_2 are the test duration in days to get 65% elongation at 120 and 130°C temperature respectively, and were obtained from ageing curves (Figure-1)

$T_1 = 120+273 = 393°K$ $T_2 = 130+273 = 403°K$

$t_1 = 131$ days $t_2 = 66$ days

ϕ = Activation energy (kCal/mole)

k = Boltzmann's constant = 2 x 10^{-3} kCal/mole/°K

$$\frac{131}{66} = \exp\left[\frac{\phi}{k}\left(\frac{1}{393} - \frac{1}{403}\right)\right]$$

$\phi = 21.71$ kCal/mole

Constant A of the Arrhenius Eqn.(1) can be determined using any one of the accelerated temperature data:

$t = Ae^{\phi/kT}$

where t= 131 days at 393°K

$A = 1.31 \times 10^{-10}$ days

After determining the values of A and ϕ, life at normal operating temperature of 60°C (333°K) can be determined as follows using Eqn. (1):

$t = Ae^{\phi/kT}$

t = 52 Years

Similar evaluations for other cables were also carried out and estimated the life-span. Test results are shown in Table 2.

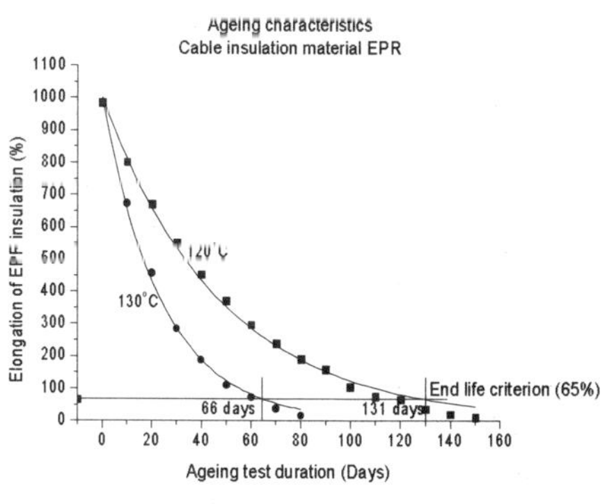

Fig. 1

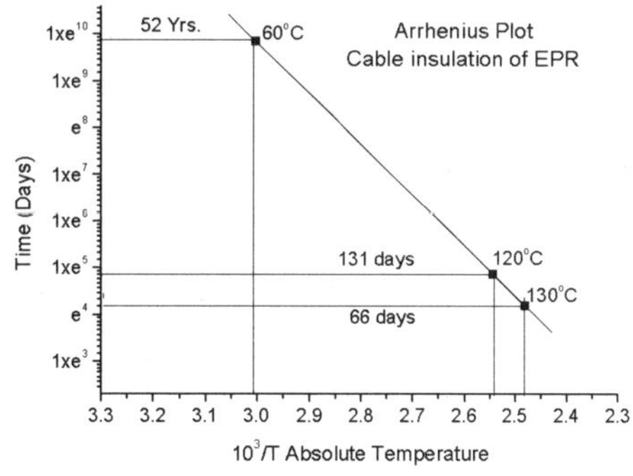

Fig. 2

Table 2 Estimation of life-span using Arrhenius Equation

Insulation Material	Test duration for 65% Elongation-at-Break		Estimated life-span at 60°C, Yrs.
	at 120°C	at 130°C	
EPR	131	66	52
LTXLPE	128	65	48
HTXLPE	140	69	55

6. CONCLUSION

The measurements of elongation-at break for the cable insulation materials is the requirement for monitoring the ageing characteristics of the cables. By measuring the elongation at break, life span of the cable insulation materials can be calculated. The estimated life can be used as a guideline for the selection and standardization of the materials. Some uncertainties are also associated due to approximations made during designing of the accelerated thermal ageing tests. Condition Monitoring (CM) techniques can be used to get an assurance of the residual life of the cables. as discussed earlier part of this paper.

7. REFERENCES

1. "IEEE Standard for Qualifying Class 1E Equipment for Nuclear Power Generating Stations", IEEE Std., 323(1974).

2. S.G. Burnay, J. Cook and N. Evans, "Round-robin testing of cable materials (IAEA Co-ordinated Research Programme 2), RCM, 8-12 June, 1998.

3. "Working Material on the Co-ordinated Research Programme on Management of Aging of In-containment I&C Cables", Report on the IAEA Research Co-ordination Meeting, 8-12 June, 1998, EDF, Bordeaux, France.

Reliability, Safety and Hazard
Editors: P.V. Varde et al.

COMPARATIVE STUDIES OF VARIOUS SOLUTION TECHNIQUES FOR DYNAMIC FAULT TREE ANALYSIS OF COMPUTER BASED SYSTEMS

Prabhudeva S.[1], *A. Srividya*[1], *A.K. Verma*[1] *and V. Gopika*[2]

[1]Reliability Engineering Group, Department of Electrical Engineering, IIT Bombay, Mumbai
[2]Reactor Safety Division, BARC, Mumbai

ABSTRACT

Fault tree analysis process is an effective method for calculating the reliability of a system. It provides a pictorial and logical framework for analyzing the reliability of systems. Also it has been used for a long time for the quantitative and qualitative analysis of the failure modes of critical systems. An important characteristic of many engineering systems is that they behave dynamically, i.e., their components interact with each other and with the environment. A fault tree is classified as Static or Dynamic depending on the types of gates used. Recently different methods have been proposed by various authors, to solve Dynamic Fault Trees. In this paper mainly two methods from literatures have been used for comparative study purpose i.e., the DIFtree (Dynamic Innovative Fault Tree) method and the Distribution method and the results are compared at the end of analysis process.

Keywords: Static Fault Tree, Dynamic Fault Tree, Markov chain, Binary Decision Diagram, Modularization, Reliability, Unreliability

1. INTRODUCTION

Complex systems are made up of components. In some cases the failure of a component may lead to the failure of the whole system and in other cases may not depend upon the functional relationship of components. In any case each component contributes to the system reliability or unreliability, so it is important to express system reliability in terms of component reliability. This makes a careful analysis of functions of all components and their failure models.

Component failure models can be obtained in two ways:

1. From the basic failure rates and working stresses using stress models and
2. From part-failure data obtained either from life-tests or failure reports of customers.

It is almost impossible to establish failure rates of components for all working conditions. However, it is possible through extrapolation or interpolation to predict reliability under specific stress conditions from the available data.

A fault tree is classified as Static or Dynamic depending on the types of gates used. Dynamic fault trees are generally solved using automatic conversion to Markov models.

The DIFtree method provides a methodology for reliability calculation of large complex fault trees. In this approach, the fault tree is divided into independent sub-trees (module) using modular approach; and different solution techniques are applied to each subtrees. If the subtree is dynamic, then the Markov chain solution is used; otherwise the Binary Decision Diagram (BDD) solution is used. The solutions of various independent subtrees are integrated to get the system reliability. The Distribution method is same in all respects to DIFtree method except in that, it solves the dynamic module without translating it into Markov chain. The DIFtree method is applicable to exponential and weibull time to failure distributions whereas the Distribution Approach handles all type of distributions but restricted to coherent systems, systems with perfect coverage and non-repairable components. This paper is intended to consolidate the work done by various researches on reliability/ unreliability estimation and a comparison of different methods and models with illustrative examples of these approaches.

2. METHOD TO SOLVE DYNAMIC FAULT TREES

Any dynamic fault tree can be solved either by translating it to Markov-chain (DIFtree method) or by distribution method to achieve the desired result.

2.1 DIFtree methodology

DIFtree [1-7]. It uses a modular approach which provides a BDD solution for static fault trees and Markov chain solution for dynamic fault trees.

A more efficient and powerful means for fault tree analysis can be provided by the use of both BDD and Markov solution methods in tandem.

A fault tree is translated to a Markov chain only if it has one or more dynamic gates. If a fault tree has only static gates it is solved by translating into a BDD.

Often a very small part of the entire fault tree is dynamic in nature. Hence independent sub trees are identified in the modular approach [5] and the decision to use a Markov or a BDD solution is made for the sub tree instead of the whole fault tree. These independent sub trees are treated separately and their solutions are integrated to get the solution for the entire fault tree. In other words several Markov models and BDDs are used for the solution of the fault tree instead of one Markov model or one BDD. The advantage of using several Markov models and BDDs is that each will be substantially smaller in size than the single Markov model or BDD.

Feature of DIFtree

1. The independent sub trees can be solved separately, thereby, allowing an exact solution in minimum time.
2. Coverage modelling capability has been included in both static and dynamic solution techniques.
3. The dynamic fault tree model allows the analysis of complex management techniques in a relatively simple manner.
4. Modular approach to dynamic models can allow the analysis of much larger systems.
5. Since this methodology allows the combination of constant probabilities and rates of occurrences, it is therefore applicable to more than just hardware. It is equally applicable to software and human operator failures as to hardware failures.

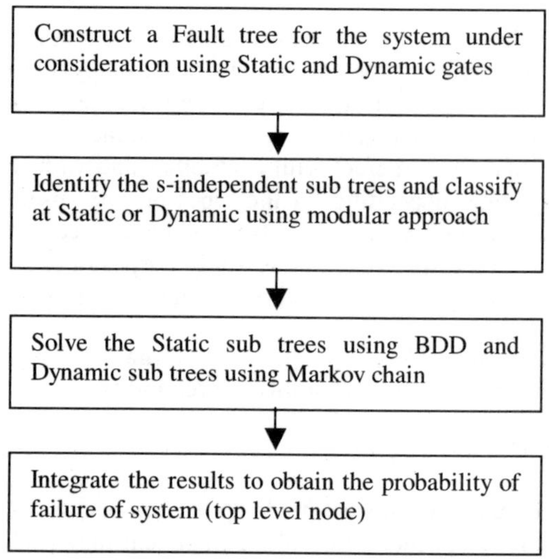

Fig. 1: Flow Chart for DIFtree Methodology

2.2 Distribution method

Distribution method is proposed [8] to analyze dynamic fault trees without translating them into Markov models. In this method, BDD or any best methods applicable to static fault tree analysis are used in solving dynamic fault trees. This method is straightforward for modular sub trees, and in general case, conditional probabilities are used to calculate the results of a dynamic module from the results of the sub trees. Only exact methods have been considered. Here the theory which is applicable for finding approximate solutions of Markov models and fault trees can be used to find the approximate solutions in this method. Generally, this method is applicable to: (a) Coherent Systems (b) Perfect Coverage Models (c) System with Non-Repairable Components (d) Fault trees without repeated constrained events.

This method concentrates only on the static - AND, OR, and Voting gates and dynamic-PAND, SEQ, and generic SPARE gates .

This approach is similar to the approach presented in [5]. *S*-independent modules are identified and classified as specified in [5]. Once these modules are solved, the results are integrated. The main focus of this method is solving dynamic modules without using Markov chains or simulation.

The basic concept of this method is that if the distributions of a set of random variables are known, then the distribution of another random variable (dependent/output), which is a function of these random variables, can be found using numerical methods. This concept can be used to find the top event probability (or distribution) of a dynamic module from the occurrence distributions of its subtrees. The method of finding the distribution of the dependent random variable is straightforward if all the input random variables are *s*-independent. If the input random variables are dependent, then we can use conditional probabilities such that the conditional distributions of input random variables become *s*-independent.

In this paper, both the methodologies are implement as discussed in previous sections and from the results obtained, a comparative analysis is carried out.

3. EXAMPLES

3.1 Example 1

Consider a PAND (Dynamic gate) with AND and OR gates as inputs [8].

DIFtree method: The top gate is a dynamic gate, the given fault tree can be divided into 3 modules as M1 (AND subtree), M2 (OR sub-tree) and M3 (PAND main module). The static module M1 and M2 solved by using BDD, and M3 is translated into Markov chain. The unreliability is calculated for mission t=1000 hrs, is as follows:

For Module M 1(AND Gate):

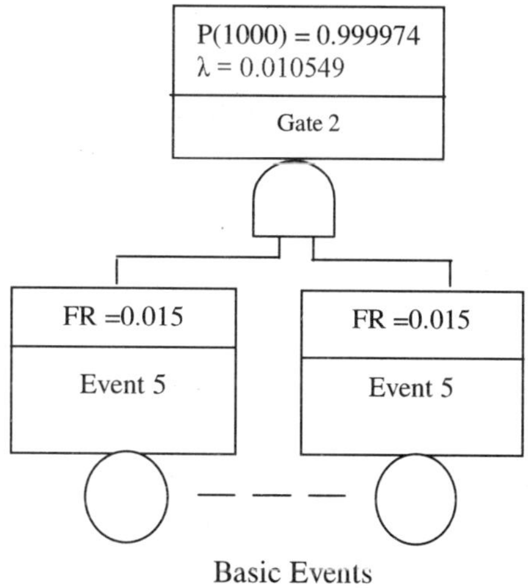

Fig. 2: Module M1

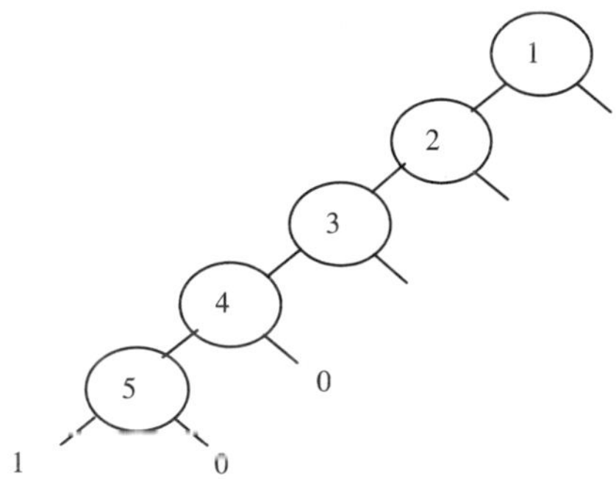

Fig. 3: BDD for M1

P(M1) = P(1) P(2) P(3) P(4) P(5) = 0.99997345

For Module M2 (OR Gate):

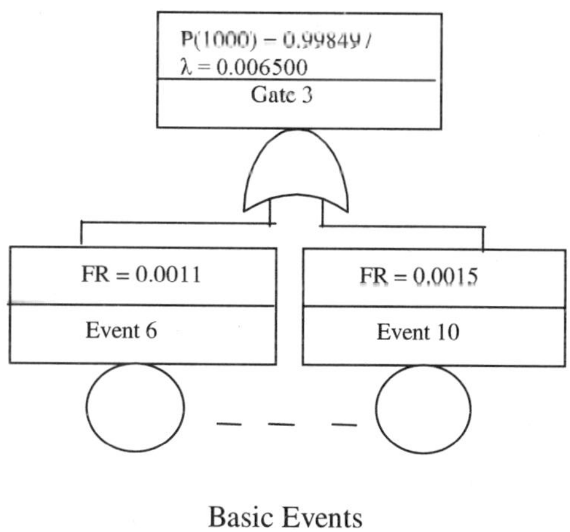

Fig. 4: Module M2

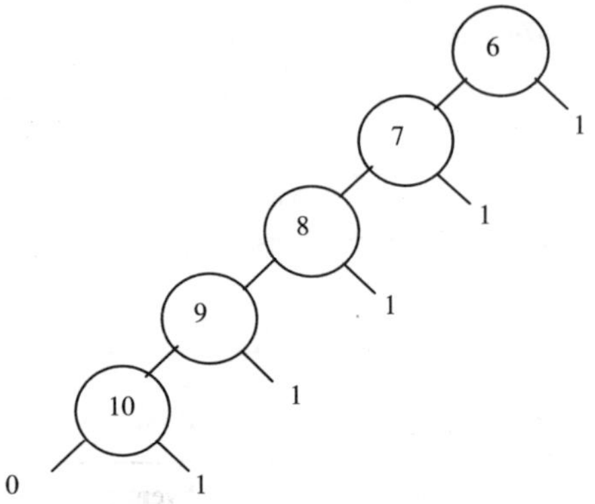

Fig. 5: BDD for M2

Module M3 includes Modules M1 and M2 as subtees. Here PAND Gate is a dynamic Gate and have two inputs G1(t) = P(M1) = 0.99997345, and G2(t) = P(M2) = 0.998494091. Hence in this DIF tree method PAND gate is solved by transforming it into Markov chain as shown in Figure 6a and 6b. "1" entry in a state indicates that the corresponding basic event has one operational replicate. Replicates are identical instances of the same component, and replication is used to reduce the size of the Markov model by lumping together basic events.

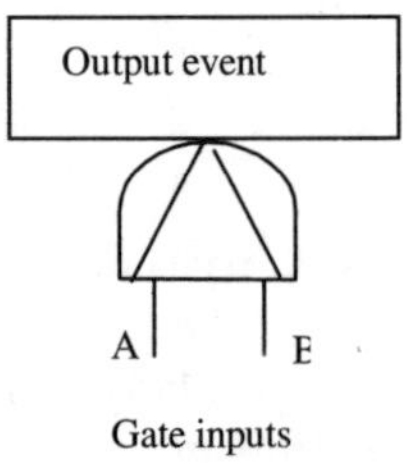

Fig. 6a: Module M3 (PAND **Gate)**

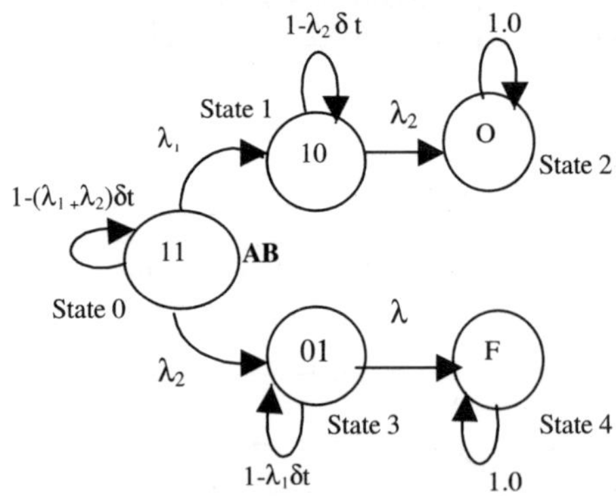

Where O: Operational State, F: Failed State

Fig. 6b: PAND to Markov Chain Translation

The set of differential equations can be obtained from the state probability equations and then taking the Laplace Transform for these equations and substituting initial values , after that solving by partial fraction method and after that by taking Inverse transform we get

$$P_0(t) = e^{-(\lambda_1+\lambda_2)t}$$

$$P_1(t) = e^{-(\lambda_2)t} - e^{-(\lambda_1+\lambda_2)t}$$

$$P_2(t) = \frac{\lambda_1}{(\lambda_1+\lambda_2)} - \left[e^{-(\lambda_1)t}\right] + \frac{\lambda_1}{(\lambda_1+\lambda_2)} e^{-(\lambda_1+\lambda_2)t}$$

$$P_3(t) = e^{-(\lambda_1)t} - e^{-(\lambda_1+\lambda_2)t}$$

$$P_4(t) = \frac{\lambda_2}{(\lambda_1+\lambda_2)t} - \left[e-(\lambda_1)t\right] + \frac{\lambda_1}{(\lambda_1+\lambda_2)} e^{-(\lambda_1+\lambda_2)t}$$

Therefore $R(t) = P_0(t) + P_1(t) + P_2(t)$ the resulting reliability equation is:

$$R(t) = \frac{\lambda_1}{(\lambda_1+\lambda_2)} + \frac{\lambda_2}{(\lambda_1+\lambda_2)} e^{-(\lambda_1+\lambda_2)t}$$

for $\lambda_1 = 0.010549, \lambda_2 = 0.006500, t = 1000$,
R(t) = 0.618768015 = G(t) and unreliability : (1-R(t)) = 0.38123.

Distribution method: In this method the fault tree is divided into modules M1, M2 and M3. M1 and M2 are static and hence the probability distribution for them is evaluated using BDD method as described above in DIFtree method. And finally the probability of the top event (Dynamic gate) is evaluated with convolution integral.

Let $G_1(t) = P(M1) = 0.999974$;

Let $G_2(t) = P(M2) = 0.998494091$ as above,

and G(t) is the probability of failure of the top event and it is calculated from the equation

$$G(t) = \sum_{i=1}^{M} \left[G_1(i \times h) - G_1((i-1) \times h)\right] \times \left[G_2(t) - G_2(i \times h)\right]$$

where $G_1(x) = G_2(x) = \exp(-\lambda x)$, G(1000) = 0.382 (unreliability). This is obtained by evaluating G1(t) and G2(t) (using numerical integration method trapezoidal integral rule).

Table 1 Reliability Comparison of Results in Example 1

Methods	Reliability	Unreliability
DIFtree	0.618768015	0.38123
Distribution	0.618	0.382

3.2 Example 2

Cardiac Assist System (CAS) [6], Cardiac Assist System is designed to treat heart failure. The structure of cardiac assist system and Fault tree are given in the reference [6] and the dynamic sub-tree (module 5) of that is as shown below:

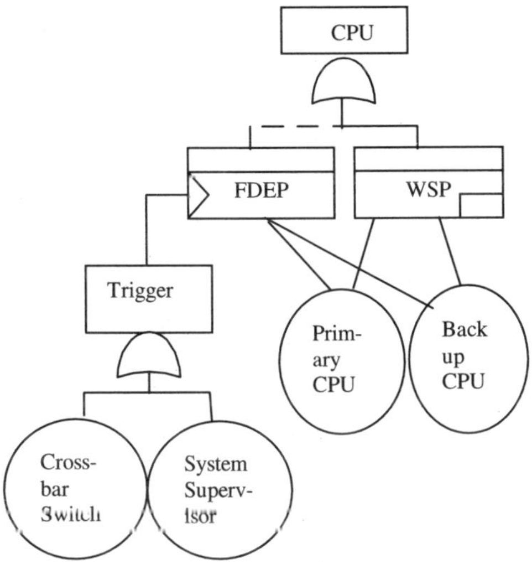

Fig. 7: Module M5 of CAS

DIF Tree Method: This is a Dynamic module, hence in this DIFtree method, first we have to transform this module into Markov chain and then solving the deferential equations from the state probability equations, using substitution method or by solving the matrix (by taking inverse of it) or any other methods to solve simultaneous equations. Here the FDEP gate cause to failure the primary (P) or secondary(S) CPUs whenever the corresponding dependent events (P) or (S) Trigger event occur. On the other hand it disenables the corresponding gate from the rest of the system. And also it will not produce any output. Hence we have to concentrate on the WSP gate (dynamic gate). The WSP gate [3] and its corresponding Markov chain are as shown in figure 8a and 8b respectively .

$\lambda_p = \lambda_S = 0.00006$, t=100, and Dormancy factor $\propto = 0.5$

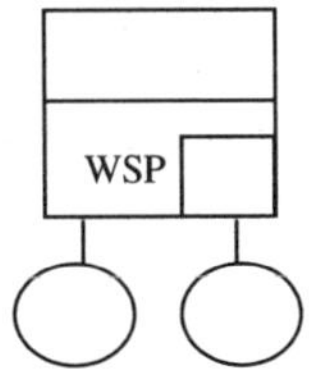

Fig. 8a: WSP Gate

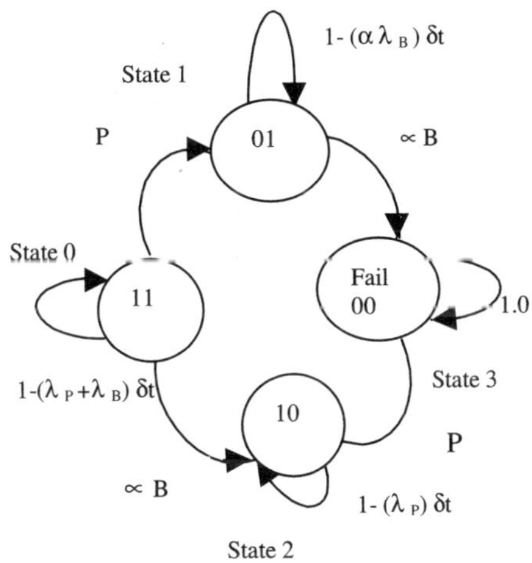

Fig. 8b: Markov Chain of WSP

from the above Markov chain, The set of differential equations can be obtained from the state probability equations and taking Laplace transform for these equation, then substituting the initial values, and by solving these equations using partial fraction method and taking inverse transform, we get

$$P_0(t) = e^{-(\lambda_P + \alpha\lambda_B)t}$$

$$P_1(t) = e^{-(\alpha\lambda_B)t} - e^{-(\lambda_P + \alpha\lambda_B)t}$$

$$P_2(t) = e^{-(\lambda_P)t} - e^{-(\lambda_P + \alpha\lambda_B)t}$$

therefore $R(t) = P_0(t) + P_1(t) + P_2(t)$, for $\lambda_P = \lambda_B = 0.00006$, $\alpha = 0.5, t = 100$ $(0 \leq \propto \leq 1)$.

$$R(t) = e^{-(\lambda_P)t} + e^{-(\alpha\lambda_B)t} - e^{-(\lambda_P + \alpha\lambda_B)t}$$

R(t) = 0.999981586, and the unreliability of WSP gate of CAS = 0.000018411, Therefore P(M5) = 0.000018411

***Distribution Method*:** In this method all the static modules are solved by using BDD or any standard methods. And for dynamic gates using distribution (exponential) and numerical methods (trapezoidal rule or Simpson rules) can be used to solve. Therefore, For Module M5 using the following expression:

$$R(t) = \exp(-\lambda_P t) + \left[\frac{\lambda_P}{\lambda_P + \alpha\lambda_B - \lambda_B}\right]$$

$$\{\exp(-\lambda_P t) - \exp(-(\lambda_P + \alpha\lambda_B)t)\}$$

R(t) = 0.999907964

P(M5) =1-0.999907964 = 0.000092036

Similarly other modules (static) are solved by BDD method and the final results are as given in Table 2.

Table 2 Reliability Comparison of Results in Example 2

Methods	Reliability	Unreliability
DIFtree	0.976503406	0.023496594
Distribution	0.96135	0.03864

4. CONCLUSIONS AND SCOPE OF THE WORK

1. Markov chain models the stochastic behaviour of the system. But the drawback of Markov chain conversion is that a huge state-space explosion results even for a moderate size fault tree. Hence conversion of fault tree containing only static gates to Markov model is very inefficient.
2. DIFtree methodology that uses a modular approach to find the independent modules classifies them as static and dynamic sub trees. The static modules are then solved using BDDs and Dynamic modules are solved using Markov model. The main feature of this methodology is that several Markov models and BDDs are used for the solution of the fault tree instead of one Markov model or one BDD that greatly reduces the complexity.
3. DIFtree method is applicable only to exponential distribution and for Weibull distribution semi-markov chain is used.
4. Distribution method is proposed to analyze dynamic fault trees without translating them into Markov models, it gives more accurate results, in less time compared to DIFtree method.
5. The two methods DIFtree and Distribution method have some common features but in some respects they are complementary for each other. Therefore a software module can be developed which will include the features of both the methods for dynamic fault tree analysis. The proposed algorithm is as shown in figure7below:

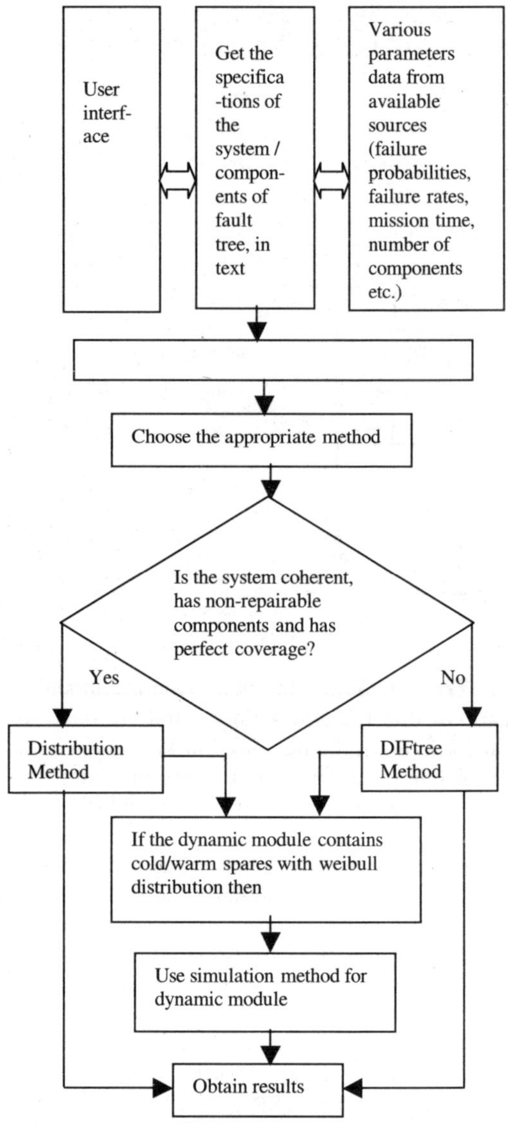

Fig. 9: Flow chart for the proposed software algorithm

5. REFERENCES

1. J. B. Dugan, Bharath Venkataramanan, Rohit Gulati, "*DIF Tree : A software packge for the analysis of Dynamic Fault tree Models*", In Proceedings of the Annual Reliability and Maintainability Symposium, **64-70**, (1997).

2. Raghavan Manian, David W. Coppit, Kevin J. Sullivan, J.B. Dugan. "*Bridging the gap between systems and dynamic the fault tree models*", In proceedings of the annual Reliability and Maintainability Symposium, (1999).

3. J. B. Dugan, Salvatore J. Bavuso, Mark A. Boyd, "*Dynamic fault tree models for fault tolerant computer systems*", IEEE Transactions on Reliability, **41**,.**3**, (1992).

4. Raghavan Manian, David W.Coppit, Kevin J. Sullivan, J.B. Dugan. "*Combining various solution techniques for dynamic fault tree analysis of computer system*", In proceedings Third IEEE International High assurance Systems Engineering Symposium, (2000).

5. J. B. Dugan, Rohit Gulati, "*A modular approach for analyzing static and dynamic fault trees*", *in* Proceedings of the Reliability and Maintainability Symposium, (1997).

6. Yansong J. Ren, J.B. Dugan, "*Design of reliable systems using static and dynamic fault trees*", IEE Transactions on Reliability **47**,**3**, (1998).

7. Rauzy, Yves Dutuit,"*A linear time algorithm to find modules of fault trees*", IEEE Transactions on Reliability, **45**, **3**, (1996).

8. Suprasad Amari, Glenn Dill, Eileen Howald, "*A new approach to solve dynamic fault trees*", In proceedings of the annual Reliability and Maintainnability Symposium (2003).

Reliability, Safety and Hazard
Editors: P.V. Varde et al.

EVOLUTION OF STOCHASTIC PETRI NETS AND ITS APPLICATION IN DEPENDABILITY MODELING

Manoj Kumar[1], *A.K. Verma*[2], *A. Srividya*[2] *and G. Govindarajan*[1]

[1]Control Instrumentation Division, BARC, Mumbai 400 085
[2]Reliability Engineering, IIT Bombay, Mumbai 400 076

ABSTRACT

Petri nets (PN) are a graphical and mathematical modeling tool. Due to the generality and permissiveness inherent in PNs, they are applicable to wide variety of applications. Many structural and modeling extensions have been proposed by researchers in various areas to address application specific issues. In this paper, we have reviewed stochastic extensions of PN for dependability modeling.

Dependability modeling plays an important role in our lives, as we are getting more and more dependant on the working of various systems. In this paper we have discussed PN based models, such SPN, GSPN, SRN etc, for dependability modeling and analysis

Keywords: Petri nets, Stochastic Petri Nets, Generalized Stochastic Petri Net, Markov Regenerative Stochastic Petri Nets

1. INTRODUCTION

Dependability includes, survivability (i.e. quality, reliability, maintainability and availability) and safety. In today's world, our lives are getting more dependant on various systems, such as, computer and computer based systems, communication networks, railway and aviation services etc. failure of these systems could result in loss of time, effort, money or even life in some critical applications. As these systems plays an increasingly important in our lives, assessment of system dependability becomes a key step in design, analysis and tuning of such systems. Measuring dependability with actual systems is very costly, time-consuming affair. A model can help in this reference, while being cost effective, it takes much less time and provide vital information to improve the system. Models also help to build a framework for gathering, organizing, evaluating, and understanding information about these systems [1]. Researchers have developed both simulation and analytical models for this purpose, e.g. discrete-event simulation models, combinatorial models, and Markov models [1].

In many cases, as the complexity of the system increases the size of the system model may become too large to handle. The size and complexity of real systems is always a major obstacle for an analyst in obtaining accurate model. Because if all the details of the system are included the model will become intractable, i.e. it will require too long simulation time to get accurate results in discrete-event simulation models, very large storage in combinatorial models, and too many states in Markov Model. For Markov models, largeness even makes it impossible to describe the system using Markov chain.

To overcome the difficulties in describing the system, a modeling technique is required that is more concise in specification and whose form is closer to designer's or analyst's intuition about what a model should look like.

In this regard, we will have a look at Petri nets (PN). The concept of Petri nets has its origin in C.A. Petri's dissertation, submitted in 1962. Petri nets are a graphical and mathematical modeling. Due to the generality and permissiveness inherent in PN, they can be applied informally to any area or system that can be described graphically like flow charts [2]. In recent years considerable efforts has been devoted to enrich PN formalism in order to improve their capability to easily capture system behavior [4]. In this paper we review the main structural and stochastic extensions of Petri nets and their application in dependability modeling.

The paper is organized as follows. Section 2 defines the basic Petri net model, its basic properties and structural extension. Section 3 introduces stochastic extensions, Stochastic Petri net (SPN) and its isomorphism with Markov chains. Section 4 introduces GSPN, SRN, and MRSPN models. In section 5 an example is taken and discussed its dependability model. In section 6 the analysis tools have been reviewed. Concluding remarks follow in section 6.

2. PETRI NETS, PROPERTIES AND STRUCTURAL EXTENSIONS

2.1 Petri Net

A Petri net is an abstract, formal model of information flow [3]. Graphically Petri net is shown as a particular kind of directed graph. The graph is a bipartite graph consisting of two kinds of nodes, called places and transitions, where arcs are either from a place to transition or from a transition to place. In graphical representation, places are drawn as circles and transitions as bar or boxes. Formally a Petri net is defined as a 5-tuple [2],

$PN = (P, T, F, W, M_0)$ where:

- $P = \{p_1, p_2, \cdots, p_n\}$ is a finite set of places
- $T = \{t_1, t_2, \cdots, t_m\}$ is a finite set of transitions
- $F \subseteq (P \times T) \cup (T \times P)$ set of arcs
- $W: F \rightarrow \{1, 2, 3, \cdots\}$ is a weight function
- $M_0: P \rightarrow \{0, 1, 2, \cdots\}$ is the initial marking

A positive directed arc connecting a place to a transition is called an input arc of the transition, while arc from transition to place is called output arc of the transition. With each arc an integer is associated, it is called the weight or multiplicity of the arc. Places connected to a transition by input arcs are called the input places of this transition, and those connected by means of output arcs are called the output places. Each place may contain zero or more tokens. A transition is enabled if each of its input places has at least as many tokens as the weight (or multiplicity) of the corresponding input arc. A transition can fire when it is enabled. The firing redistributes the tokens. A number equal to the multiplicity of the input arc is removed from each of the input places, and a number of tokens equal to the multiplicity of the output arc is deposited in each of its output places.

The analysis of Petri net revolves around investigating the possible markings. Petri net semantic does not state which of multiple simultaneously enabled transitions fires first, hence every possible firing order needs to be analyzed.

PN can be used to capture the behavior of many real-world situations including sequencing, synchronization, concurrency, and conflict. Application areas widely using PN are analysis of communication protocols, analysis of distributed-software systems, distributed database systems, concurrent and parallel programs, asynchronous system design and logic programs etc [2].

2.2 Structural extension to PN

Many extensions have been proposed to increase either the class of problems that can be represented or their capability to deal with the common behavior of real systems. Extensions are generally meant either to increase modeling power or modeling convenience. Modeling convenience is the ability to represent common behavior by the introduction of structural extensions to the basic Petri net formalism. These extensions provide a powerful way to improve the ability of PNs to model real problems. Some extensions of this type have proven so effective that they are now considered part of the standard PN definition. They are:

- Inhibitor arcs
- Transition priorities
- Marking-dependent arc multiplicity

Inhibitor arcs are useful extension to standard PN formalism. An inhibitor arc from place p to transition t disables t for any marking where p is not empty. It is also possible to use the arc multiplicity in inhibitor arcs. In this case a transition t is disabled whenever the place p contains at least as many tokens as the multiplicity of the inhibitor arc. Inhibitor arcs are used to model contention of limited resources to represent situations in which one activity must have precedence over another.

An integer "priority level" is assigned to each transition. a transition is enabled only if no higher priority transition is enabled. However, the convenience of priorities comes at a price. If this extension is introduced, the standard PNs ability to capture the entire system behavior graphically is partially lost [4].

Practical situations often arise where the number of tokens to be transferred depends upon the state of the system state. These kinds of situations can be easily managed adopting marking-dependent arc multiplicity, which allows the multiplicity of an arc to vary according to the marking of the net. Marking dependent arc multiplicities allow simpler and more compact PNs. When exhaustive state space exploration techniques are employed, their use can dramatically reduce the state space.

The extensions meant to increase the modeling power of the Petri net are covered in subsequent sections

3. STOCHASTIC PETRI NETS (SPN)

Transitions of the PN do not have timing information associated with them. Associating random (geometric or exponential) firing times with transitions of the PN extend it to stochastic PN (SPN), it also increases the modeling power of PN. When waiting times (time to fire) are associated with transitions, an enabled transition fires only after the waiting time.

A transition's firing time represents the amount of time required by the activity associated with the transition. it is counted from the instant the transition is enabled to the instant it actually fires, assuming that no other transition firing effects it.

These new SPNs are isomorphic to homogenous Markov processes [5]. Means SPNs can be transformed to discrete or continuous time Markov chains depending upon whether the waiting times associated with transitions are geometrically distributed or exponentially distributed.

Now we will take an example to show the isomorphism form Molloy's [5]. The SPN is shown in Fig. 1. The SPN displays sequential operation (t_1, t_5), parallel operation (t_2, t_3), forking (t_1), joining (t_5), and contention (t_4, t_5). Assuming an initial marking of one token in place p_1 and no tokens in the remaining places, then solving for the reachability set. The reachability set is shown in Table 1.

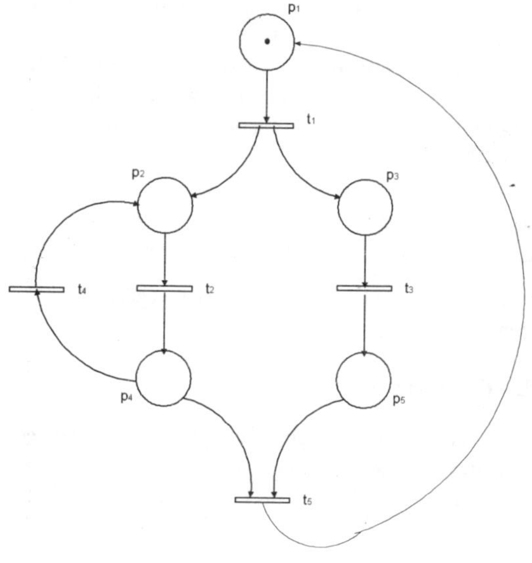

Fig. 1: Example of a SPN

Table 1 Reachability set of the SPN

	P_1	P_2	P_3	P_4	P_5
M_1	1	0	0	0	0
M_2	0	1	1	0	0
M_3	0	0	1	1	0
M_4	0	1	0	0	1
M_5	0	0	0	0	1

Using these markings its equivalent continuous time Markov chain (CTMC) is shown Fig. 2.

The CTMC shown in Fig. 2 can be solved for steady-state probabilities using the established methods.

If we assume a different initial marking in Fig. 1 then we obtain a different Markov chain. The more tokens we have in the system the larger the Markov chain. For example, if we start with two tokens in place p_1 we would find 14 states in the reachability set. Similarly, if we start with 3 tokens in place p_1 we would find 30 states in the reachability set. This shows the modeling convenience of SPN. Other stochastic modeling formalisms are discussed in the next section

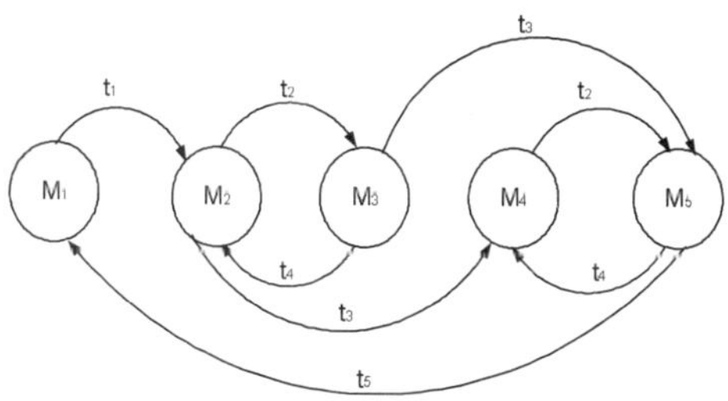

Fig. 2: The equivalent CTMC

4. OTHER STOCHASTIC MODELING EXTENSIONS TO PN

4.1 Generalized Stochastic PN (GSPN)

GSPNs are an extension to stochastic Petri nets obtained by allowing the transitions of the underlying PN to immediate as well as timed. Immediate transitions are drawn as thin bars, while timed transitions by rectangular boxes or thick bars. Immediate transitions are assumed to fire in zero time once enabled. Time transitions are associated with a random time to transition as in the case of SPNs. Immediate transitions always have priority over timed transitions to fire. If several immediate transitions are enabled together in a marking, a specified probability mass function is used to break the tie.

A marking of a GSPN is called vanishing if at least one immediate transition is enabled in the marking and tangible otherwise. While mapping reachability set of GSPN to CTMC vanishing states are eliminated.

4.2 Stochastic Reward Net (SRN)

SRNs introduce a stochastic extension into SPNs consisting of possibility to associate reward rates with the markings [4]. The reward rate definitions are specified at the net level as a function of net primitives like the number of tokens in a place or rate of a transition. The underlying Markov model is then transformed into a Markov Reward model thus permitting evaluation of not only performance and availability but also the combination of the two. The tool such as SPNP [X] implements this extension of Petri nets, allows the reachability graph to be generated automatically and easily provides dependability, and performance measure.

4.3 Markov Regenerative Stochastic PN (MRSPN)

A system in which transition times are not exponentially distributed, but have some embedded time epochs at which system have Markov (memory less) property. This kind of systems can be modeled by MRSONs. MRSPN is formally defined [4] as follows:

"A MRSPN is a SPN whose marking process is a MRGP"

MRGPs (or Semi Regenerative Processes) are discrete state continuous time stochastic processes with embedded Regenerative Time Points, at which the system have the Markov (memory less) property Based on the concept of memory of general transitions [4], Regenerative Time Points (RTPs) can be defined as follows:

"A regenerative time point in the marking process is an instance of the time when all the active and sampled time intervals of the general transitions are concluded"

The importance of this definition comes form the fact that MRSPNs can be studied by the results available for MRGPs. The analysis methods of the MRSPNs published in the literature so far are based on one of the following approaches [4]:

- Markov renewal theory
- Method of supplementary variable
- Approximate analysis by phase type expansion

5. EXAMPLE

Here we take an example of a Triple Modular Redundant (TMR) system, and describe its dependability model using GSPN.

TMR configuration is widely used in critical applications. TMR has several advantages over simplex configuration, mainly fault tolerance due to fault masking.

Each unit of the system in example has an exponentially distributed failure times. The system has built-in test and fault detection mechanism, which detects failure with a given probability. On detection of fault repair and restoration activity starts. The time of competition of repair activity also follows exponential distribution. The undetected fault may result in system failure. The failures could be safe with given probability otherwise unsafe.

The GSPN model in Fig. 3 models the system just described. The meaning of the places and transitions is shown in Table 2.

With this model various dependability parameters such as reliability, availability and safety can be analyzed. Reliability is the state probability of not being two or more tokens in failure state (safe or unsafe). Similarly, safety is the state probability of being two or more tokens in state fail-safe.

As the system size increases, manual derivation of Markov chain and solution is difficult. So, concurrent with the stochastic evolution of Petri nets, efforts have been devoted to defining and implementing automatic solution tools for dependability and performance analysis of PN based model. In the next section we briefly indicate the well-known tools for analytical evaluation.

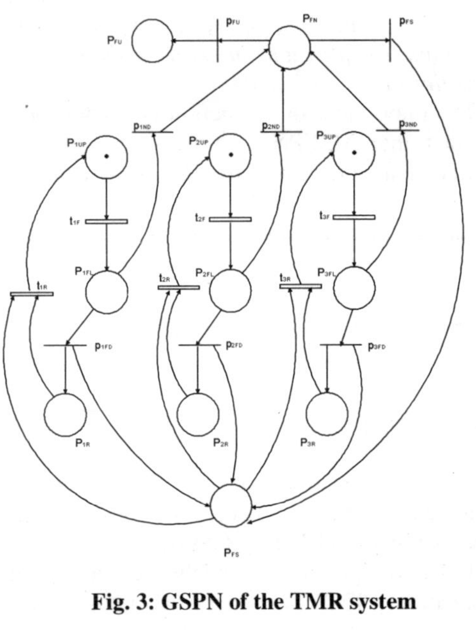

Fig. 3: GSPN of the TMR system

Table 2 Meaning of places and transitions

Places	Meaning
P_{1UP}, P_{2UP}, P_{3UP}	Unit of TMR in working state
P_{1FL}, P_{2FL}, P_{3FL}	Unit in failed state
P_{1R}, P_{2R}, P_{3R}	Unit under repair
P_{FS}	Safe failure of units
P_{ND}	Undetected failure
P_{UF}	Unsafe failure of units
Places	**Meaning**
t_{1FL}, t_{2FL}, t_{3FL}	Failure times of units
t_{1R}, t_{2R}, t_{3R}	Repair times of units
p_{1FD}, p_{2FD}, p_{3FD}	Prob. of fault detection of units
p_{1ND}, p_{2ND}, p_{3ND}	Prob. of undetected faults of units
p_{FU}	Prob. of unsafe failure
p_{FS}	Prob. of safe failure

6. ANALYSIS TOOLS

This list of tools from literature [4] is as follows:

6.1.1 SHARPE: It provides a variety of probabilistic, discrete-state models used to assess the reliability and performance of computer and communication systems. Originally provided with a textual input, now graphical interface is also available.

6.1.2 GreatSPN: It is based on graphical input and provides steady state as well as transient analysis of GSPNs.

6.1.3 SPNP: It is a C-based tool for the analysis of stochastic Reward Nets, an extension of PNs which allows specification of rewards directly at the net level. Steady state, transient and cumulative measures can be computed.

6.1.4 DSPNexpress: It deals with deterministic and stochastic Petri nets. Under the assumption that at most one deterministic transition is enabled in a given time instant, and steady state probabilities are computed.

7. CONCLUSION

Petri nets represent a powerful tool for specifying and analyzing dependability measures of computer and communication systems. In this paper we surveyed the Petri net literature and discussed structural and stochastic extensions.

We took an example, discussed its GSPN model for dependability measures. Non-Markovian PNs were also reviewed.

8. ACKNOWLEDGEMENT

The authors are thankful to Shri S. Bhattacharya (Head, CnID, BARC) and U. Mahapatra and P.P. Marathe for their encouragement and support.

9. REFERENCES

1. X. Zang, "Dependability Modeling of Computer Systems and Networks", Ph.D. Thesis, Duke University, 1999.
2. T. Murata, "Petri Nets: Properties, Analysis and Applications", *Proc. of IEEE*, Vol. 77 (4), pp. 541-560, 1989.
3. J.L. Peterson, "Petri Nets", *ACM Computing Surveys*, Vol 9 (3), pp. 223-252, Sept. 1977.
4. A. Puliafito et al., "The Evolution of Stochastic Petri Nets", *Proc. of WCSS' 97*, Singapore, pp 1-3, 1997.
5. M.K. Molloy, "Performance Analysis Using Stochastic Petri Nets", *IEEE Transaction on Computer*, Vol. C-31(9), pp. 913-917, Sept. 1982.

Reliability, Safety and Hazard
Editors: P.V. Varde et al.

SOFTWARE V&V METHODS FOR A SAFETY-GRADE PROGRAMMABLE LOGIC CONTROLLER

Jang Yeol Kim, Young Jun Lee, Kyung Ho Cha, Se Woo Cheon, Jang Soo Lee and Kee Choon Kwon

Korea Atomic Energy Research Institute (KAERI)
P.O. Box 105, Yuseong, Daejeon, 305-600, Republic of Korea

ABSTRACT

This paper addresses the Verification and Validation(V&V) process and the methodology for an embedded real time software of a safety-grade Programmable Logic Controller(PLC). This safety-grade PLC is being developed as one of the Korean Nuclear Instrumentation and Control System(KNICS) projects. KNICS projects are developing a Reactor Protection System(RPS) and an Engineered Safety Feature-Component Control System(ESF-CCS) as well as a safety-grade PLC. The safety-grade PLC will be a major component that encomposes the RPS systems and the ESF-CCS systems as nuclear instruments and control equipment. This paper describes the V&V guidelines and procedures, V&V environment, V&V process and methodology, and the V&V tools in the KNICS projects. Specifically, it describes the real-time operating system V&V experience which corresponds to the requirement analysis phase, design phase and the implementation and testing phase of the software development life cycle. Main activities of the V&V for the PLC system software are a technical evaluation, licensing suitability evaluation, inspection and traceability analysis, formal verification, software safety analysis, and a software configuration management. The proposed V&V methodology satisfies the Standard Review Plan(SRP)/Branch Technical Position(BTP)-14 criteria for the safety software in nuclear power plants. The proposed V&V methodology is going to be used to verify the upcoming software life cycle in the KNICS projects.

1. INTRODUCTION

A safety-grade Programmable Logic Controller (PLC) can be applied to a Bistable Processor (BP), Coincident Processor (CP), and an Automatic Test Interface Processor(ATIP) of a reactor protection system in nuclear power plants. Also, a Group Controller (GC) and Local Controller (LC) of an Engineered Safety Feature(ESF)-Component Control System (CCS) are used in a Plant Protection System(PPS). Testing environment of the safety-grade PLC consists of a host system and a target system. The safety-grade PLC target system has been linked with the MPSD510PP emulator using a parallel port. It will be downloaded by the Code Composer including the CPU unit's operating system, Operating System Software(OSS)/ Agents and an executable application program. The safety-grade PLC's schematic diagram is shown in Figure 1(KAERI. 5, 2003).

This paper addresses the Verification and Validation(V&V) process and the methodology for an embedded real time software of a safety-grade Programmable Logic Controller(PLC). This safety-grade PLC is being developed as one of the Korean Nuclear Instrumentation and Control System(KNICS) projects. KNICS projects are developing a Reactor Protection System(RPS) and an Engineered Safety Feature-Component Control System(ESF-CCS) as well as a PLC. The safety-grade PLC will be a major component that encomposes the RPS systems and the ESF-CCS systems as nuclear instruments and control equipment. This paper describes the V&V guidelines and procedure, V&V environment, V&V process and methodology, and the V&V tools in the KNICS projects.

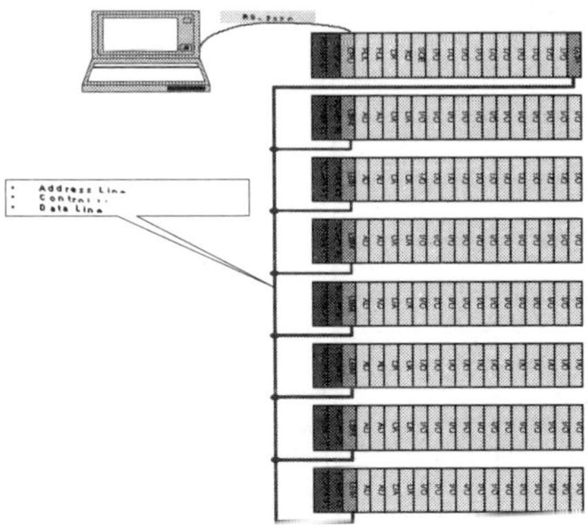

Fig. 1: Safety-grade Programmable Logic Controller(PLC)

Secifically, it describes the real-time operating system V&V experience which corresponds to the requirement analysis phase, design phase, implementation phase and the testing phase of the software development life cycle. Main V&V activities of the real-time operating system for the designed output of the PLC are a technical evaluation, licensing suitability evaluation, inspection and traceability analysis, formal verification, software safety analysis, and a software configuration management(Figure 2). First, the technology evaluation performs a technical review which is based on an engineering decision making by verifiers. Second, the licensing suitability evaluation is performed according to the Code & Standard criteria by analyzing the designed outputs based on the functional characteristics and process characteristics defined in the V&V guidelines. Third, the inspection and traceability analysis uses the System Inspection Support(SIS)-Requirement Traceability(SIS-RT) of a self-developed tool after a functional analysis and decomposition as an input and output, behavioral entity, interface etc. The inspection covers the correctness, consistency, completeness, properties of the inputs and outputs, behavioral entities, interfaces, etc. with a three-fold checklist topology. Fourth, the formal verification was performed for the statechart based SRS by a Graphical Back Animation(GBA) and a simulation. Fifth, the software safety analysis uses the HAZard OPerability(HAZOP) method and analyzes the in failure mode effect. Finally, the software configuration management is estimated by using the Nuclear Software Configuration Management(Nu-SCM) tool developed by the KNICS projects. We believe that we can achieve the functionality, performance, reliability and safety that are the V&V objective goals of a safety-grade PLC system at the SRS, SDS, implementation and testing phase during the software development life cycle through the suggested software V&V methods for a PLC.

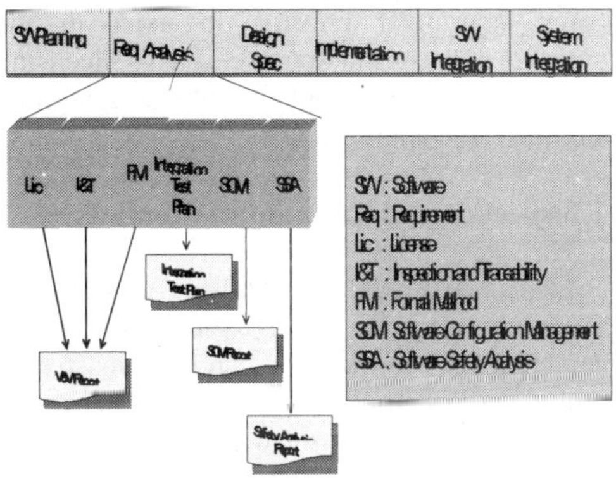

Fig. 2: V&V activities for SRS of PLC

2. METHODS OF THE SOFTWARE VERIFICATION AND VALIDATION

2.1 Review of the licensing suitability

The purpose of the licensing suitability review confirms whether or not the software requirements which coincide with the criteria of the software, performance and safety requirements defined in the Safety-grade PLC software requirement statement are suitable from a Code & Standard and technological viewpoint. According to the SRP/BTP-14 criteria(USNRC, 1997), it must satisfy all the functionality characteristics and process characteristics such as those in Table 1.

Table 1. Criteria of Licensing suitability

Functional Characteristics	Process Characteristics
Safety	Completeness
Robustness	Consistency
Security	Correctness
	Style
	Traceability
	Verifiability

Whether or not the contents described in the requirement specification from the viewpoint of the functional characteristics and the process characteristics are the correctness, consistency and completeness in the software requirement phase. It also verifies whether or not the formula, latency error and priority do exist in the algorithm and control logic etc. This licensing suitability can also be applied to the design phase using the same approach.

2.2 Inspection and Traceability Analysis

Inspection and traceability analysis is performed via the three properties of a completeness, correctness and consistency. Accuracy of a definition, input/output accuracy, accuracy of a software's behavioral characteristics, and the accuracy of an interface of a software function are very important. Inspection and traceability analysis are classified by basis reviews & comments and the latter details the review criteria as shown in Table 2.

Traceability analysis for the requirements between the Software Requirement Specification(SRS) and the Software Design Specification(SDS) using a SIS-RT tool for scheduling, task synchronization, ISR processing and a time management that are major requirements of a real time kernel was undertaken. Traceability analysis between the requirement specification and design specification has also been performed for the scheduling and task management as shown in Figure 3.

Table 2 Checklist example on the correctness of a software's behavioral characteristics

Items	Cri teria	Detailed Review Criteria for RTOS		ST
Correctness of SW behavioral	SRP/BTP-14	How is the clock tick defined? Is there a definition of interrupt latency, response, return time? Each ISR execution time required How is the context switching time defined? How is the Task Response time defined?	Is it periodic and stable for real timeclock tick? Interrupt disable time, Interrupt entering time Context switching time Dispatching time from Ready Queue to CPU	

Notes : ST : Status

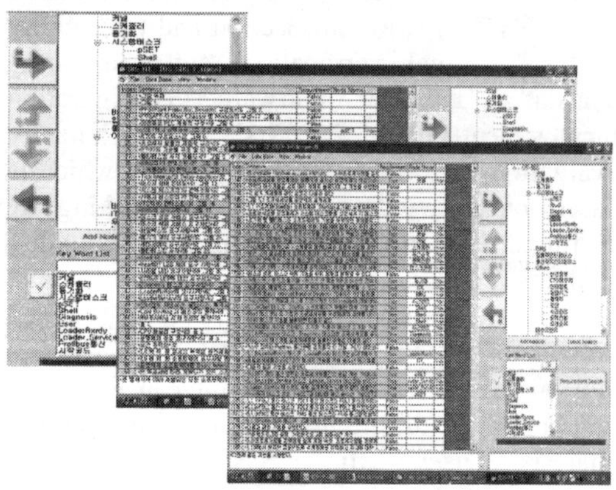

Fig. 3: Traceability View of the Scheduling and Task Management between the SRS and SDS

Through the traceability analysis and an inspection of the matrix style using a SIS-RT the missing requirements from the SRS to the SDS were found.

2.3 Formal Verification

Formal verification in this paper means the use of verification techniques for the formal logic and discrete mathematics in the specification, design, and implementation of the PLC system and software. There are three types of formal verification techniques which are an equivalence checking, theorem proving, and a model checking. Among these formal verification techniques, we used the model checking method. Model checking is a technique for verifying a finite state of a machine. Verification can be performed automatically by the Model Checker/Model Certifier. Specifications for the model checking are usually based on the finite state of a machine which is a model of the system. The procedure normally uses an exhaustive search of the state of the space of a system to determine if some specifications are true or not.

As a consistency details the verification of the Safety-grade PLC operating system, the software requirement specifications, are verified by using commercially available Statemate Model Checker/Model Certifier tool supports. Verified with the following six properties on a Statechart based requirement specification that appear as follows.

a. Non-Determinism
b. Write/Write Race Condition
c. Range Violation
d. Drive to State
e. Drive to Config
f. Drive to Property

An error trace can be used as a counter example for the six checked properties and it can help the verifier to track down where the error has occurred.

2.4 Preparation for the Integration Test Plan

Several test cases are necessary for testing the safety-grade PLC software. In the software requirement phase, a test plan should be prepared to satisfy the functional requirements and performance requirements of the safety-grade PLC. A test plan includes the following table of contents.

Integration Test Plan (Table of Contents)

A. Functional Item
A.1 Scheduling
A.2 Inter-Task Communication
A.3 Interrupt Handling
A.4 Click Tick/Time Management
A.5 Communication
A.6 Operating System Software(OSS)/Agents
A.7 Basic Input and Output Handling System/Board Support Package(BSP)
A.8 Hardware Interface
B. Performance Item
B.1 Kernel
B.2 Communication
B.3 OSS/Agents
B.4 Basic Input and Output Handling System/Board Support Package(BSP)
C. Safety and Reliability Items
C.1 Kernel
C.2 Communication

We made a unit/module test plan, integration test plan and a system test plan A schematic diagram of the safety-grade PLC's testing environment is shown in Figure 4 . The real-time operating system and OSS/agents included the safety-grade PLC verified by using a Cantata++ commercially available testing tool with a Code Composer Debugger. The application program which is downloaded for the safety-grade PLC will use pSET. pSET also checks the integrity of the memory value, and the status of the registered and specific values inside the PLC.

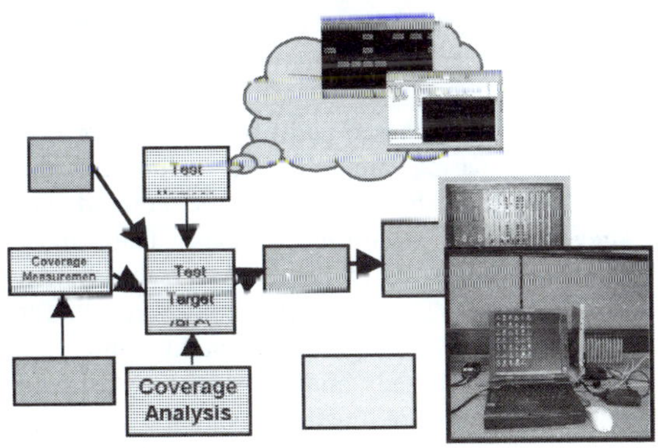

Fig. 4:Test Environment Configuration for the PLC

2.5 HAZOP Safety Analysis

At the requirement specification phase for the software development life cycle, we will use the HAZOP (Hazard and Operability) plan to achieve a hazard analysis. HAZOP is a technique that has been used successfully for a hazard analysis in industry such as a chemical factory. Once we have achieved the software hazard analysis for the developed guidelines and checklist, we will take advantage of this HAZOP concept in the hazard analysis of the safety-grade PLC software requirement specification. After the hazards and risks are collected, they will be used in the software verification and the hazard analysis in the next step. Level of significance that is imposed on the various software requirements becomes necessary information to determine the resources to medicate in the development, verification, test etc.

2.6 Software Quality Measurement

After the system testing is completed, an inside peer review should be performed in the same way as that of the unit testing and integration testing. The defects, problems and corrective actions list should be submitted to the Software Quality Assurance(SQA) organization if any. A review meeting should be held according to the quality assurance procedure. It should be focused on an objective quality goal as a quality management in the system testing phase. After a review of the system's test results, a SQA team should inspect and follow-up the test results with a checklist as to whether a system's testing will achieve enough of the original objectives, or not. Also, whether the system testing level was proper for the physical configuration audit and the performance based audit by the SQA or not. After comparing the expected value and the resulting value by the SQA, it is required that they take the necessary actions. They must confirm the results of the SQA for the system testing results. They can use the SVV results for the physical configuration audit and the performance based audit by the SQA. In summary, software quality assurance activities are recommended during the software life cycle as shown in Table 3.

Table 3 Software Quality Activities under a Nuclear QA Program

SWLC	Software Development Baselines	Recommended Software QA Activities
Requirements	SQAP, SVVP, SRS	. SRS review . In-process audit . SW Test Plan Review . Managerial Review
Design	Preliminary Design Description	. Preliminary Design Review, . In-process Audit
	Software Design Description	. Detailed Design Review
Implementation	Code Listing, Other Documentation(Code Implementation Specification)	. In-Process Audit
Test	Test Documentation	. Functional Configuration Audit
Installation and Checkout	Deliverable items, Installation Report etc, SVVR	. Physical Configuration Audit . Performance-based Audit
	User Documentation	. User Documentation Review

Notes : SWLC : Software Development Life Cycle, SW: Software, SRS : Software Requirement Specification, SVVP : Software Verification and Validation Plan, SQAP : Software Quality Assurance Plan, SVVR : Software Verification and Validation Report. QA : Quality Assurance.

3. CONCLUSION

A safety-grade PLC V&V's Systematic approach is a methodology, a tool and a technique. Through the KNICS projects a V&V methodology is established. The toolset used was a self-developed one and a commercially available one. The technique took advantage of the V&V techniques that use the formal verification technique. In this paper, at the requirement specification analysis phase, software design phase, implementation phase and testing phase, the suggested safety-grade PLC's V&V methodology, techniques and experiences were used. We introduced the software V&V experience at the requirement, design, implementation and testing phase using the proposed V&V methodology. V&V experiences are the licensing suitability evaluation, inspection and traceability analysis, formal verification, integration testing plan preparation, hazard analysis etc. The proposed V&V methodology satisfies the SRP/BTP-14 criteria for the safety software in nuclear power plants. The proposed V&V methodology is going to be used to verify the upcoming software development life cycle in the KNICS projects.

4. REFERENCES

1. USNRC, "*BTP-14: Guidance on Software Reviews for Digital Computer-based I&C Systems*," NUREG-0800 (1997).
2. IEC 61131-3, "*Programmable controllers - Part 3: Programming languages*," (1993).
3. IEEE Std-1012, "*IEEE Standard for Software Verification and Validation Plans*" (1986)
4. IEEE Std-1028, "IEEE *Standard for Software Reviews*," (1997).
5. Korea Atomic Energy Research Institute(KAERI), "*Software Verification and Validation Plan(SVVP) for Software Requirement Specification(SRS) of Safety-grade Programmable Logic Controller (PLC)*," KNICS-PLC-(SRS) SVP121 (Rev. 00), Jan. 14, 2003. (In Korean)
6. KAERI, "*SVVP for Software Design Specification(SDS) of PLC*," KNICS-PLC-(SDS) SVP131 (Rev. 00), Jan. 14, 2003. (In Korean)
7. KAERI, "*SVVP for Software Implementation of PLC*," KNICS-PLC-(IMP) SVP141 (Rev. 00), Jan. 14, 2003. (In Korean)
8. KAERI, "*SVVP for Software Integration of PLC*," KNICS-PLC-(INT) SVP161 Jan. 14, 2003. (In Korean)
9. KAERI, "*Software Safety Plan for PLC*," KNICS-PLC-(SSP) SVP107 (Rev.00) Jan. 14, 2003. (In Korean)
10. KAERI, "*Software Configuration Management Plan for PLC*", KNICS-PLC-SEP103 (Rev. 02), Jan. 14, 2003. (In Korean)

Reliability, Safety and Hazard
Editors: P.V. Varde et al.

RELIABILITY ANALYSIS AND COMPONENT FUNCTIONAL ALLOCATIONS FOR THE ESF MULTI-LOOP CONTROLLER DESIGN

Seop Hur, D.H. Kim, J.K. Choi, J.C. Park, S.H. Seong and D.Y. Lee

Korea Atomic Energy Research Institute, Dukjin-dong 150 yusung-gu Daejeon, Korea

ABSTRACT

This paper deals with the reliability analysis and component functional allocations to ensure the enhanced system reliability and availability. In the Engineered Safety Features, functionally dependent components are controlled by a multi-loop controller. The system reliability of the Engineered Safety Features—Component Control System, especially, the multi-loop controller which is changed comparing to the conventional controllers is an important factor for the Probability Safety Assessment in the nuclear field. To evaluate the multi-loop controller's failure rate of the k-out-of-m redundant system, the binomial process is used. In addition, the component functional allocation is performed to tolerate a single multi-loop controller failure without the loss of vital operation within the constraints of the piping and component configuration, and ensure that mechanically redundant components remain functional.

1. INTRODUCTION

The ESF-CCS (Engineered Safety Features-Component Control System) is a Class 1E system is a very important system in the nuclear power plants because the system automatically actuate the ESF system equipment.

The system reliability is defined to be the probability that a system will perform a required function without failure under the stated conditions during the stated period of time. To develop a highly reliable system, it is imperative that the reliability analysis is performed over the system's design life cycle. The reliability analysis methods applied in the nuclear industry are the Probabilistic Safety Assessment (PSA), Failure Mode Effect Analysis (FMEA), Fault Tree Analysis (FTA) and Mean Time between Forced Outage (MTBFO). The MTBFO is required as follows. 'Mean time between forced outage caused by failures of the equipment shall be greater than fifty reactor operating years over the entire design life of the equipment. The meaning of forced outage includes shutdowns that result directly from the failure, and shutdowns the operators must perform to avoid violation of a plant Technical Specification due to these failures. [1]

The safety related component in nuclear power plants are traditionally controlled by single-loop controllers. However, a review of the plant components and the process system in which they reside shows that a majority of the components do not posses a unique functional identity, in that they are not individually important to the plant but are collectively important as a part of a subgroup or group. In the new design of ESF-CCS, the multi-loop controllers with data networks are widely used. The one of key issues for design of multi-loop controller is component functional allocation to the each multi-loop controller through plant and function analysis and grouping.

In this paper, the reliability analysis of multi-loop controllers in ESF-CCS using the binomial process [2] is performed to calculate the availability of the ESF-CCS with a fault diagnosis function. In addition, to ensure the enhanced system reliability and availability, ESF component functional allocation is performed through allocation criteria and fault analysis

2. RELIABILITY ANALYSIS OF MULTI-LOOP CONTROLLERS

ESF-CCS Architecture

The ESF-CCS through-out the world usually consists of the 2-out-of-3 or the 2-out-of-4 architecture to prevent a single failure. The 2-out-of -4 ESF-CCS system consists of four channels, and each channel is implemented with the same architecture. A single channel of the ESF-CCS consists of three Group Controllers (GC), twelve Loop Controller (LC), an ESF Test & Interface Processor (ETIP), a Cabinet Operator Module (COM), a Control Channel Gateway (CCG), two Control Panel Multiplexer (CPM), and Soft Controllers as shown in the Figure 1. The GC acquires the logic level ESF actuation signals transferred from the four Coincidence Processors in the Reactor Protection System (RPS). When two or more logic-level actuation signals from the RPS are activated, the GC will activate the ESF function-level initiation signals. The LC generates a control signal to control the grouped ESF components which belong to each LC through performing the two-out-of-three logic using the ESF function-level initiation information from the three GCs within a channel.

Availability Analysis with Binomial Process

All the channels of the *2*-out-of-*4* GC are electrically and physically separated from the other channels to prevent a common cause failure, so the failure of each channel is independent. As the channel failure is an independent event, each channel can be observed as either a 'good' or 'fail' status. This is similar to the experiment of flipping four coins and observing the result of each coin, so a *2*-out-of-*4* GC architecture can be presented with the binomial process.

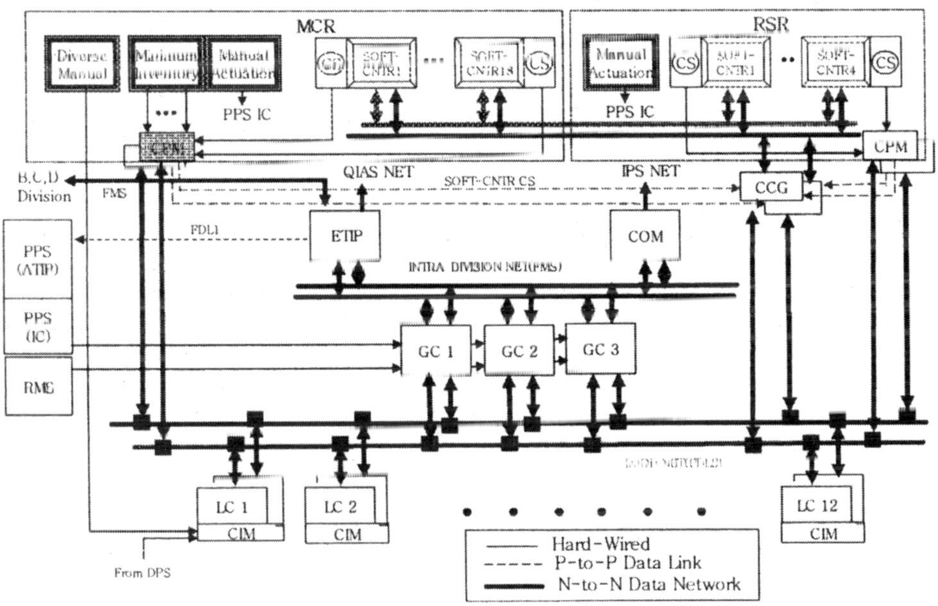

Fig. 1: KNICS ECCS-CCS Configuration (single channel)

Without Fault Diagnosis

The *2*-out-of-*4* architecture can provide a reliable operation in the case of any single channel failure, but it might cause a spurious reactor trip when the failures happen in more than two channels at the same time. If failures happen in more than the *k* channels in the *k*-out-of-*m* RPS architecture at the same time, the RPS system may generate the reactor trip signal and shutdown the reactor. The failure rate of the *k* channels out of the *m* channels without maintenance is as follows:

$$\lambda_S = \sum_{j=k}^{m} {}_mC_j \lambda^j (1-\lambda)^{m-j} \quad (1)$$

where, λ_s is the failure rate of the GC system, λ is the failure rate of each channel and assumed exponential distribution, and ${}_mC_j$ is the binomial coefficient.

With Fault Diagnosis

There are many kinds of fault diagnosis functions to detect the miss-operation status of the GC. The GC sets the alarm signal and the trouble/disable warning for the plant operator whenever faults are detected, and it starts the maintenance activity. So the GC having the fault detection function can reduce the spurious reactor shutdown frequency by the reduction of the time interval for simultaneous failures in more than two channels. The *2*-out-of-*4* architecture with the repair activity generates a spurious actuation only when the other channel fails again during the maintenance period after any initial channel failure. Therefore the system level failure rate of the *2*-out-of-*4* architecture with a repair action is given by

$$\lambda_s = {}_4C_1 \times \lambda \times {}_3C_1 \times \lambda \times MTTR = 12\lambda^2 \times MTTR \quad (2)$$

Where, the *MTTR* is the mean time to repair. The system level failure rate of the *2*-out-of-*3* architecture with a repair action is given by

$$\lambda_s = {}_3C_1 \times \lambda \times {}_2C_1 \times \lambda \times MTTR = 6\lambda^2 \times MTTR \quad (3)$$

The selective *2*-out-of-*4* architecture creates a spurious reactor trip when the failures happen in channel A or C, and in channel B or D at the same time. The system level failure rate of the selective *2*-out-of-*4* architecture with a repair action is given by

$$\lambda_s = {}_2C_1 \times 2\lambda \times 2\lambda \times MTTR = 8\lambda^2 \times MTTR \quad (4)$$

The Mean Time between Failure (*MTBF*) is expressed by the system level failure rate as follows, and has the same representation as MTBFO.

$$MTBF = \frac{1}{\lambda_s} \text{ [hours]} \quad (5)$$

Failure Rate of the GC and LC

The *2*-out-of-*4* ESF-CCS architecture can provide a reliable operation the case of any single channel failure. But the ESF-CCS might generate a spurious actuation signal when failures happen in more than two channels. In the simplified block diagram of the ESF-CCS channel in Fig. 1, the failures of the single GC in a channel do not create an important effect for the reliable operation of the ESF-CCS because of three (3) redundant GC in a channel. But In the case of LC the failures of the single LC in a channel could create an important effect for the reliable operation of the ESF components.

The ESF-CCS consists of the PLC hardware. The number of the PLC modules used to implement the GC and LC, the failure rate for each module, and the failure rate for a single GC and LC are shown in Table 1. In this evaluation, the failures of the sensor and transmitter, Intra-Channel Network (ICN), and Inter-Channel Data Network (ICDN) are not included. Using the PLC module data [4] the failure rates of GC and LC are calculated. For the single GC channel the failure rate is 5.02*exp(-05), and 5.45*exp(-05) for a single LC channel. The GC in a channel consists of triple independent PLC system. During normal operation LC receives the actuation status from the three PLCs, and then performs the 2-out-of-3 voting logics to finally determine the actuation status. If a GC PLC is fails, LC performs 2-out-of-2 voting logics. If two GC PLCs in a division, the A-division is fail. The failure rate of GC in a channel is therefore 2.5*exp (-09).

Table 1 Failure rate of single GC & LC

module	Quantity	Failure rate/ Module	Failure rate
Power supply	2(dual)	2.15E-05	4.6E-10
Processor	1	7.75E-06	7.75E-06
Comm. Card	2	8.24E-06	1.65E-05
Digital Input	4	6.25E-06	2.5E-05
Base Board	1	0.98E-06	0.98E-06
Failure rate of single GC			5.02E-05
Power supply	2(dual)	2.15E-05	4.6E-10
Processor	2 (dual)	7.75E-06	6.0E-11
Comm. Card	1	8.24E-06	8.24E-06
Analog input	1	4.06E-06	4.06E-06
Analog output	1	4.06E-06	4.06E-06
Digital input	2	6.25E-06	1.25E-05
Digital output	2	5.93E-06	1.69E-05
Base board	2(dual)	0.98E-06	0.96E-12
Failure rate of single LC			4.57E-05

System-level Failure Rate except Manual Devices

The system level failure rate and mean time between forced outage for the several kinds of ESF-CCS structures are calculated. 8 hours for the MTTR time is used to calculate the system level failure rate. Mean time between forced outage is determined from Equation (5), and has the unit with *hours*. The MTBFO of an automatic ESF-CCS is about 2.5*104 hours in the case of 2-out-of-3 structures

3. ESF COMPONENT FUNCTIONAL ALLOCATION

Problem Issued

Comparison to the Traditional Design

Comparing the ESF-CCS functional group control to traditional Interposing Logic System designs which use dedicated logic circuits for each plant component, it can be seen that although the ESF-CCS does not achieve the same degree of independence for each component. In fact, additional independence within the traditional single loop controller system beyond which exist in the process system in a detriment to system reliability.[3]

Another benefit of the ESF-CCS functional group configuration can be seen in failure modes and effects analysis. Since components are assigned to ESF-CCS functional groups in a manner consistent with their process relationship, the effects of failures are predictable and manageable. Traditional single-loop controller systems utilize dedicated processors for each component but that component independence is compromised through sharing of power supplies, auxiliary logic modules and auxiliary I/O cards. Failures in shared devices can affect large number of unrelated plant components, requiring difficult failure modes analysis often with unacceptable results.

Comparison to the Centralized Control

The centralized control type of system reduces the number of electronic components even further that achieved in the ESF-CCS, therefore, long MTBF can be achieved. However, when failures occur that cannot be accommodated by the built-in redundancy the effects can be catastrophic. Depending on the number of components and their dispersement in the process, the effects are usually worse that complete cabinet failures in a traditional design.

KAERI has adopted the functional group configuration of the ESF-CCS design as the right compromise between the centralized-redundant design and the single processor to component configuration of the traditional system design.

Allocation Methodology and Criteria

The following procedure is a general procedure for assignment of the ESF components to LCs.

- Identify the plant safety system including the systems which are not controlled by ESF-CCS.
- Define the system functions and functional independence.
- Identify the system trains and the supplied electric power divisions
- Group & allocate the system function.
- Group and allocate the safety plant components to the LCs.
- Perform failure modes and effects analysis.

The following criteria are general criteria for assignment of the ESF components to LCs.

- Safety criteria
  - Defence-in-Depth criterion
  - Single failure criterion
  - Fluid independency
  - Component redundancy

- Electric power independency
  - Functional Diversity
  - Diversity means
- Performance criteria
  - Number of the controls in a LC
  - Minimization of interface among LCs
  - Response time
  - Processing load of a LC
  - Consideration of local installation of LCs; location of the component.
  - Minimization of unexpected operation of the ESF systems
  - Testability.

Application Examples

There are approximately 300 components which are related to the safety injection actuation signal (SIAS), and the several process system such as safety injection system (SIS), shutdown cooling system (SCS), several HVAC systems, component cooling water system (CCWS), etc., are controlled by SIAS. All safety injection components are on the four safety-related divisions AE, BE, CE, and DE. There are approximately 110 components about equally divided between the four divisions. The CCWS, essential service water system (ESWS) components are also on the four safety-related divisions. While, SCS, load sequencers, HVAC systems are on the two safety related divisions AE and BE. Safety depressurized system (SDS) which is performed to depressurize the primary pressure during event of total loss of feedwaters, is not related ESF actuation signals, but we allocated these system components to the ESF-CCS multi-loop controller. The ESF component functional allocation results related to SIAS are shown in the table 1 and 2. Table 1 represents the functional allocation results to each ESF-CCS division. Table 2 represents the component allocation results to each loop controller.

Fault Analysis

The component assignment is made with the final aim of being able to tolerate a multi-loop controller failure without loss of redundancy provided by the plant components and fluid paths. Majority of the components are valves, diaphragm-air-operated, motor-operated and solenoid-operated. Solenoid-operated valves can be fail-open or fail-close but the motor operated valves are fail-as-is. The valve fail status is selected to provide a fail safe valve position from plant safety considerations and sometimes operational considerations. For critical applications, two fail-close valves may be used in a series path if the safety considerations demand valve closure. Table 3 shows the samples of the fault analysis due to multi-loop controller.

Table 2 Function allocation to the ESF-CCS division

ESF-CCS Division	Allocated system/train
Division A	SIS train-1 SCS train-1 SDS train-1 CR HVAC Train-1 CTNT HVAC train-1 Aux BLD HVAC train-1 CCWS train-1 Load Sequencer Division-A
Division B	SIS train-2 SCS train-2 SDS train-2 CR HVAC Train-2 CTNT HVAC train-2 Aux BLD HVAC train-2 CCWS train-2 Load Sequencer Division-B
Division C	SIS train-3 SDS train-3 ECCS Equip. RM HVAC Train-1 ESF SG RM HVAC train-1 CCWS train-3
Division D	SIS train-4 SDS train-4 ECCS Equip. RM HVAC Train-2 ESF SG RM HVAC train-2 CCWS train-4

In the case of LC-A-1 controller fails one active SI-train are faced with the loss of function, but the remains of 4-redundancy active SI-trains have a capability to perform their function. In the case of LC-A-2 fail, when plant is under normal operation, there is no impact, but under SIAS actuation condition, the injection flow from affected passive SI train would be reduced. But There is no impact to perform the required function of passive SI due to 4-redundant passive SI-trains.

Table 3 Component allocation to the LC

LC	Allocated Components
LC-A-1	Active SI train-1 components SC train-1 components
LC- A -2	Passive SI train-1 components CTMT HVAC/AUX BLD HVAC train-1 components
LC-A-3	CCWS train 1 components SW/TSS train-1 components
LC-A-4	CR HVAC train-1 component (CREVAS related)
LC-A-5	CR HVAC train-1 components
LC-A-6	SDS train 1 components
LCA-7	Load sequencer A division ESWS train-1 components
LC-B-1	Active SI train-2 components SC train-2 components
LC-B-2	Passive SI train-2 components CTMT HVAC/AUX BLD HVAC train-2 components
LC-B-3	CCWS train 2 components SW/TSS train-1 components
LC-B-4	CR HVAC train-2 component (CREVAS related)
LC-B-5	CR HVAC train-2 components
LC-B-6	SDS train 2 components
LC-B-7	Load sequencer B division ESWS train-2 components
LC-C-1	Active SI train-3 components SI common header related components (AE)
LC-C-2	Passive SI train-3 components ECCS equip. RM HVAC train-1 components
LC-C-3	CCWS train 3 components ESWS train-3 components SW/TSS train-3 components
LC-C-4	ECCS equip. RM HVAC train-1 components (SIAS related)
LC-C-5	ESF SG Room HVAC train-1 components
LC-C-6	SDS train 3 components
LC-D-1	Active SI train-4 components SI common header related components (BE)
LC-D-2	Passive SI train-4 components ECCS equip. RM HVAC train-1 components
LC-D-3	CCWS train 4 components ESWS train-4 components SW/TSS train-4 components
LC-D-4	ECCS equip. RM HVAC train-2 components (SIAS related)
LC D 5	ESF SG Room HVAC train-2 components
LC-D-6	SDS train 4 components

Table 4 Examples of the fault analysis during LC failure

(a) LC-A-1 failure

Affected components	Normal / Affected status	Compensating Provision
Nitrogen supply valve	Closed/ Closed	Provide Inherent isolation due to fail-close and series valve (BE division)
Atmospheric vent valve	Closed/ Closed	Provide inherent depressurization of SIT due to redundant valve (BE division)
SIT discharge iso. valve	Closed/Closed	No change in the valve line-up Not allow non-safety function of fill & drain of SIT
SIT fill and drain iso. valve	Closed/Open Closed/Closed	Not perform safety function. Not allow non-safety function of fill & drain of SIT 3 SITs perform the injection

(b) LC-A-2 failure

SI pump #1	Start/Stop Stop/Stop	4- redundant pumps No impact
SC pump #1	Start/Stop Stop/Start	2- redundant pumps No impact
IRWST iso valve	Open/Close Close/Close	Redundant SI trains No impact
SI injection line iso valve	Open/Close Close/Close	Redundant SI trains No impact

4. CONCLUSION

The reliability analysis and component functional allocations have been performed to ensure the enhanced system reliability and availability of ESF-CCS design. As a result of the availability analysis using the binomial process, the automatic parts of ESF-CCS, such as group controller and Loop controller has the MTBT of about 2.5E+4 hours.

The ESF component functional allocation has been performed to design the multi-loop controller such as loop controller of ESF-CCS. As the results of the fault analysis, the inherent functions of ESF system are maintained under a multi-loop controller failure.

5. REFERENCES

1 ALWR utility requirement document, Ch.10, vol. II, EPRI, 1990.

2. D.Y.Lee, J.B.Han and J, Lyou, Reliability analysis of the reactor protection system with fault diagnosis, Key Engineering Materials Vols. 270-273, pp1749-1754, 2004.

3. Design Procedure for Component Control System Component Functional Grouping for NUPLEX 80+, NPX80-IC-DP640-01, 1990.

Reliability, Safety and Hazard
Editors: P.V. Varde et al.

SUBCRITICALITY ANNUNCIATION FOR NUCLEAR POWER PLANTS & RESEARCH FACILITIES

Shantanu Das and B.B. Biswas

Reactor Control Division, Babha Atomic Research Centre, Mumbai 400 085

ABSTRACT

Sub-criticality, criticality, reactivity are the terms which are not very clear to the reactor operators. These parameters are prime movers defining Nuclear Reactor and should annunciate in the control room by very simple means. More so the need for this simple and important annunciation is called for in wake of recent criticality accidents and criticality excursion in the nuclear facilities worldwide and here. The reactivity measurements are done primarily by two methods deterministic and statistical, depending on the measurement fidelity. Reactivity meters are under operation since several decades based on inverse point kinetic calculations perhaps can be utilized to annunciate sub-criticality percentage, based on neutron detector signals. When the detector is not working in the current mode then based on the count mode and statistical mathematics the sub-criticality annunciation may be realized. The embedded solution for realizing this type of equipment is possible through use of parallel processing digital signal processing modules developed and was proposed for modified APSARA reactor regulating system. This simple annunciation will aid the operator confidence and remove hazards while managing reactivity specially, in cold and shut down states of reactor

1. INTRODUCTION

Reactivity management is not well indicated to the reactor operator, nor does quantitative annunciation exists in the control room. When the reactor is cold or when put under long shut down periods the annunciation of the sub-critical level should and must aid the operator for his reactivity management. Since reactivity and sub-criticality is not truly indicated, a black box approach prevails, by controlling the reactor by the reactor power and reactor power rate, the criticality is thus controlled indirectly.

Criticality or sub-criticality is the parameter what is to be managed. Reactivity i.e. deviation from criticality is reactivity when expressed in percentage is easier to understand. For instance the reactivity indication showing sub-criticality as 20% and value 0.5% to operator clearly indicates that the second value is close to reactor criticality $k=1$. Therefore this percentage indication of sub-criticality gives annunciation to the operator rather than displaying criticality values in absolute terms of numbers or in other units as say milli-K.

Reactivity is the physical property of the reactor core based on composition, geometry, temperature, pressure, ability of core to produce fission neutrons and external disturbances arising out of control/safety mechanism perturbations fuel loading and other various operator activities and poisons. The derivation of reactivity is based on average neutron life cycle model and is put in Point Kinetic Equations with various delayed neutrons energy group. From this life cycle model reactivity can be used as measure of deviation of reactor from criticality. Operationally, reactivity is the prime mover of all power change in nuclear reactors. During certain reactivity management operations such as reactor start up reactor operator introduces enormous positive reactivity change, while generating a small and sluggish power increase. Yet once criticality is approached and attained a further relatively minute change can cause a power excursion to escalate rapidly that all reactor power and rate meters pegs at full scale before operator can blink. The avoidance of such an event is only through an annunciation system for sub-criticality level.

World wide study show that it is possible to have sub-criticality annunciation system based on inverse point kinetics when detector are in current mode and by employing statistical methods when detectors are in counting mode. However applicability of inverse point kinetics to a sub-critical core is still point of debate though few practical assumptions do make this method possible to annunciate sub-criticality. The statistical method based on Feynman method is widely used to study the sub-criticality of pulsed reactors and getting popularity in arriving at sub-critical annunciation of ADS recently, though mathematically complex yet realizable through the parallel embedded architecture with digital signal processors.

2. APPLICABILITY OF INVERSE POINT KINETICS (IPK) TO SUB-CRITICAL SYSTEMS

Digital reactivity meters can give continuously give in real time reactivity values based on simple one point kinetic equations. To use this concept as sub-criticality annunciation system following points need be addressed.

a) Applicability of the inverse point kinetics (IPK) to a system where neutron flux distribution is dependent on the sub-criticality is not confirmed.
b) In the environment of the sub-criticality measurement expected neutron flux is very low, and fluctuation of signal is quite large, making IPK solver difficult.
c) The IPK requires that source strength be known accurately is difficult.

These above problems when solved the IPK based digital reactivity meter will be apt for annunciating the sub-criticality. The basic idea of the IPK solver is to monitor reactivity change from known initial stable condition. The sub-criticality annunciation system will be based on following assumption.

a) A certain measurable level of neutron flux of the target system should be available, else no measurement is possible.
b) Unknown reactivity can be expected by any operator action of say withdrawal of rods for service, insertion of new fuel bundle/element. These will be as step/ramp disturbances in the IPK. Therefore an alarm must be given when sub-criticality reaches a set point or certain rate of reactivity change is declared.

The sub-criticality is defined as $(1-k)/k$ in percentage. When this term is close to 0% means reactor is approaching criticality. 20% is more sub-critical than 15% displayed value. Simply reactor kinetics equations are expressed as the one-point reactor model, assuming that the spatial dependency of neutron density can be described by a single spatial mode shape.

$$dn/dt=[(1-\beta)k-1]*n/l+\lambda C+S$$

$$dC/dt=\beta\, k\, n/l-\lambda C$$

All notations are standard.

Sub-criticality is calculated by solving the above equations with the neutron density as input. The neutron density is the detector signal located at certain position. In this method the assumption is that neutron spectrum is constant having effective neutron velocity v. Therefore following is holding good:

$$n=K\Phi/v$$

Where K is constant based on detector cross-section and detection probability per neutron and constant attenuation effects. Φ is neutron flux at the detector. In the sub-criticality calculations absolute value of the neutron density is not required but only the relative value is sufficient. In the above reactor equations where only two equations are used the temperature feed back terms are not shown as for cold conditions those are not required. From the above equations in the steady state $dn/dt=dC/dt=0$ the following is derived :

$$(1-k)n=lS$$

This implies that if one can find the position where this relation holds in the target system, IPK can be used as sub-critical monitoring logic. The above relation says that the *normalized flux* is constant, and if so then one can use IPK. *The normalized flux is the neutron flux multiplied by (1-k), is a good barometer of sub-critical monitor equipment.*

3. NORMALIZED FLUX

The above section used a term called *normalized flux* in sub-criticality evaluation by IPK method. To have a feel of the same let us consider a homogeneous neutron source through out the core. Example is photo-neutron as source in PHWR. The second case will be a point source kept outside the core. In the case of homogeneous source the sub-criticality when changed from 20%, to 5% and then 0.5% the neutron flux peak is changed from 0.5E00 to 1E01 and then three decades change to 1E04. Therefore the fact is the absolute flux is *not stationary* with the sub-critical level. However if one tries to plot the normalized flux distribution i.e. multiplying the absolute flux with *(1-k)* through out the core area the normalized flux does remain constant at about 3.5 E03 at all the sub-critical values In other words the *normalized flux is stationary* for all sub-critical levels. For the case of point source placed at the outside of the active core, in the reflector area, the absolute flux plot in the reactor is peaked at the source end and gradually decrease as we go towards periphery of the reactor at higher sub-criticality levels. As the sub-criticality approaches 0% there is peaking inside the core but the flux is not stationary. Same thing is done to normalize the flux i.e. to multiply absolute flux with *(1-k)* and the plot in this case, show that there exist, one *stationary point* where there is constant normalized neutron flux at every sub-critical level.

Therefore the cases above show that the assumption that the normalized flux stays constant stationary for all points in homogeneous source and at particular point for the point source. So if the detector be placed at the stationary point then by solving IPK value of sub-criticality be enunciated.

1. 1.For the homogeneous source, the normalized neutron flux remains stationary even if sub-criticality varies.
2. 2.For system with point source normalized neutron flux varies with sub-criticality but a stationary point does exist.

Therefore placing detector at the stationary point gives IPK a steady state nature and can be solved for annunciating the sub-criticality.

4. FILTERING THE NEUTRON FLUCTUATION READINGS.

The neutron signal fluctuates around the average. Such fluctuations can be easily filtered with low-pass first order filter. When using the low pass filter with first order, the response of reactivity is also delayed, but the delay is about 4.5 times the time constant of the filter. Say for filter delay of 2s, the reactivity response is delayed by 10s, and for sub-criticality instrument it is perhaps acceptable.

The studies in this instrumentation development worldwide, show this by inserting plus minus 50% of the neutron source strength to generate white noise in the detector signal. Sampling rate of 1s and Filter Time constant of 1s gave very stable acceptable sub-criticality annunciation for the monitoring application.

5. MONITORING ACCURACY & SENSITIVITY AND ERRORS

Source strength not accurately known:

When the neutron source is not known but stable initial neutron flux *no* at the detector position is known and the corresponding *k* is accurately calculated by physics then neutron source is

$$S=(1-k)^* no/ l$$

In reality value of initial reference multiplication factor is not difficult using the state of the art nuclear calculation codes.

Sensitivity to initial reference multiplication factor k:

In above section we stated that initial *k* i.e. reference multiplication factor is accurate. Now if that is inaccurate then solving IPK will yield errors in the sub-criticality annunciation. When a step reactivity is added and system is still sub critical, then the neutron flux reaches a new stable level *n2*, from initial level *n*1. The new value is given by: $n2=lS/(1-k2)$, *k2*is final multiplication factor and *k1* is initial value. Obviously if *k1* is erroneous the final sub-critical level is also erroneous. The two sub-critical values are related as

$$(1-k2)=(1-k1)\, n1/n2$$

Let the $k2$ be away from the true value $k'2$, then error

$$e=(1-k2)-(1-k'2)=\{(1-k1)-(1-k'1)\}n1/n2.$$

Where $k'1$ is the true value of multiplication factor at the initial reference condition. Substituting on the error term from the relationship of the neutron flux levels i.e. $n1/n2=(1-k'2)/(1-k'1)$ we get relative error as $r=e/(1-k'2)=\{(1-k1)-(1-k'1)\}/(1-k'1)$, in simple terms the relative error of the calculated value of $(1-k2)$ is equal to the relative error of the initial reference value $(1-k1)$. This expression is based on point kinetics and without measured errors in n.

Physically what this mean is elucidated by the examples. Say a sub critical system is actually at 15% sub-criticality and at this we enter in the system initial state of $(1-k)/k$ as 10%, then at this due to disturbance of step input the sub-criticality approaches towards 0% then this 10% initial absolute estimate error would close to zero. The reason is obvious about inverse proportionality of sub-criticality to neutron flux and as above discussed the relative error remains thus constant. The same argument will hold if initial value is entered as 20% as against true value of 15% sub-criticality. As the reactor is closing towards criticality the initial error in the value of k would also close in towards true value. Therefore sensitiveness towards initial error in the assessment of sub-criticality will not hamper the safety of the sub-criticality annunciation system, or system is not too sensitive to initial error in k.

Position dependency of detectors and sensitivity:

As discussed in the section about the normalized flux section for a point source there is need to have detector placed in the stationary point where the normalized flux stays constant, in order to have the applicability of the IPK. However as seen for homogeneous source the detector can be placed anywhere as the normalized flux is stationary through out always. For point source however, the normalized flux is non stationary through out but is having one stationary point. In reality this stationary point may not be always the convenient point of observation of neutron flux. A place be thus chosen where the normalized flux is not stationary and the effect of source is small. Then there may be calculation errors as at very low value of sub-criticality but as the disturbance make the reactor towards critical zone or sub-criticality tends to zero the calculated value converges to true value. This is due to fact that the neutron flux becomes higher than value expected by the assumption that flux distribution is independent on sub-criticality. Therefore from preliminary calculations of normalized flux distributions, it is possible to find a detector position where the calculated sub-criticality results are on the conservative side i.e. on safer towards zero side.

6. STATISTICAL METHOD

The statistical method though laborious and too computation intensive can be ported into embedded electronics. The proven methods of Feynman may be a good method to start with as sub-criticality annunciation system. This method will be useful when the reactor detectors are kept in count mode. Monitoring of criticality in low-power sub critical systems have gained renewed interest specially for ADS. Among the well established statistical methods Feynman method is widely used and its complex statistics is well established as solutions to give representation of the criticality levels. Feynman method was first established with stationary sources for criticality evaluation and now being modified to take care of stochastic pulsing sources operating the reactors in pulsed mode. The statistical principles which is measurement of neutron detection counts for few measurement time intervals and then plotting of the variance to mean statistics for different time window. The curve fit obtained is used to calculate the neutron time constant and thus the value of criticality. The following relation gives the relation between variance to mean ratio and the reactor criticality parameter.

$$ratio = [\sigma^2/(\overline{N}) - 1] = Y(t)$$

$$Y(t)=(\varepsilon\lambda_f^2/\alpha^2)[\,(\nu(\nu-1))\{(1-(1-e^{-\alpha t})/\alpha t)\}]$$

t : Measurement time period
α: Neutron time constant=β-ρ)/l
ν: Fission neutrons per fission.
λ_f : probability of fission ,
ε : Detector efficiency

$\overline{\sigma}$: Standard deviation of neutron counts

$\overline{N}$ = Mean of neutron detected at detector

The above expression is for stationary neutron sources. With the parallel computing embedded DSP techniques this method can be realized for obtaining sub-criticality on line in real time. However, these statistical methods require analysis of certain length of time sequence neutron counts to estimate the sub-criticality using complex statistical theory. These presently are good for steady conditions and for detectors in count mode. The time require to display the sub-criticality level is comfortable for the application where this instrument shall give value at extremely low power. There are various other statistical methods and neutron noise analysis statistics. The Feynman method is recently regained popularity world wide for the pulse reactor systems and ADS sub-criticality monitoring functions. For the pulse source the Feynman statistic, as expressed by above relation gets modified ,by converging oscillating terms based on neutron pulse source width and time period applied in MUSE ADS. For continuous neutron source the formula is well established as expressed above.

7. CONCLUSION

The annunciation of sub-criticality to the reactor operator will aid the safety of operation and avoid any hazards or ambiguity in reactivity management. To realize this instrumentation application of classical inverse point kinetic method with the suitable simple filter and the same instrument used in count mode with statistical method will provide annunciation when sub-criticality exceeds or rate of sub-criticality is in excess. The requirement of this instrument in Indian PHWR plants and other research facility is must to aid the reactor operator from the abstract concept of reactivity. Worldwide reactivity meters are on operation since long, but to use the similar method for sub-criticality annunciation should be encouraged. However with the developed parallel processing digital signal processing modules the sub-criticality annunciation system is realizable from both the described method. These parallel processing modules were developed for employing them in APSARA reactor regulating bins for the modified control system. These methods to be able to be less sensitive to detector position, source strength value and initial criticality value a state of the art nuclear calculation codes for each type of configurations be aimed at, which is possible in today's technology.

8. ACKNOWLEDGEMENT

The authors acknowledge initiative taken by Dr. Kallol Roy and Sri Parag Punekar RRMD/BARC and Sri M.B.Patil for development of parallel processing digital signal processing embedded design for using the same in APSARA reactor regulation system bin for the application of on line transfer function analysis with sub-criticality annunciation.

9. REFERENCES

1. Development of compact digital reactivity meter and reactor physics data processor. Y.Shimazu et al *"Nuclear Technology (1986)*
2. Measurement of sub-criticality by Feynman method at KUCA *"Koyoto University" (1994)*
3. Analytical Solution for the Feynman formula for ADS with pulsed neutron sources Ceder Pazsit" *Progress in Nuclear Energy"(2003)*

Reliability, Safety and Hazard
Editors: P.V. Varde et al.

RELIABLE STABILITY ANALYSIS OF UNCERTAIN NON-LINEAR FRACTIONAL ORDER SYSTEMS

Mangala Deshpande, Nataraj S.V. Paluri and Jayesh J. Barve

Systems and Control Engineering, Room 101, ACRE Building
Indian Institute of Technology Bombay, Mumbai 400 076

ABSTRACT

We propose a reliable algorithmic tool to analyse the stability of non-linear fractional order systems with parametric uncertainties in both the fractional order linear and non-linear elements. The proposed algorithm is developed in the framework of interval analysis and guarantees that the limit cycle locus is reliably computed to a prescribed accuracy. Two examples are given to demonstrate the reliability of the proposed algorithm.

1. INTRODUCTION

Inspired by fractional calculus [1], which is a natural extension of the classical calculus, many real world systems can now be adequately modelled using fractional calculus in the Laplace domain through transfer functions that are not quotients of polynomials in s but in s^{α}, α being a positive real number.

In the last decade, a significant amount of research has been reported for the analysis and design of linear fractional order control systems. However, methods for the analysis and design of non-linear fractional order control systems are yet sparse in the literature. This motivates the need for development of new tools for the analysis of non-linear fractional order systems.

The stability analysis of autonomous non-linear fractional order control systems can often be converted into a problem of investigating the existence of sustained oscillations known as limit cycles, perhaps the most prominent feature of non-linear control systems.

Mainly, the describing function analysis technique is employed to predict the existence of limit cycles. If limit cycles are predicted, then it is also of interest to know number of limit cycle points, their frequencies and amplitudes and other key characteristics such as stability or instability.

The objective of this paper is to propose an efficient tool to construct the limit cycle locus of uncertain non-linear fractional order systems with separable nonlinearities. The uncertainties can be in the parametric values of the linear and non-linear elements. The proposed tool is in the form of an interval analysis based algorithm (see [2] for the essentials of interval analysis required in this work) and has several useful features: it guarantees that all points on limit cycle locus in the region of interest are found; if no limit cycle point exists in the region of interest, then this fact is established with a mathematical and computational guarantee (such guarantees are crucial in stability analysis)

The rest of the paper is organized as follows. Section 2 presents the proposed algorithm for computing the limit cycle locus. Section 3 contains one non-linear fractional order example to demonstrate the two-fold capability of the proposed algorithm. Section 4 gives the conclusions of the work.

2. PROPOSED ALGORITHM

Consider the closed loop system of Fig. 1, where $G_{fro}(s^{\alpha}, \mathbf{q}_g)$ denotes the transfer function of the fractional order linear element with parameter vector $\mathbf{q}_g$, $N(a, \omega, \mathbf{q}_n)$ denotes the describing function of the non-linear element with parameter vector $\mathbf{q}_n$, and a, ω

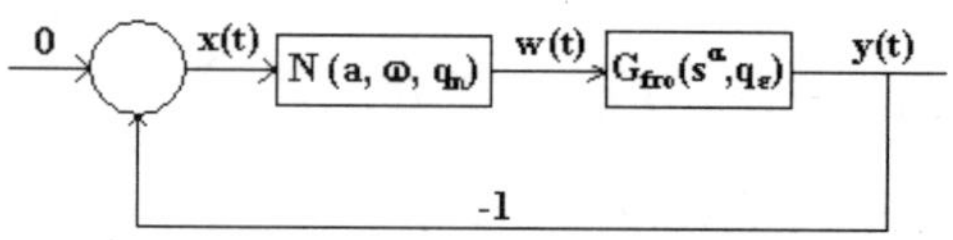

Fig. 1: Non-linear Fractional Order System

denote the amplitude and frequency of the periodic input signal to non-linear element. Let $y = (a, \omega) \in \Re^2$, $\mathbf{q} = (\mathbf{q}_g, \mathbf{q}_n) \in \Re^n$. Then, under certain assumptions stated in [3], the non-linear system exhibits a limit cycle if there exists a solution $y*$ to the characteristic equation $f(y, q) = 0$, where $f(y, q)$ is defined as

$$f(\mathbf{y},\mathbf{q}) = (f_{\mathrm{Re}}(\mathbf{y},\mathbf{q}), f_{\mathrm{Im}}(\mathbf{y},\mathbf{q}))$$
$$f_{\mathrm{Re}}(\mathbf{y},\mathbf{q}) = \mathrm{Re}\left\{1 + N(a,\omega,\mathbf{q}_n) G_{fro}(s^{\alpha},\mathbf{q}_g)\right\} \quad (1)$$
$$f_{\mathrm{Im}}(\mathbf{y},\mathbf{q}) = \mathrm{Im}\left\{1 + N(a,\omega,\mathbf{q}_n) G_{fro}(s^{\alpha},\mathbf{q}_g)\right\}$$

The solution y* is called a limit cycle point, and the corresponding a* and ω* are called the limit cycle amplitude and frequency, respectively.

Now, suppose there is parametric uncertainty in the system such that the parameter vector q varies over a bounding box $\mathbf{Q}^0 \in I(\Re^n)$. The parametric uncertainty gives rise to an uncertain non-linear system. The problem of computing the limit cycle locus for this uncertain system is defined as

$$L_c(f, \mathbf{Q}^0) = \{ \mathbf{y} \in \Re^2 : f(\mathbf{y},\mathbf{q}) \} = 0 \quad (2)$$

for some $\mathbf{q} \in \mathbf{Q}^0$. The following assumption is made in this work.

The transfer function G_{fro} and the describing function N are continuous in **q** and continuously differentiable in y.

The proposed algorithm can be briefly described as a binary search algorithm based on a geometrical subdivision process, root exclusion test and the Generalised Krawczyk operator. It consists of three parts: an initialisation part, an iterative part, and a termination part.

In the initialisation part, one first constructs a natural inclusion function F for f, and, an initial box $\mathbf{Y}^0$ that is guaranteed to contain the limit cycle locus by applying the procedure in [2]. Next, the two lists that are needed in the algorithm are initialised here: a working list L that contains boxes for processing is initialised with the box $\mathbf{X}^0 = (\mathbf{Y}^0, \mathbf{Q}^0)$ and the solution list L_{sol} that contains boxes having limit cycle points is initialised with the empty list. The algorithm then branches to the iterative part.

In the iterative part, all boxes $\mathbf{X}_{(i)}$ present in the working list L are chosen, and a branch and bound strategy is applied to discard irrelevant parts (or the whole) of these boxes using the zero exclusion test, the Generalized Krawczyk test, and subdivisions.

A box is accepted as a solution box if its width does not exceed a prescribed domain accuracy tolerance ε_x. For all such boxes, $\mathbf{Y}_{(i)}$ is deposited in the solution list L_{sol}. The remaining boxes are subdivided along their longest directions, and the resulting sub boxes are put in L. The entire iterative part is repeated till no more boxes are left in L for processing. Then, the algorithm branches to the termination part.

In the termination part, the obtained solution list L_{sol} is first examined. If L_{sol} is found empty, then no limit cycle exists, and the algorithm exits. Else, the limit cycle set L_c^{alg} is constructed as the union of all boxes $\mathbf{Y}_{(i)}$ present in L_{sol}, and following [3], the limit cycle stability for each solution box $\mathbf{Y}_{(i)}$ is assessed, and the obtained information displayed. The algorithm now exits.

The entire process is greatly speeded up by concurrently (rather than sequentially) processing all the sub boxes present in a given iteration. Concurrent processing is possible through the use of vectorized interval operations for function evaluations, subdivisions, width checks, etc.

The proposed algorithm is presented next.

Algorithm to compute Limit cycle locus for fractional order non-linear systems.

Inputs: An expression for the function f in (1), the parameter box $\mathbf{Q}^0$, and the prescribed domain accuracy tolerance ε_x.

Output: The message "no limit cycle point exists" or the computed limit cycle locus L_c^{alg}.

Note: The algorithm is to be executed in the order given below, except when otherwise indicated.

BEGIN Algorithm

1. (Initialisation part) From the expression for f, find a natural inclusion function F.
2. Construct initial boxes and lists:
   a) Construct an initial search box $\mathbf{Y}^0$ and then construct $\mathbf{X}^0 = (\mathbf{Y}^0, \mathbf{Q}^0)$.
   b) Set $k \leftarrow 0$ and initialise lists $L^{sol} \leftarrow \{\}$,
3. (Iterative part) Start a new iteration:
   a) Set $k \leftarrow k+1$ and $l_r \leftarrow$ length of L.
   b) Pick all boxes $\mathbf{X}_{(i)} = (\mathbf{Y}_{(i)}, \mathbf{Q}_{(i)})$, $i = 1, 2, ..., l_r$ from L and delete their entries from L.
4. Test phase using an interval zero exclusion test and Generalized Krawczyk test:
   a) Evaluate $F(\mathbf{X}_{(i)})$, $i = 1, 2, ..., l_r$.
   b) (Interval zero exclusion test): IF $0 \notin F(\mathbf{X}_{(i)})$, THEN discard $i = 1, 2, ..., l_r$.
   c) Generalized Krawczyk test): For each remaining $\mathbf{X}_{(i)} = (\mathbf{Y}_{(i)}, \mathbf{Q}_{(i)})$ do the following:
      i) Find a natural inclusion function $F_y^{'}$ for the derivative $f_y^{'}$, and set $C \leftarrow \{m(F_y^{'})\}^{-1}$. Evaluate the Generalized Krawczyk operator $K(X_{(i)})$ defined as,
      ii) $K(\mathbf{X}_{(i)}) = \hat{\mathbf{y}}_{(i)} - CF(\mathbf{X}_{(i)}) + \{I - CF^{'}_y(\mathbf{X}_{(i)})\}(\mathbf{Y}_{(i)} - \hat{\mathbf{y}}_{(i)})$, where $\hat{\mathbf{y}}_{(i)} = m(\mathbf{Y}_{(i)})$.
      iii) Find $\mathbf{Y}_{(i)}^{'} = K(\mathbf{X}_{(i)}) \cap \mathbf{Y}_{(i)}$.
      iv) IF $\mathbf{Y}_{(i)}^{'} = 0$, THEN discard $\mathbf{X}_{(i)}$. ELSE set $\mathbf{X}_{(i)} \leftarrow (\mathbf{Y}_{(i)}^{'}, \mathbf{Q}_{(i)})$.
   d) If no more $\mathbf{X}_{(i)}$ remain, go to step 8.
5. Solution set:
   a) For each remaining $\mathbf{X}_{(i)}$, evaluate $w(\mathbf{X}_{(i)})$. IF $w(\mathbf{X}_{(i)}) < \varepsilon_x$, THEN enter $\mathbf{Y}_{(i)}$ in L^{sol} and discard $\mathbf{X}_{(i)}$ from further processing.
   a) If no more $\mathbf{X}_{(i)}$ remain, go to step 8.
6. Subdivision phase:
   For each remaining box $\mathbf{X}_{(i)}$, find a coordinate direction $k_{(i)}$ parallel to which $\mathbf{X}_{(i)}$ has an edge of greater length. Subdivide $\mathbf{X}_{(i)}$ in direction $k_{(i)}$ getting sub boxes such that $\mathbf{X}_{(i)} = \mathbf{X}_{(i)}^1 \cup \mathbf{X}_{(i)}^2$. Enter the sub boxes $\mathbf{X}_{(i)}^1 \cup \mathbf{X}_{(i)}^2$ in L.
7. End current iteration: Return to Step 3.
8. (Termination part) Determine non-existence of limit cycles or construct limit cycle locus:
   a) IF $L^{sol} = 0$, THEN print, "no limit cycle point exists" and go to step 10.
   b) Construct $\mathrm{L}_c^{alg} = \bigcup_{\mathrm{L}^{sol}} \mathbf{Y}_{(i)}$ and output L_c^{alg}.
9. Determine stability of limit cycle boxes: For each $\mathbf{Y}_{(i)} \in L^{sol}$ do the following: Determine $\mathbf{Y}_{(i)}$ as a stable or unstable limit cycle box as per known methods for fixed parameter systems [3], and display the obtained stability information.
10. 10. EXIT Algorithm.

END Algorithm.

3. ILLUSTRATIVE EXAMPLE

The proposed algorithm is implemented using the interval arithmetic toolbox INTLAB [4] on a Pentium 1.5 GHz machine with 128 MB RAM, and demonstrated on a fractional order example.

The proposed algorithm does not require rational approximation of any fractional order term in the transfer function.

Example: The linear element is a fractional order system described in [5].

$$G\left(s^{\alpha},\mathbf{q}_g\right)=\frac{q_1 s^{1.648}-q_2 s^{0.824}+q_3}{q_4 s^{2.472}+q_5 s^{1.648}+q_6 s^{0.824}+q_7}$$

while, the non-linear element is chosen as a memory type nonlinearity comprising of relay with hysteresis. The describing function of this non-linear element is

$$N(a,\omega,\mathrm{q})=\frac{4H}{\pi\,\alpha}\left[\sqrt{1-\left(\frac{D}{a}\right)^2}-j\left(\frac{D}{a}\right)\right]$$

The nominal parameter values of the linear and non-linear elements are

$$q_1=0.054, q_2=0.271, q_3=0.466, q_4=1.0,$$
$$q_5=3.15, q_6=2.84, q_7=0.367, H=5, D=1$$

Parametric uncertainty of 10% around the nominal values is introduced in q.

(a) The objective is to construct the limit cycle locus for each uncertain parameter to a specified accuracy $\varepsilon_x=0.001$.

Solution: The proposed algorithm is applied to construct the limit cycle locus for each uncertain parameter. The initial search box for the limit cycle is taken as $Y^0=((0,100],[0.01,100])$.

The results obtained from the algorithm are plotted in Fig. 2 and Fig. 3. In these plots, the limit cycle points corresponding to the nominal parameter values are shown as cross-marked circles.

Before proceeding further, it is desirable to first verify the limit cycle results generated by the proposed algorithm versus those of the conventional graphical method at least for the nominal values of the uncertain parameters.

The conventional graphical describing function Method [6] is therefore applied for the nominal parameter values with $\pm 10\%$ uncertainty.

The limit cycle amplitude and frequency obtained from the intersection point of the polar plot and the $-\frac{1}{N}$ plot with nominal values of the parameters are 1.2 and 0.6 rad/sec., respectively (see Fig. 4). On the other hand, for the nominal parameter values, the proposed algorithm gives the limit cycle point as

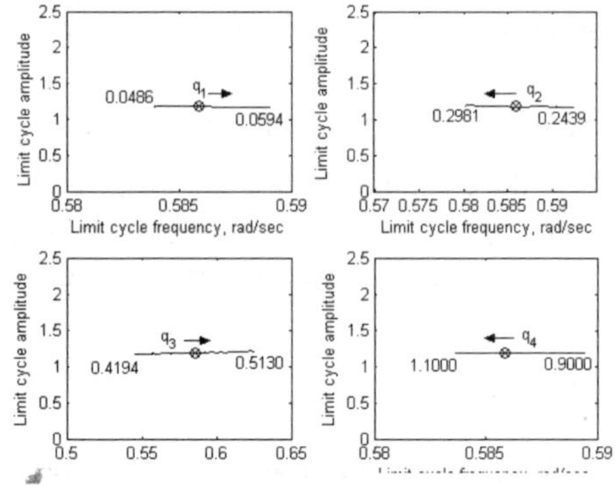

Fig. 2: Limit Cycle Locus varying q_1 to q_4.

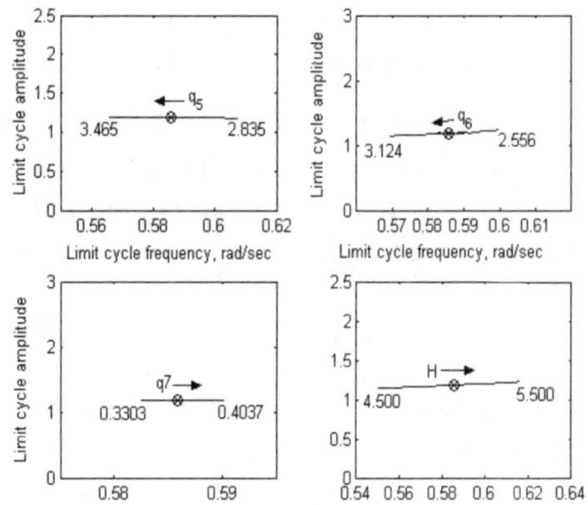

Fig. 3: Limit Cycle Locus varying q_5 to q_7 & H.

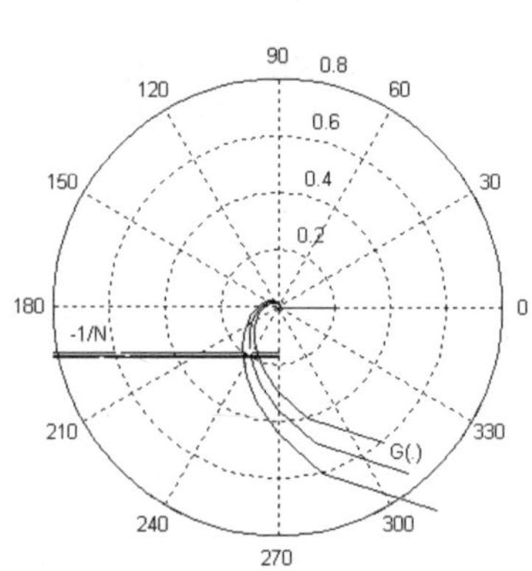

Fig. 4: Polar plot of $G\left(s^{\alpha}\right)$ and $-\left(\frac{1}{N}\right)$ plot.

drawn for $\pm 10\%$ uncertainty, showing locus of their intersection points i.e. limit cycle points. $a = [1.1822, 1.1855]$, $\omega = [0.5846, 0.5873]$, which agrees well with the results of the graphical describing function approach.

Next, from the limit cycle plots in Fig. 2 and Fig. 3, the following observations can be made: For each parameter, there is a single branch of the limit cycle locus corresponding to stable limit cycle points. The limit cycle frequency decreases with q_2, q_4, q_5, q_6 but increases with q_3, q_7 and H. The limit cycle amplitude decreases with q_6 but increases with q_3, H. Both the limit cycle frequency and amplitude hardly vary with q_1, whereas both decrease with q_6 and both increase with q_3, H.

Table 1 Performance of the Proposed Algorithm.

Parameter	Boxes	Iterations	Tsec.
q_1	4	23	22
q_2	21	26	24
q_3	48	31	32
q_4	16	25	24
q_5	37	26	34
q_6	43	26	19
q_7	09	23	21
H	84	28	27

Table 1 gives the performance of the algorithm in terms of the number of solution boxes generated to enclose the limit cycle locus for each uncertain parameter, the number of algorithmic iterations required, and the computational time taken by the algorithm.

(b) The objective is to tune a controller to achieve a given limit cycle point.

Solution: Consider again the same fractional order linear element and memory type non linear element comprising of a relay with hysteresis, but now including a fractional order PD^{β} controller in the feedback system given by

$$C(s) = c_1 + c_2 s^{\beta} = 5.5 + 3.7343 s^{1.15}$$

First, the proposed algorithm is applied to construct the limit cycle locus with each controller parameter varying in some band, say 10%, around the above nominal controller parameter values. The initial search box for the limit cycle is taken as

$$Y^0 = ((0,100], [0.01,100])$$

The proposed algorithm generates sets of 92 and 478 boxes enclosing the limit cycle locus, in 27 and 32 iterations and computational time of 42 and 52 seconds, respectively for c_1 and c_2 controller parameters. The obtained results are plotted in the top row of Fig. 5. In these plots, the limit cycle points corresponding to the nominal parameter values are shown as cross-marked circles. The limit cycle amplitude and frequency corresponding to this closed-loop non-linear fractional order system for the nominal values of the parameters are $a = 1.2$, $\omega = 0.6$ i.e., the limit cycle point is $(1.2, 0.6)$.

Suppose for any operational reason, the limit cycle behaviour is required to be shifted to some other point, say (1.86, 2.034) and (1.95, 2.1). Then, the next task is to find all controller parameter values (in the same band) that can produce the new limit cycle behaviour.

For the first task, the initial search box for limit cycle points is simply set as (first new limit cycle)

$$Y^0 = ([1.86, 1.86], [2.034, 2.034]),$$

whereas for the second task, the initial search box for limit cycle points is set as (second new limit cycle)

$$Y^0 = ([1.95, 1.95], [2.1, 2.1]).$$

The results generated by the proposed algorithm for the above two newly specified limit cycle points are plotted in the bottom row of Fig. 5. We find that the proportional controller parameter c_1 remains fixed at its nominal tuning value of 5.5 for both the two new limit cycles. Whereas, the fractional derivative controller parameter c_2 can take any value in $[3.4544, 4.0146]$ to achieve the first new limit cycle of $(1.86, 2.034)$, and to achieve the second limit cycle of $(1.95, 2.1)$, c_2 can assume any value in $[3.9213, 4.1080]$.

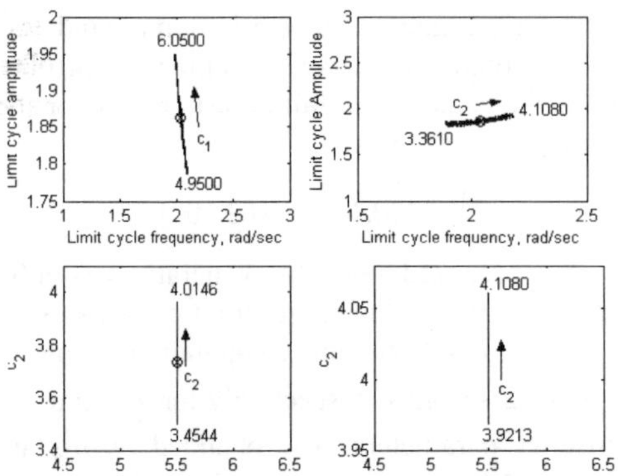

Fig. 5: Non-linear fractional order system with a fractional order PD^{β} controller. Top row: original Limit cycle locus for variations in the PD^{β} controller parameters. Bottom row: locus of controller parameters to achieve prescribed limit cycle behaviour (left bottom: new limit cycle 1, right bottom: new limit cycle 2).

Further, the new limit cycle $(1.95, 2.1)$ cannot be achieved by the nominal values of the fractional PD^{β} controller parameters. Thus, the limit cycle amplitude and frequency can be controlled by appropriately retuning only the fractional derivative term. Moreover, all limit cycles are stable.

6. CONCLUSIONS

The tool of limit cycle locus is clearly useful in gaining insight into the effect of each uncertain parameter on the limit cycle behaviour. The proposed interval-based algorithm computes this locus reliably and accurately, for a very wide class of linear and non-linear elements of fractional order and nonlinearities of the type frequency dependent / independent, with memory / without memory. The algorithm aids in tuning a controller to either suppress an existing limit cycle or achieve a new one. It must be noted that any errors in the limit cycle locus results due to the approximate nature of the describing function method cannot be avoided.

7. REFERENCES

1. I. Podlubny, "Fractional Differential Equations", Academic Press, San Diego, 1999.
2. R. E. Moore, "Methods and applications of interval analysis", SIAM, Philadelphia, PA, 1979.
3. D. P. Atherton, "Non-linear control engineering", Van Nostrand, 1975.
4. S. M. Rump, "INTLAB - interval laboratory" in Csendes, T. (Eds.): Developments in reliable computing, Kluwer Academic Publishers, Dodrecht, 1999.
5. O. Cois, A. Oustaloup, E. Battaglia, and J. L. Battaglia, "Non integer model from model decomposition for time domain system identification", Proc. 41st IEEE conference on Decision and Control, Tutorial Workshop, Las Vegas, U.S.A., pp. 200-206, 2002.
6. A. Gelb and W. E. Van der Velde, "Multiple-Input Describing Functions and Nonlinear System Design", McGraw-Hill, 1968.

Reliability, Safety and Hazard
Editors: P.V. Varde et al.

RELIABLE SIMULATION OF AERO GAS TURBINE PLANT USING FRACTIONAL ORDER MODEL: A COMPARATIVE STUDY

Vishwesh Vyawahare, Rambabu K., Nandkishor Kubal and Nataraj S.V. Paluri

Systems and Control Engineering Group, Indian Institute of Technology Bombay
Mumbai 400 076

ABSTRACT

Owing to the fact that many real-world systems are well characterized by fractional-order (FO) models, the use of fractional calculus for modeling physical systems has been widely considered in last decades. Fractional-order system (FOS) is the system, which is better described by a fractional-order differential equation. Simulation of a system represented by a FO model is a key step in its analysis. Various methods for simulating an FO system are available in the literature. In this paper, the FO model of a twin-spool aero gas turbine plant, obtained by output-error identification method is simulated using three different techniques: the Grunwald-Letnikov (GL) Approximation Method, Continued Fraction Expansion (CFE) Method and Podlubny's analytical formula involving infinite series expression. The reliability of these simulation techniques is analyzed in terms of computational complexity, simulation time and accuracy. The simulation results indicate that a convergence problem indeed occurs in using Podlubny's infinite series expression and CFE method. The study suggests that GL method can be considered as a reliable method for the simulation of FO systems.

Keywords: Discrete approximation, Fractional calculus, Fractional-order system, Gas turbine, Inverse Laplace transform, Step response, Time-domain simulation

1. INTRODUCTION

Many important real-world physical systems are well characterized by fractional order differential equations, i.e., equations involving noninteger-order derivatives. In particular, it has been shown that materials having memory and hereditary effects, and dynamical processes, such as mass diffusion, heat conduction, and long transmission lines [1], [2] can be more adequately modelled by FO models than integer order (IO) models. Thus FO systems are basically distributed parameter systems characterized by long memory transients and infinite dimensional structure. Because of the higher complexity and the absence of adequate mathematical tools, FO dynamical systems were only treated marginally in the theory and practice of control systems. Their analysis requires familiarity of work with FO derivatives and integrals [3], [4]. For geometrical and physical interpretation of fractional integration and differentiation, see [5], [6].

With the success in the synthesis of real noninteger differentiator and the emergence of new electrical element called "fractance" [7], fractional-order controllers, including fractional-order PID controllers [8], have been designed and applied to control a variety of dynamical processes, including IO and FO systems. The main advantage of using FO controllers for a linear control system is that the time and frequency responses can be shaped using functions rather than of exponential type and, as a consequence, the performance of the feedback control loop can be improved over the use of IO controllers.

2. REVIEW OF SIMULATION METHODS

Consider a FO system represented by the general transfer function (FOTF)

$$F(s)=\frac{Y(s)}{U(s)}=\frac{b_p s^{\beta_p}+b_{p-1}s^{\beta_{p-1}}+\ldots+b_1 s^{\beta_1}+b_0 s^{\beta_0}}{a_n s^{\alpha_n}+a_{n-1}s^{\alpha_{n-1}}+\ldots+a_1 s^{\alpha_1}+a_0 s^{\alpha_0}} \quad (1)$$

where a_i, α_i, $(i = 0,1,2,\ldots,n)$ and b_j, β_j, $(j = 0,1,2,\ldots,p)$ are real numbers and $\alpha_n > \alpha_{n-1} > \ldots > \alpha_1 > \alpha_0 > 0$ with $\beta_j < \alpha_n$, $(j = 0,1,2,\ldots,p)$.

Careful examination of (1) reveals the fact that each term in the numerator and denominator is an infinite dimensional linear filter due to the FO differentiator [7], which poses a major difficulty with fractional models for its time-domain simulation. Often, the analytical solution of such model output is not simple to compute. There are various approaches and methods proposed in the literature for the simulation of FO systems. See [9], [10]. In this paper, we have carried out a comparative study of the three main simulation methods, namely

1) Inverse Laplace transform (ILT) using Podlubny's analytical method [8]
2) Grunwald-Letnikov (GL) Approximation method [13], and
3) Continued Fraction Expansion (CFE) method [13].

2.1 Inverse Laplace transform (ILT) using Podlubny's analytical method

Podlubny [8] derived an analytic expression for the inverse Laplace transform (ILT) of the FOTF represented by (1). It is given below. This method basically gives the impulse response and step response of the FOTF.

$$f(t)=\sum_{r=1}^{p} b_p \frac{1}{a_n}\sum_{m=0}^{\infty}\frac{(-1)^m}{m!}$$

$$\times \sum_{\substack{k_0+k_1+\ldots+k_{n-2}=m \\ k_0\geq 0, k_1\geq 0,\ldots,k_{n-2}\geq 0}} (m;k_0,k_1,\ldots,k_{n-2})$$

$$\times\left(\prod_{i=1}^{n-2}\left(\frac{a_i}{a_n}\right)^{k_i}\right)$$

$$\varepsilon_m\left(t,\frac{-a_{n-1}}{a_n};\alpha_n-\alpha_{n-1},\alpha_n-\beta_r+\sum_{j=1}^{n-2}(\alpha_{n-1}-\alpha_j)k_j\right) \quad (2)$$

where $(m;k_0, k_1, \ldots, k_{n-2})$ are multinomial coefficients and

$$\varepsilon_m\left(t,\frac{-a_{n-1}}{a_n};\alpha_n-\alpha_{n-1},\alpha_n-\beta_r+\sum_{j=1}^{n-2}(\alpha_{n-1}-\alpha_j)k_j\right)$$

$$=\sum_{q=0}^{\infty}\frac{(q+m)!\left(\frac{-a_{n-1}}{a_n}\right)^q t^{\left[(\alpha_n-\alpha_{n-1})(q+m)+\alpha_n-\beta_r+\left(\sum_{j=1}^{n-2}(\alpha_{n-1}-\alpha_j)k_j\right)-1\right]}}{q!\Gamma\left[(\alpha_n-\alpha_{n-1})(q+m)+\alpha_n-\beta_r+\left(\sum_{j=1}^{n-2}(\alpha_{n-1}-\alpha_j)k_j\right)\right]} \quad (3)$$

Γ is the Euler's Gamma function.

Equation (2) also gives the impulse response of the FO system given by (1). The step response is obtained by multiplying (1) by 1/s and then applying (2). It is noted that (2) is in the form of double infinite series. The truncation of these series is a very vital part of in the calculation. The method may fail to converge to the true solution when t is larger than a certain value [11], as shown in the results in following sections.

2.2 Grunwald-Letnikov (GL) Approximation method

The discretization [13] of the continuous FOTF (1) can be obtained using the generating function

$$s=\omega(z^{-1}) \quad (4)$$

where z^{-1} is the shift operator. Thus (1) becomes,

$$F(\omega(z^{-1}))=\frac{Y(\omega(z^{-1}))}{U(\omega(z^{-1}))}$$

$$=\frac{\sum_{i=0}^{p} b_i (\omega(z^{-1}))^{\beta i_i}}{\sum_{j=0}^{n} a_j (\omega(z^{-1}))^{\alpha_j}} \quad (5)$$

For discrete approximation of s^{α}, we can use the generating function corresponding to the Z-transform of backward difference rule as

$$\omega(z^{-1}) = \frac{(1-z^{-1})}{T} \quad (6)$$

where T is the time step. Performing the power series expansion (PSE) of $(1-z^{-1})^{\alpha}$ we obtain the GL approximation

$$(\omega(z^{-1}))^{\alpha} = \frac{1}{T^{\alpha}} \sum_{j=0}^{N} \left(\frac{\Gamma(\alpha+1)}{[\Gamma(j+1)\Gamma(\alpha-j+1)]} \right) z^{-j} \quad (7)$$

where N = [L/T]. L is the memory length, and [x] means the integer part of x, see [4].

2.3 Continued Fraction Expansion (CFE) method

The CFE method is for evaluation of functions that are basically irrational in nature [12]. Consider an irrational function G(s) having the approximation using CFE as

$$G(s) = a_0(s) + \cfrac{b_1(s)}{a_1(s) + \cfrac{b_2(s)}{a_2(s) + \cfrac{b_3(s)}{a_3(s)+\dots}}} \quad (8)$$

where a_i's and b_i's are rational functions of the variable s, or are constants. The application of the method yields a rational function, $G^*(s)$, which is an approximation of the irrational function.

The FO operator s^{α} can be realized based on direct discretization using the Tustin and CFE [13] as

$$\omega(z^{-1})^{\alpha} = \left(\frac{2}{T}\right)^{\alpha} CFE\left\{ \left(\frac{1-z^{-1}}{1+z^{-1}} \right)^{\alpha} \right\}_{p,q} = \left(\frac{2}{T}\right)^{\alpha} \frac{P_p^{\alpha}(z^{-1})}{Q_q^{\alpha}(z^{-1})} \quad (9)$$

where P and Q are polynomials of degrees p and q respectively, in the variable z^{-1}.

3. SIMULATION RESULTS

A comparative study of three methods mentioned in the Section II is performed by simulating the FO model of a twin-spool aero gas turbine plant. The FO model was estimated using the output-error identification technique [14] with fuel rate to the gas turbine and high pressure spool speed as the input and output variables respectively [15]. The methods were tested for various combinations.

3.1 Open Loop Step Response of Gas Turbine

The FO model obtained by an output-error identification technique is

$$P(s) = \frac{107.44715}{0.01017 s^{1.64345} + 0.1587 s^{0.7755} + 1} \quad (10)$$

We obtain the unit step response of the above open loop FOTF by all the three methods, as shown in Fig. 1. In case of Podlubny's ILT method, the upper limit used for both the infinite summations involved in analytic expression was selected as 60. For simulation using GL method, each fractional power term in (10) was replaced by (7) with T = 0.01 sec. and N = 25. This resulted into the discrete filter approximation of (10) with degree equal to 25. For CFE method, we selected q = p = 9. As a result each fractional power term in (10) was replaced by

$$(\omega(z^{-1}))^{\alpha} = \frac{P(z^{-1})}{Q(z^{-1})} = \frac{P_9 z^{-9} + P_8 z^{-8} + \dots + P_0}{Q_9 z^{-9} + Q_8 z^{-8} + \dots + Q_0} \quad (11)$$

where the coefficients $P_9, \dots, P_0$ and $Q_9, \dots, Q_0$ are given in [13].

Simulation by all the three methods give good results except that after t > 0.92 seconds, Podlubny's ILT method fails to converge. Both GL and CFE method give correct results.

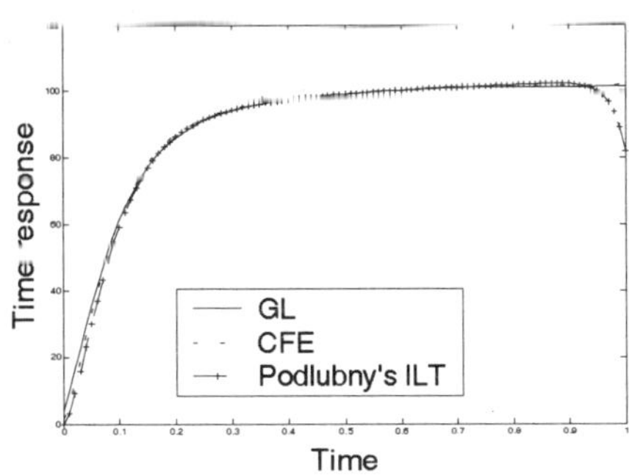

Fig. 1: Comparison of the unit-step responses of the open loop FO gas turbine model computed using various methods

3.2 Closed Loop Step Response of FO Gas Turbine with Integer Order (IO) Controller

Now we simulate the unit step response of the closed loop system comprising of Bejkar's IO controller (12) (see [16]) in cascade with the gas turbine. Due to convergence problem, Podlubny's ILT method was unable to give solution.

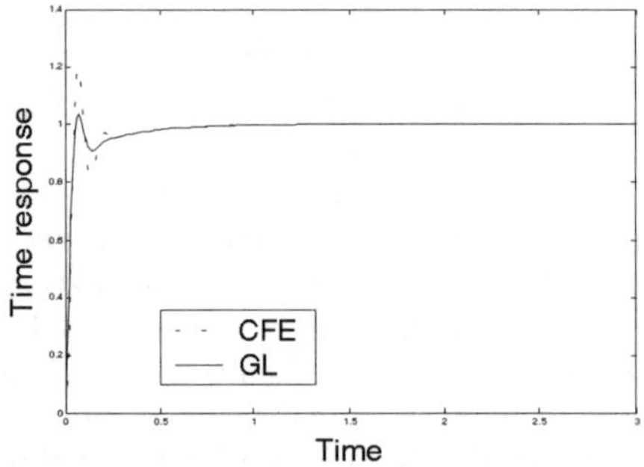

Fig 2: Comparison of the unit-step responses of the closed loop TF of IO controller in cascade with FO gas turbine model computed using GL and CFE methods

$$G(s)=\frac{1.837\times10^{10}s^2+1.313\times10^{12}s+5.455\times10^{12}}{s^3+6.344\times10^{7}s^2+2.778\times10^{13}s} \tag{12}$$

Both CFE and GL gave good results, see Fig. 2. It can be observed that the responses differ in the high frequency region. This is because of the fact that GL method gives erroneous results during the transient period.

3.3 Closed Loop Step Response of FO Gas Turbine with FO Controller

We consider another possibility wherein the FO plant is controlled by a FO controller. The closed loop step response of gas turbine with Deshpande's [17] FO controller was obtained using various methods. The FO controller is given by

$$G(s)=\frac{\begin{array}{c}3.6682\times10^{-5}s^{2.6435}+5.7241\times10^{-4}s^{1.7755}\\+5.8864\times10^{-3}s^{1.6435}+3.6069\times10^{-3}s\\+0.0916s^{0.7755}+0.5771\end{array}}{5.0865\times10^{-8}s^3+5.0965\times10^{-3}s^2+s} \tag{13}$$

In this case also, the Podlubny's ILT method failed to converge for very small values of t. Simulation results using the other two methods are given in Fig. 3. Also after t > 0.29 seconds, convergence problem was observed with the CFE method.

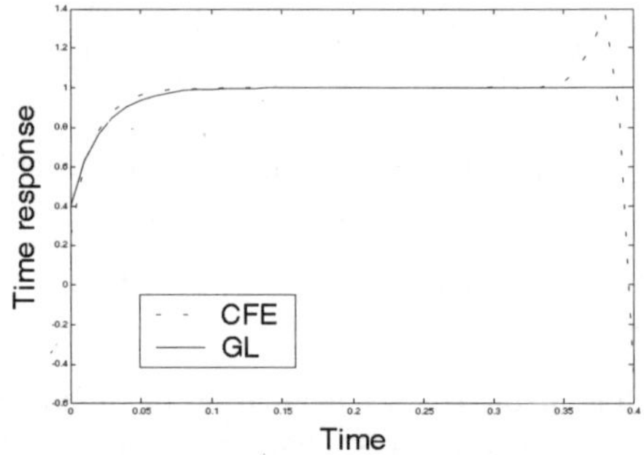

Fig. 3: Comparison of the unit-step responses of the closed loop TF of FO controller in cascade with FO gas turbine model computed using GL and CFE methods

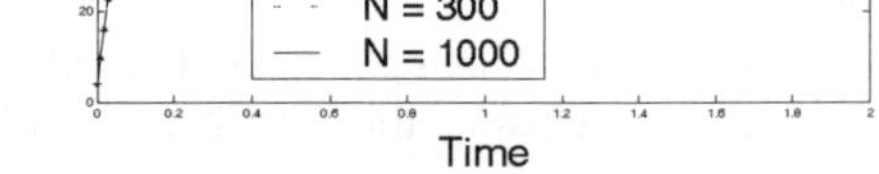

Fig. 4: Step responses using GL method for various values of N

Due to the fact that simulation results obtained using GL method are good and the method does always converge for any FOTF, we focussed our attention to this particular method and carried out further analysis. For all previous results, the value of N in (7) was fixed to 25. Fig. (4) shows the unit step response of the FOTF (11) for various values of N. It can be clearly noticed that the responses start deviating from t = 0.5 seconds (approx.), with a change in steady state value from 99.0 for N = 25 to 104.8 for N = 1000.

Thus it can be said that the value of N plays a key role in the simulation of a FOTF using GL method. Basically, by increasing the value of N, we increase the "memory" of the FO system. For instance, when N=1000, the output of the FO system at the instant t=1001 depends on all the previous 1000 values of input and output.

Table 1 Comparison of various simulation methods

Method	Computational Complexity	Simulation time	Accuracy
ILT	Very High	Very Large	High
GL	Very Less	Smallest	Depends on N
CFE	Medium	Medium	Fairly accurate

4. CONCLUSIONS

We carried out a comparative study of various methods for the simulation of the FO system. Three generalized methods given in the literature are considered, namely, Podlubny's analytical method giving inverse Laplace transform, the Grunwald-Letnikov method, and the Continued Fraction Expansion method. The performance of these methods was tested for the simulation of FO model of gas turbine. Various closed loop and open loop combinations of both fractional order and integer order controllers were simulated using the given methods.

It is observed that Podlubny's method has serious problem of convergence. There is a significant deviation from the true solution after a certain value of t. Also, computational time required for this method was very large, of the order of 1 hour. These facts imply that Podlubny's analytic method has its inherent limitation for the time-domain simulation and analysis of FO systems.

The CFE method, though, gave good results for certain FO systems, also does not guarantee convergence for all systems.

The GL method seems to converge always for all types of FO systems, both open loop and closed loop. But its performance during transient period is poor than that of the CFE method. See Table (1).

Also it is seen that the output of the FO system changes considerably for the gas turbine system when N is changed from 25 to 1000.

Thus, based on the above analysis, it can be concluded that Grunwald-Letnikov method can be considered as a reliable method for the simulation of FO systems over the Podlubny's and CFE method.

5. REFERENCES

1. A. Oustaloup, F. Levron, B. Mathieu, and F. M. Nanot. Frequency band complex noninteger differentiator: Characterization and synthesis. *IEEE Transactions on Circuits and Systems-I: Fundamental Theory and Applications*, 47(1):25-39, January, 2000.
2. S. Manabe. The non-integer integral and its application to control systems. *JIEE (Japanese Institute of Electrical Engineering)*, 80(860):589-597, 1960.
3. K. B. Oldham and J. Spanier. *The Fractional*
4. *Calculus*. Academic Press, New York, 1974.
5. I. Podlubny. *Fractional Differential Equations*. Academic Press, New York, 1998.
6. I. Podlubny. Geometric and physical interpretation of fractional integration and fractional differentiation. *Fractional Calculus and Applied Analysis*, 5(4):367-386, 2002.
7. M. Moshrefi-Torbati and J. K. Hammond. Physical and geometrical interpretation of fractional operators. *Franklin Institute*, 335B(6):1077-1086, 1998.
8. I. Petras, I. Podlubny, P. Leary, L. Dorcak, and B. M. Vinagre. *Analog realization of fractional order controllers*. Fakulta BERG TU, Kosice, 2002.
9. I. Podlubny. Fractional-order systems and $PI^{\lambda}D^{\mu}$-controller. *IEEE Transactions on Automatic Control*, 44(1):208-214, January, 1999.
10. T. Poinot and J. –C. Trigeassou. A method for modelling and simulation of fractional systems. *Signal Processing*, 83:2319-2333, 2002.
11. L. Dorcak. Numerical models for the simulation of the fractional-order control systems. *Slovak Academy of Sciences*,

Institute of Experimental Physics, Kosice, Slovak Republic, November, 1994.

12. C. Hwang, J. F. Leu, and S. Y. Tsay. A note on time-domain simulation of feedback fractional order systems. *IEEE Transactions on Automatic Control*,47(4):625-631, April, 2002.
13. B. M. Vinagre, I. Podlubny, A Hernandez, and V. Feliu. Some approximations of fractional order operators used in control theory and applications. *Fractional Calculus and Applied Analysis*, 3(3):231-248, 2000.
14. L. Dorcak, I. Petras, I. Kostial, and J.Terpak. Fractional-order state space models. *Proc. International Carpathian Control Conference*, Malenovice, Czech Republic, May 27-30, 2002, 193-198.
15. P. Thierry and T. Jean-Claude. Identification of fractional systems using output-error technique. *Nonlinear Dynamics*, 38(1-2):133-154, December, 2004.
16. A. Khan. Nonlinear mathematical modelling, simulation, and robust control of marine gas turbine. *M. Tech Thesis*, Indian Institute of Technology Bombay, 2004.
17. S. R. Bejkar. Algorithms for QFT controller and prefilter design using interval analysis. *M. Tech Thesis*, Indian Institute of Technology Bombay, 2005.
18. M. Deshpande, Analysis and synthesis of fractional order systems. *Annual Ph. D. progress seminar report*, Indian Institute of Technology Bombay, 2005.

Reliability, Safety and Hazard
Editors: P.V. Varde et al.

RELIABILITY ANALYSIS OF FAULT TOLERANT COMPUTER BASED CORE TEMPERATURE MONITORING SYSTEM OF PFBR

L. Srivani[a], B. Sasidhar Rao[a], A. John Arul[b], C. Senthil Kumar[c], S. Ilango Sambasivan[a] and P. Swaminathan[a]

[a]Electronics and Instrumentation Group, [b]Reactor Physics Division,
Indira Gandhi Centre for Atomic Research, Kalpakkam
[c]AERB-Safety Research Institute, Kalpakkam

ABSTRACT

The architecture of fault tolerant computer based core temperature monitoring system (CTMS) of Prototype Fast Breeder Reactor (PFBR) and its hardware Reliability evaluation are presented in this paper. In PFBR, Protection against individual fuel subassembly blockage is provided by Real time computer based system (RTCS). This process involves computations in every cycle and hence, a Real Time Computer is used. This paper elaborates the importance of RTCS and how the fault-tolerant architecture (Triple modular redundancy) is arrived with the help of Reliability evaluation. The operational and safety requirements, i.e., both safe and unsafe failures are analyzed. Common Cause Failure analysis has also been carried out to arrive at the results.

Keywords: CCF, CTMS, Fault tolerant, FSA, PFBR, Reliability analysis, Reliability target, RTC, RTCS, Safe failure, SDS, TCs, Unsafe failure

1. INTRODUCTION

The Prototype Fast Breeder reactor (PFBR) is a 500 MWe sodium cooled pool type Fast Reactor. The reactor core, with no moderator, is very compact and therefore has a high power density when compared to thermal reactors like Pressurized Heavy Water Reactor or Boiling Water Reactor. If coolant flow is blocked, the clad temperature increases beyond design limits and may lead to clad rupture and further to fuel melt down. Hence, continuous monitoring of the core cooling and initiation of safety actions in case of any abnormal temperature rise in the core are essential. Core Temperature Monitoring System (CTMS) is designed to detect and protect against local and global faults in reactor core. CTMS consists of both hardware-based system and Real Time Computer (RTC) based system. Protection against global faults is achieved by hardware-based system. Where as Protection against local faults like individual Fuel Sub Assembly (FSA) blockages is provided by Real Time Computer System (RTCS). This paper elaborates the importance of RTCS and how the fault-tolerant architecture is arrived with the help of Reliability evaluation.

2. FAULT TOLERANT

To attain high system reliability, methods like Fault-avoidance and Fault-Tolerance are followed. Fault-avoidance is how to prevent occurrence of faults, by good design and careful component screening, design reviews and by following quality control techniques. Fault-tolerance is the extent to which a system will continue to operate at a defined performance level even though one or more of its components are malfunctioning. Fault can be detected by comparing similar system outputs. In a comparison-based system, fault masking is achievable by voting in turn allowing the correct value to pass on. Fault containment is to contain the effects of faults as locally as possible. Among these fault tolerance is required for continued operation of any system. Fault tolerance is achieved by redundancy, i.e. the existence of more than one means for accomplishing a given function. In H/W redundancy, more than one component is used which can vote out erroneous signal or switch on a standby system to replace the failed one. Since RTCS is a safety critical system, the fault tolerant architecture of computers is used. The number of fault tolerant RTCs required is arrived with the help of reliability calculation with respect to the target.

3. CORE TEMPERATURE MONITORING SYSTEM

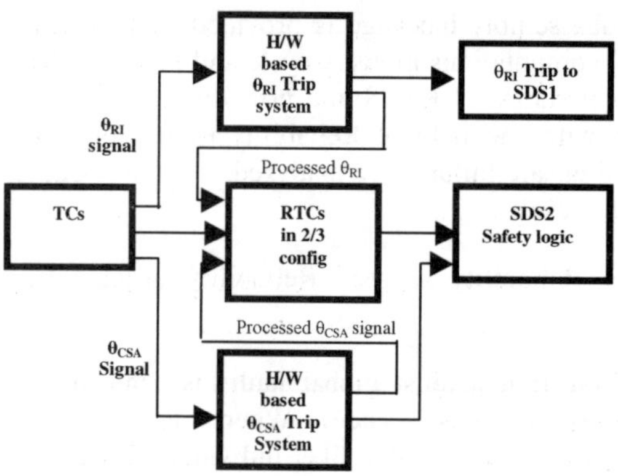

Fig. 1: Core Temperature Monitoring System

CTMS is part of diversified shutdown system (SDS) of PFBR. The primary function of CTMS is to detect flow blockage and coolant boiling [1] in addition to online calculation of the fuel burn up, reactivity balance and thermal balance. The simplified block diagram of CTMS is shown in Figure-1. Along with RTCS, CTMS consists two independent hardware based systems for Reactor Inlet Temperature measurement (H/W based θ_{RI}) and Central Subassembly temperature measurement (H/W based θ_{CSA}). These two systems take care of the global faults. For calculating the trip parameters, RTCS accepts the conditioned output of the hardware-based systems.

4. REAL TIME COMPUTER SYSTEM

RTCS consists of thermocouples, signal conditioners and RTCs in 2/3 fault tolerant configuration. The fault tolerant 2/3 voting logic of RTCS is shown in Figure-2. Except the central FSA, RTC monitors the healthiness of all FSAs outlet temperature with thermocouples (TCs) as sensor. Each FSA is provided with 2 TCs.

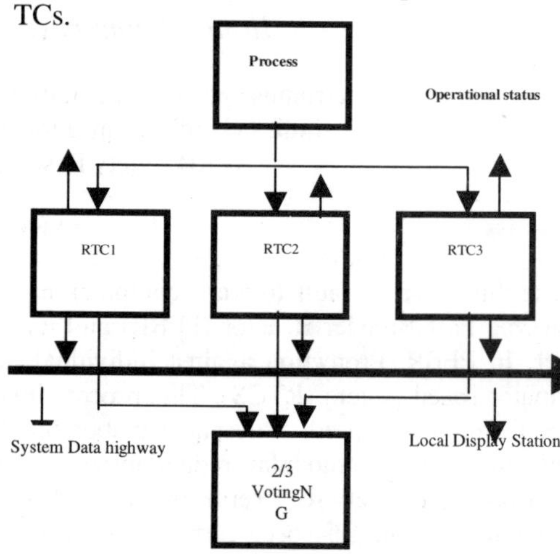

Fig. 2: Real Time Computer System

RTCS calculates Mean core outlet temperature (θ_M), Mean temperature rise across the core ($\Delta\theta_M$) and deviation in individual FSA coolant outlet temperature over the expected value ($\delta\theta_i$). RTCS compares the calculated values of θ_M, $\Delta\theta_M$ and $\delta\theta_i$ against their threshold limits, and generates respective trip or alarm signals. In addition, this system has to send acquired and computed data to main control room & backup control room for operator aid. θ_i is measured by two TCs in each of the FSAs. θ_{RI} is measured by four thermocouples in each of the two primary coolant pump side.

$\theta_M = (1/N) \sum \theta_i$, where

N, represent the number of valid FSA TC signals.

$\Delta\theta_M = (\theta_M - \theta_{RI})$;

$\theta_i = \theta\theta_i - (a_i \, . \, \Delta\theta_M + \theta_{RI})$,

where a_i is the ratio of temperature rise of the ith FSA to the mean temperature rise in the core.

Each RTC consists of subsystems like Central processing unit, analog input, digital input and digital output processing cards, power supply and signal conditioners. The hardware reliability of the RTC has been calculated using the "parts count" method and the values obtained were used to arrive at the optimum configuration.

4.1 RTCS Success Criteria

RTCS is said to function successfully if the system generates a trip signal whenever
i) Average temperature rise across all FSA outlet temperature ($\Delta\theta_M$) exceeds specified threshold.
ii) the difference between actual rise in temperature and expected rise in temperature across any FSA ($\delta\theta_i$) exceeds specified threshold.

5. RELIABILITY ANALYSIS

CTMS is classified as safety critical system and this system must be demonstrated to be safe and reliable with appropriate degree of confidence. Hence theoretical reliability analysis is essential for acceptance of the proposed design. RTCs employed in the safety critical and safety related systems have stringent reliability targets. The probability of failure on demand i.e. probability of unsafe failure must be very low of the order of 10^{-4}/demand (as per AERB guidelines). Simultaneously the probability of safe failure also must be kept low to improve the availability factor.

For a single RTC, the hardware reliability can be predicted and demonstrated where as the software reliability cannot be quantified. It can only be verified and validated to assure performance as per the requirements. Even the Hardware description languages used to program, programmable logic devices also have to be verified and validated.

For the reliability analysis, the basic failure rates of RTCS subsystems like, CPU card, analog input, digital input & digital output processing cards and Signal Conditioner are calculated using parts count method with the help of RELTEC software which is based on Handbook of Reliability Data (HRD-5) of British Telecom, UK. In each board, all the components are assumed in series. This is a conservative approach suitable for systems with very large number of reliable components. At system level reliability analysis is done using reliability block diagram method. In this way single RTC failure rate value is calculated. This value is used for various architecture validations.

6. RELIABILITY TARGET FOR RTCS

The reliability targets for safe and unsafe failures of RTCS are specified separately to meet operational and safety requirements respectively. Since RTCS is part of SDS, the reliability target for RTCS is arrived at after analyzing and allocating reliability to other sub-systems of SDS. The target for unsafe failure probability of RTCS is obtained by subtracting the unreliability of Safety Logic (SL), Power Gates (PG), and Reactor Actuation System (RAS) from the allocated value for SDS2.

The overall target for unsafe failure frequency of SDS stipulated by the regulatory authority is 1E-6/y. The failure frequency of SDS due to local fault, which is of interest, is calculated as follows.

If λ_F denote the flow blockage frequency per reactor year, the flow blockage frequency per FSA is λ_F/N, where N= total the number of FSA. There are N pairs of thermocouples to measure the coolant outlet temperature. Therefore, the failure frequency of SDS due to local fault (λ_{LF}) anywhere in the reactor if the thermocouples are used in 2/2:Success voting is

$\lambda_{LF} = N * \lambda_F/N *(2\lambda_{TC}T)+ \lambda_F*(P_{RTC} +P_{SL} + P_{PG} + P_{RAS})$

i.e.,$\lambda_{LF} = \lambda_F *(2\lambda_{TC}T + P_{RTC} +P_{SL} + P_{PG} + P_{RAS})$

where P stands for failure probability of system denoted in suffix and λ stands for failure rate of the component denoted in suffix.

After evaluating the above equation through fault tree method and based on 1 demand per year [2], the probability of failure of SDS on demand for local FSA faults is 1E-6/de. In PFBR, there are two SDSs. Hence the target for unsafe failure probability of each SDS is 1E-3/de without considering Common Cause Failure (CCF). With a Common cause failure (CCF) of 1%, the unsafe failure probability of individual SDS is 1E-4 /de. A low value of 1% CCF is used because with respect to the local fault event detection the two shut down systems have good amount of diversity.

The SDS with RTCS failure probability is written as,

$P_{SDS} = (P_{TC} + P_{RTCS}) + (P_{RI} + P_{CSA}) + (P_{SL} + P_{PG} + P_{RAS})$

Rearranging the terms,

$(P_{TC} + P_{RTCS}) + (P_{RI} + P_{CSA}) = P_{SDS} - (P_{SL}+P_{PG}+P_{RAS})$

$(P_{TC} + P_{RTCS}) + (P_{RI} + P_{CSA}) =$

$1 \times 10^{-4} - (1x10^{-7} + 1x10^{-6} + 3x10^{-6})$

1×10^{-4}/de

If $P_{TC} = 2*\lambda_{TC}T \approx 1 \times 10^{-5}$ /de and

$(P_{RI} + P_{CSA}) = 2*(\lambda\lambda_{TC}T + P_{HW})$

1×10^{-5} /de,

then the target for RTCS failure probability on demand is 1×10^{-4} /de. If CCF is ignored the target is $1x10^{-3}$ /de.

Here, β = Contribution of CCF (in terms of % of failure rate of the system).

For safe failures, the spurious trip per year for each SDS is 0.5. Of this 0.2 /y is allocated for SL, PG and RAS. The remaining 0.3/y is divided among the two hardware based systems (H/W based θ_{RI} and H/W based θ_{CSA}) and Delayed Neutron Detection system (DND) and RTCS. The spurious trip allowed for RTCS and sensors is 0.15/y, which includes all FSA TCs and RTCs. Allocating 0.05/y for TCs, the safe failure frequency permitted for RTCS is 0.1/y.

7. RELIABILITY OF RTCS IN VARIOUS CONFIGURATIONS

Four different configurations, namely hot stand by system, dual system architecture with 1/2 voting logic & 2/2 voting logic and triplicated system architecture with 2/3 voting logic are studied and compared for reliability.

7.1 Assumptions and Limitations

All hardware components of RTCS have been assumed to be following constant failure rate model. This is justified due to the fact that stringent screening procedures have been specified for the systems to eliminate early failures. In some of the RTCS subsystems, PLDs have been used. For programming PLDs hardware description languages are used. These HDL codes have to be verified and validated (V & V). For completeness of hardware reliability, this is under process. The procedure for V & V are presented in other paper (Reliability analysis of Diverse Safety Logic Systems for FBR.......by Bh. RaviKumar et all).

Software component of computer-based systems is not amenable to quantitative assessment of reliability. Hence assessment of software in the computer-based systems has to be based on evidence that the software is correct with respect to specifications, and meets all the requirements. The various steps of software life cycle are having defined IEEE guidelines[4]. Like System Requirements Specifications (IEEE 1233), Software Requirements Specifications (IEEE 830), Software Design Description (IEEE 1016), Software Test Documentation (IEEE 829), Software Configuration Management (IEEE 1042), Software_Quality Assurance Plan (IEEE 730) and V&V (IEEE 1012) are all followed for software development. All the steps are followed based on AERB/SG/D25.

Hence for analysis of failure rate, software module can be assumed to be free of bugs. Human error in configuration/maintenance of hardware and software is not modeled

7.2 Common Cause Failures

Table 1 lists the independent features built into the design of the system to take care of most of the CCFs. CCF is modeled using β factor model [3]. In this model, CCF rate λ_c is defined as $\lambda_c = \lambda_\tau \beta$, where, λ_τ is the total failure rate and β is a small fraction to account for any dependence between components. Based on the redundancy and diversity measures provided, a β of 5 % is used in the analysis.

Table 1 Independent measures between RTCs

Independence Measure	RTC Hardware	RTC Software	Signal Conditioner
Physical Separation	Yes	Yes	Yes
Electrical Isolation	Yes	NA	Yes
Component Technology	No	No	No
Design Team	No	No	No
EMI & EMC	Yes	NA	Yes
Maintenance and Testing	No	No	No

7.3 Results

The results for the different configuration studied are given in Table 2. Although hot standby configuration meets the target for safe failures it falls short with respect to unsafe failures. Hot standby system reduces the spurious trips arising from RTC, compared to 1/2 configuration, by switching over to the second system. However unsafe failure probability of hot standby system is comparable to that of 2/2 configuration, as unsafe failures may not lead to a switch over to RTC2, even if it were to work successfully. As expected, for the 1/2 voting dual system architecture spurious trips are more. The triple modular redundancy (i.e 2/3 voting configuration) satisfies both safe and unsafe reliability targets as shown in the last column of Table 2.

Table 2 Comparison of System Reliability for Different Options

Config	With CCF		Without CCF	
	Safe Failure Freq.(/y) Target 1E-1 /y	Unsafe Failure Prob. (/de) Target 1E-4 /de	Safe Failure Freq. (/y) Target 1E-1/y	Unsafe Failure Prob. (/de) Target 1E-3 /de
Hot-Standby	9.1E-3	1.3E-3	5.6E-3	1.3E-3
1/2	1.5E-1	5.6E-5	1.5E-1	1.2E-6
2/2	3.9E-3	2.2E-3	1.5E-3	2.2E-3
2/3	**4.2E-3**	**5.9E-5**	**4.5E-4**	**3.6E-6**

The failure rate value of one RTC is arrived by taking the individual failure rate values of analog input, digital input, digital output and CPU cards, which are designed indigenously and used already in various project sites like RAPS-II, Isotope separation plant etc. These cards were designed using commercial grade components for which probability of failure is high when compared to industrial and military grade components. By using industrial grade components the reliability can still be improved.

8. CONCLUSION

The reliability target for fault tolerant RTCS for safe and unsafe failure is arrived based on reliability allocation to various subsystems of SDS2. Four different options were studied with respect to safe failure rates and unsafe failure probability. The Triple modular redundancy configuration has a safe failure rate of 4.2E-3/y and unsafe failure probability on demand of 5.9E-5/de (with CCF), which meets both the safety and operational reliability targets. CCF is included in arriving at the target as well as reliability evaluation.

9. REFERENCES

1. Vijakumaran P.M et al, Conceptual design of Core Temperature Monitoring System, PFBR/ 63110 / DN / 1000.
2. Bhuvana V, John Arul A, Sasidhar Rao B, Om Pal Singh, Reliability allocation for Core Temperature Monitoring System.
3. Mosleh, A, Felming, K.N., et al, 1988, Procedures for treating Common Cause Failures in Safety and Reliability studies, NUREG/CR-4780, EPRI NP-5613, PLG-0547.
4. S. Ilango Sambasivan, Real Time Computers for Safety Critical and Safety Related Systems of Prototype Fast Breeder Reactor, SACI 2005

10. BIBLIOGRAPHY

1. Reliability Prediction of Electronic Equipment, MIL-HDBK-217
2. Design Safety Guide on Computer based systems, AERB/SG-D25

Reliability, Safety and Hazard
Editors: P.V. Varde et al.

DESIGN OF FPGA BASED SAFETY CRITICAL SYSTEM, ITS TESTING AND RELIABILITY ASSESSMENT

Nilanjan Das

Interdisciplinary Program in Reliability Engineering, IIT Bombay, Mumbai 400 076

ABSTRACT

With the importance of FPGAs increasing in the electronics field, issues relating to their integrity and reliability are becoming more important by the day. A feasibility study of an FPGA based safety critical system was carried out and compared with the performance of the present relay based logic. The failure rates of the two systems are compared and a formal conclusion was drawn based on the results obtained.

1. INTRODUCTION

In today's world FPGAs play an important role in our lives as well as the technology market. With growing demands for these programmable devices, FPGAs are becoming more and more complex and are being used in varied fields from simple daily life gadgets to highly complex and critical systems. These devices being cheaper than the ASICs and also being more flexible than ASICs have drawn a lot of interest. The ease of reconfiguring the FPGAs has made it an attractive substitute to ASICs which lack this flexibility. With the field of usage and complexity of FPGAs increasing, various issues regarding the reliability of these devices have aroused. This aspect is becoming more important as FPGAs are being regularly used as a part of the safety critical systems. With the improvement of FPGA technology the fields in which they are being implemented is expanding rapidly. But one of the major issues related to FPGA reliability, especially the SRAM based ones is their susceptibility to failure, namely, high temperature, neutron induced failures, high energy radiation, time dependent dielectric breakdown, etc. Of these the cause of failure due to high energy neutrons is the most damaging. But the good news is that with improvement of FPGA technology these failure mechanisms are rapidly being overcome and they are being utilized in many fields where reliability is one of the major constraints. This paper does a formal study of the feasibility of implementing the interlock logic of the AC/Ventilation system of a safety critical system on an FPGA. The present interlock logic is relay based and a comparative study of the two systems was performed to draw the conclusions.

2. FPGA FAILURE MECHANISMS

Semiconductor technology continues to evolve at an aggressive rate. The main elements of this evolution focus on continual miniaturization and higher integration. The failure [3] modes are also becoming more complex. As designers are increasingly turning to FPGAs for highly integrated next generation products, there are a number of potential failure mechanisms impacting overall device reliability that need to be understood. Some of the common failure modes are briefly discussed in the following section.

Temperature

As plastic encapsulated semiconductor devices, the reliability problems most frequently encountered by FPGAs are typically due to one of four root causes: the packaging technology, assembly technology, environmental overstress, or ESD. These failure mechanisms are all accelerated by exposure to high temperatures. Temperature is the single dominant stress factor affecting the reliability of electronic devices. With the potential to create failures at the device, package, and board level, it is critical to allow for margin at each level of the automotive system.

Table 1 FPGAs and their Recommended Operating Conditions [2].

Supplier	Product Family	Specified Automotive Temp. Range	Maximum Junction Temp.
Actel	SX-A, eX, MX	-40°C to + 125°C (Amb)	+ 150°C
Altera	Cyclone, ACEX 1K	-40°C to + 125°C (Jnc.)	+ 125°C
Xilinx	Spartan XI, II, IIE	-40°C to + 125°C (Jnc.)	+ 125°C

Neutron Induced Errors

Most neutrons striking an integrated circuit will pass through without interacting with the circuit, but some will pass close enough to a silicon or dopant atom to disturb the atom. The result of this interaction is the creation of secondary particles, which in turn create a trail of electron-hole pairs up to several tens of microns in length. If this occurs near the junction of a reverse-biased device in a memory cell or flip flop, then a voltage spike can occur, resulting in the memory cell or flip-flop changing state. This change of state is frequently referred to as a **Single-Event Upset, (SEU)**.

Consequences of Neutron Impact

Single-Event Upset: May affect the FPGA in two ways. They may alter the memory elements the design embeds or may alter the content of the memory storing the device configuration information.

Total Ionizing Dose: This is the long term ionizing damage to a semiconductor device caused by high energy protons and electrons.

Single-Event Latchup: This is a potentially destructive condition in which a single charged particle induces latchup within a CMOS device. With enough energy, a charged particle may trigger the parasitic npn-pnp circuit found within CMOS circuits. Once in latchup, high currents will flow through the parasitic bi-polar transistors and destroy the device.

Time Dependent Dielectric Breakdown (TDDB)

Even at normal voltages, continuously applying stress to the gate oxide film of an FPGA will eventually produce insulating film breakdown. This breakdown of the insulating film over time is called time-dependent dielectric breakdown.

3. RELIABILITY ANALYSIS METHODS

FPGAs are being regularly used as a part of the safety critical systems, e.g., in aircraft controllers, in control systems of nuclear reactors, as a part of the electronic circuitry in space crafts, etc. Thus there is a growing need for the proper assessment of the reliability of these devices. There are two methods of reliability analysis of electronic components as proposed by the U.S. Department of Defense, namely, Parts Count Method, and Parts Stress Method. These methods were defined by the U.S. Military and the complete reference is compiled into a single handbook the MIL-HDBK-217F [4]. This handbook is updated from time to time in accordance with the improving technology.

In this paper the Parts Stress method is applied to calculate the failure rate of the FPGA used in the study. This method requires detailed information of the ambient environment and the different stress levels to which the circuit or equipment will be subjected to during its normal operational life and is applied at the later stages of the design when a detailed knowledge about the equipments operational environmental conditions is known. The failure rate is calculated as:

$$\lambda_p = \sum_{i=1}^{i=n} N_i \lambda_i \quad (1)$$

where:

λ_p = total equipment failure rate.

λ_i = generic failure rate of the ith generic part.

N_i = quantity of ith generic part.

n = number of different generic part categories in the equipment.

The failure rate of the FPGA was calculated by referring to the MIL-HDBK-217F. There are two sections which cater to this category of electronic devices, the section 5.3, which caters to devices with gate counts of > 60,000, and section 5.1, which caters to gate counts > 60,000. The FPGA considered here has 200,000 gates.

Failure Rate of FPGA Using Section 5.3

The FPGA used here is the XC2S200 PQFP-208 0.22/0.18 μm Si Gate CMOS device manufactured by Xilinx. It has a Gate capacity of 200,000 and has a Plastic Quad-Flat Pack with 208 pins. The failure rate is given by:

$$\lambda_p = \left(\lambda_b \pi_t \pi_{mfg} \pi_{cd} + \lambda_{bp} \pi_e \pi_q \pi_{pt} + \lambda_{eos}\right) \times 10^{-6} \quad (2)$$

where:

λ_p = part failure rate.

λ_{eos} = electrical overstress failure rate.

λ_b = die base failure rate.

λ_{bp} = package base failure rate.

π_t = temperature factor.

π_{mfg} = manufacturing process corr. factor.

π_{cd} = die complexity corr. factor.

π_e = environment factor.

π_q = quality factor.

π_{pt} = package type corr. factor.

Substituting the appropriate values for each of the parameters the failure rate of the device was obtained as: **λ_p= 8.6977 x 10^{-6}.**

The above result was also verified with the *Reliability Data and Information Management System (REDIMS)* software developed at BARC, Mumbai for the better collection and estimation of data from the operational field. The failure rate of the FPGA calculated is

λ_p = 8.69598 x 10^{-6}.

Failure Rate of FPGA Using Section 5.1

The section 5.1 of the MIL-HDBK-217F caters to devices with gate counts > 60,000. The failure rate was calculated up to 60,000 gates and then using MATLAB it was extrapolated up to 250,000 gates. The extrapolation was done using 2nd order curve fitting. The plot below shows the extrapolated data.

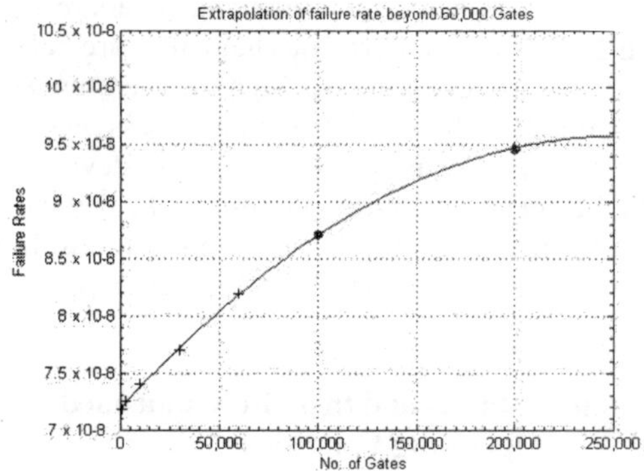

Fig. 1: Failure rate for 200,000 gates using 2nd order fit.

On the plot is shown the failure rates for 100,000 and 200,000 gate devices as two dark circles.

Analysis of data from the Xilinx Reliability Report

The data obtained from the Xilinx Reliability report was analyzed based on the two relations given below.

$$A_T = \exp\left[\frac{E_a}{k}\left(\frac{1}{T_O} - \frac{1}{T_H}\right)\right] \quad (3)$$

where

A_T = Temp. acceleration factor;

E_a = Activation energy (assumed = 0.7eV);

k = Boltzmann const;

T_o = Operating temp. in °Kelvin;

T_H = High test temp. in °Kelvin;

Using this acceleration factor the failure rate of the FPGAs can be found as:

$$\lambda_f = \frac{\chi^2 \times 10^9}{2 \times N_d N_h A_T} \quad (4)$$

where:

χ^2 = Chi-squared value at a desired confidence level and (2f +2) degrees of freedom, and f is the no. of failures.

N_d = No. of devices under test.

N_h = No. of device hours of testing.

λ_f = failure rate in FIT.

The data given by Xilinx are interpreted at a confidence level of 60%. The failure rates of the devices were hand calculated at confidence levels of 90% and 50%. The table below shows the FIT's at different levels of confidence and different junction temperatures.

Table 2 Comparison of Xilinx data with hand calculation

Device type	No. of failures	Xilinx FIT @ 60% C.L.	Calculation of FIT			
			$T_j = 55°C$		$T_j = 30°C$	
			@90% C.L.	@50% C.L.	@90% C.L.	@50% C.L.
XC2SXXX	1	15	29	13	4	2

The table below, Table 3, shows the comparison of the failure rates of the three methods of FIT calculation for the FPGA at different junction temperatures.

Table 3 Comparison of Xilinx data with MIL-HDBK-217F data

T_j	Failure rates		
	XILINX data	Section 5.1	Section 5.3
25°C	3×10^{-9}	8.6841×10^{-8}	7.6841×10^{-6}
30°C	4×10^{-9}	9.4743×10^{-8}	8.6977×10^{-6}
35°C	6×10^{-9}	10.739×10^{-8}	9.739×10^{-6}
40°C	9×10^{-9}	12.477×10^{-8}	11.739×10^{-6}
45°C	13×10^{-9}	14.689×10^{-8}	13.739×10^{-6}

4. THE SAFETY CRITICAL SYSTEM

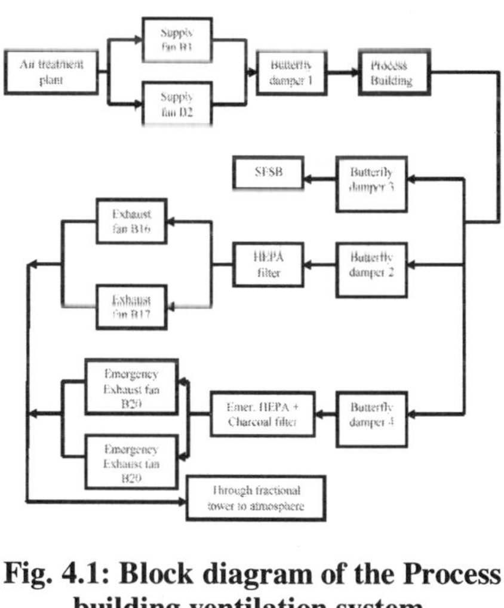

Fig. 4.1: Block diagram of the Process building ventilation system.

The interlock logic of the A/C and Ventilation system that was modeled on an FPGA was a very safety critical system and required a highly reliable operation to meet the necessary system specifications. This interlock logic is at present implemented using mechanical relays. The simplified block diagram of the system is given below in **Fig. 4**.

General Description of the A/C and Ventilation System

The A/C and Ventilation system consists of two supply fans, two exhaust fans and two emergency exhaust fans for redundancy. Of these only one supply fan and one exhaust fan operates under normal conditions. The main aim of the A/C and Ventilation system is to maintain the pressure and the chemical composition of the air inside the process building within certain specified levels. When under normal conditions the operating fan fails the other fan automatically turns on after a certain delay. The emergency exhaust turns on automatically if the pressure inside the process building or the chemical composition of the air increases beyond a limit and when normal conditions return they turn off automatically. The fans can also be turned on and off manually by the operator. The main controlling component of the Ventilation system is the Butterfly dampers. They control the air flow rate and if at any time one of them closes the whole ventilation system shuts down.

The programming was done in VHDL and the functionality was checked using the Symphony simulation software *VHDL SIMILI 2.3*.

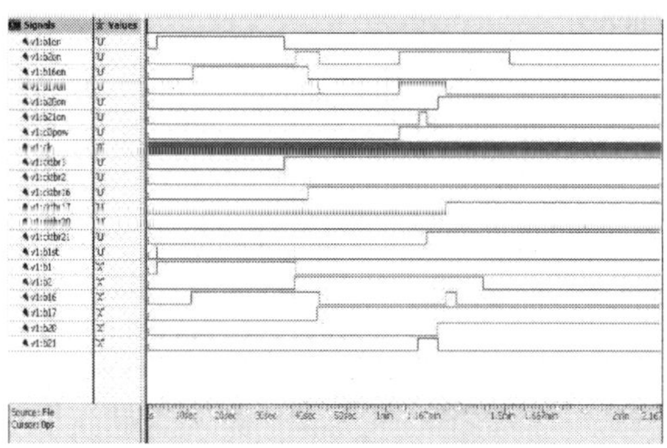

Fig. 5: Simulation result of the Ventilation System.

The design was then synthesized using the *Synplifypro v7.3.1* and the synthesized file was then again simulated for functional conformity using *VHDL SIMILI 2.3*.

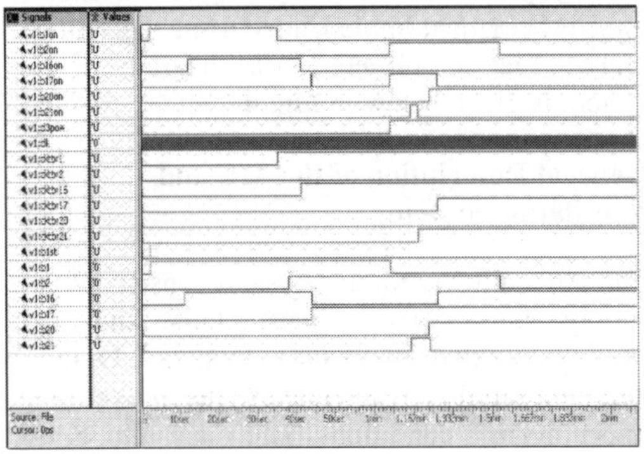

Fig. 6: Post Synthesis Simulation result of the Ventilation System.

The results obtained were verified using *Modelsim6.1* simulator software. The results matched each other.

5. TEST PLAN

The test carried out on the FPGA configured with the interlock logic was planned in accordance with the test plan implemented by Xilinx. The testing was done under accelerated conditions of temperature. The operating temperature of the FPGA device was considered to be 25°C. The model used for the accelerated testing was the Arrhenius [5] model, given by:

$$\tau = A \times \exp\left(\frac{E_a}{kT}\right) \quad (5)$$

where:

τ = mean life time;

E_a = Activation energy (assumed = 0.7eV);

A = proportionality const.;

k = Boltzmann const.(8.617x 10^{-5});

T = temperature in Kelvin;

From this relationship and using equations 3 & 4 the failure rate of the FPGA was found out from the experiment carried out. Since there was only one very costly component, the acceleration factors employed were not very high. The experiment was carried out at environmental test setup at BARC, Mumbai. Three temperature stress levels were employed for different time durations. The results of the experiment are tabulated below.

Table 4 Test Results

Test Temp	Hours of testing	Acceleration factor	Equivalent device Hrs. at 25°C	No. of failures
30°C	72	1.57	113.04	0
35°C	72	2.42	174.24	0
40°C	24	3.71	89.04	0
Total device hours at 25°C = 376.32 Hrs.				

The failure rate of the FPGA device based on the experimental results at 95% confidence level is:

λ_f = **8.348 x 10^{-4} /hr.**

6. FAULT TREE ANALYSIS OF THE RELAY BASED SYSTEM

The fault tree analysis for the existing relay based interlock logic was performed with the failure rate for each relay contact taken from the handbook of components for safety operations. Since field data was not available, generic data was considered. The fault tree analysis was carried out using the *ISOGRAPH software* ***FaultTree+ V10.1***. The values for the failure rate considered for each relay contact was 10^{-4} /demand, the repair time was considered as 3hrs., and the failure mode was considered as relay contact not remaining in position. Conservative data was considered as actual field data was not available. The overall failure rate of the system was generated by the software.

7. RESULTS

1. The failure rate obtained from the Xilinx reliability report at a confidence level of 90% at 25°C is **3 x 10^{-9} per hr**.
2. The failure rate obtained from MIL-HDBK-217F section 5.1 at 25°C is **8.6841x10^{-8}/hr**.
3. The failure rate obtained from MIL-HDBK-217F section 5.3 at 25°C is **7.6841 x 10^{-6}/hr**.
4. The failure rate obtained from the fault tree analysis of the present relay based interlock logic is **8.07x10^{-4} per hr**.
5. The failure rate obtained from the experimental results is **8.348 x 10^{-4} /hr**.

6. It is seen that the result obtained from the experiment is almost equal to the present failure rates, in spite of there being no failures observed during the experiment. This is due to the fact that there was only one device under test. So the device-hours was low and also the acceleration factors were low, leading to a high failure rate computation.

8. CONCLUSION

The study was conducted to do a feasibility study of the FPGA implementation of the relay based interlock logic of a safety critical system with the aim to improve its reliability. It can be concluded from the results obtained that even after considering the most conservative method, the MIL-HDBK-217F section 5.3, the failure rate of the FPGA based system is far better, as far as reliability is concerned, than the present relay based interlock system. There is an improvement in the failure rate by a factor of 100. A better judgment can be made regarding this if the experiment is carried out on a larger sample size at higher acceleration factors. Then the result could also be experimentally verified. Also from the field data obtained from the Xilinx Reliability Report it can be seen that there is a vast improvement in the failure rate. So, from the results obtained, it can be concluded with confidence that the FPGA implementation of the interlock logic is feasible.

9. ACKNOWLEDGEMENTS

The author wishes to thank Prof. (Mrs.) A. Srividya, IIT Bombay, and Dr. P.V. Varde of B.A.R.C. Mumbai for their invaluable support and suggestions.

A special note of thanks to Prof. A. K. Verma, for the invaluable suggestions given during the experiment.

10. REFERENCES

1. Stephen Brown and Jonathan Rose "Architecture of FPGAs and CPLDs: A Tutorial" *Department of Electrical and Computer Engineering University of Toronto.*
2. Jonathan Green, Esmat Hamdy, and Sam Beal "Antifuse Field Programmable Gate Arrays," *Proceedings of the IEEE*, Vol.81, No.7, 1993.
3. Michael Wirthlin, Eric Jhonson and Nathan Rollins, "The Reliability of FPGA Circuit Designs in the Presence of Radiation Induced Configuration Upsets," *Proceedings of the 11th Annual IEEE Symposium on Field-Programmable Custom Computing Machines (FCCM'03)*, 2003.
4. U.S. Department of Defense, 1991 "*MILITARY HANDBOOK MIL-HDBK-217F,"* Washington D.C.
5. W. Nelson 1990 "*Accelerated Testing-Statistical Models, Test Plans and Data Analysis,"* New York: John Wiley and Sons.
6. J .J. Wang, W. Wong, S. Wolday, B. Cronquist, J. McCollum, R. Katz and I. Kleyner, "Single Event Upset and Hardening in 0.15μm Antifuse-Based FPGA," *IEEE Transactions on Nuclear Science,* Vol. 6, No. 6, Dec. 2003.

Reliability, Safety and Hazard
Editors: P.V. Varde et al.

ANALYSIS AND MANAGEMENT OF RISK FOR DISTRIBUTION SYSTEM RELIABILITY

Ashok M. Salsingikar

Reliance Energy Ltd., Mumbai

ABSTRACT

Electric power systems in India have expanded immensely during the past few decades. However, the progress has been quantitative and not qualitative. The frequency and duration of outages is very high. For ensuring reliable and continuous power supply at consumer premises, the distribution systems have to be designed and operated keeping reliability aspects in view.

Keywords: Reliability indices (SAIDI and SAIFI), Outage response, Outage isolation

1. INTRODUCTION

Electricity undertakings are under a constant pressure from the State Energy Regulatory Commissions and the threat of privatisation for rendering their operation more efficient, fair structuring of tariffs, and improving the quality of service to the consumers.

The distribution system of an electricity supply undertaking is the target for criticism by consumers who consider the electric supply either poor or unreliable. Quality and reliability are recognized as being essential to the commercial success of a supply undertaking. The quality in electricity distribution requires the consumer's voltage to be maintained as near to the declared values as possible.

Considerable improvements in reliability are possible by simplifying the design. The design of the protective scheme will have a considerable effect on reliability. Its primary function is to guard the supply from the effects of faults on the system it controls and, although this object must be the first consideration, it should not seriously reduce the reliability of its own local system by light settings and consequent unstable operation.

Risk analysis regarding frequency and duration of interruptions for individual customers can be readily applied for actual Distribution Systems based on network attributes, on the utility's general network protection practices and on current system average reliability indices.

The paper also shows how existing models for investment planning can be extended to consider individual customer reliability indices. It outlines a model for the analysis and management of Risk for Distribution System Reliability.

Customers' perception of Risk associated with the safety, financial loss and other factors depicted in the model provide a way of valuing service and determining service reliability targets for consumers.

The paper addresses the issues of assessment of reliability indices for electric power distribution utility and provides a methodology for the analysis and management of risk for distribution system reliability.

2. TRADITIONAL RELIABILITY INDICES

Traditionally reliability indices are basically monitored on average figures i.e. the duration and frequency of yearly interruption for an average customer of the system. However, in view of threatened privatisation of utilities, reliability indices have to be evaluated by considering each individual customer so that utilities are subject to penalties when specific reliability targets are violated.

The paper describes a comprehensive approach for providing improved service reliability. In addition to considering the actual system performance for service reliability, customer perception of quality of service and public opinion regarding outage frequency and response has become important factors as well. Finally safety and business risks are being addressed to come up with the best business solution to meet customer and public expectations.

For the purpose of this paper, risk will be broken into two components. The first is public and employee safety, which would include normal and emergency operation of the system during any catastrophic situation. The second risk is business risk associated with the financial investments and customer's perception of risk related to service reliability performance.

The elements, which must be considered as a total system to develop design that will produce the desired result for the customer, are failure modes, outage isolation, response to customer during an outage and customer expectations.

One of the more common ways to measure service reliability on a distribution system is to monitor System Average Interruption Duration Index (SAIDI) and System Average Interruption Frequency Index (SAIFI). There are many other ways to measure service reliability that are used less frequently [1]. Although these methods are adequate measures of system trends they are too broad to be useful to the engineer who is designing the system, to meet very specific customer expectations. Measure that take into account customer specific needs, either on a density or targeted basis, are more appropriate for targeting investment in the system to reduce overall costs while satisfying customer retention requirements and meeting specific customer expectations.

3. RISK ASSESSMENT CONCEPTS

Risk management, as it applies to service reliability in this paper, includes both safety and business risks. The two types of risk have some common characteristics, which are: a probability of a negative even occurring, the consequences of the event, and the individual perception based on a combination of factors. The factor includes things such as: previous event experiences, controllability of the event, catastrophic nature of the event, and prior communications about the possibility of the event and its management. The linkage between the negative event and the consequences also heavily influence the individual's perception of the risk and its acceptance [2].

By anticipating the perception of risk, the distribution system designer can factor in the unique expected customer response to safety and service reliability performance. There are various approaches to describing customer perception of risk and system performance, which will be described in subsequent sections of this paper.

4. SYSTEM RELIABILITY DESIGN

System Reliability Design must include element to prevent outages and to have a system configuration that provides redundancies and rapid restoration capabilities. In special cases network service and automatic transfer capability can be employed to provide superior service reliability. The total number of outages that will occur is defined by the following relationship.

Outages = (Σ Component failures + Σ Exposure events)

An integrated system design should include four main approaches: component failure prevention, equipment exposure mitigation, fault isolation, and outage response capability. The four approaches have their corresponding cost-to-benefit ratios with respect to each of the service reliability performance objectives. The engineer must exercise judgment to keep the various approaches in balance and must implement the solutions in the proper sequence to achieve the optimum service reliability improvement. Figure 1 illustrates the integration for a complex situation.

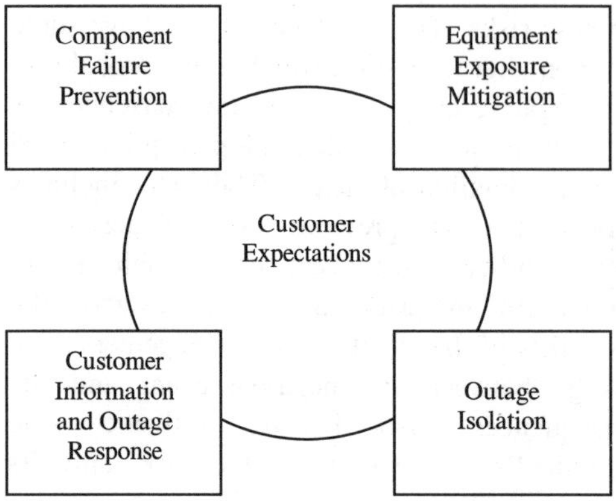

Fig. 1: Integrated Service Reliability

Component failure prevention requires an active program to record not only the outages that occur but also the component failure that precipitated the failure. By thoroughly analysing and recording each failure, it is possible to develop a strategy for the initial selection of distribution components, predicting potential failures in advance, and replacing components prior to their failure in some cases. Obviously, one can only reduce the failure rate and the design must recognize some failures will occur.

Equipment exposure mitigation is primarily accomplished through tree trimming and insulator washing programs, although, animal guards, tree wire and equipment protection from cars is also employed.

The impact of the outages on customers is a function of the duration and the isolation of the outage by system protection design.

Once the fault is isolated, it is critically important to the customer to find out as soon as possible when service will be restored. The utility is currently deploying outage detectors in homes and linking that information into an outage response system that permits the operators to dispatch crews more effectively and to provide more timely information to the customers regarding the outage.

5. RISK ASSESSMENT MODEL

Traditionally utilities have taken a supply side view of service reliability performance. They have decided what level of service should he provided to customers based on experience with little input from the customer except for complaints to the utility about service reliability performance. This traditional approach has been challenged more recently due to rising customer expectations and the more unique service requirements resulting from home offices and the power quality demands of modern commercial business.

The risk assessment model that can be used by the utility is designed to strike a balance between customer expectations and perceived performance of the system. This is described in the following relationship.

Customer expectations = Customer perceptions

Customer expectations are a function of service experience and the customer's assessment of risk. Customer perceptions of service reliability are a function of actual system performance, sensitivity level driven by recent experience, and the types of customer use of the electric service.

Customers consider a variety of factors when determining their tolerance for risk and what level of service is acceptable for their unique situation. Of course, personal safety is a top priority for them and for the utility.

Financial implications are particularly important to commercial customers and the risk of losing business revenues weighs heavily in the assessment of service expectations.

Customers also have unique views related to risk associated with loss of security, loss of food, and simply loss of control of the outage event.

5.1 Customer Expectations

Customers are different in their expectations about service quality and service response and information. Electric utility serves a wide variation of customer segments. Each of these customer segments has different expectations based on their business or personal requirements. Their expectations are also influenced by previous service experience from the utility and previous experience from other service providers, such as banks and telephone companies.

5.2 Customer Segments

The segments that are identified for determination of their expectations with respect to electric service reliability included: high technology, office buildings, food processing, retail, as well as, the traditional customer categories of industrial, commercial, agricultural, and residential.

5.3 Customer Density Segments

Since customer segments are rarely grouped so that the system can be designed for a specific segment, a practical approach is needed to target electric distribution service improvements. A good proxy for segment expectations and customer perceived risk turned out to be customer density. The densities that are chosen best reflect customer segments with similar expectations. They are shown in table 1.

Table 1 Customer Density Segments

Density Segment	Population per sq. km	Customers	% of total customers
Low	0-25	950,000	20 %
Medium	25-1350	1,950,000	42 %
High	1350+	1,775,000	38 %
Total		**4,675,000**	**100 %**

In addition to the segments identified in Table 1, it is decided to target specific customers with unique solutions in special cases.

6. SEGMENTED RELIABILITY TARGETS

The service reliability attributes considered are number of sustained outages, number of momentary outages, and duration of outages. Other factors being considered are alarm time, customer information time, and restoration time.

Table 2 Reliability Targets by Customer Density

Density	Population per sq. km	SAIFI
Low	0 25	8
Medium	25-1350	2
High	1350+	1

Customer Average Interruption Duration Index (CAIDI), along with SAIDI and SAIFI, are used to establish targets for different customer density segments that reflected the expectations of the customers in those segments. The targets are based on the judgment of experienced engineers. Table 2 shows the different targets.

By targeting in this way, it is expected to result in customers' service reliability expectations being met more frequently without having to spend the extra amount to bring all customers up to an artificially high performance level. As time goes on, the targets will he adjusted based on actual customer feedback, primarily through surveys. Further customer feedback analysis is required to provide targets for SAIDI and CAIDI.

6.1 Comparison to Traditional Methods

Traditional methods of service reliability improvement focus on the worst performing circuits and on customers with the most complaints. This reactive approach to service reliability is rarely an optimum solution from a customer or a utility standpoint.

The traditional approach tends to over design for the less sensitive customers and under design for those customers that assess the risk to be high for loss of business.

The segmented approach is designed to approximate equal perceived benefit-to-cost for all customers. That is to say customers with the highest expectations will have the greatest perceived benefit to service reliability when improvements are made.

Table 3 compares the outcomes of the two approaches for one attribute, SAIFI.

Table 3 Traditional Versus Segmented

Density	Traditional SAIFI	Segmented SAIFI
Low	5	8
Medium	5	2
High	4	1

It is evident from Table 3 that, although the system average performance is the same, the ability to target service reliability improvements to the high-density customers will significantly improve service to the customers with the highest expectations.

7. CONCLUSIONS

The paper presented the Risk Assessment Model, which is useful for establishing customer expectations. Customer's perception of risk associated with safety, financial loss among other factors provides a way of valuing service and determining service reliability targets for the specific customer segments.

8. REFERENCES

1. Warren C (1995), *Distribution Reliability - What is it?*, IEEE Conference Paper, 95-132, 132-3.
2. Royal Society (1992), *Risk*, Royal Society, London, England, p. 101.

Reliability, Safety and Hazard
Editors: P.V. Varde et al.

RISK CONSIDERATIONS IN THE IAEA's SAFETY STANDARDS AND PSA-RELATED DEVELOPMENTS

M. El-Shanawany and I. Kouzmina

International Atomic Energy Agency (IAEA), Wagramer Strasse 5,
P.O.Box 100, Vienna, Austria, A-1400

ABSTRACT

Probabilistic safety assessment (PSA) of a nuclear power plant (NPP) that is complementing the traditional deterministic analysis is widely recognized as a comprehensive, structured approach to identifying accident scenarios and deriving numerical estimates of risks dealing with NPP operation and associated plant vulnerabilities. Increasingly, during the last years, PSA has been broadly applied to support numerous applications and risk-informed decisions on various operational and regulatory matters. The expanded use of PSA in the risk-informed decision-making process requires that PSA possess certain features to ensure its technical consistency and quality. The issue of PSA quality is in the focus of current activities of the International Atomic Energy Agency (IAEA). The paper draws on risk considerations in the IAEA Safety Standards and the IAEA's work in the area of PSA quality and highlights the achievements to date and plans for the future.

1. INTRODUCTION

Consideration of risks associated with nuclear activities and requirements on risk control are included in high-level IAEA documents, e.g. Refs. [1]-[4]. Very often these requirements, while characterizing the potential of nuclear accidents, are expressed in qualitative terms like "very low probability" or "extremely low likelihood" of a significant accident. The basic principles of the traditional deterministic approach to nuclear safety (e.g. providing for defence-in-depth, compliance with the single failure criterion, provision of adequate safety margins, preventing common cause failure, and limiting the claims made on the plant operators) are aimed to ensure that nuclear accidents are very unlikely. However, the qualitative criteria being applied to the risk characterization in the deterministic safety analyses can only be reliably confirmed if the risk is quantified and the derived estimates demonstrate that the risk is indeed below the established acceptance criteria.

A tool that allows assessing and quantifying the risk is PSA. Today, in addition to the results of the traditional deterministic analysis, PSA insights are widely used in the decision-making process to assess the NPP safety level and to support various risk-informed applications such as risk-informed changes to technical specifications, risk-based plant configuration control (Risk Monitor), maintenance program optimization, etc. Regulatory bodies in many countries require that a PSA be performed for NPP licensing purposes. PSA has reached the point where, if performed to acceptable standards, it can notably influence the design and operation of NPPs.

2. IAEA SAFETY STANDARDS AND RISK CONSIDERATIONS

Since its creation in 1957, in accordance with the Statute [5], the IAEA has been establishing safety standards to protect health and minimize danger to life and property and to provide for the application of these standards to peaceful nuclear activities.

2.1. Status and Structure of IAEA's Safety Standards

IAEA Safety Standards are being published in the Safety Standards Series. They formulate what should be done in relation to various activities aimed at providing nuclear safety rather than how it should be done.

The IAEA's Safety Standards are not legally binding on Member States, but may be adopted by them, at their own discretion, for use in national regulations in respect of their own activities. However, the Safety Standards are binding on the IAEA for application in relation to its own operations and on States in relation to operations assisted by the IAEA.

The Safety Standards Series addresses the following five application areas[1]:

Safety of nuclear facilities	- NS
Radiation protection and safety of radiation sources	- RS
Safe management of radioactive waste	- WS
Safe transport of radioactive material	- TS
General safety (cross-cutting themes).	- GS

Some of the IAEA's Safety Standards address matters that are relevant to safety in a range of different facilities and activities (thematic areas); others address different aspects of safety for a specific type of facility or activity.

Regulating nuclear and radiation safety is a national responsibility, and many Member States have decided to adopt the IAEA's Safety Standards for use in their national regulations. For the contracting parties to the various international safety conventions, IAEA standards have been recognized as providing a consistent, reliable means of ensuring the effective fulfilment of obligations under the conventions. The standards are also applied by designers, manufacturers and operators around the world to enhance nuclear and radiation safety.

The IAEA Safety Standards represent a three-level hierarchy as shown in Figure 1. The IAEA Safety Fundamentals present basic objectives, concepts and principles of safety and protection in the development and application of nuclear energy for peaceful purposes. The Safety Requirements establish the requirements that must be met to ensure safety. These requirements are governed by the objectives and principles presented in the Safety Fundamentals. Finally, Safety Guides recommend actions, conditions or procedures for meeting safety requirements. Recommendations in Safety Guides are expressed with the implication that it is necessary to take the measures recommended or equivalent measures to comply with the requirements.

There are several committees supervising the development of Safety Standards. Member States are widely represented on these committees. IAEA's Safety Standards are periodically updated to ensure they can provide guidance on contemporary means of achieving a high level of safety. Five years after publication, the Standards are reviewed to determine whether a revision is necessary. There is an ongoing programme on updating and developing new publications in the Safety Standards Series.

2.2. Application of IAEA's Safety Standards

Safety Standards are only effective, however, if they are properly applied in practice. The IAEA's safety services - which range in scope from engineering safety, operational safety, and radiation, transport and waste safety to regulatory matters and safety culture in organizations - assist Member States in applying the Standards and appraise their effectiveness. The core of IAEA's safety services are peer review missions, and one of those is the International Probabilistic Safety Assessment Review Team (IPSART) mission providing an independent review of PSAs and PSA applications upon a request of a Member State. The IAEA's safety services enable valuable insights to be shared.

2.3. Risk Considerations

Risk considerations and requirements on PSA are included in IAEA Safety Standards at all levels (Safety Fundamentals, Safety

[1] Each application area has an identifier assigned to it that is used for numbering of the publications in the Safety Standards Series.

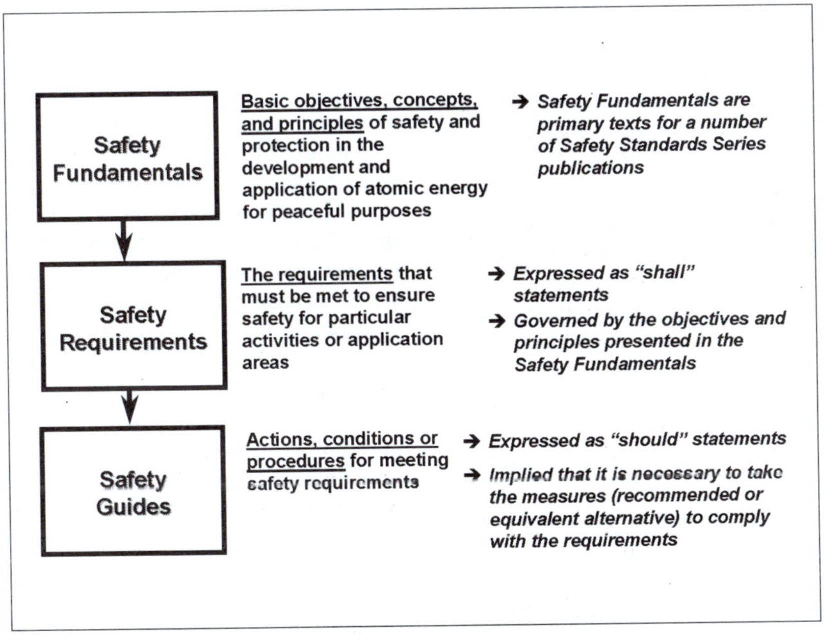

Fig. 1: Structure of IAEA Safety Standards

Requirements, Safety Guides) to the extent on which there is a consensus amongst Member States. The discussion below highlights specific points considered in the documents.

Safety Fundamentals

In the publication of the highest level, the Safety Fundamentals [1], a General Nuclear Safety Objective is set that is "to protect individuals, society and the environment from harm by establishing and maintaining in nuclear installations effective defences against radiological hazards." The General Nuclear Safety Objective is supported by two complementary Safety Objectives:

"<u>Radiation Protection Objective</u>: To ensure that in all operational states radiation exposure within the installation or due to any planned release of radioactive material from the installation is kept below prescribed limits and as low as reasonably achievable, and to ensure mitigation of the radiological consequences of any accidents.

<u>Technical Safety Objective</u>:

To take all reasonably practicable measures to prevent accidents in nuclear installations and to mitigate their consequences should they occur;

To ensure with a *high level of confidence* that, for all possible accidents taken into account in the design of the installation, including those of *very low probability*, any radiological consequences would be minor and below prescribed limits; and

To ensure that the *likelihood of accidents with serious radiological consequences is extremely low*."

The above statements imply that it is necessary to be aware of the estimate of risk of radiological releases associated with potential accidents of different severities at nuclear installations. It is also emphasized in Ref. [1] that in order to achieve the Safety Objectives, measures need to be taken "to control radiation exposure in all operational states to levels as low as reasonably achievable and to *minimize the likelihood* of an accident that might lead to the loss of normal control of the source of radiation." However, it is recognized in Ref. [1] that accidents can happen and, therefore, measures are needed to mitigate the radiological consequences, such as on-site accident management procedures and off-site intervention. In conclusion, one of the main viewpoints expressed in Ref. [1] is that "the *greater the potential hazard* from an uncontrolled release of radioactive material, the *lower the likelihood* must be of its occurrence."

Safety Requirements

The Safety Requirement publication dealing with NPP design safety [2] provides the design requirements for structures, systems and components important to safety, as well as requirements for a comprehensive safety assessment. It is stated that the safety assessment process includes the complementary techniques of deterministic and probabilistic safety analyses. It is emphasized that PSA is used in order to:

Demonstrate the design compliance with general safety objectives;

Demonstrate that design is balanced and there are no "cliff edge effects";

Assess the risk associated with external hazards;

Assess the risks of major off-site releases necessitating a short term off-site response;

Identify improvements/modifications (systems and procedures) that could reduce the probabilities of severe accidents or mitigate their consequences;

Verify compliance with Probabilistic Safety Criteria (PSC), if set.

It should be mentioned that the Safety Requirement dealing with NPP operational safety [3] also emphasizes that in order to complement the deterministic assessment, "consideration shall be given to the use of PSA for input to *the periodic safety review* to provide insight into the relative contributions to safety of different aspects of the plant."

In the Safety Requirement on site evaluation of nuclear installations [4], it is stated that "for the analysis of plant *protection against external events*, preferential consideration should be given to applicable *probabilistic methodologies*."

Safety Guides

General guidance on performance and independent verification of the safety assessments for NPPs, both deterministic and probabilistic, is provided in the Safety Guide on safety assessment and verification for NPPs [6]. This document provides recommendations to:

designers for carrying out a safety assessment during the initial design process and design modifications, and

operating organizations in carrying out independent verification of the safety assessment of new NPPs with a new or already existing design.

However, the methods and the recommendations of the Safety Guide [6] can also be used by regulatory bodies to conduct the regulatory review and assessment.

It is emphasized in the Safety Guide [6] that the achievement of a high level of safety should be demonstrated primarily in a deterministic way. However, the safety analysis should incorporate both deterministic and probabilistic approaches. These approaches have been shown to complement each other and both should be used in the decision making process on the safety and ability of the plant to be licensed. The probabilistic approach provides insights into plant performance, defence in depth and risk that are not available in the deterministic approach. In particular, the PSA can help to determine if the safety systems contain an adequate level of redundancy and diversity, if there is sufficient defence in depth and if the overall design is balanced. In a balanced design, the PSA should show that:

No particular feature of the design makes a disproportionately large contribution to risk;

No group of initiating events makes a disproportionately large contribution to risk;

The achievement of an overall low level of risk does not rely on contributors which have a significant uncertainty;

The first two levels of defence carry the primary burden of safety;

Within each level of defence, none of the safety systems is disproportionately more significant than the others.

It is stated in Safety Guide [6] that the PSA should be used throughout the design and operation of the plant to assist in the decision making process on the safety of the plant. For a new plant, the PSA should ideally be started during the conceptual design, continued through the more detailed design phase to assess more detailed design issues, and used to support the operation of the plant. For an existing plant, the PSA should be carried out either as part of a periodic safety assessment or to support the safety case for proposed modifications. It is important that PSA addresses the contributions to risk arising from all modes of plant operation. If the PSA is only carried out to Level-1, then the reactor core is, by definition, the focus of the analysis. If the PSA is carried out to Level-2 or Level-3, then the scope of the PSA may include contributions to risk arising from other sources of radioactive material on the site, such as used fuel and radioactive waste. In addition, the scope of PSA should include both internal and external initiating events.

The Safety Guide [6] provides a set of high-level specific recommendations on the development of Level-1, Level-2, and Level-3 PSAs. In addition, the use of PSAs and PSC are discussed. In the part of PSC, a reference is provided to an INSAG publication [7].

3. LATEST DEVELOPMENTS IN THE AREA OF PSA QUALITY

To promote developing the capabilities for PSA in Member States, the IAEA produced a comprehensive set of guidance documents aimed to support PSA development, review, and application. Some of the recent publications are indicated in Refs. [8]-[11].

It has been widely recognized that PSA techniques, especially in the area of Level-1 PSA, are well matured today. Increasingly, during the last years, PSA has been broadly applied to support numerous applications, such as risk-informed changes to technical specifications, risk-based plant configuration control, maintenance program optimization, regulatory inspections, etc. At several international meetings as mentioned in [12], Member States recommended that the IAEA consider writing guidance on a process to review a PSA to determine its technical adequacy for addressing specific applications, or in other words, to provide guidance to assure that a PSA is of sufficient quality to support the application. The quality was meant to be commensurate with the intended use of the PSA.

In order to further promote the use and application of PSA techniques in the Member States, in 2002 the IAEA launched a project on development of a technical document (TECDOC) "PSA Quality for Applications" that would take into consideration the advanced worldwide experience in the area of PSA quality assessment and verification, and in particular the ASME PRA Standard [13].

Four consultants meetings of small groups of experts and two technical meetings with a wide representation of PSA specialists from IAEA Member States have been held during the course of development of the TECDOC "PSA Quality for Applications". The draft document had been widely circulated amongst Member States. The comments received had been taken into account in the final version of the TECDOC [14]. This publication is aimed at providing an approach for achieving the quality of PSA in order to support reliably various PSA applications. The approach involves consideration of a set of technical features, called "attributes", of the major PSA elements relevant for specific applications, including a "base case PSA" that is performed with the aim to assess the NPP safety level. A comprehensive list of PSA applications is provided that have been categorized in the TECDOC [14] in the following way according to their purpose:

(1) *Safety Assessment:* to assess the overall safety of the plant and to develop an understanding of the main contributors to risk (this is the original motivation for the development of PSAs);
(2) *Design Evaluation*: to provide support for design evaluation;
(3) *NPP Operation:* to provide support for day-to-day operation of the plant (not including permanent changes to design or operational practices);
(4) *Permanent Changes to the Operating Plant:* to assess the safety significance of proposed permanent changes to the plant design, hardware, or administrative controls (e.g., operating procedures, the licensing basis) as an aid to decision making;
(5) *Oversight Activities:* to support plant performance monitoring and assessment (both regulatory and industry);
(6) *Evaluation of Safety Issues:* to evaluate safety issues.

The TECDOC [14] can be used by PSA practitioners for appropriate planning of PSA projects taking into account possible uses of the PSA in the future. The document can also be used by reviewers as an aid in assessing the quality of PSAs and judging the adequacy of a PSA for particular applications.

4. PLANS FOR NEW SAFETY STANDARDS ON SAFETY ASSESSMENT

The advanced experience accumulated worldwide in the area of safety assessment techniques, as well as the comprehensive corpus of methodological and guidance documents on deterministic and probabilistic safety assessments issued by the IAEA, have led to a logical decision to develop a new set of Safety Standards supporting the NPP safety assessment and verification. This set of documents will consist of one Safety Requirement publication [15] and several Safety Guides supporting it. In particular, it is planned to produce in the near future two new Safety Guides on PSA:

1. Development and Application of Level-1 Internal Events PSA for Nuclear Power Plants;
2. Development and Application of Level-2 PSA for Nuclear Power Plants.

These documents are intended to provide guidance and recommendations to plant licensees and technical support organizations in performing or managing a Level-1 or Level-2 PSA project, and use it to support the safe design and operation of NPPs, based on internationally recognized good practices. The guides will also address the PSA applications and the technical attributes and quality aspects for their sound development. An additional aim is to promote a standard framework that can facilitate the external peer or regulatory review of PSAs and their applications.

5. CONCLUSIONS

There is general international agreement, as reflected in various IAEA Safety Standards, that both deterministic and probabilistic approaches should be used to provide insight into the relative contributions to safety of different aspects of NPPs design and operation. Risk considerations are included in IAEA Safety Standards in the manner that implies the necessity for comprehensive PSAs. To assist Member States to take full benefit from PSA applications, the IAEA provides international peer review services, training courses and workshops, and publishes a wide range of technical reports on PSA and its applications.

The matter of PSA quality is important from the viewpoint of risk-informed decision making on various aspects of NPP operation and licensing. The IAEA finalizes a technical document [14] which will provide an approach for achieving the quality of PSA in order to support more reliably various PSA applications.

The IAEA has extensive plans to develop a new set of Safety Standards on NPP safety assessment and verification addressing both deterministic and probabilistic analyses.

6. ACKNOWLEDGMENTS

The authors would like to thank the members of the IAEA's Division of Nuclear Installation Safety for their active support and effective contribution to the IAEA projects on development of a wide range of technical publications.

7. REFERENCES

1. INTERNATIONAL ATOMIC ENERGY AGENCY, Safety Fundamentals: The Safety of Nuclear Installations, Safety Series No. 111-F, Vienna: IAEA, 1993.
2. INTERNATIONAL ATOMIC ENERGY AGENCY, Safety of Nuclear Power Plants: Design Requirements, Safety Standards Series, No. NS-R-1, Vienna: IAEA, 2000.
3. INTERNATIONAL ATOMIC ENERGY AGENCY, Safety of Nuclear Power Plants: Operation Requirements, Safety Standards Series, No. NS-R-2, Vienna: IAEA, 2000.
4. INTERNATIONAL ATOMIC ENERGY AGENCY, Site evaluation for Nuclear Installations, Safety Standards Series No. NS-R-3, Vienna: IAEA, 2003.
5. INTERNATIONAL ATOMIC ENERGY AGENCY, Statute, Vienna: IAEA, 1957.
6. INTERNATIONAL ATOMIC ENERGY AGENCY, Safety Assessment and Verification for Nuclear Power Plants, Safety Guide, Safety Standards Series No. NS-G-1.2, Vienna: IAEA, 2002.
7. INTERNATIONAL ATOMIC ENERGY AGENCY, Basic Safety Principles for Nuclear Power Plants 75-INSAG-3 Rev. 1, INSAG Series No. 12, Vienna: IAEA, 1999.
8. INTERNATIONAL ATOMIC ENERGY AGENCY, Risk Informed Regulation of

Nuclear Facilities: Overview of the Current Status, TECDOC-1436, Vienna: IAEA, 2005.
9. INTERNATIONAL ATOMIC ENERGY AGENCY, Review of Probabilistic Safety Assessments by Regulatory Bodies, Safety Report Series No. 25, Vienna: IAEA, 2002.
10. INTERNATIONAL ATOMIC ENERGY AGENCY, Applications of Probabilistic Safety Assessment for Nuclear Power Plants, TECDOC-1200, Vienna: IAEA, 2001.
11. INTERNATIONAL ATOMIC ENERGY Agencies, Living Probabilistic Safety Assessment, TECDOC-1106, Vienna: IAEA, 1999.
12. I. KOUZMINA, F. NIEHAUS, V. RANGUELOVA, Development of Guidance for PSA Quality for Applications, Proceedings of PSAM 7 - ESREL'04 Conference on PSA and Management, Springer-Verlag London Limited, 2004, Vol. 4, pp. 2137-2142.
13. THE AMERICAN SOCIETY OF MECHANICAL ENGINEERS, Standard for Probabilistic Risk Assessment for Nuclear Power Plant Applications, ASME RA-S-2002, New York: ASME, 2002.
14. INTERNATIONAL ATOMIC ENERGY AGENCY, PSA Quality for Applications, Final Draft TECDOC, June 2005.
15. INTERNATIONAL ATOMIC ENERGY AGENCY, Safety Assessment and Verification for Nuclear Facilities, Draft April 2005.

Reliability, Safety and Hazard
Editors: P.V. Varde et al.

INSIGHTS FROM THE CASE STUDY OF COMMON CAUSE FAILURE ANALYSIS OF SPECIAL SAFETY SYSTEMS OF QINSHAN CANDU NUCLEAR POWER PLANT

Song Minghai

Engineering Master of Shanghai Jiao Tong University
Third Qinshan Nuclear Power Co. Ltd., Haiyan, Zhejiang, P.R. China, 314300

ABSTRACT

This paper presents the practical process of case study of Unified Partial Method (UPM) in Qinshan CANDU NPP Common Cause Failure (CCF) analysis, and discusses the insights of the important technical points in the analysis. Through these insights, suggestions are raised on future development strategy of CCF analysis in Qinshan CANDU NPP Probabilistic Safety Assessment (PSA).

Keywords: Common Cause Failure (CCF), Probabilistic Safety Assessment (PSA), CANDU, Unified Partial Method (UPM)

1. INTRODUCTION

Common cause failure (CCF) is a dependent failure event where simultaneous or near simultaneous multiple failures result from a single shared cause, e.g. system maintenance error, internal flooding, common design faults. Common cause failures are infrequent, but can lead, when they occur, to severe consequences: to fail several redundant trains or systems simultaneously or near simultaneously. CCFs are difficult to predict and model. The most common methods to model CCF are Multiple Greek Letters (MGL) method, β factor method and Alpha factor method.

In the design phase of Qinshan CANDU NPP, the parametric CCF analysis were not performed in the system models of Probabilistic Safety Assessment (PSA) due to previous Canadian CANDU PSA practice. In order to show the impact of CCFs on system reliability and to perform the pilot study of CCF analysis of Qinshan CANDU PSA, separate case studies of CCF analysis on Special Safety Systems of Qinshan CANDU NPP have been performed using Unified Partial Method (UPM).

2. SUMMARY OF UNIFIED PARTIAL METHOD (UPM)

The approach of UPM[1] utilizes a refinement of the structure of the Rolls Royce and Associates (RRA) Partial β-Factor Model developed in the mid-1980s and represents a practical approach to dependent failure assessment.

The Partial Beta Factor method was originally proposed as a way of decomposing the overall assessment of a simple beta factor into a series of judgements relating to identifiable topics, which all have an impact on redundant component vulnerabilities to CCF. A total of 19 such topics were included. The analyst was required to assess a partial beta factor for each of these topics. The value that was selected had to be between a specified minimum and unity, based on the features of the set of components under review. An overall beta factor was then calculated, by multiplying all 19 partial beta factors together.

UPM represents the development and simplification of the original partial beta method. The 19 topics requiring assessment in the original model have been consolidated into eight causal groups (it is called eight sub-factors). Each of the beta sub-factors relates to a different aspect of system design, operation and working environment, including its effectiveness in defending against CCF, and is described as follows:

1) Redundancy, itpresents the status of redundancy and diversity of common cause components group.
2) Separation, it presents the isolation status and layout of common cause components group.
3) Understanding, it presents the maturity status of the technology used in common cause components group.
4) Analysis, it presents the status of previous analysis experience of fault analysis and operation experience feedback.
5) Man machine interface, it presents the status of procedure and human error likelihood.
6) Safety culture, it presents the status of training and safety culture.
7) Environmental control, it presents the status of access control to common cause components group.
8) Environmental testing, it presents the status of environmental qualification of common cause components group.

Each of the above eight sub-factors normally has A to E five categories of definitions for different levels. The analyst needs to choose one of five categories that most closely matches the components under review. Then the β factor and CCF probability can be calculated, and which can be used in fault tree quantification process.

3. ANALYSIS PROCESS

There are five major steps to perform the CCF analysis by UPM in the case study [2][3],

1) Selection of Common Cause Component Groups and Fault Tree Reconstruction. The definition of the common cause component group is considered as the most important and perhaps the most difficult task in the component-level CCF analysis as it is related to the final results of the analysis. It is the general step to perform CCF analysis compared to other CCF methods. Once the common cause component groups have been identified, it is, of course, necessary to incorporate the appropriate basic events into the fault tree model to reflect the failure dependency of common cause component groups. Note that all CCF events have assigned probabilities. The probabilities of CCF basic events are obtained by beta factor multiplying component total random failure probability:

$$P_{CCF} = \beta P_{TOTAL};$$

Where: $P_{TOTAL} = \lambda T$(running failure);
$P_{TOTAL} = \lambda T/2$ (dormant failure).

As shown, the total random failure probability is a function of the total random failure rate, λ, and a time parameter. The time T is equal to the mission time for running failures, and for dormant failures, it is the test interval.

2) Screening Analysis. Having identified a CCF group, the beta factor and CCF basic event probability must then be calculated. Since it is unnecessary to calculate every beta factor by using the in-depth qualitative assessment. of Unified Partial Method, a quantitative screening, using a conservative beta factor (0.1) shall be performed, before applying the UPM method, to identify the CCF groups that contribute most to the fault tree top event unavailability.
3) Calculation of Beta Factors by UPM. The calculation of beta factors for the identified CCF groups consists of the following five main steps:
   a. Define system boundary. It is the general step of CCF analysis which is discussed in 1) of section 3.
   b. Determine the level of assessment. A component level of assessment should be performed for fault tree quantification of common cause component groups.
   c. Consult the UPM estimation table to determine the values of each beta sub-factors. See Table-1 for the estimation

table [1]. The analyst must choose the description that most closely matches the system under consideration to determine each of the values of beta sub-factors.

d. The justification for the choices of beta sub-factors must be recorded in judgement table (see Table-2 for the practical process of beta factor calculation for Emergency Core Cooling system PV1 and PV2 fail closed common cause failure event [3]), which summarizes the judgement made in the previous step and is used to calculate the beta factor. This step can be combined with the above step, and these two steps constitute the bulk of the analysis.

e. Finally, the value of beta factor can be calculated.

4) After obtaining the beta factor, the CCF basic event probability should be calculated according to the formulas described in 1) of section 3, and the value should be incorporated into the fault tree model in order to replace the screening values for these selected CCF basic events.

5) Then fault tree should be re-evaluated to obtain the final result.

Table 1 Partial Beta Factor Estimation Table

	A	A+	B	B+	C	D	E
Design							
Redundancy	1750	875	425	213	100	25	6
Separation	2400		580		140	35	8
Understanding	1750		425		100	25	6
Analysis	1750		425		100	25	6
Operation							
Man machine interface	3000		720		175	40	10
Safety culture	1500		360		90	20	5
Environment							
Environmental control	1750		425		100	25	6
Environmental testing	1200		290		70	15	4

4. INSIGHTS FORM THE CASE STUDY OF UPM ANALYSIS

During the case study of CCF analysis on Special Safety Systems of Qinshan CANDU plant, some insights of important technical points of UPM were realized:

1) When the sub-factor of redundancy is estimated, it can be mapped of high levels of redundancy m out of n for success to 1 out of (n-m+1) for success for n<=5. At higher n, it may be optimistic, especially if n-m is large. When the situations of n>5 occurs, it should be analyzed case-by-case. When n is slightly exceeding 5, e.g., 6 or 7, this mapping can still be used. This will induce a little optimism in the result and can be considered as acceptable.

2) When the sub-factor of separation is estimated, the actual location of common cause component groups must be checked for the judgement of the sub-factor. See Table-2 as an example.

3) The estimation of the sub-factor of Man Machine Interface is not required, as pre-accident human errors should have been explicitly modeled in fault tree model with conservative probabilities. The sub-factor shall be assigned a value of zero to avoid double counting.

4) The consolidation and simplification of UPM reduces the amount of analysis and time required. Estimation table (Table-1) and judgement table (Table-2) make the consistency between different analysts is easier to achieve, although specific guidance on the interpretation of the system descriptions may still be required for particular project.

5) It gives the analyst opportunities to check the real design, location and operation of common cause component groups, and it can be a useful tool for engineering practice to estimate CCF and their defences.

6) The source of the values in estimation table (Table-1) is not clear, which may be based on expert judgement or operation experience. The future development of UPM needs to incorporate real operation data into the UPM process.

It is showed in the analysis results that the contribution of CCF to system unavailability is above 50%. As the fault tree analysis of CANDU NPP is detailed to the level of relay and contacts for instrumentation & control equipments, the detail level of fault tree structure may need

Table 2 Judgement Table for ECC System PV1 and PV2 Fail Closed Common Cause Failure Event

Applied CCF Basic Events	**3432PV1/2—FCCCF**		
Sub-Factor	**Judgement Decision**	**Category/ Numerical Value**	**Comments**
Redundancy	1 out of 2 for success.	A/1750	
Separation	The two valves are located in two rooms (S-004, PL1057 and S-019, PL1251).	C/140	
Understanding	The design Experience is over 10 years; novelty, complexity and misfit can be considered as small.	E/6	As CANDU 6 has more than 200 reactor years of operation experience and the design has passed long time of verification, further more there is no very important design change or new technology applied in the 2 Qinshan CANDU units. This is the general consideration in all of the ECC analysis, and no software is involved in the design of ECC.
Analysis	General Consideration of Special Safety Systems of CANDU reactor.	D/25	This is the general consideration in all of the ECC analysis, as reliability assessment of Special Safety Systems is mandatory for regulation in Canada and the reliability performance of Special Safety Systems are monitored in each CANDU units. Designers apply Safety Design Guides in their design work and can get feedback form reliability and PSA analysis.
Man Machine Interface	N/A	N/A	The estimation of the sub-factor of Man Machine Interface is not required, as pre-accident human errors have been explicitly modeled in Fault tree model with conservative probability. The sub-factor shall be assigned a value of zero to avoid double counting.
Safety Culture	Short time of operation experience in the 2 Qinshan CANDU units, the general safety level is appropriate to assign.	C/90	The sub-factor of Safety Culture can be assigned a category of C, as the 2 Qinshan CANDU units have not come through very long time of operation experience even with operators have got sufficient simulator training of normal and emergency conditions. The treatment is conservative and it is acceptable.
Environmental Control	The equipments are located in service building, and it is limited access area.	D/25	According to the location of ECC equipments, the sub-factor of environmental control can be assigned to category of E when equipment of common cause component groups are located in Main Control Room as it's strict access control. For ECC equipment located in Service Building, Reactor Building or E-101, the sub-factor of Environmental Control can be assigned as category of D as for special safety systems they are very important and only trained personnel are allowed to work in this limited access area.
Environmental Testing	Environmental qualified according to the requirements in Safety Design Guide.	E/4	ECC is one of the four special safety systems and it is strictly designed according to the requirements in Safety Design Guide. The compliance verification with the Safety Design Guide has been performed (see 98-03650-ASD-001); the sub-factor of Environmental Testing can be assigned to category of E.
Total Numerical Value (Summation of Sub-factors)			2040
Beta Factor (β)=Total Numerical Value /50000			0.0408
P_{CCF} of 3432PV1/2--FCCCF			7.079E-5

to be adjusted to make it more consistent with engineering practice. On the other hand, the conservation in probability estimation needs to be reduced.

5. SUGGESTIONS TO FUTURE DEVELOPMENT OF CCF ANALYSIS IN QINSHAN CANDU NPP PSA

For the future successful application of PSA in Qinshan CANDU NPP, the CCF analysis needs to be performed for all of the fault tree models, and the Unified Partial Method can be a useful engineering tool to estimate CCF and their defences.

For the future development of CCF analysis in Qinshan CANDU NPP PSA, technical suggestions are raised as follows:

1) The UPM manual [1] is the general description to the method, and it is limited description of sub-factor estimation to specified equipments. The specified guidance or procedure of UPM for CANDU PSA needs to be developed.
2) The detailed equipment category is not involved in the value estimation of sub-factors, e.g. pumps and valves of same sub-factors categories have same beta factor. The comparison to statistical operation data needs to be performed to determine the conservation of the method.
3) The detail level of fault tree structure of Qinshan CANDU NPP PSA may need to be adjusted to make it more consistent with engineering practice and the conservation of probability estimation needs to be reduced.

6. REFERENCES

1. P V Brand, UPM 3.1: A pragmatic approach to dependent failures assessment for standard systems, Cheshire, UK, AEA Technology plc, ISBN: 085 356 4337, 1996.
2. P.Santamaura, et al, CANDU Generic Probabilistic Safety Assessment-Methodology, Chapter 5 Dependent Failure Analysis, 91-03660-AR-001, Rev.0, Mississauga, Ontario, Canada, AECL, July, 2002.
3. Song Minghai, Emergency Core Cooling System Reliability Analysis with Common Cause Failure, Qinshan CANDU Project, 98-03600-ASD-003, Rev. 0, Mississauga, Ontario, Canada, AECL, Nov. 2003.

Reliability, Safety and Hazard
Editors: P.V. Varde et al.

ASSURED PROBABILISTIC MARGINS OF RISK MEASURES VS. BEST ESTIMATES: THE STATE OF THE ART AND PERSPECTIVES

Vladimir Morozov

ATOMENERGOPROEKT, 107005, Bakuninskaya str.7 Moscow, Russia

ABSTRACT

PSAs, which assume the best estimate approach that comprises CDF or LER point estimates on the basis of Bayesian technique and expert judgements may have only limited potential in the light of using them in regulation practice as latter assumes reliable checking resulting values against specified targets. The way to meet this requirement is accepting the position that the PSA quantitative results should explicitly account for uncertainty, which, in turn, should as much as possible reflect operating experience. This problem could be resolved by using assured probabilistic margins for CDF or LERF instead of their means. The paper describes the issues related to implementation of this method in PSA practice.

1. INTRODUCTION

For many years PSA was considered as a tool that can be applied concurrently to deterministic safety analyses - to extend their conclusions and findings. Such a look at the PSA role was supported by the following arguments:

- PSA resulting values are much uncertain.
- Internal PSAs are generally not full-scale i.e. developed for the limited number internal initiators
- Site specific and external hazards sometimes are beyond PSA scope and therefore, do not contribute to the plant risk profile.

As the result of above, the main PSA purpose was to provide a coherent balance in risk profile rather than to give an accurate representation of quantitative risk measures, like CDF or LERF. Hence, it is clear that the point estimates of these measures, which are appropriate for the summation of different cut-sets contributors to the integrated risk profile, are considered first to represent PSA results. Following this, uncertainty indices provide accessorial information that usually not carefully analyzed in decision-making process.

However, PSAs, which are carried out in such a way may have only limited potential in the light of using them in regulation practice as latter assumes that resulting values should be reliably checked and assured against specified targets.

The only way to meet this requirement is accepting the position that the PSA quantitative results should not be expressed solely in terms of point estimates or means, but also should account for uncertainty, which is entirely inherent for the PSA. This problem could be resolved by using assured probabilistic margins for CDF or LERF instead of their means or similar estimates.

The concept of assured probabilistic margins however, requires response to many specific problems related to PSA methodology itself, computative methods used and data treatment procedure.

The paper presents the results of the analysis of some important issues, related to uncertainty nature:

- Methods for obtaining assured probabilistic margins.

- Adequacy of model assumptions and estimating associated errors.
- Decreasing the role of subjective factors in data treatment.

2. CALCULATING ASSURED PROBABILISTIC MARGINS

The problem description and methods to be used

The following example illustrates the situation. Consider the hypothetic CDFs for two NPPs being presented by their means end error factors assuming log-normal distribution. Say that the CDF1 mean value is of 1E-4 /a and error factor E =3, whereas mean of CDF2 = 5.0E-5 /a and E = 7. The question is: Does CDF2 really lower than CDF1 and if it does what degree of confidence supports this conclusion?

If both values are log-normal the answer can be obtained analytically. For specified above input data the formula yields P = 0.695. The result shows that there is a chance of 30% that conclusion is wrong in spite of the fact that CDF1 two times exceeds CDF2!

The lessons learned from this example show that for the task of comparison of PSA results and accordingly, checking the results against targets specified in regulations point estimates of CDF or LER do not provide the appropriate measures as uncertainty which is inherent for the PSA methodology is not taken into account in this measures. Let us define assured probabilistic margin $H\gamma$ of level γ: $0<\gamma<1$ for PSA quantitative measure H (say, for example CDF) as dependent on experience and subjective analyst assumptions and judgments value that provide probability of at least γ for random variable H not exceed $H\gamma$. Mathematically this can be expressed by the following formula:

$$P\{H(\vec{\theta}) \leq H(\gamma, \tilde{\theta})\} \geq \gamma \,. \qquad (1)$$

where vector $\vec{\theta}$ represents parameters of the model and $\tilde{\theta}$ their estimates that are obtained using operating experience and subjective expert judgements. For Bayesian approach the former are random whereas the latter are constants that depend on above types of input data.

The set of random parameters typically used in PSA is presented by Poison component failure rates λ_i, Bernoulli failure probabilities per demand θ_i, parameters of CCF models, unavailability values due to component (or train) repairs υ_i.

Generally speaking the problem of obtaining assured probabilistic margins comprises two basic questions:

- Using conservative computation methods and procedures to estimate multi-parameter function $H(\vec{\theta})$ given fixed input variables
- Finding an appropriate tools to calculate $H(\gamma, \tilde{\theta})$ that provides $P\{H(\vec{\theta}) \leq H(\gamma, \tilde{\theta})\} \geq \gamma$

The first issue really not at all a problem for simple functions of limited number variables. However, full-scale PSA model used to estimate CDF typically comprises more than hundred parameters and therefore the special methods for this case to be applied to assure in appropriate accuracy or conservatism of results obtained. It is assumed here that Minimal Cut Set approach that allows calculating so-called updated upper bounds together with step-wise procedure of modularisation of logic model and screening MCS of negligible probability level provide a good basis for the response to the first question. Therefore, the problem computation of $H(\vec{\theta})$ is no more a subject for discussion in this paper except for the fact that $H(\vec{\theta})$ is a function of a large number variables and can be calculated by running a special PSA software.

Regarding the second issue it is widely adopted that the distribution of $H(\vec{\theta})$ that is need for calculating desired probabilistic margins can be obtained by Monte-Carlo simulations of input random variables. However, this provides correct results only if the distribution lows of above-mentioned variables used in Monte-Carlo procedure are exactly known. For the variables of simple nature like λ_i, these distributions can be obtained by Bayesian updating process that integrates priors, constructed by analyst with available operating failure statistics. In some cases, however, the input parameter of interest is a complex parameter as well, and its distribution is not known neither analytically nor by computation. Instead, the probabilistic margin of the parameter that corresponds to a certain level of credibility is available. This often takes place in PSA models developed for external or site-specific hazards (for example seismic hazard curve may be derived by seismologists for the certain number originally specified levels of credibility). In internal PSA model parameters of this type may represent structural component failure probabilities or frequencies of rare initiating events caused by rupture of mechanical components and piping.

In all cases of this type applying the the substitution method of calculating CDF margins for the originally specified level of credibility would be desirable. However, the method itself and its using for a set of model parameters together with Monte-Carlo simulations for the others needs to be verified.

The mathematical basis for substitution method

The two basic functions of a large number of random parameters are under consideration now:

$$H_1(\vec{\theta}) = \alpha_1\theta_1 + ...\alpha_n\theta_n \quad (2)$$

$$H_2(\vec{\theta}) = \theta_1 \cdot ... \cdot \theta_n \quad (3)$$

The first one is a linear combination, whereas the second one is a product of parameters of interest. The following theorem have been proven:

Theorem 1

If random independent variables $\theta_1,..., \theta_n$ have aging distributions, than there is a specific $\gamma_0 \approx 0.714$ that for all $\gamma_0 \le \gamma < 1$

$$H_1(\gamma,\tilde{\theta}) = \sum_{i=1}^{n} \alpha_i\tilde{\theta}_i(\gamma)$$

and

$$H_2(\gamma,\tilde{\theta}) = \tilde{\theta}_1(\gamma) \cdot ... \cdot \tilde{\theta}_n(\gamma)$$

where $\tilde{\theta}_1,..., \tilde{\theta}_n$, have been calculated as margins for each parameter at the level of credibility γ, are margins for a level of credibility at least γ for (2) and (3) accordingly. In mathematical notation this means that for k=1,2:

$$P\{H_k(\vec{\theta}) \le H_k(\gamma,\tilde{\theta})\} \ge \gamma \quad (4)$$

The aging distribution mentioned in the theorem is defined as a distribution for which intensity rate

$$\lambda(x) = \frac{f(x)}{1 - F(x)}$$

is a monotonous non-decreasing function.

It should be noted that for basic uncertainty related distributions used for parameter models (Weibull, Gamma, Gauss) this requirements has been met.

As a consequence from Theorem1 it should be pointed that the substitution method can be applied only for the part of parameters in (1) or (2) concurrently to simulations used for others. The two forms of the function in Theorem1 allows to apply the substitution approach for the different types of probabilistic logic models, like train-based structures and redundant blocks.

3. MODEL ASSUMPTIONS AND THEIR INFLUENCE ON PSA RESULTS

Typical classes of model assumptions

There are two basic classes of model assumptions used in PSA related to data representation:

- Assumptions related to distribution types of random values of different nature (time between failures, repair times, number of outcomes, etc.)
- Assumptions applied for statistical data treatment

Each class from mentioned above model assumptions has its own nature and needs to be considered separately in spite of the fact that they are strongly correlated. In this Section the first is considered, whereas the second is discussed in Section 3.

Poisson model for failure events and associated errors

The two models for failure events widely shared in PSAs are time-dependent Poisson model that represents random time between component failures and Bernoulli model that characterizes number of outcomes when failures per demand are considered. Here only the former is discussed because of its importance and a number of possible alternatives. The problems of interest is 1) an area, where using exponential distribution would be desirable and 2)estimation of possible errors in PSA results related to its adoption. As appropriate alternative to Poisson-exponential

model a family of aging distributions can be assumed.
With respect to the question the following theorems have been proven:

Theorem 2

This theorem touches the questions related to modeling so-called run-back transients initiated as deviations from normal conditions followed by protection system failure.

Suppose $S = \sum_{i=1}^{k} \xi_i$, where ξ_i are mutually independent and have the same aging distribution.

The upper limit in the sum k is a random number that has, in turn geometric distribution $P_k = qp^{k-1}, p + q = 1$.

Then distribution of $S_q = qS$, also belongs to the class of aging distributions and at $q \to 0$ converges to exponential distribution with parameter $\lambda = \frac{1}{T}$, where T designates the mean time of ξ_i.

Theorem 3

Let us assume that X_i i =1, 2, k... - are regenerative flows of events induced by corresponding aging random variables $\xi_1 \cdots \xi_k \ldots$, with limited and smooth density functions. Designate $T_i = M[\xi_i]$. Suggest $1/T_i$ values to be sampling from a specific hyper-distribution with a mean λ. Then for the steady conditions the random value $n\eta$, where η the time interval between consequent events in integrated flow converges AS to exponential distribution with parameter λ.

It should be noted however, that for the finite n $n\eta$ still has aging distribution, and estimation for $n\eta$ obtained in the theorem 3 would be optimistic.

The mean of the rest time for the nearest anticipated event of each X_i flow at steady conditions is

$$Tr_i = \frac{(T_i^2 + \sigma_i^2)}{2T_i} \qquad (5)$$

This fact will be used later.

Theorem 4

Now consider random variable ξ that has aging distribution with the mean $T = M[\xi]$. Designate $\lambda = \frac{1}{T}$ Then exponential distribution function $F(t) = 1 - exp(-\lambda t)$ provides conservative estimation (exceeds) for distribution function of ξ at each t from the interval $[0,T]$.

Theorems from the second to fourth allow to get response to the question what would be deviations in PSA results if the hypothesis of Poisson model is wrong.

The Theorems 2 and 4 state that that for many groups of initiating events the Poisson model seems to be natural and at the same time provide a certain level of conservatism regarding the influence on PSA results when aging takes place.

The Theorem3 states that if statistical data treatment method for on the basis No of events per exposure time is used to obtain estimations of model parameters, representing component failures the Poisson model has a good chance to be adopted. However, in this case it results in optimistic estimations. To estimate the level of underestimation the following reasoning can be applied.

Let a system train comprises a number of components with mean times between failures equal to $T_i = M[\xi_i]$. When "child-disease" period at the beginning plant operation passed, an exponential vs. slow-aging alternatives related to times between failures of system train components can be verified. If aging takes place, than at steady time interval (5) can be used for each component and for the sequential circuit the formula:

$$\lambda = \sum_{i=1}^{n} \frac{1}{T_i\left(\frac{1+v_i^2}{2}\right)} \qquad (6)$$

where v_i designates variation coefficient provides the upper bound for the failure rate of train as a whole. Therefore, assuming v be the maximum value of these, and taking into account that

$$\frac{1+v_i^2}{2} \geq v_i$$

the possible relative underestimation in assessment of the train reliability characteristic due to adoption of Poisson assumption is limited by v.

In the case of k 100% redundant trains this will be v^k.

4. PARAMETER ESTIMATION FOR POISSON CASE: BASIC APPROACHES TO ASSURED MARGINS

Assumptions used for Poisson data model

In the paper most frequent Poisson model for component failure rates and initiating event frequencies, for which the total number of failures or events divided by the total operating time represents the sufficient statistics, is considered.

The Poisson flow is a random process of event occurrence, which meets three basic conditions:

- Stationarity (steady state condition);
- No after effects (independence condition);
- No multiple events in infinitesimally small time intervals (ordinarity condition).

It should be noted here that Poisson model is the simplest model that could be used for time-dependent event representation It characterized by a single parameter λ known as a event occurrence rate or frequency expressed in reciprocal time.

The Bayes-non informative and frequentist approaches

Here we consider the cases when the number of events observed is statistically enough to utilize solely the plant-specific data, or at least data from sister plants can be used in statistically enough volume to get objective, i.e. based on experience, parameter estimations.

For the first case the posterior density can be obtained applying the technique of a Bayesian update with a non-informative Jeffrey's prior distribution [1,3], which "allows the specific data to speak for themselves".

With Poisson data, the Jeffrey's prior is obtained if the shape parameter of a gamma distribution is taken to be ½ and the second parameter is taken to zero.

Therefore, the posterior density distribution function as well as mean and accordingly, upper and lower credible bounds that correspond to probability of q for Bayes procedure can be obtained as follows:

$$f_1(\lambda) = \frac{T_\Sigma^{\,d+0.5}\lambda^{d-0.5}}{\Gamma(d+0.5)} e^{-\lambda T_\Sigma}$$

$$M[\lambda] = \frac{2d+1}{2T_\Sigma} \qquad \lambda_q = \frac{\chi^2_{1-q}(2d+1)}{2T_\Sigma}$$

The comparison of these expressions with those, obtained by using classic frequentist approach, leads to the following conclusions.

Formally the procedure can be interpreted in classic terms assuming that additional 1/2 failure at the average occurs during an infinitesimally small time interval in addition to observation period, i.e. just after the end of it. The 90% posterior credible interval is

$$\lambda_{0.95} = \frac{\chi^2_{0.05}(2d+1)}{2T_\Sigma} \qquad \lambda_{0.05} = \frac{\chi^2_{0.95}(2d+1)}{2T_\Sigma}$$

These may be compared with the 90% confidence interval:

$$\bar{\lambda} = \frac{\chi^2_{0.95}(2d+2)}{2T_\Sigma} \qquad \underline{\lambda} = \frac{\chi^2_{0.95}(2d)}{2T_\Sigma}$$

The confidence intervals differ from the Bayes credible intervals only in the degrees of freedom, and moreover, only slightly. The lower and upper confidence limits have degrees of freedom 2d and 2d+ 2, respectively The two Bayesian limits each use the average, 2d + 1.

Methods used pooling the data from different plants

Here there are n different objects of analysis, each corresponds to a specific data set. It is also assumed that the data from each data set may be represented as (δi, Ti ,), where δi is the number of failures, and Ti is the whole time of observation in Poisson case.

It is assumed that parameters Li are sampling from the same hyper-distribution. The parameters of sets in this case are independent random variables with some mean M and variance V.[2,4]. The estimation of M is searching in the following form:

$$\bar{M} = \sum_{i=1}^{n} w_i \frac{d_i}{T_i}$$

The method of estimating M described here is MLE method and it requires the knowledge about type of distribution that generates parameter values for each data subset. The natural ways is the choice of a conjugate distribution families for parameter λ, that in our case Gamma with PDF:

$$g_\lambda(s,\tau) = \frac{1}{\Gamma(s)}\tau^s \lambda^{s-1} e^{-\lambda\tau}, \lambda \in (0,\infty)$$

For this distribution:

$$M = \frac{s}{\tau}, V = \frac{s}{\tau^2}$$

Along with this the likelihood expression for the polled sample is:

$$P_I(d_1,...d_n;T_1,...T_n;s,\tau) = \prod_{i=1}^{n}\int_0^\infty P_I(d_i,T_i|\lambda)g_\lambda(s,\tau)d\lambda$$

On integrating the expression for the likelihood functions are the follows:

$$L_I(s,\tau) = \frac{\prod_{i=1}^{n}\Gamma(s+d_i)}{[\Gamma(s)]^n \prod_{i=1}^{n}\Gamma(d_i+1)} \prod_{i=1}^{n}\left(\frac{T_i}{T_i+\tau}\right)^{d_i}\left(\frac{\tau}{T_i+\tau}\right)^s$$

The following theorem has been proved [4].

Theorem 5

The maximisation of the likelihood function allows obtaining the optimal (finite or infinite) values s and | resulting in unbiased estimations of M. With this, the minimal variance will correspond to:

$$w_{iopt} = \frac{1}{V+\frac{M}{T_i}}\left(\sum_{j=1}^{n}\frac{1}{V+\frac{M}{T_j}}\right)^{-1} = \frac{P_i}{\sum_{i=1}^{n}P_i}$$

$$P_i = \frac{T_i}{T_i+\tau}.$$

The variance is given by the formula:

$$V_{ar}^{(min)}[\lambda] = \frac{\lambda}{\tau}\frac{1}{\sum_{i=1}^{n}P_i} = \frac{\lambda}{\sum_{i=1}^{n}T_i(1-P_i)}$$

and it always exceeds the variance value for homogeneous groups of data.

As a consequence from the theorem it is also important that the estimation for L can be re-written by the formula:

$$\hat{\lambda} = [\sum_{i=1}^{n} d_i(1-P_i)]/[\sum_{i=1}^{n} T_i(1-P_i)]$$

Comparison with what frequentist simple summation approach yields, shows that the result can be interpreted in similar ways in terms of equivalent number of counts *de* and equivalent total exposure time *Te*, where *de* and *Te*, in turn, can be obtained by formulas:

$$d_e = \sum_{i=1}^{n} d_i(1-P_i) \qquad T_e = \sum_{i=1}^{n} T_i(1-P_i)$$

These formulas can be used to derive margins in the same way like for single data set, described above.

5. CONCLUSION

In the paper some part of specific methodological problems related to obtaining assured margins of system reliability characteristics and probabilistic safety targets have been discussed. The results obtained allow concluding that this approach rather than best estimate method has a potential to be used in a regulation process as it naturally and in a correct manner accounts for uncertainty that is inherent for PSA and reliability study.

6. REFERENCES

1. NUREG/CR-6823 SAND2003-3348P Handbook of Parameter Estimation for Probabilistic Risk Assessment Washington, DC 20555-0001
2. Vaurio, J.K., On Analytical Empirical Bayes Estimation of Failure Rates, Risk analysis, Vol.7, No.3, 1987
3. Siu, N. O., and D. L. Kelly, 1998, Bayesian Parameter Estimation in Probabilistic Risk Assessment, in Reliability Engineering and System Safety, Volume 62, Issue 1-2, pp. 89-116.
4. Morozov, V.B., A treatment of uncertainties for component reliability or initiator frequency estimates based on combining data sources with the potential of non-homogeneity, PSA'99 International Conference, 1999

Reliability, Safety and Hazard
Editors: P.V. Varde et al.

NEW METHODOLOGIES FOR STATION BLACKOUT STUDIES IN NUCLEAR POWER PLANTS

C. Senthil Kumar[1], A. John Arul[2], S. Marimuthu[2] and Om Pal Singh[3]

[1]AERB-Safety Research Institute, Kalpakkam
[2]Indira Gandhi Centre for Atomic Research, Kalpakkam
[3]Atomic Energy Regulatory Board, Mumbai

ABSTRACT

Station Blackout (SBO) events in Nuclear Power plants are recognized to be significant contributors to risk of core damage and consequences to public. Station blackout accident scenario is initiated by loss of offsite power, followed by loss of power from onsite emergency AC power sources. The further progression of the accident sequence is determined by the capability of the plant to cope up with loss of AC power, for maintaining core cooling over long periods of time. In this paper, we are concerned with the first part of the analysis, which is the estimation of Station blackout frequency as a function of non-recovery time. First we present a brief review of the methods currently used for the estimation of SBO frequency. Then we present the improved method developed at IGCAR for SBO frequency estimation, viz., use of power law statistical model for representing loss of offsite power failure and non-recovery characteristics and application of Markov technique for the time dependent reliability modeling of onsite AC power system and combining these two to get SBO frequency. We also discuss common cause failure modeling issues in the onsite power distribution system. Finally we bring out the merits and problems associated with conventional as well as the new method.

Keywords: Station Blackout Frequency, Loss of offsite power, Power law, Diesel generator unavailability, Markov Model, Fault Tree, Time dependent cut set, Common cause failures, Nuclear power plant

1. INTRODUCTION

Station blackout (SBO) is defined as the complete loss of alternating current (AC) electric power to the safety buses in nuclear power plant. This means the loss of offsite electric power (LOSP) and unavailability of the onsite emergency AC power. Onsite emergency AC power system usually consists of one or more diesel generators (DGs). A simple schematic of the power distribution system is shown in Fig. 1. The switch gear bus which, receive power from the grid is known as class IV bus and that which receive power from class IV and DG is known as class III bus. Estimation of SBO frequency and its duration provide important safety related information for decay heat removal system design, arriving at optimal DG configuration and to determine the station battery capacity.

SBO calculation proceeds in three steps. First is the estimation of LOSP frequency duration relation. Next is the calculation of onsite power system (DG) unavailability. The final step is the combination of LOSP with DG unavailability to obtain SBO frequency as a function of blackout duration.

In this paper, we review the standard methods used for LOSP, DG unavailability and SBO estimation and then discuss the new methods developed at IGCAR for SBO frequency estimation, viz., use of power law statistical model for representing loss of offsite power failure and non-recovery characteristics and application of Markov technique for the time dependent reliability modeling of onsite AC power system (DG), combining these two to get SBO frequency. We also discuss common cause failure modeling issues in the onsite power distribution system. Finally we bring out the merits and problems associated with conventional as well as the new method.

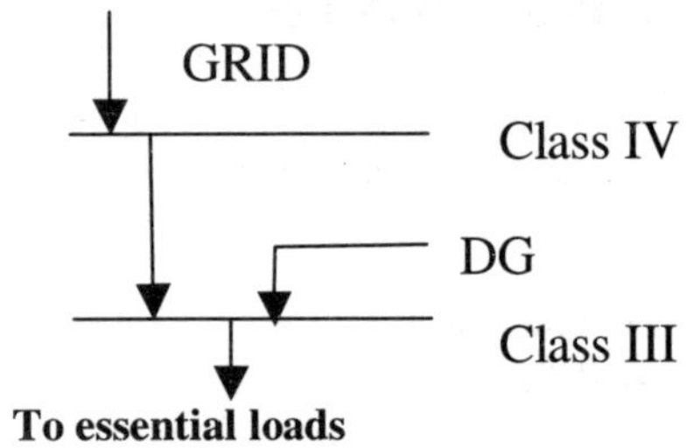

Fig 1: Schematic of AC power distribution system in a Nuclear Power Plant

2. LOSS OF OFFSITE POWER (LOSP)

The main parameters of interest in LOSP are its duration and frequency, i.e., how long a failure can occur and how frequent it is. In (Baranowsky, 1988), LOSP is categorized as plant-centered faults, utility grid blackouts and failures due to severe and very severe weather. In this report, data collected from experience is fitted to two-parameter weibull functions of the form:

$$\lambda_{LOP}(t) = \sum_{i} \lambda_{LOP_i} e^{-(\alpha_i t^{\beta_i})} \tag{1}$$

where $\lambda_{LOP_i}(t)$ is the frequency of loss of offsite power of type i, whose durations are equal to or greater than duration t. In IAEA-TECDOC-593, LOSP frequency is calculated using Bayesian estimation technique using the LOSP history at the site and assuming gamma as prior distribution for LOSP frequency. The study (Iman, 1988) reports a composite model of Weibull and Gamma to represent LOSP frequency and its duration. Most of the earlier studies on the offsite power failure analysis (Atwood, 1998; Wyckoff, 1996) have pointed out the existence of a mixture of distributions with respect to grid failures, plant-centered failures and failures due to severe weather conditions and a single probability model may not be appropriate to represent the overall distribution. Though a few studies of different modelling techniques have been carried out for offsite power failure analysis, to date, published work has mostly focused on the distributions such as Exponential, Lognormal, Gamma and Weibull. Since LOSP is an important safety parameter for a NPP, it is attempted to find a suitable model to estimate the frequency-duration correlation based on the failure rate observed at an Indian nuclear power plant.

2.1 Power Law Character of LOSP

Although the mixture model fits well, it is better if some physical basis could be established for the fitted model. Further, considering the scarcity of data, instead of a series of stretched exponentials, a single function would be preferable. As discussed in the previous section, there is a wide range of components involved in the offsite power supply scheme. For instance, grid, onsite equipment like transformers, switch gear etc., each with widely differing complexity. We propose that components belonging to each class have a specific failure rate λ_i and specific repair rate μ_i. Then the probability of observing a failure of duration $t' \geq t$, is,

$$P(t' \geq t) = \int g(\lambda) e^{(-\mu . t)} d\lambda \tag{2}$$

where $g(\lambda)$ is a normalized distribution function for component class. Since more frequent failures are generally repairable in shorter times compared to rare failures, which take more time to set right, it is assumed that $\mu \propto \lambda$. Therefore

$$P(t' \geq t) = \int g(\lambda) e^{(-c\lambda t)} d\lambda \tag{3}$$

where $c > 0$ is a constant of proportionality. The form of g is not known, however if gamma distribution is assumed for g (Takahashi, 1999), then Pareto distribution is obtained as the cumulative distribution function from equation (3). Thus, the probability of observing a failure of certain duration, is represented as,

$$P(t' \geq t) = \frac{1}{\left(1 + \frac{c}{\alpha} t\right)^k}, \tag{4}$$

where constant α, k > 0.

For large t,

$$P(t) \approx \left(\frac{\theta}{t}\right)^k, \ k > 0;\ \theta > 0;\ t \geq 0,$$

where θ = α/c and

$$F(t) = 1 - P(t)$$
$$\approx 1 - \left(\frac{\theta}{t}\right)^k \qquad (5)$$

The function, P(t) is plotted for data observed at two plant sites, over a period of 15-20 years, in Figures 2 and 3. The plots show P(t) versus LOSP time, t in Log-Log scale. The linear fits are shown alongside, which confirm the power law character of the distribution. In Figure 2 there is considerable deviation from straight line for the first few points. This is due to the fact that for small values of t, i.e., for $\frac{c}{\alpha}t \ll 1$, P(t) is virtually constant, and the power law dependence manifests fully for large t.

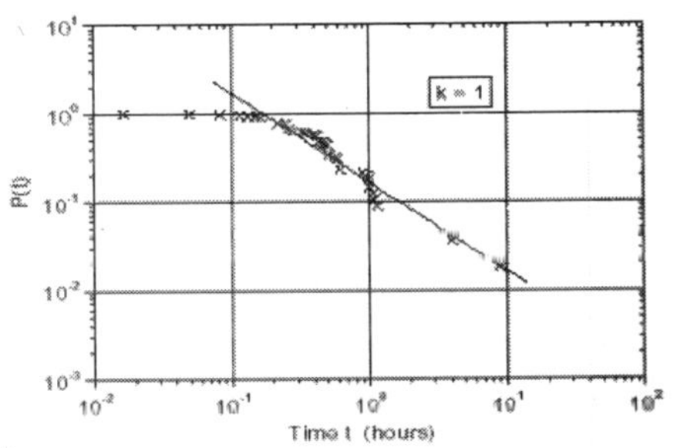

Fig. 2: P(t) Versus LOSP duration (PFBR)

The LOSP frequency-duration correlation function has the same time dependence as it will be different from P(t), only by a constant factor.

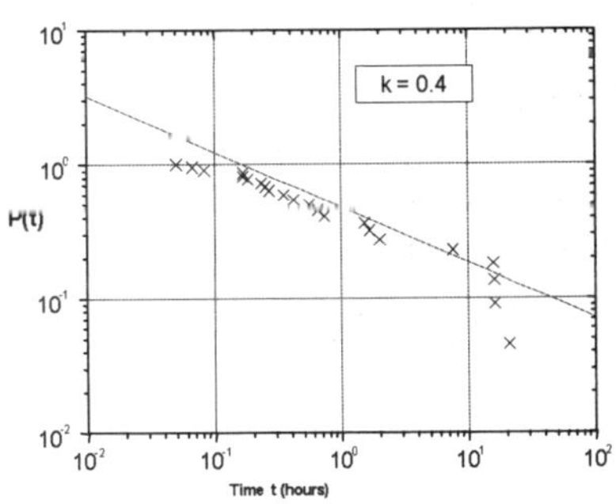

Fig 3: P(t) Versus LOSP duration (FBTR)

It is found from this analysis that power law model provides an elegant description of offsite power failure characteristics. Only two parameters need to be estimated from the observation.

3. DG UNAVAILABILITY

On failure of offsite power (class IV) supply, the backup power to the essential loads is provided by Diesel Generators (DG). Often DG unavailability is studied using Fault Tree analysis (Babar, 1988) where the minimal cut sets are quantified with a fixed mission time Tm. The resulting DG unavailability is multiplied with a (exponential) non-recovery probability as

$$U_{DG}(t \geq T) = U_{DG} \times e^{-T/T_R}$$

To get the DG unavailability as a function as a of time. Here T_R is the repair time.

There is a better approach [IAEA-TECDOC-593], where the minimal cut sets are evaluated by introducing appropriate time dependence for each of the terms. We call this approach as Time-Dependent Cut set Evaluation (TDCE) method.

3.1 Simple Fault Tree Model

Fault tree analysis is a well-established analytical tool commonly used in reliability studies. In a Fault Tree model, basic events are connected through a series of logic gates to a terminal event that usually represents the failure of the system. Most systems are represented by the basic 'AND' and 'OR' gates. An 'AND' gate represents a so-called parallel system. All of the components must fail for the system to fail. An 'OR' gate corresponds to a series system. The system fails if any of the components fail.

The fault tree for a 1/2 DG configuration is shown in Figure 4.

For the fault tree model, the probability that a DG in 1/2 mode is not available for a duration T or more, is

$$U_{DG12}(T) = U_{DG} \times e^{-2\mu T}$$
$$\approx \left[\begin{array}{c} p^2 + 2p\left(1 - e^{-\lambda Tm}\right) \\ + \left(1 - e^{-\lambda Tm}\right)^2 \end{array} \right] e^{-2\mu T} \qquad (6)$$

where U_{DG12} is the probability that the DG system will fail to start or run during a mission time, T_m.

Similarly, for 1/3 DG configuration

$$U_{DG13}(T) \approx \begin{bmatrix} p^3 + 3p^2\left(1-e^{-\lambda Tm}\right) \\ +3p\left(1-e^{-\lambda Tm}\right)^2 \\ +\left(1-e^{-\lambda Tm}\right)^3 \end{bmatrix} e^{-3\mu T} \quad (7)$$

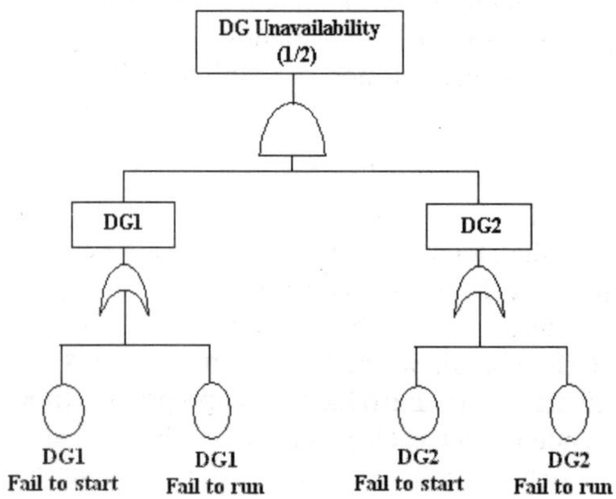

Fig. 4: Fault Tree for 1/2 DG unavailability

3.2 Time-Dependent Cut Set Evaluation Method

In this method, DG unavailability of a duration T or more is obtained from the mathematical integration of the prominent cut sets obtained from FT analysis. In 1/2 DG configuration, we have the following three events:

- Both DGs failed to start and NOT repaired in time T.
- One DG failed to start and one started successfully, but subsequently failed while running at time t and was not repaired within time T after failure. And the first DG was also not repaired in time t +T.
- Both started, but one failed while running at time t1, and second DG failed at time t2 and not repaired within time T after second failure. And the first is not repaired until t2-t1+T.

That is,

$$U_{DG12}(T) \approx p^2 e^{-2\mu T} + 2p.\lambda.e^{-2\mu T}.\int_0^{Tl-T} e^{-(\lambda+\mu)t}dt + 2\lambda^2 e^{-2\mu T}.\int_0^{Tl-T}\int_{t1}^{Tl-T} e^{-(\lambda-\mu)t1}.e^{-(\lambda+\mu)t2}dt1.dt2 \quad (8)$$

For 1/3 configuration, DG unavailability for duration T or more is

$$U_{DG13}(T) \approx p^3 e^{-3\mu T} + 3p\lambda^2 e^{-3\mu T}.\int_0^{Tl-T}\int_{t1}^{Tl-T} e^{-(\lambda-\mu)t1}.e^{-(\lambda+2\mu)t2}dt1.dt2 + 3p^2\lambda.e^{-3\mu T}\int_0^{Tl-T} e^{-(\lambda+2\mu)t}dt + 3\lambda^3 e^{-3\mu T}\int_0^{Tl-T}\int_{t1}^{Tl-T}\int_{t2}^{Tl-T} e^{-(\lambda-\mu)t1}.e^{-(\lambda-\mu)t2}.e^{-(\lambda+2\mu)t3}dt1.dt2dt3 \quad (9)$$

3.3 Comparison of Fault Tree and TDCE Method

It is observed that in fault tree, mission time (Tm) plays a significant role in the estimation of SBO frequency. The value of mission time to be chosen is not known a priori. A lower value of mission time will under predict the SBO frequency for shorter black out durations and a higher value of mission time may over predict at longer black out durations. To get accurate results an effective mission time should be used in the fault tree model. SBO frequency estimated for the PFBR site (without accounting for CCF) is shown in Figure 5. It can be seen that if the mission time (Tm) is equal to the average LOSP time, the fault tree model produces correct result consistent with the time dependent cutset evaluation model [Senthil Kumar et. al, 2003].

4. MARKOV METHOD

We present here the application of Markov model for the calculation of DG unavailability as a function of the unavailability duration. The advantages of using Markov model are easy representation of failure states and accounting for time dependence with only single to integrals.

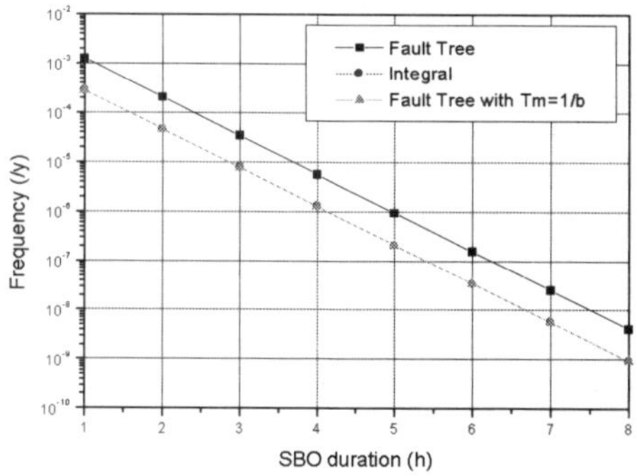

Fig. 5: Comparison of SBO Frequency Estimation For 1/2 DG Configuration

For a specific DG configuration, say, 1/2, or 2/4 DG, there will be many possibilities such as DG's failed to start, failed while running, not repaired in a given interval, maintenance etc., that will lead to the unavailability of DG. In Markov model, each such possibility is represented as states with corresponding repair rates or failure rates represented by transition from one state to another. For 1/2 DG configurations, the Markov representation is given in Figure 6. The number inside the circle indicates the number of DG that has not failed while the number in square brackets denotes the system state. This is a diagram with four states.

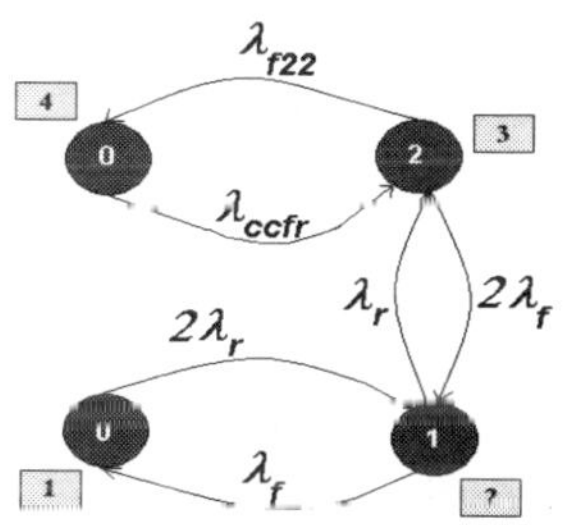

Fig. 6: Markov State Transition Diagram of 1/2 DG Configuration

where,

λ_f = Failure rate of a DG

λ_r = Repair rate of a DG.

λ_{ccfr} = CCF repair rate.

λ_{fij} = CCF rate of i out of j DG.

The probability that a system is in state, i at time t, is represented in the form of P(i, t).

$P(1, 0) = P_{fts}$^2 ; $P(2, 0) = 2\,(1-P_{fts})\,P_{fts}$;

$P(3, 0) = [1- \sum_{i\neq 3} P(i,0)\,]$; $P(4, 0) = P_{fts22}$.

where P_{fts} is DG fail to start probability and P_{ftsij} is the probability of failure to start i out of j DG due to CCF.

The time evolutions of the state probabilities are governed by the following differential equation (Daniel Gillespie, 1992).

$$\frac{d\mathbf{P}}{dt} = \mathbf{A} \times \mathbf{P} \qquad (10)$$

where A is the transition matrix,

$$A := \begin{bmatrix} -2\lambda_r & \lambda_f & 0 & 0 \\ 2\lambda_r & -\lambda_r-\lambda_f & 2\lambda_f & 0 \\ 0 & \lambda_r & -2\lambda_f-\lambda_{f22} & \lambda_{ccfr} \\ 0 & 0 & \lambda_{f22} & -\lambda_{ccfr} \end{bmatrix}$$

Once P(i, t) is obtained by solving eqn. (10), many reliability parameters of interest can be computed (Husseiny, 1982) based on the knowledge of P(i,t). For example, the DG unavailability in 1/2 mode at any time, t is computed as, $U_{DG}(t) = P(1, t) + P(4, t)$.

The probability that a DG in 1/2 mode is not available for a duration of T_{sb} or more, given a power failure interval of T_{lop} is given by,

$$U_{DG12}(T_{sb}) := e^{(-2\lambda_r T_{sb})}\,P(1,0) + P(4,0)\,e^{(-\lambda_{ccfr} T_{sb})} + \int_0^{T_{lop}-T_{sb}} P(2,t)\,\lambda_f\,e^{(-2\lambda_r T_{sb})} + P(3,t)\,\lambda_{f22}\,e^{(-\lambda_{ccfr} T_{sb})}\,dt \qquad (11)$$

Here [P(i,t) λ] conservatively approximates the pdf of system failure from the critical states i (=2 and 3). Better estimates could be obtained by making the failed states absorbing. Figure 7 gives the state transition diagram for 2/4 DG configuration. The starting state for 2/4 DG configuration is not indicated in the diagram. The states involving common cause failures (e.g., 2CF, 1RF means 2 DGs have failed due to a common cause and 1 due to independent cause) are indicated.

The initial conditions, P(i, 0) for 2/4 configuration are as follows:

$P(1,0) = P_{fts}$^4; $P(2,0)=4 \times P_{fts}$^3 $\times (1-P_{fts})$;

$P(3,0) = 6 \times P_{fts}$^2 $\times (1-P_{fts})$^2;

$P(4,0) = 4 \times P_{fts} \times (1-P_{fts})^3$;

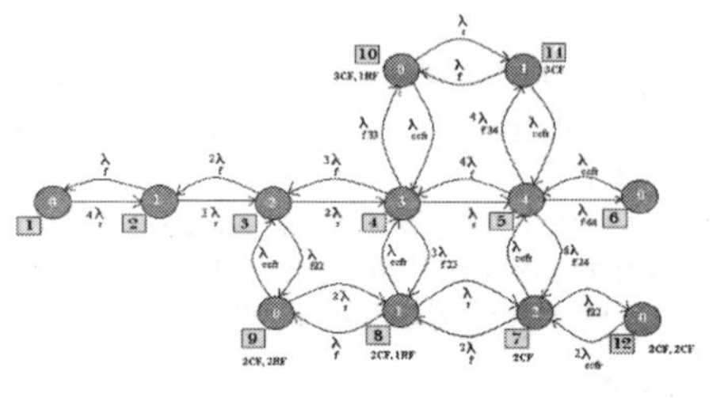

Fig. 7: Markov State Transition Diagram of 2/4 DG Configuration

$P(5,0) = (1- \sum_{i \neq 5} P(i,0))$ $P(6,0) = P_{fts44}$;

$P(7,0) = 6 \times P_{fts24} \times (1-P_{fts})^2$;

$P(8,0) = 12 \times P_{fts24} \times P_{fts} \times (1-P_{fts})$;

$P(9,0) = 6 \times P_{fts24} \times P_{fts}^2$;

$P(10,0) = P_{fts} \times P_{fts34}$;

$P(11,0) = 4 \times P_{fts34} \times (1-P_{fts})$;

$P(12,0) = 6 \times P_{fts24} \times P_{fts24}$;

Once the transition matrix is known, one can proceed in a similar fashion as in 1/2 DG case and obtain P(i, t). The unavailability at time t is computed as,

$$U_{2/4}(t) = \sum_{i \neq 3,4,5,7} P(i,t) \quad (12)$$

The probability that DG in 2/4 mode is not available for a duration more than T_{sb}, given a power failure interval of T_{lop}, is obtained in a similar way as in eqn. (11).

$$\begin{aligned} U_{DG24}(T_{sb}) := {} & e^{(-3\lambda_r T_{sb})} P(2,0) + P(8,0)\, e^{(-(\lambda_{ccfr}+\lambda_r) T_{sb})} \\ & + (P(6,0)+P(10,0)+P(11,0)+P(9,0))\, e^{(-\lambda_{ccfr} T_{sb})} + P(12,0)\, e^{(-2\lambda_{ccfr} T_{sb})} + \\ & \int_0^{T_{lop}-T_{sb}} P(3,t)\left(2\lambda_f e^{(-3\lambda_r T_{sb})} + \lambda_{f22} e^{(-\lambda_{ccfr} T_{sb})}\right) \\ & + P(4,t)\left(3\lambda_{f23} e^{(-(\lambda_{ccfr}+\lambda_r) T_{sb})} + \lambda_{f33} e^{(-\lambda_{ccfr} T_{sb})}\right) \\ & + P(5,t)\left(\lambda_{f44} e^{(-\lambda_{ccfr} T_{sb})} + 4\lambda_{f34} e^{(-\lambda_{ccfr} T_{sb})}\right) \\ & + P(7,t)\left(2\lambda_f e^{(-(\lambda_{ccfr}+\lambda_r) T_{sb})} + \lambda_{f22} e^{(-2\lambda_{ccfr} T_{sb})}\right) dt \end{aligned} \quad (13)$$

5. COMMON CAUSE FAILURE MODEL AND FAILURE DATA

While evaluating the reliability of a system with redundant components, Common Cause Failure (CCF) is important. The most widely used parametric CCF model is Multiple Greek Letter (MGL) method (IAEA-TECDOC-648, 1992). It is used in this study to estimate the CCF rates. Whenever there is CCF of equipment it is assumed that the failed components will be repaired as if it is a single component.

Many SBO studies consider only DG reliability and their configuration. Hence SBO frequency could always be computed as the joint failure of class IV and DG systems. However if class III system distribution components are to be considered especially with the fault tree model, then events like class III bus / switch gear failure directly leads to SBO.

Failure Data

Data on DG failures observed in MAPS for a period of 12 years and data observed in FBTR for a period of 10 years is used in following analysis. The average failure rate while running is found to be 1.1E-3/h and the average probability of failure on demand is 2.6E-3/d. These estimates to be used in the computation of the unavailability of the DG in 1/2 and 2/4 mode are given as follows:

Q_{tr} = 1.1E-3/h Q_{ts} = 2.6E-3/d

β = 0.147 γ = 0.528

δ = 0.505 β_2 = 0.035

β_3 = 0.087 γ_3 = 0.351

$\lambda_f = (1-\beta) \times Qtr$ $\lambda_{f24} = (1-\gamma) \times \beta \times Qtr/3$

$\lambda_{f34} = (1-\delta) \times \gamma \times \beta \times Q_{tr}/3$

$\lambda_{f44} = \beta \times \gamma \times \delta \times Qtr$ $\lambda_{f22} = \beta_2 \times Q_{tr}$

$\lambda_{f33} = \gamma_3 \times \beta_3 \times Qtr$

$\lambda_{f23} = (1-\gamma_3) \times \beta_3 \times Q_{tr}/2$

$P_{fts} = (1-\beta) \times Qts$

$P_{fts24} = (1-\gamma) \times \beta \times Qts/3$

$P_{fts34} = (1-\delta) \times \gamma \times \beta \times Q_{ts}/3$

$P_{fts44} = \beta \times \gamma \times \delta \times Qts$

$P_{fts23} = (1-\gamma_3) \times \beta_3 \times Q_{ts}/2$

$P_{fts22} = \beta_2 \times Qts$

λ_r = 0.125 / h

λ_{ccfr} = 0.125 / h and 0.25 / h.

where, β, γ and δ are the CCF parameters, and

Q_{tr} = Total failure rate of a DG to run.

Q_{ts} = Total failure probability of a DG to start.

T_{sb} = SBO duration.

T_{lop} = Loss of power duration.

The probability that DG in 1/2 and 2/4 mode are not available for duration more than T_{sb} for PFBR, given a power failure interval of T_{lop} is given in Figure 8. It is plotted for CCF repair time of 4 h and T_{lop} = 24 h.

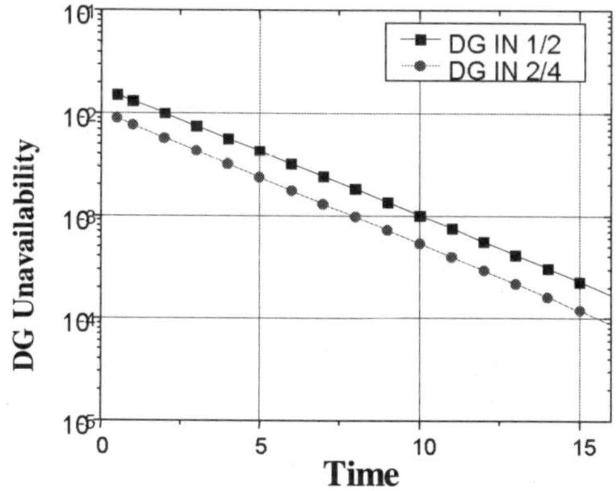

Fig. 8: Unavailability of DG Exceeding Specified Duration

6. STATION BLACKOUT FREQUENCY

Simple Fault Tree Method: As mentioned earlier, station blackout means the simultaneous failure of offsite (class IV) and onsite (class III) power systems of the plant.

> **Frequency of station blackout of (duration ≥ t)** = Frequency of class IV power failure (of duration ≥ t) * Unavailability of class III power * Probability of non recovery of class III power in time, t.

In the simple fault tree method station blackout frequency is determined approximately by combining the model for the frequency of LOSP with the unavailability of DG over a duration of interest (Arul, et al., 2001) i.e.,

$$\lambda_{SBO}(T) = U_{DG}(T) \times \lambda_{LOP}(T).$$

$$= U_{DG} \times e^{-T/T_R} \lambda_{LOP}(T) \quad (14)$$

Time-Dependent Cut set Evaluation Method: However, in the Time-Dependent Cut set Evaluation (Integral) method the SBO frequency as a function of blackout duration is estimated is obtained more accurately, by combining the expression for $\lambda_{LOSP}(t)$ with the DG unavailability equation (8) for 1/2 DG configuration,

$$\lambda_{SBO^2}(T) \approx p^2 e^{-2\mu T} \lambda_{LOP}(T) + 2p\lambda.e^{-2\mu T}.\int_0^{Tl-T} e^{-(\lambda+\mu)t} \lambda_{LOP}(t+T)dt + 2\lambda^2.e^{-2\mu T}.\int_0^{Tl-T}\int_0^{Tl-T} e^{-(\lambda-\mu)t1}.e^{-(\lambda+\mu)t2} \lambda_{LOP}(t2+T)dt1.dt2 \quad (15)$$

Markov Method: SBO frequency is obtained by combining the expression for $\lambda_{LOSP}(t)$ with the DG unavailability equation (11). It is simpler than that of the time dependent cutset evaluation model as only single integrals are always involved for numerical computations.

$$\lambda_{SBO12}(T_{sb}) := \left(e^{(-2\lambda_r T_{sb})} P(1,0) + P(4,0)\, e^{(-\lambda_{ccfr} T_{sb})}\right)\lambda_{lop}(T_{sb}) + \int_0^{T_{lop}-T_{sb}} \left(P(2,t)\lambda_f e^{(-2\lambda_r T_{sb})} + P(3,t)\lambda_{f22} e^{(-\lambda_{ccfr} T_{sb})}\right)\lambda_{lop}(t+T_{sb})\,dt \quad (16)$$

Similarly for 2/4 DG system $\lambda_{SBO2/4}(T_{sb})$ is obtained as

$$\lambda_{SBO24}(T_{sb}) := \Big(e^{(-3\lambda_r T_{sb})} P(2,0) + P(8,0)\, e^{(-(\lambda_{ccfr}+\lambda_r) T_{sb})} + (P(6,0) + P(10,0) + P(11,0) + P(9,0))\, e^{(-\lambda_{ccfr} T_{sb})} + P(12,0)\, e^{(-2\lambda_{ccfr} T_{sb})}\Big)\lambda_{lop}(T_{sb}) + \int_0^{T_{lop}-T_{sb}} \Big(P(3,t)\left(2\lambda_f e^{(-3\lambda_r T_{sb})} + \lambda_{f22} e^{(-\lambda_{ccfr} T_{sb})}\right) + P(4,t)\left(3\lambda_{f23} e^{(-(\lambda_{ccfr}+\lambda_r) T_{sb})} + \lambda_{f33} e^{(-\lambda_{ccfr} T_{sb})}\right) + P(5,t)\left(\lambda_{f44} e^{(-\lambda_{ccfr} T_{sb})} + 4\lambda_{f34} e^{(-\lambda_{ccfr} T_{sb})}\right) + P(7,t)\left(2\lambda_f e^{(-(\lambda_{ccfr}+\lambda_r) T_{sb})} + \lambda_{f22} e^{(-2\lambda_{ccfr} T_{sb})}\right)\Big)\lambda_{lop}(t+T_{sb})\,dt \quad (17)$$

The SBO frequencies obtained for 2/4 and 1/2 DG configuration are depicted in Figure 9.

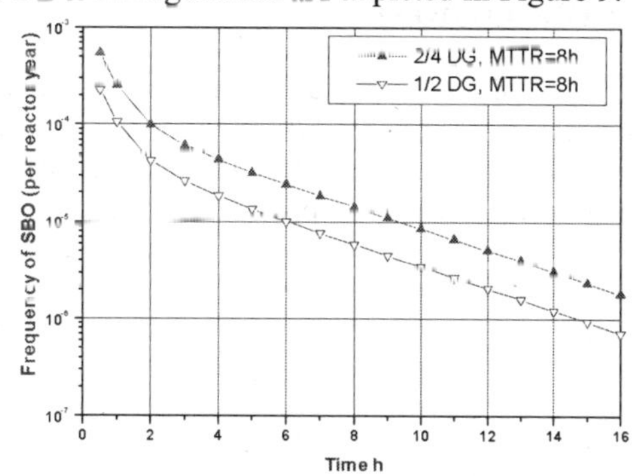

Fig. 9: Frequency of SBO Events Exceeding Specified Duration

A comparison of SBO frequency estimation for a 1/2 DG configuration using the simple fault tree method, time dependent cut set evaluation method and Markov method for the actual data collected from FBTR is shown in Figure 10.

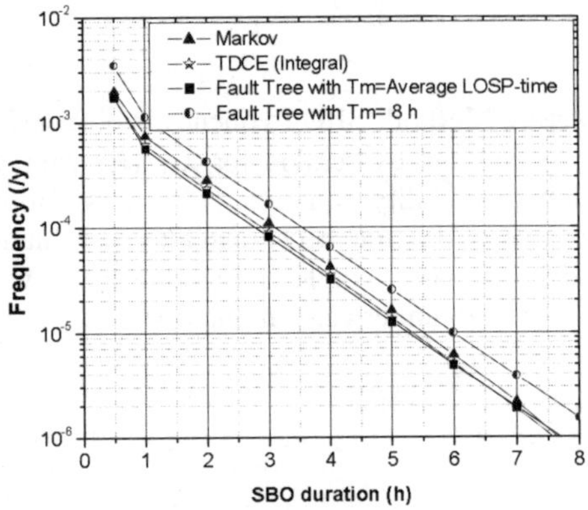

Fig. 10: Frequency of SBO for Time Exceeding Specified Duration

From the figure it can be seen that the simple fault tree method over predicts at least for 1/2 DG configuration compared to the other two methods. In terms of accuracy, the TDCE and Markov methods are comparable. The simple FT method also gives good result for the 1/2 DG case, when the mission time is chosen as average LOSP duration. The difference between simple fault tree and Markov method for higher DG redundancies is expected to be similar. The simple fault tree approach could be used only where approximate SBO frequency estimates are needed.

7. CONCLUSIONS

In this study, a power law model with adequate physical basis is proposed for describing offsite power failure characteristics. The Markov model (state space method) applied for the calculation of onsite power system (DG) time dependent unavailability is a versatile approach compared to the time dependent cut set evaluation method and it is more accurate compared to the simple fault tree approach. The disadvantage with Markov model is that it cannot be used for systems with large number of components. However in emergency DG power supply case, it has been observed that only DG reliability and their redundancy configuration have large influence on the likelihood of SBO events. Therefore we conclude that Markov method is better suited for SBO applications.

8. REFERENCES

1. Arul A.J., Sanyasi Rao, V.V.S., C. Senthil Kumar, C., Saraf, R.K. and Venkataratnam, M. (2001). Estimation of Station Blackout Frequency, IGCAR Internal Report, PFBR/01160/DN/1000.
2. Atwood C. L. et al. (1998). Evaluation of Loss of Offsite Power Events at Nuclear Power Plants: 1980 – 1996, NUREG/CR-5496, INEEL/EXT-97-00887.
3. Babar, A.K., et al. (1988). "Probabilistic Safety Assessment of Narora Atomic Power Station", BARC report.
4. Baranowsky, P.W. (1988). "Evaluation of Station Blackout Accidents at Nuclear Power Plants", NUREG-1032, USNRC.
5. Daniel Gillespie T. (1992), Markov Process, Introduction for Physical Scientists, California: Academic Press.
6. IAEA-TECDOC-593 (1991). "Case study on the use of PSA methods: Station Blackout risk at Millstone Unit 3".
7. IAEA-TECDOC-648 (1992). Procedures for Conducting Common Cause Failure Analysis in Probabilistic Safety Assessment.
8. Iman, R. L., Stephen C. Hora (1988). Modelling Time to Recovery and Initiating Event Frequency for Loss of Off-Site Power Incidents at Nuclear Power Plants, NUREG/CR-5032.
9. Senthil Kumar, C., John Arul, A.and Om Pal Singh. (2003). Comparison of Fault Tree and Integral Models for the Estimation of Nuclear Power Plant Blackout Frequency, Eleventh International Conference on Nuclear Engineering, Tokyo, Japan, April 20-23.
10. Takahashi, H., Fukuda, M., Inoue S. and Hirano, M. (1999). "Level 1 PSA for Japanese PWR Plant During Low Power and Shutdown Operation", 5th Korea-Japan PSA Workshop, Korea.
11. Wyckoff, H. (1996). "Losses of Offsite Power at U.S. Nuclear Power Plants – Through 1995", EPRI TR-106306.

Reliability, Safety and Hazard
Editors: P.V. Varde et al.

PROBABILISTIC SAFETY ASSESSMENT OF KK-NPP

Alok Mishra and Ashok Chauhan

Nuclear Power Corporation of India Limited, Mumbai 400 094

ABSTRACT

The risk of operation of nuclear power plants are quantitatively analysed by probabilistic safety analysis (PSA). PSA has become a standard nuclear safety tool, and while it does not replace the deterministic analysis, PSA results are being considered in various phases from plant design to operational practices. Safety functions for preventing or mitigating accidents and the associated systems necessary to carry out the safety functions are evaluated by these analyses. This paper presents the methodology and steps involved in the KK-PSA analysis. The analysis results for the cumulative core damage frequency, the uncertainty analysis, the sensitivity analysis are also discussed in the paper. The distribution of contributions to the core damage frequencies for different initiating events is presented and the reason for some of the highest contributors is discussed.

1. INTRODUCTION

The risk of operation of nuclear power plants are quantitatively analysed by probabilistic safety analysis (PSA). The probabilistic approaches to evaluate the safety of a nuclear power plants have been used since the mid-seventies and today it is a mature technology that can provide a quantitative assessment of the risk from accidents in nuclear power plants. It involves the development of models that delineate the response of the systems and operators to accident initiating events. In PSA additional models are also generated to identify the component failure models required to cause the accident mitigating systems to fail and also evaluate system failure probability/ unavailability. Each component failure mode is represented as an individual "basic event" in the system models. Estimates of uncertainty in risk are obtained by propagating the uncertainty distributions for each of the parameters through the PSA models.

Today plant specific PSA is a regulatory requirement in most of the countries with commercial nuclear power. PSA has become a standard nuclear safety tool, and while it does not replace the deterministic analysis, PSA results are being considered in various phases from plant design to operational practices. Safety functions for preventing or mitigating accidents and the associated systems necessary to carry out the safety functions are evaluated by these analyses.

Thus it is generally recognised that the primary objective of PSA studies is to evaluate the robustness of the actual safety status of a plant, and to identify priorities in safety upgrading measures.

In international practice three levels of PSA have evolved [4]:

Level 1: The assessment of plant failures leading to core damage and determination of core damage frequency.

Level 2: The assessment of containment response leading to the determination of release and release frequencies.

Level 3: The assessment of off-site consequences leading, together with the results of level 2 analysis, to estimate of public risks.

This paper discusses the results of the PSA Level-1analysis for the KudanKulam Nuclear Power Plant (KK-NPP). This work is carried out in Russian Federation (RF) with active participation from the Indian side.

2. DESCRIPTION OF KK-NPP

A nuclear power project is being set-up at KudanKulam (KK) in Tamil Nadu State of India in collaboration with the Russian federation. The project comprises two units each of 1000 MWe Water Water Energy Reactor, popularly known as VVER type reactor. The VVER (version V-412) reactor that is under construction at Kudankulam site is an advanced PWR. The two units of KudanKulam Nuclear Power Project (KKNPP) are scheduled to achieve first criticality in the year 2007-08.

V V E R is an acronym for *"Voda Voda Energo Reactor"* meaning water-cooled, water moderated energy reactor. This type of reactor uses 3.92% enriched uranium as fuel. The VVER reactors belong to the family of the Pressurized Water Reactors (PWRs), which is the predominant type in operation, world over. The advanced 1000 MWe design of VVER (VVER-1000) has many variants in different countries, which are derived from the basic VVER model V-392. The models of these plants have some modifications based on the client - country requirements [8]. The VVER reactor that is under construction at KudanKulam site is an advanced PWR i.e. VVER NSSS model Version V-412, which incorporates all the features of a modern PWR as per the current Russian, Western, and IAEA standards [12]. The KK-VVER has a three-year fuel cycle. This reactor requires annual refueling of one third of the core i.e., approx. 55 fuel assemblies.

The reactor plant consists of four circulating loops and a pressurizing system connected to the reactor with each loop containing a horizontal steam generator, a main circulating pump and passive part of Emergency Core Cooling System (Accumulators). The loops are connected with the Reactor Pressure Vessel Assembly by interconnecting piping. The reactor plant also consists of a reactor protection and regulation system, engineered safety features, auxiliary system, fuel handling and storage system etc [9].

3. SAFETY FEATURES OF KK-NPP

The engineered safety features are provided to mitigate and limit the consequences of abnormal conditions including accident conditions. This task is accomplished by making the reactor sub-critical, cooling and maintaining the water level in the core, limiting the rise of fuel temperature, containing the radioactivity release from the core and safeguarding various system from over pressure. The KudanKulam VVER-1000 incorporates advance engineered safety features apart from standard safety features of PWRS. The active safety systems have high redundancies and are backed by passive safety systems.

The general design concepts of the safety systems in the KK-VVER are as follows.

a. The safety systems are designed on fail-safe principle and are capable to function under loss of power supply.
b. Each active safety system has four trains 4 x 100 %, i.e. each capable to perform completely the intended safety function except a few like emergency boron injection system which has 4 x 50 % capacity.
c. The numbers of the safety trains are chosen proceeding from the consideration that one train may not be available due to single failure criteria, another may not be available due to maintenance and the third may be knocked out because of a postulated initiating event.
d. Each active safety system is backed up by passive safety system and each is capable to perform the intended safety function.
e. The safety system trains are completely segregated in terms of physical location, layout and source of motive force.
f. The design of control safety systems are such that a failure in the system causes actions directed to ensure safety.
g. The safety system actuates on demand automatically and operator actions to disable an automatically actuated safety system are blocked in the first 30 minutes to exclude human error.
h. All safety systems are supplied with power from independent sources (Dedicated Diesel Generators), designed in accordance with

the requirements of the protective safety systems.

4. APPROACH

The comprehensive Level-1 PSA for KK-NPP for internal initiating events are based on the Worldwide applicable methodology of event trees and fault trees.

Guidelines for the analysis are adopted suitably from the IAEA Safety Guidelines, IAEA TECDOCS and NUREG documents [1,2,4]. The main steps of the PSA methodology are discussed in the following sections.

Selection of the Initiating Events [6,7,9]

An essential requirement for the PSA is the availability of a comprehensive list of initiating events that have the potential to lead to core damage. A comprehensive list of the initiating events was prepared by (i) engineering evaluation of the plant (ii) Review of the design documentation for "Kudankulam" NPP available until now (PSAR), (iii) Review of the operating documents on incidents on Balakovo nuclear power plants with WWER-1000 reactors (V-320) (reference NPP for "Kudankulam" NPP). (iv) reference list available in literature. Subsequently the initiating events are grouped based on the demands they place on safety functions, frontline systems and support systems. For KK-NPP the IEs are grouped into 17 IEs.

Event Sequence Modelling

Event trees are used for event sequence modelling. Small event tree large fault tree approach is used in this analysis. Event trees with front line systems as headings are developed based on knowledge of plant responses. For each event tree handling suitable boundary conditions are incorporated whenever necessary to reflect the effect of the conditions imposed by the initiating events on the mitigating action and any other condition that developed during the event progression. Event trees are expressed in-terms of initiating events and failures of mitigating systems, in various combinations. The following are the end states considered in the event trees:

- **CD** – core damage state;
- **CSS** – cold shutdown state – no damage to the reactor core;
- **HSS** – hot shutdown state – no damage to the reactor core;
- **RVS** – reactor vessel overcooling.

However, only CD has been considered in the analysis.

Initiating Event Data Sources and Frequency Estimation

The following data sources are used for consideration of initiating event data:

a. Russian Federation (R.F) National database on incidents in NPP operation that contains information on events at Balakovo NPP and other NPPs of VVER-1000 type operating in Russia and Ukraine [11];
b. Statistics collated by OKB Gidropress on all the VVER-1000 reactor facilities [6] .
c. Results of fracture mechanics calculations performed by OKB Gidropress [7].
d. Evaluation of logic models such as fault trees developed within system analysis of NPP "Kudankulam".

For IE frequency estimation the following steps are followed:

- For derivation of the frequencies of such groups of rare initiating events as reactor pressure vessel rupture or large LOCA, fracture mechanics studies for estimation of probabilities of vessel failure and pipe breaks are used.
- Estimation of frequencies of other initiating events is performed using Bayes updating procedure based on operating experience of reference and other NPPs with VVER-1000 reactors.
- Frequency of frequent initiating events (more frequently than once per year) is estimated based on operating experience of Balakovo and Zaporizhe NPPs for last ten calendar years.
- Frequency of other initiating events is derived from available operating experience of VVER-1000s provided with reliable data except for initial period of Unit commissioning.

- Besides, for frequency estimation of initiating events caused by such system failures as failure of two trains of service water system, fault tree analysis is used.

However, beyond design-basis events characterized by catastrophic damage of the reactor vessel or steam generator headers are excluded from the analysis.

Data Base For Component Reliability

The component failure data used are a combination of both plant specific and generic data. The various data sources used are as follows:

a. Plant specific information containing event records that reflects VVER NPP operating experience and has been gathered at Balakovo NPP [2].
b. Data from Russian Reliability Data Base (VNIIAES) [11] on failures at VVER NPPs within boundaries of the former Soviet Union (Ukraine, Zaporozjhe NPP).
c. IAEA-TECDOC-478 is used to cover PSA needs in the areas where the lack of plant specific information has been identified (mainly electrical items).

The parameter estimation is used for both the cases of data i.e.,

- when the plant specific data sample is representative
- when plant specific data sample is not representative

Common Cause Failure Analysis

In KK-PSA the common cause failure analysis is carried out by using α factor model. As no specific information related to CCF is available from VVER-1000 (V-320) plant experience, a source of generic data, namely NUREG/CR-5497 [1] is used.

Human Reliability Analysis

Human reliability analysis is generally performed to quantify the human error probability and this is considered in three ways in the PSA model.

- Pre initiators/latent human actions – Actions that cause equipment or systems to be unavailable when required post-fault due to errors that occurred pre-fault.
- Initiators that either by themselves or in the combination with equipment failures lead to IEs.
- Post initiators/dynamic human actions

The pre-initiators are generally implicit in the quantification of the IEs and system un-availabilities contribute as part of their total frequency. The latent errors are quantified using screening values. The post accident personnel error is considered by using human cognitive reliability model for quantifying dynamic actions giving due credit to performance shaping factors.

Importance, Sensitivity and Uncertainty Analysis

In the PSA analysis the Importance, Sensitivity and Uncertainty Analysis are carried out with the following objective

- The importance analysis is performed for the ten most basic events and all groups of IE. The analysis of importance is performed with account for the Fussel-Vesely criteria.
- Sensitivity analysis is performed for the effect of basic events on total CDF.
- The analysis of uncertainty of the final result is performed for the whole design with respect to consequences of CD type. Mean values of reliability indicators for separate components and the respective values of the error factor is taken as the input data for the analysis of uncertainty and for all components log-normal distribution for reliability indicators is used.

5. RESULTS AND DISCUSSION

The quantitative estimation of the core damage frequency and analysis of importance and uncertainties are performed by using the computer code RISK SPECTRUM [10]. The calculations are made for the group of the final states characterized by the consequence of core damage.

The result of estimation of contributions to total core damage frequency from separate initiating event groups are given in Table-1. It is seen that the total value of cumulative core damage frequency from all IE groups considered in this analysis is 2.38E -7 /Year.

The distribution of contributions to total CDF value from separate IEs are given in the Table-1 and is depicted in Figure-1. The contributions to total CDF value from IE at power operation and shutdown modes are 87.1% and 12.9 % respectively.

The highest contributions into total CDF value give large (37.84 %) and small (17.5 %) LOCAs inside the containment, leakages from the primary to the secondary circuit (14.9 %), small leakage from the primary circuit outside the containment (6.14 %), leakage of the steam line in unisolable (7.29 %) and isolable (1.32 %) parts and loss of offsite power at power operation (1 %) and shutdown mode (12.9 %). The contributions from other IE groups are less then 1 %. The main cause of high contribution from LOCA is due to the assumption of high IE frequencies for such event and the conservative assumption that the failure of 1st stage Hydro Accumulators leads to core damage.

The sensitivity analysis is reported for the ten basic events in Table-2. The sensitivity is defined as the ratio of total CDF values estimated for basic event values increased ten times to total CDF values estimated for basic event values decreased ten times. The Fussel-Vesely importance measures for the ten most important basic events, including events corresponding to initiating events is given in Figure-2.

Table 1 Core Damage Frequency For Different IEs

Initiating events	**Initiating events frequency**	**Core damage frequency**	**Contribu-tion, %**
Power operation			
Small leakage of the primary circuit inside containment	3.82E-03	4.25E-08	17.50 %
Small leakage of the primary circuit outside containment	3.60E-03	1.49E-08	6.14 %
Inadvertent opening of the safety valve of the pressurizer	3.60E-03	4.23E-12	0.00 %
Large leakage of the primary circuit	5.91E-04	9.19E-08	37.84 %
Leakage from the primary circuit to the secondary circuit. Conventional diameter is 13 mm	3.60E-03	1.81E-08	7.45 %
Leakage from the primary circuit to the secondary circuit. Equivalent diameter is 100 mm	3.60E-03	1.81E-08	7.45 %
Leakage of the steam line in a section isolated from the steam generator	3.90E-02	3.21E-09	1.32 %
Leakage of the steam line in a section not isolated from the steam generator	3.90E-02	1.77E-08	7.29 %
Leakage of feed water pipelines in a section isolated from the steam generator	4.60E-02	3.47E-10	0.14 %
Loss of normal heat removal through the turbine condenser	3.90E-01	6.23E-10	0.26 %
Loss of offsite power	1.00E+00	2.40E-09	0.99 %
General transients	1.10E+00	1.67E-09	0.69 %
Inadvertent closure of a one fast-acting isolating valve	1.20E-01	1.69E-10	0.07 %
Failure in the service water system (PE) and the intermediate circuit for users of the reactor compartment (KAA)	5.30E-02	7.48E-11	0.03 %
Core damage frequency for power operation		**2.12E-07**	**87.11 %**
Shut down			
Loss of normal heat removal from the core with the reactor depressurized	2.60E-06	3.90E-11	0.02 %
Loss of off-site power with reactor depressurized	2.30E-03	3.13E-08	12.89 %
Core damage frequency (shut down)		**3.13E-08**	**12.89 %**
Total core damage frequency		2.38E-07	100 %

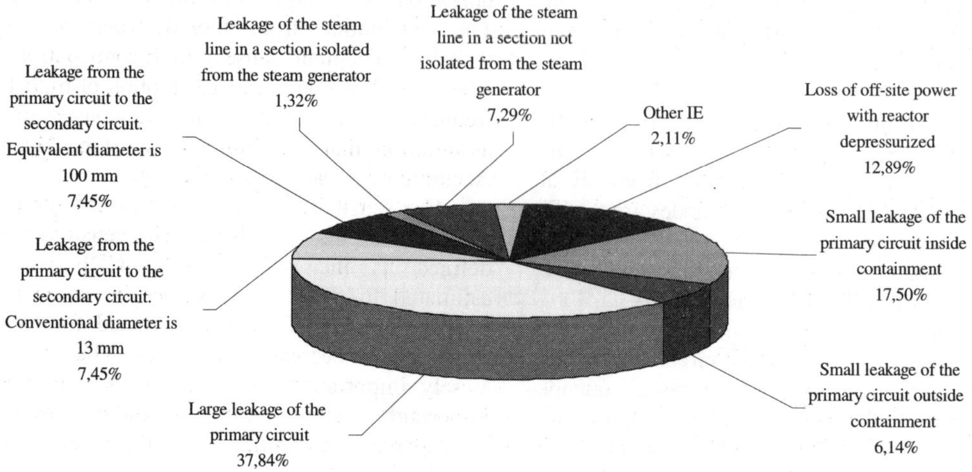

Fig. 1: Distribution of contributions to the value of core damage frequency over initiating events

Table 2 Sensitivity Analysis

Sl. No	Basic Event code	Description	Sensitivity
1	L3.	IE	3.082E+00
2	SL.	IE	2.978E+00
3	L5.	IE	2.508E+00
4	L4.	IE	2.503E+00
5	HUM-JEF	Operator does not open SVP	2.478E+00
6	C&I-LBA40AA003	Failure of control system	2.172E+00
7	JNG10-40-1-ALL	Common cause failure of CVs JNG10-40AA601-open	2.055E+00
8	JNG10-40-2-ALL	Common cause failure of CVs JNG10-40AA602-open	2.055E+00
9	LBA40AA004VMC	MOV fails to close	2.043E+00
10	SGCL.	IE	1.933E+00

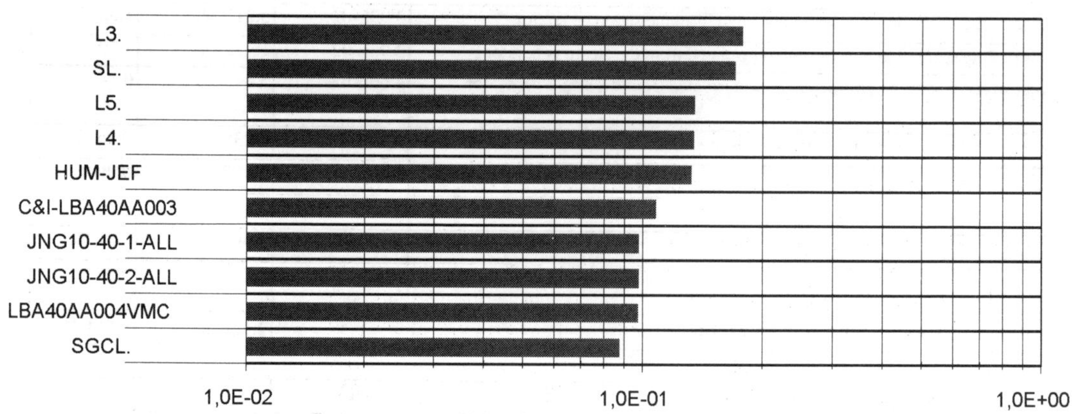

Fig. 2: Fussel-Vesely importance measure for the ten most important basic events

The analysis of uncertainty of the final result is performed for the whole design with respect to consequences of CD type. Mean values of reliability indicators for separate components and the respective values of the error factor will be taken as the input data for the analysis of uncertainty. For all components log-normal distribution of reliability indicators is used. The mean value of the CDF is found to be 2.38E-7/yearr, the median is 1.24E-07/year, the lower boundary 90% of the confidence interval is 3.09E-08 /year and the upper boundary 90% confidence interval is 8.26E-07 /Year.

6. CONCLUSION

The low CDF 2.38E-7 /Year computed by KK-NPP Level-1 PSA indicates that the KK-NPP is an advanced pressurised water reactor. The features of 4-train frontline safety systems and the features of active and passive system has resulted in low safety function unavailability. The high level of redundancy has also resulted in lower contribution of CCF events towards the core damage frequency. Different modes of operation and different states (operational or standby) of equipment provide additional protection from CCF.

7. REFERENCES

1. Common-cause failure parameter estimations. NUREG/CR-5497, October 1998, USA.
2. Data categorization and initiating event frequency calculation for NPP with VVER-1000 type reactors for the period from 1986 to 01.01.2000, VNIIAES report, Moscow, 2000
3. Guidelines on modelling Common-cause Failures in Probabilistic Risk Assessment NUREG/CR-5485, November 1998, USA.
4. IAEA Safety Series 50-P-, "Procedures for Conducting Probabilistic Safety Assessments of Nuclear Power Plants (Level 1)".
5. IAEA TECDOC-478, Component reliability Data for use in Probabilistic Safety Assessment, Vienna, Austria.
6. NPP "Kudankulam" Unit 1,2. Preliminary Safety Analysis Report. Probabilistic safety assessment report. Frequency of initiating events Working document. EDO "Gidropress", Package Number St-2.18, Number of the Document KK. UJA.JA.TM.VAB.PR013 412-Пр-120, 2001.
7. PP "Kudankulam" Unit 1,2. Preliminary Safety Analysis Report. Probabilistic safety assessment report. Data on initiating event frequencies on the basis of analyses by fracture mechanics methods. Working document. EDO "Gidropress", Package Number St-2.18, Number of the Document KK. UJA.JA.TM.VAB.PR013 412-Пр-285, 2000.
8. KudanKulam Project Technical Assignment, NPCIL, Mumbai, 1998.
9. Preliminary Safety Analysis Report for KudanKulam Nuclear Power Plant, AEP, Moscow, 2002.
10. SK SPECTRUM, User's Manual, Version 2.1, Relkon Teknik AB, Box 1288, S-172 25 Sundbyberg, Sweden, 2002.
11. Reliability data collection on safety related equipment at VVER-1000 NPPs for the period 1990…1999 years. VNIIAES, 2000.
12. Seminar on Safety of VVER Reactors, Chennai, 5th-6th November 1998.

Reliability, Safety and Hazard
Editors: P.V. Varde et al.

EXPERIENCES WITH APPLICATIONS OF PSA RESULTS FOR OPTIMIZATION/REVISION OF TECHNICAL SPECIFICATIONS FOR OPERATION OF NUCLEAR POWER PLANTS

A.K. Vijaya, Rajee Guptan, Nalini Mohan, S.G. Ghadge and S.S. Bajaj

Directorate of Safety, Nuclear Power Corporation of India Limited,
Nabhikiya Urja Bhavan, Anushakti Nagar, Mumbai 400 094

ABSTRACT

Over the past few years, NPCIL has performed comprehensive Level-1 Probabilistic Safety Assessment (PSA) for a 220 MWe Pressurised Heavy Water Reactor (PHWR) at Kakrapar Atomic Power Station (KAPS) and for the first 540 MWe PHWR at Tarapur Atomic Power Project (TAPP-3&4). The major objective of these PSAs was to present an integrated picture of the safety of the plant to identify and understand key plant vulnerabilities. As a result of the availability of these PSAs, there is a desire to use them to operate the plants in the most efficient manner practicable. In recent years, the operation of Indian Nuclear Power Plants has been characterized by improved availability/capacity factors and reduced forced outages. Frequency of planned outages is also being reduced. In order to achieve this, the PSAs are now being used as an engineering tool for optimization of Technical Specifications with regard to Allowed Outage Time (AOT) and Surveillance Test Intervals (STIs).

USNRC Regulatory Guides have established Probabilistic Safety Criteria (PSC) which can be used as guidelines for optimization of technical specification, taking into consideration Instantaneous (conditional) Risk, Cumulative Risk over the AOT period, and Average Risk over a long period. This paper discusses the approaches followed for such optimization. Case studies are also presented.

Keywords: PSA, Allowed Outage Time, Surveillance Test Interval

1. INTRODUCTION

Probabilistic Safety Assessment (PSA) is a standard tool in safety evaluation of Nuclear Power Plants which provides a useful, consistent and structural framework for evaluation of safety. PSA provides an integrated model of the overall safety of the plant, which puts the safety issues in perspective, and help in safety decisions related to both design and operations.

The insights derived from PSA can be used to enhance plant safety while operating the nuclear power stations in most efficient manner. PSA explores the risk significance of various aspects of plant design or operation and to explore the risk impact of changes in plant design or operation. The PSA can be used to evaluate the risk significance of each change or to define a risk measure as the basis of prioritizing various changes. PSA assist plant management to target resources where the target benefit for plant safety can be attended. A high level of confidence in the credibility of data, models and the results is an essential requirement for using PSA for various applications.

2. PSA APPLICATIONS IN INDIAN NUCLEAR POWER PLANTS

Comprehensive Level-1 PSA have been carried out for Indian Nuclear Power Plants such as KAPS-1&2 and TAPP-3&4. These PSAs were conducted for Internal Initiating Event during full power operation considering the reactor core as the main source of radioactivity.

In recent years, the operation of Indian NPPs has been characteriszed by progressively improving availability/ capacity factor due to reduced forced outages. Frequency of planned outages is also being reduced. Extensive efforts are also taken up for reducing manrem consumption. In order to achieve these, revision of some of the Limiting Conditions (LCOs) are being considered based on PSA. KAPS PSA is being used as an engineering tool for optimization of Technical Specification with regard to Allowed Outage Time (AOT) and Surveillance Test Interval (SIT) for equipment/components important to ensure safe operation of Indian PHWRs having design similar to KAPS.

Similarly TAPP-3&4 PSA was used for Optimization of AOTs and STIs to be specified in Technical Specification being prepared for TAPP-3&4

Quality Assurance

A Quality Assurance of various procedural tasks of PSA was recognized in this PSA project. Methods adopted in these, KAPS PSA ensured an end product which adequately meets the objectives.

The engineering functions in NPCIL are governed by ISO 9001. By virtue of this, the Quality Assurance aspects of these Level-1 PSA were covered by the Engineering Directorate procedures applicable for design and analysis. The completeness of the list of Initiating Events was assured through co-ordinated review and discussion meetings with experts in system design, plant operation and deterministic analysis. Small Event Tree-Large Fault Tree approach was used in this analysis. The event trees were cross-checked by independent experts to ensure that all scenarios are appropriately modeled and end states correctly defined. System modeling acquires great importance in this kind of approach (Large Fault Trees). A procedural guide RSA PROC-1 (5)was prepared to establish the guidelines for system modeling, system failure, criteria, assumptions, boundaries, basic event coding practices, review process, interface control, documentation control of inputs and results. The system Reliability Analyses were independently reviewed by design experts. For operational aspects, plant personnel were consulted. Consistent and standardized event coding (consisting of component tag no., failure mode and system identification), for representation of basic events human errors and dependent failures were adopted. Data analysis in this PSA was generally attempted in two phases for all the areas of data requirement; the first phase covering reference to generic data sources and a database compilation and a second phase covering references to plant specific data and a database development. Finally, a judicious mixture of generic and plant specific data was used to ensure realistic input data. Swedish Risk and Reliability Software Risk Spectrum was used for quantitative analysis of this PSA which ensured the correct processing of the inputs and minimal cutest analysis. Outputs of minimal cutest analysis were critically checked against physical understanding of expected outcomes.

Atomic Energy Regulatory Board (AERB), India, conducted a 2-tier review of KAPS Level-1 PSA. AERB has a standing committee for PSA constituted by PSA experts from AERB, Bhabha Atomic Research Centre (BARC) and NPCIL. A Working Group constituted by the Standing Committee conducted a detailed review of KAPS PSA Level-1. The Updated PSA based on review was the reference PSA for TAPP Level-1 PSA. This ensures the QA aspects of this PSA.

3. CASE STUDIES ON OPTIMIZATION OF ALLOWED OUTAGE TIME (AOT)

Approach

Probabilistic Safety Criteria are used to decide a Technical Specification change alongwith other traditional deterministic considerations, operating experience, etc. Evaluation of proposed Technical Specification changes are evaluated based on the guidelines provided in USNRC Guidelines [3,4].

The guidelines for increase in CDF as a function of CDF is given in USNRC 1.174 [3] is shown below:

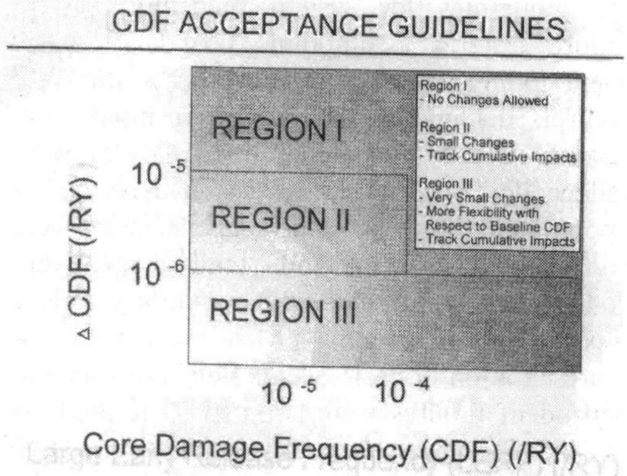

Region I:
- ❑ No changes allowed

Region II:
- ❑ Small changes
- ❑ Track Cumulative Impacts

Region III:
- ❑ Very small changes
- ❑ More Flexibility with respect to baseline CDF
- ❑ Track Cumulative Impacts

USNRC guideline 1.177 [4] gives the acceptance guidelines on Incremental Conditional Core Damage Probability (ICCDP) . ICCDP is defined as below:

ICCDP = [(Conditional CDF with the subject equipment out of service) – (Baseline CDF with nominal expected equipment vulnerability)] x Duration of AOT under consideration

An ICCDP less than 5.0 E-7 is considered acceptable for an AOT change.

These approaches were followed for the Technical Specification modifications for Indian NPPs.

The following case studies were done in NPCIL.

AOT for Unit Transformer of Kaiga-1&2

Kaiga Generating Station (KGS-1&2) are 220 MWe PHWRs of Standard Indian Design. Class-IV power supply for this station is derived by two routes: one from 220 KV grid through Start-up Transformer and another from Main Generator through Unit Transformer (UT). There is also a provision to manually connect the SUT of the other unit. The Station Auxiliary power supply is backed up by 3 x 100% Diesel Generator Sets. The failure of Current Transformers (CTs) associated with the Unit Transformers, resulted in the outage of the UTs. The technical specification calls for availability of Auto-transfer scheme and availability of two separate power sources before the startup of the unit. A proposal was considered for operation of KGS without UT. The design of Kaiga being similar to KAPS, the KAPS PSA was used to support the proposal from Probabilistic Approach. The effect of UT unavailability on Core Damage Frequency was studied. The unavailability of UT results in higher likelihood of occurrence of Class-IV power supply failure. This will increase the Class-IV failure Initiating Event Frequency. The increased Class-IV failure frequency was evaluated based on the following plant specific data:

i) Grid Failure
ii) SUT Failure
iii) Circuit Breaker Failure
iv) Failure in Fast Transfer Scheme

The Class-IV frequency hence calculated was used in the PSA model and CDF recalculated. Incremental CDF was calculated. Guidelines underlined in USNRC 1.174 and 1.177 were used for assessing the limits on increased CDF and to arrive at a suitable Allowed Outage Time (AOT) for the UTs. The increase in CDF values were marginal (0.43%). This falls in Region III as specified in USNRC 1.174. The ICCDP value was calculated as 2.48×10^{-8} and within the limits specified in USNRC 1.177. The proposal was reviewed internally by Safety Review Committee of NPCIL and was approved by regulatory authorities.

Optimum AOT for ECCS Valves

Technical Specification calls for testing of ECCS Header Injection Valves once in a month as a part of periodic surveillance. These are 2x100% valves for each Reactor inlet/outlet header. The technical specification calls that all motorized valves in ECCS shall be operable in auto mode. The failure of any one valve calls for reactor shutdown. These values are located in Shutdown access areas, and cannot be repaired when reactor is on power. KAPS PSA [1] was used to analyse the maximum Allowed Outage Time for one of the redundant ECCS Valves; the other remaining healthy. Outage of one of the Header Injection Valves for one month with increased test frequency for the redundant valve, was analysed. The incremental CDF was calculated for the changed configuration as 0.26×10^{-6}. This was judged based on Regulatory Guide 1.174. The ICCDP was also calculated. The ICCDP was calculated as 2.5×10^{-7} and compared with the criteria of USNRC 1.177. The corresponding system unavailabilities, was also examined to arrive at the optimum AOT. These values were found to be within acceptance guidelines and hence, recommended.

Fire Water Backup to Process Water

In 220 MWe Indian PHWRs, Fire Water System is providing backup to Process Water System in Shutdown Cooling Heat Exchangers, Moderator H/X and Bleed Cooler Heat Exchanger. The technical specification calls for quarterly testing of these valves. Due to the reduction in forced outages there was a requirement for increased test interval for these valves. This was examined with reference to KAPS PSA. It was found that impact on CDF was negligible and hence this was recommended for incorporation in Technical Specification.

Preventive Maintenance Duration of DG Sets in TAPP-3&4

The AOTs are generally defined for corrective maintenance. Planned preventive maintenance of DG sets during power operation are taken up for improvement equipment reliability. In addition to this Annual Preventive Maintenance jobs are taken up as recommended by DG suppliers. This involves jobs which are extensive and require duration more than AOTs specified for DG sets. A maximum period of 15 days are required for those activities. This requirement was reviewed based on TAPP-3&4 study. It was found that once in a year one DG set being out for PM activities can be permitted based on the ICCDP and ΔCDF criteria specified in USNRC [3].

Optimization of Surveillance Test Intervals (STI) for TAPP-3&4

Optimization of STI is essential to reduce unnecessary equipment wear, excessive radiation exposure to plant personnel, reduce unnecessary administrative burdens on plant personnel that are not justified by the safety significance of the surveillance requirement. The change provides operational flexibility and allows allocation of the plant personnel's time to more safety-significant aspects. An assessment of the surveillance test intervals of plant safety systems was conducted and an optimum interval could be arrived at. The surveillance test intervals of the following pumps and valves were studied for optimization. A test interval of one month was found acceptable for ECCS Pumps and Valves considering the change in CDF as well as the system unavailability. A weekly test interval was found to be acceptable for Auxiliary Boiler Feed Water Pumps. Similarly, the test interval of the shutdown system – 2 i.e. the Liquid Poison Injection System Valves and Atmospheric Steam Discharge Valves were optimized to one month. A reduction in test frequency was acceptable in this case. The DGs will be tested once in a month, if the number of failures in the last 100 valid tests is not more than 1; once in a fortnight, if the no. of failures is 2; and, once in a week, if it is more than 2. DG Sets were otherwise tested weekly. These test intervals were used in the Level-1 PSA which was being done at design stage and the CDF calculated was within the IAEA guidelines of 1×10^{-5}/yr. The testing frequency of DG Sets for new plants are being revised as per USNRC guidelines [6].

4. CONCLUSION

PSA is beginning to be employed in NPCIL to develop risk informed safety programme strategies. NPCIL with the availability of Level-1 PSA for 220 MWe and 540 MWe plants has goals for efficient use of these PSAs for applications related to Technical Specification. Optimization and Reliability Centred Maintenance. As a beginning, the above cases were attempted, giving importance to safety aspects. In all such novel applications, PSA is used as another important technique to complement engineering judgement experience, defence-in-depth concepts and design basis considerations. Also, as the uses of PSA increase, it is important that the models, the database and the calculations of PSA be kept updated over the lifetime of the plants.

5. REFERENCES

1. Probabilistic Safety Assessment Level-1 Internal Events for KAPS (April 2002).
2. Probabilistic Safety Assessment Level-1 Internal Events for TAPP-3&4 (March 2005).
3. USNRC Regulatory Guide 1.174 - An approach for using Probabilistic Risk Assessment in Risk Informed Decisions on Plant Specific Changes to the Licensing Plant (1998).
4. USNRC Regulatory Guide 1.177 - An approach for Plant Specific Risk Informed Decision Making (1998).
5. RSA PROC-1 – Preparation and Approval of Reliability Analysis Reports.
6. USNRC Regulatory Guide 1.108.

Reliability, Safety and Hazard
Editors: P.V. Varde et al.

QUALITATIVE SAFETY ANALYSIS IN ACCELERATOR BASED SYSTEMS

P.K. Sarkar and Lekha M. Chowdhury

Health Physics Unit (VECC), ARSS, RSSD, BARC
1/AF, Bidhannagar, Kolkata 700 064

ABSTRACT

In recent developments connected to high energy and high current accelerators, the accelerator driven systems (ADS) and the Radioactive Ion Beam (RIB) facilities come in the forefront of application. For medical and industrial applications high current accelerators often need to be located in populated areas. These facilities pose significant radiological hazard during their operation and accidental situations. We have done a qualitative evaluation of radiological safety analysis using the probabilistic safety analysis (PSA) methods for accelerator-based systems. The major contribution to hazard comes from a target rupture scenario in both ADS and RIB facilities. Other significant contributors to hazard in the facilities are also discussed using fault tree and event tree methodologies.

Keywords: Radiological hazard evaluation, PSA, Fault tree, Event tree, ADSS, RIB

1. INTRODUCTION

Present day high power accelerators particularly those used for accelerator driven systems (ADS) and the Radioactive Ion Beam (RIB) facilities no longer remain benign research tools for the physicists. Moreover, high current accelerators for medical and industrial purposes are required to be located in densely populated areas. There is a need for evaluating the reliability of these new accelerator systems requiring a systematic approach. At present, although a significant history of accelerator operation has been accumulated over the past 50 or so years, there is a deficiency in reliability and risk estimates of accelerator systems[1]. At the moment, despite the fact that standard reliability and risk assessment tools are suitable for accelerators, no formal database for major accelerator components is available. In the present work, we have attempted to use qualitative methods like Failure Mode Effect and Criticality Analysis (FMECA) as well as Hazards and Operability Analysis (HAZOP) for analysing accelerator based systems. Fault Tree and Event tree methods have been used to depict system failure logic, identify different initiating events and analyse the safety systems required under such circumstances.

2. SYSTEM DETAILS

For the ADS, molten Lead-Bismuth Eutectic (LBE) and Mercury are mostly used as targets. Molten LBE has certain distinct advantages and therefore, considered as a potential target to be developed for use in ADS. On the other hand, RIB targets are made of various materials depending on the required radioactive ion beam specifications. However, irrespective of the material used for target design and the actual mechanical design there exist some common issues like target cooling, containment of radioactive nuclides released from the target, reducing the leakages and reducing the dose rate received by the personnel during operation and maintenance. Typically the increase of the projectile ion beam intensities goes along with a drastic increase of the radioactive inventory in the production target area and thereby enhances other radioprotection related problems. The radioactive inventory to be handled is close, in certain cases, to that of a research reactor fuel element. We have carried out a safety analysis with ADS target as the system.

2.1 ADS Target Details

The target module consists of window, proton beam pipe, annular riser pipe, Separator, Mixer (Gas injector), Downcomer pipe, Heat-exchanger and Dump tank. In addition, there will be closed nitrogen gas loop and heating system to operate the lead-bismuth-eutectic (LBE, 45% lead and 55% bismuth) at ~ 225^0C.

Initially the LBE is melted and raised to the operating temperature (i.e. 225^0C). By pressurizing the cover gas in the dump tank, the liquid metal is transported to the loop and filled to the required level. Nitrogen gas is injected in to the loop through the mixer at appropriate pressure through a compressor. Due to this, a two-phase flow is created in the riser and this gives rise to circulation of the LBE in the loop. Liquid metal flows through the down comer, heat exchanger and window flow region. The flowing LBE extracts the heat that is deposited in the window due to proton beam and the same is rejected in the heat exchanger.

3. SYSTEM ANALYSIS

A qualitative analysis covering detailed FMECA (Failure Mode Effect and Criticality Analysis) of ADS with molten LBE target was done. In Table 1 we have shown some of the important FMECA results. In this we have found that the consequence of beam window rupture, malfunctioning of the flow meter and leak in the gas injector are likely to lead to severe consequences.

Table 2 contains some of the important results of HAZOP. We find that the flow rate near the window is to be maintained within the stipulated range; otherwise, it may lead to significant hazard. There is a very high possibility of beam misalignment. There should be some control to ensure that the beam is right on the window. If beam window ruptures, beam tube will be contaminated, leading to the possibility of entire cyclotron contamination. Polonium generated due to nuclear reactions in

Table 1 FMECA – Failure of Components, their mode of Failure, Effects of Failure and Criticality Analysis

Component	Failure Mode	Failure effect	Criticality
Beam Window / Proton beam pipe inside the target loop	Leak / rupture	Direct activation of LBE, contaminates Beam tubes	Severe
Proton beam pipe outside the target module	Leak / rupture	Vacuum loss Beam loss	Low
Separator	Leak	Air borne activity released to the atmosphere	Moderate
Gas injector	Leak	Circulation LBE decreases Release of nitrogen / LBE (containing polonium) in the cave	High
Dump tank	Leak / rupture	Spread of contamination	Moderate
Flow meters	Malfunction	Interlocks will not work	Severe

LBE, if released in air will cause significant radiological hazard. This release can occur due to any leak in the LBE loop or in the nitrogen gas loop.

The important initiating events for the LBE system are listed in Table 3., where criticality and likelihood of the events have been assessed. The most important initiating events envisaged (both criticality and likelihood high) are Window meltdown, Small crack in the window, Station blackout and Fire due to metal oxide. When the likelihood of an event is high, it affects availability of the system, so it should be ensured that the frequency of occurrence of the event is reduced. For highly critical events, safety is affected and so it should be ensured that their likelihood is very low.

Table 2 HAZOP - Hazard and Operability Analysis

Hazard	Cause	Hazard effect	Criticality
Low flow near window	Nitrogen flow inadequate	Temperature increase, possibility of window rupture	Severe
High flow near window	Excess flow of Nitrogen	Higher LBE flow rate leading to crack in window	Low
Vacuum deterioration in beam tube	Vacuum Pump Failure Or leak	High beam loss and high radiation field. Nil / reduced beam	Depends on the location
Proton beam misalignment	Faulty beam tuning/ beam transport mechanism Rupture of beam pipe.	High radiation field where beam strikes beam tube	Severe
Fire	High concentration of lead dust or powder	Highly toxic lead oxide fumes and radioactivity released in the environment	High
Explosion	Exposure to heat or incompatible materials.	Release of Radioactive elements and system damage	High
Proton beam power > 15 kW	Unlikely event restricted by machine capacity	Window crack	Moderate
Lead vapour in the ambient beyond limit	Leakage in LBE loop and vaporisation	Toxicity high	High
Radiation level in cave > limit	Leakage of active LBE	Inaccessible for a longer period	High
Polonium concentration in air > limit	Leakage in LBE or Nitrogen System	Entry restricted for a longer period	Very high

The most critical event identified i.e. Target window rupture was analysed in detail. Fault tree for that event is shown in Figure1

Table 3 List of Identified Initiating Events

	Initiating event	Criticality	Likelihood
1.	Window meltdown	Very high	Moderate
2.	Small crack in Window	High	High
3.	Hot LBE leak	Moderate	Moderate
4.	Fire	High	Low
5.	Dump tank system leak	Low	Low
6.	Dump tank over pressurization	Low	Low
7.	Lead concentration in air exceeding limits	High	Low
8.	Station blackout	High	Moderate
9.	Beam power high	Moderate	Very low
10.	Explosion	Very High	Low
11.	LBE frozen in pipe	High	Moderate
12.	Proton Beam pipe crack	Moderate	Low
13.	Nitrogen system failure	Low	High
15.	Cooling Air system failure	Very Low	High
16.	Vacuum system failure	Low	Very high
17.	Ventilation system	high	Moderate

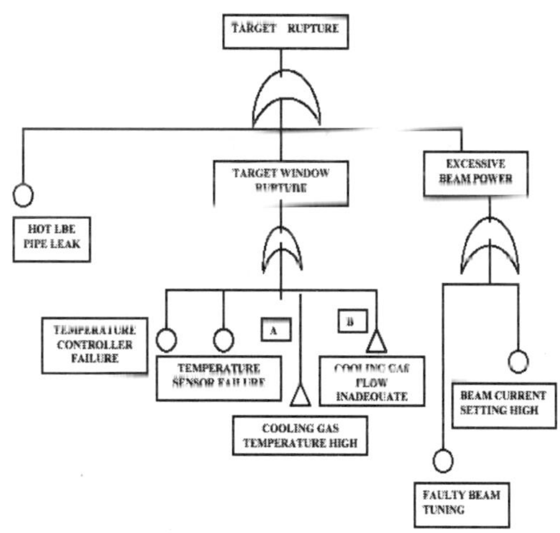

Fig. 1: Fault tree for target rupture

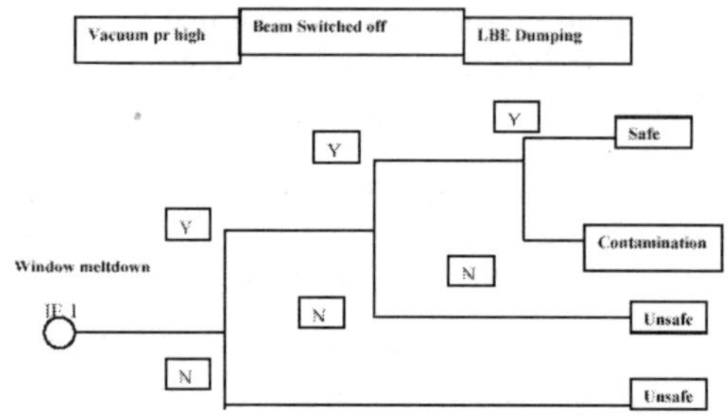

Fig. 2: Event tree for target window rupture

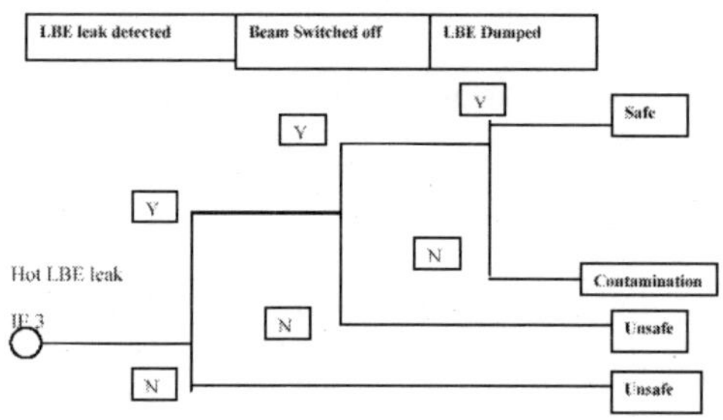

Fig. 3: Event tree for LBE leak

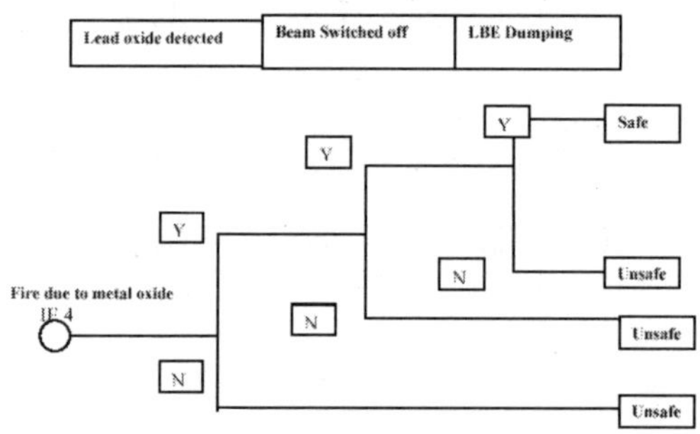

Fig. 4: Event tree for fire due to lead oxide

4. RESULTS

Event trees in case of target window rupture, LBE leak and fire due to lead oxide are given in the Figures 2-4. As can be seen the available safety systems include beam switch off and LBE dumping system. It is imperative that beam should be switched or diverted off even if LBE is dumped safely for these initiating events. LBE may be dumped automatically when the window ruptures and high pressure is sensed in the proton beam pipe. The proton beam needs to be switched off or diverted automatically at the same time. The proton beam if stopped by a closing gate valve will lead to irradiation of the valve, which consequently would give a high radiation dose during maintenance. Alternatively, we suggest the design of a beam dumping system so that the beam is automatically dumped when LBE is dumped (same signal). The beam dumping system should be shielded as required. The beam can then be manually switched off, after dumping, if required.

5. REFERENCE

1. Chowdhury L.M. and Sarkar P.K.., "Probabilistic Safety Analysis in particle accelerators", International Conference on Topical Issues in Nuclear Installation Safety, Beijing, China, 18-22 October, 2004.

Reliability, Safety and Hazard
Editors: P.V. Varde et al.

SIGNIFICANCE OF COMMON CAUSE FAILURES IN LEVEL-1 PSA AND TECHNIQUES FOR REDUCING ITS IMPACT ON CORE DAMAGE FREQUENCY

G. Srinivas, Rajee Guptan, Nalini Mohan and S.S. Bajaj

Safety Directorate, NPCIL, Nabhikiya Urja Bhavan, Anushaktinagar, Mumbai 400 094

ABSTRACT

Failure of multiple components due to a common cause represents one of the most important issues in evaluation of system reliability or unavailability. The frequency of such events has relatively low expectancy, when compared to random failures, which affect individual components. However, in many cases the consequence is a direct loss of safety system or mitigative safety function. For this reason, the modeling of a common cause failure (CCF) and its presentation in fault tree structure is of the uttermost importance in probabilistic safety analyses (PSA).

Amongst the different techniques available to reduce the impact of common cause failures, the most important operations related technique, is that of staggered testing.

In the TAPP 3, 4 Level-1 PSA it was found that the contribution of Common Cause Failures to the Core Damage Frequency (CDF) was significant. The Common Cause Failures were modeled using the Alpha Factor Model. This model is capable of evaluating the impact of staggered testing of components of the redundant standby systems. The NUREG/CR 5801 prescribes a different calculation methodology to evaluate the staggered testing on the Common Cause Failure component group. The proprietary software used for the Level-1 PSA, had a built-in non-staggered testing Alpha factor model. Hence, the factors were modified to introduce the effect of the staggered testing. There was a significant reduction in the CDF due to the introduction of staggered testing.

Keywords: Common Cause Failures, Staggered Testing, Non-staggered Testing, Core Damage Frequency, Probabilistic Safety Assessment

1. BACKGROUND

All modern technological systems are highly redundant but still fail because of common cause failures. Failure of multiple components due to a common cause represents one of the most important issues in evaluation of system reliability or unavailability. The frequency of such events has relatively low expectancy, as compared to random failures, which affect individual components. However, in many cases the consequence is a direct loss of safety system or mitigative safety function. The treatment of this kind of failures is therefore very significant to PSA (Probabilistic Safety Assessment). For this reason, the modeling of a common cause failure (CCF) and its presentation in fault tree structure is of the uttermost importance in probabilistic safety analyses (PSA).

Common Cause Failures (Definition)

These are multiple failures, which are a direct result of a common or shared root cause. The root cause may be extreme environmental conditions (fire, flood, earthquake, lightning, etc.), failure of a piece of hardware external to the system, or a human error.

Another type of failures, which are significant while discussing dependent failures are the Cascading failures. These are multiple failures initiated by the failure of one component in the system (chain reaction or domino effect.)

When several components share a common load, failure of one component may lead to increased load on the remaining ones and, thus, to an increased likelihood of failure. Such functionally dependent failures are best modeled directly into the fault tree.

One of the best-known accidents resulting from a common cause failure is the fire at Browns Ferry nuclear plant (Alabama, USA, 1975). Two operators used a candle to check for air leaks between the cable room and one of the reactor buildings, which was kept at a negative air pressure. The candle's flame was sucked in along the conduct, and the methane seal, used where the cables penetrate the wall, caught fire. The fire went on to damage 2000 cables, including those of automatic Emergency Shutdown systems (ESD's), and of all manually operated valves, except for four relief valves. With these four valves, it was possible to close down the reactor and avoid nuclear meltdown. As a feedback from the accident, the cables of the different ESDs were placed in separate conduits and no combustible filling (e.g. urethane foam) was used.

Definition of Dependent failures

From a probabilistic point of view, two events A and B are said to be independent if:

$$P(A,B)=P(A)\cdot P(A|B) \neq P(A)\cdot P(B)$$

We distinguish between Common Cause initiating events (external events) and Intersystem dependences.

External Events are those that have the potential for initiating a plant transient and increase the probability of failure in multiple systems. These types of events require a full, dedicated risk analysis. e.g. fires, floods, earthquakes, loss of offsite power

Intersystem System dependences

Functional dependences (from plant design): System 2 functions only if system 1 fails.

Shared-equipment dependences: Dependences of multiple systems on the same components, subsystems, or auxiliary components. E.g. components in different systems fed from the same electrical bus.

Physicals interactions: Failures of some system create extreme environmental stresses, which increase the probability of multiple-system failures. E.g. failure of one system to provide cooling results in excessive temperature, which cause the failure of a set of sensors.

Human-interaction dependences: Dependences introduced by human actions, including errors of omission and commission. E.g. an operator turns off a system after failing to correctly diagnose the conditions of a plant (Emergency Core Cooling System, ECCS, in Three Mile Island).

Intercomponent dependences

Events or failure causes that result in dependence among the probabilities of failure for multiple components or subsystems. The same cases *A-D* of intersystem dependences exist for intercomponent dependences as well.

2. METHODS FOR DEPENDENT FAILURE ANALYSIS

There are essentially two methods, as described below.

EXPLICIT methods: Involve the identification and treatment of specific root causes of dependent failures at the system level, in the event-and fault-tree logic. Examples are earthquakes, fires and floods, which are treated explicitly as initiating events in the risk analysis.

IMPLICIT methods: Multiple failure events, for which no clear root cause event can be identified, can be modeled using implicit, parametric models. In these methods, new reliability parameters are added to the usual list to account for dependent failures. If there were a common cause component group of three components, it would be represented as shown below.

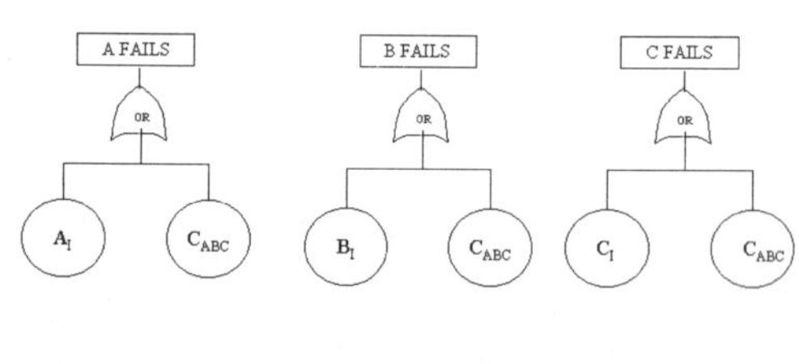

A_I, B_I, C_I = independent failure events

C_{ABC}=common cause failure event

The common cause component groups considered in this study are IDENTICAL, REDUNDANT components within a system.
The methods for modeling such inter component dependences causing common cause failures has now been fairly standardized through the various guides [1-4].

3. STAGES OF COMMON CAUSE FAILURE ANALYSIS

The most commonly accepted procedural framework of CCF consists of four major stages. These stages were followed in TAPP3, 4 Level-1 PSA study [7].

Stage 1 - System Logic Model Development - is covered in the general guidelines for PSA [5]. This stage is a prerequisite for common cause failure analysis.

Stage 2 - Identification of Common Cause Component Groups - focuses on the screening process and is critical for definition of the scope of the detailed analysis.

Stage 3 - Common Cause Modeling and Data Analysis - The incorporation of common cause events in the logic model is achieved by a straightforward modification of its structure. The selection of models for quantification of CCF contributions is the most significant part of this stage.

Stage 4 - System Quantification and Interpretation of Results - synthesizes the key output of the previous stages leading to quantification of system failure probability. In addition, uncertainty and sensitivity analyses provide additional perspective for interpretation of results.

Of the above-mentioned four stages the Stage-2 was given the highest importance in the TAPP3, 4 studies. The details followed for this stage were as follows:

- Identifying the groups of system components to be included in, or eliminated from, the CCF analysis;
- Prioritizing the groups of system components identified for further analysis so that time and resources can be best allocated during the CCF analysis;
- Providing engineering arguments to formulate defence alternatives and stipulate recommendations in stage 4 (interpretation of results) of the CCF analysis.

Based on previous experience worldwide some components are particularly susceptible to common cause failures and/or are critical for the level of plant safety. Such components were subject to relatively detailed common cause failure analysis. The focus in this context is on active redundant components and intra system contributions. The review of KAPS PSA Level – 1 study [6] indicated that the most important CCF contributors are circuit breakers, motor-operated valves and pumps. Other important CCF contributors are diesel generators, batteries, pressure relief valves, air-operated valves and sump screens. These findings were used in selecting the CCF groups for TAPP Level-1 PSA Study.

The Stage 3 of CCF modeling and Data Analysis had to be restricted to selection of a suitable Parametric model and the use of generic data. This was necessitated due to the small operating history for Indian Reactors and the absence of sufficient data for quantification of CCF parameters. As there are several systems where the system failure criteria involves 2/3 or 3/4 component failures, it was decided to use a multiparameter model. The Alpha factor model [2] was found suitable for the requirement. The generic alpha factors based on the CCF data generated as a part of the EPRI-3967 [3] were used in our assessment.

The Alpha Factor Model

The alpha factor model defines common cause failure probabilities from a set of failure frequency ratios and the total component failure probability Q_t. In terms of the basic event probabilities, the alpha factor parameters are defined as,

$$\alpha_k^{(m)} = \frac{\binom{m}{k} Q_k^{(m)}}{\sum_{k=1}^{m} \binom{m}{k} Q_k^{(m)}} \qquad (1)$$

where

$\binom{m}{k} Q_K^{(m)}$

is the probability of events involving k component failures in a common cause group of m components, and the denominator is the sum of such probabilities. In other words,

$\alpha_k^{(m)}$ = ratio of the probability of failure events involving any k components over the total probability of all failure events in a group of m components.

For example, for a group of three similar components we have

$$\alpha_1^{(3)} = \frac{3Q_1^3}{3Q_1^{(3)} + 3Q_2^{(3)} + Q_3^{(3)}} \quad \textbf{(2)}$$

$$\alpha_1^{(3)} = \frac{3Q_2^3}{3Q_1^{(3)} + 3Q_2^{(3)} + Q_3^{(3)}} \quad \textbf{(3)}$$

$$\alpha_1^{(3)} = \frac{Q_3^3}{3Q_1^{(3)} + 3Q_2^{(3)} + Q_3^{(3)}} \quad \textbf{(4)}$$

and $\alpha_1^{(3)} + \alpha_2^{(3)} + \alpha_3^{(3)} = 1$ as expected.

The ratio of $Q_{(k,m)}/Q_T$ for non-staggered tested components is given by:

$$\frac{Q_k^{(m)}}{Q_T} = \frac{k\alpha_k}{{}^{m-1}_{k-1}C\alpha_t} \quad \textbf{(5)}$$

where

$$\alpha_t = \sum_{i=1}^{m} k\alpha_k$$

The ratio of $Q_{(k,m)}/Q_T$ for staggered tested components is given by:

$$\frac{Q_K^{(m)}}{Q_T} = \frac{k\alpha_k}{{}^{m-1}_{k-1}C} \quad \textbf{(6)}$$

Some general defensive strategies to avoid dependent failures are:

- Barriers (physical impediments that tend to confine and/or restrict a potentially damaging condition)
- Personnel training (ensure that procedures are followed in all operation conditions)
- Quality control (ensure the product is conforming with the design and that its operation and maintenance follow the approved procedures and norms)
- Redundancy
- Preventive maintenance
- Monitoring, testing and inspection (including dedicated tests performed on redundant components following observed failures)
- Diversity (equipment diversity as for manufacturing, functional diversity as for the physical principle of operation)

Testing and Maintenance Policy

Staggering test and maintenance activities offers some advantages over performing these activities simultaneously or sequentially. First, it reduces the coupling associated with certain human-related failures-those that are introduced during test and maintenance activities. The probability that an operator or technician repeats an incorrect action is lower when test or maintenance activities are performed months, weeks, or even days apart than when they are performed a few minutes or a few hours apart.

A second potential advantage of staggering test and maintenance activities relates to the exposure time for the CCF events. If multiple components are indeed failed because of a CCF event and if this type of failure is detectable by testing and inspecting, then evenly staggering these activities reduces the maximum time that the multiple components would be failed because of that CCF event; that is, the exposure time is reduced for the CCF event. On the other hand, staggering of test and maintenance activities is unlikely to affect errors associated with programmatic or procedural deficiencies.

4. RESULTS

In TAPP3,4 Level-1 PSA study we had considered about 75 CCF component groups and the major contributor to the Core Damage frequency is shown in the table below:

Table 1

S.No	CCF group Description	Nominal Value of Unavaila-bility	Fractional Contri-bution	Sensitivity
1	ECCS Recirculation Valves CCF	8.161×10^{-5}	.074	1.785
2	CCF of Pump Suction MV's of ECCS	8.161×10^{-5}	.074	1.785
3	CCF of ECCS ROH1 MV's Fails to open	7.984×10^{-5}	.0739	1.767
4	CCF of ECCS ROH2 MV's Fails to open	7.984×10^{-5}	.0739	1.767
5	CCF of ECCS ROH3 MV's Fails to open	7.984×10^{-5}	.0739	1.767
6	DG's Fail to Start due to CCF	2.8×10^{-4}	.0786	1.749
7	CCF of ECCS pumps fail to start	6.964×10^{-5}	.06314	1.663
8	CCF of POSRV fail to open	1.468×10^{-4}	.00747	1.074
9	CCF of Service Water Pumps fail to start	2.5×10^{-5}	.00631	1.063

The Generic Alpha factors used are shown in Table-2.

In the study it was found that the quantification model used in the proprietary software Risk Spectrum Version 2.1 the effect of staggered testing was not available. Hence based on the ratio of $Q_{(k,m)}/Q_T$ the effect of staggered testing was obtained by modifying the Alpha factors. The alpha factors were modified such that using the same quantification formula for non-staggered testing (equation 5) the ratios shown in Table-4 are met. The ratios of $Q_{(k,m)}/Q_T$ for non-staggered and staggered testing are shown in Table 3 and 4 were obtained by using the generic alpha factors shown in Table-2 and equations (5) and (6) respectively. The Core Damage frequency was found to be 30 percent less due to the incorporation of staggered testing for the ECCS header injection valves in the TAPP 3,4 level-1 PSA study [7].

Table 2 Generic Alpha Factors

Alpha → / Group size (m) ↓	Alpha-1	Alpha-2	Alpha-3	Alpha-4
2	0.95	0.05		
3	0.95	0.04	.01	
4	0.95	.035	.01	.005

Table 3 Failure On Demand, Non-Staggered Testing $Q_{(K,M)}/Q_t$

K → / Group size (m) ↓	1	2	3	4
2	0.904	0.095		
3	0.896	0.037	.028	
4	0.887	0.021	0.009	0.018

Table 4 Failure On Demand, Staggered Testing $Q_{(K,M)}/Q_t$

K → / Group size (m) ↓	1	2	3	4
2	0.95	0.1		
3	0.95	0.04	0.03	
4	0.95	0.023	0.03	0.001

5. CONCLUSIONS

Amongst the different techniques available to reduce the impact of common cause failures, the most important operations related technique, is that of staggered testing. It not only reduces the human error that may get introduced in sequential or continuous testing but also reduces the exposure to a potential common cause event.

6. REFERENCES

1. Mosleh, A., Fleming, K.N., Parry G.W., Paula H.M., Rasmuson D.M., and Worledge D.H. Procedures for Treating Common Cause Failures in Safety and Reliability Studies, NUREG/CR-4780, EPRI NP-5613, Electric Power Research Institute, Palo Alto, CA, Vol. 1 (1988), Vol. 2 (1989).
2. Mosleh, A. Procedure for Analysis of Common- Cause Failures in Probabilistic Safety Analysis, NUREG/CR-5801, SAND91-7087,RG (1993).
3. Parry, G.W. et al, Procedures for Conducting Common-Cause Failure Analysis in Probabilistic Safety Assessment, IAEA Tecdoc-648 (1992).
4. Flemming, K.N. and Mosleh, A, Classsification and Analysis of Reactor Operating Experience Involving Dependent Events, EPRI NP-3967 (1985).
5. Procedures for Conducting Probabilistic Safety Assessment of Nuclear Power Plants (Level-1), IAEA Safety Series 50-P-4 (1992).
6. Bajaj, S.S. et al Probabilistic Safety Assessment Level-1 Internal Events for KAPS, NPCIL (2002).
7. Bajaj, S.S. et al Probabilistic Safety Assessment Level-1 Internal Events for TAPP 3&4, NPCIL (2005).

Reliability, Safety and Hazard
Editors: P.V. Varde et al.

SAFETY CONSIDERATIONS OF ACTIVE PROCESS WATER SYSTEM SHUTDOWN FOR TAPP-3&4

P. Krishna Kumar, S. Hajela, P.K. Malhotra, S.G. Ghadge and S.S. Bajaj

Directorate of Safety, Nuclear Power Corporation of India Limited, Nabhikiya Urja Bhavan, Anushakti Nagar, Mumbai 400 094

ABSTRACT

Active Process Water (APW) System, provided as unitized closed loop system in Tarapur Atomic Power Project Units-3&4, serves to remove heat from various heat exchangers. One of the important loads served by APW system is shutdown cooling heat exchangers and if APW shutdown is taken then reactor cannot be maintained in cold shutdown condition. It is estimated that after 7 days of reactor shutdown, if about 20% of the normal cooling flow to shutdown cooling heat exchangers is provided then along with keeping PHT in cold shutdown state, reactor components, moderator, end shield water, calandria vault water and calandria vault concrete temperature can be maintained within technical specification limits for extended duration.

1. INTRODUCTION

The maiden 2 x 540 MWe Pressurized Heavy Water Reactor (PHWR) power station, Tarapur Atomic Power Project-3&4 (TAPP-3&4) of which unit # 4 attained criticality in March 2005 is situated at Tarapur in Maharastra.

TAPP-3&4 is an indigenous design, which has evolved from the 220 MWe PHWR. The reactor is of pressurized heavy water type using heavy water moderator, heavy water coolant and natural uranium dioxide as fuel. The stainless steel reactor vessel (calandria) has integral end-shields at either end filled with carbon steel balls and water to limit radiation dose in Fuelling Machine (F/M) vaults. Calandria is a horizontal cylindrical vessel and is submerged in water filled calandria vault, which provides shielding and cooling to calandria. The reactor is having 392 coolant channel assemblies. A coolant channel consists of Pressure Tube (PT), Calandria Tube (CT) and end fitting assemblies. The uranium dioxide fuel bundles reside in the pressure tubes and around each pressure tube, a concentric calandria tube is provided with an annular gap. Carbon dioxide gas filled in this gap serves as thermal insulation between the high temperature Primary Heat Transport (PHT) system and low temperature moderator. PHT system has two independent loops and each has two Steam Generators (SGs) and two pumps.

For a shutdown reactor, two modes are possible. In first mode PHT temperature is kept close to normal operating temperature with SGs providing heat sink. Another mode is a planned cool down to maintain the PHT system at or below 55°C on a continuous basis by removing the decay heat generated in the core by another system called shutdown cooling system. In TAPP-3&4 this system has two independent loops (corresponding to independent PHT loops) and each loop has one set of heat exchanger and pump operating and another set of heat exchanger and pump as standby. Primary coolant flowing in the shutdown heat exchangers is cooled by APW.

APW takes heat from heavy water heat exchangers/ equipment and potentially active light water heat exchanger/ equipment and dissipate to service water (SW) through active process water heat exchangers. APW system has been considered unit wise for TAPP-3 and TAPP-4. APW system of each unit consists of six pumps and nine heat exchangers, out of which four pumps and eight heat exchangers are required for normal full power operation. The APW system shutdown is needed for the maintenance of the supply and discharge valves of the APW pumps, inlet and outlet valves of the APW heat exchangers. The APW system shutdown may also be required for bringing down the activity build up in APW system. The APW system provides cooling water to essential loads like shutdown cooling pumps and Shut Down Heat Exchangers (SDHXs), end shield cooling water heat exchangers, calandria vault cooling water heat exchangers, bleed cooler, Spent Fuel Storage Bay Heat Exchangers (SFSBHXs) etc. Thus if APW shutdown is taken, it will effectively mean that all the above systems will simultaneously be deprived of their cooling. As Service Water is heat sink for APW, any shutdown on SW system will also make APW system ineffective.

As mentioned above one of the important loads served by APW system is shutdown cooling heat exchangers and if APW shutdown is taken then reactor cannot be kept in cold shutdown condition. As APW shutdown will affect cooling of all the serviced systems; the scenario of APW system shutdown needs to be analysed w.r.t acceptable temperature of moderator, end shield water, end shield tube sheets, calandria vault water and concrete etc in addition to maintenance of cold shutdown condition.

The analysis results indicated that even after 30 days of reactor shutdown the APW system cannot be kept shutdown for more than 5 hours due to end shield temperature limitation. This period is too short to carryout any maintenance of APW/SW system.

Therefore, to enhance shutdown safety, a design provision is implemented in TAPP-3&4, interconnecting APW systems of both units in such a way that APW system of one unit in addition to its own full power or shutdown requirements can provide limited supply of cooling water to one shutdown cooling pump and shutdown cooling heat exchanger in each of the PHT loops of the other unit in which APW shutdown is taken up. The requirement of cooling water from the other unit APW system was quantified with an in house code 'COOLTMP-500'. This code is used to predict PHT, moderator, end shield water and calandria vault water temperature under reactor shutdown condition. This code is evolved from similar code for 220 MWe reactor which has earlier been validated against similar scenario, wherein APW shutdown after three days of reactor shutdown was taken up at Naroroa Atomic Power Station, Unit-2 (NAPS-2) with Non Active High Pressure Process Water (NAHPPW) system providing cooling to shutdown cooling heat exchanger only.

This paper describes the code COOLTMP-500 in brief, validation of COOLTMP-220 and results of TAPP-3&4 APW shutdown scenario.

2. COOLTMP MODELLING

The code 'COOLTMP-500' is developed to find out temperatures of PHT, moderator, end shield water, calandria vault water, end shield components and calandria vault concrete with reactor in shutdown condition.

Model

The domain of analysis involves solid as well as fluid volumes. The Calandria shell, end shield, Calandria tube, Calandria Side Tube Sheet (CSTS), Fuelling Machine Tube Sheet (FSTS), end fitting, calandria vault concrete are the solid volume. Where as, PHT, moderator, end shield cooling water and calandria vault cooling water are the fluid volume. For the purpose of thermal analysis, convection has been neglected in the fluid domain. Except for CSTS & calandria vault concrete, single node has been considered. In CSTS & calandria vault concrete region, multiple nodes have been considered, as detailed temperature distribution is required. For the single-node regions, 1-D energy balance equation of the form given below has been applied,

$$\rho Cp\left(\tfrac{dT}{dt}\right) = q'''_{gen} + q'''_{in} - q'''_{out} \qquad (1)$$

For multiple –nodes region the governing equation used is,

$$\rho Cp\left(\frac{\partial T}{\partial t}\right) = k\left(\frac{\partial^2 T}{\partial x^2}\right) + q'''_{gen} + q'''_{in} + q'''_{out} \qquad (2)$$

where 'ρ' is density (kg/m^3), q''' is Volumetric heat generation (W/m^3), 'T' is Temperature (^{o}C), 'Cp' is Specific heat of node(J/kg^{o}K), 'dt' is

Time step (sec) and 'k' is the thermal condutivity (W/mK).

Calculation Methodology

The code is integrated transient thermal model of the following system and components, which are considered, connected with each other by thermal resistances shown in Fig. 1.

i) PHT shut down cooling system
ii) Moderator system
iii) End Shield Cooling System (ESCS)
iv) Calandria Vault Cooling System (CVCS)
v) CSTS, FSTS and End fitting
vi) Calandria shell & Calandria Tube
vii) Calandria Vault concrete and liner

The code utilizes solution of transient energy balance to obtain the temperatures of different components represented as nodes in each module. Large number of components are considered and huge amount of inventory is involved. (eg: Moderator inventory is 260 tons) So considering the running time and accuracy the following methodology is used.

The heat accumulated in node 'i' is obtained in the following way,

$$E_{acci}^{(n)} = E_{ini}^{(n)} - E_{outi}^{(n)} + E_{geni}^{(n)} \qquad (3)$$

$E_{ini}^{(n)}$ =Heat input to node i for n^{th} time step, J

$E_{outi}^{(n)}$ =Heat output from node i for n^{th} time step,J

$E_{geni}^{(n)}$ =Heat generated in node i for n^{th} time step, J

A general form of the equation for calculating E_{acci} is

$$E_{acci} = m_i Cp_i \left(\frac{dT_i}{dt_i} \right) dt \qquad (4)$$

Where,
E_{acci}= Heat accumulated in node i, J
m_i = Mass of node i, kg
dT_i =Temperature difference between node 'i' and neighbouring nodes

Heat input and output to and from the nodes are calculated on the basis of the temperature differences with the neighbouring nodes and the heat transfer coefficients between the nodes. Since the temperature differences, heat transfer coefficients and the heat generation rates which are time dependent; transient E_{acci} are obtained at every time step. A typical E_{ini} or E_{outi} can be represented as,

$$E_{ini}^{(n)} (or) E_{outi}^{(n)} = h_{ij}^{(n)} A_{ij} (T_i^{(n)} - T_j^{(n)}) dt \qquad (5)$$

Where $h_{ij}^{(n)}$=heat transfer coefficient between i^{th} & j^{th} node for n^{th} time step

Explicit finite difference scheme marching on time steps is used to obtain the temperature of each node. The convergence of the result in this scheme is dependent on time step 'dt'. Proper choice of 'dt' was done to obtain an error margin less than $1x10^{-5}$. Temperature of ith node for (n+1) time step is obtained as,

$$T_i^{(n+1)} = T_i^{(n)} + \left(\frac{E_{acci}^{(n)}}{m_i Cp_i} \right) \qquad (6)$$

And in the solid the E_{ini} or E_{outi} are calculated with conduction equations.

The components and fluids are connected with each other by considering the resistance between them. The resistances are calculated by all possible modes conduction, convection and radiation and some complicated resistance paths are simplified.

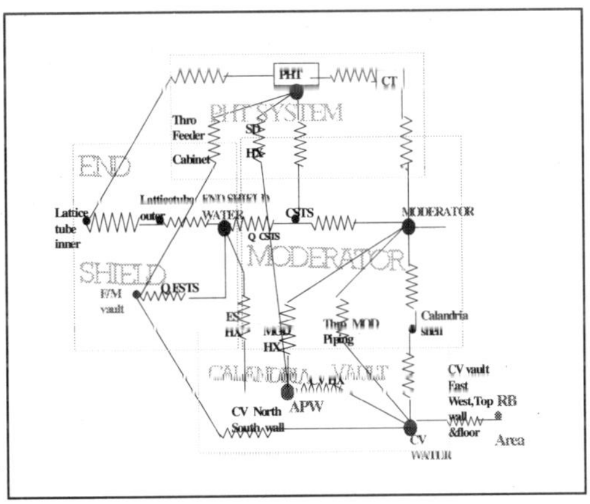

Fig. 1: Thermal resistance diagram of fluids/ components

Using electrical analogy these resistances are simplified. In the end shield region the ball water mixture is modelled based on packed bed model [1]. Theoretical evaluation of heat transfer coefficient between end shield tube sheets and ball water medium are carried out with the model of an infinite plate to ball water medium [2].

The code includes heat exchanger modelling of PHT Shutdown cooling system moderator, end shield cooling system and calandria vault water-cooling system. The Kern's Method [3] is used for thermal analysis of heat exchangers.

Sources, Sinks

i) Sources

Decay heat of fuel bundle[4] in a average rated channel is the main heat source. Other sources like nuclear heat in calandria side tube sheet, end fitting, calandria shell, calandria vault concrete, calandria vault water, moderator and end shield ball & water are also considered. Shutdown cooling pump heat is also added as source.

ii) Sinks

For PHT system heat loss through shutdown heat exchanger is the main heat sink and heat loss through feeder cabinet from PHT is also considered. Reactor building area is another heat sink.

3. VALIDAION OF CODE 'COOLTMP'

Initially the code was developed for 220 MWe reactors and the same methodology is used for 540 MWe reactors. This code is validated with data collected conducted at Narora Atomic Power Station-2 (NAPS-2) in September 1998. In that exercise temperatures of moderator, end shield water and calandria vault water temperatures after stopping APW flow to their respective heat exchangers were monitored. Limited cooling water was provided by Non Active High Pressure Process Water (NAHPPW) to shutdown cooling Heat exchangers. [5]

The scenario was simulated in the code using the following inputs.

1. Initial temperature of PHT =45°C
2. Initial outlet temperature of Moderator=37°C.
3. Initial outlet temperature of end shield cooling water =34.8°C
4. Initial outlet temperature of calandria vault cooling water=39°C.
5. Total Reactor Shut down time= 81hrs.
6. Decay heat of Average rated channel=9.17 kW.
7. NAHPPW flow to shutdown cooling heat exchanger =85 m^3/hr.

A comparison between predicted and observed temperature of PHT, moderator, end shield water and calandria vault water is given in Fig-2 which shows the predicted results are within ±5% of the experimental data.

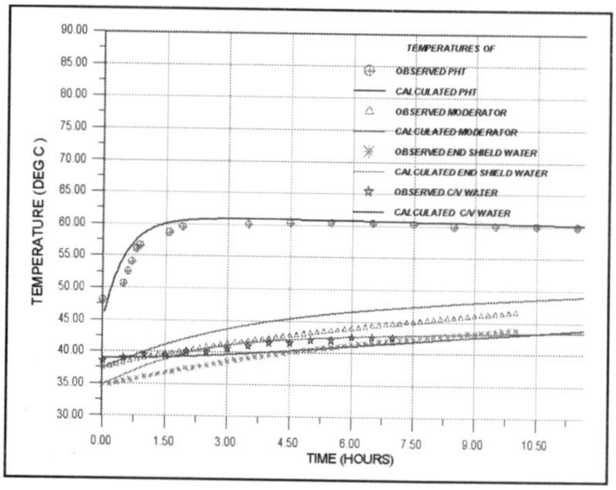

Fig. 2: Validation of COOLTMP-220 for APW shutdown scenario at NAPS-2.

4. ANALYSIS OF APW SHUTDOWN IN TAPP-3&4

Using the code an analysis was done to get the temperatures of PHT, moderator, end shield water and calandria vault water, with 7 days of core decay heat as heat source and with 20% of normal process water flow to shutdown heat exchangers. (i.e. One heat exchanger in each PHT loop)

These analyses are done for average rated channel. As per the procedure proposed to be adopted (see next section), first APW from the operating unit will be valved in to one shutdown cooling pump and one shutdown cooling heat exchanger per coolant loop of the unit in which APW shutdown is planned and then APW shutdown will be taken. With this configuration there is no interruption of cooling, only the quantity of APW supply to Shutdown cooling heat exchanger will get reduced from 100% to 20% of normal APW flow.

Due to APW shutdown, no cooling will be available to moderator, end shield water and calandria vault water. Since operation of circulation pumps in these systems will add heat in the respective system, no circulation in the primary sides of these systems is considered in the analysis.

The calculated temperature profiles of fluids and components are shown in Fig–3 &4.

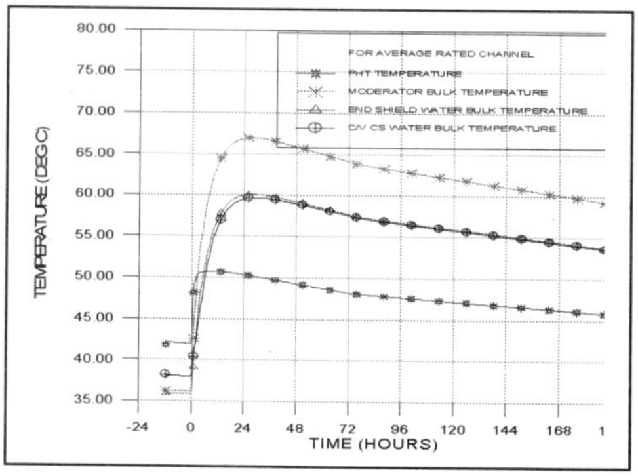

Fig. 3: Temperature profiles of PHT , moderator, end shield water and calandria vault water.

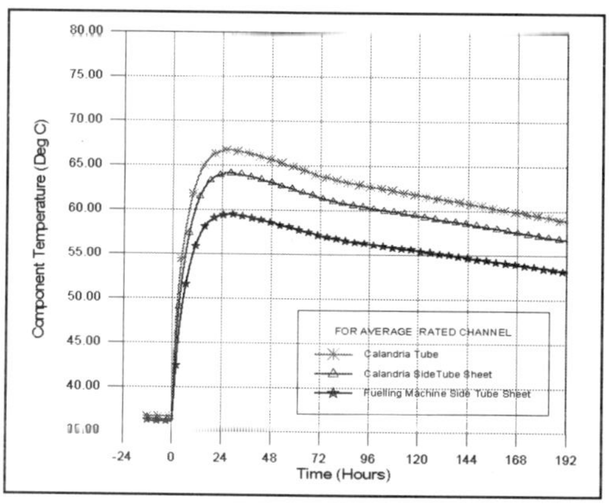

Fig. 4: Temperature profiles of calandria tube CSTS &FSTS

From the figures it is seen that initially there is rise in PHTS temperature due to reduction in APW supply. As the decay heat is quite low after 7 days of reactor shutdown; 20% of normal heat flow provides sufficient heat sink to bring down PHT temperature to less than 55°C. Just after APW supply isolation, heat sink to moderator, end shields and calandria vault water becomes unavailable, which causes rise in their respective temperatures. Subsequently, with PHTS cooling maintained and further falling of decay heat, temperature of moderator, end shields and calandria vault water also starts falling.

In this scenario, there is a reverse heat transfer from moderator to PHTS, moderator heat flows to calandria vault and end shields also. The decay heat in end shields and heat coming from moderator is mainly getting removed through PHT circulation. RB atmosphere provides heat sink to calandria vault water.

The main heat rejection path for PHT system is through the shutdown cooling heat exchangers. The heat rejection through feeder cabinet is insignificant. From the calculated temperatures it is seen that during APW shutdown moderator temperature and end shield water temperature remain below their respective Technical Specification Limiting Condition for Operation (LCO) of 83 °C and 64°C respectively. Calandria vault water temperature crosses its LCO of 59°C marginally but comes down subsequently. End shield components temperatures are within their respective design/acceptable temperatures i.e. 85 °C for CSTS and FSTS and 125 °C for rolled joint. Similarly, calandria vault concrete temperature remains below acceptable value of 65 °C.

5. MODIFICATION FOR APW SHUT DOWN

From the analysis results it is concluded that the APW system can be shutdown after seven days of reactor shutdown with about 20% of normal APW supply provided from other unit to one shutdown cooler and pump in each loop. Before taking the shutdown of the APW system a reconfiguration of the APW system needs to be done to establish the flow of APW to the shutdown pumps and coolers from the other unit.[6]

The total APW flow required for the shutdown unit (APW system under shutdown) from other unit is about 830 m^3/hr i.e. 280 m^3/hr for two shutdown coolers, 13 m^3/hr for two shutdown pumps and 535 m^3/hr for SFSB cooling heat exchanger. The normal APW flow to each shutdown heat exchangers is 650 m^3/hr. A separate analysis shows that cooling water interruption to SFSB can be tolerated for 33 hrs.[7] As APW shutdown is required for extended period; APW supply for SFSB cooling is also included. Thus the total APW requirement for TAPP-4 (if APW of TAPP-3 is under shutdown) and TAPP-3 (if APW of TAPP-4 is under shutdown) during this case is 9300 m^3/hr and 9100 m^3/hr respectively. As the pump motor is designed for the flow of 130% of the normal flow, the pump motor are able to deliver this additional flow, which is only 10% more than the normal flow. Due to the interconnection of both the units of APW system the resultant APW system resistance will decrease and will result into more flow from the pumps. In view of the above, four APW pumps in operation can meet this additional requirement. Eight active process water heat exchangers, which were in service during normal operation, will continue to be in service with one standby.

In the unit where APW shutdown is taken, all the process loads except for one shutdown pump and SDHX in each PHT loop are isolated by closing their inlet and outlet valves. A separate operating procedure will be followed to carry out all necessary reconfiguration and isolation of all the loads, which are not required. Hence the chances that any loads left in service in the flow path of shutdown coolers and pumps is rare. The isolation of the loads are done by closing both inlet and outlet valves. Hence the leakage flow through these loads is also ruled out. In case if any load (of shutdown unit) left in service, the normal flow to the various loads in operating unit will reduce and will be reflected through low flow or high temperature signals. This will warn the operator to initiate the corrective action to isolate the loads properly.

6. CONCLUSIONS

As the calculated temperatures of PHTS, moderator, end shield water, calandria vault water and core & end shield components are within the acceptable/design limits, the APW shutdown in one unit does not cause any safety concern provided,

1. APW shutdown is initiated at least 7 days after reactor shutdown.
2. PHTS cooling is maintained through shutdown cooling circuit with one shutdown cooling pump and heat exchanger in service in each loop and around 20% of normal APW flow supplied to shutdown cooling heat exchangers from the other unit APW system.

With this design modification implemented for which proper analytic base is provided, it is possible to take APW and/or SW system shutdown ensuring core cooling through shutdown heat exchangers. This provision is included in the Technical Specifications for operation of TAPP-3&4 units to ensure safety under plant shutdown condition.

7. REFERENCES

1. Daizokuni and J.M.Smith, Heat Transfer characteristics of porous rocks, A.I.Ch.E Journal, 1960, Vol.6, No.1.
2. S.Yagi and D.Kunni, Studies on heat transfer near wall surfaces in Packed Beds, A.I.Ch.E Journal, 1960, Vol.6, No.1.
3. D.Q.Kern, Process Heat Transfer, 1950, Mc Graw-Hill Book Co, New York.
4. A.G.Croff, A user's manual for the ORIGEN2 computer code, 1980, OAK Ridge National Laboratory, Oak Ridge, Tennessee.
5. P.Krishna Kumar, S.Hajela, & S.G. Ghadge, Design note on modification of 'COOLTMP' code standard PHWR, 2005, (NPC/DN/01500/05).
6. SanjeevKumar, M.C.Tripathi, U.C.Muktibodh & K.N.Gupta, Design note on modification in the active process water system, 2005, (TAPP-3&4 /DN/ 71310 /02).
7. TAPP-3&4, Safety Report-Volume-II, 2004, (Accident Analysis).

Reliability, Safety and Hazard
Editors: P.V. Varde et al.

OPERATIONAL SAFETY PERFORMANCE INDICATOR SYSTEM - A MANAGEMENT TOOL FOR THE SELF ASSESSMENT OF SAFETY AND RELIABILITY OF NUCLEAR POWER PLANTS

Anil Kumar, S.L. Mandowara and S. Mittal

Nuclear Power Corporation of India Limited, Vikram Sarabhai Bhavan,
Anushaktinagar, Mumbai 400 094

ABSTRACT

Operational Safety Performance Indicator system is one of the self assessment tools for station management to monitor safety and reliability of nuclear power plants. It provides information to station management about the performance of various areas of the plants by means of different colours of relevant performance indicators. Such systems have been implemented at many nuclear power plants in the world and have been considered as strength during WANO Peer Review. IAEA had a Coordinated Research Programme (CRP) on this with several countries participating including India. In NPCIL this system has been implemented in KAPS about a year back and found very useful in identifying areas which needs to be given more attention. Based on the KAPS feedback implementation of this system has been taken up in RAPS-3&4 and KGS-1&2.

Keywords: Operational Safety Performance Indicator, Self Assessment Safety and Reliability, Specific Indicator, Strategic Indicator, Overall Indicator, Attributes

1. INTRODUCTION

The world over, it is realized that safety assurance of Nuclear Power Plants (NPPs) is best achieved through self assessment and self motivation apart from independent regulation. Many qualitative methods have been in vogue like internal audit, peer review, record review etc. However, these could only provide qualitative feedbacks and had limitation of inadequate attention. The quantified feedbacks are vital to improve the management attention. The quantification further have advantages of trend monitoring and thus proactive alertness could be achieved, so that speedy corrective actions could be implemented before the safety and reliability aspects approach alarming level in a NPP.

Often, it looked upon safety and production reliability as conflicting objectives. However, the operating experience of nuclear industry led to understand that this is not so. In fact, plant with excellent safety records also tends to be good performer. Therefore, a complete set of performance indicators for monitoring safety and reliability have been included. As the objective is to monitor safety and reliability of NPPs and hence pure economical indicators have not been included in this system.

In late 1990s, observing international trends and the advantages of quantified feedbacks; NPCIL realized the need of development of a computer based Operational Safety Performance Indicator (OSPI) System, which could continually evaluate the station performance quantitatively and prompt station management to take proactive actions.

The OSPI system has been developed for KAPS and after initial trial operation for about six months at KAPS, since April 2004 it is under regular use in this station.

2. WHAT IS OSPI SYSTEM?

OSPI System is a web based multi user computer based system, which provides quantified feedbacks to station management to assess safety and reliability of an operating nuclear power station by means of trending and providing timely alert on each individual areas as well as at sectional and over all level.

In this system, NPP's operational safety and reliability performance is regularly monitored using well identified measurable indicators in the areas of concern for giving trends and alerts. Using software, this is synthesized and output is displayed on the screen in user friendly manner.

Figure-1

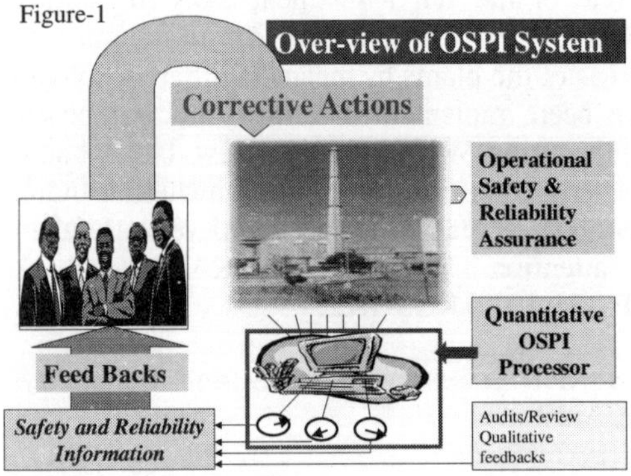

As illustrated in figure-1, the OSPI system processor takes various inputs related to plant operation and maintenance and gives output to plant management.

3. HOW IS SYSTEM STRUCTURED?

The system is structured in hierarchical way with bottom up approach. The directly measurable indicators of plant concern areas are called specific indicators and are placed at lowest level in hierarchy and overall plant level safety and reliability is represented by three attributes, which are at the top of hierarchy namely 'smoothness in plant operation', 'nuclear power plant safety assessment' and 'attitude towards safety and environment'. As these attributes cannot be measured directly, the indicator structure is expanded at different levels. Each attribute is calculated through overall indicators (functional level). Each of the overall indicators is calculated using strategic indicators (sectional level), which in turn, are calculated using specific indicators. The OSPI system structure is illustrated in figure-21:

Figure-2

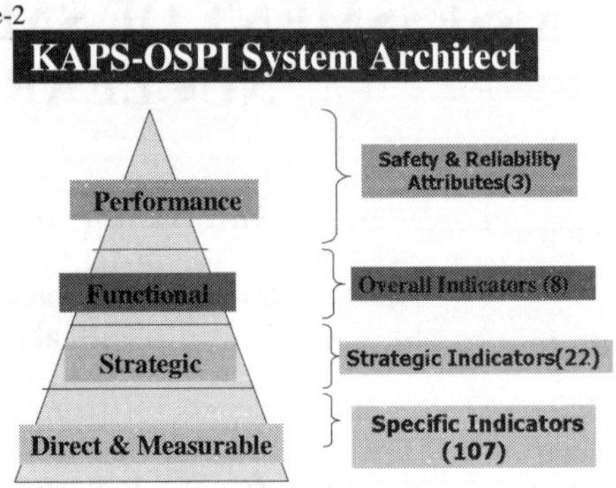

4. DEFINITIONS:

Attributes:
Key elements that are associated with plant that operates safely

Overall Indicators:
Provide overall performance evaluation of relevant functions of safety and reliability. Combinations of these provide the value of attributes.

Strategic Indicators:
Provide performance evaluation of sections supporting the function. Combinations of these provide the value of overall indicators.

Specific Indicators:
These represent quantifiable basic elements that provide physical appreciation for quick understanding and proper investigation that would help in effective corrective actions in performance improvement. Combinations of these provide the value of strategic indicators.

Weightage Factor:
The weightage factor is used for getting the higher-level indicators by the method of aggregation. A weightage factor of 1, 2 and 3 is assigned to indicators on the basis of their safety and reliability significance.

Threshold Values:

These are the quantitative values of the indicators for a year and are classified into five categories i.e. Excellent, Good, Satisfactory, Caution and Unsatisfactory. Actual yearly value of the indicators is compared with these threshold values to classify them into any of the above five categories. Indicator display changes when it crosses any of a threshold.

Aggregation:

A method to derive higher level indicators from their set of lower level indicators by using their weightage factors and category points (based on five categories).

Data Input Frequency:

This is the frequency at which data for the specific indicators are required to be collected and inputted manually or fetched through connectivity into the OSPI system software. The frequency of data collection is generally quarterly, but in some case it could be daily, monthly, half yearly or yearly. Suitable grace period are allowed in computer programme for data collection.

5. DEVELOPMENT STAGES OF OSPI SYSTEM:

Stage I: Design and develop OSPI system framework

Stage II: Collect data at KAPS and validate OSPI framework

Stage III: Define software requirements and develop software

Stage IV: Install OSPI system at KAPS and trial run

Stage V: Improve software based on feedbacks

Stage VI: Regular use of OSPI system at KAPS

For design, development, installation and testing of OSPI system at KAPS, teams were formed both at NPCIL,HQ and KAPS, which worked in close coordination.

6. FEATURES OF OSPI SYSTEM:

1. Overall display of indicators in five different colors based on its performance up to the level of attribute.
2. Trend graphs (line, bar) of yearly performance as well as performance in the period of data entry frequency
3. Section and department wise display
4. Summary about the no. of indicators in excellent, good, satisfactory, caution or unsatisfactory category at a given time.
5. Historical information about no. of times an indicator had been in excellent, good, satisfactory, caution or unsatisfactory category
6. Status of schedule adherence for data entry

7. DESCRIPTION OF OSPI SYSTEM:

1. The frame work of OSPI system consists of 3 Attributes, 8 Overall Indicators, 22 Strategic Indicators and 107 Specific Indicators (refer Annexure- 1, 2 and 3). It was finalized in consultation with officials from Operations Directorate, Health & Safety Directorate, Reactor Safety Directorate, QA Directorate in HQ and KAPS. While finalizing the frame work, due consideration was given to the usefulness of each of the specific indicators and its relationship with plant performance and also easy retrieval of data for its computation. (see reference-1, 2, 3 and 4).
2. For each specific indicator a table consisting of the following information is prepared:

   - Relationship with Attributes, Overall Indicator and Strategic Indicators
   - Rational of the indicator
   - Definition, Unit and Formula
   - Time period for reporting/ displaying the indicator
   - Source of data
   - Responsible agency for data collection and corrective action
   - Threshold values and criteria for deciding threshold values

3. Threshold values were decided based on any one of the following criteria:

   - Standard Deviation Method
   - KAPS Experience / Design basis
   - Engineering Judgment
   - Memorandum of Understanding
   - World Median
   - NPCIL Median

However, in most of the cases threshold values were decided by Standard Deviation Method using historical data of 3 to 5 years.

4. Specific Indicators are either unit Specific or Station Specific. Unit specific means indicators can be calculated separately for each unit. Whereas, station specific means indicator represent the station value. Further, some of the specific indicators are directly measurable i.e. data itself represent an specific indicator, whereas, others are to be computed using an appropriate formula.
5. Data collection frequency for the specific indicators is any one out of daily, monthly, quarterly, half yearly and yearly. About 70% specific indicators have quarterly data collection frequency.
6. To ensure prompt entry of the data by the respective section, following grace periods have been assigned:

Daily	- One day
Monthly	- One Week
Quarterly	- Two Weeks
Half Yearly	- Three Weeks
Yearly	- Four Weeks

Provision exists to find out the specific indicators for which data is not entered as per the schedule or within the grace period.

8. HOW OSPI SYSTEM WORKS:

1. Each Indicator is assessed into five categories viz. excellent, good, satisfactory, caution and un-satisfactory. Each such category is represented in one colour for ease to invite attention viz. excellent-green, good-blue, satisfactory-white, caution-yellow and un-satisfactory-red colour. This categorization is achieved by comparing the indicator value for a given period with the predefined threshold values those are worked out based on historical data. Thus each attribute, specific indicator, strategic indicator, overall indicator and specific indicator is displayed in one of the colours according to its prevailing value.
2. One year data of each specific indicator is compared with its threshold value.
3. Based on the value, specific indicators are categorized into any one of the five categories i.e. excellent, good, satisfactory, caution and unsatisfactory, assigned points +2, +1, 0, -1, -2 respectively and displayed in colour windows of green, blue, white, yellow, and red respectively.
4. From the specific indicators, strategic indicators are derived by the method of aggregation using weightage factor and points for specific indicators (i.e. +2, +1, 0, -1, -2).
5. Based on the value (between > 1.5 to < -1.5) strategic indicators are categorized into five categories i.e. Excellent (>1.5), Good (> + 0.5 to + 1.5), Satisfactory (- 0.5 to + 0.5), Caution (-1.5 to < - 0.5) and Unsatisfactory (< - 1.5), assigned points +2, +1, 0, -1, -2 respectively and are displayed in the applicable colours (green, blue, while, yellow, red).
6. In the same manner overall indicator and Attributes are derived.
7. Thus, just by seeing the hierarchical structure of OSPI system on the computer screen, the station management will get the information about the areas that need to be improved.

9. RESULTS AND DISCUSSIONS:

The results from the OSPI system are periodically discussed in the station management meeting and areas where performance is under caution or unsatisfactory category as shown by this system corrective action is taken.

As on 21/9/05 in KAPS-1, all 3 attributes and 8 overall indicators are either under good or excellent category. Out of 22 strategic indicators 18 are either good or excellent and 4 are satisfactory. Out of 107 specific indicators 81 are either good or excellent. In September 2004 only 74 specific indicators were under good or excellent category.

As on 21/9/05 in KAPS-2, all 3 attributes and 8 overall indicators are either under good or excellent category. Out of 22 strategic indicators 19 are either good or excellent and 3 are satisfactory. Out of 107 specific indicators 78 are either good or excellent. In September 2004 only 72 specific indicators were under good or excellent category.

Annexure-1

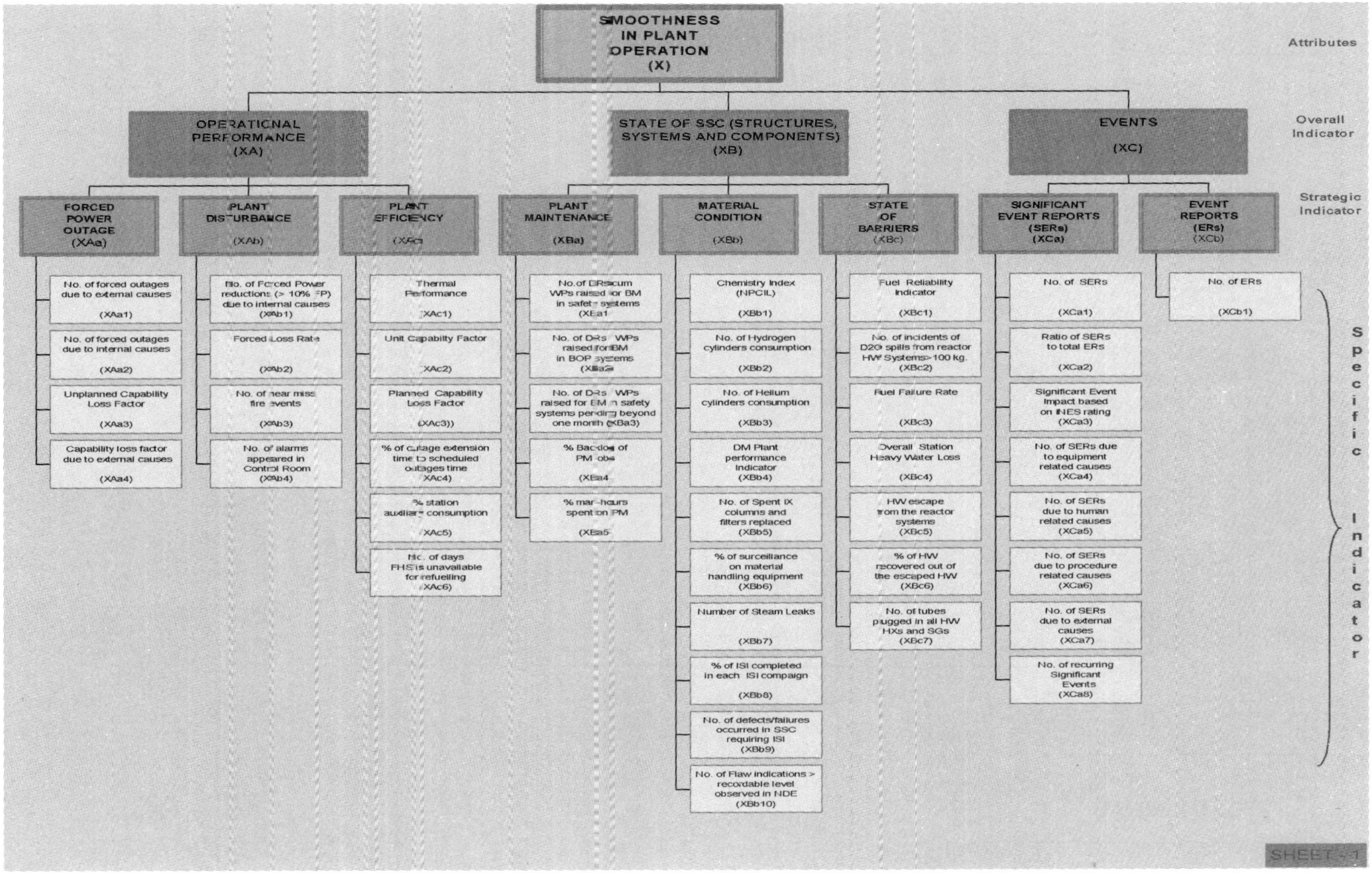
SMOOTHNESS IN PLANT OPERATION (X)
Attributes
OPERATIONAL PERFORMANCE (XA)
STATE OF SSC (STRUCTURES, SYSTEMS AND COMPONENTS) (XB)
EVENTS (XC)
Overall Indicator
FORCED POWER OUTAGE (XAa)
PLANT DISTURBANCE (XAb)
PLANT EFFICIENCY (XAc)
PLANT MAINTENANCE (XBa)
MATERIAL CONDITION (XBb)
STATE OF BARRIERS (XBc)
SIGNIFICANT EVENT REPORTS (SERs) (XCa)
EVENT REPORTS (ERs) (XCb)
Strategic Indicator
No. of forced outages due to external causes (XAa1)
No. of forced outages due to internal causes (XAa2)
Unplanned Capability Loss Factor (XAa3)
Capability loss factor due to external causes (XAa4)
No. of Forced Power reductions (> 10% FP) due to internal causes (XAb1)
Forced Loss Rate (XAb2)
No. of near miss fire events (XAb3)
No. of alarms appeared in Control Room (XAb4)
Thermal Performance (XAc1)
Unit Capability Factor (XAc2)
Planned Capability Loss Factor (XAc3))
% of outage extension time to scheduled outages time (XAc4)
% station auxiliary consumption (XAc5)
No. of days FHS is unavailable for refuelling (XAc6)
No. of DRs cum WPs raised for BM in safety systems (XBa1)
No. of DRs / WPs raised for BM in BOP systems (XBa2)
No. of DRs / WPs raised for BM in safety systems pending beyond one month (XBa3)
% Backlog of PM jobs (XBa4)
% man-hours spent on PM (XBa5)
Chemistry Index (NPCIL) (XBb1)
No. of Hydrogen cylinders consumption (XBb2)
No. of Helium cylinders consumption (XBb3)
DM Plant performance Indicator (XBb4)
No. of Spent IX columns and filters replaced (XBb5)
% of surveillance on material handling equipment (XBb6)
Number of Steam Leaks (XBb7)
% of ISI completed in each ISI compaign (XBb8)
No. of defects/failures occurred in SSC requiring ISI (XBb9)
No. of Flaw indications > recordable level observed in NDE (XBb10)
Fuel Reliability Indicator (XBc1)
No. of incidents of D2O spills from reactor HW Systems>100 kg. (XBc2)
Fuel Failure Rate (XBc3)
Overall Station Heavy Water Loss (XBc4)
HW escape from the reactor systems (XBc5)
% of HW recovered out of the escaped HW (XBc6)
No. of tubes plugged in all HW HXs and SGs (XBc7)
No. of SERs (XCa1)
Ratio of SERs to total ERs (XCa2)
Significant Event Impact based on INES rating (XCa3)
No. of SERs due to equipment related causes (XCa4)
No. of SERs due to human related causes (XCa5)
No. of SERs due to procedure related causes (XCa6)
No. of SERs due to external causes (XCa7)
No. of recurring Significant Events (XCa8)
No. of ERs (XCb1)
Specific Indicator
SHEET - 1

Annexure-2

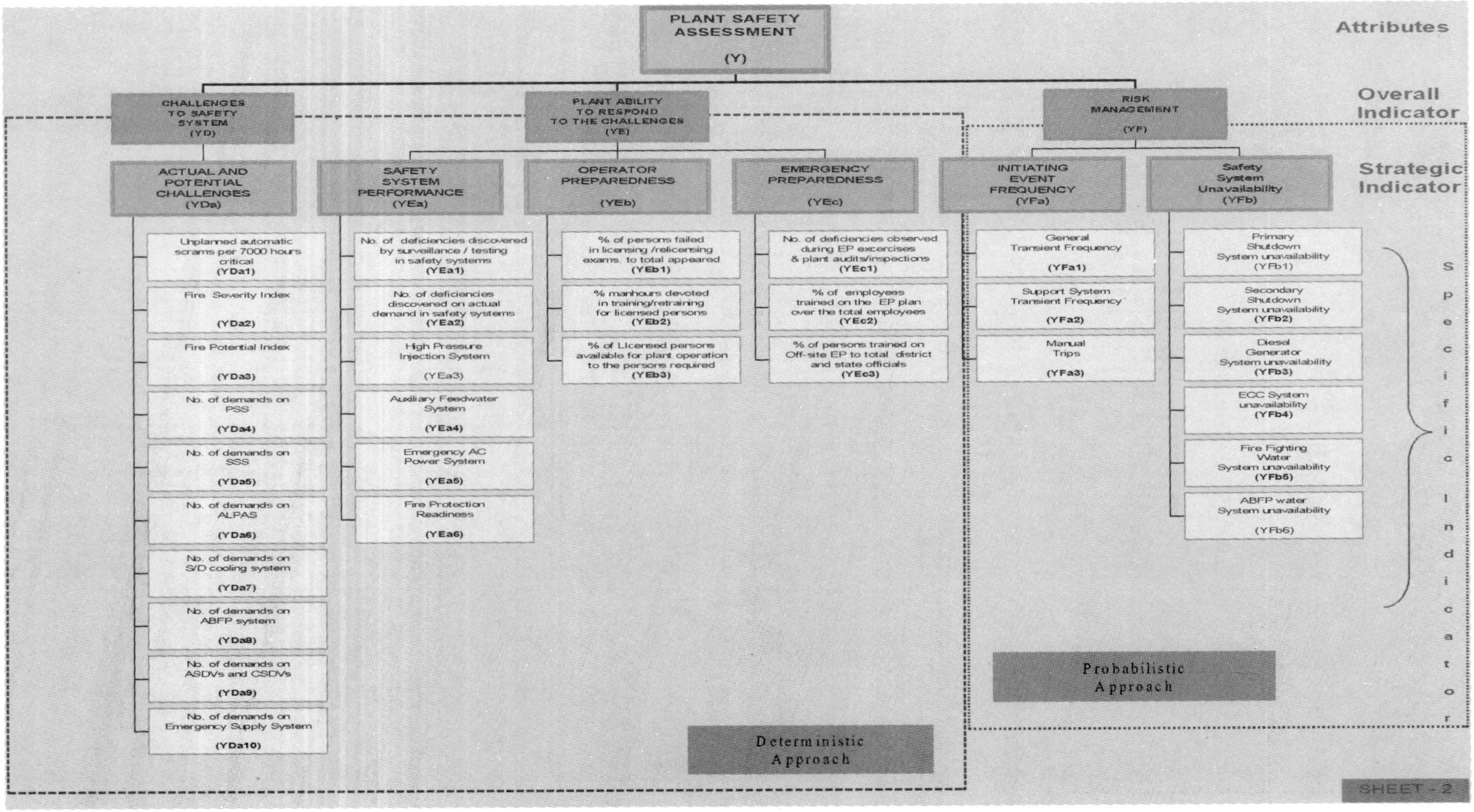

Attributes
Overall Indicator
Strategic Indicator
Specific Indicator
PLANT SAFETY ASSESSMENT (Y)
CHALLENGES TO SAFETY SYSTEM (YD)
PLANT ABILITY TO RESPOND TO THE CHALLENGES (YE)
RISK MANAGEMENT (YF)
ACTUAL AND POTENTIAL CHALLENGES (YDa)
SAFETY SYSTEM PERFORMANCE (YEa)
OPERATOR PREPAREDNESS (YEb)
EMERGENCY PREPAREDNESS (YEc)
INITIATING EVENT FREQUENCY (YFa)
Safety System Unavailability (YFb)
Unplanned automatic scrams per 7000 hours critical (YDa1)
Fire Severity Index (YDa2)
Fire Potential Index (YDa3)
No. of demands on PSS (YDa4)
No. of demands on SSS (YDa5)
No. of demands on ALPAS (YDa6)
No. of demands on S/D cooling system (YDa7)
No. of demands on ABFP system (YDa8)
No. of demands on ASDVs and CSDVs (YDa9)
No. of demands on Emergency Supply System (YDa10)
No. of deficiencies discovered by surveillance / testing in safety systems (YEa1)
No. of deficiencies discovered on actual demand in safety systems (YEa2)
High Pressure Injection System (YEa3)
Auxiliary Feedwater System (YEa4)
Emergency AC Power System (YEa5)
Fire Protection Readiness (YEa6)
% of persons failed in licensing /relicensing exams. to total appeared (YEb1)
% manhours devoted in training/retraining for licensed persons (YEb2)
% of Licensed persons available for plant operation to the persons required (YEb3)
No. of deficiencies observed during EP excercises & plant audits/inspections (YEc1)
% of employees trained on the EP plan over the total employees (YEc2)
% of persons trained on Off-site EP to total district and state officials (YEc3)
General Transient Frequency (YFa1)
Support System Transient Frequency (YFa2)
Manual Trips (YFa3)
Primary Shutdown System unavailability (YFb1)
Secondary Shutdown System unavailability (YFb2)
Diesel Generator System unavailability (YFb3)
ECC System unavailability (YFb4)
Fire Fighting Water System unavailability (YFb5)
ABFP water System unavailability (YFb6)
Deterministic Approach
Probabilistic Approach
SHEET - 2

Annexure-3

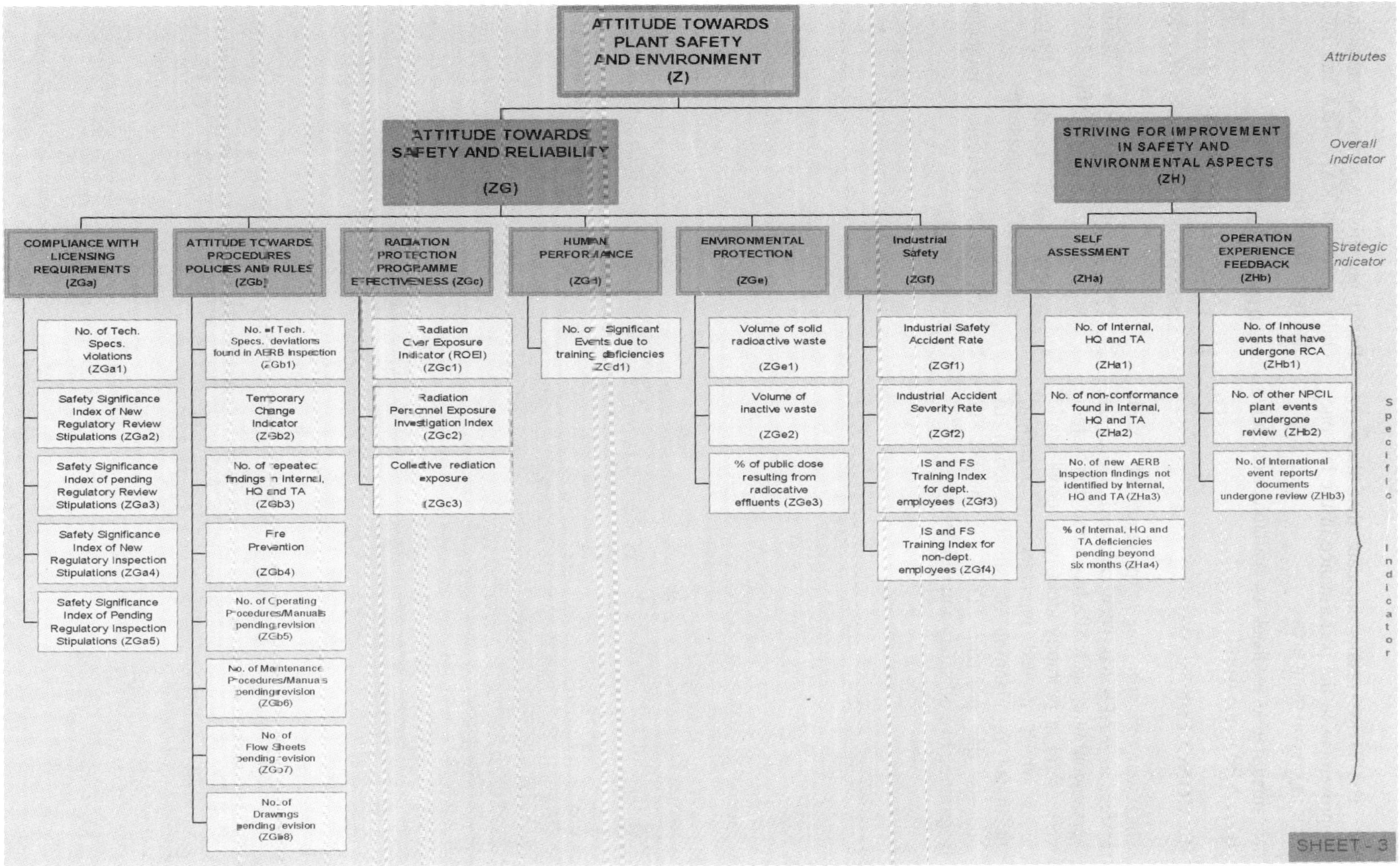
ATTITUDE TOWARDS PLANT SAFETY AND ENVIRONMENT (Z)
Attributes
ATTITUDE TOWARDS SAFETY AND RELIABILITY (ZG)
STRIVING FOR IMPROVEMENT IN SAFETY AND ENVIRONMENTAL ASPECTS (ZH)
Overall Indicator
COMPLIANCE WITH LICENSING REQUIREMENTS (ZGa)
ATTITUDE TOWARDS PROCEDURES POLICIES AND RULES (ZGb)
RADIATION PROTECTION PROGRAMME EFFECTIVENESS (ZGc)
HUMAN PERFORMANCE (ZGd)
ENVIRONMENTAL PROTECTION (ZGe)
Industrial Safety (ZGf)
SELF ASSESSMENT (ZHa)
OPERATION EXPERIENCE FEEDBACK (ZHb)
Strategic Indicator
No. of Tech. Specs. violations (ZGa1)
Safety Significance Index of New Regulatory Review Stipulations (ZGa2)
Safety Significance Index of pending Regulatory Review Stipulations (ZGa3)
Safety Significance Index of New Regulatory Inspection Stipulations (ZGa4)
Safety Significance Index of Pending Regulatory Inspection Stipulations (ZGa5)
No. of Tech. Specs. deviations found in AERB Inspection (ZGb1)
Temporary Change Indicator (ZGb2)
No. of repeated findings in Internal, HQ and TA (ZGb3)
Fire Prevention (ZGb4)
No. of Operating Procedures/Manuals pending revision (ZGb5)
No. of Maintenance Procedures/Manuals pending revision (ZGb6)
No. of Flow Sheets pending revision (ZGb7)
No. of Drawings pending revision (ZGb8)
Radiation Over Exposure Indicator (ROEI) (ZGc1)
Radiation Personnel Exposure Investigation Index (ZGc2)
Collective radiation exposure (ZGc3)
No. of Significant Events due to training deficiencies (ZGd1)
Volume of solid radioactive waste (ZGe1)
Volume of inactive waste (ZGe2)
% of public dose resulting from radioactive effluents (ZGe3)
Industrial Safety Accident Rate (ZGf1)
Industrial Accident Severity Rate (ZGf2)
IS and FS Training Index for dept. employees (ZGf3)
IS and FS Training Index for non-dept. employees (ZGf4)
No. of Internal, HQ and TA (ZHa1)
No. of non-conformance found in Internal, HQ and TA (ZHa2)
No. of new AERB inspection findings not identified by Internal, HQ and TA (ZHa3)
% of Internal, HQ and TA deficiencies pending beyond six months (ZHa4)
No. of Inhouse events that have undergone RCA (ZHb1)
No. of other NPCIL plant events undergone review (ZHb2)
No. of International event reports/ documents undergone review (ZHb3)
Specific Indicator
SHEET - 3

The above results indicate that there is improvement in the performance in the last one year in several plant areas.

No quantitative feedback is available from any other NPP in the world.

10. CONCLUSION:

One year of operating experience of OSPI system at KAPS has been obtained, which is very encouraging as KAPS has found this system very useful in promptly identifying problematic areas which needed management attention, facilitated in planning of suitable action to address the problem. Thus, it helped in improving overall safety and reliability.

Based on this experience, implementation of OSPI system at RAPS-3&4 and KGS-1&2 has also been taken up.

11. REFERENCES:

1. IAEA-TECDOC-1141: Operational Safety performance Indicators for nuclear power plants.
2. Performance Indicator Monitor: INDI version 2.03 – User's Manual
3. AERB Safety Guide AERB/SG/26 for renewal of authorization
4. WANO Performance Indicator Manual version 2001

Reliability, Safety and Hazard
Editors: P.V. Varde et al.

FIRE PROBABILISTIC SAFETY ASSESSMENT FOR KUDANKULAM NPP IN INDIA

G. Tokmachev

Atomenergoproekt, Bakuninskaya 7, Moscow, 107005 Russia

ABSTRACT

Fire Probabilistic Safety Assessment (PSA) was carried out by Atomenergoproekt, Moscow for Kudankulam NPP in design. The PSA addressed fire events initiated during plant operation at full power. Fire hazard analysis was an essential part of the fire PSA that resulted in design changes. Estimation of the fire frequency for all critical plant locations and identification of the resulting fire-induced initiating events and dependent failures were performed. The overall result of 9.1E-9 per reactor year represents approximately 17% of the frequency core damage caused by internal events. The methodology used in the study, results and findings found are described.

1. INTRODUCTION

A fire Probabilistic Safety Assessment (PSA) is a useful tool for supplementing deterministic analyses on which the Kudankulam plant design and fire protection are based. The first phase of the fire PSA was performed at the Preliminary Safety Analysis Report (PSAR) stage according to contractual terms. Since that time, Russian fire protection standards have been raised. In 2004, the Russian Regulatory Authority published the document [1] that requires a fire PSA to be carried out at the PSAR stage. Thus, now this is the regulatory requirement.

2. PLANT DESIGN FEATURES

Kudankulam design features ensure that the necessary safety functions can be maintained during and after a fire. These include substantial physical separation of cables and systems providing redundant safety functions, a high degree of fire resistance of structural elements of safety and normal operation systems, implementation of weakly combustible cable insulation, and use of fire resistant control circuits at control rooms, where redundant equipment and cables of the systems required to achieve and maintain stable shutdown conditions are located in proximity to each other. The established criteria specify fire barriers between redundant trains to have a minimum fire resistive rating of 3 hours. Each of four safety trains has the capability to provide heat removal from the reactor and maintain control and monitoring activities should a fire occur in another train.

The most important feature of the Kudankulam design is associated with the fact that safety systems comprise new passive technologies. For instance, emergency residual heat removal is fulfilled by two diversified redundant systems, one of which is operated in passive mode. From a fire protection point of view a passive heat removal system (PHRS) is of ultimate importance. This system consisting of four independent circuits of natural secondary coolant circulation is intended for long-term residual heat removal from the reactor both with sealed and leaked primary circuit in the case of loss of all auxiliary power supply sources (including emergency ones). The PHRS is designed in such a way that it is able to operate in all NPP operation modes, both independently and in combination with other normal operation and safety systems.

3. METHODOLOGY

The methodology applied is based on the IAEA Guidelines [2, 3] and Russian experience in similar activities [4-8] being carried out since the mid-1990s.

The general approach for conducting a fire PSA was little changed from that used in the studies for operating plants. Preparation of the necessary data for the fire PSA project was very difficult due to incomplete information and documentation, e.g. lack of operating procedures. Some typical tasks were impossible to perform, e.g. plant walkdowns. Therefore, the methodology typically used for performing a fire PSA of operating plants needed to be revised.

On the one hand the uncertainty associated with the results of the fire PSA performed at the PSAR stage are relatively high, on the other hand the fire PSA was recognized as a valuable tool that can provide insights into plant design. Carrying out the fire PSA from the very beginning of the design development gave the benefit of modifying plant design easily to mitigate fire consequences. Therefore, a through fire analysis could provide an extremely cost-effective approach to fire protection improvement at the design stage. Such a benefit is impossible when performing a fire PSA for operating plants.

The study was performed in a highly iterative manner, i.e. certain tasks needed refinement after conducting one or more of the subsequent tasks or incorporating changes to the plant design.

The fire PSA involved building and room analyses (including a fire hazards analysis), screening technique, determination of fire-induced events and systems with reduced functionalities due to a fire, definition of fire related event sequences, quantification of the core damage frequency (CDF), uncertainty, sensitivity and importance analyses.

It is important to note that a more comprehensive PSA for internal fires will be performed during the detailed design, providing extensive documentation for the plant operating staff for their fire protection program.

4. FIRE HAZARDS ASSESSMENT

Fire hazard analysis was a comprehensive assessment of potential fire hazards throughout the plant and the effect of potential fires on the safety-related plant areas. This was an essential fundamental part of the fire PSA. It included determination of fire compartments, identification of potential fire hazards, determination of impacts of a fire in a compartment, and identification of fire induced initiating events. A fire containment approach was accepted as the design basis, which assumes that all combustibles within a fire compartment can be consumed during a fire without a failure of the compartment boundaries regardless of operation of fire suppression systems.

The plant was divided into the fire compartments for purpose of evaluation of fire consequences, which coincide with structural components and fire resistance barriers. Judgment and experience were used to determine the adequacy of the fire protection measures for each compartment, and the possibility of propagation to other compartments. All the boundaries of a fire compartment credited as fire barriers (wall, floor, ceiling, doors, ventilation dampers, and penetrations) were evaluated carefully. A fire was supposed to propagate via a fire barrier if the duration of the standard fire would be longer than the fire resistance of the barrier expressed in hours.

In the case of high cable concentrations additional protection for the area of concern was provided, i.e. the deterministic analysis resulted in widespread physical changes to the plant such as additional fire barriers.

Examples of the plant modifications incorporated in the design are: (a) constructing fire barrier walls between redundant cables to divide the containment annulus into smaller compartments; (b) providing cooling for those compartments; (c) replacing surplus fire doors by fire-rated walls in cable shafts; (d) reinforcing fire-rated doors in certain compartments of the auxiliary reactor building; and (e) enclosing certain cables in conduit to prevent cables from contacting other equipment in the fire compartment. In the last case, the fire resistance rating of the fire-retardant wrap was designed to be equivalent to that of the fire compartment barrier walls.

The fire PSA followed a two-phase approach. In phase 1, a screening analysis was performed to identify the critical or important fire locations and screen out those areas that are not risk significant. Fire areas, which have either safety-related equipment or associated cables, were identified as requiring further analysis. The process, electrical, control and cable compartments were included into the analysis. The compartments not containing vital equipment were screened from consideration. In phase 2, a detailed analysis was performed for the important fire scenarios.

Many efforts were made in the deterministic fire hazard analysis to establish the failure modes of equipment and especially electrical circuits because of a cable failure. The circuit analysis was performed to identify the power, control, and instrumentation cables that are necessary to support the operational requirements for active safety-related equipment that is electrically controlled and/or powered. The types of cable failures considered were short circuit, short to ground, open circuit, and hot short. A fault of the last type may lead to actuation of non-energized circuits. The lack of detailed design information on cable routing was resolved by means of addressing the worst-case situation.

Fire consequences were defined in terms of initiating events and mitigating equipment failures caused by a fire. The fire initiated event sequences that need to be addressed were identified. Only a single, independent fire was assumed to occur in any plant location.

Based on the results of the study done all the fire compartments under consideration were categorized as the following groups:

- Compartments in which a fire may lead to a design basis initiating event in combination with a failure of a safety system train.
- Compartments in which a fire may cause a design basis-initiating event without failures in safety systems.
- Compartments in which a fire may not lead to a design basis initiating event, but failures in safety systems are possible.
- Compartments in which a fire cannot affect safety related systems.

The fire hazard assessment validated that adequate plant system response is available to achieve a safe shutdown state following any postulated fire.

5. FIRE FREQUENCY

Insights gained from operational experience of all the 18 units with VVER-1000 reactors in Russia and Ukraine were used in the fire PSA as the statistical basis for establishing frequencies of fires as a function of the initiating fire source. The fire sources listed below were considered:

- turbine & generator facility;
- pumps;
- fans, compressors and HVAC units;
- diesel generators;
- motor operated valves;
- cables;
- board devices such as I&C cabinets and control panels;
- electrical equipment such as transformers, inverters, and rectifiers;
- circuit breakers;
- oil tanks;
- transient fuels.

Information on fire incidents that occurred during the period 1986 through 1999 was collected. A total of 201.8 calendar years, including 140.4 years of power operation, pertain to the observation period. The available information for each event was reviewed to estimate the frequency of future fires involving a given fire ignition source based on the occurrence rate of similar fires in the past. The frequencies presented in Table 1 for certain groups of components were calculated on a "per plant" basis.

Table 1 Frequency of fires at operating NPPs with VVER-1000

Equipment becoming an ignition source	Frequency of large fires per plant-year
Turbine, generator, exciter	2.9E-3
Turbine driven pump	8.3E-4
Motor driven pump, including oil ignitions	3.5E-3
Switchgear, including circuit breakers	1.4E-3
Electrical components (transformers and others)	1.4E-3
Cable	1.1E-2
Transient fuel	2.7E-4

It is important to note that since the mid-1980s substantial effort has been made to upgrade fire safety at VVER plants. This includes the application of fire retardant coatings on cables, upgrading of fire barriers, improved fire suppression systems, improved plant procedures and operator training, etc. Improvements to the VVER-1000 plants have likely contributed to a significant reduction in the potential for fires [9]. Therefore, the estimates presented in Table 1 can be considered as conservative ones for the current state.

Fire incident information gathered confirms the fact that fires at VVER-1000 plants are relatively rare, and those occurred were usually minor and had little or no safety significance. Therefore, the fire occurrence frequency in compartments was estimated involving data on both fires and ignitions at VVER-1000 plants, which may be considered as precursors of fires. The probability of the ignition-fire transition was derived from operational experience.

The fire frequency estimation for each particular compartment was conducted on the basis of counting the number of individual ignition sources (a portion of the total mass – for cables). The resulting frequencies obtained for the particular equipment types from operational experience were assigned to the Kudankulam NPP compartments proportionally to the number of pieces of equipment in these compartments, i.e. these frequencies were weighted by the quantity and type of combustibles located in the compartment. Transient combustibles were also addressed.

6. PROBABILISTIC MODEL

The fire PSA relied on the plant response probabilistic model developed for the internal initiating events, which was modified to take into account fire-induced failures and impact of a fire on human actions. The fire risk model includes all equipment and cables required to support the functionality of the selected systems.

The event trees and fault trees were solved using the computer code RISKSPECTRUM, as used in the internal PSA. The methodology of this code is based on the method of small event trees / large fault trees widely used in the world practice.

The methodology for re-development of the event/tree fault tree models to represent fire scenarios includes the following steps:

- Review of any simplifications or approximations incorporated during the internal event PSA development to confirm their applicability for the fire risk assessment.
- Modification of the fault trees to include logic keys (house events) to enable fire induced plant failures to be modelled correctly.
- Incorporation of additional fire-induced events in the fire risk model.
- Review of the common cause failure modelling since the degree of redundancy in safety systems may be reduced because of the fire.
- Re-evaluation of human error probabilities modelled in the event trees and fault trees taking into account the impact of the fire on the operational scenarios represented.
- The following assumptions were used in developing the fire probabilistic model:
- The plant is operating in its normal operating mode at power at the time of the fire.
- Fire need not be postulated to be coincident with independent, low frequency incidents in the plant, unless they are caused as a consequence of the fire.
- Components that fail safe after a damage by the fire can be credited as fulfilling their safety functions.
- Additional random failures are considered together with those directly attributable to the fire.
- Passive mechanical components, such as heat exchangers, valves, and pipes that may be exposed to a fire, remain structurally intact as a pressure barrier.
- The fire PSA used conservative assumptions and excluded credit for operator recovery actions for modelling the subject fire area. Any actions that require operators to enter to affected compartment were neglected. In the case of components not affected by the fire, the recovery of failed components could continue to be claimed.
- To account for the factor of the fire impact to operators, the post-accident human error probabilities were increased by a factor of 10 for any action in comparison with the PSA for internal initiating events (except for the neglected actions).
- A mission time of 24 hours was used.

7. RESULTS

The overall fire-induced CDF for Kudankulam NPP was quantified to be 9.1E-9 per reactor year that is six times lower than the CDF addressed in the internal event PSA. Figure 1 shows the contributions to the total CDF from different fire-induced initiating events. Inadvertent openings of safety/relief valves were calculated to be dominant contributors. It was found that a fire-induced electrical fault (hot short) along the path of the connected power/control cables might cause the valve inadvertently to change position to the undesired open condition leading to the initiating event in the case of a fire. However, since the mechanisms leading to spurious actuation of equipment are not well understood, many uncertainties are associated with scenarios involving such failure modes. A detailed circuit analysis to identify the specific circuit faults causing the valve to change position will be performed at the detailed design stage.

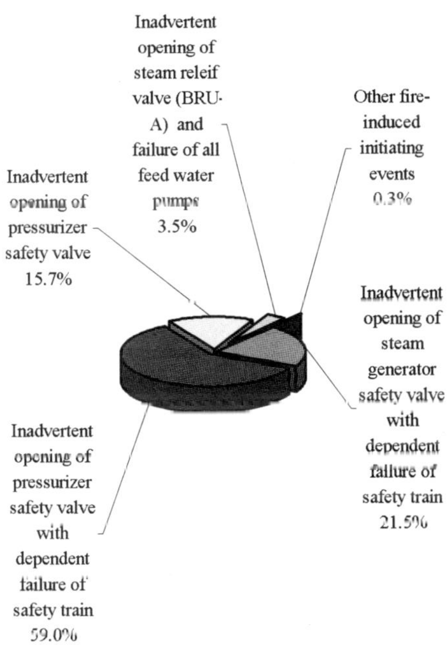

Fig. 1: Proportional contribution to core damage frequency from fire-induced initiating events at Kudankulam NPP

Contributions into the CDF from fires in different locations were also estimated. The largest contribution to the total CDF comes from fires in diesel generator buildings UKD and relevant cable tunnels UKZ. Accident sequences combining the inadvertent opening of a pressurizer safety valve with a failure of a safety system train contribute approximately 59% to the overall fire-induced CDF. Accident sequences involving inadvertent opening of a steam generator safety valve with its subsequent non-closing and a failure of one safety system train contribute 21.5% to the total fire-induced CDF. Although the fire PSA was performed in a conservative manner, the extremely low value of the CDF obtained shows that passive safety features incorporated in the design of Kudankulam NPP assure reliable fire resistance of the plant.

8. CONCLUSIONS

Results of the fire PSA demonstrate a strong possibility to achieve in the Kudankulam design the probabilistic safety goals established for future plants [10].

Although the numerical results of the fire PSA are associated with large uncertainties at the PSAR design stage, the fire PSA including the fire hazard analysis can provide a cost-effective approach to fire protection improvement.

9. REFERENCES

1. Federal Authority of Nuclear Regulation. Main recommendations for performing PSAs for NPPs, RB-032-02, Russian Federation, Moscow, 2004 (in Russian).
2. International Atomic Energy Agency. Treatment of internal fires in probabilistic safety assessment for Nuclear Power Plants, Vienna, 1998.
3. International Atomic Energy Agency. Procedures for Conducting Probabilistic Safety Assessment of Nuclear Power Plants (Level 1), Vienna, 1992.
4. V.Morozov and G.Tokmachev. PSA for fires at VVER-1000/320 nuclear power plants. In: Proceedings of the TOPSAFE'98 International Conference held by the European Nuclear Society and the Spanish Nuclear Society in Valencia, Spain, April 15-17, 1998. Senda Editorial,S.A: CD-Rom "TOPSAFE'98 Papers".
5. J A Carretero, Y Shvyryaev, G Tokmachev, V Morozov, P Fernandez. Results of

Novovoronezh-3 PSA performed within TACIS-91 Programme In: Proceedings of the TOPSAFE'98 International Conference held by the European Nuclear Society and the Spanish Nuclear Society in Valencia, Spain, April 15-17, 1998. Senda Editorial,S.A: CD-Rom "TOPSAFE'98 Papers".

6. V.Morozov, G.Tokmachev, M.A.Azarm, A.Minister. Probabilistic study of fire ifluence on Smolensk NPP safe shutdown capability. In: Proceedings of International Conference Fire&Safety 2001, held in London on 12-14 February, 2001: CD-Rom "Nuclear Engineering International, Conference documents, FIRE&SAFETY 2001, 12-14 February London"
7. V.B.Morozov, G.V.Tokmachev, A.Yu.Zhuravlev. Probabilistic and Deterministic Safety Assessments of NPPs with VVER-1000 reactors given a fire, All-Russian Conference «Practice of PSA development and their application for NPPs with VVER-type reactors both operating and under design", 18-20 November 2002, Atomenergoproekt, Moscow
8. V.B.Morozov and G.V.Tokmachev. Analysis of impact of possible fires and their consequences on safe shutdown of VVER-1000 reactor unit. Teploenergetika (Rus) - Thermal Engineering (Engl), Moscow, 2005, No 1, Pages 47-50
9. S.P.Nowlen, M.Kazarians, N.Siu, H.W.Woods. Fire Risk Insights from Nuclear Plant Fire Incidents. In: Proceedings of International Conference Fire&Safety 2001, held in London on 12-14 February, 2001: CD-Rom "Nuclear Engineering International, Conference documents, FIRE&SAFETY 2001, 12-14 February London"
10. International Atomic Energy Agency. Basic Safety Principles for Nuclear Power Plants 75-INSAG-3 Rev.1 INSAG-12, Vienna, 1999

Reliability, Safety and Hazard
Editors: P.V. Varde et al.

INSIGHTS FROM THE SEISMIC PROBABILISTIC SAFETY ANALYSIS OF PAKS NUCLEAR POWER PLANT

József Elter

Paks Nuclear Power Plant Ltd., P.o.box 71, H-7031 Paks, Hungary

ABSTRACT

As an integral part of the seismic safety program of the Paks NPP a level 1 seismic PSA was prepared. Currently unit specific level 1 PSA models and results are available for internal initiating events, internal fires and flooding at full power. Internal initiating events for off-power operating modes of the plant have also been analyzed by level 1 PSA. The recent quantification of seismic risk was a substantial extension to the existing level 1 PSAs, by this the majority of important initiating events have already been taken into account since earthquakes are considered to be the only risk significant external hazard.

The major steps of the Paks seismic PSA model development were the analysis of seismic response, assessment of failure modes and seismic fragility, accident sequences delineation and the quantification of core damage risk from earthquakes.

The PSA models were used to evaluate the progressive risk improvements that may be achieved from implementation of seven potential seismic upgrades. An overview of the core damage frequency from the baseline PSA results and from each upgrade analysis is also provided.

Keywords: PSA for NPPs, Seismic Safety, External Hazard, Quantification of Seismic Risk

1. INTRODUCTION

The earthquake occurred on 4 March 1977 in Vrancea attracted attention to the seismic safety problems of the NPPs of VVER-440 V230 and V213 types. Seismic problem of the VVER type units is derived from the fact that during their design the seismic hazard of the sites was underestimated and safety aspects related to the external events were neglected. In the case of Paks NPP the site seismic hazard was also underestimated, so the units of VVER 440/V213 type of Paks NPP were not designed for any earthquake loads. At the end of 1980s it was clear that seismic hazard of the Paks site can be much greater, than it was assumed during the design. The NPP launched a comprehensive program for seismic assessment and upgrading of the plant years ago.

Preparation of the seismic PSA for the units was an integral part of the program. The objective of the study was to quantitatively show the current level of plant safety with respect to the seismic hazard representative for the Paks site. The PSA took into consideration all the improvements implemented so far in support of increasing the seismic resistance of the plant. The study was supposed to partly rely on results of existing analyses performed in the seismic safety enhancement program. However, additional analyses were also needed in areas like seismic response and fragility assessment.
The models of the existing level 1 PSA for internal initiating events were utilized.

2. OBJECTIVES AND CONTENTS OF THE SEISMIC SAFETY PROGRAM

The essential safety requirement for the nuclear power plants is to ensure the shutdown of the reactor, its cooling down and decay heat removal function, and limitation of radioactive releases. According to the international practice, the specific features of Paks NPP and the site, the seismic safety program included the following tasks:

- re-evaluation of seismic hazard of the site, which includes the geotechnical survey of the site, analysis of soil liquefaction, etc.
- establishing the safe shutdown and heat removal technology, and elaboration of the list of the structures, systems and equipment, which are essential to ensure seismic safety,
- installation of seismic instrumentation and elaboration of pre-earthquake preparedness and post-earthquake actions,
- evaluation of the seismic capacity of systems and structures relevant for safety,
- performing the necessary upgrading measures taking into account the priorities and the feasibility of the necessary corrective actions.

Seismic re-evaluation of the NPP has been carried out for the design base earthquake (SSE or SL-2 level) according to the international practice.

In the design of Paks NPP an operational base earthquake (OBE or SL-1 level) was not determined and considered. The earthquake level for safe continuous operation could be determined on the basis of the experiences of re-qualification of safety relevant systems.

A level 1 seismic PSA has become an integral part of the seismic safety program of the Paks NPP.

3. LEVEL 1 SEISMIC PSA

Currently unit specific level 1 PSA models and results are available for internal initiating events, internal fires and flooding at full power [2], [3], [4]. Internal initiating events for off-power operating modes of the plant have also been analyzed by level 1 PSA [5]. Quantification of seismic risk now is a substantial extension to the existing level 1 PSAs for Paks, and the majority of important initiating events have been taken into account because earthquakes are considered to be the only risk significant external hazard.

The following objectives of the seismic PSA have been defined in accordance with the international practice and the available PSA for other types of initiating events.

- Quantification of frequency of core damage from seismic events.
- Identification of important contributors to seismic risk (initiating events, accident sequences, component failures, human errors).
- Identification and ranking of plant vulnerabilities based on qualitative insights.
- Development of recommendations for safety improvement including feedback to the seismic safety program.

4. SITE SEISMIC HAZARD REVALUATION

Probabilistic seismic hazard has been assessed for Paks NPP for very low probability events (return period of 10^2-10^8 years) for purpose of determining the probabilities of seismic induced failure of structures and components.

Peak ground acceleration (PGA) and uniform hazard response spectrum (UHRS) at bedrock have been calculated for three return periods. Non linear effective stress methodology was then used in site response analysis to propagate the bedrock spectra to the surface and to assess liquefaction potential. Logic tree formulation has been used to incorporate input parameter uncertainties.

All available historical data and instrumental earthquake information have been considered for the probabilistic seismic hazard assessment. A comprehensive earthquake catalogue with more than 20,000 entries has been compiled and homogenized from different regional databases. Three alternative area source models have been created with 20, 25 and 15 source zones respectively and earthquake recurrence estimated particularly within each of these areas.

During re-evaluation of the site the most discussed question was, whether the tectonic structures on the site had been active during the recent tectonic regime. The conclusion was that the probability of recent activity and the existence of a capable fault are very low. In order to prove this fact we have established and operated a micro-seismic monitoring network. Since then no micro-seismic event has been registered around the site. However, neither historical seismicity nor recent local monitoring data support any fault activity an additional 10% probability has been assigned to possible local fault sources.

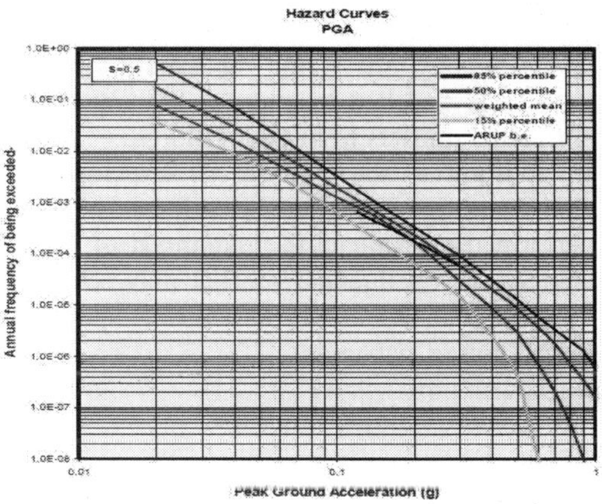

Fig. 1: Hazard curves: annual frequencies of peak ground accelerations

The PSHA has resulted PGA (see Figure 1.) and spectral acceleration values for all the assessed return periods for the bedrock outcrop at the site. The results of the study match pretty well with those given an earlier by an other study [8] namely PGA=0.25 g at around 10,000 years return period.

Site response analysis has been carried out to propagate the bedrock spectra to the surface taking account the effect of site-specific geotechnical and soil parameters.

5. METHODOLOGY OVERVIEW

A detailed seismic PSA was necessary so that the above objectives and the corresponding regulatory requirements could be met and a solid safety upgrading strategy proposed. The Paks seismic PSA has been performed in a depth and detail that is not yet typical for most VVER-440/213 reactors. In this respect experience from the seismic PSA for Paks is expected to yield insights which can be usefully applied to similar reactors in Central and Eastern Europe.

Presently earthquakes only during full power operation were analysed. A reference unit has been selected for the target of seismic PSA. Also, the state of the reference unit (e.g. design basis, plant modifications and safety improvements made) for which the analysis is to be performed has been defined as reference state of the reference unit.

The analysis has been performed in a cooperation of internationally recognized experts, domestic research organizations, plant engineers and other personnel.

The major steps of the Paks seismic PSA have been decided in accordance with most common international practices [6] and methodologies applied recently in most OECD countries [7]. These steps are as follows:

- assessment of seismic hazard,
- analysis of seismic response,
- assessment of failure modes and seismic fragility,
- PSA model development for accident sequences from seismic events,
- quantification of core damage risk from earthquakes.

5.1 Seismic Acceleration Ranges

It was not practical to quantify the PSA models using continuous families of seismic hazard curves and associated equipment fragility distributions. Instead, seven acceleration ranges as it shown in Table 1 were selected to define seismic initiating events. The lower bound is 0.07g, this corresponds to the lowest seismic capacity for all structures and equipment. The upper bound 1.0g is the highest acceleration evaluated in the seismic hazard analysis. The intervals in the ranges are progressively larger to account for the fact that the frequencies change more slowly at higher accelerations.

Table 1 Initiating event acceleration ranges

Initiating Event	Acceleration Range (g)	Main Frequency (event/year)
SEIS1	0.07 – 0.10	$2.69 \cdot 10^{-3}$
SEIS2	0.10 – 0.15	$1.08 \cdot 10^{-3}$
SEIS3	0.15 – 0.22	$3.16 \cdot 10^{-4}$
SEIS4	0.22 – 0.32	$8.71 \cdot 10^{-5}$
SEIS5	0.32 – 0.48	$2.35 \cdot 10^{-5}$
SEIS6	0.48 – 0.70	$4.76 \cdot 10^{-6}$
SEIS7	0.70 – 1.0	$8.99 \cdot 10^{-7}$

5.2 Failure Modes and Seismic Fragility

The analysis of seismic response based on the results of finite element evaluations of structures and floor response spectra were already available for practically all levels of interest within the buildings of safe shutdown components.

Comprehensive walk-downs were conducted to examine all structures and plant locations that contain mechanical equipment, piping, switchgear, electrical equipment, instrumentation and control cabinets, and cables that may affect any systems or functions that are analyzed in the PSA. Systematic screening criteria were applied during the walk-downs to determine whether the seismic capacity of an examined item is sufficiently high to justify no need for additional evaluation.

As a result of the initial evaluation process, many structures and components were assigned to the following screening groups, based on their assessed capacities [9].

- Structures, mechanical equipment, electrical equipment, relays, and cabinets that satisfy the predefined higher screening capacity criteria (HCLPF= 0.41g pga).
- Mechanical equipment that satisfy the lower screening capacity criteria (HCLPF= 0.27g pga).
- Relays that have been tested and satisfy the relay screening capacity criteria.

The three screening groups are included in the PSA models to ensure completeness of the seismic risk quantification process. The failure fractions for each group are determined by their respective screening fragility distributions. The functional impacts from each group are determined by the most limiting combination of failures, based on the specific structures and equipment that are included in the group. In this manner, the PSA quantification process completely accounts for seismic failures of all relevant structures and equipment in the plant, and no items are omitted from the analysis process.

More detailed analyses were performed to determine the capacities of specific structures, mechanical equipment, electrical equipment, cabinets, and relays that did not satisfy the initial screening criteria. The fragility distributions determine the likelihood of failure for each structure or component in the PSA models, as a function of the seismic acceleration. In some cases, similar types of equipment are combined in a single group when the PSA models are quantified because they have similar seismic capacities and are expected to fail at approximately the same acceleration. These items are modelled together to account for the correlated impacts from their failures.

In addition to the analyses of structural and equipment capacities, site-specific analyses were performed to evaluate the likelihood of various degrees of soil liquefaction that may occur and the potential damage to structures, equipment, buried piping, cables, etc.

5.3 Seismic Failure Fractions

Having the information about the seismic hazard and component fragilities, failure fractions of the seismic induced failures were calculated by convoluting the seismic hazard and fragility curves. In addition to calculating the failure fraction point estimates the uncertainties of the failure fractions using Monte Carlo method were calculated.

Each failure fraction represents the mean conditional likelihood for the given seismic induced failure at the designated seismic acceleration.

5.4 PSA Model Development

A baseline seismic PSA model was developed in order to construct an event tree structure that integrates event trees developed earlier within the internal initiator PSA study and specific earthquake-induced transients into a generic model that reflects the specifics of an earthquake and the reference state of the plant. The potential seismic-induced transients were identified and evaluated with respect to the needs of their modelling in the seismic PSA. An event tree representation of the special seismic safety cool down procedure was produced in support of further modelling. Finally, a generic event tree was developed as an integrated event sequence model for seismic-induced transients, seismic and non-seismic equipment failures. In addition, containment functions and systems were examined to identify vulnerabilities that involve early failure of containment functions.

All the failure modes that are consequences of an earthquake, and all the seismic related dependent failures were transferred into an OR gate in the fault trees as shown on the Figure 2.

Extensive use was made of the existing level 1 PSA models for internal initiators throughout the model development to ensure consistency in terms of modelling approaches, depth, and level of detail.

The baseline seismic PSA model thus was developed in the following major steps:

- selection (grouping) of equipment level failures that can be caused by different seismic-induced failures (groups),
- identification of transient initiating failures and additional system, train or component level failures and degradations that can be caused by any combination of equipment failures selected for a group, establishment of a list of transient initiating failures that can be caused by an earthquake,
- development of functional event trees for single transient initiating failures,
- development of a generic event tree for modelling plant responses to an earthquake with combinations of single and multiple transient initiating failures,
- modelling containment performance.

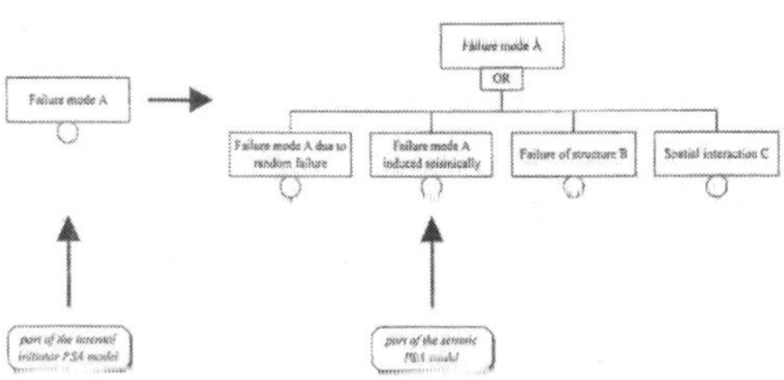

Fig. 2: Transfer of failure modes to include seismic component failures

6. QUANTIFICATION OF CORE DAMAGE RISK FROM EARTHQUAKES

The baseline seismic PSA model was developed using the Risk Spectrum PSA Professional computer code. It is not able, however, to convolute seismic hazard and fragility curves, and so seismic failure fractions calculated by a different code were put into it manually. Point estimate calculations were performed by Risk Spectrum that required a new type of calculation as compared to the internal initiator PSA: in many cases, especially for high acceleration ranges the conditional core damage probability was calculated (given the seismic initiator occurred) instead of the core damage frequency. The reason for that was that the frequency calculation gives incorrect results for high failure probabilities that frequently appear in the high acceleration ranges. In those ranges the conditional core damage probability given the seismic initiator occurred is close to unity. Final results for core damage frequency were derived by multiplying the conditional core damage probabilities with the corresponding frequency of the seismic initiator.

The baseline analyses include credit for previously identified seismic upgrades that the plant has either implemented or the detailed designs were already available.

The baseline results seen in Table 2 are strongly dominated by structural failures of bolted connections in the Turbine Building. Eventual collapse of the turbine building is assumed to cause steam and feedwater header ruptures that result in total loss of main and auxiliary feedwater and disables closed loop heat removal through the secondary side. Emergency feedwater may remain available; however it can be used only for open loop secondary side heat removal that is assumed to be insufficient to ensure long term heat removal. The only means of heat removal from the core in such a situation would be the service water system through emergency core cooling system heat exchangers using primary feed-and-bleed.

Table 2 Calculated core damage frequencies

Initiating Event	Base line CDF (event/year)	Upgrades CDF (event/year)
SEIS1	$4.22 \cdot 10^{-6}$	$2.13 \cdot 10^{-10}$
SEIS2	$7.18 \cdot 10^{-5}$	$1.91 \cdot 10^{-7}$
SEIS3	$1.17 \cdot 10^{-4}$	$8.67 \cdot 10^{-7}$
SEIS4	$7.26 \cdot 10^{-5}$	$3.91 \cdot 10^{-6}$
SEIS5	$2.34 \cdot 10^{-5}$	$1.89 \cdot 10^{-5}$
SEIS6	$4.76 \cdot 10^{-6}$	$4.74 \cdot 10^{-6}$
SEIS7	$8.99 \cdot 10^{-7}$	$8.99 \cdot 10^{-7}$

The PSA models were used to evaluate the progressive risk improvements that may be achieved from implementation of seven potential seismic upgrades. The most effective upgrade that was considered in these analyses increases the structural capacity of the Turbine Building bolted connections. The revised PSA analysis is based on the assumption that the seismic capacities of these connections are increased to the extent that failures of these structures are now governed by buckling of the vertical frames. The combined effects from these upgrades reduce the total CDF by a factor of 10, compared with the baseline results. The most important risk contributors are now determined by failures of offsite power at high accelerations that also disable onsite operator performance and may prevent effective response from offsite personnel, and extensive soil liquefaction that damages the underground service water piping. Approximately only 4 % of the revised CDF is caused by failures that occur below the 0.25g design basis earthquake.

7. REFERENCES

1. L. Tóth, E.B. Győri, P. Mónus, T. Zsíros: Seismic hazard assessment at Paks NPP site, GeoRisk Institute, November 2000.
2. A. Bareith, E. Holló et al. Level 1 PSA of Paks NPP, Units 1-4. Summary Report and Appendices, VEIKI Report No. 20.21-235/1-4, October 1998
3. Cz. Sándor, Z. Karsa, I. Neubauer, J. Nigicser, P. Siklóssy: Probabilistic Assessment of Fire Risk in NPP Paks, Final Report, VEIKI Report No. 22.11-814, November 1998
4. Z. Karsa, I. Neubauer, J. Nigicser, P. Siklóssy: Probabilistic Risk Assessment of Internal Flooding in NPP Paks, Final Report, VEIKI Report No. 22.11-813, November 1998
5. Bareith et al. Level 1 PSA for Low Power and Shutdown Operating Modes of Paks NPP, Main Report and Appendices, VEIKI Report No. 22.21-816, June 1998
6. Probabilistic Safety Assessment for Seismic Events, IAEA-TECDOC-724, IAEA, Vienna, Austria, October 1993
7. R. J. Budnitz: State-of -the-Art Report on the Current Status of Methodologies for Seismic PSA, NEA/CSNI/R(97)22, OECD Nuclear Energy Agency, Paris, March 1998
8. Ove Arup & Partners, PHARE Program, Seismic Hazard Re-evaluation, Final Report, No 94-0600, 1996.
9. Walkdown Screening for Conducting Seismic PSA of Paks NPP. Prepared by EQE International, Inc. and Westinghouse Electric Europe. 250997-R-003, October 2000.

Reliability, Safety and Hazard
Editors: P.V. Varde et al.

APPROACH TO FIRE-PSA STUDY OF KAPS IPHWR

S.C. Maiti[1], D. Chatterjee, Rajee Guptan, Nalini Mohan, S.G. Ghadge and S.S. Bajaj

Directorate of Safety, NPCIL, NUB, Anushaktinagar, Mumbai 400 094

ABSTRACT

Some safety significant fire events in nuclear power plants like Browns Ferry (U.S.A, 1975), NAPS (Narora, India, 1993) had serious implication on nuclear safety. Some of them had direct or indirect potential to cause sever reactor core damage. Therefore fire-induced core damage frequencies (FCDF) can be a significant contributor to the overall risk for a plant depending upon the design and operational characteristics of the NPP. Fire probabilistic safety assessment (Fire-PSA) studies for KAPS has been initiated using probabilistic and deterministic analysis like 'Fire Hazard Analysis' to identify the main contributing 'Fire Zone' and 'fire scenarios' to FCDF. The study is being done based on the IAEA guideline and as per international practice for carrying out Fire-PSA. Fire zone are screened out qualitatively and quantitatively based on their direct or indirect potential to Conditional Core Damage frequency (CCDF). Fire database used for the analysis was for 25 Reactor-years of commercial operation of KAPS. Fire frequency was very low in most of the 'Fire Zone' and most of the fire were detected & suppressed at incipient stage. In most of the cases safety systems were not challenged, and all safety functions were fully available. It gives an insight to fire risk related strength-weaknesses of design and operation of IPHWR. Major contributor to fire risk ie vulnerable area, vulnerable nuclear safety items, vulnerable fire risk action, and important fire safety components can be identified. It helps the designer and plant management to prioritize their effort in fire risk reduction.

Keywords: Fire Hazard Analysis, Probabilistic Safety Assessment, Fires, Fires Frequency, Fire-PSA, Core Damage Frequency, Fire induced Core Damage Frequency, Event Tree, Fault Tree, Initiating Event

1. INTRODUCTION

KAPS-1&2 is a twin unit 235 Mwe Indian Pressurized Heavy Water Reactor (IPHWR) with heavy water as coolant and moderator. Since it's commercial operation it is operating with a very high availability factor. PSA Level-1 has been completed with internal initiating events (LOCAs and transients) and found to have CDF= 6.7 E(-6). The result of this detailed analysis had established the high degree of defence-in-depth of IPHWR against Internal IEs. IPHWRs are planned, designed, constructed with the best quality standard and operated at per with international safety practice to ensure defence against external events like Fire, Flood and Seismic. However a quantitative risk analysis has been started to quantify CDF due to fire as an external event. It will give an insight about the strength - weakness of IPHWR and priority areas required improvement. A study on Fire-PSA has been started with this basic objective along with other objectives.

KAPS 1&2 plant had been divided into fire compartments. Each compartment has been evaluated qualitatively based on it's potential for fire which can directly or indirectly contribute to CDF. Fire data is collected from the date of commercial operation (for last 25 Reactor-years of operation) and fire frequency of each compartment are calculated. Then the conditional CDF is calculated based on fire frequency, safety system inventories, and fire detection-suppression probability. If the CCDF due to fire in the compartment is less than certain percentage of combined CDF (due to all internal IEs) that compartment is screened out. Remaining fire compartments are required to be analysed in detail using deterministic and probabilistic analysis to assess the risk potential. Result of the analysis is reviewed to find strength-weaknesses and areas required management attention.

2. FIRE PSA METHODOLOGY

Fire-PSA of KAPS is being done as par IAEA guideline [2,4] in following steps:

- Division of KAPS-1&2 into 'Fire Zones'
- Fire data collection of each zones
- Fire Hazard Analysis of each zone.
- Qualitative screening of zones based on safety system & fire hazard inventories.
- Quantitative screening of Zones based on CCDF
- Results and insights after above step
- Detailed analysis of the remaining vulnerable zones.
- Results and insights.
- Report preparation as per standard format of LEVEL-I PSA of KAPS.

2.1 Division of KAPS–1&2 into 'Fire Zones'

KAPS-1&2 have been divided into 56 fire zones based on IAEA [3] standard definition of fire compartment. Fire zone is a compartment to which fire cannot practically spread from neighbouring zone unless otherwise fire barrier is damaged, or the fire will not spread even after complete burning of all combustible due to distance. Some of the zones are as listed below:

- start
- RB – 100 MEL,
- RB – Pump Room,
- RB – 103 MEL,
- TB – 111 MEL,
- etc

2.2 Fire data collection of each zones

Fire data are collected by the Fire Protection Engineer for each zone with special emphasis to following input data:

- List of KAPS-1&2 'Fire Zones'
- Fire Zone data
- fire compartment geometry
- list of ignition sources
- list & amount of combustible
- location/distribution of combustible
- location of safety systems / items,
- list of redundant safety item within the compartment, if any
- ventilation details,
- list of fire suppression items
- list of compartment to which it can spread fire and barrier rating,
- ventilation details,
- etc
- Fire Event data of each zones
- plant status before and after fire,
- ignition source and material burnt,
- suppression means,
- time taken to suppress the fire,
- root cause of fire,
- amount of smoke and means used to remove smoke,
- list of equipment damaged and mode of damaged,
- list of fire barriers and their rating,
- list of compartment to which it can spread fire and barrier rating provided,
- etc

2.3 Walkthrough to verify plant data

A team of experts consisting Analyst, and plant personnel visited the plant to verify the plant specific data and also to get more insight about zones. Walkthrough was done for five zones to identify and assess important parameters. Following zones were visited:

- Control Equipment Room,
- DG Room
- DG Control panel room
- 415 V class –III Switchgear room
- 6.6 KV Class –III Switchgear room
- etc

2.4 Fire Hazard Analysis (FHA) of zone.

As par the practice adopted by NPCIL, station carried out FHA using CALFIRE software package to assess following:

- fire hazard class of the compartment
- barrier rating required as per Indian Standard,
- duration of fire (it was assumed that all inventories were lumped at a single place),
- equipment damage (for the worst case)
- possibility at all to spread fire to the safety related zone (for the worst case)

This information was used for further analysis.

2.5 Qualitative screening of zones based on safety system & fire hazard inventories.

Qualitative assessment is done for each zones based on fire load and safety system inventories. If the compartment has no safety significant component and the fire inventories is limited to such a extent that it will completely burn before any possibility of fire spread, the compartment is screened out.

After thorough analysis 9 zones out of 56 zones were screened out. Some of the zones screened out are WMP, GT, etc.

2.6 Quantitative screening based on CCDF

The quantitative screening was done in two sub-stages as indicated below:

- In the first stage, all safety equipment in the room are assumed to be damaged,
- in the second stage, equipment damage or vulnerability of safety component was taken based on analysis output.

The conditional core damage frequency (CCDF) is calculated with the assumption that all safety components in the zone-compartment had failed after failure of suppression of potential big fire.

First Stage is done in two steps. In first step, following parameters were calculated for each fire zones:

- Frequency of fire, (ω_i), see Table 1.
- Probability of fire not on a safety system, p_{NOS}
- detection probability, p_D
- suppression probability, p_S
- CFF, Critical fire frequency with potential to affect safety system

Then critical fire frequency, CFF is calculated based on the equation:

$$CFF = \omega_i (1 - p_D * p_S * p_{NOS}) \quad ......(1)$$

Table 1 Location-Wise Fire Frequency for KAPS

Fire Zone Area	No. of fires	Oprtion Yrs.	Fire Frecy
WMP	1	13	0.077
TG Air Cooler	1	25	0.040
Cable Vault 95	0	25	0.000
Cable Vault 103	0	25	0.000
DG	0	50	0.000
Battery	0	25	0.000
6.6 KV Class-IV Bus E	0	13	0.000
6.6 KV Class-III Bus D	0	13	0.000
SUT	0	25	0.000
UT	0	25	0.000
GT	0	25	0.000
BFP	0	25	0.000
UPS	0	13	0.000
415 MCC	0	25	0.000
415 Class II Bus S	0	13	0.000
415 Class III Bus J	0	13	0.000
MCR	0	13	0.000
CER	0	13	0.000
RB Pump Room	2	25	0.080
Moderator Room	0	25	0.000
95 RB	0	25	0.000
103 RB	1	25	0.040
100 RB	1	25	0.040
SI Yard – Class-IV	1	13	0.040
Chiller Room	1	25	0.040
TB-111 MEL	4	25	0.160

Then the worst case CCDF is calculated from existing Level-1 PSA of KAPS [1] using above frequency as fire induced IE's frequency of that zone with all safety components in the zone had failed (ie q=1.0).

Such analysis was carried out for balance 46 zones. If the CCDF is less than certain percentage of combined CDF (due to all internal IEs) that compartment is screened out. Remaining fire compartments were taken for further analysis

In the second stage screening is being done in three steps with assumption that (1) all fire materials and sources are accumulated at one place; (2) safety component in the zone/compartment had failed after a time as found in vulnerability analysis; (3) fire spread to another zone after failure of fire suppression and damage of fire barrier.

In the first step, Fire Hazard Analysis of each zone is done using COMPBRN IIIe to assess:

- dynamics of heat and smoke production,
- Vulnerability analysis of electrical, mechanical and electro-mechanical safety components including fire barriers,
- damage of electrical or electronics related safety items due to effect of heat-smoke.
- CFF is calculated using equation (1),

In second step, fire spread to next zone, equipment damage in that zone, etc are calculated. Critical fire spread frequency, CFSF, is calculated as:

$$CFSF = \omega_i(1 - p_D * p_S) * P_{FBij} \qquad (2)$$

- Where, CFSF = Fire frequency with potential for affecting safety system in neighbouring zone,
- P_{FBij}, Probability that fire will spread from zone-i to safety significant zone-j due to failure of barrier between (i – j).

Vulnerability and damage study of zone j is done based on smoke production and time temperature output of FHA.

CCDF is calculated using the existing Level-1 PSA of KAPS with unavailability of only damaged components equals to unity. IE frequency was modified taking CFF as IE's frequency of that zone and CFSF as IE's frequency of next zone. If the CCDF is less than certain percentage of combined CDF (due to all internal IEs) that compartment is screened out.

2.7 Results and insights after above step

Fire frequency thus calculated was very low in most of the 'Fire Zone'. It is also observed that most of the fire were detected & suppressed at incipient stage and safety systems were not challenged, and safety functions were fully available. Hence it is expected that the final calculated CCDF will be low.

3. FIRE-PSA OF VULNERABLE ZONES

After quantitatively analysing the base case study and obtaining the CCDF, vulnerable zones were identified and detailed analysis was done in three steps. In the first step, deterministic and probalistic parameter of fire model were calculated for all possible potential fire scenario in a zone. Fire propagation, vulnerability and equipment damage probability study shall be done in this step. In the second step, fire non-suppression and propagation ET shall be developed. Event Tree (ET) of one fire zone / compartment as given in figure-1.

LOCAs, transients or any new IEs that can be initiated due to fire, fire fighting activity or fire propagation will be listed for further analysis. Sequence of events after fire that can cause a particular IE shall be assessed. Fire specific ET's shall be developed with that sequence to find out combined IE frequencies (ie frequency of transient event caused by fire). In the last steps, sequence of safety systems required responding to IE under fire condition for prolonged cooling and long-term shutdown shall be checked. Especially, validity of existing success criterion of system and component under fire condition also will be checked. IEs will be grouped based on same sequence of safety system response and same consequence criteria. Then final fire induced CDF shall be calculated using existing Level-1 PSA of KAPS-1&2 for all above group of IEs with IE's group frequency and unavailability of damaged components.

Plan of all these studies shall be carried out as indicated below:

- detailed fire hazard analysis using COMPBRN IIIe to study
  - Vulnerability,
  - damage of electrical and electronics component due to smoke and heat,
- initiating events caused by fire,
- fire propagation analysis,
- analysis for effect on HEPs after fire,
- importance & Severity Analysis,
- uncertainty analysis,
- uncertainty analysis,
- report preparation and

areas of further study

4. CONCLUDING REMARKS

After the preliminary quantitative screening analysis (refer Table 1.), it was found that frequency of fire with potential to damage safety system and their function in various identified fire zones was extremely low. It has given a

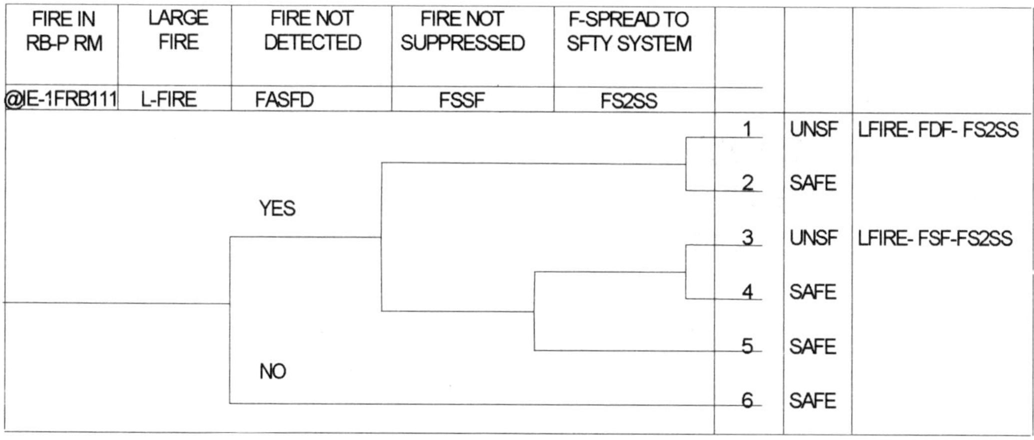

Fig. 1: Sample ET for RB Pump Room Fire.

confidence about defence-in-depth of philosophy adopted against fire. Different case studies will be prepared to get more insight about fire vulnerability of zone, system, safety components, and will be extensively used for completing level – 1 Fire PSA of KAPS.

5. REFERENCES

1. Bajaj, S.S., et al Probabilistic Safety Assessment Level-1 Internal Events for KAPS, NPCIL(2002).
2. IAEA, Treatment of External Hazards in Probabilistic Safety Assessment for Nuclear Power Plants, Safety Series No. 50-P-7, IAEA, Viena(1995),
3. IAEA, Evaluation of Fire Hazard Analysis for Nuclear Power Plants, Safety Series No. 50-P-9, IAEA, Viena(1995),
4. IAEA, Treatment of Internal Fires in Probabilistic Safety Assessment for Nuclear Power Plants, Safety Report Series No. 10, IAEA, Viena(1998).

Reliability, Safety and Hazard
Editors: P.V. Varde et al.

FIRE RATINGS OF CIVIL ENGINEERING STRUCTURES OF KKNPP USING RUSSIAN FIRE SAFETY REGULATIONS

Abhijit Harshan, V.R. Rajagopalan and M.N. Ray

Kudankulam Project, Nuclear Power Corporation of India Ltd, Mumbai

ABSTRACT

This paper explains the design approach, methods and principles adopted by the Russian Designers for estimation of the fire loads, the classification of building and structures into different fire categories conforming to the various Russian Codes and Standards. Civil Engineering Structures are categorized into different fire resistance categories. Within the structures, the rooms housing the combustible materials are graded as per the degree of fire hazard into sub categories. Besides these categorizations, the various components of the building such as walls, slabs, roofs etc., are rated depending upon its functionality, importance of the component in the total building system and limits of fire rating. The fire rating of various rooms in the NPP buildings are estimated by calculating the specific fire load contained in each room, together with the data such as area of the room, total fire load in the rooms, and heat of combustion of the materials in the room and duration of fire from standard fire curves. These aspects are discussed and a few examples have been worked out to estimate the specific fire load (g) in MJ/m^2 of a few rooms of KKNPP as per the Russian standards.

Keywords: Fire Resistance, Fire rating, Explosion and fire hazard category, Specific fire load, Fire load, Loss of Bearing Capability, Loss of Integrity, Loss of Heat Insulation Capability

1. INTRODUCTION

To cater to the increasing power needs of Southern India, 2 Units of 1000MWe Pressurized Water Reactors (PWR) is being constructed in Kudankulam near Kanyakumari. The Project is conceived and is being developed as an Indo-Russian Joint Venture. The Russian side is responsible for providing the technical expertise in the form of working drawings and supply of equipment along with expert supervision. The Indian side, through the public sector enterprise Nuclear Power Corporation of India Ltd (NPCIL), bears the responsibility of Construction, Erection and Commissioning of the necessary infrastructure and the Nuclear Power Plant (NPP).

In the design of civil engineering structures, the estimation of fire hazard is an important aspect. This paper attempts to explain the principles adopted by the Russian Designers in the estimation of fire loads and the classification of various buildings and structures from fire hazard point of view. This is carried out by adopting the provisions of various Russian Codes and Standards [1] [2] [3] related to design for fire loads. A few example problems have been worked out to estimate the specific fire load (g) in MJ/m^2 of few rooms of the KKNPP buildings to demonstrate the calculation aspects as per the Russian practice.

Fire protection of KKNPP is based on the principle of passive fire localization by following the IAEA guidelines 50-SG–D2 [4]. In this connection, fire barriers and safe (limiting) distances ensures non-spreading of fire beyond the zone during full burn out of the fire load.

2. CATEGORIZATION OF BUILDINGS AND STRUCTURES

The categorization of the building and structures with respect to their explosion and fire safety are based on the analysis and conforms to Russian standard NPB 105-95 [2]. Explosion and fire hazard categories of rooms and buildings are determined for the most unfavorable duration with respect to the fire or explosion based on the type of combustible substances and materials in the equipment and rooms, their quantity and their fire hazardous properties and peculiarities of the processes.

2.1 Categorization of Rooms

The categorization of the rooms are performed by successive check of the rooms for the possibility to classify them to different categories starting from the highest i.e., A to the lowest i.e., E.

2.1.1 Category A

These are rooms in which combustible gases, easily flammable fluids with a flash point not more than 28^0C are circulated or handled in such an amount that explosive steam-gas-air mixtures can be formed and which, at ignition, develops in the room a design explosive overpressure of more than 5 kPa.

2.1.2 Category B

These are rooms in which combustible dust or fiber, easily flammable fluids with a flash point more than 28^0C are circulated or handled in such an amount that explosive air-dust or steam-air mixtures can be formed, which, during ignition, develops in the room a design overpressure of more than 5 kPa

2.1.3 Category C (C1 to C4)

These are rooms in which combustible and low-combustible fluids, hard combustible and low-combustible substances and materials (including dust and fiber), substances and materials which burn only in case of interaction with water, air, oxygen or with each other, are stored, circulated or handled and are referred to categories C1 to C4 provided that the rooms where they are stored, circulated or handled are not referred to Categories A or B. Each sub classification is as listed below:

Rooms in which the specific fire load (g) is more than 2,200 MJ/m^2 are referred to Category C 1.
Rooms in which the specific fire load (g) is between 1,401 and 2,200 MJ/m^2, are referred to Category C 2.
Rooms in which the specific fire load (g) is between 181 and 1,400 MJ/m^2, are referred to Category C 3.
Rooms in which the specific fire load (g) is between 1 and 180 MJ/m^2, are referred to Category C 4.
Specific fire load (g) in MJ/m^2 is determined by the relationship:

$$g = \frac{\sum_{i=1}^{n} G_i * Q_{Hi}^{p}}{S} \qquad (1)$$

where G_i - quantity of the ith material of fire load, in kg;

Q_{Hi}^{p} - lowest heat of combustion of the ith material of fire load, in MJ/kg;

S – area of the fire load location, m^2

2.1.4 Category D (Hot non-combustible substances / burning of combustible fuels)

Rooms in which non-combustible substances and materials in hot, incandescent or melted state are circulated or handled, the processing of which is accompanied with release of radiation heat, sparkles and flame as well as combustible gases, fluids and hard substances which are burned or utilized as a fuel, are referred to Category D.

2.1.5 Category E (Non Combustible materials in cold state)

Rooms in which non-combustible substances and materials in cold state are circulated or handled are referred to category E.

2.2 Categorization of Buildings

Categorization of buildings by explosion and fire hazard is carried out in five categories A to E as follows:

2.2.1 Category A

A building is referred to Category A if the total area of Category A rooms in this building is either more than 5% of the area of all rooms or 200 m^2.

It is not allowed to refer a building to category A, if total area of Category A rooms in the building is not more than 25% of the total area of all rooms in this building (but not more than 1,000 m^2), and these rooms are equipped with automatic fire fighting facilities.

2.2.2 Category B

A building is referred to Category B, if two conditions are met simultaneously:

a) The building is not referred to Category A
b) The total area of Category A and B rooms is more than 5% of the total area of all rooms or 200 m^2.

It is not allowed to refer a building to category B if total area of Category A and B rooms in the building is not more than 25% of the total area of all rooms in this building (but not more than 1,000 m^2), and these rooms are equipped with automatic fire fighting facilities.

2.2.3 Category C

A building is referred to category C, if two conditions are met simultaneously:

a) The building is not referred to Categories A or B;
b) The total area of Category A, B and C rooms is more than 5% (10%, if this building does not have Category A and B rooms) of the total area of all rooms.

It is not allowed to refer a building to Category C if total area of Category A, B and C rooms in the building is not more than 25% of the total area of all rooms in this building (but not more than 3,500 m^2), and these rooms are equipped with automatic fire fighting facilities.

2.2.4 Category D

A building is referred to Category D if two conditions are met simultaneously:

a) The building is not referred to Categories A, B or C;
b) The total area of Category A, B, C and D rooms is more than 5% of the total area of all rooms.

It is not allowed to refer a building to Category D, if total area of Category A, B, C and D rooms in the building is not more than 25% of the total area of all rooms in this building (but not more than 5,000 m^2), and these rooms are equipped with automatic fire fighting facilities.

2.2.5 Category E

A building is referred to Category E, if it is not referred to Categories A, B, C or D.

3. ESTIMATION OF FIRE LOAD

The fire load of a space is defined as the total amount of combustible material in the space divided by the gross floor area. It is expressed in terms of kilograms / square meter (kg/m^2). The fire load for each confined space shall be calculated separately on a space-by-space basis by totaling the weight of all the combustible items in a space and dividing it by the gross floor area.

All combustible materials inside of a space contribute to the fire load for that space. The only items which do not have to be considered as part of the fire load are those which are inherently non combustible, such as glass, metals, ceramics etc.

4. FIRE LOAD DISTRIBUTION

The distribution of combustible materials in a space within a room is a significant factor in determining the intensity and duration of a fire. Although the fire load for the entire space may be within the acceptable limits, excessively uneven distributions of combustible materials can lead to barrier attack in the areas of high concentration and ultimately a breakdown in the structural fire protection of the entire space.

Fire classification of construction materials, structures, rooms, buildings, components and parts of the buildings are based on their properties that lead to the development of hazardous fire factors and thus the fire development.

Fire classification is provided to establish the necessary requirements regarding fire protection of structures, rooms, buildings, components and parts of the buildings depending on their fire-resistance and/or fire safety.

5. FIRE RESISTANT STRUCTURES AND COMPONENTS

Buildings as well as building parts enclosed by fire-resistant walls or fire compartments are classified based on fire-resistance rating, constructional and functional fire hazard classes and are governed by Russian Federation (RF) Fire Safety Code for Buildings and Works – SNiP 21-01-97 [1].

Fire resistance is characterized by a fire resistance rating. Fire resistance rating of civil structures is characterized by period of time (minutes) after which one or several signs of the limit states rated for this structure appear. These are:

i. Loss of bearing capability (R)
ii. Loss of integrity (E)
iii. Loss of heat insulation capability (I)

Fire hazard of a structure is characterized by a fire hazard class. Civil structures are divided into four fire hazard Classes. These are:

i. K 0 (non fire-hazardous)
ii. K 1 (low fire hazard)
iii. K 2 (medium fire hazard)
iv. K 3 (fire hazardous)

The degree of fire resistance of a building is determined by the fire resistance of its components (Type I, II up to IV).

The class of construction fire hazard of a building is determined by the extent to which the building structure contributes to the development of a fire and the formation of fire factors determine the constructional fire hazard class of a building (K 0 to K 3).

A building of high degree of fire resistant such as type I should be constructed with a structural component made of non fire hazardous class such as K 0.

As from the fire safety point of view, all the buildings of a NPP are important; therefore all the structural elements such as load bearing columns, walls, trusses, roofs, outside walls, stair case etc are of fire hazard class K 0.

Civil engineering structures provided for fire resistance are called fire barriers. These structures help to prevent fire and products of combustion from propagating from a room or fire compartment where the fire has occurred to other rooms. Fire barriers include firewalls, partitions and ceilings.

Fire barriers based on the fire resistance requirement of the structures are divided into different types as shown in Table 1.

Fire rating of doors and windows are given in Table 2.

Table 1 Fire resistance type of fire barriers

Fire-resistant structures	Type of fire-resistant structures	Fire rating of a fire-resistant structure, not less than	Type of Doors, Windows for openings, not less than	Type of air lock, not less than
Walls	1	REI 150	1	1
	2	REI 45	2	2
Partitions	1	EI 45	2	1
	2	EI 15	3	2
Floors	1	REI 150	1	1
	2	REI 60	2	1
	3	REI 45	2	1
	4	REI 15	3	2

Table 2 Fire rating of doors and windows

Closure of openings in Fire walls	Type of Closure in the opening of the fire wall	Fire resistance rating, not lower than
Doors, gates, hatches, valves	1	EI 60
	2	EI 30*
	3	EI 15
Windows	1	EI 60
	2	EI 30
	3	EI 15
Curtains	1	EI 60

* Fire resistance of the lift well doors is allowed to be taken not less than E 30.

Table 3 Fire resistance rating of civil structural elements

Degree of Fire resistance of the building	Fire resistance rating of civil structural elements, Minimum						
	Building load bearing elements	Outside walls non load bearing	Structural Floors between stories	Elements of roof structures without false ceilings		Staircases	
				Roof slab with heat insulation	Trusses, beams, and girders	Indoor walls	Flights and landings
I	R 120	E 30	REI 60	RE 30	R 30	REI 120	R 60
II	R 90	E15	REI 45	RE 15	R 15	REI 90	R 60
III	R 45	E15	REI 45	RE 15	R 15	REI 60	R 45
IV	R 15	E15	REI 15	RE 15	R 15	REI 45	R 15
V	Not rated						

Buildings and fire compartments are divided by fire resistance as per Table 3.

For example, the turbine building of KKNPP is a category II fire resistant structure whose fire ratings of various structural members match with the values given in Table 3.

6. ESTIMATION OF FIRE RATING OF ROOMS AND STRUCTURAL MEMBERS

In the procedure for comparative determination of the fire resistance of civil structures, a standard fire is considered wherein when the relationship of average volumetric temperature variation in a room versus time is expressed as

Follows, which is indicated in curve A of Figure -1 [5]:

$$t = 345 \lg (8\tau + 1) \qquad (2)$$

where *t – average volumetric temperature during fire in a room, in degree Centigrade*

τ – time since the start of fire, in minutes

The above relation is obtained experimentally by considering combustion of a fire load evenly distributed throughout the room. As the fire load, cellulose pulp was taken in Germany and wood in Russia.

The curve A of the standard fire adopted from EDF (Electricite de France) is expressed in the form of a graph.

Another Curve B (Figure-1) was generated experimentally using heat release during a fire in MJ/m^2 with respect to time of full burnout.

With the help of these two curves A & B, the time of full burnout of the combustible materials and the maximum temperature built up within the confinement during the fire from the expected heat release from the combustible materials can be estimated.

Example 1: [5]
To evaluate the fire rating of a safety tunnel room in the Reactor Auxiliary building (UKC)

Structure /Room : UKC FC4
Structural component : Tunnel in Reactor
: Auxiliary Building
Floor Elevation : -8.00 m
Area : 75.80 m^2
Height : 3.80 m
Type of Fire load : Cable insulations
Fire load in the specified area : 4000 kg
Lowest heat of combustion of
Cable / insulation
Materials : 25 MJ/m^2

$$\text{Specific fire load} = \frac{\text{Fire load in the area} \times \text{heat of combustion load}}{\text{Area of the Room}}$$
$$= 1319\ MJ/m^2$$

As the specific fire load (g) is between 181 to 1400 MJ/m^2, category of fire hazard is C3. Also by referring to the fire rating graph (see Figure-1), the fire rating of the structural members shall be a minimum of 80 minutes.

However in accordance with the RF standard VSN 01-87 [3], fire prevention regulations for design of NPPs, the minimum fire resistance limits for the building structural elements for Category I structure for the internal walls is REI 120. But based on the Technical

Assignment of KKNPP, the walls separating the safety system channels and the other systems shall have a minimum fire resistance rating of REI 180. Therefore, though the time of full burnout is 80 minutes as per the fire load, the structural components have been designed for a fire rating of REI 180 which covers the requirement of SNiP [1] mentioned in Table 3.

For the above example, for a heat release of 1319 MJ/m^2, the time of full burn-out of the fire load is 80 minutes, and the maximum value of average volumetric temperature in the room reaches ~ 950 ^{0}C by extending the abscissa to curve A and reading the temperature.

Similar calculations for estimation of time of burnout for additional rooms are given below.

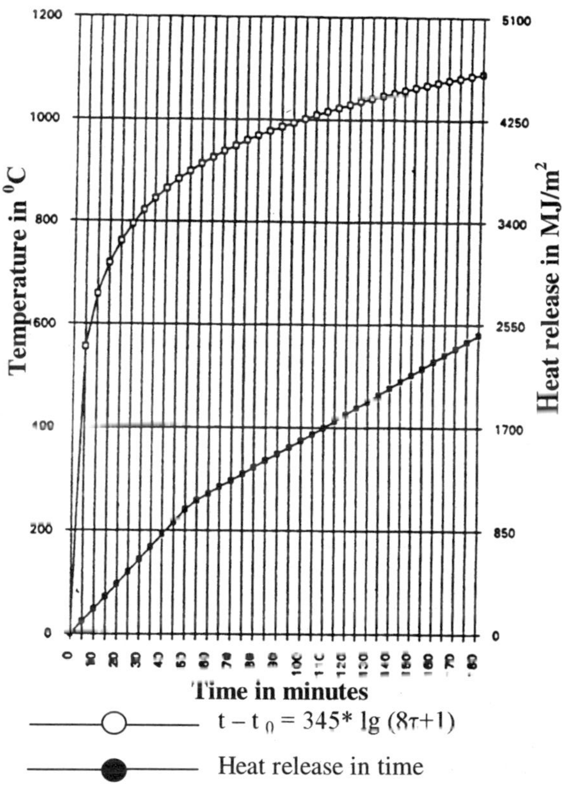

Fig. 1: Curves of heat release with time and maximum temperature during full burn out of fire [5]

Example 2: [5]

Structure/Room : Normal power supply building UBA/ UBAFC1

Structural component : Cable chambers and corridors for communication with reactor compartment

Floor elevation : -6.0 m up to +13.0 m

Area : 4 m^2

Height : 17.2 m

Type of fire load : Cable insulation

Fire load in the specified area : 1500 Kg

Lowest heat of combustion of cable insulation : 10 MJ/Kg

$$\text{Specific fire load (g)} = \frac{\text{Fire load in the area} \times \text{heat of combustion}}{\text{Area of the room}}$$

$$= 3750 \text{ MJ/m}^2$$

As the specific fire load (g) is more than 2,200 MJ/m^2, category of fire hazard is C1. Also by referring to the fire rating graph (see Figure-1), the fire rating of the structural members shall be a minimum of 180 minutes.

Example 3: [5]

Structure/Room : Fire area of annulus UJB/ UJBFC14

Structural component : Annulus between containments

Floor elevation : +5.40 m up to +9.0 m

Area : 40 m^2

Height : 3.6 m

Type of fire load : Cable insulation

Fire load in the Specified area : 200 Kg

Lowest heat of combustion of cable insulation : 25 MJ/Kg

$$\text{Specific fire load (g)} = \frac{\text{Fire load in the area} \times \text{heat of combustion}}{\text{Area of the room}}$$

$$= 125 \text{ MJ/m}^2$$

As the specific fire load (g) is between 1 and 180 MJ/m^2, the category of fire hazard is C4. Also by referring to the fire rating graph (see Figure-1), the fire rating of the structural members shall be a minimum of 10 minutes.

7. CONCLUSION

The principles involved in the categorization of buildings in accordance with their fire hazard degree, the method of estimation of specific fire load for particular areas and the structural requirement of civil engineering structures capable of withstanding the estimated fire loads, in accordance with the Russian fire safety regulations, have been brought out in this paper. The design method followed in the RF is comprehensive and a quick assessment of structural fire resistance can be arrived at from the estimated fire loads using curves indicated in Figure-1. The method compares well with the international practice of assessing the fire severity based on fire test performance of structural materials under defined conditions with respect to quantity and type of combustible load.

8. REFERENCES

1. SNiP 21-01-97 -Fire Safety of Buildings and Works-1997.
2. NPB 105-95 -Determination of Categories of rooms and buildings ranked by Fire explosion and Fire hazard, 1995.
3. VSN-01-87-Fire protection design norms at Nuclear Power Plants, 1987.
4. IAEA 50-SG-D2- Rev1- Fire protection in Nuclear Power Plants, 1992.
5. Detailed Project Report 4/4: Fire protection measures, Explanatory Note - Atomoenergoproekt, NPP Kudankulam, Unit 1 & 2, 2001.

Reliability, Safety and Hazard
Editors: P.V. Varde et al.

SAFETY IN SHUT DOWN

K.R. Mohan Ram, K.M. Somasundaram, V.V. Sanat Kumar and G. Nageswara Rao

Nuclear Power Corporation of India Limited, Kaiga Generating Station, Karnataka

ABSTRACT

Safety in nuclear power plant at all times is ensured right from design stage to plant construction and operation. Adhering to regulatory limits or criteria ensures safety in shut down, such that safety margins are not reduced. In shut down conditions, sufficient redundancy, reliability and capacity in equipment are provided such that safety margins are not lowered. There is provision of two fast acting independent shut down systems of diverse design, each independently capable of safely terminating any reactivity transient at all circumstances. The depth required would be governed by appropriate combination of reactivity effects, which include total voiding in the core; cool down of fuel, coolant and moderator as well as xenon decay. Depth of the shut down system required during prolonged shutdown is maintained by neutron poison in moderator In shut down condition, reactivity control and cooling of fuel requirements are met at all times.

1. INTRODUCTION

Nuclear Power is today a well-established source of electricity, with about 16% of the total electricity generation in the world coming from this source. In India, nuclear power is about 3% of our installed electric capacity. The nuclear power industry has traditionally given the utmost attention to safety, right from the time of the earliest reactors some 50 years ago. In all activities connected with nuclear power plants, including their design as well as in operation, a major part of the total effort is devoted to ensure safety of operating personnel as well as the environment.

2. SAFETY DESIGN PRINCIPLES

The objective of the nuclear power plant safety is to ensure and demonstrate that the risk from plant to public and plant personnel is acceptably low. This safety philosophy is based on deterministic approach that includes defence-in-depth principle, safety design criteria, and the concept of design basis accidents. There are well-established safety design principles, which are applied in nuclear reactor technology to ensure that safety functions will be performed with a very high degree of reliability. The defence-in-depth approach, the heart of safety philosophy, where there are several lines of defence, one backing another. The radioactive fission products, which are produced as a result of the nuclear reaction process, are generated within the solid uranium dioxide fuel. A series of physical barriers are provided between the radioactive fission products and the environment. Most of the fission products stay embedded in the uranium dioxide fuel pellets, where they are produced. These pellets are in turn sealed in metallic clad and contained within a high integrity reactor coolant system. Standards in design, construction and operation ensure that these systems maintain their integrity with high degree of reliability. In addition, there are several features and systems built into the design to prevent deviations from normal mode of operation that could adversely affect the integrity of these barriers [1]. Still unlikely failures are postulated and systems are provided to cope with such postulated failures. One such system is the containment system, which is a massive concrete building housing the reactor, and designed to remain leak tight even under pressure. Further, an exclusion zone around the nuclear power reactors is also established as a measure of abundant caution.

3. SAFETY REVIEW PROCESS

The Probabilistic Safety Assessment (PSA) complements the traditional deterministic approach. PSA is a standard tool in safety evaluation of the nuclear power plants and it provides a useful, consistent and structured framework for the evaluation of safety and helps in safety decisions related to design and operations [2]. Of the three levels of PSA, level-1 addresses the identification of plant failures leading to core damage and their frequencies of occurrences. Level-2 addresses the assessment of containment response. Level-3 addresses the assessment of off-site consequences leading to estimates of public risk[3]. A comprehensive Level-1 PSA for standard Indian PHWRs. indicates a fairly high level of defence-in-depth.

A multi-tier safety review process for nuclear power projects right from the site selection stage through design, construction, commissioning and operation exists. Based on the reviews by the different safety committees, AERB makes stipulations, which is a mandatory binding on the operating organization.

A system to track all significant events at operating plants ensures that significant events are systematically documented, reviewed and analyzed to find the cause of the incident and for taking corrective action by way of design modification, revision of operating procedure or replacement of a component with one having much higher reliability and lower failure probability. This approach of analyzing all unusual occurrences has contributed greatly producing nuclear power safely.

4. SAFETY DURING OPERATION

A combination of sound design and conservative operating practices makes a reactor safe. The policies and practices for operation of nuclear power stations in India recognize this fact. Technical Specifications for Plant Operation –a document, approved by the regulatory authorities defines the safety limits for various systems parameters, requirements for minimum, periodic surveillance and testing requirements and administrative controls are strictly adhered to. Trained and qualified personnel operate and maintain operating stations. Training simulators tailored to provide training on all aspects of operation of nuclear plant including handling of malfunctions and accidents have been set up. In-service inspection program assesses the healthiness of various components and equipments during their service life of the plant to ensure that no incipient problem results in an actual failure of the component.

5. SAFETY DURING SHUT DOWN

Shut down state of the reactor is that state of the reactor when it is maintained sub critical with specified negative sub criticality margin in any of the modes as under:

1. The Primary Shutdown System (PSS) rods, regulating rods, absorber rods and shim rods are fully "IN" the reactor core and 45 mk (38 mk after 140 FPD) (mode-P) equivalent of boron is added through LPIS to the moderator.
2. Secondary Shutdown System (SSS) has actuated 45 mk (38mk after 140 FPD) has been added to the moderator equivalent of boron in mode 'S' and the regulating rods, shim rods and the absorber rods are fully "IN" the reactor core.

The reactor is said to be in cold shutdown condition when in addition to any state above, the temperature of the primary heat transport system is less than 55 degree centigrade and the core is cooled by shutdown cooling system. However before achieving cold shutdown state a total of 45 mk (38 mk after 140 FPD) equivalent of boron shall have been added to the moderator [4].

Safety in shutdown is ensured by safety systems that carry out the functions such as fast reduction in power, compensating all positive reactivity additions, maintaining sub-criticality margins are capable of completely terminating any of the postulated reactivity transients in the most reactive state of the core. These frontline safety systems are supported by safety support systems [1].

A specific shutdown state Guaranteed Shutdown State (GSS) of the reactor with sufficiently large reactivity shutdown margin, established by the addition of liquid poison into moderator to provide positive assurance that an inadvertent increase in reactivity, by withdrawal of all other Reactivity devices cannot lead to criticality [5]. This state is provided to enable maintenance of shutdown system and reactivity devices. During shut down state of the reactor, the neutronic over power trip set points are automatically driven down to lower value (6% FP and 8 % FP) for the two shut down systems. Normally only protective system channels are taken for maintenance ensuring that the protective action is available. During maintenance of safety support systems it is ensured that no maintenance is carried out on shut down systems and vice versa. How ever two events namely Loss of Regulation Accident (LORA) and Loss of Coolant Accident (LOCA) are envisaged in design as initiators for single failures that may impair safety in shut down. These initiating events are briefly described below.

The design of the regulating system is such that the malfunction of any single component will leave the system still safe and operating. Failure or malfunction can be due to either mechanical or regulating system input channels misbehaving. The control from the regulating system and the mechanical components like servomotors etc. are provided individually for each rod and thus their failure can cause only a single rod failure. The processing is done by a dual processor hot standby (DPHS), microprocessor-based system.

The regulating system consists of two parts, system A and system B. Each system has two processors. One processor does the control function and other supervisory function. (The signals from the triplicate sensors are input to each of these processors). The two processors perform the inter-comparison of the corresponding signals at various processing stages. In case of detection of any fault based on this comparison, the control of the reactivity devices is transferred to the healthy system. Adjustment, maintenance or testing of defective system can be done even with the reactor operating.

The system A controls seven control rods, while system B controls the other seven rods. In case of failure of system A, the rods connected to this system are transferred to the system B and vice versa. Therefore, a failure of a single system will not result in a malfunction of the control rods. However, if one further assumes the simultaneous failure of the transfer circuit, then a maximum of 4 absorbers, 1 shim and 2 regulating rods can malfunction from either system. During shutdown, the absorbers are locked 'IN' by mechanical ratchets and can be withdrawn only by deliberate manual action. As an accident, it may be postulated that on seeing the rods traveling out of the reactor core during LORA, the operator unlocks the ratchets on absorbers.

For the incident of LORA at power is low (10^{-5} FP) two cases are considered - one where in the reactor is at low power with a shutdown cooling pump in service. The coolant flow is taken to be 6% of full flow and coolant inlet temperature as 25°C. In the other case, the reactor is at low power and in hot standby condition with PCPs (primary coolant pumps) pumps in service and coolant inlet temperature of 249°C.

In the initial core, for the actual rate of addition of the reactivity during LORA, On PSS actuation with the over power trip set at 6%FP, the maximum power reached is 7.2 %FP in 84.4 sec, and the maximum fuel centre line temperature is around 42°C. The log rate trip is encountered much earlier at 40.6 sec and the maximum power reached is 7.2E-05 FP. For all cases of rate log trips, all temperatures remain at around 25°C itself. On failure of PSS and SSS actuation on over power trip set at 6%FP, the maximum power reached is 9.7 %FP in 87.7 sec and the maximum fuel centerline temperature is 51°C [4].

For the case of LORA in hot-standby condition of the fresh core, a maximum power of 10.3 %FP is reached when SSS actuates on failure of PSS . the fuel central line temperature is well with in designed limits[4].

6. LOSS OF COOLANT ACCIDENT - LOCA

The design of primary heat transport system conforms to the relevant codes for high pressure systems. Integrity of the PHT system is maintained during the life of the plant and ensured by in service inspections. Nevertheless, pipe failures in the PHT system are still postulated and analysed to demonstrate the capability of safety systems to cope with such severe accidents. During Shut down state LOCA due to failure of pressure tube and calandria tube is assumed as a design basis accident.

In the event of failure of calandria tube as a consequence of failure of pressure tube, the PHT coolant will discharge into calandria. Additional effects to be considered in this situation will be the pressurisation and loading of calandria and its components inside calandria as well as possible dilution of boron in moderator.

In this case, as the PHT is in cold condition there will not be any flashing of PHT into the moderator. So there will not be any bubble formation. For the largest break within the capacity of PHT pressurizing pump, so that light water ECCS does not actuate. The discharge through OPRD will match the flow from the PHT, which could be around 900 Kg/min. If this process continues then the boron in the moderator could get diluted drastically when all the PHT inventory including inventory in steam generator goes into the calandria resulting in a gradual gain in reactivity. However, shutdown systems including LPIS would actuate and keep the reactor sub critical. LPIS system injects negative reactivity into the moderator increasing the boron concentration in calandria . Part of this boron could also escape through OPRD due to incoming PHT. To evaluate this aspect, two cases enveloping the scenario as given below have been analysed:

- Poison added through LPIS is s discharged through OPRD after mixing with moderator in calandria and then expelled..
- Part of ALPAS poison is directly expelled through OPRD before mixing with bulk moderator

As the reactor is in cold and xenon free condition, there will not be any addition of positive reactivity to power coefficient, channel temperature coefficient and the reactivity transient in this case is very slow and is adequately compensated by boron addition through LPIS. PSS or SSS also would have actuated, the reactivity worth provided by this would provide additional sub criticality margins [4].

7. CONCLUSIONS

It is of paramount importance that health and safety of plant personnel and members of the public are fully ensured. The current generation of nuclear power stations in India has several inherent safety features. The designs provide redundancy in protective and safety systems and adopt the concept of defence-in-depth. Operation of nuclear power plants is characterized by strict adherence to a set of prescribed limits and guidelines. Periodic regulatory inspections, WANO peer reviews, information exchange programmes, pre job briefings, post job critiques together reinforce safety of nuclear power plants.

8. REFERENCES

1. NPCIL KAIGA-1&2 Safety Report Vol II
2. IAEA - INSAG – 12, Basic Safety Principles for NPPs
3. IAEA – Safety Guide No. NS-G-1.2, Safety Assessment and Verification for NPPs
4. Design Basis Report of Reactivity Control Scheme for NPCIL KGS-1&2
5. Technical Specifications for NPCIL KGS-1&2

Reliability, Safety and Hazard
Editors: P.V. Varde et al.

DESIGN SOLUTIONS ON SAFETY FOR NPP UNITS WITH WWER REACTORS OF NEW GENERATION—SHORT DESCRIPTION OF NPP POWER UNITS WITH NEW GENERATION WWER REACTORS

V.M. Berkovich, I.I. Kopytov and Y.V. Shvyryaev

FSUE «Atomenergoproekt», Moscow

1. INTRODUCTION

The design of new generation NPP power units with rector plant VVER-1000/V-392 (NPP-92) has been developed and licensed in RF Gosatomnadzor. The construction of the VVER-1000/V-393 has already commenced at Novovoronezh (NVN-NPP-2), in Russia and at Kudankulam (KK-NPP-2), in India. The conceptual design of the 1500Mwe (VVER-1500) is in progress. The design concepts of new projects of NPP with VVER is aimed at two main objectives:

- Increase of safety level;
- Increase of economic efficiency of nuclear power production and reduction of costs of construction and operation.

This paper discusses the design improvements, which ensures achieving the above objectives.

2. DESIGN SOLUTIONS FOR INCREASING THE SAFETY LEVEL

Design solutions for safety of NPP-92 are directed at development of NPP with reactors VVER of increased safety level, to provide, that the total risk of NPP operation is as low as it's reasonably achieved (ALARA principle). At that, there should be strictly fulfilled the requirements of effective Russian safety regulations, as well as recommendations of IAEA and International Commission on Radiation protection.

Development of design solutions for safety level increase in NPP-92 design is based on realization of basic principles of defense-in-depth, which include erection of physical barriers on the way of radioactive substances release into the environment and ensuring the high level of safety systems efficiency and reliability, designated for protection of these physical barriers against damage.

Assurance of high reliability level of safety systems (SS) is provided by setting the requirements to safety probabilistic indices, which are the base for quantitative assessment of NPP safety level.

During the development of design solutions for NPP-92, consideration was given to experience of designing and operation of serial NPP unit with VVER-1000/V-320. At present, in Russia, Ukraine and Bulgaria, 17 power units with V-320 reactors are in operation, including power units at NPP "Temelin" in Czech Republic and at NPP Rostov in Russia, commissioned in 2000 2001. Operation of NPP with reactors V-320 proved the possibility of ensuring high safety level, since during the total operation time of these units more than 250 reactor-years, there were not registered any serious accidents.

The safety concept of NPP-92 design includes the following engineering principles /1, 2/:

- Single failure principle for analysis of design basis accident;
- Diversity principle for the systems, which perform safety functions;
- Physical separation (by distance or physical barriers);
- Ensuring protection against common-cause failures;
- Ensuring protection against in-site effects (fires, flooding, steam-water jets, missiles, pipe-lines whipping in the NPP premises);
- Ensuring protection against (characteristical for the NPP site) external natural and man-made effects (earthquakes, tornadoes, hurricanes, high or low temperatures, accidents at air, ground or water transport, etc.);
- Ensuring protection against personnel erroneous actions;
- Expanded application of passive systems for performing of safety functions;
- Ensuring increased level of reliability during performing of safety functions with high frequency of requirements for their implementation;
- Reduction (down to minimal levels) of frequencies or probabilities of equipment and pipe-lines failures;
- Implementation of analyses, which contain assessment of ways of Ultimate Design Basis accidents (UDBA) development, including accidents with severe core damage, and development of technical measures for their mitigation.

Setting the requirements to targeted values of probabilistic safety indices is based on requirements of pars.1.2.17, «General provisions for ensuring nuclear plants safety » (OPB-88/97) /1/, in accordance with which, mean value of Limit Accident Release (LAR) frequency should not exceed 1,0E-7 per reactor per year. Limit Accident Release is defined as the release of such quantity of radioactive products, at which there may be reached or exceeded the dose criteria (set by RF Minzdrav – Health Ministry) at the zone of protective measures planning.

The second requirement are the requirements of par.4.2.2 of OPB -88/97, in accordance with which the frequencies of severe accidents with the reactor core damage (CD) should not exceed 1,0E-5 per reactor per year.

Generation and making solutions for increasing the safety level in NPP-92 design were based on application of results of PSA, implemented for operating power units 1-4 of Balakovo NPP with reactor VVER-1000 (project V-320).

The main contribution into total CD frequency is from IEs with transient processes without primary circuit leaks (~80%) and leak from primary to secondary circuit (~14%).

The main reasons of dominating contribution into CD frequency from these groups of IEs for NPP with V-320 are as follows:

- Relatively high frequencies of realization of such IEs in comparison with primary circuit leaks;
- Relatively low level of reliability of systems for heat removal from the reactor plant via the secondary circuit. This is explained by the fact, that the design of NPP with V-320 includes three-channel safety systems, based on using of similar components (diesel-generators, BRU-A, pumps, valves, check-valves, etc.) in separate safety systems channels. The main contributions into unavailability give common-cause failures of similar components;
- Inadequate level of protection against human erroneous actions.

Protection of safety physical barriers, which for NPP with VVER are the nuclear fuel (fuel matrix and fuel elements cladding), boundary of the reactor cooling circuit and containment, is provided by engineering systems, designated for performing the following safety functions:

- Reactivity control, i.e. putting the reactor into subcritical state maintaining it in this state in all the ranges of operating parameters;

- Heat removal from nuclear fuel in the core and from spent nuclear fuel in the fuel pool;
- Maintaining the coolant reserve in the core in case of accident with reactor coolant loss; limitation of release of radioactive substances into the environment .

Providing the high level of safety functions reliability in the NPP-92design is based on the actual realization of the above safety engineering principles and requirements to the structure and design of safety systems, which are supported by the results of quantitative reliability analyses and probabilistic safety assessments (PSA). During the development of safety systems structures, special attention was paid to the following basic engineering principles:

- Active emergency core cooling system (ECCS) and stages 1 and 2 hydroaccumulator systems for maintaining the coolant reserve in the core in case of primary circuit leaks. At that, stage 2 hydroaccumulators together with stage 1 hydroaccumulators are redundant to ECCS with respect to functions of maintaining the coolant reserve in the core during 24 after the accident start. This time may be used for restoring the operability of active ECCS in case of its failure. ECCS trains are used for the functions of maintaining the coolant reserve in the core at high and low pressure and the functions of heat removal from the containment volume (spray system function);

Table 1.1 Core damage frequency values for V-320 and V-392

	Core damage frequency, per reactor year		
Initiating event	V-320 (24 h)*	V-392 (24 h)*)	V-392 (72 h)*)
Transients without LOCA's	2,30E-05	2,20E-08	2,26E-08
Loss of coolant accident (LOCA)	2,10E-06	5,60E-10	2,80E-08
Leakage from the primary to the secondary circuit	9,20E-06	3,50E-09	3,80E-09
Leakage from the primary circuit outside the containment	1,30E-06	5,50E-09	5,50E-09
All power operation	3,60E-05	3,1E-08	6,00E-08
Shutdown mode	-	3,80E-09	7,70E-09
All	3,60E-05	3,5E-08	6,8E-08

*) In the brackets is given the mission time.

- Ensuring protection against common-cause failures;
- Ensuring protection against personnel erroneous actions;
- Expanded application of passive systems;
- using of functional and design diversity;
- Ensuring protection against internal and external hazards.

Safety concept of NPP with RP V-392 is realized by application of the following design solutions:

Application of mutually-redundant passive and active safety systems for performing safety functions, including the following systems (safety functions and SS structure are presented in table 1.1 and in figure 1):

- Modernized reactor emergency protection system with double-increased (in comparison with V-320) number of control rods and fast boron injection system (FBIS) for putting the reactor into subcritical state and maintaining it in this state in wide range of operating parameters (FPS provides maintaining of subcritical state up to temperature below 100°C);
- Active (EHRS) and passive (PHRS) emergency heat removal system via the secondary circuit. At that, both systems are capable to remove heat during unlimited time, while for NPP with V-320, EHRS may operate during limited time (about 30-40 hours), which is determined by the coolant reserve in the tanks of these systems;

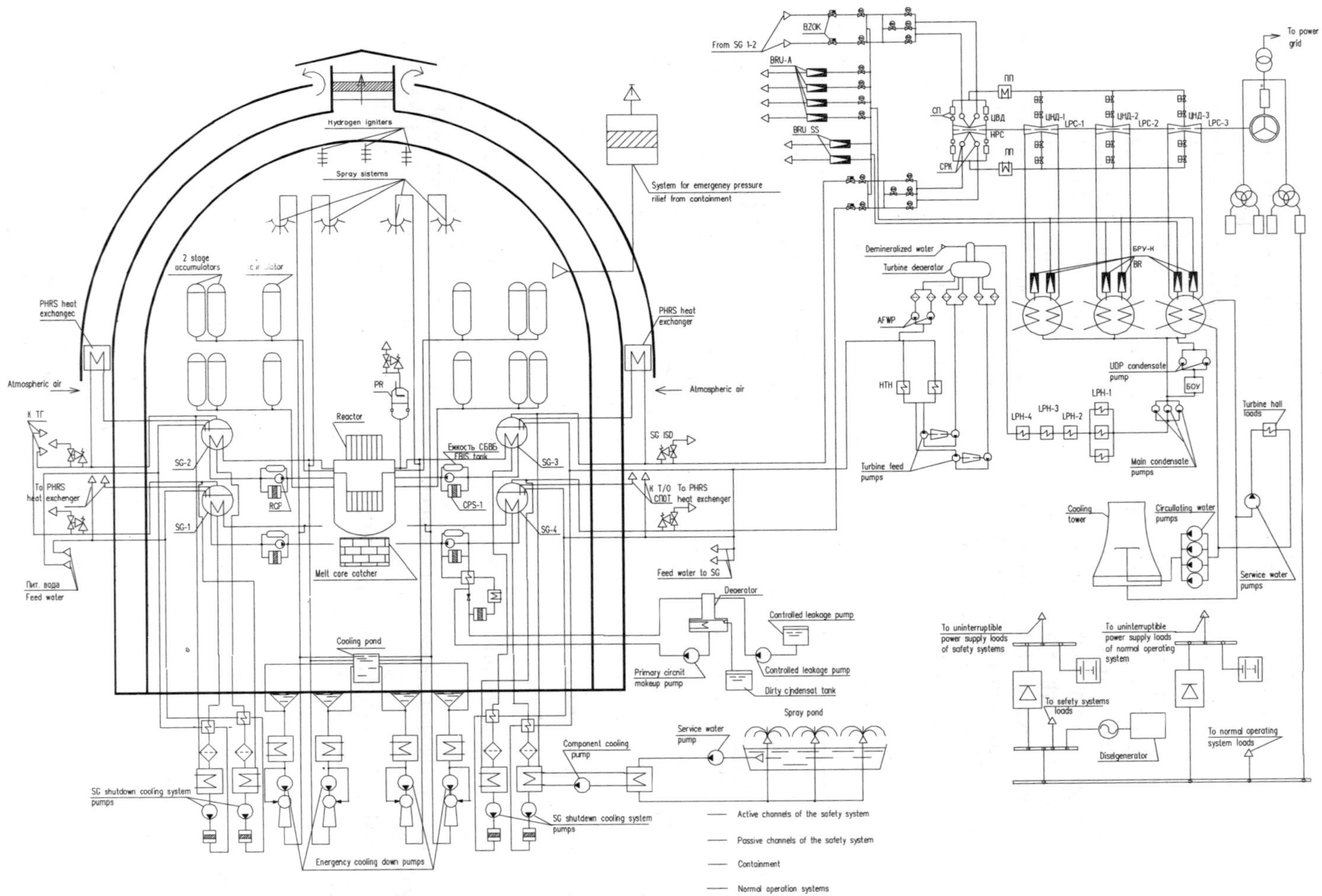

Fig. 1: Schematic diagram of power unit of NPP-92.

- Using the design diversity for the components (fast-acting cut-off valves, isolating gate valves with electric or pneumatic drives) in systems, which perform functions of isolating of steam generators and the containment from the environment.

Application of mutually active and passive systems and active systems with components of different design allows to provide high level of reliability of performing the safety functions by reducing the effects of common-cause failures (application of functional and design diversity) and by reducing the effects of personnel erroneous actions (operation of passive systems do not require any personnel actions).

Besides, the design of active safety systems envisages additional measures for reducing the effects of common-cause failures and of personnel erroneous actions. As an additional measure for protection against common-cause failures, design envisages using of separate channels of emergency core cooling systems and emergency heat removal systems via the secondary circuit for the purpose of normal operation. Two of four channels of these systems are in constant operation, two remaining are in standby mode during unit power operation.

It should be noted that major part of the equipment (pumps, valves, etc.) of operating channels are in the same states, which are required for implementation of assigned safety functions in case of emergency situations. Such solution allows to increase the level of readiness of safety systems due to exclusion of hidden failures of operating components and provide additional protection against common-cause failures due to use of diversity of modes of the components using.

2. Protection against personnel erroneous actions for active safety systems is provided by increased automation of these systems control, when there is necessity of their operation in transients and emergency modes, and by using of passive systems, which do not require operator actions.

3. Application of the concept of «leak before break» reduces the values of frequencies of large leaks from the primary circuit and frequencies of damages of vessel equipment and SGs headers down to negligibly low values.

4. Exclusion of containment sumps pollution by using the block structure of thermal insulation of the primary circuit pipe-lines and using of new design of filtering devices. It should be noted that in the last times newly designed filtering devices are installed at operating NPP units with reactor V-320.

5. Using of double reinforced concrete containment with passive system for hydrogen removal, spray system and molten core retention system (catcher for molten nuclear fuel) in the volume of primary containment and systems for ventilation and clean-up of the medium in the annular gap between primary and secondary containments, provides reduction of releases and reduction of the size of sanitary-protection area for design-basis accidents and prevents the exceeding of limit accident release for ultimate design basis accidents, including severe accidents with complete melting of the nuclear fuel.

It should be noted that by this time design and effectiveness of new components and equipment of safety systems (PHRS heat exchangers and gates, ECCS injectors, passive recombiners in the hydrogen removal system and other components) is fully justified by experimental and analytical studies.

3. ASSESSMENT OF SAFETY LEVEL

Quantitative assessments of NPP safety level are conducted on the basis of comparison between frequencies or probabilities of core damage and limit accident releases, received during PSA implementation, and the targeted values, presented above.

Table 1.1 present the results of evaluation of contribution into total core damage frequencies (CD) value from internal initiating events, received during implementation of PSA for NPP-92 (V-392) and NPP (V-320).

Basing on these results review, the following conclusions may be made:

1. Application of basic design solutions in accordance with safety concept for NPP-92, allowed to reduce the mean CDF value for this NPP by approximately 1000 times for 24 hours mission time and 500 times for 72 hours mission time in comparison with NPP with V-320.

2. Application of mutually-redundant EHRS and PHRS and realization of diversity principle in SG isolation systems, allowed to reduce the contribution from IE with transients by more than 1000 times in comparison with NPP with V-320.

3. Use of HA-2, block structure of thermal insulation of the primary circuit pipe-lines and using of new design of filtering devices of containment sumps allowed to reduce the contribution from the primary circuit leaks inside the containment by approximately 75 times in comparison with NPP with V-320.

4. Contribution into total CD frequencies from personnel erroneous actions reduced by approximately 500 times in comparison with NPP with V-320. At that, the main contribution is from personnel erroneous actions in sutdown modes, which are characterized by relatively low automation level and for which the PHRS is ineffective .

5. Contribution from the primary to secondary circuit leaks was reduced by approximately 2500 times in comparison with NPP with V-320.

4. REFERENCES

1. Vesely, W., Principles of resource-effectiveness and regulatory-effectiveness for risk-informed applications; reducing burdens by improving effectivenessel. *Engng and System Safety* **63** (1999), p. 283.
2. Painton, L, and Campbell, J., Genetic algorithms in optimization of system reliability. *IEEE Transactions on Reliability* **44** 2 (1995), p. 172.
3. Vaurio, J.K., Optimization of test and maintenance intervals based on risk and cost. *Rel. Engng and System Safety* **49** (1995), p. 23.
4. Harunuzzaman, M., and Aldemir, T., Optimization of standby safety system maintenance schedules in nuclear power plants. *Nuclear Technology* **113** (1996), p. 354.
5. Mutloz, A., Martorell, S., and Serradell, V., Numerical absolute and constrained optimization of maintenance based on risk and cost criteria using genetic algorithms, Proceedings of Advances in Safety and Reliability, ESREL, 97, Lisbon, p. 1749.
6. Levitin, G., and Lisnianski, A., Join redundancy and maintenance optimization for multistage series-parallel systems. *Rel. Engng and System Safety* **64** (1999), p. 33.
7. Martorell, S., Sanchez, A., Carlos, S., and Serradell, V., Simultaneous and Multi Criteria Optimization of TS requirements and Maintenance at NPPs, Annals of Nuclear energy 29 (2002), pp. 147-168.
8. Yang, J., Sung, T., and Jin, Y., Optimization of the surveillance test interval of the safety systems at the plant level. *Nuclear Technology* **132** (2000), p. 352.
9. Henley, J., Kumamoto, H. reliability engineering and risk assessment. Englewood Cliffs, NJ: Printice-Hall; 1981.
10. Risk Spectrum Manual
11. Martorell, S., Serradell, V., Verdu, G., Fragio, R., Sanchez, A., Martinez, M.L., Gil, S., and zquierdo, J.A., Maintenance optimization by using PSA results in the RCM context., 1996, IAEA-TECDOC-873, 257.
12. Horn, J., Nafploltis, N., and Goldberg, D.E., (1994). A niched pareto genetic algorithm for multi-objective optimization. Michalewiez, Z, editor, proceedings of the 1* IEEE Conference on Evolutionary Computation, pp82-87, Piscataway, Newjersy.
13. Srinivas, N., and Deb, K., Multi-objective function optimization using non-dominated sorting genetic algorithms. Evolutionary Computation, 2(3), 1995, 221-248.
14. Deb, K., Multiobjective optimization using evolutionary algorithms. Wiley, Chlchester, UK.

15. Deb, K., Agrawal, S., Pratap, A., and Meyarivan, T., A fast and elitist multi-objective genetic algorithm: NSGA-II, IEEE Transactions on Evolutionary Computation, 6(2), 2002, 182-197.

16. Deb, K., and Agarwal, R.B., Simulated binary crossover for continuous search space, Complex Systems 9(1995), 115-148.

Reliability, Safety and Hazard
Editors: P.V. Varde et al.

EVOLUTION OF PSA ACTIVITIES IN DAE, INDIA

R.K. Saraf

Reactor Safety Division, Bhabha Atomic Research Centre, Mumbai 400 085

ABSTRACT

Probabilistic Safety Assessment (PSA) is an analytic method for identifying, analysing and evaluating risk and show evidence that the public health and safety is protected. The world wide use of PSA within operating and regulating organizations is increasing. A substantial amount of effort has been applied in many countries to use PSA in a beneficial manner. In DAE, India regulatory body, viz., Atomic Energy Regulatory Body is now insisting on PSA results for licensing and renewal of licence though not in formal manner. Nuclear Power Corporation is also convinced that the results of PSA study are useful for decision making regarding design evaluation, back fitting, and technical specification modification. This talk covers evaluation of PSA activity in DAE, India right from its inception to the present state of art, giving emphasis on problems faced, data collection & analysis, modelling of human errors, and common cause failures. The talk also includes evaluation of external event PSA, viz., fire PSA, seismic PSA, and application of PSA towards development of Risk Monitor, Risk informed in-service inspection. The talk also outlines the future course of action like Dynamic Fault Tree Analysis, inclusion of ageing effects in PSA, online implementation of Risk Monitor, Software Reliability evaluation, and Passive System reliability. The areas where efforts are needed for effective utilization of PSA are also highlighted.

Keywords: Probabilistic Safety Assessment, Internal/External events, Applications

1. INTRODUCTION

Probabilistic Safety Assessment (PSA) is an analytical method for identifying, analyzing and evaluating risk of a plant (nuclear/ chemical) and can be used to show that the public health and safety are ensured. In addition to this ultimate goal, PSA can also be used for identifying the strengths and weaknesses of the plant so that remedial measures, if needed, can be taken.

In October 1975, the Nuclear Regulatory Commission released the Reactor Safety Study also known as WASH-1400 [1]. This study was the first comprehensive Probabilistic Risk Analysis of a nuclear power plant and represented a step forward in the risk analysis of a complicated engineering system. A number of publications have reviewed this effort and described the methodology developed by the Reactor Safety Study team. Several text books were published with detailed descriptions of the event tree- fault tree methodology in a form suitable for helping those interested in pursuing these techniques. Based on the event tree-fault tree methodology several PSA studies have been carried out in USA and other countries. Results of PSA study are increasingly used for optimizing a variety of operational tasks and procedures in nuclear power plants. A plant specific PSA is very useful tool for safety control optimization since it models both the design and the operation of a plant. The most common applications of the PSA are the optimization of technical specifications and risk based configuration control. Other applications include optimization of operating procedures and maintenance activities and plant modifications/

back fittings. The main objectives for the assessment of risk with respect to nuclear power plants are

(i) Identify initiating events and accident sequences that might contribute significantly to risk
(ii) Provide realistic quantitative measures of the likelihood of these risk contributors
(iii) Provide a realistic evaluation of the potential consequences associated with the identified accident sequences
(iv) Provide a reasonable risk-based framework for making decisions regarding nuclear plant design, operation and siting.
(v) Design of emergency operating procedures to cater to the accident sequences that were identified as significant.

2. PSA ACTIVITIES IN DAE

Earlier nuclear power plants were designed based on deterministic methods and the licensing procedures were also essentially based on the deterministic safety analysis. In this analysis, important safety features are derived from the requirement that the consequences of postulated design basis accidents are within the specified regulatory guidelines.

In DAE, India, the PSA activities have started sometimes in late 1985 and the first level-1 PSA report was released in 1989. Prior to these PSA activities, reliability analysis of reactor systems, viz., safety and safety related systems were carried out for research reactors and the early nuclear power plants. The reliability analysis of reactor systems, though not mandatory, was carried out and was included in preliminary safety analysis reports for licensing proposals. For all the existing power plants which are operating and under construction, one will find the chapter on system reliability analysis in safety report volume II (appendix I). The reliability analysis of a system is based on fault tree technique using generic component data and operating power plant experience. The reliability analysis in terms of unavailability of a system was earlier done by hand calculations for single value (median value) and later on by using latest software packages like PSAPACK, RISK SPECTRUM and ISOGRAPH. The uncertainty and sensitive analyses have become easier by using these softwares. When the reliability analysis of reactor systems for NAPS was completed the BARC team extended these activities for carrying out Level 1 PSA study for internal initiating events. NPCIL provided the required design inputs for carrying out these activities. At that time, only WASH-1400 PSA study and NUREG-2300[2] reports were available. Based on the guidelines of these reports, report on Level 1 PSA study for NAPS[3] was released in 1989. The results of the study were based on hand calculations and Monte Carlo simulation code for uncertainty analysis. Many design modifications were suggested based on the results. The report was reviewed by NPCIL and the comments on the report were encouraging which lead us to continue these activities confidently. Later on BARC-NPCIL team carried out Level 1 PSA study for Kaiga power plant [4] for obtaining clearance from Ministry of Environment, Government of India for licensing purpose. This report was released sometimes in the year 2000. In Kaiga PSA study latest procedures for carrying out human reliability analysis and common cause failure analysis have been considered. Also attempt for carrying out uncertainty analysis using fuzzy set theory was employed for the first time.

During carrying out these studies, BARC team has encouraged and trained other concern groups working in research reactors in BARC and power reactors in NPCIL. This regular training helped research reactor group to come up with first level 1 PSA study report on Dhruva Reactor, IGCAR to come up with reliability analysis reports of various systems of Prototype Fast Breeder Reactor (PFBR) and Fast Breeder Test Reactor (FBTR). This training has helped NPCIL to come up with KAPS & TAPS 1&2, 3&4 PSA study reports. In fact, NPCIL has extended these activities to Level 2 PSA and has come up with Level 2 PSA for KAPS which will be shortly reviewed. Health Physics Division of BARC is well equipped to carry out Level 3 PSA methodology. All these studies have helped the concerned agencies to modify and retrofit the systems. The major outcome of all these studies is the systematic efforts for data collection and analysis for plant components/ equipment/ systems and human errors at almost all power stations. Atomic Energy Regulatory Board

(AERB) also found these reports quite supportive to them for taking decisions in licensing and renewal of license. AERB now insists on Level 1 PSA study though not in formal manner. At present Level 1 PSA of MAPS based on operating experience is nearing completion and PSA activities on AHWR is in progress.

The above PSA studies were confined to internal initiating events and full power operation of the reactors. International studies, especially French PSA studies, have indicated a significant contribution to risk from low power and shutdown states of the reactor. A PSA study for shutdown state of an operating reactor has also been carried out by BARC. The experience of the PSA activities has also helped us to develop methodology for external event PSA. In the case of external event PSA, methodology has been developed for Fire PSA and for seismic PSA is also being developed.

Atomic Energy Regulatory Board also started taking initiative in the field of PSA. As early as in 1989, it constituted a committee for an action plan for PSA activities. The terms of reference of this committee were [5]:

i. to prepare a status report on the activities being carried out by the various units of DAE on PSA
ii. to identify the gaps in the activity
iii. to suggest manpower requirements for such activities at the various DAE units/ AERB and also an organizational structure for the purpose
iv. to identify areas where computer software have to be developed and also the agencies for carrying out this work
v. to develop methodologies to validate PSA models
vi. to carry out uncertainty analysis in the PSA program
vii. the necessity for conducting human reliability exercise, man machine interface and also to identify agency/agencies to develop this area

This action plan committee has recommended the formation of a standing committee for overseeing the development of PSA methodologies and their applications. The committee has also suggested for the constitution of working groups for development of PSA methodologies, data collection and analysis, human reliability analysis and software development. PSA standing committee and the working groups were constituted by AERB. These have played a major role in promoting PSA activities in various units of DAE.

Later on to meet the day to day review requirements of AERB, Chairman, AERB, constituted PSA committee consisting of experts from BARC, NPCIL, IGCAR and AERB. This committee has a key role in reviewing the PSA reports of TAPS 1&2, Kakrapar 1&2, Kaiga 1&2 and the reliability analysis of various systems of PHWRs, Fast Breeder Test Reactor and Prototype Fast Breeder Reactor. Review of PSA report on Kudankulam reactor project is in progress.

AERB has identified the need for standardizing the procedures for the conduct of PSA as a consequence of the large number of PSA reports it is likely to receive shortly. In view of this it has prepared a safety manual for Probabilistic Safety Assessment for Nuclear Power Plants and Research Reactors (Draft). In spite of the operating experience, there is a limitation of data. To overcome the limitation of availability of data, AERB has compiled a generic data base to be used in situations where plant specific data is not available.

3. APPLICATION OF PSA

The confidence in carrying out PSA study for nuclear reactors has helped us to use the methodology and results of the study for different applications.

(i) Risk informed In-Service Inspection procedures have been formulated for Kaiga Reactor Primary Heat Transport System using Kaiga PSA results.
(ii) Risk Monitor software has been developed for KAPS based on the PSA model of KAPS. The software is useful to control room operators for taking decisions regarding maintenance priority, allowable outage time and technical specification optimization based on current state of risk.

This software will be distributed shortly to AERB, NPCIL for field trials.

(iii) Symptom Based Diagnostic System using Artificial Neural Networks to identify selected initiating events (and to suggest the necessary corrective actions) depending up on the combination of the values of numerous operating parameters of the power plant is being developed for Kaiga atomic power station. One such system has been developed for NAPS.

4. CURRENT ISSUES

PSA activities are now well established but the procedures are still evolving. The results of PSA study solely depend on modeling the systems and component/ human failure data. The present models in the following areas have limitations which can introduce significant uncertainties and hence additional efforts are required to reduce these uncertainties and standardize the methodologies.

(i) Human Reliability Analysis: Presently Technique for Human Error Rate Prediction (THERP) is being applied. Accident Sequence Evaluation Programme (ASEP) methodology is used for screening purpose. Lot of uncertainties exist in HRA results due to scanty data on human error (especially for cognitive tasks) and modeling human as component. Currently expert elicitation techniques are being tried. The debate arises when the human error is dominant contributor to Core Damage Frequency. (Application of performance shaping factors on operator crew performance have to be investigated more thoroughly)

(ii) Common Cause Failure: Most of the NPPs are characterized by high degree of redundancy in safety systems. For such systems, common cause failures can pose a major threat to system reliability. Based on experience β factor method is used conservatively. For other methods like multiple Greek letter/ alpha factor models there is no adequate data available. Also there is no recommendation of special CCF models in any PSA guidelines. Hence the state of art is not satisfactory for high redundancy systems with a limited amount of observed failures. Thus using conservative estimates leads to debatable situations when CCF is dominant contributor to CDF.

(iii) Dynamic Fault Tree Analysis: Quantification of the fault tree in the present system reliability analysis uses component failure data in terms of failure/hazard rate, which is assumed to be independent of time. In many of the components especially mechanical components even though they are periodically maintained this assumption may not be true. Mechanical components failure rate does depend upon time of usage and increases with the time as the parts start wearing out. Such phenomena need to be considered in reliability analysis in terms of dynamic fault tree analysis.

(iv) Software Reliability: In the design of complex systems, today an increasing amount of functionality is dedicated to the software. Since software always operates in the context of a larger system, the dependability requirements of the system, filter down to the software components and become the desired software dependability requirements. software reliability analysis models need to be incorporated in PSA models in a consistent manner.

(v) Passive System Reliability: Recently development of advanced reactors with passive safety features has been very prevalent. In this situation, conventional PSA may not be enough. Attention must be paid to checking the applicability of the conventional PSA as a safety measure of advanced reactor design. Passive systems rely upon natural physical laws such as gravity, natural circulation etc. Passive systems may fail its mission not only as a consequence of mechanical failures of components but also for deviation from expected behavior due to physical phenomena mainly related to thermal hydraulics or due to different initial conditions.

5. CONCLUSIONS

1. Various units in DAE are well equipped to carry out full scope PSA for research and power reactors.

2. Level 1 PSA studies have helped designers and regulators to a great extent for design modifications/ backfittings and decision making.
3. Results of PSA are applied to develop systems like risk monitor and risk informed inservice inspection procedures.
4. Further efforts are required to develop/ standardize methodologies in the area of human reliability analysis, common cause failures, dynamic fault tree analysis, software reliability and passive system reliability.

6. REFERENCES

1. N.C.Rasmussen, *Reactor Safety Study*, WASH-1400 (NUREG-75/014) Appendix II, U.S. Nuclear Regulatory Commission, 1975.
2. NUREG-2300 PSA Procedures Guide
3. PSA of NAPP, BARC/1445/1989, Internal Report
4. Kaiga PSA Internal Report
5. AERB Report of the committee to prepare an action plan for PSA Activities, 1989.

Reliability, Safety and Hazard
Editors: P.V. Varde et al.

SAFETY MANAGEMENT IN NPPs : A MULTI-OBJECTIVE OPTIMIZATION

Alok Mishra[1], Anand Patwardhan[2], Ashok Chauhan[1] and A.K. Verma[3]

[1]Nuclear Power Corporation of India Limited, Mumbai 400 094
[2]S.J.M. School of Management, IIT Bombay, Mumbai 400 076
[3]Reliability Engineering Programme, IIT Bombay, Mumbai 400 076

ABSTRACT

Technical specification and maintenance (TS&M) activities in a plant are associated with controlling risk or with satisfying requirements, and are candidates to be evaluated for their resource effectiveness in risk-informed applications. Accordingly, the Risk-Based analysis of Technical Specification (RBTS) is being considered in evaluating current Technical Specification (TS). The multi objective optimization of the TS&M requirements of a Nuclear Power Plant (NPP) based on risk and cost, gives the pareto-optimal solutions, from which the utility can pick its decision variables suiting its interest. In this paper a multi objective Genetic Algorithm (GA) technique has been used to make a trade-off between risks and cost both at the system level and at the plant level for Loss of Coolant Accident (LOCA) as an initiating event.

Keywords: Safety Management, Multi Objective Optimization, and Probabilistic Safety Assessment

1. INTRODUCTION

It is widely recognized that there is a growing interest, from both regulatory authorities and the nuclear industry, to stimulate the use of Probabilistic Safety Analysis (PSA) for risk-informed application aimed to emphasize both effective risk control and effective resource expenditures at Nuclear Power Plants (NPPs). The general goal of these risk informed applications is to make requirements and activities more risk-effective and less costly by using the risk information to focus better on what is risk-important for better safety management of the nuclear power plants [1].

The resources versus risk-control effectiveness principles formally enter in optimization problems where the cost or the burden is to be minimized simultaneously with the risk or performance. This relationship applies when TS&M activities (Surveillance testing Interval (STI), Allowed Outage time (AOT), Preventive Maintenance (PM) frequency) are optimized to minimize both the cost and the unavailability of safety-related systems/ plant risk taking TS&M-related parameters as the decision variables.

In the past, normally, the resolution of such a kind of optimization problem has been limited to focus only on a single parameter (STI, AOT, PM frequency, etc.) and/or adopting a single optimization criterion (availability, costs, plant risks, etc.) [2,3,4,5,6,7]. However, the interaction that exists between TS&M requirements, and the similar purpose of the different TS&M tasks on increasing the system availability, which in some cases may be redundant, have brought a growing interest during the last years to focus on the simultaneous and multi-criteria optimization of TS&M at NPPs. However, the optimization problems were restricted to minimizing the cost or the burden while restricting the risk / performance to be constrained at a given level.

In the simultaneous optimization of TS&M-related parameters based on risk (or unavailabi') and cost, like in many engineering optimization problems (i.e. design, reliability, etc), one normally faces multi-modal and non-linear objective functions and a variety of both linear and non-linear constraints. This results in a complex and discrete search space with regions of feasible and unfeasible solutions for a discontinuous objective function that eventually presents local optima. For such problems, it is desired to find global optima not violating any constraint. Nowadays, new methods based on Simulated Annealing (SAs) and Genetic Algorithms (GAs) are considered in order to try to solve this kind of problems. Both in general and GA in particular possess many characteristics desirable in such a kind of optimizer [2,3,4,5,6,7,8].

In this paper a multi objective optimization case study for a Nuclear Power Plant (NPP) at the plant level is presented. A PIE (Postulated Initiating Event) i.e. LOCA of a Pressurised Water Reactor (PWR) has been considered for the analysis. And an optimization study has been done to decide the surveillance test interval STI), allowed outage time (AOT), the preventive maintenance (PM) period and the replacement/ overhaul time of components, such that the risk (CDF due to the initiating event) and the cost of maintaining the required availability of the systems are optimized. A multi objective Genetic Algorithm technique has been used to arrive at the Pareto optimal solution, for optimizing both risk and cost.

2. PROBLEM FORMULATION

The multi-objective optimization of TS&M will be formulated by adopting the model of Risk and Cost as the objective functions to be minimized. TS&M are represented through appropriate parameters included within the models of risk and cost, which will be adopted as decision variables for the optimization process.

2.1 Risk Function

In a nuclear power plant the occurrence of a postulated Initiating Event (IE) with the unavailability of the mitigating safety systems may lead to Core Damage (CD) sate. The Core Damage Frequency (CDF) can be defined as the risk associated with the operation of a Nuclear Power Plant 'NPP). Thus the total such risk can be defined as

$$\sum Risk = \sum_{m=1}^{n} \left\{ \begin{matrix} [I.E.\ frequency]_m \times \\ \sum_{p=1}^{q} [C.D.EventSequences]_p \end{matrix} \right\} \quad (1)$$

And the risk associated with a particular initiating event 'r'can be defined as

$$Risk_r = I.E.frequency_r \times \sum_{p=1}^{q} \begin{bmatrix} C.D.Event \\ Sequences \end{bmatrix}_p \quad (2)$$

The CD event sequences results due to the occurrence of an I.E. and unavailability of a single or a group of mitigating safety systems. Thus the $Risk_r$ can be defined as a function of the I.E. frequency and the unavailability of the mitigating systems i.e.

$$Risk_r = I.E.\ frequency_r \times [U_a(x) \times U_b(x)...... + U_a(x) \times U_c(x)........] \quad (3)$$

$U(x)$ – unavailability of mitigating system and sub-script a,b,c are for different mitigating systems.

The system unavailability model (which can also be modeled as a risk function at a system level) is often formulated into the PRA adopting the rare event approximation, which is an upper bound, as follows [9]:

$$U(x) \approx \sum_j \prod_k u_{jk}(x) \quad (4)$$

Where the sum in j extents to the number of Minimal Cut sets (MCS) after the reduction of the system structure function and the product in k extents to the number of basic events relevant to the corresponding MCS. In addition, $u_{jk}(x)$ is the unavailability associated with the basic event k belonging to MCS j, which quantifies a given unavailability state of a safety component that depends on the vector of decision variables x. Consequently, quantification of the different contributions to the component unavailability is needed to derive the final risk function using.

Table 1 Components of risk function

No.	**Description**	Formulations Standby System	On-line System
1.	Random failures	$\rho + \frac{1}{2}\lambda_s^* T$	$\lambda_r^* T_m$
2.	Testing	$\frac{t}{T}$	
3.	Corrective Maintenance	$\left(\frac{1}{T}\right)(\rho + \lambda_s^* T)\left(1 - e^{-\left(\frac{D}{\mu}\right)}\right)\mu$	$\lambda_r^* \mu\left(1 - e^{-\left(\frac{D}{\mu}\right)}\right)$
4.	Preventive Maintenance	$\frac{m}{M}$	$\frac{m}{M}$

$\lambda^* = \lambda_0 + \frac{1}{2}\alpha.M.(\psi(z))^2.(2-\varepsilon).(\frac{1}{\varepsilon})$ PAS Model

$\lambda^* = \lambda_0 + \frac{1}{2}\alpha.M.(\psi(z))^2.[1+(1-\varepsilon)(\frac{L}{M}-3)]$. PAR Model

ρ demand failure probability, λ_s standby failure rate, λ_r run-time failure rate, D allowed outage time, T surveillance test interval, t testing time, M preventive maintenance period, m preventive maintenance time, μ mean time to repair, α linear aging factor, $\psi(z)$ working condition ε maintenance effectiveness, L overhaul period, T_m is the mission time.

Eq (2-4), which, in turn, will be used as one of the objective functions. Thus by using only Eq.(4), we can write the risk function at the system level and using Eq.(2 or 3), we can formulate the risk function at the plant level for a particular initiating event.

The unavailability contributions of a component normally in standby are divided into two categories: 1) unavailability due to random failures, named the reliability effect, and 2) unavailability due to testing and maintenance downtimes, named the downtime effect. In modern NPPs some of the active safety system functions are combined with normal operation functions, with a view to increase functional reliability. In such cases the design provides full-fillment of safety functions with minimum number of changes of valves position. These models are presented in detail in Ref [7] and are summarized here in Table-1.

2.2 Cost Function

The relevant costs in analyzing TS&M optimization of safety-related equipment include the contributions of on-line systems as well as the standby components, which undertake surveillance testing with period T,

Preventive maintenance with period M and corrective maintenance to restore their operability after a failure has been discovered during a test. Consequently, summing up the corresponding cost contributions of the relevant components to yield the cost model:

$$C(x) = \sum_i c_i(x) \quad (5)$$

which will be used as the other objective function in the optimization problem. These models are presented in detail in Ref [7] and are summarized here in Table-2.

Table 2 Components of COST function

No.	Description	Formulations Standby System	On-line System
1.	Testing:	$\left(\frac{t}{T}\right)\cdot C_{ht}$	
2.	Preventive Maintenance:	$\left(\frac{m}{M}\right)\cdot C_{hm}$	$\left(\frac{m}{M}\right)\cdot C_{hm}$
3.	Corrective Maintenance:	$\left(\frac{1}{T}\right)\cdot\left(\rho+\lambda_s^* T\right)\cdot\mu\cdot\left(1-e^{-\left(\frac{D}{\mu}\right)}\right)\cdot C_{hc}$	$\lambda_r^*\mu\left(1-e^{-\left(\frac{D}{\mu}\right)}\right)\cdot C_{hc}$
4.	Plant Shut-down (Repair time exceeds AOT):	$\left(\frac{1}{T}\right)\cdot\left(\rho+\lambda_s^* T\right)\cdot\left(e^{-\left(\frac{D}{\mu}\right)}\right)\cdot C_u$	$\lambda_r^*\cdot\left(e^{-\left(\frac{D}{\mu}\right)}\right)\cdot C_u$
5.	Overhaul Maintenance:	$\left(\frac{1}{L}\right)\cdot C_r$	$\left(\frac{1}{L}\right)\cdot C_r$

C_{ht} hourly cost for surveillance testing, C_{hm} hourly cost for preventive maintenance, C_{hc} hourly cost for corrective maintenance, C_u cost for plant shut-down and C_r cost for overhaul, cost of train maintenance, CHM cost of preventive maintenance of a train.

3. EXAMPLE OF APPLICATIONS

In a nuclear power plant the emergency core cooling system provides core cooling under postulated accidents involving leaks of primary coolant. In the case of loss of coolant accident, borated cooling water is injected into the reactor core to remove the decay heat and to preserve core integrity. This borated cooling water is stored in concentrated boron solution emergency supply tank, in accumulators, or is recirculated from the reactor building sump.

In a typical modern Pressurised Water Reactor the emergency core cooling system comprises of four sub-systems known as (1) High Pressure Active Safety Injection (HPSI) System, (2) High Pressure 1st stage Passive Safety Injection System (HA-I), (3) Low Pressure 2nd stage Passive Safety Injection System (HA-II) and (4) Low Pressure Active Safety Injection (LPSI) system. These are all quadruplet train safety systems. Failure of these safety systems during LOCA leads to core damage and hence the availability of these systems is one of the main concerns for the TS&M requirements.

The multi-objective optimization study is carried out to under-stand the trade-off between the cost and unavailability of these safety systems individually and also the cost and risk trade-off, while considering the effect at the plant level. Thus for the system level the objective functions are formulated using the Eq.(4&5) and for the plant level using Eq.(3&5). For our case study, the risk function at the plant level has the following form:

$$Risk_{LOCA}=\left[I.E.frequency_{LOCA}\right]\times\begin{bmatrix}\{U_{HPSI}\times U_{HAI}\}+\\ \{U_{HPSI}\times U_{LPSI}\times U_{HAII}\}+\\ \{U_{HPSI}\times U_{HAII}\}+\{U_{HAI}\}\end{bmatrix} \quad (6)$$

4. SYSTEM MODELING AND DATA

The system modeling consists of the active and passive safety injection systems and the underlying safety support systems. A train is assumed to consist of the front-line safety system and its support system. For the active systems the surveillance test interval, the allowed outage time and the preventive maintenance period are train specific and are chosen same for all the components in a train. It is also considered that during the preventive maintenance the entire train with all its components are under preventive maintenance. The over-haul periods of the components are for the individual components. For a passive system the surveillance test interval and the preventive maintenance are to coincide with the annual scheduled plant outage, and hence these values has to be forced as once per year.

The success/failure criteria of the above systems as defined by the safety analysis report are used and the fault tree analysis for generation of the Minimal Cut Sets (MCS) of these systems have been done using the computer software "Risk Spectrum" [10]. The cut-off value for the Minimal Cut Sets (MCS) for the unavailability of the High Pressure Injection System and the Low Pressure Injection System are taken as $3.85*10^{-11}$ and $3.78*10^{-11}$ respectively. The frequency of a Loss of Coolant Accident is considered to be $5.91*10^{-4}$.

The obtained Minimal Cut-sets have been used to define the two objective function based on risk and cost. The various component data required for the fault tree analysis are considered as per the Probabilistic Safety Analysis for Nuclear Power Plant guidelines. For considering the effect of ageing for all the components Proportional Age Reduction (PAR) model is used, though option exists for Proportional Age Set-back (PAS) model.

The component unavailability and the cost parameters as required in the formulations for the risk and cost functions modeling are given in Reference [7,11].

4.1 Decision Variables

Any solution to the optimization problem can be encoded using the following array of decision variables: $X = \{T, D, M, L_{1\ n}\}$ In this analysis it is considered that the T i.e. Surveillance test interval, D i.e. Allowed Outage Time and M i.e. Preventive Maintenance Period. L_{1-n} are the over-haul time of a components in the mitigating front-line systems and support systems. The above decision variables can be bounded by lower and upper bound to get a solution in the desired feasible domain. The bounding conditions lower bound and upper bound for the decision variables are considered as follows *T(168,4032), D(72,720), M(4000, 16000)* and $L_{1-n}(8000, 80000)$, all the above mentioned values are in hours. The current TS&M requirement for the decision variables are as follows : $X = \{$ **T=672, D=168, M=8000, L1-N =32000}.**

5. GENETIC ALGORITHM FOR MULTI OBJECTIVE OPTIMIZATION

Most classical multi-objective methods convert multiple objectives into a single objective either by aggregating them or by converting all but one objective into suitable constraints. Although some of these classical methods can be applied repeatedly for finding multiple pareto-optimal solutions, each method usually requires the setting of a number of parameters (aggregating weight vectors or constraint limits). Recently, increasing interest has been shown in the use of evolutionary optimization algorithm for finding multiple pareto-optimal solutions.

Evolutionary Algorithms (EAs) mimic natural evolutionary principles to constitute search and optimization procedures in a variety of ways. Genetic Algorithms (GAs) are search and optimization algorithms inspired by the principles of natural evolutionary principles. Conceived by John Holland in the early sixties, GAs have enjoyed a number of applications and developments. One of the niches of GA's application domains is in the field of multi-objective optimization, where instead of one objective the problem is faced with number of conflicting objectives.

Over the past decade, a number of multi-objective evolutionary algorithms (MOEAs) have been suggested [12,13]. The primary reason for this is their ability to find multiple pareto-optimal solutions in a single simulation. In this study one such multi-objective EA – a real coded "Non-dominated Sorting Genetic Algorithm with Elitism (NSGA-II) [13,14,15] is used and is discussed in the following sections.

5.1 NSGA-II Solution Algorithm

The Non-dominated Sorting Genetic Algorithm or NSGA-II procedure [15] for finding multiple pareto-optimal solutions in a multi-objective optimization problem has the following three features:

- It emphasis the non-dominated solutions

- It uses an explicit diversity preserving mechanism
- It uses an elitist principle

In NSGA-II, the initial population (P_t) of size N is randomly created within the feasible boundary for each decision variables. Then the offspring population Q_t is created by using the parent population P_t. The parent and the child population are then combined to form R_t of size $2N$. Then, a non-dominated sorting is used to classify the entire population R_t. Although this requires more effort compared to performing a non-dominated sorting on Q_t alone, it allows a global non-domination check among both the parent and child solutions. Meaning the elitism in the solution set is preserved. Once the non-dominated sorting is over, the new population is filled by solutions of different non-dominated fronts, one at a time. The filling starts with the best non-dominated front and continues with the solutions of the second non-dominated front, followed by the third non-dominated front, and so on. Since the over-all size of R_t is $2N$, not all fronts may be accommodated in N slots available in new population. All fronts which could not be accommodated are simply deleted. When the last front is being considered there, may exist more solutions in the last front than the remaining slots in the new population. Then instead of arbitrarily discarding some members from the last front, it would be wise to use niching strategy to choose the members of the last front, which reside in the least crowded region in that front.

5.1.1 Crowded Tournament Selection Operator

A solution I wins a tournament with solution j if any of the following conditions are true:

- If solution I has a better rank, that is, $r_i < r_j$.
- If they have the same rank but solution I has a better crowding distance than solution j, that is , $r_i = r_j$ and $d_i > d_j$.

The first condition makes sure that the chosen solution lies on a better non-dominated front. And the second condition resolves the tie of both solutions being on the same non-dominated front by deciding on their crowded distance. The one residing in a less crowded area (with a large crowding distance) wins.

5.1.2 Crowding Distance

To get an estimate of the density of solutions surrounding a particular solution I in the population, we take the average distance of two solutions on either side of solution I along each of the objectives. This quantity d_i (the crowding distance) serves as an estimate of the perimeter of the cuboid formed by using the nearest neighbors as the vertices. The crowding distance d_i can be calculated by several ways, in this study a crowding distance metric [14] , which requires $O(MN \log N)$ computations is used. The other methods have $O(MN^2)$ complexity.

5.1.3 Cross-over Operator

Since the design variables take any real number, a real-coded genetic algorithm is used in this study. In a real coded GA, variables are not coded in binary strings instead GA operators are directly applied on the real numbers. Although any selection operators can be used, specialized cross-over and mutation operators are used to effectively create off-spring solutions from parent solutions. In this study, we use Simulated Binary Crossover (SBX) and a parameter-based mutation operator [16] . The cross-over rate is fixed at 0.9 and the mutation rate is fixed at 0.01. All the simulations are run for 500 generations.

6. RESULTS AND DISCUSSION

The multi objective optimization at the system level has been studied for the unavailability vs the cost/hr of maintaining the system . The decision variables are the surveillance test interval, the allowed outage time, the preventive maintenance period and the overhaul time of various components appearing in the minimal cut sets (mcs). The pareto-optimal solution for the active HPSI system and LPSI systems are given in figure-1 and 2 respectively. The implications of the cost and unavailability for meeting the current Technical Specification requirement are also plotted in the same figures. From the figures it is seen that the current Technical Specification requirements for the maintenance of the systems are not cost and risk consistent. It is observed that the present requirement is very much away from the pareto optimal front and lot of scope exists for making the resource allocation more risk consistent. The decisions variables can be chosen by the utility from the set of pareto-optimal solutions for the given minimum unavailability requirement. The optimal solution for the passive HA-I and HA-II systems are single points as the decision variables are imposed because of their configuration in the NPP.

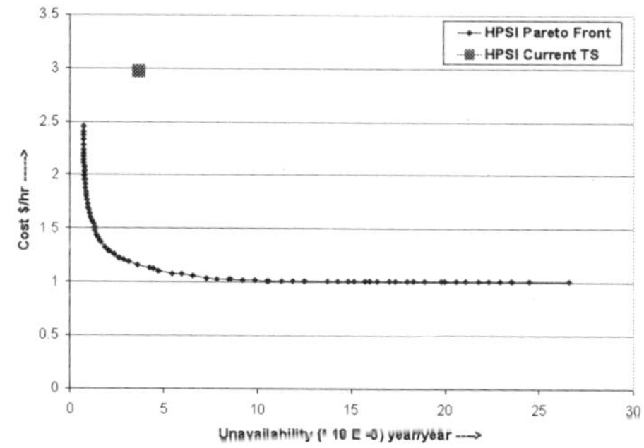

Fig. 1: Pareto optimal solution for HPSI and current TS results

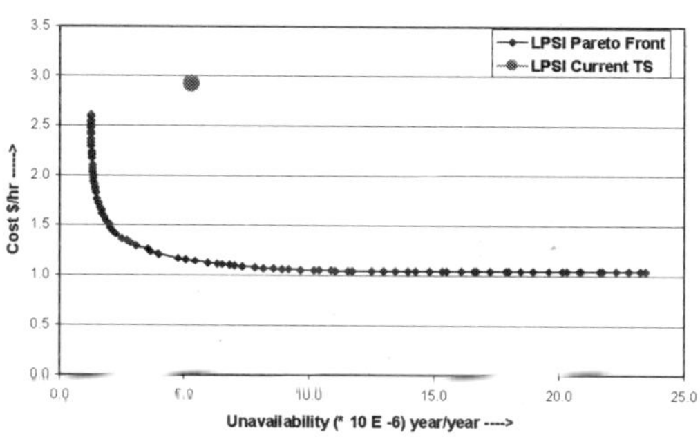

Fig. 2: Pareto optimal solution for LPSI and current TS results

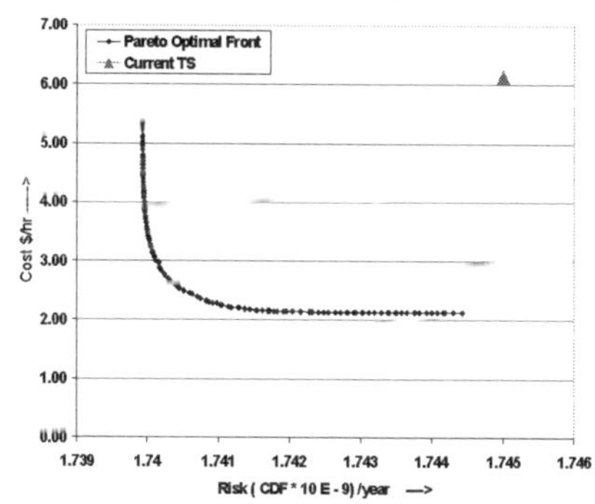

Fig. 3: Pareto optimal solution for Risk vs Cost and current TS results for LOCA

The multi objective optimization at the plant level for the risk vs cost/hr of maintaining the risk level due to a particular initiating event LOCA is given in figure-3. The implications of the cost and risk for meeting the current technical specification requirement are also plotted in the same figure. It is seen that the current TS requirement are not risk consistent and is far off from the pareto optimal solutions. Thus the plant level risk vs cost optimization suggests redistribution of allocation of resources (Decision Variables) in terms of STI, AOT, PM frequency for different systems and overhaul time of equipment to be risk consistent.

7. CONCLUSION

The TS&M requirements for increasing the system availability and for minimizing the risk due to the probable postulated initiating events have brought a growing interest to focus on multi objective optimization in the context of the use of PRA for risk-informed applications and better safety management of NPPs.. The multi objective optimization of the TS&M requirements of a NPP based on risk and cost, gives the pareto-optimal solutions, from which the utility can pick its decision variables suiting its interest. All of the optimized TS&M parameters in the example of application represent practical alternatives that can be implemented in a NPP under a real situation. The decision variables of STI, AOT, PM and the over-haul period for different systems and equipment chosen from the pareto-optimal front will be risk consistent. The Genetic Algorithm has proved to be very efficient in arriving at the optimal solutions in a reasonably small number of generations.

8. REFERENCES

1. Vesely, W., Principles of resource-effectiveness and regulatory-effectiveness for risk-informed applications: reducing burdens by improving effectiveness. *Rel. Engng and System Safety* **63** (1999), p. 283.
2. Painton, L., and Campbell. J., Genetic algorithms in optimization of system reliability. *IEEE Transactions on Reliability* **44** 2 (1995), p. 172.
3. V aurio, J.K., Optimization of test and maintenance intervals based on risk and

cost. *Rel. Engng and System Safety* **49** (1995), p. 23.
4. Harunuzzaman, M., and Aldemir, T., Optimization of standby safety system maintenance schedules in nuclear power plants. *Nuclear Technology* **113** (1996), p. 354
5. Muñoz, A., Martorell, S., and Serradell, V., Numerical absolute and constrained optimization of maintenance based on risk and cost criteria using genetic algorithms. Proceedings of Advances in Safety and Reliability, ESREL'97. Lisbon, p. 1749.
6. Levitin, G., and Lisnianski, A., Join redundancy and maintenance optimization for multistage series-parallel systems. *Rel. Engng and System Safety* **64** (1999), p. 33.
7. Martorell, S., Sanchez, A., Carlos, S., and Serradell, V., Simultaneous and Multi Criteria Optimization of TS requirements and Maintenance at NPPs, Annals of Nuclear energy 29 (2002), p. 147-168.
8. Yang, J., Sung, T..and Jin, Y., Optimization of the surveillance test interval of the safety systems at the plant level. *Nuclear Technology* **132** (2000), p. 352.
9. Henley J, Kumamoto H. reliability engineering and risk assessment. Englewood Cliffs, NJ:Printice-Hall;1981.
10. Risk Spectrum Manual
11. Martorell, S., Serradell, V., Verdu, G., Fragio, R., Sanchez, A., Martinez, M.L., Gil, S., and Izquierdo, J.A., Maintenance optimization by using PSA results in the RCM context. 1996,IAEA-TECDOC-873, 257.
12. Horn, J.,Nafploltis, N., and Goldberg, D.E., (1994). A niched pareto genetic algorithm for multi-objective optimization. Michalewicz, Z, editor, proceedings of the 1st IEEE Conference on Evolutionary Computation, P 82-87, Piscataway, Newjersy.
13. Srinivas, N., and Deb, K., Multi-objective function optimization using non-dominated sorting genetic algorithms. Evolutionary Computation, 2(3), 1995,221-248.
14. Deb, K., Multiobjective optimization using evolutionary algorithms. Wiley, Chlchester, UK.
15. Deb, K., Agarwal, S., Pratap, A., and Meyarivan, T., A fast and elitist multi-objective genetic algorithm: NSGA-II, IEEE Transactions on Evolutionary Computation, 6(2), 2002, 182-197.
16. Deb, K., and Agarwal, R.B., Simulated binary crossover for continuous search space, Complex Systems 9(1995), 115-148.

Reliability, Safety and Hazard
Editors: P.V. Varde et al.

AN APPROACH TO ACCIDENT MANAGEMENT FOR INDIAN PRESSURIZED HEAVY WATER REACTORS

S. Hajela, R. Grover, S.G. Ghadge and S.S. Bajaj

Directorate of Safety, Nuclear Power Corporation of India Limited
Nabhikiya Urja Bhawan, Anushakti Nagar, Mumbai 400 094

ABSTRACT

As part of accident management for Indian Pressurized Heavy Water Reactors (IPHWRs), the entire regime from transients to beyond design basis accidents is proposed to be covered in two distinct phases. In the first phase the existing event based EOPs are proposed to be supplemented by symptom-oriented event handling by keeping identified safety functions satisfied so that the barriers to radioactivity release remain intact. In the second phase, it is proposed to prepare severe accident management guidelines for selected set of severe accident event sequences. This paper describes the concept, under active consideration for accident management for IPHWRs.

1. INTRODUCTION

The objective of nuclear power plants is to generate electrical power in a safe and environmentally benign way. To achieve this objective, various stages viz. design, construction, commissioning, operation and subsequent decommissioning of nuclear power plant (NPP) are carried out by the utility's own commitment to safety and to conform to the national regulatory requirements.

Public perception of nuclear safety hinges primarily on release of radioactivity from nuclear power plant and to address this issue, a well established principle, defence-in-depth, to ensure nuclear safety, is achieved by means of four successive barriers which prevent release of radioactivity (i.e. fuel matrix, fuel cladding, primary coolant boundary and containment). The defence-in-depth is structured in five distinct levels and the means to achieve the objectives at each level are clearly defined (Table-1). In this structuring it is ensured that should one level fail, the subsequent level comes into play [1].

In accordance with defence in depth philosophy, each of the above levels should be protected individually, independent of other levels. This implies that accident management provisions should be in place even if all provisions within the design basis are adequate.

As of now, Indian nuclear power programme is based on Pressurized Heavy Water Reactors (PHWRs) and these nuclear power stations have procedures to deal with normal operation and anticipated operational occurrences. Event based Emergency Operating procedures (EOPs) are also in place for these nuclear power stations.

In addition, emergency planning of Indian NPPs include On-site and Off-site emergency plans. Recalling the levels of defence in depth in order to secure barriers of radioactivity release; it is seen that procedures exist for the first three levels and the fifth level.

Table 1 Levels of Defence in Depth

Level	*Objective*	*Essential Means*
1	Prevention of abnormal operation and failures	Conservative design and high quality in construction and operation
2	Control of abnormal operation and detection of failures	Control, limiting and protection systems and other surveillance features
3	Control of accidents within the design basis	Engineered safety features and accident procedures
4	Control of severe plant conditions including prevention of accident progression and mitigation of the consequences of severe accidents	Complementary measures and accident management
5	Mitigation of radiological consequences of significant releases of radioactive materials	Off-site emergency response

The International scenario in the area of accident management was studied and IAEA, WANO, US-NRC documents were referred. Based on this study, following two-pronged accident management approach for Indian PHWRs (to address the fourth level of defence in depth) is proposed.

- Management of accident within design basis, which is oriented towards safety parameters monitoring and symptom oriented EOPs.
- Severe accident management with emphasis on mitigation strategies.

This paper covers concepts with respect to above two areas of work, which are under active discussion to finalize the strategy for accident management for IPHWRs, both within design basis and beyond, in line with the best practices developed for NPPs elsewhere.

2. SYMPTOM ORIENTED EVENT HANDLING

As brought out above, IPHWRs have event based EOPs for handling deteriorated plant conditions within design basis. Event based EOPs are developed on the premise that before or during the recovery process operator is able to identify the specific event causing the transient or accident. These types of procedures facilitate the optimum response to mitigate the consequences of the transient or accident. In this approach, the number of EOPs tends to become more as new events are postulated based on experience and safety analysis guidelines. For an IPHWR, the typical number of EOPs for the main plant is 60. EOPs dealing with fuel handling system are not included in this count.

Another approach is to use either symptom, function or state based EOPs , all of which basically use similar principle but differ in the way the scheme is implemented. In these approaches, following a plant upset, plant safety is ensured by keeping a certain number of safety related plant parameters satisfied for 'symptom oriented', safety functions for 'function based', and plant states for 'state based' EOPs.

Subsequent to Three Mile Island (TMI) accident, worldwide trend is to opt for symptom oriented procedures where the emphasis is to ensure the safety of the plant and diagnosing the event is given a lesser priority following plant upset condition.

The approach proposed for IPHWRs is based on fundamental principles of nuclear safety wherein it is considered that radiation release from an NPP is prevented through a series of barriers i.e. fuel matrix, fuel cladding, primary coolant boundary and containment. This approach is in line with similar practices followed in other countries. Figure-1 illustrates this concept. It is seen that integrity of these barriers is ensured if corresponding 'Safety Functions' are satisfied during and after the event.

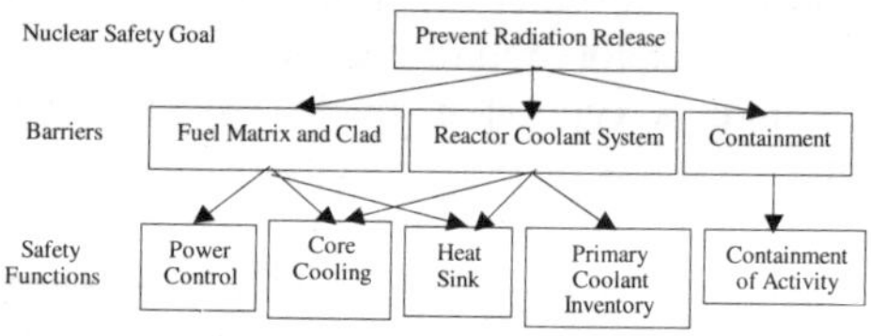

Fig. 1: Safety Function Relationship

These safety functions are identified by certain plant parameters termed as 'Core Safety Parameters'. These parameters are listed in Table-2.

The proposed approach includes event based EOPs also for those events which can be unambiguously identified thus supplementing Symptom oriented EOPs. During event handling, the Core Safety Parameters will continuously be monitored to ensure that the event is being handled in such a way that the plant safety is ensured or to take care of situations where the actual event progression differs from the event considered in the EOP in force or if other (multiple) failures occur subsequent to initiating event. This concept is explained though a flow chart in Figure-2.

Table 2 Core Safety Parameters

Sr. No.	Safety Function	Core Safety Parameter
1	Power Control	Actual Power
2	Core Cooling	ROH* sub-cooling
3	Heat Sink	RIH# sub-cooling
4	Primary Coolant Inventory	PHT Storage Tank Level
5	Containment of Activity	Containment Pressure RB Exhaust Activity

*: Reactor Outlet Header
#: Reactor Inlet Header

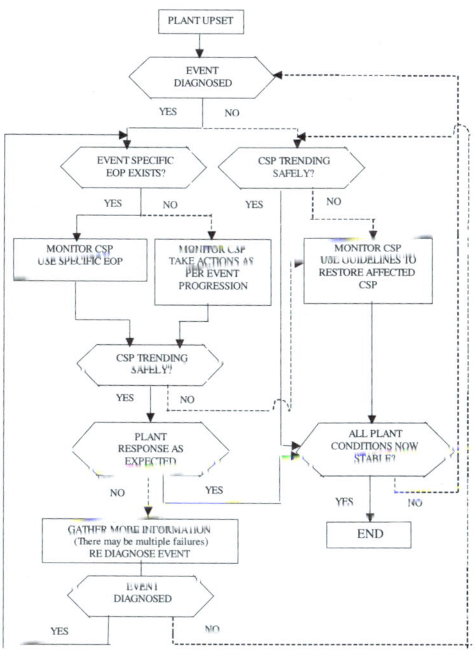

Fig. 2: Accident Management Concept

Continuous monitoring of safety functions is proposed through safety function status tree, one status tree is prepared for each safety function. The entry for a safety function status tree is through its corresponding Core Safety Parameter. At each branch of status tree, values of other identified parameters determine the further path on the status tree. In this way through multiple branches, different end states of the status tree can be reached and at each end state, status of safety function is known. The end state could be 'Satisfied', in which case no operator action is required and on the other extreme, the end state could be under 'Severe Challenge', where an immediate action is required to bring back the safety function to an acceptable state. The other end states for a status tree are classified as 'Off-Normal' or 'Challenging State', calling for graded operator response. Thus except for 'Satisfied' end state, all the other end states of the status tree will have certain guidelines for operator to bring back the affected safety function to 'Satisfied' state, irrespective of the initiating event; thereby ensuring safety of the NPP. The safety function status trees for all the five safety functions are prepared and are being deliberated further for their finalization. The concept of safety function status tree is shown in Figure-3. In order to draw operator attention and comprehend the severity of the situation, the status tree end states of 'Severe Challenge', 'Challenging State', 'Off-Normal' and 'Satisfied' are colour coded as 'Red', 'Orange', 'Yellow' and 'Green' respectively.

As seen, safety function status through status trees is determined by corresponding core safety parameters and some other related parameters. These additional parameters are termed as 'Main Safety Parameters'. The guidelines to operator at each end state for each safety function status tree, will call for operation of certain systems and therefore healthiness of these systems is to be ensured. For this purpose, some more parameters required for system operation as intended are also proposed to be monitored continuously and are termed as 'System Healthiness Indicators'.

Thus, in all, the proposed approach identifies three sets of plant parameters or symptoms. They are 'Core Safety Parameters' (Table-2), 'Main Safety Parameters' (viz. Primary Heat Transport System Pressure, Steam Generator Pressure, Steam Generator Blow Down Activity etc.) and 'System Healthiness Indicators' (viz. Secondary Shutdown System Poison Tanks Level and Pressure, Steam Generator Level, Calandria Level etc.).

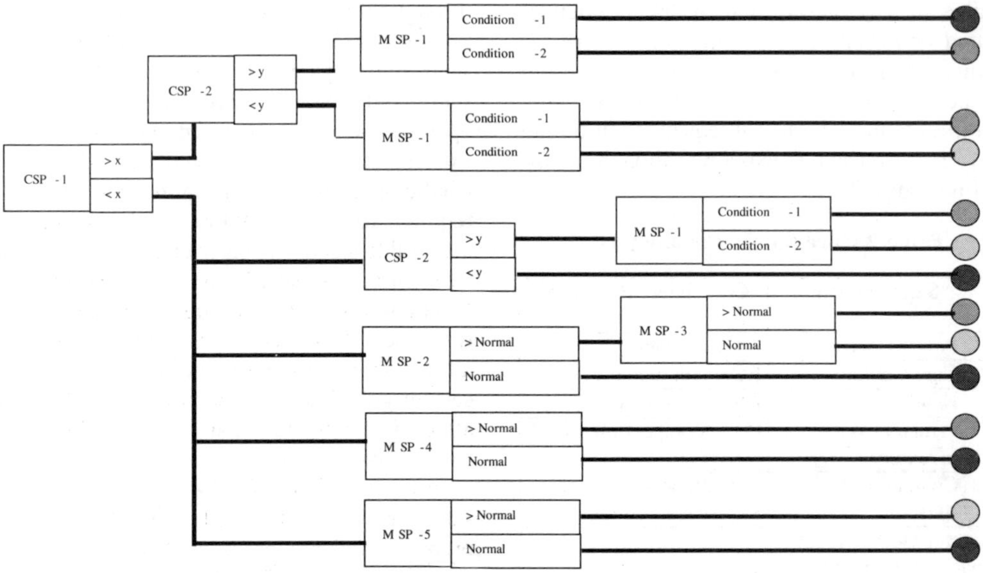

<u>Legend</u>
CSP: Core Safety Parameter
MSP: Main Safety Parameter

Fig. 3: Conceptual Safety Function Status Tree

These parameters are identified as part of development of this concept of accident management. There are a total of 6 core safety parameters, 10 main safety parameters and 24 system healthiness indicators.

It is conceptualised that safety function monitoring will be computerized , in which the various identified plant parameters will be input to a safety function display and monitoring system [2]and whenever any of the Core Safety Parameters will deviate from the normal expected value within a predetermined range , the corresponding status tree will appear on the screen. For each of the end state reached based on Main Safety Parameters; appropriate guideline to the operator to restore back the safety function to the 'Satisfied' state will also be displayed on the screen to aid operator online and in real time.

Presently validation of this concept is being attempted through safety analysis results for various PIEs and also by simulating various plant upset conditions in the simulator of Kaiga Generating Station.

3. SEVERE ACCIDENT MANAGEMENT

In the regime of accident management, the term 'Severe Accident', in practice has come to be synonymous with core melt accident with severity of accident depending on:-

- Degree of fuel degradation
- Potential loss of containment integrity and
- Resultant radioactivity release to environment

PHWRs have multiple, separate volume of water surrounding fuel (i.e. moderator and calandria vault water). These large inventories of water can remove decay heat for varying lengths of time, upto hours or more. Therefore, an accident scenario like Loss Of Coolant Accident (LOCA) followed by failure of Emergency Core Cooling (ECC) which falls in the category of severe accident for Light Water Reactors (LWRs); is a Design Basis Severe Accident (DBSA) for PHWRs. In order to reach a beyond Design Basis Severe Accident scenario for PHWRs, failures of moderator cooling and calandria vault water-cooling will have to be postulated in addition to LOCA and failure of ECCS. Such accidents in PHWRs, where core damage occurs are termed as Severe Core Damage Accidents or Beyond Design Basis Severe Accidents.

Severe accidents are addressed in IAEA Safety Standards Series [3]. Though this report discusses about the design activities for addressing severe accidents; the following can be taken into account while preparing accident management procedures for operating reactors.

- Important event sequences that may lead to severe accidents shall be identified using a combination of probabilistic method, deterministic methods and sound engineering judgment.
- Potential design or procedural changes that could either reduce the likelihood of the selected events or mitigate their consequences, should the selected event occur, shall be evaluated and shall be implemented if reasonably practicable.
- Consideration shall be given to the plants full design capabilities, including possible use of some system (safety and non-safety system) beyond their originally intended function and anticipated operating conditions, and the use of additional temporary systems to return the plant to a controlled state and/or to mitigate the consequences of a severe accident provided that it can be shown that the system are able to function in the environmental conditions to be expected.

  For multiple unit plants, consideration shall be given to the use of available means and/or support from other units, provided that the safety operation of the other units is not compromised.

  Accident management procedures shall be established, taking into account representative and dominant severe accident scenarios.

For arriving at Severe Accident Management Guidelines (SAMGs) for IPHWRs, the available analytical information, both probabilistic and deterministic will be used. First based on PSA Level-1 and Level-2 studies for IPHWRs , dominant severe core damage sequences will be identified which require mitigation.

After finalizing the severe accident event sequences needing Severe Accident Management Guidelines; plant capabilities will be assessed to mitigate the consequences. This evaluation will include monitoring needs, identification of instruments required for generating information and need to provide additional qualified instruments, identification of systems required to operate beyond their design range and assurance for their operation in harsher environment. As part of this, strategies w.r.t. augmentation of heat sinks, drawing power from other unit, temporary systems supplying water, temporary power arrangements etc. will be deliberated.

Subsequently, a decision whether Severe Accident Management Guidelines need to be a separate set of document or the selected existing EOPs can be extended to include steps for severe accident mitigation will be taken.

4. CONCLUSION

In line with international practice, the work on accident management upto design basis through symptom oriented EOPs and beyond design basis through Severe Accident Management Guidelines is underway in NPCIL for IPHWRs and the final concept of accident management will be ready by end 2005. This work is being done by a task group comprising of personnel having expertise in relevant disciplines. The entire accident management is being covered in two distinct phases.

Firstly, the present practice of handling plant upsets is proposed to be supplemented by continuous monitoring of Safety Parameters to ensure plant safety and to take care of multiple failures. Computer based event identification will also be a part of this job. After event identification through the suggested methodology, optimum recovery through existing event based emergency operating procedures will be continued.

In the second step Severe Accident Management Guidelines are proposed to be developed for selected severe accident event sequences. This selection will be based on postulated events frequency and consequences.

5. REFERENCES

1. International Atomic Energy Agency, Vienna INSAG-10 (1996) *Defence In Depth In Nuclear Safety.*

2. International Atomic Energy Agency, Vienna Technical Reports Series No. 387 (1999) *Modern Instrumentation and Control For Nuclear Power Plants-A Guidebook.*

3. International Atomic Energy Agency, Vienna Safety Standards Series NS-R-1 (2000) *Safety of Nuclear Power Plants: Design.*

Reliability, Safety and Hazard
Editors: P.V. Varde et al.

THE PAST, PRESENT AND FUTURE DEVELOPMENT OF PROBABILISTIC SAFETY ANALYSIS SOFTWARE

Stephen Flanagan

Isograph Ltd, The Malt Building, Wilderspool Park, Warrington, Cheshire, WA4 6HL, UK

ABSTRACT

This paper traces the development of probabilistic safety analysis software from its beginnings in the 1960's to the present day and also looks into trends for the future.

Fault tree analysis is closely linked to probabilistic safety analysis and the first fault tree analysis software was conceived in the 1960s for a specific defence project in the USA. In the following few years fault tree analysis was further refined by the Boeing Aircraft Company.

The modern revolution in fault tree analysis software occurred in the mid 1980s onwards with the advent of desktop workstations and personal computers (PCs).

1. INTRODUCTION

The aim of this paper is to follow the mainstream history of fault and event tree analyses software used in Probabilistic Safety Analysis (PSA) over the past forty years or so and to look at the current challenges and opportunities for the future. In addition to the discussion of software developed solely for fault and event tree analysis the paper examines some related activities such as risk monitors and integration with other software tools involved in the system engineering process.

The development of fault tree analysis software has gone through several phases since its inception in the early 1960s. When fault tree analysis was first used the software was far different to that we are familiar with today. Back in those days software programs were run on large central computers using punched tape or cards as the input medium. Output appeared in printed form on continuous stationary of on punched media. The use of these early programs was limited to a few computer experts and sometimes to just the programmer. They concept of 'user friendly' software was indeed a long way away! The early computers were quite limited in both their processing speed and memory capacity which limited the effective size of problems that could be solved in a reasonable time. Often the computers were being used by many other analysts for a variety of purposes and the time slots on the machines were often expensive and rare.

As computer hardware and associated devices were invented and improved upon the power and user-friendliness of software products began to improve. Plotting devices, Cathode Ray Tube (CRT) screen displays, tape storage, disk storage and increased processing speed and memory availability were improving rapidly.

By the late seventies high speed mainframes were connected to colour graphic displays and colour plotters, and featured increased memory and disk space. By then programs were developed on screen and stored on disk in both source and executable form. These computers were often linked by networks across the world and were available on a commercial pay-as-you-use basis

Minicomputers were also widespread and were capable of carrying out fault tree analysis and were even used for risk monitors. The minicomputers were not as powerful as the mainframes but they were generally dedicated to a particular task and were not required to services tens or hundreds of simultaneous users as were the mainframes. An example of the early development of a risk monitor on a minicomputer will be reviewed later. The subsequent use us this technique in a more modern environment on a PC will be described.

A step change in the computer environment came in the early 1980s with the advent of workstations and Personal Computers (PCs). The workstations had extensive colour graphics facilities and high-speed processors which combined into a desktop computing platform with powerful capabilities.

As fault and event tree analysis software has matured, there has been greater interest in integrating this software with other software commonly used in system engineering. This integration has not been achieved at the present time but there is continuing work especially in data transfer between products using international data definition standards and also in the use of common databases.

One of the important considerations in recent years has been the position of PSA in the design process. Traditionally a PSA has been carried out by a group separated from the design of a system and has been undertaken either at the end of a design or at specific 'Freeze points' during the design. This has led to expensive and time-consuming design changes. A later section will describe some recent work to overcome this problem.

2. PROBABALISTIC SAFETY ANALYSIS OFTWARE HISTORY

The first use of probabilistic modeling as a modern research tool stems from work on the atomic bomb. In the 1950s, Hermann Kahn of the RAND Corporation used it to evaluate the "what if" scenarios of nuclear proliferation. By 1960, probabilistic scenario analysis (PSA) was used in financial analysis, engineering applications, and general economic evaluations

Fault tree analysis was first conceived by H A Watson of Bell Laboratories in connection with a US Air Force contract to study the launch control system for the Minuteman Intercontinental Ballistic Missile (ICBM). The successful use of this new technique was taken up by Dave Hassl, then at the Boeing Company, who led a team to analyse the entire Minuteman Missile.

The Boeing Company subsequently utilised the experience of the Minuteman analysis on their commercial aircraft. In 1965 Boeing and the University of Washington sponsored the first System Safety Conference at which several papers were presented on Fault Tree Analysis leading to a worldwide interest in this field.

In 1966 Boeing developed a program called BACSIM for the analysis of multi-phase fault trees. This software could handle twelve phases and included repair and K-factor adjustment of failure rates. This in-house software, developed by a team led by Bob Schroeder, ran on an IBM 370 mainframe and had the capability to plot the fault trees on a 26 inch wide Calcomp drum plotter.

The Apollo 1 launch pad fire in January 1967 led NASA to contract Boeing to implement a completely new, comprehensive safety assessment programme. This helped to bring fault tree analysis to the notice of the general public.

The next few years saw a rapid development of the theory and computational solution of fault trees. Here are some of the main contributions to the field. In 1970 W Vesely developed the Kinetic Tree Theory (KITT) and the PREPP/KITT computer program [1], [2]. Fussell and Vesely developed the top down cut set generation algorithm called MOCUS (Method of obtaining Cut Sets) [3], [4]. A bottom up cut set algorithm call MICSUP (Minimal Cut Sets Upward) was developed by P. Pande, M. Spector and P. Chatterjee [5]. The FATRAM algorithm was developed by D. Rasmuson and N. Marshall to improve upon the MOCUS algorithm [6]. S. Semanderes developed a cut set algorithm using prime numbers in his ELRAFT program, which efficiently stored cut sets and eliminated super sets [7]. Randall Willie developed the ever-popular computer program call FTAP [8]. Dick Worrell developed the SETS computer program, which is still in usage in various versions [9]. Howard Lambert extensively developed importance measures, and developed the program called IMPORTANCE [10], [11].

In 1975 a landmark PSA study, conducted for the USNRC, headed by Prof Norman Rasmussen from MIT was published under the title WASH-1400, '(The) Reactor Safety Study'. It is thus often referred to as the Rasmussen Report. It considered the course of events which might arise during a serious accident at a (then) large modern Light Water Reactor. It estimated the radiological consequences of these events, and the probability of their occurrence, using a fault tree/event tree approach.

The report concluded that the risks to the individual posed by such stations were acceptably small, compared with other tolerable risks. Specifically, the report concluded that the probability of a complete core meltdown is about 1 in 20,000 per reactor per year.

The study was intended to estimate the probabilities of occurrences of accidents involving radioactivity release and to assess the risk of such accidents relative to other risks. The study involved (1) a list of potential accidents in nuclear reactors, (2) estimation of the likelihood of accidents resulting in radioactivity release, (3) estimation of health effects associated with each accident, and (4) comparison of nuclear accident risk with other accident risks. The study determined that the nuclear accident risk was small - almost negligible compared with more common risks.

The WASH 1400 risk assessment was subsequently reviewed by a Risk Assessment Review Group in 1977 (the Lewis Report) that concluded that "they were unable to determine whether the absolute probabilities of accident sequences in WASH 1400 are high or low, but believes that the error bounds on those estimates are, in general, greatly understated." They went on however to say:

WASH-1400 was largely successful in at least three ways: in making the study of reactor safety more rational, in establishing the topology of many accident sequences, and in delineating procedures through which quantitative estimates of the risk can be derived for those sequences for which a data base exists.

Despite its shortcomings, WASH-1400 provides at this time the most complete single picture of accident probabilities associated with nuclear reactors. The fault tree/event tree approach coupled with an adequate database is the best available tool with which to quantify these probabilities.

WASH-1400 made clear the importance to reactor safety discussions of accident consequences other than early fatalities.

The NRC accepted the findings of the Risk Assessment Review Group and issued a statement which said in part:

It is important to note that the Risk Assessment Review Group, while criticizing (sic) the risk assessment of WASH 1400 per se, commended the description of accident sequences and the "fault tree/event tree" approach as an analytic tool for quantifying probabilities of accidents.

The next major developments in fault and event tree analysis software came in the early 1980s with the advent of new computer hardware in the form of workstation and Personal Computers (PCs). Combining both local access to powerful processing facilities and on-screen graphics these new platforms generated a new generation of fault and event tree analysis software. One such software product was FaultTree+ which has been developed by Isograph Ltd under the leadership of Dr Richard Pullen. The next two sections describe some of the applications derived from FaultTree+ and its antecedents.

3. THE ESSM

In 1987 the Essential Systems Status Monitor (ESSM) was installed on a Honeywell DPS6/92 mini-computer at the Heysham 2 AGR in England [12], [13]. The ESSM was directly accessed by control room staff and was the earliest application of a nuclear safety risk monitor in an operational environment.

Before the ESSM was introduced to Heysham 2, operating instructions had been used to stipulate the minimum levels of redundancy required in the plant's safety systems. These documented instructions were concise and unambiguous and, due to the complexity of the safety systems, interpret requirements for minimum levels of plant in a conservative manner. The introduction of the ESSM Risk Monitor provided the ability to perform probabilistic safety assessments and allowed the plant to be operated in a less restrictive manner whilst preserving the same safety objectives.

The ESSM provided risk assessments based on fault tree methodology. The underlying fault trees modeled the effectiveness of essential safety systems to protect against a variety of initiating events leading to serious consequences.

The fault trees were similar to those used in off-line probabilistic safety assessments and represented systems such as post-trip sequencing equipment, the pressure support system, the start/standby boiler feed system, the emergency boiler feed system, the essential electrical system, the decay heat boiler system, the reactor seawater system, the inlet guide vane system, the gas circulators, the circulator auxiliaries cooling system and the circulation diverse cooling system.

The analysis of fault trees representing complex nuclear safety systems had traditionally required long computing times. Reducing these computing times to provide on-line assessments in one or two minutes was probably the greatest challenge for the software development team. The fault trees representing AGR systems are particularly complex due to the high degree of system redundancy combined with the effect of common failures. In fact, common basic events were so extensive that no appreciable modularization of the trees (automatic or manual) was possible. Reducing the computing time relied heavily on developing efficient algorithms to process the system minimal cut sets using traditional Boolean Algebra methods.

ESSM Functionality

One of the principal functions of the ESSM was to perform probabilistic safety assessments for control room operators within one or two minutes. Assessments took into account current plant status (component outages due to faults or scheduled maintenance) as well as the current plant configuration. Although the ESSM performed a probabilistic assessment, the results of the assessments were provided in terms of maintenance categories that correspond to bands of failure frequencies.

Deterministic rules could also be monitored in parallel to probabilistic assessments using a special set of fault trees. Violation of such rules could also be displayed to the operator. During normal operation of the plant, however, advice displayed to the operator was provided using probabilistic criteria.

The ESSM fault trees could also be manipulated by the computer software to provide advice on how to improve the availability of the essential systems. Single items, or combinations of items, would be listed as priority maintenance candidates. These items, when restored to service, would substantially improve the availability of the essential systems.

The ESSM could also be accessed from the maintenance planning office. What-if scenarios could be investigated in an off-line environment. This allowed the effect of planned maintenance to be determined on the availability of the essential systems. The future availability of these systems could be optimized taking into account existing faults.

Adapting Fault Trees To An Operational Environment

A fault tree graphically represents the interaction of failures and other events within a system. Basic events at the bottom of a fault tree are linked via logic symbols (known as gates) to one or more TOP events. These TOP events generally represent system failure modes or hazards for which predicted reliability data is required. Basic events generally represent component failures for which a probability of failure is provided based on historical data. The traditional analysis process is to produce the system minimal cut sets, apply the basic event probabilistic data, and then determine the probability of the TOP event.

By using special events known as 'House Events' we can interactively modify the logic of a fault tree to take account of the real time changes in a system's configuration and status. House events are events that have a probability of one or zero (status true or false). By switching their status we can effectively re-configure the fault tree. In addition component failures may be represented by temporarily converting a basic event to a 'true' house event.

Re-analysis of the fault tree, taking into account the current house event settings, provides the current availability status of the system.

4. THE RISKVU MONITOR

RisVu V3 is an example of a modern risk monitor that may be adapted to provide on-line safety assessments or off-line comparisons of different design options. In effect it may be used as an operational aid or a management tool.

RiskVu is a 'commercial off-the-shelf' product that operates in parallel to the FaultTree+ computer program. FaultTree+ is widely used throughout all the major engineering industries to perform integrated fault tree, event tree and Markov analyses.

Using RiskVu As An Operational Tool

ARiskVu project is connected to one or more FaultTree+ project files. These project files might represent risk models for nuclear, aerospace, railway, process and other types of engineering systems. These project files may be temporarily disabled for any given risk assessment. This is particularly useful when modeling multiple system configurations that may vary at different times. For example, if RiskVu was being used to model the boats within a dockyard and the connected on-shore auxiliary systems, the project disable feature would allow a user to quickly indicate which classes of boats were currently in the dockyard.

RiskVu allows users to provide a custom high-level interface to the underlying fault tree, event tree and Markov models constructed in FaultTree+. Component and configuration events are created by the RiskVu administrator. These events are linked to primary or house events in the fault or event trees. Modification of an event's status in RiskVu will modify the status of the primary or house events in the underlying FaultTree+ projects. The status of a component event indicates whether it is currently in service or out of service due to a fault or scheduled maintenance. A configuration event (corresponding to a house event in the fault or event tree model) has a true or false status. Component and configuration events may be grouped by an administrator.

System structures may be displayed in the form of hierarchical or schematic diagrams. Schematic diagrams may be 'activated' allowing users to navigate between systems, modify the status of plant items and view system degradation levels using colour codes.

After indicating a change of status for a component or configuration event the user may ask RiskVu to reassess the current risk level for the monitored systems and indicate any operating restrictions. Current risk levels may be compared to previous assessments in a history profile.

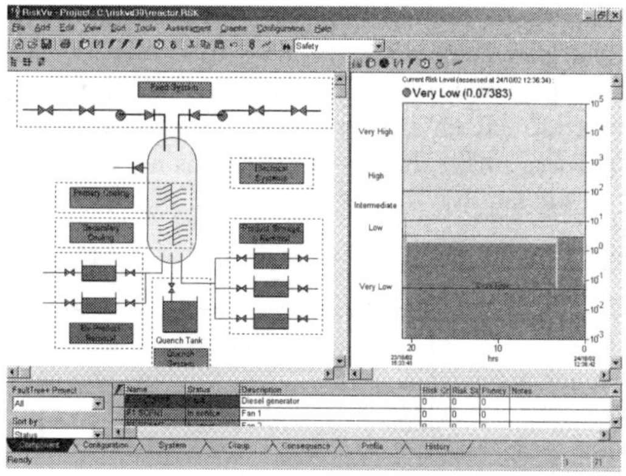

Fig. 1: RiskVu screen display showing the current status of systems together with the history profile

Once an assessment is performed RiskVu provides priority restoration rankings for components that are currently out of service. These rankings indicate which components will result in the largest reduction in risk if they are restored to service. RiskVu also provides importance rankings indicating which operating components would result in the highest increase in risk if they were to fail.

RiskVu also allows the effects of planned maintenance to be assessed. A time profile may be quickly constructed by a RiskVu user indicating at what times in the future individual components will be removed from service for scheduled maintenance. RiskVu will then assess the resultant risk levels taking into account any current outages. Alternative time profiles may be compared.

An important feature of the RiskVu risk monitor is its ability to be customized to adopt the functionality required by the user. A RiskVu project administrator (with password protection) may customize the screen interface to display only the information required by a user or operator of the system.

Using RiskVu For Design Comparisons

RskVu may also be used as an off-line tool allowing users to assess the effects of design changes on risk or comparing the risk associated with alternative designs. Configuration events may be used to switch between different levels of redundancy and the failure rates and test intervals of component events may be temporarily changed.

In addition, the effect of different operational environments over the course of a mission may also be assessed using the time profile facility.

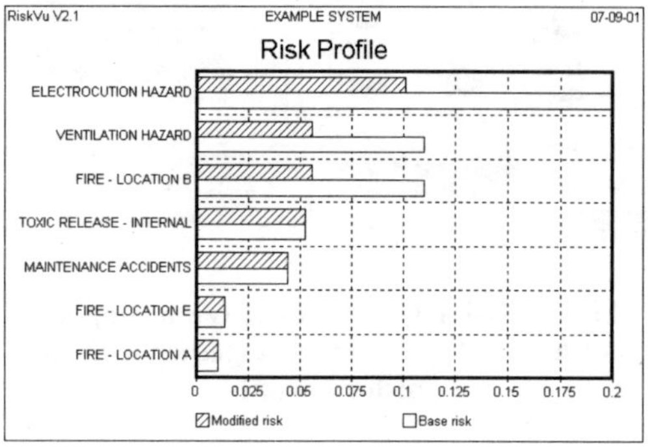

Fig. 2: Risk profile comparing 'base' risk with 'modified' risk

5. RBOT

Wen applied at an early stage of a project, PRA can be a valuable design tool. Current PRAs however, are generally performed to demonstrate safety and are often unsuited for applications aimed at making design or operating decisions. In making design decisions, if comparisons are made using simplified fault trees, the models may not accurately represent the system's functioning or malfunctioning: Valid design or operation comparisons require detailed assessments. Developing a detailed PRA during the design phase of a project is an iterative process where the models are developed at design "freeze" points. However, as the design matures, the PRA results continue to reflect a previous design and not the current design. An example of the process is

- Development of a PRA model
- Use of the model to identify weaknesses,
- Quantification of PRA benefits of alternate design and operational strategies, and
- Adoption of selected design and operational improvements.

This process was followed by Westinghouse during the design of its AP600. As expected, the scope and detail of the AP600 PRA model increased as the plant design matured. Although the iterative process resulted in a number of design and operational improvements, the process spanned five stages and six revisions.

This iterative process can continue throughout the design cycle and into the certification process. For example, when considering the final design approval for the CE System 80+ reactor containment design, the Nuclear Regulatory Commission (NRC) requested that the designers evaluate design alternatives that would help mitigate the consequences of severe accidents. Sixty-three design alternatives were considered, and twenty-seven were quantified. This back-end approach is expensive and time-consuming. A much better approach would be to use a dynamic PRA coupled to the design to reduce, control, or eliminate risk by quantitatively identifying risk drivers as the design develops.

Westinghouse, in its design of the International Reactor Innovative and Secure (IRIS) plant, brought the use of a dynamic PRA to reality by using a safety-by-design approach to cover all aspects of safety at the very beginning of the design process. Rather than evaluate design alternatives on a system-by-system basis, one change at a time, and discarding previous evaluations, the different system and component configurations were evaluated in various combinations for the entire range of initiating events. Thus, unlike other design "optimization" programs, Westinghouse used its PRA program to evaluate different types of plant systems and components, in different operational states, that were interchangeably evaluated as the design progressed. Because each "new" plant configuration may or may not improve plant safety or may marginally improve safety at great cost or operational flexibility, all previous design alternatives were retained to maximize the benefits of a risk-based design. More importantly, because of the ease in evaluating alternate component or system arrangements, dramatic increases in reliability were observed with atypical, unusual, or simply different design configurations compared to the designs using "proven" reliability design practices. This contributed greatly in a two order-of-magnitude reduction in the base-case core damage frequency for IRIS. The design lessons learned from the dynamic PRA for IRIS were then applied by Westinghouse to the AP 1000 even though it was undergoing design certification review by the NRC.

Innovative Approaches to Probabilistically Evaluating Alternative Designs

Aspart of its safety-by-design approach and commitment to reduce the core damage frequency to as low as realistically possible for its IRIS plant, Westinghouse had many team members focused on safety and reliability. One of the team members was the Oak Ridge National Laboratory (ORNL); its task was to develop a one-button architecture that would allow components, modules, and data to be interchanged at will with the probabilistic effect immediately apparent. ORNL's program, called risk-based design optimization tool (RBOT), is coupled to FaultTree+ and allows users to evaluate design changes; new modeling approaches, methods, or theories; modeling uncertainties and completeness; physical assumptions; and data changes on component, cabinet, train, system, and facility bases.

With RBOT, changing the design is as simple as picking one of the design alternatives from a scroll-down menu that have been developed by PRA analysts. The default (base case design) and currently selected design options are highlighted; graphical displays provide additional indicators of the option(s) chosen. With the scroll-down menu, many design alternatives, including unconventional alternatives, can be easily evaluated. If additional design alternatives are desired and are not available in the scroll-down list of options modeled in RBOT, they can be easily created by a PRA analyst and added to the scroll-down list. The basic RBOT computer program in development automatically inserts the correct fault tree module(s) for the design configuration chosen into the PRA model in FaultTree+, maintains any links to event trees (if used), re-links the correct support systems, adjusts the common cause failure probabilities on the train and system levels, and accounts for changes in recovery actions and human error probabilities. FaultTree+ then recalculates the parameters of interest for the entire spectrum of initiating events, and transmits the results back to RBOT. The resulting fault tree is the same as that which would result if those choices were the original base case. Any or all of the design options can be made in any order using RBOT. Because all choices remain available, no information is lost and the design can be returned to any previous state.

Because all of the current and previous design, modeling, and data sets are available via the one-button architecture, the safety ramifications of design options are evaluated very early, feedback on design alternatives is immediate, and true optimization of the plant design can be achieved. Thus, even with complicated systems and alternatives, designers can not only observe if the system is more reliable, but can also understand why the system is better or worse.

Selected Examples of Design Optimization Using BOT and FaultTree+

The IRIS design is aimed at achieving four major objectives: proliferation resistance, enhanced safety, economic competitiveness, and reduced waste. The safety approach of Generation II reactors [current nuclear power plants (NPPs)] can be defined as coping with accidents/consequences by active means, while the improved approach of Generation III reactors (e.g., AP600) is coping by inherently safer passive means. The design of IRIS, a next-generation NPP, promotes safety to the next level through safety-by-design that eliminates most accident initiators while the consequences of other accidents are rendered acceptable.

Figure 3 shows the safety systems for the IRIS reactor.

The two Systems that were evaluated using RBOT were the automatic depressurisation system (ADS) and the emergency heat removal system (EHRS). The ADS is a pressure relief system with two parallel trains for blowing down into a suppression pool. There are our trains of emergency heat removal. The number of trains required for operation is dependent on the initiating event and varies from 2 to 4. The EHRS is a natural circulation system that removes heat via the steam generator. The flow path is form the steam generator, to the refuelling water tank, through a main and/or bypass leg, back to the steam generator.

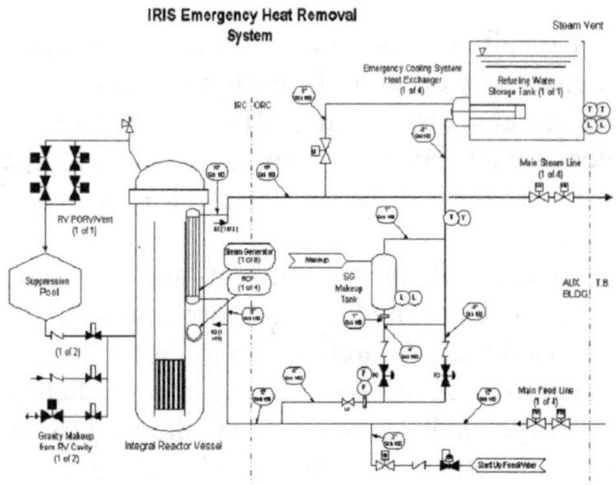

Fig. 3: Safety systems for the IRIS reactor

A collection of design alternatives for various systems for IRIS were identified through a review of current and Generation III NPP designs. For example, a review of operating NPPs showed that typical relief systems (similar to IRIS's ADS) consist of one or two trains. Generation III NPPs have more complex systems such as two trains with four stages of pressure relief. Further review shows that the block valves for the power-operated relief valves (PORVs) may be initially open or closed. (The block valves are used to prevent a loss-of-coolant accident given a stuck open PORV. However, a block valve that fails to open will prevent depressurization of the primary system. Thus, a trade-off exists between having the block valves open or closed.) When a system arrangement, component type, component position is chosen from the menu, RBOT automatically changes the fault tree in FaultTree+ to reflect that choice. The following choices were made available in RBOT for optimizing the ADS:

- The choice of block valves and PORVs were air-operated valves (AOVs) and motor-operated valves (MOVs) (other type valves could have been added at any time),
- The initial position of the block valves (AOVs and MOVs) could be open or closed,
- The number of trains for pressure relief could be 1 or 2 and
- The capacity of each valve, whether AOV or MOV, could be 50% or 100% (i.e., able to accommodate 50% or 100% of the maximum steam flow required for successful depressurization of the primary system for the design basis accidents).

The level-1 PRA for IRIS evaluated using RBOT included 4 event trees with the following initiating events (IEs): a primary system loss-of-coolant accident, a steam generator tube rupture, a secondary system pipe break, and a loss of offsite power. The mitigating systems necessary to prevent core damage for these IEs are represented by II fault trees that cover sub criticality of the reactor, primary system depressurization, and long- and short-term heat removal. Choosing any of the options developed and made available via RBOT provides users with a PRA that is the same that would be developed if that were the base case model. Again, because all previous choices are still available, the design can be returned to any state, at any time, in any order.

In this example, the one-button architecture allowed designers to evaluate 40 unique design alternatives of the ADS for the 8 design choices given above. However, simply providing design alternatives is not beneficial unless various metrics of interest can be compared, improved, or optimized. Using this example, evaluating the failure modes for the different design options provides clues on how the system could be improved. By clicking a single "button" to change the design of the system, designers learned that for those events requiring pressure relief that:

- With only one block valve, the initial position of the block valve does not matter-failures of the PORV(s) dominate the reliability of the system;
- 2 PORVs, each with a 50% capacity rating (i.e., 2-50% capacity), do not provide sufficient redundancy and are in fact worse than having just one PORV because the failure of either PORV /block valve arrangement causes the system to fail;
- True redundancy (2- vs 1-100% capacity valve) improves reliability by an order of magnitude and not the square of the paR V's failure probability because of shared systems and

components. For those events requiring pressure relief, it is 4-6 times better to have the block valve initially open, where it has to spuriously transfer to the closed position than to have it initially closed and having to open. (For many system arrangements, the dominant contributor to system failure is the CCF probability of the block valves to open. The simplest way to reduce the CCF probability of the block valves' failing to open is to have the block valves already open.), and

- For the ADS, it is observed that having the block valves originally open improves system reliability 20% and reduces the core damage frequency 3.5%. Once this CCF probability of the block valves failing to open is eliminated, the type of block valve (e.g., MOV or AOV) is inconsequential.

The other safety system evaluated using RBOT was the ERRS. Identifying design alternatives for the ERRS was a challenge because it was already a pretty elegant system in its simplicity. The original design used natural circulation cooling (i.e., no pumps), and only 1 of the 2 AOVs needed to open to start the system. In fact, the designers purposefully used the most reliable valves they could-the AOVs. As such, identifying design alternatives for the ERRS was not originally considered because of the simplicity of the existing system design.

With RBOT, the number of trains in a system, or even component type or position-alternatives that would not normally be evaluated-were evaluated because of the ease of changing out components, adding trains, etc. This is contrary to typical alternatives that focus on big changes such as number of trains and cross-ties between trains (although these can also be easily evaluated using RBOT).

Because AOVs have the best reliability characteristics for active valves, it was expected that their use would result in the most reliable system. With the ease of evaluating different components and their arrangement, RBOT was used to change one or both of the AOVs to MOVs for each train of the ERRS. Changes automatically made by RBOT to the FaultTree+ model for the valve options included

- Changing the selected component(s) from AOV to MOV (or vice-versa if changing from MOV to AOV),
- Changing the support system for those components from instrument air to electric power (or vice-versa if changing back to AOVs),
- Changing the CCF probabilities on the train level, and
- Changing the CCF probability on the system level (4 trains).

Surprisingly, using the "better" valves actually made the system (and thus the plant) worse. Using the more unreliable MOVs actually improved system performance. This is because the CCF probability for the AOVs failing to open is larger than that for the MOVs, and the CCF probability of the valves was the largest contributor to system failure. Although the MOVs significantly improved the system, using a mix of AOVs and MOVs provided the most reliable system. In fact, a mixed-valve system improved system reliability by 42%; the core damage frequency was reduced by 23%.

Surprisingly, the relatively few component choices for the ADS and EHRS resulted in a very large number of design alternatives. The 11 component choices-component type, position, and capacity-resulted in 160 design alternatives. The one-button architecture allowed all of these alternatives to be evaluated probabilistically with ease and provided designers with insights into system behavior.

With the wealth of information available through the one-button architecture, design alternatives can now begin to consider production, operation, and maintenance costs. For example, if the use of a mixed-valve system significantly increases operation and maintenance costs, a decision could be made to maintain all MOVs in the EHRS because system unreliability would not increase appreciably. Thus, for the first time, plant reliability and other attributes, such as cost, can be optimized simultaneously.

6. SUMMARY

The development of software for the Probabilistic Safety Analysis (PSA) has developed rapidly in the past 45 years. Most of the basic theory was developed in the 1960s and early 1970s by pioneers in the aerospace and nuclear power industries.

The development of computer hardware, operating systems and software development products have enabled the software for PSA to move out of the narrow confines of a few specialists in government laboratories and major industrial companies to the desktop of thousands of safety engineers around the world.

These trends have extended the scope of PSA software into both the design process and the operational environment.

The next few years will see elements of PSA software become integrated with the other software tools commonly used by design and operational engineers. Several international collaborations are under way to investigate methods of common data exchange and storage to ensure that data is input once and used by many times by different software tools, improving both productivity and minimizing data errors.

7. REFERENCES

1. *Analysis of Fault Trees by Kinetic Tree Theory,* Idaho Nuclear Corp IN-1330, 1969, W. E. Vesely
2. *PREPP & KITT: Computer Codes For The Automatic Evaluation Of A Fault Tree,* IN-1349; Idaho Nuclear Corp, 1970, W. E. Vesely & R. E. Narum
3. *A New Methodology For Obtaining Cut Sets For Fault Trees,* Transactions American Nuclear Society (Vol 15, No 1), 1972, p262-263, J. B. Fussell & W. E. Vesely
4. *MOCUS – A Computer To Obtain Minimal Sets From Fault Trees,* Aerojet Nuclear Corp; ANCR-1156, 1974, J. B. Fuseell & E. B. Henry & N. H. Marshall
5. *Computerized Fault Tree Analysis: TREEL and MICSUP,* 1975, University of California ORC-75-3, P. K. Pande & M. E. Spector & P. Chatterjee
6. *FATRAM – a Core Efficient Cut Set Algorithm,* 1978, IEEE Transactions On Reliability (R-27 No 4), p250-253, D. M. Rasmuson & N. H. Marshall
7. *ELRAFT: A Computer Program For The Efficient Logic Reduction Analysis Of Fault Trees,* 1971, IEE Transactions On Nuclear Science (NS-18 No. 1), p481-487, S. N. Semanderes
8. *Computer Aided Fault Tree Analysis: FTAP,* 1978, University of California Operations Research Center; OC 78-14, R. R. Willie
9. *Set Equation Transformation System (SETS),* 1974, SLA-73-0028A Sandia National Laboratories, R. B. Worrell
10. *Measures Of Importance Of Events And Cut Sets In Fault Trees,* Reliability And Fault Tree Analysis; SIAM, 1975, p77-100, H. E. Lambert
11. *The IIMPORTANCE Computer Code,* UCRL-79269; Lawrence Livermore Lab, 1977, H. E. Lambert & F. M. Gilman
12. *ESSM Gives On-Line Real Time Probabilistic Reliability Assessment,* Nuclear Engineering International, May 1989, R. A. Pullen
13. *Employing PRA Techniques In An Operational Environment,* Nuclear Engineering International, January 1986, R. A. Pullen

Reliability, Safety and Hazard
Editors: P.V. Varde et al.

APPLICATION OF PSA IN RISK INFORMED DECISION MAKING

M. Hari Prasad, Gopika Vinod, R.K. Saraf, A.K. Ghosh and H.S. Kushwaha

Health, Safety & Environment Group, Bhabha Atomic Research Centre, Mumbai 400 085

ABSTRACT

Probabilistic Safety Assessment (PSA) models have been successfully employed during design evaluation to assess weak links and carry out design modifications to improve system reliability and safety. Recently, studies are directed towards applying PSA in various decision making issues concerned with plant operations and safety regulations. This necessitates development of software tools like Living PSA, Risk Monitor etc. Risk Monitor is a PC based tool developed to assess the risk, based on the actual status of systems and components. Such tools find wide application with plant personnel and regulatory authorities since they can provide solutions to various plant issues and regulatory decision making issues respectively.

Keywords: PSA, Risk Monitor, RAW, RRW, AOT, STI, What-If analysis

1. INTRODUCTION

In Nuclear Power Plants (NPPs), safety is the major concern. PSA analysis leads insight into plant processes, mechanisms and possible interaction between plant systems, both for existing plants with operating histories and for plants still in the design stage. In view of this, on line safety has received lot of attention from operation and maintenance personnel.

Risk Monitoring can be defined as being the process whereby a complex technical facility is continuously monitored as regards the functioning or non-functioning of its different subsystems and the risk emanating from the facility is evaluated on the basis of this information. In the widest sense it can be regarded as being part of the risk management of a plant. Risk Monitoring provides safety status information for a plant and thus aids decision making about whether continued plant operation is tolerable under certain system function outages. It may also support operations and be of help in deciding on maintenance strategies allowing immediate assessment of different plant configurations

Reactor Safety Division of Bhabha Atomic Research Centre (India) has developed a PC based tool, which can assess the risk profile. This package can be used to optimise the operation in NPPs with respect to a minimum risk level over the operating time. Risk Monitor is user friendly and can re-evaluate Core Damage Frequency (CDF) for changes in component status, test interval, initiating event frequency etc.,. Plant restoration advice, when the plant is in high risk configuration, current status of all plant equipments and equipment prioritisation are also provided in the package. What-If analysis can be done by using this package, which provides guidance to plant operators to choose among various options.

2. WHY RISK MONITOR?

Plant configuration undergoes changes due to changes in component status and/or operating /

maintenance procedures. Some components are randomly down and/or others can be planned for test, maintenance and repair. This results in a variation of the risk level over operating time, which is termed as risk profile, and indicates the trends which could lead to deviation from desired CDF. PSA models can be used to quantify risk due to changes in components status, system design and operations consequent to changes in plant configuration.

3. PSA METHODOLOGY

Probabilistic Safety Assessment (PSA) is an analytical technique for assessing the risk by integrating diverse aspects of design and operation of NPP. Risk can be defined as the product of the probability of an accident and the consequences from that accident. In the context of a NPP, core damage of the reactor represents an accident and release of radioactivity in the public domain and its effects on them will be the consequence. In order to prevent the occurrence of an accident, various engineered safety features are designed. An accident situation occurs when an initiating event is coupled with the unavailability of one or more Engineered Safety Features. The accident frequency can be minimised by reducing the frequency of initiating event and improving the availability of safety systems. This can be achieved by incorporating good design, operating practises and selection of reliable components. The consequence of an accident that may occur can be minimised by providing barriers (e.g. reactor containment and exclusion zone) for arresting the release of radioactivity and reaching the public. In general PSA aims at (1) identifying the Initiating Events that may lead to severe accidents in conjunction with the unavailability of associated engineered safety features and (2) evaluating the Accident Sequence Frequency and resulting CDF. PSA analysis can be extended to evaluation of importance measures for the identification of critical components, which in turn assists in maintenance planning of Nuclear Power Plant.

4. ISSUES RELATED TO DECISION MAKING IN NPPS

It is very much essential to ensure that a Nuclear Power Plant operates in a safe manner at all times. Whenever any modifications or change of configuration are required to be incorporated in an operating plant, Risk Monitor can assist plant operators as well as regulators to assess the risk and ensure that plant is in the prescribed safe domain. Various such applications are given below:

(i) Decision making in operations: If CDF value exceeds the prescribed probabilistic safety criterion, efforts should always be made to lower the CDF through proper configuration control, different test and maintenance policies.

(ii) Maintenance Strategies: Risk Achievement Worth (RAW) is the well suited indicator for deciding maintenance policies. Maintenance actions need to be planned according to the order in which RAWs of components are ranked, i.e. components having higher RAWs need to be maintained with higher priority. Similarly, component having higher Risk Reduction Worth (RRW) should be given attention from the design point of view, since these can enhance the reliability of the system. This type of decision is less sensitive to the absolute values of the component failure parameters; however, the relative values of failure parameters influence the values of RAW and RRW.

(iii) Risk Informed In-Service Inspection: The Risk Informed In-Service Inspection (RI-ISI) program aims at integrating traditional engineering evaluations with insights gained from PSA. The PSA can be used to obtain an estimate of risk and relegate it to various systems and down to components to obtain an idea of their importance in terms of contribution to the Risk. Risk Monitor can be effectively employed for analysing the change in CDF whenever there is a change in Inspection plans and thereby analyse for an optimum scheduling plan. Risk importance measures like Fussel-Vesely and Birnbaum Importance are evaluated for various components and systems in the Risk Monitor for risk informed inspection planning.

(iv) Review of Technical Specifications: Technical Specifications are usually based on deterministic assessment and engineering judgement. Technical specifications based on probabilistic considerations can be evolved to optimise the Allowable Outage Times (AOT) and Surveillance Test Intervals (STI) for various Systems [1] [2].

(v) Emergency Operating Procedures and Risk Management: The Emergency Operating Procedures (EOPs) are generally based on the considerations of failures in process systems. From the event tree relevant to a particular process system failure, safety systems can be identified and their availability can be ensured so as to maintain the plant in a safe domain. EOPs based on dominating accident sequences as identified in PSA can be effectively used in risk management.

The decision making issues described herein have guided in the design and development of the Risk Monitor.

5. SOFTWARE DEVELOPMENTAL ASPECTS OF RISK MONITOR

The Software has been developed in Visual Basic. The Data Flow Diagram of Risk Monitor is shown in the Fig. 1. The various modules developed in the package are as follows.

a) System Modelling Options
b) Main Summary & On-Line Risk
c) Component data base
d) Component Out-of-Service & Restore
e) What-If Analysis

5.1 System Modelling Options

In this module user can choose the different options like plant operation (Normal power or low power operation), initiating events frequency (Direct or through Minimal Cut Sets (MCS)), core damage frequency (Through accident sequences or MCS) and can set the risk levels (Acceptable risk, unacceptable risk, moderately safe, totally safe and design level risk). The software uses plant specific PSA models. However, option is provided for inclusion of PSA models for different NPP also.

5.2 Main Summary & On-Line Risk

This module summarises status of the safety systems (available, degraded or unavailable) based on the status of the components (Available, out of service), list of components which have been taken out from the service and risk profile (CDF vs Time) on day basis, monthly basis or year basis. Logs on the CDF values on time basis are shown in the risk profile module. If the user double clicks on this table it will give the status of the plant of any day. Risk Monitor presents the graphical display of Risk profile with respect to CDF value. Eventhough, coloured bands are defined for risk levels, standardisation is required for defining these level. The user interface of this module is shown in the Fig. 2.

5.3 Component Database

A Reliability Data Base is used for the management of data which is designed using MS ACCESS which also stores the PSA models and analysis results. The package provides database for basic events probability, initiating events frequency and human errors probability. Different component reliability models [3] [4] such as repairable, non-repairable, tested, mission time, probability and frequency models have been incorporated in calculation of the component unavailability. Risk Monitor can re-evaluate the CDF depending on the change in the value of initiating event frequency or safety system unavailability. In case of Safety System, the unavailability will be affected when component unavailability changes. Depending on the mode in which the component is functioning in the safety system, parameters like failure rates, test intervals, mission time, repair time etc. will alter its unavailability value. Risk Monitor has provided the option to alter the necessary parameters depending on the mode of the component functioning, thereby computes its unavailability. This change is propagated to the Core Damage Frequency and user can see its implication on Risk profile also. The user interface of this module is shown in the Fig. 3.

5.4 Component Out-of-Service & Restore

This module shows all the components which are in service and out of service as of today system wise. The user can take any component out of service (for maintenance, testing or inspection) or can restore by checking the relevant components in the in-service list or out of service list respectively. This operation changes the configuration of the systems and in turn affects the CDF. CDF is re evaluated from the minimal cut sets provided in the database. The user interface of this module is shown in the Fig. 4.

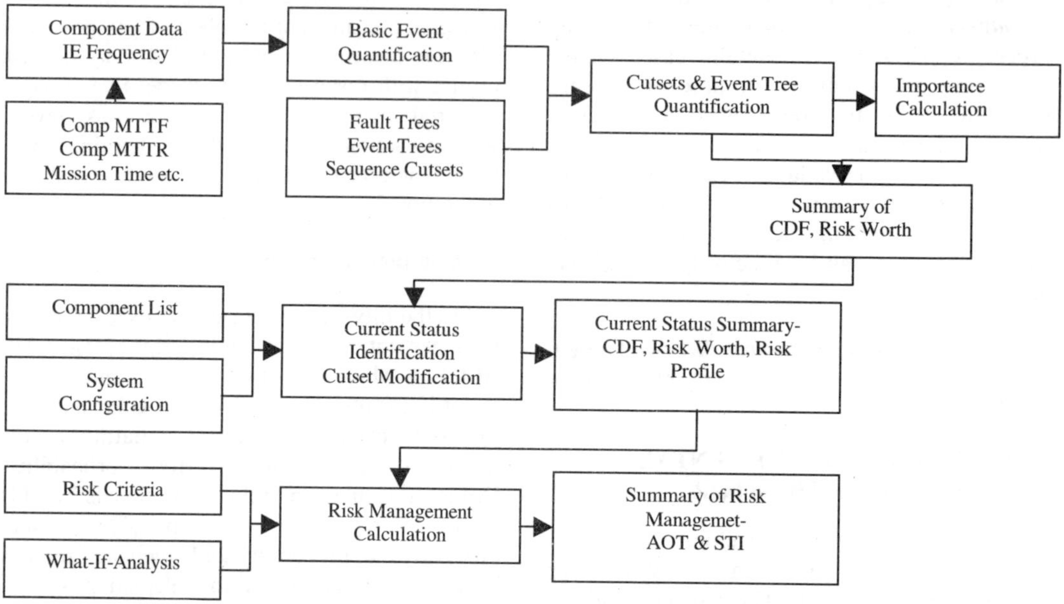

Fig. 1: Data Flow Diagram of Risk Monitor

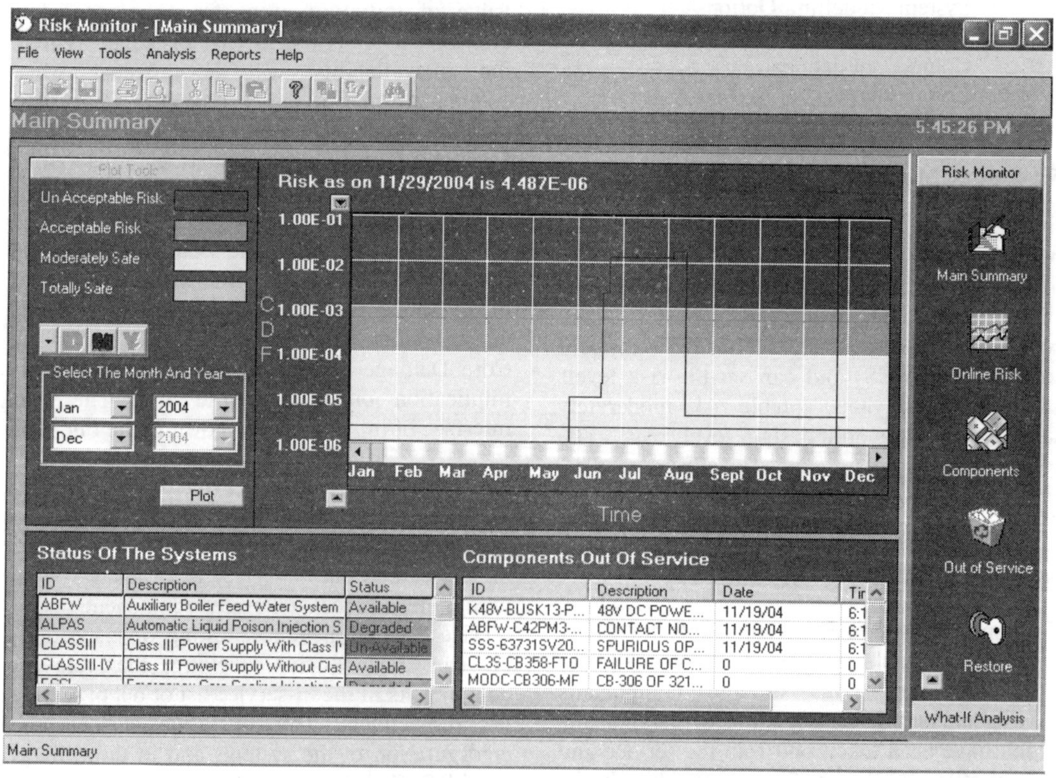

Fig. 2: Main Summary & On-Line Risk

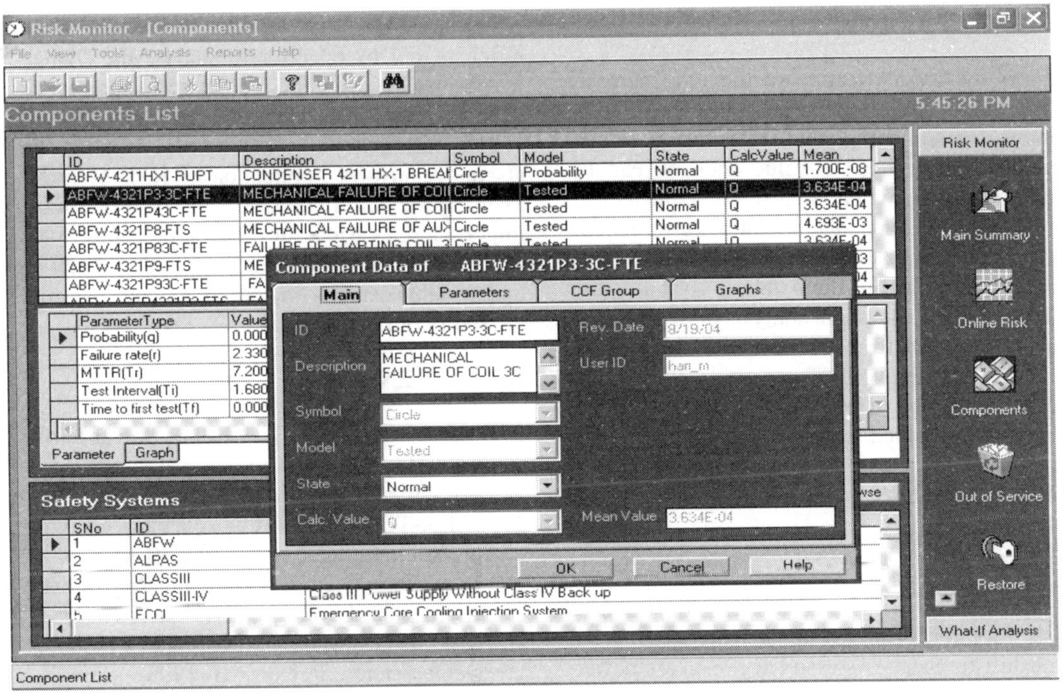

Fig. 3: Component Database

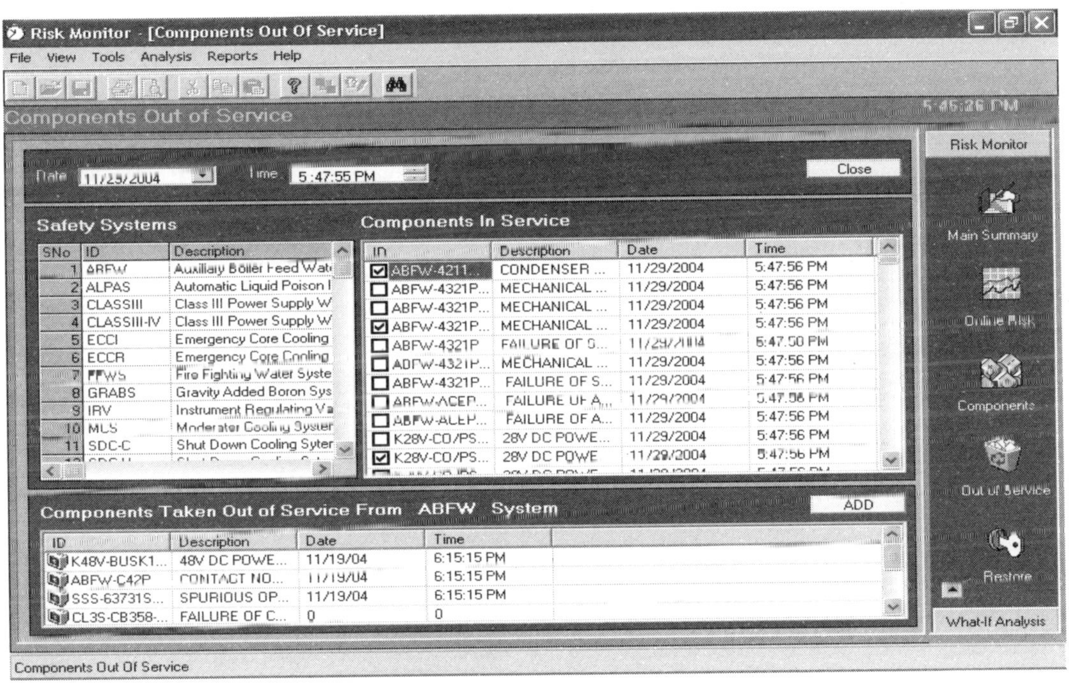

Fig. 4: Component Out-of-Service & Restore

5.5 What -If Analysis

This is the unique feature of the risk monitor. With this analysis user can analyse different combinations of component states and based on the change in the CDF value decision can be made on which combination of components can be taken for maintenance or can be restored. All the cases which have been analysed by this analysis are stored in the database so that if the user wants to do the same analysis in the future it can be retrieved from the database and can be applied, this will avoid the repetition of the same analysis. On-line "Help" and report generation have been provided with the software package, so as to assist the user in navigating through the software as well as to get familiarised with the PSA terminologies.

6. CONCLUSIONS

Risk Monitors backed up by the results of Probabilistic Safety Assessment (PSA) are power plant operations. Risk monitor can be used as an efficient tool by operator to analyse the change in Risk whenever the component parameters change. Also it can be considered as an efficient system to track plant history. Envisaging the above scenario of application of Risk Monitor, assuring the quality of Risk Monitor is of utmost importance. Various testing approaches have been adopted to expose the detectable as well as undetectable errors in order to improve the reliability of risk Monitor.

7. REFERENCES

1. IAEA-TECDOC-737; (1994), Advances in reliability analysis and probabilistic safety assessment for nuclear power reactors.
2. IAEA-TECDOC-1106; (1999), Living probabilistic safety assessment (LPSA).
3. PSA Pack 4.2, A Code for Probabilistic Assessment Level 1, IAEA, Vienna (1993).
4. Risk Spectrum, PSA Professional 1.00.05: A Demo version, Relcon AB (1998).

Reliability, Safety and Hazard
Editors: P.V. Varde et al.

ADVANTAGES AND DISADVANTAGES OF A RISK-BASED REGULATORY REQUIREMENT (the experience in Argentina)

Jorge Barón

Instituto CEDIAC – Facultad de Ingeniería
Universidad Nacional de Cuyo – Mendoza – Argentina

ABSTRACT

Argentina has its own nuclear regulations, which include a risk-based criterion curve for the licensing of nuclear installations. This requirement, established in the early '70s, must be fulfilled with a PSA study.

It has been applied to several installations, and the advantages and disadvantages of this approach are discussed in the paper through several examples. The main disadvantage is a somehow large amount of PSA work that needs to be performed for the licensing of a nuclear installation. The main advantage is the effective risk reduction that can be achieved by retrofitting the risk-based lessons learned into the design teams (not only for design of systems and components, but also for design of operation, testing and maintenance schemes).

1. INTRODUCTION

Argentina started its nuclear activities in the early 1950s, with the establishment of the Atomic Energy Commision, devoted exclusively to peaceful applications. Several research reactors were designed and built, and the first Nuclear Power Plant of Latin America (Atucha I) started its commercial operation in Argentina in 1974.

Accompanying this development, a national regulatory approach was started in the early 1970s, with the establishment of a national regulation, which includes probabilistic features. In particular, regulation AR10.1.1 [1] (formerly AR3.1.3 [2]) was enforced, which requires the fullfillment of a risk criterion curve for the acceptance of nuclear installations.

This criterion curve is based on an iso-risk acceptability line, and embeds the concept of putting a negligible contribution to the public risk due to the existence of the nuclear facility under consideration. The basis for the iso-risk come from the works by Farmer regarding the individual biological risk of iodine accidental emissions, and are consistent with ICRP recommendations (at present, adapted for ICRP-60) [3] [4].

This approach was initially established for the public, and then it was extended also to those accidents involving the operating personnel.

The objective of this criterion is to limit the individual risk associated to potential exposures to values comparable to the risk associated to normal exposures, during normal operations. For risk estimations, the set of accidental sequences that imply radiological consequences for either the public or the workers should be identified and the annual probability of such sequences must be calculated, as well as the effective individual doses in the most exposed group of the public or the most exposed worker. For the public, the weather probabilities may be taken into account, but not the application of eventual counter-measures. If sequence grouping is performed, the worst sequence (from the dose point of view) should be taken, with the probability resultant from the addition of those sequences that compose the group. PSA tools are used to estimate yearly probabilities.

The criterion establishes that all the resulting values of probability and dose should result in points in the *acceptable* region of a criterion curve. This criterion curve is indicated in fig. 1 for members of the public, and in fig. 2 for workers. If the case is that more than 10 sequences are plotted, the probability value of each point must be divided by N/10, being N the number of sequences.

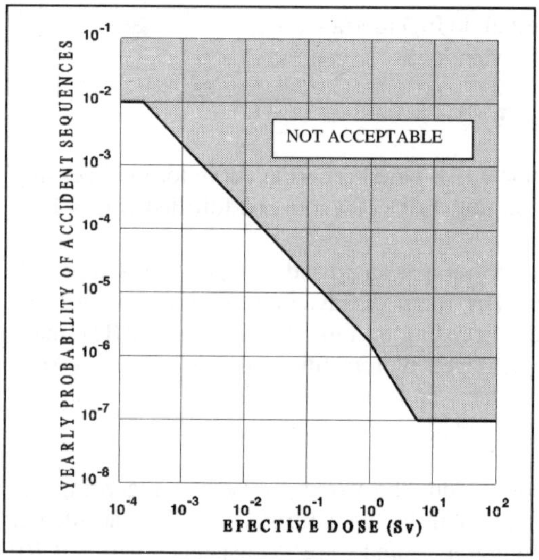

Fig. 1: Criterion for members of the public

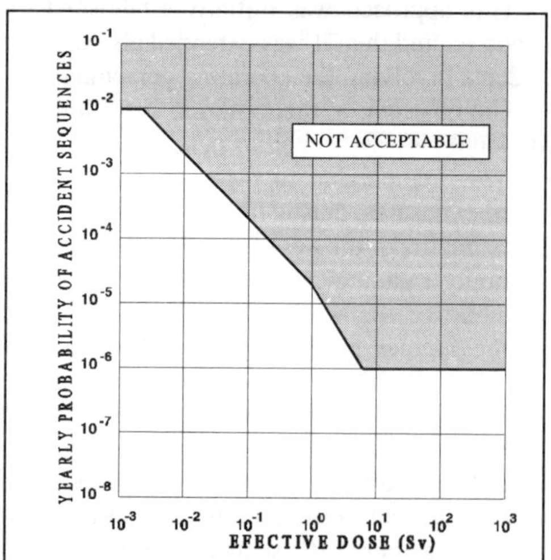

Fig. 2: Criterion for operators

2. CRITERION FUNDAMENTALS

The individual risk concept is used in the derivation of the criterion curves, posing a limit to that risk. The individual risk is defined as the probability that a certain individual, during a certain period of time is accidentally exposed to ionizing radiation and dies due to that exposure. Being X= "exposure" and F = "death due to that exposure", the individual risk R is defined as:

$$R = P(X \bullet F) = P(X) \cdot P(F / X)$$

where P(X) is the exposure probability of the individual, incurring on an effective dose E, and P(F / X) is the death probability of the individual after exposed to an effective dose E. P(X) depends on many factors related to the practice or accident itself, the siting, the individual location, etc., while P(F / X), instead, depends on the incurred dose E. This dependence increases in a linear relation up to approximately 1 Sv, with a slope of $\cong 5 \times 10^{-2}$ Sv^{-1}, and then with a steep sigmoid to reach values close to unity for doses of 6 Sv or higher. The individual risk can therefore be expressed as:

$$R = P(X) \cdot P(F / X) = P(X) \cdot f(E)$$

It is assumed that for a certain installation N accident sequences are possible to occur, each one with a given individual risk. Being so, the sequence *j* has a related risk $R_j = P(X_j).f(E_j)$. Assuming that the accident sequences are disjointed, the total individual risk R_T results

$$R_T = \sum_N R_j$$

Following the ICRP recommendations, and considering the limitations of the PSA methodology, the risk is limited for the public individuals to $R_T = 10^{-6}$ per year, and for the workers to $R_T = 10^{-5}$ per year. It is also assumed that with N=10 accident sequences, all the significant potential accidents are included (if they are less than 10, the situation is conservative, and if they are more than 10, the probability values must be scaled by N/10). Under this assumption, the contribution of each accident to the total risk R_T must be $R_j = 10^{-7}$ per year for the public, and $R_j = 10^{-6}$ per year for the workers, in order to not exceed the respective imposed limits.

Now, coming back to the expression of R = P(X).f(E), considering P(X) and E as variables, and R as a fixed parameter, it results, for the public and for the workers:

$$P(X) = \frac{10^{-7}}{f(E)} \text{ and } P(X) = \frac{10^{-6}}{f(E)}$$

Plotting these functions in a *probability vs. dose* graph (log-log) the criterion curves of figs. 1 and 2 are obtained. These criterion curves have four segments:

- a first horizontal segment, which results from the truncation of the indicated functions at the value of $P(X) = 10^{-2}$ – this indicates that *high probability* accidents with radiological consequences are not acceptable, independently of the doses involved
- a first inclined segment (with a 45° slope in the log-log graph), that corresponds to the domain where f(E) is linear with a slope $\alpha \cong 5 \times 10^{-2}$ Sv^{-1}.
- a second inclined segment, which corresponds to the domain where f(E) is sigmoid (which should imply a sigmoid representation on the log-log graph). For simplicity, this has been conservatively approximated with a linear behavior (in the log-log graph)
- a second horizontal segment, which corresponds to the domain where f(E) ≅ 1 (6 Sv and higher doses)

These "criterion curves" can therefore be considered *iso-risk curves* in all their segments, with the exception of the initial horizontal segments. The iso-risk values involved are 10^{-7} for the public and 10^{-6} for workers. The criterion curves limit two regions, identified as *acceptable* and *not acceptable* in the figures, being that all the points in the acceptable region have risk values lower to the limit indicated in the iso-risk curve.

3. CRITERION USAGE

To make use of the criterion curves, each radiological potential accident for a particular practice or plant is represented by a point A_j (j ≤≤ N) which coordinates are[$P(X_j)$ - E_j]. For each criterion curve, a *constellation* of points should be obtained from PSA studies. These points must be represented in the criterion curves graphs, and all of them should fall in the *acceptable* region. Being this the case, it means that for each of the points the associated risk is lower than or equal to the iso-risk regulatory requirement, and the practice or plant is considered acceptable.

In the case that one or more of these points fall into the *not acceptable* region, the licensing requirements are not fulfilled and the practice or plant is considered not acceptable. To become acceptable, modifications must be made to the practice or the plant, in order to *introduce* the points in the acceptable region. These modification may either reduce the likelihood (probability) by improving the safety systems, operator training, etc, reduce the consequence (dose) by providing mitigation systems or other means, or both. In this sense, the criterion curves can be used not only for the acceptance or rejection of a certain practice or plant, but also as a reference tool on how to turn it acceptable in case of initial rejection.

Furthermore, the need to ellaborate a full PSA for licensing requirements, allows for risk-ranking of different accident sequences, and its contributors, as a decision-making tool to improve the overall safety of the plant or practice, from the *risk point of view*.

Atucha II NPP example

Atucha II is a project developed by KWU and built in the Atucha site of Argentina, near the Parana river. It has a 700MW(e) PHWR of the german design, with a pressure vessel and vertical fuel channels.

Its construction begun in 1982 and was stopped due to budgetary reasons in a state very close to completion. At the time the Preliminary Safety Report [5] was written, a licensing PSA was performed, according to the licensing requirements of the moment [2].

These requirements had the same philosophy stated in these papers, but the criterion curve was more permissive, as it was based in ICRP-26 limits. When the limits were reviewed by the ICRP, so did the regulation in Argentina.

The results of this case are indicated in Fig. 3.

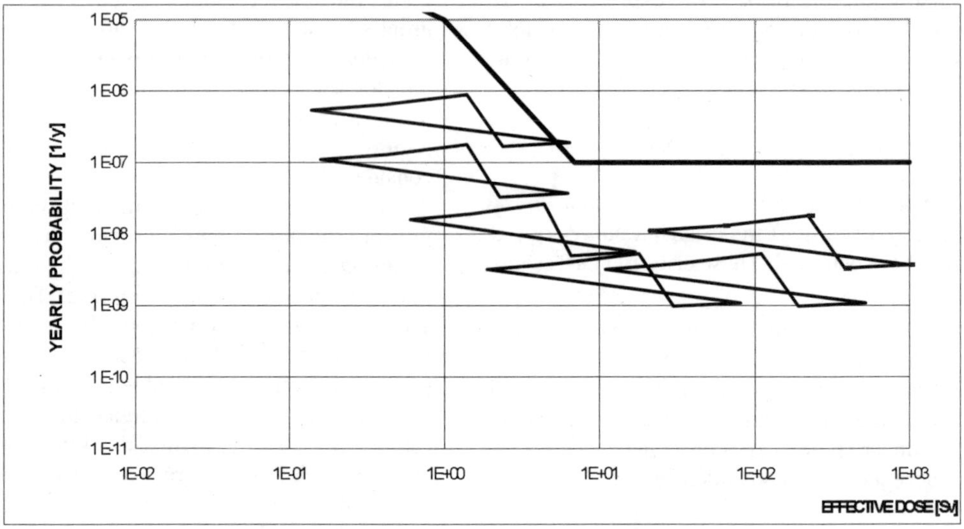

Fig. 3: Application to Atucha II (public)

In this case, six *release cathegories* were obtained from the PSA, and they are represented by a set of points that correspond to different atmospheric stability classes, each having a different probability at the site, and also different associated doses due to the local population distribution. Almost every point fits in the *acceptable* region. However, some effort is needed to improve the plant response to specific scenarios.

If the more restrictive requirements of the present regulation should be applied [1], in case the project is continued and the plant finished, further effort should be made to fit in the *acceptable* region, from the *risk point of view.*

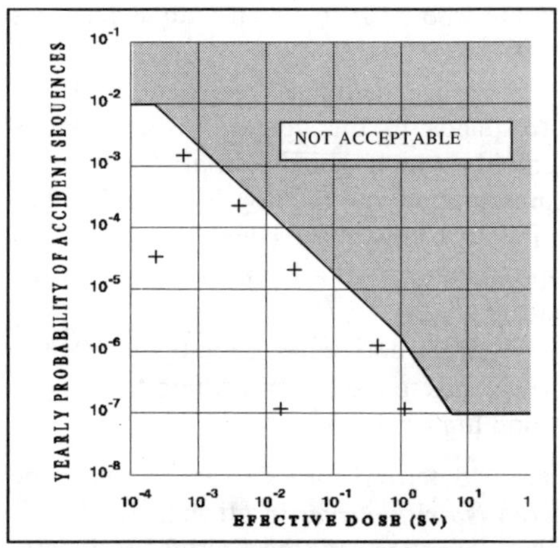

Fig. 4: Application to the MPR (public)

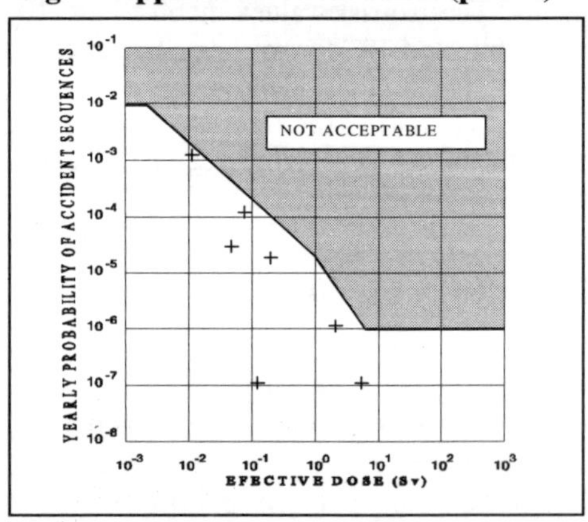

Fig. 5: Application to the MPR (workers)

Multi Purpose Reactor (MPR) example

Applications of the criterion curves were performed for research reactors, as indicated in [6]. Under the same framework, a PSA was performed [7] for a specific pool-type research reactor (MPR). This is a 22 Mw, argentine design, with some distinctive characteristics, such as separate cooling circuits for the core and the pool, pneumatic driven control rods, gadolinium chambers as a second shutdown system, and distributed software-based safety.

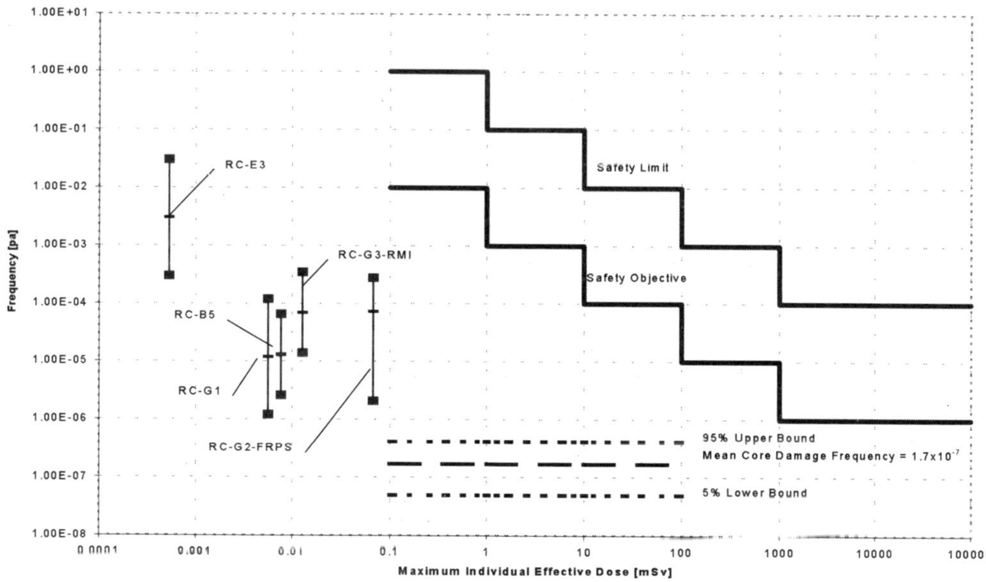

Fig. 6: OPAL PSA results vs. ARPANSA criteria

The PSA considered a series of initiating events (i.e. internal events, seism, shutdown radiation beams, etc) and both doses to the public and the workers were estimated.

The PSA results are represented against the criterion curves in figs. 4 and 5. All the points (both for the public and the workers) fall in the *acceptable* region. From the *risk point of view*, they show a well-balanced design. Although they are too close to the limit, it should be noticed that the basic design was contrasted to the old argentine criteria without the ICRP 60 correction and here results are contrasted to the actual criteria.

OPAL Reactor example

In July 2000, ANSTO signed a contract with the Argentinean company INVAP S.E. for the design, construction and commissioning of a replacement research reactor (RRR). INVAP contracted CEDIAC to prepare the Probabilistic Safety Assessment (PSA) for the RRR in support of the ANSTO application for the construction licence.

At that moment, Australia did not have a risk-based regulation, but mainly based on the same principles of iso-risk acceptance, soon the Australia Regulatory Authority (ARPANSA) developed risk-acceptance criteria [8] as indicated in Fig. 6. Within these criteria, there are two curves that are approximately iso-risk curves, indicated as "Safety Limit" and "Safety Objective". The regulation imposes that the PSA results of individual doses to the public, and their associated annual frequencies, including uncerta-inties, must be lower that the safety limits. It is desirable that these risk contributors are also lower than the safety objectives.

The PSA performed for this reactor provided results which are also indicated in Fig. 6.

The overall CDF with its uncertainty band is indicated with dotted line, and the doses were not calculated, as the frequency alone fulfilled the most stringent safety objective.

This means that those accidents that imply damage to the core are not significant contributors to the overall risk of OPAL reactor.

There are, however, some accidents that do not involve core damage (i.e. manipulation of fuel elements during shutdown, experiments, etc.) that contribute to the risk and for which its frequencies and related doses were calculated, and are indicated in Fig. 6. The overall results from the risk point of view for this reactor are very low, which indicates a high degree of redundancy and quality.

This degree was achieved via an extensive interaction between the PSA performers and the designers, as will be discussed later.

4. ADVANTAGES & DISADVANTAGES

The use of a risk-based criterion for licensing imposes an extra effort during the usually excruciating licensing process. This is the main disadvantage that has been identified. As PSA is not a common practice for all nuclear facilities, and there is a lack of experts for its performance and/or revision, this somehow complicates the whole licensing process.

On the other hand, it is extremely useful to provide the *risk point of view* in the early stages of development of a nuclear project, as this point of view is not usually present in the designers. When a PSA is required during the design phase, it is possible to retrofit the lessons learned during its performance (i.e., major risk contributors) and to modify the design, the operation schemes, the maintenance and test schemes, the human/machine interfaces, etc. As an overall result, an optimisation from the risk point of view is obtained. This means not only that the overall risk figures are low, but also that a *balanced* design can be obtained, therefore optimising the use of resources.

5. REFERENCES

1. ARN; Regulación AR10.1.1 "Norma Básica de Seguridad Radiológica", Buenos Aires, 1997.
2. CALIN, "Norma 3.1.3, Criterios Radiológicos Relativos a Accidentes", Buenos Aires, 1979.
3. ICRP, "1990 Recommendations of the International Commission on Radiological Protection", ICRP-60, Pergamon Press, 1991.
4. ICRP, "Protection from Potential Exposure: A Conceptual Framework", ICRP-64, Pergamon Press, 1993.
5. CNEA, "Preliminary Safety Analysis Report – NPP Atucha II", App B, 1981.
6. Chiossi, C, "An Application of the Argentine Criterion Curves to Research Reactors", Use of Probabilistic Safety Criteria TCM, IAEA, Vienna, 1988.
7. INVAP, "Preliminary Safety Report of the MPR Reactor", 076753203IBLI00110, Bariloche, 1994.
8. ARPANSA, "Regulatory Assessment Principles for Controlled Facilities", RB-STD-42-00 rev 1, Sydney, 2001.

Reliability, Safety and Hazard
Editors: P.V. Varde et al.

INTEGRATED RISK ASSESSMENT

Dan Serbanescu

PBMR, Gordon Hood ave.1267 Lake Buena Vista Building, Centurion PO Box 9396, 0046 South Africa

ABSTRACT

Risk analyses are now at the point of moving to new horizons and solve new challenges. The search for new methods and / or integrated ones to cope with these tasks is a high challenge. The Integrated Risk Assessment (INTGRPRA) is part of this process and one of the possible solutions to many of the identified problems

1. INTRODUCTION

The issues related to risk evaluation are becoming more and more important for an extended area of topics. However there is a series of paradoxes identified, which are generated by various myths circulating in the risk theories. They are related to the definition of the object for the risk analyses in complex systems (RACS), the use of appropriate to the model methods in order to support extending the area of application to topics like ageing, commission errors (which could be called non linear tasks) and the communication of results, connected in fact with the first two problems.

One possible approach proposed to solve the problems generated by these paradoxes and requiring a solution is to perform with priority the following (see also Appendix):

- Clarify the representation of the whole model of RACS
- Improve methods able to use the combined deterministic and probabilistic approaches
- Clarify definitions of risk for various components of the RACS (public, regulator, industry)

2. MODEL AND METHODS

Consideration of risk as being associated with the theory of complex systems will contain in itself reference to the notion of randomness and entropy. On the other side this process is also related to the need to identify the boundaries of the system evaluated, the method to be used in finding out truth statements about this system and the subject performing this knowledge quest. A representation of the complex system - the nuclear power plant for the purpose of the evaluation of their safety was refercenced in (Serbanescu, 2004). The complex system was represented as a hierarchical structured one on three levels, of which the first one is related to the generic society targets and objects. The next levels go to subsystems related to the regulatory, design and concept subsystems and to the operation and detailed evaluation. The system is represented with a feedback connection to the previous levels, assuring self-regulation. (Figure 1(a)) (See also Appendix).

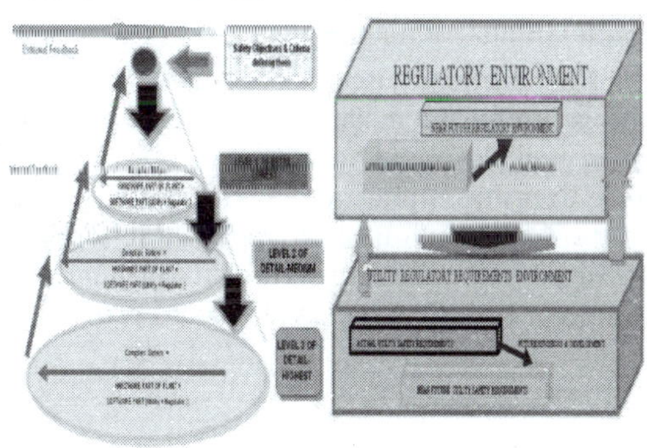

(a) (b)

Fig. 1: RACS model

Each level is defined by elements of the subsystems composing the whole. The elements of the sets at each level are connected on vertical (from level to level and on horizontal (at a given level). The vertical structure is defined by a relationship of a type "Element X is requiring element(s) Y with the degree of certainty W". Figure 1(b) shows that at each level there are uncertainties connected to the rules environment, which indicates on the high importance of the type of inferences to be used, usually grouped in a conventional manner into deterministic and probabilistic ones. The three levels with the defined relation between levels generate on the created structure a sigma algebra, for which a norm can be used. For the third level of detail, risk can be used as a norm for the sigma algebras defined as per (Serbanescu, 2004). The method is using the standard tasks for Probabilistic Risk Assessment (PRA), with the structure as per Figure 2. In case of new designs, or for special type of reactors (gas cooled ones for instance) important adjustments have to be made to PRA. In addition to the specifics of some new designs (for which a full PRA level 3 is needed due to the absence of the Core Damage) there is also the need to perform extensive sensitivity model runs, which require an integrated model of internal and external events (INTGRPRA) for all sources of radioactivity and all modes of operation (vanGraan et al, 2005)

Integration of external and internal events into one PRA model will also be done considering the fact that various tasks use various sub-models and computer codes as shown in Figure 3 for a model used for licensing purposes and in Figure 4 for a model used for the design optimization purposes.

The methods used to evaluate the risk criteria are for the models above are as per (van Graan et al, 2005 and Serbanescu, 2004).
In the licensing application the risk norm is used. For the design optimization application however the use of more refined criteria as for instance the Aggregate Risk Criteria. (ARC) could provide better resolution. ARC could be also used during the sensitivity runs for licensing applications. The use of such criteria is part of the issue of the risk metrics, which as it results from above is highly dependent of the application of INTGRPRA (Serbanescu, 2004).

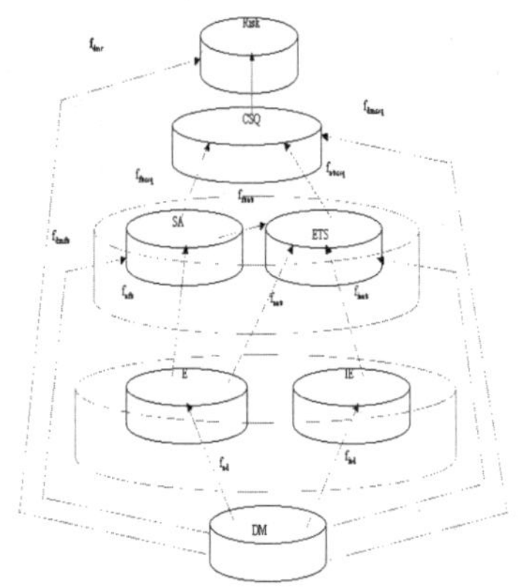

Fig. 2: PRA model to support the definition of sigma algebras

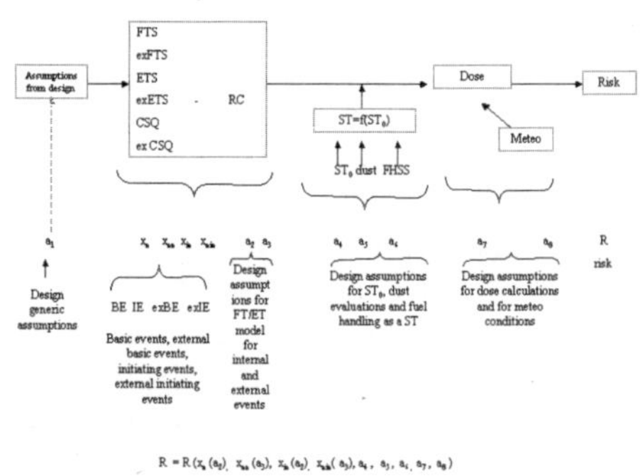

Fig. 3: Simplified flow-path of PRA tasks

The applications could also be supported to a high degree by using other methods indicating on areas of potential high risk even at initial design phase (Serbanescu and van Graan, 2005). The risk norms chosen have to comply with the boundaries as defined by licensing requirements or as targets for design optimization

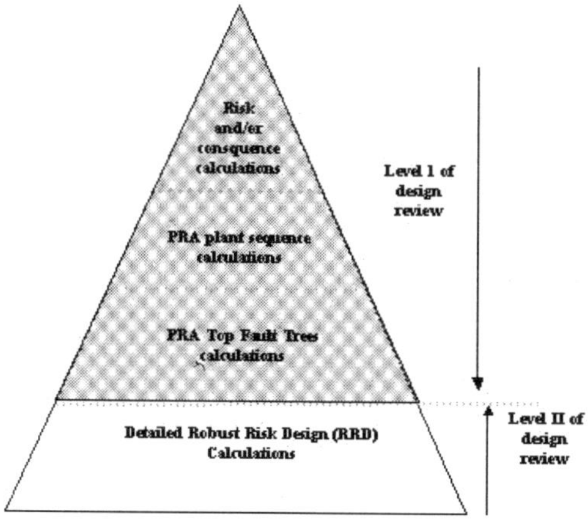

Fig. 4: Simplified PRA tasks for design optimization application

The limits are usually in the form of a Pareto type as in Figure 5 for which extensive sensitivity cases are needed to confirm the credibility of the results and their uncertainty in order to be able to use them for the decision process. INTGRPRA also allows the transformation of the non-linear tasks into linear ones, once the judgments are done in the risk space and not in the space of the frequency of events.

Fig 5: The Pareto set for a certain criteria for design adequacy and/or licensing requirements

The application of the superposition principle for the modeling of combined effects from various random variables as translated into the risk space is one possible way to be used for linearization of other PRA tasks apparently non linear (ageing effect for instance). In order to implement such an approach it would be possible to use the linearization methods applied to integrate seismic into internal PRA model. For seismic issues this was shown in (van Graan et al, 2005) A PRA model is based on the results of graph theory in building Fault Trees (FT) and Event Trees (ET), which are connected between them by a well-defined set of rules, using a grammar in which the statements have a certain degree of credibility (probability to be true).

Figure 6 shows an example of the results presented in (Serbanescu and van Graan, 2005) related to the impact and areas of use in the decision process for various types of approaches (deterministic or probabilistic). A decision process might lead to various results depending on the type of problem we want to decide on. This could be an illustration of the expected solutions for one of the well-known RACS paradoxes generated by the tendencies to use deterministic type of thinking looking for certainty in statements in randomness governed models (see also Appendix).

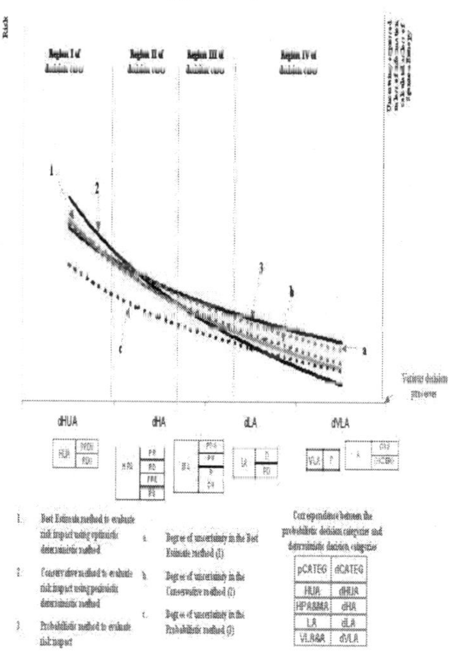

Fig. 6: Illustration of decisions using deterministic and probabilistic approaches

If the decision is aimed at evaluating high foreseen risk situations above the acceptable limits, then the deterministic pessimistic statements will lead to the most conservative decision, even if that will happen under less credibility than for the probabilistic ones. But by other reasons than technical ones, the deterministic based decisions could be expected.

If the decision is aimed at evaluating high or moderate foreseen risk situations below the acceptable limits, then there will be no difference between the very pessimistic way of thinking, an optimistic one or a probabilistic one except the fact that the probabilistic one will have more credibility, which could make it the most probable choice for the decision.

If the decision is aimed at evaluating low and very low foreseen risk situations below the acceptable limits, then it will be based on the probabilistic approach, giving the fact that it generates the most conservative results with highest credibility. Evaluation of risk impact using extensive sensitivity cases is one of the key issues to support the probabilistic type of thinking and its more extensive use in decision process. This is integrated in the verification and validation process, of which independent review and benchmarking play a very important role in confirming the truth value of probabilistic statements. INTGRPRA is very well fit for extensive and detailed reviews of this type.

3. RESULTS AND CONCLUSIONS

The INTGRPRA results are represented as in Figures 7 and 8. They will indicate the lines of isorisks for a two dimensional case and the surfaces of isorisk for the three dimensional case. The results indicate the acceptance areas and could support the identification of the main issues for further risk improvement and compliance. The results as shown in Figure 8(b) also support the need for refined risk criteria for applications like design optimization (Serbanescu and van Graan, 2005)

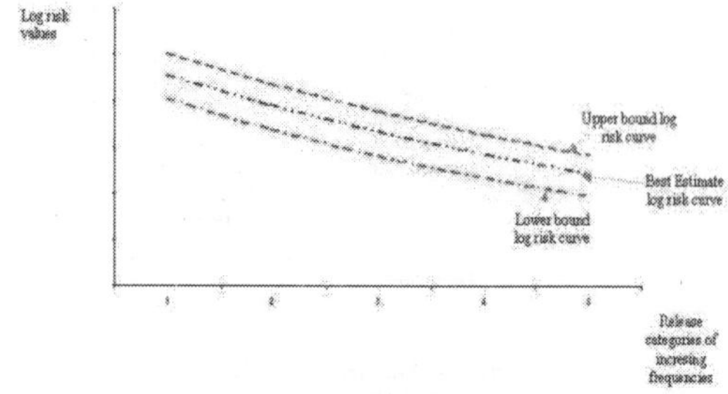

Fig. 7: Sample two dimension results of the risk curves obtained for an ITGRPRA model

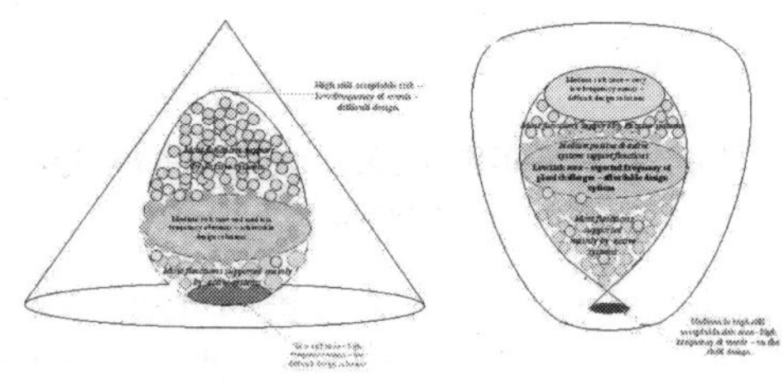

(a) (b)

Fig. 8: Sample 3D results of the risk curves obtained for an ITGRPRA model

4. APPENDIX - SOME ISSUES ON PARADOXES OF THE THEORY OF RISK IN COMPLEX SYSTEMS

One possible theory on the philosophy of science is that every science evolves in phases governed by principles as listed in Table 1.

The phases follow one after another in principle; however they could overlap, lead to jumps from one science to another

For a given phase the principles cluster around one dominant, which is driven mainly by the isomorphism between a specific set of mythologies. Each phase is characterised by deep believes creating paradoxes. The solutions of the paradoxes are looked for initially in the science environment itself, but for the crucial ones the solutions come in final analysis from other sources, basically using the isomorphism with the mythical thinking.

Table 1 Phases of science evolution, driving principles, paradoxes and believes

Phase	Main characteristic of the Phases in science evolution	Main governing principle of a given phase	Main features of the interface of governing principle, paradoxes and believes
1	Define the need	P1-An unique source	An isomorphism exists between the mythical thinking in general and the scientific thinking acting to solve paradoxes
2	Clarify the differences	P2- Basic science features as a dual approach	Instinctive impulse exists to bring the issues to a duality in defining the truth, applicable criteria and methods for its evaluation and value of results
3	Develop main tools	P3-Creation of new in science governed by a triple way approach	Creating the new in science has a triple facets in order to try to overpass the dual approach of the basics and look for the third option, starting from paradoxes
4	Check the stability of the theories	P4-Durability of a theory shown as robustness to paradoxes	Continue quest from phase 3 and check the science robustness on the embedded in it paradoxes
5	Identify main believes preventing evolution	P5-Each historical stage is an engine powered by intuition and believes	Identify the believes behind the paradoxes identified in previous phases
6	Refine and increase usefulness	P6-For any theory has to be useful	Clarify and refine the value of the acquired set of knowledge
7	Attempt to merge tools	P7-God is not playing dice- Absolute and relative knowledge	Support the need for creating the new with merged tools in the attempt to make a synthesis of those ones appearing as fundamentally opposed
8	Manage the built up	P8-Hierarchy and science theories structures involve the need for management	Manage the gained corpus of knowledge in consolidation of the tools and results obtained so far.
9	Attempt to solve the unsolved yet issues and expand the object	P9- Irrepressible need to reach perfection as a cause for restarting the whole cycle	Start systematic challenge of the existing paradoxes in a final attempt to eliminate them from the built up achieved and clean the corpus of knowledge.

The solution is found in a process of unspoken use of a "truth switch" applied as part of a coincident search for paradoxes and believes behind them as a set of solutions to the paradoxes, which blocked the situation. (Figure 9). Some results of application of this interpretation to the evaluation of risk of complex systems are as follows:

- The safety system representation with a feedback connection to the previous levels, assuring self-regulation., which is a 3 + 1 structure (where the 1 is denoting the real object and the feedback to it from the model built to represent that object). It is in accordance with mythical understanding of solving similar type of paradoxes. This model is actually used for all the risk regulation purposes at national and international levels and it is also supported by results from decision and game theory, as the best adopted so far solution for the paradoxes of defining the risk object and the RACS model.
- The belief in the deterministic way of thinking and its use in the method of this science leads to conclusions about risk and its use in applications, which cannot be inferred using deterministic tools without acknowledging the limits of this approach,

since they are profoundly probabilistic ones. The semantic of the method, as in every myth has to reflect the model and the real object, so to distort it less while studying it, by applying correctly the rules to the created structure and understanding its limits. The method is built using a semantic aimed to solve problems in the increasing order of complexity

- The moment of understanding the RACS results is an individual one and subject to holistic approaches, coming for instance in illuminations even for the most rational statements, even if the recognition comes in the societal acceptance / myths built around this topic

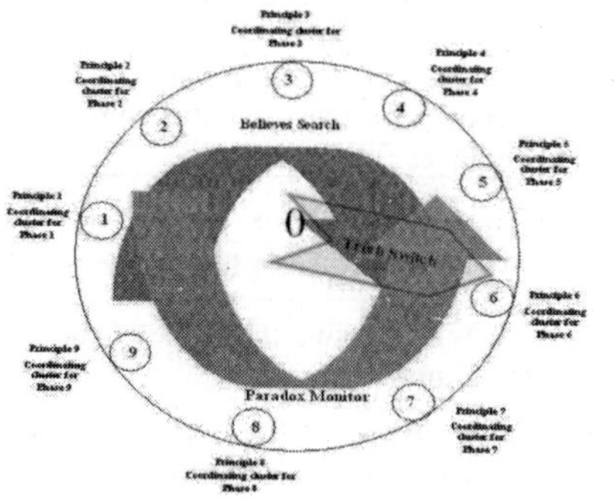

Fig. 9: The Paradox and Believes switches

5. REFERENCES

1. Van Graan, H., Serbanescu D, Combrink Y.,2005, *A Methodology for the Integrated development of Internal and External Events in the Probabilistic Risk Analysis for PBMR Plant,* ESREL 2005, Gdansk 2005
2. Serbanescu D. 2004, *Some Insights on Issues related to Specifics of the use of Probability, Risk, Uncertainty and Logic in PRA Studies,* IJCI, 1,2, 2004
3. Serbanescu D., van Graan H, 2005, *"Some specifics on the integration of the sensitivity and uncertainty analyses in pra for a gas reactor in the process of demonstration of the plant safety margins",* IAEA TM on Safety Margins, Pisa, 2005

6. BIBLIOGRAPHY

4. Kuhn, T.S, 1962, *The structure of scientific revolutions*, University of Chicago, 1962.
5. Dubuisson, D., 1993, *Mythologies du XX-e siecle* Presses Universitaire du Septentrion, 1993.
6. Ferdinand de Sausure, 1965,*"Course in General Linguistics"* in Adams and Searle, ed., Critical Theory Science, 1965.
7. René Descartes, 1966, *Discours de la méthode,* Paris : Garnier - Flammarion, 1966
8. Guenon, R.,1985, *La Methaphysique Orientale, Editions Traditionelles,* Paris, 1985.
9. Rossi, P.2000, *La nascita della scienza moderna in Europa*, Gius, Laterza&Figli, 2000.

Reliability, Safety and Hazard
Editors: P.V. Varde et al.

DECISION-MAKING IN INTEGRATED ASSET MANAGEMENT FOR SAFETY OPTIMISATION

Rudolph Frederick Stapelberg

The Cooperative Research Centre for Integrated Engineering Asset Management, Australia

ABSTRACT

Integrated Asset Management (IAM) involves integrated processes of managing physical assets and requires a certain level of Knowledge Management (KM) insight and decision-making from diverse organisational disciplines. IAM is a systematic, structured process covering the whole life of an asset whereby the underlying assumption is that an organisation's assets exist to support the organisation's delivery strategies. The principal KM-based decision-making in IAM is thus to achieve the best possible match of assets within an organisation's delivery strategies. This could however, pose a *risk* on *reliability* and *safety* in the implementation of such delivery strategies.

Keywords: Asset management, Knowledge Management, Decision-making, Delivery strategies, Risk, Safety

1. INTRODUCTION

The principles of Integrated Asset Management (IAM), representing current thinking with asset owners and asset management organisations, can be listed as:

- An effective integrated framework can be established for asset management;
- Asset management decisions are integrated with strategic and tactical planning;
- Asset planning decisions are based on the 'life-cycle' and integrity of assets;
- Asset ownership/stewardship bears risks that can be assessed and evaluated;
- The appropriate tactical planning considers asset condition and safety of use;
- Financial accountability is established for asset performance and residual life.

These principles are not definitive but are consistent with trends in improving decision-making in Integrated Asset Management in the public and private sectors. Each principle reflects a fundamental notion of Knowledge Management-based decision-making for an organization's delivery strategies. Within an IAM context, a practical measure for the outcome of such strategies would be the extent of continual improvement in reliability and safety in the management of risk-based assets. As organizations are being driven toward KM-based decision-making to meet organizational delivery strategies and create value, they are increasingly finding that the structured use of existing knowledge and the creation of new knowledge allows for better decision-making. Furthermore, Knowledge Management (KM) technology is ideally suited for organisational focus groups such as Asset Management Groups (AMG's). The challenge of research in this field therefore, is to determine through a series of surveys that target public and private sector asset owners as well as asset management organisations, whether the principles are applied in an *integrated framework*, and to exercise judgement in their *integrated application* from the perspective of *knowledge-based decision-making*. This is done through individual interviews with selected personnel and through *action research* participation in various targeted Asset Management Groups that are directed to achieving reliability and safety in the implementation of organisational delivery strategies. Action research in KM-based decision-making is an iterative process involving researchers and practitioners acting together on a particular cycle of activities, problem diagnosis, action intervention, and reflective learning (Avison *et al* 1999).

Action research represents an ideal research method as it can address complex real-life problems and the immediate concerns of practitioners. Yet, paradoxically, the academic community has almost totally ignored action research. Only one of the 155 articles in a survey of 19 journals in 1991 describing research methods used in information systems covered action research (Lau, F. A. 1997).

2. KNOWLEDGE MANAGEMENT IN INTEGRATED ASSET MANAGEMENT

Knowledge Management in IAM is concerned with the exploitation and development of knowledge for informed decision-making with a view to furthering the organization's objectives. The knowledge used for decision-making includes explicit, documented knowledge, and tacit, subjective knowledge. Organizations that successfully apply KM-based decision-making are likely to view knowledge as an asset and to develop organisational norms and values which support the creation, and sharing of knowledge (Rowley, 2000).

Historically, KM has been aimed at managers through what has been generally referred to as an *executive information system* aimed at supporting managerial decision-making. More recently, however, KM systems are increasingly designed for entire organizations. If executives need access to information and knowledge, their employees are also likely to have an interest in and need for that information (O'Leary and Watkins, 1992).

Several frameworks have been developed to understand knowledge management phenomena. These frameworks can be broadly classified into two categories: descriptive frameworks and prescriptive frameworks. The descriptive frameworks attempt to characterize the nature of KM phenomena, whereas prescriptive frameworks prescribe methodologies to follow in the knowledge management process. Each framework contributes to an understanding of KM phenomena. A comparison of these frameworks by Holsapple and Joshi (1999) yields several notable observations: First, the dimension of knowledge resources has received comparatively little attention. Second, there is no common or standard method of characterizing knowledge management activities. Third, there is no common or standard method of characterizing influences on the conduct of knowledge management. Fourth, there is no overall KM framework that subsumes the others (Holsapple and Joshi 1999).

Choo's (1996) KM framework identifies knowledge management activities in which an organization uses information strategically for sense-making, knowledge creation, and decision-making. The model takes a view of decision-making as a process concerned with understanding how an organization processes information to resolve task uncertainty for the delivery of its strategies (Choo, 1996).

Nonaka's (1994) KM framework focuses on knowledge creation that happens through human interaction and knowledge conversions and incorporates knowledge modes (i.e., tacit vs. explicit) with respect to human knowledge resources. The model identifies four kinds of knowledge conversion that drive knowledge creation: socialization, externalization, internalization, and combination. These conversions are based on a dichotomy between tacit versus explicit modes of knowledge. Tacit knowledge refers to knowledge that cannot be easily verbalized and articulated, whereas explicit knowledge refers to knowledge that can be readily verbalized in a formal, systematic language. Organizational knowledge is created by the interactions among these four conversion processes, and through transferal of tacit/explicit knowledge from individual to group to organizational levels (Nonaka, 1994).

Szulanski's (1996) framework focuses on impediments to transferring best practices (i.e., knowledge) within an organization and is designed for the purpose of exploring knowledge transfer, and thus to the activity of transferring best practice. The framework identifies four stages involved in knowledge transfer: initiation, implementation, ramp-up, and integration (Szulanski, 1996).

The KM framework of van der Spek and Spijkervet (1997) focuses on a cycle of stages that governs knowledge management activities in an organization. The framework identifies a cycle of four knowledge management stages: conceptualize, reflect, act, and retrospect, that are oriented toward problem-solving and can be viewed as coordinating knowledge manipulation activities within a problem-solving environment (Van der Spek and Spijkervet, 1997).

Wiig's (1993) KM framework focuses on management issues that affect the conduct of KM in an organization, and involves three KM pillars which are based on a broad understanding of knowledge creation, manifestation, use, and transfer (Wiig, 1993).

3. DECISION THEORY IN INTEGRATED ASSET MANAGEMENT

As the principal management decision objective of IAM is to achieve the best possible match of assets within an organization's delivery strategies, a fundamental theory related to the topic of IAM is that of *Decision Analysis*. Decision Analysis refers to the broad quantitative field, overlapping *operations research* and *statistics* that deals with modelling, optimizing and analyzing decisions made by individuals, groups and organizations. The purpose of Decision Analysis is to assist decision-makers in making better decisions in complex situations, usually under uncertainty, typically such as the required *reliability and safety* of risk-based assets. The significance of decision-making for reliability and safety is measured by the *objectives and consequences* of the decision, and *adherence to statutory compliances* of the resulting action.

Decision Theory has applicability only when (a), there is more than one choice of action possible, and (b), it is uncertain which course of action will provide more gains in terms of the desired objectives. The decision may be to make no change or take no action, a form of null response, but it still should be considered a positive action from a decision theoretic perspective. What action to take, will depend upon the decision-maker's expectations or *preferences*.

The decision analytic framework helps the decision-maker think systematically about objectives and preferences, and the uncertainty and consequences of a problematic situation within the confines of risk; then model quantitatively these and other important aspects of the problem and their interrelationships to either minimize losses or to maximize gains such as reliability and safety in the management of risk-based assets. A schematic decision analytic framework is proposed by the author in Figure 1 which represents a two-tier KM-based decision-making regime. The figure depicts three different types of knowledge flow, namely: 1) Principal two-way knowledge creation and knowledge transfer flows between KM-based decision-making elements and knowledge verification and validation (dark double arrows); 2) Secondary two-way knowledge flows between KM-based decision-making elements and knowledge verification and validation (light double arrows); and 3) Unidirectional and sequential knowledge manipulation flows between initial decisions, outcomes, response decisions, and consequences of the decision-making process (light single arrows).

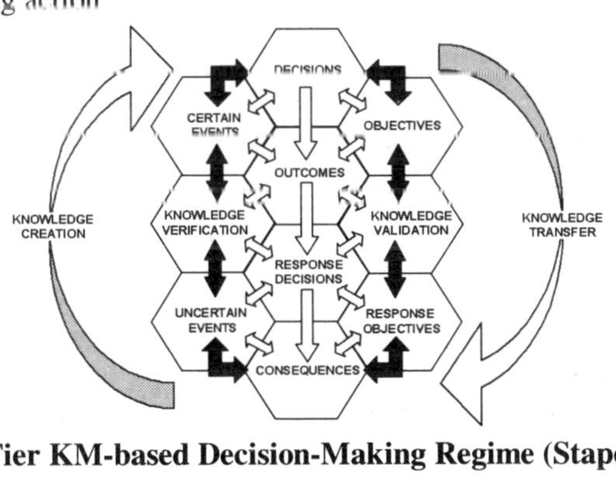

Fig. 1: Two-Tier KM-based Decision-Making Regime (Stapelberg, 2005)

4. MAIN ELEMENTS OF AN INTEGRATED KM-BASED DECISION MODEL

Complex decision problems in IAM normally follow a two-tier KM-based decision-making regime as indicated in Figure 1, and include some or all of the following elements:

- *Decisions* to achieve the best possible *match* of assets for an organisation's delivery strategies;
- *Objectives* which determine how the decision-makers' *preferences* impact upon certain events;
- *Certain Events* which occur through *implementation* of an organisation's delivery strategies;
- *Outcomes* of certain events that determine the *state* of an organisation's delivery strategies;
- *Response Decisions* in the management of *integrity* of an organisation's delivery strategies;
- *Response Objectives* which determine the *effect* of response decisions upon uncertain events;
- *Uncertain events* occurring from response decisions to an organisation's delivery strategies;
- *Consequences* of uncertain events that determine *risk* of an organisation's delivery strategies;

Decisions

A *decision*, ***Di, i* = 1, 2,...** , refers to a point in time when the decision-maker has to choose one *alternative*, ***di***, out of a domain of available alternatives, that could be discrete or continuous, simple or combined. An *alternative* can be a combination of values relating to a number of *decision variables* for the best possible match of assets for an organisation's delivery strategies.

Objectives

An *objective*, ***Xj, j* = 1, 2,...** , in solving a decision problem relates to the decision *verification* process of a decision-maker's *preference*, ***xj***, in achieving an organisation's delivery strategies.

Certain events

Certain events, ***Cj, j* = 1, 2,...** , typically occur interspaced between subsequent decisions. If a certain event realizes before a decision is made, through implementation of an organisation's delivery strategies, its *occurrence*, ***cj***, will typically be observed by the decision-maker before that decision is made.

Outcomes

The *outcomes*, ***Ok, k* = 1, 2,...** , of certain events are the results of the alternatives chosen that determine the *state*, ***ok***, an organisation's delivery strategies.

Response Decisions

A *response decision*, ***Ri, i* = 1, 2,...** , refers to a point in time when the decision-maker has to choose one *response alternative*, ***ri***, out of a domain of available response alternatives. A response alternative can be a combination of values relating to a number of *response decision variables* to ensure *integrity* of an organisation's delivery strategies, based on certain outcomes.

Response Objectives

A *response objective*, ***Yj, j* = 1, 2,...** , in solving a response decision problem relates to the decision *validation* of the *effect*, ***yi***, in achieving an organisation's delivery strategies.

Uncertain events

Uncertain events, ***Uj, j* = 1, 2,...** , typically occur interspaced between subsequent response decisions. If an uncertainty realizes before a response decision is made, its *influence*, ***uj***, will typically be observed by the decision-maker before that response decision is made. Together with the decisions already made, such uncertainties determine the *consequences* of the event.

Consequences

The *consequences*, ***Zk, k* = 1, 2,...** , of uncertain events are the results of the response alternatives chosen that determine the *risk*, ***zk***, of an organisation's delivery strategies.

Development of an Integrated Decision Model

From the main elements of an Integrated KM-based Decision Model, the following algorithms can be developed to evaluate the best *action* to take for an organisation's delivery strategies:

Let the *initial decision* criteria be designated by **A**, and the *response decision* criteria by **B**.

$A \equiv \{Di, i = 1, 2,... ,\} : Ok\ (Xj, Cj), j, k = 1, 2,...,$
$B \equiv \{Ri, i = 1, 2,... ,\} : Zk\ (Yj, Uj), j, k = 1, 2,...,$

Each outcome, ***Ok*** is a function of a decision objective ***Xj***, and certain event **Cj**, ***i*** **= 1, 2,... ,** and consequence, ***Zk*** is a function of a response objective ***Yj***, and uncertain event **Uj**, ***i*** **= 1, 2,... .**

The most appropriate strategic and/or tactical planning action to take will depend on these decision criteria, which might all be subsumed into a single parameter, **θ**, where **θ = A . B**, such that they have a density function:

$$f\,(i; \theta) \qquad (1)$$

If **θ** and the form of the function (e.g. **Γ** or the log-normal) were known, the density function would be known and therefore the best *action* to take would also be known.

5. RISK IN INTEGRATED ASSET MANAGEMENT DECISION-MAKING

The concept of *risk* has appeared frequently in management literature as a means for gauging the adoption of a precautionary approach to management of an organisation's delivery strategies. Though terminology varies, there are basically two stages in analyzing risk in KM-based decision-making. The first stage, *risk assessment*, is the formulation of *initial decision* criteria in a way that conveys possible *outcomes* of *certain events*. Such criteria provide input to knowledge creation that enables determination of the *preferences* of alternative decision options. The second stage, *risk management*, is the formulation of *response decision* criteria in a way that conveys possible *consequences* of *uncertainty*. Such criteria receive input from knowledge transfer that enables identification of the *effects* of alternative response decision options. This stage describes the way decision-makers take *uncertainty* into account in their decision-making. In this context, uncertainty, effect, and risk are concepts that go together - where risk is the quantitative aspect of effect and uncertainty. Even among technocrats the word *risk* is commonly understood to mean that something *detrimental could happen* (i.e. effect, uncertainty). In this sense, the term implies:
i) a *loss* is incurred when the event does happen;
ii) there is a possibility of the event happening, i.e. it has a finite, if unknown, event probability.

Loss in Integrated Asset Management Decision-Making

An important element in determining the best action to take in a situation of uncertain risk is the concept of *loss*. The *consequence* of a given action is validated through a *loss function*:

$$l\,(i; \theta) \qquad (2)$$

This represents the loss in taking action where **θ** represents the decision criteria, or state space. The best action for **θ** implies zero or negative loss. As indicated before, all that might be known is that **θ** has a particular distribution and not its parametric value. One form of a very robust loss function is the *squared error loss function*:

$$L\,(i; \theta) = f\,(\theta i - \theta)^2,\ i = 1, 2,... , \qquad (3)$$

This function is used in fitting a least-squares regression model. Unlike the parameter fitting situation, losses arising from decisions about the value of **θ**, which can be taken to represent the state equation, will not be symmetrical about the parametric mean.

Utility in Integrated Asset Management Decision-Making

Decision Theory has an explicit method of dealing with situations where stakeholders, who may agree completely on the facts of the situation (**A**, **B**, **θ** or **L**), may insist upon different *actions*. This is done through the concept of *utility* which measures the value of a particular delivery strategy.

The concepts of *risk* and *utility* are linked, and their relation depends on the attitude to risk of the decision-maker. For increasing utility of some delivery strategy, the risk declines for the risk-averse decision-maker, while the reverse if true for the risk-prone decision-maker.

A linear relation holds for the risk-neutral decision-maker. Decision-makers may not be consistent in their attitude to risk as an organization's delivery strategies change. The risk tolerance of the decision-maker may also depend on the stake that is involved. This is especially so where risk relates to *reliability* and *safety* in the implementation of such delivery strategies.

6. CRITICAL EVALUATION OF ESTABLISHED DECISION THEORY

Comprehensive test: Decision Theory is usually in the form of an algorithm representing a mathematical dynamic programming solution that evaluates all combinations of decisions, each with uncertain consequences that will yield a value for the lowest risk of an organization's delivery strategies.

The constraints placed upon these decisions may include capital availability and financial objectives or restrictions such as earnings or cash flows, production, asset replacement or acquisition, but most common are regulated requirements for *reliability* and *safety* in the implementation of such delivery strategies.

Parsimonious test: The Decision Analysis method is conceptually simple but computationally complex. However, it is preferred by some analysts over deterministic evaluations because it yields more realistic values, utilizes available information more thoroughly, and quantifies risk. Expansion of the use of Decision Analysis is limited, however, by several factors. Among them is generating distributions for decision criteria, $\boldsymbol{f}\ (\mathbf{i};\ \boldsymbol{\theta})$, and loss functions, $\boldsymbol{L}\ (\mathbf{i};\ \boldsymbol{\theta})$.

Substituting expected values for deterministic values ignores the distribution of risk and the diversification effects and interdependencies among the chosen alternatives. Decision-makers in general, simply have not understood the use of these distributions and related risk statistics.

Clearly defined constructs: In the Decision Theory concept of utility, the locus of the minimum risk points in a utility-versus-risk plot, called here the *decision frontier*, is the set of all strategies that satisfies the criteria that, given certain constraints, no increase in value is possible without greater risk, and no reduction in risk is possible without loss in value – the definition of rational choices in portfolio theory (Markowitz, 1952). Any point *off* this frontier has alternatives that are higher value, lower risk, or both, under the same constraints.

Internally consistent test: Decision Analysis helps the decision-maker think systematically about alternatives, preferences, and outcomes, and the structure of response decisions, uncertainty and consequences of a decision-making problem as well as the important concepts of *risk*, *utility* and *loss* and their interrelationships. The elements of the algorithms fit sensibly together to determine a value at each consequence or risk level to the extent that Genetic Algorithms can be employed to approximate this optimization calculation. The results are the same.

Testable hypotheses: The process can be repeated for all other specific target value predictions, each time finding a specific decision frontier with the lowest risk under specific constraints.

Consistent with well known data: In developing Decision Analysis, David Hertz (1964) extended Markowitz's (1952) portfolio theory concepts from investments in financial assets to investments in physical assets marked by uncertainty. Newendorp (1975) recognized Hertz's work and speculated about a two-dimensional display (expected value and expected loss) to illuminate *decision-making* selections, but did not propose an optimization approach. Ball (1983) applied Hertz's insights specifically to the best possible match of industrial assets within an organization's delivery strategies. This idea was later advanced by Hightower and David (1991) and Edwards and Hewitt (1995) and updated by Ball *et al.* (1999), among others.

7. DECISION-MAKING FOR SAFETY OPTIMISATION

An extensive and systematic examination of *reliability* and *safety* aspects has to be carried out carefully and at the earliest possible opportunity in the management of risk-based assets. To meet these essential demands, a thorough Safety and Hazards Analysis is compulsory during the engineering design and development stages of such assets, for official approval to commence with their construction, installation and operation. Steps in this analysis are achieved by a variety of methods such as What-If Analyses and Safety Checklists, usually incorporated in a more formal **Hazard and Operability Study (HazOp)** of integrated risk-based assets. The

HazOp is usually conducted in the method of *action research*, using focus groups.

Essentially the **HazOp** procedure is a KM-based decision-making procedure that involves establishing a knowledge-based description of an integrated risk-based asset and systematically analyzing every system to decide on how deviations from the asset's design intent can arise. Once identified, a *risk analysis* is conducted as to whether such deviations and their *consequences* can have a negative effect upon the safe and efficient operation of the asset. If considered necessary, *action* is then taken to remedy the situation. An essential feature in this approach of Decision Analysis is the use of keywords to focus attention on deviations and their possible causes. Primary keywords reflect both the original design intent and operational aspects of the system being studied. Typical process oriented words are very similar to the process parameters as the words employed will generally depend upon the process being studied, whether at systems level, or at a more detailed component level. However, the technique is Hazard *and* Operability studies; thus, added to the primary keywords of; No; More; Less; As Well As; Reverse; Other Than; might be relevant operational words such as isolate, drain, vent, purge, open, inspect, clean, maintain, start-up and shutdown. In simple terms, the **HazOp** study involves applying in a systematic way all relevant keyword combinations to the system in question in an effort to uncover potential problems. The results are recorded in columns under the following headings: *Node, Attributes, Deviation, Cause, Consequence, Safeguards, Decision, Action.*

The *deviation* would be described using keyword combinations. Potential *causes* are identified that would result in the deviation occurring. The *consequences* would arise both from the *effect* of the deviation and, if appropriate, from the cause itself. The recording of consequences should be explicit. An important point to note, particularly from a decision-making point of view, is that when assessing consequences, credit for protective systems or instruments which are already included in the design should *not* be considered. However, any existing protective devices that either prevent the cause or safeguard against the adverse consequences must be recorded. For example: "Local pressure gauge in discharge from pump should indicate arising problem." might be reported. Safeguards need not be restricted to hardware, but, where appropriate, credit can be taken for procedural aspects such as the use of a standard work instruction (SWI) and job safety instruction (JSI). Where a credible *cause* results in a negative *consequence*, it must be decided whether appropriate *action* should be taken. It is at this stage that consequences and associated safeguards are considered. If it is deemed that the protective measures are adequate, then no action need be taken, and words to that effect are recorded in the Decision column. Actions fall into two groups: actions that remove the cause; and actions that mitigate or eliminate the consequences. Whereas the former is to be preferred, it is not always possible, especially in dealing with equipment malfunction. However, removing the cause first should always take preference, and only where necessary, the consequences mitigated. Having gone through the steps involved in recording a single deviation, the technique can now be put into the context of a qualitative hazard and operability KM-based Decision Model. Such a model is feasible as the **HazOp** method is an iterative process, applying in a structured and systematic way the relevant knowledge for decision-making in identifying potential safety hazard problems.

8. CONCLUSIONS

Integrated Asset Management (IAM) involves integrated processes of managing physical assets and requires a certain level of Knowledge Management (KM) insight and decision-making from diverse organisational disciplines. Knowledge Management in IAM is concerned with the exploitation and development of knowledge for informed decision-making with a view to furthering the organisation's objectives. The principal of KM-based decision-making in IAM is to achieve the best possible match of assets within an organisation's delivery strategies, to minimise *risk*, and to optimise *reliability* and *safety* in the implementation of such delivery strategies. Action research as an iterative process involving researchers and practitioners acting together on a particular cycle of activities, including problem diagnosis and action intervention, represents an ideal research method in KM-based decision-making. In this context, Decision Theory provides a useful framework to explore alternatives. It allows the decision-maker to recognize that deciding *not* to take action is just as much a decision as deciding which action to take, and that a judgment error may be made either by taking an unnecessary action or by failing to take a necessary action. Decision Theory provides a method of assessing which parts of decision-making have a particular impact on the outcome of preferred decisions, and of determining how robust a selected course of action is, based on potential consequences. It also provides a construct for the development of a KM-based decision-making model for use in Hazard and Operability Studies, **HazOp** of complex integrated risk-based assets for identifying potential hazard problems and optimising safety.

9. REFERENCES

1. Avison, D., Lau, F., Myers, M., and Nielsen, P.A. (1999) "Action Research", Communications of the ACM, Vol. 42, No. 1, Jan. 1999.
2. Ball, B C. (1983) "Managing Risk in the Real World", European Journal of Operational Research, Vol. 14, 248-261, 1983.
3. Ball, B C. and Savage, S L. (1999) "Notes on Production Portfolio Optimization", Operations Research Letters, 1999.
4. Choo, C. (1996) "An Integrated Information Model of the Organization: The Knowing Organization". http://www.fis.utoronto.ca/people/faculty/choo/
5. Edwards, R A. and Hewitt, T A. (1995) "Applying Financial Portfolio Theory to the Analysis of Producing Properties", SPE Paper 30669, 70th Annual Technical Conference and Exhibition, Dallas, TX, 22-25, October 1995.
6. Hertz, David B. (1964) "Risk Analysis in Capital Investment", Harvard Business Review, 42: 9510, January/February 1964).
7. Hightower, M L and David, A. (1991) "Portfolio Modeling: A Technique for Sophisticated Investors", SPE Paper 22016, Economics and Evaluation Symposium, Dallas TX, April 1991.
8. Holsapple, C. W. and Joshi, K. D. (1999) "Description and Analysis of Existing Knowledge Management Frameworks", 1999 IEEE Proceedings of the 32nd Hawaii International Conference on System Sciences.
9. Lau, F. A. (1997) "Review of action research in information systems studies". *Information Systems and Qualitative Research*, A. Lee, J. Liebenau, and J. DeGross, Eds. Chapman & Hall, London, U.K., 1997, pp. 31–68.
10. Markowitz, H. M. (1953) "Portfolio selection", The Journal of Finance, March 1952.
11. Newendorp, Paul D. (1975) "Decision Analysis for Petroleum Exploration", Petroleum Publishing Company, Tulsa, 1975.
12. Nonaka, I. A. (1994) "Dynamic Theory of Organizational Knowledge Creation". Organization Science, 5(1), 14-37, Feb. 1994.
13. O'Leary, D. and P. Watkins (1992) "Integration of Intelligent Systems and Conventional Systems". Intelligent Journal: Intelligent Systems in Accounting, Finance, and Management, Vol. 1, No. 2, 1992, pp. 135-145.
14. Rowley, J. (2000) "From learning organization to knowledge entrepreneur". Journal of Knowledge Management, **4**, 1, 2000, 7-14.
15. Szulanski, G. (1996) "Exploring Internal Stickiness: Impediments to the Transfer of Best Practice Within the Firm". Strategic Management Journal, 17, 27-43, 1996.
16. Van der Spek R. and Spijkervet A. (1997) "Knowledge Management: Dealing Intelligently with Knowledge". Knowledge Management And Its Integrative Elements, eds. Liebowitz & Wilcox, New York: CRC Press, 1997.
17. Wiig, K. (1993) "Knowledge Management Foundations". Arlington: Schema Press, 1993.

Reliability, Safety and Hazard
Editors: P.V. Varde et al.

PLANT SPECIFIC RISK INFORMED DECISION MAKING - A VISION FOR INDIAN PHWR

Rajee Guptan, Nalini Mohan, S.G. Ghadge and S.S. Bajaj

Directorate of Safety, Nuclear Power Corporation of India Limited,
Nabhikiya Urja Bhavan, Anushakti Nagar, Mumbai 400 094

ABSTRACT

The objective of Nuclear Power Plant (NPP) safety is to ensure and demonstrate that the risk from NPP to public and plant personnel is acceptably low. As a supplement to the deterministic approach, use of probabilistic techniques has been gaining grounds. Probabilistic Safety Assessment (PSA) is a popular international practice to calculate plant risk to the public, environment and plant personnel, in the event of an accident taking place in the plant.

The risk posed by Nuclear Reactors to the Public at large is a very important issue in the public acceptance of a Nuclear Power Programme in any country. Risk Definition, generally accepted at present as the measure of possibility for an accidental event and severity of its effects to occur, simultaneously comprises both the probability (frequency) of potential damage occurrence and extent of this damage.

NPCIL's vision is to provide safe, reliable and cost effective energy to the nation by ensuring safe operation of the plants, limiting radiation exposure to plant personnel and public within prescribed limits.

To achieve this vision our mission is to enhance the technical competency, resources and the awareness towards improved safety culture specific to Probabilistic Safety Assessment for applying PSA Studies to Risk Informed Decision making. This mission translates into the goal of completion of full scope PSA, with the co-operation and synergetic efforts of the Head quarters and Operating Station experts. NPCIL will then be self sufficient in applying Probabilistic Studies in Risk Informed Decision making to minimize risk and back fitting design changes to optimize the existing design and operating practices, thus providing for safe, reliable and cost effective energy to the nation.

Keywords: Risk Informed Decision, Probabilistic Safety Assessment, Operating Practices

1. INTRODUCTION

The objective of Nuclear Power Plant (NPP) safety is to ensure and demonstrate that the risk from NPP to public and plant personnel is acceptably low. The safety philosophy is based on deterministic approach, which includes defense in depth principle, safety design criteria and the concept of design basis accidents. As a supplement to this approach, use of probabilistic techniques has been gaining grounds.

Probabilistic Safety Assessment (PSA) is a popular international practice to calculate plant risk to the public, environment and plant personnel, in the event of an accident taking place in the plant. It guides design and operation by identifying potential accident sequences that contribute to risk. Measures can then be taken to reduce this contribution. It provides inputs to decisions on design, design back fitting, plant operations, emergency operating procedures, safety analysis and on regulatory issues. PSA's are useful for design optimization and design verification when carried out at the project stage of the plant. PSA is considered to complement the deterministic analysis for Design Basis Events (DBE's) and for Beyond Design Basis Accidents (BDBA's). It focuses attention on the low probability and high consequence events along with high probability and low consequence events. Today the results of a PSA study are used for a wide variety of applications such as Technical Specifications (TS) optimization, Reliability Centered maintenance (RCM), Risk-Based In-service Inspection (RB-ISI), Living PSA (LPSA) and Risk Monitor (RM).

Historically from the early 70's reliability analysis of the safety systems and the safety support systems were a pre-requisites for the first criticality of the NPP's and NPCIL used to carry out these analyses in co-ordination with BARC for its various stations. In 1997, NPCIL felt the need to develop an in-house expertise in PSA activities. In this regard, technical experts from various design, analysis and operating stations were pooled together to form a PSA team. This resulted in a PSA activity of International Standards for a Level-1 PSA for Kakrapara Atomic Power Station (KAPS-PHWR) and Tarapur Atomic Power Station (TAPS-BWR) at NPCIL. KAPS and TAPS-BWR PSA's had been successfully completed in April 2002. Review by the Atomic Energy Regulatory Board (AERB) of the TAPS-BWR PSA has been completed and that for KAPS is in the final stages. Subsequent to the Level-1 analysis, substantial progress has been made in using the results of the analysis in various applications (e.g. Technical Specification Optimization, Surveillance Test Frequency optimization, Allowed Outage Time of Components etc.) for the Operating Stations. The use of PSA as a support to Plant Life extension has been used for the TAPS BWR design. Similar analysis for the Plant Life extension is being carried out for RAPS-2 (PHWR). To add to the above accomplishments a Level-2 PSA of KAPS-PHWR is under preparation, in its final stages.

For the first time in the history of NPP's in India, a comprehensive Level-1 PSA at project stage is being conducted for TAPP-3, 4 to fulfill the objective of design optimization and design verification. Work has been initiated for Shut Down PSA, Low Power PSA and External Event PSA.

Simultaneously, from 1999 onwards all the sites were technically trained in a structured manner for carrying out Reliability Analysis (RA)/PSA. Class Room and hands on training on Risk Spectrum Software package was imparted to plant personnel to perform the Reliability analysis of safety systems and safety support systems as required for Application for Renewal of Authorization (ARA). Efforts were also made to develop awareness amongst plant personnel about the collection of Plant Specific Component failure data with a vision to generate an Indian PHWR component failure data bank, which can be prudently used for all future studies.

2. RISK DEFINED

Risk Definition, generally accepted at present as the measure of possibility for an accidental event and severity of its effects to occur, simultaneously comprises both the probability (frequency) of potential damage occurrence and extent of this damage.

The risk posed by Nuclear Reactors to the Public at large is a very important issue in the public acceptance of a Nuclear Power Programme in any country. Ours is no exception. However unlike any conventional power sources technology the method of quantification for defining risk for a Nuclear power reactor is itself couched in terse terms. We attempt here to de-mystify these terms.

Risk is the combined answers to (1) what can go wrong? (2) How likely is it? And what are the consequences?

Internationally three levels of PSA and the terms for Risk are defined as follows:

Level 1 PSA

The Level - 1 PSA concerns the probabilistic analysis of internal as well as external events. Internal events are events due to random type failures of systems and equipments whereas in external events fire, flood, seismic factors are studied leading to damage to the core, which is considered as the main source of radioactivity.

Core Damage Frequency (CDF): An expression of the likelihood that, given the way a reactor is designed and operated, an accident could cause the fuel in the reactor to be damaged. (Ref: USNRC RG-1.174).

In a level-1 PSA system analysis techniques are used to define accident sequences and to arrive at their frequencies. Accident sequences are defined in terms of initiating events and a series of system functional failures. The various possible accident sequences with respect to each initiating events are identified by event tree analysis. A Level-1 PSA provides insights into design weaknesses and into ways of preventing core damage which in most cases is the precursor to accident leading to major radioactive releases with potential health and environmental consequences.

The CDF is an outcome of the Level-1 PSA

Level 2 PSA

Starting from the results of level-1 the physical processes of accident progression post core damage are analyzed. Probabilities are determined for timing and mode of containment failure as well as for the release of radioactivity within the plant as well as the source term for a release into the environment.

Large Early Release Frequency (LERF): It is defined as the frequency of those accidents leading to significant, unmitigated releases from containment in a time frame prior to effective evacuation of the close-in population such that there is a potential for early health effects. Such accidents generally include unscrubbed releases associated with early containment failure at or shortly after vessel breach, containment bypass events, and loss of containment isolation. (Ref: USNRC-RG 1.174)

The starting point of Level-2 PSA is the grouping of a large number of accident sequences, derived in a Level-1 PSA, into a smaller number of plant damage states in accordance with accident characteristics and containment response characteristic for various accident sequences. After determining the plant damage states the progression of accidents and impacts on containment behaviour are examined probabilistically with containment event trees. The various end states of the containment event trees are grouped to a more manageable set of release categories for which district source terms are estimated.

A Level-2 PSA provides additional insight into the relative importances of accident sequences leading to core damage in terms of the severity of the radioactive releases they might cause.

The LERF is an outcome of the Level-2 PSA.

Level 3 PSA

Starting from the probabilities for releases as determined in level-2, the probabilities and extent of damages in the environment of the plant are determined.

It provides insights into the relative importances of accident prevention and mitigatory measures expressed in terms of the adverse consequences for the health of both plant workers, public contamination of land, air, water and foodstuffs. It also provides relative effectiveness of various aspects of accident management related to emergency response planning.

For the distinct release categories identified through Level-2 PSA analysis the magnitude of consequences are estimated for a variety of meteorological conditions and wind directions. The resulting consequence magnitudes and probabilities are presented in the form of Complementary Cumulative Distribution Functions (CCDFs). CCDFs are often shown in a logarithmic scale with both set of axes to accommodate the wide range of frequencies and consequences.

This cumulative curve depicts the Distribution of the Likelihood with reference to measures of fatalities or damage to the public e.g. deaths, occurrences of cancer etc.

The CCDF is an outcome of the Level-3 PSA.

2. VISION, MISSION AND GOAL FOR RISK INFORMED DECISION MAKING:

3. VISION

To provide safe, reliable and cost effective energy to the nation by ensuring safe operation of the plants, limiting radiation exposure to plant personnel and public within prescribed limits.

4. MISSION

To enhance the technical competency, resources and the awareness towards improved safety culture specific to Probabilistic Safety Assessment amongst the experts at Head

Quarters and Operating Stations for applying PSA Studies to Risk Informed Decision making.

A full scope PSA combined with Risk informed decision making would help achieve this mission. A full scope PSA consists of comprehensive Level-1 PSA for both internally generated plant initiating events and externally generated initiating events i.e. fire, seismic and flood, shutdown PSA and low power operation PSA and subsequently comprehensive Level-2 PSA and finally Level-3 PSA.

A systematic structured approach was followed in creating awareness about PSA among various stations. To gather information from plant on operation practices which formed the basis for the development of plant models, frequent visits were made to each stations. Plant information was assembled using sources such as the Safety Analysis Report, DBRs/DMs, Piping and Instrumentation Diagrams, Technical Specifications, Operating Procedures and Maintenance Records.

After having completed the KAPS Level-1 study, the PSA Group has been entrusted with additional activities for Level-1 PSA for 540 MWe PHWRs at the design stage and for external events and Level-2 PSAs for 220 MWe PHWR. Also, it was decided that Level-1 PSAs for 220 MW PHWRs will be performed by individual stations taking KAPS PSA as a reference. This was intimated to the individual stations during early 2000. An action plan as per the Table-1 was accordingly evolved to impart the necessary training to plant personnel and to equip them with the necessary software. At each Station, System Performance Engineers of Technical Unit are responsible for regularly keeping track of system performance and recording all the problems in the system. These system performance engineers update their data by regular discussions with concerned O&M personnel and by regularly referring control engineer's log which is the main source of information for most of the data items.

A guideline was prepared and issued to all the stations to ensure Quality Assurance (QA) aspects (as per IAEA TEC DOC-1101) in the studies with regards to Fault Tree Development, Basic Event Coding Reliability Analysis Documentation, etc.). Additionally the systems (Process, Front line and Support) required in Level-1 PSA were listed in the guideline and the parameters required for all the major components were grossly discussed. A list of components in each system alongwith the necessary reliability parameters was passed on to the technical unit engineers for compiling the component failure data, initiating event frequency etc. required for PSA.

Table 1 Strategies and Goals

S. No.	STRATEGIES FOR HQ	STRATEGIES FOR STATIONS	GOALS
1	Plant Specific Level-1 PSA • Dissemination of information through international standards, discussion and training programmes, workshops and seminars to develop the expertise for a Plant Specific PSA. • To inculcate awareness for plant specific PSA data collection.	Development of Plant Specific Level –1 PSA • Use of State of the art techniques for Reliability analysis • Systematic collection of Plant Specific Component failure data, and Initiating events and its post processing • Event Tree analysis	Applying Probabilistic Studies in Risk Informed Decision making to minimize risk, quantifying the accident consequences and back fitting design changes to optimize the existing design and operating practices.
2	PSA Applications: Specific Procedures to integrate the PSA results for Applications such as – • TS optimization • AOT & STI evaluation • RCM • Risk informed ISI • Re-licensing • Assessment of Design Configuration change • Severe Accident Management Strategies	To become self sufficient in the long run to perform all the PSA applications exclusively for the respective Stations.	• Quantification of safety margins, and specific changes to Licensing Basis • Adequacy of Severe Accident Management Strategies • Optimization of maintenance strategy including ISI • Evaluation of Risk Reduction Worth (RRW) of proposed changes to Structures, Systems and Components (SSC)

3	Full Scope PSA: a) External Events PSA • Acquisition of Resources and expertise • Plant Walk downs to identify the vulnerabilities of SSC's to external events and to make engineering judgment on the relative resistance capacity of them. b) Low Power and Shut down PSA c) Level-2 PSA d) Level-3 PSA	To perform Synergetic functions to fulfill the strategy of HQ.	• To quantify the limits of radiation exposure to plant personnel and public in line with the ALARA principle. • For the development of advanced designs and for promoting the expansion NPP's design objectives.
4	To formulate Probabilistic Safety Criteria		For periodic safety reviews encompassing a PSA, Probabilistic safety criteria are always used for evaluation of results.

5. CONCLUSION

With the completion of full scope PSA, with the co-operation and synergetic efforts of the Head quarters and Operating Station experts, NPCIL will be self sufficient in applying Probabilistic Studies in Risk Informed Decision making to minimize risk and back fitting design changes to optimize the existing design and operating practices, thus providing for safe, reliable and cost effective energy to the nation.

Though the PSA fraternity will agree that it is a very daunting vision, but with a well-defined mission backed with prudent strategies and SMART goals, we are confident of achieving it.

6. REFERENCES

1. KAPS Level-1 PSA.
2. RSA Proc for conducting Reliability Analysis.
3. IAEA Safety Guide-50-P4.
4. USNRC RG-1.174.
5. IAEA TEC DOC – 1101.

Reliability, Safety and Hazard
Editors: P.V. Varde et al.

RISK INFORMED REGULATORY DECISION-MAKING IN ASSESSMENT OF CHANGES IN TECHNICAL SPECIFICATIONS

R.B. Solanki, R.S. Rao, U.K. Paul, Om Pal Singh and S.K. Gupta

Atomic Energy Regulatory Board Anushaktinagar, Mumbai 400 094

ABSTRACT

Probabilistic Safety Assessment (PSA) is increasingly being used in Nuclear Power Plants (NPPs) worldwide for assessment of core damage frequency, radioactivity release frequency to the environment and assessment of risk to the public. PSA methodology has matured over the years and is also finding applications in areas such as on-line risk monitoring and performing living PSA, Optimisation of Technical Specification, Risk Informed In-Service Inspection and Reliability Centered Maintenance. PSA provides a systematic approach for integrating the risk concept into safety assessment and regulatory decision-making. A worldwide consensus is emerging for the use of PSA as a complement to the deterministic analysis and in arriving at more balanced decision-making. This approach is normally referred to as "Risk Informed" decision-making.

PSA models should meet certain technical requirements for being useful in such applications and the quality of PSA should be consistent and commensurate with the intended application. Effective use of PSA results requires that comparison be made against a set of either established acceptance criteria or targets.

In this paper, the technical requirements for adequacy of PSA for use in Risk Informed decision-making in different areas are presented. The methodology for application of PSA for technical specification optimisation is described. The acceptance criteria for the assessment of the submittals on technical specification change proposals are discussed. The illustrative example for application of this method of technical specification optimisation is presented.

Keywords: Risk-Informed, Technical Specifications, Probabilistic Safety Assessment, Regulation, Pressurized Heavy Water Reactor, Decision-making.

1. INTRODUCTION

PSA if performed with adequate care and detail, provides more detailed and realistic picture of the [1]safety importance of plant design features, operating procedures and operational practices. PSA studies are of interest not only in determining the absolute value of the risk of damage to the reactor core, release frequency and the risk to the public, but also the information on the various components of the risk and their relative importance. As on today, PSA has been carried out for more than 200 NPPs in the world and for most of the remaining ones, it is in various stages of development and completion [1].

The risk considerations have been used, implicitly, in making safety related decisions along with deterministic regulatory requirements. However, increased maturity of the PSA methodology in the recent years has replaced the implicit part by explicit quantitative data and provided additional insights that would not otherwise be available. This move from a traditional, deterministic approach of regulatory decision-making to one that also takes into account the risk insights derived from PSA in an integrated way is known as "Risk-Informed (RI)" approach.

Technical Specification (TS) requirements are generally based on deterministic analysis and engineering judgements. However, in some cases the requirements may be unduly restrictive or not conducive to safety, and thus changes may be desirable. At the same time, there are certain requirements, which can improve safety if strengthened. The use of PSA to support TS optimisation is an important element for "Risk-Informed" approach.

Considering the benefits of the PSA, many countries like USA, UK, Germany, France, Canada, Japan and Korea [2] have started using PSA insights in areas like finding weak links in the design, operating procedures and technical specifications, prioritisation of operational tasks, operator training and regulatory decision with risk informed/based approach. Atomic Energy Regulatory Board (AERB) is a competent authority for the regulation of nuclear and radiation facilities in India. Keeping in view the international trend and benefits it accrues, AERB also plans to adopt RI approach in the regulatory process.

In this paper, the methodology and acceptance criteria for risk impact evaluation on the TS due to Allowed Outage Times (AOTs) and Surveillance Test Intervals (STIs) change proposal is presented. A case study is also presented to provide the practical utility.

2. TECHNICAL REQUIREMNTS OF PSA

In spite of the recognized strength of PSA, the views that PSA results cannot be relied upon because there are possible limitations arising from (a) difficulty in ensuring completeness of initiating event identification, (b) unavailability of reliable plant component failure data, (c) difficulties in modelling and quantification of human errors, (d) difficulties in modelling and quantification of common mode/cause failures and uncertainties associated with various models and analysis steps. However, these aspects, with present state-of-the-art PSA methodology, can be treated sufficiently/adequately and each source of limitation/uncertainty can be meticulously quantified to improve the quality of PSA. For this, PSA should meet the following technical requirements.

(a) The scope of the PSA should be well defined keeping in view the objective of the study.

(b) Detailed QA programme should be established and made effective in every PSA project.

(c) Failure Mode and Effect Analysis (FMEA) of each component should be performed.

(d) All the risk significant components should be modeled in the PSA.

(e) As far as possible, plant-specific data should be used. If plant-specific data are not adequate, a Bayesian updating technique should be used to update the data.

(f) Success criteria used for system reliability assessment should be based on the deterministic accident analysis.

(g) Appropriate human actions should be incorporated in the Fault Trees (FTs) as well as Event Trees (ETs).

(h) While choosing particular model for any component, its true status in the system such as operating, stand-by, repairable and non-repairable should be identified and accounted.

(i) Multiple failure of events for which a clear cause-effect relationship can be identified should be explicitly modelled in the FTs. Multiple failures for which no clear root cause event can be identified can be modelled using Common Cause Failure (CCF) models such as β-factor model, α-factor model, Multiple Greek Parameter model or Binomial Failure Rate models.

(j) PSA studies should be supported by the uncertainty analysis and sensitivity studies.

3. METHODOLOGY FOR TECHNICAL SPECIFICATION EVALUATION

The TS optimisation methodology involves the following steps:

(a) Carry out the Consequence analysis for Core Damage Frequency and identify the Risk Critical Components (RCC) for which, Surveillance Test Intervals (STIs) and

Allowed Outage Times (AOTs) are to be optimised.

(b) Determine the risk contribution associated with the change in STIs and AOTs

(c) Ascertain the acceptability or otherwise of the risk impact

(d) Perform sensitivity and uncertainty analysis.

Various risk measures need to be considered for risk based TS evaluations with regards to AOT and STI [3]. For AOT evaluation, instantaneous (conditional) risk while the component is in the failed or repair state, Incremental Conditional Core Damage Probability (ICCDP) and Yearly AOT Risk should be considered. For STI evaluation, Test Limited Risk and Test caused Risk should be considered.

Instantaneous (conditional) risk while the component is in the failed or repair state has been considered for demonstrating the procedure for evaluation of AOT change proposal. The graphical interpretation of ICCDP is shown in Fig. 1.

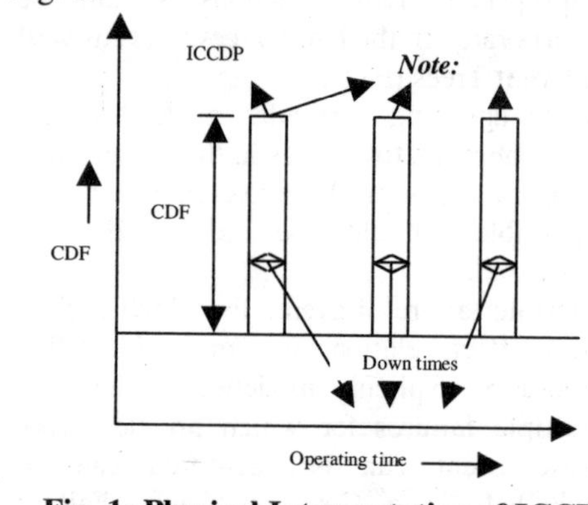

Fig. 1: Physical Interpretation of ICCDP

Note: Cumulative Downtime Risk (accumulated CDF increase over some time period)

The risk-based AOT can be calculated as follows:

As per the guidelines given in RG1.177, Incremental ICCDP needs to be calculated for the equipment under consideration.

ICCDP = (ΔCDF)*duration of single AOT (1)

where, ΔCDF = (CDF with subject equipment out of service – baseline CDF with nominal expected equipment unavailability).

The CDF with subject equipment out of service can be obtained by making the unavailability of the subject component as 1.0 (total unavailability). This ΔCDF is a risk contribution of the subject component.

4. ACCEPTANCE CRITERIA

For the acceptability of the proposed AOT, acceptance guidelines discussed in Regulatory Guide RG 1.174 [4] should be considered. The guidelines are provided for two levels i.e. (i) at Core Damage Frequency (CDF) level for level-1 PSA studies, and (ii) Large early Release Frequency (LERF) level for Level-2 PSA studies as given in Figs. 2 and 3 respectively.

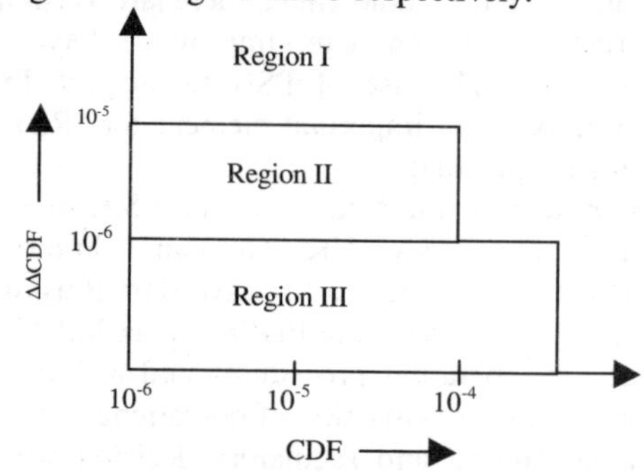

Fig. 2 Acceptance Guidelines for CDF

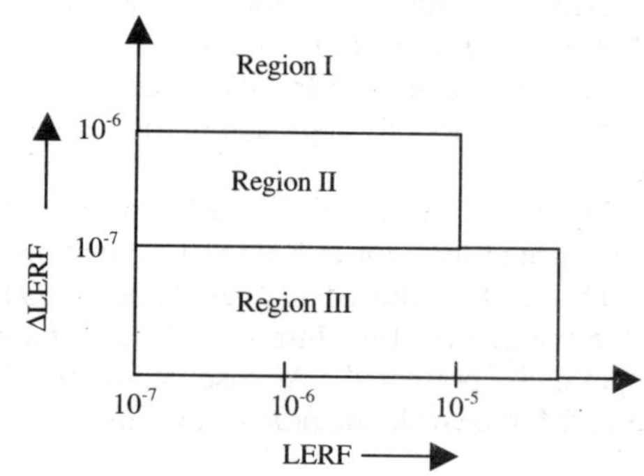

Fig. 3: Acceptance Guidelines for LERF

If the application for proposed change indicates that there is a reduction in the risk (i.e. CDF or LERF), the change should be considered to have satisfied the relative principle of risk-informed regulation with respect to CDF or LERF. The proposed change should be acceptable. If the ΔCDF falls in region I, no change is permitted. In region II small changes are permitted and the cumulative risk impact of the change needs to be monitored and appropriate measures should be incorporated to reduce the overall risk. In region III, change in the overall risk is very small and hence the change can be acceptable. However, the cumulative risk impact of the proposed change should be monitored.

These guidelines are given for the full scope PSA (i.e. including internal events, external events, full power, low power, and shutdown). However, same guidelines can be used for limited scope PSA allowing margin for the contribution from the out-of-scope items.

Apart from the acceptance guidelines given in RG-1.174, the guidelines given in RG-1.177 [3] specific to PSA application for TS optimisation should be considered while assessing the TS change proposal.

RG1.177 provides the acceptance guidelines with respect to ICCDP. An ICCDP of less than 5.0E-07 is considered small for a single AOT. This criterion is based on the hypothetical situation in which the subject equipment at a representative plant is out for five hours causing the CDF of the plant, with an assumed baseline CDF of 1.0E-04/reactor year, to conditionally increase to 1.0E-03/reactor year.

It is to be noted that these numerical guidelines should be used to decide the acceptability of the proposed TS change along with the other traditional considerations such as operating experience, feasibility of the proposed change and consideration of risk-critical configuration due to proposed change, etc.

5. ILLUSTARTIVE EXAMPLE

In this paper, as an illustration, the above methodology is applied to evaluate the AOTs for risk critical components for a typical NPP.

AERB has yet not established acceptance criteria for PSA applications. For demonstration purpose, the criteria given in the RG1.174 and RG1.177 have been used.

Now consider a Pressurised Heavy Water Reactor (PHWR) in operation with the baseline CDF calculated to be 5.75E-06/reactor year from a limited scope Level-1 PSA performed for internal initiators. For optimisation of the AOTs in the present Technical Specifications, the RCCs need to be identified. This can be done by reviewing the dominant Minimal Cut Sets of the consequence analysis of CDF. For an Illustration, let us assume that these include Diesel Generators (DGs) of emergency power supply system. The following are the results of the Level-1 PSA study (say).

- The baseline CDF is calculated to be 5.75E-06/Reactor year.
- The CDF increases to 1.15E-05/reactor year if one out of three DGs becomes unavailable.

Let us assume that the Allowed Outage Time (AOT) and Test intervals for DGs, as per the current TS requirements, are 72 hours and 7 days respectively.

The PSA study has been performed at Level-1 and hence the risk should be considered at CDF level and not for LERF. Also, the level-1 PSA study is a limited scope study and hence appropriate margins should be considered for the out-of-scope items while comparing the results with the acceptable guidelines.

Now, from equation no. 1, for AOT of 7 days, the ICCDP can be calculated as follows:

ICCDP = (1.15E-05 – 5.75E-06) * (7/365)
= 1.102E-07

Similarly, for different AOTs, ICCDPs can be evaluated. These are calculated for different AOTs and given in Table 1.

Table ICCDP Results for Proposed AOTs

Sr. No	AOT duration (days)	ICCDP
1	7	1.102E-07
2	10	1.157E-07
3	15	2.363E-07
4	20	3.150E-07
5	25	3.938E-07
6	30	4.726E-07
7	35	5.513E 07

6. RESULTS AND DISCUSSIONS

From Table 1, it can be seen that with the current TS requirements (i.e. AOT = 7 days), the ICCDP is less than the acceptable value of 5.0E-07. Hence, there is a scope for relaxing the TS requirements on AOT for DGs. It can also be noted that the proposed AOT of 30 days meets the acceptance criteria given in RG1.177.

As stated earlier, the acceptance guidelines of RG1.174 also need to be considered for the decision-making. These guidelines suggest to ensure that the increase in the overall risk (i.e. CDF) due the proposed change is "acceptable".

In the present example, the increase in CDF (i.e. ΔCDF) is found to be 5.76E-06/reactor year. This falls in the region II of the Fig. 2. Here, the increase in the CDF due to proposed AOT is within the acceptable range (i.e. 1.0E-06/reactor year to 1.0E-05/reactor year). Hence, the proposed AOT of 30 days instead of current requirement of 7 days, can be considered as "acceptable". However, it needs to be ensured that the total CDF is less than the 1.0E-04/reactor year. This is the INSAG-12 [5] safety goal established to be demonstrated by the existing NPPs at every stage of operation during the lifetime of NPP.

The new CDF need to be evaluated by modifying the DG maintenance unavailability taking into consideration of proposed AOTs. These new CDF values should be compared against INSAG-12 recommended value of 1.0E-04/reactor year. It is reiterated that guidelines given in Fig. 2 is for full scope PSA. One needs to consider the risk contribution from the out-of-scope items when the TS evaluation is being done with a limited scope PSA.

7. REFERENCES

1. ATOMIC ENERGY REGULATORY BOARD, "Probabilistic Safety Assessment Guidelines", AERB Safety Manual, AERB/NF/SM-1(draft), June 2002.
2. OECD NEA, "Review Procedures and Criteria for different Regulatory Applications to PSA", report on the survey of national practices, Paris, 1998.
3. UNITED STATES NUCLEAR REGULATORY COMMISSION, "An approach for plant-specific, Risk-Informed Decision-making: Technical Specifications", RG-1.177, August 1998.
4. UNITED STATES NUCLEAR REGULATORY COMMISSION, "An approach for using Probabilistic Safety assessment in Risk-Informed Decisions on plant specific changes to the licensing basis", RG-1.174, November 2002.
5. INTERNATIONAL ATOMIC ENERGY AGENCY, Basic Safety Principles for Nuclear Power Plants, Rev. 1., Safety Series No. 75-INSAG-12, IAEA, Vienna, 1992.

Reliability, Safety and Hazard
Editors: P.V. Varde et al.

APPLICATION OF PROBABILISTIC SAFETY ASSESSMENT TO A HIGH POWER RESEARCH REACTOR

P.V. Varde, U.L Sharma, S.K. Marik and A.C. Tikku

Reactor Group, BARC, Mumbai 400 085

ABSTRACT

Probabilistic Safety Assessment of 100 MWt Dhruva research reactor has been carried out. The scope of this work extends beyond Level 1 PSA study and includes the limited scope Level 2 PSA study to give the likelihood of radioactivity release, during the postulated LOCA scenario.
The work involved under this project includes the reliability analysis of safety systems and safety support systems, estimation of the frequency of failure of initiating events and modeling of accident sequences towards giving a statement on *'Core Damage Frequency'* (CDF) of the plant. The 'Uncertainty' analysis has been carried out at system level as well as at CDF level to account for possible errors in data and modeling. The 'Sensitivity' analysis has been performed to check the affect of major assumptions and critical system parameters on the result of this analysis. Though LOCA contribution to the CDF is small the consequences of LOCA have potential of release of radioactivity into the reactor containment. There exists potential, however small it may be, for release of radioactivity outside the plant depending on the performance of containment isolation and emergency exhaust system. Based on the results & findings of the deterministic safety analysis of the plant, further study was carried out to estimate the frequency of release of radioactivity during LOCA scenario. The likelihood of release of radioactivity was found to be 'very low'.

Keywords: Probabilistic Safety Assessment, Research Reactor, Performance data, Loss of Coolant Accident, Reliability Analysis of Safety Systems, Initiating Event Frequency

1. INTRODUCTION

Dhruva is a 100 MW(th), high neutron flux (maximum ~ 1.8 x 10^{14} n/cm^2/Sec) research reactor located at BARC and has the distinction of being one of the high flux, high power research reactor in the world. Dhruva was commissioned in 1985 and achieved full power operation in 1988. In this reactor, natural uranium is used as fuel and heavy water as coolant, moderator and reflector. The safety analysis of the plant was carried out using the deterministic methods. Now-a-days the Probabilistic Safety Assessment (PSA) methodologies are increasingly being used in safety assessment of nuclear plants. The PSA technique complements the conventional deterministic methodology in safety assessments of design basis accident (DBA) scenarios. At the same time the insight obtained using PSA methods on 'beyond design basis' scenarios supplements the deterministic findings. The strength of PSA lies in offering a framework which is relatively more systematic and enables the quantification of potential risk level for a plant.

PSAs are performed at three Levels as given below:
i) ***Level 1*** *PSA*: The assessment of plant failure events leading to the determination of reactor **core damage frequency,** ii) ***Level 2 PSA*** : The assessment of reactor containment response,

together with Level 1 PSA results, leading to the determination of radioactivity **release frequency** and iii) ***Level 3 PSA*** : The assessment of off-site consequences, together with the results of Level 2 PSA analysis, leading to estimates of **public risk**.

The scope of this work includes development of a Level 1 PSA and limited scope Level 2 PSA model of the plant. Therefore this study has been referred as *Level 1 + PSA study for Dhruva* [1]. The results of this study mainly comprises of identification of accident sequences and the statement on Core Damage Frequency (CDF). Further, as part of limited scope Level 2 PSA study the estimation of radioactivity release frequencies from the containment considering Minor LOCA scenario has been performed.

To account for possible errors in data and modeling, 'uncertainty' analysis was performed at system as well as CDF level. Detailed 'sensitivity' analysis was also performed with a view to have assessment of the impact of various assumptions made during the study. In the absence of a standard definition for CDF, this study uses the definition given by AERB (Atomic Energy Regulatory Board) in its draft document entitled "An Approach paper for PSA - AERB Policy for PSA"[2].

The objectives for development of this study were;

- identification of dominant accident sequences,
- identification of systems, components and human action important to safety
- assessment of important dependencies,
- identification of new safety issues,
- analysis of severe accidents,
- suggest design modification, if any,
- comparison with the target values,
- determination of release frequency from containment for dominant LOCA accident sequence, i.e. Minor LOCA (as part of limited scope Level 2 PSA).

In addition, this study is expected to be useful for better understanding of the performance of the systems and in generating valuable inputs for decisions making on, a) plant operation, b) design modification and back-fitting c) safety analysis and d) regulatory issues.

As part of Level 1 PSA study the work envisages estimation of *'core damage frequency'* (CDF) considering i) full power operating condition ii) internal initiating events and iii) reactor core as the source of radioactivity release. The special issues like assessment of human factor, common cause failures, uncertainty in data and model & sensitivity of assumptions have been addressed wherever applicable. The aspects like treatment of external events and the estimation of risk from the shutdown conditions of the reactor are beyond the scope of this study.

2. SALIENT FEATURES OF THE STUDY

In this PSA study, attempts have been made to follows the current international standards existing in the field of PSA. The existing international codes and standards like IAEA Safety Series 50-P-4:Procedures for Level 1 & 2 PSA of NPPs [3] and various IAEA-TECDOCs have been followed as guideline documents for performing this study. This study can be considered as an important step towards standardization of PSA methodology. The salient features of this study include,

i) Extension of Level 1 PSA to include the Limited Scope Level 2 PSA
ii) Use of plant specific data for quantification,
iii) Discussion on important safety aspects related to performance of safety systems,
iv) System performance analysis is integral part of this study. Where possible, comparison of the results of this study with the similar systems have been done to get a better feel on performance of the systems,
v) Considerations related to taking credit for the availability of backup shutdown device on case to case basis for protection action for various accident sequences,
vi) Evaluation of human reliability analysis for safety significant human actions,
vii) Performance of uncertainty analysis at CDF and System level to account for variations in data and possible modeling errors,
viii) Sensitivity analysis for various assumptions and critical data,
ix) Formatting of the document using IAEA specified format.

3. OVERVIEW OF METHODS AND APPROACH

Major procedural steps followed for performing this study are i) Identification of initiating events, ii) Accident sequence modelling, iii) Data assessment and parameter estimation, iv) Accident sequence quantification and v) documentation of the analysis and interpretation of results. The quality assurance aspects in various area of this study, like organisation, technical work which includes independent review and documentation have been addressed throughout this study.

The main objective of task associated with selection of initiating events was to generate a complete list of Initiating Events (IEs), to the extent possible. However, it should be recognised, that a judicious consideration is required to identify IEs which have only small contribution to the total risk and hence need not be included in this PSA study. Accordingly, 27 initiating events were identified during the initial stage of PSA study. Based on engineering & operational judgement and reference to the list of initiating events from generic sources, 17 initiating events were considered for the final modeling & analysis.

There are various approaches for the selection of initiating events like i) Engineering evaluation, ii) Reference to previous list of initiating events, iii) Deductive analysis and iv) Operational experience. To prepare a list, as complete as possible, all the approaches were followed. Engineering evaluation approach was particularly useful for qualitative and quantitative analysis of events, which have potential for common cause failures. Most of the initiating events for this study were identified based on the references to already existing lists of initiating events. This include plant specific documents like safety analysis report for Dhruva [4] and generic source like, Canadian NRU reactor [5], ORNL High Flux Research Reactor(HIFR)[6], CANDU Power reactor IEs list [3], etc. Operation experience approach was particularly useful for identifying IEs having relatively high frequency of occurrence. It may be noted that the generation and review and classification of list of IEs was a continues and iterative process. The final list of IEs could be consolidated only during the accident sequence quantification phase. Keeping in view the scope of this analysis only internal initiating events have been considered in this study.

For each initiating event, the safety functions that need to be performed in order to mitigate the possibility of core damage were identified. Accordingly, the front line systems identified for analysis were i) Emergency Cooling System or Shutdown Cooling System, ii) Protection System (Primary Shutdown & Backup Shutdown System), iii) Emergency Core Cooling System. The Emergency Power Supply System (Class III, Class II and Class I System), Compressed systems and other systems which are required for the operation safety systems were considered as safety support systems. Apart from this analysis of 'Containment Isolation and Emergency Exhaust System' was carried out as part of Level 2 PSA analysis. For the major part of this study Fault tree analysis methodology has been used for modeling various systems of the plant. For complex scenarios other methods like Markov analysis, Reliability Block Diagram, etc. have also been used for estimating system unavailability. Event tree methodology was applied to model accident sequences for 17 initiating events.

The emphasis had been to use plant specific data for quantification of component and sub-system failure rates. In case the data from plant specific source are not adequate then the data from generic sources, like IAEA-TECDOC-478 [7], IAEA-TECDOC-900 [8], WASH-1400 [9], etc. have been used. For the estimation of electronic components failure rates, MIL-217: American Military Standard, 1991 [10] methodology has been used. The estimation of *Common Cause Failures* for various systems has been carried out using empirical 'B-factor model'. A value of β=0.1 was assumed as the highest value that characterize a system which is highly susceptible to CCFs. A judgment was used to assign lower value of β for the system where the redundant system operate on diversified principle or the system which over the years of operations have demonstrated very low dependent behaviour.

Human error analysis was carried out for important human actions as part of accident sequence evaluation program. The NUREG-CR-1278 [11], IAEA-TECDOC-618 [12], IAEA-TECDOC-592 [13], etc. were the main source of data and information on human reliability analysis for this study.

The assessment of frequencies of various accident sequences involves assigning the quantified estimates to the safety functions (which forms headers of the event tree) and the initiating events. The final estimate of frequency of accident sequences associated with an initiating event has been arrived by propagating the accident scenario on the event tree by considering the unavailability estimates of the applicable safety function.

Sensitivity and *Uncertainty* analysis was carried out as part of the accident sequence evaluation programme. There were nine cases identified for sensitivity analysis. The objective was to have an assessment of sensitivity of the important assumptions and critical data on the core damage frequency. The uncertainty analysis was performed using the Monte Carlo simulation. For most of the cases ~ 1200 iterations have given good results. The error factor used in the uncertainty analysis was ~ 2.0 for plant specific data and 3.0 for generic data. For piping system in primary and secondary coolant system higher error factor of ~ 10 was considered as the scatter in the data obtained from the generic source was relatively wide.

Definition of Core Damage Frequency

A Level 1 PSA usually implies assessing plant failures events leading to severe core damage and the corresponding frequencies. In the absence of a standard definition of **'Core Damage'**, the definition given in AERB's draft document entitled "An Approach paper for PSA" has been used in this study. This document proposes the definition of Core Damage as follows:

"The Core Damage (CD) is defined as the accident conditions which involve exceeding design limits for more than one parameter and result into any (or in combination) of the following:

a) *Fuel failure more than 10 % in the core*
b) *Damage of critical core components (e.g., Reactor Vessel / Calandria, End Shield, etc.)*
c) *Rupture of more than one fuel bundle housing (e.g. Pressure Tube)*
d) *Prolonged clad oxidation and /or H_2 generation from metal water reaction leading to global volumetric hydrogen concentration in the containment exceeding 10%.*

Above definition was adopted to define different core damage states in this study as follows: i) *Reduction in Safety Margin:* Cases where failure of one of the redundant safety system has occurred. This need not mean violation of safety limit, ii) *Safety Limit Violation:* where safety limits are violated but it need not necessarily results in to bulk fuel failure, iii) *Delayed Core Damage* : to cover those accident sequences where the core damage is not immediate but over a period of time ranging from 1/2 hrs to couple of days, iv) *Core Damage State:* Indicative of scenario as given in item a) or b) or c) or d) in the above list or any combination of item a, b, c, and d. and v) *Severe Core Damage*: Occurs when failure of primary and secondary or back up safety system takes place. In some cases where the scenario is fuzzy and no conclusion could be drawn about type and severity of damage, the unsuccessful termination of the scenario itself has been considered as core damage state on conservative basis.

4. LIMITED SCOPE LEVEL 2 PSA

The next natural step was to extend the Level PSA 1 study to Level 2 PSA study. Though this study does not deal with the complete aspects related to Level 2 PSA study, it attempts to perform a limited scope containment analysis using most credible accident - LOCA for Dhruva. This study integrates the results of level 1 study, the containment failure modes & its probabilities and the available deterministic insights on release of radioactivity from the containment, towards giving a estimate of different *'release frequencies'* for LOCA scenario. Accordingly, this requires i) estimation of failure probability of Containment Isolation and Emergency Exhaust System, ii) determination of accident sequences using Containment Event Tree (CET) for LOCA group of CDFs, for different modes of containment failures and finally iii) estimate of release frequency for different release categories.

The choice of selection of only LOCA related accident sequences was due to following factors, i) The list of accident sequences in the level 1 PSA study shows that accident sequences associated with LOCA (major or minor) are the only scenarios which have potential for direct release of radioactivity to the containment, ii) Minor LOCA contributes ~ 8 % to the total CDF and iii) the data available from deterministic study, like thermal hydraulic study and estimation of release of radioactivity, only deals with LOCA condition.

4.1 Containment Performance

Dhruva containment is tested after every six month interval by pressurizing it to a pressure of 150 mm of WG. Leakage rate from the containment should be less than the stipulated limit for declaring the containment 'available'.
The Containment performance evaluation has been carried out in terms of estimation of containment isolation and emergency exhaust system failure probability. The Various containment failure modes and the associated failure probabilities are as follows:

i) Containment leakage probability
ii) Containment Isolation failure
iii) Containment direct bypass
iv) Emergency Exhaust System failure

Other mode of containment failure, viz., Vapour explosion, combustion, over temperature have not been considered as Dhruva being a low pressure low temperature reactor these mechanisms which could lead to above type of failures have low possibility. The study indicates that in relative term, the containment leakage and containment isolation failure contributes more to containment failure probability.

4.2 Scenario During Postulated LOCA Condition

Dhruva containment is a Pressure-vent type containment with a design intent to ensure that during the low probability event of LOCA, the exposures are not likely to exceed the stipulated limit for accident conditions. Release to the environment comprise of, i) release from the stack through emergency exhaust system, ii) ground level leakage as a result of overpressure within the containment and iii) direct bypass route.

During postulated LOCA scenario, the overpressure signal in the containment and /or the presence of high activity in the exhaust air causes the automatic closure of containment isolation dampers, tripping of main ventilation system fans and starting of the emergency exhaust system. The emergency exhaust system has a set of combined HEPA and charcoal filters and has the capability to steadily reduce the over pressure to slightly negative pressure of around 30 mm of WG within 12 minutes following LOCA.

4.3 Source Term For Postulated LOCA Scenario

For the postulated accident scenario of Dhruva reactor, i.e. LOCA with Emergency Core Cooling System available to perform its intended function, only limited coolant boiling in the channel is expected and no clad failures are indicated in the LOCA study [4]. The results of Level 1 study shows that contribution from Major LOCA to the total CDF is very small while Minor LOCA contributes ~ 8 % to the CDF. Hence, for this analysis, a 'Minor LOCA' scenario has been considered with failure of ECCS to come on demand, while the light water injection to Reactor Core has been actuated successfully. For this scenario there may exist a threat of delayed core damage if ECCS could not be recovered in 7.0 hrs which makes this scenario a very low probability event.

The definition of core damage considered for this analysis assumes core damage as a condition when fuel failure occurs for more than 10% in the core. For LOCA scenario considered in this analysis it is assumed that light water injection caters to core cooling and the flow and pressure conditions are similar to the one when ECCS comes on demand. For this, core cooling conditions it is assumed that certain fraction of total fuel experiencing boiling leads to fuel failure. Accordingly, the affected fuel leads to release of 100% noble gases, 25% of radio-iodines and 1% of particulate into the immediate coolant medium. This release gives rise to a 'Source Term' for calculating the radiation doses at a distances of 300 Meter, 1600 Meter and 2000 Meter from the reactor.

4.4 Containment Event Tree (CET)

The four failure modes of the containment, as discussed in the previous sub-section, forms the heading of the CET. CET was drawn for the

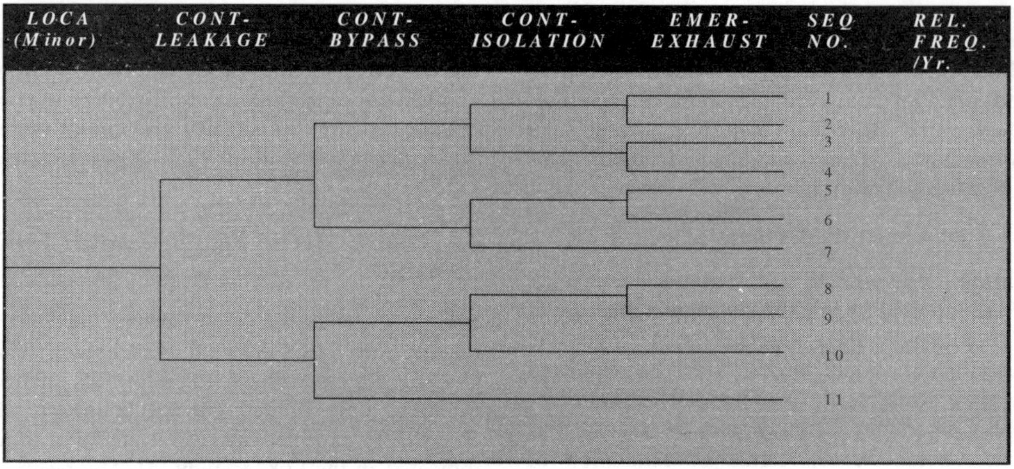

Fig. 1: Simplified representation of Containment Event Tree

Minor LOCA category of CDF and the scenario was propagated inducting status (availability or non availability) of containment sub-systems / failure modes. Figure 1 shows the CET for Minor LOCA scenario.

Though provision exists in some of the sub-systems to resume the system function or use alternate option to carry out the demanded function with human actions, no credit has been taken in this analysis for 'recovery functions'. These actions include manual closing of isolation dampers, manual opening of emergency exhaust line dampers, closing of Air Locks, etc. The reason for not taking credit of manual recovery action is that one of the assumptions of this analysis that reactor building pressure is brought from over-positive to negative condition well within 12 minutes, hence, i) available time is not enough to perform recovery actions and ii) these actions are not included in plant EOPs.

The quantification of the four headers of the CET has been carried out based on the results of the fault tree analysis of containment system. If the leakage is stipulated value of leakage from the containment then it is treated as success and one moves up from the nodal point on the CET and vice versa. For estimating containment bypass probabilities no data were available either for conditional probability of failures or for component failures, hence operational judgement has been used for estimating these values.

5. RESULTS & DISCUSSION

The outcome of the accident sequence evaluation program was spectrum of various degree of plant damage states. Accordingly, the overall Core Damage Frequency estimate for Dhruva was 3.8 * 10^{-5} /yr, which is much below the value of 10^{-4} / yr. applicable for old plants.

5.1 Dominant Sequences Contributing to CDF

Out of the total number of accident sequences identified using probabilistic criteria and analytical assessment only few accident sequences are likely to result in varying degree of core damage. The accident sequences, which were identified as contributing significantly to the CDF are referred as the *dominant accident sequences.* Sequences wherein the clad failure does not result in a significant way (<10%) have not been included in CDF calculation. Similarly, a judgment has been used in choosing those sequences, which results into 'delayed core damage' after a couple of hours. An engineering and operational insight was used to postulate that the possibility of recovering a failed system for mitigating the consequences.

The initiating event, normal Class IV power failure, Failure of Emergency Cooling Water line in some specific locations, Reactor Trip and Minor LOCA contribute to the CDF.

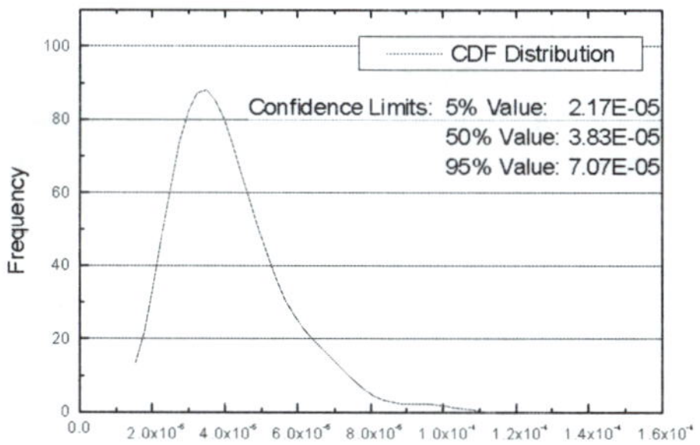

Fig. 2: Distribution of CDF

5.2 Results of Uncertainty & sensitivity Analysis

Uncertainty analysis was carried out at two levels, at system level and CDF level. It may be noted that the error factor for all the system is relatively low. This is due to the fact that plant specific data, particularly for critical equipment, were used for this analysis. Generally at component level an error factor of ~ 2.0 is used if the data is plant specific and a factor of > 3.0 was considered adequate for giving variability in generic data. The Monte-Carlo simulation was carried out for propagating uncertainty in the fault tree from component level to system level and subsequently from system level to CDF level. Normally 1200 iterations were performed for one simulation. After obtaining the uncertainty bounds at system level, the uncertainty bounds were obtained for CDF. Figure 2 shows the frequency distribution for the CDF. The lower bound median value, and upper bound for CDF has been estimated as $2.17 * 10^{-5}$ /yr, $3.83 * 10^{-5}$ /yr, and $7.07 * 10^{-5}$ /yr, respectively. Accordingly, the error factor in CDF estimation works out as 1.80.

The sensitivity analysis was performed to i) determine the sensitivity of core damage frequency to possible dependencies among component / sub-systems and human failures and ii) to address those modeling assumptions suspected of having a potentially significant impact on the results. Accordingly, nine cases which were considered as significant from the sensitivity point of view are analysed. A typical result of sensitivity analysis is shown in Fig. 3. Here a sensitivity analysis was performed to see the affect of variation in frequency of class IV power failure on CDF.It can be seen that even by considering a value as high as 10 / yr , the CDF value increases to only $2.25 * 10^{-4}$ / yr.

5.3 Interpretation of Level 1 PSA Results

The overall CDF value for Dhruva is $3.8*10^{-5}$ / **yr.** Median Value (50 %), Lower bound (5%) and Upper bound (95 %) of the Core Damage Frequency of Dhruva is $3.83 * 10^{-5}$ /yr, $2.17 * 10^{-5}$ /yr, and $7.07 * 10^{-5}$ /yr, respectively.

The results of this analysis should be seen in the light of the well established safety principles which include i) provision of redundancy in safety related equipment / components such that minimum safety functions can be performed even in the event of failure of a single active component, ii) there exists more than one means of achieving reactor shutdown (shut off rod and moderator dumping) and core cooling. Even through defence in depth is the basic design philosophy adopted for designing the system, the potential common cause element may affect more than one safety systems. This was the main consideration in performing the CCF analysis for finding out the dependencies among various redundant components in the system and also at independent safety systems levels. The extent of the sensitivity of this dependency of these systems, viz., variation of CDF with different degree of dependency was analysed by carrying out sensitivity analysis.

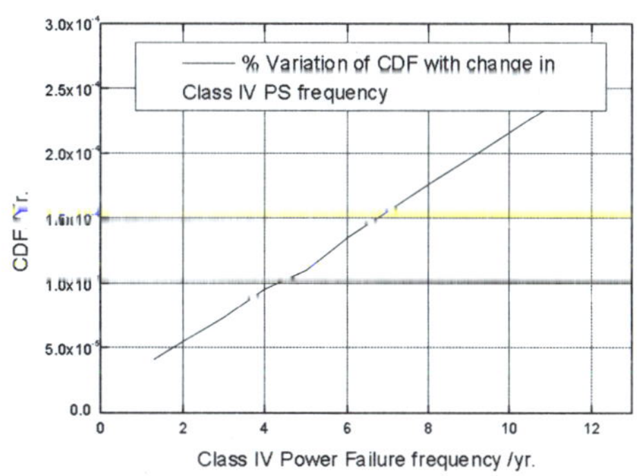

Fig. 3: Sensitivity Analysis - effect of Class IV power frequency on CDF

The major contribution comes from the Accident sequence (Class IV Power Failure and Failure of Emergency Cooling System. Even though the ECS system is complex by design, failure probability of this system is one of the lowest. In spite of this the above accident sequence becomes prominent because it was assumed, conservatively, that after a reactor trip minimum two out of three shutdown cooling pumps are required to come automatically.

The quantified estimates obtained using PSA methodology does provide a vital mean for knowing safety level of the systems in general and plant in particular. However, these results are better utilized for comparing safety status of different systems in the plant and similar systems in different plant for obtaining the relative safety status of the plant under considerations. There is an assumption that methodology adopted and input data used for system / system are similar. With this assumption in background PSA results provide a yard stick for comparing different systems and plants. Different plants of similar type can be compared using the CDF estimate. A survey was carried out on PSA studies on research / test reactor. The PSA report provides a list compiled from the available literature on the CDF estimates for different research reactors world over. The objective was to compare the results of Dhruva Level 1 PSA in terms of CDF with other PSA studies. Though CDF statement for Dhruva does not include the contribution from external events like Fire, Flood, Seismic, etc, a general look on the table shows that Dhruva CDF of $3.8 * 10^{-5}$ / yr. is much lower (in some cases almost an order of magnitude lower) compared to other nuclear plants.

5.4 Results of Limited Scope Level 2 PSA

As shown in the CET, totally there are eleven accident sequences. These sequences has been binned into four broad release categories. i) *Low Probability* accident sequence ($3.4*10^{-6}$ / yr) but with the consequences in terms of expected dose to the public to be less than the estimated value of ~ 7 mSv. Accident sequence -1 forms this release category. II) The second bin contains those sequences with *Very Low Probability* - of the order of 10^{-8} / yr, e.g. accident sequence no. 3 and 8, but could lead to consequences more than the value of 7 mSv depending on the mode of containment isolation failure as applicable for seq. 3 and magnitude of leakage in case of seq. 8. III) The third bin contains those sequences having the *Very Very probability* of occurrence (of the order of ~ 10^{-11} or less) but could result into higher doses depending on the severity of the accident scenarios and type and extent of containment sub-system failures. IV) The fourth category of accident sequences (having the likelihood of occurrence of < 10^{-13}) have been categorised as *hypothetical events.*
Based on above discussion it can be concluded that likelihood of a scenario when the expected dose to the public will be less than or equal to 7.33 mSv is $3.4*10^{-6}$ /yr. For higher order public dose the likelihood could be given as less than or equal to $4*10^{-8}$ / yr.

6. CONCLUSIONS

The contribution to CDF from major LOCA has been estimated to be less than 1% to the total CDF. This study shows that the probability of Major LOCA for Dhruva is very low. Though Dhruva design philosophy follows single failure criteria, the PSA methods provides a an efficient mechanism for the analysis of various accident scenarios, including the beyond design basis and scenario involving multiple system failures. Considering that performance of a PSA study for a plant is resource intensive (in terms of money, man power and infrastructure) and the experience & insight obtained from this study, it can be argued that for research reactors the Level 1+ PSA study provides, to a large extent, a holistic and effective framework for carrying out the safety analysis.

7. REFERENCES

1. Level 1+ Probabilistic Safety Assessment of Dhruva, A BARC internal report, 2002.
2. AERB, "Approach Paper on AERB Policy on PSA", Atomic Energy Regulatory Board, AERB/PSA/61090/01/52, Jan. 31, 2001.
3. International Atomic Energy Agency,."Procedures for conducting probabilistic safety Assessment of Nuclear Power plants (level 1) safety series No. 50-P-4, Vienna 1992.
4. BARC, "Dhruva Safety Report, Volume-3" A BARC internal document, 1994.
5. Winfield, D.J.,, "Safety Reassessment of AECL Test and Research Reactors", Proc International Symposium on Research Reactor Safety, Operation and Modifications" AECL(CRNL), Canada, Oct 1989 (IAEA-SM-310/94).

6. Flangen, G.F., etal., “Risk and Safety Analysis in Support of the Operation at the High Flux Isotope Reactor at the Oak Ridge”, Proc International Symposium on Research Reactor Safety, Operation and Modifications” AECL(CRNL), Canada, Oct 1989 (IAEA-SM-310/16).
7. International Atomic Energy Agency, “Reliability Data for Probabilistic Safety Assessment of NPP” , IAEA-TECDOC-478, , Vienna 1986.
8. International Atomic Energy Agency, “Reliability Data for Probabilistic Safety Assessment of Research Reactors”, IAEA-TECDOC-900, , Vienna.
9. US NRC Safety Study ((WASH-1400), NUREG-75/104, 1975.
10. MIL-HDBK-217:Military Standardization Handbook for Reliability Prediction of Electronic Equipment, 1991.
11. Swain, A.D. & Guttmann, H.E., "Handbook of Human Reliability Analysis with Emphasis on Nuclear Power Plant Applications", Final Report, NUREG-CR-1278, Sandia National Laboratories, August 1983.
12. International Atomic Energy Agency, “Human Reliability Data Collection and Modelling', IAEA-TECDOC-618, Vienna 1991.
13. International Atomic Energy Agency, “Case Study on the use of PSA Methods: Human Reliability Analysis, IAEA-TECDOC-592, , Vienna 1991.

Reliability, Safety and Hazard
Editors: P.V. Varde et al.

RISK INFORMED PLANT MODIFICATIONS AND UPGRADES FOR LIFE EXTENSION OF INDIAN BWR

Vibha Hari, Rajee Guptan, Nalini Mohan, S.G. Ghadge and S.S. Bajaj

Directorate of Safety, Nuclear Power Corporation of India Limited
Nabhikiya Urja Bhavan, Anushakti Nagar, Mumbai 400 094

ABSTRACT

Probabilistic Safety Assessment techniques can be used to determine how ageing is affecting component and system unavailability; to identify plant components and systems that have a propensity to ageing and are risk significant. An essential requirement for PSA is the failure data for various components. An analysis of the failure data TAPS operation revealed that they were still in their useful life period. Hence, age related component modelling were not called for. With the objective of TAPS life extension, NPCIL initiated Level-1 Probabilistic safety Assessment (PSA) studies for Tarapur Atomic Power Station –1 & 2. This Plant specific PSA Level – 1 study resulted in identifying and understanding the key plant vulnerabilities and in making recommendations for improvements by identifying risk significant systems and component and . In particular, in case of TAPS, PSA has proved to be a valuable tool to understand the implications of non- conformance to the current standards and in making decisions on upgrades. The same methodology was used for trying out the different configurations of the system under consideration for upgrade and for optimizing the final decision on the same. These results are also useful for prioritizing the implementation of the upgrades/modifications, technical specification optimisation, where the maximum benefit for plant safety can be obtained. Insights available based on this risk informed study for TAPS life extension are discussed in this paper.

Keywords: PSA, Life Extension, Ageing, Risk Informed Decision

1. INTRODUCTION

A Level 1 PSA provides an assessment of plant design and operation, focusing on events that could lead to core damage. It can provide major insights into design strengths and weaknesses and into ways of preventing core damage that would be a precursor to a large release of radioactive material. Indian Nuclear Power Station - a BWR, built at Tarapur (TAPS) is a 200 MWe twin unit station. It was commissioned in 1969 and today also it continues to operate successfully. The station has been the first and the only BWR type Indian Nuclear Power Station. Since these reactors are of the older design, while seeking their life extension, a fresh assessment of their safety has been attempted by NPCIL.

As a part of this assessment, a Level-1 PSA study of TAPS-1 & 2 was taken up for internal events where PSA was considered to complement the deterministic analysis for Design Basis Events (DBE's) and for Beyond Design Basis Accidents (BDBA's). focusing attention on the low probability and high consequence events along with high probability and low consequence events.

Insights from this risk informed PSA study provided inputs to decisions on design back fitting, plant operations, emergency operating procedures, safety analysis and on regulatory issues.

This paper discusses the salient features of the study. The results obtained and the insights derived from the study are discussed in details. The recommendations for modifications to certain systems in order to improve the overall safety of TAPS are discussed in the paper.

2. BRIEF STATUS OF PSA IN INDIAN NPPS

India has 13 PHWRs and 2 BWRs in operation with 3260 MWe installed capacity. Several other PHWRs, VVERs and Fast Breeder Reactor (FBR) type nuclear power plants are in different stages of construction and planning/approval. The first BWR was commissioned in 1969 and the first PHWR was commissioned in 1971. Several modifications have been carried out in each of these nuclear power plants based on in-house operating experience gathered so far, periodic safety assessments carried out and international feedback based on accidents such as TMI-2, Chernobyl etc. A systematic safety assessment based on PSA results has been a more recent development. The designing and licensing of nuclear power plants have, so far, been based on the deterministic criteria (which include defense-in-depth principle, safety design criteria and the concept of design basis accidents) to a large extent with some element of probabilistic considerations in terms of reliability targets for safety systems. In recent years NPCIL has taken up actions to extend this application of probabilistic approach to full-fledged PSA of nuclear power plants in a progressive manner in its endeavor to complement deterministic approach to nuclear safety by a probabilistic approach. Three Level-1 PSA studies, for TAPS (BWR) and the other for PHWRs KAPS and TAPP3&4 are completed.

3. BRIEF DESCRIPTION OF TAPS (BWR)

Tarapur Atomic Power Station (TAPS) is a twin unit BWR type nuclear power station designed and built by GEC, USA. Brief summary of design features of TAPS is given in Table0 -1. Both the units were originally dual cycle type, designed to generate 220 MWe power each. In 1984 after isolation of their secondary steam generators due to excessive tube leakage, the reactors are being operated in a single cycle mode at a load of 160 MWe each. Low Pressure Core Spray System, Auto Blow Down System, Post Incident System, Liquid Poison Injection System and Emergency Condenser System are the engineered safety systems of TAPS. A general plant schematic is provided in Fig-1.
This PSA study was conducted for Internal Initiating Events during full power operation considering the reactor core as the main source of radioactivity. Small Event Tree-Large Fault Tree approach was adopted in this analysis

4. QUALITY ASSURANCE

A Quality Assurance of various procedural tasks of PSA was recognized in this PSA project. Methods adopted in this study ensured an end product which adequately meets the objectives.

5. KEY PLANT VULNERABILITIES AND UPGRADATIONS PROPOSED

Key plant vulnerabilities emerged based on minimal cut sets. Also, after analysing the event sequences, dominant ones contributing significantly to plant core damage frequency were identified. Based on the minimal cutset for these dominant sequences, certain plant vulnerabilities emerged. Based on the results the proposed upgradation listed below have been arrived at in the order of priority and also based on the safety impact, targeting the resources, cost benefit and other considerations.

Modifications and upgrades proposed

Emergency Power Supply System : Emergency Power Supply System (EPSS) is considered first for up gradation. It is obvious that upgrading EPSS would have direct impact on availability of many frontline systems such as Low Pressure Core Spray System, Emergency Feed Water System, Emergency Condenser System, Shut Down Cooling System, Liquid Poison Injection System and so on. Emergency Diesel Generators have been located in the same room and in the vicinity of each other. They exhibit vulnerability for common cause failure (CCF) Needless to say that improving the availability of EPSS would improve the Core Damage Frequency to a great extent. Following modifications are proposed:

Existing emergency DGs (3 x 50% capacity) are proposed to be replaced with full capacity (3 x 100% capacity) DGs qualified to IEEE and other relevant standards.

Table 1 Brief summary of design features : TAPS

1.	Low Pressure Coolant Spray System	1. Low pressure (14.04 kg/cm^2) core spray system provides coolant to the core during accidents in which vessel pressure is low. Two independent trains and Four motor driven (Class III) pumps have been provided. This system is shared between two units. 2. Automatic depressurisation system has been provided to depressurise the reactor to 14.04 kg/cm^2 at which the low pressure core spray system can spray the coolant on the core.
		3. Three relief valves with capacity of about 150 MT/hr each have been provided. In addition there are five non auto blow down relief valves which relieve steam pressure at progressively higher pressures with similar relieving capacities. Control Rod Drive system provides back up source of high pressure injection with two trains of one pump each with 265 LPM capacity at 110 Kg/cm^2
2.	Reactor Core Isolation cooling system	This is also called as Emergency Condenser system. It is a passive system and can cool the reactor from 285 degree C to about 120 degree C after scram when reactor is isolated from main condenser. There are two tube bundles in the condenser.
3.	Key support systems	1. D.C. Power with 1200 AH station batteries. Two banks (8 hours duration). 2.Emergency AC power from three DGs is shared between two units. 3.R.B. cooling water system provides cooling water to safety systems. This system is shared between two units. R.B. cooling water system provides cooling water to safety systems. This system is shared between two units. 4. Salt service water system acts as the ultimate heat sink and cools rector building cooling water system. Instrument Air system is shared between two units.
4.	Heat removal system :	1. Residual heat removal / Post Incident system to reduce pressure/ remove heat from the containment during accidents. Two independent trains and four motor driven (class III) pumps have been provided. This system is shared between two units. 2.Residual heat removal (Class IV) shut down cooling system to remove decay heat, maintain reactor vessel integrity and reactor at low pressure. This system is shared between two units. It consists of one heat exchanger and three pumps out of which only one pump is required for operation.
5.	Reactivity control system	1. Control Rods 2. Liquid Poison Injection System with two parallel positive displacement pumps rated at 50 gpm per pump. This system is partially shared between two units. (Poison tank and reciprocating pumps are shared)
6.	Containment structure	1. A steel enclosure. 2. Volume is 1,32,880 cubic feet. (45 psig design pressure)

- The concept of forming a single bus when Emergency Power is needed, is proposed to be dispensed with. The new concept proposed would consist of a dedicated bus for each emergency DG, one DG each supplying emergency loads of Unit-1 and Unit-2 and the 3rd DG catering to the redundant loads for the normal emergency loads of both the units. This arrangement would provide adequate redundancy for the emergency loads of both the units in case any one DG or any emergency bus becomes unavailable.
- It is proposed to redistribute the emergency loads suitably.
- It is also proposed to redistribute these loads such that all the associated loads are on the same bus. (e.g. core spray pumps, their associated motor operated valves and their control logic circuits would be connected to the same bus). Presently, this aspect is not always taken care of.
  - It is further proposed to provide tie lines between the three buses so that in case any DG fails, its loads can be fed from one of the other two DGs.
  - It is proposed to relocate the Emergency DGs to a separate location for each with proper barriers in between them.
  - It is also proposed to relocate the associated emergency buses suitably.
  - It is further proposed to segregate emergency buses and non-emergency buses properly. (Today they are very near to each other.)

Shut Down Cooling System : Though Shut Down Cooling System (SDCS) plays a low key role for TAPS from the PSA considerations, it is the only system to cool the reactor enough to permit personnel entry into the reactor dry well for any inspection / maintenance work after the Emergency Condenser (EC) has cooled it to about 120 degree Celsius. Being a shared system between two units, demands for its operation were such that, it was thought best to take this system for upgradation and segregate the system for each unit. This decision was also influenced by the consideration that due to spent fuel getting accumulated in the fuel pool over the past several years, the redundant heat exchanger from SDCS had to be diverted to fuel pool cooling system to cope up with the increased heat load and SDCS is left with only one heat exchanger. Following modifications are proposed :

- It is proposed to provide independent heat exchanger(s) for Fuel Pool Cooling System. The upgraded SDCS would consist of 3 pumps (3×100% capacity), 3 heat exchangers (3×100% capacity) and adequate isolation valves, making two independent SDCS for two units with one heat exchanger (100% capacity) and one pump (100% capacity) as common redundant for both units.
- The control logic circuits also are proposed to be modified suitably.

Compressors : Compressors for Instrument Air System have been located in the same room as that of emergency DGs and in the vicinity of each other. They exhibit vulnerability for CCF. Following modifications are proposed:

- It is proposed to relocate the three compressors, keeping them in the same room but utilizing the empty space created by shifting of the emergency DGs.
- It is also proposed to provide barriers for each compressor.

Equipment/Components which need improvement in test and maintenance : Critical components from Primary Steam Isolation Valve System, Liquid Poison Injection System, Emergency Feed Water System, Emergency Condenser System, Shut Down Cooling System and Emergency Core Cooling System were identified for consideration of improvement in test and maintenance.

The test and maintenance procedures and its frequency being followed presently is based on several considerations such as MTBF, MTTR, accessibility of components, criticalness of the function the component and so on. USNRC guide is followed as the guideline to arrive at the optimum AOT in the Technical Specification. TAPS PSA is being used as an engineering tool for optimization of Technical Specification with regard to Allowed Outage Time (AOT) and Surveillance Test Interval (SIT) for equipment/components important to ensure safe operation of the plant.

6. CONCLUSIONS

The conventional safety philosophy. is based on deterministic approach, which includes defense in depth principle, safety design criteria and the

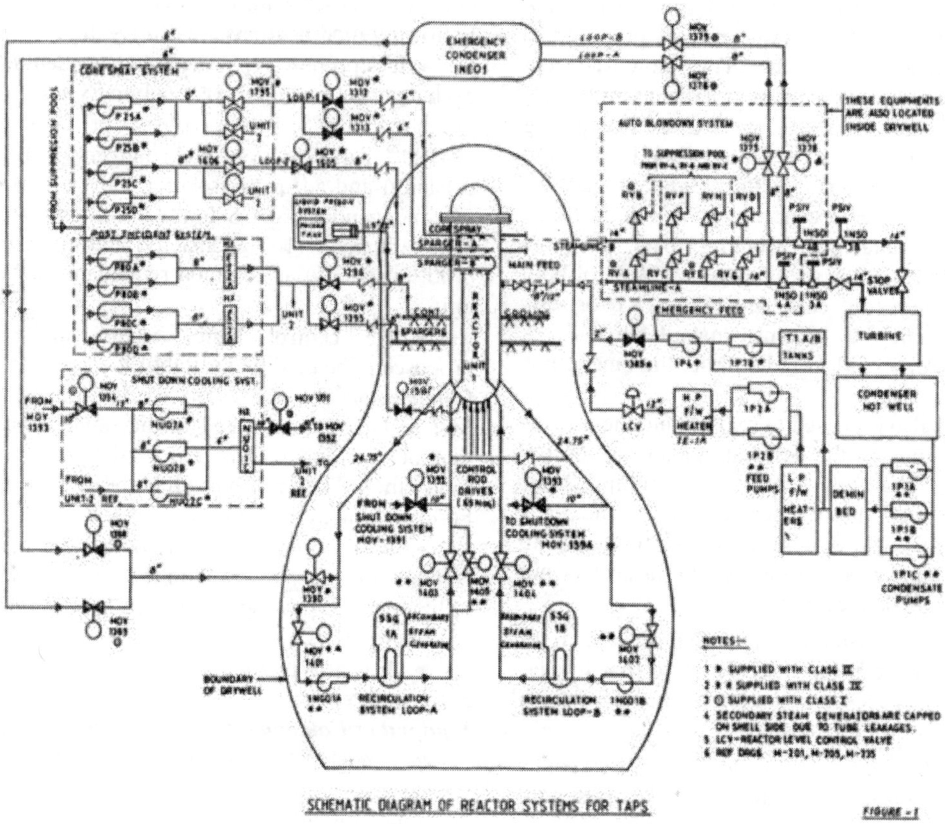

Fig. 1: Schematic Diagram of Reactor Systems for TAPS

concept of design basis accidents. PSA is used to complement, enhance, and validate conclusions that are based on these deterministic design principles.

In case of TAPS , this modern risk informed approach to safety regulation was adopted in which the PSA is used to provide one of the inputs to decisions concerning safety.

PSA, which considers a much wider range of faults, takes an integrated look at the plant as a whole (system inter-dependencies), and uses realistic criteria for the performance of the plant and systems. The PSA, therefore, is a powerful tool for safety management and its use can increase the level of safety by providing information on design and operation by identifying potential plant vulnerabilities that contribute to risk. Measures can then be taken to reduce this contribution. To achieve the desired end effect, there can be several ways the modifications can be thought of. Analysing the impact of each proposal on the plant safety enables to select the optimum proposal for implementation. Indeed, for life extension upgradation of TAPS, PSA results have played an important role.

7. REFERENCES

1. Design and Analysis Reports of TAPS – Vol-1 and Vol-2 (1966).
2. TAPS Design Basis Reports Vol-I Through Vol-VI (1974).
3. Level-1 PSA report for TAPS Vol. 1-4.
4. International Atomic Energy Agency, A frame work for a quality assurance programme for PSA, IAEA -TECDOC-1101, IAEA, Vienna (1999).
5. International Atomic Energy Agency, Applications of Probabilistic Safety Assessment (PSA) for Nuclear Power Plants, IAEA-TECDOC-1200, IAEA, Vienna (2001)

Reliability, Safety and Hazard
Editors: P.V. Varde et al.

FRAMEWORK FOR APPLYING RI-ISI METHODOLOGY FOR INDIAN PHWRS

Gopika Vinod, R.K. Saraf, A.K. Ghosh and H.S. Kushwaha

Health, Safety & Environment Group, Bhabha Atomic Research Centre, Mumbai 400085

ABSTRACT

Risk Informed In-Service Inspection (RI-ISI) aims at categorizing the components for In-Service inspection based on their contribution to Risk. For defining the contribution of risk from components, their failure probabilities and its subsequent effect on Core Damage Frequency (CDF) needs to be evaluated using Probabilistic Safety Assessment methodology. During the last several years, both the U.S. Nuclear Regulatory Commission (NRC) and the nuclear industry have recognized that Probabilistic Safety Assessment (PSA) has evolved to be more useful in supplementing traditional engineering approaches in reactor regulation. The paper highlights the various stages involved in applying RI-ISI and then compares the findings with existing ISI practices.

Keywords: Risk Informed In-Service Inspection, CanStandards, Conditional Core Damage Probability

1. AN INTRODUCTION TO RI-ISI

The objective of the In-Service Inspection (ISI) program is to identify service-induced degradation that might lead to pipe leaks and ruptures, thereby meeting, in part, the requirements set in the General Design Criteria and applicable standards. ISI programs are intended to address all piping locations that are subject to degradation. Incorporating risk insights into the programs can focus inspections on the more important locations and reduce personnel exposure, while at the same time maintaining or improving public health and safety. The justification for any reduction in the number of inspections should address the issue that an increase in leakage frequency or a loss of defense in depth should not result from decreases in the numbers of inspections. Keeping in view of this requirement, risk informed technologies have also established their strength in determining an optimum inspection plan.

International experience has shown that degradations detected in nuclear power plants were mainly due to causes which are not anticipated during design. Obviously, there is need for setting up an ISI program that has the capability of more accurately finding the components where the probability of degradation and its consequence will be greatest. Last decade has seen development of Risk informed procedures for plant maintenance activities, where focus is set on the risk which is measured in terms of CDF [1]. American Society of Mechanical Engineers (ASME) in co-operation with Westinghouse Owner's Group (WOG) has developed a procedure based on importance measures, where failure probability is combined with consequence to introduce a ranking among the components for ISI selection. Electrical Power Research Institute (EPRI) has defined another procedure which uses a risk matrix including the degradation mechanisms and consequence of failure. ASME and EPRI were able to chart out a proper methodology for Risk

Informed In-Service Inspection (RI-ISI), which has resulted in ASME code cases 560 & 578 respectively [2,3].

Definite improvements in ISI programme has been observed in implementing RI-ISI procedures over ASME section IX in terms of reduced plant down time, safety significant focused inspection and reduced radiation exposure. This paper is an attempt to study the improvement of RI-ISI programme over Can standards [4] followed in Indian Pressurised Heavy Water Reactors (PHWRs) and introduction of risk informed concepts in inspection planning in Indian PHWRs. Typical case study was conducted by applying RI-ISI on components appearing in Primary Heat Transport System (PHTS) & Shut Down Cooling System (SDCS).

2. CURRENT INSPECTION PHILOSOPHY IN INDIAN PHWRS

The inspection currently followed at Indian PHWR is partly based on CAN standards, ASME Section IX and expert judgment derived from operating experience. Current ISI philosophy considers the determination of inspection areas and the degree of inspection by taking into account the safety margins, magnitude, and type of failure. Inspection categories are function of size of failure, stress ratio and fatigue usage factor. The size of failure is expressed as the ratio of the maximum energy release rate from the failure being considered to maximum energy release rate from the most severe failure considered during design of the system that directly transports heat from the nuclear fuel. This ratio is denoted by RE. The failures are categorized as large, medium and small depending on calculated value of RE. Fatigue usage factors shall be determined by the rules given in ASME Boiler Pressure Vessel Code Section III. Stress Ratio is defined as the ratio of calculated stress to allowable stress as per PHWR practices. The criteria for developing inspection programme are derived from stress ratio and fatigue usage factor. According to the above criteria, four categories are defined viz., A, B, C1 & C2 signifying decreasing order of severity. Fig. 1.1 and 1.2 represent four categories for the matrices of fatigue usage factor and stress ratio for medium and large size failures respectively.

		FATIGUE USAGE FACTOR		
		0.01 (Low)	0.1 (Medium)	1.0 (High)
STRESS RATIO	High (1-2/3)	C1	B	A
	Medium (2/3-1/3)	C1	B	B
	Low (1/3-0)	C2	C1	C1

Fig. 1.1: Determination of category for medium failure size

		FATIGUE USAGE FACTOR		
		0.01 (Low)	0.1 (Medium)	1.0 (High)
STRESS RATIO	High (1-2/3)	B	A	A
	Medium (2/3-1/3)	C1	B	A
	Low (1/3-0)	C2	C1	C1

Fig. 1.2: Determination of category for large failure size

Depending on the category in which each component is falling, the inspection method, quantum of inspection and interval are determined. Regarding Inspection interval, first inspection will be carried within first 5 years of reactor operation and subsequent inspections are to be done for time intervals that do not exceed 1/3 of the design operational life of the plant or 10 years, whichever is less.

3. APPLYING RI-ISI

RI-ISI involves following program tasks, which are to be carried out for successful application.

(i) Identify the systems to be included in RI-ISI program using importance measures

(ii) Estimate the component failure probabilities for leak and rupture cases and identify the applicable degradation mechanisms for each component.

(iii) Estimate the consequence of failure

(iv) Apply these estimates in EPRI matrix for risk categorisation

(v) Conduct a Risk impact analysis

Upcoming sections describe how each of these tasks are handled in this case study.

3.1 System Identification

ASME & WOG has suggested importance measure based approach for system identification. They have used combination of Risk Reduction Worth (RRW) and Risk Achievement Worth (RAW) for prioritization as shown in Table 1 [5].

Table 1 ASME-WOG RI-ISI categorisation

Risk Category	Criterion
Potentially high (1)	RRW >1.005
High (2)	1.001>RRW > 1.005 & RAW >2
Low (3)	RRW< 1.001

This scheme has been applied to process and safety system of a typical Indian PHWR. For applying RAW to process system, a value of 1/yr has been taken in the case study since the failure frequencies considered for the process systems are much less than 1/yr.

Table 2 Results of Ranking Systems using ASME-WOG scheme

System Name - Type	RRW	RAW	ASME Rank Cate.
Fire Fighting System	1.248	25.59	1
Moderator System	1.14	63.96	1
Active Process Water Cooling System (IE)	1.628	1485	1
Secondary Steam Relief System	1.049	126.62	1
Shut-Down Cooling System	1.036	37.44	1
Main Steam Line Break (IE)	1.085	88.14	1
Emergency Core Cooling System (Injection)	1.014	5.35	1
Active Process Water System (IE)	1.001	1.96	2
Small LOCA (IE)	1.00006	1.009	3

Table 2 presents the results of applying this ranking scheme to some systems in an Indian PHWR. Various drawbacks were found in this ranking scheme such as (i) difficulty in prioritising the systems/ segments having the same rank (ii) RAW and RRW are not additive, which is a prerequisite for ranking group of structures, systems and components and (iii) these measures cannot be extended to the parameter level.

Since the conventional importance measures cannot bring out a relative ranking of process and safety system, Differential Importance Measure has been applied system wise. Differential Importance Measure (DIM) can be defined as the fraction of total change in Risk or Core Damage Frequency that is due to a change in parameter x_i [6].

$$DIM_{x_i} = \frac{\delta CDF_{x_i}}{\delta CDF}$$

$$\delta CDF_{x_i} = CDF(x_i + \delta x_i) - CDF_0$$

$$\delta CDF = \sum_i \delta CDF_i$$

Table 3 Ranking of systems using DIM

System Name - Category	DIM	ASME Rank
Fire Fighting System	0.355	1
Moderator System	0.218	1
Active Process Water Cooling System (IE)	0.184	1
Secondary Steam Relief System	0.083	1
Shut-Down Cooling System	0.063	1
Main Steam Line Break (IE)	0.037	1
Emergency Core Cooling System (Injection)	0.025	1
Active Process Water System (IE)	5.9E-4	2
Small LOCA (IE)	3E-5	3

DIM has overcome the limitations found in ASME-WOG scheme. The advantages found in applying DIM are (i) it is a additive measure, hence suitable to handle ranking of group of Systems, Structures and Components, (ii) importance of changes affecting multiple pipe segments can be analyzed and finally, effect of parameters such as different applicable degradation mechanisms in a single pipe segment can be analyzed. Table 3 shows the list of ranking obtained from applying DIM to various process and safety systems appearing in PSA model which includes components subjected for ISI

3.2 Estimation of Failure Frequency of Components

Indian PHWRs have not so far observed the presence of any degradation mechanism in any of the components in critical systems. But from the international experience, it has been found that components are subjected to various damage mechanisms which are not anticipated in the design stage. The presence of these mechanisms is mainly attributed to the material of the components, its functional requirement as well as due to the environmental conditions it is subjected to. An attempt has been made to associate possible damage mechanisms to each component that are considered in this case study. For components like pumps, heat exchangers etc., statistical estimates were considered for failure probability analysis. Estimation of piping failure probabilities has always been a debatable topic from the point of view of PSA and Structural reliability analysis. Structural reliability analysis requires lot of input data as well as models to come up with a realistic estimate. The next best option is the recently developed Statistical estimates, which are popular under Thomas model. Thomas model has been applied to piping segments to derive a representative failure probability value. H.M. Thomas [7] has suggested an approach by which the base failure probability is multiplied with factors derived from service experience. Since these factors were derived way back in 1981, B.O.Lydell [8] has modified these factors in 2000, with the reactor experience till then. A correlation model has been proposed between piping design and operational parameters and frequency of leakage.

$\lambda_{F\text{-}TOT} = \lambda_{BASE}\, Q_E\, FB$

where,

$\lambda_{F\text{-}TOT}$ = Plant specific total leakage frequency

λ_{BASE} = Base line generic frequency (in the range of 1E-9 to 1E-7/ year : recommended value of 1e-8/year)

F = plant age factor B = design learning curve

Q_E = Multiplier representing the change in reliability by piping size and shape differences

$= Q_P + (AQ_W)$

$Q_P = L(D/t^2)$ $Q_W = N1.75(D/t)$

A = Weld Penalty factor (suggested value 50)

L = Length of pipe D = Pipe Diameter

t = Pipe wall thickness

N = Number of circumferential welds in the system

For calculating Age factor F, cumulative number of failures in first ten years as unity and then plotted the number of failures per year as a fraction of unity against age. F factor is then plotted as a function of pipe size and degradation mechanism. As far as design learning curve factor B is concerned, Thomas remarks that unless one is estimating the leak or rupture frequency for a new design there is no need to consider the B factor. It may be assumed to 1. Finally, Thomas defined relationship between frequency of catastrophic rupture (λ_C) and frequency of leakage (λ_L).

$$\lambda_C = \lambda_L \times P(C|L)$$

where, P(C|L) = conditional probability of rupture given leakage

Another alternative is SKI published values which gives failure probabilities depending on the dominant degradation mechanism acting on it. Conditional rupture probability has also been derived for all degradation mechanisms from SKI-PIPE database [9] as shown in Table 4.

Table 4 Statistical data on degradation mechanisms (SKI database)

Failure mechanism	Fail	Rup	$\lambda_j(F)$	P(C|L)
Corrosion Fatigue	14	0	6.8×10^{-3}	<0.071
Thermal Fatigue	38	0	1.8×10^{-2}	<0.026
Stress Corrosion Cracking	166	0	8×10^{-2}	<0.006
Erosion Corrosion	15	0	7.3×10^{-3}	<0.067
Corrosion	72	3	3.5×10^{-2}	0.042
Water Hammer	35	15	1.7×10^{-2}	0.43
Flow Accelerated Corrosion	280	19	1.4×10^{-1}	0.068
Vibration Fatigue	364	25	1.8×10^{-1}	0.069
OTHERS	43	8	2.1×10^{-2}	0.19
Design & Construction	192	13	9.3×10^{-2}	0.068
UNKNOWN	177	11	8.6×10^{-2}	0.062

3.3 Estimate the Consequence of Failure

In RI-ISI, consequence of failure of a component is expressed in terms of Conditional Core Damage Probability (CCDP). Depending on the function and state of the component, various expressions are available for finding the CCDP. A component failure can either initiate an initiating event or impair a safety system function.

In the systems considered, any failure in the components present in PHT will be an initiating event, viz., Loss of Coolant Accident (LOCA). Depending upon the area of leak or rupture, they are classified either as small or large LOCA events. In the case of Small LOCA, CCDP comes from the accident sequence involving failure of Reactor Protective system, where as in the case of Large LOCA, CCDP comes from accident sequence involving failure in injection and recirculation of Emergency Core Cooling System (ECCS).

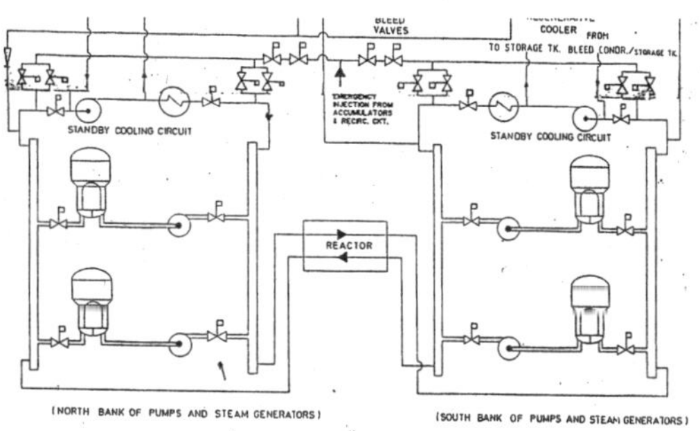

Fig. 2: Schematic of PHTS & SDCS

For the shut down cooling system, pipelines upto the first isolation valves (MV1, MV2, MV3 & MV4) are connected with PHTS. So any leak or rupture in these pipelines will be equivalent to a LOCA event and treated likewise. After these isolation valves, components will be in an on-demand configuration. Here, mean time between the tests/ inspection interval or demand is to be considered. The expression can be written as

$$CCDP_i = [CDF_{(qi=1)} - CDF_{(BASE)}] * T_t$$

Where,

$CDF_{(qi=1)}$ = CDF given the component failure due to joint i

$CDF_{(BASE)}$ = BASE CDF

q_i = Pipe break unavailability

T_t = Mean time between tests or demands

For SDCS components, weekly test interval is followed. Also for calculating the piping unavailability online non-repair model is assumed with mission time taken as 2 months.

3.4 Application of Estimates in EPRI Matrix for Risk Categorisation

EPRI has designed a Risk matrix with different inspection categories, depending on the CDF values and degradation mechanism for determining the inspection interval. Each segment is assigned the appropriate category depending on it Conditional Core Damage Probability (CCDP) and degradation mechanism. Risk matrix can be defined as a Decision matrix that is used to categorise the pipe segments into high, medium, and low importance, based on degradation mechanism and consequence of its failure. By examining the service data, a basis has been established for ranking pipe segment rupture potential as High, Medium, or Low simply by understanding the type of degradation mechanism present. Consequence can be quantified through the estimation of CCDP.

Potential	Degradation Mechanism
High	Flow Accelerated Corrosion, Vibration Fatigue, Water Hammer
Medium	Thermal Fatigue, Corrosion Fatigue, Stress Corrosion Cracking, Pitting, Erosion Corrosion
Low	No degradation mechanism

likelihood frequency		None	Low	Medium	High
	Low	Low 7	Medium 5	High 3	High 1
	Medium	Low 7	Low 6	Medium 5	High 2
	High	Low 7	Low 7	Low 7	Medium 4

Fig 5: Risk Matrix

Table 5 presents the risk category for some of the components in the case study as per the Risk matrix suggested by EPRI (as shown in Figure 3).

Table 5 Risk Category from EPRI Risk Matrix

Component Description	Fail –Freq (/Yr)	CCDP	Risk Cate.
Piping from Outlet Header – Steam Generator	9.54E-06	8.74E-06	5
Piping from Steam Generator – Primary Coolant Pump	2.50E-06	8.74E-06	5
Piping from Primary Coolant Pump Valve	7.93E-07	2.03E-07	6
Piping from Primary Coolant Pump – Inlet Header	4.66E-06	8.74E-06	5
Steam Generator	8.76E-06	2.03E-07	6
Primary Coolant Pump	1.05E-05	8.74E-06	5
Reactor headers	3.00E-04	8.74E-06	3

3.5 Risk Impact Analysis

Risk impact analysis essentially carries out a comparison between the existing inspection programme and risk based inspection programme. For the components considered in the case study, these aspects were compared and summarized in the table 6 given below. Code has been given to various inspection methods such as: 1 -Visual examination, 2-Dimensional examination, 3A-Surface examination (Liquid Penetration Test, Magnetic Particle Test, Eddy current Test), 3B-Volumetric examination (Ultrasonic Test, Radiography Test, Eddy current Test) and 4. - Integrative examination

Table 6 Risk Impact Analysis

Component Description	Current Cate.	Insp Method	Risk Cate.
Piping from Outlet Header – Steam Generator	B	1, 3A, 3B	5
Piping from Steam Generator – Primary Coolant Pump	B	1, 3A, 3B	5
Piping from Primary Coolant Pump Valve	B	1, 3A, 3B	6
Piping from Primary Coolant Pump – Inlet Header	B	1, 3A, 3B	5
Steam Generator	A	1, 3A, 3B	6
Primary Coolant Pump	A	1, 3A, 3B	5
Reactor headers	A	1, 3A, 3B UT	

4. CONCLUSIONS

Can Standards has categorized Steam Generator and pump in high category which demands a 5 year inspection period. From RI-ISI consideration, they can be put in low and medium category, which allows for an inspection interval greater than 5 years. Similar reduction in inspection activities were observed for other components which are not listed in Table 6. Risk based inspection methodology has demonstrated substantial reduction in inspection efforts. It should be noted that risk to the plant has not increased by reducing these inspection efforts, since the high safety significant components are still following the higher frequency of inspection. Risk based inspection has organized the category of inspection for each component thereby changing its method of inspection. Also, it provides a technical basis for fixing the inspection category. It is very much essential to understand the prevailing degradation mechanism in each component so that a suitable inspection method can be suggested accordingly. Given below is a suggested inspection method as per the possible degradation mechanisms acting in the respective component [5].

Degradation Mechanisms	Method
SCC or Low Cycle Fatigue	Ultrasonic Testing (UT)
Erosion Corrosion , Pitting	Radiography Test/ UT
Flaws in non-ferro magnetic materials	Eddy current Testing
Flaws in Ferro magnetic materials	Magnetic Particle Test
Surface degradation or cracking	Visual Test

5. REFERENCES

1. Syed ali, Bagchi, "*Risk Informed In-Service Inspection*", Nuclear Engineering Design, 181, pp221-224 (1998).
2. ASME Code CASE N-560, *Alternative Examination Requirements for Class1, Category B-J Piping welds*
3. ASME Code CASE N-578, *Risk Informed methods for In-Service Inspection of pipe welds*
4. CAN/CSA-N-285.4, "*Periodic Inspection of CANDU Nuclear Power Plant components*".

5. RIBA PROJECT, "*Risk Informed Approach for In-Service Inspection of Nuclear Power Plant components*", EUR 20164 EN, Project Summary, December 2001.
6. Borgonovo et.al., "*A new importance measure for risk informed decision making*", Reliability Engineering and System Safety, 72 pp. 193-212 (2000)..
7. H. M. Thomas., "*Pipe and Vessel Failure Probability*", Reliability Engineering, Vol.2, (1981), pp. 83-124.
8. B.O.Y.Lydell, "*Pipe Failure Probability-the Thomas paper revisited*", Reliability Engineering and System Safety, Vol.68, (2000), pp. 207-217.
9. Gosselin, Fleming, "*Evaluation of Pipe failure potential via degradation mechanism assessment*", ICONE 5, pap 2641, May 26-30, Nice, France (1997).

Reliability, Safety and Hazard
Editors: P.V. Varde et al.

AN OVERVIEW OF EROSION CORROSION MODELS AND RELIABILITY ASSESSMENT FOR CORROSION DEFECTS IN PIPING SYSTEM

A. Srividya, H.N. Suresh, A.K. Verma, V. Gopika[1] and Santosh[1]

Reliability Engineering Group, Indian Institute of Technology Bombay, Mumbai 400 076
[1] Reactor Safety Division, BARC, Mumbai 400 085

ABSTRACT

Piping systems are part of passive structural elements in power plants. The analysis of the piping systems and their quantification in terms of failure probability is of utmost importance. The piping systems may fail due to various degradation mechanisms like thermal fatigue, erosion-corrosion, stress corrosion cracking and vibration fatigue. On examination of previous results, erosion corrosion was more prevalent and wall thinning is a time dependent phenomenon. The paper is intended to consolidate the work done by various investigators on erosion corrosion in estimating the erosion corrosion rate and reliability predictions. A comparison of various erosion corrosion models is made. The reliability predictions based on remaining strength of corroded pipelines by wall thinning is also attempted. Variables in the limit state functions are modelled using normal distributions and Reliability assessment is carried out using some of the existing failure pressure models. A steady state corrosion rate is assumed to estimate the corrosion defect and First Order Reliability Method (FORM) is used to find the probability of failure associated with corrosion defects over time using the software for Component Reliability evaluation (COMREL).

Keywords: Piping System, Reliability, Erosion Corrosion, FORM

1. INTRODUCTION

Large numbers of existing structures deteriorate with time due to exposure to environment. These stressors affect the strength of piping system. Erosion Corrosion is one of the major causes of degradation in steel pipelines. The major effect of corrosion is the loss of metal cross-section, which reduces the load carrying capacity. If this is not detected and repaired, it will reduce the remaining strength of the pipelines. It is observed that the corrosion increases with time. The erosion corrosion rate can be found using suitable model from which the thickness of metal corroded and hence the thickness of pipe can be determined.

Many investigations have been carried out to determine the remaining strength of corroded pipelines [1-5]. B31G model is found to be the most widely used among all the failure pressure models. But because of its conservative estimates for longer defect length, it is modified by Keifner and Veith [3] and has given rise to modified B31G model, which overcomes the draw back of B31G model and is used in this work for reliability analysis using COMREL software.

The failure pressure can be estimated by using the available techniques [1] and a comparative study on various models available to evaluate failure pressures to estimate reliability is attempted in this paper.

2. ESTIMATION OF EROSION CORROSION RATES

Erosion corrosion also called Flow Assisted Corrosion (FAC) rate can be found by using KWU-KR [6] model to determine the thickness of pipe. The method estimates the erosion corrosion rate, which is a function of flow velocity, temperature, material chemical composition, fluid chemistry and exposure time.

The model KWU-KR FAC [6] is used to calculate corrosion rate as a function of Keller's geometry factor, Flow velocity, Fluid temperature, Material composition, Fluid Chemistry (pH at 25 ^{0}C and dissolved oxygen), Exposure time and in case of 2 phase flow, steam quality. The pipe thickness can be calculated as a function of time.

The wall corrosion W_C (t) is the thickness of pipe corroded and is given by,

$$W_C(t) = \frac{\Delta\phi_R\, t}{\rho_{St}} \tag{1}$$

$\Delta\phi_R$ = FAC rate (μg/cm^2 hr)

t = exposure time (hr)

ρ_{St} = density of steel (μg/cm^3)

Once FAC rate is found, the thickness can be calculated by

$$W_{Pipe}(t) = W_{Original} - W_{Calculated}(t) \tag{2}$$

W_{pipe} = pipe wall thickness at time t (cm)

$W_{Original}$ = original nominal pipe thickness (cm)

$W_{Calculated}$ (t)= calculated thickness of pipe corroded at time t (cm)

$$\Delta\phi_R = 6.35\times K_C(B\cdot e^{(N-W)}[1-0.175\times (pH-7)^2]\cdot 1.8\cdot e^{0.118g}+1)\times f(t) \tag{3}$$

$$B=-10.5\sqrt{h}-(9.375*10^{-4}T^2)+(0.79T)-1325 \tag{4}$$

$$N=-0.0875h-(1.275\times10^{-5}T^2)+(1.078\times10^{-2}T)-2.15 \tag{5}$$

for 0=h=0.5%

$$N=(-1.29\times10^{-4}T^2+0.109T-22.07)\times 0.154e^{-1.2h} \tag{6}$$

for 0.5%≤h≤5%

$\Delta\phi_R$ = calculated specific rate of material loss (μg/cm^2 hr)

K_C=geometrical factor (0.04 for straight pipes)

w = flow velocity (m/s)

pH = pH value (9.5 to 10.5)

g = oxygen content ((μg/kg)

h = content of chromium and molybdenum in steel (is taken as < 0.5%)

T = temperature (^{O}K)

f(t) = a time correction factor (0.79 for periods >9.6x10^4 hrs.)

3. FAILURE PRESSURE MODELS

It is evident that volume defects are generated by a corrosion process. In the estimation of failure pressure, the critical parameters are depth of metal loss and axial length. It is believed that the corrosion damage accumulates with an increase in exposure period. Hence it can be assumed that the critical parameters of a corrosion defect are depending on the exposure period and corrosion rate. To evaluate the pipeline failure pressure, various pressure models which are available in the literature [1] are given in the subsequent sections using the following notations:

P_f = failure pressure

T = pipeline elapsed time

t = pipeline wall thickness

L = Length of defect

D = Diameter of pipe

d = depth of defect

M – Folio's factor

UTS = ultimate tensile strength of pipe

YS – yield strength of pipe

i) Model B31G

$$G=0.893\frac{L(T)}{\sqrt{Dt}} \text{ and } M=\sqrt{1+0.893\frac{L(T)^2}{Dt}} \tag{7}$$

$$Pf=1.11\frac{2YS\,t}{D}\left[\frac{1-\frac{2d(T)}{3t}}{1-\frac{2d(T)}{3tM}}\right] \text{ For G<4} \tag{8}$$

$$P_f = 1.11\frac{2YSt}{D}\left[1-\frac{d(T)}{t}\right] \text{ for } G \geq 4 \quad (9)$$

ii) Modified B31G

$$Pf = \frac{2(YS+68.95)t}{D}\left[\frac{1-0.85\frac{d(T)}{t}}{1-0.85\frac{d(T)}{tM}}\right] \text{ for } G< 4 \quad (10)$$

$$M = \sqrt{1+0.6275\frac{L(T)^2}{Dt}-0.003375\frac{L(T)^4}{D^2t^2}} \text{ for}$$

$$\frac{L^2}{Dt} \leq 50 \quad (11)$$

$$M = 0.032\frac{L(T)^2}{Dt}+3.3 \text{ for } \frac{L^2}{Dt} > 50 \quad (12)$$

iii) Battelle Model

$$Pf = \frac{2UTS\,t}{D}\left[\frac{1-d(T)}{t}M\right] \quad (13)$$

$$M = 1-\exp\left[-0.157\frac{L(T)}{\frac{\sqrt{D(t-d(T))}}{2}}\right] \quad (14)$$

iv) DNV 99 Model

$$P_f = \frac{2UTS\,t}{(D-t)}\left[\frac{1-\frac{d(T)}{t}}{1-\frac{d(T)}{tM}}\right] \text{ for } G<4 \quad (15)$$

$$M = \sqrt{1+0.31\frac{L(T)^2}{Dt}} \quad (16)$$

v) Shell - 92 Model

$$P_f = \frac{1.8UTS\,t}{D}\left[\frac{1-\frac{d(T)}{t}}{\frac{1-d(T)}{tM}}\right] \quad (17)$$

$$M = \sqrt{1+0.805\frac{L(T)^2}{Dt}} \quad (18)$$

4. RELIABILITY ANALYSIS

Deterministic approach has been described above. The probabilistic reliability analysis was given by Ahammed [5]. Pipe parameters are treated as random variables following Gaussian distribution. For reliability analysis, a failure function is defined which expresses the criterion for failure. The failure function can be considered to be equal to the difference between failure pressure P_f and applied pressure P_a.

$$Z=P_f - P_a \quad (19)$$

The limit state function is given by

$$Z = 2(S_Y+68.95)\frac{t}{D}\times \left[\frac{1-[d_O+R_d(T-T_O)]/t}{1-[d_O+R_d(T-T_O)]/tM}\right]-P_a \quad (20)$$

Positive value of Z indicates that it is safe otherwise indicates failure state. The method involves linearizing the limit state function and determining the reliability index (β) which is a measure of pipeline safety. β is found using the iterative procedure adopted in COMREL [8] and the starting value of reliability index can be determined by Cornell's β,

$$\beta = \mu_{Z}/\sigma_{Z} \quad (21)$$

μ_Z and σ_Z are the mean and standard deviation of Z respectively.

5. EXAMPLE

In order to demonstrate, PHWR outlet feeder piping system [7] is considered as a case study. There are 306 number of small diameter pipes ranging from 32 mm to 70 mm and length 2 meter to 22 meter that connects outlet header to the steam generator. The feeder pipe is made of Carbon steel, A106GrB. It is assumed that erosion corrosion is present in the outlet feeder. Only straight pipes are considered for study.

Table 1.1 FAC values for different flow velocity

K_c	W m/sec	pH	g μg/kg	h %	T °K	f(t)	ΔΦR μg/cm²hr
0.04	1.5	9.5	15	0.5	373	0.79	0.16808
0.04	2.0	9.5	15	0.5	373	0.79	0.18090
0.04	2.5	9.5	15	0.5	373	0.79	0.18868
0.04	3.5	9.5	15	0.5	373	0.79	0.19625
0.04	4.0	9.5	15	0.5	373	0.79	0.19799
0.04	4.5	9.5	15	0.5	373	0.79	0.19904
0.04	5.0	9.5	15	0.5	373	0.79	0.19968

Table 1.2: FAC values for different oxygen content

K_c	W m/sec	pH	g μg/kg	h %	T °K	f(t)	ΔΦR μg/cm²hr
0.04	1.5	9.5	0	0.50	373	0.79	0.00941
0.04	1.5	9.5	5	0.50	373	0.79	0.09464
0.04	1.5	9.5	10	0.50	373	0.79	0.14189
0.04	1.5	9.5	15	0.50	373	0.79	0.16808
0.04	1.5	9.5	20	0.50	373	0.79	0.18260
0.04	1.5	9.5	25	0.50	373	0.79	0.19065
0.04	1.5	9.5	30	0.50	373	0.79	0.19511

Table 1.3: FAC values for Different pH values

K_c	W m/sec	pH	g μg/kg	h %	T °K	f(t)	ΔΦR μg/cm²hr
0.04	1.5	7.5	0	0.5	373	0.79	2.15144
0.04	1.5	8	5	0.5	373	0.79	1.13361
0.04	1.5	8.5	10	0.5	373	0.79	0.58069
0.04	1.5	9	15	0.5	373	0.79	0.30491
0.04	1.5	9.5	20	0.5	373	0.79	0.18260
0.04	1.5	10	25	0.5	373	0.79	0.13926
0.04	1.5	10.5	30	0.5	373	0.79	0.13296

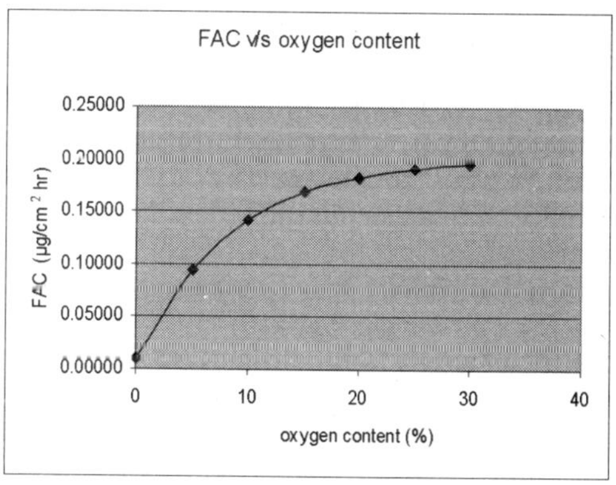

Fig. 1: FAC with Oxygen content

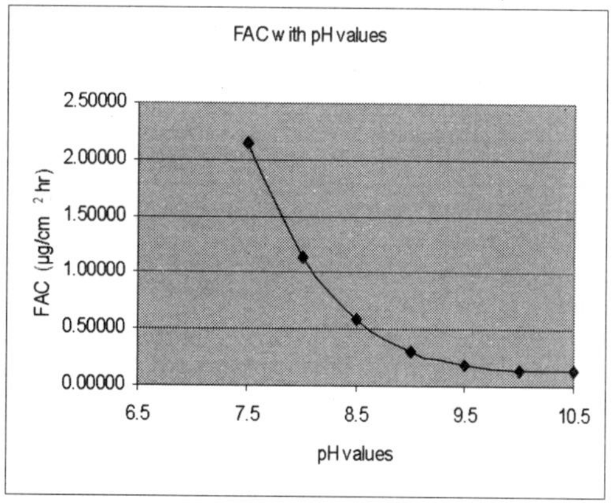

Fig. 2: FAC with pH value

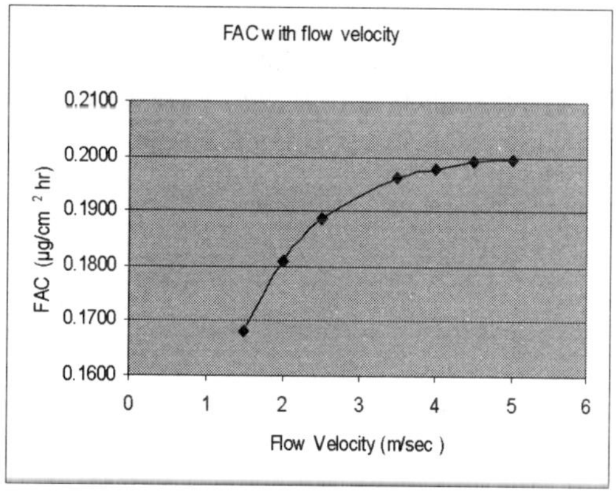

Fig. 3: FAC with Flow velocity

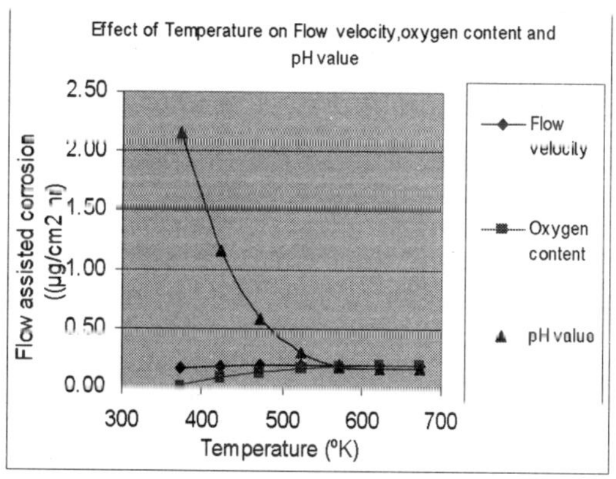

Fig. 4: Effect of temperature on flow velocity, oxygen content and pH value

Table 2 Data for random variables used in example with mean and Co- variance

Parameter	Mean value	Co Variance
Depth of defect (mm) d_o	2	0.1
Yield strength (MPa) YS	358	0.069832
Ultimate tensile strength (MPa) UTS	455	0.07329
Diameter of pipe (mm) D	72	1.5
Radial corrosion rate (mm/year) R_d	0.051	0.015
Fluid pressure (MPa) P_a	8.7	0.9
Service life (Years) T	40	-
Inspection interval (years) T_O	10	-
Length of defect (mm) L	300	0.05
Thickness of pipe (mm)	7	0.148

To illustrate the concepts discussed above, a pipeline from an outlet feeder of PHWR plant was analyzed with known defects. The variables considered are shown in table 2 [7].
The above data is solved with COMREL demo version [8] and the results are given in table 3 and the plot of reliability index with exposure period in figure 5.

Table 3 Effect of Exposure time

Parameter T	Reliability Index	Parameter sensitivity
10	4.683	-0.1827
15	1.568	-0.3354
20	0.6367	-0.1012
25	0.2925	-0.04661
30	0.1169	-0.02652
35	0.01034	-0.01706
40	-0.06088	-0.01188

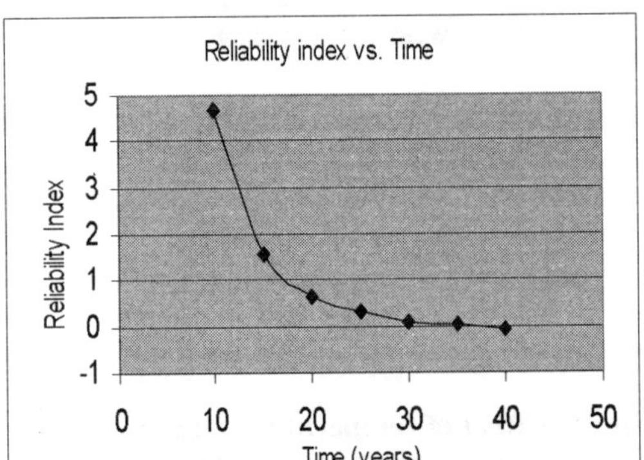

Fig. 5 Reliability index with time

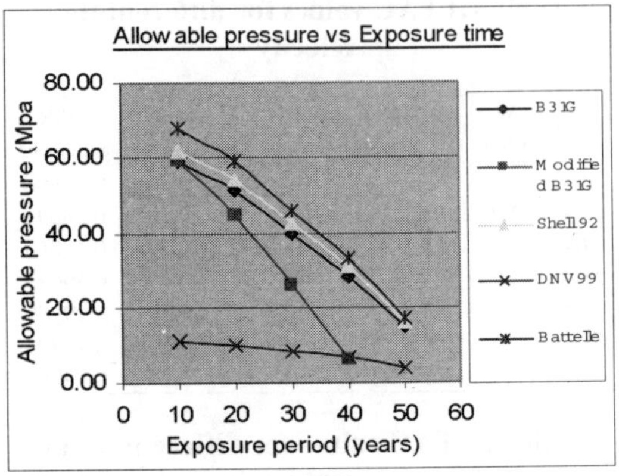

Fig. 6 Comparative study on various failure pressure models

6. RESULTS AND DISCUSSIONS

A PHWR pipeline is considered for the study and the results obtained for FAC are shown in Tables 1.1-1.3. It is observed in Figure 1 that FAC increases from 0.00941 to 0.19511μg/cm^{2}hr with increase in oxygen content from 0 to 30%. FAC decreases from 2.15144 to 0.13296μg/cm^{2}hr with increase in pH from 7.5 to 10.5 as shown in figure 2. As the flow velocity increases from 1.5 to 5 m/s, the FAC increases from 0.16808 to 0.19968μg/cm^{2}hr. With increase in temperature from 373 to 673°K FAC increases with respect to oxygen content and flow velocity and decreases with respect to pH. A comparison of allowable pressure with exposure period by different failure pressure models is given in figure 6.

The reliability analysis for the data given in Table 2 is carried out using COMREL demo version and the results are shown in table 3 and figure 5 gives an insight of the effect of reliability with respect to life.

7. CONCLUSIONS

From this study it is concluded that the FAC is dependant on the parameters like oxygen content, pH value, flow velocity and operating temperature. These parameters are to be controlled carefully. The defect depth increases with time and the length of pipe doesn't have significant effect on the erosion corrosion. The reliability index decreases from 4.683 at 10 years to -0.06088 at 40 years. Hence, whenever a defect in the piping system is detected it can be repaired if possible else the component is replaced to achieve the desired reliability level in the piping system.

8. REFERENCES

1. Caleyo F, Gonzalez J. L, Hallen J. M "*A study on the Reliability Assessment Methodology for Pipeline with Active Corrosion Defects*", Pressure Vessel and Piping, **79**, 77-86, 2002.
2. ASME-B31G. "*Manual for determining the remaining strength of corroded pipelines, A supplement to ASME B31G code for pressure piping*", New York, American Society for Mechanical Engineers, 1991.
3. Kiefner F, Vieth P.H. "*New method corrects criterion for evaluating corroded pipe*", Oil and Gas journal, **6**, 56-59, 1990.
4. Ahammed M, Melchers R.E, "*Reliability estimation of pressurized pipelines subject to localized corrosion defects*", Int. J. of pressure vessel and piping, **69**, 267-272, 1996.
5. Ahammed M, "*Probabilistic estimation of remaining life of a pipeline in the presence of active corrosion defects*", **75**, 321-329, 1998.
6. Smith C. L, Shah V. N., Kao.T, Apostolakis. G, "*Incorporating Aging Effects into Probabilistic Risk Assessment- A Feasibility Study Utilizing Reliability Physics Models*", **NUREG/Cr-5632** report, 2001.
7. Gopika V, "*Risk informed studies on in-service inspection*", Ph.D., thesis, 2005.
8. COMREL, www.strurel.de, demo version for reliability analysis.

Reliability, Safety and Hazard
Editors: P.V. Varde et al.

ROOT CAUSE ANALYSIS OF SECONDARY COOLANT PUMP FAILURE AT DHRUVA: A CASE STUDY

N.S. Joshi, C. Satheesh, P. Ganeshan and S.K. Marik

Research Reactor Services Division, Bhabha Atomic Research Centre Mumbai 400 085

ABSTRACT

Efficient and reliable operation of secondary coolant pumps in a nuclear reactor is very important as heat generated in the reactor core is transferred through heat exchangers to the ultimate heat sink, with the help of these pumps. At Dhruva, (A 100 MWt natural uranium fuelled, nuclear research reactor), the reactor core is cooled by heavy water, the primary coolant in a closed loop, which, in turn, is cooled by secondary coolant (Process water). The secondary coolant is recirculated in closed loop by a set of single stage, double suction centrifugal pumps operating in parallel. Frequent bearing failures at pump driven end (DE side), pump-motor coupling failures, pump- seal leakages etc. were observed. These failures were considered, initially, to be of random nature. A close condition monitoring programme, undertaken subsequently, could identify the existence of severe cavitations in one of the pumps due to various reasons like increased wear ring clearances, changes in operating condition of the pumps etc. This paper describes the observations and various methods employed for identifying the root cause for frequent failures of secondary coolant pumps. By addressing the root cause, the conditions promoting failures are either eliminated or reduced to minimum.

1. INTRODUCTION

Root cause analysis is a technique used to identify the conditions that initiate the occurrence of an undesirable activity or state. It is a process designed for use in investigating and categorizing the root causes of events with safety, health, environmental, quality, reliability and production impacts.

Parallel pumps operation, with redundant pump kept as an active-standby, is a common feature in nuclear reactors. The stand-by pumps would be pressed into service promptly in case of any problem with the operating pumps without interrupting the system operation. Depending on the safety significance and other maintenance requirements, the number of active stand-byes can be more than one in some cases. Such a case where two active stand-by pumps are provided for three pumps operating system is presented here. This case study highlights methods used for detection of cavitation occurring in one such pump and subsequent identification of the root cause behind the failures.

'Dhruva' is natural Uranium fuelled, heavy water moderated and cooled thermal nuclear research reactor with maximum power rating 100MW. Three numbers of Primary Coolant Recirculation Pumps, operating in parallel, circulate the heavy water through fuel channels for removal of nuclear fission heat and then transfer it to the secondary coolant through a set of shell and tube heat exchangers. Secondary coolant recirculation is achieved through a set of three numbers of centrifugal pumps operating in parallel as shown in the layout (Fig: -1).

Individual pumps are provided with pressure gauges at the suction and the discharge sides and only gross flow from all the pumps is measured. The pumps take suction from a common header and deliver to common discharge header. An expansion tank rides on the system providing positive suction to the pumps.

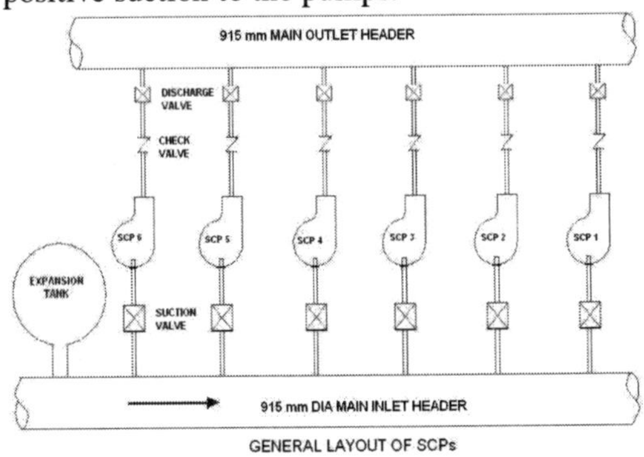

Fig. 1: General Layout of the pumps

2. SECONDARY COOLANT PUMPS (SCP)

The SCPs are of single stage, horizontally split case, double suction centrifugal pumps. Each pump is rated to provide a flow of 1284 m^3/hr / (21400 lpm) at 76 MWC. A six vane bronze impeller is supported between ball bearings. The pump is driven by a 450 Kw electrical motor through a flexible coupling, operating at 1493 rpm. A Non return Device (NRD) is keyed to the motor shaft to prevent reverse rotation. The pumps are mounted on a rigid concrete foundation.

The pumps have been provided with motorized discharge and manually operated suction valves of butterfly type. Throttling of discharge valves is not possible as the valves are of butterfly type. The suction and discharge valves of standby pumps remain in closed condition to avoid any flow bypass through check valves.

The original design intent was to operate five numbers of these pumps in parallel with a gross flow of 1,07,000 lpm. However, during commissioning trials, flow induced vibrations were observed in the heavy water heat exchangers and as a corrective measure, the total secondary coolant flow was restricted to around 84000 lpm with three-pump operation. Thus each pump was operated at a new operating point of 1640 M^3/hr. (27333 lpm) Further the pump operation at higher flow has resulted in an increase in NPSHR from 7.0 m to around 8.5 m there by reducing the NPSH available margin by $\approx$ 1.5 m. This modification was done, as the available NPSH was 12.5 MWC i.e. more than the required NPSH at the new operating conditions.

Though the discharge flow from each pump was slightly higher than the flow at BEP, the pump operation was smooth and trouble free for about 10 years.

3. OPERATION HISTORY

The system was commissioned in 1984. Operating parameters like suction, discharge pressures, motor currents were regularly monitored with calibrated instruments. The values were generally remaining normal. The secondary coolant pumps are in operation since the year 1984 and each of them have clocked about 60,000 hours.

Frequent failures of motor bearings, couplings, mechanical seals etc., were experienced in Secondary Coolant Pumps in the recent past. Condition monitoring of these pumps, largely on vibration based parameters, was able to predict these failures well in advance, on regular basis. Initially these failures were assumed to be of random nature and no hidden cause was envisaged. The following types of failures were found on the rise:

a) Bearing failures of pump especially at the drive (DE) end.
b) Breaking of Coupling and Non-Return Device (NRD) bolts.
c) Bearing housing failure on DE side with scouring marks on shaft under bearing location.
d) Frequent pump-seal leakage and noisy operation of the pump.

4. CONDITION MONITORING

Regular condition-monitoring programme was initiated on these pumps to predict impending problems. As the frequency of bearing failures in different pumps was on the rise, specifically on pumps SCP # 3, SCP # 4 and SCP # 5, a detailed vibration mapping at specific points like pumps casing, headers, suction lines, discharge lines etc. was undertaken for identifying hidden cause if any. Readings were taken with different combination of pumps in operation. The vibration values were found to be high on pump # 4 and pump # 3 with dominant frequencies 25

Hz, 150 Hz etc in the spectrum. These frequencies were observed at various points like pipelines and adjacent pumps due to transmission. Correction of minor deficiencies observed in the piping supports of pump # 4 did not make any significant improvement in the vibration characteristics. One important observation was that, vibration values increased when pumps # 4 and # 3 were in operation, highest being at pump # 3. The spectrum showed dominance of 25 Hz (1x) and 150 Hz (6x, i.e. vane passing frequency) signals on these pumps. Refer figure # 2 for vibration spectrum obtained during condition monitoring. The other associated parts of the pump and system like suction nozzle, suction

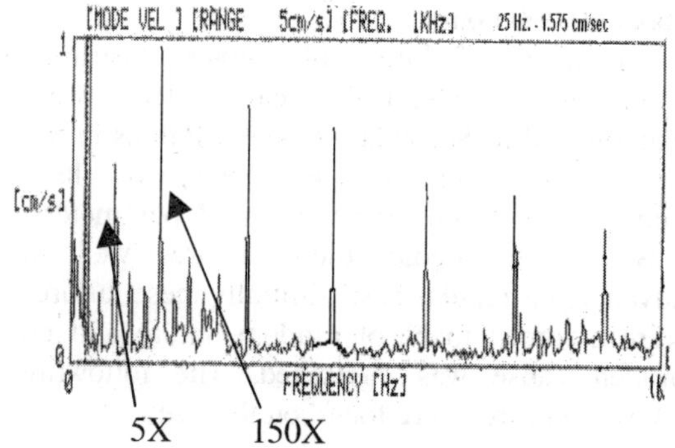

Fig. 2: Vibration Spectrum Prior to opening of the pump

header were also showing 150 Hz (6x) signals. In another combination of pump operation, excluding pump # 4, vibration values were low and 150 Hz signal (6x) was conspicuously absent. This lead to the conclusion that contribution of 150 Hz (6X) from pump # 4 could be the reason for high vibrations.

The presence of 1X, 6X, and harmonics of 6X observed in pump # 4 are, indicative of hydraulic unbalance and cavitations respectively. Hence the pump was opened for inspection of its internals.

5. OBSERVATIONS

Based on the condition monitoring signals of hydraulic unbalance, the pump # 4 was opened for inspection. It was found that the impeller was in damaged condition and most of the damage appearing on the Driving End side of the impeller eye area. Minor erosion was found on the impeller vanes at the impeller eye region on the NDE side. Following are the detail observations. Refer figure # 3 for further details.

5.1 Driving End (DE) side

- Severe pitting / erosion marks seen on the impeller neck and on the hidden side of the vanes.
- Formation of a groove was observed on the impeller neck in the region of wearing ring for the entire periphery.
- Considerable metal loss was observed from four numbers of vanes out of the six from the inlet edge. Erosion appears to have started from the hidden side of the impeller vanes.
- Formation of holes due to pitting / erosion seen on the remaining two vanes.
- Discolouration was observed on all the vanes for about half of the vane length. Remaining length of all the vanes towards the exit tip was found intact.
- Wearing ring ID was found to have similar pitting erosion marks and grooving was observed over the entire periphery.
- Both the impeller shrouds were found to be in healthy condition.
- The clearance between casing ring and impeller neck was about 5.0 mm that is more than 10 times the recommended clearance specified by the manufacturer.

Fig. 3: Mirror Images of Cavitation failures on the Hidden side of Impeller Vanes. (Pump # 4)

5.2 Non-Driving End (NDE) side

- Initiation of erosion / pitting on all the six vanes near the hub.
- Slight roughening of wearing ring ID and the impeller neck.
- The clearance between casing ring and the impeller neck was found to be 1.2 mm, which is three times more than the recommended value.

6. ANALYSIS

6.1 Analysis of the failed assembly:

Observations as mentioned above were indicative of cavitation failure in the pump impeller. Vibration analysis also indicated signals of hydraulic unbalance (1X) and vanes related problems (presence of vane passing frequency, 6X). However, by virtue of design provisions, cavitation was not envisaged in these secondary coolant pumps.

6.2 Pump Rating and NPSHR

When three-pump operation was resorted, flow delivered by each pump had increased to around 1650-m^3/ hr. This modification resulted in an increase of flow by ≈ 30 %.

With the suction piping size and layout remaining same, flow velocity at the suction pipe increased to > 15 ft / sec, which is higher than the recommended velocity and this caused higher turbulence at the impeller eye region.

6.3 Wearing Rings

In these secondary coolant pumps there are no impeller wearing rings, only casing wearing rings have been provided. A close diametrical clearance of 0.4 mm is required to be maintained between the casing ring and the impeller hub to minimise the leakage flow from the high-pressure side of the impeller to the impeller suction eye.

Excessive wearing ring clearance can impact the tendency for formation of both classical and recirculation cavitation. As the increased leakage and eddies formation as shown in figure 4 disturbs the flow pattern into the impeller eye, the tendency for pockets and bubbles to form is increased. The eddy currents at the impeller eye have effectively reduced the flow channel size, thereby increasing the liquid velocity for the rated flow. When the velocity increases, the pressure drop due to friction also increases. If the drop is large enough to cause the pressure to fall below the liquid vapour pressure, the pump will develop classical cavitation because of the initiating action of recirculation cavitation. A doubling of wear-ring clearances require 40 % to 45 % higher NPSHR to suppress these tendencies and hence at this point the wearing rings are generally replaced.

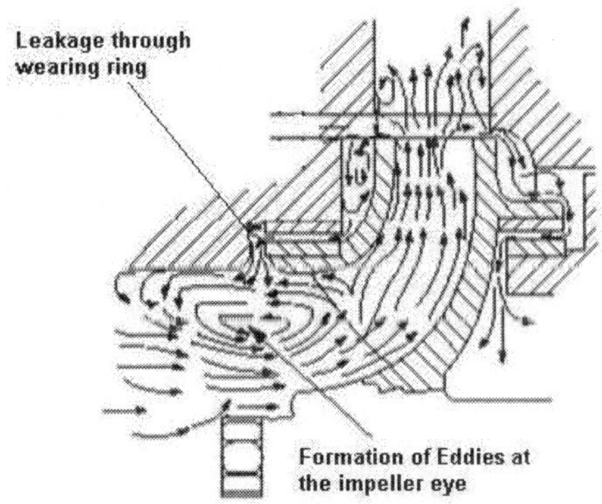

Fig. 4: Formation of vortices at impeller suction eye & at vane tips

As both types of cavitation generate noise and vibration, the only way to distinguish between them is by careful inspection of the damaged impeller. The classical cavitation damage is located on the invisible side of the vanes. It starts from the leading edge and can extend up to two-third of the vane length before the pressure implodes the bubbles. Either feeling or looking the underside of the vane with a mirror is necessary to evaluate the damage.

6.4 Cavitation at pump suction:

The abnormal wear-ring clearance between the casing ring and impeller on DE side has resulted in high leakage and formation of eddies thereby changing the flow distribution on either side of impeller vanes. The clearance on the NDE side was not abnormally high thereby drawing normal or slightly higher flow through this side of the impeller. This uneven flow on both sides of the impeller has caused hydraulic unbalance as revealed during condition monitoring.

The classical symptoms of cavitations like fluctuation in fluid parameters and metallic noise were undistinguishable due to the general background noise.

7. CONCLUSION

The combined effects of abnormally high wearing ring clearance, higher suction velocity and the reduced NPSHA / NPSHR margin due to three pumps operation has resulted in the failure of the impeller vanes on DE side due to cavitation. Since the suction velocity and the NPSHA cannot be altered due to the existing piping layout / site constraints and the operating conditions, maintaining of proper wearing ring clearance is of utmost importance for reducing the effects of cavitation. The appearance of hydraulic unbalance (dominant frequency at the operating speed and the vane pass frequency) signals can be taken as one of the indications of onset of cavitation.

8. REFERENCES

1. Hydraulic Institute Standards – 14th Edition.
2. Pump Handbook: McGraw Hill, edited by Igor Karassik, William Krutzsch, Warren Fraser, Joseph Messina
3. Stepanoff: Selected advanced topics in centrifugal pumps and blowers in two phase flow.
4. Igor Karassik: Centrifugal pumps.
5. Igor Karassik: "Centrifugal pumps and system hydraulics", Chemical Engineering, Oct-1982
6. W. E. Nelson: "Understanding pump cavitation", Chemical Processing, Feb-1997.
7. W. E. Nelson: "Pump cavitation- correction & cures", Chemical Processing, Dec-1997.
8. Mouafak Zaher: "Avoid cavitation in centrifugal pumps", Chemical Engineering, Jun-2003.
9. W. H. Fraser: "Recirculation in centrifugal pumps", Indian pumps, Mar-1983.
10. Material science and technology – Corrosion and environmental degradation Vol. II.

Reliability, Safety and Hazard
Editors: P.V. Varde et al.

EFFECTIVE MAINTENANCE OPTIONS THROUGH MULTI-OBJECTIVE MAINTENANCE MODELING

A.K. Verma[a], *A. Srividya*[a], *M.S. Loknath*[b], *P.G. Ramesh*[b] *and R. Jaikumar*[b]

[a]Indian Institute of Technology Bombay, Mumbai
[b]Institute of Armament Technology, Pune

ABSTRACT

Optimum maintenance decision-making requires satisfaction of several objectives and constraints. A number of models in literature have considered maintenance objectives and constraints in a limited way. For today's systems there exits a need for comprehensive and user-definable decision models, which could serve as sound foundations for maintenance decision support systems (MDSS). Multi-objective optimization using Evolutionary Algorithms can offer suitable alternatives in this regard. This paper aims to bring out the need for such an approach for maintenance decisions. With a case of two simple objectives of cost and reliability in preventive inspection of a system, the role of NSGA based MOOP in making effective maintenance decisions has been demonstrated.

1. INTRODUCTION

Dependability of modern complex large engineering systems is fundamental to all economic, strategic and technological endeavours. Dependability of such systems during their life cycle is enabled thorough effective maintenance. Considerable work has been done in the field of studying the need, techniques and strategies for maintenance of equipment and systems including large engineering systems such as power plants, nuclear installations and marine machinery systems. Several methods have also been implemented with varying levels of effectiveness. Maintenance effectiveness has been limited largely due to the fact that it is a complex activity involving decision making in the presence of several objectives, constraints, variation in products and their operating conditions and in the absence of reliable failure data and adequate or timely knowledge of equipment condition. Not withstanding the complexities involved, sound maintenance planning is imperative for modern systems, where down times due to system breakdowns are expensive. Reliability Centered Maintenance (RCM) has emerged as a maintenance management strategy in recent times to provide cost effective maintenance solutions. However, there is a general lack of sound foundations of decision making for claiming that the maintenance strategy derived from RCM is indeed 'optimal'. Maintenance strategies have been studied and implemented considering only a convenient portion of the issues being faced. References at [1] to [3] are some of the reviews of various types of maintenance methods/ policies in practice catering to a small set of objectives and constrained by a few important factors. But the need of the day in managing maintenance of large engineering systems is to be able to consider all possible issues associated with maintenance so as to make it effective. Therefore maintenance activity can benefit from the techniques of multiple criteria decision making (MCDM).

2. AN OVERVIEW OF SOLUTION TO MULTI CRITERIA MAINTENANCE OPTIMISATION PROBLEMS

As brought out in section 1, a multi objective maintenance optimization problem (MOOP) is as follows [4]:

Minimize/maximize $f_m(x)$, m=1,2,..M;

$$\text{Subject to } g_j(x) \geq 0,\ j = 1,2,..J;$$
$$H_k(x)=0,\ k=1,2,..K;$$
$$x_i^{(U)} \geq x_i \geq x_i^{(L)}\ i=1,2,..n.$$

where, there are M objectives, J inequality constraints, K equality constraints and n decision variables with upper and lower bounds. The objective functions for a maintenance model are manifold, such as, reliability, maintainability, availability, cost and less often considered ones such as, inventory, safety, environmental, production, and operation characteristics.

There are several common classical methods to solve MOOP such as weighted sum, value function, goal programming methods, etc. A case in example is the multi attribute value model discussed by Jiang and Ji [5]. The most common approach in classical methods is to convert the MOOP into a single objective optimization problem (SOOP) by according weights or preferences to each objective. The drawback of such a composite objective function is that the individual criteria and the corresponding risks lose their identities [6]. Further, in many of them, all objectives have to be converted into one type. Determination of the weights needs problem knowledge and is not always easy. Some of the shortcomings associated with most classical methods are: (a) Convergence to an optimal solution depends on the chosen initial solution. (b) Algorithms can get stuck to a sub-optimal solution. (c) Only one solution is obtained in each iteration. The problem will have to be solved afresh with new weights to generate a set of options.

The other technique to handle MOOP is Evolutionary Algorithm (EA), Genetic lgorithms (GA) being the most popular among them. These algorithms mimic nature's evolutionary principles to direct the search towards an optimal solution. In every iteration we get a population of solutions in the objective space. After a certain number of iterations (or generations) we get a population of optimal solutions called the Pareto optimal set (front) as shown in figure 1. (also called non-dominated or non-inferior solutions). Each solution in the set represents a particular combination of objective values. No

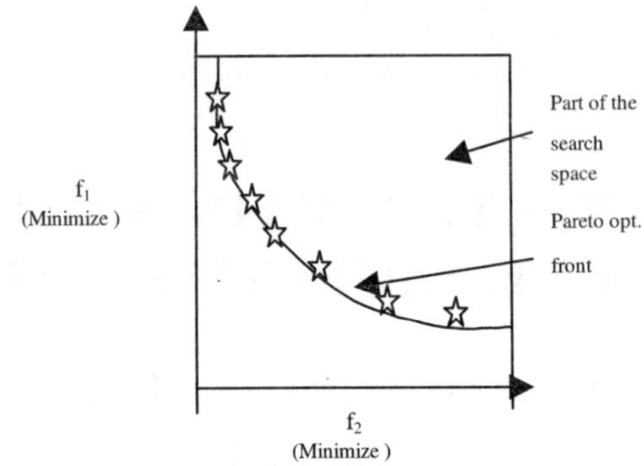

Fig. 1: Pareto optimal solutions (trade off solutions)

solution is inferior to any other solution the set. The solutions of such a set differ from one another only in the extent of trade offs between the objectives.

The multiple objectives can also be combined to give a SOOP using preferences or weights [7] as was the case in the classical methods, of course with similar limitations. Various options available in solving a MCDM problem is shown in figure 2.

3. GA BASED MULTI OBJECTIVE OPTIMIZATION.

Genetic algorithms are general-purpose search algorithms (Evolutionary Algorithms), which are computationally simple but powerful in their search for improvement [8]. The algorithms are inspired by the principles of genetics and solutions of problems evolve in a similar manner. In GAs, a population of knowledge structures called chromosomes (each made of a string of genes) is generated randomly, which represent candidate solutions to the problem. In every generation of the algorithm, the solutions that contribute to better function values (alternatively, fitness values) survive and are passed on (or have a greater probability of being passed on) to the next generation. The evolution of the population over time is also influenced by controlled variations brought about by means of cross over and mutation in the chromosomes.

GAs as also other EAs, work with a population of solutions instead of a single solution. In each generation a population of solutions is obtained.

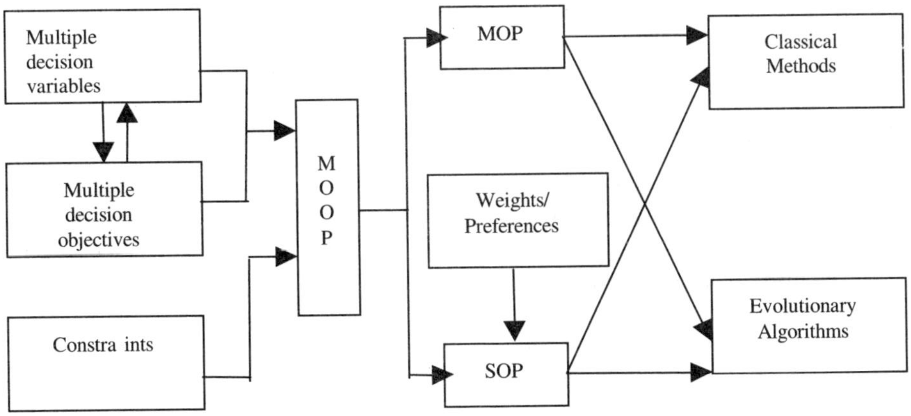

Fig 2: MCDM solution

The population of solutions approaches a set of optimal solutions, called the Pareto optimal solution set.

The two requirements of a multi objective optimization problem (MOOP) arc (a) to find as many Pareto optimal solutions in as less a time as possible and (b) to obtain a good diversity of solutions in the in the Pareto optimal front. EAs are particularly suited for the above two requirements. The effectiveness with which the former is achieved is improved by suitable measures in the algorithm, called elite operators to ensure that good solutions are carried forward to the next generation without being lost (Elitist Multi Objective EAs.). The latter is accomplished using diversity preserving operators, such as mutation, sharing function model etc. Thus a number of algorithms are found in literature attempting to achieve the above-mentioned objectives of MOOPs [4]. Some of the recent work on application of EAs in maintenance related fields use powerful algorithms such as, Strength Pareto Evolutionary Algorithm (SPEA) [9,10], Multi Objective Genetic Algorithm (MOGA) [11].

4. APPLICATION OF GA (NSGA-2) FOR PROBLEM SOLUTION

Non-Dominated Sorting Genetic Algorithm (NSGA-2) is one of the efficient evolutionary algorithms to aid MCDM. Comparatively, the algorithm has been rated as one of the best available [12]. Goldberg [8] proposed the idea of using non-dominated sorting concept in GAs. A detailed coverage of the technique can be found in [4]. The algorithm aims to retain the twin objectives of MOOP mentioned earlier. The algorithm begins with the generation of a random population of decision variables. The population is sorted according to an ascending level of non-domination. The best non-dominated solutions are called non-dominated solutions of level 1. They are temporarily removed from the population of solutions. Non-dominated solutions out of the remaining population are again obtained, which form the non-dominated solutions of level 2, and so on. The algorithm is summarized as follows:

(i) A random population of decision variables is generated and corresponding values of objective functions obtained.

(ii) Now, to obtain the fitness values, a non-dominated sorting method is employed. In this method, the population of solutions in the objective functions space P is sorted according to non-domination into several mutually exclusive non-dominated sets P_i (i = 1,2,,...r) till all population members are classified so that

$$P - P_1 \cup P_2 \cup \ldots P_r$$

(iii) All solutions in the first non-dominated set P_1 are the best in terms of closeness to the Pareto-optimal set and hence are assigned the highest fitness value. Generally the value is numerically equal to the population size. The successive sets are assigned progressively worse fitness values.

(iv) In order to ensure the second objective of MOOP, namely, that of achieving diversity among the solutions, a diversity preserving operation is used. In NSGA the required diversity is maintained by degrading the fitness obtained in the step (iii) by an amount, which depends on the crowding of the solutions in the front. This is done by 'sharing function method'.

(v) The process is repeated for other non-dominated sets likewise. It will be seen that no element of one non-dominated set is inferior to an element of a non-dominated set with lesser fitness value.

(vi) Cross over and mutation are then carried out as per the assumed cross over and mutation probabilities.

(vii) The algorithm is repeated for a predetermined number of generations to get the Pareto-optimal set.

5. PROBLEM STATEMENT

In certain electronic systems, there are costs associated with inspecting the systems, C_i to ensure that they are in a state to be operated successfully. The probability of surviving till the inspection is R(T). In case the system suffers a break down, the average repair contract cost is C_r. The model by Barlow and Hunter [13] can be adopted for the present case to give the cost per unit time as follows:

$$C(T) = \frac{[C_i R(T) + C_r(1 - R(T))]}{\int_0^T R(t)dt} \quad (1)$$

The reliability of the system is assumed to follow an exponential pattern:

$$R(T) = \exp[-(\lambda T)] \quad (2)$$

The failure probability is therefore

$$F(T) = 1 - \exp[-(\lambda T)] \quad (3)$$

The multi-objective optimization problem will therefore be framed as follows:

$$Minimise \left\{ \begin{matrix} C(T) \\ F(T) \\ 1 \le T \le 10 \end{matrix} \right\} \quad (4)$$

It is to be noted that in the case of inspection decisions, the restrictive assumptions of increasing hazard rate that are necessary for replacement decisions are not required. Hazard rate need not be increasing to justify an inspection. The equipment need to be attended to by means inspections even though the hazard rate is constant, failing which, there may be catastrophic failure of the equipment. All the same, failures can still occur despite inspections [14].

Table 1 Details of the GA for the subject case

1.	C_i	1
2.	C_r	4
3.	λ	0.1
4.	Type of GA	NSGA2
5.	Population size	32
6.	Number of generations	10
7.	Cross over probability	0.6
8.	Mutation probability	0.05

In the above problem there are two objectives, namely cost and failure probability, to be minimized. The details of NSGA2 which was run for the above problem are enumerated in table 1.

6. DISCUSSION OF RESULTS

The NSGA 2 for the above multi (twin) objective problem was executed using the C programme from reference [15] after suitable modification of the source code for the objective functions of the present problem. The above two objectives were taken up for optimization to demonstrate the process and the benefits of MOOP. It can be seen from the plots in figures 3 to 5 that the solutions form a single optimal front due to the inherent simplicity assumed for the purpose of the subject problem.

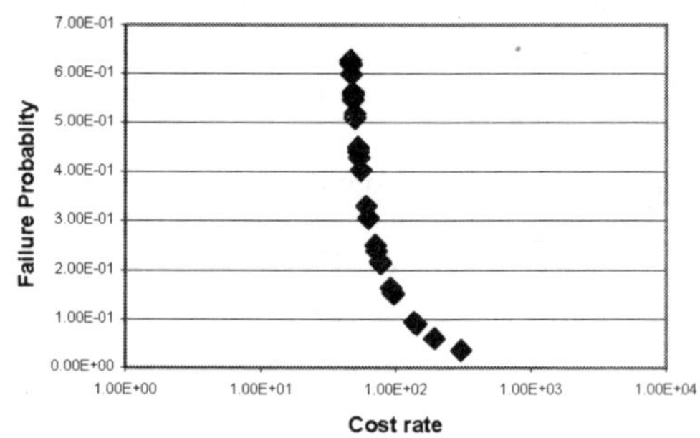

Fig. 3: Initial population

In the case of problems with multiple decision variables and objective functions having very

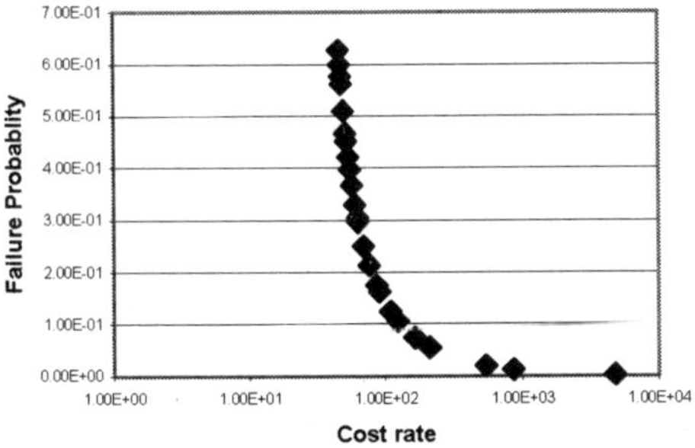

Fig. 4: Objective space after 5 generations showing greater spread of solutions.

little correlation, the initial population of solutions would be scattered in the objective space. Assisted by the sorting mechanism, as the GA progresses in generations, the solutions would converge towards the Pareto optimal front. However, the effect of sharing function method can be seen clearly.

The solutions after the 10th generation are well dispersed compared to those after the 5th generation. Each solution in the Pareto optimal front is an optimal solution and one is not inferior to the other. The solutions in the front differ from one another in the extent of trade off between cost and probability of failure. Thus there are multiple options to choose from each of which would correspond to a particular maintenance strategy. Further, all solutions in the P-optimal front are superior to the solutions in the rest of the feasible space.

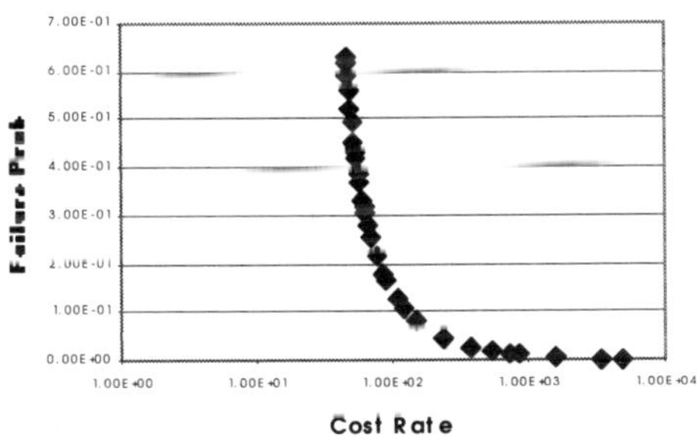

Fig. 5: Objective space after 10 generations showing further spread of solutions.

7. CONCLUSIONS

In this paper, the need for multi-criteria decision making (MCDM) in the area of maintenance has been highlighted. The use of classical and evolutionary optimization methods have been discussed and shown that multi-objective optimization problem (MOOP) approach using Evolutionary Algorithms (EAs) like genetic algorithms are well suited for maintenance decisions. The several Pareto optimal (non-dominated) solutions available to the user enable choice of the right maintenance option in the light of technical and non-technical considerations under prevailing circumstances. Such dynamic selection of maintenance options can render the maintenance plan effective.

8. REFERENCES

1. Gertsbakh, I.B. (1977) *Models of Preventive Maintenance.* North Holland Publishing Company.
2. Cho, D.I. and Parlar, M. (1991) *A survey of Maintenance Models for Multi Unit Systems.* Eur Jnl of Oper res. **51** 1-23.
3. Wang, H. (2002) *A survey of Maintenance Policies of Deteriorating Systems.* Eur Jnl of Oper res. **139** 469-489.
4. Deb,K. (2001) *Multi-Objective Optimization Using Evolutionary Algorithms.* John Wiley & Sons, Inc.
5. Jiang, R. and Ji, P. (2002) *Age replacement Policy: a Multi-Attribute Value Model.* Reliability Engineering and System Safety. **76** 311-318.
6. Chareonsuk, C., et.al. (1997) *A Multi Criteria Approach to the Selection of Preventive Maintenance Intervals.* Int Jnl of Prod Econ. **49** 55-64.
7. Lapa, C.M.F. (2005) *A model for preventive maintenance planning by Genetic Algorithms based in cost and reliability.* Reliability Engineering and System Safety) 1 8.
8. Goldberg, D.E. (1989) *Genetic Algorithms in search, Optimization and Machine Learning.* Pearson Education (Singapore) Pte Ltd. India.

9. Martorell, S., et. al. (2004) *Alternatives and Challenges in Optimising Industrial Safety Using Genetic Algorithms.* Reliability Engineering and System Safety. **86** 25-38.

10. Martorell, S., et. al. (2005) *RAMC+C Informed Decision Making With Application to Multi-Objective Optimisation of Technical Specifications and Maintenance Using Genetic Algorithms.* Reliability Engineering and System Safety. **87** 65-75.

11. Anderson, S.R., et. al. (2005) *Multi-Objective Optimization of Operational Variables in Waste Incineration Plant.* Computers and Chem Engg. **29** 1121-1130.

12. Baran, B., Lucken, C.V., Sotelo, A. (2005) *Multi-Objective Pump Scheduling Optimization Using Evolutionary Strategies.* Advances in Engineering Software. **36** 39-47.

13. Barlow, R.E. and Hunter, L.C. (1960) *Optimum Preventive Maintenance Policies.* Oper Res. **8** 90-100.

14. Bhadury, B. and Basu, S.K. (2003). *Terotechnology: Reliability Engineering & Maintenance Management.* Asian Books Pvt. Ltd.

15. Deb, K. (2005) *nsga2-v1.1.* KanGAL. IIT Kanpur.

Reliability, Safety and Hazard
Editors: P.V. Varde et al.

RELIABILITY CENTRED MAINTENANCE – AN OVERVIEW IN INDIAN PHWR CONTEXT

D. Chatterjee, Rajee Guptan, Nalini Mohan, S.G. Ghadge and S.S. Bajaj

Directorate of Safety, Nuclear Power Corporation of India Limited
Nabhikiya Urja Bhavan, Anushakti Nagar, Mumbai 400 094

ABSTRACT

Over the last few decades, attitude and approach to maintenance has changed very rapidly together with new maintenance techniques, growing awareness of effect of equipment failure on safety and environment and increasing pressure to achieve high plant availability and to contain cost. Broadly, the approach and technique for maintenance are categorized in four groups viz. reactive maintenance, preventive maintenance, predictive maintenance and finally, the latest concept of Reliability Centered Maintenance (RCM). While the reactive maintenance is basically 'run it till it breaks' mode is an old concept for maintenance. In preventive maintenance either a time based schedule or a schedule based on that of machine run time. The goal here is to minimize system and component degradation and thus, sustain or extend the useful life of the equipment. A predictive maintenance approach strives to detect the onset of equipment degradation and differs from preventive by basing maintenance needs on the actual condition of equipment rather than some predetermined schedule.

Finally, the Reliability Centered Maintenance (RCM) is the maintenance approach used when following a process that assesses equipment condition and determines the maintenance requirement in operating context.

This paper brings out a case study of Realiability Centred Maintenance Technique carried out for Indian PHWR highlighting various aspects of RCM and application of the same. Though this technique of maintenance is not strictly followed presently. This can improve maintenance approach in future.

1. INTRODUCTION

Over the past few decades, the awareness, technique and approach to maintenance have changed very rapidly and to a great extent. The change has been more due to greater awareness in terms of the extent of equipment failure affecting safety and environment and increasing pressure to achieve high plant availability and to contain cost prudently. Broadly, the approach and technique for maintenance are categorized in four groups, viz. reactive maintenance, preventive maintenance, predictive maintenance and finally, the latest concept of Reliability Centred Maintenance (RCM).

NPCIL as a utility is responsible for design, commissioning, operating and maintaining Nuclear Power Plants. In addition to better plant availability for better revenue, it also has a commitment to environment and society by ensuring highest standard of safety. So, better and improved approach to maintenance and safety are always strived and attempted. NPCIL also has successfully completed Level-1 Probabilistic Safety Assessment study for its plants at both design level and operating level. Thus, a detailed PSA study is available for different internal Initiating Events. This also makes available the detailed system analysis (through Fault Tree Technique) for different safety and process systems. The detailed fault trees are developed by including all components/equipment which are required for system success and failure of which can lead to system failure/unavailability. After qualitative and quantitative analysis of these systems, the relative importance of various components are established.

2. SYSTEM LEVEL STUDY

Though the plant level PSA study is available with CDF calculated, it was decided to carry out the study at system level. The reason behind this is, individual system level study gives the basis of corresponding change in maintenance planning and finally when such changes are made, on a number of systems, its combined impact on the plant CDF. The plant PSA study helps to identify the system important to safety of the plant and can be a large contributor to CDF if fails. Thus, DG was identified as an important safety system.

3. PRESENT PILOT STUDY

Present study is carried out to understand the impact of change in maintenance for a Diesel Generator. Diesel Generator is the main equipment/component for Class-III Electrical Power Supply System. There are three Diesel Generator for Kakrapara Atomic Power Station (KAPS) Class-III Emergency Power Supply System. Any single DG available will make Class-III supply available as far as DG availability is concerned. Also, the Technical Specification restricts the outage of more than one DG at any particular time due to preventive maintenance, Hence, single Diesel Generator unavailability has been studied in detail using Fault Tree Technique.

4. WHY FAULT TREE TECHNIQUE?

As per Reliability Centred Maintenace standard tool, the Failure Mode and Effect Analysis study is used to understand the component role in the system failure. However, Fault Tree Technique being an advanced tool for handling bigger systems (in NPP, the systems are consisting of large number of components) also uses the failure mode (type of failure), effect (through logical tree) and criticality (through importance analysis).

The RCM process entails asking seven questions about the system under review, as follows:

1. What are the functions and associated performance standard of the assessment in present operating context?
2. In what ways does it fail to fulfill its functions?
3. What causes each functional failure?
4. What happens when each failure occurs?
5. In what way does each failure matter?
6. What can be done to predict or prevent each failure?
7. What should be done if a suitable pro-active task cannot be found?

All the above seven questions are addressed by Fault Tree Analysis (FTA) Technique and hence FTA Method has been used in the present study.

4. MAIN OUTCOME OF THE STUDY

After developing the fault tree using different gates and basic events for the Diesel Generator failure as a top event, the qualitative and quantitative analysis was done using "RISK SPECTRUM" computer code. This code generates the list of basic events/causes responsible for system failure alongwith their relative importance to the top event failure. The fault trees developed for DG are shown in Figures 1.1 upto Figure 1.4 and the Minimal Cut Set List alongwith the relative importance is shown in Table-1.

5. RESULTS

After analyzing the DG Failure (both with failure to start and run), the cutsets were generated. The contribution of different cutsets are shown in Table-1. It is decided to group the cutsets in three groups. The first group consists of cutsets with contribution upto 10% (first 5 cutsets in this case), the second group consists of cutsets with contribution from 1% to 10% (from cutset no. 6 upto cutset no. 11) and for the third group consists of cutsets with contribution less than 1% (from cutset no. 12 upto cutest no. 20). After obtaining these categories, it is suggested that most critical cutsets (i.e. first group) can be taken

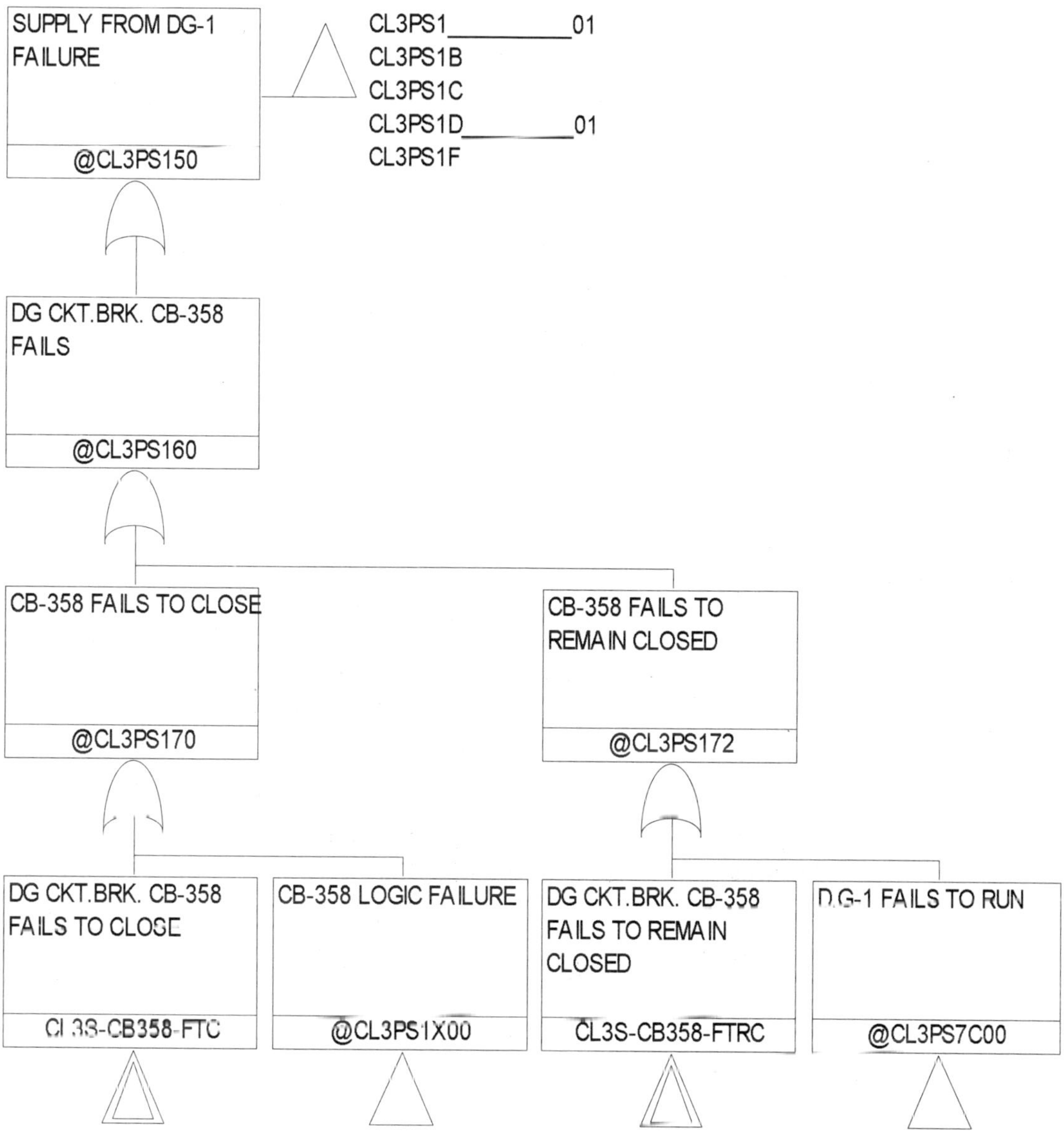

Fig. 1: Class III Electrical Power Supply Not Available at 6.6 KV Level

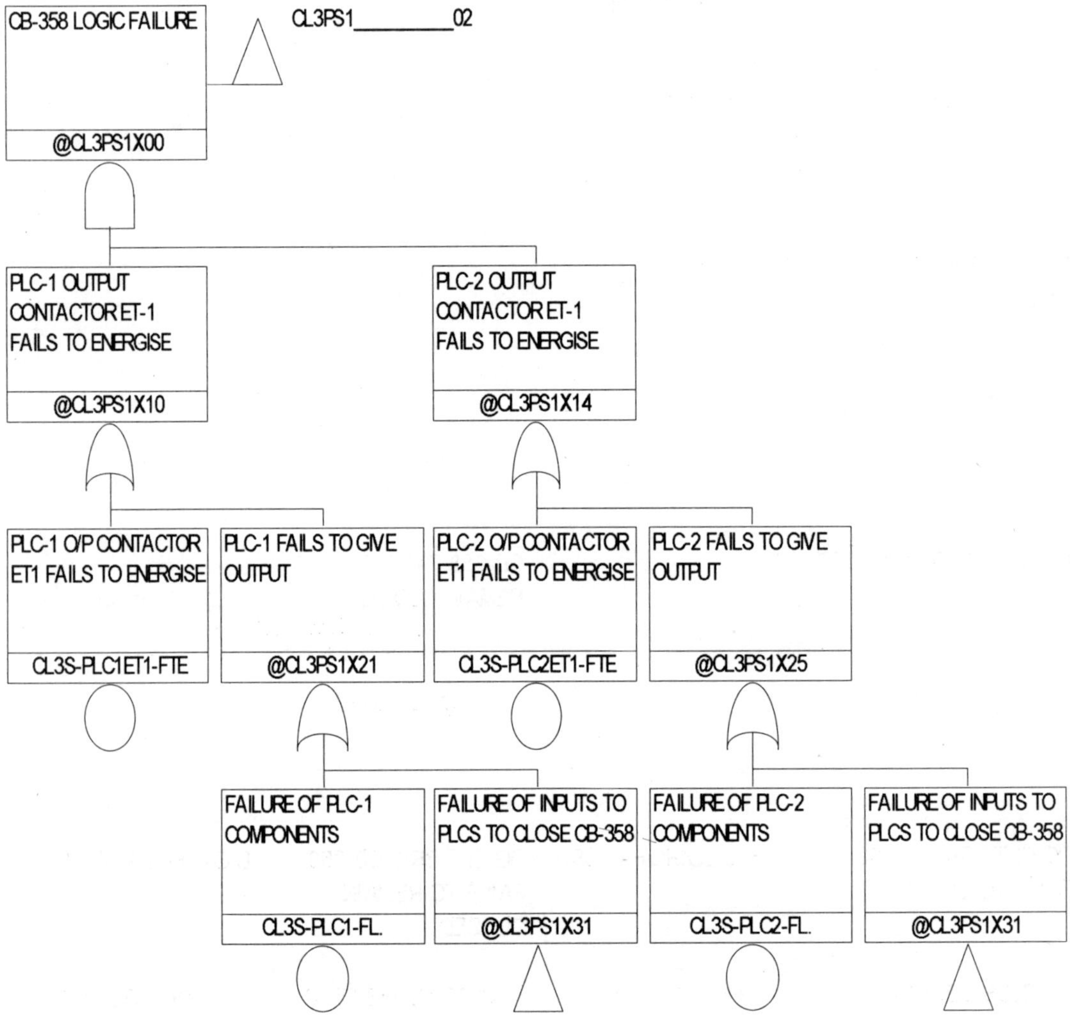

Fig. 2: CB-358 Logic Failure

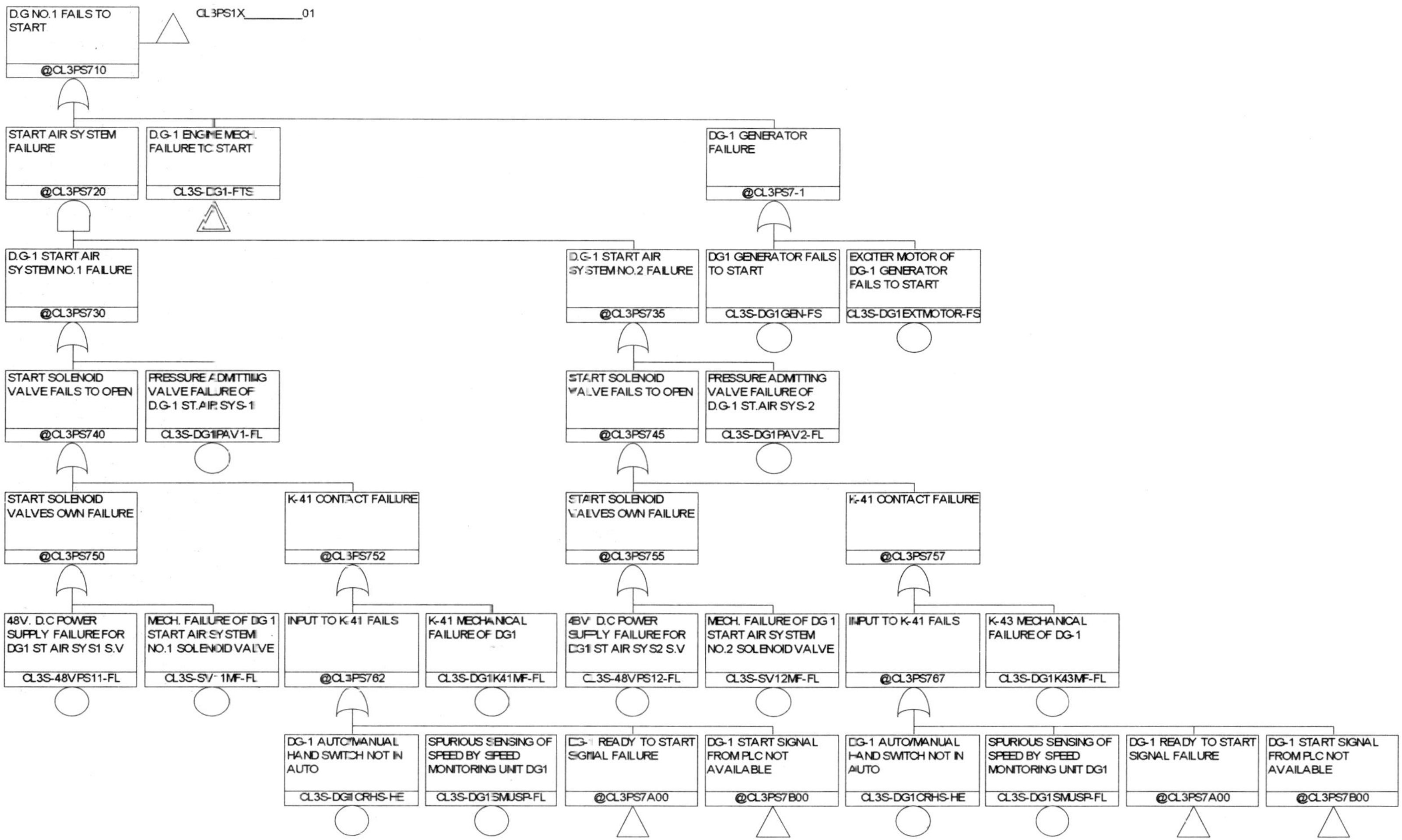

Fig. 3: Diesel Generator D.G-1 Fails to Srart

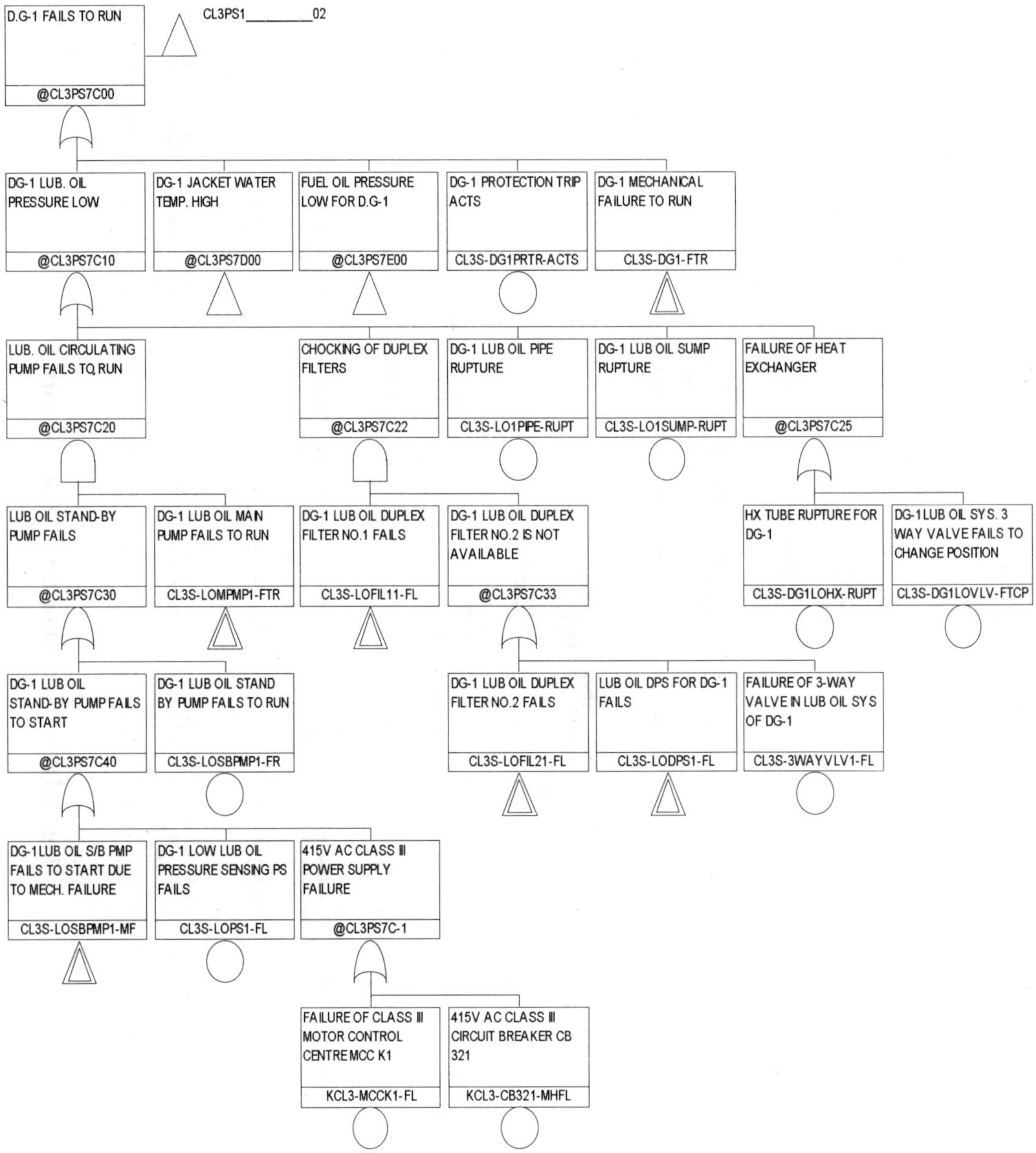

Fig. 4: Diesel Generator D.G-1 Fails to Run

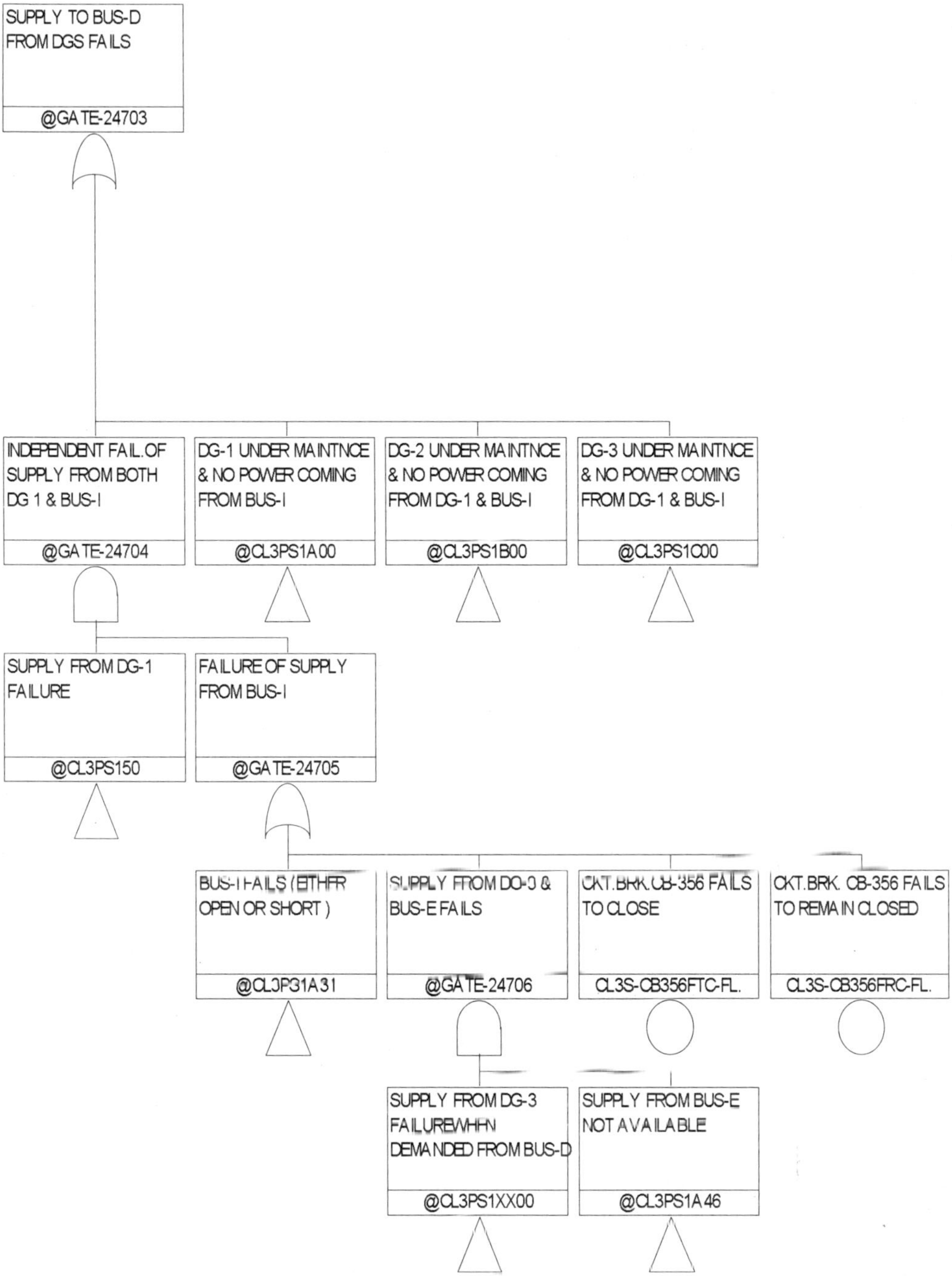

Fig. 5: Fault Tree for Single DG Maintenance UNAV. With Other DG Independent Failure

Table 1 List of Minimal Cutsets for D.G. Failure

Top event probability Q = 2.249E-02

No.	Prob.	%	Q	Event	Description
1	4.547E-03	20.22	4.547E-03	CL3S-DG1-FTS	D.G-1 ENGINE MECH. FAILURE TO START
2	3.591E-03	15.97	3.591E-03	CL3S-CB356-FTO	FAILURE OF CB-356 CONNECTING BUS D.G 1 TO TRIP
3	3.591E-03	15.97	3.591E-03	CL3S-HS7521-FL	CB-358 HS-7521 FAILS TO BE IN AUTO
4	3.591E-03	15.97	3.591E-03	CL3S-HS7518-FL	DG-1 HS-7518 FAILS TO BE IN AUTO
5	3.232E-03	14.37	3.232E-03	CL3S-ISOCB351-FTO	FAILURE OF CL-IV SIDE CB-351 TO TRIP
6	7.947E-04	3.53	7.947E-04	CL3S-CB358-FTC	DG CKT.BRK. CB-358 FAILS TO CLOSE
7	7.570E-04	3.37	7.570E-04	CL3S-DG1-FTR	DG-1 MECHANICAL FAILURE TO RUN
8	4.563E-04	2.03	4.563E-04	CL3S-UVD2794-FL	FAIL. OF UNDER VOLT LOCKOUT RELAY BU-D 2794SENSING CL-4 FL
9	3.599E-04	1.60	3.599E-04	CL3S-DG1PLC-FL	DG1 LOCAL PLC FAILURE
10	3.591E-04	1.60	3.591E-04	CL3S-CBISO-CCF-ALL	COMMON CAUSE FAILURE OF ISOLATION CKT.BKRS TO OPEN
11	2.272E-04	1.01	2.272E-04	CL3S-CB358-FTRC	DG CKT.BRK. CB-358 FAILS TO REMAIN CLOSED
12	1.666E-04	0.74	1.666E-04	CL3-DGFTS-CCF-13	CCF OF DGS FAILING TO START
13	1.666E-04	0.74	1.666E-04	CL3-DGFTS-CCF-12	CCF OF DGS FAILING TO START
14	1.428E-04	0.64	1.428E-04	CL3-DGFTS-CCF-ALL	CCF OF DGS FAILING TO START
15	1.176E-04	0.52	1.176E-04	CL3S-DG1SMU48-FL	DG-1 SPEED MEASUR. UNIT 48V. SUPPLY FAILURE
16	1.000E-04	0.44	1.000E-04	CL3S-DG1CRHS-HE	DG-1 AUTO/MANUAL HAND SWITCH NOT IN AUTO
17	1.000E-04	0.44	1.000E-04	CL3S-DG1SDRLY-HE	DG-1 S/D RELAY NOT RESET AFTER MAINT - HUMAN ERROR
18	8.400E-05	0.37	8.400E-05	CL3S-DG1SMU-FL	FAILURE OF DG-1 SPEED MONITORING UNIT
19	8.400E-05	0.37	8.400E-05	CL3S-DG1SMUSP-FL	SPURIOUS SENSING OF SPEED BY SPEED MONITORING UNIT DG1
20	3.360E-05	0.15	3.360E-05	CL3S-DG1LOHX-RUPT	HX TUBE RUPTURE FOR DG-1

for maintenance in each cycle, second group cutsets can be taken in each alternate cycle and the third group can be taken by skipping two cycles and for each third cycle. Thus, the average maintenance time can be brought down from the existing followed uniform maintenance time. The presently followed maintenance duration is 7 days in every 6 months and proposed for a change to 5 days in every 6 months.

Next study is done to see the impact of the change in average maintenance time to system unavailability. As described earlier, the Class-III supply is available and fails when all three DGs fail together. Figure 2.0 shows the possible combinations of one DG under maintenance and simultaneous independent failure of other two DGs. By changing maintenance induced unavailability to 70%, the top failure probability of Class-III power supply system changes from 1.79×10^{-4} to 1.36×10^{-4}, which shows an improvement of 24% at unavailability level (reduction in top value).

6. CONCLUSION

This study shows that the effective maintenance down time can be reduced by prudently planning a better maintenance management backed by the reliability analysis. Though at present, this study is limited to single system, the same can be extended to other systems of the plant and then a single integrated study can be done on plant CDF to see the impact of changed maintenance duration.

This study also helps in better planning of spare component management and inventory control which is an added advantage.

7. REFERENCES

1. KAPS Level-1 PSA (with internal event)
2. Federal Energy Management Program on Operations and Maintenance
3. Plant Engineering Maintenance and Reliability Resources at Maintenance Resources.Com
4. Reliability Centered Maintenance by John Moubray.
5. Reliability Centered Mainenance (RCM) by Alan Pride, Associate Director, System Reliability, Smithsonian Institution.

Reliability, Safety and Hazard
Editors: P.V. Varde et al.

RELIABILITY CENTERED MAINTENANCE

Rajesh Nair, H.N. Bhat and G. Nageswara Rao

Kaiga Generating Station 1&2, Kaiga 581 400

ABSTRACT

RCM is a systematic way of establishing a preventive maintenance (PM) program or improving an existing PM program. RCM uses a structured methodology that establishes the maintenance requirements for equipment based on the consequences of failure. Equipment failures that may impact system functions are identified and PM tasks are then identified for those equipment failures. The intent of RCM is to establish maintenance practices which focus more on the functional importance of a piece of equipment, its failure/repair history and less on vendor PM recommendations, traditional "time-directed" overhaul tasks.

Reliability Centered Maintenance (RCM) is an efficient maintenance plan. Structures, systems and equipment that directly supports safe, reliable, cost-effective plant operations, strongly influence production reliability. All RCM-based production facilities see lower operating costs from reduced failures and less maintenance. RCM is found to be an efficient scheduled maintenance and condition monitoring plan development process.

Effective adoption of RCM to enhance safety and reliability to nuclear power stations are outlined.

1. INTRODUCTION

With few exceptions, preventive maintenance has been considered the most advanced and effective maintenance technique available for use by industrial and facility maintenance organizations. A Preventive Maintenance (PM) program is based on the assumption of a "fundamental cause-and-effect relationship between scheduled maintenance and operating reliability. This assumption was based on the intuitive belief that because mechanical parts wear out, the reliability of any equipment is directly related to operating age. It therefore followed that the more frequently equipment was overhauled, the better protected it was against the likelihood of failure. The only problem was in determining what age limit was necessary to assure reliable operation."

A maintenance policy based exclusively on some maximum operating age would, *no matter what the age limit, has little or no effect on the failure rate.*

In separate independent studies, it was noted that a difference existed between the perceived and the intrinsic design life for the majority of equipment and components. In fact, it was discovered that in many cases equipment greatly exceeded the perceived or stated design life

Reliability-Centered Maintenance (RCM) is the optimum mix of reactive, time- or interval-based, condition-based, and proactive maintenance practices. The basic application of each strategy is shown in Fig. 1. These principal maintenance strategies, rather than being applied independently, are integrated to take advantage of their respective strengths in order to maximize facility and equipment reliability while minimizing life-cycle costs RCM includes reactive, time-based, condition-based, and proactive tasks. In addition, a user should understand system boundaries and facility envelopes, system/equipment functions, functional failures, and failure modes, all of which are critical components of the RCM program.

2. DESCRIPTION

Preventive Maintenance (PM) assumes that failure probabilities can be determined statistically for individual machines and components, and parts can be replaced or adjustments can be performed in time to preclude failure. For example, a common practice has been to replace or renew bearings after so many operating hours assuming that bearing failure rate increases with time in service. Fig. 2, *Bearing Life Scatter*, shows the failure distribution of a group of thirty identical 6309 deep groove ball bearings installed on bearing life test machines and run to failure. The wide variation in bearing life is obvious and precludes the use of any effective time-based maintenance strategy. Fortunately, computer advances in the 1990s have made it possible in many cases to identify the precursors of failure, quantify equipment condition, and schedule the appropriate repair with a higher degree of confidence than was possible when performing strictly interval-based maintenance relying upon usually erroneous estimates of when a component might fail. Also, it has been discovered recently that there are many different equipment failure characteristics, only a small number of which are age- or use-related. This new knowledge has increased the emphasis on Condition Monitoring (CM), often referred to as Condition-Based Maintenance, which has caused a reduction in the reliance upon time-based PM.

It should not be inferred from the above that all interval-based maintenance should be replaced by condition-based maintenance. In fact, interval-based maintenance is appropriate for those instances where abrasive, erosive, or corrosive wear takes place, material properties change due to fatigue, embrittlement, etc. and/or a clear correlation between age and functional reliability exists.

In addition, for those systems or components where no failure consequences in terms of mission, environment, safety, security, or Life-Cycle Cost (LCC) exist, maintenance should not be performed, i.e., the equipment should be run to failure and replaced The concept of RCM has been adopted across several government and industry operations as a strategy for performing maintenance. RCM applies maintenance strategies based on consequence and cost of failure. In addition, RCM seeks to minimize maintenance and improve reliability throughout the life-cycle by using proactive techniques such as improved design specifications, integration of condition monitoring in the commissioning process

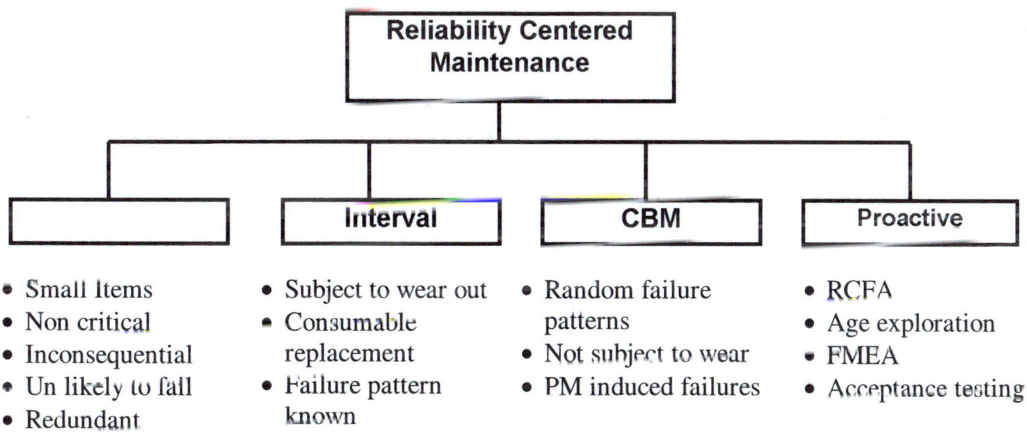

Fig. 1: Components of a RCM program

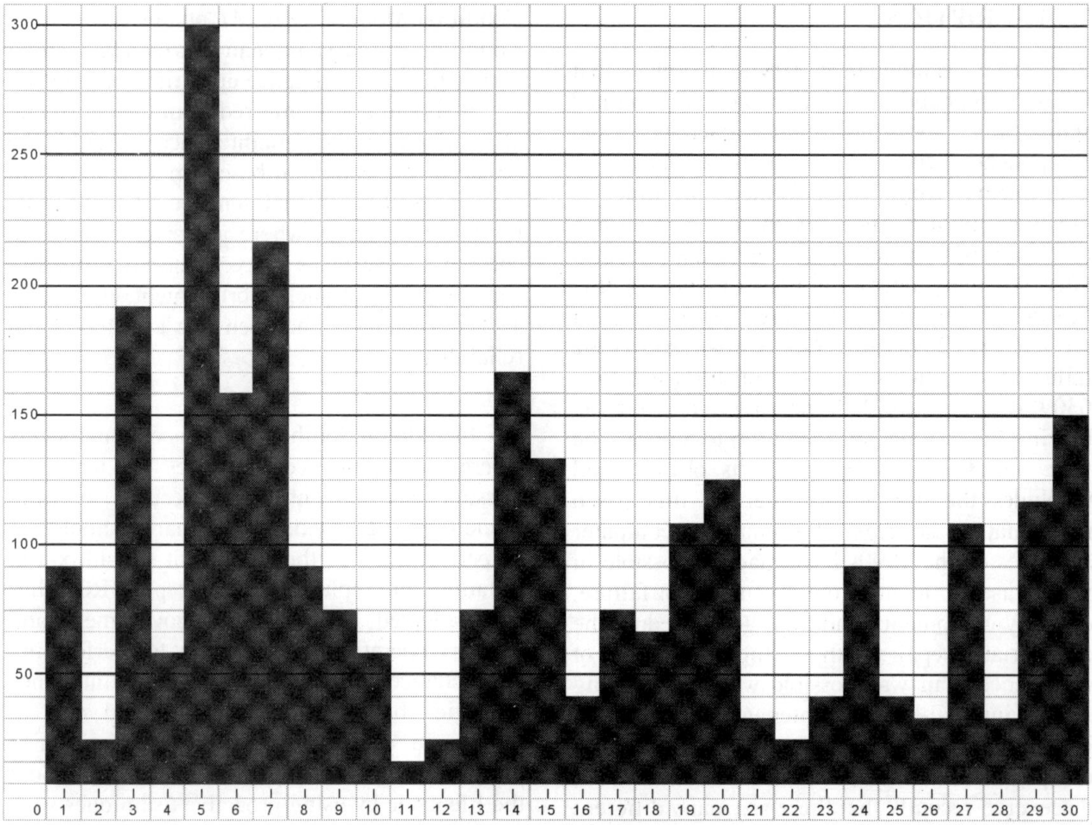

Fig. 2: Bearing Life Scatter

2.1 RCM Principles

The primary RCM principles are:

- ***RCM is Function Oriented***—RCM seeks to preserve system or equipment function, not just operability for operability's sake. Redundancy of function, through multiple pieces of equipment, improves functional reliability but increases life-cycle cost in terms of procurement and operating costs.
- ***RCM is System Focused***—RCM is more concerned with maintaining system function than with individual component function.
- ***RCM is Reliability Centered***—RCM treats failure statistics in an actuarial manner. The relationship between operating age and the failures experienced is important. RCM is not overly concerned with simple failure rate; it seeks to know the conditional probability of failure at specific ages (the probability that failure will occur in each given operating age bracket).
- ***RCM Acknowledges Design Limitations***—RCM objective is to maintain the inherent reliability of the equipment design, recognizing that changes in inherent reliability are the province of design rather than of maintenance. Maintenance can, at best, only achieve and maintain the level of reliability for equipment that was provided for by design. However, RCM recognizes that maintenance feedback can improve on the original design. In addition, RCM recognizes that a difference often exists between the perceived design life and the intrinsic or actual design life and addresses this through the Age Exploration (AE) process.
- ***RCM is Driven by Safety, Security, and Economics***—Safety and security must be ensured at any cost; thereafter, cost-effectiveness becomes the criterion.
- ***RCM Defines Failure as "Any Unsatisfactory Condition"***—Therefore, failure

may be either a loss of function (operation ceases) or a loss of acceptable quality (operation continues but impacts quality).

- ***RCM Uses a Logic Tree to Screen Maintenance Tasks***—This provides a consistent approach to the maintenance of all kinds of equipment.
- ***RCM Tasks Must Be Applicable***—The tasks must address the failure mode and consider the failure mode characteristics.
- ***RCM Tasks Must Be Effective***—The tasks must reduce the probability of failure and be cost-effective.
- ***RCM Acknowledges Three Types of Maintenance Tasks***—These tasks are time-directed (PM), condition-directed (CM), and failure finding (one of several aspects of Proactive Maintenance). Time-directed tasks are scheduled when appropriate. Condition-directed tasks are performed when conditions indicate they are needed. Failure-finding tasks detect hidden functions that have failed without giving evidence of pending failure. Additionally, performing no maintenance, Run-to-Failure, is a conscious decision and is acceptable for some equipment.
- ***RCM is a Living System***—RCM gathers data from the results achieved and feeds this data back to improve design and future maintenance. This feedback is an important part of the Proactive Maintenance element of the RCM program.

2.2 Types of RCM

There are several ways to conduct and implement an RCM program. The program can be based on rigorous Failure Modes and Effects Analysis (FMEA), complete with mathematically-calculated probabilities of failure based on design or historical data, intuition or common-sense, and/or experimental data and modeling. These approaches may be called Classical, Rigorous, Intuitive, Streamlined, or Abbreviated. Other terms sometimes used for these same approaches include Concise, Preventive Maintenance (PM) Optimization, Reliability Based, and Reliability Enhanced. All are applicable. The decision of what technique to use should be left to the end user and be based on:

- Consequences of failure
- Probability of failure
- Historical data available
- Risk tolerance
- Resource availability

Classical/Rigorous RCM

a. *Benefits*: Classical or rigorous RCM provides the most knowledge and data concerning system functions, failure modes, and maintenance actions addressing functional failures of any of the RCM approaches.

b. *Concerns*: Classical or rigorous RCM historically has been based primarily on the FMEA with little, if any, analysis of historical performance data. In addition, rigorous RCM analysis is extremely labor intensive and often postpones the implementation of obvious condition monitoring tasks.

c. *Applications*: This approach should be limited to the following three situations:

- The consequences of failure result in catastrophic risk in terms of environment, health, or safety, and/or complete economic failure of the business unit.
- The resultant reliability and associated maintenance cost is still unacceptable after performing and implementing a streamlined type FMEA.
- The system/equipment is new to the organization and insufficient corporate maintenance and operational knowledge exists on function and functional failures.

Abbreviated/Intuitive/Streamlined RCM

a. *Benefits*: The intuitive approach identifies and implements the obvious, usually condition-based, tasks with minimal analysis. In addition, it eliminates low value maintenance tasks based on historical data and Maintenance and Operations (O&M) personnel input. The intent is to minimize the initial analysis time in order to realize early-wins that help offset the cost of the FMEA and condition monitoring capabilities development.

b. *Concerns*: Reliance on historical records and personnel knowledge can introduce errors into the process that may lead to missing hidden failures where a low probability of occurrence exists. In addition, the intuitive process requires that at least one individual has a thorough understanding of the various condition monitoring techniques.

c. *Applications*: This approach should be utilized when:

- The function of the system/equipment is well understood.
- Functional failure of the system/equipment will not result in loss of life or catastrophic impact on the environment or business unit. For these reasons, the streamlined or intuitive approach has been recommended for both discrete and continuous manufacturing facilities

2.3 RCM Analysis

The RCM analysis should carefully consider and answer the following questions:

- What does the system or equipment do; what are the functions?
- What functional failures are likely to occur?
- What are the likely consequences of these functional failures?
- What can be done to reduce the probability of the failure(s), identify the onset of failure(s), or reduce the consequences of the failure(s)? Answers to these four questions can be used with the decision logic tree depicted in Fig. 3, *Reliability-Centered Maintenance (RCM) Decision Logic Tree*, to determine the maintenance approach for the equipment item or system.

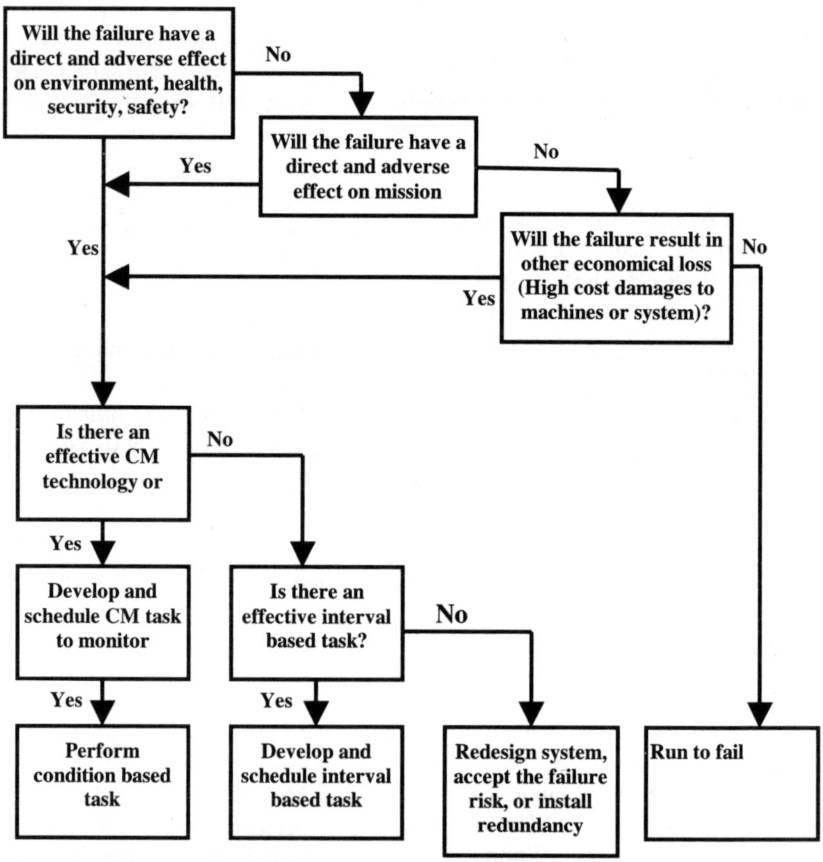

Fig. 3: Reliability Centered Maintenance RCM Logic Tree

Note that the analysis process as depicted in Fig. 3 has only four possible outcomes:

- Perform Condition-Based actions (CM).
- Perform Interval (Time- or Cycle-) Based actions (PM).
- Determine that redesign will solve the problem and accept the failure risk, or determine that no maintenance action will reduce the probability of failure install redundancy.
- Perform no action and choose to repair following failure (Run-to-Failure).

2.4 Failure

Failure is the cessation of proper function or performance. RCM examines failure at several levels: the system level, sub-system level, component level, and sometimes even the parts level. The goal of an effective maintenance organization is to provide the required system performance at the lowest cost. This means that the maintenance approach must be based on a clear understanding of failure at each of the system levels. System components can be degraded or even failed and still not cause a system failure. A simple example is the failed headlamp on an automobile. That failed component has little effect on the overall system performance. Conversely, several degraded components may combine to cause the system to have failed, even though no individual component has itself failed.

System and System Boundary

A system is any user-defined group of components, equipment, or facilities that support an operational requirement. These operational requirements are defined by mission criticality or by environmental, health, safety, regulatory, quality, or other agency/business defined requirements. Most systems can be divided into unique sub-systems along user defined boundaries. The boundaries are selected as a method of dividing a system into subsystems when its complexity makes an analysis by other means difficult:

- A system boundary or interface definition contains a description of the inputs and outputs that cross each boundary.
- The facility envelope is the physical barrier created by a building, enclosure, or other structure; e.g., a cooling tower or tank.
- Standardize on selecting boundaries. For example, a pump could include the first upstream/downstream isolation valve, the coupling, and associated gauges. The motor would include the electrical circuit from the load side of the motor control center but not the coupling.

The intent is to develop a series modular FMEAs and assemble them as if they were Lego Blocks and select the maintenance actions based on the consequences of risk determined by the criticality and probability factors.

Function and Functional Failure

The function defines the performance expectation and can have many elements. Elements include physical properties, operation performance including output tolerances, and time requirements such as continuous operation or limited required availability.

Functional failures are descriptions of the various ways in which a system or subsystem can fail to meet the functional requirements designed into the equipment. A system or subsystem that is operating in a degraded state but does not impact any of the requirements addressed in *System and System Boundary*, has not experienced a functional failure.

It is important to determine all of the functions of an item that are significant in a given operational context. By clearly defining the functions' non-performance, the functional failure becomes clearly defined. For example, it is not enough to define the function of a pump to move water. The function of the pump must be specific and defined in such terms flow rate, discharge pressure, vibration levels, efficiency, etcFailure Modes

Failure modes are equipment- and component-specific failures that result in the functional failure of the system or subsystem. For example, a machinery train composed of a motor and pump can fail catastrophically due to the complete failure of the windings, bearings, shaft, impeller, controller, or seals. In addition, a functional failure also occurs if the pump performance degrades such that there is insufficient discharge pressure or flow to meet operating requirements. These operational requirements should be considered when developing maintenance tasks.

Dominant failure modes are those failure modes responsible for a significant proportion of all the failures of the item. They are the most common modes of failure.

Not all failure modes or causes warrant preventive or conditioned based maintenance because the likelihood of their occurring is remote or their effect is inconsequential.

Reliability

Reliability is the probability that an item will survive a given operating period, under specified operating conditions, without failure usually expressed as Mean Time to Failure (MTTF) or Mean Time Between Failure (MTBF). The *conditional probability of failure* measures the probability that an item entering a given age interval will fail during that interval. If the conditional probability of failure increases with age, the item shows wear-out characteristics. The conditional probability of failure reflects the overall adverse effect of age on reliability. It is not a measure of the change in an individual equipment item.

Failure rate or frequency plays a relatively minor role in maintenance programs because it is too simple a measure. Failure frequency is useful in making cost decisions and determining maintenance intervals, but it tells nothing about which maintenance tasks are appropriate or about the consequences of failure. A maintenance solution should be evaluated in terms of the safety, security, or economic consequences it is intended to prevent. A maintenance task must be applicable (i.e., prevent failures or ameliorate failure consequences) in order to be effective.

Failure Characteristics

Studies show that random failures accounted for 77-92% of the total failures and age related failure characteristics for the remaining 8-23%. The basic difference between the failure patterns of complex and simple items has important implications for maintenance. Single-piece and simple items frequently demonstrate a direct relationship between reliability and age. This is particularly true where factors such as metal fatigue or mechanical wear are present or where the items are designed as consumables (short or predictable life spans). In these cases an age limit based on operating time or stress cycles may be effective in improving the overall reliability of the complex item of which they are a part.

Complex items frequently demonstrate some infant mortality, after which their failure probability increases gradually or remains constant. A marked wear-out age is not common. *In many cases scheduled overhaul increases the overall failure rate by introducing a high infant mortality rate into an otherwise stable system.*

Preventing Failure

Every equipment item has a characteristic that can be called *resistance to* or *margin to* failure. Using equipment subjects it to stress that can result in failure when the stress exceeds the resistance to failure. Fig. 6, *Preventing Failure*, depicts this concept graphically. The figure shows that failures may be prevented or item life extended by:

- Decreasing the amount of stress applied to the item. The life of the item is extended for the period f_0-f_1 by the stress reduction shown.
- Increasing or restoring the item's resistance to failure. The life of the item is extended for the period f_1-f_2 by the resistance increase shown.

Decreasing the rate of degradation of the item's resistance to or margin to failure. The life of the item is extended for the period f_2-f_3 by the decreased rate of resistance degradation shown Stress is dependent on use and may be highly variable. It may increase, decrease, or remain constant with use or time. A review of the failures of a large number of nominally identical simple items would disclose that the majority had about the same age at failure, subject to statistical variation, and that these failures occurred for the same reason. If one is considering preventive maintenance for some simple item and can find a way to measure its resistance to failure, he or she can use that information to help select a preventive task.

Adding excess material or changing the type of material that wears away or is consumed can increase resistance to failure or the rate of degradation. Excess strength may be provided to compensate for loss from corrosion or fatigue. The most common method of restoring resistance is by replacing the item. The resistance to failure of a simple item decreases with use or time (age), but a complex unit consists of hundreds of interacting simple items (parts) and has a considerable number of failure modes. In the complex case, the mechanisms of failure are the same, but they are operating on many simple component parts simultaneously and interactively so that failures no longer occur for the same reason at the same age. For these complex units, it is unlikely that one can design a maintenance task unless there are a few dominant or critical failure modes.

Failure Modes and Effects Analysis (FMEA)

FMEA is applied to each system, sub-system, and component identified in the boundary definition. For every function identified, there can be multiple failure modes. The FMEA addresses each system function (and, since failure is the loss of function, all possible failures) and the dominant failure modes associated with each failure, and then examines the consequences of the failure. What effect did the failure have on the mission or operation, the system, and on the machine?

Even though there are multiple failure modes, often the effects of the failure are the same or very similar in nature. That is, from a system function perspective, the outcome of any component failure may result in the system function being degraded.

Likewise, similar systems and machines will often have the same failure modes. However, the system use will determine the failure consequences. For example, the failure modes of a ball bearing will be the same regardless of the machine. However, the dominate failure mode will often change from machine to machine, the cause of the failure may change, and the effects of the failure will differ.

2.5 Criticality and Probability of Occurrence

Criticality assessment provides the means for quantifying how important a system function is relative to the identified Mission. Table 1, Criticality/Severity Categories, provides a method for ranking system criticality. This system, adapted from the automotive industry, provides 10 categories of Criticality/Severity. It is not the only method available. The categories can be expanded or contracted to produce a site-specific listingThe Probability of Occurrence (of Failure) is also based on work in the automotive industry. Table 2, *Probability of Occurrence Categories*, provides one possible method of quantifying the probability of failure. If there is historical data available, it will provide a powerful tool in establishing the ranking. If the historical data is not available, a ranking may be estimated based on experience with similar systems in the facilities area. The statistical ("Effect") column in Table 2 can be based on operating hours, day, cycles, or other unit that provides a consistent measurement approach. The statistical bases ("Comment") may be adjusted to account for local conditions. For example, one organization changed the statistical approach for ranking 1 through 5 to better reflect the number of cycles of the system being analyzedF. RCM Implementation

There is no one set path for successfully implementing RCM because RCM is more than just performing a Failure Modes and Effects Analysis (FMEA), adopting condition monitoring techniques, and/or optimizing a maintenance and overhaul program through the application of an Age Exploration (AE) process. A successful RCM implementation process first must recognize what and where the source of return on investment (ROI) resides. The source(s) of ROI may be tangible and/or intangible. For the former, a quantifiable business case may be developed based on financial benefit (savings, cost avoidance, reduced Work in Progress (WIP) and/or reduced liability) to the organization while for the latter, the benefit may be unquantifiable (employee skills, morale, customer relations, etc.) In either case, a baseline and goal must be established through some mechanism such as internal or external benchmarking, which results in a defined gap between the "As-Is" and the "To-Be" state and the ROI identified for closing all or a portion of the gap.

RCM is not for everyone and very few organizations will benefit from implementing all elements of a classical RCM program. RCM like all tools/processes has an element of diminishing return. Not all the elements of RCM which are applicable to a nuclear power plant, the aircraft industry, and/or a 24/7 continuous process plant in a sold out condition, will be applicable to a batch process operation or a non-production facility. However, there are a few truths everyone should follow and there is no need to pilot or perform an FMEA analysis. They are:

1. Key performance indicators are essential for establishing the baseline, goal, and the gap. Progress cannot be measured or sustained without KPIs.
2. Thermography works for electrical distribution, boilers, couplings, roofing systems and building façadesIf your specifications for alignment, imbalance, motor circuit phase impedance, oil condition and cleanliness, and vibration are not quantified, the product you receive will have latent defects 80% of the time.
3. If you do not commission and check the sequence of operation of your equipment to a predetermined quantifiable specification, you will not get what you expect.
4. Pareto analysis is the best tool for determining where to start your RCM process. Look for the bottlenecks, the recurring failures, and follow the money.
5. RCM implementation in a team environment works better.
6. Failure modes for identical equipment are the same. It is only the consequence and probability of failure that changes.
7. The impact of poor water chemistry is underestimated in terms of energy consumption and life-cycle cost.
8. The majority of failures are random. Very few machines understand how a calendar works. Age Exploration can reveal hidden assets.
9. Celebrate and advertise your successes and address your failures. Credibility is a key to building support for long-term success.

2.6. Key Performance Indicators (KPI) Selection

Significant thought must go into the process of selecting KPIs to support the maintenance program. The value of meaningful KPIs cannot be overstated; however the significance of KPIs that are inaccurate or inapplicable cannot be understated. First identify the goals and objectives of the organization because they will have an impact on the selection of KPIs at all levels of maintenance activity. KPIs that cannot possibly be obtained should not be chosen, and only those that may be controlled should be selected. Issues of concern should also be identified so that they will be considered in the selection of KPIs. All processes owners who are key to the implementation of the overall effort should have a self-selected metric to indicate goals and progress in meeting those goals. This will foster the acceptance of collecting data to support the KPIs and will also promote the use of the KPIs for continuous improvement. Also one must consider the capabilities of the organization to collect the data for KPIs, i.e. the process used for collecting and storing the data and the ease of extracting and reporting the KPIs. In doing this, the cost of obtaining data for the KPIs and the relative value they add to the overall program must be calculated. While advocating doing the right things within the maintenance program with life-cycle cost as a driver, the cost of the capturing supporting KPIs must also be watched closely.

1. *Benchmark Selection*

After selection of the appropriate KPIs is complete, benchmarks should be established. These characterize the organization's goals and/or progress points for using KPIs as a tool for maintenance optimization and continuous improvement. Benchmarks may be derived from the organizational goals and objectives or they may be selected from a survey performed with similar organizations. These benchmarks will be used as a target for growth and to evaluate risk associated with non-achievement of progress.

2. *Utilization of KPIs*

After benchmarks are established and data collection has begun, the information must be acted on in a timely manner to maintain continuity within all of the processes that are counting on KPIs as a performance enhancement tool. In order to take full advantage of the benefits of KPIs they should be displayed in public areas.

(a) *The tracking and publication of KPIs inform the people of what is important, what are the goals, and where they stand with respect to performance expectations. The impact of displaying KPIs often has an immediate effect on the workers in the functional area being measured. In addition, KPIs are an integral part of any Team Charter as they allow the Team and Management to determine Team priorities and measure productivity CONCLUSION*

Today, cutting operations and maintenance cost and preserving reliability of Nuclear Power plant are top priorities for Managers of our plant. The concept of RCM is sound and provides with a structured approach to maintenance program with an optimum balance between cost of maintenance and reliability improvement. RCM and predictive maintenance analyses complement each other and when performed simultaneously offer an excellent approach to maintenance optimization.

3. REFERENCES

1. http://www.wbdg.org
2. Reliability Centered Maintenance Proceedings at NPCIL Kaiga conducted by HIMER, Chennai.

Reliability, Safety and Hazard
Editors: P.V. Varde et al.

IMPLEMENTATION OF RCM (RELIABILITY CENTERED MAINTENANCE) AT INDIAN NUCLEAR POWER PLANTS (INPP)

A.K. Sinha and S. Mittal

Nuclear Power Corporation of India Limited (NPCIL), V S Bhavan , Anushaktinagar, Mumbai

1. INTRODUCTION

NPCIL's primary objective is to continue to operate its nuclear power plants safely and reliably into the new century. All efforts are geared to this objective, covering all areas of plant operations. The experience gained at individual NPCIL plants and other International plants is reviewed at corporate HQ and disseminated to all plants for action at plant levels. Developments of state of art techniques and technology for maintenance as well as the changing views on maintenance organization and responsibilities have also affected the philosophy of maintenance management in recent days. Our whole aim is to ensure a positive move towards attaining World Class Maintenance i.e. World Class Standards in maintenance.

RCM (Reliability Centered Maintenance) is being considered as one of the strongest tool to develop world class maintenance strategies.

Plant machinery and system, structure and Component (SSC) in an industry are known as physical assets. These assets must be maintained in a proper working condition to get the maximum output from them. These assets must continue to achieve their in-built capacity throughout their service. RCM ensures the continued reliable performance of asset as its users want it to perform. RCM paves the way for this achievement by maintaining their inherent reliability.

FMECA is a technique of conducting RCM that evaluates system reliability by examining the potential failure modes and determines the effect of the failure modes on system performance. It is a design analysis tool to identify critical parts/subsystem, study effect on individual part/subsystem failure on next higher assembly.

RCM is an approach to optimise maintenance and provides the basis for PM (Preventive Maintenance) and its frequency. By focusing on failure modes and effects, streamlined RCM identifies components critical to safety and production and provides the best options for preventive maintenance for the failure modes. Analysis of failure data on a regular basis is essential to improve the reliability of plant and equipment on a sustained basis.

2. IMPLEMENTATION

As a first step towards implementation of RCM in INPP (Indian Nuclear Power Plants) a training programme for 70 maintenance engineers from all operating plants was conducted at KGS, Kaiga and KAPS for familiarising them with latest technology of RCM- II. Task groups have been constituted at the individual plants to identify systems facing consistent problem where RCM can be applied.

RCM is mainly done to redefine and fine tune the existing PM programme/schedules/ maintenance strategies by verifying and suggesting suitable modifications:

RCM is implemented in following two phases.

Phase - I

i) Whether presently followed strategy/schedule is correct and appropriate for the equipment and suggest schedule/strategy modifications.
ii) Whether PM planned and carried out are optimum
iii) Whether any design modification is required as a solution to recurring/unsolved problems, if any.

Detailed analysis is carried out using FMEA (Failure Mode Effective Analysis) by pin pointing and assigning Root-Cause for failure modes of each component in equipments.

Analysis is carried out as per RCM I/RCM-II i.e. by assigning and DETECTION (1 to 10), and obtaining the RISK PRIORITY NUMBER (RPN) - a numerical value out of 1000, and then concentrating our exercise to attack the higher and non-acceptable RPNs to suggest relevant changes of strategies/scheduled maintenance / modifications. Then suggested modifications and changes in PM frequency is implemented.

Phase - II

After the successful implementation of the first phase recommendations, again the FMEA is carried out and a new set of RPN derived and then compared with the first phase RPN to see for improvements/scope for further improvements by phase-II recommendations for fine tuning of the PM schedules. These recommendations are implemented and process repeated, if desired.

Thus the cycle continues till a new improved maintenance strategy is established for the equipment so that the equipment availability can be maximised.

Methodology:

We focus on

i. Function of a component
ii. Functional failure (loss of function)
iii. Failure mode
iv. Failure Effect

Based on above we draw decision tree for selecting the suitable technique of Preventive maintenance and finalising the frequency of the particular technique to be used for particular component.

3. CASE STUDY

RCM has been successfully implemented at TAPS on large number of Motor Operated Valves (MOVs), at KAPS for Helium Blowers, at MAPS for Emergency Process Sea Water Pump (EPSWP) on pilot study basis. Findings, recommendations and implementation of design modifications are discussed below.

Case study - I

KAPS Moderator Helium Cover Gas Circulation

Kakrapar Atomic Power Station has been provided with total 4 X 100% capacity Helium Blowers of M/s. Aerzen Machines Ltd., UK out of which 2 Blowers are installed in Unit # 1 and the other 2 Blowers are installed in Unit # 2

Total 19 failures were reported during a span of five years on four blowers.

Based on number of failures, helium Blower # 1 has been taken up for RCM as per RCM-II and calculated the **criticality index.**

Criticality Index was calculated considering following parameters.

i. Consequence of failure
ii. Consequence of failure factor
iii. Consequence of providing redundancy
iv. Criticality score
v. Consequence of frequency of failure
vi. Consequence of Down Time
vii. Consequence of cost of spares replaced on failure
viii. Consequence of cost of manpower to attend the failure
ix. Priority Score

***Ranking of system and equipment*:** Based on the Criticality Index , the priority class of the equipment is fixed as **Most critical (if, Criticality Index > 100)**

FMEA WORK SHEET

Function	Functional Failure (loss of function)	Failure Mode (cause of failure)	Failure effect
Moderator Helium cover gas circulation	Cover gas circulation stopped	1. Oil seal failed	Oil leak Gas leak, DAC
		2. Coupling failed	Blower halt
		3. Gear failed	Blower halt
		4. Lubrication failed due to water ingress	Blower jammed
		5. Foundation frame broken	Blower hault/fault

RCM ANALYSIS WORK SHEET

Failure mode (cause of failure)	Failure cause	Task	Frequency
Oil seal failed	Poor material and improper seal design	Condition based maintenance (CBM)	Weekly
Blower failed	Bearing failed Coupling failed	CBM CBM	Monthly Yearly
Blower jammed	Lubrication failed Timing gear failed	CBM Replace gears	Weekly As found
Blower discharge flow low	Internal leakages are more/wear out more	Inspect internal parts, lobes, casing, and clearances. Repair it.	As found

Modifications carried out on Helium Blowers.

1. Foundation grouting with special material
2. Gas seal chamber oil seals material was modified from Neoprene rubber to Viton rubber
3. Motor and Blower coupling was replaced with Meta-Flex coupling.

PM and CBM program were modified accordingly.

Case study - II

RCM ON MAPS # 1 EMERGENCY PROCESS SEA WATER PUMP (EPSWP)

Problems causing unavailability of MAPS # 1 PSWP (3 numbers)

FAILURE DATA

Sl. No.	Description	Frequency
1.	Motor Coupling	3
2.	Conical barrel coupling	2
3.	Cutless bearing	7
4.	Journal bearing	9
5.	Thrust bearing	6
6.	Shaft	5
7.	Shaft Protection Pipe	5
8.	Impeller	5
9.	EP (Embedded Part) bolt	2
10.	Motor	4
11.	Impeller guide piece	1
12.	Breaking holder	2

Based on above failure data. Failure Mode Effect Analysis (FMEA) was carried out for EPSWP.

Recommendation for EPSWP:

1. Material of shaft protection pipe was changed from Ni Casting to SS on an experimental basis and after changing the material, no breakdown has been reported so far.
   Hence it is recommended the changing of material of the shaft protection pipe for all other pumps.
2. Regarding the cutless bearing failure it has been changed on experimental basis to thordon bearing which is having hard nylon instead of soft rubber leading to better life of the bearing.
   Hence it is recommended the replacement of all cutless rubber bearings with thordon bearings.
3. Shaft dia was changed from 35 mm to 45 mm (since failure frequency was more) on an experimental basis. It has been found successful.
   Hence shaft dia should be changed for all pumps.
4. EP (Embedded Part) bolts and other fasteners checks to be added to Quarterly PM check list.
5. Thrust bearing lock nut tightness to be checked in PM.
6. Hydraulic induced vibration and its effect may be studied for pump no 2 and 3 in detail for identifying the root cause of the higher failures of pump no. 2.
7. Hydraulic induced vibration and its effect may be studied for pump no 5 and 6 in detail for identifying the root cause of the higher failures of pump no. 5.
8. TIR (Total Indicated Reading) check feasibility for the assembled shaft may be looked into to avert any possible failure due to misalignment wrong thread etc. before pump assembly and installation.
9. Overhauling Schedule may be reworked for optimum maintenance based on running hrs. clocked.

Case study - III

On large number of Motor Operated Valves (MOVs) at TAPS -1 & 2:

Large number of MOV's are in operation at our oldest Boiling Water Reactor, TAPS-1&2 which are under operation since 1969 and 1970 respectively.

Most of the MOV's are part of the safety system. Frequent failure of the MOV's has promoted us to carry out RCM analysis's in some what different ways.

The structured defects which could not be eliminated in MOV's by the improvement of the maintenance skill were searched and analysed through the ten years maintenance history and failure data.

Thirteen years failure data of nine problematic MOV's was collected and total 51 failures was reported. Failure mechanism was then categorised.

(i)	Drift in Torque Switch (TS) and setting in Limit Switch (LS)	- 38%
(ii)	Loosening of parts	- 9%
(iii)	Breaking of parts	- 9%
(iv)	Insulation breakdown	- 9%
(v)	Grease hardening or lack of lubrication	- 9%
(vi)	Gear Wear	- 8%
(vii)	Clutch mechanism wear	- 8%
(viii)	Slipping of motor brakes	- 8%
(ix)	Leaking of Bonnet	- 0%

As most dominant failure mechanism was found drift in Torque Switch (TS) and setting in Limit Switch (LS), appropriate action was taken to rectify the problem and inspection frequency of these item has been increased.

Now most of MOV's are operating smoothly without any problem.

4. USE OF DIFFERENT PM TECHNIQUES

The use of different Predictive Maintenance Techniques like Vibration Analysis, Ferrography (Wear Particle Analysis), Thermography, Signature Analysis of MOV's (Motor Opereated Valves), SOAP (Spectrography Oil Analysis Procedure), Shock Pulse technique, EMPATH

(Electric Motor Performance Analysis Trending Hardware), Accoustic Emmission etc. provides gradual foundation to proactive approach to maintenance. Proactive tasks means tasks undertaken before a failure occurs, traditionally known as predictive and preventive maintenance. A proactive task is worth doing if it reduces the risk of the associated multiple failure to an acceptable low level such an approach with application of Information Technology (IT) to RCM shall go a long way in economic reliable power generation.

Now most of MOV's are operating smoothly without any problem.

5. CONCLUSION

It is estimated that after plant wide implementation of RCM, Indian NPPs will be able to reschedule the maintenance activities and thereby achieve more economic and reliable plant operation through a better maintenance management.

Implementation of RCM Programme is expected to reduce the corrective maintenance hours and to achieve the highest reliability of the system and equipment.

RCM is essential to implement for achieving zero breakdown of the equipment.

It is also necessary to incorporate lesson learnt into day-to-day maintenance programme activities to enhance the value of the RCM programme.

6. ACKNOWLEDGEMENT

We are very much thankful to Shri D. Subba Rao, SO/F, leader of RCM team at KAPS, Maintenance Group at MAPS and Shri Babar, SO/H (Rtd.) of BARC for their contribution in conducting RCM at KAPS, MAPS and TAPS respectively on pilot study basis.

Reliability, Safety and Hazard
Editors: P.V. Varde et al.

THERMAL ANALYSIS OF MODERATOR AND ASSOCIATED SYSTEMS IN CASE OF FAILURE OF APWC AND FOLLOWED BY UNAVAILABILITY OF FFS

I. Thangamani, Vishnu Verma, S.G. Markandeya, G. Chakraborty, A.K. Ghosh and H.S. Kushwaha

Health, Safety and Environment Group, Bhabha Atomic Research Centre, Mumbai 400 085

ABSTRACT

In Indian Pressurized Heavy Water Reactors (PHWR), under normal operating conditions the heat generated in moderator and other systems and equipments is transferred by Active Process Water (APW) system to Active Process Water Cooling (APWC) system, which in turn dissipated to atmosphere via Induced draft cooling tower. Fire Fighting System (FFS) water has been provided as a backup to APW in case of its failure. Based on Probabilistic Safety Analysis (level-1) for Kaiga Nuclear Power Plant (NPP), it has been identified that failure of APWC system together with unavailability of FFS water is one of the dominant accident sequence having significant core damage frequency. In order to understand the consequences of such an accident and also as a part of overall level-2 analysis for Kaiga NPP, a lumped parameter model was developed to study the consequences of the above accident scenario to the moderator and associated systems. During the accident scenario, on account of loss of APWC, the temperatures of all associated systems increase. This leads to moderator swelling, spillage, flashing, boiling, OPRD (Over Pressure Rupture Disk) rupture and cover gas pressure transients. The developed code is capable of analysing the above transients. A brief description of the model and typical results of the analysis are presented in this paper.

Keywords: PSA, Moderator analysis, PHWR

1. INTRODUCTION

The design of PHWR based Indian NPPs has adopted defence-in-depth approach. Under the normal operating conditions of the plant, most of the heat generated in the core is transferred via PHT system to the steam generators for power production. The low grade thermal energy produced in other systems (such as moderator, calandria vault, end shield, etc.) and components (pumps, glands, etc.) is transferred to APW system which is ultimately dissipated to atmosphere via the APWC system. To mitigate the consequences of failure in APW and APWC systems, the use of FFS water is also provided. Thus, the use of FFS water as back up to APW and APWC systems besides its intended fire fighting function is a unique characteristic of Indian PHWR's. Since moderator system of the PHWR acts as a major heat sink during accident conditions and under shut-down conditions, the behaviour of APWC can directly influence the reactor core performance. In one of the earlier PSA level-1 studies [1] carried out for the Indian PHWRs, the contribution to core damage frequency (CDF) was noted to be significant for three postulated accident scenarios. Of these, one of the postulation comprised of failure of APWC followed by unavailability of FFS water, and its percentage contribution to CDF was 9.5%. Similarly, for the sequence APWC Failure - Human Error-Fire Fighting System unavailability, it was 29.77%.

In order to assess the consequences and ascertain quantitatively the damage potential of the above accident sequence, the analytical model has been developed for predicting the steady state and transient behaviours of the affected systems. The present paper describes these studies and presents some of the salient features of analysis carried out.

2. NOMENCLATURE

Cp	Specific Heat (J/kg/K)
h	Enthalpy (J/kg)
H	Enthalpy (J/kg)
m	Mass (kg)
M	Mass (kg)
Q	Rate of heat generated or Heat transferred (W)
t	Time (sec)
T	Temperature (K)
V	Volume (m^3)
x	Steam Quality
ρ	Density (kg/m^3)

Subscripts

apw	Active Process Water system
apwc	Active Process Water Cooling system
core	Reactor core
cov	Cover gas
cv	Calandria Vault System
es	End Shield System
fg	Latent
flash	Flashing
he	Helium
hx	Moderator Heat Exchanger Tube
L	Liquid
loss	Energy loss
mod	Moderator system
pht	Primary Heat Transport system
pump	Energy addition pump
sg	Steam Generator
stm	Steam
v	Vapor

3. SYSTEM DESCRIPTION

The calandria of the PHWR has 306 pressure tube assemblies containing fuel bundles through which high pressure coolant (D2O) circulates and removes the core heat and transfers it via the PHT system to the secondary side of the steam generator. Besides the reactor PHT system, as depicted in Fig-1, the moderator system, end shield system, calandria vault system and other miscellaneous systems (clubbed as other systems in Fig.1) such as Spent Fuel Bay coolers, D2O upgradation plant and D2O evaporation and clean up system, are the other heat generating systems which are cooled by the APW system. The APW system, in turn, transfers all the heat through intermediate heat exchangers to APWC system which finally dissipates all the heat to the atmosphere via cooling towers [2,3].

The calandria vessel is filled with the moderator (D2O) in which some fraction of the reactor core heat gets dissipated [4]. Besides, there is also volumetric heat generation in the moderator due to neutron moderation and other radiation interactions. All this heat is removed by APW system through moderator heat exchangers. Usually the moderator level in the calandria is 96%. Remaining portion of calandria is filled with inert gas (Helium) called as cover gas. Cover gas is maintained at a slightly higher pressure (0.3 kg/cm2 (g)) and is circulated, through OPRD (Over Pressure Rupture Disk) ducts as shown in figure-2, for purging of deuterium, generated due to radiolysis of the moderator. The other function of cover gas is to accommodate the moderator level variations on account of its temperature.

A set of four OPRDs provided on top of the calandria restricts the over pressure build up in calandria during accident. The OPRD burst pressure is 1.4 kg/cm2 (g).

Calandria vault is filled with de-mineralized water. This serves dual purpose of radiation shielding and effective removal of heat generated in the biological and water shields. Calandria vault cooling system transfers heat generated in calandria vault to APWC via APW [5].

End shields are provided at both the ends of calandria, which support and locate calandria tubes and pressure tube assemblies. End shields form part of the calandria vault enclosure and are filled with steel balls and water medium to provide shielding. End shield cooling system serves to remove heat generated in end shields and transfers it to APWC via APW [6].

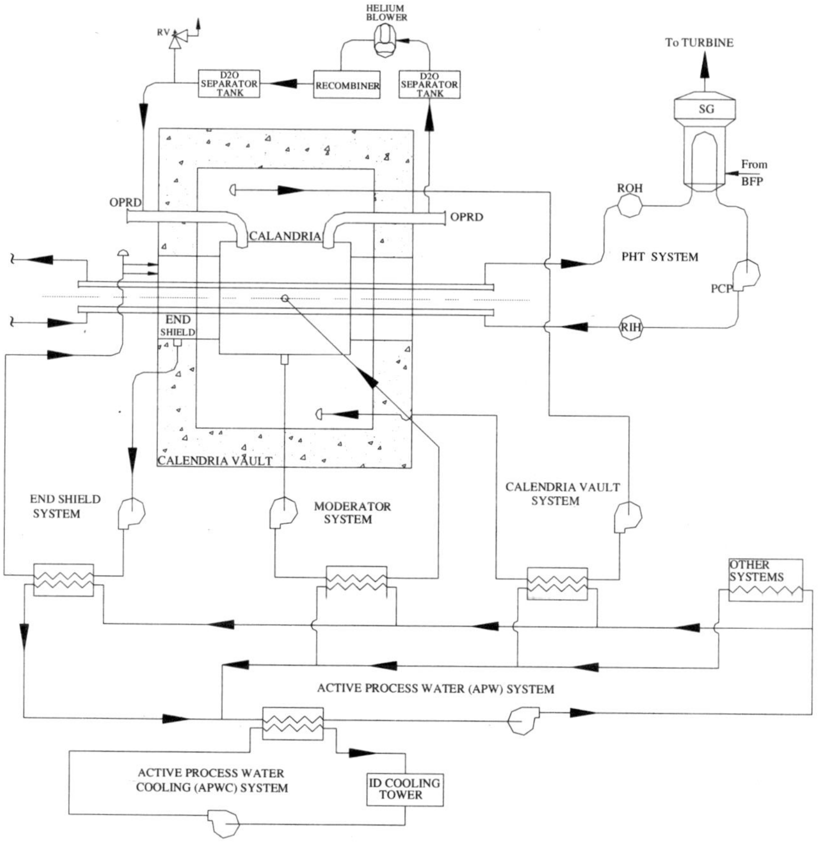

Fig. 1: Schematic diagram of Indian PHWRs

4. MATHEMATICAL MODELLING

In order to analyze the above accident scenario, the complex network of piping systems and components interfacing between, reactor systems, APW, and APWC systems were modeled using a lumped mass approach. Each process system is considered as a single volume and the volumes are interconnected by heat exchangers which are represented by heat transfer coefficient and area. Heat transfer coefficients were calculated by using appropriate empirical correlations.

A block diagram of various systems that are modeled and their heat exchange paths are shown in figure-3.

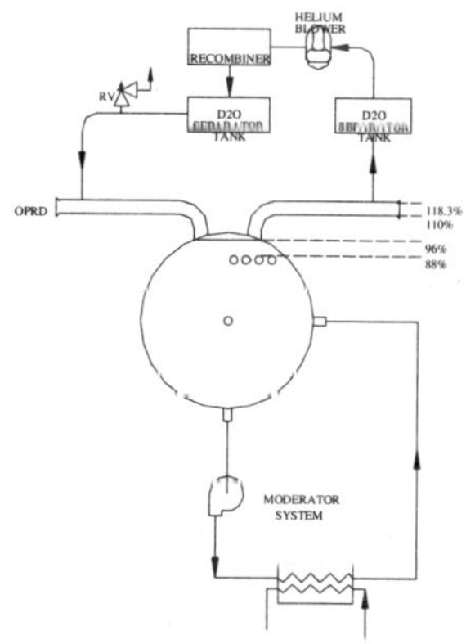

Fig. 2: Schematic diagram of Moderator system

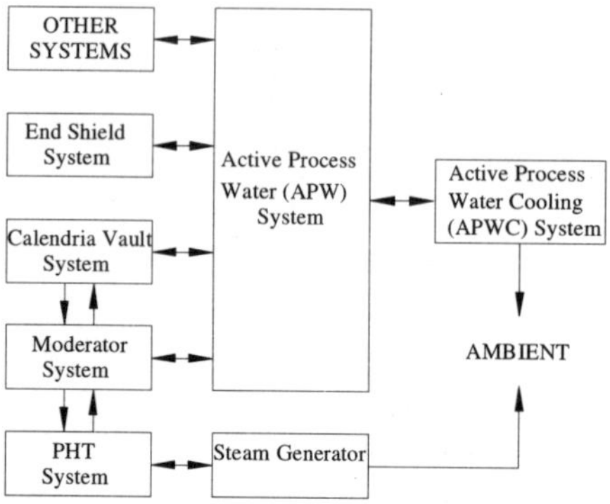

Fig. 3: Block diagram of mathematical model

4.1. PHT System

The maximum heat generated in the core is transferred by PHT system to steam generator [7]. The lumped energy equation for PHT system can be given as

$$M_{pht}Cp_{pht}\frac{dT_{pht}}{dt} = Q_{core} + Q_{pump} - Q_{sg} - Q_{loss} - Q_{pht\text{-}mod} \quad (1)$$

where Q_{core} is the rate of core heat addition to PHT system, Q_{pump} is rate of heat added by the PHT pumps, Q_{sg} is the rate of heat transferred to steam generator, Q_{loss} is the rate of heat loss to ambient by convection and radiation and $Q_{pht\text{-mod}}$ is the rate of heat transferred from PHT system to the moderator .

Similarly for other systems energy equations are given below.

4.2. Moderator System

The energy equation for the moderator system can be given as

$$M_{mod}Cp_{mod}\frac{dT_{mod}}{dt} = Q_{mod} + Q_{pump} + Q_{pht\text{-}mod} - Q_{mod\text{-}hx} - Q_{loss} \quad (2)$$

where $Q_{\text{mod-}hx}$ is rate of heat transferred from moderator to moderator heat exchanger tubes. The heat exchanger tube material is a large metallic mass and has a large thermal inertia. Hence, the energy equation for the heat exchanger tube mass can be as,

$$M_{hx}Cp_{hx}\frac{dT_{hx}}{dt} = Q_{\text{mod}\_hx} - Q_{\text{mod}-apw} \quad (3)$$

where $Q_{\text{mod-}apw}$ is the rate of heat transferred from moderator heat exchanger tube to APW system.

There will be moderator swelling and contraction due to change in density caused by temperature changes. Density (ρ) is dependent on moderator average temperature and cover gas pressure [5].

4.3. End shield System

The energy equation for the end shield cooling system is

$$M_{es}Cp_{es}\frac{dT_{es}}{dt} = Q_{es} + Q_{pump} - Q_{loss} - Q_{es-apw} \quad (4)$$

while Q_{es} is the rate of heat generated in end shield.

4.4. Calandria vault System

The energy equation for the calandria vault system can be given as,

$$M_{cv}Cp_{cv}\frac{dT_{cv}}{dt} = Q_{cv} + Q_{pump} + Q_{\text{mod}-cv} - Q_{loss} - Q_{cv-apw} \quad (5)$$

where Q_{cv} is the rate of heat generated in calandria vault.

4.5. APW System

The energy equation for the APW system can be given as,

$$M_{apw}Cp_{apw}\frac{dT_{apw}}{dt} = \sum Q_{in} - \sum Q_{out} \quad (6)$$

where

$$\sum Q_{in} = Q_{\text{mod}-apw} + Q_{cv-apw} + Q_{es-apw} + Q_{pump} + Q_{others}$$

$$\sum Q_{out} = Q_{loss} + Q_{apw-apwc}$$

$Q_{apw\text{-}apwc}$ is heat transferred from APW to APWC.

4.6. APWC System

$$M_{apwc}Cp_{apwc}\frac{dT_{apwc}}{dt} = Q_{apw\text{-}apwc} + Q_{pump} - Q_{loss} - Q_{apw\text{-}los} \quad (7)$$

$$Q_{apw-los} = 2\dot{m}_{apwc}Cp_{apwc}\left(T_{apwc} - T_{amb}\right).$$

Where $\dot{m}_{apwc}$ is mass flow rate of APWC and T_{apwc} is the average temperature of APWC system.

4.7. Steam generator

For steam generator portion [8],

$$M_{sg}Cp_{sg}\frac{dT_{sg}}{dt} = Q_{pht-sg} + Q_{pump} - Q_{sg-los} - Q_{loss} \quad (8)$$

where

$$Q_{sg-los} = \dot{m}_{sg}\left[Cp\left(T_{sat} - T_{feed}\right) + xH_{fg}\right]$$

$\dot{m}_{sg}$ is feed flow to steam generator, T_{sat} is the saturation temperature corresponding to steam generator pressure, and T_{feed} is feed water temperature. T_{sat} and H_{fg} are taken at steam generator pressure.
For each system, heat loss (Q_{loss}) is the loss due to convection and radiation mode.

4.8. Boundary Condition

APWC system failure can be of two types; one is active component failure and other is passive component failure. In order to do conservative analysis, passive component failure (i.e. header rupture) was considered. By considering passive component failure, APWC system will not have any water and hence temperature in all associated systems will increase. It is assumed that the water in APWC is not available immediately after failure. In case of failure of APWC the boundary condition can be given as

$Q_{apw-apwc} = 0$, $M_{apwc} = 0$

4.9. Cover gas

Moderator level change affects the cover gas pressure which can be calculated using equation of state.
The Mass balance for cover gas can be given as,

$$M_{cov} = M_{he} + M_{stm} \tag{9}$$

where M_{he} and M_{stm} are helium mass and steam mass respectively which can be evaluated by following equations.

$$\frac{dM_{he}}{dt} = M_{he,in} - M_{he,out} \tag{10}$$

$$\frac{dM_{stm}}{dt} = M_{stm,in} - M_{stm,out} \tag{11}$$

Helium is injected into the cover gas circuit in case cover gas pressure falls below the specified set pressure. Steam (D_2O) is added into the cover gas volume due to evaporation/boiling of moderator. If the cover gas pressure rises above the relief valve set point it relieves (by relief valve) both the helium and steam out to the atmosphere. Cover gas pressure is calculated using equation of state. Once the cover gas pressure reaches the OPRD rupture pressure the disk ruptures and results in moderator flashing followed by boiling. Mass of moderator flashed was calculated in a simplified manner,

$$m_v = \frac{M_{mod}\left(H_L - h_L\right)}{\left(h_v - h_L\right)} \tag{12}$$

where H_L is Liquid enthalpy of moderator at OPRD rupture pressure and h_v & h_L are vapor and liquid enthalpies of moderator at atmospheric pressure.

4.10. Protective Actions

There are several alarm [3,9] and trip settings for various parameters of the individual systems modeled. All these are modeled suitably in the program. The few of these which are more important in the present analysis are stated below.

a. Reactor setback at moderator outlet temperature of 70°C [10].
b. Reactor trip at calandria vault water outlet temperature high.
c. OPRD burst pressure at 1.4 kg/cm^2(g).

5. ASSUMPTIONS

Some of the assumptions required to be made to simplify the analysis are stated below,

1. Active process water keeps circulating in the APW system all the time.
2. Pumps for circulating moderator, end shield cooling water, and calandria vault cooling water keep running.
3. PHT pumps and PHT feed pumps (Pressurising pumps) keep running.
4. Heat transfer coefficients for various heat exchangers are calculated by using appropriate empirical correlations.
5. A constant overall heat transfer coefficient was assumed for Steam generator.

6. NUMERICAL ANALYSIS

In order to solve the above set of equations various data in respect of system geometry, flow parameters, system mass inventories, heat loads in all the systems at full power condition, decay heat generation rates, initial temperature and ambient conditions were given as input. Reactor trip, set back and annunciations and other inputs were given for the analysis.

Initial temperatures of all the systems were given same as ambient temperature. All the first order equations were solved by fourth order Runge-Kutta method. Steady state conditions of all the systems were reached first and then initiating condition for accident scenario was introduced to carry out the analysis.

Since the heat sink is not available during transients, temperatures start rising in all associated systems. Due to temperature rise in moderator system, moderator level rises, thereby cover gas pressure rises. Moderator level, cover gas pressure, cover gas mass, moderator mass, etc. were calculated and reactor set back, reactor trip and other annunciation actions were implemented whenever the concerned parameters exceed the permissible values.

7. RESULTS AND DISCUSSION

Steady state operational conditions for all the systems were reached after quasi-steady state iterative calculations over a process time of about 3×10^{5} seconds. The maximum rate of change in steady state parameter, among all chosen parameters, was ensured to be less than 10^{-11} between two consecutive iterations. Table-1 shows comparison of system temperatures as obtained by the code with the values in design manuals.

Figures 4 and 5 show the average temperature variation in all the systems until reaching the steady state. Average temperatures are found to be in close agreement with those provided in design manuals.

Some of the larger deviations between the compared values were noted to be on account of difficulties in getting precise input data in respect of geometry and hold up capacities.

Table 2 shows the results obtained and sequence of events after initiation of the accident. It was found that at 1230 seconds moderator outlet temperature was very high and reactor set back gets initiated. At around 3060 seconds reactor trip gets initiated due to high temperature of the calandria vault water at the outlet. Temperatures in all the systems continue to rise gradually due to decay heat and at around 70435 seconds (approx 19.56 hrs) OPRD is seen to get ruptured.

Figures 6 and 7 represent the temperature variations in all the systems after initiation of accident analysis.

From Fig 6 it has been found that temperatures in all the systems start rising initially at slower rate and later at faster rate. At 1230 seconds temperature of moderator starts decreasing due to initiation of reactor set back signal. Thereafter the temperatures keep rising in all the systems due to addition of decay heat. In CV system, rate of heat generation is less as compared to its huge water inventory; hence the temperature rises at a slower rate.

The steam generator temperature varies according to steam generator (SG) pressure, which is controlled by an independent pressure control module. The SG pressure varies from 40 kg/cm^2(g) at full power to 48 kg/cm^2(g) at zero power. The reactor set back signal at 1230 seconds causes power reduction at a rate of 0.5% full power per second. Due to this, as depicted in Fig-7, SG temperature rises and the PHT temperature decreases and at near zero power both the PHT and SG temperatures get equalized [11] beyond about 1430 seconds.

In Fig 8 moderator level variation (i.e. moderator swelling/shrinking) with respect to time has been plotted. At around 1230 seconds, the moderator level starts decreasing due to set back and again starts rising slowly due to addition of decay heat. At 18759 seconds the moderator level starts rising steeply because of its expansion in the vertical portion of the OPRD pipe (see Fig.2 i.e. from 100% to 110%). Thereafter it rises (i.e. swells) gradually and at 53157 seconds moderator starts to boil. Around 57000 seconds onward the moderator swelling rate is less than the evaporation rate; this causes the moderator level to decrease. The final sharp decrease in moderator level is due to moderator flashing followed by OPRD rupture.

Fig 9 shows that the cover gas pressure rises at faster rate until 1230 seconds and decreases thereafter due to set back and after that it rises gradually. Once pressure reaches the relief valve set pressure (0.5 kg/cm^2(g)), relief valve opens until the pressure comes down to valve reseating pressure (0.45 kg/cm^2(g)). As the moderator evaporation starts after 53000 seconds the cover gas pressure starts rising gradually until OPRD ruptures.

Fig 10 shows the moderator mass variation with respect to time. Moderator evaporation starts when its temperature reaches the saturation temperature at the cover gas pressure. After 53157 seconds moderator starts to evaporate gradually. Sharp decrease in moderator mass is due to moderator flashing that takes place after OPRD rupture at 70435 seconds.

Once the OPRD rupture occurs, the evaporation of moderator will result into loss of moderator heat sink following which the core heat-up will be at a higher rate. The estimated time as per the present analysis for the initiation of moderator heat loss event is about 70,000 seconds i.e. approximately 20 hours. Due to several assumptions made in the analysis and uncertainties involved in the mathematical modeling of the systems, it is quite likely that the actual time to initiation of moderator heat sink loss may be actually larger than the above estimate. In order to mitigate the consequences of the postulated accident, 20 hours time may be considered as adequate for taking corrective actions such as organizing manual provision of fire fighting water in the moderator circuit.

Table 1 Comparison of Steady State Values

	Values obtained from Analysis			Design Manual Values		
	Average Temperature	Temp at inlet of system (°C)	Temp at outlet of system (°C)	Average Temperature	Temp at inlet of system (°C)	Temp at outlet of system (°C)
Moderator System	51.37	42.35	60.40	52.85	43.70	62.00
Calendria System	43.72	39.84	47.61	44.35	42.50	46.20
End Shield System	44.39	39.74	49.04	44.95	43.30	46.60
APW System	39.53	34.64	44.43	39.38	36.00	42.76
APWC System	34.00	29.85	38.15	35.49	31.28	39.70
PHT system	270.46	248.11	292.81	271.20	249.00	293.40
Secondary Side	250.59			210.80	171.00	250.60

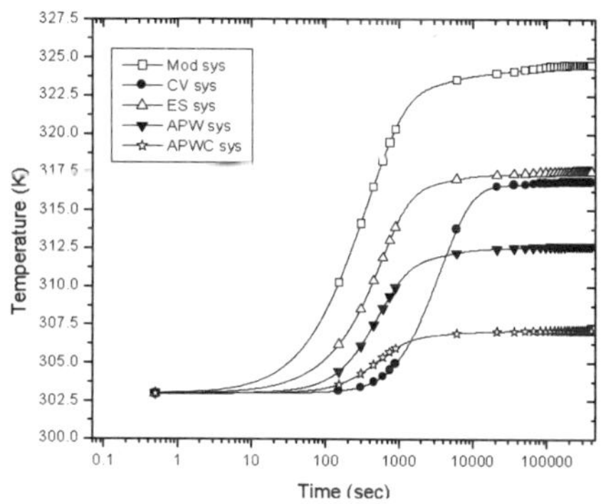

Fig. 4: Ave. Temp variation before initiation of accident conditions (to get steady state)

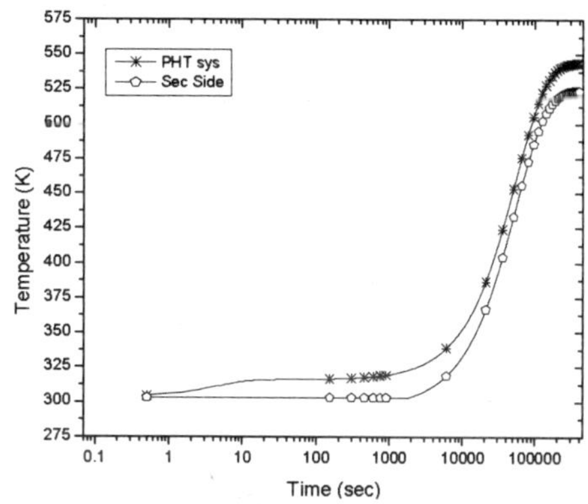

Fig. 5: Ave. Temp variation before initiation of accident conditions (to get steady state)

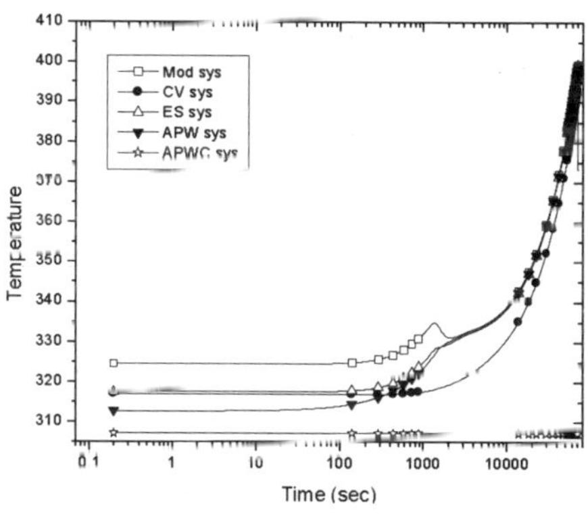

Fig. 6: Ave. Temp variation after initiation of accident conditions

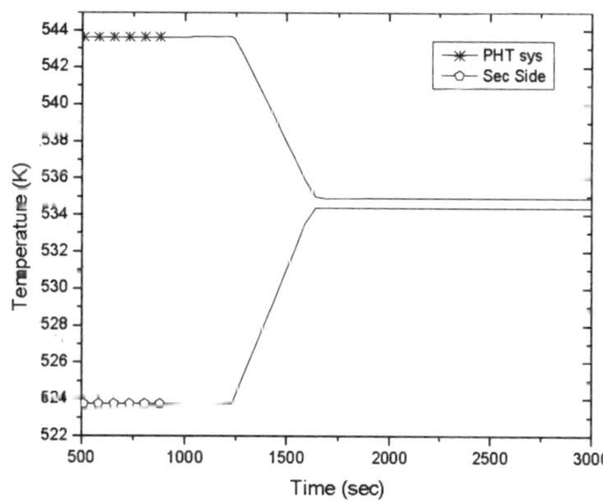

Fig. 7: Ave. Temp variation after initiation of accident conditions

Table 2 Results and sequence of events after initiation of accident conditions

Time (sec)	Mod Temp (K)	CV	ES	APW	PHT	SG	Cover gas Pressure (Pa)	Mod Level (%)	Mod Mass (kg)	Command Actions
398.4	326.5472	317.01	319.5	317.45	543.61	523.74	132835.74	96.1995	138000	End Shield Water Outlet Temp 51 °C: High Alarm
440.9	326.8958	317.04	319.84	317.9	543.61	523.74	133200.15	96.2344	138000	APW Inlet Temperature 41°C: High Alarm
739.4	329.5764	317.32	322.5	320.97	543.62	523.75	136122.42	96.5207	138000	End Shield Water Outlet Temp 54 °C: Very High Alarm
1017.9	332.2178	317.69	325.14	323.7	543.62	523.76	139154.78	96.8261	138000	Mod Outlet temp. 68 °C, Alarm; High Temperature
1229.9	334.2353	318.03	327.18	325.73	543.63	523.77	141585.92	97.0784	138000	Mod Outlet temp. 70 °C, Reactor Setback Initiated
3064.9	332.3086	321.17	331.65	332.02	534.91	534.38	139260.59	96.83689	138000	Calandria Vault Inlet Temp 46.5 °C: Reactor Trip
18759.32	348.1036	340.67	347.53	347.73	534.84	534.48	145807.12	100.00003	138000	Moderator lvel 100% : Alarm High Level
20452.32	349.8962	342.5	349.32	349.53	534.83	534.48	148571.63	103.00031	138000	Moderator lvel 103% : Alarm Very High Level
70434.9	399.5647	393.29	399.31	399.53	534.8	534.53	242542.97	115.11951	137556.4	OPRD Ruptured

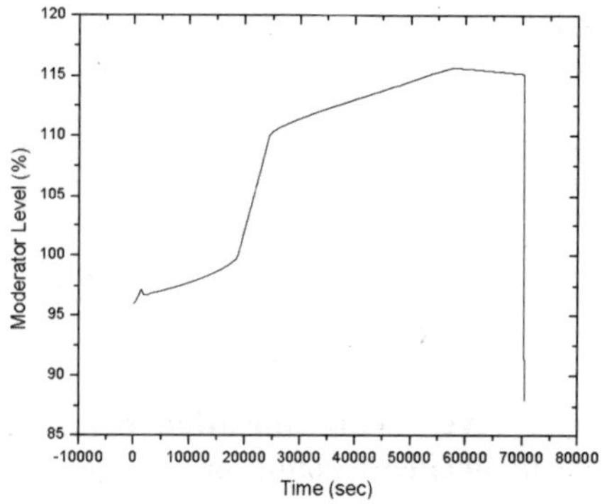

Fig. 8: Moderator level variation after initiation of accident conditions

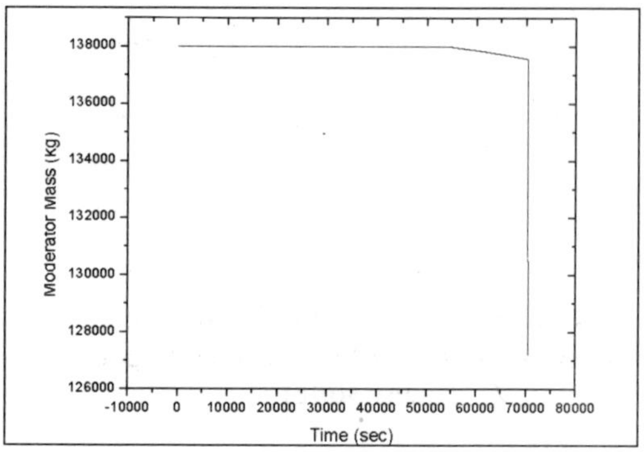

Fig. 10: Moderator mass variation after initiation of accident conditions

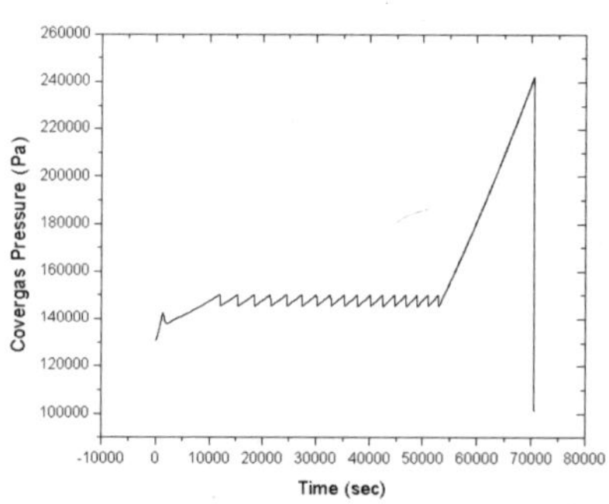

Fig. 9: Cover gas pressure variation after initiation of accident conditions

8. CONCLUSION

The model has been developed based on lumped parameter approach to analyse the transients arising out of the postulated accident scenario comprising of APWC failure followed by unavailability of FFS water for Indian PHWRs. It is observed that about 20 hours after the accident, the moderator heat sink is likely to get lost. The analysis assumes that the PHT system's decay heat removal capability is intact, which needs to be further investigated. Further studies are to be continued to improve the present model.

9. REFERENCES

1. Babar,A.K, Saraf,R.K, et al "*Probabilistic safety assessment of Kaiga Atomic Power Project*", Internal Divisional Report.
2. Kaiga-1,2 Preliminary Safety Analysis Report Vol-II: Technical Details.
3. Design Basis Report Kaiga-1,2/DBR/71300/106; "*Process Water Systems*"
4. Design Manual; "*Moderator System*"
5. Design Manual Rajasthan Atomic Power Project-3,4/34160; "*Calandria Vault Water Cooling System Process and Instrumentation*"
6. Design Manual, Kaiga-34110; "*End Shield Cooling System Process and*
7. Design Manual Kaiga-33110; "*Primary Heat Transport system Main circuit*"
8. Kaiga-3,4 Preliminary Safety Analysis Report Vol-I, Section-6; "*Turbine Generator and Secondary System*"
9. Design Manual Kaiga-1,2/DM/67130/145; "*Active Process Water System Control and Instrumentation*"
10. Design Manual Kaiga-1,2/DM/63722/144; "*Reactor Set Back System*"
11. Design Manual Kaiga-333000; "*Feed, Bleed, Storage and Gland Supply Systems*"

Reliability, Safety and Hazard
Editors: P.V. Varde et al.

PRESSURISER SAFETY VALVE STUCK OPEN SEVERE ACCIDENT ANALYSIS FOR VVER 1000 NPP

U.S. Chikkanagoudar, S.K. Dubey, S.K. Gupta and R.I. Gujrathi

Atomic Energy Regulatory Board, Anushaktinagar, Mumbai 400 094

ABSTRACT

Kudankulam Nuclear Power Project is a 1000 MWe VVER type reactor incorporating special design features. In AERB analysis was performed for this reactor using RELAP5/SCDAP/Mod 3.2 for an accident, which is initiated due to inadvertent opening of a pressuriser safety valve and remaining in stuck open position. Without Passive Heat Removal System (not taking credit of PHRS), the accident turns into severe accident due to active safety systems failure. Continued heating and loss of primary coolant lead fuel failures and melting. This paper gives the brief description of system and its modeling using RELAP5/SCDAP/Mod 3.2 along with the Initial and Boundary conditions. Finally the analysis results and event sequence based on variation of important parameters such as core maximum temperature, oxidation and hydrogen generation, fuel melting and relocation etc. are discussed.

Keywords: Severe Accident Analysis, VVER, SCDAP, RELAP5, Pressuriser, PSD, Safety Valve

1. INTRODUCTION

The Kudankulam VVER is a 1000 MWe light water cooled, light water moderated reactor of Russian design. Reactor core consists of 163 hexahedral Fuel Assemblies (FAs) and each FA comprises of 311 fuel elements [1]. This reactor incorporates many advanced safety features to mitigate Design Basis Accidents and Beyond Design Basis Accidents, to avoid severe core damage [3].

The primary circuit consists of reactor downcomer, lower plenum, core, upper plenum and four reactor coolant loops [2]. Each loop is having a hot leg, primary (tube) side of steam generator, cold leg and Reactor Coolant Pump (RCP) set. Pressuriser surge line is connected to hot leg of 3rd loop and spray lines to cold leg of 4th loop. Three Pulse Safety Devices (PSDs) (one control and two safety) are mounted on the pressuriser relief line. Passive first stage ECCS consists of four accumulators, of which two are connected to reactor upper plenum and two to reactor downcomer.

The secondary circuit consists of four loops and each loop is having a horizontal Steam Generator (SG), steam line and main steam header connecting all four steam lines [4]. On each steam line one Main Steam Isolation Valve (MSIV) is mounted. Governor system controls the supply of steam to turbine from main steam header. There are two over pressure protection devices mounted on main steam header, i.e. BRU-A (Atmospheric Steam Discharge Valve) and BRU-K (Condenser Steam Discharge Valve). SG overpressure protection is done only by SG PSDs. One advanced feature called Passive Heat Removal System, controls the SG pressure and removes decay heat through SG in case of total power loss.

2. MODELING AND INITIAL CONDITIONS

This section describes the complete Nodalisation of KK NPP. Nodalisation diagram of the reactor vessel and core is given in Figures 2.1. The core is divided into five parallel channels (410 to 414), each of ten sub-volumes. Lateral flow between adjacent core channels is simulated using the cross flow model. Downcomer is represented by Annulus component (505) and core bypass is represented by pipe component (508). Branch component (506) represents the lower plenum. The upper plenum is divided into two single volumes 510 and 511. Heat structures are used to model the reactor vessel metal structures. The core is divided into five regions by grouping similarly powered fuel assemblies (FA). Each FA consists of 311 fuel elements and 18 control rod elements. Table 2.1 lists the FAs, control rod (CR) bundles and burnable absorber rod (BAR) bundles in each core region.

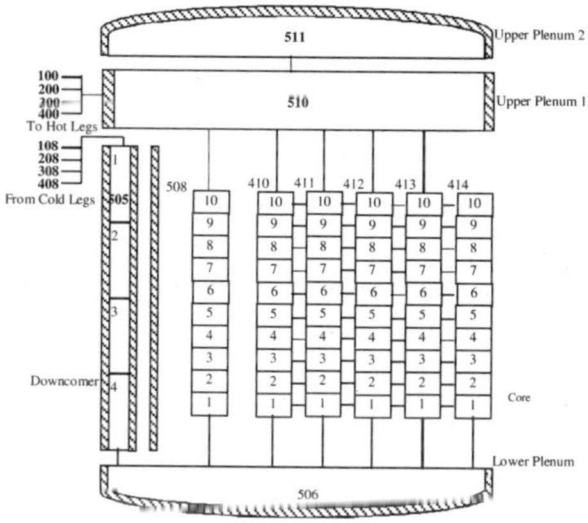

Fig. 2.1: Nodalisation of reactor vessel

Primary system is modeled as four loops connected to reactor pressure vessel. Nodalisation of one primary coolant loop (Figure 2.2) consists of hot leg (pipe component 100), hot collector (pipe component 101), primary side of steam generator (3 parallel pipe components 102, 103 and 104), cold collector (pipe component 105), pump suction leg (pipe component 106), and cold leg (pipe component 108). The pump is modeled as the pump component (107). Pump trip is simulated by specified coast down speed as input. The other three loops are identical except for some differences in pressuriser and ECCS connections. Heat structures are used to represent the thermal behavior of the primary piping.

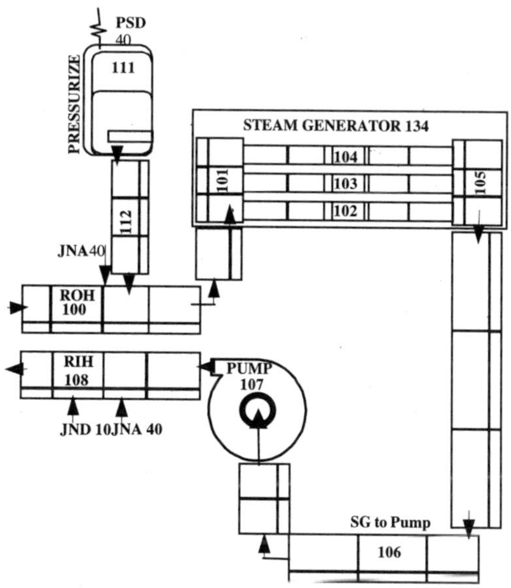

Fig. 2.2: Nodalisation of reactor vessel

The pressuriser is modelled with pipe component 111. The surge line is represented with 3 sub-volumes (pipe component 112). Three pulse safety devices (PSD) are represented using motorised valve components. Surge line is connected to hot leg of the 3rd loop. Pressuriser heaters are simulated by the four heat structure components.

Nodalisation of secondary involves horizontal steam generators modelled as pipe component with five volumes. The steamlines are modelled using pipe and single volume components. Main steam header is modelled as branch component. Atmospheric Steam Discharge Valves (BRU-A) located on MSH and are modelled with motor valve component. Feed water is supplied by using boundary conditions. The other three loops are identical to the first loop.

Table 2.1 Core region distribution for SCADAP

Region	Fas	CR Bundles	BAR Bundles
1 (inner)	19	13	6
2	36	18	18
3	36	36	-----
4	36	18	18
5 (outer)	36	-----	-----
Total	163	85	42

2.1 Assumptions

Following assumptions are considered;

- Station Black Out (Total Power Loss) occurs 1.9 s before the start of reactor shutdown operation.
- The first signal for reactor trip (in this case low primary pressure) is ignored.
- Emergency Protection action starts after delay of 3.5 s after signal generation for EP.
- One out of four BRU-A is not available.
- For accident to progress to severe core damage, the Passive Heat Removal System (PHRS) is not considered in the analysis.
- High Pressure Emergency Core Cooling system is assumed failed due to power loss
- SG Emergency Blowdown and Cooldown is not available due to power loss.
- One accumulator connected to reactor lower plenum fails to actuate.

2.2 Initial Conditions

The steady state has been obtained after modelling of the various reactor components. The steady state values are used as the initial conditions for the accident scenario considered and are given in Table-2.2 to Table-2.4 [7]. Table 2.5 lists the various SCDAP parameters used in the analysis.

Table 2.2 Initial Conditions [5]

Parameter	Value
Reactor thermal power, MW	3120.0
Coolant temperature at reactor inlet, °C	296.0
Coolant pressure at the core outlet, Mpa	15.4
Coolant flow rate through reactor, m^3/h	82200.0
Maximum linear heat flux, W/cm	448.0
Level in steam generator, m	2.2
Pressure in SG main steam header, Mpa	6.27
Steam rate to turbine kg/s	1696
Feedwater flow rate into one SG, kg/s	424.0
Feedwater temperature, °C	220.0
Maximum radial peaking factor	1.86
Effective fraction of delayed neutrons, %	0.77
Fraction of coolant core bypass %	3.0

Table 2.3 Core axial power distribution [7]

Height%	5	15	25	35	45
Value (Relative unit)	0.36	0.84	1.13	1.30	1.39
Height, %	55	65	75	85	95
Value (Relative unit)	1.39	1.27	1.12	0.84	0.36

Table 2.4 Decay heat (relative unit) after shutdown

Time, s	0.01	0.1	1	10
Decay Heat	0.07446	0.07445	0.07439	0.05746
Time, s	100	1000	10000	50000
Decay Heat	0.03846	0.02279	0.01182	----

Table 2.5 RCP speed during coast-down [7]

Time, s	0.0	0.5	1.0	2.0	5.0	10.0
Speed rps	16.25	15.8	15.4	14.6	12.7	10.5
Time, s	20.0	30.0	50.0	80.0	100.0	104.0
Speed rps	7.6	6.1	3.5	2.01	0.601	0.00

Table 2.6 SCDAP input parameters

Parameter	Value
Fuel Rods	
Active Length (m)	3.53
Rod Pitch (m)	0.01275
Pellet Radius (m)	0.00385
Cladding Inner Radius (m)	0.003935
Cladding Outer Radius (m)	0.00457
Fuel density (% T.D.)	95.0
Mass of He Fill Gas (kg)	1.8×10^{-5}
Inert Gas Pressure (bar)	20
Upper & Lower Plenum Void Volume (m^3)	1.77×10^{-5}
Plenum Length (m)	0.1689
Burn-up	0.0
Control Rods and Burnable Absorber Rods	
Guide Tube Inner Radius (m)	0.0063
Guide Tube Outer Radius (m)	0.0055
Cladding Inner Radius (m)	0.0041
Cladding Outer Radius (m)	0.0046
Grid Spacers	
Grid Spacer Mass (kg)	0.55
Grid Spacer Height (m)	0.02
Grid Spacer Thickness (m)	8.0×10^{-4}

3. RESULTS AND DISCUSSION

Power reduction begins when absorber rods start dropping after 3.5 s delay. Whole scram process takes 4.0 s to complete. Turbine Generator stop valve closes following the reactor trip, resulting in pressurisation of SG and opening of BRU-A. Figures 3.1 to 3.8 give variation of some of the important parameters along with description for each parameter.

Table 3.1 Chronological Sequence of Events

Time (s)	Event
0.0	PRZ PSD opening and fails to seat (Initiating Event)
17.1	Generation of first scram signal ($P_{reactor} \leq 14.7$ MPa)
26.0	Generation of the 2nd scram signal (DNBR ≤ 1.35)
27.6	SBO (Total power loss) - Trip of all RCP sets - Disconnection of feedwater system - Disconnection of primary makeup system - BRU-K disconnection - Pressuriser heater power off - TG stop valve closing
29.5	CPS AR rods starts dropping (after 3.5 s delay)
30.1	BRU-A opening and operation in pressure control mode (Steam header pressure more than 7.2 MPa)
33.4	Signal for ECCS pumps connection (Decrease of difference in saturation temperature of any hot leg). Pump fail to start on loss of power.
131.6	RCP set coast down complete (for 104 s)
2000.0	Upper part of core exposed as Level drops below top of core
4087.1	First failure of cladding at high temperature under compressive load at axial node-7 & 6 of inner ring
4261.8	Debris formation caused by phase change (at node 6, 7 & 8 within 100s range)
4365.7	Melt pool forms at node 6 of inner ring and spreads. Blocked by freezing of relocating cladding.
4500.0	Coolant level drops below Lower Core Support Plate. Table 3.2 gives variation of major core parameters during core degradation phase (4500 s to 8000 s).
8000.0	Calculation is terminated, due to property error after complete core blockage. The error is being investigated. Final state of core is represented in Figure 3.1

The primary coolant leak through PSD results in core uncovery. The fuel temperature rises and cladding oxidation takes place due to Zircaloy and steam interaction. The upper part of the core is exposed at 2000 s, leading to gradual heating of fuel rods. First failure of cladding was observed in upper half region of innermost ring. In about 2300 s, the fuel starts melting and forms a molten pool, which spreads radially as well as axially. At the same time control rod material starts melting, but is confined inside the guiding tube. At about 4500 s, the coolant level drops below lower core support plate, exposing the complete core. From this stage, successive stages of important parameters of core degradation are listed in table 3.2. The final stage of core at about 8000 s is presented in pictorial form in Figure 3.1. Actually, the calculation was terminated at 8600 s, because of thermophysical property error observed after complete blockage in core due to crust formation, the results in this paper are presented up to the stage of formation crust in lower part of core.

Table 3.2 Core degradation parameters

Core degradation state	Time (s)				
	4500 s	5500 s	6500 s	7400 s	8000 s
Volume of molten pool	2.52	8.25	10.66	10.96	12.160
Total mass of UO_2 in pool	15243	51311	66450	68329	75846
Total mass of oxidic Zr	2291	5681	5809	5817	5832.4
Total mass of metallic Zr	1866	6833	1027	1071	1245.7
Mass of partially liquefied porous debris	6699	1924	1743	1746	0.0

I	I	I	I	V	I	I	I	I
I	I	V	V	V	V	V	I	I
V	V	V	V	V	V	V	V	V
V	V	V	V	V	V	V	V	V
V	V	V	V	V	V	V	V	V
V	V	V	V	V	V	V	V	V
Mx	Mx	Mx	Mx	Mx	Mx	Mx	Mx	Mx
Mx	Mx	Mx	Mx	Mx	Mx	Mx	Mx	Mx
Mx	Mx	Mx	Mx	Mx	Mx	Mx	Mx	Mx
Mx	Mx	Mx	Mx	Mx	Mx	Mx	Mx	Mx

Fig. 3.1: Degraded core at 8000 s (I-Intact, V-Molten & slumped, Mx – Frozen melt to form crust)

3.1 Primary Pressure

Due to break of primary pressure boundary, initially there is sharp drop in primary pressure (Figure 3.2). Afterwards till 1900 s, the pressure is almost constant at about 8 MPa due to increased void formation. Pressure starts to rise on excessive void formation and then reduces at low core inventory. Water supply from ECCS accumulator is actuated when core pressure reduces to 5.89 MPa at about 4920.0 s (accumulator water flow was intermittent).

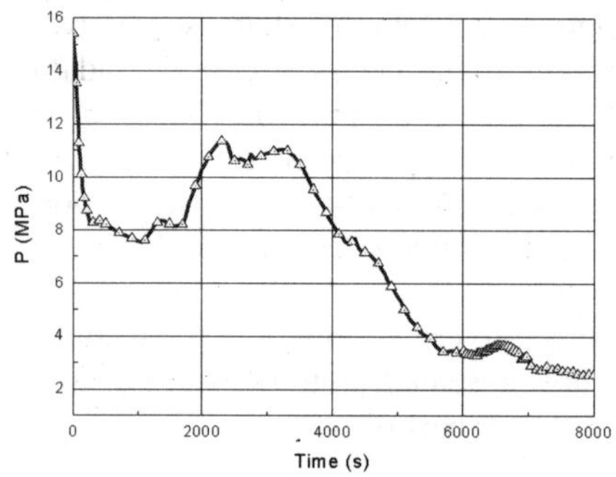

Fig. 3.2: Primary Pressure

3.2 Primary Level

Level zero corresponds to the lowest point in the lower plenum (Figure 3.3). Initial level is 10.9 m, (lower plenum + core + upper plenum). Initially the level continuously decreases due to loss of inventory, and then starts fluctuating up to 4920 s and gradually increases after accumulator injection. Finally the level then is maintained at about cold leg elevation.

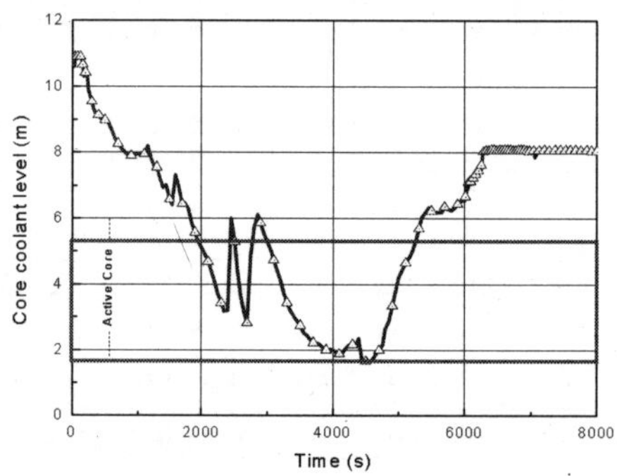

Fig. 3.3: Primary Coolant Level in Vessel

3.3 Leak Flow

Leak calculations were performed considering the leak flow area equal to 0.00196 m^2 (corresponds to critical steam flow of 50 kg/s at 18.11 Mpa). During initial phase of transient, single-phase steam flows at about 45 kg/s and it keeps reducing due to reduction in primary pressure (Figure 3.4). Later on due to void swelling, pressuriser starts refilling with steam water mixture, which leads to higher discharge rates under two-phase flow.

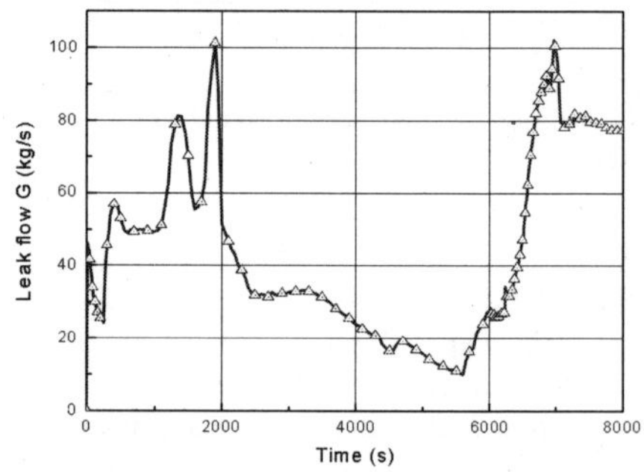

Fig. 3.4: Leak flow from pressuriser PSD

3.4 Coolant Flow

RCP trips at 27.6 s on loss of NPP off-site power. This leads to gradual reduction in core flow for about 104 s due to coast down of RCPs (Figure 3.5). After RCP coast down, natural circulation flow persist for considerable time due to cooling by secondary inventory present in SG. After SG dryout, natural circulation flow ceases and is almost stagnant (fluctuating with very low mass flow) at 1600 s.

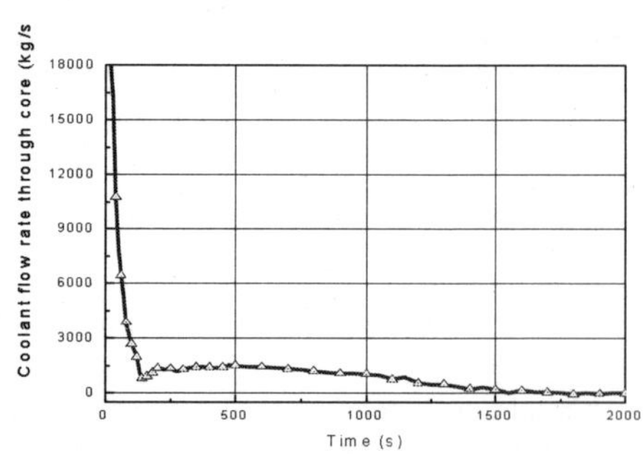

Fig. 3.5: Primary Coolant Flow

3.5 Core Max Surface Temperature

Figure 3.6 shows the core maximum surface temperature during the accident scenario. Initially the temperature is almost constant at saturation temperature of coolant. As core uncovery begins, the temperature rises gradually, and the increase is sharp after 4000 s i.e. when most of the core is uncovered. The sharp increase is due to large amount of heat addition from zircaloy oxidation. Later on the temperature is around Fuel melting point (≈3000 K).

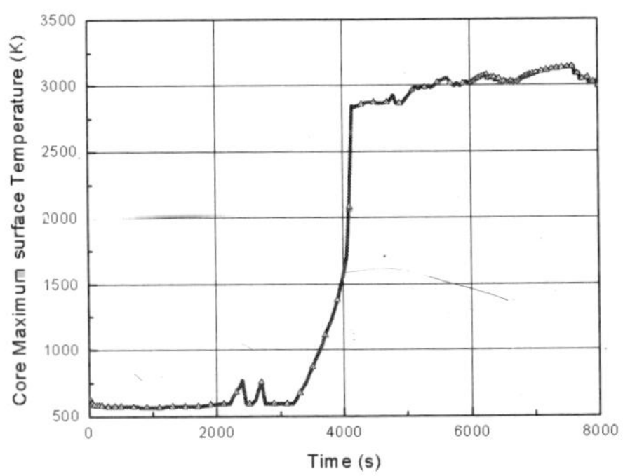

Fig. 3.6: Core maximum surface temperature

3.6 Oxidation Heat

The total oxidation heat generation during the accident scenario is shown in the Figure 3.7. The maximum oxidation heat of 150 MW is being liberated at about 4000 s.

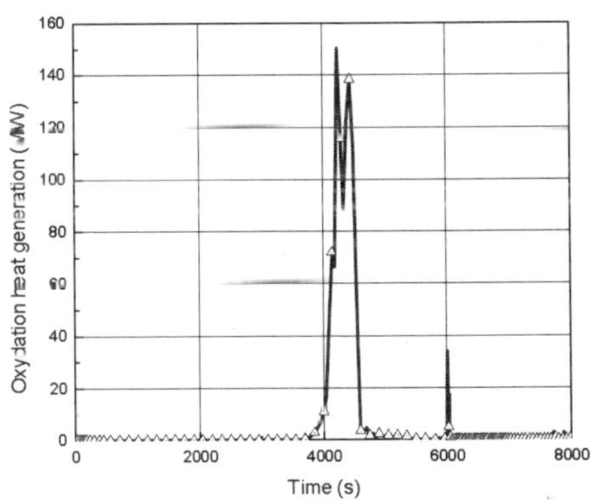

Fig. 3.7: Core total oxidation heat generation

3.7 Hydrogen Generation

Figure 3.8 shows the hydrogen generation rate during the accident scenario. The hydrogen generation depends on the oxidation and the trend is similar to the oxidation heat generation. Maximum hydrogen generation rate is about 1.1 kg/s and total hydrogen generated is about 300 kg.

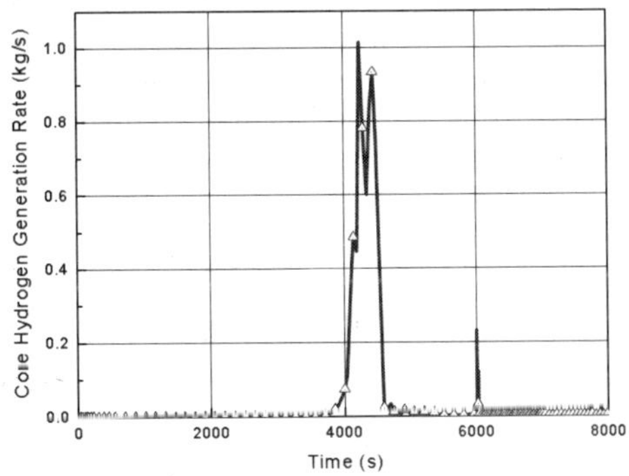

Fig. 3.8: Core total hydrogen generation rate

4. CONCLUSIONS

By analysis it is seen that;

- In about 34 min of accident, the coolant level drops below upper part of core
- After 75 min, the coolant level drops below lower core support plate, exposing the complete core
- After 2.22 hr, most of the central zone of core is in molten and slumped downward where; crust of frozen melt is formed after coming in contact with coolant. Crust formed was about 1 m above the LCP.
- Equivalent radius of melt pool was 1.77 m, which same as that of core radius (In core central region, melt has spread completely along radial direction).
- The maximum oxidation heat generation rate observed was 150 MW. Corresponding maximum hydrogen generation rate observed was 1.1 Kg/s.
- After 2.22 hr of accident, the total hydrogen generated in core is about 300 Kg.

5. REFERENCES

1. Bocharov Yu. L, Preliminary Safety Analysis Report, Reactor, S-4, Sep. 2000.
2. Krushelnitsky, V. N., Preliminary Safety Analysis Report, Reactor Coolant System, S-10, Feb. 2002.
3. Preliminary Safety Analysis Report, Revision 2, Chapter S-6/2, Engineered Safety Features, Kudankulam Project, February 2002
4. Shmygin, A. I., Preliminary Safety Analysis Report, Steam Conversion System, S-10, Feb. 2002.
5. Borisov L. N, Preliminary Safety Analysis Report, Primary Coolant Inventory Decrease, S-15/6, Jan 2001.
6. Code Manual I-V, RELAP5/MOD 3.2, Idaho National Laboratory, IdahoFalls, USA, 2001
7. Preliminary Safety Analysis Report, Revision 1, Chapter S-15/0, Accident Analysis, Introduction, Kudankulam Project, March 2001.

Reliability, Safety and Hazard
Editors: P.V. Varde et al.

PLANT TRANSIENT ANALYSIS ON ONE PUMP TRIP FOR 540MWe INDIAN PHWR

Santanu Pahari, P. Pramod, H.P. Rammohan, P.K. Malhotra, S.G. Ghadge and S.S. Bajaj

Directorate of Safety, Nabhikiya Urja Bhavan, NPCIL, Anushakti Nagar, Mumbai 400 094

ABSTRACT

In 540MWe Indian Pressurised Heavy Water Reactor (PHWR), in case of single Primary Circulating Pump (PCP) trip, Reactor Outlet Header (ROH) pressure at the suction side of the tripped pump increases and the other ROH pressure in the same loop decreases because of redistribution of pressure along the loop. Pressuriser has significant effect on the pressure imbalance during the transient as it is only on the north side of the core and provides interloop connection. The analysis is done by the plant transient analysis thermal-hydraulic nutronic computer code "ATMIKA.T" with the modelling of all the associated control systems. This paper deals with the two scenarios of single pump trip in north side and south side with a number of schemes to reduce the pressure imbalance and to check whether the reactor survives without trip in such scenarios.

1. INTRODUCTION

In 540MWe Indian Pressurised Heavy Water Reactor (IPHWR), Primary Heat Transport System (PHT) is divided into two loops. In each loop there are two Primary Circulating Pumps (PCPs) and they are kept in either side of the reactor core. These two loops are independent of each other, except the connection through pressurizer, which is at north side of the core. So the transient behaviour will be different in case of a North side pump trip and a South side pump trip. This is basically due to the interloop transfer of coolant, due to the pressure imbalance between the North-ROHs of the two loops which are connected to Pressurizer.

In case of single PCP trip, ROH pressure at the suction side of the tripped pump will increase and the other ROH pressure in the same loop will decrease because of redistribution of pressure along the loop. There is a stepback which initiates immediately with the single PCP trip to drop the power down to 65%.

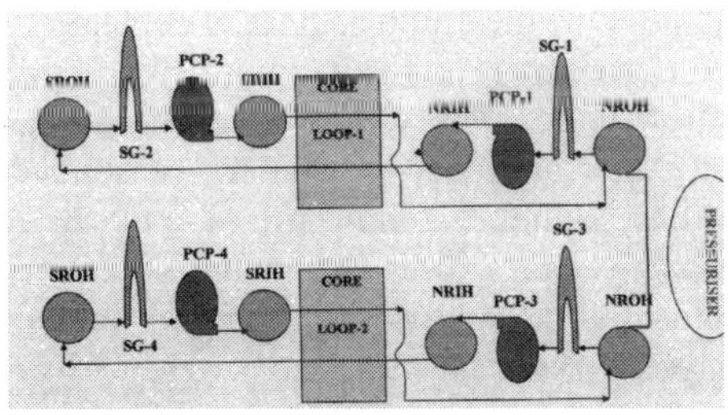

Fig. 1: Schematic diagram of two loop in IPHWR with pressuriser

Here, the two scenarios of single pump trip on North side and single pump trip on south side have been analysed by plant transient analysis code ATMIKA.T with different schemes and the result is presented below.

2. BRIEF DESCRIPTION OF THE CODE

ATMIKA.T [1] is a thermal hydraulic nutronic computer code for simulation of Loss of Coolant Accident (LOCA) and other transients. The code has been used for the Plant Transient Analysis of Kaiga Generating Station-1&2 [2], Rajastan Atomic Power Station-3&4 [3] and Tarapur Atomic Power Project -3&4 [4]. To assess performance of the code, it has been validated against the phase-C test results of Kakrapara Atomic Power Station-1 [5]. Thermal-hydraulics of the system is simulated by dividing the entire system into a number of lumped control volumes called nodes. Thermal-hydraulic part of ATMIKA is based on unequal velocity equal temperature (UVET) model. Governing equations of UVET model consist of 3 conservation equations of mass, energy and momentum. A staggered mesh arrangement where pressure, density and enthalpy are defined at the node and flow is defined along the flow path at the junction of two control volumes is used. Mass and energy conservation equations are applied on lumped control volume and momentum equation is applied on flow paths. The code has been modified to predict flow stratification and counter current flow in a vertical channel [6].

The pumps are placed at the junction of flow paths. With the given boundary conditions and input variables, the pump model calculates the pressure differential [7]. It also calculates energy transmitted by the pump as heat to coolant.

Pressurizer is considered as two separate volumes of liquid + vapour phase and vapour phase. Mass and energy balance are applied to both the volumes assuming pressure is same for both. The steam flow from the interface is calculated as a function of the voids in the liquid + vapour zone, surface area, etc. using a correlation derived from the experimental data. In this calculation, total volume of the pressurizer is maintained constant. Level tracking is done using the liquid + vapour zone volume and its cross section. This gives the swell level. As far as instrumentation is considered, it is the collapsed level in the range of the instrument.

All the process controls (e.g. PHT pressure inventory control, steam generator pressure and level control, Bleed Condenser (BCD) pressure and level control, turbine speed control) are modelled. Reactor regulating system along with setback and stepback function is modelled separately with point kinetics.

3. PCP TRIP ON SOUTH SIDE (i.e. AWAY FROM THE PRESSURISER)

In case of South side pump trip the pressure in the south side ROH (ROH-2) of the affected loop was found to rise and the north side ROH (ROH-1) pressure was found to dip (Figure 2). As the ROH-1 pressure of the affected loop (i.e.loop-1) was below the ROH-1 pressure of other loop (Figure 3), inter loop inventory transfer to the affected loop was found to occur through the pressuriser connection. This also contributes to the pressure rise in the affected loop and the reactor was found to trip on high pressure even though stepback is actuated on PCP trip.

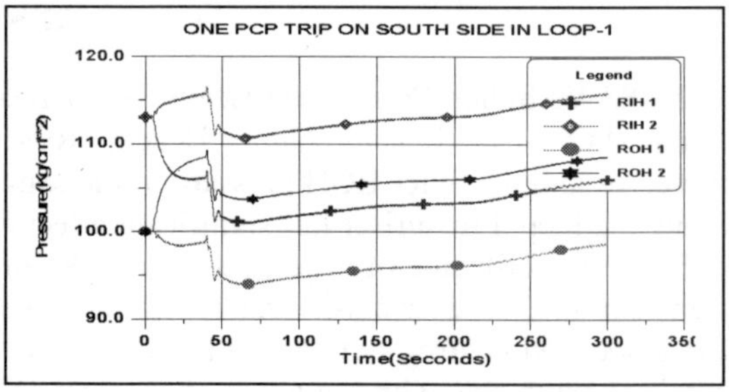

Fig. 2: PHT pressure variation in Loop -1

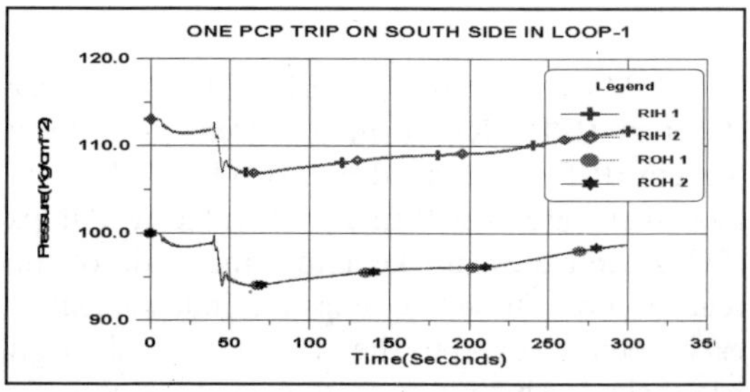

Fig. 3: PHT pressure variation in Loop -2

In another case, to avoid the interloop transfer, auto tripping of the same side (South) PCP on the other loop within 5.0 sec, was considered with power step back initiated on PCP trip (Figure 4&5).

There are four atmospheric steam discharge valves (ASDVs), one in each steam line. These transients were also analysed with the ASDVs kept fully opened manually for 10 seconds to reduce the ROH pressure by reducing corresponding steam generator (SG) pressure.

Subsequently these remain closed as SG pressure is below the ASDV opening setpoint.

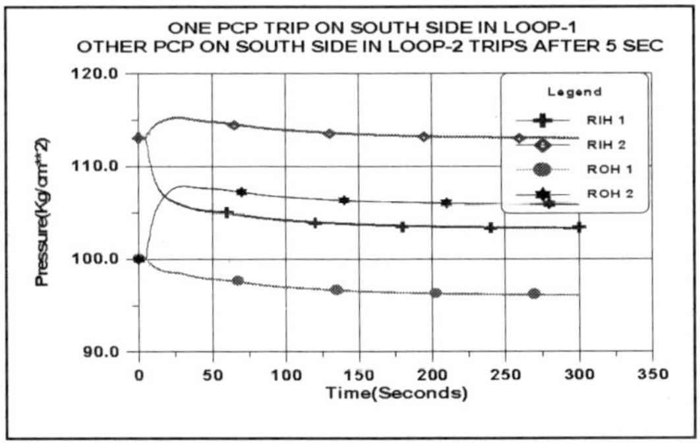

Fig. 4: PHT pressure variation in Loop -1

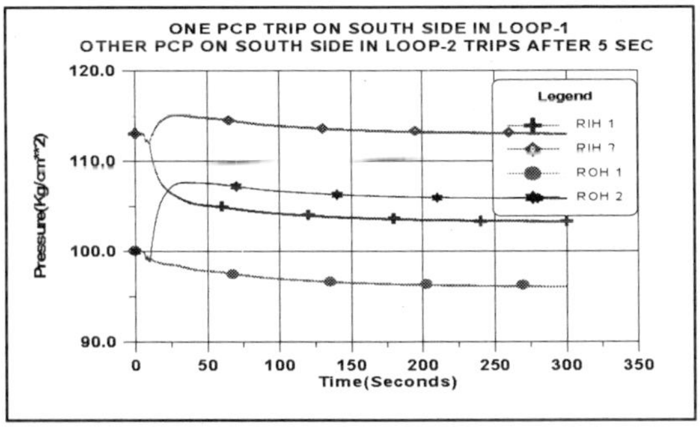

Fig. 5: PHT pressure variation in Loop -2

In case of the same side pump is tripped after 5.0 seconds of the first pump trip, the highest pressure found in the ROH was around 107.9 Kg./cm^2(g) which is 0.6Kg./cm^2 lower than the high pressure trip set point. As the pump in the other loop is also tripped, the north side ROH pressure in the other loop also got lowered. This has reduced the inventory transferred to the first loop, thereby reducing the peak pressure. Effect of opening of ASDVs on highest pressure is found to be negligible [~107.7 Kg./cm^2(g)].

In the case with the other loop pump trips simultaneously with the first pump the maximum pressure was about 107.7 Kg./cm^2(g).

The effect of delay in tripping of same side pump of the other loop was investigated. It is found that a delay of more than 25 sec will result in high PHT pressure trip.

Initiation of setback instead of stepback with one PCP trip will result in faster pressure rise because of swelling due to power and flow mismatch and reactor will trip on high pressure.

Another case where instead of tripping the other pump, ASDVs of the north bank was kept opened for 10.0 seconds was tried out. In this case the gain in reducing the peak pressure was found to be very minimal, and the reactor was found to trip on high pressure.

4. PCP TRIP ON NORTH SIDE (i.e. NEAR THE PRESSURISER)

In case of single PCP trip on north side, the pressure in the north side ROH (ROH-1) of the affected loop was found to rise and south side ROH (ROH-2) pressure was found to dip (Figure 6&7). In the affected loop the pressure of ROH (ROH-1) to which pressurizer is attached was above the pressure of the corresponding ROH (ROH-1) in the other loop. It causes the inventory transfer from the affected loop to the other loop. It causes the south side ROH (ROH-2) pressure to reach low pressure trip setting [92 Kg./cm^2(g)]. If this trip is ignored the pressure was found to dip upto 89.6 Kg./cm^2(g).

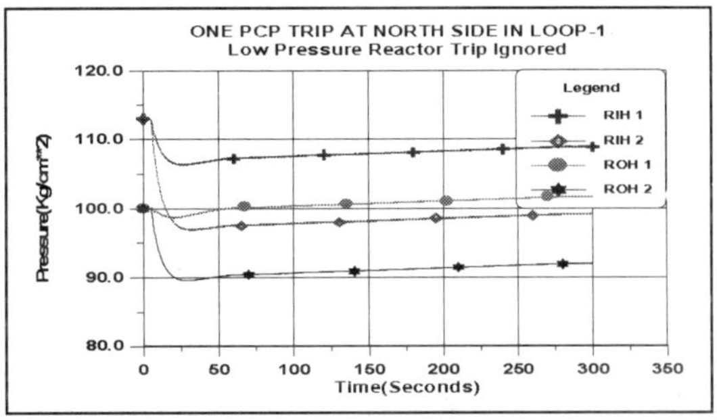

Fig. 6: PHT pressure variation in Loop -1

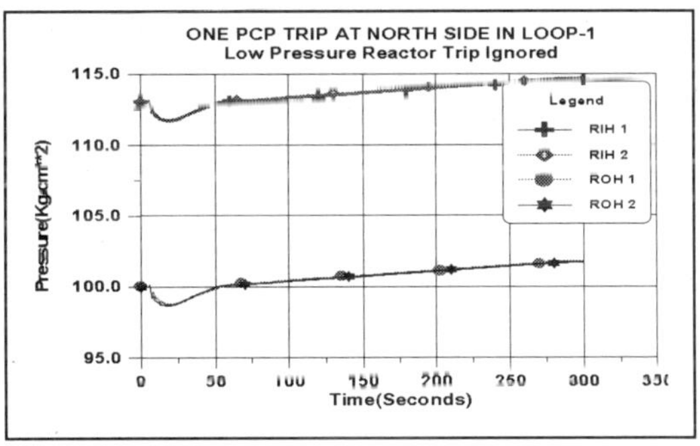

Fig. 7: PHT pressure variation in Loop -2

Here also auto tripping the corresponding pump at the other loop was considered with step back on PCP trip. In this case the minimum pressure reached was around 90.2 Kg./cm^2(g). Here no void is seen in the ROH as the coolant temperature in ROH also decreases because of reduction in boiler delta-T. One PCP trip in one loop results in flow reduction to ~74.5% in that loop but stepback reduces power to 65%FP [Figure 8]. As core flow remains higher than the power, boiler delta-T gets reduced by ~4^{0}C [Figure 9].

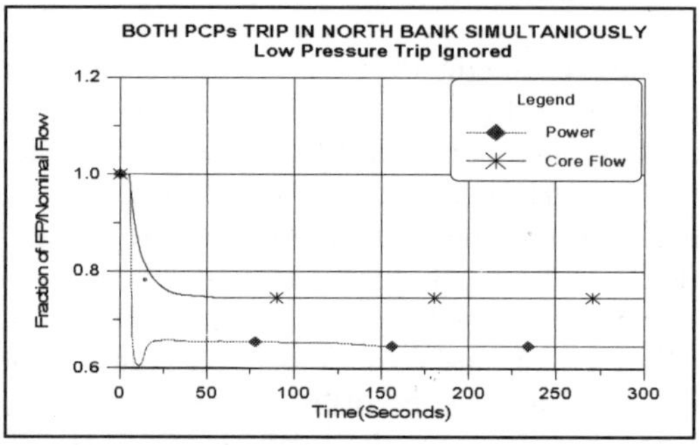

Fig. 8: Core flow and Power Variation

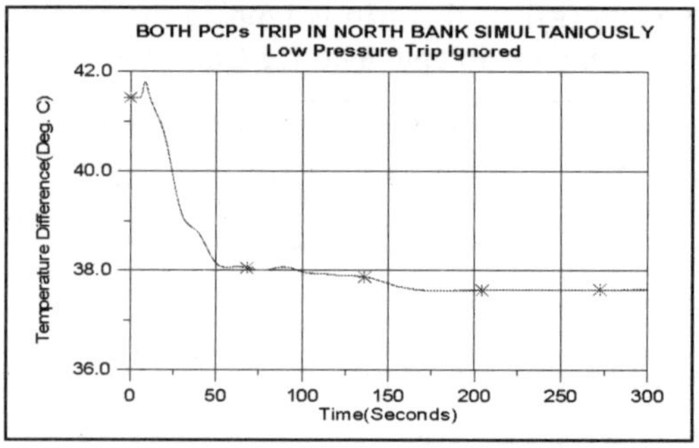

Fig. 9: Average delta-T variation across SGs

Both the above cases were also analysed using setback (at the rate of 0.5% FP/s) instead of stepback, in order to reduce the pressure dip. But in both the cases north side ROH pressure increases because of power-flow mismatch and reaches the high-pressure trip setting [Figure 10].

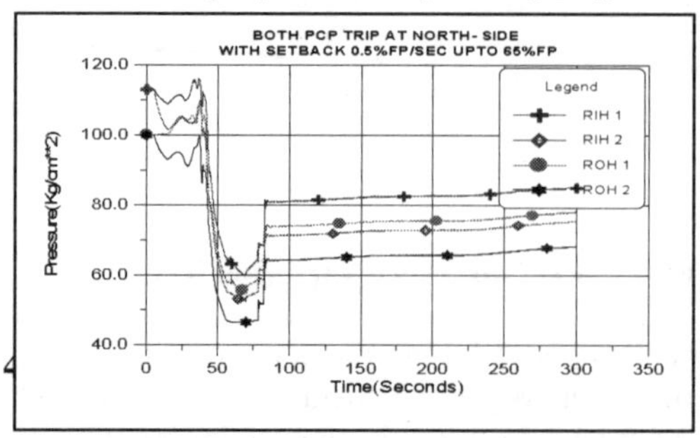

Fig. 10: PHT pressure variation in Loop-1

5. CONCLUSION

In case of a south side PCP trip, reactor trips on high PHT pressure. Reactor trip in this case can be avoided by tripping the same side PCP of the other loop within 25sec of the first pump trip.

In case of north side PCP trip, reactor trips on low PHT pressure. Tripping of same side PCP of the other loop helps in reduction in pressure dip but can not avoid reactor trip on low PHT pressure.

6. REFERENCES

1. Rammohan H.P. and Bajaj, S.S., 1996 .ATMIKA . A System Thermal-Hydraulic Computer Code Model Description.,Internal Report, NPCIL.
2. Safety Report Vol. II (Accident Analysis) part B, Kaiga Generating Station-1&2, Rev 1, June 2001.
3. Safety Report Vol. II (Accident Analysis) part B, Rajastan Atomic Power Station-3&4, Rev 1, June 2001.
4. Safety Report Vol. II (Accident Analysis) part B, Tarapur Atomic Power Station- 3&4, Rev 1, June 2001.
5. Pramod et al., Validation of computer code ATMIKA.T- National Heat and Mass Transfer Conference, 2002, January 2-5, Kolkata.
6. Pahari S., Rammohan H.P., and Bajaj S.S., 2002 .Modification in code ATMIKA.,Internal Report, NPCIL.
7. TRAC-P1, "An advanced best estimate computer program for PWR-LOCA Analysis", Los Alamos Scientific Laboratory, 1979.

Reliability, Safety and Hazard
Editors: P.V. Varde et al.

DEVELOPMENT OF COMPUTER CODE CRIFLAN FOR CRITICAL FLOW ANALYSIS IN REACTOR SAFETY

D. Kumar, S.T. Panchal, Debjit Rakshit and S.M. Chaudhry

Light Water Reactor Division, Bhabha Atomic Research Centre, Mumbai 400 085

ABSTRACT

The computer code **Cri**tical **Fl**ow **An**alysis (CRIFLAN) has been developed for critical flow analysis in reactor safety. The code employs a heuristic model to describe critical discharges of water for both sub cooled and saturated conditions. Unlike many large thermal-hydraulic codes (RELAP5/MOD3, ATHLETE, and ASTEC) currently in use, CRIFLAN is a stand alone code and is not system specific. The proposed CRIFLAN model is validated against correlation based models such as H-F, Moody, and HEM and against the published experimental data. The CRIFLAN predictions are in good agreement with H-F results. It is concluded that the CRIFLAN code offers a reliable, user-friendly analytical tool for critical flow analysis in the field of reactor safety.

1. INTRODUCTION

The critical flow rate is the maximum flow rate that can be attained by a compressible fluid as it passes from a high pressure region to a low pressure region. In the case of a LOCA in a water cooled nuclear power reactor, coolant is lost through circuit ruptures or inadvertently opened valves. As long as the system pressure is larger than a certain critical pressure compared with the environment to which the coolant is released, sound velocity limits the mass discharge rate through the phenomenon of critical flow. Flow area and critical mass flow rate determine the energy and the mass loss from the system. The determination of critical discharges governs the plant behaviour during transients as well as under accidents conditions. Quite often either due to the limitations of the various models in terms of applicability of the range of parameters, or limited experimental database, analysts had resorted to conservative analysis in the past. Even for the best estimate critical flow models of current generation, quite often some correction factors are recommended by the code developers (RELAP5/MOD 3.2, [1995]) to get an agreement with the experimental data. Computer code CRIFLAN has been developed to provide the ease of handling critical flow analysis with flexibility to the user to select any models of choice and base the decision on the knowledge of the deviations vis a vis other models.

2. LITERATURE SURVEY OF THE IMPORTANT CRITICAL FLOW MODELS

Many works concerning the description of critical flow models for a large variety of fluids is available in the literature. The range of parameters of the applicability of these models and the associated uncertainties make the selection difficult of a particular model for the given application. Equally complex is the experimental data base available for critical flow with diverse and varied range of flow parameters, flow geometries and the nature of the fluid itself. For water-steam system, Table 1.0 summarizes the experimental data base reported on critical flows. Salient features of some of these models are listed in Table 2.0.

Table 1 Critical Flow Experimental Data Base (For Water)

Source & Date	Size mm	Pressure bars	Comments (*)
Burnell, 1947	5-38	1-12	P, Sat. l/d=0.656
Semiscale Project, 1975-77	4-25	3-103	P, N, Sat., SC l/d=4
Marviken CFTProject,1978-79	12-50	40-50	P, Sat., SC l/d=0.3-0.7
Hutcherson,1975	108	1-18	P, Sat l/d=3
Edwards & Jones, 1974	32	2-54	P, Sat. l/d=28
Henry,1970	8	10-20	P, SC l/d=115
Fauske,1965	6	7-124	P, Sat. l/d=0-40
Uchida & Narai,1966	4	0.2-0.8	P, Sat., SC l/d=25-625
Zaloudek,1963-65	13 13	8	P, O, Sat., SC l/d=1-20
Sozzi,Suther-land,1975	13	30-71	P, Sat., SC l/d=0.4-140
Dryndrozhik1975	6,11	2-5	N, Sat.
Simoneau, 1975	4	5-66	N, SC
Zimmer et al, 1979	25	1-10	N, SC
Morrison, 1976	28	58-67	N, Sat., SC

* P – Pipe, N – Nozzle, O – Orifice,
Sat. – Saturated, SC – Subcooled, regime

3. CRITICAL FLOW MODELS

3.1 Moody Model (Moody, 1965)

This model assumes the flow to be in thermodynamic equilibrium, but with the phases flowing at different rates. The fluid is assumed to undergo isentropic change starting from stagnation conditions. Thus, at a given stagnation regime the flow can be expressed as a function G (p, γ) of the local values of pressure p and slip γ in the fluid at the assumed flow. Critical conditions are defined as the values of pressure and slip maximizing the function G (p, γ). These values apply solely to stagnation conditions. It is assumed that the critical values of pressure and slip are reached simultaneously in the critical section.

3.2 Henry-Fauske Model (Henry and Fauske, 1971)

The Henry-Fauske model uses a large number of assumptions and approximations. It is assumed that the fluid changes between stagnation conditions and the critical section as follows:

- Isentropic flow
- Slip equal to one
- No mass and heat transfer between phases
- Incompressible liquid

As a result, each phase retains the same mass fraction and entropy while the fluid changes proceed.

3.3 Homogeneous Equilibrium Model(HEM) (Hutcherson, 1975)

The HEM model assumes that the two phases are in thermodynamic and mechanical (no slip) equilibrium. The two-phase mixture is considered to be a compressible single-fluid, whose average density is a function of the qualities and the densities of each phase. Single-phase flow data can be applied to this fluid. The HEM model assumes isentropic expansion occurs between the centre of the upstream volume and the critical section. This is similar to the approach used for the isentropic flow of a perfect gas.

4. CRIFLAN MODEL: A HEURISTIC APPROACH

Empirical models based on pioneering works of Burnell (Tong, 1988) and Zaloudek (Table1.0) continue to evince interest for the estimation of the critical flow, primarily due to their simplicity and ease in implementation in the code. Taking clue from this a heuristic approach has been followed in the CRIFLAN model. The model applies both for sub cooled and saturated fluid conditions for a water-steam system. The model is based on the pressure critical ratio (ε), defined in terms of length to diameter (l/d) as follows

$$\varepsilon' = 0.56[\frac{(l/d)}{8}]^{0.36}$$

For sub cooled water flow, the degree of sub cooling (ΔT_H) is related to the pressure critical ratio, as follows

$$\varepsilon_H = \varepsilon'[1-(0.14+0.66\frac{\Delta T_H}{100})(\frac{\Delta T_H}{100})]$$

$$\varepsilon = 1.44.\varepsilon_H \ln[1+\frac{P}{(94-545(\varepsilon_H - 0.25)^2)}]$$

$$\varepsilon z = \min(\varepsilon_H, \varepsilon)$$

For Steam-Water mixture,

$$\varepsilon z = \varepsilon'(1-\alpha)+0.56\alpha$$

Finally the Mass Flux G is computed by

$$G = C_d A\{2\rho P(1-\varepsilon z)\}^{0.5}$$

The theoretical value of discharge coefficient (C_d) for ideal flow through an opening is given by (Ghosh, 2001)

$$\pi/(2+\pi) = 0.611$$

4.1 Salient Features of the Code CRIFLAN

The CRIFLAN computer program is written in FORTRAN 90 language and follows a modular approach. The CRIFLAN code consists of a suite of modules based on Moody, HEM, and Henry-Fauske models and are built into the same code so that user can specify any of the models or in any combination thereof. The code follows a user dialogue interface to navigate through various thermodynamic parameters so that appropriate region of fluid state e.g. 1Φ liquid, 1Φ vapour 2Φ mixture can be identified. The Thermodynamic properties are derived from the WASP (Water And Steam Properties Package program). The program has been verified for its correctness before hand. The program issues the error messages for any conflicting or unphysical parameters, if user specifies an erroneous data. Although program interpolates linearly for the points in between, however, it issues a caution to the user if any time it encounters the limits of applicability of the correlation. The user can specify the input data in terms of any of the thermodynamic properties viz. pressure, temperature, enthalpy, quality etc. besides specifying the choice of a particular critical flow model to be used for the analysis. The input can be specified in any system of units, however the results will always be generated in SI system of units. The output results will be available both in tabulated data form and graphical plot form.

4.2 Validation of the CRIFLAN Model

The CRIFLAN model is subjected to extensive comparison with various available models like Moody, HEM, Henry-Fauske, RELAP5/MOD3 and with some published experimental data too. The modified Henry-Fauske(H-F) model, widely recommended for the validation of leak rate calculations (IAEA-TECDOC-774, 1998) due to the fact it covers both the sub cooled and the saturated two phase flow regimes. Also, it predicts the one of the best fit against the experimental data. The inter-comparison of the code against H-F is reported in this paper. The H-F data used for comparison with CRIFLAN is independent of parameter l/d and is equal to unity. The code automatically generates the percentage deviations for any user specified models in tabulated form. This feature helps the user to take a wholistic decision to use critical flow data in any further engineering applications with known bounds of uncertainty.

5. RESULTS AND DISCUSSION

Some results obtained by using CRIFLAN are illustrated as variation of mass flux with pressure, in figure 1.0 to 4.0. For sub cooled water, at lower degree of subcooling (delTH=10 ^{0}C) and for low value of l/d ratio (0.1), CRIFLAN predicts very well against the Henry-Fauske(H-F) data over the entire range of pressure from 2-16 MPa. The maximum error is less than 2%, with CRIFLAN code under predicting against H-F(figure 1.0). For the same degree of sub cooling, with higher value of l/d (=1.0) the results still show a good agreement with CRIFLAN under prediction not exceeding 10%(figure 1.0). Figure 2.0 reveals that with increasing degree of sub cooling (delTH=40 ^{0}C) and for different values of parameter l/d equal to 0.1 and 1.0, the results continue to match quite well for lower pressure ranges. However, the maximum under prediction does not exceed 22% at pressure 16 MPa (figure 2.0). In case of critical flow in the two-phase regime comprising of steam-water mixture, larger value of l/d (=3.5) as against smaller values of l/d (=0.1) exhibit a very excellent match, for mixture quality of X=0.1, with the H-F model. The maximum deviation does not exceed 7%. Although the

TABLE 2 Literature Survey of Major Critical Flow Models and Their Characteristics

Model		Name of author and publication Date	Main Model Characteristics & Observations
Perfect Fluid		Perfect gas & Incompressible liquid	Early models, diabatic, adiabatic, with and without friction, Flow through DeLaval Nozzle, Cylindrical duct, Isentropic flow,
Equilibrium Theories	Homogeneous theories	HEM,1977	Homogeneous equilibrium model, isentropic flow, the most simple two-phase flow theory
		Babitskiy, 1973	Equilibrium Scheme, Substitution of momentum equation with isentropic conditions
		Lahey etal., 1977	Fluid dynamic approach
	Non-Homogeneous Theories	Moody,1965	Slip equilibrium model(energy model)
		Moody,1966	Analysis of whole depressurization process
		Fauske,1964	Slip equilibrium model (momentum model)
		Levy,1965	Lumped model, functional link between α and χ obtained
		Cruver et al,1967	Unified theory of 1-D isentropic equilibrium separated two phase flow
		Moody,1975	Consistent slip model
		Ransom et al. 1978	Method of characteristics, RELAP5/MOD0 onward
Non-Equilibrium Theories	Frozen Theories	Burnell, 1974	Semi-empirical correlation, scheme very easy to use
		Zaloudek, 1963	Semi-empirical correlation, scheme very easy to use
		Starkman,1964	Classical Frozen theory, based on experimental data
		Moody,1969	Pressure Pulse model, equations for moving CV
		Henry et al, 1971	Non-equilibrium accounted, used in RELAP4 code
		Ransom et al,1978	Methods of characteristics to solve used RELAP5/MOD 3.2
	General Theories	Henry1970	Model valid in low quality
		Malne,1975	Release of dissolved gases taken into account
		Porter,1975	Moody theory as reference, covers whole Mollier diagram
		Winters et al,1979	Lumped non-equilibrium model based on bubble growth
Theories Described by Giot		Giot et al1968,1972	Method analogous to Fauske and Moody
		Katto et al,1974	Equilibrium isenthalpic flow model
		Flinta et al,1975	Effect of nucleation on critical flow rate
		Bauer et al,1976	Non-equilibrium model, Pseudo-criticality
		Stadtke,1977	Irreversible thermodynamic analysis,

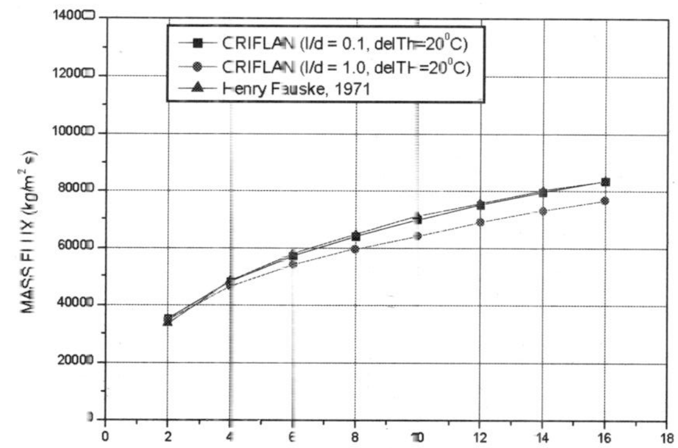

Figure 1.0 : Critical Flow of Sub-Cooled Water

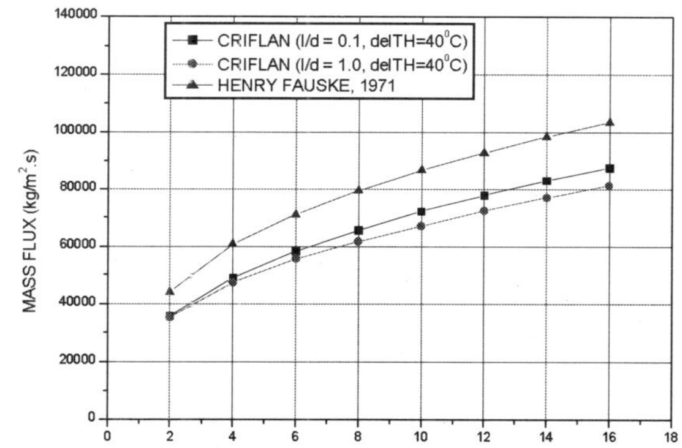

Figure 2.0 : Critical Flow for Sub-Cooled Water

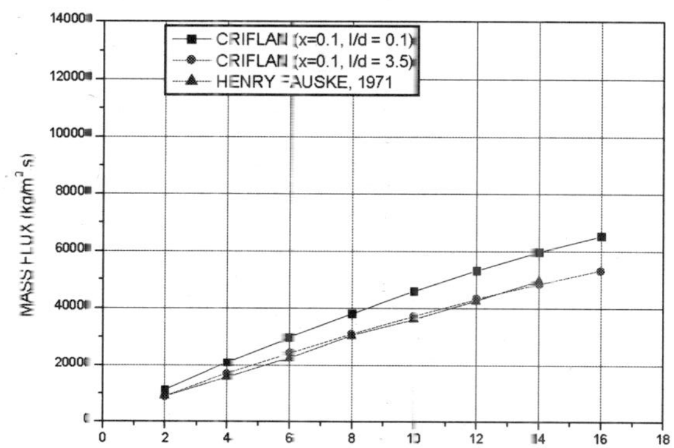

Figure 3.0 Critical Flow for Steam-Water Mixture

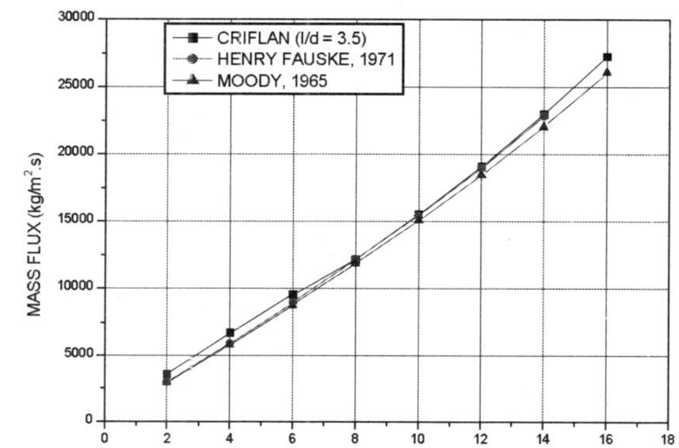

Figure 4.0 : Critical Flow for Steam

agreement at lower pressure does not have significant deviations, with smaller value of l/d (=0.1) and quality (x=0.1), CRIFLAN code over predicts the flow with mismatch not exceeding 24% at the highest pressure limit of 16 MPa (figure 3.0). Figure depicts the critical mass flux calculated by CRIFLAN code for pure steam(X=1.0). The inter-comparison of results of CRIFLAN has been drawn both against the H-F and the Moody models. This is due to the fact that moody model is applicable only in the saturated range. The results show that the predictions of CRIFLAN with H-F show an excellent match with deviations not exceeding 12% over prediction. In case of comparison with Moody model, the over all agreement is excellent. However, CRIFLAN over predicts Moody slightly higher than against H-F model.

6. CONCLUSIONS

A general purpose, simple, reliable and a practical analytical tool has been developed for the analyzing the critical flow system under wide range of the parameters. In the subcooled regime, CRIFLAN code predicts the critical discharges consistently well especially for lower value of l/d with lower degrees of sub cooling. This is also true for the single phase steam critical flows predictions with Moody and H-F models. For both 1-phase and 2-phase water-steam system, CRIFLAN prediction at low pressure range is found to be better than at higher pressures. However, the deviations in results are comparable to those possible with most of the widely used computer codes. Due to its simple architecture and user friendliness, the code is a potent tool for analyzing the critical flows which is central to the assessment of nuclear safety under accident conditions.

7. FUTURE DEVELOPMENT

The provision in the code to take into account the effect of non-condensable and incorporating crack morphology for irregular and narrow gaps is being pursued.

8. ACKNOWLEDGEMENT

Thanks are due to Shri R.S. Yadav, Head LWRD for his encouragement and support in this work.

9. REFERENCES

1. "RELAP5/MOD3.2 Code Manual: Models and Correlations, Volume 4", NUREG/CR-5535, (1995).
2. Moody F. J. (1965), Maximum Flow Rate of a Single Component, Two-phase mixture, Journal of Heat Transfer, ASME, 87, *pp134-142.*
3. Henry, R.E. and Fauske, H.K.(1971), The Two-Phase critical flow of One component Mixtures in Nozzles, Orifices and Short Tubes, *ASME Transactions of the Journal of Heat Transfer, pp179-187.*
4. Hutcherson, M.N. (1975) contribution to the two-phase blowdown Phenomenon,ANL/RAS.
5. Ghosh, B.etal. "LBB Analysis of Indian PHWRs and BWRs, 2. Leak Flow Rate", *Proc. 1 st Natl. Conf. Nucl. Reactor Safety*, Paper No. 47-B, Nov. 25-27(2002), BARC, Mumbai, INDIA.
6. "Guidance for the application of the Leak before Break Concept", IAEA-TECDOC-774(1998).
7. Tong, L.S.,(1988):*Principles Of Design Improvement For Light Water Reactors,* Hemisphere Publishing Corporation.

Reliability, Safety and Hazard
Editors: P.V. Varde et al.

ESTIMATION OF CHF RATIO AT VARIOUS POWER LEVELS IN TAPP-3&4 REACTORS

H. Kalra, P. Pramod, H. P. Rammohan, P. K. Malhotra, S.G. Ghadge and S.S. Bajaj

Directorate of Safety, Nuclear Power Corporation of India Limited
Anushakti Nagar, Mumbai 400 094

ABSTRACT

TAPP-3&4 are the 540 MWe PHWRs having horizontal fuel channel. At normal 100% FP operation there is no boiling in the channel. However, when the power increases due to any transient, the boiling may start in the channel. The main application for critical heat flux (CHF) prediction is to set the operating power with a comfortable margin to avoid CHF occurrence. This margin of CHF can be expressed in terms of minimum critical heat flux ratio (MCHFR), which is the ratio of CHF to local heat flux for the same pressure, mass flux and quality. The CHF depends on power, coolant flow rate as well as coolant condition in the channel. As the power increases the flow reduces in the channel and cooling is degraded. The thermal hydraulic code is developed for present analysis. The output of analysis are CHF prediction quality calculation at axial locations of the maximum rated channel at various power levels and channel flow reduction with increase in power.

1. INTRODUCTION

TAPP-3&4 are the 540 MWe PHWRs having horizontal fuel channel. At normal 100% FP operation there is no boiling in the channel. However, when the power increases due to any transient, the boiling may start in the channel.
In forced convective boiling the boiling crisis occurs when the heat flux is raised to such a high level that the heated surface can no longer support continuous liquid contact. This heat flux is usually referred to as the critical heat flux (CHF). It is characterized by a sudden rise in surface temperature caused by blanketing of the heated surface by a stable vapor layer.

CHF ratio is the ratio of critical heat flux to the actual heat flux. This is defined at the local point. The critical heat flux is a function of pressure,flow and relative enthalpy at that local point. Thus CHF ratio depends on the heat flux and coolant condition at that point. The coolant condition at any point in the core depends on the whole core power and its distribution in the core, secondary side condition which determines the inlet temperature; pump characteristic and pressure distribution within the system. Hence an attempt is made to relate the channel power with Minimum Critical Heat Flux Ratio (MCHF ratio) with other conditions assuming to be normal.

As per the safety guide, acceptance limit on MCHF ratio is 1.1 in case of Normal Operation Condition (NOC) and Anticipated Operational Occurrences (AOO).

2. METHODOLOGY OF ANALYSIS

The maximum rated channel (5.5 MWth) of TAPP-3&4 reactors has been chosen for the analysis. Analysis carried out to estimate the steady state parameters in the channel at various powers with the header pressures remaining same. The D2O properties are used for the analysis. The Reactor inlet header and outlet header pressures are used as boundary condition. It is assumed that these pressures remain constant for various powers.

RIH temperature is assumed to be 260°C at all powers and the flow at full power of the channel is calculated to be 24.849 kg/sec.

The whole channel is divided into 24 axial nodes and critical heat flux is calculated along channel length for various powers. The CHF is calculated using Groenevold CHF look up table, which provides CHF values for discrete values of pressure, mass flux and quality. These are valid in the range of 0.1-20 MPa pressure, 0-7500 Kg/m2-sec mass flux and -50-100% quality. Linear interpolation between tabulated CHF values is used for determining the CHF. A total of 22946 CHF data were used in deriving the look up table [1]. As the data are from random experiments the CHF will follow normal distribution.The scatter of CHF data with respect to the values presented in look up table is defined by root mean square error (standard deviation) as 7.82%[1].

⑨ = 7.82%

As per the international practice it is necessary to find value of CHF such that the 95% chance the actual value exist above this value. Such a value of CHF is called CHF with 95% probability. Thus CHF with 95% probability is 1.65⑨ (90% of data falls within ± 1.65⑨)

$$CHF_{95\%probability} = CHF\ (1-1.65⑨\ /100) = 0.87097\ CHF$$

Mean value of CHF on which the look-up table is built is based on a finite number of experimental data The variation of mean CHF is again defined by std. deviation of mean $⑨_m$.

$⑨_m$ = ⑨ / {n = 7.82/ { 22946 = 0.0516%

Hence the certainty with which mean is predicted is called confidence limit. CHF with 95% confidence can be written as

$$CHF_{95\%confidence} = CHF\ (1-1.65\ ⑨_m\ /100) = 0.999\ CHF$$

Hence the factor to account 95% probability is estimated to be 0.87097 and to account 95% confidence is around 0.999. Hence the total factor to define 95% probability and 95% confidence limit is estimated to be 0.87. The CHF look up table is valid only for vertical upward flow inside a uniform heated 8-mm tube. To facilitate the application of CHF table method to our geometry and heat flux distribution etc., various correction factors are used. The corrected CHF value is expressed as,

CHFa=CHF8 (P, G, Xe) K1K2K3K4K5K6Kpro
K1= Tube cross section factor =0.957
K2= Bundle factor = min (0.8, 0.8exp (-0.5X1/3)
K3 = Grid spacer factor = 1.0
K4 = Heated length factor = 1.0
K5 =Axial flux distribution= 1.0
K6 =Flow factor horizontal flow = 1.0
Kpro=Factor to define 95% probability and 95% confidence limit= 0.87

The flow condition depends largely on 2-phase friction drop model used. The Friedel correlation[2] is used for 2-phase friction drop calculation.

3. RESULTS

Figure-1 to Figure-8 show the variation of critical heat flux with the length of channel at different operating powers. It can be seen from Figure-8 that at 165% FP operation the minimum critical heat flux ratio becomes less than unity. It can be seen that at ode no.19 (which represents the 10th axial bundle location in the channel) MCHFR reaches to unity.

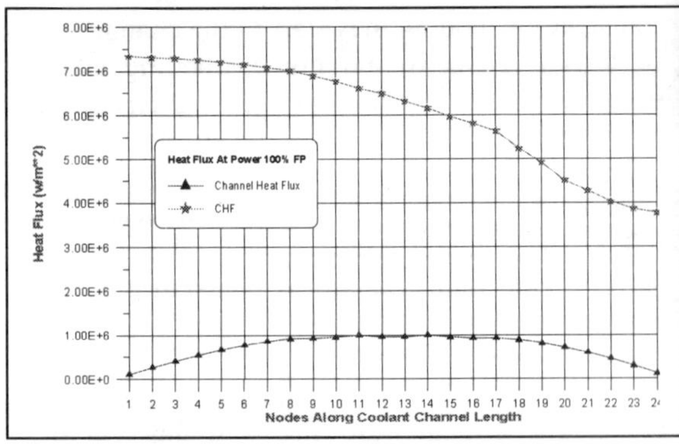

Fig. 1: Heat flux at 100% FP

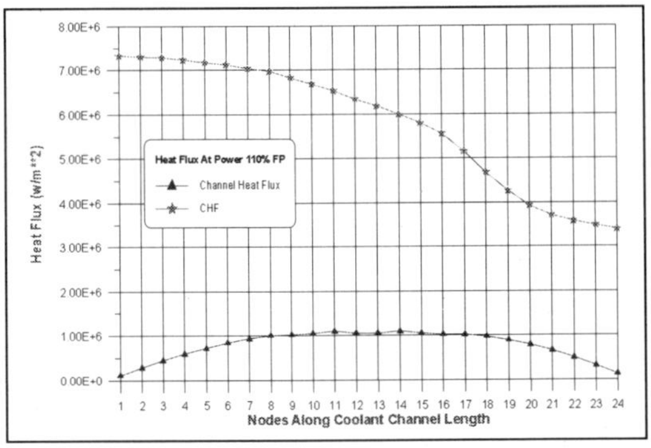

Fig. 2: Heat flux at 110%FP.

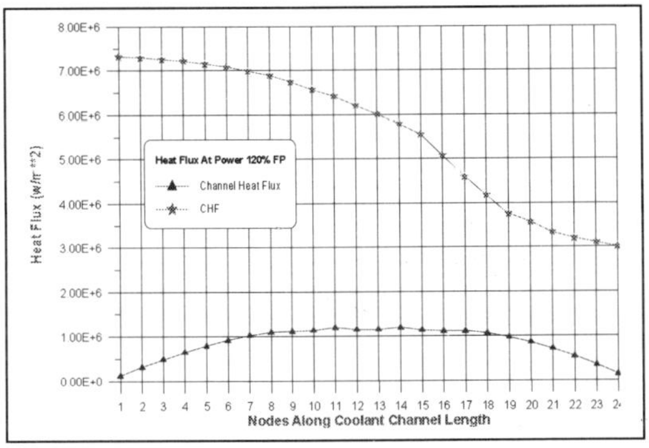

Fig. 3: Heat flux at 120%FP.

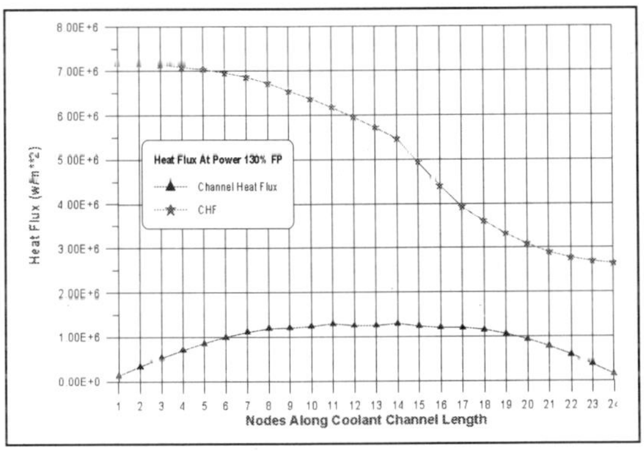

Fig. 4: Heat flux at 130%FP.

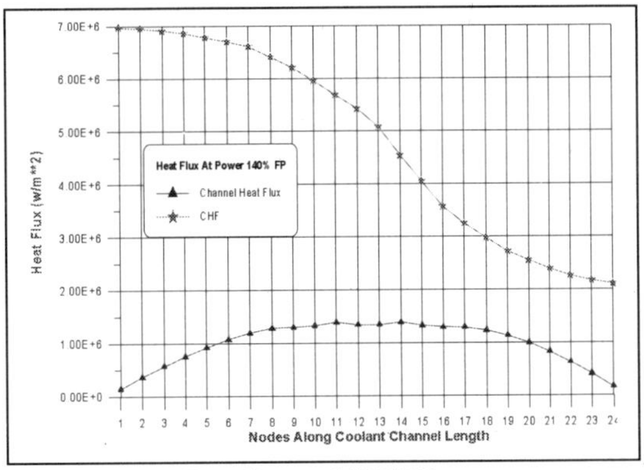

Fig. 5: Heat flux at 140%FP

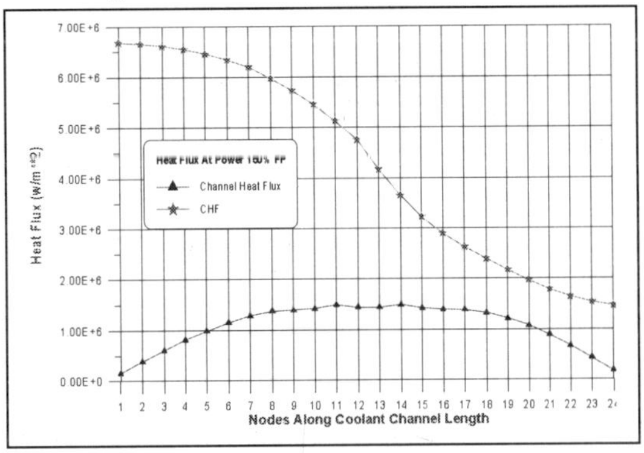

Fig. 6: Heat flux at 150%FP.

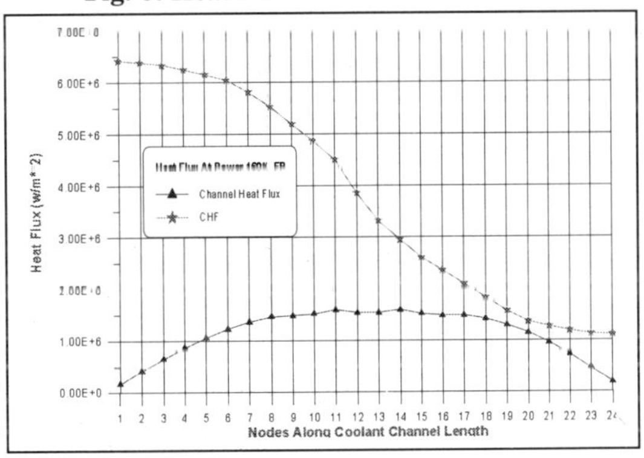

Fig. 7: Heat flux at 160%FP

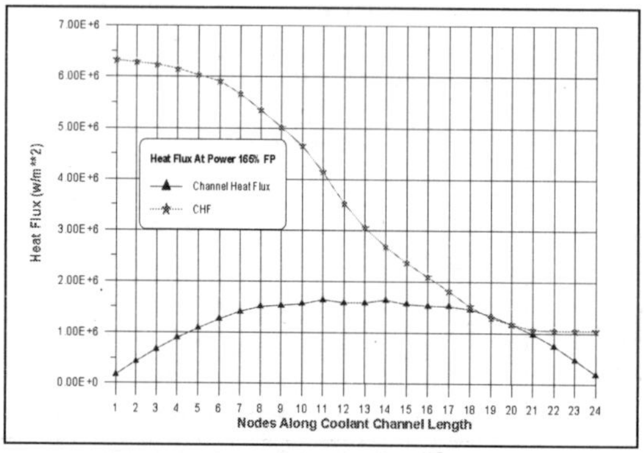

Fig. 8: Heat flux at 165%FP.

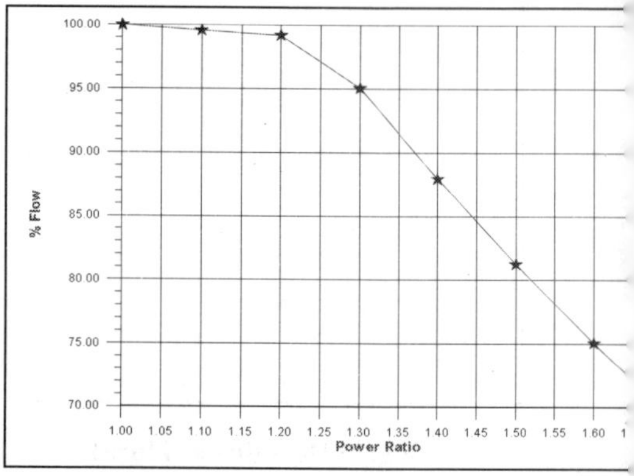

Fig. 10: Quality at the exit of Channel.

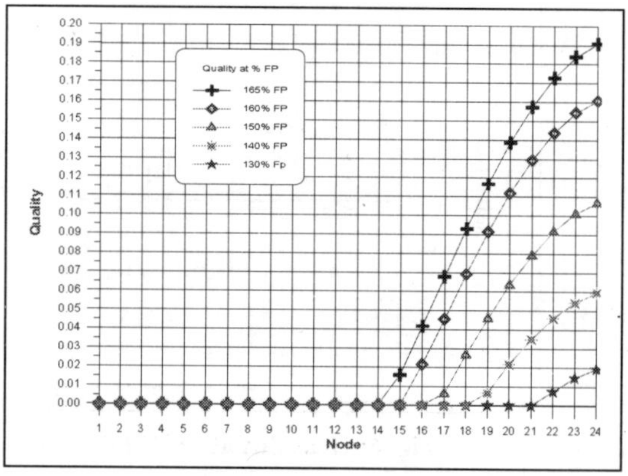

Fig. 9: Heat flux at 170%FP.

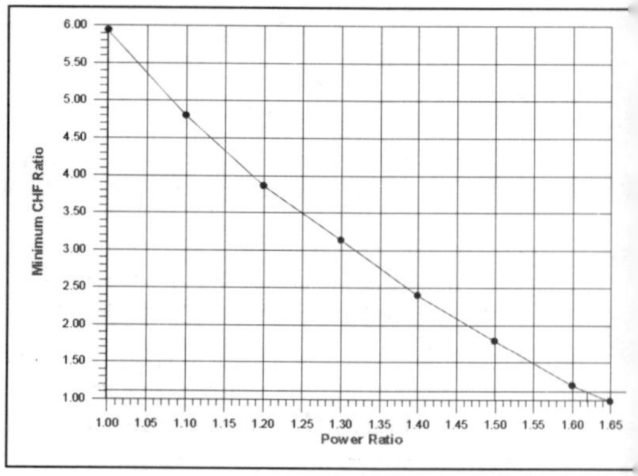

Fig. 11: MCHFR v/s Power Ratio.

Fig. 9: shows the variation of equilibrium quality along the length of channel at different operating powers. It can be seen from Figure-9, as the power increases by 30% the voiding starts in the channel. When reactor is operating at 165% FP the quality at the exit of **channel is 19%.**

Figure-10 shows the variation of flow with increase in power. It is seen that there exist a significant flow reduction as two phase boiling starts in the channel. At 162% FP operation the flow reduces to 73% flow.

Figure-11 shows the variation of MCHFR with increase in power. It can be seen from Figure-8 that MCHFR reaches to unity at 165% FP. The operating limit on MCHFR is 1.1. At about 162% FP the MCHFR reaches to 1.1 at 10th bundle location.

4. CONCLUSION

A limit of 162% FP can be fixed for this type of fuel configuration. Under this condition it is observed that MCHFR remains above 1.10. If any change in power profile the limit is to be recalculated.

5. REFERENCES

1. GROENEVOLD,D.C.,et al.,"*The 1995 look-up table for heat flux in tubes*", Nuclear Engg. and Design, vol 163, pp 1-23, 1996.
2. FRIEDEL,L.,"*Improved friction pressure drop correlations for horizontal and vertical two phase flow*",European two phase flow group mtg, ISPRA 1979.

3. IAEA-TECDOC-1203,"*Thermohydraulic relationships for advanced water cooled reactors*" 2001.
4. RAY,S.,"*Axial power profile for Tapp 3&4*", NPCIL Internal Communication, 2004.
5. CHISLOM,D.,"*Two Phase Flow in Pipelines and Heat Exchangers*", Longman Inc., New York,pp 92-120, 1983.

Reliability, Safety and Hazard
Editors: P.V. Varde et al.

MULTIPLE SHAKER CONTROL (MIMO) FOR RELIABILITY TESTS OF LARGE STRUCTURES

Deepak Jariwala

Regional Manager (SE Asia), Spectral Dynamics, Inc. – USA

ABSTRACT

The conventional method of vibration testing large Device Under Test (DUT) is to employ a large Electrodynamic Shaker with a costly fixture. This is MISO i.e. Multiple In and Single Out meaning many input channels but just one output or drive channel. Spectral Dynamics emphasises on MIMO Control where we employ more than one shaker and a less simpler fixture which means more economical. Again, conventional vibration testing calls for single axis testing at a time but State-of-the-Art for better reliability tests calls for 3 or 6 axes simultaneously being tested. The paper explains a Seismic Test done using shakers & a MIMO Controller.

Keywords: MIMO, Vibration Tests, Large Dut, Multiple Shakers

1. WHY ARE WE CONCERNED ABOUT RELIABILITY?

In the present day and age when things are getting more automatic and electromechanical , it is extremely important that products involved directly or indirectly are reliable and able to face environmental abuse. One severe abuse is Vibration which normally cannot be avoided. It can be suppressed to a certain extent . Where it cannot be suppressed, we need to make our product in question more reliable. Let us think of important equipment which have to survive in case of earthquakes e.g. Communication Equipment at Fire Stations, Police Stations or Control Equipment at Nuclear Plants or Navigational Equipment in Aircrafts or even your modern car electronics etc. Vibration Testing has to be done in design, developmental and production stages for rugged products. Resonance is a very important parameter to be considered in Vibration Testing.

Conventional Methods of Vibration Testing

It was more of a **shake and bake** approach. The shake also was done by crude methods using a motor and a platform. These were the Electromechanical shakers. In the 1950's the Electrodynamic Shaker was invented and this was and is still being used extensively for higher accelerations and frequency ranges beyond 2000 Hz. The Electrohydraulic shaker is also being used but for low frequency and high displacement requirements. However, the Electrohydraulic shaker has frequency limitations (500 Hz maximum) for Random Testing. All these testing approaches were single axis. Different fixtures were used like L or Cube to achieve other axes. The Slip Table is a very common arrangement which is used for Horizontal Testing. The Z axis is achieved by keeping the Shaker in vertical mode. The X and Y axes are achieved using the Slip Table and then later changing the alignment of the DUT.

Multiple axes of DUT tested simultaneously

In the real world situation, vibration abuse on the DUT does not come just on one axis rather experts have defined 6 minimum axes on which the abuse comes i.e. X, Y, Z, Pitch, Roll and Yaw. We are normally familiar with the X, Y and Z. Pitch, Roll and Yaw are well understood when we think of an Aircraft or ship in water in motion.

Thus for a more realistic simulation in the lab, we need simultaneous excitation in multiple axes.

This calls for motion of the DUT in multiple degree of freedom. We need multiple shakers and a compatible fixture and a Controller which can produce such motion with good accuracy, phase control etc.

Spectral Dynamics, Inc. – USA has several US patents in the area of MIMO Control and this paper discusses this application.

Recent advances in digital control of multi-shaker and multi-axis systems have made Rectangular Control an achievable, if not a simple, reality. Most current multi-shaker, MIMO (Multiple Input Multiple Output), tests are performed on the basis of a "square" arrangement, in which the number of control transducers equals the number of controlled exciters. However, cases exist in which it is desirable to have more control transducers than exciters. This has been termed "Rectangular" control.

Aerospace Applications : In the past 4 years tremendous advances have been made in both theoretical development and practical implementation of multiple exciter and multiple axis testing techniques. The work done in this area has followed a rigorous, closed-loop, system matrix formulation with no open loop approximations attempted or permitted. The result has been both a validation of the mathematical developments and a continual learning process as ever more imaginative schemes are implemented. One of the most recent developments has been the formulation and implementation of a control technique termed Rectangular control. Much of the impetus for working in this direction has come from customers facing ever more difficult testing requirements, but not having the budgets necessary for procurement of larger shakers or expensive/complex fixtures.

Whereas there are many tests where multi-axis or even 6 DOF simulation are necessary, perhaps the largest current desire for MIMO testing is where 2 electrodynamic shakers will be used in a single, usually vertical, axis. This is a realistic simulation, for example, for many missile transportation or airborne applications. In this type of test the traditional control technique was to use 2 control accelerometers which were placed either on the two shaker heads, the two attachment points or 2 locations on the missile near the attachment points. With this type of testing, many response points could also be measured giving a good picture of the dynamics of the entire test system. It is also possible, in this scenario, to define several or many Limit channels, in conjunction with the 2 Control channels, to avoid severe overtesting of the test article. However, as with any Limit Channel test scenario, the use of Limit control will often reduce the excitation level dramatically and create uneven motion on the test article. One reason for attempting Rectangular control, rather than Limit control, has been to create a more uniform motion on the shaker head/head expander and on the test article. Recent testing experience on 2 shaker loaded missile container tests has verified this concept.

Random Testing (MIMO) : Multi-shaker Random testing is performed for many reasons including both dynamic simulation and multi-reference Frequency Response Function measurements. In Multiple-Exciter-Single-Axis (MESA) testing for example, it may be necessary to drive all shakers in Phase with high Coherence values to create "Platform Motion". But it may also be desirable to create uncorrelated Random excitations with very low Coherence values, for the purpose of revealing Modal characteristics of the test fixture or test article. Also some test requirements may include the requirement of various Phase relationships between shakers. Each of these and more are possible with JAGUAR MIMO (Multiple Input/Multiple Output) Control.

For Random control with a high Coherence value of 0.9 or 0.95 specified, control of Phase is typically set to zero degrees. However, in the 2 shaker case, it might also be specified as 180° or 90°. Note that Phase is controlled in the sense of an averaged Cross Spectrum as part of a specified Spectral Density Matrix. Nevertheless, actual control accelerometer signals will often exhibit the desired Phase conditions in real time, if the test system is well designed, balanced and matched in key components.

Seismic Simulation : There are two preferred MIMO techniques available with JAGUAR,

when the desired simulation is to recreate a given time waveform. These are:

A. Multiple-Exciter Transient Waveform Control (TWC) - Used when the desired time waveform can fit in one memory block. For example if a seismic simulation to 100 Hz is planned, and the waveform is sampled at 1,024 samples per second (Sampling Multiplier = 10.24), then by specifying 32,768 samples as the effective frame size, a transient of 32 seconds will be reproduced.

B. Multiple-Exciter Waveform Replication - Used when the desired time waveform exceeds a single block or frame in length. For example, this is the preferred method for road simulation studies where the record length may be 45 minutes, an hour or more. In this application, JAGUAR MIMO Replication will seamlessly connect as many frames as necessary to complete the specified road simulation interval. In the test described here, the objective was to first synthesize 3 uncorrelated "broadband" burst-random pulses, each of 32 seconds duration. Then these pulses were to be used to drive the 3 axes of the seismic simulation table in such a way as to equal or exceed a given Shock Response Spectrum. Note that the Shock Response Spectra were slightly different in each of the 3 major axes.

Fig 1: Seismic Table controlled by JAGUAR TWC

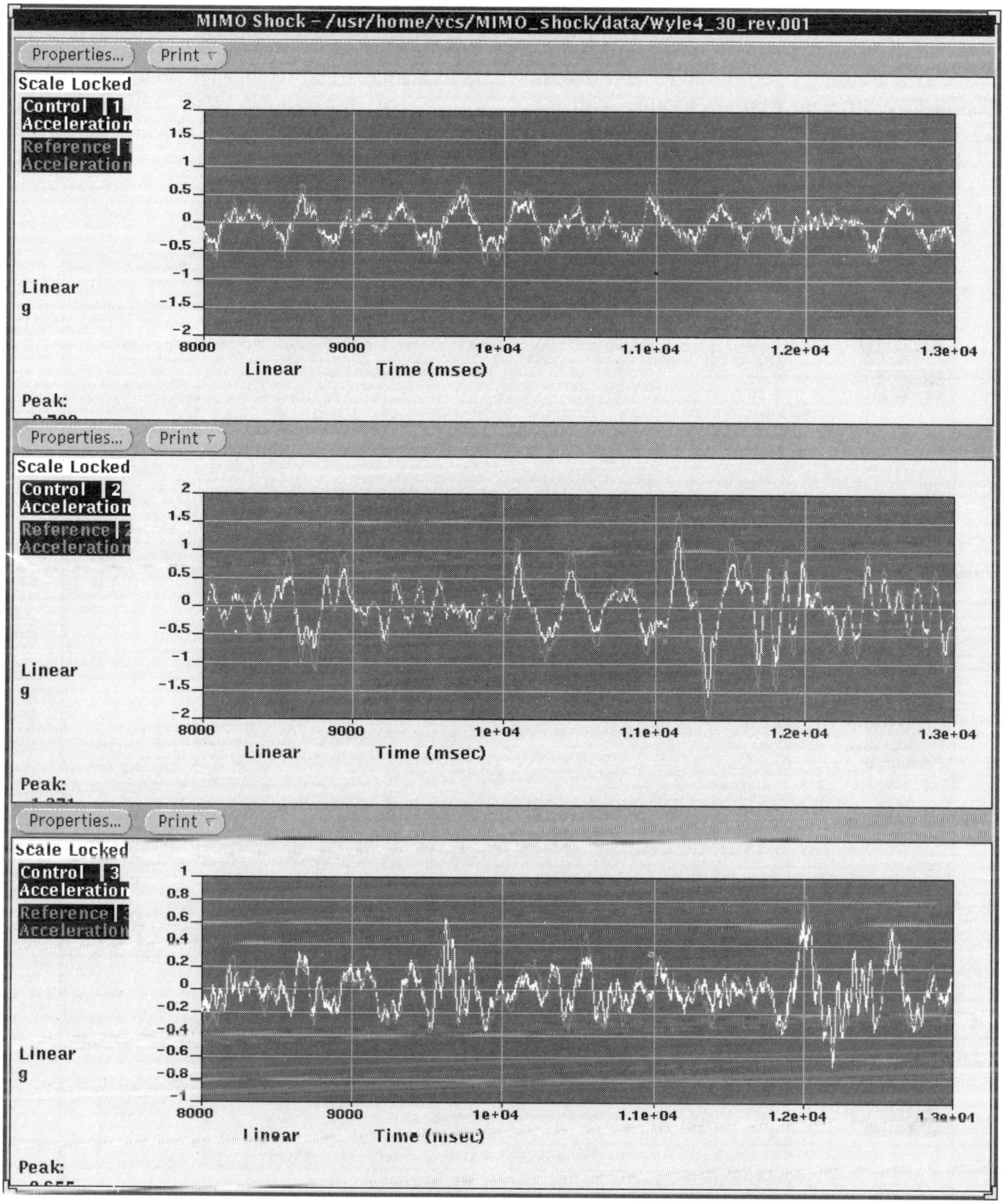

Fig. 2: Reference and Control Acceleration Waveforms used to create desired SRS

Note in Fig 2 that the higher frequency perturbations present in the Control (White) trace but not in the Reference (Green) trace are due to oil column resonances in the hydraulic excitation system. In this case there are 4 electro-hydraulic exciters with one used for front-to-back motion, one for side-to-side motion and two for the vertical axis. The two vertical actuators work in concert and were driven by a common JAGUAR output signal. In the following Fig 3 the Reference Response Spectra (RRS) and the resulting measured Shock Response Spectra

(SRS) for each of the 3 axes are shown overlaid. Note that the peak response in the Maxi-Max for the X-axis (top), Y-axis (middle) and Z-axis (bottom) are different from each other with the Z-axis showing over 21g's.

Note that the RRS and SRS show excellent matching from 0.1 Hz to about 7 Hz at which point the measured SRS begins to exceed the RRS achieving successful test response performance. The complete reference pulses which are not shown had zero to peak Displacements of about 4 inches and the resulting tests were near the 40 ips capabilities of the excitation system.

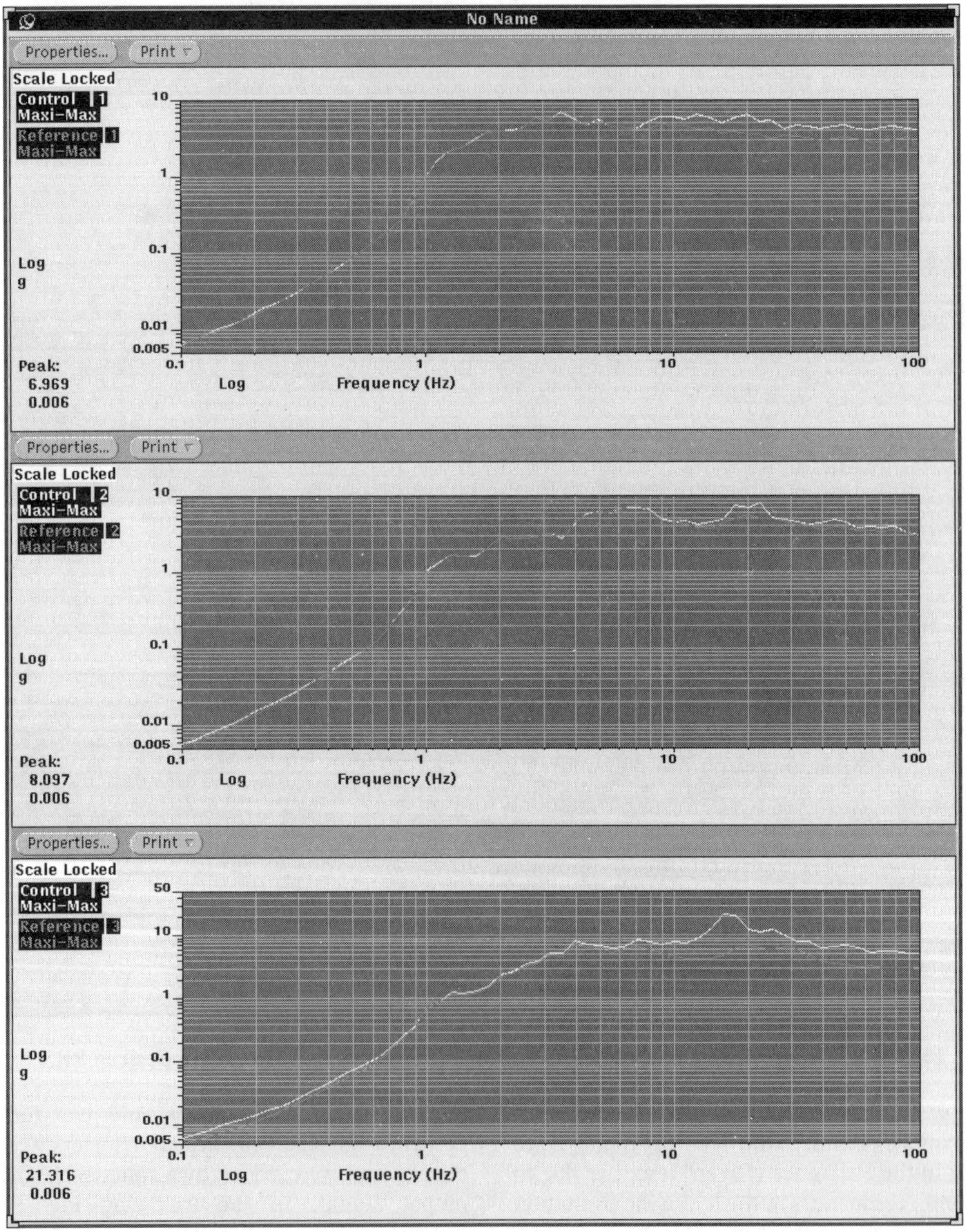

Fig.3: Reference Response Spectra (RRS) and Control Maxi-Max SRS of X, Y and Z axes

2. REFERENCES

1. 'Seismic Simulation with JAGUAR' (paper) by Tony Keller, International Sales Director, Spectral Dynamics, Inc. , San Jose, California.

2. 'A Minimal-Mathematics Introduction to the fundamentals of Random Vibration & Shock Testing' by Wayne Tustin, Year 2005.

Reliability, Safety and Hazard
Editors: P.V. Varde et al.

ROBUST OPTIMIZATION UNDER UNCERTAINTY

Subrata Chakraborty[1] and Achintya Haldar[2]

[1]Department of Civil Engineering, Bengal Engineering and Science University, Shibpur Howrah
[2]University of Arizona, Tucson, Arizona, U.S.A.

ABSTRACT

A robust optimization procedure is proposed when the design variables are uncertain but only limited statistical information on them is available. The procedure is particularly desirable at the initial phase of complicated design problems. The method deviates from the current practice of considering the worst propagation of uncertainties with the aid of nonlinear programming. The efficiency and robustness are achieved by not providing equal importance to the gradients of the objective function and constraints with respect to uncertain variables. In the proposed formulation, the robustness of the objective function is improved by using a new gradient index defined as a function of the weighted sum of the gradients of the performance function. To enhance the robustness of the associated constraints, the penalty terms are logically adjusted by introducing the weight factors proportional to importance of the associated gradient of the constrains. The desirability of the proposed formulation is demonstrated with the help of an example. The method is found to be more effective in achieving robustness in the design without significantly adversely influencing the performance requirements. The weight factors used in represent the relative importance of the variables in the optimization process.

1. INTRODUCTION

The state-of-the-art in the deterministic structural optimization is quite advanced. The available optimization techniques enable us to design efficient and economical structures, in most cases. However, a major limitation of the deterministic optimization techniques is that they are unable to incorporate uncertainties in the structural parameters even when the information is available. It is now well-established in the profession that engineering analysis and design cannot be completed without considering the presence of uncertainty. Conceptually, the development of optimum design procedures for structures under uncertainty can be broadly divided in to three categories: (i) the performance based optimization where the objective is to minimize the expected value of the performance function, (ii) the reliability based optimization by incorporating statistical information of all the decision variables in the objective function and by minimizing the probability of failure, and (iii) the robust optimization where the design is the least sensitive to the changes in input uncertain parameters. The reliability based optimization of the cost function brings specified reliability for a particular limit state. But it may be sensitive to variations of specific design variables. A robust optimization, as a measure of the performance, can be considered as a design procedure that is insensitive (or less sensitive) to the changes in the input decision variables within a range of interest while satisfying the safety requirements. Moreover, when statistical information on the decision variables are not sufficient or complete, the robust optimization is also an attractive alternative.

The robust design method was originally proposed by Taguchi (Phadake 1989), almost two decades ago. Subsequently, the progress in the area of robust design optimization method has been taken place in various forms through nonlinear programming-based optimization procedures. These are primarily based on the probabilistic and the sensitivity based approaches. Two most commonly used measures in probabilistic approaches are the mean and variance of the performance function. Generally, the variance is minimized by compromising the original performance requirements. A more balanced approach of cost minimization and satisfying the performance requirements at the same time is expected to be very desirable. The various approaches that utilize probabilistic information reported in the literatures are: Weighted-Sum method (Lee and Park 2001, Doltsinis et al. 2005), Compromise Programming approach (Chen et al 2000), and Physical Programming method (Messac and Ismail 2000). In these probabilistic approaches, statistical information on the design variables is required. The sensitivity based approaches are applied in a deterministic framework and do not require detailed probabilistic information on the parameters. These approaches utilize the sensitivity information and uncertainty ranges Therefore, it is expected to be cost effective and can be used at the early design stage. The worst case estimate of propagated uncertainty provides the designer the maximum possible deviation of the uncertain variables from their nominal values. This practice provides an alternative when complete statistical information is not available. However, the limited information on uncertainty when integrated with a nondeterministic optimization framework leads to the robust design (Belegundu and Zhang 992, Su and Renand 1997, Gu et al. 2000). Gunawan and Azaram (2004) proposed a sensitivity measure for single objective robust optimization problem. They extended the work to consider multi-objective optimization problems (Gunawan and Azaram, 2005). Han and Kwak (2004) obtained robustness in optimization by minimizing an index calculated from the maximum gradient of the performance function with respect to uncertain variables. The robustness of the constraints feasibility was introduced through an additional penalty term. Widyan and Angles (2005) formulated the robust design problems as the minimization of the sensitivity matrix of the variations in single and multiple objectives performance functions.

Most of the abovementioned sensitivity based formulations use equal importance to each individual gradient of the objective function and the constraints with respect to the uncertain variables. But it is well established in the structural reliability community (Gupta and Manohar 2004) that all the gradients are not of equal importance. It is expected to be much more elegant and efficient to use the weighted value of gradients instead of using equal weightage. The basic idea of the proposed robust optimization formulation is to improve the robustness of the objective function using a new gradient index. It is defined as a function of the weighted sum of the gradients of the performance function. The same concept can be extended to the constraints. To enhance the robustness of the constraints, instead of using simple gradient based penalty function, a weighted gradient function can be incorporated. In the present study, the penalty terms are logically adjusted by introducing weight factors proportional to the importance of the associated gradients of the constraints with respect to corresponding design variables. Essentially, the proposed robust optimization formulation becomes two-criterion equivalent deterministic optimization problem where the weighted sums of the mean and gradient index are optimized. It provides a robust design; i.e., the performance function becomes less sensitive to the variation in the input variables compared with the conventional sensitivity based approaches. The procedure is clarified further with the help of an example. The weight of a two-bar truss is optimized and the robustness is demonstrated by observing that the performance function becomes flatter over the range of variables of interest. The results are compared with the usual sensitivity based robust optimization procedure as proposed by Gu et al. (2000).

2. SENSITIVITY BASED ROBUST OPTIMIZATION

A typical design optimization problem of finding design variables that minimizes an objective function such that a set of constraints is satisfied can be expressed as,

$$\begin{aligned} &\textit{Minimize} && f(\mathbf{x},\mathbf{z}) \\ &\textit{Subjected to} && g_j(\mathbf{x},\mathbf{z}) \le 0 \quad j=1,2,......,m \qquad (1) \\ & && x_i^L \le x_i \le x_i^U \quad i=1,2,.....,n \end{aligned}$$

where, $\mathbf{x}$ is an n-dimensional vector containing design variables and $\mathbf{z}$ is an l-dimensional vector containing the system parameters, and x_i^L and x_i^U are the lower and upper bounds of the design variables, respectively. In a deterministic optimization procedure, the problem is defined with the above information.

The proposed robust optimization formulation improves the robustness of the objective function by using a gradient index (GI), defined as the function of the weighted sum of the gradients of the performance function as,

$$GI = \sum_{i=1}^{n} \frac{\partial f}{\partial u_i} I_{fi} \Delta u_i \quad (2)$$

Where, u_i denotes the uncertain variables in **x** and/or **z**, and I_{fi} is the weight factor of the ith variable (u_i) defined as,

$$I_{fi} = n \frac{\left(\frac{\partial f}{\partial u_i}\right)^2}{\sum_{i=1}^{n}\left(\frac{\partial f}{\partial u_i}\right)^2} \quad (3)$$

To enhance the robustness of the associated constraints, an additional term is normally introduced as a penalty as defined below,

$$k\sum_{i=1}^{n} \frac{\partial g_j}{\partial x_i} \Delta x_i \le 0 \quad (4)$$

The penalty term needed to be added for a greater than type constraint and subtracted for a less than type constraint to ensure safety. However, it seems to be unrealistic to put equal weight to all the sensitivity gradients. It will be more elegant to modify the penalty term as below,

$$k\sum_{i=1}^{n} \frac{\partial g_j}{\partial u_i} I_{gi} \Delta u_i \le 0 \quad (5)$$

Where, I_{gi} is the weight factor of the ith variable (u_i) defined as,

$$I_{gi} = n \frac{\left(\frac{\partial g}{\partial u_i}\right)^2}{\sum_{i=1}^{n}\left(\frac{\partial g}{\partial u_i}\right)^2} \quad (6)$$

Several other forms for the weight factor and the corresponding penalty function can be considered. But at this time, the weight factor proposed in this paper appeared to be the most advantageous. However, it needs further investigations. It is to be noted that the weight factor defined as a function of the sensitivity gradients is multiplied by a factor n (number of random variables) in Eq. 6 so that the results can be compared and are consistent with the conventional sensitivity based optimization procedures where equal weight of the gradients is used.

The robust optimization problem is finally represented as a two-criterion equivalent deterministic optimization problem and can be mathematically expressed as:

$$\begin{aligned} &Min. \quad (1-\alpha) f(\overline{\mathbf{x}}, \overline{\mathbf{z}}) + \alpha GI \\ &Sub.\ to \quad g_j(\overline{\mathbf{x}}, \overline{\mathbf{z}}) + k\sum_{i=1}^{n} \frac{\partial g_j}{\partial u_i} I_{gi} \Delta u_i \le 0 \\ &x_i^L \le x_i \le x_i^U \quad i = 1, 2, .., n,\ j = 1, 2, .., m \end{aligned} \quad (7)$$

Here, the parameter α serves as a weighting factor in the bi-objective optimization problem. The maximum robustness will be achieved when α becomes 1.0.

To assure variations in the design variables within the original design space, the upper and lower bounds of the design variables need to be shifted as,

$$x_i^{L^R} = x_i^L + \Delta x_i, \quad x_i^{U^R} = x_i^U + \Delta x_i \quad (8)$$

3. NUMERICAL STUDY

A two-bar truss, as shown in figure 1, is considered to clarify the proposed robust optimization technique. The truss is subjected to an inclined static load of 200kN as shown in figure 1 with $tan\,\psi = 0.8$. The optimization problem involves minimizing the weight of the truss. The design variable vector (**x**) is composed of the cross-section area of the truss members and half of the horizontal distance between nodes 1 and 2. The design parameters are the density, applied force and maximum allowable tensile strength. The design variables are considered to be uncertain with a spread given by certain specified percentage about their nominal values. The design parameters are considered to be deterministic in present study. However, the formulation developed here can be easily extended when both the design variable vector (**x**) and the design parameter vector (**z**) are uncertain. This extension is now under development. The two constraints are the limiting axial stress in the truss members. The optimization problem can be mathematically expressed as,

$$\text{Min}\, f = \rho_{bar} x_1 \sqrt{1+x_2^2}$$

$$\text{s.t.}\quad g_1 = [1 - \frac{|F|}{2\sqrt{65}s_{max}}\sqrt{1+x_2^2}\left(\frac{8}{x_1}+\frac{1}{x_1 x_2}\right) \geq 0$$

$$\text{and,}\, g_2 = [1 - \frac{|F|}{2\sqrt{65}\sigma_{max}}\sqrt{1+x_2^2}\left(\frac{8}{x_1}-\frac{1}{x_1 x_2}\right) \geq 0$$

$$0.2 \leq x_1 \leq 4.0 \quad and \quad 0.1 \leq x_2 \leq 1.6 \tag{9}$$

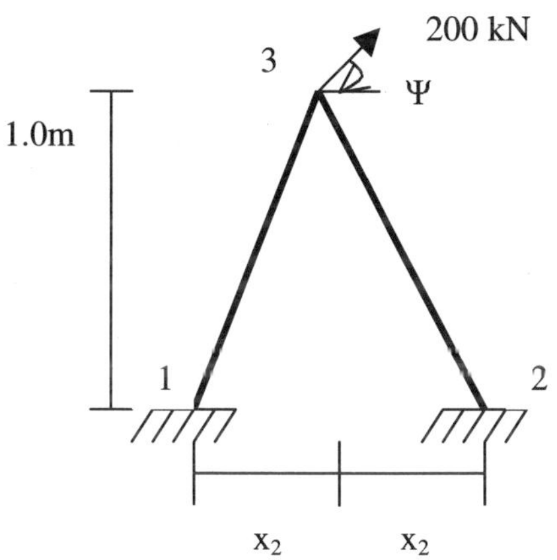

Fig. 1:The two bar truss

The deterministic design parameter vector (**z**) are taken as,

$$\{\mathbf{z}\} = [\rho_{bar}\ F\ \sigma_{max}]^T = \left[10^4 kg/m^3\ \ 20kN\ \ 1000MPa\right]^T \tag{10}$$

The problem is solved by sequential quadratic programming using standard IMSL subroutines. For deterministic optimization, x_1 is found to be 0.2000 cm^2 and x_2 is found to be 0.1249 meters. The optimum design variables and corresponding optimum weight of the truss obtained by the proposed formulation is shown in table 1 for different settings of the parameter α. The optimum results obtained by the conventional sensitivity based robust optimization as proposed in Gu et al. (2000) are shown in bracket in the same table for comparism. The penalty term k is taken as 3.0 and the variability in the design variables due to uncertainties as defined by Δx_i are considered as 10% of the nominal values.

Table 1 The robust design result

α	Optimum x_1	Optimum x_2	Optimum Weight
0.0	1.897 (1.834)	0.266 (0.441)	1.963 (2.003)
0.5	1.895 (1.847)	0.273 (0.423)	1.964 (2.005)
1.0	2.020 (1.949)	0.373 (0.330)	2.052 (2.156)

The optimum weight of the truss versus the uncertainty defined as percentage of the nominal values of the design variables (both area and the horizontal distance) are plotted in figure 2. The corresponding variations in the truss weight are shown in figure 3. In both figures 2 and 3, k and α is assumed to be 3.0 and 0.5, respectively. The optimum weight of the truss with increasing value of the penalty term is shown in figure 4. The corresponding variations in the truss weight are shown in figure 5. It can be observed from figures 3 and 5 that the variations of the weight of the truss are less in proposed approach i.e. design is less sensitive to the changes in the design variables than that obtained by using the conventional sensitivity based robust design. The proposed method also produces the optimal weight for a particular level of robustness as can be observed from figures 1 and 4, indicating its efficiency.

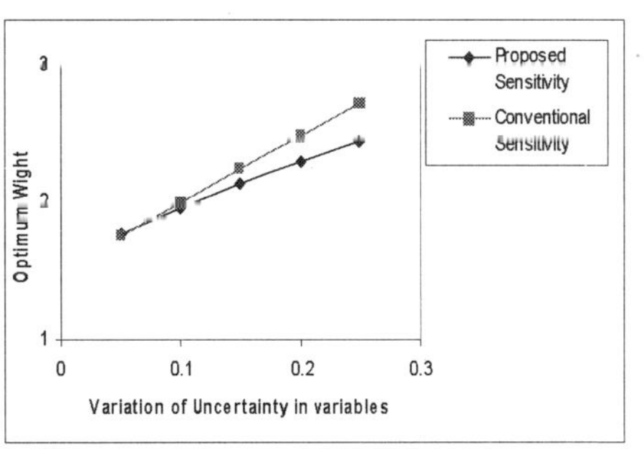

Fig. 2: Optimum weight of the truss with varying uncertainty in design variables

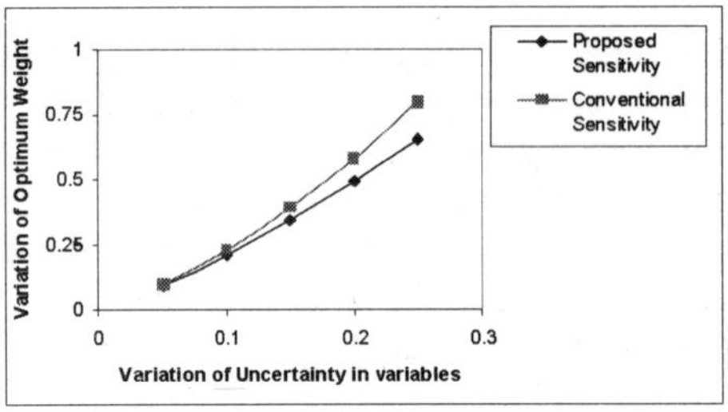

Fig. 3: Variation of optimum weight of the truss with varying uncertainty in design variables

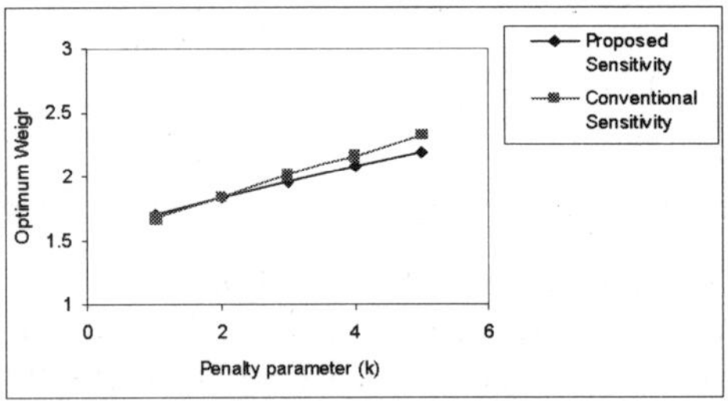

Fig. 4: Optimum weight of the truss with varying penalty parameter (k)

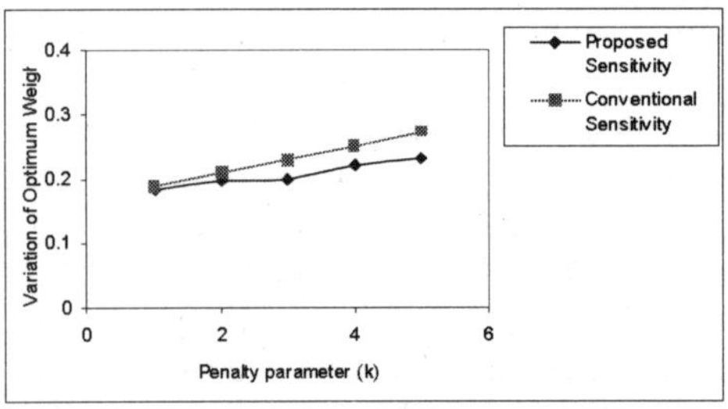

Figure 5 Variation of optimum weight of the truss with varying penalty parameter (k)

4. CONCLUSIONS

A new gradient index-based robust optimization method is proposed. The gradient index is defined as a function of the weighted sum of the gradients of performance function with respect to design variables. The robustness in the constraints is introduced by modifying the penalty term obtained as a function of the weighted sum of the gradient of the constraints. The weight factors are defined in such a way that the results can be compared and are consistent with the conventional sensitivity based optimization procedures where equal weight of the gradients is used. The method is found to be more effective in achieving robustness in the design without significantly adversely influencing the performance requirements. In the proposed robust optimization approach, the mean value and the variation of the performance function are controlled simultaneously. This is a very desirable feature. At the same time, it puts more importance to the uncertain variables that contribute more to the variations of the objective function and violations of the constraints. The weight factors computed in the proposed approach represent the relative importance of the variables in the optimization process. Thus, when there are large numbers of variables in a design problem, the factors can be used in selecting the critical variables making the optimization problem more efficient and robust.

5. ACKNOWLEDGEMENT

The first author gratefully acknowledged the financial support of DST, Govt. of India in BOYSCAST Fellowship scheme to pursue this research work and also CSIR in sponsored research scheme.

6. REFERENCES

1. Belegundu A.D.and Zhang S.(1992), *Robustness of design through minimum sensitivity, ASME,* J. of Mech. Design, **114**, 213-217.
2. Chen, W. Sahai, A., Messac, A. and Sundararaj G. J. (2000), *Exploration of the Effectiveness of Physical Programming in Robust Design*, ASME J. of Mech. Design, **122**, 155-162.
3. Doltsinis, I., Kang, Z. and Cheng G. (2005), *Robust design of non-linear structures using optimization methods,* J. of Computer Methods Appl. Mech. Engnrg., **194,** 1779-1795.
4. Gu, X., Renaud, J. E., Batill, S. M., Brach, R. M. and Budhiraja A. S. (2000), *Worst case propagated uncertainty of multidisciplinary systems in robust design optimization*, Struct. Multidisc. Optim, **20**, 190–213.
5. Gunawan, S. and Azarm S. (2004), *Non-Gradient Based Parameter Sensitivity Estimation for Single Objective Robust*

Design Optimization, ASME, J. of Mech. Design, **126**, 395-402.

6. Gunawan S. and Azarm S. (2005), *Multi-objective robust optimization using a sensitivity region concept*, Struct. Multidisc. Optim., **29**, 50–60.
7. Gupta, S and Manohar C. S. (2004), *An improved response surface method for the determination of failure probability and importance measures,* Structural Safety, **26**,123-139.
8. Han, J. S. and Kwak, B. M. (2004), *Robust optimization using a gradient index: MEMS applications*, Struct. Multidisc. Optim., **27**, 469–478.
9. Lee, K. and Park, G. (2001), *Robust optimization considering tolerances of design variables*, Comp. and Struct., **79**, 77-86.
10. Phadke, M.S. (1989), *Quality Engineering using robust design*, NJ, Prentice Halls, Englewood Cliffs.
11. Su, J. and Renand, J. E. (1997), *Automatic differentiation in robust optimization*, AIAA J., **35**, 1072-1079.
12. Messac, A. and Yahaya I. (2002), *Multi objective robust design using physical programming,* Struct. Multidisc. Optim., **23(5)**,357-371.

Reliability, Safety and Hazard
Editors: P.V. Varde et al.

GIRSANOV LINEARIZATION AND LEVEL CROSSING STATISTICS OF STOCHASTICALLY DRIVEN NONLINEAR OSCILLATORS

Nilanjan Saha and D. Roy

Indian Institute of Science, Bangalore 560 012

ABSTRACT

Unlike deterministic engineering systems, simulations of stochastically driven oscillators of engineering interest present several computational difficulties. First, numerical integration schemes for oscillators under stochastic excitations are generally far less accurate and derivations of higher order methods are not straightforward. Next, for simulations of extremely unlikely (rare) events corresponding to the tail of the associated probability distribution, one generally has either to use a prohibitively large ensemble size during Monte Carlo simulations or use an importance sampling technique. While a few importance sampling techniques are presently available in the literature for being used with simulations of stochastically driven oscillators (Macke and Bucher 2003), all of them require an educated guess on the choice of a new probability distribution function which converts a rare event into a highly probable event. Such difficulties naturally make the importance-sampled Monte Carlo simulation, which is possibly the only practicable choice for most non-linear and large dimensional engineering dynamical systems, a numerically incorrect and somewhat heuristic procedure.

The objective of this paper is to take a step towards the resolutions of such difficulties using a Girsanov transformation of the probability measure. To start with, a Girsanov linearization is proposed for derivations of higher order methods that can locally linearize a given non-linear oscillator irrespective of its dimensionality. It is thus possible to obtain weak stochastic solutions of higher accuracy, not achievable with most other available methods in the literature. Next, a Girsanov transformation of the underlying probability measure is again exploited to compute the level crossing statistics of non-linear oscillators. Finally it is demonstrated that such a transformation may also be used to arrive at an importance-sampled version of Monte Carlo method for effective simulations of rare events.

1. THE GIRSANOV TRANSFORMATION

Consider an n-DOF oscillator under additive noise processes as:

$$[M]\ddot{X} + C(X,\dot{X})\dot{X} + K(X,\dot{X})X = \sum_{r=1}^{q} G_r(t)\dot{W}_r + F(t)$$

where $X = \{x^{(1)}, x^{(2)}, \ldots. x^{(n)}\}^T \in R^n$ is the displacement vector, $[M]$ is a constant mass matrix, $C(X,\dot{X}),\ K(X,\dot{X}) \in R^{n\times n}$ are (state-dependent for non-linear systems) damping and stiffness matrices respectively, $\{G_r(t) : R \to R^n\}$ is the r-th element of a set of n diffusion vectors,

$$\{W_r(t) \mid r \in [1,q]\} \tag{2.1}$$

constitutes a q-dimensional vector of independently evolving zero-mean Wiener processes with $W_r(0) = 0$, $\mathbf{E}[| W_r(t) - W_r(s) |^2] = (t-s), t > s$

$\forall r \in [1,q]$ and $F(t) = \{F^{(j)}(t) \mid j = 1,\ldots,n\}$ is the externally applied (non-parametric) and deterministic force vector. $\mathbf{E}$ denotes the expectation operator. The description of the oscillating system as in Eq. (2.1) being entirely formal (due to non-differentiability of Wiener processes, which implies that $\dot{W}_r(t)$ exists merely as a valid measure, but not as a mathematical function), they may more appropriately be recast as the following system of $2n$ first order SDE-s in an incremental form:

$$dx_1^{(j)} = x_2^{(j)} dt$$

$$dx_2^{(j)} = a^{(j)}(X,\dot{X},t)dt + \sum_{r=1}^{q} b_r^{(j)}(t) dW_r(t), \; j - 1,2,\ldots,n \tag{2.2}$$

where,

$$a^{(j)}(X,\dot{X},t) = -\sum_{k=1}^{n} \bar{C}_{jk}(X,\dot{X})\dot{x}^{(k)} - \sum_{k=1}^{n} \bar{K}_{jk}(X,\dot{X})x^{(k)} + \bar{F}^{(j)}(t)$$

$$b_r^{(j)}(t) = \bar{G}_r^{(j)}(t)$$

$$[\bar{C}] = [M^{-1}][C], \quad [\bar{K}] = [M^{-1}][K], \quad \{\bar{F}\} = [M^{-1}]\{F\}, \quad \{\bar{G}\} = [M^{-1}]\{G\} \tag{2.3}$$

Moreover, we assume that the drift vector may be decomposed into its constituent linear and nonlinear parts as: $a^{(j)} = a_l^{(j)} + a_{nl}^{(j)}$, where the subscripts '$l$' and '$nl$' are respectively used to denote the linear and nonlinear parts. To ensure boundedness of the solution vector

$$\bar{X} \stackrel{\Delta}{=} \{X^T, \dot{X}^T\}^T \subset R^{2n}$$

(where $X \stackrel{\Delta}{=} \{x_1^{(j)}\} \in R^n$ and $\dot{X} \stackrel{\Delta}{=} \{x_2^{(j)}\} \in R^n$, j=1…n), it is assumed that the drift and diffusion vectors, $\mathbf{a} = \{a^{(j)}\}$ and $\mathbf{b_r} = \{b_r^{(j)}\}$ are measurable, continuous, satisfying Lipschitz growth bound:

$$\| \mathbf{a}(\bar{X},t) - \mathbf{a}(\bar{Y},t) \| + \sum_{r=1}^{q} \| \mathbf{b_r}(\bar{X},t) - \mathbf{b_r}(\bar{Y},t) \| \le Q \| \bar{X} - \bar{Y} \| \tag{2.4}$$

where $\bar{Y} \in R^{2n}$, $Q \in R^+$ and $\|.\|$ denotes the Euclidean norm (i.e., the L_2 norm in the associated probability space $[\Omega, \mathrm{F}, P]$). Let the initial conditions be mean square bounded, i.e., $\mathbf{E} \| \bar{X}(t_0) \|^2 < \infty$ (for most practical purposes the initial condition vector may be treated as deterministic and this is followed in this study too). The time interval $[0,T]$ of interest is so ordered that $0 = t_0 < t_1 \ldots < t_i < \ldots < t_L = T$ and $h_i = t_i - t_{i-1}$ where $i \in Z^+$. For further simplification of the rest of the presentation, we assume a uniform time step $h_i = h \, \forall i$.

Obtaining Non-Linear Response-Moments:

We are in presently interested in computing the weak response i.e., expectations of the form:

$$v = E[f(\bar{X},t)] \tag{2.5}$$

where $f(\bullet)$ denotes some real-valued function. When we utilize Monte Carlo simulation to evaluate the moment, v is replaced by its sample-mean formula (Rubinstein 1981)

$$v \approx \hat{v} = \frac{1}{N} \sum_{i=1}^{N} \left[f(\bar{X}^{(i)},t) \right] \tag{2.6}$$

for large N

where $\bar{X}^{(i)}(t)$ denotes the ith sample path of $\bar{X}(t)$ and N is the number of sample paths. Since evaluating $\hat{v}$ requires weak solutions of Eq. (2.1) only, the measure transformation method, based on Girsanov's theorem (Oksendal 2004), may be applied to linearize Eq. (2.1) by removing the nonlinear parts $a_{nl}^{(j)}(\bar{X},t)$ of the drift field. Accordingly, the linearized SDE-s take the form:

$$d\tilde{x}_1^{(j)} = \tilde{x}_2^{(j)} dt$$

$$d\tilde{x}_2^{(j)} = a_l^{(j)}(\tilde{X},\dot{\tilde{X}},t)dt + \sum_{r=1}^{q} \underbrace{b_r^{(j)}(t)\{[b_r^{(j)}(t)]^{-1} a_{nl}^{(j)}(\tilde{X},\dot{\tilde{X}},t) + 1\} dW_r(t)}_{d\hat{W}_r(t)}, \quad j = 1,2,\ldots,n \tag{2.7}$$

$$dy = \sum_{j=1}^{n} \sum_{r=1}^{q} [b_r^{(j)}(t)]^{-1} a_{nl}^{(j)}(\tilde{X},\dot{\tilde{X}},t) \; y \, dW_r(t)$$

where $\hat{W}_r(t)$ is another set of q-dimensional Brownian motion processes and initial conditions are such that $\tilde{X}(t_0) = \bar{X}(t_0)$. In order to compensate for the removal of non-linear drift terms, we have augmented the linearized SDE-s by the one-dimensional correction process $y(t)$ (called the Radon-Nikodym derivative) with initial condition $y(0) = 1$. Moreover, an exact solution to the linearized SDE-s corresponding to the first two of Eq. (2.7) may be readily constructed independent of $y(t)$. If $\hat{W}_r(t)$ replaces $W_r(t)$ for each r in Eq. (2.7), the resulting SDE-s are obviously linear in $\tilde{X}, \dot{\tilde{X}}$. Under Novikov's restrictions on $a_{nl}^{(j)}(\tilde{X}, \dot{\tilde{X}}, t)$, which are usually met in most practical applications, the functional $\hat{v}$ of Eq. (2.6) may be evaluated according to Girsanov's theorem as:

$$v = \mathbf{E}\left[f\left(X, \dot{X}, t\right)\right] = \mathbf{E}[y(t) f(\tilde{X}, \dot{\tilde{X}}, t)] \quad (2.8)$$

The Radon-Nikodym derivative is a strictly positive random process (an exponential martingale), computable as:

$$y(t) = \exp\left[-\int_0^t \sum_{j=1}^{n} a_{nl}^{(j)}(\tilde{X}, \dot{\tilde{X}}, t) dW(t) - \frac{1}{2}\int_0^t (a_{nl}^{(j)}(\tilde{X}, \dot{\tilde{X}}, t))^2 ds\right]$$

(2.9)

During numerical implementation, the Radon-Nikodym derivative is actually computed over each time step and then appropriately concatenated to obtain the weak response. Finally, the functional $\hat{v}$ of Eq. (2.6) can be replaced again by its linearized Monte Carlo estimator:

$$\hat{v} = \frac{1}{N}\sum_{i=1}^{N} y^{(i)}(t)\left[f(\tilde{X}^{(i)}, \dot{\tilde{X}}^{(i)}, t)\right] \quad (2.10)$$

Simulation to Determine the Level-Crossing Statistics:

For simplicity in describing the rest of the procedure, we restrict our attention to $n = 1$, i.e., to SDOF oscillators. We now try to obtain the first moment of the number of up-crossings, v, of the stochastic response process X (the displacement process, for instance) with respect to a level α which can be written as the functional:

$$v(t_j) = \mathbf{E}[\sum_{i=1}^{j>i} \mathrm{I}(X(t_i) \geq \alpha \text{ and } X(t_{i-1}) < \alpha)]$$

(2.11)

where $I(\bullet)$ denotes indicator function and $\{t_i\}$ is a discretization of the time axis. Using the Monte Carlo simulation to evaluate Eq. (2.11), v is replaced by its sample-mean formula:

$$v(t_j) \approx \hat{v}(t_j) = \frac{1}{N}\sum_{i=1}^{N}\left[\sum_{k=1}^{j>k} \mathrm{I}(X^{(i)}(t_k) \geq \alpha \text{ and } X^{(i)}(t_{k-1}) < \alpha)\right]$$

for large N (2.12)

where $X^{(i)}(t)$ denotes the i-th sample path of $X(t)$ and N is the ensemble size. Since evaluating $\hat{v}$ requires weak solutions of Eq. (2.11) only, the measure transformation method, based on Girsanov's theorem, may be applied to linearize the Eq. (2.1) by removing the nonlinear parts $a_{nl}^{(j)}(\overline{X}, t)$. Accordingly, the SDE-s take the form of equation (2.7).

Similarly restricting $a_{nl}^{(j)}(\tilde{X}, \dot{\tilde{X}}, t)$ to satisfy Novikov's condition, the functional $\hat{v}$ of Eq. (2.12) is evaluated according to Girsanov's theorem as:

$$v(t_j) = \mathbf{E}[\sum_{i=1}^{j>i} \mathrm{I}(X(t_i) \geq \alpha \text{ and } X(t_{i-1}) < \alpha)] = \mathbf{E}[y(t_j)\sum_{i=1}^{j>i} \mathrm{I}(X(t_i) \geq \alpha \text{ and } X(t_{i-1}) < \alpha)]$$

(2.13)

As before, the Radon-Nikodym derivative is a strictly positive random process (an exponential martingale), computable as in equation (2.9). Finally, the functional $\hat{v}$ of Eq. (2.12) can be replaced again by its Monte Carlo estimator

$$\hat{v}(t_j) = \frac{1}{N}\sum_{i=1}^{N} y^{(i)}(t_j)\left[\sum_{k=1}^{j>k} \mathrm{I}(X^{(i)}(t_k) \geq \alpha \text{ and } X^{(i)}(t_{k-1}) < \alpha)\right]$$

(2.14)

Importance Sampling Formulation

Now our interest shifts towards finding the expected number of up-crossings v of a level α so that v crossing is very low, or, equivalently,

α is high. Here, we introduce a constant or time-dependent forcing β such that the response displacement is driven to a level very close to the level α.

$$
\begin{aligned}
d\widetilde{x} &= \widetilde{x}\,dt \\
d\widetilde{x} &= [a(\widetilde{X},\dot{\widetilde{X}},t)+\beta\ ]dt + \sum_{r=1}^{q} \underbrace{b_r(t)\{[b_r(t)]^{-1}\beta + 1\}dW_r(t)}_{d\hat{W}_r(t)}, \qquad j=1,2,\ldots,n \\
dy &= \sum_{r=1}^{q} [b_r(t)]^{-1} a_{nl}(\widetilde{X},\dot{\widetilde{X}},t)\ y\,dW_r(t)
\end{aligned}
\tag{2.15}
$$

In order to find β we write a differential form of equation (2.1) by removing the additive noise and putting a forcing function β in the form:

$$\ddot{x}_2 + a(x,\dot{x},t)dt = \beta, \qquad x_0 = 0,\ \dot{x}_0 = 0 \tag{2.16}$$

Recasting the equation in this form actually shifts the fixed point close to the level α so that such a system, under an added noise process, frequently u-crosses the level α. For instance, if $(0,0)$ is a stable fixed point of equation (2.18) with $\beta = 0$ and if it is intended to shift this fixed point to $(0.9\alpha,0)$, then it is required to choose β as:

$$\beta = a(0.9\alpha,0,t) \tag{2.17}$$

Note that since equation (2.15) is non-linear; hence a strong stochastic Newmark method (SNM) is used to numerically solve it and this is followed by a multiplication of the SNM-based solution by the Radon-Nikodym derivative to finally obtain the weak response. Finally, the functional $\hat{v}$ of Eq. (2.12) can be replaced again by its Monte Carlo estimator given by equation (2.14).

2. NUMERICAL IMPLEMENTATIONS

In order to keep the numerical illustration simple and focussed, consider the non-linear second order SDE corresponding to an SDOF hardening Duffing oscillator under an additive white noise, described by the following Eq.:

$$\ddot{x} + C\dot{x} + K_1 x + K_2 x^3 = \sigma \dot{W}(t) \tag{3.1}$$

The incremental state space representation for the oscillator is:

$$
\begin{aligned}
dx(t) &= \dot{x}(t)dt \\
d\dot{x}(t) &= a(x,\dot{x},t)dt + \sigma\, dW_1(t)
\end{aligned}
\tag{3.2}
$$

where the velocity drift coefficient is given by $a(x,\dot{x},t) = -C\dot{x} - K_1 x - K_2 x^3$, under initial conditions $X_0 = \begin{Bmatrix} x_0 \\ \dot{x}_0 \end{Bmatrix}$.

Obtaining Response-Moments:

For this part, we write the drift term as $a_l(\widetilde{X},\dot{\widetilde{X}},t) = -C\dot{x} - K_1 x$ and $a_{nl}(\widetilde{X},\dot{\widetilde{X}},t) = -K_2 x_1^3$. Now writing Eq. (3.1) according to Eq. (2.4), we get:

$$
\begin{aligned}
d\widetilde{x}_1(t) &= \widetilde{x}_2(t)dt \\
d\widetilde{x}_2(t) &= (-C\widetilde{x}_2 - K_1\widetilde{x}_1)dt + \sigma\underbrace{(-\sigma^{-1}K_2\widetilde{x}_1^3 + dW(t))}_{d\hat{W}(t)}
\end{aligned}
\tag{3.3}
$$

with the augmented equation as:

$$dy(t) = -\sigma^{-1}K_2\widetilde{x}_1^3\ y\,d\hat{W}(t) \tag{3.4}$$

Hence, the Radon-Nikodym derivative turns out to be:

$$y(t) = \exp\left[-\int_0^t \left(-\sigma^{-1}K_2\widetilde{x}_1^3\right)d\hat{W}(t) - \frac{1}{2}\int_0^t (-\sigma^{-1}K_2\widetilde{x}_1^3)^2 ds\right] \tag{3.5}$$

Note that both the integrals of Eq. (3.5) are stochastic and hence, expansions of integrands in them are done through Ito-Taylor expansion up to $O(h^3)$ and then they are modelled as multiple stochastic integrals (MSI-s). Under the new probability law, the first two equations of

(3.3) become linear (in new variable $\widetilde{X} = \{\widetilde{x}_1, \widetilde{x}_2\}^T$ (say)):

$$d\begin{Bmatrix} \widetilde{x}_1 \\ \widetilde{x}_2 \end{Bmatrix} = \begin{bmatrix} 0 & 1 \\ -K_1 & -C \end{bmatrix} \begin{Bmatrix} \widetilde{x}_1 \\ \widetilde{x}_2 \end{Bmatrix} dt + \begin{Bmatrix} 0 \\ \sigma \end{Bmatrix} d\hat{W}(t) \quad (3.6)$$

and its solution with initial conditions $\widetilde{X}_0 = \begin{Bmatrix} \widetilde{x}_{10} \\ \widetilde{x}_{20} \end{Bmatrix} = \begin{Bmatrix} x(t_0) \\ \dot{x}(t_0) \end{Bmatrix}$ is of the form:

$$\widetilde{X} = \exp(A\int_0^t ds)\widetilde{X}_o + \exp(A\int_0^t ds)\int_0^t \exp(-A\int_0^s ds_1)\begin{Bmatrix} 0 \\ \sigma \end{Bmatrix} d\hat{W}(s) \quad (3.7)$$

where, $A = \begin{bmatrix} 0 & 1 \\ -K_1 & -C \end{bmatrix}$ is the system matrix. Now using the Eq. (2.6) we get the non-linear weak solution of the function $f(x_1, x_2, t)$ as:

$$E[f(x_1, x_2, t)] = E[y(\widetilde{x}_1, \widetilde{x}_2, t)\ f(\widetilde{x}_1, \widetilde{x}_2, t)] \quad (3.8)$$

Expectation of the Number of up-crossings:

The numerical formulation of out-crossing rate is precisely similar to previous case of moment formulation with the difference lying in the functional to be computed. Hence, using eq. (2.14),

$$\hat{v}(t_j) = \frac{1}{N}\sum_{i=1}^{N} y^{(i)}(t_j)\left[\sum_{k=1}^{j>k} \mathrm{I}(\widetilde{x}_1^{(i)}(t_k) \geq \alpha \text{ and } \widetilde{x}_1^{(i)}(t_{k-1}) < \alpha)\right] \quad (3.9)$$

Importance Sampling:

Here we write the equations in this form with the introduction of the forcing β :

$$d\widetilde{x}_1(t) = \widetilde{x}_2(t)dt$$
$$d\widetilde{x}_2(t) = (-C\widetilde{x}_2 - K_1\widetilde{x}_1 - K_2\widetilde{x}_1^3 + \beta)dt + \sigma\underbrace{(-\sigma^{-1}\beta + dW(t))}_{d\hat{W}(t)} \quad (3.10)$$

with the augmented equation as:

$$dy(t) = -\sigma^{-1}\beta\ y\, d\hat{W}(t) \quad (3.11)$$

so that $\beta = K_1(0.9\alpha) + K_2(0.9\alpha)^3$. (3.12)

The explicit solution for the Radon-Nikodym derivative is:

$$y(t) = \exp\left[-\int_0^t (-\sigma^{-1}\beta)d\hat{W}(t) - \frac{1}{2}\int_0^t (-\sigma^{-1}\beta)^2 ds\right] \quad (3.13)$$

Using strong SNM once again to solve equation (3.10) followed by the computation of the Radon-Nikodym derivative over each time step, one may readily obtain the weak response and thus the up-crossing statistics as in equation (2.14).

3. NUMERICAL RESULTS:

Moment Repsponse: We have observed that expectations of the first few powers of the displacement and velocity components, simulated through this linearization and through the conventional form of the strong stochastic Newmark method (SNM) (Roy 2005), match very well. The hardening Duffing oscillator under additive white noise is particularly suited to cross-check the validity of numerical solutions as exact stationary solution of the probability density function is available for this case (Wang and Zhang 1998). Figures 1 through 3 plot time-histories of expectations of up to second powers of displacement and velocity components with a uniformly chosen ensemble size of 500. For obtaining these figures, we have chosen the system parameters as $C = 2.0,\ K_1 = 100,\ K_2 = 100$. We have chosen a rather high intensity, $\sigma = 5$, for the additive noise and the step size is uniform at $h = 0.001$. Unlike the SNM (which is an implicit method), the Girsanov linearization provides a semi-analytical, explicit formulation of the stochastic time marching map. However, it may lead to a higher variance for the estimation of the required expectation (especially for lower ensemble sizes). This is a problem that may be well addressed through a variance reduction procedure.

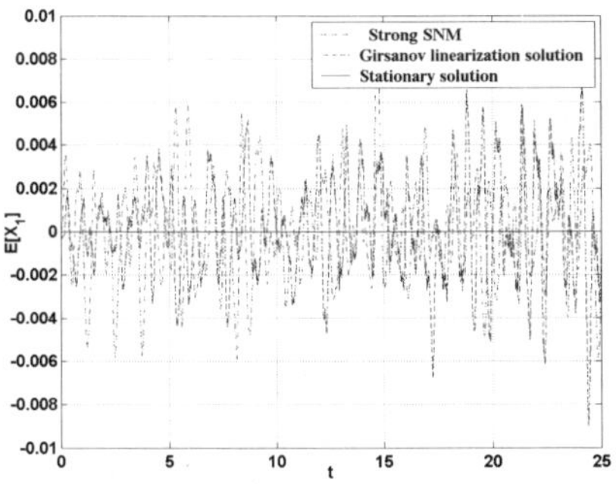

Fig. 1: First moment histories of Duffing equation under additive noise -mean of displacement - $C = 2.0$, $K_1 = 100,\ K_2 = 100,\ \sigma = 5.0$, ***h*=0.001, number of samples: 500**

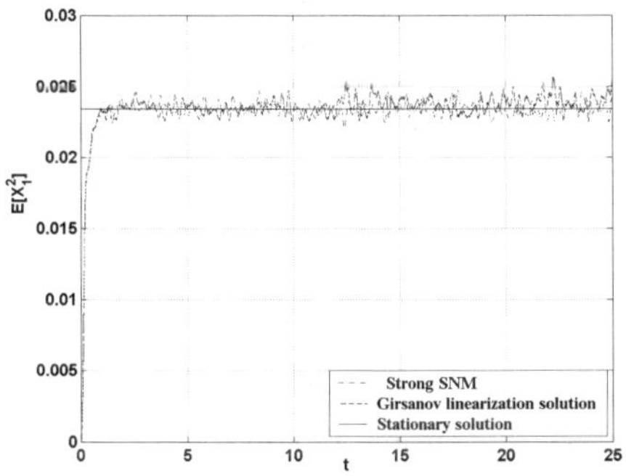

Fig. 2: Second moment histories of Duffing equation under additive noise -mean of displacement squared - $C = 2.0, K_1 = 100,\ K_2 = 100, \sigma = 5.0$, ***h*=0.001, number of samples: 500**

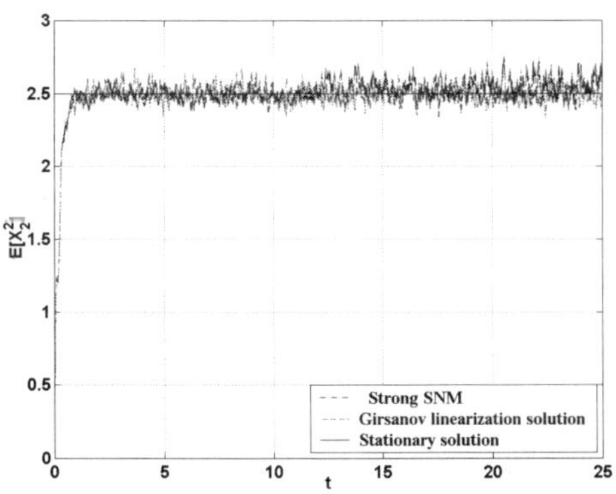

Fig. 3: Cross-moment histories of Duffing equation under additive noise -mean of velocity squared -

$C = 2.0, K_1 = 100,\ K_2 = 100, \sigma = 5.0$, *h*= 0.001,number of samples: 500

Up-crossing Statistics:

We have observed that up-crossing statistics of displacement, simulated through Girsanov linearization and through the conventional form of the strong stochastic Newmark method (SNM) (Roy 2005), match very well. Figure 4 gives the time-histories of expectations of out-crossing of displacement component with a uniformly chosen ensemble size of 500. For obtaining these figures, we have chosen the system parameters as $C = 2.0,\ K_1 = 100,\ K_2 = 100$. We have chosen, $\sigma = 5$, for the additive noise and the step size is uniform at $h = 0.001$.

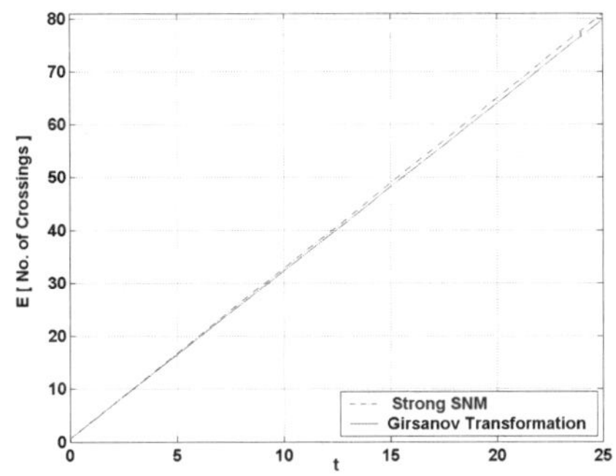

Fig. 4: Mean no. of up-crossings history of Duffing equation under additive noise – $\alpha = 0.020, C = 2.0$,

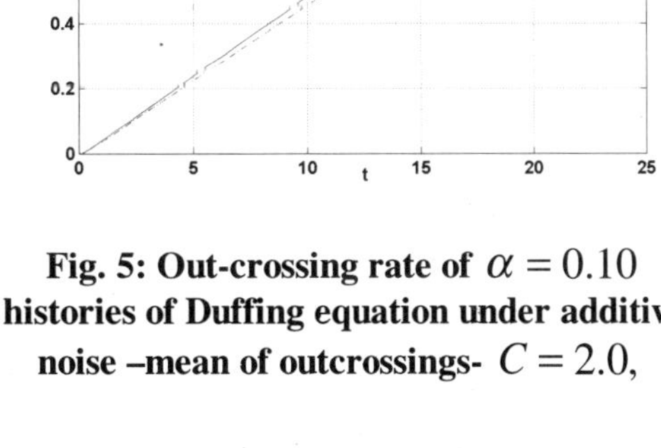

Fig. 5: Out-crossing rate of $\alpha = 0.10$ **histories of Duffing equation under additive noise –mean of outcrossings-** $C = 2.0$,

$K_1 = 100,\ K_2 = 100,\ \sigma = 5.0$, h=0.001,

number of samples: 500 (Girsanov's Method)
5000 (Strong SNM)

$K_1 = 100,\ K_2 = 100,\ \sigma = 5.0$, h=0.001,
number of samples: 500

Importance Sampling

We have observed that on forcing the equations close to the level α through the forcing parameter β (as in equation (3.12)) allows us to drastically reduce number of samples required for a satisfactory computation of the up-crossing statistics as compared with a direct application of the SNM that requires a considerably more number of samples. Figure 5 provides the time-histories of expectations of numbers of up-crossings of the displacement component. For obtaining these figures, we have chosen the system parameters as $C = 2.0,\ K_1 = 100,$ $K_2 = 100$, $\alpha = 0.10$. We have chosen $\sigma = 5$ for the additive noise intensity and the step size is uniformly kept at $h = 0.001$.

4. ACKNOWLEDGEMENTS

The authors thankfully acknowledge the financial support from the Board of Research in Nuclear Sciences (BRNS), Government of India, through grant no. DAEO/MCV/DR/112.

5. REFERENCES

1. Macke, M. and Bucher, C., 2003, "Importance sampling for randomly excited dynamical systems", Jl. Sound and Vibration, Vol. 268, pp. 269-290.
2. Oksendal, B., 2004, "*Stochastic differential equations - An introduction with applications*", Springer, Sixth Edition.
3. Roy, D. and Dash, M K, 2005, "Explorations of a family of stochastic Newmark methods in engineering dynamics", *Comp. Meth. in App. Mech. and Engg.* (in the press, ref. No. CMA-5859).
4. Rubinstein, R. Y., 1981, "*Simulation and the Monte Carlo Method*", Wiley, New York
5. Socha, L, 2005, "Linearization in Analysis of Nonlinear Stochastic Systems: Recent Results—Part I: Theory" *Applied Mech. Review*, Vol. 58, pp. 178-205.
6. Socha, L, 2005, "*Linearization in Analysis of Nonlinear Stochastic Systems: Recent Results—Part II: Applications" Applied Mech. Review*, Vol. 58, pp. 303-315.
7. Wang R and Zhang Z 1998 "Exact stationary response solutions of six classes of nonlinear stochastic systems under stochastic parametric and external excitations" *J. Eng. Mech. ASCE* **18** pp. 18-23.

Reliability, Safety and Hazard
Editors: P.V. Varde et al.

CONTAINMENT TEMPERATURE, PRESSURE AND ACTIVITY RELEASE DURING LIMITING DESIGN BASIS ACCIDENT IN TAPP 3&4 REACTOR

Sanjeev Kumar, D.K. Bhartia, Manoj Kansal, N. Mohan, R.N. Bhawal, S.G. Ghadge and S.S. Bajaj

Directorate of Safety, Nuclear Power Corporation of India Limited,
Nabhikiya Urja Bhavan, Anushakti Nagar, Mumbai 400 094

ABSTRACT

Containment is considered as ultimate safety system and is designed to enclose whole reactor system and prevent the spread of active air-borne fission products. For Pressure and Temperature calculation, Design Basis Accident (DBA) is double ended break of reactor inlet header or main steam line break but activity release studies are done to access its performance following limiting design basis accident i.e. Loss of Coolant Accident (LOCA) and Emergency Core Cooling System (ECCS). In such accident scenario, the core is severely damaged and results in production of steam and hydrogen along with release of activity to containment environment. Containment functions are maintained in such accident, and radiological consequences are within the prescribed limits.

Keywords: Containment Temperature, Pressure, Activity, Consequence of accident

1. INTRODUCTION

Indian Pressurized Heavy Water Reactor (IPHWR) of 540 MWe capacity is a pressure tube type reactor with heavy water as moderator and coolant and natural uranium dioxide as fuel. It consists of 392 horizontal fuel channel assemblies each surrounded by heavy water moderator within the calandria vessel. The moderator is insulated from hot pressure tubes by carbon dioxide filled annulus between pressure tube and its surrounding calandria tube.

The moderator is cooled by process water in an independent circuit. The whole reactor is enveloped in the reactor containment building. Double containment philosophy is adopted for current design of IPHWRs. Containment is considered as ultimate barrier to safety to prevent the spread of active air-borne fission products. Containment building of Indian nuclear power plants is designed for the pressure and temperature that could result from the Design Basis Accident (DBA) which is defined as double ended guillotine break of reactor inlet header or main steam line break. In IPHWR containment, while suppression pool system acts as a major heat sink for limiting peak pressure, Reactor Building coolers contribute in the pressure reduction during post accident condition. In addition to these, there are other systems within primary containment and secondary containment for fission products management.

This paper presents the results of the studies carried out to assess the performance of containment system of IPHWR under limiting design basis accident scenarios i.e. LOCA with coincident failure of ECCS. Following the postulated accident, prolonged degraded cooling conditions prevail in reactor core, fuel temperature rises, and pressure tubes deform and touch the calandria tube. Heat transfer to cold moderator limits the maximum temperatures reached in fuel and terminate the scenario. It is seen that under such scenario, pressure tube integrity is maintained and no fuel meltdown is predicted.

The consequences of such postulated accident depend greatly on containment safety features and containment performance. If the containment functions were maintained in such accident, the radiological consequences would be within the prescribed limits.

2. CONTAINMENT PRESSURE TEMPERATURE TRANSIENT ANALYSES FOLLOWING SEVERE ACCIDENT

2.1 Modelling, Inputs and Assumptions

The PHT system of TAPP-3&4 consists of two independent loops connected by a small (about 100 mm OD) pressure balancing line. In case of a postulated break in piping of one loop, blow down of the reactor coolant will be limited to half of the core (one loop) due to automatic closure of isolation valves present in the pressure balancing line. This is considered as Design Basis Accident (DBA).

In Case of DBA, the coolant will flash into steam at a very high rate. This causes a rapid pressure rise in the Reactor Building. The discharge rate (Figures 1) calculated by computer code ATMIKA[1] have been used for this analysis. The resulting pressure rise drives the steam air mixture through the vent shafts and distribution headers into the suppression pool, where 100% condensation [2] of steam takes place while air passes and gets cooled by suppression pool water and released into volume V2 atmosphere, raising its pressure.

The calculation of the pressure-temperature transient has been done by computer code `PACSR' [3].

PACSR [3] a lumped parameter code for containment thermal hydraulic calculation is used for the analysis. In this code containment can be modeled by various compartments connected by junctions. PACSR code can model containment atmosphere having two phase i.e. liquid and vapor (atmosphere/fluid) and can handle saturated and superheated atmospheric conditions and can also calculate free and forced convection heat transfer. This code can handle liquid , steam and three non-condensable gases.

The calculation method involves breaking up the whole duration of the event into small time steps and then considering the effect of energy and mass addition in the small time interval for arriving at the pressure/temperature at the end of each time step using the applicable thermodynamic laws.

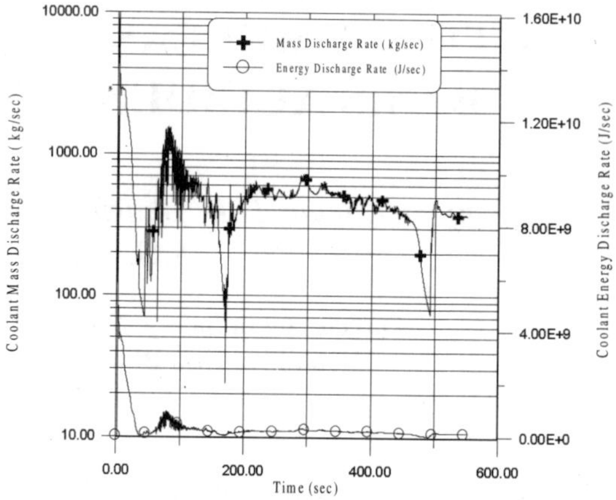

Fig. 1: Reactor Coolant mass and energy discharge rate following LOCA with coincident failure of ECCS.

This code has been developed to simulate multi compartments. In the analysis reported here, the containment has been simulated by two accident-based volumes namely, volume V1 (the area containing high enthalpy piping) and volume V2 (other areas of primary containment) connected by vent system via suppression pool.

Volumes V1 is further divided into five compartments. Every compartment of containment Volume V1 and Volume V2 is divided into two regions: a liquid region containing water, and a vapor region containing uniform mixture of air, water vapor and liquid droplets. Vapor region is assumed to be in thermo-dynamic equilibrium, and have uniform temperature and density throughout the region. The flashed coolant/(steam) is assumed to mix instantaneously and uniformly in the vapor region of volume V1.

The code can also simulate the containment walls and structures for the heat absorption and conduction calculations. A number of correlations are incorporated in the program to determine Heat Transfer-Coefficient (HTC) between containment atmosphere and structures such as Tagami[4], Uchida [5], Diffusion based heat transfer coefficient [6]. For calculations presented in this paper diffusion based HTC is used.

Computer code 'PACSR' is verified against various international experiments for containment thermal-hydraulics and hydrogen distribution [7,8]. Verification of the code was also done against experimental data obtained from the containment model test facility at Madras Atomic Power Station (MAPS)[9].

The free-air volume used in calculation for various compartments are given in Table 1.

Table 1 Primary containment free air volumes

Room No.	Name of Compartment	Free air volume (m^3)
R-301	Calandria Vault	125.2
R-403-501	Fuelling Machine Vault (N)	3233.8
R-701-005-103	Pump Room	22671.3
R-404-502	Fuelling Machine Vault (S)	3233.8
R-801-705	Dome Region	25362.9
	V-2 Volume	30994.7

3. RELEASE OF ACTIVITY TO ATMOSPHERE

The calculation of activity release for postulated accident i.e. limiting design basis accident(LOCA +ECCS failure) performed by using computer code 'ACTREL: ACTivity RELease'. The computer code 'ACTREL' has been developed to model the behavior of fission products in primary and secondary containment and to calculate its release due to pressure resulting from postulated accident. This also models various engineered safety features to reduce the concentration of fission products in the containment. For the calculation of activity release, it is assumed that all released activity is instantaneously mixed uniformly in primary containment environment and is available for leakage from the containment right from the beginning of the postulated accident. For the computation of activity releases from the containment, the detailed methodology is given in reference 12. Figure-2 shows schematically the composite picture of containment activity release barriers and removal mechanisms.

For the purpose of the present evaluation, 100% equilibrium core inventory of iodine and noble gases is considered as released from the fuel. However an attenuation factor of 2 for iodine is considered for water trapping. For calculating the release from containment to the atmosphere, a leakage rate of 0.4% containment volume/hr at 1.25kg/cm^2(g) and 0.6% at 200 mm of W.C. containment volume/hr stipulated in the technical specification at specified test pressure of primary and secondary containment respectively are used. 10% leakage from primary containment is assumed to bypass the secondary containment.

Based on the calculated release of activity into the atmosphere, the doses at exclusion distance of 1.6 km are worked out using dose conversion factors [13] For initial two hrs into the accident, the worst meteorological parameters are assumed to maximize the dose. For subsequent releases, time averaged dispersion parameters are used, still conservatively assuming the wind direction to be same throughout.

4. RESULTS & DISCUSSION

The pressure and temperature transient in the containment following the accident is shown in Figures 3 and 4. It can be observed that, maximum pressures and temperatures reached in volumes V1 (break compartment) and V2 (suppression pool chamber) at around 30 sec.

The predicted peak pressure and temperature are 0.56 kg/cm^2(g) (at ~29.7 sec) and 113.72^{o}C (at ~29.7 sec) in the accident. In this case, FM Vault has been considered as break compartment.

It is observed that initially the rate of pressure fall in V1 is faster than that in V2. The reason is that V1 is full of steam, which is condensing, and also most of the Reactor Building. air coolers are located in V1. After some time when V1 pressure falls below the V2 pressure by 0.035 kg/cm^2(g) specially provided blowout panels open out (from V2 to V1) ,and pressure in the two volumes get equalized.

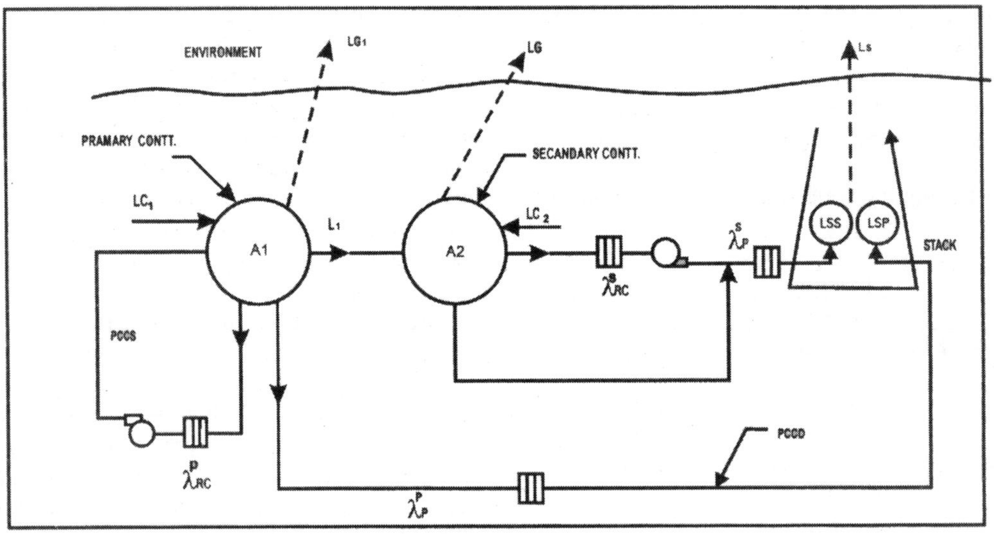

LC_1 = IN LEACKAGE OF COMPRESSED AIR TO PRIMARY CONTAINMENT

LC_2 = INLEACKAGE OF COMPRESSED AIR TO SECONDARY CONTAINMENT

L_1 = INLEACKAGE FROM PRIMARY CONTT. TO SEC CONTAINMENT

LG_1 = GROUND LEVEL LEAKAGE FROM PRIMARY CONTT. TO ENVIROMENT (SC BY PASS ASSUMED)

LG = GROUND LEVEL LEAKAGE FROM SECONDARY CONTT. TO ENVIRONMENT

LSP = STACK RELEASED FROM PRIMARY CONTT. THROUCH PCCD

λ_P^p = IODINE RETAINED BY FILTERS OF PCCD (Primary Containment Control Discharge System)

λ_{RC}^p = IODINE RETAINED BY FILTERS OF PCCS (Primary Containment Cleanup System)

λ_{RC}^s = IODINE RETAINED BY FILTERS OF SECONDARY CONTT. RECIRCULATION SYSTEM

λ_P^s = IODINE RETAINED BY FILTERS OF SECONDARY CONTT. PURGE SYSTE.

LSS = STACK LEVEL RELEASE FROM SECONDARY CONTAINMENT

Fig. 2: Schematic Diagram of Double Containment Activity Release Model

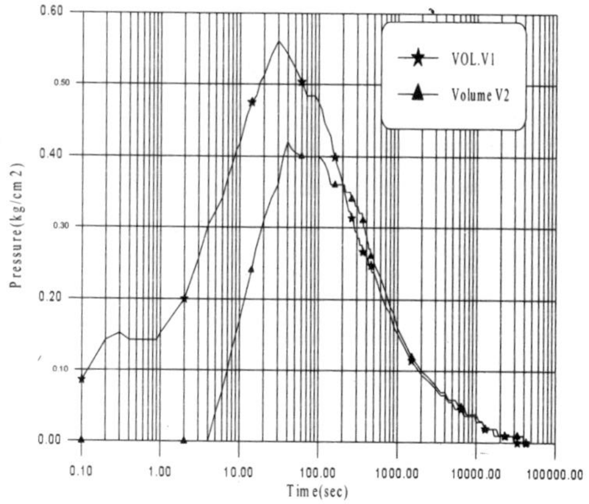

Fig. 3: Pressure in Break Compartment following accident in TAPP 3&4

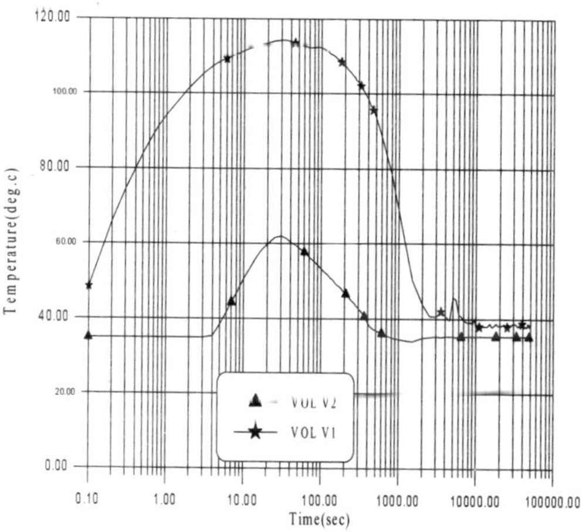

Fig. 4: Temperature in Break Compartment Following accident in TAPP 3&4

The calculated releases to the atmosphere and corresponding doses are depicted in Table 2.

5. CONCLUSION

Based on the analysis carried out for postulated accident, the following conclusion can be drawn.

1. The calculated peak pressure under postulated accident condition is 0.56 $kg/cm^2(g)$ (at ~29.7 sec) and temperature is 113.72^{O}C (at ~29.7 sec), which are within the design pressure (1.44 $kg/cm^2(g)$) and temperature(125°C).
2. The doses calculated for the postulated accident are within the Acceptable Emergency Doses (AED) of 100 mSv for whole body dose and 500 mSv for thyroid dose at exclusion distance of 1.6 km.

Table 2 Activity Releases and Corresponding Doses

Fraction of Fuel Failure	100%	
Fission Product Release from Core	I-131	23.8x10^{06} Ci
	Noble Gases	209 x10^{06} Ci
Release to Containment	I-131	11.9 x10^{06} Ci
	Noble Gases	209 x10^{06} Ci
Release From Containment	Ground Level I-131 Noble gases	258 Ci 2.48 x10^{05} Ci
	Stack Level I-131 Noble Gases	174 Ci 4.11 x10^{06} Ci
Doses at Exclusion Distance of 1.6 km.	Thyroid	45.0 mSv
	Whole Body	21.0 mSv

6. REFERENCES

1. H. P. Rammohan. ATMIKA-a system thermal hydraulic computer code, 1996.
2. Kakodkar, and V.K. Mehra. Vapor suppression pool experiments (phase 1). BARC/I-171. 1972.
3. S. Kumar, M., Kansal, N.Mohan, R.N. Bhawal and S.S.Bajaj. *PACSR:* A computer code for post accident containment system response. No. NPC/01550 / 2005/ M/ 2005.
4. J. J. Carbajo. Heat transfer coefficients under LOCA conditions in containment building. Nuclear engineering and design, 1981,**65**, 369-386.
5. J. Green and K. Almenas. An overview of the primary parameters and methods for determining condensation heat transfer to

containment structures. Nuclear Safety 1996.,**37, (1)**, 26-47.

6. Description of the Mod2/ 85 Versions of RALOC 1988, Part-2. Physical modeling of thermal hydraulics and integration methods. GRS-A-1426.
7. S.S. Bajaj. Comparison of Code Calculation with Experiments on containment Pressure Response during LOCA Condition, The 2nd International Conference on Containment Design and Operation, October 14-17, 1990. Toronto,Canada.
8. M., Kansal, N.Mohan, R.N. Bhawal and S.S.Bajaj..Verification of Computer Code PACSR against experiment D15 on Battelle Frankfurt Containment Model, Report No. NPC/01540/96/B/8131 dated September 5.1996
9. M., Kansal, N.Mohan and S.S.Bajaj. Verification of Computer Code PACSR against experiment ISP-23 on HDR Containment Model, (Internal Report) 1994
10. Hillard, and Postma.Large-scale Fission Product Containment Tests. Nuclear Technology,1981, **53**.
11. Head HPD . Review of exclusion zone. Head HPD, BARC, Internal Report,2000.

Reliability, Safety and Hazard
Editors: P.V. Varde et al.

CONTAINMENT INTEGRATED LEAKAGE RATE TEST (ILRT) OF INDIAN PHWR

D.K. Bhartia, Nalini Mohan, R.N. Bhawal and S.S. Bajaj

Directorate of Safety, Nuclear Power Corporation of India Limited
Nabhikiya Urja Bhawan, Anushakti Nagar, Mumbai 400 094

ABSTRACT

Integrated Leakage Rate Test (ILRT) of containment system plays a very important role in safety of a Nuclear Power Plant. Containment system constitutes the last physical barrier to release of radioactivity from the core and is called upon to mitigate the consequences of not only accidents within the design basis, but also some of the highly unlikely severe accidents. Hence, leak tightness of containment becomes uttermost priority for the safety of plant personnel & public. The containment and associated ESFs are tested before the first criticality and there after periodically during service. The pre-operational integrated leakage rate is carried out at LOCA based design pressure, at periodic test pressure and at some intermediate pressure points to assess the leakage characteristics. This paper summarizes the various requirements and activities relevant to the ILRT of the Indian Pressurized Heavy Water Reactor (PHWR) containment system.

1. INTRODUCTION

Indian PHWR Containment system design incorporates a primary containment constructed from Prestressed Cement Concrete (PCC) enveloped in a Reinforced Cement Concrete (RCC) secondary containment on a common base raft. A passive suppression system is also incorporated in the design. Figure-1 shows the section of Reactor Building of 540 MWe PHWR at Tarapur Atomic Power Station . The primary containment of 540MWe is designed for a nominal uniform overpressure of 1.44 kg/cm^{2}g (144 kPag) which is based on the calculated peak pressure in the event of the postulated design basis accident of double ended rupture in main steam line. However, the peak pressure calculated for postulated design basis accident of double ended rupture in the reactor outlet header is 0.8 kg/cm^{2}g (80 kPag), which has been used for integrated leakage rate test.

Primary Containment (PC) houses reactor and its various associated systems. It also provides shielding to Gamma Radiation from reactor. In case of postulated loss of coolant accident, the containment holds the radioactivity released from the reactor and/or its coolant in such a way that the release of radioactivity to the environment is within the acceptable and prescribed limits. The secondary containment does not house any plant system. The high enthalpy steam lines passing through the secondary containment (SC) space are designed to pass through pipe sleeve to eliminate the need for designing the secondary containment for postulated high enthalpy pipe break of steam line resulting in internal pressure and temperature loading. However, small increase in SC pressure could be expected under accident condition involving pressure rise in PC, due to the air leakage into the SC atmosphere under the condition when SC purge is not available. The purge damper is fail safe (i.e. fail to open) with respect to air and power failure to ensure that SC is not pressurized even if such failures occur.

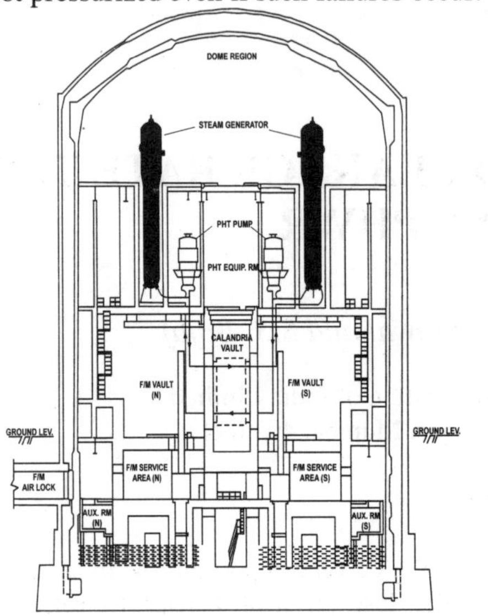

Fig. 1: Typical Cross Section of 540 Mwe Reactor Building

2. TEST OBJECTIVES

The objectives of conducting leakage rate tests on the primary containment and its various accessories and engineered features is to demonstrate the leaktightness of the containment so that in the event of an accident, the consequences of an off site release of radioactive fission products are reduced to within the acceptable limits.

The leakage rate tests on the barriers between the two accident-based volumes V1 & V2 is conducted to ensure that the flow of air steam mixture following DBA bypassing the suppression pool is restricted to within the acceptable limit so that the peak internal overpressure in the Primary Containment remains limited to design value.

The leakage rate test with negative pressure in Volume V1 is conducted to ensure maintenance of sufficient negative pressure in Volume V1 with respect to areas in Volume V2 with the design purge rate so as to restrict the escape of tritium and other activities from Volume V1 to V2 and to ensure better recovery of D2O vapour.

The leakage rate tests on secondary containment is conducted to ensure low leakage under post DBA condition so that adequate vacuum can be maintained with the secondary containment purge operating.

3. TEST SEQUENCE

3.1 Pre-Test Inspection

Before commencement of ILRT , it is ensured that all pre-requisites have been completed and the containment system is as close as possible to that during Reactor operation . A visual inspection is made of the passive as well as active parts of the containment to ensure that no defect exists in the containment boundary that may jeopardize the safety of the structure or successful completion of test.

3.2 Pre-Operational Tests

This involves demonstration of automatic fast isolation of the containment when demanded. The test is carried out by actuation of the containment isolation logic with verification of closure of barriers. The time elapsed between onset of closure signal and completion of isolation are verified for individual devices to be within the design specification of 5 secs. The active devices falling under this category are the various valves in the airlines open to containment atmosphere and penetrating the containment boundary.

Local Leakage rates tests is carried out on individual penetrations, closure devices, access airlocks, various joints, etc. before commencement of integrated leakage rate test. Individual leakage rate tests are done by method of absolute pressure rundown .

This is followed by preliminary leakage rate test on the primary containment envelope at low pressure generally 0.2 to 0.3 kg/cm^2g in order to assess and identify leakages through various paths, and to verify operability of the test instrumentation.

3.3 Secondary Containment Leakage Rate Test (Sc-Lrt)

Secondary containment space is also leak tested to establish a leakage correlation for assessing the ground level release from secondary containment for the duration when the interspace between primary and secondary containment remains at little positive pressure during LOCA. It includes a Positive pressure test which is conducted at 200mm of wc and a negative pressure test to demonstrate the maintenance of adequate negative pressure (lower than minus 12mm of water column) in the secondary containment with purge system operational.It also includes demonstration of interception of leakages from primary containment.

3.4 Volume V1 To V2 Interity Test

Effectiveness of pressure suppression system requires integrity of the barriers between the two accident-based volumes V1 & V2. Also, during normal plant operation Volume V1 area is maintained at some negative pressure compared to V2 by maintaining a purge via dryers and particulate filters. The tests include a positive pressure leakage rate test to evaluate the effective suppression pool bypass area between V1 & V2 and a test to demonstrate, maintenance of measurable negative pressure in V1 with respect to V2, with normal ventilation & purge system in V1 operating.

4. TEST PROCEDURES

The procedures for testing 540MWe PHWR containments were prepared based on the experience of testing two units of RAPS, MAPS, NAPS ,KAPS, RAPP-3&4 & Kaiga each and the requirements of the various international standards, codes and guides viz. ASME Sec.III, 10CFR50 App-j, ANS56.8, KTA Standard 3401 etc. The current procedures for 540MWe PHWR broadly incorporate the essential requirements of these guides and standards. A review of the Canadian standard CAN3-N287.6-M80 also indicate that all requirements of this standard have generally been met.

4.1 Leak Detection

During the leakage rate tests, leak search is carried out on individual penetrations and on the containment envelope from both inside and outside. The inspection from inside under hyper atmospheric conditions (upto 0.14 kg/cm^2g) necessitate appropriate selection and training of the personnel under pressurized atmosphere and use of various protection measure to safeguard the health of the crew. Both Indian and International standards are followed in this respect. Also the experience available elsewhere are utilised for selection and training. The leak detection and estimation during the test is done using soap solution (equal parts of corn syrup, liquid detergent and glycerin) , smoke (useful while carrying out leak checks from inside of the pressurized containment), Audible noise& thin paper strips.

4.2 Leakage Rate Determination

The integrated leakage from the containment is determined from estimation of mass of air in the containment as a function of time during the pressure rundown. Determination of mass of air is based on use of ideal gas law of PV =mRT. The pressure of dry air is used for the above computation.The contained mass is calculated at different time inteval (5 to 10 minutes) during the run down.The air mass calculated by using Least Square Method is plotted against time. The slope of this line will give the rate of change of air mass with time,which represent the leakage rate.

The mass of dry air is worked out from the correlation

$m = 9.81\ E{+}4(Pa + Pb - Pv)V/RT$

Where,

m = Mass of dry air, kg

Pa = Containment pressure, kg/cm^2g

Pb = Barometric pressure, kg/cm^2a

Pv = Partial pressure of water vapour, kg/cm^2

V = Containment free volume m^3

R = Gas constant for air 288.7 J/kg^oK

T= Containment average air temperature, ^oK

The air mass determined at successive points of time is fitted using the methodology of least squares to form a correlation of

$m = A\phi + B$ for constant leakage rate with respect to time.

Where m = mass of containment air at time ϕ

A and B = Constants

The slope of this correlation

'A' indicate the integrated leakage rate.

4.3 Verification Of Leakage Test Accuracy

In order to verify the accuracy of leakage rate test measurement, superimposition test is done. For this purpose, a calibrated known inleakage of compressed air is imposed through a 100 mm NB line provided with a flowmeter on the existing leakage in the containment at leak test pressure for 3 hrs. The flowmeter readings are then recorded at least hourly. Consequently, the readings of containment pressure is recorded in order to estimate the composite leakage rate (Lc) of both the containment system leakage rate (Lam) and superimposed leakage rate (Lo).

The superimposed leakage rate (Lo) shall be between 75 and 125 percent of maximum allowable leakage rate (La).

$$(Lo + Lam - 0.25La) <= Lc <= (Lo + Lam + 0.25La)$$

4.4 Data Rejection And Error Analysis

The accuracy of the results are dependent not only on the instrument accuracies but also on the total number of observations, and test duration. Based on the error analysis in TAPP-4 rundown period and frequency of data collection have specified as 8 hrs. with data collection frequency of 5 mins. during rundown.

Data rejection for the computed air mass in the containment at specified time interval is carried out in accordance with ANSI Standard ANS 56.8

An upper confidence level of 95% is used to account for the error in the estimation of leakage rate.

5. TEST EXPERIENCE AT TAPP-4

The leakage test conducted on the primary containment indicated that there was no appreciable leakage though construction joints. However leakage were detected around the Steam Generator Embeded Parts (EP's) in dome region from outside PC. Minor leaks were observed through electrical canister EP's. Leakage were observed through FM Airlock and Main Air lock EP to concreate interface.

The leak test results which are summarized in Table-1 indicate that observed leak rate for Kaiga , RAPP-3&4 & TAPP-4 are between 0.06% to 0.154% of contained volume/hr. However, the leakage rate for TAPP-4 is 0.11% which lies between Kaiga & RAPP-3&4.

The combined contribution of leakage through all the local penetrations, isolation valves and airlock seals to the overall leakage of the containment was observed to be quite less (< 10% of total leakage).

A simultaneous test conducted on the secondary containment enveloping the primary containment confirmed interception of the leakages. This is an important observation as the overall containment performance with regard to ground level release following an accident would significantly depend on the extent of interception. The interception of leakage in SC were observed to be between 75% to 89%.The secondary containment interception for TAPP-4 is 87% which lies between Kaiga & RAPP-3&4.

Table 1 Summary of leakage test result

	Leakage Rate (Sm^3/hr)*	% of contained vol./hr	Secondary Containment Interception	
			Sm^3/hr	%
Kaiga-1	63.07	0.059%	46.52	75%
Kaiga-2	96.71	0.097%	81.24	84%
RAPP-3	163.00	0.154%	143.80	88%
RAPP-4	150.20	0.150%	142.0	89%
TAPP-4	155.60	0.110%	135.0	87%

* Sm^3/hr. - Standard cubic meter per hour

The preoperational tests carried out to determine the leaktightness between Volumes V1 andV2 of the pressure suppression system indicated that the bypass area were less than 0.02 m2 of equivalent opening. This parameter is also monitored indirectly during normal operation by the negative pressure in Volume V1 with respect to V2. Minor leaks were seen through FM vault and service area doors.

6. CONCLUSIONS

With regard to the experience gained from ILRT of Indian PHWR containment following conclusions could be drawn.

1. The test procedures and specifications applicable to TAPP-4 containment are commensurate with the design objectives relevant to normal operation of the plant as well as post accident radiological safety considerations. These procedures are

generally in line with the relevant safety codes, guides and standards applicable in Canada, USA and FRG and sufficiently adequate for demonstration of design objectives.

2. In most of the cases leaks were observed through EP to concrete joints. Hence proper precaution is required to be taken during the construction stage in order to avoid this leak. This needs to be tackled by suitable design modification, QA procedures implemented in the issuance of work permits, review of work reports etc. along with inspection and testing following the maintenance action in relevant areas.
3. The summary of leakage test result shows that there is a scope to reduce the leakage rate through primary containment as observed in Kaiga –1 in which the observed leakage rate was at lowest value. Comparative study of the construction procedure of the three representative primary containment i.e. RAPP, Kaiga & TAPP will help to bring down the leakage rate through primary containment in future reactors.
4. The summary of leakage test result also shows that there is a scope to reduce the leakage rate through secondary containment. Similarly, as observed in RAPP-4 the interception of leakage from primary containment could be improved in future reactors. Comparative study of the construction procedure of the three representative secondary containment i.e. RAPP, Kaiga & TAPP will help in a similar manner to bring down the leakage rate through secondary containment in future reactors.

7. REFERENCES

1. AERB/SG/CON-TEST
   "Proof and leakage rate testing of Reactor Containments", 1999.
2. Nalini Mohan, D.K. Bhartia, A.K.Shit, R.N.Bhawal,R.K.Sharma,V.C.Agrawal
   "Report on TAPP-4 Primary Containment Integrated Leakage Rate Test", 2004.
3. O.P.Arora,K.S.Rao,A.K.Pal
   "Procedure for R.B.Proof test & ILRT of TAPP3&4", 2004.

Reliability, Safety and Hazard
Editors: P.V. Varde et al.

AN EFFICIENT METHOD OF IMPROVING THE PERFORMANCE OF MODIFIED HL-RF OPTIMIZATION ALGORITHM

Santosh, A.K. Ghosh and H.S. Kushwaha

Health, Safety and Environment Group, Bhabha Atomic Research Centre, Mumbai 400 085

ABSTRACT

Several optimization algorithms in structural reliability have been studied in order to determine their suitability and performance measures to various types of structural reliability problems. It is seen from the study that the modified HL-RF method is one of the most widely being used First Order Reliability Methods (FORMs) for obtaining the safety index of a structural component. The main objective of this study was to improve the performance of modified HL-RF method by implementing an efficient step length selection rule. Several step length selection rules were studied and implemented in order to make the algorithm more robust and efficient. It is clear from the study that the variant of Goldstein's rule which is popularly known as Armijo rule is one of the most efficient step length selection rule. It is observed from the study that a proper step length selection rule will not only make the algorithm more robust but also improves the other performance parameters such as efficiency, capacity etc. This paper describes the concepts of the implemented step length selection rule and its implementation in the modified HL-RF algorithm. Several case studies have been taken from the literature in order to demonstrate and validate the algorithm. The results of the implemented algorithm have been verified with the existing results presented in the literature.

Keywords: Structural reliability, Optimization algorithms, Limit state function, Goldstein rule, Armijo rule, Reliability index, Modified HL-RF method

1. INTRODUCTION

In any engineering problem, several parameters are involved and some of these parameters may be either deterministic or probabilistic. Similarly in structural engineering problems, the geometric parameters of the section (dimensional variations), the physical properties of the materials (cube strength of the concrete, yield strength of the steel, the Young's modulus of the steel and the concrete, etc.) and the loads (live loads on floors, wind loads, etc.) coming on structures are subjected to random variations. If the coefficient of variation of a random variable is very small, probably this may be ignored and the variable may be considered as deterministic. Hence in any engineering problem the parameters, which are to be considered as random variables are initially well-defined [1].

A typical structural reliability analysis deals with models that are mathematical idealizations of the physical processes [2]. The idealization requires the definition of basic variables describing the geometry, loads and material properties. The mathematical model then establishes the relationship between the load and the resistance of a structural component under consideration. A structural reliability analysis then aims at estimating the probability of exceeding the structural limit states imposed on a structural component. Any equation that represents a particular limit state condition (or failure condition) of the structural component will be interconnecting these basic random variables and is a function of these variables. Hence this equation is known as Limit State Function (LSF) or performance function.

In the simplest case of a component reliability analysis, the main goal of structural reliability analysis is to estimate the probability of failure of a structural component by evaluating the following integral:

$$p_f = \int_{g(x)\leq 0} f(x)dx \quad (1)$$

where p_f is the probability of failure, x is the vector of random variables, $g(x)$ is the limit state function or the performance function such that $g(x) \leq 0$ defines the failure domain, and $f(x)$ is the joint probability density function for the basic random variables.

The integration of Equation 1 is very difficult since it involves multiple integrals in addition to evaluating the joint probability density functions (PDFs) of the random variables [3]. Further more, a closed form of solutions for general cases is not available. Therefore a transformation is required to transform the basic random variables from their original space into the standard normal space where the solution of the above equation can be obtained easily. Approximate solutions for this type of problems are found either by simulation or by approximating the limit state surface at the so-called design point or checking point which is a point in the failure region that has the highest probability density in the standard normal space. For this reason the point is sometimes called as the "Most Probable Point of failure" (MPP). Hence determination of this point constitutes a constrained optimization problem in the context of structural reliability analysis.

The objective of this work was to study several optimization algorithms in structural reliability and improve the performance of modified HL-RF method. In the Mean Value First Order Second Moment (MVFOSM) method, the LSF is linearized about its mean values to obtain the safety index of a given structural component. But evaluation of reliability by linearizing the LSF about the mean values leads to erroneous estimate for performance function with highly non-linear or for large co-efficient of variation. Also the reliability index is distinct for different formulations of the same limit state condition [3]. In order to over come these drawbacks Hasofer and Lind proposed a new approach in which the linearization is done about a point which lies on the failure surface and it corresponds to a maximum likelihood of failure occurrence. The reliability in this approach is measured through safety index which is defined as the minimum distance from the origin to the failure surface in the standard normal space with zero mean and unit standard deviation [3, 4].

In general, Advanced First Order Second Moment (AFOSM) method is widely being used in structural reliability analysis. It is an approximate method that estimates the safety index of given structural component at the design point rather than at the mean values. The AFOSM method belongs to the family of FORMs which includes various approaches such as Hasofer Lind (HL), Hasofer Lind & Rackwitz Fiessler (HL-RF), modified HL-RF etc [3]. The HL method is computationally intensive but more accurate than the MVFOSM method. Rackwitz and Fiessler extended the HL method to include the information on the distribution of the basic random variables and the method is popularly known as HL-RF method. This method in particular requires least amount of storage and less number of iterations. But when the limit state function is highly complex and nonlinear this method may converge slowly or even sometime result in divergence [6].

In order to overcome these limitations, there has been an increasing trend towards analyzing the performance of these algorithms. Many researchers have made several modifications to HL-RF method by implementing various step length selection rules to make the method more efficient and robust [6]. The implemented step length selection rule in this work is one of such modifications to the modified HL-RF method. A variant of Goldstein's step length selection rule which is popularly known as Armijo rule has been implemented in this study [6, 7]. It is seen from the experience that the step length selection rule not only makes the algorithm efficient but also makes more robust by finding a global solution to the problem. The proposed methodology has been demonstrated with the help of several numerical examples taken from the literature.

2. BASIC CONCEPTS IN STRUCTURAL RELIABILITY ANALYSIS

When introducing the formulation of the structural reliability problem, it is quite logical to start with an idealized component problem. In this problem there are two random variables, a load variable s and a resistance variable r and each of them is characterized by their probability density function. The entire goal of the structural reliability analysis is to find the probability of failure, i.e., the probability that the load s exceeds the resistance r. This relationship between the load and the resistance can be represented in the mathematical form, i.e.

$$g(r, s) = r - s \tag{2}$$

The failure surface or the limit state can then be defined as $g(r, s) = r - s = 0$. This defines the boundary between the desired and undesired performance of a structural component and it also represents a state beyond which a structural component can no longer fulfill the required function for which it was designed. The failure, safe and unsafe regions of linear and non-linear performance functions are shown in Figure 1.

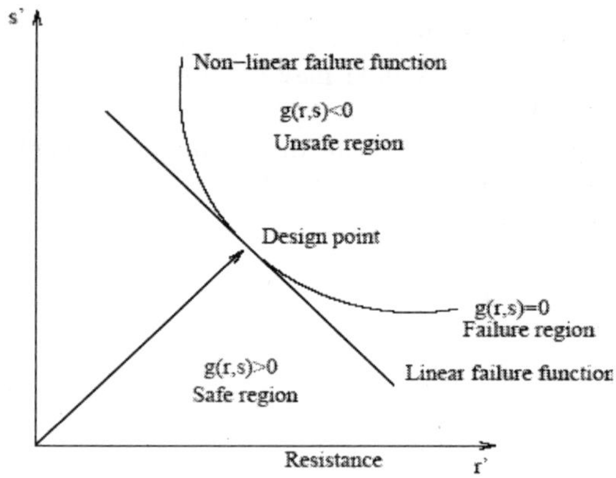

Figure 1: Limit state concept

From Equation 2, one can observe that the failure occurs when $g(r, s) < 0$. From this condition, the probability of failure p_f is evaluated by the integral

$$p_f = \int \cdots \int_{g() < 0} f_x(x_1, x_2, \ldots, x_n) dx_1 dx_2 \ldots dx_n \tag{3}$$

where $f_x(x_1, x_2, \ldots, x_n)$ is the joint probability density function for the basic random variables. The computation of p_f from Equation 3 is almost impossible since evaluating the multiple integral is extremely a difficult task. Therefore this problem will be solved by analytical methods known as first and second order reliability methods. The basic principle of these methods is that they estimate the reliability index or safety index in the reduced co-ordinate system by linearizing the LSF about a point [3–5, 8]. The probability of failure p_f can then be obtained from the relation:

$$p_f = \Phi(-\beta) = 1 - \Phi(\beta) \tag{4}$$

where

$$\Phi(\beta) = \int_{-\infty}^{\beta} \frac{1}{\sqrt{2\pi}} e^{-\frac{1}{2}z^2} dz \tag{5}$$

In structural reliability analysis, the use of approximation concepts is quite common and necessary because of difficulty in directly computing the multi-dimensional integration over the irregular domain in probabilistic space. In general, approximation techniques are used in the transformation of non-normal distributions into standard normal distributions for calculating the safety index denoted by β.

2.1 Transformation of Non-normal Variables

The non-normal random variables x can be transformed to standard normal variables z using the popularly known Rosenblatt transformation [9]. When the variables are mutually independent, the transformation takes the following form:

$$z = \Phi^{-1}[F_x(x^*)] \tag{6}$$

The approximate normal or equivalent normal distribution is derived by equating the cumulative distribution functions and probability density functions of the original non-normal and the approximate or equivalent normal random variables at the design point on the failure surface [3, 5]. It can be considered as a result of Taylor's series expansion of Equation 6 at MPP denoted by X*, neglecting higher order terms, we have

$$z = \Phi^{-1}[F_x(x^*)] + \frac{\partial}{\partial x}(\Phi^{-1}[F_x(x^*)])|_{x^*}(x - x^*) \quad (7)$$

where

$$\frac{\partial}{\partial x}(\Phi^{-1}[F_x(x^*)]) = \frac{f_x(x^*)}{\phi(\Phi^{-1}[F_x(x^*)])} \quad (8)$$

substituting Equation 8 in Equation 7 and rearranging we have

$$z = \frac{x^* - \Phi^{-1}[f_x(x^*)]\phi(\Phi^{-1}[F_x(x^*)])/f_x(x^*)]}{\phi(\Phi^{-1}[F_x(x^*)])/f_x(x^*)} \quad (9)$$

this can be written in the following form:

$$z = \frac{x - \mu'_x}{\sigma'_x} \quad (10)$$

where $F_x(x)$ is the marginal cumulative distribution function, $f_x(x)$ is the probability density function, $\Phi(.)$ is the standard normal cumulative distribution function, $\phi(.)$ is the standard normal probability density function, and μ'_x and σ'_x are the equivalent means and standard deviations, and they are given by the expressions [3, 9, 10]:

$$\sigma'_x = \frac{\phi(\Phi^{-1}[F_x(x^*)])}{f_x(x^*)} \quad (11)$$

$$\mu'_x = x^* - \Phi^{-1}[F_x(x^*)]\sigma'_x \quad (12)$$

using Equation 10, Equation 11 and Equation 12, the transformation of the random variables from the original x-space to the standard normal z-space can be easily performed, and the performance function $g(z)$ in the z-space is approximately obtained [3, 4]. The main effort in the reliability analysis is to compute the safety. In general, the calculation of the safety index can be formulated as a constrained optimization problem in finding the minimum distance point [4, 6, 11]. There are many algorithms available in the literature that can solve this type of problem. However, this paper describes only the modified HL-RF method.

3. MODIFIED HL-RF METHOD

Since the modified HL-RF method is the extension of the original HL-RF method, it is logical to describe it from the basics of the original HL-RF method. First consider the special case in which $g(z) = (b+a^T z)$ is a linear function of z. The optimization problem then becomes,

$$Minimize \qquad \beta = |z| = \sqrt{z^T z} \quad (13)$$

$$Subjected\ to \qquad g(z) = 0 \quad (14)$$

where β is the objective function and $g(z)$ is the limit state function in standard normal space, is now equivalent to computing the shortest distance from the hyper plane $(b + a^T z) = 0$ to the origin. Using the conditions that the solution z^* lies on the hyper plane, the optimal point z^* is expressed as:

$$z^* = \frac{1}{|a|^2}[a^T z_0^* - g(z_0^*)]\{a\} \quad (15)$$

where z^*_0 is the starting point. Now extend this concept to a more general case in which $g(z)$ is nonlinear. The recursive formula of the nonlinear version of the HL-RF method is expressed as follows:

$$z_{k+1}^* = \frac{1}{|\nabla g(z_k^*)|^2}[\nabla g(z_k^*) z_k^* - g(z_k)]\nabla g(z_k^*) \quad (16)$$

where $\nabla g(z^*_k)$ is the gradient vector of the performance function at z^*_k which is the kth iteration point. The detailed formulation of this recursive formula can be found elsewhere [3].
The above formula is applied repeatedly until the sequence z_k converges. Compared to other nonlinear optimization algorithms, this algorithm requires the least amount of storage and computation in each step. For most situations this method not only converges, but also converges faster [3, 4, 6]. However, under certain conditions this method may fail to converge and when the limit state function is complex and highly nonlinear, this method may converge slowly or even sometime result in divergence.

In order to improve the robustness of the HL-RF method, certain modifications have been suggested by several investigators. In one of the modified procedures, the robustness of HL-RF method can be improved by introducing a merit function, $m(z)$ which monitors the convergence of the sequence and at each step the new iteration point is selected by a line search along the direction vector [2,6,12]. The merit function should have a global minimum at the solution point of limit state equation and it should decrease in each iteration step. Here, a local minimum point of limit state equation must satisfy:

$$g(z^*) = 0 \qquad and \qquad z^* + \lambda^* \nabla g(z^*) = 0 \quad (17)$$

where

$$\lambda^* = -\frac{g(z^*)z^*}{|g(z^*)|^2} \quad (18)$$

From these conditions a non-negative merit function can be formulated as

$$m(z) = \frac{1}{2}|z - \frac{\nabla g(z) z}{|\nabla g(z)^2|^2} \nabla g(z)|^2 + \frac{1}{2} c g(z)^2 \quad (19)$$

where c is a positive constant. Obviously, all minimum points of limit state equation are global minimum points of $m(z)$. In the modified HL-RF method, the new iteration point is selected by a line search along the direction vector d_k given by the expression:

$$d_k = \frac{1}{|\nabla g(z_k)|^2} [\nabla g(z_k) z_k - g(z_k)] \nabla g(z_k)^T - z_k \quad (20)$$

until a sufficient decrease in $m(z)$ is achieved. Hence the merit function is a convenient guide for selecting the step size, since it is in terms of quantities that are already known [6].

4. STEP LENGTH SELECTION RULE

Various selection rules based on much less restrictive conditions have been proposed by several investigators. These rules allow us to obtain a point with the desired characteristics with a very restricted number of function evaluations [7, 13, 14]. The primary difference between the most algorithms rests in the rule by which the successive directions of descent are selected. The secondary difference lies in the selection of step length also known as learning rate denoted by α. In this paper we discuss the basic rules and the implementation concepts of a step length selection algorithm in the modified HL-RF method.

The common principle of all these search techniques is that:

Condition (1). *the step length, α must not be chosen too large (otherwise there is a risk that the algorithm exhibits an oscillatory behavior)*

Condition (2). *the step length, α must not be chosen too small (otherwise there is a risk of the algorithm converging prematurely)*

There are certain conditions which ensure that the selected α is an optimum value. In Goldstein's rule, one ensures the above condition (1) by satisfying the relation

$$g(\alpha) \leq g(0) + m_1 \alpha g'(0) \quad (21)$$

where m_1 is a constant chosen in interval [0, 1]; and the above condition (2) is ensured by the relation

$$g(\alpha) \geq g(0) + m_2 \alpha g'(0) \quad (22)$$

where m_2 is a constant chosen in the interval [m_1, 1].

A variant of Goldstein's rule proposed by Armijo consists of choosing α such that

1. α *satisfies the Equation 21*
2. *M*α *does not satisfy the Equation 21*

where *M* being a constant larger than 1 (in practice *M* is chosen in the range from 5 to 10). The typical procedural steps are given below:

1. Choose $\alpha_{min} = 0$ and $\alpha_{max} = \infty$
2. Compute $g'(0) = \nabla g^T(x) d_k$
3. Compute $g(\alpha) = g(x + \alpha d_k)$
   (a) If $g(\alpha) \leq g(0) + m_1 \alpha g'(0)$, go to step 4
   (b) Otherwise set $\alpha_{max} = \alpha$ and go to step 6
4. Compare $g(\alpha)$ and $g(0) + m_1 M \alpha g'(0)$
   (a) If $g(\alpha) > g(0) + m_1 M \alpha g'(0)$:END
   (b) Otherwise go to step 5
5. Set $\alpha_{min} = \alpha$
6. Look for a new value of α in the interval $[\alpha_{min}, \alpha_{max}]$ and return to step 3

Note that it is important to test these two relations in order to ensure the property of descent. It is seen from the experience that the starting values of α will improve the efficiency of the algorithm.

5. RESULTS

In order to demonstrate and validate the proposed algorithm, several numerical examples have been taken from the literature. A few of them have been discussed here.

Example (1). The following failure function is taken from Ref. [15] for testing the algorithm for a nonlinear performance function.

$$g(x) = x_1^3 + x_2^3 - 18$$

where x_1 and x_2 are normally distributed random variables. Here, two cases are considered. In case 1, the mean and standard deviations are set to *10* and *5* respectively for both the variables where as in case 2, the mean value of x_2 is set to 9.9 and all other parameters remain unaltered.

Example (2). In another example, a failure function takes the following form

$$g(x) = x_1^2 + x_2 - 8.757$$

where x_1 and x_2 are normally distributed random variables with mean values and standard deviations $\mu_1 = \mu_2 = 10.0$ and $cov_1 = cov_2 = 0.3$ respectively.

Example (3). Another example related to a cantilever beam taken from Ref. [1] defines the performance function as

$$g(x) = x_1 x_2 - 2000 x_3$$

with x_1 and x_2 are normally distributed random variables and x_3 is a lognormal random variable. The mean and standard deviation of the variables are *(0.32, 0.032)*, *(1400000, 70000)* and *(100, 40)* respectively.

Example (4). In order to test the algorithm for nonnormal distributions, another example has been taken from [4]. The limit state function takes the following form

$$g(x) = x_1 x_2 - 78.12 x_3$$

in which x_1, x_2 and x_3 are random variables with mean values $x_1 = 2.0 \times 10^7$, $x_2 = 1.0 \times 10^{-4}$, $x_3 = 4.0$ and standard deviations $\sigma_1 = 0.5 \times 10^7$, $\sigma_2 = 0.2 \times 10^{-4}$, $\sigma_3 = 1.0$. Here two cases have been considered. In case 1, x_1 and x_2 follows normal distributions where as x_3 follows Type-I extreme value distribution. In case 2, x_1 and x_2 follows lognormal distributions where as x_3 follows the same Type-I extreme value distribution.

The results of these examples have been summarized in Table 1.

Table 1: Comparison of results

Examples	Present algorithm		HL-RF method		Literature	
	Reliability Index	Iterations	Reliability Index	Iterations	Reliability Index	Iterations
Example1 & case1	2.24009	7	2.239	49	2.2401 [15]	7
Example1 & case2	2.22588	7	2.225	49	2.2260 [15]	8
Example2	3.05857	6	3.059	33	3.0588 [15]	5
Example3	2.19109	4	2.191	6	2.1920 [1]	4
Example4 & case1	3.31819	13	3.322	12	3.3214 [4]	9
Example4 & case2	4.45272	8	4.428	8	4.4271 [4]	4

6. CONCLUSIONS

Several optimization algorithms have been studied in order to determine their suitability and performance measures to various types of structural reliability problems. A brief review of these methods has been given in this paper. In order to improve the performance of the modified HL-RF method several step length selection rules were studied and implemented in this work. It is observed that the variant of Goldstein's rule which is popularly known as Armijo rule is one of the most efficient optimal step length selection rule. The proposed algorithm has been validated from several examples taken from the literature.

It is clear from the study that an appropriate step length selection rule will not only improve the efficiency of the algorithm but also makes it more robust. Also the initial value of step length has a major effect on the performance of the algorithm. It is found from the experience that a step length of 0.1 is a good initial value for this algorithm. However, a value of 0.5 makes the algorithm converge within a few iterations for problems with less complexity. Most of the problems were converged in about less than 10 iterations. However, only a few cases where some perturbation was imposed took more than 10 iterations. In comparison with the HL-RF method, the modified HL-RF method is more robust. The tolerance imposed on the merit function and the $g(\ )$ function is 0.0001. In general, the overall performance of the modified HL-RF algorithm depends on the step length selection rule and initial value of the step length.

7. REFERENCES

1. R. Ranganathan, *"Structural Reliability: Analysis and Design"*, Jico Publishing House, 2000.
2. Terje Haukaas, *"Structural Reliability Analysis in OpenSees"*, August 21, 2001.
3. Achintya Haldar & Sankaran Mahadevan, *"Reliability Assessment using Stochastic Finite Element Analysis"*, John Wiley & Sons Inc., 2000.
4. Liping Wang & R. V. Grandhi, *"Safety Index Calculations using Intervening Variables for Structural Reliability"*, Computers & Structures, Vol. 59, No. 6, pp 1139-1148, 1996.
5. Jae-Ohk Lee, Young-Soon Yang & Won-Sun Ruy, *"A Comparative Study on Reliability Index and Target-Performance-based Probabilistic Structural Design Optimization"*, Computers & Structures, Vol. 80, (2002), pp 257-269.
6. Pei-Ling Liu & Armen Der Kiureghian, *"Optimization Algorithms for Structural Reliability"*, Structural Safety, 9 (1991) 161-177.
7. Jean-Pierre Dussault, *"Convergence of Implementable Descent Algorithms for Unconstrained Optimization"*, University of Sherbrooke, Canada, April 4, 2000.
8. Manolis Papadrakakis & Nikos D. Lagaros, *"Reliability-based Structural Optimization using Neural Networks and Monte Carlo Simulation"*, Computer Methods in Applied Mechanics and Engineering, Vol. 191, pp. 3491-3501, 2002.
9. R. E. Melchers, *"Structural Reliability Analysis and Prediction"*, Ellis Horwood, 1987.
10. Rudiger Rackwitz & Bernd Fiessler, *"Structural Reliability under Combined Random Load Sequences"*, Computer & Structures, Vol.9, pp 489- 494,1978.
11. Kyung K. Choi & Byeng D. Youn, *"Hybrid Analysis Method for Reliability-based Design Optimization"*, Proceedings of DETC'01 ASME 2001 Design Engineering Technical Conference and Computers and Information in Engineering Conference, Pittsburgh, Pennsylvania, September 9- 12, 2001.
12. Sharif Rahman, *"Probabilistic Fracture Mechanics: J-estimation and Finite Element Methods"*, Engineering Fracture Mechanics, Vol. 68, pp. 107-125, 2001.
13. Chee Keong Kwoh, Gul Nawaz Khan & Duncan Fyfe Gillies, *"Automatated Endoscope Navigation and Advisory System from Medical Imaging"*, International Proceedings of SPIE, Vol. 3660, 1999.
14. Ioannis Akrotirianakis & Berc Rustem, *"A Globally Convergent Interior Point Algorithm for General Non-linear Programming Problems"*, Technical Report 97/14, November 1997.
15. X. Yu, K. K. Choi & K. H. Chang, *"Reliability and Durability Based Design Sensitivity Analysis and Optimization"*, Technical Report, R97- 01, January 1997.

Reliability, Safety and Hazard
Editors: P.V. Varde et al.

SERVICE LIFE PREDICTION OF EXISTING REINFORCED CONCRETE STRUCTURES IN MARINE ENVIRONMENT

A. Srividya, and Satish B. Allampallewar

Reliability Engineering Group and Indian Institute of Technology Bombay
Mumbai 400 076

ABSTRACT

Problem related to the fundamentals of durability caused by chloride corrosion of reinforced concrete structures is discussed. The corrosion process has two stages, the initiation time (t_i) including depassivation time and the corrosion propagation time (t_p). The total service life of existing RC structure can be expressed as $t_s = t_i + t_p$. Ramp-type surface concentration is the idealization for exponential variation of surface chloride concentration with time. In the literature, exponential variation is obtained from regression analysis of field data of bridge decks during biennial inspections over a period of 15 years in snow-belt region. A simple mathematical model to predict the corrosion initiation time of reinforced concrete structure for ramp type surface concentration is proposed in this paper. The results found by this model compare well with those published using the complicated mathematical model requiring the use of software to solve numerical algorithms. Standard deviation of service life is determined based on previous studies to account for variation in cover depth and diffusion coefficient and reliability index using FORM is calculated.

Keywords: Corrosion, Chlorides, Concrete, Service life

1. INTRODUCTION

Reinforced concrete structures are most widely used through out the world. These structures may require to withstand adverse environmental conditions such as marine and chemical like building in the chemical industry, foundations in the chemically aggressive ground water, or due to use of contaminated construction materials.

As time elapses concrete structures deteriorate naturally affecting aesthetic, functional and structural aspects in service. The rate of deterioration depends on factors such as planning, design, construction, operation/use, and maintenance. Degradation of concrete is related to it's durability aspect and of reinforcing steel to it's corrosion, which leads to irrecoverable losses and collapse of structures. Hence, the safety of concrete structures in service should be considered.

Reinforcement corrosion starts once the concentration of chloride ions at the reinforcing steel surface exceed a threshold value (critical chloride content). Literature survey have shown that this value can vary widely and is a probabilistic function rather than being fixed or constant. Corrosion has been observed to initiate at chloride concentrations from 0.1 to 1 % by weight of cement. Such a wide variation causes service life estimates to differ, depending upon model input parameters. This variation or uncertainty affects the validity of the choice and project scheduling of maintenance, repair, rehabilitation alternatives. For example, if the chloride concentration at the steel depth already approaches or exceeds the critical value, then any maintenance or repair action to restrict the chloride ingress is less effective or ineffective, and the associated resources are wasted. Hence, a complete evaluation method for effective calculation of service life is necessary. Service life or reliability index is useful for owner, designer, or to a organization to take decision in time of priority of repairs, repair strategy, corrosion protection, sell, purchase, or demolish. In order to optimize maintenance planning and budgeting, planned maintenance at the optimum time is the safest and most cost effective approach. The approach needs up to date information when the structure will need maintenance and repair and hence required to predict service life.

In literature [1] measured core data (Weyers et al. 1994) for surface chloride accumulation during bridge biennial inspections over a period of 15 years is used. With regression data has been modeled by exponential variation of surface chloride accumulation with time. The same measured data for the surface chloride is modeled by a ramp-type variation and a transcendental equation to predict the corrosion initiation time (t_i) is presented [2], and for numerical solution use of MATLAB 6.1 is suggested.

In this paper, a simple mathematical equation is proposed to predict corrosion initiation time (t_i) for ramp-type surface chloride accumulation. Corrosion initiation time is calculated and compared with the results of the literature. Standard deviation of Corrosion initiation time to account for variation in cover depth and diffusion coefficient based on previous studies is presented by using FORM.

2. CORROSION OF STEEL IN CONCRETE

The corrosion process that takes place in concrete is electrochemical in nature. Corrosion will result in flow of electrons between anodic (where corrosion occurs and current flows from) and cathodic (where no corrosion occurs and current flows to) sites on the steel reinforcement. Concrete is capable of conducting electric current by ionic current flow and act as electrolyte. Connection between the anode and cathode, allows current return and completes the circuit. Reinforcing steel in concrete normally does not corrode because of the formation of a passive oxide film on the surface of the steel due to the initial corrosion reaction. The hydration process of cement in freshly poured concrete develops a high alkalinity, which in the presence of oxygen stabilizes the film on the surface of embedded steel, ensuring continued protection while the alkalinity is retained. Normally, concrete exhibits a pH above 12 because of the presence of calcium hydroxide, potassium hydroxide, and sodium. Although the precise nature of this passive film is unknown, it isolates the steel from the environment and slows further corrosion as long as the film is intact. However Carbonation and chloride contamination are two major situations in which corrosion of reinforcing steel can occur.

2.1 Chloride corrosion of steel

Chloride ingress in the concrete can be from de-icing salts that are applied to the concrete surface or from seawater in marine environments. If chlorides are present in sufficient quantity, they disrupt the passive film and subject the reinforcing steel to corrosion. The levels of chloride required to initiate corrosion are extremely low. The process of deterioration of concrete structures [3] due to chloride corrosion consist of:

Incubation stage: Chloride ion ingress into cover concrete and are accumulated around the reinforcing steel. It's duration is determined mainly by the diffusion rate of chloride ion within concrete. This stage is absent when content of chlorides attributable to concrete materials is high.

Development stage: In this stage, the reinforcing steel begins to corrode. Corrosion products accumulate and cracks are generated in cover concrete due to the expansion pressure of the product. Dominant factors are the quantity of oxygen in the dissolved state, the moisture present, and the electrical resistance of the concrete.

Acceleration stage: The rate of corrosion is accelerated by the generation of cracks along the reinforcing steel (longitudinal cracks). Delaminating and peeling-off of cover concrete start. The dominant factors are same as in the development stage, but loading actions also influence the process. While the static strength does not decrease even when longitudinal cracks have been generated, the strength and ductility begin to be reduced if repeated loading at high stress is applied.

Deterioration stage: Corrosion of reinforcing steel advances and the reduction in cross-sectional area and strength becomes conspicuous. The dominant factors are almost the same as in the acceleration stage.

2.2 Predicting the corrosion initiation time:

The initiation period, t_i, defines the time it takes for sufficient chlorides to penetrate the concrete cover and accumulate in sufficient quantity at the depth of the embedded steel to initiate corrosion i.e. the time taken for the critical threshold concentration, c_{th} , to reach the depth of cover, d.

The diffusion of the chloride ions in the uncracked RC concrete is governed by the equation,

$$\frac{dC}{dt} = D * \frac{d^2C}{dx^2} \quad (1)$$

2.2.1 Constant surface chloride concentration:

Chloride accumulation at the surface is assumed same and equal to it's maximum value, $C_s = C_o$. The t_i is given as:

$$t_i = \frac{d^2}{4*D} * \frac{1}{\left[erf^{-1}\left(1 - \frac{C_{th}}{C_o}\right)\right]^2} \quad (2)$$

2.2.2 Ramp-type surface chloride concentration:

The ramp-type is the idealization for exponential variation (derived from regression of field data) of surface concentration with time. In this type, surface chloride concentration is assumed to increase uniformly with age from zero (t=0) to a maximum value at some time t_o ; and then remaining constant for $t \geq t_o$. A transcendental equation [2] to predict the corrosion initiation time is given as:

$$g(x,t) = \frac{t}{to}\left[\begin{array}{c} -\frac{x}{\sqrt{\pi * Do * t}} \exp\left(-\frac{x^2}{4*t*Do}\right) + \\ \left(1 + \frac{x^2}{2*Do*t}\right) erfc\left(\frac{x}{\sqrt{4*Do*t}}\right) \end{array}\right] \quad (3)$$

The time (year) predicted to initiate corrosion of the reinforcement steel at depth x = d is computed by finding the roots of Eq.3 for available data i.e. C_o, C_{th}, t_o, d, D_o

$$\frac{C_{th}}{C_o} = g(x,t), \; t \leq t_o$$

$$\frac{C_{th}}{C_o} = g(x,t) - g(x, t - t_0), \; t \geq t_o \quad (4)$$

The roots of the Eq. 4 can be obtained numerically using *Mat Lab 6.1* with the bisection method [2]. It is suggested that, in extracting the roots of Eq. 4, the search process is required to be terminated when the calculated function is small enough to ensure accurate roots.

2.2.3 Proposed mathematical model

Proposed simple equation to predict the corrosion initiation time for ramp-type surface chloride concentration ($t_{i,ramp}$) is given as the sum of constant corrosion initiation time (Cc) and half the time required to reach the maximum surface chloride concentration (t_o)

$$t_{i,ramp} = (Cc + 0.5 * t_o) \quad (5)$$

Results obtained for the data, C_{th} = 52.362, and C_o = 5.818, (both in N/m^3) [2] by using Eq. 2 for constant chloride concentration, and, by using Eq. 4 (results previously obtained [2] and (5) are tabulated in Table 1.

Variation of corrosion initiation time (Rc-P) with diffusion coefficient for various values of cover depth, from the set of data in Table 1, is shown in Figure 1.

2.2.4 Reliability index (β): Reliability index is obtained for the data x = 44.5 mm, Do=38.7 mm^2/year, C_{th} = 5.818 N/mm^2, Co=52.362 N/mm^2, and for range of t_i equal to five to fourteen years by using software COMREL 7.1. Results are tabulated in Table 2 and a plot of variation of ß with t_i is shown in Figure 2

2.2.5 Combination of statistical and deterministic models: First order second moment method [5] is used to predict mean corrosion initiation time of RC structure. Mean values are used for the parameters in deterministic model to calculate mean corrosion initiation time. The standard deviation of the corrosion initiation time (σ) to account for variation in cover depth and diffusion coefficient only, is determined. Standard deviation value for cover depth and diffusion coefficient considered are 8.9 mm and 9.675mm^2/year (i.e. 20 % and 25 %) respectively. Values of (σ) are shown in the Table 1.

2.3 Propagation time (t_p):

The propagation period, t_p is the time necessary for sufficient corrosion to occur to cause an unacceptable level of damage to the structure or element under consideration. The length of this period depends on unacceptable damage level and rate of corrosion process.

Table 1 Time Predicted to Initiate Corrosion (Year) – Constant Diffusivity (Do)

Do	X	Time to initiate corrosion			
		Deterministic model			FORM
		Cc	Rc-L	Rc-P	σ
38.7	44.5	10.08	12.6	12.579	4.754178
	50.8	13.13	15.7	15.635	5.653627
	57.1	16.59	19.1	19.094	6.631093
	63.5	20.52	23	23.023	7.708417
	69.8	24.8	27.3	27.297	8.855465
	76.2	29.55	32.1	32.053	10.11155
51.6	44.5	7.559	10.1	10.059	3.339358
	50.8	9.851	12.4	12.351	3.914841
	57.1	12.45	15	14.946	4.527516
	63.5	15.39	17.9	17.892	5.19089
	69.8	18.6	21.1	21.098	5.886696
	76.2	22.16	24.7	24.665	6.639175
64.5	44.5	6.047	8.6	8.5473	2.583409
	50.8	7.881	10.4	10.381	3.003763
	57.1	9.957	12.5	12.457	3.444459
	63.5	12.31	14.8	14.814	3.914841
	69.8	14.88	17.4	17.378	4.401865
	76.2	17.73	20.2	20.232	4.922501
77.4	44.5	5.039	7.6	7.5394	2.111901
	50.8	6.567	9.1	9.0673	2.443189
	57.1	8.297	10.8	10.797	2.786717
	63.5	10.26	12.8	12.761	3.149465
	69.8	12.4	14.9	14.899	3.52122
	76.2	14.78	17.3	17.276	3.914841
90.3	44.5	4.32	6.8	6.8195	1.78871
	50.8	5.629	8.2	8.1291	2.062561
	57.1	7.112	9.7	9.6119	2.344314
	63.5	8.796	11.3	11.296	2.639475
	69.8	10.63	13.2	13.127	2.939605
	76.2	12.67	15.2	15.166	3.254983
103.2	44.5	3.78	6.3	6.2796	1.552794
	50.8	4.925	7.5	7.4255	1.786566
	57.1	6.223	8.8	8.7229	2.025719
	63.5	7.696	10.2	10.196	2.274779
	69.8	9.299	11.8	11.799	2.526525
	76.2	11.08	13.6	13.582	2.789496
116.1	44.5	3.36	5.9	5.8596	1.372699
	50.8	4.378	6.9	6.8782	1.576886
	57.1	5.531	8.1	8.0315	1.784898
	63.5	6.841	9.4	9.341	2.000566
	69.8	8.266	10.8	10.766	2.217564
	76.2	9.851	12.4	12.351	2.443189
129	44.5	3.024	5.5	5.5237	1.230537
	50.8	3.94	6.5	6.4404	1.411962
	57.1	4.978	7.5	7.4783	1.596198
	63.5	6.157	8.7	8.6569	1.786566
	69.8	7.439	10	9.9391	1.97743
	76.2	8.866	11.4	11.366	2.175159

Unacceptable damage level varies with the requirements of owner and the nature of the structure. Corrosion rate will be influenced by nature of the embedded metal, properties of the surrounding concrete and the composition of the pore solution within the concrete, and the environmental conditions (particularly temperature and moisture availability).

When the total amount of corrosion product reaches the critical amount of rust products, the internal expansion force will exceed the tensile strength of concrete and cause the cracking of the cover concrete. The time measured from corrosion initiation to the cracking is termed as propagation time. Tensile strength of concrete depends on the tensile stress and area of cracked surface. Tensile stress varies with concrete grade whereas area of cracked surface is dependant on location and spacing of reinforcement, and cover concrete. In a structure placing of reinforcement not only varies with member but also it may vary within a member at different locations. The propagation period which is minimum in a structure is required to be used. In view of the complexity of the corrosion process computer program Life-365 [6] specifies a fixed value of 6 year for black steel and 20 years for epoxy coated steel. Cady and Weyers [7] specifies 2 to 5 years for black steel.

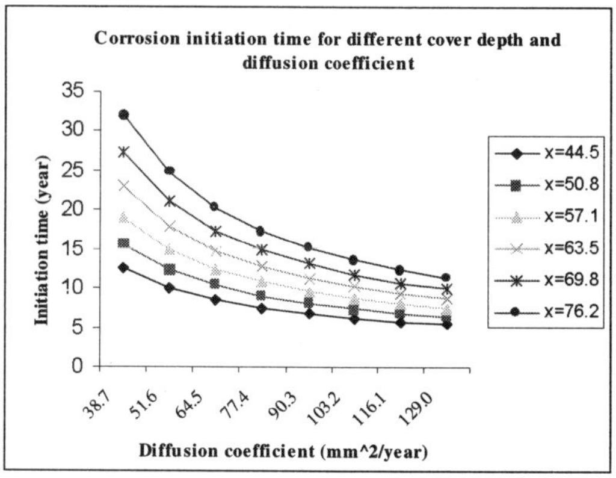

Fig. 1: Variation of initiation time with different diffusion coefficient for different cover depth.

Service life (t_s)

Total service life of existing RC structure is given as,

$$t_s = t_i + t_p \quad (6)$$

Residual life (t_r)

Residual life of existing RC structure is given as,

$$t_r = (t_i + t_p) - a \quad (7)$$

Discussion of results

Comparing the results in table 1, of corrosion initiation time for ramp-type surface chloride concentration, it is observed that the proposed simple method gives accurate results for the same data. Figure 1 shows decrease in initiation time, with increase in diffusion coefficient and with decrease in cover depth. Rate of decrease of t_i with increasing diffusion coefficient is more when higher cover depth is provided.
From Table 2 and Figure 2 it is observed that reliability index reduces with increasing time. Change in the sign of ß corresponds to corrosion initiation time and set in of corrosion propagation.

Table 2 Reliability index (ß) for different Time to Initiate Corrosion (Year) – Constant Diffusivity (D_o)

t_i	β
5.000	1.360
6.000	1.034
7.000	0.7420
8.000	0.4780
9.000	0.2374
10.00	0.1662E-01
11.00	-0.1868
12.00	-0.3749
13.00	-0.5494
14.00	-0.7117

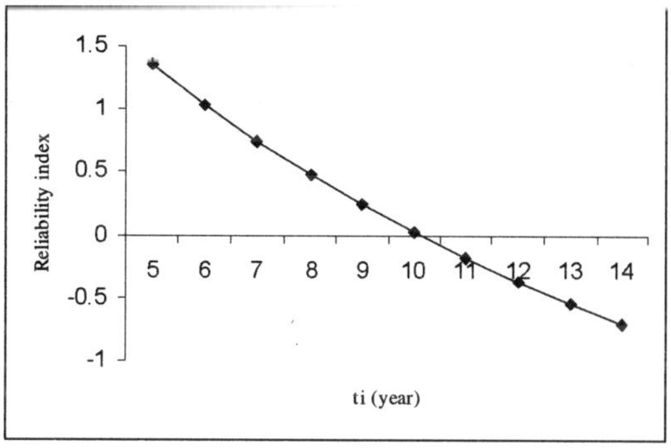

Fig. 2: Variation of ß with t_i (year)

Nomenclature:

t_s	*service life, year*
C	*chloride content*
D, D_o	*apparent diffusion coefficient,* $mm^2/year$
x	*depth , from the exposed surface, mm*
t	*time, year*
C_s	*surface chloride concentration*
t_i	*corrosion initiation time ,year*
t_p	*propagation time*
d	*cover depth, mm*
C_{th}	*chloride threshold*
C_o	*max. Surface chloride concentration*

Erf( ) *error function[4]*
erf^{-1}( ) *inverse standard error function*
erfc( ) *complement of error function erf( )*
t_o *time to reach the max. Surface chloride concentration (measured from t=0)*
$t_{i,ramp}$ *ti, for ramp-type chloride concentration*
σ *standard deviation of t_i*
Cc *t_i, for constant chloride concentration*
Rc-L *$t_{i,ramp}$, from literature[2]*
Rc-P *$t_{i,ramp}$, obtained from proposed equation (5)*
t_r *residual life*
a *age of RC structure*

3. REFERENCES

1. Mumtaz K. Kassir, Michel Ghosn, (2002) *Chloride-induced corrosion of reinforced concrete bridge decks*, Cement and concrete research, 32 , 139-143.
2. P. Phurkhao and M. K. Kassir, (2005), *Note on Chloride induced corrosion of reinforced concrete bridge decks*, Journal of engineering mechanics, ASCE, **131** , 97-99.
3. T. Miyagawa, (1991) *Durability design and repair of concrete structures: Chloride corrosion of reinforcing steel and alkali-aggregate reaction*, Magazine of concrete research, 43, no. 156, 155-170.
4. Korn A. and Korn *M.*, (1968) *Mathematical handbook for scientists and engineers*, (2nd edition, Mc Graw Hill).
5. Andrzej S. Nowak, Kelvin R. Collins, (2000) *Reliability of structures*, 5, 101, (International edition, McGraw Hill Book Co. Singapore, 101).
6. Thoms, M. D. A. and Bentz E. C., (2000), manul *Computer program for predicting the service life and life cycle costs of reinforced concrete exposed to chlorides, 13* .
7. Cady, P. D. and Weyers, R. E., (1984) *Deterioration rate of concrete bridge decks; J.* Transp. Eng., 110(1), 34-44.

Reliability, Safety and Hazard
Editors: P.V. Varde et al.

SYSTEM RELIABILITY IN FUZZY ENVIRONMENT - AN APPLICATION TO SLOPES

Vageesha S. Mathada[1], G. Venkatachalam[2], and A. Srividya[1]

[1]Reliability Engg,. IIT Bombay, Mumbai 400 076
[2]Emeritus Fellow, Civil Engg., IIT Bombay, Mumbai 400 076

ABSTRACT

Landslides and slope failures represent a challenge to geotechnical engineers, as the failure is associated with devastating damage to property, transportation, communication network, and even loss of life. Slope stability assessment is a daunting task as it encounters various uncertainties. Therefore, possibilistic reliability approach can be used to compute fuzzy failure probabilitites (possibilities) for determining safety of the slope in such circumstances. The slope stability problem can be considered as a system with many failure surfaces, calculation of system reliability in fuzzy environment is also important by considering correlation between different failure surfaces. This paper addresses the issue of calculation of system failure possibility bounds by extending the philosophy of random variables to fuzzy variables. The proposed method has been applied to calculate the system failure possibility bounds for the congress street cut. The results show that $\beta_{Possibilistic}$ is greater than $\beta_{Probabilistic}$ on different surfaces and the range of system bounds also increases when the uncertain parameters are fuzzy.

Keywords: Possibilistic Reliability, Fuzzy sets, System Reliability, Slope Stability

1. INTRODUCTION

The instability of both man-made and natural slopes constitute majority of failures in geotechnical engineering. These catastrophic slope failures are associated with devastating damage to property, transportation, communication network, other utilities, and even loss of life. The assessment of stability of a slope is an important issue, based on which, design and mitigation works are undertaken. There is increasing awareness among geotechnical engineers, after almost two decades of assessment, of the fact that significant uncertainties are associated with slope stability problem and that these uncertainties often cannot be removed by reasonable investigative effort or expenditure. Analysis has to be performed with limited or imperfect information. The probabilistic approaches consider uncertain parameters to follow the characteristics of random uncertainty. However, not all uncertainties are random and objectively quantifiable. Some uncertainties, especially those based on incomplete information are due to cognitive sources (Zimmermann, 1991). Accordingly, fuzzy sets / fuzzy numbers were used to represent uncertainty in geotechnical engineering (Nawari & Liang, 2000; Juang et al., 1998; Dodagoudar & Venkatachalam, 2000; Giasi et al., 2003). Recently, Mathada et al. (2005) proposed a stable and convergent method for possibilistic reliability analysis for slopes.

2. RELEVANCE OF SYSTEM RELIABILITY TO SLOPE STABILITY

The overall reliability of any geotechnical or structural system with several components will depend not only on the reliabilities of the individual components but also on other factors including correlations between different components or elements of the system. Similarly, the overall reliability of a geotechnical system with several modes of failure will depend on reliability considering individual modes of failure as well as the correlations between the different failure modes. The importance of 'combinatorial' reliability in geotechnical engineering was shown earlier by Harr (1977). The calculation of system reliability in geotechnical engineering is not that easy. However, one may simplify the problem by considering it as a simple series system or a parallel system or a combination of both. The assumption of series system leads to significantly high failure probabilities, whereas the assumption of a parallel system leads to extremely low failure probabilities.

The characterization of a complex geotechnical problem as a series or parallel system is not justified. It is difficult to obtain exact mathematical solutions to the probability of failure or reliability of a system. This can be tackled appropriately by considering the 'Upper bound' and the 'Lower bound' to the mathematically exact probability of failure or reliability. Cornell (1967) and Moses & Kinser (1967) originally proposed solutions for such bounds for structural systems. The usefulness of these bounds depends on how close they are to each other and the bounds presented by Ditlevson (1979) gives the solution for narrow reliability bounds.

In soil slope stability problems the search for the critical slip surface on the basis of the conventional factor of safety is done by locating the slip surface with least factor of safety. A similar search is done to locate the critical probabilistic surface with high probability of failure. The same concept can be extended to get the critical possibilistic surface with highest possibility of failure. Mathada et al. (2005) have suggested a method to calculate possibilistic reliability ($\beta_{Possibilistic}$) for slopes.

The minimum factor of safety or the minimum possibilistic reliability based on a single slip surface means that only one mode of potential failure has been considered. There are many possible failure surfaces along which the slope can fail in reality, and considering it as a system with several potential failure modes is appropriate. The overall probability of a slope with many potential slip surfaces can be greater than the probability of failure considering any individual slip surface (Cornell, 1971; Chowdhury & Xu, 1995). This is true even with possibilities. The difference depends on the correlation between the failure possibilities of different slip surfaces, for which no general formulation is available for fuzzy uncertain parameters. This paper extends the concepts of system probability of failure to system possibility of failure with fuzzy uncertainties.

3. PROBABILISTIC SYSTEM RELIABILITY

3.1 Unimodal (Simple) Bounds

Cornell (1967) has established simple bounds on the reliability of structural systems subjected to k failure modes. The assumption that all failure modes are perfectly correlated yields the upper bound reliability (or safety) p_s as

$$p_s = 1 - \max_{i(i=1,k)} (p_{f_i}) \tag{1}$$

where p_{fi} is the probability of failure of ith failure mode. The assumption of all the failure modes to be statistically independent yields the lower bound as

$$p_s = \prod_{i=1}^{k} (1 - p_{f_i}) \tag{2}$$

Hence, the reliability of the system is bounded by

$$\prod_{i=1}^{k} (1 - p_{f_i}) \le p_s \le 1 - \max_{i=1,k} (p_{f_i}) \tag{3}$$

If $p_{fi} << 1$, then Eqn (3) can be approximated by

$$\max_{i=1,k} (p_{f_i}) \le p_{fs} \le \sum_{i=1}^{k} p_{f_i} \tag{4}$$

3.2 Bimodal (Narrow) Bounds

The assumption of perfect correlation or no correlation between failure modes is not proper. The modes are usually are positively correlated. Ditlevson (1979) has developed narrow bounds for structural system failure probability through indicator function algebra. Suppose that the performance function in each of the k failure modes is represented by

$$G_j(X) = G_j(X_1, X_2, \ldots X_n); \; j = 1,2,\ldots\ldots, k \quad (5)$$

such that the individual failure events are

$$E_j = [G_j(X) < 0] \quad (6)$$

The lower bound on p_{fs} is

$$p_{f_s} \geq p_{f_1} + \sum_{i=2}^{k} max\left[p_{f_i} - \sum_{j=1}^{i-1} P(E_i \cap E_j), 0\right] \quad (7)$$

and the upper bound on p_{fs} is

$$p_{fs} \leq \sum_{i=1}^{k} p_{f_i} - \sum_{i=2, j<i}^{n} max \; P(E_i \cap E_j) \quad (8)$$

where $P(E_i \cap E_j)$ in equations (7) and (8) is the joint probability of events E_i and E_j.

The joint probability $P(E_i \cap E_j)$, may be approximated as follows:

For the lower bound (Eqn.(7))

$$P(E_i \cap E_j) = P(A) + P(B) \quad (9)$$

and for the upper bound (Eqn.(8))

$$P(E_i \cap E_j) = max[P(A), P(B)] \quad (10)$$

where $P(A)$ and $P(B)$ in equations (9) and (10) are given by

$$P(A) = \Phi(-\beta_i)\Phi\left[-\left(\frac{\beta_j - \rho_{ij}\beta_i}{\sqrt{1-\rho^2_{ij}}}\right)\right] \quad (11)$$

$$P(B) = \Phi(-\beta_j)\Phi\left[-\left(\frac{\beta_i - \rho_{ij}\beta_j}{\sqrt{1-\rho^2_{ij}}}\right)\right] \quad (12)$$

In equations (11) and (12), β_i and β_j are the reliability indexes of the events E_i and E_j, respectively and ρ_{ij} is the correlation between these failure modes.

SYSTEM RELIABILITY USING $\beta_{POSSIBILISTIC}$

The concept given in section 3 can be easily extended to get the system reliability when fuzzy uncertainty is present. For unimodal (simple) bounds, the values of p_f in equations (1) – (4) can be replaced by p_f obtained from possibilistic reliability analysis. For narrow bounds, additional effort is required to get the correlation between different failure modes ρ_{ij}. The correlation coefficient between the slip surfaces (potential failure modes) k and l, based on linearization of the performance function using Taylor's series and truncating at second order terms is given by

$$\rho_{kl} = \frac{\sum_{i=1}^{m}\left(\frac{\partial G_k}{\partial X_i}\right)_{X=\bar{X}}\left(\frac{\partial G_l}{\partial X_i}\right)_{X=\bar{X}}\sigma^2_{X_i}}{\sqrt{\left(\sum_{i=1}^{m}\left(\frac{\partial G_k}{\partial X_i}\right)^2_{X=\bar{X}}\sigma^2_{X_i}\right)\left(\sum_{i=1}^{m}\left(\frac{\partial G_l}{\partial X_i}\right)^2_{X=\bar{X}}\sigma^2_{X_i}\right)}} \quad (13)$$

where m is the number of uncertain variables. There are no methods to evaluate partial derivatives of the performance function when uncertain quantities are modelled as fuzzy numbers. A method has been proposed to compute the partial differentials which finally can be used to get the correlation between different slip surfaces. The method of evaluating the partial derivatives with respect to certain fuzzy variable is explained below:

Let $G = f(x_1, x_2, \ldots, x_m)$. We are interested in calculating the partial derivative $\partial G/\partial X_1$ (say). The partial derivative with respect to a variable at any α level is calculated using numerical differentiation technique. The superior and inferior value of the performance function at any α-level is calculated by using the extreme values of the variable X_1 and all other variables are kept at the mean value at that α-level.

$$G(X_{1\alpha+}) = f(x_{1\alpha+}, x_{2\alpha mean}, \cdots, x_{m\alpha mean}) \quad (14)$$

$$G(X_{1\alpha-}) = f(x_{1\alpha-}, x_{2\alpha mean}, \cdots, x_{m\alpha mean}) \quad (15)$$

where

$$x_{i\alpha mean} = \frac{x_{i\alpha+} + x_{i\alpha-}}{2} \qquad i = 1,m \quad (16)$$

Then by numerical differentiation $\left(\frac{\partial G}{\partial X_1}\right)_\alpha$ at any αlevel is given by

$$\left(\frac{\partial G}{\partial X_1}\right)_\alpha = \frac{G_{x1\alpha+} - G_{x1\alpha-}}{x_{1\alpha+} - x_{1\alpha-}} \qquad (17)$$

Then the partial differential $\left(\frac{\partial G}{\partial X_1}\right)$ to be used in Eqn.(13) is calculated with the formula (Mathada et al., 2005) as

$$E\left[\frac{\partial G}{\partial X_1}\right] = \frac{1}{N}\left[\left(\frac{\partial G}{\partial X_1}\right)_{\alpha_N} + 2\sum_{i=1}^{N-1}\alpha_i\left(\frac{\partial G}{\partial X_1}\right)_{\alpha_i}\right] \quad (18)$$

where N is the number of α-cuts and

$$\alpha_i = \frac{i}{N}, \qquad i = 1, N \qquad (19)$$

The same procedure can be applied to get partial derivatives with respect to other variables. When once these partial derivatives are obtained correlation between different failure surfaces can be obtained using Eqn. (13). The narrow bound system possibilities are calculated subsequently using equations (7) – (12). The procedure is explained with the help of an illustrative example.

4. ILLUSTRATIVE EXAMPLE

The geometry of the slope is as shown in Fig. 1 and this example is the well known congress street cut (Ireland, 1954). It is also the case history analyzed by Chowdhury & Xu (1995). The clay deposit is divided into three layers following Ireland and others. One set of mean values and standard deviations of the undrained shear strength are shown in Table 1.

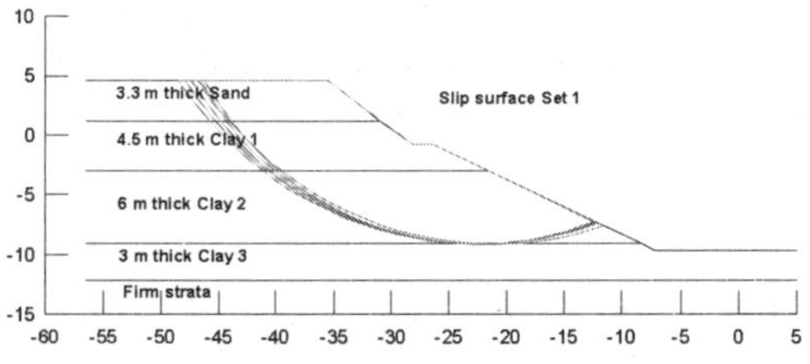

Fig. 1: Cross section of Congress street cut with Set 1 slip surfaces.

Table 1 Shear Strength Properties for soils in the Congress street cut

	Mean (kPa)	Standard Deviation(kPa)	CoV(%)
Clay 1	55.0	20.4	37
Clay 2	43.0	8.2	19
Clay 3	56.0	13.2	24

The set of slip surface tangential to bottom of clay 2 is denoted as Set 1 (Fig. 1) and the set of slip surfaces tangent to bottom of clay 3 is denoted as Set 2 (Fig.2) similar to Chowdhury & Xu (1995). The top layer of sand has negligible influence on stability because of zero cohesion and low normal stresses and has therefore, been neglected as in the analysis of Ireland (1954) and Chowdhury & Xu (1995).

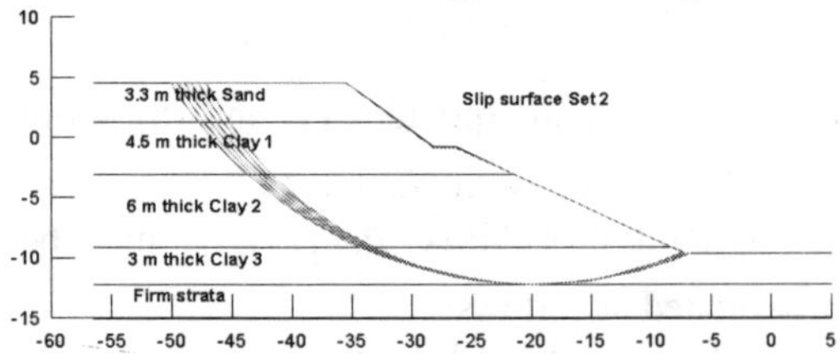

Fig. 2: Cross section of Congress street cut with Set 2 slip surfaces.

Gaussian fuzzy numbers with the characteristics given in Table 1 are used to characterize the uncertain properties as fuzzy uncertainties. The construction of Gaussian fuzzy numbers is done as given in Fig. 3. $\beta_{Possibilistic}$ for a given slip surface is calculated using the method suggested by Mathada et al. (2005). No correlation between the uncertain parameters is considered.

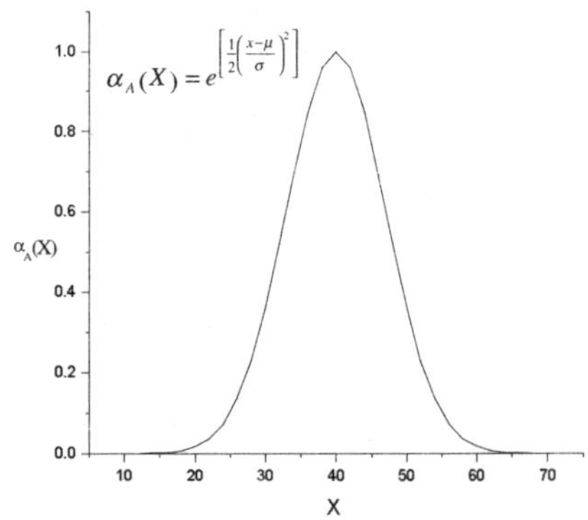

Fig. 3: Gaussian fuzzy number

5. RESULTS

The simple bounds for failure possibility obtained are shown in Table 2. Table 3 gives the narrow bounds for failure possibility.

Table 2 System failure possibilities (Simple Bounds)

	Condition	Failure Possibility
Layer 2 only (Set 1, 6 surfaces)	Critical surface	0.266465
	Lower bound	0.266787
	Upper bound	0.842587
Layer 3 only (Set 2, 6 surfaces)	Critical surface	0.318735
	Lower bound	0.318735
	Upper bound	0.897350
Layer 2 and 3 (12 surfaces)	Critical surface	0.318735
	Lower bound	0.318735
	Upper bound	0.983842

Table 3 System failure possibilities (Narrow Bounds)

	Condition	Failure Possibility
Layer 2 only (Set 1, 6 slip surfaces)	Critical slip surface	0.266465 (0.26592)[a]
	Lower bound	0.266787 (0.26592)[a]
	Upper bound	0.312215 (0.26592)[a]
Layer 3 only (Set 2, 6 slip surfaces)	Critical slip surface	0.318735 (0.27389)[a]
	Lower bound	0.318735 (0.27389)[a]
	Upper bound	0.320652 (0.27389)[a]
Layer 2 and 3 (12 slip surfaces)	Critical slip surface	0.318735 (0.27389)[a]
	Lower bound	0.318735 (0.27389)[a]
	Upper bound	0.528416 (0.44733)[a]

[a] Failure probabilities reported by Chowdhury & Xu (1995) considering the same variables as random on a similar set of surfaces.

7. DISCUSSION AND CONCLUSIONS

The difference between the upper and lower bounds in case of simple bounds is so large to make any judgement as can be seen in Table 2. The lower bound is matching with failure possibility of the critical slip surface. When it comes to narrow bounds, the range between upper and lower bound possibilities is less in all the three cases. When Set 1 and Set 2 are considered separately the difference between the upper and lower bound possibilities is found to be 4.7 % and 0.2 %, respectively. When all the slip surfaces (12 surfaces) are considered the range is around 21%. Similar results for system probabilities were reported earlier by Chowdhury & Xu (1995) considering the uncertainties as random variables on a similar set of surfaces.

The conventional approach of finding reliability for slopes either on a deterministic critical surface or on a critical probabilistic/ possibilistic surface may not lead to a realistic estimate of slope reliability. The slope stability problem must be treated as a system reliability problem with several modes of failures. The correlation between different slip surfaces plays an important role to get realistic estimates on bounds. As many of the uncertainties connected with geomaterials cannot be represented precisely, characterizing them as fuzzy variables appears to be a viable means. The results obtained for the system possibility by the proposed method to get the correlation between slip surfaces with fuzzy uncertainty are in conformity with the consistency principle.

8. REFERENCES

1. Chowdhury, R.N. and Xu, D. W. (1995). "Geotechnical System Reliability of Slopes." *Reliability Engrg. and System Safety*, Vol. 47, pp. 141-151.
2. Cornell, C. A. (1967). " Bounds on the Reliability of Structural Systems." *J. Struct. Div.*,ASCE, Vol. 93, pp. 171-200.
3. Cornell, C. A. (1971). "First-order Uncertainty Analysis of Soils Deformation and Stability." *Proc. of the 1st Int. Conf. Applications of Statistics and Probability in Soil and Struct. Engrg.*, Hong Kong University Press, Hong Kong, pp. 126-144.

4. Ditlevsen, O. (1979). "Narrow Reliability Bounds for Structural Systems." *J. Struct. Mech.* , ASCE, Vol.7, pp. 453-472.
5. Dodagoudar, G. R. and Venkatachalam, G. (2000c). "Reliability Analysis of Slopes using Fuzzy Sets Theory." *Computers and Geotechnics,* Vol. 27, pp. 101-115.
6. Giasi, C. I., Masi, P. and Cherubini, C. (2003). "Probabilistic and Fuzzy Reliability Analysis of a Sample Slope near Aliano." *Engineering Geol.,* Vol. 67, pp. 391-402
7. Harr, M. E. (1977). *Mechanics of Particulate Media – A Probabilistic Approach*, McGraw Hill, New York.
8. Ireland, H. O. (1954). "Stability Analysis of the Congress Street Open Cut in Chicago." *Geotechnique*, Vol 4, pp. 163-168.
9. Juang, C. H., Jhi, Y, -Y. and Lee, D. H. (1998). "Stability Analysis of Existing Slopes Considering Uncertainty." *Engrg. Geol.,* Vol. 49, pp. 111-112.
10. Mathada, V. S., Venkatachalam, G. and Srividya, A. (2005). "A Method for Possibilistic Analysis of Slopes," *Proc. of 9th Int. Conf. on Str. Safety and Reliability (ICOSSAR 05),* MilPress, Rotterdam, Netherlands (CD-ROM).
11. Moses, F. and Kinser, E. (1967). "Analysis of Structural Reliability", *J.Struct Div.*, ASCE, Vol. 1967, pp. 147-164.
12. Nawari, N. O. and Liang, R. (2000). "Fuzzy-based approach for determination of characteristic values of measured geotechnical parameters." *Can. Geotech. J.,* Vol. 37, pp. 1131-1140.
13. Zimmermann, H. J. (1991). *Fuzzy Sets Theory and its Applications,* Kluwer Academic Publishers, Boston.

Reliability, Safety and Hazard
Editors: P.V. Varde et al.

STRUCTURAL PROB ABILISTIC METHODOLOGY FOR MECHANICAL COMPONENTS

Mariappan V.[a], Sakhardande M.J.[a], and Kakodkar U. V.[b]

[a]Mechanical Engineering, Government Engineering College, Farmagudi, Goa
[b]Satyam Computer Services Ltd., Secunderabad, Andhra Pradesh

ABSTRACT

Probabilistic methods incorporated with the various other user friendly tools, unlike traditional methods, provide a means to quantify the inherent risk of a design and to quantify the sensitivities of design variables. This paper explains shortcomings of the current analysis approach and the potential for improvement via incorporation of probabilistic analysis methods. Various user-friendly tools like the COMREL and IDEAS are applied to solve and analyze the design problems. Stress concentrations of the physical object and the optimality of the design variable using FEM are presented.

Keywords: Probabilistic design, CAD tools, Stress analysis

1. INTRODUCTION

Knowing the inherent risk of failure in any design is becoming increasingly important to both the manufacturer and the customer. Designers and management must concern themselves with the ability to assess risk, identify parameters, which drive risk, and minimize the risk given other program constraints. Analysis using probabilistic methods provides a tool for meeting these needs. This section provides an overview of the current structural analysis approach, expounding on its shortcomings and listing potential dilemmas in applying the approach to future designs.

1.1 Need for a Different Approach

Current system design analysis methods [1] do not directly account for the random nature of most input parameters. The result of treating parameters such as material properties, geometry, environment, and loads as singly determined (deterministic) values is a design of unknown reliability, or conversely, unknown risk. Risk is defined as the chance of encountering harm or loss. Virtually all activities contain risk, but we are willing to accept the risk levels, given our life experience. When we elect to participate in these activities, we accept the risk, either overtly or tacitly.

Design process yielding unknown risk poses the following problems for future designs:

1. As designs grow more critical and competitive, there is a need to quantitatively assess and optimize reliability (or minimize risk), given other program constraints.
2. New system developments departing dramatically from traditional environments. Application of historical uncertainty factors may not be sufficient to provide adequate safety. Conversely, the trend to design to all possible unfavorable events occurring simultaneously could produce an unacceptable weight.

2. RELIABILITY BASED APPROACH

All design parameters are treated as variables and the basic result from the analysis is a probability of failure, or risk. The effect of realistic variability of the design input parameters can thus be obtained. Specifically, given the structural analysis methodology in conjunction with statistical characterization of applied loading, geometry, material behavior, and expected environment, the probabilistic structural analysis methodology is capable of producing safety (risk) quantification, design variable sensitivity analysis, cost/weight reduction scenarios and optimum inspection intervals. A coupling equation [2] is to be established according to the situation for the desired reliability level. The typical coupling equation for normally distributed strength S and normally distributed stress, s is given in Equation (1). Equations (1) and (2) are for tensional and torsional elements respectively.

$$R = 1 - \phi\left[-\frac{\pi \bar{r}^{2}\bar{S} - \bar{P}}{\pi \bar{r}^{2}\sqrt{\sigma_S{}^2 + \left\{\frac{\sigma_P{}^2 + 4.44\alpha^2 \bar{P}^2}{\pi^2 \bar{r}^4}\right\}}}\right] \quad (1)$$

$$R = 1 - \phi\left[-\frac{\pi \bar{r}^{3}\bar{S} - 2\bar{T}}{\pi \bar{r}^{3}\sqrt{\sigma_S{}^2 + \left\{\frac{4\sigma_T{}^2}{\pi^2 \bar{r}^6} + \frac{36\bar{T}^2\sigma_r{}^2}{\pi^2 \bar{r}^8}\right\}}}\right] \quad (2)$$

where, α is tolerance fraction, T is applied torque and P is the applied load.

This equation will be for the statistical distributions of stress and strength. Solving the equation will yield the value of the design variable and applying statistical tolerancing on it gives the complete design dimension

2.1 Using Software 1

Here a limit state equation [3] is identified for the given situation involving the distribution s of resistance and load to indicate a failure. The limit state is attained for g(X) = 0 and the failure domain is F: {g(X) < 0}. Reliability Index (RI) obtained form the given desired reliability level. The software also provides a facility to use various statistical distributions. Though this is primarily meant for structural design, an attempt is made to use it for mechanical components

3. DESIGN OF MECHANICAL COMONENTS

Mechanical elements circular shaft under tension and torsion are considered in this paper. They are designed for a specified reliability analytically i.e, using interference theory and COMREL [4]. Following is the data:

1. **For tension:**
   Load:
   Mean : 4000 N
   Standard deviation: 100 N
   Distribution of the tensile stress: Normal
   Strength:
   Mean: 100000 Pa
   Standard deviation: 5000 Pa
   Distribution : Normal
   Reliability: 0.99
2. **For torsion**
   Torque:
   Mean: 100000 N
   Standard deviation: 10000 N
   Distribution of the shear stress: Normal
   Strength:
   Mean: 50000 Pa
   Standard deviation: 5000 Pa
   Distribution: Normal

Fixing the limiting state equations for both the cases in the software COMREL is shown in the Equation (3).

$$g(X) = X_s - X_p * A \quad (3)$$

where A is the parameter.

The design data that we used in the probabilistic design approach (analytical) was again used to find a solution using COMREL and the following results were generated by the software. The parameter was calculated using the Software COMREL and then the diameter was computed as shown in Tables 1 and 2.

Table 1 COMREL result of tensional element

Xp	Xs	A	μ_p	μ_s	d^2= $(4\,\mu_p)/(\pi * A * \mu_s)$	d
Nor	Nor	0.75	4k	10^5	0.0679	0.26

Table 2 COMREL result of tensional element

Xp	Xs	A	μ_p	μ_s	d^3 = $(16\,\mu_p)/(\pi * A * \mu_s)$	d
Nor	Nor	0.75	10^5	50k	13.5	2.3

The various plots are also automatically generated by the software out of which plots of reliability index for both the cases are shown in Figure 1 and Figure 2 respectively.

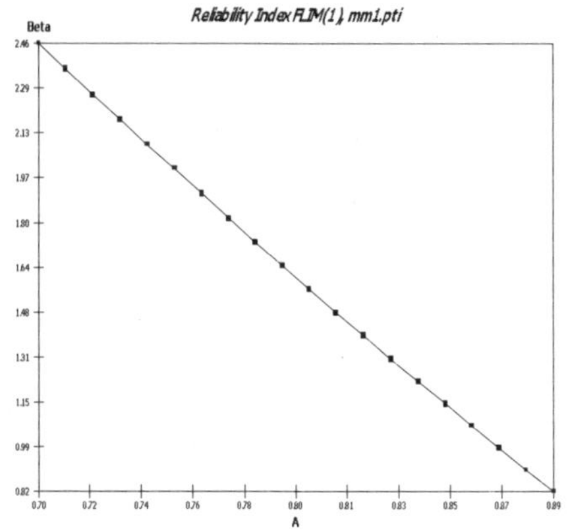

Fig. 1: Reliability Index by COMREL

The consolidated results using the deterministic, reliability based and structural probabilistic methods for tensional and tensional cases are provided in Table 3.

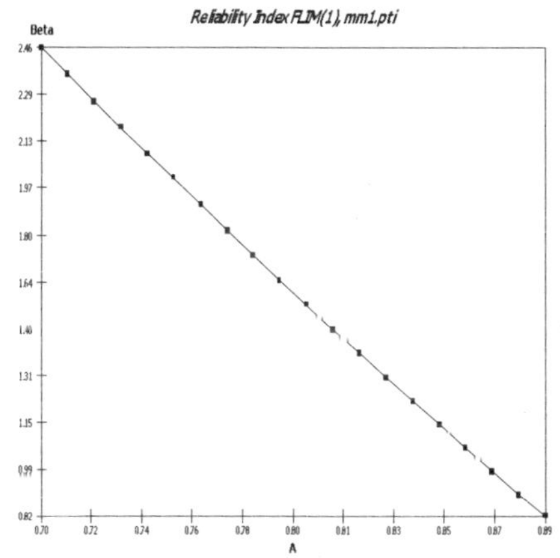

Fig. 2: Reliability Index by COMREL

Table 3 Consolidated results

Approach	Torsion	Tension
Deterministic	d = 0.39 m	d =3.10 m
Reliability based	d = 0.26 m	d = 1.90 m
Structural probabilistic (COMREL)	d = 0.23 m	d = 2.3 m

3. ANALYSIS USING SOFTWARE I-DEAS

Once the design parameter i.e., the diameter a 3-D model of the object, the shaft can be created using the software IDEAS [5]. Among the various modules provided in the software, Finite Element Analysis (FEA) wherein the physical model is converted to mathematical model to analyze the part for various modes of failure. Applying the required forces on the part with tools provided in software and the built in solver one can come to the final model where the stress distribution is clearly shown using colour codes. One has to compare the various configurations of stress distributions to find the best possible solution i.e., the one that results in the least stress concentration at a single point. The results of the analysis carried out on the design parameters that we computed in the illustration mentioned earlier are shown in the Figure 3

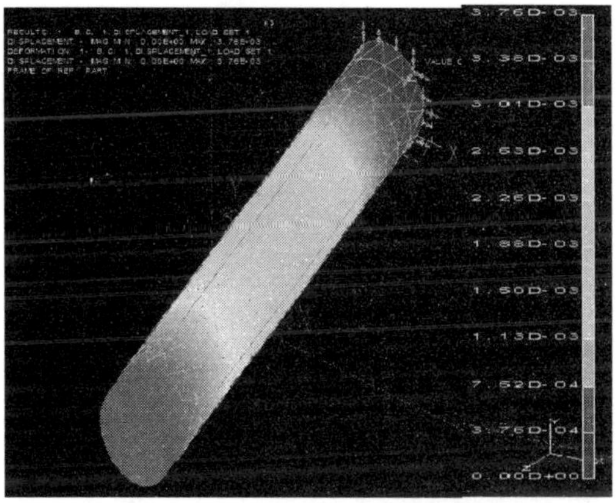

Fig. 3: Analysis using I-DEAS

4. CONCLUSION

It is evident from the results that the deterministic approach leads to over design in both the cases of tension and torsion. Using software IDEAS an in depth analysis on the selection of the parameter for an optimal stress concentration. Thus it can give the actual distribution stress concentration in various finitely segmented regions. By doing so, it attaches physical significance to the design. Therefore, the results obtained would be economic, optimum and with of course less risk involvement Though very simple elements are considered in this study, the approach can be extended to any complex situations too.

5. REFERENCES

1. Shigley, Mischke (2003), Mechanical Engineering Design (SI Units), Tata McGraw Hill Publishing Company Limited, India.

2. Kapur, Lamberson (1977), Reliability in Engineering Design, John Wiley & Sons, Singapore.
3. Ranganathan R.,"Structural Reliability Analysis & Design" Jaico Publishing House, 2000.
4. www.strurel.com
5. www.ugsplm.com

Reliability, Safety and Hazard
Editors: P.V. Varde et al.

INTEGRITY ASSESSMENT OF 37 ELEMENT FUEL BUNDLE OF TAPS 3&4 REACTOR UNDER BEYOND DESIGN BASIS ACCIDENT CONDITION

S.K. Tripathi and S.K. Gupta

Atomic Energy Regulatory Board, Anushaktinagar, Mumbai 400 094

ABSTRACT

Tarapur Atomic Power Station 3 & 4 is a 540 MWe Pressurized Heavy Water Reactor. It uses 37 element natural Uranium dioxide (UO_2) fuel pallets encapsulated inside the cylindrical sheath and are welded to the end plate at each end. During an postulated accident in which part of the fuel bundle are exposed to very high temperature (no means of heat removal) and other are at lower temperature (coolant temperature) possibility of failure of end plate weld due of thermal stresses developed by these relative temperature cannot be ruled out. In this report an attempt is made to study behaviour of fuel bundle under different temperature loading. Modelling of 37 element fuel bundle was done in ANSYS FEM. System was analysed for various sets of temperature loading. The system was analysised for plasticity and creep as material nonlinearity. The total strain, creep strain and stress increase as the temperature increases in upper portion of fuel bundle and decrease with increase in temperature in lower region. In earlier case, the total strain exceeds fracture strain after 45 sec. This Gross deformation leads to failure of end plate.

Keywords: Integrity Assessment, Finite Element Modelling, Fuel Bundle, Regulation, Pressurized Heavy Water Reactor, ANSYS

1. INTRODUCTION

Tarapur Atomic Power Station 3&4 is 540 MWe Pressurized Heavy Water Reactor. It uses natural Uranium dioxide (UO_2) as a fissile material. The fuel bundle is designed for neutron economy hence the content of parasitic neutron absorbing material is kept to minimum. The fuel bundle consists of 37 cylindrical fuel element joined together by welding the element to the end plates at both ends. The fuel bundle has one element at the centre and 6, 12 and 18 elements arranged in concentric rings. Each element contains long stack of about 35 sintered natural UO_2 pellets in a thin Zr-4 sheath with end plugs welded at both the ends. Reactor Core of TAPS 3&4 has 392 coolant channels. 13 fuel bundles are contained in each coolant channel. Out of the 13 fuel bundles, 11 full bundles and half a bundle on each end of the coolant channel are in the core and participate in nuclear heat generation. A total of 5096 fuel bundles are in the reactor core.

2. ACCIDENT SCENARIO

During a beyond design basis accident scenario like LOCA and partial/complete LOECC the cool ability of core will not maintained. As a result of flow stratification this some portion of fuel element may be expose to coolant temperature (coolant is available) and other portions may exposed to no coolant thus see high temperature. The end plate of fuel bundle may see higher thermal stresses. The material properties of sheath & endplate (Zr-4) degrade drastically at higher temperature [1].

The creep plays important role at higher temperature. The fuel bundle behaviour under creep and degraded material properties may impose high thermal strain at end plate welds. The excessive deformation of end plate under creep & high temperature loading is a matter of concerns for Regulatory bodies. Various criteria exist for fuel failure but failure of fuel bundle end plate because of temperature loading is not studied so far for Indian 540 MWe reactor.

Since the temperature history of such accident is not readily available it was assumed those 12 fuel elements are expose to lower temperature. The remaining 25 element will have higher temperature. Following cases were considered.

Case 1: The lower 12 elements were at 300^{0}C and other elements were at 1000^{0}C, 1100^{0}C, 1300^{0}C and 1600^{0}C respectively.

Case 2: The lower 12 elements were at 400^{0}C and other elements were at 1000^{0}C, 1100^{0}C, 1300^{0}C and 1600^{0}C respectively.

Case 3: The lower 12 elements were at 500^{0}C and other elements were at 800^{0}C, 1000^{0}C, 1100^{0}C, 1300^{0}C and 1600^{0}C respectively.

3. MODELLING DESCRIPTION OF FUEL BUNDLE

The fuel bundle was modelled in ANSYS Finite Element Software. The fuel element was modelled with SHELL 181 element. SHELL181 is suitable for analyzing thin to moderately-thick shell structures. It is a four-noded element with six degrees of freedom at each node. It is well-suited for linear, large rotation, and/or large strain nonlinear applications. The modelling of end plate was done with beam 44 element. The complete model consists of 11920 elements and 12276 nodes. The 3 D model of fuel bundle system was shown in Figure 1. The 3 D model of end plate was shown in Figure 2.

4. LOADING AND BOUNDARY CONDITION

The system was analysised for temperature loading. The lower 12 fuel elements were loaded with 300^{0}C, 400^{0}C and 500^{0}C for the three cases given above. The upper 25 fuel elements were loaded with 800^{0}C, 1000^{0}C, 1300C and1600^{0}C for the three cases given above. The temperatures were given at each node level. A typical temperature loading was shown in Figure 3. The material properties of Zr-4 was taken upto 850^{0}C and True Stress Strain curve was shown in Figure 4.

The fuel bundle system was constraint at all DOF at one end plate and allowed to expand at other end.

The system was analysised for plasticity and creep as material nonlinearity. The creep law used for analysis are given by [2]

$$\varepsilon^{\cdot} = 10.4\sigma^{3.4}\exp\left(\frac{-19600}{T}\right) + \frac{3.5\times10^{4}\sigma^{1.4}\exp\left(\frac{-19600}{T}\right)}{1+274\int_{t_2}^{t}\exp\left(\frac{-19600}{T}\right)(T-1105)^{3.72}\,dt}$$

Where

$\varepsilon^{\cdot}$: Strain rate

σ: Stress

T: Temperature

T_2: time when temperature is 850^{0}C

Since the second term diminishes very quickly after a short time, only the first term was used.

5. ASSUMPTIONS

Following assumption was taken during analysis.

1. The material properties were used up to 850^{0}C. The properties of fuel bundle beyond 850^{0}C had taken same as the properties at 850^{0}C.
2. The temperature distribution across the thickness was neglected.
3. The temperature along the fuel element was not varying and considered to be constant value.
4. The pressure & weight loading was not considered.

6. RESULTS AND DISCUSSIONS

The system was analysised for cases 1, 2 and 3.

Case 1: Four set of static nonlinear analysis was carried out. In all the set the lower 12 elements were at 300^{0}C and remaining elements were at 1000^{0}C, 1100^{0}C, 1300^{0}C and 1600^{0}C

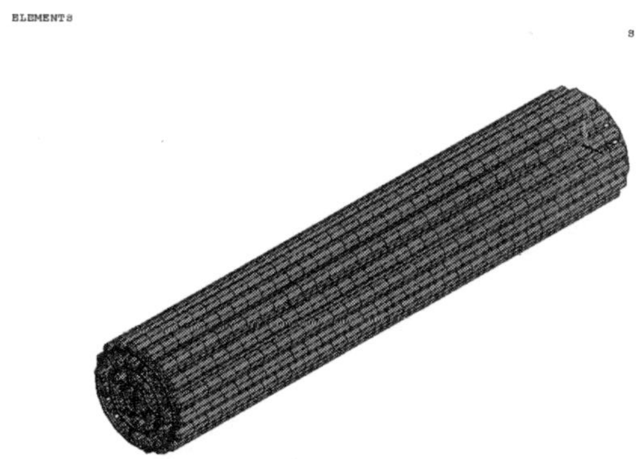

Fig. 1: 3D model of fuel bundle in AYSYS

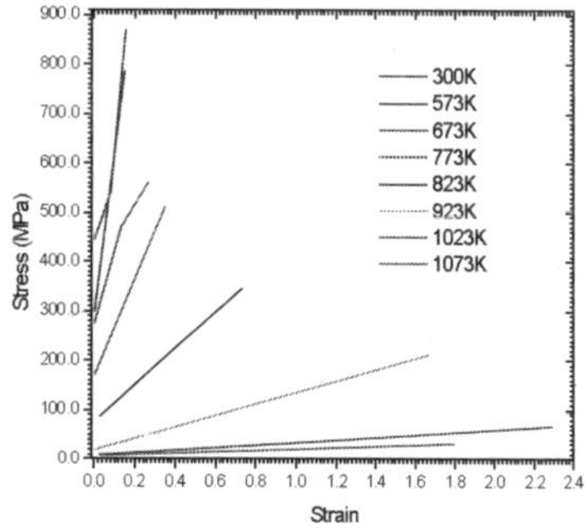

Fig. 4: True Material Stress Strain Curve of Zr-

Fig. 2: 3D Model of end plate in ANSYS.

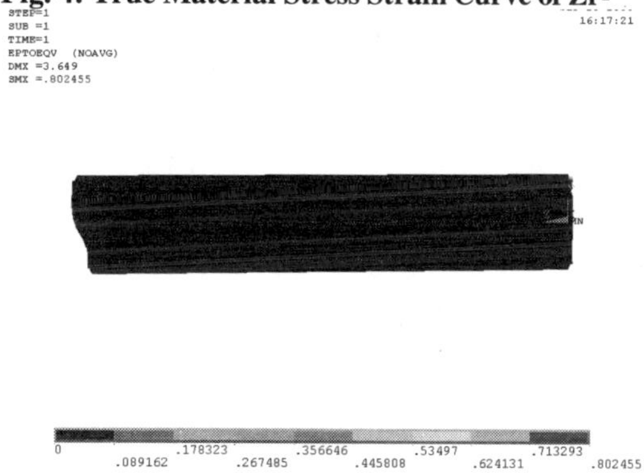

Fig. 5: Total Strain contour plot for case 1

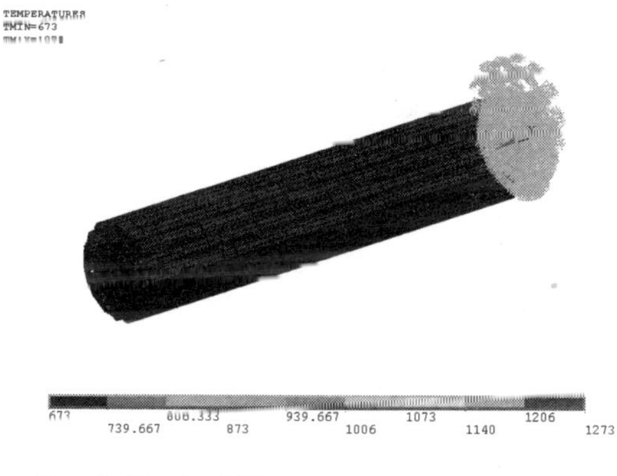

Fig. 3: Typical Temperature loading with Boundary condition

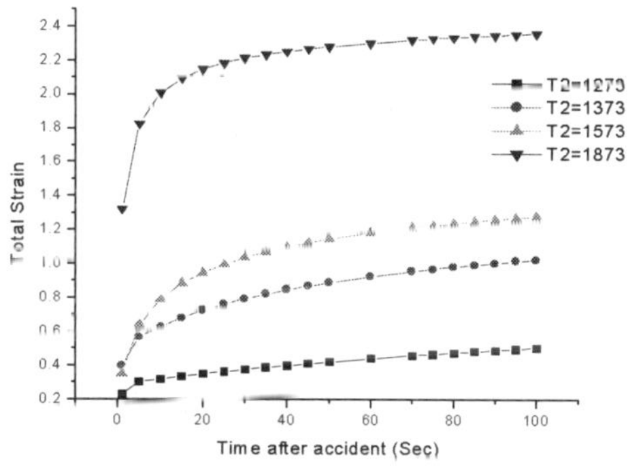

Fig. 6: Total Strain variation with time after accident for case 1

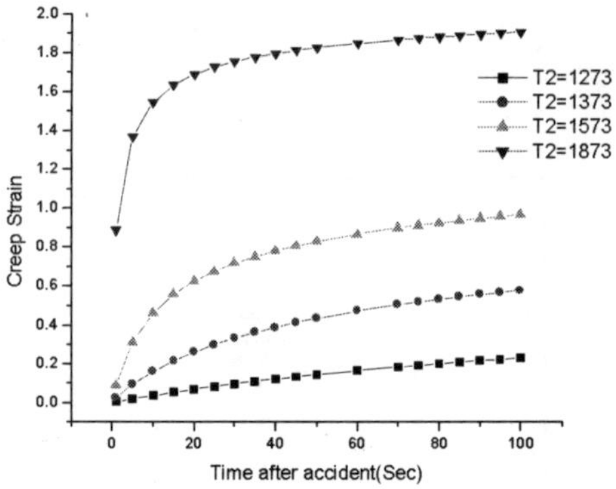

Fig. 7: Creep Strain variation with time after accident for case 1

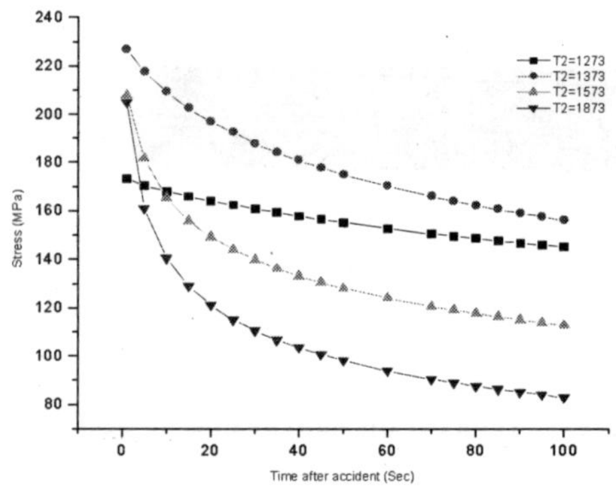

Fig. 8: Stress variation with time after accident for case 1

Fig. 9: Deformed shape of end plate for case 1

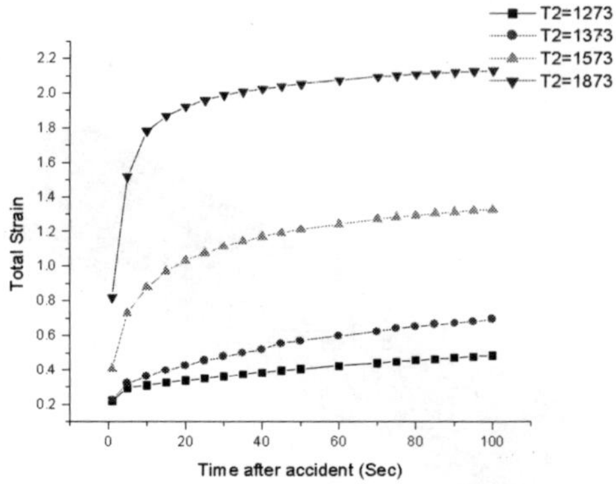

Fig. 10: Total Strain variation with time after accident for case 2

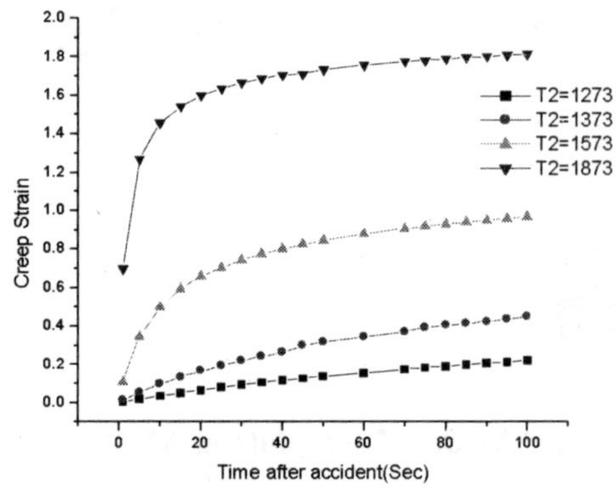

Fig. 11: Creep Strain variation with time after accident for case 2

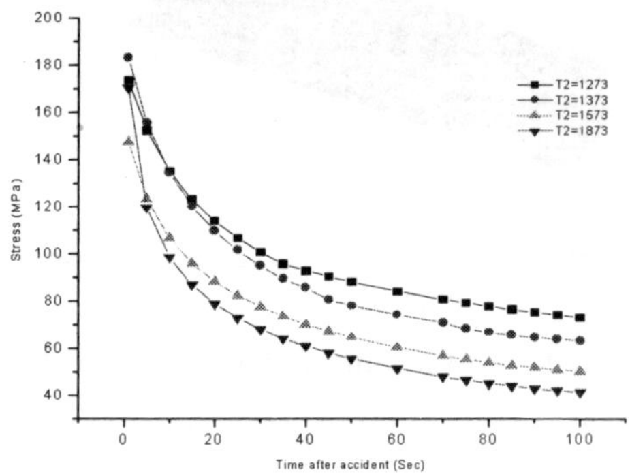

Fig. 12: Stress variation with time after accident for case 2

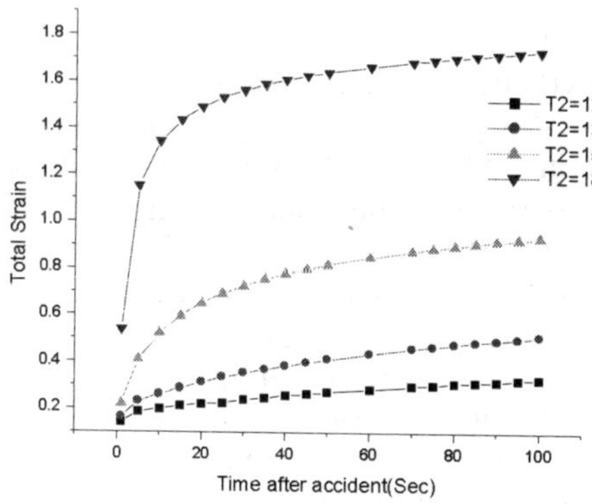

Fig.13: Total Strain variation with time after accident for case 3

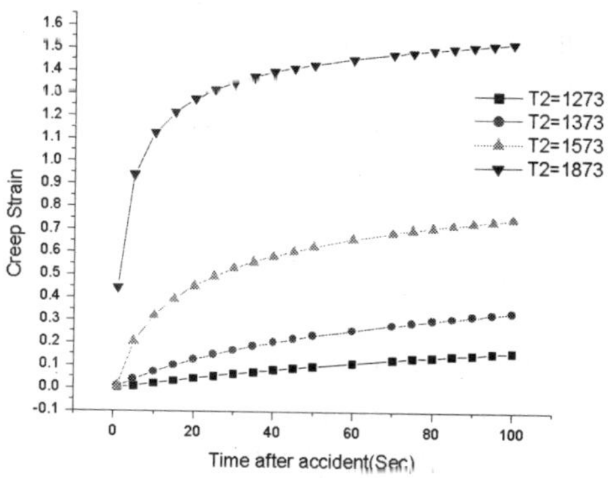

Fig. 14: Creep Strain variation with time after accident for case 3

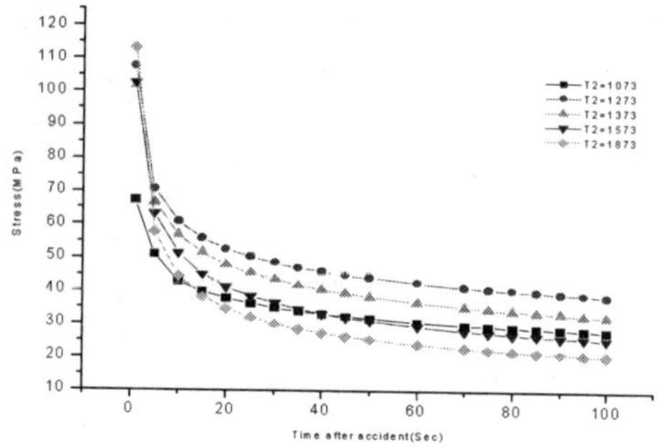

Fig. 15: Stress variation with time after accident for case 3

respectively. The strain contour of a typical set (lower elements 300^0C and remaining elements at 1600^0C) is shown in Figure 5. The plot of total strain, creep strain and stress with time after accident were shown in Figure 6, 7 and 8 respectively. It can be observed that for 1000^0C temperature loading case the total strain, creep strain and stress at 100 sec was 0.5024, 0.2316 and 145.099 MPa respectively and these values are through more then yield values but within the limit of fracture. The total strain, creep strain and stress are increasing as the temperature increases in upper portion of fuel bundle and for 1600^0C the total strain exceeds fracture strain after 45 sec and continues to increase. The total strain, creep strain and stress values are 2.357, 1.906 and 82.964 respectively after 100 sec. The deformed end plate was shown in Figure9.

Case 2: In this case the lower 12 elements were at 400^0C and remaining elements were at 1000^0C, 1100^0C, 1300^0C and 1600^0C respectively. The plot of Total strain, creep strain and stress with time after accident were shown in Figure 10, 11 and 12 respectively. The total strain, creep strain and stress values are 2.132, 1.812 and 41.151 MPa respectively after 100 sec for 1600^0C. It can be noted that the maximum strains are lower than the strains obtained in case 1. The maximum strain is less than fracture strain.

Case 3: In this case the lower 12 elements were at 500^0C and remaining elements were at 800^0C 1000C, 1100^0C, 1300^0C and 1600^0C respectively. The plot of Total strain, creep strain and stress with time after accident were shown in Figure 13, 14 and 15 respectively. The total strain, creep strain and stress values are1.731, 1523 and 20.535 MPa respectively after 100 sec for 1600^0C. It can be noted that the maximum strains are lower than the strains obtained in case 1&2. The maximum strain is less than fracture strain.

7. CONCLUSIONS

Integrity of fuel bundle under various temperatures loading condition were analysised in present paper. Modelling of fuel bundle was done in ANSYS FEM. System was analysed for various sets of temperature loading. For case 1, the total strain, creep strain and stress are increasing as the temperature increases in upper portion of fuel bundle and for 1600^0C the total strain exceeds fracture strain after 45 sec and continues to increase. The total strain, creep strain and stress values are 2.357, 1.906 and 82.964 respectively. Whereas these values were 2.132, 1.812 and 41.151 for case 2 and .731, 1.523 and 20.535 for case 3. The strains exceed fracture strain in case 1 and within the limit in case 2 and 3. The maximum strain increases with increases of temperature at upper portion of fuel bundle and decreases with increase in temperature in lower region.

Gross deformation exceeding fracture strains were observed in case1 leading to failure of end plate.

8. REFERENCES

1. SCDAP/RELAP5/Mod 3.2 Code manual Volume IV: MATPRO: A library of materials properties for Light Water Reactor Accident analysis, October 1997, NUREG/CR-6150 INEL-96/0422, Revision 1.
2. "Severe core damage experiments and analysis for CANDU applications", P.M. Mathew, A.J.White, V.G.Snell and M. Bonechi, AECL, Transactions of the 17th international conference on Structural Mechanics in Reactor Technology (SMiRT 17), Prague, Czech Republic, August 17-22, 2003.

Reliability, Safety and Hazard
Editors: P.V. Varde et al.

ESTIMATIMATION OF RELIABILITY BASED PARTIAL SAFETY FACTORS FOR PRESSURE VESSELS

Siddhartha Pramanick

Reactor Operation Division (Dhruva), Bhabha Atomic Research Centre, Mumbai 400 085

ABSTRACT

This paper evaluates the partial safety factors of applied stress, crack depth and fracture-toughness for a plate having partial through crack, under tensile load. In this report FORM (First Order Reliability Method)[1] has been used for probabilistic analysis. Partial safety factors, for both in ductile and plastic collapse region for shallow crack as well as deep crack with respect to the target failure probability, has been presented. A comparison of the PSF estimates has been made with the available data from American Petroleum Institute (API)[2]. The partial safety factors are useful to qualify components with flaws discovered during in-service inspection (ISI) and a very easy tool to assess the reliability of the structural components in the plant.

1. INTRODUCTION

Reliability based design and analysis technique are increasingly being adopted internationally by civil, aerospace, offshore, petroleum and nuclear industry. This is primarily because a lot of scatter in the design variables such as material properties is observed. Traditional design methods use a high factor of safety to accommodate this scatter. The probabilistic methods provide a better insight in the design of the structure

2. NEED OF PROBABILISTIC ANALYSIS AND PARTIAL SAFETY FACTORS

Deterministic assessment of crack-like flaws requires information that are assumed to be known. However in many instances all of the important independent variables are not known with a high degree of accuracy. In such cases, conservative estimates of the independent variables can be made to ensure an acceptable safety margin; however, this approach can lead to overly conservative results. Hence, we go for probabilistic approach.

Deterministic analysis of fracture mechanics is well established, where we assume that the material properties, geometries and loading conditions are deterministic in nature. But in actual practice they are not deterministic rather random in nature. Their randomness does not always follow normal distribution. Mostly they follow extreme value distribution Type-I, Log-Normal distribution, Weibull distributions. Probabilistic study not only takes care of changes in properties, geometries and loading condition, but also failure probability with safety margin.

The dependence of safety margin on uncertainty of the independent variable can also be evaluated using a probabilistic analysis. This type of analysis can be used to combine a deterministic FFS (Failure For Service) assessment model with the distribution prescribed for the independent variables to calculate the failure probability. However this type of analysis requires specialized technical skill and is not amenable for general use in standard.

2.1 Partial Safety Factors [3]

Individual safety factors that are applied to the independent variable in the assessment procedure can be used to model uncertainty. These factors or partial safety factors (PSF) are probabilistically calibrated to reflect the effect that each of the independent variables has on the probability of failure. The advantage of the approach is that uncertainty can be introduced in an assessment by separately combining the partial safety factors with the independent variables in the deterministic analysis model (R-6 Method has been used as deterministic model).

2.2 Derivation Of Partial Safety Factor

In a probabilistic approach, the basic design criterion is either a maximum allowable probability of failure P_f (target reliability) or a minimum allowable safety index β (target safety index). The two factors are related by the following equation:

$$P_f = \phi(-\beta) \tag{1}$$

In general, for cases where there are several variables and the limit state function are complex, determination of the PSF can be complicated.

As an example, consider a component subject to a stress state composed of primary stress (S_p) and secondary stress (S_S). The strength of the component is AR, where R is a strength parameter of the material (for example a combination of yield and ultimate tensile stress) and A is a factor that is pro optional to the geometry of the component. The limit state function for this component is given by the following equation

$$G(R, S_p, S_s) = A \times R - (S_p + S_s) \tag{2}$$

And the limit state is

$$G(R, S_p, S_s) = 0 \tag{3}$$

By adjusting the value of the parameter (or tuning factor) in an iterative process, the result from a FORM analysis for each iteration can be used to produce a solution where the converged safety index is equal to target safety index. The corresponding Most Probable Point (MPP) is computed in the basic coordinates, R^*, S_p^* and S_s^*. Because the MPP is on the limit state, it follows that

$$A \times R^* = S_p^* + S_S^* \tag{4}$$

Therefore, if a design is made so that

$$A \times R^* > S_p^* + S_s^* \tag{5}$$

Then the implication is that the actual safety index exceeds the target safety index and actual probability is less than the target probability of failure.

The partial safety factors are defined for variable R, Sp, Ss as follows respectively

$$\gamma_R = R_n \div R^*$$
$$\gamma_{S_p} = S_p^* \div S_{p,n} \tag{6}$$
$$\gamma_{S_s} = S_s^* \div S_{s,n}$$

Where R_n, S_{pn}, S_{sn} is the nominal value of R, S_p & S_s.

3 FORMULATION OF LIMIT STATE FUNCTION (USING R-6 METHOD)

The resistance to fracture in failure assessment diagram (FAD) can be described using the following equation

$$K(L_r) = (1 - L_r^{2.5})^{0.2} \tag{7}$$

Where L_r=load ratio, K_r= toughness ratio in the failure assessment diagram (FAD). The driving force for the crack –like flaw is defined by the value K_r; therefore the limit state function is

$$G = K_r^*(L_r) - K_r \tag{8}$$

Where K_r^* is the limiting value of K_r on facture assessment line in FAD. At this point a reliability program can be used with this limit state to derive the PSFs.

However, numerical difficulties motivated the development of an alternate approximate approach that provided to be robust while maintaining the accuracy. In this approach, the limit state function is defined by a radius on the Failure Assessment Diagram (FAD). To develop this alternate limit state function, the FAD point (L_r, K_r) is computed based on nominal values of the design factors denoted as (L_m, K_m). This point defines a vector in the L_r – K_r space that makes an angle θ with the abscissa of the FAD. The resistance to fracture is defined as the distance to FAD envelope (Failure assessment line, FAL), r, as shown in figure *. The radius, r is a function of θ.The driving force for the crack, R_o, is defined as

$$R_0 = (L_r^{\,2} + K_r^{\,2})^{0.5} \qquad (9)$$

R_o is also a random variable having probability density function, f_r. However, from geometrical considerations, the variability in *r* is small. Therefore, for analytical simplification, it is assumed that both θ and r are constant. Now, the limit state equation is defined as

$$G = r - R_0 \qquad (10)$$

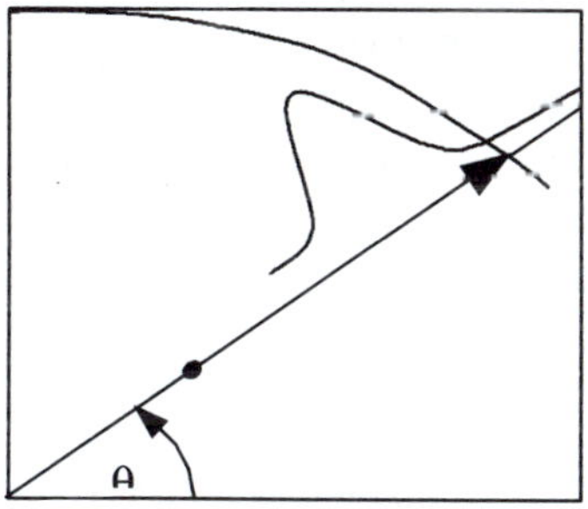

Fig. 1: Failure assessment diagram (FAD)

3.1 Reformulation of the Limit State Function

In order to facilitate the use of a reliability computer program to solve for the PSF, the limit state function is re-derived in terms of normalized random variables.

The fracture toughness of the material is a random variable having a mean of μ_K and standard deviation of σ_K.

$$K_{mat} \Rightarrow WeibullDist[\mu_K, \sigma_K] \qquad (11)$$

A new random variable is described using following equation

$$Y = K_{mat} / \mu_K \qquad (12)$$

Therefore,

$$Y \Rightarrow W[1, COV_K] \qquad (13)$$

Again, for primary applied stress,

$$S_p \Rightarrow ExtremeVal\,ueDist\lfloor \mu_{S_p}, \sigma_{S_p} \rfloor \qquad (14)$$

$$X = S_{S_p} / \mu_{S_p} \qquad (15)$$

$$X \Rightarrow E.V.Dist - 1[1.0, COV_S] \qquad (16)$$

Now, we define the constants *A* and *B* as follows

$$A = \mu_{S_p} / R_y \qquad (17)$$

$$B = \mu_K / R_y \qquad (18)$$

Where, R_y is yield stress.
Then, using above equations we have

$$K_{mat} / R_y = BY \qquad (19)$$
$$S_p / R_y = AX$$

Now, the re-worked limit state function becomes

$$G = r - (AX)[\pi a/(BY)^2 + 1.0]^{0.5} \qquad (20)$$
$$where, a = crack - size$$

The slop of the vector r is defined as
For shallow crack

$$Slop = 0.56 / B \qquad (21)$$

For deep crack

$$Slop = 1.12 / B \qquad (22)$$

Table 1 Input data for determination of PSF

Parameter (Distribution)	Normal Value (μ)	*COV*
Stress (type 1 Extreme Value Distribution)	$\mu_S = 1.0$	0.10, 0.20, 0.30
Fracture Toughness (Two parameter Weibull)	$\mu_K = 1.0$	0.25
Flaw Size ($a <$ 0.2 inches)	μ_a=1.0	0.30
Flaw Size ($a \geq$ 0.2 inches)	μ_a=1.0	0.20

Table 2 Definition of 18 different cases required to derive PSFs

Case No	Target Probability (P_f) /Safety Index (β)	Coefficient Of variance for Stress ($COVs$)
Short Crack (a<0.2 inches)		
1	$2.3(10)^{-2}/2.00$	0.10
2	$(10)^{-3}/3.09$	0.20
3	$(10)^{-6}/4.75$	0.30
4	$2.3(10)^{-2}/2.00$	0.10
5	$(10)^{-3}/3.09$	0.20
6	$(10)^{-6}/4.75$	0.30
7	$2.3(10)^{-2}/2.00$	0.10
8	$(10)^{-3}/3.09$	0.20
9	$(10)^{-6}/4.75$	0.30
Long Crack (a ≥0.20inches.)		
10	$2.3(10)^{-2}/2.00$	0.10
11	$(10)^{-3}/3.09$	0.20
12	$(10)^{-6}/4.75$	0.30
13	$2.3(10)^{-2}/2.00$	0.10
14	$(10)^{-3}/3.09$	0.20
15	$(10)^{-6}/4.75$	0.30
16	$2.3(10)^{-2}/2.00$	0.10
17	$(10)^{-3}/3.09$	0.20
18	$(10)^{-6}/4.75$	0.30

Table 3 Calculated PSF values for stress, fracture toughness and crack size (γ_S, γ_K, γ_a) for brittle fracture for all 18 cases as in above table 1. (values in brackets are from API 579-2000).

γ_S	γ_K	γ_a
Short Crack (a<0.2 inches)		
1.2(1.20)	1.43(1.43)	1.08(1.08)
1.3(1.30)	1.429(1.43	1.07(1.08)
1.54(1.55)	1.45(1.43)	1.00(1.08)
1.41(1.40)	1.41(1.43)	1.2(1.20)
1.60(1.50)	1.90(1.82)	1.03(1.10)
2.00(2.00)	2.00(2.0)	1.05(1.05)
1.75(1.75)	2.0(2.0)	1.35(1.35)
2.07(2.5)	2.21(2.0)	1.26(1.50)
2.7(2.6)	2.20(2.0)	1.25(1.5)
Long Crack (a ≥0.20inches.)		
1.1(1.20)	1.34(1.43)	1.10(1.10)
1.40(1.40)	1.55(1.54)	1.10(1.10)
1.60(1.60)	1.68(1.67)	1.10(1.10)
1.30(1.40)	1.79(1.67)	1.26(1.15)
1.77(1.80)	1.41(1.43)	1.12(1.10)
2.3(2.30)	1.33(1.43)	1.20(1.10)
1.65(1.70)	2.00(2.0)	1.24(1.25)
2.45(2.60)	2.20(1.82)	1.22(1.25)
3.42(3.50)	2.0(1.67)	1.24(1.25)

Table 4 Calculated PSF values for stress, fracture toughness and crack size (γ_S, γ_K, γ_a) for plastic collapse for all 18 cases as in above table 1.(values in brackets are from API 579-2000).

γ_S	γ_K	γ_a
Short Crack (a<0.2 inches)		
1.25(1.25)	1.0(1.0)	1.0(1.0)
1.50(1.50)	1.0(1.0)	1.0(1.0)
1.75(1.75)	1.0(1.0)	1.0(1.0)
1.5(1.50)	1.0(1.0)	1.0(1.0)
2.0(2.0)	1.0(1.0)	1.0(1.0)
2.50(2.50)	1.0(1.0)	1.0(1.0)
2.03(2.00)	1.0(1.0)	1.0(1.0)
3.06(3.10)	1.0(1.0)	1.0(1.0)
4.10(4.10)	1.0(1.0)	1.0(1.0)
Long Crack (a ≥0.20inches.)		
1.25(1.25)	1.0(1.0)	1.0(1.0)
1.50(1.50)	1.0(1.0)	1.0(1.0)
1.75(1.75)	1.0(1.0)	1.0(1.0)
1.5(1.50)	1.0(1.0)	1.0(1.0)
2.0(2.0)	1.0(1.0)	1.0(1.0)
2.5(2.50)	1.0(1.0)	1.0(1.0)
2.00(2.00)	1.0(1.0)	1.0(1.0)
3.10(3.10)	1.0(1.0)	1.0(1.0)
4.10(4.10)	1.0(1.0)	1.0(1.0)

4. CONCLUSION

Structural reliability methods (FORM /SORM, Simulation methods, etc.) allow us to account for the uncertain part of data in order to quantify the structural integrity. However this quantification is relative to an imperfect knowledge and the formulation of general structural safety problem involves then different kind of uncertainties: the inherent variability which concerns or its environment (geometry, material, loading…), imperfections in deterministic mechanical model which arises from the use of simplified mathematical and numerical models to describe complex phenomena and imperfection in the probabilistic data formulation which arises from the incompleteness of statistical data. As a consequence, partial safety factors which accounts separately for these different sources of uncertainty are formulated in order to be used in codification as a function of the information on the studied structure.

5. REFERENCES

1. Hasofer, A.M. and Lind, N.C.,"Exat and Invarient Second Moment Code Format," 1947, *Journal of Engineering Mechanics,* ASCE, Vol. 100, pp. 357-366.
2. API,*Fitness-for-Service API Recommended Practice 579,*American Petroleum Institute, Wasington D.C., 2000. Report.
3. Wirsching, P.H. and Mansour, A.E, May 1998, *"Incorporation of Structural Reliability Methods into Fitness for Service,"* the Material Properties Council, Inc.

Reliability, Safety and Hazard
Editors: P.V. Varde et al.

PROBABILISTIC FRACTURE ASSESSMENT OF TAPP 3-4 PHT PIPING

P.A. Jadhav, Rohit Rastogi, Vivek Bhasin, K.K. Vaze, A.K. Ghosh and H.S. Kushwaha

Health, Safety and Environment Group, Bhabha Atomic Research Centre, Mumbai 400 085

ABSTRACT

Methodology based on probabilistic fracture mechanics (PFM) is finding increasing acceptability in demonstrating safety of Nuclear Power Plant (NPP) piping. In PFM, the methods of fracture mechanics and reliability theory are combined for assessing the reliability of components, which contain cracks. In this work, reliability assessment of Tarapur Atomic Power Plant (TAPP) 3-4 Primary Heat Transport (PHT) piping is done using PFM. Monte Carlo simulation with stratified sampling is used as a variance reduction technique. PFM model assumes a pre-existing circumferential surface crack before the start of plant operation. The crack grows in size during the lifetime of the plant due to the fatigue loading. This part-through wall crack having escaped hydro-test and pre-service inspection, may result in either a through wall flaw (leak) or may lead to the rupture of the piping. R6 method is used as failure criteria. Steam generator inlet (SGI), steam generator outlet (SGO) and pump discharge line (PDL) are considered for piping reliability analysis. Study shows that probability of occurrence of break before leak in piping is very less as compared to that of leak before break.

Keywords: Probabilistic Fracture Mechanics, Leak before break, Piping reliability, Monte Carlo Method

1. INTRODUCTION

PFM is widely used to show fitness for purpose and residual life estimation of nuclear power plant components [4]. The objective is to ascertain the safety of the components under postulated or observed flaws. However there is large uncertainty in the input parameters needed for this calculation (e.g. Applied loads, material tensile and fracture properties, flaw size etc.). In deterministic design methods safety of power plant is assured by taking enough safety margins for loads against critical load. However this type of approach may become unrealistic. PFM method is most rational tool, which can be used in design and structural integrity assessment of NPP components. The PFM results are given out in terms of the reliability or the probability of failure of the component. In this work leak and break probabilities of Indian 540 MWe PHWR PHT piping is calculated using PFM. The analysis has been done using Code developed in BARC, which uses Monte Carlo simulation with stratified sampling.

2. PROBABILISTIC FRACTURE MECHANICS MODEL

In PFM model, the fracture mechanics and probabilistic methods are combined. PFM model is based on the assumption that failure in the piping occurs due to the sub-critical and catastrophic growth of crack like defects introduced during welding. These cracks exist in the weld with certain probability and can be detected during pre-service and in-service inspection with certain probability, which is dependent on their size. Cracks, which get detected during inspection, are assumed to be removed without introducing new cracks into the material. Only pre-service inspection is modeled in this analysis. Uncertainty in crack detection is modeled by using probability of detection curve (POD). POD curve for carbon steel pipe is taken from Gosselin [3]. Sub-critical growth of these cracks place due to fatigue. These crack growth is assessed using fracture mechanics. Assessment of safety of these cracks is done using R6 [5] method. Parameters involved in this assessment like crack size, tensile and fracture properties of material are assumed to be randomly distributed.

Initial Crack Size Distribution

Part-through semi-elliptical inside surface crack having crack depth a and length $2b$ is considered. The crack depth, a, and aspect ratio, $\beta = b/a$, are two parameters used to define the crack size and assumed to be independent. Literature survey suggests that these two parameters are exponentially distributed [4]. Crack depth probability density function is

$$p(a) = \frac{\lambda_a \exp(-\lambda_a a)}{1-\exp(-\lambda_a 0.8t)} \quad (1)$$

where $1/\lambda_a$, the mean crack depth=6.25 mm.

Aspect ratio probability density function is

$$p(\beta) = \lambda_\beta \exp(-\lambda_\beta (\beta - 1)) \quad (2)$$

where $(1 + 1/\lambda_\beta)$ is the mean aspect ratio and λ_β =1.15

Material Properties

The pipes are made of SA333 Gr.6 carbon steel. All material properties are log-normally distributed.

Table 1 Mean and coefficient of variation for different material properties

Material property	Mean	% COV*
Yield strength N/m^2	2.4E8	7
Ultimate strength N/m^2	4.49E8	5
Fracture toughness N/m$^{3/2}$	2.78E8	40
Ramberg-Osgood coefficient	1.948	40
Ramberg-Osgood exponent	4.898	13

* Coefficient of variation

Mean and coefficient of variation of material properties is given table 1. Mean of material properties is taken from BARC report [7] and their variation is taken from Rahman [6]. Rahman has given this variation for 304 SS material, which has similar tensile and fracture properties as that of SA 333 Gr-6. Correlation matrix of these material properties is

$$\begin{array}{c|ccccc|} & \sigma_y & \sigma_u & K_{IC} & \alpha & n \\ \sigma_y & 1 & 0.30 & 0 & 0 & 0 \\ \sigma_u & 0.30 & 1 & 0 & 0 & 0 \\ K_{IC} & 0 & 0 & 1 & 0 & 0 \\ \alpha & 0 & 0 & 0 & 1 & -0.80 \\ n & 0 & 0 & 0 & -0.80 & 1 \end{array}$$

Failure Criteria: R6 Method

R6 method is used for failure assessment. The R6 procedure [5] involves plotting the coordinates of the normalized crack driving force $K_r' = K/K_{IC}$ (K =applied stress intensity factor, K_{IC} is fracture toughness) and the normalized applied stress, $L_r' = L/L_C$ (L=applied load, L_C is net section collapse load), both being function of crack size, on a failure assessment diagram (figure 1). Crack initiation takes place if plotted point lies above the FAD. This evaluation is termed as Category 1 evaluation in R6 method. The equation of material specific FAD is

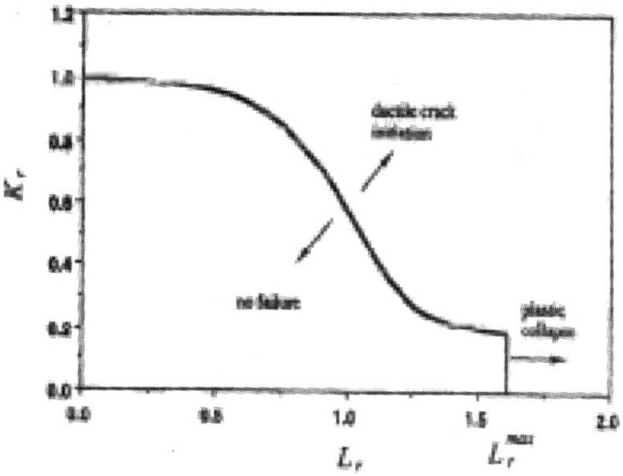

Fig. 1: R6 Failure assessment Diagram

$$K_r = \left\{1 + \alpha L_r^{n-1} + \frac{L_r^2}{2\left(1 + \alpha L_r^{n-1}\right)}\right\}^{-\frac{1}{2}} \quad (3)$$

$$K_r = 0 \qquad for \qquad L_r > L_r^{max}$$

$$L_r^{max} = \sigma_f / \sigma_y$$

σ_{flow} = flow stress and σ_y =yield stress

α =Ramberg-Osgood coefficient and n =Ramberg-Osgood exponent

Initiation criteria:

$$K_r' \geq Kr$$

Limit load and Stress Intensity Factor (SIF) solutions are used from A16 [1].

In this work R6 option 2 and category 1 method is used for the analysis.

Leak criteria used in this analysis is

$$a \geq \min(0.8t, a_{crit})$$

where a_{crit} is limiting half crack length calculated using R6 method. Break criteria used is

$$a \geq \min(0.8t, a_{crit}) \text{ and } b \geq b_{crit}$$

where b_{crit} is limiting crack length for the through-wall crack calculated using R6 method

Crack Growth Characteristics

Fatigue crack growth due to different transients is modeled using Paris' law. Arrival time of transients is assumed constant. It is assumed that semi-elliptical as-fabricated cracks continue to grow as semi-elliptical during the operation of plant. Crack growth in both directions is controlled by local-SIF associated with that direction. Paris law constants C and m for the piping material are taken from ASME pressure vessel and boiler code, section XI [2].

Loads Acting On Different Pipes

Three critical locations considered are steam generator line, steam generator outlet line and pump discharge line. Dimensions of these pipes are given in table 2.

Table 2 PHT piping dimensions

Pipe	Outer dia. mm	Thickness (mm)
SGI	508	40
SGO	510	50
PDL	457	35

Different transients considered for fatigue analysis and their frequency is given in table 3. Maximum and minimum stress during the above transients is given in table 4.

Table 3 Transients and their frequencies

Stage No.	Events	No. of cycles
1	CSD to HSB + to Full power	25 /year
2	HSB to Full power	88 /year
3	Power Manoeuvring at Full Power	375 /year
4	OBE	10 /earthquake
5	SSE	10 /earthquake

OBE-Operating basis earthquake
SSE-Safe shutdown earthquake
CSD-Cold shutdown
HSB-Hot standby

Table 4 Maximum and minimum stresses (in MPa) during transients.

Transient NO.			1	2	3	4	5
SGI	Axial	σ_{max}	27.43	27.43	27.43	28.65	29.88
		σ_{min}	0	25.14	26.50	0	0
	Bending	σ_{max}	61.23	61.23	61.23	87.77	114.3
		σ_{min}	0	56.128	59.14	0	0
SGO	Axial	σ_{max}	24.83	24.83	24.83	27.24	29.65
		σ_{min}	0	22.76	23.98	0	0
	Bending	σ_{max}	49.49	49.49	49.49	67.21	84.94
		σ_{min}	0	45.37	47.80	0	0
PDL	Axial	σ_{max}	30.08	30.08	30.08	31.14	32.21
		σ_{min}	0	27.57	29.05	0	0
	Bending	σ_{max}	53.17	53.17	53.17	62.57	71.98
		σ_{min}	0	48.74	51.36	0	0

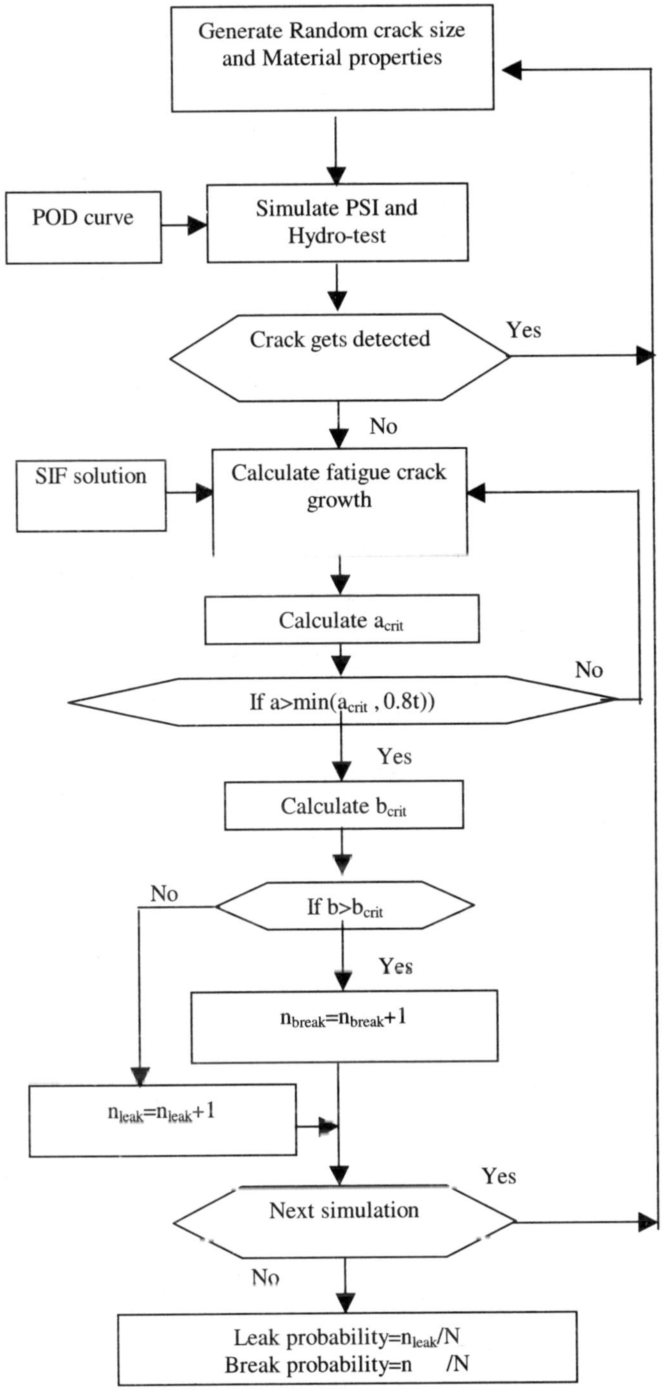

Fig. 2: Monte Carlo Algorithm for PFM analysis.

3. PFM ANALYSIS

Algorithm of Monte Carlo method used for the PFM analysis is shown in figure 2. Random crack size, material properties are generated so that they follow given distribution. Hydro-test and pre-service inspection is simulated. Those cracks, which get undetected during hydro-test and PSI grow due to different transients and may cause leak or break in the piping. Leak or break in the piping is assessed by using R6 method. Leak and break probabilities are given by

$$P_{leak} = n_{leak} / N$$

$$P_{break} = n_{break} / N$$

where N is total number of simulations, n_{leak} is number of simulations in which leak is observed and n_{break} is the number of simulations in which break is observed.

So as to increase the computational efficiency Stratified Monte Carlo method is used. The distribution of initial crack size is one of the most influential and most uncertain quantities [4] and hence only crack size domain is stratified. Cracks, which are initially deeper and longer, contribute in failure.

4. RESULTS AND DISCUSSION

Piping reliability results for SGI, SGO and PDL obtained using PFM model are shown in figure 3, 4 and 5. Both leak before break and break before leak probabilities are shown. Maximum leak probability for the piping is of the order of 10^{-5} and maximum break probability is of the order of 10^{-7}. Leak probability in nuclear piping is about hundred times more than break probability. Which means that piping is likely to leak before break and thus the LBB criteria is satisfied. Conditional leak and break probabilities in OBE and SSE are calculated. Leak probabilities under different levels of earthquake are comparable but break probabilities increase by about one order under SSE level earthquake. It is also observed that, in three pipes SGI has maximum leak probability and PDL has maximum break probability. SGO has least leak and break probability.

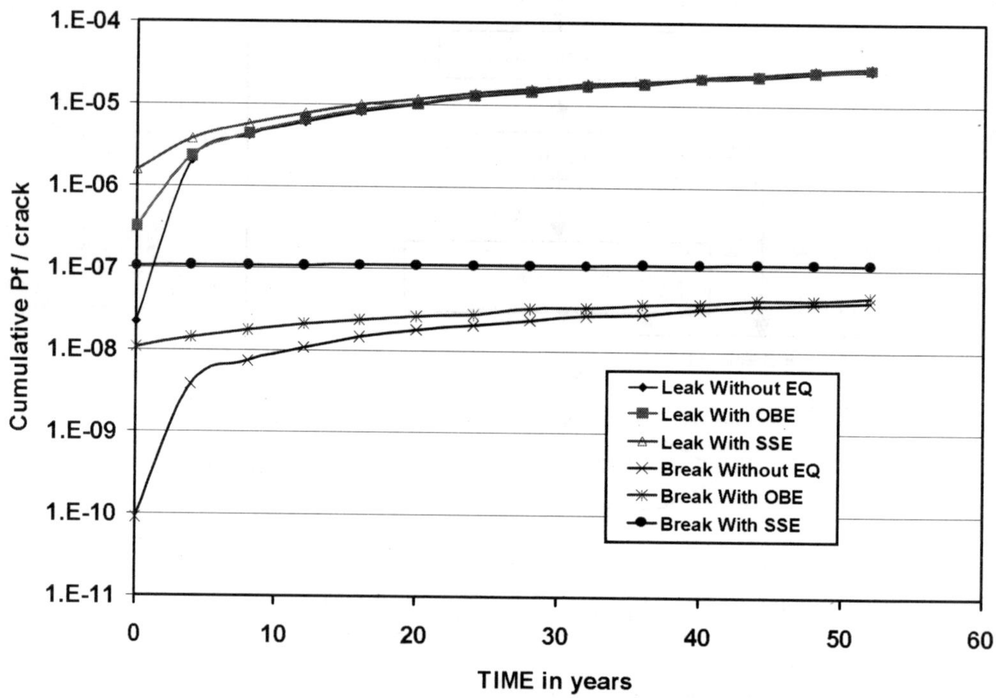

Fig. 3: Leak and Break probabilities of steam generator inlet line.

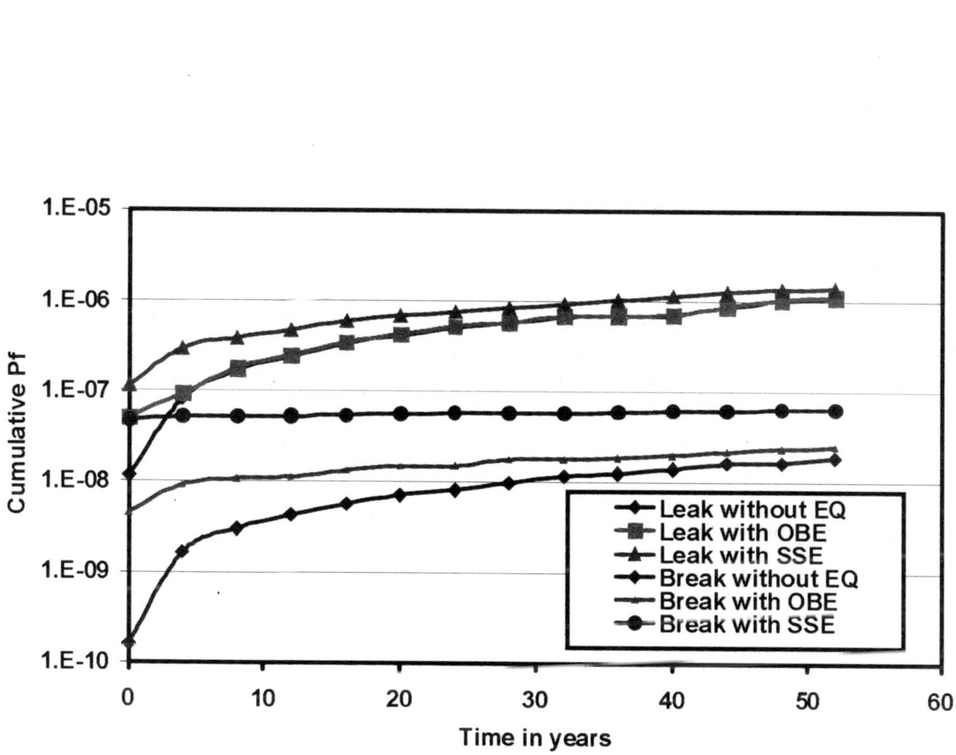

Fig. 4: Leak and Break probabilities of steam generator outlet line.

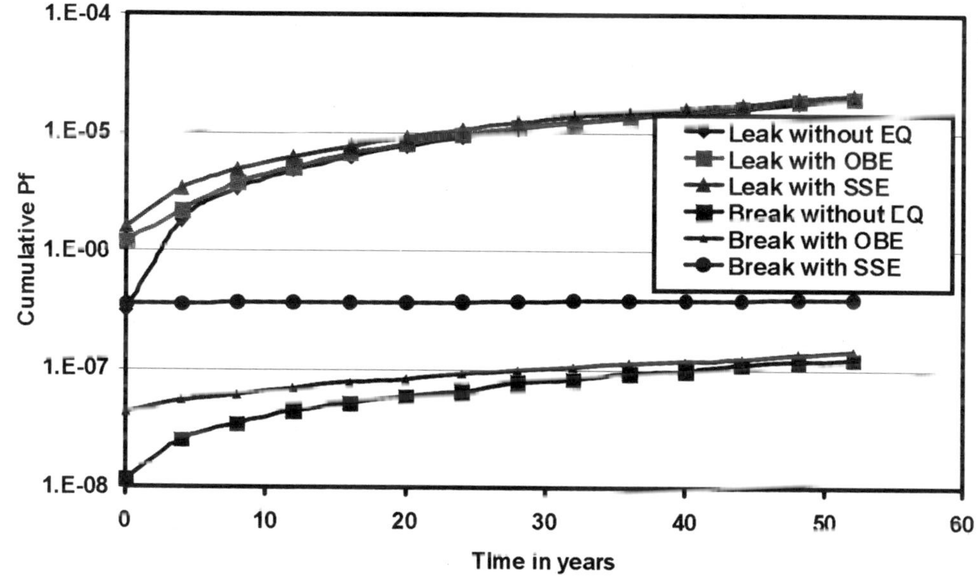

Fig. 5: Leak and Break probabilities of pump discharge line.

5. REFERENCES

1. "Appendix A16 – Guide for leak and break analysis and defect assessment", from RCC-MR Edition 2002.
2. ASME Boiler and Pressure Vessel Code, Section XI.
3. Gosselin, S. R., Simonen, F. A., Heasler, P. G., Doctor, S. R., Becker, F. L., "NDE performance (POD) curves for fatigue cracks in piping based on industry performance demonstration data", Proceedings of ICONE 10, 10th international Conference on Nuclear Engineering, Arlington, VA, April 14-18, 2002.
4. Harris, D. O., Lim, E. Y., Dedhiya D. D., "Probability of pipe fracture in the primary coolant loop of a PWR plant" NURG/CR-2189, vol. 5,1982.
5. Milne, I, Anisworth, R. A., Dowling A. R., Stewart A.J. 1986. Assessment of the integrity of structures containing defects. CEGB Report, Central Electricity Generating Board, U. K., R/H/R6-Rev3.
6. Rahman, S., "Probabilistic Fracture Analysis of cracked pipes with circumferential flaws", Int J. Pres. Ves. & Piping 70 (1997) 223-236.
7. Singh, P. K., Vaze, K. K., Kushwaha, H. S., "Characterisation of girth pipe weld for primary heat transport system of pressurized heavy water reactors", BARC/2002/E/004.

Reliability, Safety and Hazard
Editors: P.V. Varde et al.

EFFECT OF METEOROLOGICAL SAMPLING ON THE PROBABILISTIC SAFETY ANALYSIS LEVEL-3 RESULTS

*Pradeep Bhargava and V.V.S. Sanyasi Rao**
Health Physics Division
* Reactor Safety Division
Bhabha Atomic Research Centre, Mumbai 400 085

ABSTRACT

COSYMA Code is used to find out the effect of meteorological sampling in assessment of off-site radiological consequences of accidental release of radioactive material to the environment caused by meteorological sampling. Postulated release due to triple failures in Kakarapar Atomic Power Station (KAPS) is considered in this study. Individual risk of early health effect due to postulated accidental release of radionuclides is considered as an endpoint to find out the uncertainty. Results of the analyses show that ranges of predicted risk of health effect arising from different sampling scheme for 95th percentile are within the factor of two.

Keywords: Risk, Health effect, COSYMA, Radioactivity, Meteorology

1. INTRODUCTION

COSYMA (Code system from MARIA) [Hasemann et. al 1995] is applicable for probabilistic Safety Analysis (PSA) level III as well as deterministic calculations of the off-site consequences due to hypothetical accidental releases of radioactive material to atmosphere from nuclear facilities. The health effects, impact of countermeasures and economic costs of the releases can also be evaluated.
Effect of meteorological sampling on the PSA Level III [IAEA Safety Series, 1966] result is considered as an endpoint of this study. Aim of this study was to find out the uncertainty caused by meteorological sampling scheme for Kakarapar Atomic Power Station (KAPS) located at Kakarapar in Gujarat state of India.

2. DESCRIPTION OF COSYMA

Subsystem of COSYMA is Near Early (NE), Near Late (NL) and Far Late (FL). The NE sub-system is limited to calculate the early health effects and the influence of emergency actions to reduce those effects and is intended for use in the region near to the site. The NL subsystem is limited to calculate the late health effects and the impact of associated countermeasures, and is intended mainly for use in the region near to the site. The FL sub-system is concerned with calculating late health effects and effect of the appropriate countermeasures at larger distances from the site.

3. ACCIDENT SEQUENCE

In this study, triple failure i.e. LOCA + ECCS failure + Containment failure (Inlet damper failure) is considered as the accident scenario for 235 MWe Indian PHWR. This is one of the accident sequence considered for source term estimation in Chernobyl Committee Report [DAE SRC, December, 1987]. Noble gases and Iodines are the two-release groups of radionuclides considered for estimation of uncertainty in the risk of deterministic health effects in the population zone around the KAPS site at 3.3 km. Ground level release without plume rise has been considered. This provides a conservative estimate of the doses received by the off-site personnel.

The core inventories of KAPS reactor are based on the results from ORIGEN-2 [A G Croff, 1980]. The inventories are shown in Table 1. These inventories correspond to an average discharge burn up of 6400 MWD per ton of Uranium.

Methodology adopted for the source term estimation resulting from the accident sequence considered above is summarized below.

I. 100% of the core inventory of iodine and noble gases are assumed to be released from the core and all fission products (FPs) are assumed to be released instantaneously into the reactor containment after the accident.

II. For accident scenarios involving Loss of Coolant Accident (LOCA) with failure of the ECCS, a water-trapping factor of two for iodine release is considered in the PHT circuit.

III. The iodine plate-out halftime for deposition in the containment is 1.5 hr in primary envelope and 2.0 hr in secondary envelope. This assumption will have very little impact as the release duration is only six minutes, which is substantially smaller than the iodine plate out halftime.

IV. The radioactive decay of FPs in the containment has been considered.

On the basis of above methodologies as adopted in Chernobyl Committee Report [DAE SRC, December, 1987], overall release to the environment for Noble gases and Iodine are estimated to be 55% and 13.75% of the core inventories respectively. The duration of release has been taken as one hour in this study as required in COSYMA. The cure inventory and releases to the environment are shown in Table 1.

Table 1 Core inventory and Source term for KAPS

S/ No.	Radio nuclides	Inventory in Core (Bq)	Release to environment(Bq)
1	Kr-85	2.95E15	1.6225E15
2	Kr-85M	1.93E17	1.0615E17
3	Kr-87	3.63E17	1.9965E17
4	Kr-88	5.11E17	2.8105E17
5	Xe-133	1.60E18	8.8550E17
6	Xe-135	1.08E17	5.9400E17
7	I-131	8.11E17	1.0973E17
8	I-132	1.16E18	1.5695E17
9	I-133	1.62E18	2.1919E17
10	I-134	1.75E18	2.3678E17
11	I-135	1.50E18	2.0295E17

Atmospheric Dispersion

A Segmented Gaussian plume model [Hasemann et. al.,1995], which assumes that the atmospheric conditions affecting the plume are determined from those at a single meteorological station representing the site of release is considered. It assumes that the wind direction changes every hour throughout the release and the subsequent travel. In this case dispersing material assumed to follow a trajectory, which consist of a series of straight segments.

Dry deposition is calculated using the deposition velocity, with the amount of material remaining in the plume calculated using the source depletion model, in which the vertical profile of the plume is assumed to remain Gaussian. Wet deposition is included using the washout coefficient, which increases with increasing rainfall rate.

The effects of down wash and mixing in the wake of a reactor building have been considered using a virtual source model based on the work of Hanna [Hanna et. al., DOE, 1982].

4. METEOROLOGICAL SAMPLING

PRA predicts the probability distribution of consequences, if an accident occur in any of the wide range of atmospheric conditions which might occur at the site of interest during the period in which site operates. The sequences of conditions are obtained from the hourly spaced meteorological data over a period of few years, and assuming that the conditions during the future operation of the site will be similar to those observed in the past. It is impossible to undertake the calculations for every sequence of conditions over the operating period of the site in advance. Therefore a representative sample of starting times or weather sequences must be used. A meteorological sampling scheme derived for use with a trajectory model must explicitly consider wind direction as well as the variation of atmospheric conditions with distance from the site.

There are three ways in which the starting times can be selected. The simplest systems are random and cyclic sampling. In random sampling the random sample are simply chosen at random from the Meteorological data of the whole year. In cyclic sampling the starting time are taken to be every nth hour on the meteorological data.

Stratified sampling [Jones J A et. al., EUR 16239, 1995], which is widely used for sampling weather sequences, was used to select the starting time or weather sequence from the KAPS meteorological data for the year 2000. Sequences of conditions from the meteorological data, for which the consequences are likely to be similar, are combined into groups. Sequences are chosen at random from those within each group and assigned a probability based on number of sequences allocated to each group and the number selected from within each group.

The grouping scheme distinguishes two main type of sequences; 1) Dry (without rain), 2) Wet (with rain) occurring within about 100 km of the site. Dry Sequences were further divided into groups depending upon travel time of material to two distances (20 km and 60km) from release point, and the stability category in the first hour of release. Wet sequences can be defined by quantities related to the wet deposition density rather than just to rainfall rate. The distance at which high deposition occurs is also important. The sequences with rain in the scheme adopted here were grouped in terms of ratio of washout coefficient (Λ) to wind speed (u). Meteorological sampling scheme adopted in this study are shown in Tables 2 and 3.

Two different sampling schemes: cyclic and stratified sampling scheme are considered to carry out the uncertainty analysis.

1. 8760, 400 and 300 weather sequences are considered in cyclic sampling
2. Two binning scheme are considered in stratified sampling. 147 weather sequences in stratified 1 (ST-1) and 114 weather sequences in stratified 2 (ST-2) are considered

Table 2 The sampling scheme (ST-1) used in this study

Group	Definition		
Groups are based on ratio of washout coefficient to wind speed			
1	$\Lambda/u > 5.10^{-5}$ within 10 km		
2	$\Lambda/u > 5.10^{-5}$ within 25 km		
3	$\Lambda/u > 5.10^{-5}$ within 50 km		
4	$\Lambda/u > 5.10^{-5}$ within 75 km		
5	$\Lambda/u > 5.10^{-5}$ within 100 km		
6	$\Lambda/u > 5.10^{-6}$ within 10 km		
7	$\Lambda/u > 5.10^{-6}$ within 25 km		
8	$\Lambda/u > 5.10^{-6}$ within 50 km		
9	$\Lambda/u > 5.10^{-6}$ within 75 km		
10	$\Lambda/u > 5.10^{-6}$ within 100 km		
Groups defined using initial stability category and travel time in hour for 20 km and 60 km			
	Category	Travel time (hr)	
		20 km	60 km
11	A/B	Any	Any
12	E/F	<2	<6
13	E/F	<2	>6
14	E/F	>2	<6
15	E/F	>2	>6
16	C/D	<1	<3
17	C/D	<1	>3
18	C/D	>1	<3
19	C/D	Any	<5
20	C/D	Any	Any

Table 3 The sampling scheme (ST-2) used in this study

Group	Definition		
Group are based on ratio of washout coefficient to wind speed			
1	$\Lambda/u > 5.10^{-5}$ within 10 km		
2	$\Lambda/u > 5.10^{-5}$ within 25 km		
3	$\Lambda/u > 5.10^{-5}$ within 50 km		
4	$\Lambda/u > 5.10^{-5}$ within 75 km		
5	$\Lambda/u > 5.10^{-5}$ within 100 km		
6	$\Lambda/u > 5.10^{-6}$ within 10 km		
7	$\Lambda/u > 5.10^{-6}$ within 25 km		
8	$\Lambda/u > 5.10^{-6}$ within 50 km		
9	$\Lambda/u > 5.10^{-6}$ within 75 km		
10	$\Lambda/u > 5.10^{-6}$ within 100 km		
Groups defined using initial stability category and travel time in hour for 20 km and 60 km			
	Category	Travel time (hr)	
		20 km	60 km
11	A/B	<5	<8
12	A/B	<5	Any
13	A/B	Any	<13
14	A/B	Any	Any
15	E/F	<2	<4
16	E/F	<2	<6
17	E/F	<2	Any
18	E/F	Any	Any
19	C/D	<2	<4
20	C/D	<2	Any
21	C/D	<3	<4
22	C/D	<4	<6
23	C/D	<4	<7
24	C/D	<4	<8
25	C/D	<5	<9
26	C/D	Any	Any

5. PATHWAYS OF EXPOSURES

Exposure pathways considered for estimating acute risk for adults are cloud shine, inhalation, skin doses from material deposited on skin, ground shine, and resuspension of ground deposited activity. The first three routes of irradiation only affect those people who are exposed to the plume of material as it passes overhead.

External irradiation from the cloud only occurs while the plume is present. Material is only inhaled while the plume is present, but some material can be retained in the body for long periods. Irradiation from third route stops once material is removed from the skin and clothing. The other two routes can lead to the source of exposure being present for long periods of time.

6. SHORT-TERM INDIVIDUAL RISKS

The risk of suffering from early health effects following irradiation increases rapidly as the dose increase above the threshold value. Short-term individual risk [KfK and NRPB, COSYMA: EUR-13028, 1991] is estimated using the formula,

$$R = 1 - \exp(-H)$$

The hazard function H is defined as,

$$H = \ln 2 \, (D/D_{50})^{S}$$

Where, D is the dose received in the appropriate period, D_{50} is the dose which cause the effect in 50% of the population and S is the shape parameter, which characterizes the dose-risk relationship. D_{50} and S are specific to the health effect considered. Doses that are accumulated at low dose rate are much less effective in causing early health effects than dose delivered at high dose rate. This include in the model by summing over in a number of different time periods.

7. RESULTS AND CONCLUSION

Results of COSYMA runs are presented as complementary cumulative distribution function in figure 1, 2 and 3. It shows the probability of exceeding particular levels of the consequences allowing for the range atmospheric conditions that could occur at the time of accident. The ratio of maximum to minimum value for 95th percentile is presented in Table 4.

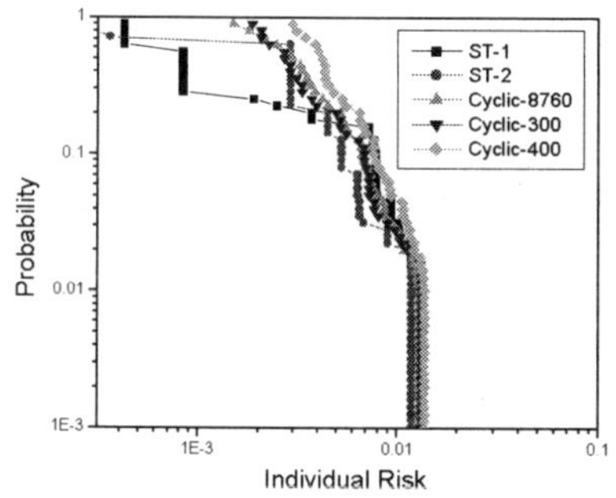

Figure 1: Total Risk of morbidity at 3.3Km

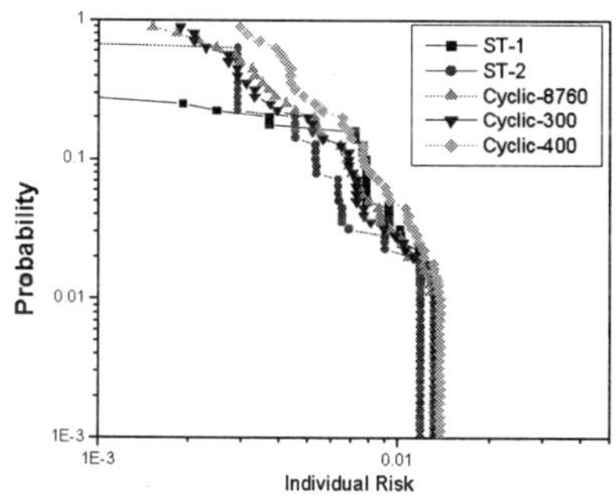

Figure 2: Risk of Skin Morbidity at 3.3Km

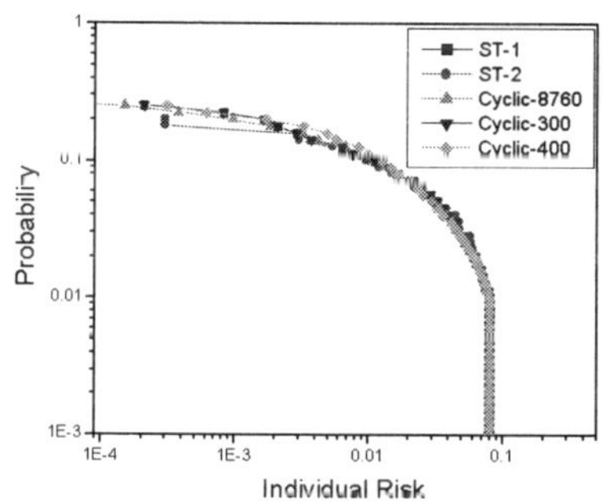

Figure 3: Risk of Mental Retardation at 3.3 Km

Table 4 Ratio of maximum to minimum values for 95th percentile of the risk at 3.3 Km

Effect (Risk of morbidity)	Ratio
Mental Retardation	1.30
Skin	1.47
Total	1.47

Cyclic sampling tends to select those conditions that occur more frequently, and so give poor predictions of high percentiles of the probability distribution. The main difficulties with stratified sampling is to identify a sampling scheme where the variation between the consequences corresponding to different sequences allocated to a group is small compared to the differences between the consequences for sequences in different group. A further difficulty is the possible mis-allocation of a sequence with high-predicted consequences, and low correct probability, to a group with much higher probability. Stratified sampling introduces uncertainty [L.H.J. Goossens et. al., EUR 18825EN] into the final predictions for two closely related reasons, namely the choice of the sampling scheme and selection of the representative starting times from within each group of the scheme. Random and cyclic sampling introduces uncertainty into the overall predictions from the choice of starting times.

8. REFERENCES

1. A G Croff, A user's manual for the ORIGEN2 computer code, ORNL/TM-7175, July 1980.
2. Hanna, S R, Briggs, G A and Hosker, R P. Handbook on atmospheric diffusion. Technical Information Center, US Department of Energy (1982).
3. Hasemann I, Jones J A: COSYMA User Guide (Version 95/1) ; A joint report by NRPB and KfK, EUR 13045, KfK 4331 B
4. Jones J A, Mansfield P A, Haywood S M, Hasemann I, Steinhaur C, Ehrhardt J and Faude D. PC-COSYMA (Version 2); an accident consequences assessment package for use on PC. EUR report 16239 (1995).
5. KfK and NRPB. COSYMA: A new program package for accident consequences assessment. CEC. Brussels, EUR-13028 (1991)
6. L.H.J. Goossens, R.M. Cooke, B.C. Kraan, J. Ehrhardt, F.Fischer, I. Haseman, J.A. Jones, J.Brown, A. Khuesheed, A. Phipps, Probabilistic accident consequences uncertainty assessment using COSYMA, Report: EUR 18825EN
7. Procedure for conducting Probabilistic Safety Assessments of Nuclear Power Plants (Level 3); Safety Series No. 50 P-12, IAEA Vienna (1996)
8. Report Of the Task Force On Review Of Indian NPPs In The Light Of The Accident At Chernobyl Reviewed And Revised By DAE SRC

Reliability, Safety and Hazard
Editors: P.V. Varde et al.

PROBABILISTIC AND POSSIBILISTIC APPROACH FOR ASSESSMENT OF RADIOLOGICAL RISK DUE TO ORGANICALLY BOUND AND TISSUE FREE WATER TRITIUM

Sudhir Dahiya, A.G. Hegde, P.C. Verma[#], M.L. Joshi and H.S. Kushwaha[*]

ESS, Health Physics Division,
Environmental Surveillance Laboratory, RAPS, Kota
*Health, Safety and Environment Group
Bhabha Atomic Research Centre, Mumbai 400 085

ABSTRACT

This study illustrates use of two approaches namely probabilistic using Monte Carlo simulation (MCS) and possibilistic using fuzzy α-cut (FAC) to estimate the radiological cancer risk to the population from ingestion of organically bound tritium (OBT) and tissue free water tritium (TFWT) from fish consumption from the Rana Pratap Sagar Lake (RPSL), Kota. Using FAC technique, radiological cancer risk rate (year^{-1}) at $A_{\alpha 1.0}$ level were 1.15E-08 and 1.50E-09 for OBT and TFWT respectively from fish ingestion pathway. The radiological cancer risk rate (year^{-1}) using MCS approach at 50th percentile (median) level is 1.14E-08 and 1.49E-09 for OBT and HTO respectively from ingestion of fresh water fish.

Keywords: Radiological Cancer Risk, Organically Bound Tritium, Tissue Free Water Tritium, Rana Pratap Sagar Lake, Monte Carlo Simulation and Fuzzy **α**-Cut Technique

1. INTRODUCTION

Rajasthan Atomic Power station (RAPS) was the site, where two units of Pressurised Heavy Water Reactor (PHWR) of 220 $MW_{(e)}$ each were commissioned during mid seventies. This station is situated on the eastern bank of RPSL. Presently there are four units (PHWR) are under operation at Rajasthan Atomic Power Station (RAPS) in which natural uranium oxide is used as fuel and heavy water as moderator and coolant. Tritium is mainly produced due to activation of the deuterium present in heavy water. The production rate of tritium in a PHWR system has been estimated to be about 6.29 TBq per $MW_{(e)}$.Y (Pillai et al., 1990). RAPS discharges low level liquid waste to RPSL in the form of tritiated water (HTO) through aquatic and atmospheric routes within the authorised limit prescribed by Atomic Energy Regulatory Board, India.

In aquatic system, tritium gets incorporated in body tissue and gets equilibrated with the body water of aquatic fauna and flora within a time span of less than one hour by exchange of tritium present in water with labile hydrogen present in tissues. A part of HTO/TFWT can get incorporated in body tissue of organism during formation of tissues or by exchange of tritium with exchangeable hydrogen present in the body tissues. Tritium occurs in biological samples in three forms (Belot and Guenot, 1984; Sweet and Murphy, 1984; Brown, 1988; Brown, 1995):

1) in the free water (HTO molecules) associated with the material;

2) bound to oxygen and nitrogen atoms in compounds of the material;

3) bound to carbon atoms in the material.

Component (1) is easily removed by evaporation at low temperature or by freeze-drying and is in the form of tritiated water (HTO). Tritium of component (2) is exchangeable with water of the system to varying degrees depending on the molecular structure. It is referred to as 'exchangeable OBT (organically bound tritium)'. It rapidly reaches equilibrium with the overall aqueous component of the system, and the free water and the exchangeable OBT therefore tends to equilibrate rapidly with local environment. In contrast, tritium of component (3) is tightly bound to the organic structure. It is thus no longer related to the local environmental hydrogen isotope composition after its incorporation, but represents an integration of the environmental tritium levels during the growing periods and, more generally, during the functioning periods of the biological organism. It is often referred to as 'nonexchangeable OBT', which can only be released on decomposition of the organic compound.

Tritium working group in IAEA, CRP-BIOMASS concluded that for routine release of tritium, present work on HTO/TFWT monitoring are sufficient and reliable, but work on OBT must be enhanced in order to make defendable radiological dose and risk assessment to the population in vicinity of PHWRs or any other facility handling tritium (IAEA, 2001). Radiological cancer slope factor and dose coefficients are more than twice for OBT in comparison to HTO through ingestion pathway due to its longer biological half life.

The wide spread use of sophisticated instrumentation in sample processing and analysis often fails to retain the uncertainty at various steps during data generation. In present study methods based monte-carlo simulation (probabilistic) and fuzzy set theory based alpha cut technique (possibilistic approach) are used to assess the health risk from TFWT and OBT from fresh water fish consumption from RPSL, Kota.

2. MATERIALS AND METHODS

Sampling and Analysis

A total of thirteen types of the fresh water fish species were collected from the RPSL and analysed for TFWT and OBT (Roy et al., 2003).

Radiological Cancer risk assessment Carcinogenic risk was calculated by multiplying the cancer slope Factor (ingestion) of the TFWT and OBT by the total ingestion of type of tritium by an individual as a linear relationship which can be written as:

$$\textbf{RISK} = \boldsymbol{CDI(Ing) \times CRC(Ing)} \text{ --- (1)}$$

where CDI = Chronic daily intake of the TFWT/OBT (Bq/day) through ingestion pathway. CRC is the cancer risk coefficient for TFWT and OBT (USEPA FGR-13, 1999).

Fuzzy set theory: The concept of a fuzzy set theory was introduced by Zadeh (1965) and has been used in various fields in dealing the uncertainty handling. In fuzzy set theory the degree of belonging of any parameters to a well defined set is described in term of membership functions. The set A is defined in term of its membership function by $A = \{\mu_A(x), x \, \varepsilon \, X\}$, these values of the membership function are also called as membership degree as given below

$$A = \begin{pmatrix} = 1 & \text{x is full member of A} \\ \varepsilon\,(0,1) & \text{x ia a partial member of A} \\ = 0 & \text{x is not a member of A} \end{pmatrix}$$

Fuzzy α-Cut Technique: Fuzzy alpha cut (FAC) technique uses above discussed model to represent imprecision in the parameters. Uncertain parameters are considered to be fuzzy numbers with some membership functions. Figure 1 shows a parameter represented as a triangular fuzzy number with support of A_0. Wider the support of membership function, higher is the uncertainty. At a resolution level of α, it will have support of A_α. Higher the value of α, higher will be the confidence in parameter (Li and Yen, 1995). The method is based on extension principle, which implies that functional relationship can be extended to involve fuzzy argument and can be used to map the dependent variable as a fuzzy set. In simple arithmetic operation, this principle can be used analytically. For each α-cut level of all the parameters, two values, one from lower side and one from higher side of triangular function will be generated. These values generated at α cut levels were used in equation 1 to estimate the carcinogenic risk to determine the minimum and maximum possible value of the output at that α-cut level. This information was directly used to construct the corresponding fuzziness (membership function) of the output, which is used as a measure of uncertainty.

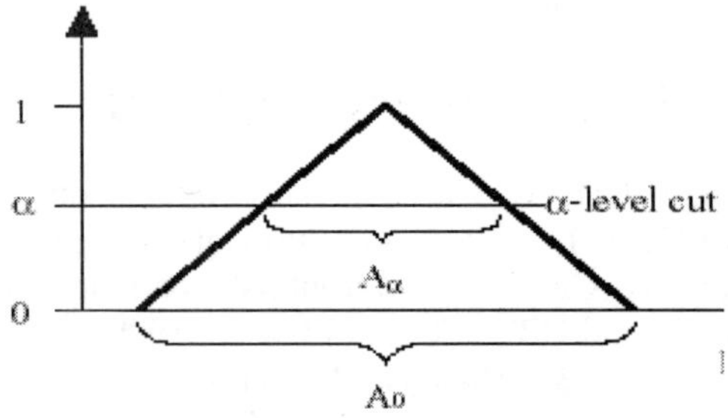

Fig. 1: Fuzzy number, its support and α-cut

Monte Carlo Simulation

The principle of MCS is illustrated by Figure 2. Data of all the input parameters (**P**) like OBT / TFWT in fish and fish intake rate of population etc subject to uncertainty were considered as a random variables and analysed to best fit a distribution. A number for each parameter from their well defined distribution was drawn and used to calculate an output for radiological risk as per the equation 1. These numbers were generated n times (n=5000 in this study) and a set of output random variable R generated. The statistical properties of **R** are computed from output of results Ri

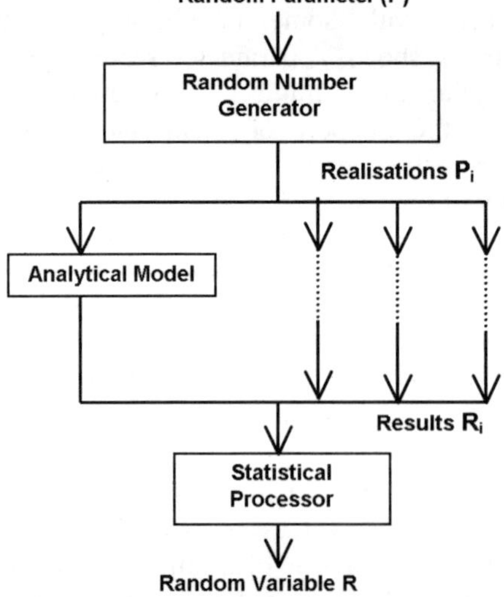

Fig. 2: Sketch of the Monte Carlo Simulation

3. RESULT AND DISCUSSION

The characteristics were used from the demographic data available in literature for population in the studied area. In the probabilistic approach the intake of fish and TFWT and OBT content in the edible part of the fish was statistically analysed to fit in the appropriate distribution and the random numbers were generated from that distribution for cancer risk estimation. Cancer Risk coefficients were used from FGR-13 (USEPA, 1999). In case of possibilistic analysis all the parameters were considered as triangular fuzzy functions for the analysis of uncertainty in cancer risk assessment. In addition to the cancer mortality risk internal dose from TFWT and OBT through fish ingestion food pathway were also estimated. An average annual dose to population through fish ingestion pathway using MCS approach worked out to be 0.03μSv/year (90% CI- 0.006, 0.09μSv/year) and 0.23μSv/year (90% CI- 0.06, 0.55μSv/year) for HTO and OBT respectively. At RAPS site, the apportioned dose to tritium through aquatic pathway stipulated by AERB is 30μSv/year (20μSv/year for RAPP 1, 2, 3&4 and 10μSv/year to heavy water plants). Using FAC technique, radiological fatal cancer risk rate ($year^{-1}$) at $A_{\alpha 0.05\ lower}$, $A_{\alpha 1.0}$ and $A_{\alpha 0.05upper}$ level are found to be 3.27E-09, 1.15E-08 and 4.45E-08 respectively from OBT via fish ingestion pathway alone (Figure 3).

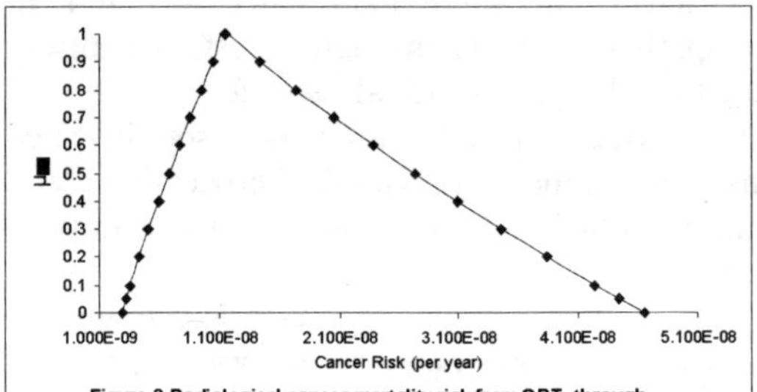

Figure 3 Radiological cancer mortality risk from OBT through Fish ingestion pathway using FAC technique

Radiological fatal cancer risk rate ($year^{-1}$) from TFWT through fresh water fish ingestion using FAC technique are found to be 3.14E-10, 1.50E-09 and 1.01E-08 at $A_{\alpha 0.05\ lower}$, $A_{\alpha 1.0}$, and $A_{\alpha 0.05upper}$ level respectively (Figure 4), which is almost eight times less than the risk from OBT ingestion at $A_{\alpha 1.0}$ level.

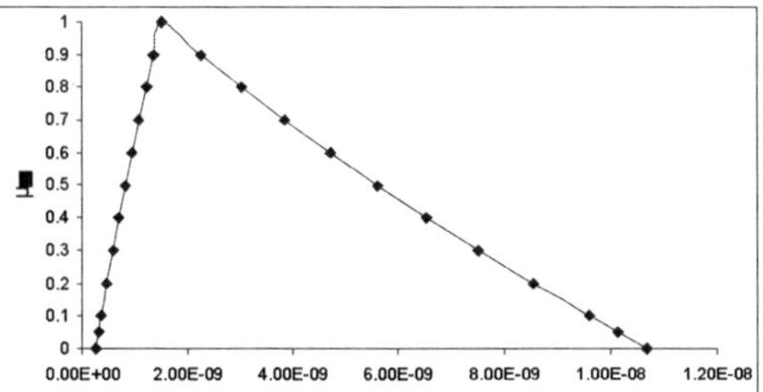

Figure 4 Radiological cancer mortality risk from TFWT through Fish ingestion pathway using FAC technique

The radiological cancer risk rate (year^{-1}) using MCS approach at 50th percentile (median) level was 1.14E-08 (90% CI- 3.85E-09, 3.48E-08) from OBT on the ingestion of fresh water fish as shown in Figure 5. There is spread of almost one order at 5th and 95th percentage levels due to the variation in OBT contents in fish and intake rate as well.

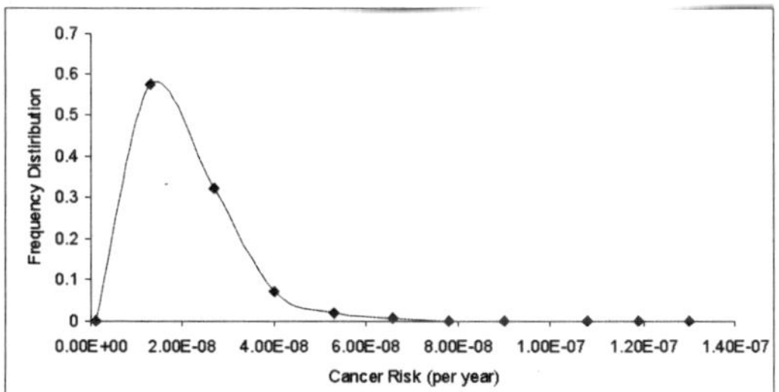

Figure 5 Frequency distribution of radiological cancer mortality risk from OBT through Fish ingestion pathway using MCS

The radiological fatal cancer risk rate (year^{-1}) in MCS approach at50th percentile (median) level was 1.49E-09 with a ninety percent confidence interval 3.73E-10 to 6.28E-09 for HTO respectively from ingestion of fresh water fish from RPSL (Figure 6).

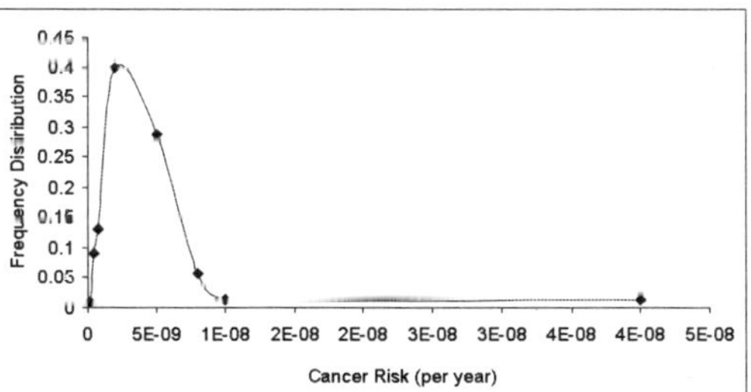

Figure 6 Frequency distribution of radiological cancer mortality risk from TFWT through Fish ingestion pathway using MCS

Both the results from MCS and FAC are not having the exactly same risk levels due to the difference in the form of inputs used. Random number from various well defined distributions were drawn for the assessment of risk in MCS technique, whereas in FAC technique, which is extension of interval calculation the values are drawn from the linear slope from triangular fuzzy membership function. In MCS assessment the analytical model given in equation 1 is applied on the random number basis whereas in FAC technique respective alpha cut values from all triangular fuzzy membership function are used for the risk assessment, which always provides space for over or under estimation of the risk in both the approaches of uncertainty analysis. FAC approach which is extension of interval calculation method always gives a conservative result which is confirmed from the results in this work as well.

The radiological cancer risk at 95 percentile level in MCS approach was 3.48E-08 and 6.28E-09 from OBT and TFWT from fish ingestion pathway are found to be negligible in comparison to the risk from medical and natural exposure at the site. It can be concluded from the results of this work that OBT monitoring in food items is equally essential as of HTO/TFWT because the radiological cancer risk rate is more than five times higher from OBT than TFWT at 90% confidence level using MCS and using FAC, the radiological risk from OBT through fish ingestion pathway is 4.4 times higher than TFWT at $A_{\alpha 0.05 upper}$ level.

4. REFERENCES

1. Belot, Y., Guenot, J., 1984. *Assimilation of 3H in photosynthesizing leaves exposed to HTO*. Health Physics, **47**, 849–855.
2. Brown, R.M., 1988. *Assessment of the significance of organically-bound tritium in environmental materials.* Report AECL-CRNL/INFO-0283.
3. Brown, R.M., 1995. *The measurement of tritium in Canadian food items.* Report AECL-CRNL/INFO- 0499 AECL Project No. 3.135.1.
4. Fisher, B. (2003). *Fuzzy environmental decision-making: applications to air pollution.* Atmospheric Environment, **37**, 1865-1877.
5. IAEA (2001) BIOMASS PROGRAMME version beta-2 CD-rom IAEA February 2001.

6. Li, H.X. and Yen, V. C. (1995). Fuzzy set theory and Fuzzy decision making. CRC Press
7. Pillai, K. C. and Iyengar, T. S. (1990) *An Overview of tritium generation from nuclear operations and its significance- symposium on tritium measurements and applications.* Bulletin of Radiation Protection, Feb 22-23. **13** (1), 129-184.
8. Roy, Alpana; Verma, P.C. and Gurg , R.P. (2003). *Organically bound tritium in Fish at Rana Pratap Sagar Lake.* National Symposium on Environment **(NSE-12)**, 535-538.
9. Sweet, C.W., Murphy, C.E., 1984. *Tritium deposition in pine trees and soil from atmospheric releases of molecular tritium.* Environmental Science Technology **18**, 358–361.
10. United State Environmental Protection Agency (1999). *Cancer Risk Coefficients for Environmental Exposure to radionuclides.* Federal Guidance Report No.13, **EPA 402-R-99-001**, 94
11. Zadeh, L. A. (1965). Fuzzy sets. Inf. Control, **8**:338-353.

Reliability, Safety and Hazard
Editors: P.V. Varde et al.

ASSESSMENT OF INHALATION RISK DUE TO RADIOACTIVITY RELEASED FROM COAL-BASED THERMAL POWER PLANT

S.K. Sahu, G.G. Pandit, V.K. Shukla, V.D. Puranik and H.S. Kushwaha[1]

Environmental Assessment Division
Health, Safety and Environment Group[1]
Bhabha Atomic Research Centre, Mumbai 400 085

ABSTRACT

In India, the coal based thermal power plants have been the major source of power generation in the past and would continue for decades to come. As the coal contains naturally occurring primordial radionuclides the burning of pulverized coal to produce energy for generation of electricity in thermal power plants will result in the emission of a variety of natural radioactive elements into the environment in the vicinity of thermal power plants. In this paper we have used two different methods for characterization of uncertainty in inhalation risk to the general public around 10 Kms radius in the neighborhood of a coal-fired thermal power plant.

Keywords: Thermal Power Plant, Radionuclides, Inhalation Risk, Fuzzy Logic, Monte Carlo Simulations

1. INTRODUCTION

In India, the coal based thermal power plants have been the major source of power generation. The burning of pulverized coal for generation of electricity in thermal power plants results in huge quantity of fly ash as the coal used in India for power generation is of high ash content. Although particle emission controls in power stations may reach high retention efficiency (99% by using electrostatic precipitators), considerable amounts of fine fly ash particles are still emitted from the stack to the atmosphere because of the high rate of coal consumption [Mishra, 2004]. As the coal contains naturally occurring primordial radionuclide the use of coal as a heat source for electric power generation will result in the emission of a variety of natural radioactive elements into the environment in the vicinity of thermal power plants through fly-ash emissions and is one of the sources of technologically enhanced exposure to population. The collective doses to the population arise primarily through inhalation of radioactivity during the passage of the cloud containing fly-ash emitted from the stack [Tso and Leung, 1996]. Since many of the thermal power plants in India are situated in densely populated areas, the estimation of radiological risk to the neighborhood population may be of significance.

This paper deals with assessment of radiological risk to the general public around 10 Kms radius in the neighborhood of a 1000 MWe coal-based thermal power plant. We have used two different methods for characterization of uncertainty in inhalation risk due to radionuclide escaping from the stack of thermal power plant. The first method used is fuzzy logic-based alpha-cut analysis in which uncertain parameters are treated as fuzzy numbers with given membership functions. In the second method Monte Carlo simulation that treats parameters as random variables bound to a given probabilistic distribution was used to evaluate the distribution of the resulting output.

2. SOURCE TERM

In India, even though coal is available in abundance, its quality is poor. The ash content of Indian coal is in the range of 23–48% (average 40%). The consumption rate of coal for electricity generation is about 12.5 tonnes per MW per day. For a typical 1000 MWe coal based thermal power plant the daily consumption of coal is 12500 tonnes. If we assume the efficiency of the electrostatic precipitator for discharge control of fly ash is 99% then only 1 % of the fine ash is released into the ambient air through the stack of the plant. Therefore the emission rate of the fly ash from the stack is work out to be 0.58 Kg/sec.

The radioactivity content in fly ash and the emission rate of various radionuclides from the stack of a typical 1000 MWe coal based thermal power plant is given in Table-1. The emission rate of radionuclide was estimated by multiplying the emission rate of fly ash from the stack with the concentration of the respective radionuclide in fly ash.

Table 1 Radioactivity content in fly ash and emission rate of various radionuclide (mean values are given in parentheses)

Radionuclide	Concentration (Bq/Kg)	Emission Rate (Bq/Sec)
^{226}Ra	40.7-151.7 (77.7)	23.6-88.0 (45.0)
^{228}Th	95.1-230.5 (125.8)	55.2-133.7 (73.0)
^{210}Pb	25.7-69.6 (40.5)	14.9-40.4 (23.5)
^{238}U	18.8-121.4 (60.3)	10.9-70.4 (35.0)

3. RISK ANALYSIS

Risk assessment is the process that estimates the likelihood of occurrence of adverse effects to humans and ecological receptors as a result of exposure to hazardous chemical, radiation, and/or biological agents. Quantitative risk characterization involves evaluating exposure estimates against a benchmark of toxicity, such as a cancer slope factor. Risk is calculated by multiplying the carcinogenic slope factor of the radionuclide by the dose an individual receives and therefore can be written as

$$\text{Risk} = \text{Dose} * \text{SF} \quad (1)$$

The collective effective dose to the population living around 10 Kms radius in the neighborhood of coal-based thermal power plant were calculated following the method outlined by Papastefanou [Papastefanou, 1996].

$$\text{Dose} = \frac{Ei * IR * DCF}{vd * S} \quad (2)$$

where E_i (Bq/Sec) is the emission rate of radionuclide from the stack, IR is the inhalation rate (Bq/m^3), DCF is the dose conversion factor (Sv/Bq), v_d is the deposition velocity (m/Sec), and S is the surface area (m^2). 0.01 m/Sec is taken as deposition velocity [UNSCEAR, 1982]

Two different methods have been used to evaluate the inhalation risk due to radionuclide release from thermal power plant. Monte Carlo Analysis (MCA) is the most widely used probabilistic method in risk assessment. The MCA technique treats any uncertain parameter as random variable that obeys a given probabilistic distribution. This technique is widely used for analyzing probabilistic uncertainty. In MCA computer simulation are use to combine multiple probability distributions associated with the risk equation. Thus we get a probabilistic distribution for the risk.

Fuzzy set theory provides the basis for a method to describe uncertainty and vagueness. Fuzzy set theory was first introduced by L. A. Zadeh [Zadeh, 1965] and has been applied in various fields. Objects of fuzzy set theory are fuzzy variables whose boundaries are imprecise. Thus the fuzzy variables facilitate gradual transition between states and, consequently possess a natural capability to express and deal with observation and measurement uncertainties. One of the most widely used concepts of fuzzy set theory is the fuzzy alpha cut (FAC) technique. Given a fuzzy set A defined on X and any number $\alpha \in [0, 1]$, the α-cut, A^α, is the crisp subset of X consisting of all of the elements of X for which membership grades in A are greater than or equal to the specified value of α. To summarize, the membership value of a real number reflects the ''likeliness'' of occurrence of that number; the alpha-cut level reflects the different sets of numbers with a given minimum likeliness [Kaufmann, 1985].

4. RESULTS AND DISCUSSION:

The α-cut values for 0.05 ($A_{\alpha\text{-}0.05\ LHS}$ and $A_{\alpha\text{-}0.05\ RHS}$) and 1.0 ($A_{\alpha\text{-}1.0}$) arises from FAC method for inhalation risk due to ^{226}Ra, ^{228}Th, ^{210}Pb and ^{238}U was given in Table-2. The value corresponding to a membership function of 1.0 ($A_{\alpha\text{-}1.0}$) corresponds to the most possible value. The tabulated result show the inhalation risk arises due to ^{228}Th is much higher compare to other radionuclide.

Table 2 Inhalation risk arises due to variousbv radionuclides from FAC method

Radionuclide	$A_{\alpha\text{-}0.05\ LHS}$	$A_{\alpha\text{-}0.05\ RHS}$	$A_{\alpha\text{-}1.0}$
^{226}Ra	$6.7X10^{-10}$	$2.2X10^{-8}$	$6.4X10^{-9}$
^{228}Th	$1.5X10^{-9}$	$3.3X10^{-8}$	$1.0X10^{-8}$
^{210}Pb	$4.2X10^{-10}$	$1.0X10^{-8}$	$3.3X10^{-9}$
^{238}U	$3.3X10^{-9}$	$1.7X10^{-8}$	$5.0X10^{-9}$

Fig. 1 shows the membership function of total inhalation risk arises from radionuclides (^{226}Ra, ^{228}Th, ^{210}Pb and ^{238}U) emitted from thermal power plant. The α-cut values for 0.05 ($A_{\alpha\text{-}0.05\ LHS}$ and $A_{\alpha\text{-}0.05\ RHS}$) and 1.0 ($A_{\alpha\text{-}1.0}$) arises from FAC method was 2.94 X 10^{-9}, 8.19 X 10^{-8} and 2.50 X 10^{-8} respectively.

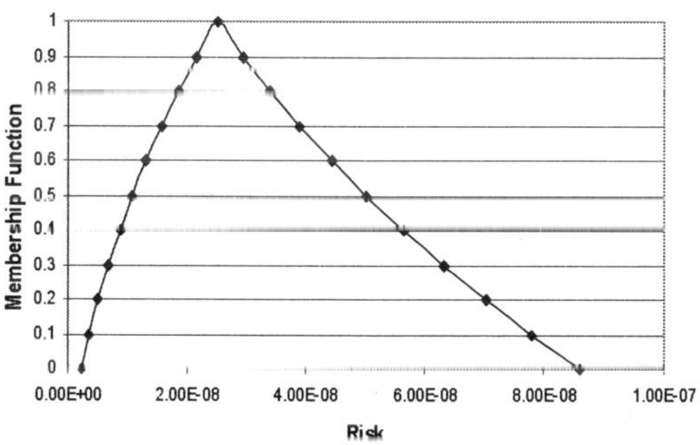

Fig. 1: Total radiological risk due to inhalation of radionuclide for FAC

Monte Carlo simulation technique has been applied to find out the probabilistic inhalation risk. The 5th percentile, 95th percentile and mean values of inhalation risk due to ^{226}Ra, ^{228}Th, ^{210}Pb and ^{238}U arises from MCA method was given in Table-3. The result show the inhalation risk arises due to ^{228}Th is significant compare to other radionuclide.

Table 3 Inhalation risk arises due to various radionuclides from MCA method

Radionuclide	5th Percentile	95th Percentile	Mean
^{226}Ra	$2.7X10^{-10}$	$1.1X10^{-8}$	$3.5X10^{-9}$
^{228}Th	$3.1X10^{-9}$	$1.6X10^{-8}$	$9.9X10^{-9}$
^{210}Pb	$8.3X10^{-10}$	$5.5X10^{-9}$	$3.2X10^{-9}$
^{238}U	$5.7X10^{-9}$	$3.2X10^{-8}$	$4.2X10^{-9}$

Fig. 2 shows the cumulative distribution of total inhalation risk arises from radionuclides (^{226}Ra, ^{228}Th, ^{210}Pb and ^{238}U) using Monte Carlo simulation. The 5th percentile, 95th percentile and mean values for total inhalation risk arises from radionuclide (^{226}Ra, ^{228}Th, ^{210}Pb and ^{238}U) was found to be 3.83 X 10^{-9}, 6.50 X 10^{-8} and 2.08 X 10^{-8} respectively using Monte Carlo simulation.

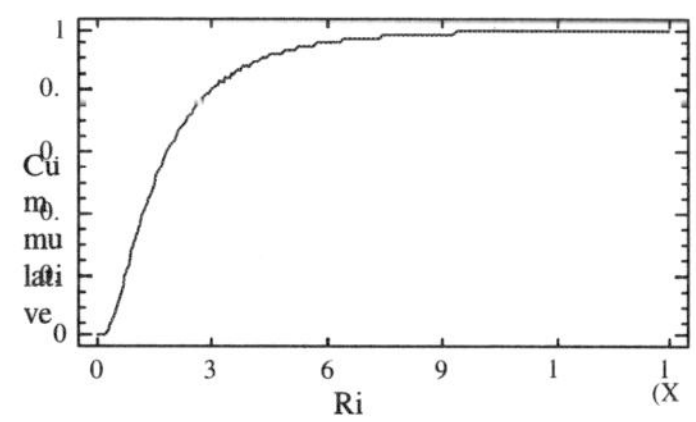

Fig. 2: Cumulative distribution of total radiological risk due to inhalation of radionuclide

Fig.3 depicted the percentage distribution of radiological risk due to inhalation of radionuclide emitted from the stack of thermal power plant. Among the radionuclide considered in this study, ^{228}Th contributed maximum to the total inhalation risk (41%) while the minimum contribution was from ^{210}Pb (13%). The higher contribution to the total risk from ^{228}Th can be attributed to its higher concentration in fly ash.

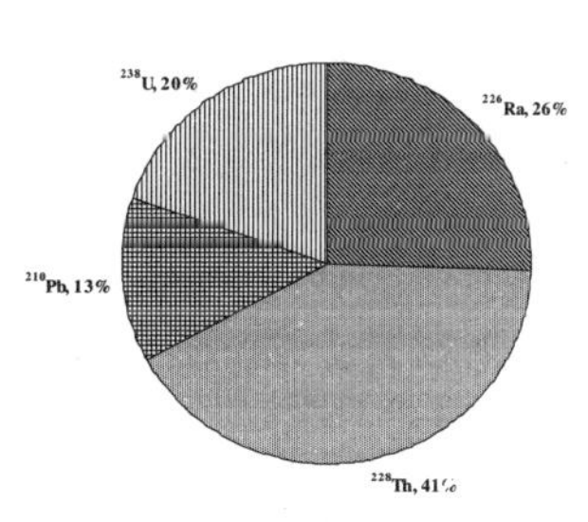

Fig. 3: Percentage distribution of radiological risk due to inhalation of fly ash.

5. REFERENCES

1. Kaufmann A, Gupta, M.M. (1985) Introduction to fuzzy arithmetic: theory and application. Van Nostrand Reinhold Co., New York.
2. Mishra, U.C. (2004) Environmental impact of coal industry and thermal power plants in India. J. Environmental Radioactivity, 72, 35-40.
3. Papastefanou, C. (1996) Radiological impact from atmospheric releases of ^{226}Ra from coal-fired power plants. J. Environmental Radioactivity, 32, 105-114.
4. Tso, Man-yin W. And Leung, K.C. (1996) Radiological impact of coal ash from the power plants in Hong Kong. J. Environmental Radioactivity, 30, 1-14.
5. UNSCEAR (1982) Ionizing radiation: sources and biological effects. United Nation Scientific Committee on the Effects of Atomic Radiation, United Nations, New York.
6. Zadeh LA (1965) Fuzzy Sets. Inf Control 8:338–353.

Reliability, Safety and Hazard
Editors: P.V. Varde et al.

ACCIDENT SOURCE TERMS AND PUBLIC DOSES FOR THE KUDANKULAM VVER-1000

Kapil Deo S. Singh[a], V.K. Sharma[a] and Anuradha V.[b]

[a]Health Physics Division, BARC, Mumbai 400 085
[b]Operating Plant Safety Division, AERB, Anushaktinagar, Mumbai 400 094

ABSTRACT

The Kudankulam VVER-1000 (Model V-392) reactor is an advanced PWR meeting Russian, Western, and IAEA standards. The safety review of this reactor was undertaken by AERB in 2001 prior to granting the construction licence. As part of this exercise, an independent evaluation of the radiological consequences of various postulated initiating events was carried out. The results of this analysis for DBAs are presented in this paper. It is found that the Russian calculations meet the AERB acceptance criteria by a good margin. This paper gives the methodology adopted for performing an independent accident safety review of the Russian reactor.

Keywords: Radiological consequences, Source term, Design basis accident, Reference level

1. INTRODUCTION

India's biggest nuclear power station is being set-up by NPCIL at Kudankulam in technical collaboration with the Russian Federation. The twin-unit VVER-1000 (Model V-392) station incorporates many advanced passive safety features [1]. The Kudankulam site was selected by the DAE Site Selection Committee in 1987-88, and an application for the construction licence, along with the Preliminary Safety Analysis Report (PSAR) was submitted to AERB in January 2001. The construction licence was granted in March 2002 after due review of the PSAR. As part of the safety review, an independent evaluation of the radiological consequences of various postulated initiating events listed in Package 15.13 of the PSAR "Radiological Consequences" [2] was carried out.

The aim of the analysis was to compare and verify the calculations using the best computational tools available and to adapt the calculations for Indian conditions. For each of the Design Basis Accidents (DBAs) studied, the accident source term was also independently evaluated. Since the critical member of the population assumed in the PSAR calculations was a European child 1-8 y of age, it was additionally required to make calculations for the critical members normally specified by AERB, viz. an Indian infant <1 y and an Indian adult. The scope of the calculation was limited to DBAs only as the Package 15.13 does not discuss events beyond the design basis.

The VVER-1000 Reactor

The Kudankulam VVER-1000 reactor [1,2,3] is a PWR of Russian design and uses light water as moderator and coolant, and enriched (upto 4.4%) uranium dioxide as fuel. The coolant system has four loops, each containing a recirculating pump and a horizontal steam generator. The reactor is housed in a Reactor Pressure Vessel (RPV) and the coolant loops are connected to the vessel by interconnecting piping. Other equipment connected to the RPV include a pressuriser floating on the hot leg of one of the four loops, and four first stage accumulators for supplying

ECCS water. The reactor plant also consists of reactor control, protection and regulation systems, engineered safety systems, auxiliary systems, fuel handling and storage system, etc.

Engineered Safety Features

Engineered Safety Features (ESFs) [3] are provided to mitigate the consequences of abnormal conditions including accident conditions. The principal emergency tasks to be carried out are: safe shut down of the reactor and maintenance of sub-criticality, uninterrupted heat removal from core to limit fuel temperature, and containment of radioactivity released from core. The VVER-1000 (Model V-392) reactor, which has been selected for Kudankulam, contains some of the most advanced active & passive safety features. The ESFs include the containment system, and those designed to meet DBAs, Beyond DBAs and Anticipated Transient Without Scram (ATWS).

Containment System

The VVER-1000 is provided with a double containment [2,3]. The primary containment is a pre-stressed concrete structure with 1.2 m thick walls and a spherical dome. It is lined with a 6 mm carbon steel plate on the walls and the dome, and with an 8 mm liner on the base slab. The secondary containment is made of 0.6 m thick RCC walls. The 2.2 m annular space between the primary and the secondary walls is maintained at negative pressure to prevent any leakage of radioactivity to the environment. The primary containment is divided into two zones:

- The maintenance-free zone (UZ areas)
- The restricted access zone (RAZ areas)

All releases take place in the UZ areas and subsequently move to the RAZ zone. The UZ areas comprise enclosures of the primary equipment such as the steam generators, the primary coolant pumps, reactor vault, valve chambers and other compartments. These areas are not accessible during reactor operation. The RAZ areas comprise the central hall, the compartments of pump motors, and other compartments. The design pressure of the containment is 0.4 MPa and the design temperature is 150°C. The design leak rate is 0.3 %/d of contained volume for DBA.

Reactor Building (RB) Ventilation System

The ventilation of the RB is shown in Fig. 1. The ESF ventilation systems influencing the consequences of accidents are:

- KLA21 (33000 m^3/h) for venting the containment air space during refuelling, and also for cleaning air after an accident (Manual start);
- KLA22 (3000 m^3/h) exhaust system equipped with aerosol and iodine filters for maintaining a vacuum of 150 Pa in the RAZ space (Trip condition: 0.003 MPa excess pressure);
- KLA02 (12000 m^3/h) re-circulation system for maintaining negative pressure within the compartments of the UZ area (Trip condition: 0.03MPa excess pressure);
- KLB22 (4000 m^3/h) annulus exhaust system with filtration capability against aerosol and iodine, for maintaining a vacuum of 100 Pa in the annulus (Operable for all reactor states).

Area Spray System

The containment spray system has two main functions: i) to remove the heat released into containment following an accident and maintain the pressure and temperature in the containment space, and ii) to scavenge radio-iodines released into the containment air, and to partially neutralize the boric acid to obviate the possibility of surface corrosion. The area spray system is actuated at a containment pressure of 0.03 MPa, and operates with a delay of 90s.

2. ACCIDENT CONSEQUNCE ANALYSIS

Release modes

Only volatile elements (Noble gases, Iodine, Caesium and Tellurium) are assumed to be released during DBAs. The release of radioactivity to the atmosphere [2] is expected to take place i) directly from the containment by way of normal or ESF ventilation, ii) from areas outside the containment where failures of primary coolant lines might occur, and iii) from the secondary system through the Steam Generator (SG) relief devices. The activity release model is schematically depicted in Fig. 1.

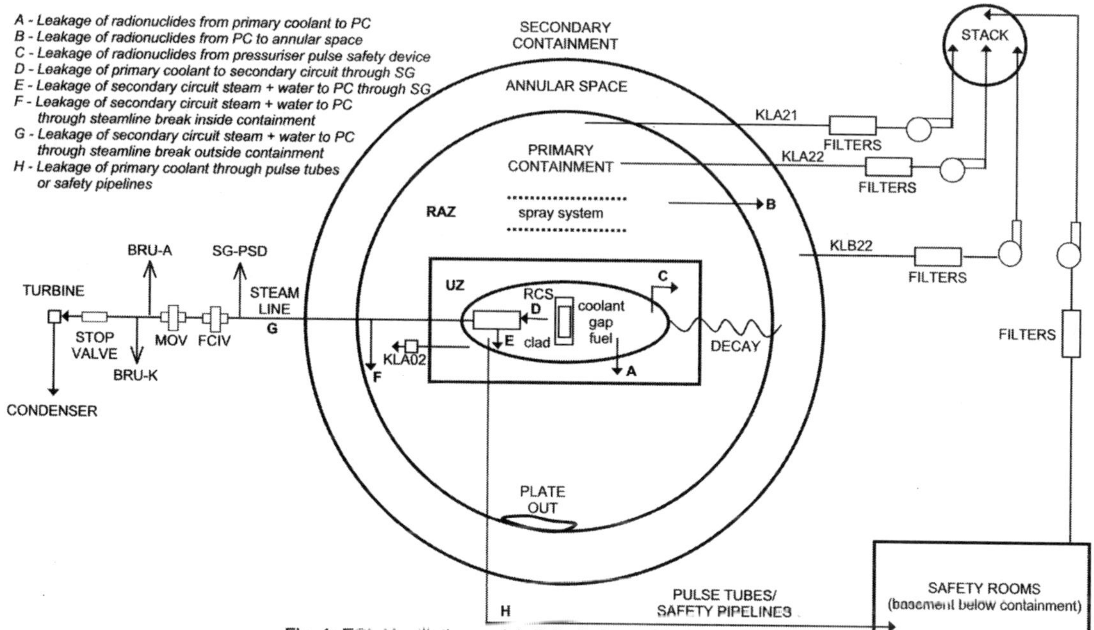

Fig. 1: ESF Ventilation and Activity Release Model of VVER-1000

The primary coolant is expected to carry a small amount of radioactivity from failed fuel (for which the design limit is 1% of fuel elements with gas-leaking defects, and 0.1% of fuel elements with substantial defects) [2]. Under conditions of fuel depressurisation, a 'spike release' may occur from the failed fuel, resulting in the coolant activity increasing several-fold. The maximum allowable concen-tration in the coolant is termed the safe limit for reactor operation. (See Table 1 [2], column 3).

In an accident sequence, additional fuel failures are expected to occur. In the present study, it is assumed that if during a transient, there occurs a departure from nucleate boiling ratio or if the clad temperature exceeds 355° C, additional fuel failure may occur. Whenever the above criteria are met, it is conservatively assumed that 100% of fuel elements fail. The assumptions for the release from failed fuel are: 100 % of gap inventory for NG & I, 30% for Cs and 1% for Te.

The gap inventory of the reactor from prolonged operation is given in Column 4 of Table 1. In addition to the release from the fuel-clad gap, a so-called heat-up release may occur during DBAs from the stored inventory in the fuel matrix. Based on the maximum temperature rise of the fuel elements during DBAs, the fraction of radioactivity released by fuel heat up is at most 0.02 % of equilibrium core inventory for NG, I, and Cs and 0.003 % for Te.

Table 1 Core Inventory of Kudankulam VVER-1000

Sr. No.	Radio-nuclide	SOL* in Coolant (Bq/kg)	Activity under HC$ (Bq)	Eql core Inventory (Bq)
1.	Kr-85m	6.2×10^7	4.2×10^{14}	8.3×10^{17}
2.	Kr-87	1.9×10^8	6.2×10^{14}	2.4×10^{18}
3.	Kr-88	2.4×10^8	1.2×10^{15}	3.1×10^{18}
4.	Xe-133	8.4×10^7	2.4×10^{16}	6.9×10^{18}
5.	Xe-135	4.6×10^7	6.7×10^{14}	8.7×10^{17}
6.	Xe-138	2.3×10^8	7.1×10^{14}	5.9×10^{18}
7.	I-131	1.6×10^8	9.5×10^{15}	2.8×10^{18}
8.	I-132	4.2×10^8	8.6×10^{13}	3.8×10^{18}
9.	I-133	3.4×10^8	7.8×10^{15}	6.9×10^{18}
10.	I-134	3.2×10^8	2.6×10^{15}	7.2×10^{18}
11.	I-135	2.6×10^8	3.2×10^{15}	5.9×10^{18}
12.	Te-132	1.8×10^3	7.3×10^{15}	3.8×10^{18}
13.	Cs-134	1.6×10^6	3.9×10^{15}	4.5×10^{17}
14.	Cs-137	2.3×10^6	2.1×10^{15}	2.9×10^{17}

*SOL: Safe Operation Limit; $HC: Hermitic Clad

Attenuation of release

The attenuation of released radioactivity [2] depends on its release form, chemical nature and radiological properties.

Noble Gases

Krypton and Xenon radioisotopes are released as non-condensable gases and are not attenuated by any natural process, except radioactive decay. Thus, the releases from fuel to coolant, from coolant to Reactor Building (RB) air, and from Steam Generator relief devices to the atmosphere may be assumed to be total.

Iodines

The iodines are volatile and highly reactive species, and are also radio-biologically important. The physical form of iodine may be i) molecular (or elemental), ii) aerosol, or iii) organic (mostly, methyl iodide). The proportion of these forms in a given environment depends on i) the concentration of the iodine in air, ii) the pH, temperature and concentrations of iodine and other additives in water solutions in contact with air, and iii) the age of the iodine mixtures. Other factors, such as radiation field and the presence of catalysts are also important.

Under normal operating conditions in VVERs, the small releases in the containment air may have Iodine forms in the ratio of 40% molecular, 30% organic and 30% aerosol [2]. The fraction of organic iodine in air decreases as the concentration of Iodine in air increases. Thus, with an air concentration corresponding to iodine released under the spike process, it is expected to be about 1%, while with a concentration corresponding to 100% gap inventory, the organic iodine is expected to be about 0.2 %. [2]. The release fractions of iodine is given in the Table 2.

Table 2 Iodine Release Fractions [2]

Release from failed fuel	100 %
Release from uncontrolled leak	100 %
Release from SG during relief device operation	0.1 %
Release from scrubbed steam carrying iodine	10 %
Release from scrubbed air containing iodine	10 %

The attenuation factors in iodine filters are: 99.9% for molecular iodine, 99.0% for organic iodine and 99.97% for aerosols.

The attenuation of iodine by natural plate-out in the RB is modelled as follows [2] :

$$\lambda = K * (S/V)$$

where S/V = surface-to-volume ratio (m^{-1}),

K = coefficient of mass transfer (m/s)

λ = plate-out rate (s^{-1}).

A desorption process on molecular iodine is also modelled. The removal of iodine forms by the area spray droplets is modelled as a rate process, R. The values of K for the natural deposition processes are given in the Table 3.

Table 3 Iodine and aerosol removal parameters [2]

Parameter	Value
Coefficient of mass transfer, K (m/s)	
♦ Molecular iodine (in humid air)	10^{-4}
♦ Aerosol (in humid air)	$5x10^{-4}$
♦ Molecular iodine (in dry air)	10^{-5}
♦ Aerosol (dry air)	10^{-5}
Coefficient of desorption of mol. iodine, s^{-1}	10^{-6}
Rate of removal by area spray system, R (s^{-1})	
♦ Molecular iodine	10^{-2}
♦ Organic iodine	$5.2x10^{-6}$
♦ Aerosol	$1.3x10^{-2}$

Caesium and Tellurium

These elements are released in air as aerosols and are retained in aerosol filters with a high efficiency (99.97 % in HEPA filters). They are also attenuated by natural and active processes to a high degree, as given above for iodine aerosols. The release fractions are given in the Table 4.

Table 4 Caesium and Tellurium Release Fractions [2]

Release from uncontrolled leak	100 %
Release of Cs from failed fuel	30 %
Release of Te from failed fuel	1 %
Release from SG during relief device operation	0.1 %

Calculation of Accident Source Terms

This paper presents results of source term calculation for the postulated DBAs in the Package 15.13 [2,4]. Computer programs were developed to compute the time-dependent releases for each DBA. All release and attenuation assumptions were in accordance with those made in the relevant section of the Package. The releases were calculated for a period of 30-d.

Public Dose Calculation

In the package PSAR 15.13, the doses are computed [2,4] for the external (cloudshine and groundshine) and internal (inhalation and ingestion) exposure pathways. The main assumptions [2] for the computation of dose are:

- All releases take place at the ground level;
- The release weather category is Pasquill F, with a wind speed of 1 m/s;
- The terrain is assumed to be smooth;
- Critical organs are: Whole body, Thyroid, and Lung;
- Down wind distances for dose calculation: [1.6 km (Exclusion boundary), 3 km, 5 km, 15 km, 25 km, 30 km]
- The critical group of population is children 1-8 y of age. The annual diets of this group and of the European adult are given in [2].

Adaptation of the Risk codes for Consequence Analysis

Some codes considered for this study (CRAC2 [5] and COSYMA [6]) are risk analysis codes. In the present case, they were used in deterministic mode for calculation of dose at various distances from the reactor under constant weather conditions. They also have as their critical member an adult Western individual. Since the critical member assumed in the VVER Package 15.13 is a child of age group 1- 8 y, the results of risk codes had to be modified suitably. External dose correction factors used were those given in Table 5 [6]. Corrections factors for inhalation and ingestion were generated using an in-house computer code CHILDULT [4].

Table 5 External Dose Correction factor (D_{ch} / D_{ad})

Age, y	Correction factor	
	Cloud- shine	Ground- shine
0 1	1.50	1.80
1 - 3	1.50	1.80
4 - 7	1.40	1.50
8 –12	1.30	1.40
13 – 15	1.15	1.15
>16	1.00	1.00

Benchmark Problem

The following computer codes were available for consequence analysis:

- CRAC2 code (1983)
- PC-COSYMA code Version 2.0 (1995)
- DOSCAL code (2000). [7]

In order to decide which of the above three codes would be best for detailed consequence analysis of all the DBAs, each of them was run for a benchmark problem, and compared with the output of the DOZA_M (VVER) Code.

The standard problem taken for this exercise was one of the DBA of the Kudankulam VVER-1000 reactor, namely "Spectrum of ruptures of steam line breaks inside the containment up to a maximum diameter of steam line". The source term for this accident was taken from Sec. 15.13.2.1 of the Package

An inter-comparison of public dose for the standard accident scenario is given in Fig. 2 & 3 for whole body and thyroid doses, respectively. The Indian consequence analysis code, DOSCAL, with its primary objective as licensing of indigenous heavy water reactor based power plants, is found to be the most conservative of all the four codes studied. It is found that the dose as calculated by VVER code DOZA_M is in line with two other international codes CRAC2 and PC-COSYMA. Among the latter codes, the former is found to be slightly more conservative than COSYMA. For this reason for all the rest of the 14 DBA scenarios analysed, only CRAC2 code is used.

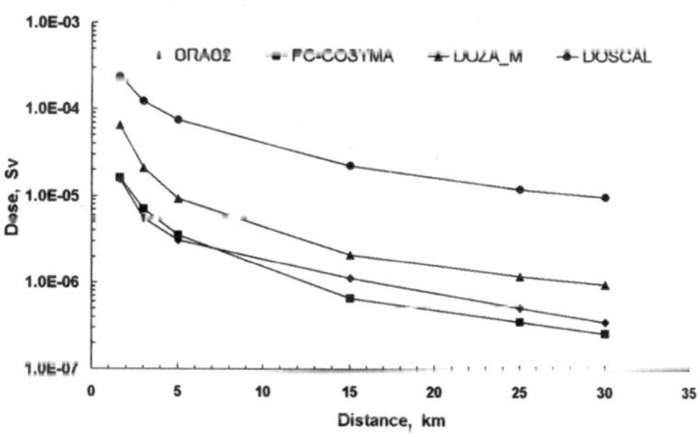

Fig. 2 : Steamline Break inside Containment. Whole Body Dose

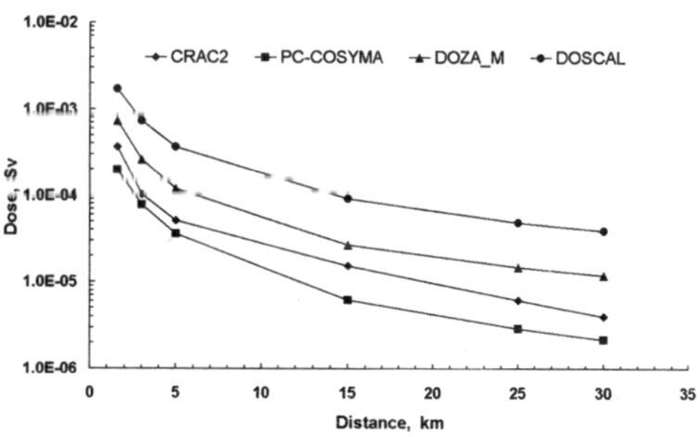

Fig. 3: Steamline Break inside Containment: Thyroid Dose

Table 6 Summary of Results of the 15 DBAs [4]

Sr. No.	Design Basis Accidents	Acceptance Criterion*	WB Dose at 1.6 km (Sv)		Thyroid Dose at 1.6 km (Sv)	
			DOZA_M	CRAC2	DOZA_M	CRAC2
1.	SLB inside containment	20a	$6.50x10^{-5}$	$1.56x10^{-5}$	$7.30x10^{-4}$	$3.66x10^{-4}$
2.	SLB outside containment	20a	$1.80x10^{-3}$	$3.90x10^{-4}$	$2.40x10^{-3}$	$4.29x10^{-4}$
3.	FW line break	20a	$1.50x10^{-4}$	$5.57x10^{-4}$	$1.30x10^{-3}$	$8.44x10^{-4}$
4.	Prompt seizure of MCP	20a	$1.50x10^{-4}$	$1.76x10^{-3}$	$1.30x10^{-3}$	$6.20x10^{-3}$
5.	Break of MCPU	20a	$5.90x10^{-5}$	$6.21x10^{-6}$	$2.00x10^{-4}$	$2.81x10^{-5}$
6.	Non-seating of PSD PC	19	$9.60x10^{-6}$	$7.77x10^{-6}$	$1.40x10^{-4}$	$7.26x10^{-5}$
7.	SB LOCA (< 100mm)	20a	$1.10x10^{-5}$	$1.04x10^{-6}$	$1.70x10^{-4}$	$6.63x10^{-5}$
8.	LB LOCA (100- 850 mm)	21	$1.60x10^{-2}$	$1.38x10^{-4}$	$2.30x10^{-2}$	$4.23x10^{-3}$
9.	DU10 break outside containment	20a	$4.00x10^{-7}$	$4.72x10^{-8}$	$7.50x10^{-6}$	$5.36x10^{-6}$
10.	DU105 break outside containment	20a	$4.40x10^{-5}$	$2.78x10^{-5}$	$4.90x10^{-4}$	$3.55x10^{-4}$
11.	SG tube rupture	20a	$5.90x10^{-5}$	$6.87x10^{-5}$	$8.20x10^{-5}$	$7.77x10^{-5}$
12.	Tearing off of SG header lid	20b	$2.50x10^{-4}$	$5.17x10^{-5}$	$8.00x10^{-4}$	$4.37x10^{-4}$
13.	Incorrect loading of FA assemblies	20a	$5.30x10^{-6}$	$5.65x10^{-6}$	$1.50x10^{-5}$	$1.06x10^{-5}$
14.	Ejection of CR (Variant-1)	21	$7.40x10^{-4}$	$2.03x10^{-4}$	$9.20x10^{-3}$	$4.50x10^{-3}$
15.	Ejection of CR (Variant-2)	21	$7.60x10^{-4}$	$5.93x10^{-4}$	$3.80x10^{-3}$	$1.99x10^{-3}$

* 19 : 0.2 mSv for Whole Body, 0.6 mSv for Thyroid and 1.2 mSv for any organ , normal operation dose limit
20a : 5 mSv for Whole Body, 25 mSv for Thyroid - < 5% of Emergency Reference Level
20b : 25 mSv for Whole Body, 125 mSv for Thyroid - < 25% of Emergency Reference Level
21 : 100 mSv for Whole Body, 500 mSv for Thyroid - Emergency Reference Level

3. RESULTS AND DISCUSSION

The summary results for the various DBAs are presented in Table 6 [4]. The calculated results are for a European child of 1-8 yrs of age. A comparison of the results with those calculated using CRAC2 code shows that:

- The results are broadly in agreement; in most cases the CRAC2 results under predict the DOZA_M results by a factor of 2 to 4.
- It is difficult to tell what the causes for this under prediction might be. Several critical data items used in DOZA_M calculations are unknown; they include deposition velocity, Dose Conversion Factors, breathing rates, age-dependent ingestion dose calculation parameters, and correction factors for external dose parameters.
- Doses were also calculated for an Indian infant and Indian adult using the best available ingestion rates and food patterns. Taking the whole body dose for the Indian adult and the thyroid dose for the Indian infant, the acceptance criteria are still met [4] for all VVER-1000 DBAs.

4. CONCLUSIONS

The calculations carried out in this paper confirm that the KudankulamVVER-1000 meets the AERB acceptance criteria for all listed DBAs by a good margin. The VVER code DOZA_M is more conservative than both CRAC2 and PC-COSYMA.

5. REFERENCES

1. A.I. Siddiqui, 'From Volga to Ganga: The Story of Kudankulam', Nu Power, Vol.16, No.1-2, 2002.
2. PSAR, NPP Kudankulam Unit1 & 2, Chapter 15 Emergency Situation Analysis, Sub-package 15.13 Radiological Consequences, 2001.
3. KK-PSAR, Package S-6, 'ESFs & Containment Systems', Kudankulam Project, 2000.
4. V.K. Sharma et al., 'A report on the accident source terms and public doses for the Kudankulam VVER-1000', BARC, November, 2002.
5. L.T. Ritchie, et al., 'CRAC2 Model Description', NUREG/CR-2552, USNRC, 1983
6. European Commission, 'PC COSYMA Version 2', User Guide, NRPB, EUR 16240 EN (NRPB-SR280), 1995.
7. V.K. Sharma, 'DOSCAL – A Consequence Analysis Code', Unpublished report, Health Physics Division, BARC, 2000.

Reliability, Safety and Hazard
Editors: P.V. Varde et al.

RISK ZONING AROUND NUCLEAR POWER PLANTS IN COMPARISON TO OTHER MAJOR HAZARDOUS INDUSTRIAL INSTALLATIONS

Christian Kirchsteiger

European Commission, DG JRC, Institute for Energy, Westerduinweg 3, 1755 LE Petten, The Netherlands

ABSTRACT

The background and current status of the information basis leading to the definition of risk and emergency zones around Nuclear Power Plants (NPPs) in different countries in Europe and beyond are analysed. Although dependable plant-specific Probabilistic Safety Assessment (PSA) of Level 2 and/or Level 3 could in principle provide sufficiently detailed input to define the geographical dimension of a NPP's risk and emergency zones, the analysis of the status in some European and other countries shows that other, "deterministic" approaches using a Reference Accident are actually used in practice. Regarding use of Level 2 PSA for emergency planning, the approach so far has been to use the Level 2 PSA information retrospectively to provide the justification for the choice of Reference Accident(s) used to define the emergency plans and Emergency Planning Zones (EPZs). There are significant differences in the EPZs that are defined in different countries, ranging from about 1 km to 30 km. Further, there is a striking contrast in the extent of using probabilistic information to define emergency zones between the nuclear and other high risk industry sectors, such as the chemical process industry, and the reasons for these differences are not entirely clear, but seem to be more related to risk perception than actual risk potential. Finally, based on consensus discussions at a recent JRC/OECD International Seminar on Risk & Emergency Zoning around NPPs, recommendations are given in the areas of more comprehensive use of available risk information for risk zoning purposes, risk communication and comparative risk assessment.

1. BACKGROUND AND OVERVIEW

Within the Institute for Energy (IE) of the Joint Research Centre (JRC) of the European Commission (EC), located at Petten, The Netherlands, a project on *Benchmarking and Harmonising Strategic Planning Practices for Risk & Emergency Zoning around Nuclear Power Plants and Information to the Public* has recently been developed with the objectives to evaluate the corresponding status in Europe and beyond, and to determine whether it would be possible to move towards a greater level of international harmonisation.

The knowledge resulting from this project should help Regulatory Authorities, Civil Protection Institutions, European Institutions such as EC Policy Directorates General, the various PSA users and developers and, last but not least, the general Public to get a clear picture on the relevance of the issue in a comparative view (e.g. versus corresponding practices of the chemical process industries), on the consistency of current approaches and on related research and development needs.

Information has been collected and analysed in a first report [1] and later on distributed in a condensed version [2] to a large audience for review and comments. As a main conclusion, significant differences have been found in the definitions of EPZ of the NPPs in different Member States of the European Union (EU) and beyond.

The original idea for this project consisted in the view that PSA is currently already mature enough to be used also for NPP Emergency and Risk Zoning (ERZ). However, at present it can be stated that not much is being done in EU Member States in application of Level 2 or 3 Probabilistic Safety Assessment (L2 or L3 PSA) results to emergency planning (EP). The approach to EP is, in general, strongly deterministic. The usual approach is that a Reference Accident is defined to be used as basis for drawing up the emergency plans. In EU Member States, the practical application of L2 PSA results for accident management is very limited and, effectively, very little risk-based information is used. In the course of this project, from the participating countries[1] only the Czech Republic and the UK informed about some cases where L2 PSA results were used in a formal way as an input to EP. The UK is the only EU Member State, which has been carrying out research on how L2 PSA outcomes could be used in a systematic way for EP purposes.

As a next step, JRC approached a large number of PSA experts on the one side and EP experts on the other side to ask whether incorporation of risk-informed support into NPP EP is currently a sufficiently relevant and mature topic to be treated by a seminar investigating prospects towards international harmonisation. JRC received a large number of very positive responses, only a few ones being reserved or skeptical. While this is certainly not an exhaustive feedback, it was nevertheless reasonably found to organise together with OECD a *JRC Seminar on Emergency & Risk Zoning around NPPs*, which was held on 26–27 April 2005 at JRC's Institute for Energy in Petten, The Netherlands. The Seminar attracted a wide participation from EU and non-EU countries. More than 20 technical presentations in the area were given by participants from Europe, USA, South Africa and India[2]. The objectives of this Seminar were:

- to provide a forum for presentation and discussion of status of EP and PSA, safety policies as well as current and possible future requirements for ERZ,
- to provide an opportunity for sharing of experience in the field on both good practice and identification of problem areas, including comparison to other major-hazardous industries, such as the chemical process industries, and thus
- to help relevant stakeholders (i.e. Regulatory Authorities, utilities, emergency response organizations as well as PSA users and developers) on both national and international levels
  - to decide on the relevance of this issue at this time,
  - to decide on related research and development needs, and
  - to consider needs for international harmonisation.

2. CONCLUSIONS AND RECOMMENDATIONS

Current Approaches to Emergency Planning

It was clear from the ERZ Seminar's presentations and discussions that there are many similarities but also some significant differences in the way that EP is drawn up and EPZs are defined in different countries.

The "standard" approach to EP is mainly deterministic and uses a Reference Accident approach (consideration of Design Basis Accident or selected reference scenarios) and almost no risk information is used. Sometimes a blend of deterministic and probabilistic approaches is used. The ERZ Seminar has gone some way to investigating these similarities and differences. However, it was clear that further work is required to provide a better understanding of the national approaches to EP and to determine whether it would be possible to move towards a greater level of harmonisation.

[1] The following countries were involved in the information collection exercise summarized in report [1]: Belgium, Czech Republic, Finland, Hungary, Netherlands, Slovakia, Spain and the United Kingdom. In addition, information was obtained from Japan and the US.

[2] Further information on the Seminar can be found under http://www.energyrisks.jrc.nl and Proceedings are available from the author.

Especially, there is a large variety in approaches and data used for EPZ in different countries. Current EPZs range from less than 10 km up to 80 km. The reasons for these differences need to be fully understood. In addition, the issue needs to be addressed on whether there is any advantage in defining larger EPZ or if there are any disbenefits from defining large EPZs. In many countries the relevant IAEA documents are used[3].

Recommendation 1: It is recommended that further work be carried out towards understanding the reasons for differences in the EP in different countries.

Recommendation 2: It is recommended that further work be carried out on how the uncertainties inherent in the L2 PSA can be taken into account in defining EP and EPZs.

Current and Future Use of L2 PSA for Emergency Planning

It was agreed that the proper response to an emergency requires understanding of the underlying hazards. The results of L2 (and L3) PSAs provide important information in this area. The standard emerging worldwide is full scope L2 PSA to be carried out for all NPPs. Such analyses are of sufficiently mature nature to be used for a wide range of applications and could also be used as one of the inputs into EP as part of an overall risk-informed approach.

However, uncertainties in PSA are large and are likely to be greater in L2 PSA (derivation of the source terms) and greater still in L3 PSA (calculation of the off-site consequences of a release of radioactive material) as compared to L1 PSA. Further consideration needs to be given as to how these uncertainties are taken into account in proper EP.

A distinction should be made between full-scope L2 (and L3) PSA and restricted scope, when not all power levels or hazards are included (e.g. it may only address internal initiating events). Further consideration needs to be given as to how a restricted scope PSA could be used as an input to EP.

[3] e.g. the October 2003 updated version of TECDOC-953 on "Methods for Developing Arrangement for Response to a Nuclear or Radiological Emergency" which gives guidance on EP and ERZ.

The approach used so far is to use L2 PSA information *retrospectively* in order to provide a justification for choice of Reference Accident(s) used to derive EP and EPZs. Where a Reference Accident approach is used based on L2 PSA information, consideration needs to be given to the number of Reference Accidents that need to be defined to characterise severe accident progression and release characteristics.

It was agreed that L2 PSA information could be used as a basis for EP and there is the potential to do this in a more *proactive* way. However, apart from the aspect of the above-mentioned uncertainties, consideration needs to be given on the methodology, i.e. how this is to be done in a risk-informed way that takes account of other factors so that it is not based on PSA information alone. L2 PSA provides an understanding of how a severe accident would occur, the accident management measures that could be used to mitigate the effects of the accident, and the resulting source terms and frequencies of releases from the plant. This provides the detailed information that would be required as input from PSA to definition of EP and EPZs.

One of the important issues is how to understand from the evolution of the accident which source term would happen. L2 PSA typically contributes to model accident management measures that can be carried out to mitigate the consequences of a severe accident. Severe Accident Management Guidelines could effectively contribute to extending time for emergency response, i.e. this would extend the period of time available before a release of radioactivity would occur from the NPP and hence increase the time available to put the EP into action.

Recommendation 3: It is recommended that further work shall be carried out to determine how the L2 PSA information could be used in a systematic way as an input to defining EP and EPZs, taking into account the corresponding uncertainties. It is recommended that JRC should undertake a pilot study to produce a scheme for use of L2 and L3 PSA information for EP. This could be within the framework of or as a spin-off from JRC's involvement in the SARNET Network of Excellence[4].

[4] http://www.sar-net.org

Full Scope L3 PSA

Full scope L3 PSA is available only for very few NPPs in the EU. However, they are being developed in a number of countries (including Netherlands, UK, USA, Japan, South Africa, India). One of the reasons could be one rather specific aspect of L3 PSA: while conducting L1 and partially also L2 PSA study is, or can be, in favour of nuclear operators / licensees due to their plant-internal character and related potential for improvement, they do not have any incentive to perform L3 PSA studies as these address exclusively plant-external effects. To have L3 PSA study available is, more or less, the concern of nuclear Regulatory Authorities and/or Radiological Protection Authorities.

PSA Quality Requirements

To be useable as one of the inputs to EP, it is recognised that a high quality PSA is required that is suitable for this specific application. As mentioned, one particular aspect here is that the source terms need to be well defined in terms of the quantities of radioactive material released and the release profile (start of the release, duration, height, energy, etc.). The PSA that is used to provide an input to EP needs to be complete and address all the contributions to the risk in terms of the hazards considered and the operational modes addressed.

However, there was no agreement on the extent to which external hazards (such as severe seismic events) and security related events (such as terrorist attacks on the plant) should be taken into account in defining EP and EPZs. Further consideration is required of these topics.

Recommendation 4: To support the current IAEA activity on PSA Quality for Applications [3] and the Proposal for a Co-ordinated Research Project on PSA of nuclear facilities in relation to external events (into which the issue of security related events could be included).

Current Leoβgal Requirements

There was a concern that the mandatory legal requirements in some countries may be over-restrictive and that the possibility of adopting a more risk-informed approach should be considered.

Future Nuclear Power Plants

The trend is to improve the level of safety for future NPPs (so-called Generation III+ - IV). This would significantly reduce the potential for severe accidents and releases of radioactive material from the plant to occur. In principle, this could be considered to reduce, or perhaps eliminate, the need for EP.

Recommendation 5: Further considerations needs to be given on how EP and the EPZs would be defined for future NPPs where the risk from the plant in terms of large off-site releases of radioactivity would be very much lower than for current plants. This needs to be reconciled with the expectations of the Regulatory Authorities and the Public. Consideration needs to be given on whether the moral obligation to provide an EP would outweigh the technical conclusion that this would not be required.

Risk Communication

There is a need to be able to communicate risk information to the Public both before and following an accident. In addition, there is a need to educate the Public on that they can understand risk information. Further consideration needs to be given to how this can be done. Care must be exercised while *communicating risk-related information and insights on potential vulnerabilities* to the Public in the view of *security issues and prevention of malevolent acts. Education of the Public on risk considerations* is important for enhancing the general understanding of risk implications and better acceptance of risk from nuclear industry. *Comparative consideration* of emergency planning and associated risks for nuclear and other industries, as well comparison with general risks from daily life may be helpful in communicating risk information to the Public.

Recommendation 6: It is recommended that further work be carried out on risk communication and how this can be done in relation to EP.

European Risk Map

There was discussion of the long-term goal of developing a European risk map for all potentially major hazardous industries, which would include both nuclear and chemical installations.

Nuclear and chemical safety have various different and various similar aspects (see reference [5] and footnote 2):

Regarding nuclear, NPPs are quite similar to each other in terms of:

- type of risk (risk of radioactive release through multiple barriers),
- steady state power production and outages,
- process system technology and supporting safety functions which are quite straightforward to analyse,
- all plant operators are used to safety and quality management systems.

Further, there are commonly accepted safety criteria, including de-facto international consensus on risk targets.

Chemical installations, on the other hand, are much larger in number and variety:

- different substances (chemical, fuels, explosives),
- different products and production technologies,
- process systems entities are often complex to analyse,
- not all plant operators are used to safety management systems nor thorough risk assessment methods (for event with very low frequencies and high consequences, in particular).

Regarding regulation, on an EU level, the so-called "Seveso Directive" 96/82/EC sets out basic safety management and risk analysis criteria, but there are considerable differences between different countries, branches (chemical, metal, refineries, mining, etc.) and operating companies. Further, there is some EU cooperation ongoing in the area of risk zoning (so-called "Land-Use Planning"), aiming at developing common databases for failure frequencies, scenarios, etc.

In summary, nuclear and chemical risks are basically similar in terms of involving a similar type and extent of consequences:

- direct consequences (fatalities, contamination, economics, etc.)
- latent consequences (cancer, genetics, etc.)
- indirect consequences (business impact, branch reputation, etc.).

Major differences between nuclear and chemical risks are:

- number and variety of facilities
- plant technology and operational features
- maturity of risk assessment and safety management practices
- resources for regulatory surveillance throughout plant life cycle
- systemic risk context (supply chains, infrastructure, clusters, etc.)
- "intangibles" (history, public concerns, branch characteristics, etc.).

However, and this is reflected in many joint discussion fora[5] and emerging risk governance initiatives[6], general views on risks are clearly converging between the nuclear and the chemical safety experts.

Recommendation 7: It is recommended to explore the possibilities as to how such an initiative could be linked to supporting implementation of some of the provisions of the EU Environmental Impact Assessment Directive or other international framework legislation in the area, such as the so-called "Aarhus Convention"[7]. On an EU level, such a support activity would have to be conditional to the request of the corresponding EC Policy Directorate General.

International Topical Working Group

It was agreed that the current EP practices for NPPs could be enhanced by moving towards a more risk-informed approach, where the process of defining the EP and EPZs takes account of the information provided by the L2 PSA. This would supplement the purely deterministic approach. However, the discussion showed that there is no guidance on how this should be done in general, though this has already been done in some countries where the source terms and frequencies

[5] see e.g. [5] and http://www.energyrisks.jrc.nl/newspopuphtml.htm

[6] see e.g. http://www.trustnetgovernance.com

[7] UN/ECE Convention on Access to Information, Public Participation in Decision-Making and Access to Justice in Environmental Matters; see also: http://www.unece.org/env/pp/welcome.html

produced by the L2 PSA have been used to check the choice of the Reference Accident(s) that have been used as the basis for EP. This is an example of a retrospective use of PSA.

Recommendation 8: It was recommended that JRC should set up an international topical working group to address risk-informed EP, incl. the above mentioned related topics of risk communication, risk mapping etc. Not only EU Member States should be involved, but also wider international framework and co-operation would be desirable, such as together with IAEA or OECD.

3. FOLLOW-UP

Based on this summary report and the therein included recommendations, feedback is currently sought from interested institutions and EC Policy Directorates General in order to develop a possible follow-up initiative. More information on these ongoing discussions and developments will shortly be added to *http://www.energyrisks.jrc.nl*

4. REFERENCES

1. Kubanyi, J., Kirchsteiger, C., Risk-Informed Support of Nuclear Power Plant Emergency Zoning, Report EUR 21580 EN, JRC Petten, 2005.
2. Kubanyi, J., Kirchsteiger, C., Information Basis to support NPP Emergency Zoning. Draft Review of Status and Possible Recommendations for Discussion at the JRC-OECD Seminar on Emergency & Risk Zoning around NPPs, JRC Petten, 26-27 April 2005, Report EUR 21593 EN, JRC Petten, 2005.
3. PSA Quality for Applications, Working Material, Draft February 2004, Rev.2.3, IAEA Vienna, 2004.
4. Shepherd, C., Kouzmina, I., Rapporteurs' Notes of the Seminar on Emergency and Risk Zoning around Nuclear Power Plants, 26-27 April 2005, JRC Petten.
5. C. Kirchsteiger (ed), International Workshop on Promotion of Technical Harmonization on Risk-Based Decision-Making, Special Issue of Safety Science, Vol. 40, Nos.1-4, 2002.

Reliability, Safety and Hazard
Editors: P.V. Varde et al.

DEVELOPMENT OF A FUZZY EXPERT SYSTEM TO STUDY THE EFFECTIVENESS OF COUNTERMEASURE DURING A NUCLEAR EMERGENCY

D. Datta and M.L. Joshi

Health Physics Division, Bhabha Atomic Research Centre, Mumbai 400 085

ABSTRACT

The primary aim of radiological or nuclear emergency preparedness is to mitigate the consequences of the accident by reducing the radiation exposure to the site personnel and members of the public. Countermeasures such as sheltering, evacuation, etc. are implemented to reduce the exposure. However, the selection of the effective countermeasure is fuzzy due to the vagueness of the countermeasure variables. Fuzzy set theoretic approach has been devised for justifying an effective countermeasure. Effectiveness of countermeasure is quantified on the basis of fuzzy compositions between averted dose and countermeasure variables. The paper describes the development of a fuzzy expert system which performs this task.

Keywords: Countermeasure; Sheltering, Evacuation, Fuzzy expert system

1. INTRODUCTION

Since the Chernobyl accident in 1986, a great deal of research has been carried out for improving the emergency response planning (ERP). The prime objective of the ERP is to reduce radiation exposure by implementing countermeasures such as 'sheltering', 'administration of stable iodine tablets', 'evacuation', 'decontamination', etc. at appropriate time. The possibility of identifying an 'optimum' set of countermeasures for a given scenario is an issue of emergency response planning. This approach has led to the development of several off-site emergency Decision Support Systems (DSS), [e.g. Ehrhardt et al, 1993, Salt and Dunsmore, 2000]. DSS often focus on short-term (days-weeks) countermeasures such as sheltering, administration of iodine tablets and evacuation. The lessons learned from the Chernobyl accident tells that the consideration should also be given to the longer term restoration strategy (months-years) to ensure that the area can be used by future generations. With a view to this fact, decision makers should have access to rational decision models which can be used to support the optimum selection of countermeasure in the event of nuclear or radiological accident. According to the recommendations of ICRP [ICRP, 1988, ICRP, 1990], the most appropriate actions in the event of an emergency can be selected on the basis of the possible benefits and detriments of all available options. This allows an assessment of whether an action is justified or not and whether it is reasonably achievable or not. However, the effect of these protective actions may be dependent on the site-specific conditions such as population, location of the sheltering place, its shielding factors and road conditions. Variables attributing these conditions have a wide distribution of values leading to an epistemic uncertainty due to inherent lack of knowledge (vagueness) about them. In addition, there exist many facts that may cause the uncertainty in the countermeasure model itself. In view of this, the real situation is very often not deterministic (crisp). Practically huge uncertainty exists in the assumed values of parameters such as the fraction of population with various shielding factors, evacuation rate, etc. Therefore a decision making process to select the appropriate countermeasure in presence of this uncertainty is the culmination of the various technical assessment process during a radiation emergency.

As vagueness or epistemic uncertainty can be well handled by fuzzy sets theory (FST) [Dubois, 1980], the latter provides the analytical formulations to address the uncertainty of parameters in the countermeasure model. Linguistic hedges and various implications of FST such as mamdani implication [Mamdani, 1981] play a very important role in designing such an effective tool applicable for decision making process during a radiation emergency. Using the powerful features of FST and associated fuzzy logic (opinion of experts), a fuzzy expert system "FEXEMER" has been developed to study the effectiveness of countermeasure during a nuclear or radiological emergency.

The paper describes the architecture of "FEXEMER" and its application to select the effective countermeasure. "FEXEMER" has been written by using Visual Basic 6.0. The firing mechanism of the fuzzy rules is justified by fuzzy reasoning.

2. FUZZY EXPERT SYSTEM - WHAT IS IT?

Fuzzy expert system (FEXS) works on the fuzzy sets and fuzzy logics. A fuzzy set A is denoted by an ordered set of pairs (x, μ(x)), where, the element x ε X of a specific universe and μ(x) denotes the degree of membership [Zimmermann, 1986]; μ(x) ε [0,1]. The translation of the crisp variable into a membership function or a linguistic variable depends on the specific problem. The crisp set of elements that belong to the fuzzy set A at least to the degree α is called the α level set or α cut given by

$$A^{\alpha} = \{x \varepsilon X \mid \mu_L(x) \geq \alpha\} \qquad (1)$$

The three basic operations on fuzzy sets are union, intersection, and complement. Union and intersection of the membership function μ(A) and μ(B) of two fuzzy sets A and B are given by

$$\mu_{A \bigcup B}(x) = \max(\mu_A(x), \mu_B(x)) \qquad (2)$$

$$\mu_{A \bigcap B}(x) = \min(\mu_A(x), \mu_B(x)) \qquad (3)$$

The 'max' and 'min' operators are one of the many S-norm functions [Dubois, 1980] and T-norm functions [Dubois, 1980] that define union and intersection and represents "most pessimistic" and "most optimistic" of them. Finally the complement can be defined by

$$\mu_A(x)^C = 1 - \mu_A(x) \qquad (4)$$

Besides these basic operations, fuzzy inference requires the association of a fuzzy membership function with each implication. For example, for the rule, "IF A THEN B", a possible fuzzy membership function can be written as,

$$\mu_{A \to B}(x, y) = \min(\mu_A(x), \mu_B(y)) \qquad (5)$$

According to Zadeh, fuzzy evidence can be propagated through a rule using what is usually referred to as a compositional rule of inference (CRI). The following CRI is common in the literature

$$\mu_B(y) = \sup\{\min[\mu_{A_\sigma}(x), \mu_{A \to B}(x, y)]\} \qquad (6)$$

Fuzzy Logic

Central to fuzzy logic is the notion of a linguistic variable, such as *very-low, low, medium, high, etc.* Linguistic variables (LVs) are groups of fuzzy sets with partially overlapping membership functions over a common (crisp) basic variable x, each describing one specific aspect. The range of numerical values spanned by a linguistic variable is called the universe of discourse for that variable. Associated with each term is a *fuzzy membership function* [Klir and Folger, 1988], whose domain is a subset of the universe of discourse. The overlaps of the membership functions define the fuzziness of the linguistic variable. A fuzzy set is generally modified using the linguistic hedges such as "very", "low", "much", "slight", "more-or-less", etc. [Zadeh, 1972]. The characteristics of the variable property is reflected by the determination of membership function (Equations 2-3) which is important to obtain the reliable results [Civanlar, 1986, Tursken, 1991].

Definition of Fuzzy Expert System (FEX)

FEXS is an expert system that uses a collection of membership functions and rules, instead of boolean logic, to reason about data. Unlike conventional expert systems [Datta, 1999, Datta et al, 2002], which are mainly symbolic reasoning engines, FEXS is oriented toward numerical processing. The rules of a fuzzy expert system are usually formatted as: *If x is low and y is high then z is medium*, where, *x* and *y* are input variable, *z* is an output variable. *Low* and high are respective membership functions defined on x and y (input variables); and *medium* is a membership function defined on z (output). The rule's premise (antecedent) describes to what degree the rule applies, while the conclusion (the rule's consequent) assigns a membership function to each of one or more output variables. The set of rules in FEXS frames its knowledge base known as fuzzy rule base. The general inference or reasoning in FEXS involves four primary processes: (a) *fuzzification*, (b) *fuzzy inference*, (c) *fuzzy composition* and (d) *defuzzification*. Under fuzzification, the membership functions defined on the input variables are applied to their actual values, to determine the degree of truth for each rule premise. Under inference, the truth value for the premise of each rule is computed, and applied to the conclusion part of each rule. Under composition, all of the fuzzy subsets assigned to each output variable are combined together to form a single fuzzy subset for each output variable. Under defuzzification, fuzzy output is converted into a crisp number.

3. ARCHITECTURE OF FEXEMER

The expert system FEXEMER consists of a database of linguistic variables associated with the input variables of the countermeasure models, a fuzzy rule base coupled with membership function selectors, and an inference engine and a defuzzifier. A schematic layout of FEXEMER is given in Figure 1.0. The inference engine of the system is built by model-based fuzzy reasoning techniques. Rule firing mechanism of the system is based on fuzzy relations and fuzzy compositions. A typical format of the RULE is: If *shielding condition is good* and *inhalation dose rate is medium*, Then *sheltering is good.* Inference engine composes the fuzzy relations under the prevalence of the fired rules and sends to the defuzzifier that finally defuzzifies and generates a crisp output. Membership function plays an important role in composing the fuzzy relations.

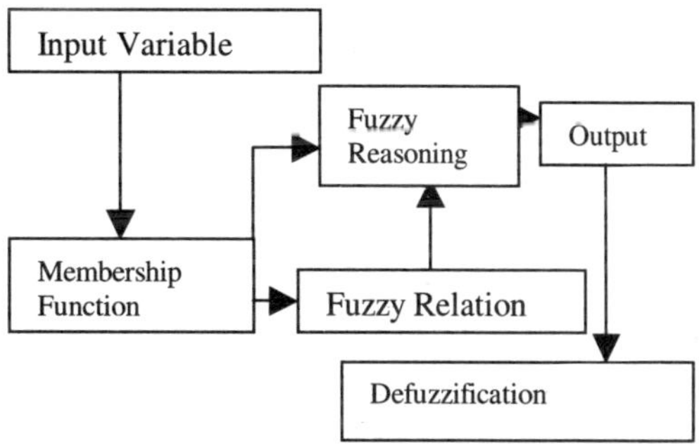

Fig. 1: Schematic layout of FEXEMER

4. FUZZY VARIABLES OF COUNTERMEASURE MODELS

Countermeasures such as Sheltering, Stable iodine tablet administration and Evacuation are considered in FEXEMER for evaluating their effectiveness during radiation emergency situation. The effectiveness of these countermeasures is computed on the basis of the fuzzy composition of averted dose [IAEA TecDoc, 2003] and input variables of countermeasures that represent the feasibility condition of implementing them. The later one is computed using fuzzy logics and the averted dose is computed using the estimated difference between the time integrated whole body dose and thyroid dose received by the public without and with countermeasure. Countermeasure models are represented mathematically by equations (5-8).

Sheltering

$$\begin{aligned} ThyD^{CM}(x) &= D_{thy}(x)*t_{cm}+S_{inh}*D_{thy}(x)*t_{eff}+ \\ &\quad D_{thy}(x)*(t_{int}-t_{eff}),\ \text{for}\ t_{cm}+t_{eff}\le t_{int}, d\le d_{cm} \\ &= D_{thy}(x)*t_{cm}+S_{inh}*D_{thy}(x)*(t_{int}-t_{cm}), \\ &\quad \text{for}\ t_{cm}+t_{eff}>t_{int},\ d\le d_{cm} \qquad (5) \\ &= D_{thy}(x)*t_{int}, \text{for}\ d>d_{cm} \end{aligned}$$

$$\begin{aligned} GC^{CM}(x) &= D_{gc}(x)*t_{cm}+S_{gc}*D_{gc}(x)*(t_{int}-t_{cm}), \\ &\quad \text{for}\ d\le d_{cm} \\ &= D_{gc}(x)*t_{int},\ \text{for}\ d>d_{cm} \end{aligned}$$

Administration of stable iodine tablets

$$\begin{aligned} ThD^{CM}(x) &= D_{thy}(x)*t_{cm}+ \\ &\quad INH*D_{thy}(x)*(t_{int}-t_{cm}), \text{for}\ d\le d_{cm} \qquad (6) \\ &= D_{thy}(x)*t_{int},\ \text{for}\ d>d_{cm} \end{aligned}$$

Evacuation

$$\begin{aligned} ThyD^{CM}(x) &= D_{thy}(x)*t_{cm},\ \text{for}\ d\le d_{cm} \\ &= D_{thy}(x)*t_{int},\ \text{for}\ d>d_{cm} \qquad (7) \\ GC^{CM}(x) &= D_{gc}(x)*t_{cm},\ \text{for}\ d\le d_{cm} \\ &= D_{gc}(x)*t_{int},\ \text{for}\ d>d_{cm} \end{aligned}$$

Averted Dose

$$\begin{aligned} &AvThyD(x)=ThyD(x)-ThyD^{CM}(x)\ \text{and} \\ &AvGC(x)=GC(x)-GC^{CM}(x) \end{aligned} \qquad (8)$$

where, the symbols in equations (5-8), i.e., tcm, t_{int}, t_{eff}, d, d_{cm}, signifies the countermeasure implementation time, dose integration time, an effective time of initiating action, distance of interest for measuring dose rate and distance at which countermeasure action is to be implemented respectively, and the symbols S_{inh}, S_{gc}, INH represent the sheltering factor for thyroid dose, sheltering factor for whole body dose and inhalation factor for prophylactic administration respectively. Thyroid dose and whole body are symbolized by $ThyD(x)$ and $GC(x)$ and the corresponding dose rate are symbolized by $D_{thy}(x)$ and $D_{gc}(x)$ respectively.

The parameters S_{inh}, S_{gc}, and INH are very much uncertain due to their vagueness in realistic situation and hence they represent as fuzzy variables. In addition, other variables such as (a) Population distribution, (b) Road condition, (c) time delay, and (d) Evacuation rate are also fuzzy due their vagueness. Membership functions of S_{inh}, S_{gc}, and INH are shown in Figures 2.0 and 3.0. The impact of their overlapping is computed with the fuzzy relation and the fuzzy reasoning [Hadipriono, 1986].

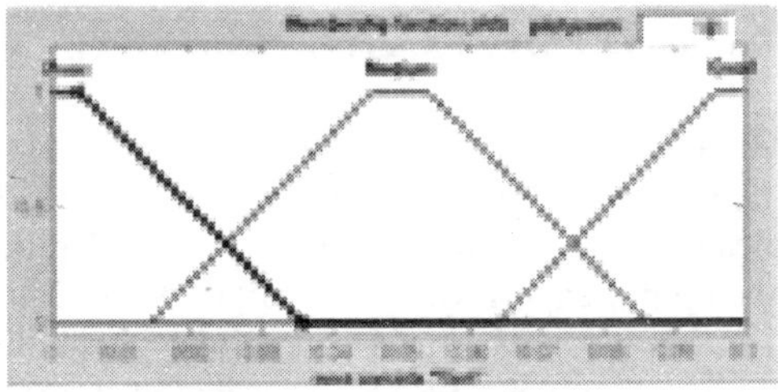

Fig. 2: Membership function of S_{inh} and S_{gc}

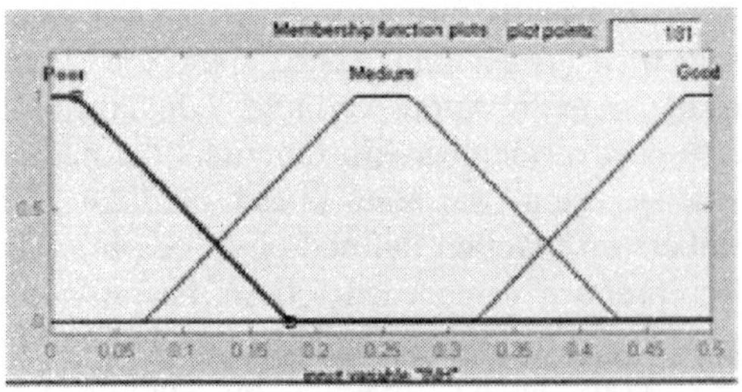

Fig. 3: Membership function of INH

Among the other variables, the percentile of population is basically derived by probabilistic treatment of human behaviour, which is not possible in realistic situation. Therefore, the uncertainty associated with the percentile of population is handled by a linguistic variable. Membership function of the sheltering is shown in Figures 4.0.

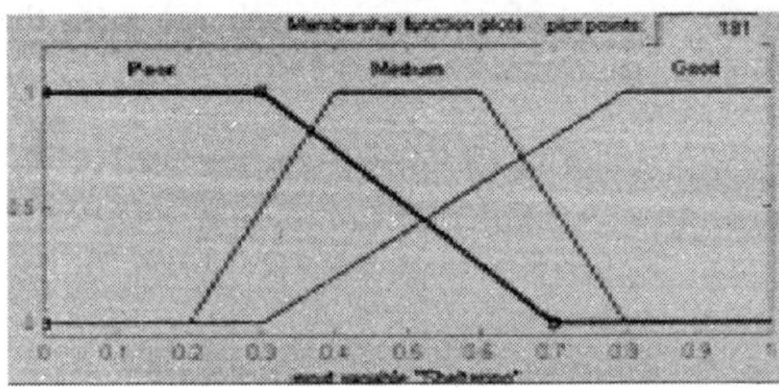

Fig. 4: Membership function of sheltering

A set of fuzzy conditional statements for these countermeasures is given in Table – 1.0.

Table. 1 Fuzzy Conditional Statements of countermeasures

Countermeasure	Fuzzy Conditional statements
Sheltering	IF S_{inh} is good and S_{gc} is good and <population> is low THEN Sheltering is good; Shielding Condition is "Good" [Effective] IF S_{inh} is poor and S_{gc} is poor and <populn> is low THEN Sheltering is medium; Shielding Condition is "Medium".
Evacuation	IF [time delay: (t_{eff})] is small and road condition is good THEN evacuation speed is "Good" [Effective]
Iodine Tablet Administration	IF INH is good and <population> is low THEN Distribution rate is "Good" [Effective]

5. RESULTS

Effective countermeasure using fuzzy logic

The code FEXEMER is applied to select the justified countermeasure. A typical scenario as an exercise is given as: Air concentration: 500 kBq/m^3, Gamma dose rate 0.06 mSv/hr, wind speed: 2m/s, stability category: F, t_{int}= 12.0 hrs, t_{cm} = 2.0 hrs, d = 1.0 km, d_{cm} = 3 km. Averted dose only for sheltering and evacuation is shown in Table 2.0 and 3.0. A typical sets of fuzzy relations between shielding conditions and shielding effect for sheltering and the same between road condition, population density and evacuation rate are given in Table 4.0.

Table 2 Thyroid and Whole body dose for Sheltering

Cm –Shelt.	2.0 km	3.0 km	4.0 km
ThyD (mSv)	2.24E+02	1.69E+02	1.08E+02
ThyD(cm)	2.24E+02	1.69E+02	1.08E+02
Av Dose	0.00E+00	0.00E+00	0.00E+00
W body Dose	5.94E-02	4.48E-02	2.87E-02
W b dose (cm)	5.94E-02	4.48E-02	2.87E-02
Av dose	0.00E+00	0.00E+00	0.00E+00

Table 3 Thyroid and Whole body dose for Evacuation

Cm –Shelt.	2.0 km	3.0 km	4.0 km
ThyD (mSv)	9.28E+02	4.95E+02	3.17E+02
ThyD(cm)	1.55E+02	8.25E+01	3.17E+02
Av Dose	7.73E+02	4.12E+02	0.00E+00
W body Dose	2.46E-01	1.31E-01	8.40E-02
Wb dose (cm)	4.10E-02	2.19E-02	8.40E-02
Av dose	2.05E-01	1.09E-01	0.00E+00

Table 4 Fuzzy Relations between A and B

Sheltering	Shielding effect (B)		
Sheltering factor (A)	Poor	Medium	Good
Poor	0.2	0.6	0.9
Medium	0.3	0.65	0.9
Good	0	0.7	1.0
	Evacuation rate (B)		
Road Condition (A)	Poor	Medium	Good
Poor	0.0	0.3	0.6
Medium	0.0	0.6	0.65
Good	0.0	0.6	1.0
Pop. Density (B)	Evacuation rate (B)		
Poor	0.01	0.02	1.0
Medium	0.01	0.6	0.6
Good	0.01	0.4	0.01

Fuzzy reasoning with the obtained fuzzy relation for input condition of sheltering and evacuation and averted dose together gives an output fuzzy set for shielding and evacuation. The operation "max-min" composition is used for fuzzy reasoning. The center of area defuzzification method is applied to obtain the crisp output for computing the effectiveness of evacuation as shown in Table 5.0.

Table 5 Effectiveness of Evacuation Countermeasure

Road Condition & Pop Density	Evacuation	Declaration of Evacuation
Very Good	0.8	Effective
Good	0.7	Effective
Medium	0.5	OK
Poor	0.4	Not effective
Very Poor	0.2	Not effective

6. CONCLUSION

The successfully development of a fuzzy expert system for the selection of an effective countermeasure with proper justification is the first achievement of this work. The final 'integrated fuzzy expert system' explicitly represented both imprecision in the input data and uncertainty in the interpretive knowledge base. Sheltering, prophylactic administration and evacuation are considered to investigate the potential applicability of fuzzy set theory. The respective membership functions and fuzzy conditional statements describing these countermeasures are formulated according to the guidelines obtained from literature [IAEA-TECDOC-1162, IAEA-TECDOC-953] on emergency preparedness (expert's opinion). The verification and validation of the present version of the software is under progress. In spite of the simple membership functions and fuzzy conditional statements, the obtained results show the potential applicability of fuzzy set theory to describe uncertain parameters in a countermeasure model.

7. REFERENCES

1. Civanlar, M.R., and Trussel, H. J., 1986, Constructing membership functions using statistical data, Fuzzy Sets and Systems, 18, pp.1-13.
2. Datta D., 1999, Development of a knowledge based system for decision making of protective measures during a radiation emergency, Proceedings of 24th IARP Conference, January 20-22, KAPS, pp. 102-107.
3. Datta D. and Hegde A.G., 2002, EEVADO – Ontology based construction of a knowledge based system for handling radiation emergency, Journal of Radiation Protection and Environment, 24(1 & 2), pp. 455- 460.
4. Dubois, D., and Prade H., 1980, Fuzzy sets and systems : Theory and applications, Academic Press, New York.
5. Ehrhardt, J. et al., 1990, COSYMA: A new program, package for accident consequence assessment, Eur 13028.
6. Hadipriono, F.C., 1986, Approximate Reasoning Models for Consequences on Structural Component due to failure Events, Civil Engineering for Practicing and Design Engineers, 5, pp. 155-169.
7. IAEA97, International Atomic Energy Agency, Method for the Development of Emergency Response Preparedness for Nuclear or Radiological Accidents, IAEA-TECDOC-953, Vienna, 1997.
8. IAEA, International Atomic Energy Agency, Generic Procedures for Assessment and Response during a Radiological Emergency, IAEA-TECDOC-1162, Vienna, 2000.
9. ICRP, International Commission on Radiological Protection, 1988, Optimization and decision-making in radiological protection, ICRP 55, Pergamon Press, Oxford.
10. ICRP, International Commission on Radiological Protection, 1990, Recommendations of the ICRP, ICRP 60, Pergamon Press, Oxford.
11. Klir, G.J., and Folger, T.A., 1988, Fuzzy Sets, Uncertainty and Information, Prentice-Hall, New Jersey, pp. 1-32.
12. Mamdani, E.H., and Gaines, R.R., (eds), 1981, Fuzzy reasoning and its applications, Academic Press, London.
13. Salt, C.A. and Dunsmore, M.C., 2000, Development of a spatial decision support system for post-emergency management of radioactivity contaminated land, Journal of Environmental Management, 58, pp. 169-178.
14. Tursken I.B., 1991, Measurement of membership functions and their acquisition, Fuzzy Sets and Systems, 40, pp. 5-38.
15. Zadeh, L., 1972, A fuzzy-set-theoretic interpretation of linguistic hedges, J. Cybern, 2(2), pp. 4-34.
16. Zimmermann, H. J., 1986, Fuzzy Set theory and its applications, pp. 47-59, Kluwer-Nijhoff Co., Boston.

Reliability, Safety and Hazard
Editors: P.V. Varde et al.

RISK-BASED APPROACH FOR RADIATION SHIELDING EVALUATION

L. Guneshwor and D. Datta

Health Physics Division, Bhabha Atomic Research Centre, Mumbai 400 085

ABSTRACT

Reduction of exposure to radiation in a nuclear facility is achieved by a shielding between the source and the radiation worker. The prime objective of the shielding design is to compute the shield thickness required to achieve the prescribed dose limit stipulated by the regulatory body. Deterministic methods of dose rate evaluation for shielding design are extended to risk based estimation by introducing the risk factor and the variability into the linear attenuation factor. A risk-based approach for estimating the shielding thickness is described in the paper. In place of a rigorous uncertainty analysis, a sensitivity analysis has been carried out in this paper.

Keywords: Risk, Effective dose, Uncertainty, Quality Factor

1. INTRODUCTION

Radiation shielding against gamma rays or neutrons is basically required for reducing the exposure to the personnel working in radioactive facilities such as nuclear power plant, fuel-reprocessing plant etc. The basic methodology of computation of the dose rate for a specified thickness of a shielding material includes parameters such as linear attenuation coefficient of the shielding material, source strength, buildup factor and particle flux to dose rate conversion factor [ANSI, 1991]. Computation of Gamma or neutron flux at the target distance in presence of the shielding material depends basically on source-shield geometry and the source strength. A large number of computational methods are available for estimating the shielding thickness even in complex source shield geometric configuration [Jaeger et al., 1968]. In order to compute the corresponding risk due to the exposure one needs to convert the estimated dose into risk by multiplying with the desired risk factor (/year) [USEPA, 1999]. In many applications, consideration of the uncertainty constitutes an integral part of the modeling process. The most evident example is the regulatory requirements established by the US Nuclear Regulatory Commission (NRC), the US Environmental Protection Agency (EPA), and the US Department of Energy (DOE) to conduct performance assessment analyses in order to consider the uncertainty associated with the predictions of the performances of a geologic repository for a high level nuclear waste [Eisenberg et al., 1987]. In exposure risk assessment problems, the probabilistic approach is equally appealing, because it makes little sense to give a deterministic value for the risk associated with a given receptor due to the large uncertainty in exposure point-kernel method. The uncertainties generally present in this case are as follows: (1) Modeling uncertainty arises due to using a simplistic relationship to describe the actual behaviours of a physical system. This can be quantified either by comparisions with other more involved models that provide a more accurate representation of reality, or by comparisions with collected data from field. Modeling uncertainty can be formally treated in a reliability context by introducing a random variable that describes the actual and predicted model response or output. (2) Statistical uncertainty [Dettinger and Wilson, 1981] arises due to the basic input data that are obtained from the shield designer's end, specially source related data (radionuclide inventory) comprising the source strength and statistical estimators (mean and higher order moments), and a probability density function (PDF) chosen to represent the distribution of each input variable.

Prediction uncertainty means that the reliability estimate depends on the state of knowledge that is available to the shield designer at the time of analyses. Various factors could affect the model response which are not included in the analysis simply due to lack of knowledge. As the state of knowledge increases, assessment of the reliability is refined.

Literature survey reveals that risk based computation is more communicative in the public domain compared to deterministic approach, for example, using high atomic mass shielding materials such as aluminium may increase risk and lower mass materials provide a limited amount of attenuation [Wilson, J.W., et al., 1991].

2. RISK BASED COMPUTATION – WHY AND WHAT?

Risk based methods are probabilistic rather than deterministic. This basis offers several immediate advantages. Deterministic methods have traditionally recognized uncertainty and reacted to it with excess conservatism, generally in the form of large safety factors. Deterministic methods in the shielding calculation have produced some seriously over-designed structures in the shielding material.

Probabilistic methods offer help in two ways. First, these methods explicitly and logically treat uncertainty. Just enough conservatism is applied to " err on the conservative side''. Further, these methods recognize that the worst case is no less a rare event than the best case, and base their conclusions on the most likely case. These methods consider: (1) just how improbable the worst case is, and (2) the consequences of the worst case occurring. This leads ultimately toward risk.

Risk is a two-dimensional measure of adversity. It combines probability and consequence. Risk-based analysis can be qualitative, quantitative or a hybrid. In qualitative analysis, one assigns the outcomes under study to broad probability and consequence bands, e.g. high, medium and low. One then uses a ranking table or risk plot to report the results. Even a qualitative risk assessment provides several benefits. Its two-dimensional nature draws attention to high risks whether they involve frequency or severity. It reduces biases by forcing systematic thinking. It presents its conclusions in a lucid graphical manner that is readily understood by others.

Quantitative risk analysis uses measured or estimated probabilities and consequences. Although moving from qualitative to quantitative risk analysis multiplies the cost, it increases the benefit. In addition to the benefits that risk-based methods provide at the qualitative level, quantitative methods calculate expected values for probability and consequence. If the consequence is measured in monetary terms, then the analyst can calculate a probable net present value (NPV) difference between existing conditions and for any proposed change.

3. RISK ESTIMATION IN SHIELDING EVALUATION

The point-kernel technique [Jaeger et al., 1968] is generally used for evaluating the thickness required of a shield material to limit the dose rates within the regulatory limit. The expression of dose rate includes parameters such as linear attenuation of the source and shielding material, and the buildup factor. All these parameters possess a specific probability distribution because their fundamental properties depend on various physical phenomenons such as Compton interaction, pair production, etc.

Considering a joint distribution of all the individual probability distributions, estimation of the distribution of dose rate can be carried out. This estimated value of the dose rate is further converted into the fatal cancer risks by taking a joint distribution of dose rate and risk factor.

The risk (/yr) due to a gamma dose accumulated slowly over a lifetime can be written as

$$R \approx K_{\gamma} QD / DDREF \qquad (1)$$

where K_{γ} is the risk coefficient (/Sv) , $DDREF$ is the dose and dose rate effectiveness factor, usually 2 or 1.5. [NCRP, 1997] and D is the lifetime accumulated dose (Sv) and Q is the quality factor. Since high LET radiations are more effective in risk paradigm than low LET gamma rays, quality factor has been introduced in equation (1). Based on observations of relative biological effectiveness (RBE), quality factor can be defined [ICRP, 1991] for use in risk estimation in radiation protection practice. The use of effective dose through Q places all radiations on an equal scale.

4. UNCERTAINTY IN WHOLE BODY RISK

Exposures to radionuclide even in the presence of shielding material produce a non-uniform distribution of dose within the body of the radiation worker. The uncertainty in risk associated with this exposure depends on the uncertainty in organ specific doses as well as the uncertainty in the risk per unit dose for each organ irradiated. A commonly encountered exposure scenario is one in which there is an approximately uniform, whole body dose of low LET radiation. For example Cs^{137} gets distributed uniformly throughout the body. In view of this and according to the guidelines published in the report [USEPA, 1994] following distribution of DDREF has been adopted. The probability density function for the DDREF is given by

$$f(x) = \begin{cases} 0.5 & ; \quad 1 \le x \le 2 \\ 0.5e^{-(x-2)} & ; \quad x > 2 \end{cases}$$

A plot of this probability distribution and its cumulative distribution are shown respectively in figures 1 and 2 below:

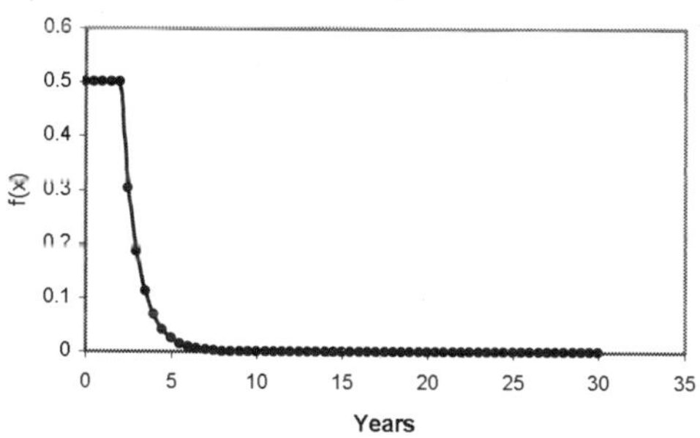

Fig. 1: Probability distribution function for DDREF

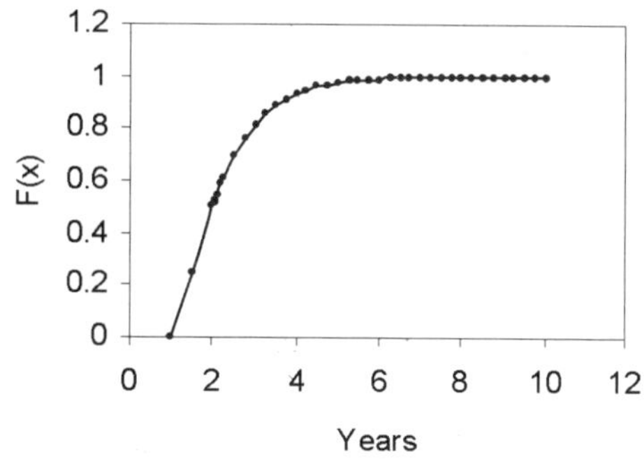

Fig. 2: Cumulative distribution function for DDREF

5. RESULTS AND DISCUSSION

Case Study

In order to focus our approach on risk based estimation of shielding evaluation a typical problem as described below is considered.

A shielded box was required for handling 5 curies (1.85×10^{11}Bq) of high-level active waste (HLW). Activity present in the waste was mainly due to Cs^{137} ($t_{1/2}$ = 30 yrs, E_γ = 0.662MeV) Shielding material chosen for this problem was either steel (density 7.89 gm/cc) or lead (11.34 gm/cc). It was required to estimate the shield thickness in presence of this source-shield configuration. Constraint of this problem was that the dose rate due to the gamma ray from Cs^{137} should be within the regulatory limit (0.1 mR/hr). The input parameters used for this problem are as shown in Table –1.

Table 1 Input Parameters

Source Strength (Ci)	Gamma Energy (Mev)	Distance of source and detector (cm)
5	0.662	50

The estimated dose rates for various thicknesses of the shielding material and the corresponding risks are given in Table 2a and 2b. Variation of risk due to variation of linear attenuation coefficient is given in Table 3a and 3b.

Table 2a Dose Rate and risk on Shielded box

Shield Material : Steel		
A	B	C
12	9.850	3.27E-11
15	2.190	7.27E-12
18	0.462	1.53E-12
20	0.164	5.44E-13
21	0.097	3.22E-13
22	0.057	1.89E-13

Table 2b Dose Rate and risk on Shielded box

Shield Material: Lead		
A	B	C
7.5	0.270	8.96E-13
8.0	0.150	4.98E-13
8.5	0.081	2.68E-13
9.0	0.045	1.49E-13
10.0	0.013	4.31E-14
*	*	

A: Shield thickness in cm
B: Dose rate in mR/hr
C: Risk (/year) without considering the variation in the buildup

Table 3a Variation of Risk and dose rate due to Variation in linear attenuation coeff

Shield Material : Steel		
A	R1	R2
12	3.71E-11	2.86E-11
15	8.27E-12	6.28E-12
18	1.78E-12	1.32E-12
20	6.26E-13	4.63E-13
21	3.69E-13	2.75E-13
22	2.16E-13	1.62E-13

Table 3 b Variation of Risk and dose rate due to Variation in linear attenuation coefficient

Shield Material: Lead		
D	R3	R4
7.5	9.34E-13	8.57E-13
8.0	5.18E-13	4.75E-13
8.5	2.80E-13	2.57E-13
9.0	1.56E-13	1.43E-13
10.0	4.51E-14	4.12E-14
*	*	*

A & D: Shield thickness in cm
R1 & R3: Risk (/y) with + 10% error in μ
R2 & R4: Risk (/y) with – 10% error in μ
μ: Linear attenuation coefficient in cm^{-1}

As per standard practice present risk factors used in the calculation are taken from the database of risk coefficients developed by FGR-13 [USEPA, 1999]. Uncertainty in this risk estimation may further arise due to the distribution in the dose and dose rate effectiveness (DDREF) factor. The value of DDREF inherent in the risk factor estimation is taken as a point estimate, which is equal to 2. But strictly speaking, the variability of DDREF is within the range 2-20. Plotting of risk against various shield thickness of steel and lead are shown in figures 3 and 4 respectively. It is observed from the Table-2 that, the risk per year corresponding to dose rate of 0.1 mR/hr is 2.68E-13 for lead and 3.22E-13 for steel.

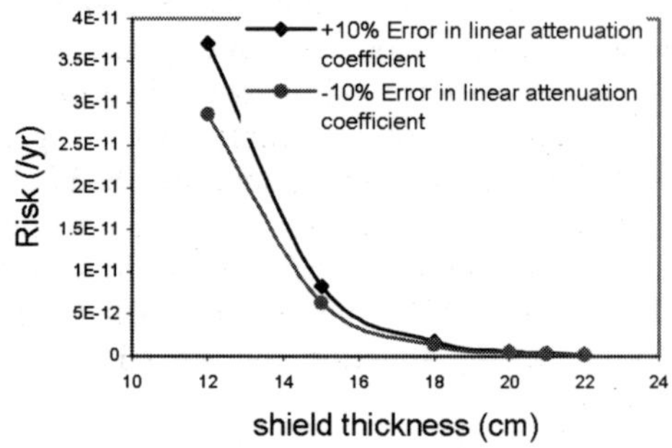

Fig. 3: Risk variation versus shield thickness (shield material: steel)

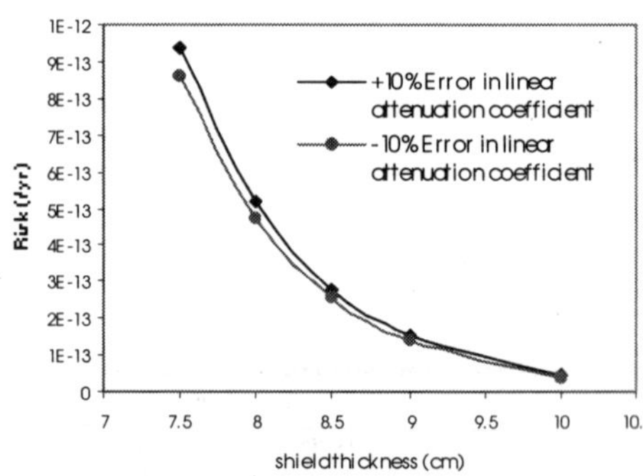

Fig. 4: Risk versus shield thickness (shield material: lead)

6. UNCERTAINTY REDUCTION

There are important prospects for reducing the uncertainties in risk projection in future. New human cancer data from exposed cohorts will be one key element in risk uncertainty reduction. The study of cancer risks in the nuclear industry is an ongoing study involving several countries and will include a large population of workers (for e.g., >100,000) (Cardis et al., 1995). One aspect that would be of great benefit from new human data is to consider any tissue dependence on the DDREF for low-LET radiation. Past observations have noted that a much larger DDREF for lung cancers than for e.g. breast cancer (NCRP, 2000). A confirmation of these results and an extension to other cancer types would be useful in reducing the uncertainties in the DDREF.

The uncertainty in the dose-rate effects may be small for high-LET particles because there is very little dose-rate dependence observed experimentally or expected from biophysical considerations. An approach that considers PDFs for a DDREF dependent on radiation quality could be more accurate than the approach used here, but at odds with conventional risk assessment approaches. Experimental studies that focus on the DDREF for photons should provide a significant reduction in the current uncertainty levels. One of the sources of uncertainty in the evaluation is the value chosen for DDREF, on which differences of view remain.

7. CONCLUSION

In this paper we have carried out a sensitivity analysis of dose by varying the linear attenuation coefficient of the shielding material as the parameter rather than a rigorous uncertainty analysis. The paper presents the approach of risk based shielding design rather than usual standard methods, which are based only on the dose rate. The advantage of this approach is to quantify the uncertainty that may arise because of the variability of the parameters such as buildup factor, self absorption coefficient of the shielding material required for computation. The dose equivalent is obtained by integrating the product of quality factor and dose over the entire linear transfer values. In addition to the quantification of uncertainties, this approach also allows the determination of dose equivalent which can be stated as reliability based shielding evaluation.

8. REFERENCES

1. American National Standard Institute (1991), Neutron and Gamma Flux-to-Dose rate factors, American Nuclear Society; ANSI/ANS-6.1.1.1997
2. Cardis, E., et al. (1995), Effects of Low Doses and Dose-rates of External Ionizing Radiation: Cancer Mortality Among Nuclear Industry workers in Three Countries. Radiat.Res.142, 117-132.
3. Dettinger, M.D. and Wilson, J.L. (1981), First order analysis of uncertainty in numerical models of groundwater flow: Part1 Mathematical development, water Resources Research, 17(1), pp.149-161
4. USEPA (1999) Eckerman, K.F, Leggett, R.W et al., "Federal Guidance Report 13: Cancer risk coefficients for environmental exposures to radionuclides" EPA (2002).
5. Eisenberg, N.A., Rickertsen, L.D., and Voss, C., (1987), "Performance assessment, site characterization, and sensitivity and uncertainty methods: Their necessity association for licensing", in B.B. Buxton (ed.), Proc. Of the conference on Geo-statistical, Sensitivity and uncertainty Methods for Ground-Water flow and Radionuclide Transport Modeling, DOE/AECL, Batelle Press, San Francisco, California, pp.9-38.
6. USEPA (1994),"Estimating Radiogenic Cancer risks" EPA402-R-93-076
7. ICRP (1991), The 1990 Recommendations of the International Commission for Radiological Protection, ICRP Report 60, Pergamon Press, Oxford, UK.
8. Jaeger, R.G., Blizzard, E.P. (1968), "Engineering Compendium on Radiation Shielding Volume 1: Shielding Fundamentals and Methods", Springer-Verlag, New York.
9. NCRP (1997), Uncertainties in Fatal Cancer Risk Estimates Used in Radiation Protection, NCRP Report No. 126.

10. NCRP (2000), Recommendations of Dose Limits for Low Earth Orbit. NCRP Report 132.Bethesda MD.

11. Wilson, J.W. et al. (1991), Transport Methods and Interaction for Space Radiation, pp.1257, NASA, Washington D.C.

Reliability, Safety and Hazard
Editors: P.V. Varde et al.

INVERSE SOURCE TERM CALCULATION FOR THE INDIAN DECISION SUPPORT SYSTEM FOR NUCLEAR EMERGENCIES

V.K. Sharma, Chitra S., S. Anand, Arti Agrawal and Manoj Kansal[#]

Health Physics Division, BARC, Mumbai 400 085
[#]Safety Analysis Group, NPCIL, Anushaktinagar, Mumbai 400 094

ABSTRACT

The paper presents details of a computer program developed for rapid estimation of the accident source term at a nuclear power plant. The software forms part of the Indian Decision Support System (DSS) for Nuclear Emergencies. It employs inverse source term estimation technique to back-calculate the release of activity into the environment, using a set of offsite dose rate measurements available on-line from gamma monitors installed around the nuclear power plant. The software has been customised for the Indian Pressurised Heavy Water Reactor (PHWR) located at the Narora site.

Keywords: Source term, Ring monitor, Nuclear emergency, Decision Support System

1. INTRODUCTION

The impact of the large-scale releases during the Chernobyl accident was felt in the erstwhile Soviet Union and in adjoining European countries through dry and wet deposition from the plume, and over rest of the world through global fallout. The accident had a profound negative impact on the development of nuclear energy worldwide. The public was dismayed by the prolonged trans-boundary release, by lack of information on the scope of contamination, and by the inadequate and inconsistent response to the emergency.

As a result of the experience gained from Chernobyl [1], the following areas were earmarked for future technical development:

- environmental monitoring systems;
- communication networks for rapid exchange of radiological information;
- guidance on universally acceptable standards of contamination in food; and
- decision support systems (DSSs) to advise emergency managers on counter-measures.

Europe has since advanced on all the fronts: Automated early warning systems now operate in most European countries, and they are now connected through efficient communication networks. Basic Safety Standards are now available for derived intervention levels in food.

In addition, several European Governments, Universities and other institutions have come together to develop decision support systems. A Real-time On-line DecisiOn Support (RODOS) system is operating in the EU countries; the Scandinavian and Baltic Countries, with Canada have developed the Accident Reporting and Guiding Operational System (ARGOS) [2]; and Russia is developing the Radio-ecological Analysis Support System (RECASS/RODOS NT). All three systems have joined hands to form the RODOS Consortium, which in turn is interacting with similar systems operating elsewhere, like the National Atmospheric Release Advisory Centre (NARAC) system operating in the US, and Worldwide version of System for Prediction of Environmental Emergency Dose Information (WSPEEDI) operating in Japan. A large part of the world is thus connected through a Decision Support System Network (DSSNET) [3].

The Indian DSS

Although India was not significantly affected by the fallout from the Chernobyl accident, it has drawn its lessons from the experience of the European countries. The Indian Environmental Radiation Monitoring Network (IERMON), with 37 on-line monitoring stations distributed throughout the country, is available at a central station located at BARC [4]. All power stations and other nuclear sites are connected through Department of Atomic Energy's ANUNET, a VSAT communication network. In addition, a Computerised On-line Plant Information System (COPIS) connects all NPPs to the Headquarters of the Nuclear Power Corporation in Mumbai.

In view of a rapidly expanding nuclear power programme and the introduction of large capacity units, it was considered imperative to put in place a DSS for nuclear emergencies at all power plant sites. The work on the Indian DSS was initiated in the latter half of year 2004 and has developed in a modest way since. It has been modelled on similar lines to EU-RODOS. The system will have near real-time and 48-hour predictive capability and will be developed for each of six NPP sites, starting with the 220 MWe PHWR site at Narora.

The DSS software comprises essentially six modules:

- the meteorological module
- the dispersion module
- the source term module
- the dose and counter-measure module
- the Geographic Information System module
- the Decision-making module.

This paper gives a detailed account of the method used to develop the source term module.

2. SOURCE TERM

Source Term (ST) is the complete description of accident release that enables its environmental consequences to be determined. Source term can be the actual release as measured off-site or the potential release. In all cases, the ST must specify the amount, rate of release, composition, and the mode of release including the environment, height and energy content of the release.

An accident source term depends on the reactor type and power, the nature of accident and its severity, and the physical and the radiological characteristics of the radionuclides. For convenience the radionuclides in the reactor core are grouped based on their volatility:

1. Noble gas (Kr, Xe)
2. Halogens (Br, I)
3. Alkali metals (Rb, Cs)
4. Tellurium & Antimony (Te, Sb)
5. Alkaline earth (Sr, Ba)
6. Ruthenium group (Ru, Rh, Mo, Tc)
7. Lanthanides (Y, La, Zr, Nb, Ce, Pr, Nd, Pu, Am, Cm)

For the present study, 64 radionuclides are selected. The radionuclides belonging to the first four groupings are the volatiles and contribute about 22% of the core radioactivity for a PHWR (see Table 1, computed using ORIGEN2 code [5]). They are important for accidents up to the design basis. The other groups contribute progressively more as the severity increases beyond the design basis.

Table 1 Inventory of a 220-MWe PHWR

Groups	Element Sub-groupings	Core Inventory Ci	Percent of core inventory
Volatiles	Noble gas	2.92×10^8	6.01
	Halogens	3.15×10^8	6.48
	Alkali metals	3.13×10^8	6.44
	Tellurium	1.72×10^8	3.54
	Total	1.092×10^9	22.47
Non-Volatiles	Fission Products	2.556×10^9	52.61
	Actinides	1.210×10^9	24.90
	Total	3.766×10^9	77.51

Types of Source Term

For the purpose of providing inputs to the DSS, the following source terms are defined:

1. Dummy source term
2. Potential source term
3. Actual source term
4. Prognostic source term

Dummy source term

It is the fictitious source term required to run the DSS in the normal mode (no-accident condition). It is used for demonstrating that all modules of the code are functioning properly.

Potential source term

The potential (or pre-release) source term is to be estimated based on plant parameters prior to the beginning of release. It may be based either on pre-calculated source terms or on the current plant conditions, based on deterministic or probabilistic methods. Expert judgement is required to relate the accident situation to a known accident type and severity.

Pre-calculated Source Term

Source terms are estimated for all postulated accident scenarios which lead to off-site releases. Source term calculation follows rule-based procedure and is highly conservative. Only Noble gases and Iodines are normally included. As an example, pre-calculated source terms for the following accident sequences were taken from the NAPS Safety Report [6]:

1. LOCA+ECCS failure (Fig. 1)
2. LOCA+Containment (inlet damper) failure (Fig. 2)

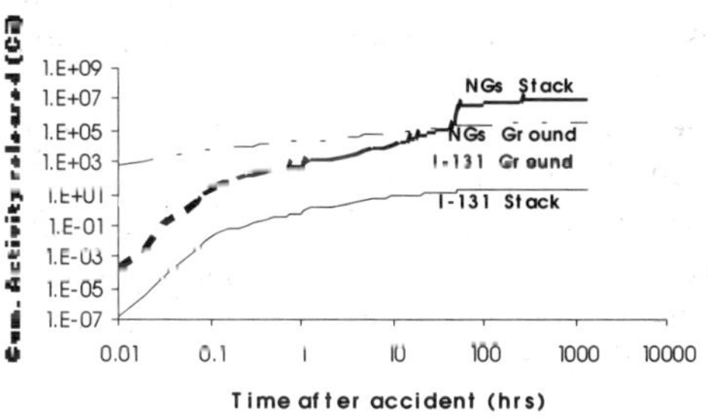

Fig. 1: Source term for LOCA+ECCS failure

Deterministic Method

In the deterministic approach, the source term is based on an actually developing accident

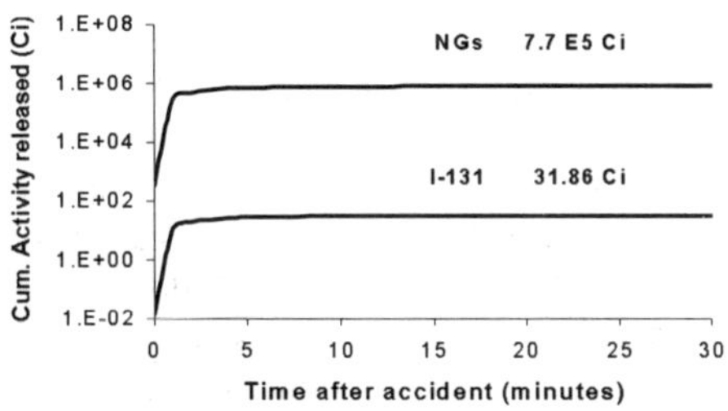

Fig. 2: Source term for LOCA+Containment (inlet damper) failure

situation. The diagnosis of plant situation must precede the prognosis of source term.

This method is used in French PWRs and encoded in the SESAME [7] code developed by IRSN, France. It is proposed to develop a similar system for the Indian DSS project.

Probabilistic Method

This method is based on Bayesian Belief Network (BBN) theory. The SPRINT code [8] being developed for RODOS is an example of this approach. A BBN model of the plant is made, with each node having discrete states, each with its probability. Nodes are connected by directed links; the dependence is given by conditional probabilities. Observations and trends of key nodes are followed up, and global changes in node state probabilities are calculated by an inference engine. Probabilities act on pre-calculated source terms (based on PSA Level-2) to generate a set of probabilistic source terms. A similar system will be developed for the Indian DSS.

Actual Source term

This refers to the real-time source term estimation. It is based on either the response of environmental monitors, or of plant stack monitors. The monitor response is used to back-calculate the released activity.

Prognostic source term

The prognostic source term is required when the DSS is to be used in the predictive mode. This source term is estimated based on the current source term and the projected status of the safety systems.

Actual (Inverse) Source Term

Inverse source term is the actual release as estimated from environmental measurements. Ideally, the installed monitor should distinguish between different radionuclides; however, since on-line gamma spectrum measurement in the field is difficult, only a gross gamma-measuring monitor is currently being installed for the Indian DSS. Supplementary information required to distinguish radionuclides will be generated by manual post-accident sampling. Since accidents in PHWRs may have simultaneous ground-level and stack-level release, it is proposed to have two rings of detectors around the plant, one at a radial distance of 800 m and the other at 1600 m. An optimisation exercise has shown that 16 monitors in each ring can resolve the ground and the elevated release.

The Gamma Monitor [4]

A typical gamma monitor (See Fig. 3) to be used for the ring monitor is similar to that developed by BARC for environmental radiation monitoring. The system is powered by a solar power pack and backed up by a battery. The data communication to the central station is achieved using a GSM Modem. The signal is transmitted to the Emergency Response Centre as an SMS message. This eliminates the dependence of the system on mains power and land-line telephone. Two detectors, one a GM counter and the other an ion chamber are used. The frequency of signal transmission is normally once every hour, changing to five-minutes on sensing an accident.

Fig. 3: On-line Gamma Monitor

Inverse Calculation

The monitor readings at two radial distances (M_{800}, M_{1600}) for simultaneous releases at two heights (10 m and 145 m) are given by the following set of equations:

$$M_{800} = D_{G,800} * R_G + D_{S,800} * R_S$$
$$M_{1600} = D_{G,1600} * R_G + D_{S,1600} * R_S$$

where

$D_{G,800}$ and $D_{S,800}$ are the dose rates at 800 m per unit activity release at ground and stack level, respectively;

$D_{G,1600}$ and $D_{S,1600}$ are the dose rates at 1600 m per unit activity release at ground and stack level, respectively;

R_G is the release rate at the ground level (10 m); and

R_S is the release rate at the stack level (145 m).

By solving these simultaneous equations, the release rates at ground and stack are estimated. The M_{800}, M_{1600} readings are transmitted on-line by the installed gamma monitors. $D_{G,800}$, $D_{S,800}$, $D_{G,1600}$, $D_{S,1600}$ values are pre-calculated for each radionuclide and for each weather category with the plume in the direction of a monitor, and stored in a data file.

On account of the uneven terrain at Narora, the inverse calculation needs to take into account the irregular arrangement of the monitors.

Three different cases

Since the ring monitor is a gross gamma monitor, it is necessary to split the source term calculations into three cases as described below:

Case 1: Only downwind gamma monitor readings are available;

Case 2: Besides on-line gamma readings, a gross gamma ground-shine reading is available;

Case 3: Besides on-line gamma readings, portable γ monitor readings and γ spectrum giving air concentrations of ^{131}I, ^{137}Cs and ^{132}Te are available.

Off-site Source Term: Case 1

This is based on on-line gamma monitors returning γ-dose rates. Since the γ-activity of the noble gases in the release is likely to be overwhelmingly large, no significant error is introduced in assuming that the response of the gamma monitors is derived from noble gases only.

Correction factors are separately computed for plume direction not coinciding with a radial line joining a monitor (see Fig. 4). Since the individual radionuclide response is pre-calculated and the gross monitor response is available on-line, the individual contributions from seven significant NG nuclides (^{85m}Kr, ^{85}Kr, ^{87}Kr, ^{88}Kr, ^{133}Xe, ^{135}Xe, and ^{138}Xe), are known.

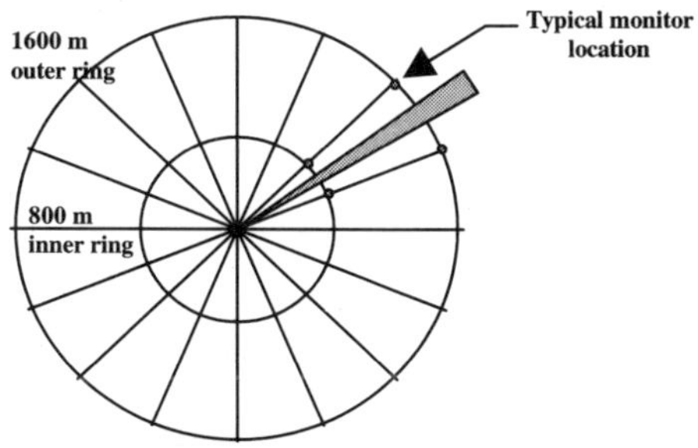

Fig. 4: Ring monitor layout around reactor

Off-site Source Term: Case 2

Two simultaneous measurements at the two rings are required for case 2:

- on-line monitors returning γ-dose rates, and
- gross gamma field on the ground.

As before, air Gamma field gives noble gas radionuclide composition. Groundshine dose rate measured using a portable instrument can give following radionuclides:

- Cs (^{134}Cs, ^{136}Cs, ^{137}Cs, ^{138}Cs)
- I (^{129}I, ^{131}I, ^{132}I, ^{133}I, ^{134}I, ^{135}I).

In the present calculation, the mass ratio of Iodine to Cesium in the equilibrium core is taken as 1:10. They are assumed to have similar release and deposition behaviour.

Off-site Source Term: Case 3

Two simultaneous measurements at the two rings are required for case 3:

- γ-dose rates, using portable gamma monitors;
  γ spectrum of representative volatile radio-nuclides in air (^{131}I, ^{137}Cs and ^{132}Te).

These measurements made beneath the plume centreline lead to a complete description of the release (27 radionuclides).

3. SOTERM CODE

The SOTERM package can handle all three cases mentioned earlier. The package (See Fig.5) takes inputs from i) the EMGST Code ii) DOSEFAC code, iii) the on-line plant parameters, iv) on-line weather data, and v) environmental measurements, including the on-line ring monitor readings.

The program is triggered by an emergency situation (a downwind monitor records >5 times the natural background). Depending upon the data available, it automatically decides on the case to be run. It subsequently prepares input for dispersion code.

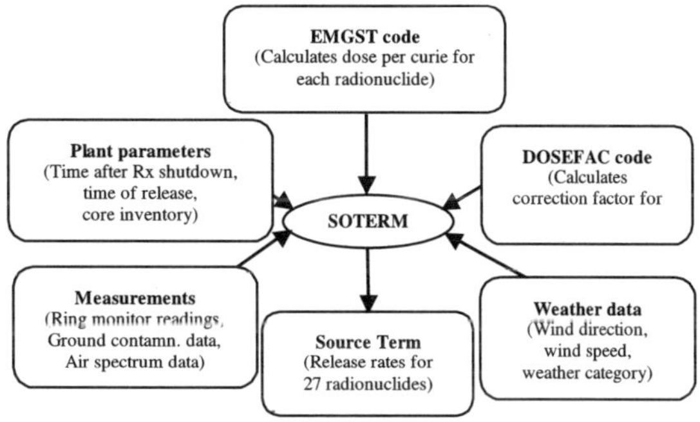

Fig. 5: Flowchart of SOTERM program

Forward dose calculation module, EMGST

EMGST Code is a forward dose calculation routine based on the Gaussian plume model. It calculates air/ground dose rate in mGy/hr for all the 64 radionuclides for both ground and stack level releases at two different downwind distances (800m/1600m). It calculates external dose for all gamma energies (150keV – 3MeV) emitted by the radionuclides and for all six weather categories. At present the output for a maximum of 27 radionuclides goes in to the SOTERM program to calculate the source term. The inputs required for running EMGST are:

- Weather category, wind speed, release height, downwind distance, and
- Dose conversion factors, deposition velocities, γ-energies and half-lives of the 64 radionuclides.

The following output for all the six weather categories at two different downwind distances due to two different release heights is generated:

- Plume/submersion dose rate (mGy/h) for Noble gases;
- Air concentration (Ci/m^3) of volatiles (Cs, I, Te);
- Groundshine dose rate (mGy/h) for unit release of iodine and Cesium.

DOSEFAC module

This routine calculates correction factor for an off-centered plume. The correction factors depend on the weather category, the gamma energy, the radial distance, the azimuthal angle, and release height.

DOSEFAC calculations are time intensive, hence are performed offline. The code output gives the ratio of the dose rate for a given plume to that of the plume in line with the nearest monitor. This is input to the SOTERM program. An example of the output of the DOSEFAC code is shown in the Fig. 6.

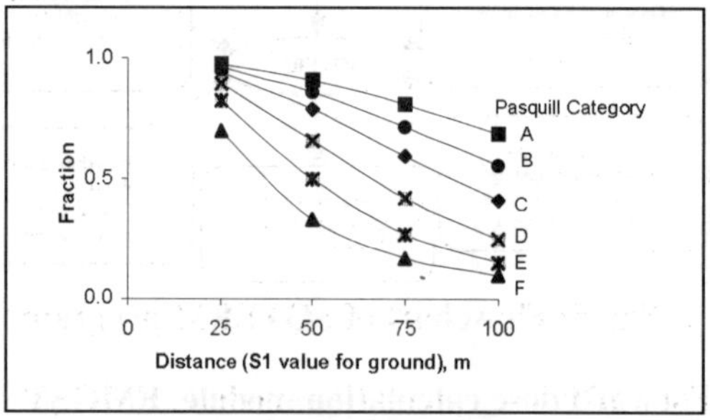

Fig. 6: Dose correction factor for 250 keV γ, ground level release

4. RESULT / DISCUSSION

Functioning of the Indian DSS was checked using the accident source terms corresponding to the accident (LOCA+ECCS failure) for Narora terrain and meteorology. The graphical representation of the calculated source term at one hour after the accident assuming case 2 conditions is given in Fig. 7.

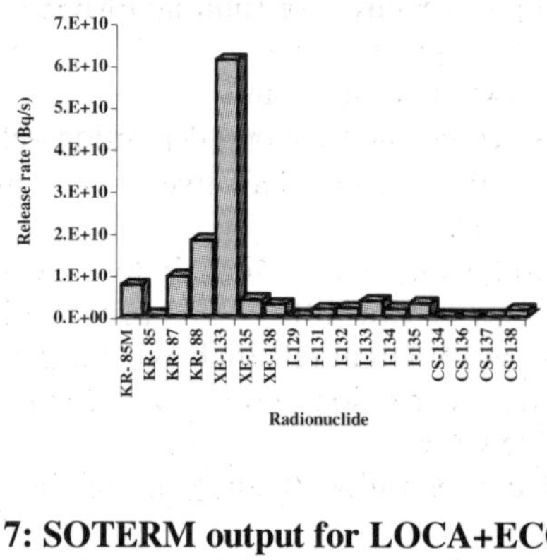

Fig. 7: SOTERM output for LOCA+ECCS failure

The time-dependent ring monitor response required for this purpose was generated by running the EMGST code for cases 1 and 2 with the above mentioned source term. It was found that the monitor readings were very sensitive to the ground level release. On this basis, it is concluded that the environmental detector should have adequate sensitivity and accuracy. An interesting observation for the Narora site is long spells of calm condition during winter when the plume hovers around the site, giving high monitor background.

5. CONCLUSION

Owing to the limitations of the on-line monitoring system, the initial source term generated for the Indian DSS comprises only Noble Gases. Subsequent manual field measurements are required to additionally calculate releases of cesium, iodine and tellurium isotopes. Future work includes development of NaI-based field monitors to give on-line spectral information on the releases, and generation of potential source terms.

6. REFERENCES

1. W. Raskob and J. Ehrhardt, 'The RODOS system: Decision Support For Nuclear Off-Site Emergency Management', 10th congress of the IRPA, 14-19 May 2000, Hiroshima, Japan.
2. The Danish Nation-wide Nuclear Emergency Preparedness Concept, Ministry of Interior, DEMA, Denmark, 1997.
3. T. J. Sullivan, 'The ARAC-RODOS-WSPEEDI Information Exchange Project', UCRL-JC-137966, USDOE, 1999.
4. IERMON website, EAD, BARC.
5. A.G. Croff, 'ORIGEN2: A Versatile computer Code for Calculating the Nuclide Compositions and Characteristics of Nuclear Materials', Nuclear Tech., Vol.62, Sept. 1983.
6. NAPS safety report, Volume 2, Accident Analysis.
7. D. Winter and J.M. Agator 'Utilisation of the SESAME System for Diagnosis and Prognosis of Plant Status during an Emergency in a French PWR', Radiation Protection Dosimetry, 73:1-4, pp273-276 (1997).
8. E. Grindon and C.G. Kinnibrgh, 'SPRINT: A Tool for Probabilistic Source Term Prediction for use with Decision support systems', Radiation Protection Dosimetry, 109:35-39 (2004).

Reliability, Safety and Hazard
Editors: P.V. Varde et al.

DETERMINISTIC SAFETY ANALYSIS OF PRODUCTION OF GASEOUS RADIOTRACERS ^{41}Ar AND ^{79}Kr IN A RESEARCH REACTOR

*Chitra S.**, *S. Anand**, *V.K. Sharma**, *V.N. Yelgaonkar*# *and Shivarudrappa V.*+

*Health Physics Division, BARC, Mumbai 400085
#Isotope Applications Division, BARC, Mumbai 400085
+Radiopharmaceuticals Division, BARC, Mumbai 400085

ABSTRACT

Gas phase radiotracers ^{41}Ar and ^{79}Kr can be produced by irradiating natural Argon and enriched ^{78}Kr gaseous targets in a nuclear reactor. While all design measures are taken to minimize the probability of failure of the gas ampoule, leakage of radioactive gas cannot be entirely ruled out. This paper presents the safety analysis of the production of ^{41}Ar/^{79}Kr radiotracers in 40-MWt Cirus reactor. Several sequences involving failure of the quartz ampoule and the consequent spread of radioactivity to working areas are analysed. Results show that even for the worst accident scenario, the on-site dose is low.

Keywords: Gaseous radiotracers, Gas irradiation, Accident Analysis, On-site impact

1. INTRODUCTION

Gas phase radioactive tracers find a large number of applications in industry and research for making diagnostic studies in gaseous process systems [1,2,3]. The applications include: air pollution studies, leakage location in underground gas pipeline, investigations in ventilation systems, study of movement of gases in oil fields, and investigations in industrial process streams. The radiotracers most commonly used include: ^{41}Ar, ^{85}Kr, ^{79}Kr, ^{133}Xe, ^{3}H, and ^{222}Rn. Many of them are amenable to production in nuclear reactors, being formed as fission products or activation products.

The production of gaseous tracers in research reactors is constrained by small sample sizes, low density of the target gas and natural dilution by other isotopes present in the target.

The primary safety concern is as regards the potential leakage of the gaseous activity during irradiation. This is addressed at three levels: design (by providing multiple containments), quality assurance (by specifying stringent qualifying tests on target holders), and operation (by strictly following operational procedures). However, excellence in design and operation can only reduce the probability of failure of the irradiation capsule, it cannot entirely rule out accidents in an operation that has to be carried out on a routine basis. It is imperative, therefore, to find out in a deterministic manner the scope of such an accident.

It is proposed to irradiate gaseous targets (^{40}Ar and ^{78}Kr) in 40-MWt research reactor Cirus to indigenously produce gaseous radiotracers ^{41}Ar and ^{79}Kr. An accident analysis of the routine production of these gaseous radiotracers was carried out by using a deterministic approach and the detailed results of the study are presented in this paper.

2. METHOD

2.1 Method of Analysis

The safety analysis of the scheme of production of ^{41}Ar and ^{79}Kr is carried out by first estimating the inventories of these isotopes and other radio-nuclides formed at the end of irradiation in the reactor and those after a brief decay period. The irradiation can containing the quartz ampoule filled with irradiated target gas is assumed to fail in a location where it will produce high consequences. For each accident scenario, the build-up of activity and dose in the working areas is integrated to yield the on-site impact. The off-site dose is not considered, as Cirus routinely releases relatively large amounts of noble gas activity through the stack and, therefore, the incremental off-site impact of a capsule failure would be negligible.

Irradiation of Argon Gas in Cirus Reactor

Natural argon has three stable isotopes; their properties are given in Table 1:

Table 1 Irradiation of Natural Argon Gas [4]

Argon isotope (atom %)	Properties of activation products			
	Thermal cross-section, b	Product nuclide	Half-life	Main emissions
^{36}Ar (0.337)	5.046	^{37}Ar	35.1 d	β+, No γ
^{38}Ar (0.063)	0.802	^{39}Ar	269 y	β-, No γ
^{40}Ar (99.60)	0.660	^{41}Ar	1.83 h	β-, γ

^{41}Ar produced with low activation cross-section of the target, which when coupled with its short half-life requires that a sufficiently large amount of argon gas be irradiated to obtain adequate activity. It is proposed to irradiate argon in a quartz ampoule (of capacity 7 cc), at a pressure of 6 kg/cm^2.a at normal temperatures, at the following two locations:

- in the horizontal self-serve facility, (max. activation flux: 1.4×10^{13} n.cm^{-2}.s^{-1})
- in a sample rod position in the reactor core (max. activation flux: 6.3×10^{13} n.cm^{-2}.s^{-1})

The purity level of the gas is taken as 99.9 % minimum, the rest being nitrogen and oxygen in the normal proportion in air. Only minor activities other than ^{41}Ar are expected to be generated. The typical dimensions of an ampoule and aluminium can are shown in Figure 1.

^{41}Ar activity reaches equilibrium in about 24 hours; however, for operational reasons, the gas will be irradiated for 7 days (in-pile irradiation) and for 1 day in a self-serve position. A cooling step of about 2 hours at the self-serve facility and 1 hour for the gas irradiated in a tray rod is envisaged. At the end of this period, the capsules will be transferred to the Tray Rod Facility (TRF) for loading and delivery.

Irradiation of ^{78}Kr in Cirus Reactor

The irradiation characteristics of natural krypton are given in the Table 2. ^{78}Kr, the target nuclide, is present in natural krypton to only 0.35%. In order to obtain sufficient ^{79}Kr activity, it is

Table 2 Irradiation of Natural Krypton Gas [4]

Krypton isotope (atom %)	Properties of Krypton activation products			
	Thermal cross-section, b	Product nuclide	Half-life	Main Emissions
^{78}Kr (0.35)	4.858	^{79}Kr	34.9 h	β+/EC, e-, X, γ
^{80}Kr (2.28)	11.76	^{81}Kr	2.1×10^5 y	β+/EC, e-, X, γ
^{82}Kr (11.58)	30.23	^{83m}Kr	1.86 h	IT, e-, x
^{83}Kr (11.49)	208.5	^{84}Kr	(Stable)	-
^{84}Kr (57.00)	0.083	^{85}Kr	10.8 y	β-, γ
^{86}Kr (17.30)	0.0616	^{87}Kr	76 m	β-, e-, X, γ

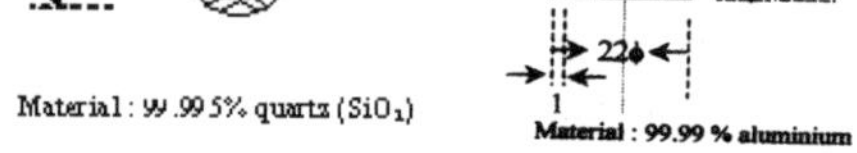

Fig. 1: Details of Ampoule and Aluminium Can

necessary to use enriched targets (^{78}Kr - 99.9%). The balance gas is natural krypton. Enriched ^{78}Kr is proposed to be irradiated in both the self-serve facility and in a tray rod position in the core, using similar ampoules. A 7-day irradiation is adequate, as it yields about 96% of the equilibrium activity of ^{79}Kr.

Qualification Tests on the Gas Container

The target gas is sealed in a pure quartz ampoule, wrapped in an aluminium foil and packed inside a cold-welded aluminium can. In the present usage, the maximum fill pressure used is 6 kg/cm^2.a, which is expected to rise to 7.1 kg/cm^2.a during irradiation, which is well within safety limits. Although the irradiation temperature is 80°C, the ampoule is tested against deformity or leakage in a furnace at 250° C for 8 hours. The capsule is also subjected to a drop-test from a height of 18m onto a 20mm steel plate. The aluminium can is tested to withstand an internal pressure of 10.2 kg/cm^2.a, thereby ensuring compliance with safety requirements.

Potential Accidents with On-site Impact

The gas capsule resides for 7 days in the reactor, for up to a few hours in the transfer flask/ fuelling machine during its transfer to the TRF, and for a brief period in the hot cell awaiting disposal.

Irradiation in the Tray Rod/ Self-serve Facility

For the capsule irradiated in the tray rod or self-serve, a failure could be postulated to occur i) during irradiation in the reactor, ii) during transit to the TRF, or iii) in the TRF cell. Both the tray rod and the self-serve facility of Cirus are air-cooled, and any leaked activity will be entrained in the cooling air, to be carried over by the ventilation stream and discharged out through the stack. As stated earlier, the off-site consequences of such a failure would be negligible. If a leakage of activity occurs in the fuelling machine or the transfer cask, the activity will be dispersed immediately in the reactor hall and will irradiate its occupants. This scenario is considered in this study. Should a release accident occur immediately after the transfer of the capsule to the TRF, on-site consequences will occur only if there is a simultaneous ventilation failure. This leads to a slow diffusion of the activity into the reactor hall. This scenario is also analysed here.

Characteristics of a Capsule Rupture Accident

Accident in the Reactor Hall

An accident occurring in the reactor hall is assumed to release activity instantaneously to the available air volume, which in the present report is restricted to the main hall, neglecting any leakage to the basement or sub-basement air spaces. The Reactor Building (RB) ventilation is treated as a parameter: it is either available at the design air change rate (0.91 air change/hour) or not at all. The radioactive decay of the radionuclides is considered throughout. Doses are calculated upto 24 hours after the release.

Accident in the TRF Hot Cell

The tray rod facility of Cirus is a cuboidal hot cell of volume 11.19 m^3. The ventilation provided in the cell ensures a negative pressure in the cell and maintains air change at the rate of 7.6 per hour. When the pressurized capsule bursts, the gaseous activity will be mixed almost instantaneously with the cell ventilation air and exhausted. The normal RB exhaust is at a flow rate of 51,000 m^3/h through a bank of HEPA filters. During emergency conditions, when the normal ventilation has been isolated, the RB is exhausted at 4,250 m^3/h through a bank of four combination filters comprising both HEPA filter and charcoal plates. Both these ventilation systems should be unavailable if the activity in the TRF is to leak out to the working areas.

Calculation of Radioactive Inventories

The accident release terms are estimated from the activities generated in the Argon/Krypton capsule at various times after end of irradiation. The computer code ORIGEN2 [5] is used for the purpose. (See Table 4)

Dose Computation for Accidents in the TRF

A postulated failure of the gas capsule coincident with the failure of the main ventilation system would eliminate the negative pressure gradient normally available and lead to a slow leakage by diffusion of the activity from the TRF to the RB air space.

The escape of radioactivity is modelled here by molecular diffusion of the particular species (krypton or argon) in air. It is assumed that only one capsule fails and that when the accident occurs, the entire radioactive inventory of the capsule is immediately released into the TRF air space, after which it slowly diffuses out. The concentration of radioactivity in an air volume is uniformly distributed. Radioactive decay is taken into account. This will affect in particular the ^{41}Ar activity in any compartment. The temperature of the system is taken as 298 K and the pressure 1 atmosphere, both of which do not change during the period of study (up to 24 h after the accident). The small isotopic effect on diffusion coefficient is considered in calculation of the leaked activity. The diffusion of the activity into the duct extending from the TRF to the isolation damper is conservatively neglected. All leakages take place into the Reactor hall.

Equation of Molecular Diffusion:

Fick's Law as applied to unidirectional diffusion of material A in a gas medium B is given as:

$$J_A = -D_{AB} \frac{dC_A}{dy}$$

Here the binary gas mixture shall be assumed to comprise of ideal gases at constant pressure and temperature where

J_A = molal flux density relative to a plane of no net volume transport, g-mol/s-cm^2

D_{AB} = Mutual diffusion coefficient of species A in a binary gas of A and B, cm^2/s

C_A = Concentration of A, g-mol/cm^3

y = Co-ordinate distance in direction of diffusion, cm.

Diffusion Coefficient, D_{AB}: In a low-pressure binary gas mixture, the mutual diffusion coefficient is given as:

$$D_{AB} = \frac{0.001858 T^{3/2} (1/M_A + 1/M_B)^{1/2}}{P\, \Omega_D\, \sigma_{AB}^2}$$

where T = Temperature, K

M_A, M_B = Molecular wts. of constituents A & B

P = Absolute pressure, atm

σ_{AB}, ε_{AB} = Lennard-Jones force constants for the binary

Ω_D = Collision integral, f (kT /ε_{AB})

k = Boltzmann's constant. The constants σ_{AB}, ε_{AB} are obtained from the corresponding values for pure substances by use of the combining rules:

$$\frac{\varepsilon_{AB}}{k} = \left(\frac{\varepsilon_A}{k} . \frac{\varepsilon_B}{k}\right)^{1/2} \qquad \sigma_{AB} = 0.5\,(\sigma_A + \sigma_B)$$

The values of the various parameters and the diffusion coefficients for the present study of argon-air and krypton-air systems are given in the Table 3.

Table 3 Parameters for calculation of the D_{AB} [6]

	σ (Å)	ε /k (°K)	σ_{AB} (Å)	(ε_{AB}/k) (°K)	Ω_D	D_{AB} (cm^2/s)
Air	3.71	78.6	-	-	-	-
Ar	3.54	93.3	-	-	-	-
Kr	3.66	178.9	-	-	-	-
Ar – Air	-	-	3.63	85.64	0.915	0.195
Kr – Air	-	-	3.68	118.6	0.998	0.153

The molecular weight of air is taken as 29. The molecular weights of the isotopes are taken based on their mass number. Ω_D is available as a function of (kT/ε_{AB}) in Ref. [6].

Leak Paths in the TRF

The TRF was inspected for potential leak paths, and it was found that the penetrations would give at most a total effective leak area of 159.18 cm^2, of which about 126 cm^2 is available in the pipe penetrations of the master-slave manipulator and about 33.18 cm^2 in the pipe connection for discharging roll-in balls into the cell. During normal ventilation, air is sucked into the cell through these paths. Following ventilation shutdown, leakage of activity by diffusion is expected to take place through these same passages.

Modeling of the Activity Leakage into the Reactor Hall

Since the TRF communicates with the Reactor hall air through a long pipe, the diffusional leakage is modeled in two stages: from the TRF to the pipe, and from the pipe volume to the Reactor hall air volume. The three air volumes are shown schematically in the figure below:

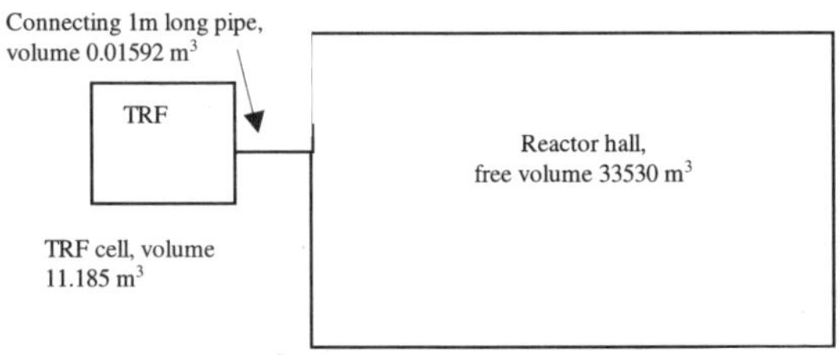

Fig. 2: TRF and the leak compartments

In the modeling of activity release into RB, as a first stage the equation for molecular diffusion is applied to the interface between the TRF & the pipe, and subsequently between the pipe and the Reactor Hall. In each case, the area of the interface is taken as 159.18 cm^2.

A computer program CIRDOSE was written to compute the concentrations of important radionuclides ^{41}Ar and ^{37}Ar (for argon gas irradiation), and ^{79}Kr and ^{83m}Kr (for krypton gas irradiation) in the TRF volume, in the pipe, and in the Reactor hall air as a function of time. In both cases, only the submersion dose by noble gas isotopes needs to be considered. The Reactor hall is modeled as an equivalent cylinder and the dose at its axis is taken as being applicable.

Accident Sequences

The various accident sequences during irradiation of argon and krypton are listed below. The parameters selected are: target gas, location of irradiation, location of accident, and status of RB ventilation. A perusal shows that the most limiting accident sequence is failure of argon or krypton capsule after in-pile irradiation with instantaneous release into the reactor hall and unavailability of ventilation (i.e. Case 6 below). The consequences for other accident sequences shall be small fractions of those of this limiting accident.

Case 1: In-pile irradiation, 1 hr decay, activity released in TRF. No ventilation;
Case 2: Self-serve irradiation, 2 hr decay, activity release in reactor hall. No ventilation;
Case 3: Self-serve irradiation, 2 hr decay, activity release in reactor hall. Ventilation available;
Case 4: Self-serve irradiation, 3 hr decay, accident in TRF. No ventilation;
Case 5: In-pile irradiation, release into reactor hall at the end of irradiation. Ventilation available;
Case 6: In-pile irradiation, released into reactor hall at the end of irradiation. No ventilation

3. RESULTS AND DISCUSSION

Gross Inventories and Radiation fields in Reactor Hall:

The gross activity available in the argon/ krypton capsule at various decay times are summarised in Table 4.

The radiation field build-up in the RB due to a capsule failure follows the same trend as activity. Calculations show that the maximum radiation field occurs, for Case 6, where values of 0.10 mSv/h for Argon and 0.04 mSv/h for krypton are noted. The field values for Argon capsule failure are high enough to set off alarm by an area gamma monitor (alarm level: 0.06 mSv/h) installed in the RB, but krypton capsule failure may go without automatic detection.

Table 4 Activity in an Argon and Krypton Capsule

Radio-nuclide	Activity (GBq)			
	In-pile irradiation		Self-serve Irradiation	
	End of Irradiation	1hr decay	2 hr decay	3 hr decay
^{41}Ar	40.7	27.79	4.218	2.893
^{37}Ar	0.134	0.134	0.005	0.005
^{79}Kr	48.470	47.360	10.323	10.101
^{83m}Kr	0.029	0.020	0.003	0.002

Cumulative Whole Body Dose

Table 5 summarises the effective whole body dose at discrete time intervals after the release transients. For the most limiting Case 6, the cumulative argon dose to a worker is observed to be 0.253 mSv in 8 h, and 0.266 mSv at 24 h. The corresponding values for krypton are 0.299 mSv at 8 h and 0.772 mSv in 24 h. (see Figures 3&4). It is worth mentioning here that Case 6 postulates ventilation failure, following which a prolonged stay in the reactor hall is unlikely, and it is more realistic to consider the next most severe accident, which turns out to be Case 5. The cumulative 8-h doses for this sequence are about 31% of those for Case 6 for argon, and only 15 % for krypton.

Table 5 Build up of whole body dose due to Argon leak and Krypton Leak

Time (hr)	Whole body Dose (μSv) Accident case					
	1	2	3	4	5	6
Argon leak						
0	0	0	0	0	0	0
4	1.170	21.50	8.050	0.122	77.70	208.0
8	2.130	26.20	8.100	0.221	78.20	253.0
12	2.490	27.30	8.100	0.259	78.20	263.0
24	2.650	27.60	8.100	0.275	78.20	266.0
Krypton Leak						
0	0	0	0	0	0	0
4	1.640	33.10	9.040	0.351	42.40	155.0
8	6.290	63.70	9.260	1.340	43.50	299.0
12	13.40	91.90	9.260	2.860	43.50	431.0
24	45.30	164.0	9.260	9.660	43.50	772.0

4. CONCLUSION

The safety analysis carried out has estimated the on-site radiological impact of rupture of an argon/ krypton gas capsule in the Cirus reactor. Significant radiological impact occurs when failure occurs in Reactor Hall or in the TRF. The worst case is when the ^{79}Kr gas capsule ruptures in the Reactor hall with ventilation unavailable. The maximum cumulative dose is 0.299 mSv in 8 h and 0.772 mSv in 24h. These on-site doses would normally be considered acceptable, given that such an event is highly improbable.

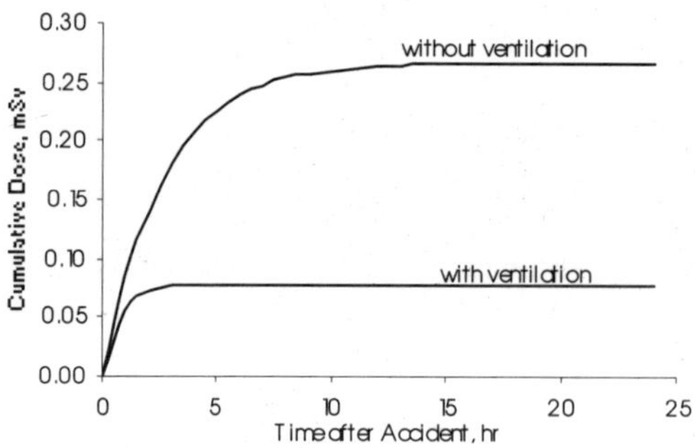

Fig. 3: Cumulative Dose due to Argon isotopes in Reactor Hall (Case 6)

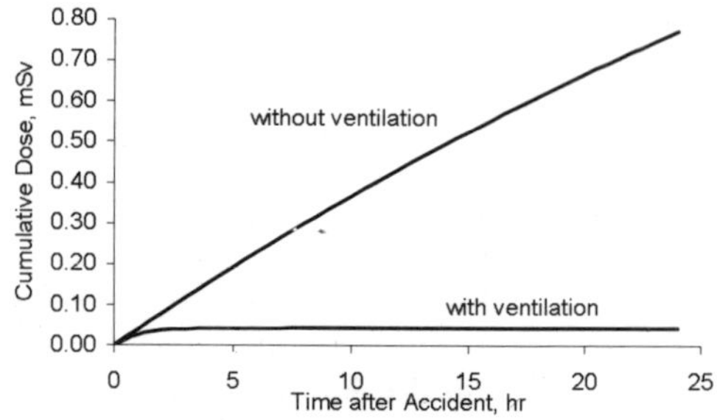

Fig. 4: Cumulative Dose due to Krypton isotopes in Reactor Hall (Case 6)

5. REFERENCES

1. Charlton, J.S. et al., "Radioisotope techniques for the Investigation of Process Problems in the Chemical Industry", Paper in Industrial Applications of Radio-isotopes and Radiation Technology, Grenoble, France, 28 Sept. – 2 Oct., 1981, IAEA, Vienna, 1982.
2. Chaim Gilath, "Ventilation and Air Pollution Studies Using Radioactive Tracers: A Critical Review", Int. Jl. Appl. Rad. Isot, Vol.28, pp 847-854, 1977.
3. Allen, L.S. et al., "Use of Nuclear Technology in Oil and Gas Exploration", Paper in Industrial Applications of Radio-isotopes and Radiation Technology, Grenoble, France, 28 Sept. – 2 Oct., 1981, IAEA, Vienna, 1982.
4. "Janis, A New Java-based Nuclear Data Display Program, NEA Data Bank, NEA, OECD.
5. Croff, A.G., "ORIGEN2: A Versatile computer Code for Calculating the Nuclide Compositions and Characteristics of Nuclear Materials", Nuclear Tech., Vol.62, Sept. 1983.
6. T.K Sherwood, R.L. Pigford, C.R. Wilke, "Mass Transfer", McGraw-Hill Kogakusha Ltd., Tokyo, Japan, 1975.

Reliability, Safety and Hazard
Editors: P.V. Varde et al.

HAZARD ANALYSIS OF ETHYLENE OXIDE DISCHARGE FROM STORAGE TANK – A CASE STUDY

S. Bhattacharya and P.K. Ghosh

Atomic Energy Regulatory Board, Niyamak Bhabhan, Anushaktinagar, Mumbai 400 094

Hazard assessment has been carried out for storage tank containing ethylene oxide, which is toxic, extremely volatile, and highly flammable substance. The consequence analysis have been carried out using discharge models, dispersion models and effect model, in case of loss of containment of storage tank and piping. The dispersion is calculated in terms of distance at various weather conditions and at predetermined conditions of toxic, thermal and overpressures and the effects are estimated in terms of human injury and environmental and equipment damage.

1. INTRODUCTION

A Hazard Assessment study of Ethylene Oxide storage tank was carried out by the Industrial Plants Safety Division (IPSD) of AERB.

The accidents that have occurred in the past involving Ethylene Oxide has been mentioned as the history of accidents in the book "Loss prevention in Process Industries" Part III - F. P. Lees as follows:

Table 1 Etheylene Oxide Accidents

Sl no.	Date/ Place	Plant/ Transport	Event	Death/ Injuries
1	1962, April 17 Doe Run, Kentucky US	Feed vessel	Vapour cloud explosion	1 death 21 injury
2	1964, June 4 Antwerp Belgium	Reactor	Internal explosion	4 death 20 injury
3	1968 January 1 Dunreith, Indiana, USA	Rail tank car	BLEVE	24 death 5 injury
4	1971, Nov 7 Morris llinois, USA	Plant	Internal explosion	4 injury
5	1973, Dec 27., Freeport, Texax, USA	Tank	Vapour cloud fire	29 injury
6	1976, May 24 Geismar, Louisiana	Reactor	Internal explosion	56 injury
7	1977, July 8 Gela, Italy	Plant	Internal explosion	1 death 1 injury
8	1987, July3 Antwerp Belgium	Distillation column	Explosion	
9	1989, March 7 Antwerp Belgium	Distillation column	Explosion	
10	1991, March 12 Seadrift Texax, USA	Plant	Vapour cloud explosion	1 death

For identified scenarios, effect of loss of containment is based on toxicity and flammability of ethylene oxide. The consequence analysis due to loss of containment and subsequent effect largely depends on three parameters i) hazardous properties of chemical ii) storage condition iii) safety measures incorporated in the storage.

2. HAZARDOUS PROPERTIES

2.1 Toxic hazard: Ethylene oxide is a respiratory, skin and eye irritant, at high concentration and it is associated with central nervous system depression. Breathing of vapour and contact with eyes and skin need to be avoided. Delayed effect of high acute exposure includes headache, nausea, vomiting shortness of breath, cyanosis and pulmonary oedema. Additional symptoms reported after acute exposure are drowsiness, fatigue, incoordination and weakness. The carcinogenic potential of ethylene oxide has been demonstrated and IARC has classified Ethylene Oxide as a Group I Carcinogen and NIOSH (US National Institute for Occupational Safety) and OSHA (Occupational Safety and Health Administration) also concluded the same, for long term exposure. It is associated with adverse reproductive effects in both animals and human being.

2.2 Fire hazard

Ethylene oxide has inflammable limit between 3% to 100% and is Very dangerous when exposed to heat or flame, can react with acids or bases. It has wide flammable limits as it supplies its own oxygen and presence of air is not a pre-requiste for fire and any small release is likely to catch fire rather than travel as toxic cloud in the atmosphere. Hence inspite of its toxic nature the primary hazard arises due to its potential for fire and explosion

3. STORAGE TANK DETAILS

Contents : Ethylene Oxide
Vessel Size : 13.5 m3 full
Vessel Dimension : 2063 mm ID x 4806 mm (full length) with 2:1 ellipsoidal dished heads
Orientation : Horizontal
Material of Construction : SS 304
Storage Temperature : 6 degree C
Storage Pressure : 2.4 kg/sq. cm of N2
Vessel Location : Above the ground
Height above ground :Bottom 1.4 m above
Location of the vessel : Open air.
Bunding : Bunded with concrete bund of 6 m3 volume. The bund drains (not overflows) into a 150 m3 containment pit.

3.1 Vessel Opening

The vessel has twenty openings in terms of flange joint for the control instruments, inlet/outlet of process liquid, nitrogen supply, manhole etc..

The large openings are two manholes of sizes 900 mm and 450 mm.

There are 80 mm nozzles at the bottom of the tank, one the drain nozzle and the other ethylene oxide outlet nozzle and also 50mm pipelines for inlet and other connections, the failure of which can lead to loss of containment of liquid ethylene oxide to the environment causing a potential hazardous condition.

3.2 Safety Measures

The ethylene oxide is stored in a tank designed for 8 kg/sq. cm. From the tanker, it enters the tank through a line, which is connected, to a nozzle to deliver ethylene oxide at top of the tank. The ethylene oxide when required for the process is pumped from the tank by a centrifugal pump, which is connected at the tank bottom. Heat leakage into the storage tank causes vapourisation and equilibrium is maintained by applying nitrogen pressure and a nitrogen blanket of about 2.4 kg/sq. cm. is maintained over it and it is stored at about 6 degree Celsius under refrigerated condition

In order to be ready for export in liquid state it is necessary to condition the liquid delivery system so that ethylene oxide can be pumped to the reactor and hence all the ethylene oxide lines are cold traced with chilled brine lines,

The storage area is also equipped with four EO detectors - The detectors are located at unloading pump, feed pump, storage tank bottom discharge valve, storage tank bottom drain valve. In addition a gas detection point at the stack has also been provided.

Ethylene oxide is stored in an area specially designated for their storage as per Explosive Department and SMPV rules the enclosure has a barbed wire fence and only authorised personnel can enter and the clearance between EO and the fence is atleast 20 m on all sides.

The vessel is provided with

- Remote isolation valve, double relief system consisting a bursting disc
- The vent of the RV discharges through a vent stack in which live steam is used.
- Alarms for high level and high temperature.

4. POSTULATION OF EVENT SCENARIO

The failure cases selected for consideration in this study were identified by "checklist" method in which failure cases are derived from knowledge of actual previous accidents. The procedure consists of first making an inventory of the hazardous materials present and then reviewing their intrinsic hazards. It has been found that escape of material constitutes the main hazard and the various possible means of escape have been identified:

(i) sudden and complete rupture of vessel;
(ii) "guillotine-breakage" of pipework connected to vessels;
(iii) small holes or cracks in piping and vessels;
(iv) flange leaks;
(v) leaks from pump glands and similar seals.

No (i) option is the worst case scenario and in this case the whole amount of ethylene oxide will come out and the consequence will be vapour formation which will travel and if it finds an ignition source before getting diluted below the lower flammable limit there will be flash fire.

Guillotine-breakage of pipework connected to vessels is credible scenario and in such cases the amount of ethylene oxide coming out will depend on the nozzle size as well as the nozzle placement on the tank and the consequences can be flash fire/explosion.

5. IDENTIFICATION OF EVENTS FOR ANALYSIS

The ethylene oxide storage tank has twenty openings of which the larger ones are the two manholes, the level indicator of 150mm diameter, a blinded relief valve, drain pipe, ethylene oxide outlet each of 80 mm diameter and ethylene oxide inlet pipe of 50 mm diameter. Four scenarios which can have considerable effect on the surrounding because of the size of opening and the characteristic properties of ethylene oxide, are selected for the consequence analysis as follows:

A) Catastrophic failure of the storage tank
B) Level indicator of 150 mm diameter guillotine rupture
C) Leakage of drain pipe 80mm diameter
D) Rupture of 50mm EO Inlet pipe dipped into the tank

The catastrophic failure of the storage tank when full will release the liquid into the bund and though the probability of this scenario is very remote, the initiating events for this scenario, which can cause such a failure are as follows:

a) External factors such as earthquakes, vehicular impact, sabotage
b) Mechanical failure due to defects, fatigues, brittle failure etc.
c) Excessive vibrations due to support damage, flow induced vibration
d) Under or overpressurisation due to operational or control failure
e) Material failure due to external fire
f) Corrosion

Full bore rupture of 80 mm and above pipelines is not probable if designed and constructed as per standard.

Whereas failure of 50 mm and less pipelines have taken place more frequently in chemical industry and an hazard assessment calls for detailed study including probability of effect on people in such cases.

Table 2 Assumptions for consequence analysis

1. Weight of liquid=12,000kg
2. Vessel is padded with nitrogen at 6 o C
3. The ambient temperature : 30oC =evaporation temperature
4. The weather condition chosen are: 2F, 3D and 1.5 A for all the three scenarios;
5. Lc50 value for Ethylene oxide is 2211ppm,= concentration for interest.
6. The explosion overpressure of interest are:
   0.3 bar – unrepairable damage to plant equipment
   structure bar – repairable damage to plant equipment and structure
   0.02 bar – shattering of glass
7. The heat radiation levels of interest are:
   kw/sq m – Sufficient to cause pain to personnel if unable to reach cover within 20 seconds, however blistering of the skin is likely,
   12.5 kW/sq. m- Minimum energy required for piloted ignition of wood, melting of plastic tubing and
   37.5 kW/sq. m- Sufficient to cause damage to the process equipment.
8. The distances upto which the effects are calculated are
   30 metre
   60 metre and
   90 metre , the basis for choosing the distances being the location of other equipment ahead is taken as 2 meters almost filled for the purpose of calculation.
10. The location of the tank is 1.4 metre above the ground.
11. Relative humidity is assumed as the average of 85% for throughout the year.
12. Surface roughness at the location of the installation is 0.1
13. The pool diameter is (equivalent diameter, assumed as circular for 27.4 m2 dyke area) 5.30 m

6. HAZARD ANALYSIS

The consequence analysis is based on the physical conditions of the storage and some predetermined weather conditions, thermal radiation levels, concentrations and overpressure values which can have adverse consequences.

6.1 Consequence Analysis

The assumptions for consequence analysis are brought together in Table 1.

The consequence analysis have been carried out with commercial validated software (*PHAST Professional*) for different scenarios for the four cases as follows:

A) Catastrophic failure of the storage tank
- Flash Fire Radiation effect

B) Level indicator of 150 mm diameter guillotine rupture
- Flash Fire radiation effect
- Jet fire radiation effect
- Explosion effect of over pressure

C) Leakage of drain pipe 80mm diameter
- Toxic effect
- Flash Fire Radiation Effect
- Jet Fire Radiation Effect
- Pool Fire Radiation Effect
- Over pressure explosion effect

D) Rupture of 50mm EO Inlet pipe dipped into the tank
- Toxic Effect
- Pool Fire Radiation Effect
- Flash Fire radiation Effect
- Explosion overpressure
- Effect model

The Table 3-6 show the results highlights of result of dispersion model and the effect model results are shown in Table 8.

The items in paranthesis in table 3-6 are the weather conditions for the maximum distance.

6.2 Event Probability

One of the method for estimating the probability of an event to occur is to refer to the relevant global data existing on failure cases and figures can be collected from various sources, which support these failure rates. These data include process plant on various types at different circumstances including extreme conditions and correction is needed while using these values. Report of the Risk Analysis of Six Potentially Hazardous Industrial Objects in the Rijnmond Area gives failure rates of tanks, vessels, pumps, pipework, valves, instrumentation, electrical equipment etc. collected from a different sources and some are reproduced below:

Table 3 Catastrophic Failure of Storage Tank

Specified condition	Effect	Max Distance(m)
50% LFL	Flash Fire	36 (2 F)

Table 4: 150 mm pipe rupture- consequence

Specified condition	Scenario	Maximum Distance (m)
4.0 kw/m2	Radiation effect jet flame	200m (3 D)
12.5 kw/m2	"	73 m (3 D)
37.5 kw/m2	"	Not reached
50% LFL	Flash Fire	263 m (3 D)
0.3 bar	Overpressure	330m (3 D)
0.14	"	374 m (3 D)
0.02	"	760 m (2 F)
Worst case- Explosion effect - 0.02 bar- 760 m		

Table 5: 80 mm pipe leakage- consequence

Specified condition	Scenario	Results
LC 50	2211ppm	450 m (2 F)
4.0 kw/m2	Radiation effect jet flame	270 m (3 D)
12.5 kw/m2	"	216 m (2 F)
37.5 kw/m2	"	Not reached
50% LFL	Flash Fire (late ignition)	98 m (3 D)
0.3 bar	Overpressure	122 m (2 F)
0.14	"	144 m (2 F)
0.02	"	244 m (2 F)
Worst case- Toxic - 450 m		

Table 6: 50 mm pipe rupture- consequence

Specified condition	Scenario	Results
LC 50	2211ppm	350 m (All)
4.0 kw/m2	Pool fire	32 m (3 D)
12.5 kw/m2	"	24 m (3 D)
37.5 kw/m2	"	17 m (3 D)
50% LFL	Flash Fire (late ignition)	166 m (2. F)
0.3 bar	Overpressure	201 m (F)
0.14	"	223 m (F)
0.02	"	388 m (F)
Worst case- Explosion effect - 0.02 bar- 388m		

Table 7 Basic Failure Rate Data of Pipelines

Pipework	Mode of failure	Failure rate
Upto 50 mm dia pipe	Catastrophic rupture	1E-9/sect-hr 1E-10/m.hr
	Significant leakage	1E-8/sect.hr 1E-9/m.hr
>50 to 150 mm dia	Catastrophic rupture	3E-10/ sect-hr 3E-11/m.-hr
	Significant leakage	6E-9/sect.hr 6E-10/m.hr
>150mm dia pipe	Catastrophic rupture	1E-10/ sect-hr 1E-11/m.-hr
	Significant leakage	3E-9/sect.hr 3E-10/m.hr
Pressure Vessel	Catastrophic Rupture	1E–6/yr

Catastrophic failure means severance of the pipe with discharge rate based on the pipe diameter, significant leakage is where a pipe develops a crack due to corrosion, erosion or fatigue and hole size is between 5-15mm diameter depending on the pipeline dia. These are base failure rates for process plant pipework where there is no vibration, corrosion/erosion or excessive thermal cyclic stresses expected otherwise correction factor of 3-10 is applicable. The unit "per section hr" is defined as average length between significant items of equipment i.e. vessels, control valves, pumps. Sections will vary 3-30 metres in length it is typically 10 metres.

6.3 Effect Model

From the Table 7 it is found that the failure rate data for the 50 mm or less diameter pipelines are more probable compared to larger diameter pipelines.

The accepted method of calculating the probability of personal injury and damage is by Probit Analysis, where Probit Pr is a parameter expressed in percentage of damage, calculated from a causative factors such as thermal intensity, overpressure and toxic concentration.

The constants used in equation are taken from literature which are validated and experimented over a wide range of situations.

The effect of this consequence from the Table 6 was calculated which gives the result in Table 8.

There are few chemicals which have information on dose response relationship that can be applied to "probit function" to quantify the number of fatalities that are likely to occur due an exposure and in this report probability analysis of four conditions were studied:

i) Probability of fatality due to dispersion cloud
ii) Probability of personal injury due to flash fire
iii) Probability of personal injury due to pool fire
iv) Probability of damage of structures due to over pressure
v) Probability of injury of human due to over pressure

Table 8 Result of Effect model for 50 mm Pipe Rupture

S No.	Event	Dis affected metre	Weather	% of damage
1	Fatality due to the toxicity of the dispersion cloud	210	A 1.5	<1% Fatality
2	"	350	F 2	<1% Fatality
3	"	300	D3	Negligible
4	Flash Fire	50-90	All	Fatality
5	2nd degree burn injury due to radiation effect of 37.5 Kw/m2 from pool fire	11-13	All	50% of population
6	1st degree burn at 25 kw/m2 for 10 secs	13-19	All	95% of population
7	Glass breaking damage for 0.0207 over-pressure.	278-388	All	3.4% of the glass structures
8	Minor damage due to 0.14 bar over pressure.	139-223	All	99% of the minor structures
9	Major structural damage due to 0.3 over pressure.	120-201	All	90.5% of the major structure

The probability of human injury due to over pressure has not been shown in the table 7. There are two types of effect of over pressure - the direct effect on human being damaging human organs and indirect effect of fragments and debris hitting people. Two major organs i.e., lung and eardrum are most affected by overpressure; however, the impact of damage depend on the location of persons. The individual organ damage has not been calculated but the population under 0.1 overpressure would have various types of injuries because of structural damage.

The distance from the ethylene oxide storage tank to the boundary would determine whether public would be affected beyond the plant premises in the most credible scenarios as mentioned i.e, rupture of 50 mm pipeline.

7. DISCUSSION

Table 3 shows that the catastrophic rupture (Case A) would cause flash fire which would extend upto 36 metres. The toxic effect of catastrophic failure has not been calculated as this failure would cause a very large mass to be released and the high range of flammability of EO would have more probability of catching fire immediately. Hence the cloud of EO is not expected to disperse further to cause toxic effect.

It has been experienced that persons present with the flash fire range are subject to fatality. The probability of large number of persons present within 36 metres are less and the fencing of the equipment being 20 metres away will not affect many other structure. But the scenario of catastrophic failure of the vessel is not much probable as the vessel has been designed based on accepted standard and 100% radiographed and maintained as per the codal requirement.

The toxic effect of 150 mm pipe rupture (Case B) would also be less as the chance of jet fire and flash fire would be more due to release of large mass which may back fire or may flash causing explosion. From the Table 4 it is seen that the maximum effect of the rupture is due to flash fire which extends upto 260 metres but the over pressure effect of 0.02 bar would cause shuttering of glass panes upto 760 metres and can extend beyond plant premises.

Leakage from 80 mm pipe(Case C) and rupture of 50 mm inlet pipelines(Case D) would cause much less quantity of ethylene oxide to be released but the duration of release would be longer. This would allow the cloud to disperse causing toxic effect and there is also chance of late ignition causing explosion when the cloud finds ignition source. Table 5 shows that the maximum distance affected from 80 mm leakage is the toxic release felt upto 450 metre(2211ppm). In the 50 mm pipe rupture the longest distance affected is around 340 metres(Table 6) due to overpressure upto 0.02 bar causing shattering of glass pane. But the probability of an explosion to occur is low because for an explosion to occur some amount of confinement of the vapour cloud is necessary and in open that may not be practicable. The chance of flash fire predominates upto 160 metre distance and also the toxicity can travel upto 350 metres.

8. CONCLUSION

If one goes by the known cases of major accidents which has taken place as given in Table 1 it is obvious that barring one all others have taken place in the process unit rather than in the storage tank since possibility of deviation from design intent is higher in process. . The basic failure rate data as given in Table 7 indicates that it is more probable for rupture of 50 mm and lesser diameter pipelines than higher diameter pipelines and catastrophic vessel failure.

The effect of dispersions of 50 mm pipe rupture have been calculated in terms of injury and damage and tabulated in Table 8. It has been seen that the effect of toxicity extending upto a distance of 350 metres is less than 1%, and hence probability of any injury due to toxicity can be totally ruled out. In case of flash fire the fatality within 50-90 metres (100% LFL and depending on weather condition) would be 100%. The distance vs radiation plot from the dispersion show that 1st degree burn (125 kJ/m2) is possible for 95% of the population within 13–19 metre distance, whereas, 2nd degree burn (250kJ/m2) would be felt between 11-13 metres where the chance of 50% of the population would be affected

9. REFERENCES

1. Guidelines for Hazard Evaluation Procedures, Centre for Chemical Process Safety, American Institute of Chemical Engineers, 1992
2. Guidelines for Chemical Process Quantitative Risk Analysis, Centre for Chemical Process Safety, American Institute of Chemical Engineers, 2000.
3. Methods for Determination of Possible Damage to People and Objects Resulting

from Release of Hazardous Materials, - Committee for the Prevention of Disasters caused by Dangerous Substances, The Hague, 1992, TNO.
4. Methods for Calculation of Physical Effects, Committee - Committee for the Prevention of Disasters caused by Dangerous Substances, The Hague, 1997, TNO.
5. Loss prevention in Process Industries Part III - F. P. Lees
6. Risk Analysis of Six Potentially Hazardous Industrial Objects in the Rijmond Area, A Pilot Study – A report to the Rijmond Public Authority.

Reliability, Safety and Hazard
Editors: P.V. Varde et al.

MODELLING EVALUATION OF SAFETY INDEX BY USING O-MAX: A CASE STUDY

Rane S.B., Mantha S.S.[1], Sayed Ahsan Mahdi and Rane Suraj[2]

Department of Mechanical Engineering, Sardar Patel College of Engg, Mumbai
[1] Department of Mechanical Engineering, V.J.T.I. Mumbai
[2] Department of Mechanical Engineering, Padre Conceicao College of Engineering, Verna, Goa

ABSTRACT

In a massive drive of product and service innovations it is necessary to focus on safety issues in concern with the products and services. Therefore HSS&E policies are rapidly developed and wide adopted. Adoption of HSSE policy suggest its continuous improvement. It is a well-known fact that if we want to improve the performance, we should have appropriate evaluation techniques. Once measurement (scale) for any output is decided, then it is easy to align the strategies/actions/initiatives to get optimum results. Therefore it is essential to develop the structured methodology for evaluation of various performance parameters in concern with HSSE. The evaluation of safety index is one of the essential processes which is under consideration. The paper introduces Structured Methodology of Evaluation of Safety Index by using O-Max with a suitable Case Study in City Gas (Natural Gas) distribution industry, providing quantitative measure for Safety Index Measurement.

Keywords: HSSE -Health, Safety, Security, Environment, O-Max- Objective Matrix

1. INTRODUCTION

In the new millennium, the managements are becoming increasingly process oriented, to meet the challenges of customers and to focus on quality, cost and agility. There are massive drives of Liberalization, Globalization & Privatization all over the world. Increasing globalization of products, services and financial markets, coupled with immensely increased rate of product and service innovations cause severe competitive pressures for the business organization. In support of the fundamental business objectives, the management is fully committed to achieve high standards of performance in HSS&E and continual improvement [6]. HSSE issues are recognized as essential parameters for the future success of the organization. Safeguarding, the health, safety & security of the employees whilst minimizing both the risk to public & the impact of activities on the environment, are the major priorities.

2. HSSE ISSUES

In support of the fundamental business objectives, the management is fully committed to achieve high standards of performance in HSS&E and continual improvement [9]. HSSE issues are recognized as essential parameters for the future success of the organization. Safeguarding, the health, safety & security of the employees whilst minimizing both the risk to public & the impact of activities on the environment, are the major priorities.

Monitoring HSSE presents one of the toughest management challenges, that industries face & success in this respect is an absolutely reliable indicator of how well industries are performing the business as a whole [13]. Success requires safety, unremitting commitment &

structured methodical approach but above all it requires genuine sense of responsibility for employees, contract partners & members of the community where we work.

Indifference or hostility towards health, safety & environmental risk can't be tolerated. The costs are too high like personal losses, loss in business performance, asset damage or loss. Losses due to accidents are like an iceberg i.e. apparently it looks very small but the impact as a whole is massive in the form of negative publicity. Today HSS&E has got amalgamated with almost all the business processes. It is almost impossible to have a business process without the help of HSS&E policy. Thus HSS&E has become the enabler of business competitiveness in the market.

3. NEED OF EVALUATION OF SAFETY INDEX

Adoption of HSSE policy also suggest its continuous improvement. Therefore it is essential to develop the structured methodology for evaluation of various performance parameters in concern with HSSE [1].

The evaluation of safety index is one of the essential processes which is under consideration. O-Max can be used for evaluation of Safety Index. It focuses on all criteria of Safety that affects its Safety Index [6]. O-max acts as a benchmark for deciding future strategies and actions to achieve better performance [3]. The assessment methodology used in this technique is easily applicable to all criteria affecting safety index and helps to quantify them.

4. INTRODUCTION & THEORY OF OBJECTIVE MATRIX (O-Max)

Objective Matrix (O-Max) Tool is developed by Jim Riggs at Oregon Productivity Center. It is suitably applied to those activities, which are intangible. A measurement method called the Objective Matrix (O-Max) is particularly appropriate to basic work units such as crews, departments & staffs, although it can also represent a complete organization [2]. It is conveniently applied to knowledge based activities that are considered difficult to measure, as well as skill based work that can be metered by more conventional measures. It has been favorably received in manufacturing, services & government sectors by both large & small organizations.

Method of Evaluation of Safety Index using O-Max

1 - Identify the system whose safety index needs to be calculated
2 - Identify the Main criteria affecting the safety index of the system
3 - Identify various sub criteria for each main criteria
4 - Assign Weights to the main criteria
5 - Assign weights to the sub criteria
6 - Make the objective matrix and map criteria on it
7 - Calculate the safety scores for each criteria
8 - Calculate safety index
9 - Calculate maximum and minimum safety index
10 - Calculate current / previous safety index [8]

5. QUANTIFYING THE SCORES WEIGHT, VALUE AND INDICATORS OF SAFETY INDEX

All the criteria of safety index do not have equal effect on overall safety of the work unit [5]. Assigned weights (100 points) are distributed over the criteria with consideration of safety objectives.

Hence the different criteria of the following strategies considered as due safety criteria. These indicators can be varied according to the Organization's policy. The weights are allotted for every main criteria.

6. CRITERIA AND POINTS

Table 1 explains the distribution of points for different criteria used for Safety Index.

Table 1 Allotment of points to different criteria

Sr. No.	Criteria	Points (%)
1	Primary Goal(Zero loss Time Injury)	40
2	Statutory Requirements	20
3	Behavior Based Safety	25
4	Risk Hazard Rating & Control	15
		100

7. DEFINING SUB CRITERIA AND THEIR WEIGHT DISTRIBUTION

The above points can be considered as 100% for the individual criteria and this 100% is again distributed as per the sub criteria importance [10]. Table-2 shows the criteria used, parameter used, nature of parameter and points distributed for this purpose [7].

8. EVALUATION OF CURRENT SAFETY INDEX:

Calculation of performance scores in Safety Index Main Criteria (Refer the Table-2) [11]:

Score for Main criteria = Sum of (Score of sub criteria x points/weights for the same)

Method of selecting the numbers is based on the Scores achieved in concern with highlighted numbers from the Table 3 for every sub-criteria as per the business organization's policy of rating the performance in that sub-criteria.

Primary Goal (Zero Loss time Injury)
(8 x 0.1) + (7 x 0.1) + (7 x 0.1) + (6 x 0.05) + (6 x 0.05) + (7 x 0.05) + (7 x 0.05) + (5 x 0.1) + (8 x 0.4) = **7.2**

Statutory Requirements:
(10 x 0.03) + (6 x 0.02) + (10x 0.05) + (9 x 0.07) + (4 x 0.03) + (9 x 0.06) + (9 x 0.04) + (9 x 0.1) + (9 x 0.25) + (10 x 0.1) + (9 x 0.15) + (10 x 0.1 = **9.07**

Behavior Based Safety:
(6 x 0.1) + (5 x 0.2) + (8 x 0.15) + (4 x 0.15) + (7 x 0.3) + (7 x 0.1) = **6.2**

Risk Assessment & Controls:
(8 x 0.5) + (7 x 0.3) + (8 x 0.2) = **7.7**

Current Safety Index = (7.2 x 40) + (9.07 x 20) + (6.2 x 25) + (7.7 x 15) = **739.9**

Maximum Safety Index = (10 x 40) + (10 x 20) + (10 x 25) + (10 x 15) = **1000**

Minimum Safety Index = (1 x 40) + (1 x 20) + (1 x 25) + (1 x 15) **=100**

Percentage Safety Index = (739.9 x 100)/1000 **=73.99%**

9. TARGETED SAFETY INDEX

Let us have a targeted Safety Index = 860 for the Next year. All the Criterions are to be improved except statutory requirements [12], [4].
Current Safety Index is **73.99%** and Targeted Safety Index is **860.** Two areas identified to improve the safety index are:

Primary Goal (need to improve 7.2 to 9)

Statutory Requirement same = 9.07

Behavior Based Safety (need to improve 6.2 to 8)

Zero Loss time Injury (need to improve 7.7 to 8)

Targeted safety Index = (9 x 40) + (9.07 x 20) + (8 x 25) + (8 x 15) = **861.4**
say **860**

Maximum Safety Index = (10 x 40) + (10 x 20) + (10 x 25) + (10 x 15) **=1000**

Minimum Safety Index = (1 x 40) + (1 x 20) + (1 x 25) + (1 x 15) **=100**

The above calculations can be utilized to monitor the performance of organization in the various sub criteria of Safety Index evaluation [14].

Table 2 Points distribution for the used criterias

Criteria Used	Parameter Used	Nature of Parameter	Points Distributed
Primary Goal:			
Zero Lost time Injury	Use of Personal Protective Equipment (PPE)	Mandatory	10%
	Level of Proactive maintenance of tools & equipments	Increasing	10%
	Quality of Safety Audits	Increasing	10%
	Near Miss reporting	Increasing	5%
	Degree of use of Safety Suggestions scheme	Increasing	5%
	Permit to Work	Mandatory	5%
	Quality of Incident Reporting	Increasing	5%

	Degree of Housekeeping	Increasing	10%
	Occurrences of accidents	Decreasing	40%
Statutory-Requirement:			
A) MC Guidelines	Application for Installation	Mandatory	3%
	Quality of Joint Inspection	Increasing	2%
	Permits	Mandatory	5%
	No of Barricades and Display boards	Increasing	7%
	Quality of Daily Progress Reporting	Increasing	3%
B) Fire Brigade	Degree of adherence to Documentation Required (NOC from Fire Brigade, Approved drawing, Test Certificate of Installation)	Increasing	6%
	Quality of Pipeline Construction	Increasing	4%
	Degree of Fire & Safety Compliance	Increasing	10%
C) Factory Act – 1948	Degree of Compliance related to Factory Act	Increasing	25%
D) Pollution Control Board	Degree of adherence to		
	Handling of Hazardous material	Increasing	10%
	Disposal of Hazardous material	Increasing	15%
	Authorization of storage	Increasing	10%
Behavior Based Safety	Feeling of Ownership Involvement of workforce	Increasing	10%
	Extent of use of Root Cause Analysis	Increasing	20%
	Measurement frequency and accuracy	Increasing	15%
	Feedback (and Goal-Setting)	Increasing	15%
	Awareness	Increasing	30%
	Management and Supervision	Increasing	10%
Risk Assessment & Control	Accuracy of Hazard identification	Increasing	50%
	Quality of hazard assessment	Increasing	30%
	Management of Health and safety	Increasing	20%

Table 3 Objective Matrix for Evaluation of Safety Index

Safety Index Criteria	Primary Goal								
Parameter	Extent of use of PPE	Level of Tools & Equip. Maint.	Quality of safety	% of Near miss Reports	Degree of use of Safety Suggestions schemes	Permit to Work	Quality of Incident Reporting	Degree of House keeping	Occurrences of accidents Per year
Weights									Decreasing nature
10	10	10	10	10	10	10	10	10	1
9	9	9	9	9	9	9	9	9	2
8	8	8	8	8	8	8	8	8	3
7	7	7	7	7	7	7	7	7	4
6	6	6	6	6	6	6	6	6	5
5	5	5	5	5	5	5	5	5	6
4	4	4	4	4	4	4	4	4	7
3	3	3	3	3	3	3	3	3	8
2	2	2	2	2	2	2	2	2	9
1	1	1	1	1	1	1	1	1	10

Table 3 contd. Objective Matrix for Evaluation of Safety Index

SIC	Statutory Requirements											
Parameter	BMC Guidelines					Fire Brigade			Maharashtra Factory Act - 1948	Degree of adherence to PCB norms		
	Application for Installation	Quality of Joint Inspect-ion	Permits	No. of Barricades and Display boards	Quality of Daily Progress Reporting	Degree of Adherence to Documentation Required	Quality of Pipeline Construction	Degree of Fire & Safety Compliance	Degree of Compliance	Handling of Hazardous Material	Disposal of Hazardous Material	Authorization of storage
Weights												
10	**10**	10	**10**	10	10	10	10	10	10	**10**	10	**10**
9	9	9	9	**9**	9	**9**	**9**	**9**	**9**	9	**9**	9
8	8	8	8	8	8	8	8	8	8	8	8	8
7	7	7	7	7	7	7	7	7	7	7	7	7
6	6	**6**	6	6	6	6	6	6	6	6	6	6
5	5	5	5	5	5	5	5	5	5	5	5	5
4	4	4	4	4	**4**	4	4	4	4	4	4	4
3	3	3	3	3	3	3	3	3	3	3	3	3
2	2	2	2	2	2	2	2	2	2	2	2	2
1	1	1	1	1	1	1	1	1	1	1	1	1

Table 3 contd. Objective Matrix for Evaluation of Safety Index

SIC	Behavior Based Safety						Risk Assessment and Control		
Parameter	Feeling of Ownership & Involvement of Workforce	Extent of use of Root Cause Analysis	Measurement of safety	Feedback & Goal setting	Awareness	Mgt. & Supervision	Accuracy of Hazard Identification	Quality of Hazard assessment	Management of Health & Safety
Weights									
10	10	10	10	10	10	10	10	10	10
9	9	9	9	9	9	9	9	9	9
8	8	8	**8**	8	8	8	**8**	8	**8**
7	7	7	7	7	**7**	**7**	7	**7**	7
6	**6**	6	6	6	6	6	6	6	6
5	5	**5**	5	5	5	5	5	5	5
4	4	4	4	**4**	4	4	4	4	4
3	3	3	3	3	3	3	3	3	3
2	2	2	2	2	2	2	2	2	2
1	1	1	1	1	1	1	1	1	1

10. CONCLUSION

This method is based on continuous improvement of Safety Index. The Safety Index is considered to be gradually improving from Level 3 to higher levels.

The challenge for the modern business organization is to sustain a holistic approach for remaining competitive on a global level. Performance evaluation for enhancing the efficiency and effectiveness of business process is crucial to remain in business and to grow. Business policy and HSSE Policy go hand in hand. Therefore evaluation of safety index in concern with adoption of HSSE policy is utmost-necessary. For keeping system under continuous improvement and to optimize (Streamlining the internal efficiency) evaluation is a must. O-max facilitates evaluation of Safety Index of any manufacturing and service organization. This process reveals present position of the system and identifies key points where focus is needed and also guides for continuous improvement in Safety Issues. Safety Index evaluation by O-Max at periodic interval can therefore lead to successfully monitoring the process of adopting HSSE Policy.

11. REFERENCES

1. Anthony V.; (1992), *Evaluating the safety function: A conceptual model* J. of Safety Research, **23-1**; 27-38.
2. Ernst A.P., Koningsveld, (2005), *Participation for understanding: An interactive method* J. of Safety Research, **36-3**; 231-236.
3. Halperin K. M. and McCann M.; (2004), *An evaluation of scaffold safety at construction sites* J. of Safety Research; **35-2**, 141-150
4. Horst R., Simon G., Graham P. and Gordon C., (1998), *Evaluating Contact Techniques: Assessing the Impact of a Regulator's Intervention on the Health and Safety Performance of Small and Medium-Sized Businesses* J. of Safety Research, **29-4**; 235-247.
5. Jayming S. F., David M. F. and M. Sam M.; (2004) *Making the business case for process safety using value-at-risk concepts,* J. of Hazardous Materials, **115-1-3**; 17-26
6. Johnston R. G., (2004), *Adversarial safety analysis: Borrowing the methods of security vulnerability assessments,* J. of Safety Research, **35-3**, 245-248.
7. Maiti J. and Bhattacherjee A., (1999), *Evaluation of Risk of Occupational Injuries Among Underground Coal Mine Workers Through Multinomial Logit Analysis* J. of Safety Research, **30-2**; 93-101.
8. O'Malley P. M. and Wagenaar A. C., (2004), *Effects of safety belt laws on safety belt use by American High School Seniors,* J. of Safety Research, **35-1**, 125-130
9. Pierson T. G. and Murphy D. J., (1996), *Safety and health educational needs of agricultural education and industry professionals,* J. of Safety Research; **27-2**; 103-116
10. Rane S. B. , Bhonsale A.V., Mantha S. S., Dalvi D. S; (2004), *Evaluation of Total Productive Maintenance Implementation Using O-Max Tool*; Proceedings of International Conference on Manufacturing And Management-2004 (GCMM-2004) December 6th-10th, 2004, Vellore Institute of Technology, Vellore; 432.
11. Rodriguez L.and Bogett W. R., (1989), *Societal considerations in scaling injury severity and effects*; J. of Safety Research, **20-2**, 73-83.
12. Rundmo T.; (1995), *Perceived risk, safety status, and job stress among injured and non injured employees on offshore petroleum installations*"; J. of Safety Research, **26-2**, 87-97.
13. Sinclair R. C., Smith R., Colligan M., Prince M., Nguyen T. and Stayner L., (2003); *Evaluation of a safety training program in three food service companies*; J. of Safety Research, **34-5**; 547-558
14. Vojtecky M. A. and Schmitz M. F. (1986) , *Program evaluation and health and safety training* ; J. of Safety Research, **17-2**, 57-63.

Reliability, Safety and Hazard
Editors: P.V. Varde et al.

HAZARD AND OPERABILITY STUDIES FOR URANIUM RECOVERY FROM SEAWATER

A.K. Saxena, T.L. Prasad and P.K. Tewari*

Desalination Division, Bhabha Atomic Research Centre, Mumbai 400 085

ABSTRACT

Based on initial success of extracting about 800 μg of uranium by harnessing the tidal wave using PAO adsorbent in June 2000, a Process flow sheet for scaled up facility to extract 100gU/year has been developed. The hazard and operability studies associated with the unit operations involving chemicals along with the marine environment encountered have been conducted. A conceptual process design and cost benefit analysis has also been carried out for a 100 Te per year 'uranium from seawater' plant. The reliability and safety studies required for extraction system are being established on prototype contactor assemblies and the possible options are presented in this paper.

Keywords: Alternate uranium resources; Uranium from seawater, Tidal energy, Extraction systems, Polyacryl Amid Oxime (PAO), Acrylonitrile handling, Preventive Environmental Management (PEM)

1. INTRODUCTION

Electricity is mankind's cleanest and most efficient energy carrier. Nuclear energy is a non-greenhouse gas emitting primary energy source for electricity production round the clock and round the year regardless of weather. Nuclear energy may have some issues that look like not fully addressed to some of the green-peace people, but it definitely has advantages besides the overwhelming one of being atmospherically very clean and most acceptable. The nuclear power generation industry is mature with half a century of experience and an ever-improving engineering behind it [1]. To ensure a more desirable work environment and a sustainable development, Preventive Environmental Management (PEM) concept has been integrated in this technology of recovery of uranium from seawater. The elements of PEM consists of, analysis of unit processes and unit operations of the basic flow sheet for identification of pollution hot spots, evolving of process related interventions to reduce wastes at source, improving quality and yield, integration of economic and environmental concerns of stakeholders and thus adoption of an integrated approach [2]. Uranium recovery from seawater has the potential of becoming a supra-clean technology. In the last century uranium has universally gained acceptance as a primary energy source. Currently it caters to ~17% of the electricity generation globally. Uranium has been projected as the main workhorse of future when the fossil energy reserves dwindle by the middle of this century [3]. The terrestrial distribution of uranium ore occurrence is grossly uneven. With a large coastline, India, Japan, Korea and a few other uranium starved nations have a larger stake in exploiting the 4.5 billion tonnes of uranium locked in seawater. Based on initial success of extracting about 800 μg of uranium by harnessing the tidal wave using PAO adsorbent, a process flow sheet for scaled up facility to extract 100gU/year has been developed [4]. The process flow scheme involving major unit

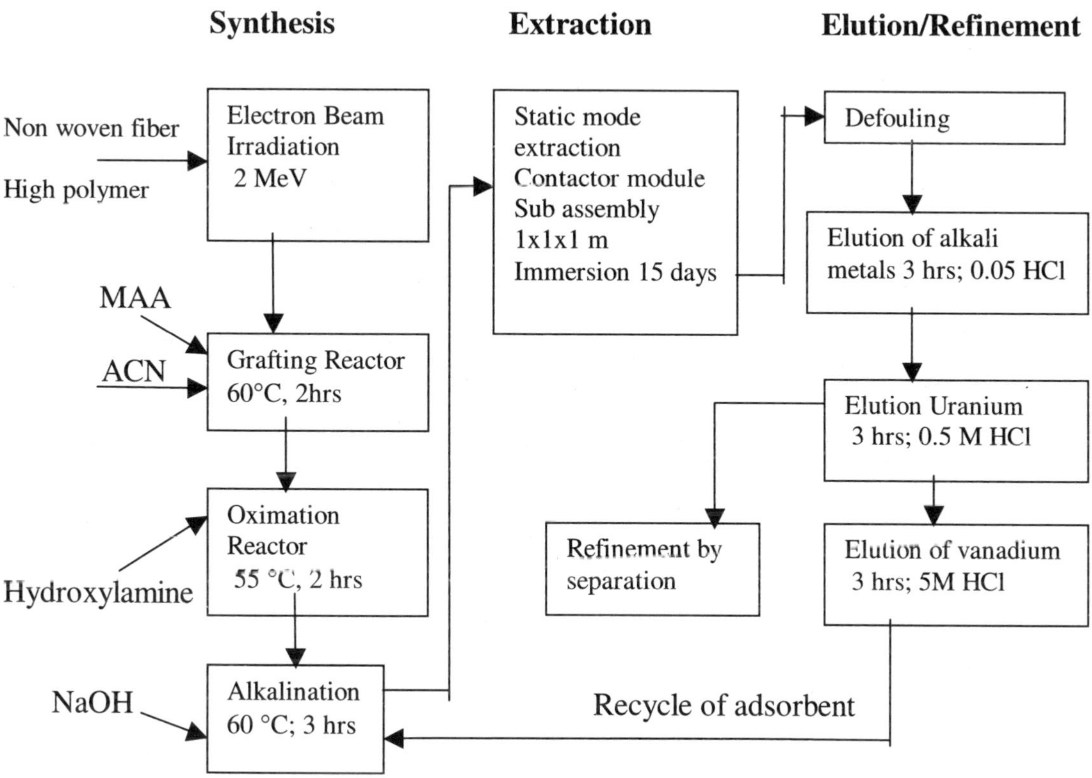

Fig. 1: Process Flow diagram

operations like radiation grafting, dissolution; drying; oximation; alkalination; extraction; defouling and elution has been evaluated on PEM principles. A conceptual process design and gross estimates of the Capital & Operational cost has also been carried out for a 100 Te per year 'uranium from seawater' plant based on the current status of uranium from seawater technology. The reliability and safety parameters required for extraction systems are being established on prototype contactor assemblies and the salient features have been brought out in this paper

2. RECOVERY OF URANIUM THROUGH A CLEANER PRODUCTION TECHNIQUE

The process flow scheme developed for recovery of uranium from seawater involves major unit operations like radiation grafting; dissolution; drying; oximation; alkalination; extraction; defouling; elution and refining [3,4]. The flow diagram of the process cycle is as shown in Fig.1 above.

The proposed process has adopted two principle techniques for achieving cleaner production and they are

- At Source Reduction
- Recycling

At Source Reduction; The first step in any waste minimization program is the reduction of wastes at source. Source reduction alleviates the problems associated with handling and disposal of wastes and ensures the preservation/protection aspects of the environment. At Source reduction is achieved by

- Good Operating Practices; GOP.
- Adopting appropriate Techniques of processing.
- Recyclables in process consumables.
- Enhanced Product yields or lower consumption norms.

Technique & Technology selection for industrial process and equipment modifications, are based on 'reduced waste generation' criteria. For example as compared to conventional mining, in this technology we have no grinding operation and no acid leaching, thus zero mill tailings are generated. There is no radon release to environment in this new resource and production method. Technology changes may involve process automation also. The process flow sheet arrived at is amenable to easy automation as well. Input consumable material substitutions are done to accomplish hazardous waste reduction.

Recycling: Recycling includes both reuse and reclamation. Recycling via reuse involves either returning a waste material to the originating process as a substitute of input material or using it in another process as an input material. The proposed process adopts recycling of the PAO adsorbent and reclamation of solvents for reuse in the plant itself.

3. PROTECTION FROM CHEMICALS

The hazards posed by the chemicals used are due to:

- Evolution of toxic & irritant fumes.
- Formation of flammable and explosive mixtures.

The hazardous properties of chemicals used in the process for recovery of uranium from seawater and their safe threshold limit values are as listed in table-1 below.

Table 1 Classification of chemicals

S N	Chemical	Hazard class for T'port	NFPA health rating	TLV (OSHA)
1	Acrylonitrile	3	4	2 ppm
2	Dimethyl formamide	3	1	10 ppm
3	Hydroxylamine hydrochloride	8	NA	NA
4	Methnaol	3	2	200 ppm
5	Caustic flakes	8	3	2 mg/m^3
6	HCl	8	3	5 ppm

Acrylonitrile exerts its toxic action by two simultaneous mechanisms namely; inhibition of the activity of cytochrome oxidase by liberation of cyanide and the inhibition of sulf-hydryl dependent enzymes by cyanoethylation. The regulations set the permissible level of 2 ppm as 8 hour weighted average concentration with a STEL (Short Term Exposure Limit) of 10 ppm for any 15-minute period. Standards require the use of protective clothing and equipment in regulated areas. During the initial trial operation of the grafting reactor at BRIT Vashi a maximum of 10ppm concentration on a few short durations was noticed. Proper ventilation system, deep design reactor concept, minimal working volumes, nitrogen blanketing and a good temperature control of reaction ensured an acceptable ambient air concentration of the toxin within TLV limits as above [5].

4. METHODS OF EXTRACTION

Recovery of uranium from seawater requires contacting of very very large amounts of seawater. Therefore only a cost free method of contacting enormous amounts of seawater and a meticulous logistics to handle the large numbers of contactor assemblies can only give an acceptably economical technology of uranium extraction from seawater. The Tidal Current based concepts have been adopted while designing the contactor assemblies for the proposed process as shown at figure 1. A Conceptual Design based Techno-economic feasibility report has been prepared for a 100 tonnes per year capacity plant [6,7]. In the report, 17000 contactor arrays (CA) of size 4m x 4m x 4m having apparent adsorption surface area of 4000 m^2 each have been conceived. These CA's are positioned in the seawaters in the path of tidal currents of the sea/oceans with the top of the top CA at about 2m depths from top of water surface and the bottom of the bottom CA at about 4 m above from the seabed with the help of appropriate size of buoys and suitable strength of fixed anchors in the seabed. The dry weight of each unit is about 4 Te. CA's are then kept submerged for about 15 days based on the seawater characteristics. The grown CA's are then taken out for desorbing the adsorbed uranium. After recovering the uranium, the CA's are again submerged in seawater. About 900 CA's of the above-mentioned design are processed everyday to achieve the target capacity. Hence at any point of time 900 x 15

Table 2 Methods of extraction systems and their operating steps

SN	Item Description	Proposal-A	Proposal-B	Proposal-C	Proposal-D
1	Catamaran dumb barges of size 100mx44m & loading 160 units	100 nos	-	-	-
2	Jettty with 10m depth	Y	Y	Y (14m wide)	Y (14m wide)
3	Shore cranes of 10 te & 50m boom	6 nos	-	6 nos	6 nos ;25m boom
4	Ship cranes of 10 te & 50m boom	-	4nos on each ship	-	-
5	Railway wagons with engines	Y	Y	Y	Y
6	15 te BP tugs	6 nos	-	-	15 nos. of 1 tonne BP tugs (on rental)
7	Moored buoys with floats	-	Y	Y	900 floating packs
8	Work boats	10 nos	10nos	10 nos	10 nos
9	Inspection boats	2 nos	2 nos	2 nos	2 nos
10	Operation – Break up of events				
10.1	Loading of CA's on to vessel at jetty	CA's on Catamaran type vessel	CA's on Geared & self propelled vessel	Wagons used to move CA's to jetty	CA's on floating pack
10.2	Moving vessel to required place	Y	Y	On to jetty	Floating pack toing
10.3	Mooring of vessel for 15 days	Y	CA's fixed by buoys	CA's into floats below jetty	Floating pack is moored
10.4	Removing of moorings	Y	Grown units removed on to vessel	Grown units removed by jetty cranes	Grown floating pack is removed

units will always be in the seawater. Various marine logistics methodologies that have been considered and the budgetary cost of each scheme are tabulated below:

5. CONCLUSIONS

The hazard and operability experience gained during pilot scale demonstrations have shown the reliability of the systems. The specially developed organic adsorbent PAO has given promising results for recovery of uranium from seawater. The in-house experience has demonstrated the safe handling capability of the engineering systems and procedures for using chemicals like acrylonitrile and DMF in large quantities. Harnessing tidal scheme at a place where good water current is coupled with lesser turbidity and lower bioactivity will give the optimum sites for locating the plants. Depending on the place selected, extraction system based on proposal-C is favored. More sites should be evaluated for their bio fouling and dirt fouling characteristics for establishing plants of similar capacity.

6. ACKNOWLEDGEMENTS

The authors thank NSDRC, especially Prof Sambandhan; Shri SC Katiyar Station Director TAPS (now retired) for the extensive help in data collection and Dr. Sunil Sabharwal, Head RTDS,BARC for electron beam irradiated substrate work. Special thanks are due to Shri DS Shukla, Director ChETG,BARC for the technical, discussions and suggestions.

Proposal-A

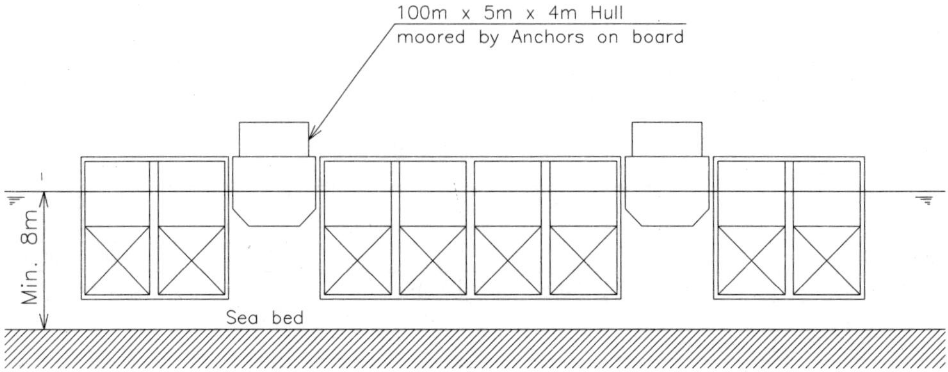

Section of the Dumb Catamaron barge of 100m x 44m, with each hull 100m x 5m x 4m.
The draft at jetty shall be more than 8m for loading/unloading.
Each vessel will carry 160 units.

Proposal-B

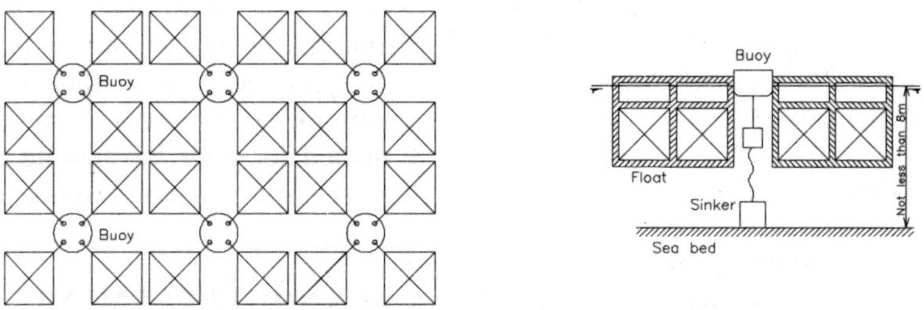

Layout of buoys with floats to hold the units at reqd. depth.

Proposal-C

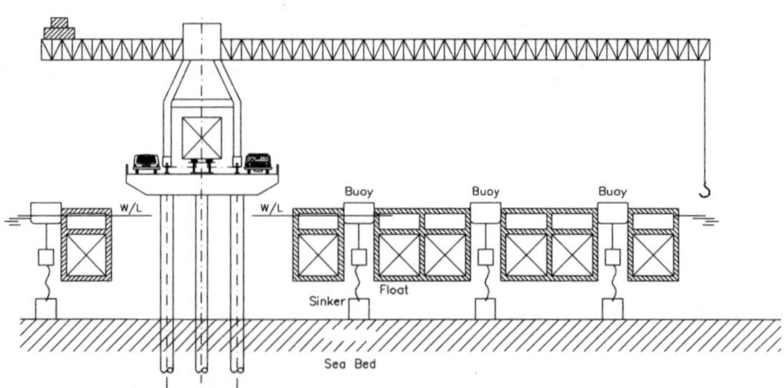

Proposal-D

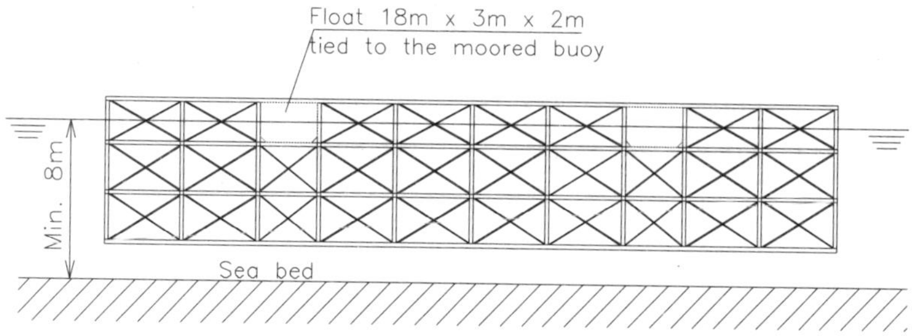

Vertical Section of 16 units Floating Pack

Table 3 Estimated Project Cost for various options (Rs. In Crores)

ITEM	PROPOSAL - A		PROPOSAL - B		PROPOSAL - C		PROPOSAL - D	
	CAPITAL	ANNUAL	CAPITAL	ANNUAL	CAPITAL	ANNUAL	CAPITAL	ANNUAL
Initial Expenses	5.0	-	5.0	-	5.0	-	5.0	-
Ships	400.0	20.0	240.0	48.0	-	-	-	-
Jetty	30.0	0.3	20.0	0.2	300.0	0.3	30.0	0.3
Cranes	30.0	3.0	80.0	8.0	30.0	3.0	20.0	2.0
Trains	20.0	5.0	20.0	5.0	20.0	5.0	20.0	5.0
Buoys / Floats	-	-	45.0	2.3	45.0	2.3	333.0	16.6
Tugs	60.0	20.0	-	-	-	-	-	10.0
Work Boats	6.0	2.0	6.0	2.0	6.0	2.0	6.0	2.0
Inspection Boats	4.0	1.0	4.0	1.0	4.0	1.0	4.0	1.0
TOTAL	555.0	51.3	420.0	66.5	410.0	13.6	418.0	36.9

NOTE:

1. The cost depends upon place selected and detailed workout is needed
2. Depending on the weather conditions proposals A, C & D can be operated for about 300 days in a year, and proposal - B can be operated for about 270 days in a year.

7. REFERENCES

1. Nuclear News Digest Volume 18 Issue 5 pp. 5to9 2005.
2. Prasad, T.L., Trivedi, R.N., Sahasranaman, K., Proceedings of seminar on Safety, Health & Environment in Indian Chemical Industry, organized by Mumbai Regional Centre of Indian Institute of Chemical Engineers at Hotel Grand Maratha Sheraton Mumbai 11th January 2003.
3. Kakodkar A, "Chemical Engineering perspectives in Atomic Energy Programme", HL Roy memorial lecture delivered at CHEMCON-2000, Organised by the Indian Institute of Chemical Engineers at Science City, Kolkata, December 2000.
4. Saxena AK, Invited Talk on progress review of 'Uranium from Seawater' on August 14, 2000 delivered in BARC.
5. Saxena AK, Prasad TL "Safety Manual for

the 100gpy RUSWapp at Trombay" February 2004

6. Saxena AK, "Uranium from Seawater: a new resource for meeting future demands of Nuclear Reactors", Indian Chemical Engineering, Section C, Vol 43, No. 3, July –September 2001.
7. Saxena AK, Prasad TL, Tewari PK & Shukla DS "Conceptual Design and Techno Economic Feasibility Report on RUSWapp ~ 100t", November 2003. Internal Report PICUS, Desalination Division BARC Mumbai.

Reliability, Safety and Hazard
Editors: P.V. Varde et al.

ENHANCEMENT OF PERSONNEL SAFETY FOR FOLDED TANDEM ION ACCELERATOR AT BARC

*Sapna Padmakumar, N.B.V. Subrahmanyam, Jignesh P. Bhatt, M.J. Kansara, A. Basu, Rajesh Kumar, A. Agarwal, S.K. Gupta, P. Singh, S. Kailas and D.S. Joshi **

Nuclear Physics Division, *Radiation Safety Systems Division, Bhabha Atomic Research Centre, Mumbai 400 085

ABSTRACT

Protection of workers, public and environment from radiation hazards is one of the key issues addressed in any accelerator project. Reviewing and assessing various safety related aspects at different stages (design, construction, commissioning, operation), to determine whether the activities comply with the applicable safety standard and requirements, is one of the principal ways to achieve high level of safety. The commissioning of FOTIA (Folded Tandem Ion Accelerator) at BARC has been done through a well-defined set of test acceptance criteria, for confirming adequacy of all safety related features of the facility. The operational limits and conditions are consistent with the regulatory requirements and an adequate level of safety is maintained. The basic objective of this paper is to highlight the enhancement of safety aspects in FOTIA.

Keywords: FOTIA, Search & Secure, SF_6

1. INTRODUCTION

The FOTIA facility at BARC [1] (Fig.1), designed and developed indigenously, has the capability of delivering accelerated ion beams of mass up to 40 and beam energy up to 66 MeV for a maximum terminal voltage of 6 MV. As per the safety requirements the accelerator shall be operated in accordance with standards specified in Code of Federal Regulations (10 CFR 20) [2]. The maximum permissible whole body dose for an operator shall be not more than 2 rem/year. The FOTIA is capable of operating under standard operating conditions without any radiation hazard. The FOTIA facility is operated using a computer based control system located in the control room where the radiation level is negligible.

Safety standards for most industrial environments are well established (OSHA 29 CFR 1910) [3], and this section merely points out certain specific industrial hazards that are often found in accelerator facilities.

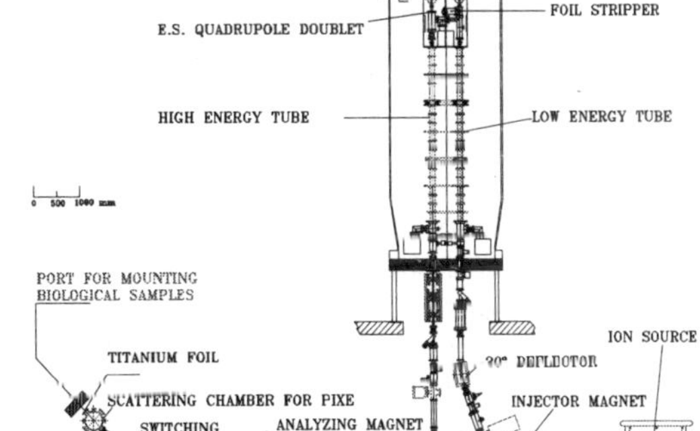

Fig. 1: Schematic diagram of FOTIA

2. RADIATION SAFETY AND SYSTEM INTERLOCKS

FOTIA interlock systems, control instrumentation and readouts on the control console is identified and made discernible. Doors to the areas of the facility where an individual could receive a radiation dose are provided with safety interlocks, which automatically stop the beam at the entry of the accelerator itself. All safety interlocks are operating as per coordinated logic. Presently, FOTIA is used for analysis work, which normally requires a terminal voltage of up to 4 MV. The radiation level measured by wall mounted and portable radiation monitors at different locations are found to be negligible. The radiation field is measured frequently and is recorded by RSO. Each location where an individual could receive a radiation dose and entrances to these locations is equipped with easily observable warning devices (both visible and audible) which will be activated for 15 seconds prior to start up of the beam. Such warning devices will clearly discernible in and near all effected areas. Prior to start up of accelerator, the FOTIA Operator will physically verify the area including access doors and interlock switches. There is a physical input by the Operator to verify that this inspection has been done before the injection of beam in to the accelerator. Personnel are evacuated from the zone by a search procedure, by pressing of search-point buttons in a pre-programmed sequence. During the search warning hooters are sounded, and on its completion warning signs are illuminated within the zone. In case of someone trapped inside the searched area, instructions are displayed prominently to press one of the emergency scram buttons. The positioning of the search-points in each area has been thoroughly reviewed, with specific emphasis to ensure that the searcher will not miss anybody. This will ensure the security of ion source and accelerator areas, to which there is currently free access. To prevent unauthorized use, a control panel key and console password is provided to operate the FOTIA system. For this purpose the operators & other members of the FOTIA team have been properly trained. TLD badges of all users and FOTIA members are validated for the duration of allocated beam-time and this information is recorded in a central database of CRPC. Radiation monitors installed in each area will gives an audible signal if the radiation level exceeds the pre-set value.

3. GENERAL SAFETY

1. Electric circuits and interconnections are wired and installed in accordance with accepted electrical building codes. Fire extinguishers of the appropriate portable type to combat electrical or solvent fires are installed around the FOTIA facility. Warning signs are displayed on the panels of high voltage equipments or at the entrance of restricted areas to warn the personnel. Entrance to the safety cage of ion source area is fitted with micro switch, which is interlocked with mains supply to high voltage and isolation transformer.
2. A fire alarm and detection system is provided in the building with smoke detectors placed at strategic locations. Independent air handling equipment and ducts are provided for different areas to prevent spread of fire smoke or air contamination to unaffected areas through these ducts.
3. The main health hazard is from SF_6 used as insulating gas, the safety system is based on oxygen detectors and emergency breathing apparatus located at several strategic points in the building. Normal oxygen content in atmosphere varies from 20 to 21%. When the local oxygen content reduces to 19%, an alarm will be raised locally with a simultaneous visual indication in the control room. At this oxygen level only some discomfort may be felt and it should be possible to continue in the area for about 1 hour. If the oxygen content falls below 16.5%, a loud siren will be sounded at all emergency centers in the building, indicating that building should be evacuated immediately. Two exhaust systems for removal of leaked SF_6 gas of suitable capacity has been commissioned in accelerator room and beam room. SF_6 monitors are installed at various locations and each monitor will be set to give an alarm at 10 ppm of SF_6 in air to detect presence of minute leak in the system.

4. THE FAULT TREE ANALYSIS SHOWING THE LIST OF PROBABILITIES

A Particle Accelerator Safety Committee for BARC facilities (PASC) has evaluated the level of safety provided in FOTIA. Suggestions provided by this committee during their visit have been implemented and testing has done, using the Fault Tree technique. The first step in this process is the 'top event' (i.e. the failure to be quantified), which was defined as "Failure to ensure that no personnel are in accelerator room or ion source room when beam is present". The technique involves defining all the possible sub events, which could independently cause the top event (in our case "Person present in the area when beam is injected" and "Person enters the area when beam is present"), then defining the sub events, which could cause these, and so on. Eventually 'base events' are encountered which can be quantified without further subdivision; examples are 'bulb fails',' relay contact fails' or 'person ignores warning sign'. Hardware failure rates can be derived from recorded failure data, or from libraries of data on similar devices.

The derivation of human failure rates is a specialization, which involves analysis of the tasks involved, and comparison of these to other tasks, which have estimated or measured failure probabilities. It is usual to draw the Event Tree, which then does Boolean reduction (Fig.2) on the resultant event logic to derive 'minimal cut sets' (MCSs). A minimal cut set is a set of base events, which together can cause the top event. The probability of each MCS is the product of the probabilities of the base events within it; while the probability of the top event is the sum of the MCS probabilities. Besides indicating a fairly good level of system safety, arguably the most useful finding is the relative contribution of the various MCSs. The complexity of the Personnel Safety system leads to a very large number of MCSs. These were analyzed, and was grouped into the following categories, for each of which the percentage contribution to the top event is given as per the statistical data available:

1. The searcher presses the buttons in sequence but makes no attempt to search for a person in that area, and that person does not respond to either the audible search warning or the visual warnings following the search (either deliberately, or due to ignorance) - 77%
2. As above, but the person in the area does not respond because the warnings are not operational - 22.5%

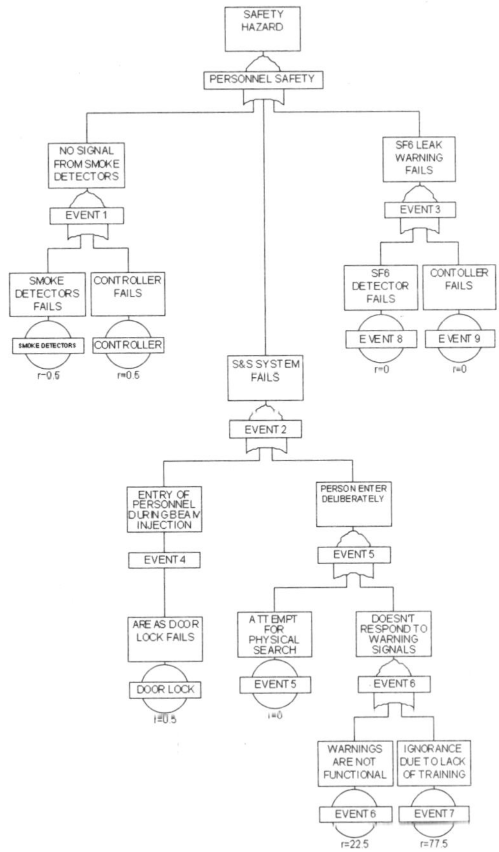

Fig. 2: Fault tree diagram with Boolean reduction

3. A person releases the interlocks and enters. This also requires that the area door's lock fail to be engaged - 0.5%

5. CONCLUSION

These possibilities clearly underline the need to ensure that operators and users are properly trained in the search & secure procedure. This indicates the enhancement in safety, which can be made by making the warnings failsafe, and reduces the probability of the second category to the same order of magnitude as the third. A copy of the current operating and the emergency procedures is supplied along with the operator's instruction manual. Radiation levels inside the accelerator rooms are monitored. All monitoring devices are electrically independent of the accelerator control and safety interlock systems.

6. ACKNOWLEDGEMENTS

We thank Dr. S. Kailas for his keen interest and encouragement during this work.

7. REFERENCES

1. Folded tandem ion accelerator facility at BARC, P. Singh, Pramana-J Phys. 57, 639 (2001)
2. IAEA Safety Guide GS-R-3 /1/,
3. "The Dares bury Personnel Safety System" DE Poole, T Ring PAC'89

Reliability, Safety and Hazard
Editors: P.V. Varde et al.

RADIATION RISK FROM POTENTIAL EXPOSURES IN GAMMA RADIATION PROCESSING PLANTS (GRPP)

A.U. Sonawane, Pravin Patil and S.P. Agarwal

Radiological Safety Division, Atomic Energy Regulatory Board, Mumbai, Niyamak Bhavan, Anushaktinagar, Mumbai 400 094

ABSTRACT

Gamma Radiation Processing Plants (GRPP) consist of Co-60 or Cs-137 sources, with typical source strength of 10^{16} Bq to 10^{18} Bq. The exposure rate inside the radiation cell when source is in exposed condition is of the order of few Gy/min to several Gy/min. The fault tree is constructed to derive logical structure for one of the new and two old GRPPs for an event leading to potential exposure in which a person trapped inside the radiation cell (radiation processing room) receives lethal dose. It is observed that the risk of an event is less than the reference risk of attributable death due to normal exposure in case of new GRPPs and more in case of old GRPPs.

1. INTRODUCTION

Gamma Radiation Processing Plants (GRPP) are increasing in India mainly for sterilization of medical or healthcare products and preservation of food. GRPPs consist of a Co-60 or Cs 137 sources, with typical source strength of 10^{16} Bq to 10^{18} Bq, mounted in a Stainless Steel source rack. Radioactive source, when not in use is stored in water pool (typical dimensions are 2.5 m x 3 m x 6 m) filled with deminaralised water and located inside the radiation cell. The water shield does not allow radiation to escape in to the radiation cell, thus permitting free access to personnel to carry out plant maintenance. The radiation processing of food is carried out inside the radiation cell shielded by 1.5 - 1.8 m thick concrete walls. For radiation processing of product, the source is brought to the exposed position above the water level after activation of all safety devices. The products in aluminum carriers or tote boxes are mechanically positioned around the source rack and are turned round their own axis, so that contents inside the product boxes are irradiated on both sides. The product, either pre-packed or in bulk, placed in suitable containers is sent into the cell with the help of automatic conveyor. The conveyor goes through a concrete wall, labyrinth, which prevents radiation from reaching the work area and operator room (control room). The operation of the plant is governed by the established operating procedures laid down by manufacturer of the GRPP. The start up of the plant is through a sequence of search operations and actuation of safety systems. Search operations require that operator of GRPP shall press, within pre-set time, all sequential push buttons provided on walls of radiation cell at strategic locations inside the cell. The source is raised from the water pool by hydraulic/pneumatic system when products are to be processed by radiation. There are 13 gamma radiation processing facilities operating in the country and two new plants are going to commence operation in a couple of weeks. The first food irradiation facility for research purposes, FIPLY (Food Irradiation and Processing Laboratory) located in BARC, commenced its routine operation in 1966 and the first commercial irradiation facility for medical product sterilization, i.e. ISOMED (Isotopes in Medicine) started operating in 1974.

2. DESIGN SAFETY

A number of safety systems and interlocks are provided between the personal access door (PAD) and source raise system in the radiation cell in order to make it impossible for any person to enter the cell when the source is in the exposed condition. Various safety features of modern GRPPs include multipurpose single key for operation of GRPP, safety systems and interlocks provided at the PAD, inside the radiation cell and labyrinth, at the product barrier doors, in the deminaralised water plant and at roof of the cell. The design has been evolved by adding improvements resulting from the operating experience and lessons learnt from previous incidents reported in the literatures [1][2].

The design of new GRPPs incorporates the defence-in-depth concept applicable to all safety systems, such that multiple levels of protection are provided and the human intervention is minimized [3]. This is mainly achieved by adhering to the design principles such as redundancy, diversity and independence in the safety systems, sub-systems and components [4].

3. EVENTS POSING POTENTIAL EXPOSURES IN GRPPs

Potential exposures are predominant threat in the operation of the GRPPs. A number of failure scenarios that could lead to potential exposures can be considered in GRPPs such as i) source is unshielded and operator enters radiation cell via personnel entry door ii) source is unshielded and operator enters via product gate iii) source pencil is transported outside the radiation cell because of its fall onto conveyor system due to its entanglement with product boxes iv) person trapped inside radiation cell fails to stop source movement and v) radiation exposure outside the room.

4. FAULT TREE ANALYSIS

Among the above scenarios, there is high probability for an event with potential exposure to a person who is trapped inside the radiation cell and fails to stop source movement and receives lethal dose. This explores the possible ways that a person could already be inside the radiation cell with source shielded and become exposed either by the actions of another person or by failure in the source movement control system. In the present studies, a fault tree analysis, for one of the new and two old GRPPs, viz., SPICE, BRIT, Vashi & FIPLY, BARC, as shown in figure-1, figure-2 and figure- 3, has been carried out for the event leading to potential exposure, in which a person trapped inside the radiation cell receives lethal radiation dose when the source is coming out of the water shield and he fails to stop source raise operations. The redundant safety systems and interlocks are provided inside the radiation cell for lowering the source into safe storage condition in case of emergency situation. For example, in case of FIPLY facility, a trip wire (emergency pull cable) runs around the wall and pulling the cable would stop the irradiator operation and sources lowers to its shielded position, in case of SPICE facility, in addition to trip wire, emergency push button is provided on one of the cell wall and pressure plate is fitted over the floor near product entry door. When a person trapped inside the cell pushes the button or steps on the pressure plate, sources lower down into the pool and aborts irradiator operations. In case of new GRPPs, in addition to the above safety features, a communication system is installed behind PAD to communicate with Control Room in case of emergency. Fault tree is constructed to derive logical structure for the above high probable event and thereby a risk from potential exposure is evaluated by identifying initiating events. The construction and evaluation of fault tree is made in such a way that the event causing potential exposure is placed at the top of the tree and other associated events are connected by "AND" and 'OR' Gates. Identification is made for "minimal cut sets" and events leading to the top event. When the tree has been constructed with minimal cut sets, suitable reasonable expected frequencies or probabilities based on the case studies have been assigned to each basic event leading to the event [5]. The frequency of occurrence of the top event is calculated by adding /multiplying the contribution from each cut set.

The risk of potential exposure is compared with the Reference Risk of Attributable Death (hereinafter called as Reference Risk) corresponding to risk owing to normal occupational exposure.

R.C = Radiation Cell
fot = fraction of time
pd = per demand
/a = per year and per installation

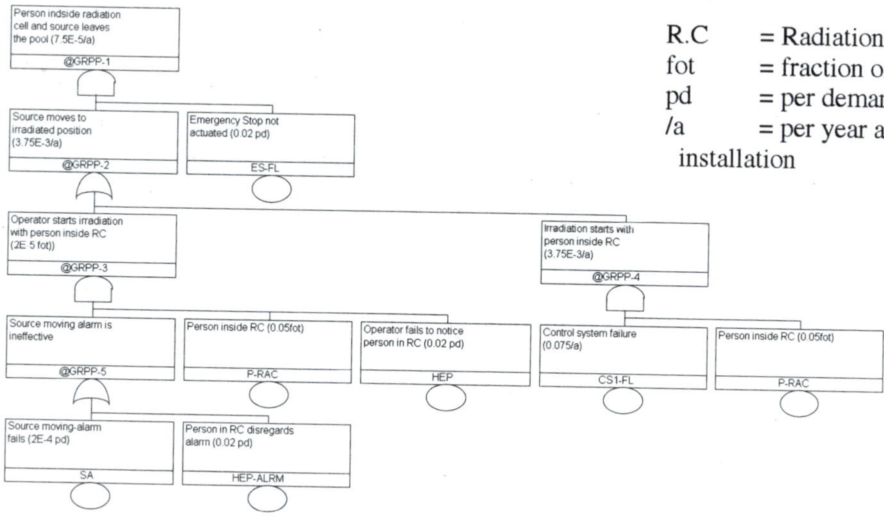

Fig.1: Fault Tree Analysis for new GRPP

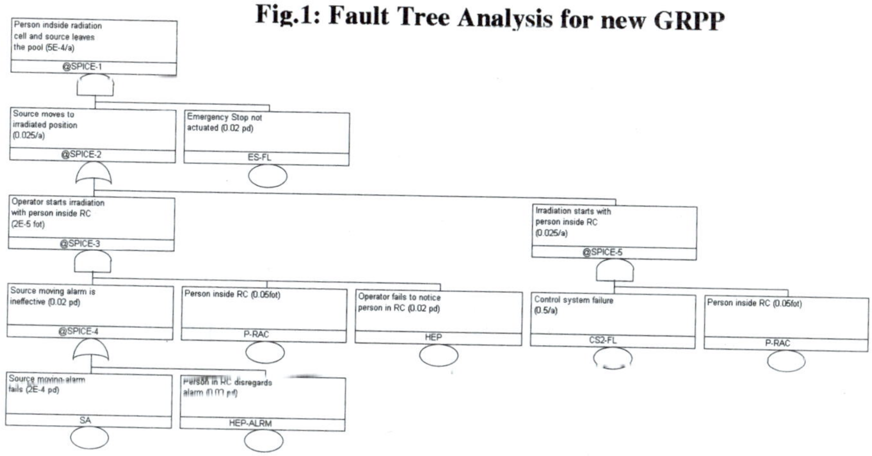

Fig.2: Fault tree analyses for SPICE, GRPP

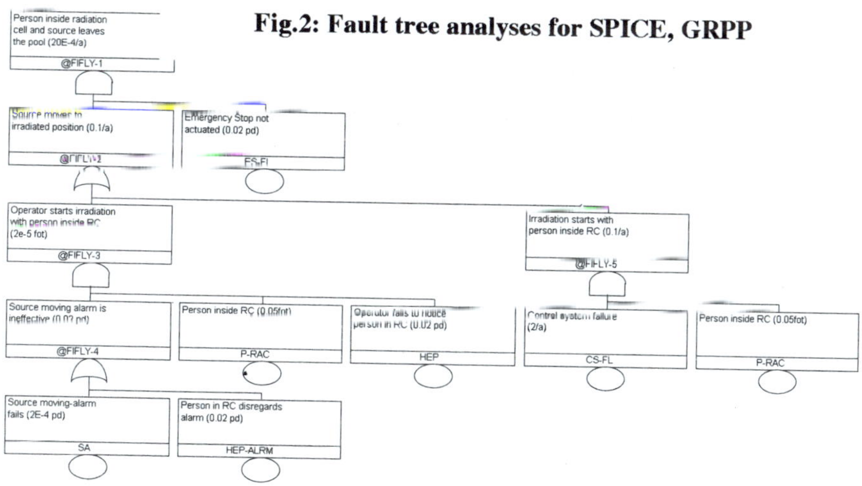

Fig. 3: Fault tree analysis for FIPLY GRPP

The component reliability data is indicated in the fault tree as shown in figure-1, figure-2 and figure- 3.

5. CRITERIA OF ACCEPTABILITY

It has been assumed that the annual occupational doses to an average individual more often is no greater than about 5 mSv. The International Commission on Radiological Protection (ICRP) has considered 4 % per Sv as cancer death risk to occupational workers. Based on the above the individual occupational risk from normal exposure is calculated as 2×10^{-4} in a year as a criteria of acceptability. The risk from the potential exposure from an event analyzed by fault tree is compared with the above reference risk.

6. RESULTS

The pathway of minimal cut set leading to maximum contribution to the value of the final expected frequency is through event involving a person present inside radiation cell (P-RAC), control system fails (CS-FL) and emergency stop not actuated (ES-FL). This gives information about the system, which requires maximum reliability for its intended operation. A comparison of the frequency of occurrence of an event involving a person trapped inside radiation cell and receiving a lethal dose is made between the new GRPPs and two old GRPPs. It is observed that the expected frequency of occurrence of event in new GRPPs is 0.75×10^{-4} which is 2.6 times less than the reference risk constraint due to normal exposure $R = 2 \times 10^{-4}$ and the same for SPICE & FIPLY is found to be 5×10^{-4} & 20×10^{-4} respectively.

7. CONCLUSION

The fault tree analysis of new GRPPs and old GRPPs indicate that the value of risk constraint, i.e. expected frequency of fatal dose to a person trapped inside the radiation cell is less by a factor of 2.6 in new GRPPs than in two old GRPPs. This is mainly because of the incorporation of more redundant and diverse safety systems and interlocks provided inside the radiation cell (e.g. emergency push button, trip wire, pressure plate and communication system) for lowering the source from its exposed position to its safe storage position.

8. ACKNOWEDGEMENT

Authors are thankful to Shri R.B. Solanki, Scientific Officer (D), AERB and Shri U.K. Paul, Scientific Officer (F), AERB for their valuable guidance and suggestions in carrying out the above studies.

9. REFERENCES

1. "Lessons Learned from Accidents in Industrial Irradiation Facilities, (1996)", STI/PUB/1015, International Atomic Energy Agency (IAEA).
2. "Review of Events at Large Pool Type Irradiators (1989), NUREG-1345, US Nuclear Regulatory Commission, DC, USA.
3. Safety series No. 107, International Atomic Energy Agency, Vienna, 1992.
4. AERB Safety Standard Specification No. 6 (1993) on "Radiological Safety for the Design &Installation of Land Based Stationary Gamma Irradiators.
5. ICRP publication No. 76, Vol No. 2, 1997, Protection from Potential Exposures: Applications to Selected Radiation Source.

Reliability, Safety and Hazard
Editors: P.V. Varde et al.

SAFETY ASPECTS OF NUCLEAR DESALINATION PLANT

A. Raha, G. Kishore, A.K. Adak, I.S. Rao, V.K. Srivastava and P.K. Tewari*

Desalination Division, Bhabha Atomic Research Centre, Mumbai 400 094

ABSTRACT

In nuclear desalination, fresh water is being produced from the saline water using nuclear energy. Desalination plant (DP) is coupled with a nuclear reactor for the required thermal and /or electrical energy. But it must always be ensured that the product water is free from any radioactive contamination under normal condition and also in anticipated operational occurrences. The desalinated water should meet the national and international (e.g. World Health Organization) drinking water standards through the application of matured and proven desalination technologies. So in all the stages of design and operational procedures, safety aspects have been considered during nuclear desalination project development to prevent any possibility of radionuclide contamination into the product water. Under non-power application of Department of Atomic Energy programs, Desalination Division (DD) of Bhabha Atomic Research Centre is setting up Nuclear Desalination Demonstration Project (NDDP) at Kalpakkam, Tamilnadu. In this case, various engineering approaches have been utilized to minimize the risk of contamination of the product water. The Isolation Loop (IL) included in the steam circuit acts as a barrier between Nuclear power Plant (NPP) and DP. Selection of proper monitoring system for contamination detection is very important. This paper presents the safety aspects of coupling a DP with a nuclear reactor and possible safety measures to ensure that the product water is safe for public distribution.

Keywords: Nuclear desalination, Radioactive contamination, Coupling, Safety, Radiation monitoring, Pressure reversal

1. INTRODUCTION

Nuclear Desalination is defined as the production of potable water from sea water in an integrated facility in which both the nuclear reactor and the desalination system are located on a common site, sharing some of common systems or facilities, and the energy used for the desalination system is supplied by the nuclear reactor [1]. The purpose of this paper is to elaborate on the safety aspects of nuclear desalination with specific reference to NDDP. Nuclear desalination involves three technologies: nuclear, desalination and the interface of coupling. Desalination processes are categorized on the basis of energy required. Thermal energy based processes, such as Multi Stage Flash (MSF), Multi Effect Distillation (MED), Low Temperature Evaporator (LTE) and mechanical energy based processes such as Reverse Osmosis (RO). Selection of a particular desalination process depends on the saline water quality, product water requirement and process economics. Now this choice of a particular desalination process also determines the manner in which the Desalination Plant (DP) is to be coupled with the nuclear reactor for supply of energy required in desalination.

2. COUPLING CONFIGURATION OF NUCLEAR REACTOR AND DP

There are three possible coupling configurations for coupling a DP with a nuclear reactor.

Thermal coupling to Thermal DP

Thermal energy can be supplied to the DP via an intermediate heat transfer loop. In such configuration, to transfer the thermal energy there is a direct fluid coupling between the reactor and DP. This introduces the risk of radioactive contamination of the product water.

Electrical coupling to RO-DP

In the case of RO DP, the only energy supplied by the NPP is electricity to operate the DP pumps. Since there is no thermal coupling between the reactor and the DP, there is no direct path for carry-over of radioactive materials from the reactor to the product water. However as the RO plant shares common resources with NPP such as common seawater intake and outfall, the effect of this on the possible contamination of the product water must be evaluated.

Thermal & Electrical coupling to Hybrid DP

Hybrid DP consists of both thermal and membrane based desalination processes. Coupling of a hybrid DP with a NPP is more complex as it involves both couplings.

3. SAFETY OBJECTIVES IN NUCLEAR DESALINATION

The safety of a nuclear DP depends mainly on the safety of the nuclear reactor and the interface between the nuclear plant and the desalination system. Adequate safety measures must be introduced to ensure near zero radioactivity release to the product water. The risk for accidental radioactivity carry-over is to be assessed. The basic objective is to prevent radioactive contamination of the DP and/or the atmosphere. It is necessary to ensure that the DP should not cause any disruption and malfunctioning in the nuclear power plant. At the same time perturbations to the DP due to the nuclear power plant needs to be analyzed.

Safety Design Requirements for the Couplings

Safety objective in nuclear desalination is translated into safety design requirements under defence in depth principle.

- Provision of multiple barriers between potentially radioactive material and product
- The provision of engineered features preventing the radioactive material from reaching the product-water even in case of any credible sequence of failure.

Engineered safety features for prevention of the radioactive contamination of product water

Assuming a three loop system (i.e. an Isolation Loop (IL) between the reactor loop and the DP loop as shown in Fig 1), pressure in these loops (P_1, P_2 & P_3) may be adjusted in such a way that transfer between loops is in the favourable direction [2].

The best way, which is difficult in practice, is for an arrangement where the reactor loop pressure is lower than the IL pressure which is in turn lowers the product loop pressure. In this case, transfer between loops is always in the safe direction. In practice, systems are usually engineered in two ways:

H-L-H Configuration

The reactor loop pressure (P_1) is higher than the IL (P_2); the desalination loop pressure (P_3) is higher than the IL (H-L-H). In this case a pressure barrier against undesired transfer exists between the product loop and the IL.
Advantages of this configuration are

a) Any leakage in the isolation heat exchanger will be directed to the IL and not to the DP loop. b) The allowed radioactive contamination for the IL is slightly higher and more easily monitored.

Disadvantage is that the operating pressure in the DP loop is higher which is difficult to use as the pressure of the DP loop is dictated by the process and it is not very high.

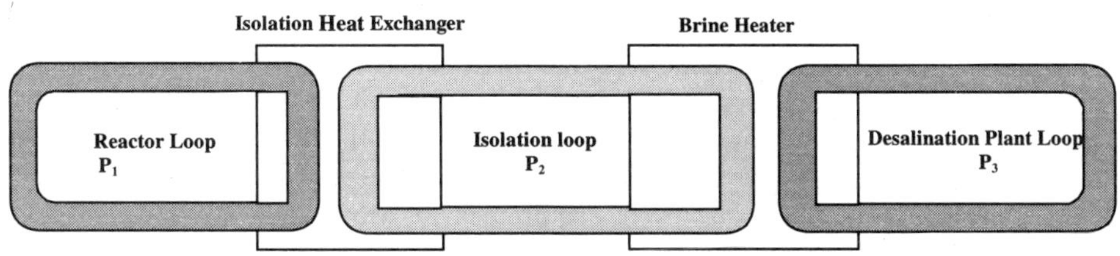

Fig. 1: Coupling of the desalination plant to nuclear reactor by an Isolation loop

L-H-L Configuration

The reactor loop pressure (P_1) is lower than the IL pressure (P_2); the DP loop pressure (P_3) is lower is lower than the IL (P_2) (L-H-L). In this case, a pressure barrier against undesired transfer exists between the reactor loop and the IL.

Advantage of this configuration is that the radio activities are enclosed inside of primary loop.

Disadvantage is that it is difficult to monitor the possible contamination of IL and to exclude the possibility of a leakage from IL to final product.

Safety analysis of a Nuclear DP

Safety analysis is the tool for confirming the adequacy and efficiency of the provisions in defence in depth. The most commonly accepted techniques of deterministic safety analysis used in the safety analysis of nuclear power plants (NPP) are adequate for NDP. In deterministic safety analysis response of an NDP is predicted and analysed for postulated initiating events (PIE's) which can be anticipated operational occurrences or a design basis accidents (DBA). The safety analysis of a NDP can be performed by [3];

(A) Identification of the defence-in-depth barriers in the NDP
(B) Construction of the postulated initiating event list (limited to DBA)
(C) Analysis of the accident scenarios and its consequence.
(D) Verification of design with respect to acceptance criteria.

There are two aspects to implement the defence in depth. One is to provide barriers that would have to be successively breached for radioactive material to escape outside the plant. The other is to prevent the barriers from being breached. As stressed, the first aspect of defence in depth in the design is the provision of a series of physical barriers to confine the radioactive material at specified locations. The integrity of these barriers is ensured by conservative design margins and by high quality in manufacture, inspection and maintenance. In the case of failure of one or more of these passive barriers, active engineered safety systems are provided to ensure the integrity of the remaining barriers. The second aspect of the defence in depth strategy is to prevent the barriers from being breached. This is accomplished by the design of the plant operational systems.

An examination of, the *defence-in-depth barriers* and a look at the NDP as a whole, six main barriers can be identified to avoid contamination of the fresh water product, these are:

1. Fuel matrix (pellet)
2. Fuel rod cladding
3. Steam generators tube walls
4. Isolation Heat exchangers
5. Desalination Plant
6. DP Isolation System (between the DP and distribution-piping grid).

DP itself also act as a barrier in preventing radioactive contamination of the fresh water product as if any radioactive isotopes did enter the desalination system feed water stream, it would be left behind in the concentrate in MED or MSF desalination system. This means that, in case there is a release of fission products due to the catastrophic damage of the fuel matrix and fuel element cladding, contamination could reach the fresh water product only by a chain of failures. The steam generators would need to break, the turbine trip would have to fail, the heat exchanger tube walls would have to break simultaneously, and then the Isolation System would have to fail to perform its function.

4. SAFETY ASPECTS OF NUCLEAR DESALINATION DEMONSTRATION PROJECT (NDDP), KALPAKKAM: A CASE STUDY

NDDP is a 6300 m^3/d hybrid MSF-RO DP coupled to MAPS 1 & 2. MSF plant is thermally coupled to MAPS as shown in Fig 2. Coolant water rejected from the MSF is used as a feed to RO system. Both the DP's are having common seawater intake and outfall with NPP.

Steam Requirement

The Multi-Stage Flash(MSF) DP needs about 21 T/hr dry saturated steam at 2.8 kg/cm^2(a) for the brine heater and 1 T/hr dry saturated steam at 8 kg/cm^2 (a) for the steam jet ejectors [4]. The steam requirement of MSF plant will be met from MAPS 1 & 2.

Source of Steam

Two sources of MAPS steam are considered for the supply of the steam. One is high pressure (HP) (41 kg/cm^2 (a)) main steam which will be used to generate steam for steam jet ejectors and another one is low pressure (LP) (3.0 kg/cm^2 (a)) cold reheat steam at the exhaust of the high pressure turbine will be used to generate steam for the brine heater.

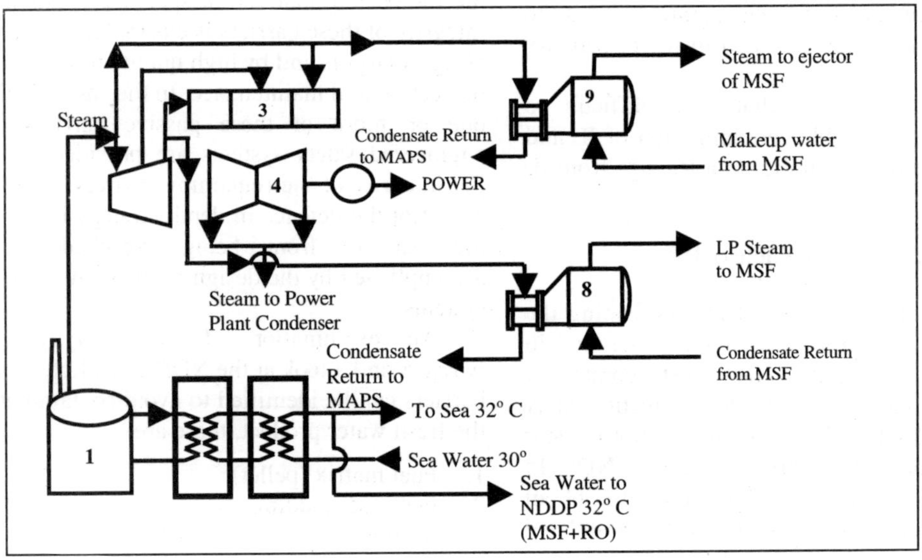

1. Nuclear Reactor
2. HP Turbine
3. Moisture Separator
4. LP Turbine
5. Moderator-DM water Cooling Loop
6. DM Water-Sea Water Cooling Loop
7. Generator
8. LP Isolation Heat Exchanger
9. HP Isolation Heat Exchanger

Fig. 2: Coupling Diagram of the Desalination Plant with MAPS

Engineered Safety features in Coupling of 4500 m3/d MSF plant of NDDP with MAPS

Process logic of steam intake from MAPS & isolation heat exchanger circuit is shown in Fig 3.

Provision of Isolation Heat Exchanger

Isolation heat exchangers are provided for generating steam for brine heater and to the ejector systems from the source steam from MAPS as shown in fig 2. Heat exchanger (HX) which generates steam for steam jet ejectors called High Pressure (HP) Isolation HX and heat exchanger which generates steam for brine heater is called Low Pressure (LP) Isolation HX.

LP Steam Supply system: The LP steam is taken from the cold reheat line before moisture separator (MS) of MAPS 1 & 2. The pipe connections are taken separately from MAPS-1 and MAPS-2 and the two lines are joined together after passing it through independent moisture separator to a common header. The common header from MAPS-1 and MAPS-2 are taken out of turbine building to NDDP area. LP isolation heat exchanger is cylindrical shell and tube, kettle type reboiler.

HP Steam Supply: The HP steam piping, which is of much smaller size, is tapped from the main steam piping within turbine building, upstream of the hogging ejector is routed through the same path for LP steam piping. The HP isolation heat exchanger is also a cylindrical shell and tube, kettle type reboiler. These isolation heat exchangers provide a physical barrier between the nuclear power plant and DP to avoid any possible ingress of radioactivity from source steam of MAPS to desalination plant.

Pressure Reversal

LP steam taken from MAPS is at higher pressure (3 kg/cm^2) than the pressure of steam (1.8 kg/cm^2) generated in the LP isolation HX. But pressure of steam generated in the LP isolation HX is lower than the brine side pressure (5 kg/cm^2) in the brine heater. So this is an H-L-H coupling configuration. Pressure reversal existing in the brine heater is the most effective design solution as any loss of integrity in the barrier would cause a leak of water toward the IL. Hence by maintaining this prevention of the ingress of radioactive contamination to DP is always ensured.

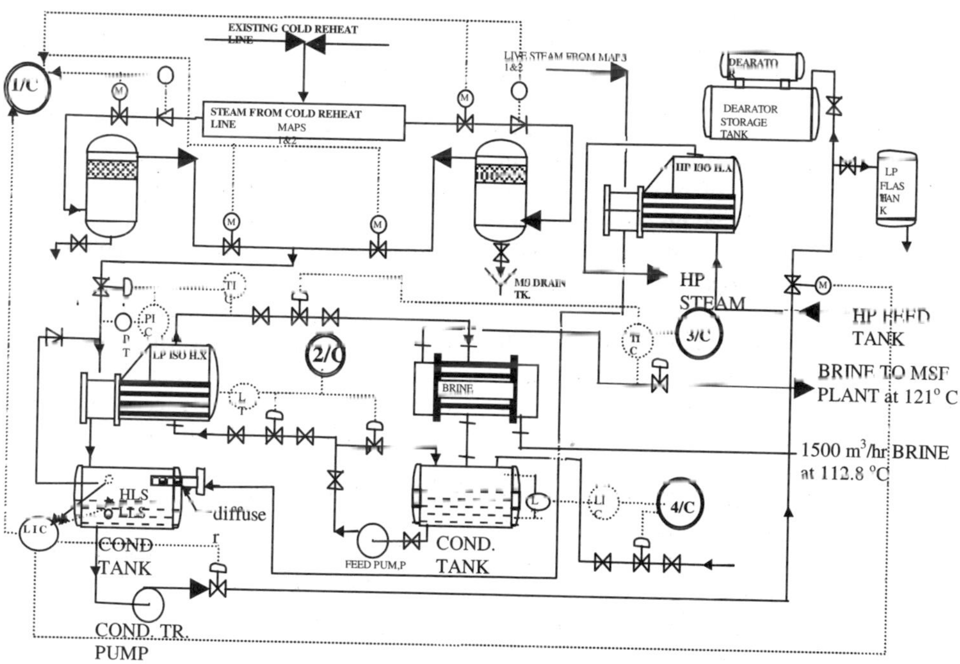

Fig. 3: Process Logic of Steam Intake & Isolation Heat Exchanger Circuit

Safety analysis of Isolation System

The Isolation System design should ensure the performance of the isolation function that is fulfilment of safety requirements during all foreseen conditions, without relying on non-categorised components, in the required time and with the required reliability [5].

Design Basis of Isolation System

The design basis of the Isolation System covers following Postulated Initiating Events (PIE's):

1. Operational transient in NDDP and MAPS in terms of flow, pressure and temperature etc
2. Failures in the IL pressure control
3. Failures in the IL circulating pump
4. Leaks in the Heat Exchangers working as barriers.

Predicted response to PIE's

1. Steam is extracted from the secondary circuit of MAPs for the MSF of NDDP. So DP also contributes to the total heat sink of the MAPS. But in this case DP represents a very small fraction (21 T/h of steam out of total 1250 T/h of steam) [4] of the load on the MAPs. The sudden decrease or cessation of steam requirement due to transients or shutdown of DP results in feedback of thermal transient in the secondary circuit of NPP. These transients are expected to be similar to the transients already considered in NPP design, such as loss of heat sink, loss of load and steam line break. So no transient more severe than those already addressed in NPP design are expected to occur as a result of addition of this DP.

Likewise transients originated in MAPS will have impact on operational stability and availability of DP but it would not cause any radiological consequences.

2. IL pressure control is necessary to maintain pressure reversal. Brine which is in the tube side of the brine heater is at a high pressure (5 kg/cm^2) and its temperature controls the IL steam temperature through temperature control valve (TCV) as shown in fig 3. IL pressure is controlled by a temperature-pressure cascade control. LP Steam coming from the MAPs is controlled by a Pressure Control Valve (PCV) which is cascaded with the temperature controller (TIC) in IL.

3. IL is provided with feed pump (2 × 100%) to prevent loss of circulation in IL due to failure in one IL feed pump.

4. In brine heater, brine side pressure is higher than the IL pressure. So any leak in brine heater will cause brine to leak into IL. This will increase the conductivity of the IL steam. Conductivity meter installed in the isolation loop is used for continuous monitoring of conductivity. Any increase in conductivity of the IL steam indicates heat exchanger leakage and operator will be informed for safe DP shutdown.

Additional Engineered Safety Features

1. A quick closing air operated non return valve is placed in the steam line in each unit between source steam and isolation heat exchanger. The non return valve will close automatically in case of turbine trip and high condensate level in condensate collection tank. In addition to quick closing non return valve, a motor operated isolation valve is also provided. Both the valves are located in the turbine building.
2. An electrically operated isolation valve is provided down stream of the moisture separator, so that if necessary, one unit of MAPS can be isolated from the MSF plant.

5. CONCLUSIONS

Nuclear energy can be safely used for desalination purpose. But in all stages of design and operational procedures, safety aspects are need to be addressed with due consideration to the coupling aspects. NDDP, Kalpakkam is designed in accordance with all required safety standards and will demonstrate that nuclear desalination is not only economically feasible but also safe for public distribution.

6. REFERENCES

1. "Introduction of Nuclear Desalination, a Guidebook", Technical Reports Series No 400, IAEA, Vienna, 2000.
2. "Safety Aspects of Nuclear Plants Coupled with Seawater Desalination Units", IAEA–TECDOC – 1235, Vienna, August 2001.

3. "Design Concepts of Nuclear Desalination Plants", IAEA–TECDOC–1326, Vienna, November 2002.
4. "Coupling Aspects of MSF Desalination Plant and Loss of Electrical Power Generation of a Nuclear Power Plant: Case Study" by A.K.Adak, P.K.Tewari and B.M.Misra, International Journal of Nuclear Desalination, Vol.1, No.3, 2004, P273-281.
5. "Pressure Reversal as a Safety Feature for Nuclear Desalination" by Nestor Masriera and Alicia Doval in IAEA Technical Meeting on Integrated Nuclear Desalination System, Chennai, December, 13-16, 2004.

Reliability, Safety and Hazard
Editors: P.V. Varde et al.

RISK ANALYSIS OF PROCESS PLANTS: APPROACH OF THE ATOMIC ENERGY REGULATORY BOARD

Lakshman Nyaya Valiveti, K. Ramprasad and P.K. Ghosh

Atomic Energy Regulatory Board, Niyamak Bhavan, Anushaktinagar, Mumbai 400 094

ABSTRACT

The analysis of the probable consequences and effects of the hazardous chemicals used in the process plants plays a vital role in regulatory decision-making. This analysis is required in the entire lifecycle of the process plant. This paper presents the detail of risk analysis required at various lifecycle stages viz. siting, design, and operation of a process plant. It also describes the regulatory methodology and safety review process adopted by Atomic Energy Regulatory Board (AERB) to ensure safety of the plant in its entire life cycle.

Keywords: Consequence Analysis, Hazard Assessment, Regulatory review, Risk Analysis

1. INTRODUCTION

The analysis of the probable consequences and effects of the hazardous chemicals used in the process plants plays a vital role in regulatory decision-making. This analysis is required in the entire lifecycle of the process plant.

The analysis in the siting stage covers the preliminary hazard analysis of the proposed process plant based on hazardous substances likely to be used, and to choose the site posing negligible/tolerable risk to the public and environment.

The analysis at the design stage involves identifying the risk contributions by various systems/equipment/materials. Any events of low but credible probability and high consequences, need to be minimised at this stage by adopting inherently safer methodologies, passive protection systems and active protection systems in the plant design.

In the operation stage the risk analysis involves the identification of the personnel exposed to maximum risk and taking adequate safety measures. The analysis at this stage is continuous since the probability of failure of systems/equipment change with time and usage The experience in operation of the plant gives further insights to the risk analysis.

2. SITING STAGE

The regulatory review of the process plant at siting stage is to ensure the acceptability of the site from the standpoint of site characteristics vis-a-vis plant design basis. The review of safety aspects is done based on the conceptual design (or actual design, if available) of the facility. The site identified for locating the facility is reviewed to verify that no adverse interaction exists between the facility and the site. In the review, the Regulatory Body takes note of the characteristics that influence the design of the facility and also the impact on the environment that may lead to specific design or operational requirements. The other factors that are considered in this stage are the demography, meteorology, geology etc. of the site and its surrounding areas.

Sitting consideration include potential external human-induced events such as earth quakes, flooding of the site, fires and explosions due to nearby process plants etc. Considering the maximum magnitude in various fields of human activities in our country, safe distances for Nuclear Power Plants (NPP) have been indicated in AERB AERB/SC/S, Code of practice on safety in NPP siting [1] for certain human activities, which are tabulated below.

Table 1 Safe distances to site a plant for some activities

Facility/Activity	Safe Distance
Facility involving storing, handling of inflammable, toxic corrosive or explosive materials (20,000 Te TNT or equivalent)	5 km
Facility involving storing, handling of toxic/ corrosive chemicals (30 Te Chlorine or equivalent)	8 km
Mining activity	5 km
Military activity of reasonable magnitude	10 km

The Site Evaluation Report (SER) forms the main document for review by the Regulatory Body in respect of siting clearance. Information relating to the site with particular emphasis on factors important to chemical / environmental safety and emphasising those site characteristics, which may influence the design and operation of the facility, should be provided. Information regarding the interaction of the facility and the environment should also be provided. The SER should specify the exact location of the process plant and provide a description of the geographical, demographic, meteorological, hydrological, geological, seismological characteristics of the site and surrounding areas. Ecological data from the site should also be turnished for assessment of the environmental/ chemical impact of the process plant. Information furnished should be adequate to permit an independent evaluation of the suitability of the site for locating the facility.

The following studies would be required to assess the influence of design and operation of facility on the environment, the population surrounding the proposed plant and the operators of the plant.

2.1 Preliminary Process Hazard Analysis

Preliminary Process Hazard Analysis is used during the conceptual, early development, early design phase, of a plant. The method is intended for use only in the preliminary phase of plant development for cases where past experience provides little or no insight into potential safety problems, e.g., a new plant with new process. The case study of the method is presented in Appendix 1

2.2 Coarse Hazard Study

The coarse hazard study is done for the plants where the design of the plant is not new and the block layouts of plant systems are available. The objective is to determine whether there are problems in areas such as information about the hazards, basic features of process design, or layout and siting of the plant. The case study of the method is presented in Appendix 1

Other methods like hazard indices, what if analysis, checklists etc can also be used in the siting stage depending on the applicability.

3. DESIGN STAGE

The regulatory review of design of the process plant is to ensure that the consented design meets the defined safety standards, operability and equipment availability when operated under design conditions.

The regulatory consents for the construction phase (approval of the design for construction of plant) of the process plants require the submission of Safety Report. This forms the principal documents for the Regulatory Body to determine whether the operation of the fuel cycle facility under review will not result in unacceptable radiological, chemical and industrial hazards/ risk to the plant/ site personnel, the public and the environment.

The safety report should give information on the site conditions, process details, Hazardous chemicals and Hazard control mechanisms, safety systems and safety analysis, waste management and mitigation of major accidents.

The important aspects reviewed by the regulatory body are

- Precautions in storage, transport and handling of hazardous substances.
  - Margin of safety in design
  - Maximum quantity that can be stored
  - Distances between storage areas
- Safety provisions for the equipment
  - Pressure relief
  - Draining arrangement
  - Isolation of equipment, vessels.
  - Alarm provisions for level, temperature, pressure as applicable
- Hazard control measures
  - Control of work environment against hazards

- o Threshold limit values
  - o Personal Protective equipment
  - o Handling loss of containment (Minor spillages, emergency conditions, off-site & on-site emergency plans)
  - o Monitoring of work environment
- Waste management (Confinement, processing & disposal)
  - o Concentration of effluents & emissions
  - o Scrubber systems, stacks for emissions
  - o Effluent treatment and disposal
  - o Solid waste disposal

The following hazard assessment studies would be required to obtain the information detailed above.

- What-if Analysis or Checklists based on
  - o Process control review
  - o Relief and blowdown review
  - o Accessibility study
  - o Operating procedures check
  - o Review of adequacy of training programme
  - o Hazardous area classification
- Hazard & Operability Study (HAZOP)
- Failure Modes Effect & Criticality Analysis (FMECA)
- Consequence Modeling

3.1 Hazard & Operability Study (HAZOP)

HAZOP study is carried out when the engineering line diagram of the plants become available. It should be carried out by multi-disciplinary team who review the process to discover hazards and operability problems using a guideword approach. Case study of HAZOP methodology is presented in Appendix 1.

3.2 Failure Modes Effect & Criticality Analysis (FMECA)

FMECA involves review of systems to discover mode of failure, which may occur, and effect & criticality of each such failure mode. The technique is oriented towards equipment rather than process parameters.

3.3 Consequence Modeling

For major process plants with high degree of hazard (Heavy water Plants etc.) the some degree of quantification of hazard shall be required to give better understanding of the likely development of accident scenarios, their identification and mitigation. Case study of consequence analysis is presented in Appendix 2

4. OPERATION STAGE

The regulatory review in operation stage of a process plants is done to ensure that the facility is being operated in accordance with the stipulations made in the consent and also the procedures outlined in the documents submitted to the Regulatory Body. Modification to the process, plant design and operating procedures, having a bearing on safety and relevant to the consenting process requires approval by the Regulatory Body. Modification may also include augmentation, expansion, life extension, etc. These reviews are periodic and done to ensure that standards of safety and reliability defined at the design stage are achieved in practice and are maintained through out the life of installation.

The important document submissions required by the regulatory body are

- Revised safety report
- Quantitative Risk Assessment report
- Technical specifications for operation of the plant identifying the safety limits, limiting safety system settings, limiting conditions for operation and surveillance requirements.
- Operation, maintenance, security and training manuals

They key elements for review are

- safety status and surveillance reports,
- personnel exposures and investigation reports,
- safety related unusual occurrences,
- violation of technical specifications,
- environmental surveillance,
- plant modification relating to safety aspects.

The risk analysis studies that would be required at this stage for regulatory review are

- Re HAZOP of the process plant based on as built design and after approved modifications
- Quantitative Risk Assessment

4.1 Quantitative Risk Assessment (QRA)

QRA provides a quantitative method to evaluate risk and to identify areas for cost-effective risk reduction. The steps involved in QRA of a process plants are detailed in the appendix 3.

5. CONCLUSION

These studies shall facilitate the regulator to arrive at a judicious decision keeping in mind the scientific and technological factors. The safety review process adopted by Atomic Energy Regulatory Board is aimed to ensure the safety of the plant in its entire life cycle.

6. APPENDIX 1: CASE STUDIES FOR PRELIMINARY PROCESS HAZARD ANALYSIS, COARSE HAZARD STUDY AND HAZOP STUDY

A1.1 Example Process Plant

Production of zirconium sponge (Zr) from the zirconium oxide (ZrO_2)

- Direct chlorination of ZrO_2 in presence of carbon at 900° – 1000°C to form anhydrous $ZrCl_4$
- Direct chlorination of off grade zirconium sponge and zircaloy scrap turnings with chlorine at 350° – 400°C to form anhydrous $ZrCl_4$
- Purification of zirconium tetra chloride ($ZrCl_4$) by re-distillation / sublimation
- Magnesio-thermic reduction pure $ZrCl_4$ at 850°C to produce the metal sponge
- Pyro-vacuum distillation at 1000°C to separate the slag magnesium chloride ($MgCl_2$) and un-reacted magnesium from zirconium sponge
- Handling of raw sponge, grading evaluation and blending

A1.2 Preliminary Process Hazard Analysis

Preliminary Process Hazard Analysis is used during the conceptual, early development, early design phase, of a plant. The method is intended for use only in the preliminary phase of plant development for cases where past experience provides little or no insight into potential safety problems, e.g., a new plant with new process. Early identification of most of the hazards could be possible resulting in effective saving in cost that could otherwise result from major plant redesigns if hazards are discovered at a later stage. It is very useful for "Site Selection". It does not preclude the need for further hazard assessment; instead it is a precursor to subsequent hazard analysis. Items for consideration consists of meticulous preparation of a list of hazards:

Table A1.1 Preliminary Process Hazard Analysis for Zirconium Sponge Plant

Hazard	Cause	Major Effects	Corrective / preventive measures
Dust explosion	Carbon dust	Injuries / fatalities Damage to equipment, structure	Ensure adequate ventilation
Chlorine release	Leak in chlorine piping	Injuries / fatalities	Selection of suitable material Reduce storage quantity Reduce piping length Provision of chlorine leak detectors Provision of chlorine gas masks for operator in case of emergency Emergency plan
Pyrophoric metals	Magnesium and zirconium in contact with water / moisture.	Injuries / fatalities	Storage and handling of metals to prevent the hazard
Rapid phase transition (metal – water reaction)	Damage of reactor walls (scrap chlorination) due to insufficient cooling	Injuries / fatalities Damage to equipment	Limit the amount of metal scrap to be chlorinated Use coolant that does not react with the metal Alarm/action on reduced coolant flow and rise in temperature
Emissions & Effluents	Emissions and effluents from various reaction steps	Damage to environment Health hazard to public and personnel	Scrubbing the emissions and treatment of effluents

A1.3 Coarse Hazard Study

The coarse hazard study is done for the plants where the design of the plant is not new and the block layouts of plant systems are available. The objective is to determine whether there are problems in areas such as information about the hazards, basic features of process design, or layout and siting of the plant. It exposes hazards due to interactions between the plant and other plants or the environment.

The coarse hazard study should identify and list the hazardous materials and their hazardous properties like toxicity & flammability, hazards like fire, explosion, corrosion & smell, and any other matters of concern caused directly by the materials and/or by the equipment and activities containing the materials. The conduct of coarse hazard Study involves compilation of lists on hazardous properties of material involved and hazards posed by the material in process in the equipment.

Table A1.2 Hazardous material and their hazardous properties list for Zirconium Sponge Plant

Hazardous materials	Hazardous properties [2]
Chlorine	Extremely Toxic
Carbon	Explosion Hazard in form of dust
Magnesium	Violent reaction with water, oxidising agents, can burn even in CO_2 atmosphere
Zirconium	Explosion Hazard in form of dust, Pyrophoric, Highly Exothermic reaction with chlorine, explosive conc. 0.16g/lit in air.
Zirconium chloride	Exothermic reaction with magnesium

- Raw materials, intermediates, by-products, final products
- Plant equipment
- Facilities
- Safety equipment
- Interfaces between system components
- Operating environment
- Operations (maintenance, testing etc.)

The guidewords like "Toxicity", "Fire", "Explosion", "Effluent" etc. should be applied to the block lay out of the equipment and the hazards and corrective measures are identified (as indicated in Preliminary Process hazard Analysis case study).

The "Toxicity" guideword applied to the fluidized bed chlorinator will address the issues like

- Probability of chlorine leak
- Reduction of leakage probability
- Reduction of chlorine inventory
- Detection of leak
- Mitigation (method, access, response time)
- Emergency action plan

A1.4 HAZOP Study

The method uses Guidewords, which are used to quantify or qualify the intention in order to guide and stimulate the hazard identification process. The Guidewords are used to generate deviations from the design intent. These guidewords are applied on parameters like flow, pressure, temperature etc.

Guidewords	***Meaning***
No	Negation of Design Intent
Less	Quantitative Decrease
More	Quantitative Increase
Part of	Qualitative Decrease
As well as	Qualitative Increase
Reverse	Logical Opposite to Intent
Other than	Complete Substitution

Scrap Chlorination step case study

Scrap chlorination is carried out in a packed bed type graphite lined reactor. This reactor is lined inside with graphite and jacketed outside for water-cooling to enable sufficient extraction of heat. The chlorination will be carried out in a continuous fashion with intermittent feeding of scrap and continuous feeding of chlorine. The exhaust gasses are scrubbed in 2-stage water and alkali scrubbing system and the scrub solutions are processed in the effluent treatment plant

Other guidewords like PART OF, REVERSE, OTHER THAN can be applied as per the system requirements.

7. APPENDIX 2: CASE STUDY OF CONSEQUENCE ANALYSIS

Caste study: Full bore rupture of 8-inch liquid discharge line from 200 MT LPG tank

Source models provide a description of the rate of discharge, the total quantity discharged, duration of discharge, and the state of discharge i.e. liquid, vapour or two-phase flow. Evaporation models are subsequently used to calculate the rate at which the material becomes air borne.

Table A1.3 Model HAZOP case study for scrap chlorination step of Zirconium Sponge Plant

Guide-word	Deviation	Causes	Consequences	Action
NO	No chlorine flow	Chlorine supply valve closed. Chlorine tank empty	No reaction	
LESS	Less chlorine flow	Leak in chlorine supply piping Low chlorine flow rate set.	Toxic gas release Batch operation delayed	Leak Detection and rectification methods to be installed Set chlorine flow accordance to batch weight.
NO	No scrap	Scrap batch not added	No reaction Excess scrubber load	
LESS	Less scrap	Full batch not added	Excess load on scrubber Unaccounted hazardous scrap	Scrubber to be designed for the load Steps to ensure full batch addition
NO LESS	No Cooling water flow Less cooling water flow	Cooling water line failure Cooling water line leak	High temperature, reactor vessel failure	Interlock chlorine flow with cooling water flow.
MORE	More chlorine flow	High chlorine flow rate set	Faster reaction, more heat generated, reactor vessel failure	Set chlorine flow accordance to batch weight.
MORE	More Scrap	Extra scrap added	Longer batch time	Administrative control to restrict the batch size
MORE	More temperature	Inadequate cooling	Reactor vessel failure	Interlock temperature and cooling water flow, chlorine supply.
AS WELL AS	Water /moisture enters the reactor	Leak from cooling water jacket into the reactor	Violent reaction, reactor vessel failure, Chlorination efficiency effected	Regular surveillance of reactor vessel, lining.

Next a dispersion model is used to describe how the material is transported downwind and dispersed to specified concentration levels. For flammable releases, fire and explosion models convert the source model information on the release into energy hazard such as thermal radiation flux and explosion overpressures.

Discharge

The discharge can be assumed to be in liquid state, since LPG has a normal boiling point of 21°C it flashes and evaporates at release.

- Estimate of flow (compressible fluid release through pipe, release duration if applicable)
- Estimate flash fraction
- Estimate liquid entrainment (conservative) and rainout (if data available)

Dispersion

The dispersion model can be assumed as continuous and that of a heavy gas dispersion. A conservative estimate of F- atmospheric stability class and wind velocity of 2m/s can be made.

Consequence

The consequence can be Jet fire (immediate ignition), Flash fire (delayed ignition) or Explosion (Delayed ignition and confinement).

The above scenario was modelled using PHAST a commercially available, validated and universally accepted software for consequence analysis and the following results were obtained.

Input to the Model

LPG stored as saturated liquid at 25°C(vapour pressure 4.5bar)

Discharge temperature 25°C

Liquid line rupture scenario, 8-inch diameter pipe

TNT explosion (TNT equivalent) model is used with 10% explosion efficiency

Weather

Wind speed 2m/s stability class F (stable night)-conservative

Surface roughness 0.2m (open area with trees and scattered houses)

Output of the Model

The jet fire radiation distances are about 240m for 4kw/m^2 radiation level and 175m for 37.5kw/m^2 radiation level.

The centre line concentration of the cloud reached the lower flammable limit of LPG at about 150m and 1/2 LFL at about 370m. These are the likely distances the flash fire may envelope.

In case of a delayed explosion the maximum distance that will be exposed to 0.02 bar over pressure level i.e. glass cracking would be 460m The repairable damage 0.1 bar over pressure levels can extend up to a maximum distance of 270m. These values represent the worst-case delayed explosion, which corresponds to the conditions that give the greatest downwind effect distance for the overpressure of interest.

8. APPENDIX 3: QUANTITATIVE RISK ASSESSMENT

The major steps of a QRA study are as follows

Define the Potential Accident Scenarios

From the hazard assessment studies carried out (HAZOP etc.), identify the events/event sequences. Screen the scenarios and identify credible accident scenarios and worst-case scenario. Identify and define the source characteristics for these scenarios and the consequences of interest.

Evaluate the event consequences

Consequence analysis is done based on the release, dispersion, weather conditions etc. The results of consequence analysis are used to identify the affected population (operator, public etc). This can be done by plotting the damage contours to scale on site/plot plan.

Estimate the potential event frequencies

The initial event frequencies can be obtained from plant history, generic data or by fault tree analysis. Event tree analysis can be used to account for the post release event frequencies.

Estimate the effects

The effects on personnel, public, property and environment are identified. The costs of these effects are also identified to enable prioritisation of risk reduction methods.

Estimate the risk

The risk is identified by combining the consequences and probabilities of the events identified. The overall risk from the process plant is obtained by summation of risks from all the events.

9. REFERENCES

1. Atomic Energy Regulatory Board (1990), *Code of practice on Safety in Nuclear Power Plant Siting*, Atomic Energy Regulatory Board, Mumbai, INDIA

2. Sax N. I. (1979), *Dangerous Properties of Industrial Materials, Fifth Edition*, Van Nostrand Reinhold Limited, New York, USA

10. BIBLIOGRAPHY

1. Atomic Energy Regulatory Board (2000), *Regulation of Nuclear and Radiation Facilities*, Atomic Energy Regulatory Board, Mumbai, INDIA

2. Lees F. P. (1996), *Loss Prevention in Process Industries, Second Edition,* Vol. 1, Butterworth – Heinemann, Oxford, UK.

3. Centre for Chemical Process Safety (2000), *Guidelines for Chemical Process Quantitative Risk Analysis, Second Edition,* American Institute of Chemical Engineers, New York, USA.

Reliability, Safety and Hazard
Editors: P.V. Varde et al.

QUANTITATIVE RISK ASSESSMENT OF AN AMMONIA BASED PLANT

R. Bhattacharya and P.K. Ghosh

Atomic Energy Regulatory Board, Mumbai

ABSTRACT

Ammonia is being handled in large quantities in a chemical plant. Since Ammonia is toxic as well as flammable a Quantitative Risk Assessment (QRA) study was carried out which included Scoping Analysis, Hazard Identification, Consequence Analysis, Probit calculations etc. to find out the Effect distances on various scenarios and the probability of persons getting affected.

Keywords: Chemical Exposure Index, Consequence Analysis, Risk Assessment

1. INTRODUCTION

Ammonia is being handled in large quantities in a process plant. There are hold ups in different sections of the plant. The portions of the plant which handles large quantity of ammonia is the static hold up in tanks and dynamic hold up in the columns. The storage is at atmospheric temperature. Ammonia being toxic and flammable has been chosen for the Quantitative Risk Assessment (QRA) to identify the effect distances and the consequences.

The Dow's Chemical Exposure Index (CEI) method has been utilised for identification and ranking of hazards and identified the scenarios for which Consequence Analysis were carried out.

For identification of Maximum Credible Accident (MCA) scenario the plant has been divided in to 27 systems, as identified in the Hazard and Operability (HAZOP) study report. From these, systems handling ammonia liquid and vapor have been chosen.

The probability of dispersion of toxic substances was obtained by calculating the Probit function (Pr). Based on this, percentage of probable fatality has been evaluated.

2. HAZARDOUS PROPERTIES

Ammonia has the hazardous properties of toxicity and flammability.

2.1 Toxic properties

Its Threshold Limt Value (TLV) is 25 ppm and Short Term Exposure Limit (STEL) is 35ppm. Its LC50 value for human is 8165 ppm and LCLO value is 5000ppm. The affect of ammonia inhalation at various concentrations on human body are as follows:

Degree of Hazard	Concentration in air
Upper Respiratory Tract irritation excluding Nose & Throat	100 ppm (Min)
Deep Lung Irritation	500 ppm (Min)
Coughing, Bronchial Spasms	1700 ppm
Dangerous, less than ½ hour exposure may be fatal	2000-3000 ppm
Serious oedema, strangulation, Asphyxia, rapidly fatal	5000 – 10000 ppm
Immediately fatal	10000 ppm

Splashes of liquid ammonia into eyes produce serious eye burns. The rapid penetration of ammonia into the ocular tissue may result in perforation of the cornea and even in death of the eyeball. Contact of liquid ammonia with skin results in chemical burns.

Industrial poisoning by ammonia is usually acute. Chronic poisoning is also possible, but is less common. Repeated exposure to the gas may result in chronic irritation to the eye, nose and upper respiratory tract.

2.2 Fire and Explosion Hazards

Ammonia gas is flammable, but the fire hazards are moderate. The explosive / flammable limits of the gas are from 16 to 25% by volume in air (i.e. it requires comparatively high concentration of ammonia in air before it catches fire). The explosive range is broadened by temperature and pressure higher than atmosphere. Though ammonia is combustible, it is very difficult to ignite and in the open air the flame is not stable.

3. SAFETY ANALYSIS

3.1 Scoping Analysis

The Scoping Analysis is to determine, under worst case assumptions, if there is a risk that can cause injury, death or financial impact to the public, the workers or the environment. The steps in Scoping Analysis involved:

- Identification of hazardous chemicals
- Quantities and location of these chemicals
- Process conditions under which these chemicals are present
- Type of hazards involved: Fire/Explosion, Toxic release
- Mitigation facilities in case of leakage
- Comparison for different chemicals with the above guidance
- Ascertain the need of detailed QRA

Ammonia is the most hazardous chemical, which is being used in this plant and ammonia hold up in a column and other storage vessels were identified as the major areas of hazard potential.

3.2 Hazard Identification Methodology

The next step of the QRA is the identification of all relevant hazards existing in a particular plant to form a basis for further analysis. Past experience has shown that in the case of leakage in the open air, there is no need to consider the risk of fire and explosion from an ammonia air mixture. Since the plant is in open air and the toxic limit of ammonia is below the flammable limit only the toxic release has been considered.

Techniques for hazard identification essentially fall into three categories. Out of the fundamental methods, Comparative methods and the Logic Diagram method, based on the HAZOP (Fundamental) study made for the systems in this plant, the Dow Chemical Exposure Index (Comparative) method has been utilised to identify the systems for which Consequence Analysis was carried out.

3.2.1 DOW's Chemical Exposure Index (CEI)

The CEI methodology utilises a linerarised expression for estimating air borne quantity released. The methodology provides an index that is consequence based and is independent of the frequency of events. This results in an index that is suitable for use as a screening tool for more sophisticated process hazard analysis.

As per the Guidelines

- Airborne Quantity was to be calculated based on liquid release with the scenario of rupture of process pipes, hoses, vessels etc.
- Airborne Quantity based on liquid release rate, total liquid released, amount flashed and the pool size formed was to be estimated.

Assumptions:

- All CEI calculations assume a wind speed of 5m/s and neutral weather conditions.
- Release from all scenarios are assumed to continue for at least 5 min. duration.

CEI was calculated based on the formula CEI = $655.1\sqrt{AQ}$/ERPG-2, where ERPG-2 (Emergency Response Planning Guidelines) is the maximum airborne concentration below which it is believed that merely all individuals could be exposed for up to 1 hour without experiencing or developing irreversible or other serious health effects or symptoms that could impair their abilities to take protective action. For ammonia it is 139 mg/m3 or 200 ppm.

As per Dow's CEI Guidelines, Consequence Analysis is required to be carried out for cases where CEI>200.

In this study therefore to identify the cases where CEI>200, the required hold up for release of ammonia for 5 mins. duration is calculated as about 3890 kg. After scrutinising all the 27 systems in the HAZOP study made earlier it was found that only the following equipment along with connected pipelines were having that hold up of ammonia.

3.2.2 Consequence Analysis

In this study Consequence analysis have been carried out with commercial validated software (PHAST *Professional*) for different scenarios. Effect distances have been calculated for the identified Maximum Credible Accident (MCA) scenarios and Worst Case scenarios with respect to Ammonia storages in columns and tanks. The generic data on frequency of failure of different systems have also been tabulated. LC50 value has been taken as the basis of our assessment of effect distances under weather stability class of 'F'. For ammonia the LC50 is 6164 mg/m^3 for human beings which is equivalent to 8865 ppm.

3.2.3 Probability of Fatality due to Dispersion loud

The probability of dispersion of toxic substances can be obtained by calculating the Probit function (Pr)

$$Pr = a + b \ln (C^n t)$$

where a, b and n are constants specific for particular substance, $C^n t$ is the causative factor for dispersion where C is the concentration and t is the time duration for the exposure.

From the Report "method of determination of possible damage" prepared by the Committee for the Prevention of Disasters caused by Dangerous Substances, Voorberg, The Director General of Labour generally referred as "green book", the constants for ammonia to affect the human are –

a = -15.8
b = 1
c = 2

The constants are found by extrapolating the effect of animal exposure data to the human population. The Pr has been calculated and chance of fatality of the exposed population was found to be 1% for 30 minute of exposure of a concentration of 2000ppm and 35% for a concentration of 5000ppm.

4. RESULTS

Results for the following cases studied are tabulated below.

Case-I: Release from bottom nozzles of a column: The hold up of the system is 14t ammonia and the system is about 17m above ground. The maximum and minimum sizes of the bottom nozzles are 50mm and 25mm respectively.

CEI SUMMARY:

Sr. no	SCENARIO	CEI value	Remarks
1	Full bore rupture of 50mm hose at pump suction (unloading from tanker to NH3 storage tank)	254.63	Consequence Analysis to be carried out
2	Release from largest (80NB) bottom most nozzle of ammonia tank	268.40	Consequence Analysis to be carried out
3	Release from largest (50NB) bottom most nozzle of column containing ammonia	379.60	Consequence Analysis to be carried out
4	Guillotine rupture of bottom most ammonia liquid line (50NB) of NH3 storage tank	423.90	Consequence Analysis to be carried out

RESULTS : Maximum Credible Accident Scenarios:

Sr. No	Scenario (considering 300 sec. release)	Amm. released	System pressure Kg/Cm^2	Nozzle size	Effect distance, M
I	Release from bottom nozzle of a column	14Te	20	25mm 50mm	16 M 22 M
II	Release from bottom nozzle of a tank	7Te	21	25mm 80mm	14 M 88 M
III	Release from bottom nozzle of storage tank	17.46Te	10	25mm 50mm	10 M 17 M
IV	Full bore rupture of 50mm hose at pump suction (unloading from tanker to ammonia storage tank)	6.3Te	10	50mm	66 M

Comment: the probability of failure from history for 25-80 NB pipelines is $2 * 10^{-6}$ per year

RESULTS : Worst Case Scenario :

Sr. No	Scenario	Weather stability class	Effect conc. LC_{50} mg/m^3	Effect distance,m (24 te case)	Probability of failure from history
I-A	Catastrophic failure of 24 te NH_3 storage tank - instantaneous release	2F	6164	730 M	10^{-6} per year
I-B	Catastrophic failure of 24 te NH_3 storage tank - instantaneous release	5D	Transition to Gaussian Dispersion	530 M	10^{-6} per year

Case-II: Release from bottom nozzles of a tank: The hold up of the system is 7t ammonia and the system is about 3m above the ground. The maximum and minimum sizes of the bottom nozzles are 80mm and 25mm respectively.

Case-III: Release from bottom nozzles of ammonia storage tanks: The hold up of each of the two tanks is 24t of ammonia and the system is about 2m above the ground. The maximum and minimum sizes of the bottom nozzles are 50mm and 25mm respectively.

Case-IV: Release from the ammonia road tanker during unloading: Full bore rupture of 50mm hose at pump suction. The hold of tank is 6.3 t.
Case-V: Catastrophic failure of ammonia storage tank: The hold up of each of the two tanks is 24t. (80% of capacity). The tanks are at atmospheric temperature and therefore maximum operating pressure would be 15 kg/cm^2.

5. DISCUSSION

The total plant was divided in to 27 loops as identified in the HAZOP study. Utilising the thumb rule that CEI values greater than 200 would be requiring Consequence Analysis to identify the Effect distance, the scenarios were identified. and The maximum effect distance calculated using international validated software for leakage through 80 mm bottom most nozzle of a tank is 88M for 2F weather condition and for catastrophic failure of ammonia storage tank, the effect distance is 730M and 530M respectively. In case of catastrophic failure of the storage vessels the effect concentration of LC50 would be felt beyond plant boundary (100M). Mitigation measures are to be adopted to reduce the probability of such happening and in case it happens the consequences are to be reduced by proper Emergency Response Planning and awareness to the public in the vicinity.

6. CONCLUSION

Though the probability of these types of failure is very remote, as mentioned in the result, the following precautions have been taken to avoid such failure.

(i) Complete HAZOP study of the system has been done and the recommendations have been incorporated.
(ii) Safety interlock logics have been provided in the system for safe operation, shutdown/start up conditions and emergency situations of the plant.
(iii) The ammonia storage tanks have been designed as per ASME Section-VIII Div.-1 and two safety valves have been provided for each ammonia storage tanks to avoid over pressurisation of the tanks.
(iv) Both the ammonia storage tanks have been kept with in the dyke and two pumps are available for inter tank transfer and also from the storage system to other sections during emergencies.
(v) Fire water sprinkler system with remote actuation has been provided in order to confine the spread of leakage.
(vi) Emergency Plan prepared and approved.

Based on the design and operation of the plant, probability of major leakage of ammonia due to containment failure would be a remote probability. However, if any such containment failure takes place, with prompt detection and actuation of emergency plan and the mitigation measures adopted to combat any emergency situation, the adverse consequences can be brought under control.

7. REFERENCES

1. DOW's Chemical Exposure Index Guide, AIChE,1994
2. Guidelines for Chemical Process Quantitative Risk Analysis, Centre for Process Safety, AIChE, 2000
3. Risk Analysis of six potentially hazardous industrial objects in the Rijmond Area, a Pilot Study- A Report to the Rijmond Public Authority, 1981

Reliability, Safety and Hazard
Editors: P.V. Varde et al.

FUZZY FAULT TREE ANALYSIS: REVISITED

Ashok Deshpande

Chair: Berkeley Initiative in Soft Computing (BISC)-Special Interest Group (SIG)
Environment Management Systems (EMS)
Hon. Adjunct Professor, Bioinformatics Centre, University of Pune
Former Deputy Director: National Environmental Engineering Research Institute (NEERI)

ABSTRACT

Fault tree has been one of the most effective tools for estimating the frequency of occurrence of a hazardous event in Probability Risk Assessment (PRA). Once the structure of the fault tree is established by the professional experts, the probability of hazardous event (known as a top event) is worked out using component failure data. Invariably the analysts are confronted with imprecise / inadequate data and work with rough estimates of probabilities of failure that is provided by the system designers. Under these circumstances, it is considered inappropriate to use conventional procedure for the estimation of Top Event Probability (TEP) of hazardous event. Fuzzy set theory offers a framework of analysis which could model imprecision in the input failure probabilities. The case study refers to the application of *fuzzy simulation concept* to a 15,000 MT ammonia storage tank in India using the available inter failure statistics of the basic events and compared these results with the conventional fault tree evaluation procedure. We believe that FTEP could be an improved representation of the frequency of occurrence of hazardous event over conventional TEP

Key words: Risk assessment, Fuzzy number, Fuzzy fault tree Ammonia storage tank, Process control instruments, Failure data, Fuzzy top event probability

1. INTRODUCTION

Environmental risks are inherent in design and operation of large and complex industrial units such as the nuclear power/ chemical plants. Any major failure or a series of minor failures in the system operation may lead to a disaster resulting in heavy toll of human life and devastating losses to ecology, environment and property. Growing concern over the risk of major chemical accidents have led to research on new techniques for identification, assessment and evaluation of risks for devising risk mitigation measures. Probabilistic Risk Assessment (PRA) is hereto the most widely employed technique for risk management.

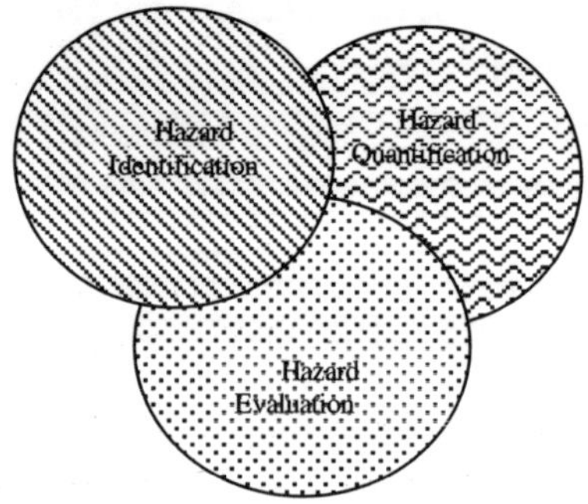

Fig. 1: Hazard Study and Risk Analysis: Conceptual Activities

The steps involved in PRA are hazard identification; hazard analysis & assessment Quantification); risk evaluation; identification of risk mitigation measures; and preparation of Disaster Management and Emergency Preparedness Plans. Fig. 1 presents the conceptual framework.

Definition of Risk

The evaluation of risk, arising out of an undesired event, comprises estimation of expected frequency of undesirable events per unit time and expected damage [1]. Then a customary definition of risk R in terms of expected frequency of occurrence and the expected damage D, is

$$R = F.D \quad (1)$$

The other possible definitions of risks include

$$R_k = F.D^k \quad (2)$$

where, k > 1 is a risk function which signifies the importance of events with larger damages. In addition to the above two, there are other possible definitions of risk [1].

Uncertainties in Probabilistic Risk Assessment (PRA)

Probabilistic Risk assessment is initiated with a qualitative phase that includes the identification of accident contributors and the consequence modeling of accidents visualizing release scenarios. Many mathematical models exist for the accidental release of hazardous chemicals using toxicity and meteorological data. Subjectivity in accidental release scenarios of hazardous chemicals can often eliminates the potential advantage offered by the existence of variety of these models Koivisto et at [2] have developed a method based on fuzzy logic for the selection of a computer model to be used in consequence analysis under specific situations. The modeling phase is then continued with the quantification of accident frequencies by using component failure and human errors data. Consequence modeling, coupled with the estimation of accident frequencies and the population at risk in turn, helps in plotting F – X and F –N curves arising out of:

- Incompleteness of accident contributors and there by the modeling of accidents and
- Inaccuracy in the quantification of risk in terms of frequency and consequences.

In the context of PRA for nuclear power plant, the estimation of the frequency of accident is termed as Level 1 and the most important phase of PRA. Table 1 presents a summary of criticisms often leveled against PRA [3].

Table 1 Criticism of PRA

Phase of Analysis	Deficiencies
1) Definition and description of system	Relevant subsystems or activities are excluded
2) Hazard identification	Description of the system does not correspond to the real life situation
3) Accident modeling	Important accident contributors or their families thereof excluded or omitted
4) Quantification of risks	Uncertainties in component failure rate or human error data Inaccuracies in accident modeling
5) Description of results	Boundaries of- and assumption in-analyses are not described Source of quantitative data are not presented

The prerequisites to risk assessment studies include event tree, development and / fault tree construction and alike wherein much of the uncertainties are introduced. Amongst the techniques used in the estimation of accident frequencies, fault tree analysis assumes a great significance. It is, therefore, essential to model uncertainties associated with Fault Tree Analysis (FTA) to enhance the credibility of risk assessment. A probabilistic approach is used in FTA and a considerable amount of uncertainty builds up in the top event probability evaluation is not adequate [4]. The probabilistic approach, as a whole lacks adequate mechanism for tackling the problem of uncertainties with respect to top event in fault tree analysis which is further demonstrated with two case studies.

Fault Tree Analysis

The technique of Fault Tree Analysis (FTA) for the estimation of the frequency of occurrence of an event was formalized in 1962 at Bell Telephone Laboratories, USA. The construction of fault tree requires professional expertise. Checking and rechecking of fault tree helps in attaining the desired accuracy. In order to describe the computational procedures, it would be appropriate to consider an illustrative example:

One of the intermediate event (Flash vessel empty) from the fault tree of atmospheric storage tank depicted in Fig.3 has been identified for describing the FTA procedure.

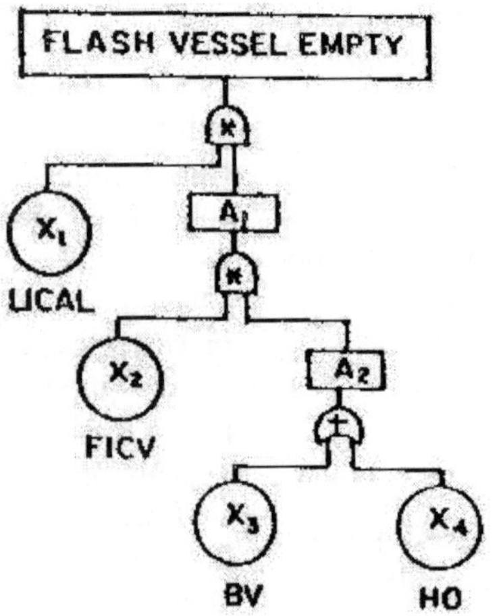

Fig. 2: Intermediate Event: Flash Vessel Empty

Minimum cut sets $\{X_1, X_2\}$, $\{X_3\}$, $\{X_4\}$ **(1)**

Event Relationship $T = X_1 \cap A_1$ **(2)**

Where $A_1 = X_2 \cap A_2$ and $A_2 = X_3 \cup X_4$ **(3)**

Top Event Probability (TEP),

$P_T = 1 - \{(1 - pX_1)\}(1\text{-}pA_1)\ (1\text{-}pA_2)\}$ **(4a)**

Where: $pA_1 = pX_2 * pA_2$ and $pA_2 = 1\text{-}\{1\text{-}pX_3\}\ \{1\text{-}pX_4\}$ **(6b)**

Fuzzy Fault Tree Analysis

The analysts are confronted with inaccurate data probability of failure of primary events occurrence of an extremely hazardous event is rare and therefore, the data rarely available. In the absence of accurate data, it becomes necessary to work with rough estimates of probabilities that are provided by the system designers. It is evident that these experts offer their subjective assessment on human error and component failure probabilities. Under these circumstances it is considered inappropriate to use conventional FTA for computing system failure probability.

Fuzzy Set Theory (FST) offers frame of analysis that could model imprecision in input failure probabilities used in FTA for the estimation of Top Event Probability (TEP)-in fuzzy set would be hereafter termed as Fuzzy Top Event Probability (FTEP) and such analysis would be termed as Fuzzy Fault Tree Analysis (FFTA). The concept of fuzzy probability in FT represents a fuzzy number, the value of which ranges between zero and one, for each primary event.

Uncertainty Number or Fuzzy Number

An interval of confidence is one way of reducing the uncertainty using lower and upper bounds. It is a practical and logical process for treating uncertainty with whatever information is available. This information can be objective or subjective [5]. Let us relate the concept of the interval of confidence to another called the level of presumption. For example, that the failure probability of certain process control instrument (eg. PSV) has been estimated from data to be in the interval: $500 * 10^{-4}$ to $5250 * 10^{-4}$ f/yr. This is termed as the interval of confidence. On the other hand, let us assume that in order to bring down the risk level to acceptable level of 10^{-6} fatality/yr, the estimated failure probability is $1000 * 10^{-4}$ failures/yr.

Two levels of confidence could be assigned to the above two situations 0 for ($500 * 10^{-4}$ – $5250 * 10^{-4}$). These two levels of confidence are in fact levels of presumption and could be represented by [0,1]. The level of presumption could also lie between 0 and 1 and could be represented as:

$$\forall \mu_1 \mu_2\ \varepsilon\ [0,1]\ (\mu_1 < \mu_2) \Rightarrow \left(\left[\alpha^{\mu 1}, \alpha^{\mu 2}\right]\right) \quad (7)$$

which means that if μ increases the interval of confidence never increases (Fig 3). The coupling between the level of presumption and interval of confidence at level will be a way defining the concept of an uncertain number or a fuzzy number. This association corresponds to the natural, often implicit, mechanism of human thinking in the subjective estimation of a value for a dimension.

Arithmetic Operations

The shape of fuzzy probabilities of primary events could take various forms. Trapezoidal representation of fuzzy number could be one such shape which has been used in this article.

Fuzzy probabilities of the basic events must be combined according to the rule of combination of fuzzy numbers to compute FFTA. Primarily, only two logical operations are required to compute FTEP: multiplication and complementation of fuzzy sets.

Multiplication

According to the extension principle [6], the computational procedure for multiplication of two fuzzy sets P_{Xi} and P_{Xj} is quite involved and therefore, Tanaka et. al. [7] have suggested an approximation of the multiplication procedure by defining as follows:

$$P_{Xi} * P_{Xi} = \{(\alpha_{i1}\ \alpha_{j1}), (\alpha_{i2}\ \alpha_{j2}), (\beta_{i2}\ \beta_{j2}), (\beta_{i1}\ \beta_{j1})\} \quad (8)$$

Multiplying the corresponding elements of fuzzy probabilities of event X_i and X_j, respectively

Complementation

The complementation of any fuzzy set P_{xi} is given as:

$$P_{\bar{X}_i} = \{(1-\alpha_{i1}), (1-\alpha_{i2}), (1-\beta_{i2}), (1-\beta_{i1})\} \quad (9)$$

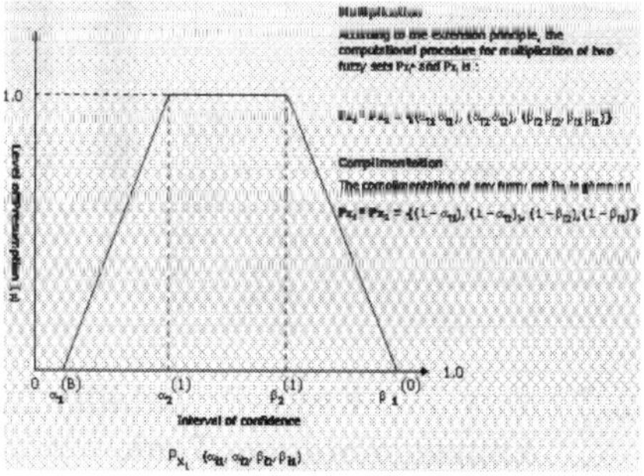

Fig. 3: Fuzzy Probability Trapezoidal Representation

2. A CASE STUDY

The chemical process industry can be broadly divided into three categories, viz. storages, process units and the transfer lines. Storage facility for hazardous chemical is by far the most vulnerable unit in any chemical process industry. Ammonia, an important intermediate in the production of fertilizer is either manufactured and stored in the pressurized storages or could be procured and then stored in atmospheric tanks near the fertilizer complexes. No item of the equipment is included in more accidents than storage tanks, probably because these are fragile and easily damaged by slight pressure and vacuum. The major failures of the storage tanks are due to over-filling, over-pressurization, sucking in, explosion/fires, floating roof accidents etc.

The case study relates to the ammonia storage tank (atmospheric storage) located at about 3 km. from the large fertilizer complex in Bombay [8]. Ammonia is stored at –33 deg. C and at atmospheric pressure. The ammonia is transferred to a tank via a 2.2 km. pipeline from jetty. The storage has been identified as the most hazardous unit as the Toxicity Index (TI.) exceeds the acceptable unit of 10 [8]. The damage distance has been computed visualizing the accident scenario of the release of ammonia from the tank. The main cause of the rupture of the tank could be attributed to over-pressurization. Majority of the storage tanks are designed to withstand guage pressure of only 8 in. (2 kpa or 0.3 psi) and will burst at about 3 times this pressure and could thus be easily damaged. Rupture of storage tanks has, therefore, been assumed as the top event in the fault tree depicted in Fig. 4.

Failure Data

To collect, store and retrieve failure information effectively for the devices of many types is an enormous task and requires an organization that is consistent with the defined objectives. In order to carry out PRA studies for chemical plants, a need for failure data of the process control instruments was recognized. The data required for failure probability computations was collected from the log books available at the fertilizer complex which maintains inter-failure statistics of the majority of process control instruments. Table 2 shows the computed values of failure probabilities of the instruments indicating upper and lower values. Data on human error probability was not available and, accordingly, international data has been used in the computation of TEP of the FTEP values.

Table 2 Failure Probabilities of Primary Events: Ammonia Plant

Instrument/Event		Failure Probability (* 10^{-4})	
	Abbreviation	Lower Bound	Upper Bound
Bypass Valve	BV	1.00	20.00
Human Operator	HO	1.00	20.00
Human Operator(overfilling)	HO	4000.00	8000.0
Flow indicator and controller valve	FICV	2500.00	7200.0
Level Indicator, controller and alarm	LICAL	3000.00	7500.0
Jetty Storage		1.00	20.0
Cooling water pump-1	CW-Pump	500.00	5250.0
Cooling water pipe	CW-Pipe	1000.00	5500.0
Heat transfer from pump		1.00	20.0
Flare fails		1.00	20.0
Compressor	Comp	8900.00	9500.0
Power Source		50.00	150.0
Breaker Switch		2500.00	4000.0
Pump		500.00	5250.0
Level Alarm	LAH	1500.00	2500.0
Level indicator	Li	5.00	20.0
Vacuum Relief Valve		500.00	5250.0
Pressure Transmitter		3500.00	7200.0
Manometer out of order		1.00	2.0
Pressure Safety Valve	PSV 2	500.00	5250.0
Pressure Alarm	PA	500.00	2500.0

A computer program was developed for FTEP in various kinds of representation of fuzzy probability of the primary event. For a particular fault tree, the program takes the lower and upper bounds of the failure probabilities of the primary events as input, and calculates FTEP. In this particular case, the nature of the membership function (μ) of the primary event is not known. The only data available is the lower and upper failure probabilities of the primary event. The membership grade function is assumed to be a trapezium as shown in Fig. 3. The two known lower and upper failure probabilities represents the points $\alpha_1^{(0)}$ and $\beta_1^{(0)}$. The other two points $\alpha_2^{(1)}$ and $\beta_2^{(1)}$ which lie on the line μ=1 are not known. Hence, they are generated as any two random numbers ranging between $\alpha_1^{(0)}$ and $\beta_1^{(0)}$ using standard pseudo-number generating technique.

In a nutshell, the program works as under:

- Generates a trapezium for each of the primary events.
- Traverses the FFT bottom up computing the intermediate results according to the **AND** and **OR** fuzzy operations (Tanaka operators) at the various interior nodes.
- Finally the trapezium (which correspond to the membership function) of the top event is generated).

3. RESULTS AND DISCUSSION

Membership grade of 1 has been considered in the study which signifies the possibility of the occurrence of the hazardous event. The Top Event Probability of hazardous event (release of ammonia from the storage tank) using FTA approach has been estimated as 1.00×10^{-5} f/yr.

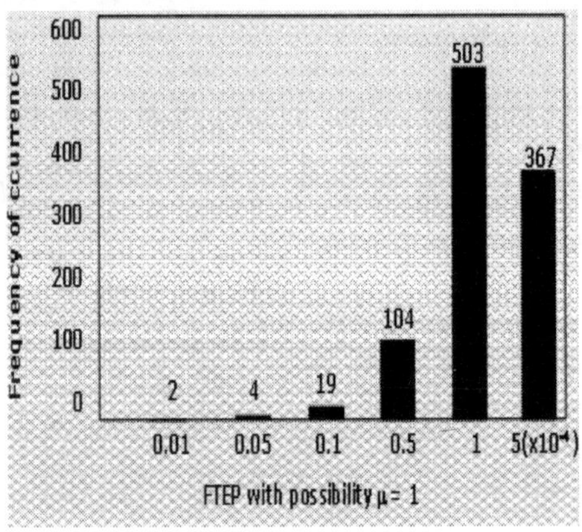

Fig. 3 Fuzzy-Top-Event-Probability for Ammonia Storage Tank

Random combinations of the fuzzy probability of primary events were used for μ = 1 and the range of TEP were estimated. The

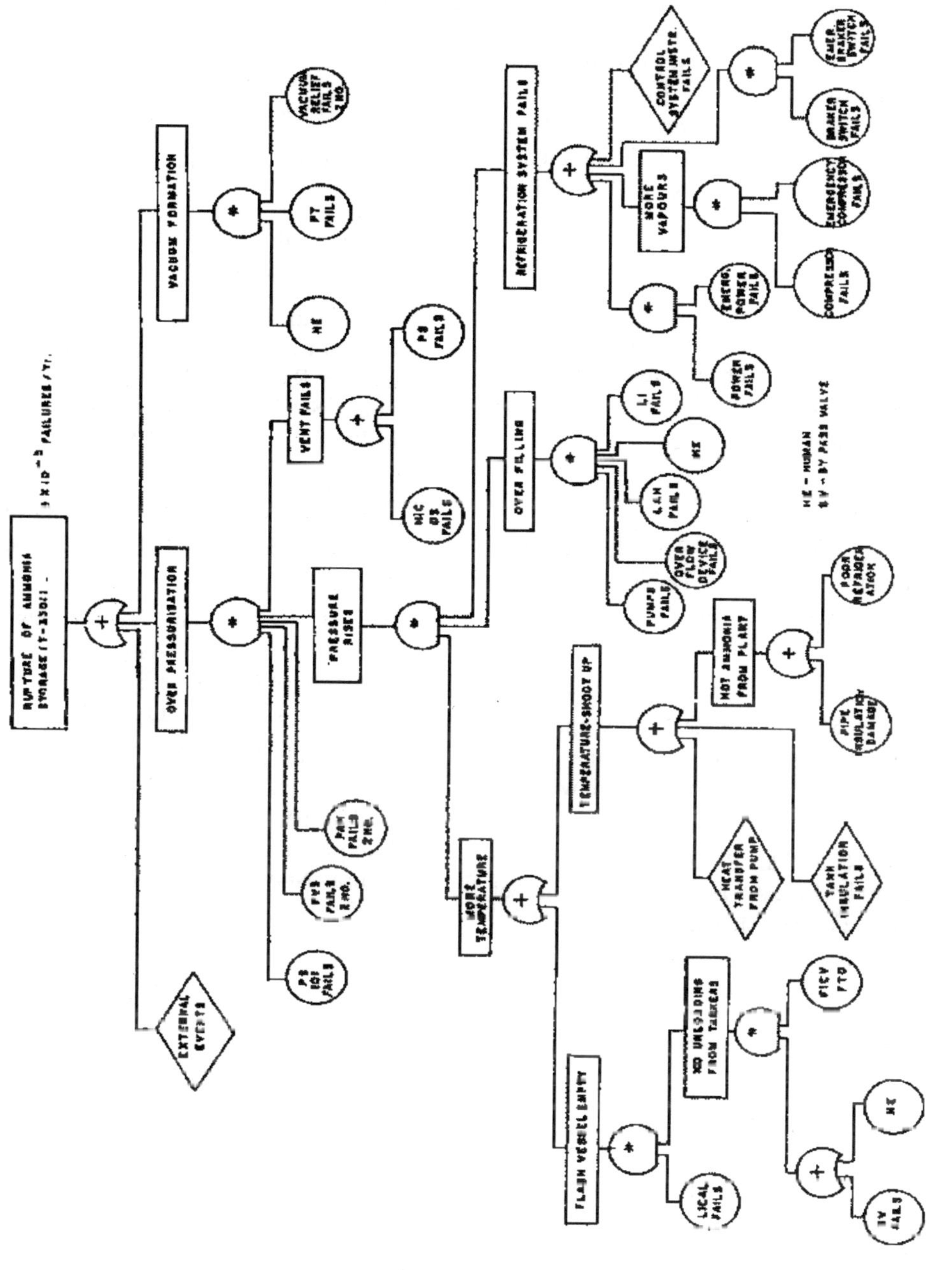

Fig. 4: Fault tree for Ammonia Storage tank

occurrence of Fuzzy Top Event Probability (FTEP) ranged between 1.75 x 10^{-6} f/yr and 7.56 x 10^{-4} f/yr. with a most likely value between 5 x 10^{-5} – 5 x 10^{-4} with possibility (μ) of 1.Frequency of occurrence of the event ranges between 104 and 503 for the most likely values with possibility one (Fig 5). Employing conventional approach, top event probability works out to 1.00 x 10-5 i.e. a value contained in the interval (1 x 10^{-6} – 5 x 10^{-4}) obtained through FFTA.

We can infer from fuzzy simulation approach that expressing possibility of occurrence of the hazardous event as a range value is a superior presentation than presenting frequency of occurrence of the event as a single value (1.00 x 10^{-5} failures/ year), using the conventional approach.

4. ACKNOWLEDGEMENT

The assistance obtained Mr. R.S. Olaniya, Mr. U.A. Deshpande and Mrs. B. Dabir is gratefully acknowledged.

5. REFERENCES

1 Norman J.,McCormik: Reliability and Risk Analysis–Methods and Nuclear Power Applications, Academic Press, New York 1981

2 Koivisto R., Kukku R. Dohral, Mand Jarv: Accidental Release–Fuzzy Selection of Adequate Formal Model.,. Journal of Loss Prevention Process., Ind 1991, Vol 4 Oct. *373-330.*

3 Suka J and Kakko., On the Problem and Future of Safety and Risk Analysis., Journal of Hazardous materials, 21 (1989), *105-124*

4 Mishra K.B. and Weber G.G: Use of Fuzzy Set Theory for Level 1 Studies in Probabilistic Risk Assessment, Journal of Fuzzy Sets and Systems 37, 1990, *139-160*

5 Kaufman and MM Gupta: Introduction to Fuzzy Arithmetic Theory and Applications, Van Nostran, New York 1984 First Edition

6 Mizumoto M. and Tanaka K.: Some Properties of Fuzzy Number in MM Gupta and RK Ragade and RR Yager, Advances in Fuzzy Set Theory and Applications, North Holland Amsterdam 1979, *153-164*

7 Tanaka H., Fan L.T. and Toguchi K.: Fault Tree Analysis by Fuzzy Probability IEEE Transactions on Reliability, 32(5), 1983. *455–457*

8 NEERI Report on Hazard Study and Quantitative Risk Assessment of RCF Complex of Chembur Bombay, Vol. 1,New Ammonia Plant (Trombay) Nov. 1990.

9 Deshpande A.W., Deshpande U. A. and P. Khanna, Fuzzy Fault Tree Analysis: A Case study, paper presented at the Second International Conference on Fuzzy logic and Neural Networks (IIZUKA 92) Japan, July 1992 17-22.

Reliability, Safety and Hazard
Editors: P.V. Varde et al.

FUZZY BASED APPROACH TO THE WELL-BEING EVALUATION OF A SUB-TRANMISSION SYSTEM

Ajit Kumar Verma[a], A. Srividya[a], and M.V. Bhatkar [b]

[a]Reliability Engineering Group, Department of Electrical Engineering,
Indian Institute of Technology Bombay, Powai, Mumbai 400 076
[b] Department of Electrical Engineering, A.C.Patil College of Engineering, Navi Mumbai 400 709

ABSTRACT

The most commonly used load point indices that are used in sub transmission system are the load point failure rate, load point outage duration, and load point annual unavailability. Reliability evaluation of generating facilities, composite generation and transmission systems and distribution systems are usually conducted independently and reliability indices obtained for each assessment are used to make decisions on the facilities concerned. The power system parameters such as failure and repair rates used in the probabilistic models basically come from historical operation records, which are necessarily limited and subject to errors. In case of new equipment, these parameters may be based on factory data or even extrapolation from other equipment types. In other words, there is considerable data uncertainty that exists in these parameters. The theory of fuzzy sets has been recognized as a potential tool to include data uncertainty in normal calculations. Methods are presented to represent data uncertainty using fuzzy numbers and fuzzy interval arithmetic. The fuzzy wellbeing of the system is quantified in terms of system health and margin state probabilities, which incorporate the agreed deterministic criteria in addition to the system risk index which describes the system inadequacy. This concept is applied to the RBTS bus 4 systems. The fuzzy wellbeing indices are determined for a supply point 2.

Keywords: Triangular fuzzy number, Load point indices, Fuzzy arithmetic, Well-being framework

1. INTRODUCTION

Almost all aspects of daily life in modern society depend on the use of electricity. It is so common that electricity consumers have come to expect electrical energy to be supplied continuously with good quality. The quality and reliability of the power supply are major concerns for an electric power utility. An electric power system in an utility firm is normally divided in to functional zones for the purpose of organizing, planning, operation and maintenance etc. The three functional zones or segments are generation, transmission and distribution. A further fundamental subdivision of an electric power system is obtained by combining the functional zones into hierarchical levels, namely hierarchical level I (HL I, for generation facilities only), level II (HL II, for composite generation and transmission) and level III, (HLIII, for all three zones are combined) [1].

The most commonly used load point indices that are used in sub transmission system are the load point failure rate, load point outage duration, and load point annual unavailability. Reliability evaluation of generating facilities, composite generation and transmission systems and distribution systems are usually conducted independently and reliability indices obtained for each assessment are used to make decisions on the facilities concerned. The power system parameters such as failure and repair rates used in the probabilistic models basically come from historical operation records, which are necessarily limited and subject to errors. In case of new equipment, these parameters may be based on factory data or even extrapolation from other equipment types. In other words, there is considerable data uncertainty that exists in these parameters. The theory of fuzzy sets has been recognized as a potential tool to include data uncertainty in normal calculations. Methods are presented to represent data uncertainty using fuzzy numbers and fuzzy interval arithmetic. The fuzzy wellbeing of the system is quantified in terms of system health and margin state probabilities, which incorporate the agreed deterministic criteria in addition to the system risk index which describes the system inadequacy. This concept is applied to the RBTS bus 4 systems. The fuzzy wellbeing indices are determined for a supply point 2.

The distribution is an important part of the total electric power system as it provides the final link between the bulk system and the customer. A distribution system can be broadly classified in to two basic areas, one that connects the bulk transmission system to the distribution substation and other that connect the distribution substation to the customer load. The first is often designated, as sub transmission system is usually the network type of distribution and used most often to supply bulk loads. Reliability indices of a distribution system indicate how adequately the different parts of the system are able to perform their intended function and satisfy customer demands. The disconnection probability of the customers can be reduced by increased investment during the planning phase, operating phase or both.

The primary customer load point reliability indices that are used in such system are the load point failure rate, the load point outage duration and load point annual unavailability (AU) [2]. These indices are fundamentally important; they do not always give a complete representation of the system behavior and the response.

2. WELL-BEING FRAMEWORK

Deterministic approaches do not and cannot account for the probabilistic or stochastic nature of the system behavior, considering the reluctance in applying the probabilistic approaches to small isolated power system. The fuzzy well being of the system is quantified in terms of system health and margin state. The fuzzy well-being analysis is described by following an analytical approach. The performance of the system is determined by using fuzzy probabilities of the system being in the healthy, marginal and at risk states. The fuzzy probabilities of these operating states can be calculated using contingency enumeration technique. The system fuzzy well-being model is shown in Figure 1 [3].

$$\tilde{P}(H) + \tilde{P}(M) + \tilde{P}(R) = 1 \quad (1)$$

Symbol "~" is used to represent the fuzzy term.

In the healthy state there is a sufficient margin with the specified criteria. In the marginal state the system is not in any difficulty but does not have sufficient margin to meet the specified criterion.

The fuzzy probability of healthy state can be evaluated using a contingency enumeration technique. Fuzzy Annual unavailability of a particular load or a supply point is an expected time in a year that loads or supply point will be on outage. It can be evaluated using conventional method, as a risk index.

The fuzzy risk index is can be evaluated by using modified equation (2)

$$\tilde{P}(R) = \text{Fuzzy (AU)}/8760 \quad (2)$$

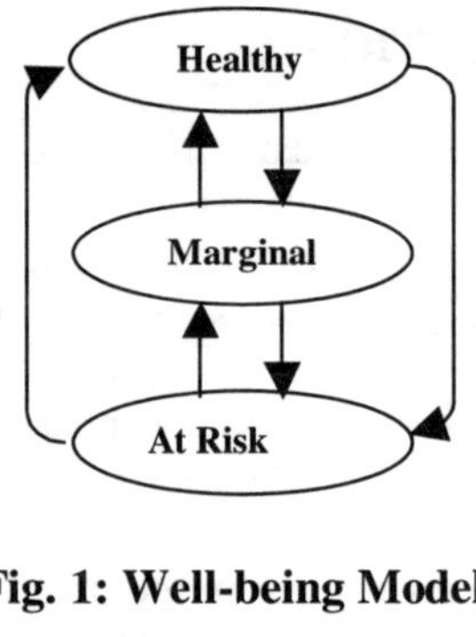

Fig. 1: Well-being Model

3. FUZZY SET THEORY

Fuzzy set theory can be used for complete and effective modeling of uncertainty and imprecision in human reasoning with the use of linguistic variables with membership functions. Fuzzification offers superior expressive power, greater generality, and an improved capability, to model complex problems at a low solution cost [4]. The use of FST in reliability engineering is gaining popularity. In order to account for aging, wear and any other contributing factors, qualitative description of condition can be used to modify original calculation of reliability using FST. In reliability analysis, it is often difficult to evaluate the failure probabilities of components from past occurrences because the environment of the systems changes. Furthermore, it might be necessary to consider possible failure of components even if they have never failed before. Therefore, FST can be used to define possibility of failure [5]. The notion of the possibility of failure is more predictive than that of the probability of failure.

3.1 Using Triangular Fuzzy Number

Triangular Fuzzy Number (TFN) is used to represent failure of the components as shown in the Figure 2 [5]. If the given failure rate is a single point estimated value based on some historical operation records, uncertainty that exists in the failure rate can be quantified using the fuzzy number. For example, in Figure 2 the given failure rate, f is quantified using a TFN with triplets (sf, f, $s'f$) where s and s′ represent certain percentage values. f is the good representation of the failure rate.

3.2 Membership functions for failure rate

$$M_f(x) = 0, \quad x < sf, x > s'f \tag{3}$$

$$= \frac{x - sf}{f - sf}, \quad sf \le x \le f$$

$$= 1, \quad x - f$$

$$= \frac{s'f - x}{s'f - f}, \quad f < x \le s'f$$

At presumption level α, interval of confidence of failure rate f is

$$f_\alpha = [sf + \alpha(f - sf), s'f - \alpha(s'f - f)]$$

$$= [sf + \alpha f(1-s), \quad s'f + \alpha f(1-s')]$$

$$= [f_{\alpha L}, f_{\alpha R}]$$

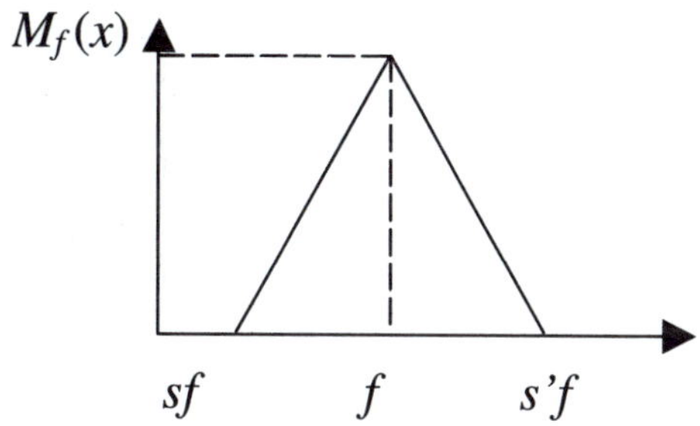

Fig. 2: Triangular Fuzzy Number

The state space diagram can be used to evaluate the reliability indices. An alternative method to the state space diagram is to use the following approximate expressions [2] can be modified evaluating the fuzzy failure rate (λ), outage duration (r), and fuzzy annual outage time or fuzzy unavailability (U) for the different contingencies.

$$\lambda_{pp} = \tilde{\lambda}_1 \tilde{\lambda}_2 (r_1 + r_2) \tag{4}$$

$$r_{pp} = r_1 r_2 / (r_1 + r_2) \tag{5}$$

$$U_{pp} = \tilde{\lambda}_{pp} * r_{pp} \tag{6}$$

4. DECRIPTIONS OF THE SUB TRANSMISSION SYSTEM

The deterministic criterion may be the presence/outage of transmission line and a transformer. The healthy states are determined using a contingency enumeration approach. In the given system B1, B2, B3, B4, are the system load buses. The fuzzy failure rate and crisp repair time of the components are given in table1. The buses B1, B3, B5 are on the high voltage side and the B2, and B5 are on the low voltage side. The single line diagram of the RBTS bus-4 sub transmission network is shown in Figure 3.

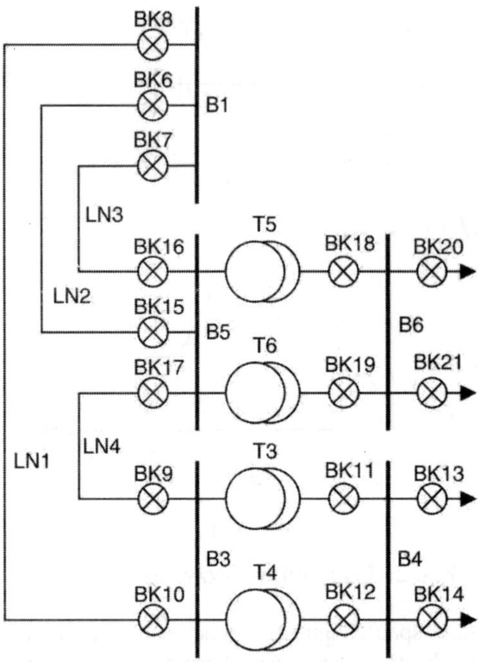

Fig. 3: Single Line Diagram

4.1 Failure rate evaluation using Fuzzy Sets

As shown in Figure 2 the given failure rate, f is quantified using a TFN with triplets (sf, f, $s'f$) where $s = 90\%$ and $s' = 110\%$ .The system lead-time is also considered as the fuzzy number and quantified using a TFN with triplets. Using the fuzzy arithmetic operations the Total failure rate is evaluated. The crisp value for the repair time is considered as given in the table in hours. Fuzzy Reliability data for the RBTS bus 4 sub transmission system is given in Table 1.

5. ILLUSTRATION AND RESULTS

The fuzzy total failure rate (λ_{total}), total outage duration (r_{total}), and fuzzy total annual outage time or fuzzy unavailability (U_{total}) can be evaluated using the following expressions with fuzzy notations.

$$\lambda_{total} = \sum \tilde{\lambda} pp \text{ Failure/year} \tag{7}$$

$$U_{total} = \sum \tilde{U}_{pp} \text{ Hours/year} \tag{8}$$

$$r_{total} = \frac{\sum \tilde{U}_{pp}}{\sum \tilde{\lambda}_{pp}} \text{ Hours} \tag{9}$$

where $\tilde{U}_{PP}$ and $\tilde{\lambda}_{pp}$ are fuzzy arithmetic addition of the fuzzy unavailability and fuzzy failure rate for all the possible contingencies.
Fuzzy Annual unavailability of the sub transmission system [6] is evaluated using the fuzzy failure rates, repair times of related contingencies and then fuzzy probability as lower and upper bounds of risk state with respect to the system data are found to be 1.12047E-05 and 1.14123E-05. Failure rate and Fuzzy annual unavailability in terms of lower and upper bound for RBTS bus 4 of supply point 2 for the different contingencies are given in Table 2 (a and b).

Table 1 Fuzzy Reliability data for the RBTS bus 4 sub transmission system

	fuzzy failure rate (TFN)		
Component	L	M	R
T 1..6	0.0135	0.015	0.0165
BR 6,..,10,15..,17	0.0018	0.002	0.0022
BR 1,..5,11..14,18…21	0.0054	0.006	0.0066
B 1,3,5	0.0009	0.001	0.0011
B 2,4,6	0.0009	0.001	0.0011
LN 1	0.414	0.46	0.506
LN 4	0.414	0.46	0.506
LN 2,3	0.621	0.69	0.759

Table 2 (a) Fuzzy annual unavailability of RBTS bus 4 of supply point 2

	Failure rate (failures/yrs)	
Contingency	Lower	Upper
B1	0.0009	0.0011
B3	0.0009	0.0011
B4	0.0009	0.0011
T4-T3	6.24E-07	9.32E-07
T4-BR11	1.58E-07	2.36E-07
BR12-T3	1.58E-07	2.36E-07
BR12-BR11	2.66E-08	3.98E-08
LN1-B5	4.25E-07	6.35E-07
LN1 -LN4	0.000313052	0.000467646
Total	0.003014444	0.003769726

Table 2 (b) Fuzzy annual unavailability of RBTS bus 4 of supply point 2

Contingency	Fuzzy AU(Hr/Yr)	
	Lower	Upper
B1	0.0018	0.0022
B3	0.0018	0.0022
B4	0.0018	0.0022
T4-T3	4.68E-06	6.99E-06
T4-BR11	4.99E-07	7.46E-07
BR12-T3	4.99E-07	7.46E-07
BR12-BR11	5.33E-08	7.96E-08
LN1-B5	6.80E-07	1.02E-06
LN1 -LN4	0.001246811	0.00186252
Total	0.006653224	0.008472099

Fuzzy probability of healthy state is evaluated using the criteria of the presence /outage of line 4. Line 4 is present. It has been observed that the fuzzy health for this criterion gives the lower and upper bound as 0.999941084 and 0.999949897. The fuzzy marginal values are estimated as 3.8691E-05 and 4.77117E-05 as shown in Table 3. Annual unavailability of the given system is evaluated using the failure rates, repair times of related contingencies and then probability of risk which corresponds to the base values of the system data is found to be 8.6231E-07. The fuzzy probability of health and margin in terms of lower and upper bound are evaluated for the different criterion. The evaluated results for the presence of T3, T3 and L4 are given in Table 4 and table 5 respectively.

Table 3 Fuzzy probability of health, margin and risk for RBTS supply point 2 for Line 4

	Lower bound	Upper bound
Total fuzzy failure rate	0.1383009	0.1487014
Fuzzy AU(H)	0.4389053	0.5161079
Fuzzy health	0.9999411	0.9999499
Fuzzy margin	3.87E-05	4.77E-05

Table 4 Fuzzy probability of health, margin and risk for RBTS supply point 2 for T3

	Lower bound	Upper bound
Total fuzzy failure rate	0.0944135	0.0993683
Fuzzy AU(H)	0.3032475	0.3548635
Fuzzy health	0.9999595	0.9999654
Fuzzy margin	4.05E-05	3.46E-05

Table 5 Fuzzy probability of health, margin and risk for RBTS supply point 2 for L4 and T3

	Lower bound	Upper bound
Total fuzzy failure rate	0.1305	0.1595
Fuzzy AU(H)	0.639	0.781
Fuzzy healthy state	0.9999108	0.9999271
Fuzzy margin state	7.22E-05	8.82E-05

6. CONCLUSION

The failures of the system components are fuzzified using fuzzy numbers. Fuzzy forced outage rates of the components are obtained from these fuzzy numbers using fuzzy interval analysis. This paper illustrates the system wellbeing analysis by fuzzy set theory. The wellbeing of the system is quantified in terms of system health and margin states, which incorporate the agreed deterministic criteria; in addition to the system risk index that describes the system inadequacy. A technique based on utilizing fuzzy system wellbeing framework for load point indices is illustrated in this paper by application to RBTS Bus-4 supply point 2. The paper has presented a fuzzy set theory to assess the adequacy of electric sub transmission system. The concept of fuzzy number and interval of confidence has been used to incorporate uncertainty in the calculation of wellbeing indices. The methodology can provide better reliability measure in terms of confidence interval at different level of fuzziness.

7. REFERENCES

1. Billinton R. and Allan R. (1988), *Reliability Assessment of Large Electric Power Systems* Kluwer Academic Publishers.
2. Billinton R. and Allan R. (1984), *Reliability evaluation of power system*, Plenum Press, New York.
3. Billinton R and Mahmud Fotuhi-Firuzabad. (1994), *A basic framework for generating system operating health analysis* , IEEE Transactions on Power System, **9(3)** , 1610-1617.
4. Zimmermann H. .J.(1991), *Fuzzy Set Theory and its Application*, Kluwer academic publishers, 2nd Edition.
5. Kaufmann A. and Gupta M. M.(1985), *Introduction to Fuzzy Arithmetic*, Van Nostrand Reinhold Company.
6. Goel L and Gupta R (2001), *Well-being indices of a sub transmission system,* Computers and Electrical Engineering, **27(4),** 145-15.
7. Billinton R. and Karki R. (1999), *Capacity reserve assessment using well-being analysis*, IEEE Transaction on Power System, **14(2)**, 433-438.
8. Goel L and Billinton R. (1995), *Pertinent factors in the adequacy evaluation of over all electric power system,* IEE proc. Generation, Transmition, Distribution , 142.

Reliability, Safety and Hazard
Editors: P.V. arde et al.

SELECTING OPTIMAL CONDITION MONITORING TECHNIQUE USING FUZZY MULTI-CRITERIA DECISION MAKING METHODS: A CASE STUDY

A.K. Verma [1], *A. Srividya* [1] *and Rajesh S. Prabhu Gaonkar* [2]

[1] Reliability Engineering Group, Indian Institute of Technology Bombay
Mumbai 400 076

[2] Mechanical Engineering Department, College of Engineering
Goa, Farmagudi, Ponda, Goa 403 401

ABSTRACT

Selecting optimal condition monitoring technique under fuzzy environment is not a trivial assignment. This paper presents a case study of condition monitoring technique selection under multiple criteria's that are fuzzy in nature. The case study is carried out on condition-based maintenance of turbine and an expert opinion of ten engineers working on turbine maintenance is incorporated. Three most significant condition monitoring techniques i.e. vibration analysis, lube oil analysis and endoscopic examination and nine maintenance decision criteria's have been considered. Fuzzy linguistic terms have been used to rate and weigh the maintenance decision criteria's. Triangular fuzzy sets/ numbers are assumed to represent linguistic terms/ variables and fuzzy set operations have been carried out using $\alpha -$ cut method. Most appropriate/ optimal technique selection procedure is demonstrated using four different fuzzy multi-criteria decision making methods. The basic method used is rating and ranking method using fuzzy set theory wherein ratings of alternatives/ strategies is determined first and then ranking is carried out to decide the optimal strategy. Other methods i.e. ranking fuzzy sets using *cardinal utilities*, by *maximizing and minimizing sets* and suitability set method are also established to confirm our choice of optimal maintenance strategy.

Keywords: Multi-criteria decision-making, Condition monitoring techniques, Fuzzy sets, Rating and ranking, Suitability set, Dominance relation

1. INTRODUCTION

Decision analysis is a technique that can be used to help decision-maker who is faced with a problem involving uncertainty. Much of the decision-making in the real world takes place in an environment in which the goals, the constraints and consequences of possible actions are not known precisely. To deal quantitatively with precision, the concepts and techniques of probability theory, decision theory and information theory are usually employed. In doing so, we consider the precision that is related to randomness only. Other major source of imprecision in many decision processes is vagueness or fuzziness. By fuzziness, we mean a type of imprecision which is associated with fuzzy sets. In many cases, it has become more and more obvious that comparing the different alternatives and determining the optimal solutions in decision problems cannot be done using single criterion. This has led to multi-criteria/ multi-attribute decision- making (under fuzziness).

In condition-based maintenance, the critical components are monitored for deterioration or health and the maintenance is carried out just before the failure occurs. The typical condition monitoring techniques are: vibration analysis, oil or debris analysis, eddy current testing, conductivity related methods, thermal monitoring technique, corrosion monitoring, endoscopic examination etc. Now the basic problem is to choose appropriate technique given the information about various criteria's that are fuzzy in nature or expressed in linguistic terms. The aim here is to obtain the optimal technique with highest degree of desirability with respect to all relevant criteria's. This problem is multi-criteria decision making problem that is tackled by various researchers [1] working in the area of decision-making in a fuzzy environment.

We use various methods from the literature to arrive at an optimal decision. These methods are: rating and ranking of fuzzy sets by Bass and Kwakernaak, ranking fuzzy sets using *cardinal utilities* by Baldwin and Guild, ranking fuzzy sets by maximizing and minimizing sets by Chen, suitability set and dominance relation etc. The rating of alternative with respect to criteria is fuzzy and is given by grade membership function. The ratings and weights have expressed by linguistic terms (variables) i.e. as fuzzy sets. These ratings and weights in linguistic terms are obtained from ten experts through a case study on turbine maintenance. The grade membership for both the variables is considered as triangular fuzzy numbers and all the mathematical operations are done using $\alpha-$cut method.

2. FUZZY MULTI-CRITERIA DECISION MAKING

The basic problem to choose between set of alternatives, given some decision criteria's. Let $A=\{a_i\};\ i=1,2,.....n$ be the set of decision alternatives and $C=\{c_j\};\ j=1,2,....m$ be the set of (fuzzy) criteria's according to which the desirability of an alternative is to be judged. The aim here is to obtain the optimal alternative with highest degree of desirability with respect to all relevant criteria's. This is multi-criteria decision making problem that is attempted by researchers [2,3] working in the area of decision-making in a fuzzy environment.

2.1 Rating and Ranking Method

Kahne [3] developed a technique of rating and ranking for stochastic models, which further on extended to fuzzy set theory by Bass and Kwakernaak. As per Kanhe's model, let the rating of the alternative *i* with respect to decision criteria *j* be r_{ij}; $i=1,2,...n$; $j=1,2,...m$. The relative importance of decision criteria *j* is called weight and is denoted as ω_j. Therefore the ranking of the alternatives is performed according to their rank that is given by:

$$R_i=\frac{\sum_{j=1}^{m}\omega_j\cdot r_{ij}}{\sum_{j}\omega_j}\quad ;i=1,2,......n \tag{1}$$

The optimal alternative is the one for which value of R_i is maximum. Bass and Kwakernaak [3,4] suggested an extension to the above model that has become classic in this area. It consists of two phases:

Phase 1:*Determination of ratings of alternatives*: Let $A=\{a_i\};\ i=1,2,.....n$ be the set of decision alternatives and $C=\{\tilde{c}_j\};\ j=1,2,.....m$ be the criteria's that are fuzzy in nature. $\tilde{R}_{ij}$ be the fuzzy rating of alternative *i* with respect to criteria *j* and $\tilde{W}_j\in R$ is the weight or importance of criteria *j*. The rating of alternative *i* with respect to criteria *j* is fuzzy and is given by grade membership function $\mu_{R_{ij}}(r_{ij})$. Also, the relative importance (weight) of criteria *j* is given by a fuzzy set $\tilde{W}_j$ with grade membership function $\mu_{W_j}(\omega_j)$. The evaluation of alternative a_i is therefore a fuzzy set, which is computed using $\tilde{R}_{ij}$ and $\tilde{W}_j$:

$$g(z)=g(\omega_1,.....\omega_n,\ r_{i1},.....r_{in})=\frac{\sum_{j=1}^{m}\omega_j\cdot r_{ij}}{\sum_{j=1}^{m}\omega_j} \tag{2}$$

Membership function of μ_{zi} is then defined as:

$$\mu_{z_i}(z) = \min\{\min_{j=1}^{n}(\mu_{W_j}(\omega_j)), \min_{k=1}^{n}(\mu_{R_{ik}}(r_{ik}))\} \quad (3)$$

The final rating is therefore $\tilde{R}_i = \{(r, \mu_{R_i})\}$ given by a membership function $\mu_{R_i}(r) = \sup_r \mu_{z_i}(z)$.

Phase 2: *Ranking*: Once final ratings have been obtained, ranking or rank ordering is done in phase 2. Bass and Kwakernaak established a measure to distinguish the 'preferable alternatives' from each other and rank them. If ratings $\tilde{R}_i$ are fuzzy, then preference set $\tilde{P}_i = \{(p, \mu_{P_i}(p))\}$, which is also a fuzzy set, can be obtained using:

$$\mu_{P_i}(p) = \sup_r \mu_R(r_1, \ldots, r_m) \quad (4)$$

This fuzzy set can be effectively used to judge the degree of preferability of an alternative over all other alternatives.

2.2 Ranking Fuzzy Sets using '*Cardinal Utilities*'

Baldwin and Guild [3,5] define a relation $\tilde{P}_{ij} = \{(r_i, r_j), \mu_{P_{ij}}(r_i, r_j)\} \ ;i \neq j$ with membership function as $\mu_{P_{ij}}(r_i, r_j) = f(r_i, r_j)$.

This function expresses the 'difference' between the ratings of two fuzzy sets. Such set is defined as $\tilde{O}(x_i) = \{x_i, \mu_O(x_i)\}$ with membership function:

$$\mu_O(x_i) = \sup_{r_i, r_j} \min\{\mu_{R_i}(r_i), \mu_{R_j}(r_j), \mu_{P_{ij}}(r_i, r_j)\}$$

This equation expresses the degree to which alternative x_i is preferable to its best rival alternative. $\tilde{O}(x_i)$ correspondes to *max-min composition* of $\tilde{R}_i$, $\tilde{R}_j$ and $\tilde{P}_{ij}$. Without going into the mathematical particulars, we present the solution of the problem as:

$$\mu_O(x_i) = \min_j \{\tilde{\mu}_j\} = \left[\frac{\delta - \alpha}{1 + (\delta - \gamma) + (\beta - \alpha)}\right] \quad (5)$$

where the parameters of the above equations are depicted in the Figure 1.

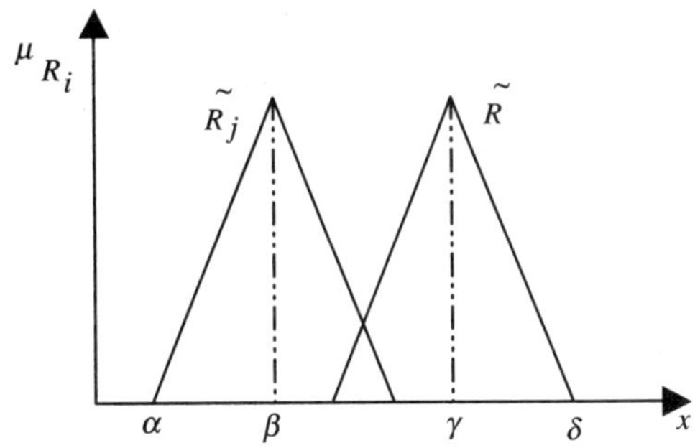

Fig. 1: Membership Functions of Ratings $\tilde{R}_i$, $\tilde{R}_j$

2.3 Ranking Fuzzy Sets by '*Maximizing and Minimizing Sets*'

A modification of ranking approach is suggested by Jain [3], which is further modified by Chen [3,6] for better discrimination of the ratings. He defines 'maximizing set ($\tilde{M}$)' and 'minimizing set ($\tilde{N}$)' whose membership functions are as follows:

$$\mu_M(r) = \left[\frac{r - r_{min}}{r_{max} - r_{min}}\right]^n \quad (6)$$

$$\mu_N(r) = \left[\frac{r_{max} - r}{r_{max} - r_{min}}\right]^n \quad (7)$$

where $r \in [r_{min}, r_{max}]$ is the real interval and n = 1 for linear, n = 2 for risk prone and n = 0.5 for risk averse membership functions. To perform the ranking and get fuzzy optimal alternative obtain:

$$O(x_i) = \frac{R(x_i) + 1 - L(x_i)}{2} \quad (8)$$

where $R(x_i) = \sup_r \min\{\mu_{R_i}(r), \mu_M(r)\}$ and $L(x_i) = \sup_r \min\{\mu_{R_i}(r), \mu_N(r)\}$.

2.4 Suitability Set and Dominance Relation

First rating matrix and weights for criteria's are determined and then the suitability set for each alternative is computed. The suitability set for each alternative *i* is assumed as fuzzy weighted sum of ratings that is considered here as an appropriate measure of suitability [7]:

$$S_i = \sum_j \omega_j \cdot r_{ij} d \tag{9}$$

For choosing an alternative, one needs to know the concept of dominance. Dominance (δ) of the normal convex fuzzy set *a* over the normal convex fuzzy set *b* is defined by:

$$\delta(A,B) = \vee_x (\mu_{\leq A}(x) \wedge \mu_B(x))$$

where $\vee$ denotes maximum operation, $\wedge$ denotes the minimum operation and $\leq A$ is the fuzzy set '*less than or equal to A*' formed from *A* by setting :

$$\mu_{\leq A}(x) = 1.0 \quad ; x < x^*$$
$$= \mu_A(x) \quad ; x \geq x^* \tag{10}$$

where x^* is the leftmost (lowest) value of *x* for which $\mu_A(x) = 1.0$. Though we get one alternative using dominance relation, this is not sufficient to make the choice since it does not take into account the shapes of suitability sets. To include the shapes of membership functions also and to confirm about the optimal technique, we define the difference function as below:

$$g_k(S_1, S_2, \ldots, S_n) = S_k - \frac{\sum_{\substack{i=1 \\ i \neq k}}^{n} v(x_i) \cdot S_i}{\sum_{\substack{i=1 \\ i \neq k}}^{n} v(x_i)} \tag{11}$$

wherein vector, $v(x_i) = min_j \{DR(i,j)\}$ and *k* corresponds to a position in $v(x_k) = 1$. The set obtained using the above difference function gives the degree of preference of the chosen alternative over the other alternatives. This procedure is particularly useful when two or more suitability sets dominate the other sets.

3. CASE STUDY

A ase study is carried out on the maintenance of turbine. The purpose of the case study is to select (also rank) the optimal technique out of three alternative techniques: vibration analysis (A_1), lube oil/ debris analysis (A_2), endoscopic examination (A_3) and nine decision criteria's namely: investment cost (C_1), operating costs (C_2), accuracy of the technique (C_3), repeatability of the instruments used (C_4), ease of use (C_5), environmental restrictions (C_6), technical expertise requirement (C_7), ease of maintenance (C_8), ease of mounting (C_9). Ten expert engineers judged the techniques as per the criteria's. The ratings and weights have expressed by linguistic terms/ variables (i.e. as fuzzy sets). The grade membership for both the variables are considered as triangular fuzzy number (TFN). Figure 2 and Figure 3 shows TFN representations of linguistic variables related to rating and weights on the scale [0,1].

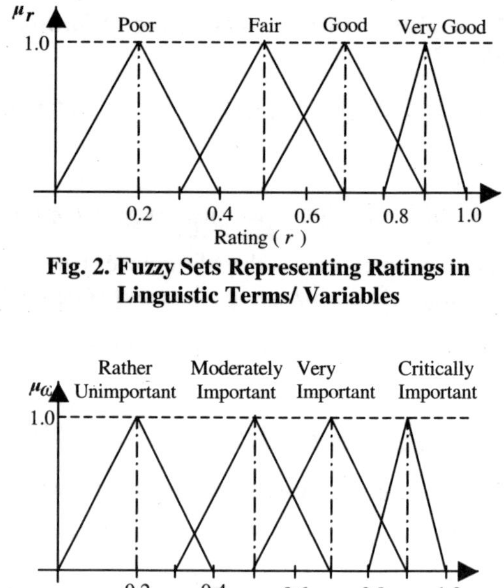

Fig. 2. Fuzzy Sets Representing Ratings in Linguistic Terms/ Variables

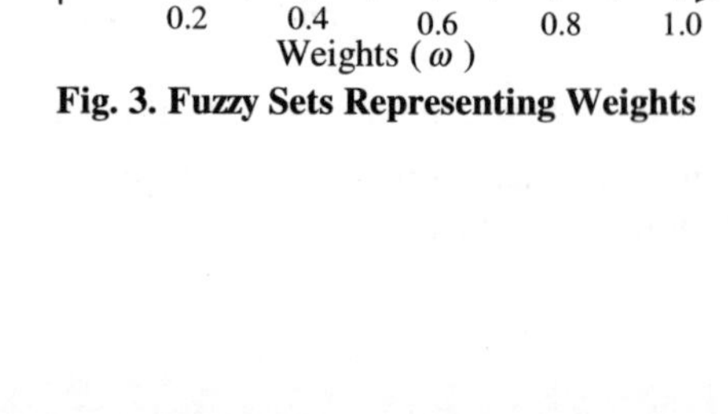

Fig. 3. Fuzzy Sets Representing Weights

Means of combining expert's judgmennt include averaging method, averaging method with feedback [8], combining method for symmetric triangular fuzzy sets [9]. The expert judgment of the 10 experts is combined using averaging method. The combined resulting fuzzy set is then required to be interpreted in terms of linguistic variable defined earlier. To get this, resultant fuzzy set is to be 'mapped' on to the nearest pre-defined fuzzy set. The resultant set is then mapped on to the nearest predefined set using distance measures (Hamming distance, Euclidean distance, distance measure proposed by Tran et al.)[10,11] and credibility score method [12]. The final ratings and weights are given in Table 1.

Table 1 Weights of Decision Criteria's and Rating of the Techniques with respect to Decision Criteria's

Decision Criteria's	Weights	Alternatives/ Techniques		
		A_1	A_2	A_3
C_1	Moderately Impo	Fair	Fair	Good
C_2	Very Important	Good	Good	Good
C_3	Critically Important	Good	Very Good	Good
C_4	Very Important	Good	Good	Very Good
C_5	Very Important	Good	Fair	Fair
C_6	Very Important	Good	Fair	Fair
C_7	Very Important	Good	Good	Good
C_8	Very Important	Good	Fair	Fair
C_9	Very Important	Fair	Good	Fair

4. RESULTS AND DISCUSSION

Ealuation of Eqn. 2 involves arithmetic operations like addition, multiplication and division of fuzzy sets [8,9,13]. These operations are to be performed with the help of interval of confidence at each presumption level α. The different grade memberships (α-cuts) of ratings of the three techniques are computed and the membership functions of the final ratings obtained using Bass and Kwakernaak method is shown in Figure 4. It is a bit evident from Figure 4 that vibration analysis technique slightly dominant over other two techniques. Still, one needs to investigate the case further to obtain the optimal technique. To do this, one more decisive factor that is given in Eqn. 4 is used to distinguish between 'preferable alternatives'. We compute the preference sets to investigate preferability of vibration monitoring alternative over the others. The preferability data for different grade membership values is obtained and plotted in Figures 5. From the results it can be observed that vibration analysis technique is the optimal technique under fuzzy decision criteria's considered in the case study under investigation.

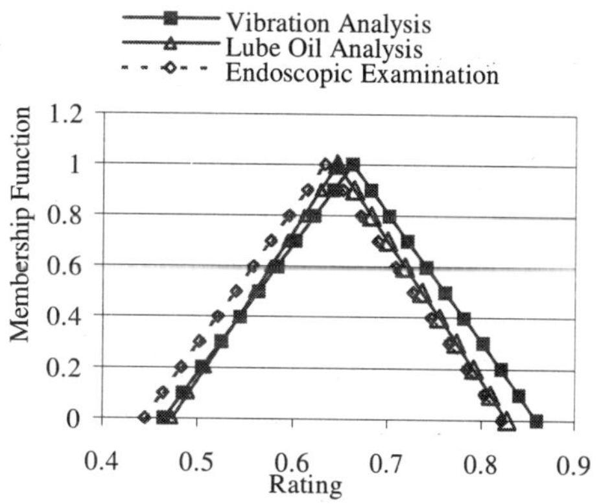

Fig. 4: Final Ratings of Techniques

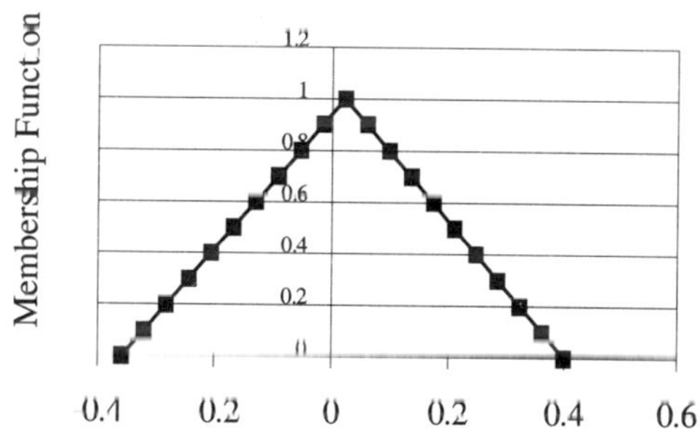

Fig. 5: Preferability of Technique 1 over 2 and 3

Now, we consider our problem of ranking the techniques using 'cardinal utilities' method by Baldwin and Guild. Optimal maintenance technique is the one that has maximum $\tilde{\mu}_j$ value obtained through Eqn. 5. We compare all three techniques i.e. vibration analysis (with rating

R_1), lube oil analysis (with rating R_2), endoscopic examination (with rating R_3) by this method considering two techniques at a time. From Table 2, one can observe that $\tilde{\mu}_j$ of technique 1 is more than $\tilde{\mu}_j$ of technique 2 and 3, i.e. rating of R_1 is greater than rating of R_2 and R_3. This means vibration analysis technique outperforms other two techniques in optimality.

Table 2. Values of $\tilde{\mu}_j$ (*Cardinal Utilities* Method)

Parameters of Eqn. 5	R_1 and R_2	
	R_1	R_2
α	0.473913	0.465217
β	0.646032	0.661905
γ	0.661905	0.646032
δ	0.86	0.8275
μ_j	**0.28177**	**0.26288**

Parameters of Eqn. 5	R_1 and R_3	
	R_1	R_3
α	0.445652	0.465217
β	0.633333	0.661905
γ	0.661905	0.633333
δ	0.86	0.82125
μ_j	**0.29900**	**0.25714**

Parameters of Eqn. 5	R_2 and R_3	
	R_2	R_3
α	0.445652	0.473913
β	0.633333	0.646032
γ	0.646032	0.633333
δ	0.8275	0.82125
μ_j	**0.27889**	**0.25539**

We assess our problem of optimal technique selection further using Chen method by considering two techniques at a time. The membership functions of the three fuzzy sets and maximizing and minimizing sets are as below and plotted in Figures 6, 7 and 8.

$$\mu_{R_1}(r) = \frac{r-0.465217}{0.196688} \quad ;0.465217 \le r \le 0.661905$$
$$= \frac{0.86-r}{0.198095} \quad ;0.661905 < r \le 0.86$$
$$= 0 \quad ;\textit{otherwise}$$

$$\mu_{R_2}(r) = \frac{r-0.473913}{0.172119} \quad ;0.473913 \le r \le 0.646032$$
$$= \frac{0.8275-r}{0.181468} \quad ;0.646032 < r \le 0.8275$$
$$= 0 \quad ;\textit{otherwise}$$

$$\mu_{R_3}(r) = \frac{r-0.445652}{0.187681} \quad ;0.445652 \le r \le 0.633333$$
$$= \frac{0.82125-r}{0.187917} \quad ;0.633333 < r \le 0.82125$$
$$= 0 \quad ;\textit{otherwise}$$

Maximizing and minimizing sets for $n = 1$:

Considering R_1 and R_2:

$$\mu_M(r) = \frac{r-0.465217}{0.394783} \quad ;0.465217 \le r \le 0.86$$
$$= 0 \quad ;\textit{otherwise}$$

$$\mu_N(r) = \frac{0.86-r}{0.394783} \quad ;0.465217 \le r \le 0.86$$
$$= 0 \quad ;\textit{otherwise}$$

Considering R_1 and R_3:

$$\mu_M(r) = \frac{r-0.445652}{0.414348} \quad ;0.445652 \le r \le 0..86$$
$$= 0 \quad ;\textit{otherwise}$$

$$\mu_N(r) = \frac{0.86-r}{0.414348} \quad ;0.445652 \le r \le 0.86$$
$$= 0 \quad ;\textit{otherwise}$$

Considering R_2 and R_3:

$$\mu_M(r) = \frac{r-0.445652}{0.381848} \quad ;0.445652 \le r \le 0.8275$$
$$= 0 \quad ;\textit{otherwise}$$

$$\mu_N(r) = \frac{0.8275-r}{0.381848} \quad ;0.445652 \le r \le 0.8275$$
$$= 0 \quad ;\textit{otherwise}$$

Ranking values for all techniques are given in Table 3. When R_1 and R_2 combination is considered, optimal ranking value $O(x_i)$ of R_1 is more than R_2 and in R_1 and R_3 combination also, $O(x_i)$ of R_1 is more than R_3. Optimal ranking value of R_2 is more than R_3 in R_2 and R_3 combination. Therefore vibration analysis is an optimal technique as per Chen method.

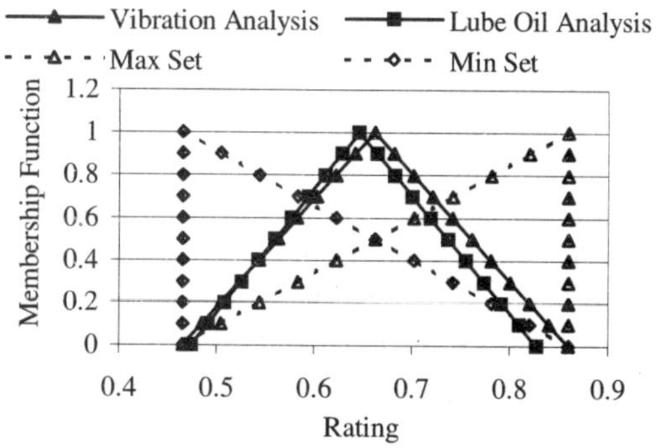

Fig. 6: Final Ratings of Technique 1 and 2 (Chen Method)

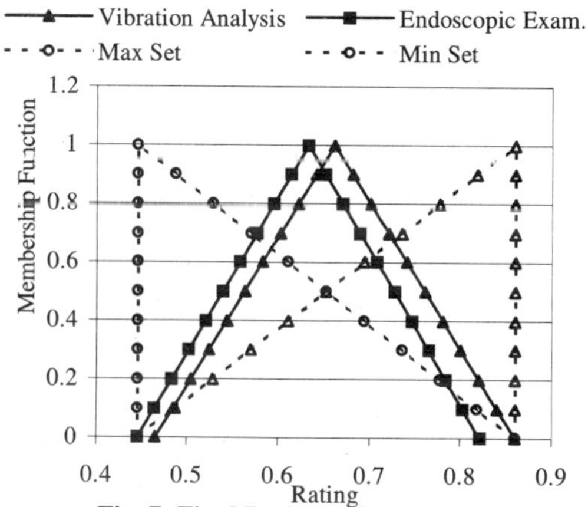

Fig. 7: Final Ratings of Technique 1 and 3 (Chen Method)

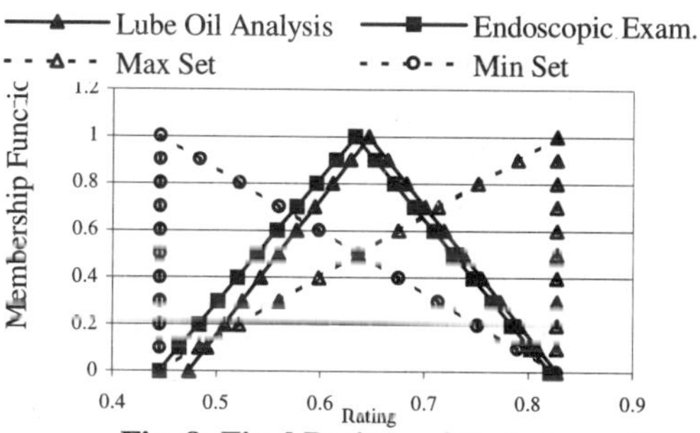

Fig. 8: Final Ratings of Techniques 2 and 3 (Chen Method)

Now, we study our problem using suitability set method. The suitability sets in the form of TFN using Eqn. 9 are: S_1 = [2.14 4.17 6.88], S_2 = [2.18 4.07 6.62] and S_3 = [2.05 3.99 6.57] and are depicted in Figure 9. Next step is to get the dominance relation $\{DR(i,j)\}_\delta$, which is the dominance of S_i over S_j. $\{DR\}_\delta$

Table 3 Values of $O(x_i)$ (Chen Method)

Parameters of Eqn. 8	R_1 and R_2	
	R_1	R_2
$R(x_i)$	0.66635	0.63017
$L(x_i)$	0.670	0.68139
$O(x_i)$	0.498175	0.47439

Parameters of Eqn. 8	R_1 and R_3	
	R_1	R_3
$R(x_i)$	0.68149	0.62395
$L(x_i)$	0.64967	0.68919
$O(x_i)$	0.51591	0.46738

Parameters of Eqn. 8	R_2 and R_3	
	R_2	R_3
$R(x_i)$	0.67919	0.6612
$L(x_i)$	0.64031	0.67173
$O(x_i)$	0.51944	0.494735

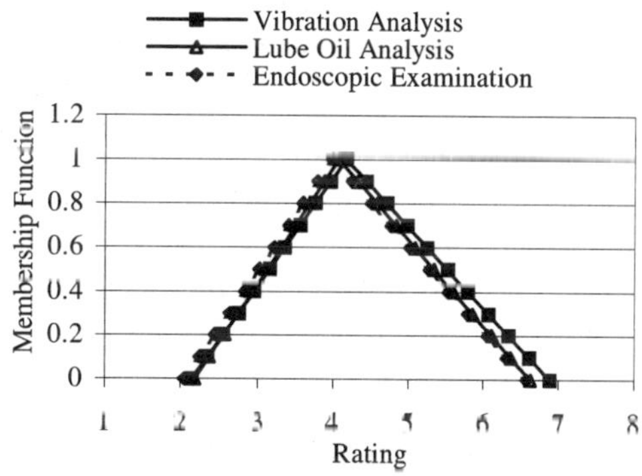

Fig. 9: Final Ratings of Techniques (Suitability Set Method)

for our problem is: $\{DR\}_\delta = \begin{bmatrix} 1 & 1 & 1 \\ 0.98 & 1 & 1 \\ 0.96 & 0.985 & 1 \end{bmatrix}$. We can observe from this relation that the first row is with all entries as 1. That means vibration analysis (with suitability set S_1) dominates over other two alternatives, with weights being [1, 0.98, 0.96]. Next to confirm our findings, we obtain the preference plot using difference function. Preferability of vibration analysis over lube oil analysis and endoscopic examination is shown in Figure 10.

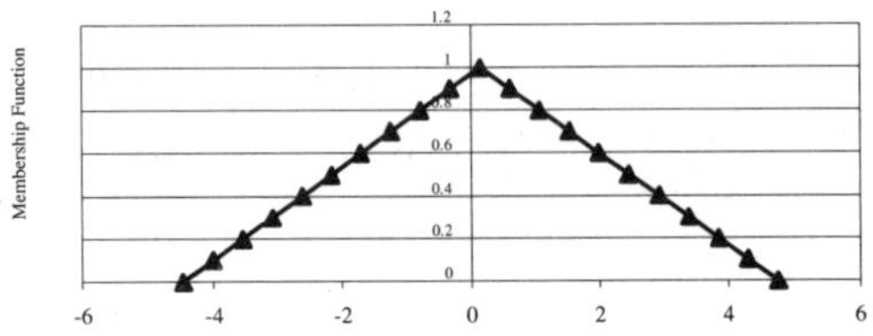

Fig. 10: Preferability of Vibration Analysis over Lube Oil Analysis and Endoscopic Examination (Suitability Set Method)

5. CONCLUSION

In this paper, four different methods have been established to select most appropriate or optimal condition monitoring technique under fuzzy decision criteria's. A case study is conducted on turbine maintenance. First condition monitoring techniques and decision criteria's have been determined and then opinions/ judgment of ten experts/ engineers is gathered in linguistic form. These opinions are combined and mapped to get the resulting rating of the techniques with respect to the decided criteria's and also weights to the decision criteria's. Optimal condition monitoring technique is then found out using rating and ranking, ranking fuzzy sets using 'cardinal utilities', by 'maximizing and minimizing sets' and suitability set method. In this case study of turbine maintenance, vibration analysis turned out to be the optimal one. It is also found that the other two techniques follow vibration analysis very closely.

6. REFERENCES

1. Mechefske C. K. , Wang Z. (2001), *Using Fuzzy Linguistics to Select Optimum Maintenance and Condition Monitoring Strategies*, Mechanical Systems and Signal Processing, **15(6)** , 1129-1140.
2. Zimmermann H.-Z. (1985), *Fuzzy Set Theory and its Applications*, Kluwer-Nijhoff Publishing, Hingham, MA.
3. Zimmermann H.-J. (1987), *Fuzzy Sets, Decision Making and Expert Systems*, Kluwer Academic Publishers, Norwell, MA.
4. Bass S. M., Kwakernaak H. (1977), *Rating and Ranking of Multiple-Aspect Alternatives using Fuzzy Set Theory*, Automatica, **13**, 47-58.
5. Baldwin J. F., Guild N. C. F. (1979), *Comparison of Fuzzy Sets on the Same Decision Space*, Fuzzy Sets and Systems, **2**, 213-231.
6. Chen S.-H. (1985), *Ranking Fuzzy Numbers with Maximizing Set and Minimizing Set*, Fuzzy Sets and Systems, **17**, 113-129.
7. Machol R. E. (Editor in Chief) (1984), *Fuzzy Sets and Decision Analysis*, Elsevier Science Publishers B. V., Amsterdam.
8. Kaufmann A., Gupta M. M. (1988), Fuzzy *Mathematical Models in Engineering and Management Science*, Elsevier Science Publishers B. V., Amsterdam.
9. Verma A. K., Srividya A., Prabhu Gaonkar Rajesh (2004), *Fuzzy Dynamic Reliability Evaluation of a Deteriorating System under Imperfect Repair*, International Journal of Reliability, Quality and Safety Engineering, **11(4)**, 387-398.
10. Klir G. J., Yuan B. (2001), *Fuzzy Sets and Fuzzy Logic*, Prentice Hall of India Pvt. Ltd.
11. Tran L., Duckstein L. (2002), *Comparison of Fuzzy Numbers using Fuzzy Distance Measure*, Fuzzy Sets and Systems, **130**, 331-341.
12. Juang C. H., Jhi Y. Y., Lee D. H. (1998), *Stability Analysis of Existing Slopes considering Uncertainty*, Engineering Geology, **49**, 111-122.
13. Kaufmann A., Gupta M. M. (1985), *Introduction to Fuzzy Arithmetic --Theory and Applications*, Van Nostrand Reinhold Company, New York.

Reliability, Safety and Hazard
Editors: P.V. Varde et al.

SAFETY STUDY OF THE TRANSPORT OF CONTAMINANT IN GROUNDWATER USING ARTIFICIAL NEURAL NETWORKS

D. Datta and B.P. Dubey+*

*Health Physics Division
+Reactor Control Division
Bhabha Atomic Research Centre, Mumbai 400 085

ABSTRACT

Groundwater flow and contaminant transport modelling has been used at many hazardous waste sites with varying degree of success. The ability to reliably predict the rate and direction of groundwater flow and contaminant transport is main issue in planning and implementing groundwater remediation. Artificial Neural Network approach has been attempted to estimate the essential components of groundwater flow and contaminant transport modelling in saturated porous media. Parameters such as hydraulic conductivity and longitudinal dispersion coefficient are estimated using sufficiently trained neural network model. A software package, called BIKAS has been adopted in this study.

Keywords: Artificial Neural Networks; Hydraulic conductivity; Groundwater flow

1. INTRODUCTION

The fate and movement of dissolved substances in groundwater has generated considerable interest out of concern for the quality of the subsurface environment. Mathematical models are often used to predict solute concentrations before management strategies are implemented. Advances in software and hardware now permit the simulation of subsurface transport using sophisticated mathematical models. However, complexity and heterogeneity of this kind of problem raise the limitation of the mathematical modelling. Basically in groundwater modelling transport of contaminants is modelled to assess the hazard associated with the contaminants so that spatial and temporal variation of the hazards associated with the contaminants can be predicted. Contaminant can be released to the groundwater through direct and indirect pathways. Contaminants after release into the groundwater move and respond to a number of interrelated natural and man made factors. In view of this fact, it can be stated that the main issue in groundwater modelling is to reliably predict the rate and direction of groundwater flow and contaminant transport [Maier et al, 2000].

Recent trend of the use of artificial intelligence technologies for assessment, modelling, and control in environmental science and engineering directs that application of artificial neural networks (ANNs) in this field will be one of the best soft computing methodology in predicting or evaluating the hazard assessment of contaminant transport in groundwater. Artificial neural networks use a system of highly interconnected processing units, or neurons, to learn patterns and relationships contained within a set of data. This ability to learn allows ANNs to be applied to a number of modelling tasks including pattern classification, clustering, function approximation, prediction and forecasting, optimization, etc. However, the presence of complexity and heterogeneity in hydrology systems raise the difficulty in solving the issue. As an alternative among the available soft computing methods, artificial neural network's (ANN) flexible structure can provide simple and reasonable solution to various problems in hydrology. Applications of ANN in groundwater related problems are as follows: Ranjithan et al. [1993] successfully used ANN to simulate pumping index for hydraulic conductivity realization to remediate groundwater under uncertainty. A similar study has been conducted by Rogers et al [1994] to simulate a regularly index for a multiple pumping realization containing multiple plumes at a contaminated site. Coulibaly et al [2001] modeled water table depth fluctuation by using three types of functionally different ANN models.

The present paper presents the usage of ANN in solving the transport of contaminant in groundwater.

2. A GROUND-WATER MODEL

A model, i.e. here a ground water system may be defined as a simplified version of a real-world system that approximately simulates the relevant excitation-response relations of the real-world system. Complexity of the real-world systems necessitates for simplification in making planning and management decisions. The simplification is introduced as a set of assumptions which expresses the nature of the system and those features of its behaviour that are relevant to the problem under investigation. These assumptions will relate, among other factors, to the geometry of the investigated domain, the way various heterogeneities will be smoothed out, the nature of the porous medium, the properties of the fluid involved, and the type of flow regime under investigation. The first step in the modelling process is the construction of a conceptual model consisting of a set of assumptions that verbally describe the system's composition, the transport processes that take place in it, the mechanisms that govern them, and the relevant medium properties [Rogers, 1994]. In the process of developing the models, properties of the system are expressed in terms of parameters. These values of the parameters are generally obtained from laboratory experiments and/or field scale experiments. However, such values may not represent the often complex patterns across a large geographic area, hence limiting the effectiveness of the model. In addition, such field scale experiments can be expensive. However, often we are interested in modelling quantities such as the depth of water table and solute concentration. The reason behind this is that they are directly relevant to hazard assessment for environmental decision making and basically these variables are measured regularly. However, there are a number of methods that have been developed for groundwater parameter estimation [Yeh, 1986]. They range from primitive trial and error techniques that are very time consuming and whose solution strongly depends on the skills of the practitioner, to advance mathematical and geostatistical methods, such as the linearised cokriging approach [Kitanidis at el., 1983]. However, usage of these methodologies is limited due to their highly theoretical nature. With a view to this detrimental groundwater contamination problem pertaining to distributed contaminant concentration values of saturated groundwater has been considered in this paper.

First, analytical methods have been used to numerically estimate the hydraulic parameters namely hydraulic conductivity (K, m/day) and longitudinal dispersion coefficient (D_x, m^2/day) and then ANN has been attempted to train the pattern to evaluate the parameters. A supervised MultiLayer Perceptron (MLP) is employed to limit the permissible parameter range and to enhance the accuracy of estimates. The paper presents role of ANN in estimating a single parameter, K. The same procedure is extended to estimate two parameters. Dataset used in this study was generated numerically by using a computer software package to simulate a 2-D groundwater aquifer. Here, in this case, development of groundwater modelling is described in a phased manner such as conceptual, mathematical and then numerical. Following sub sections describe in brief the contents of all these phases.

Content of a Conceptual Model

The assumptions that constitute a conceptual model are related to the following items:

- The geometry of the boundaries of the investigated aquifer domain;

- The kind of solid matrix comprising the aquifer;
- The mode of flow in the aquifer;
- The flow regime;
- The properties of the water (with reference to its homogeneity, compressibility, effect of dissolved solids and/or temperature on density and viscosity etc.);
- The presence of assumed sharp fluid-fluid boundaries, such as a phreatic surface;
- The relevant state variables and the area, or, volume, over which the averages of such variables are taken;

Selecting the appropriate conceptual model for a given problem is one of the most important steps in the modelling process. Over simplification may lead to a model that lacks the required information, while under simplification may result in a costly model, or in the lack of data required for model calibration and parameter estimation, or both.

The next step in the modelling process is to express the conceptual model in the form of a *mathematical model*.

Content of a Mathematical Model

The complete statement of a mathematical model used in this paper consists of the following items:

- A definition of the geometry of the considered domain and its boundaries;
- An equation (or equations) that expresses the balance of the considered extensive quantity (or quantities);
- Flux equations that relate the flux(es) of the considered extensive quantity(ies) to the relevant state variables of the problem;
- Constitute equations that define the behaviour of the fluids and solids involved;
- An equation (or equations) that expresses initial conditions that describe the known state of the considered system at some initial time; and
- An equation (or equations) that describe the interaction of the considered domain with its environment.

Content of a Numerical Model

Once the conceptual model is translated into a mathematical model in the form of governing equations, with associated boundary and initial conditions, a solution can be obtained by transforming it into a numerical model and writing a computer program (code) for solving it using a digital computer.

The main features of numerical model adopted for the present work are:

1. The solution is sought for the numerical values of state variable only at specified points in the space and time domains defined for the problem.
2. The partial differential equations that represent balances of the considered extensive quantities are replaced by a set of algebraic equations (written in terms of sought, discrete values of the state variables at the discrete points in space and time).
3. The solution is obtained for a specified set of numerical values of the various model coefficients.
4. Because of the large number of equations that must be solved simultaneously, a computer program is written using C++ language.

Horizontal Two-Dimensional Modeling

A second fundamental approximation often employed in dealing with regional problems of flow and contaminant transport is that ground water flow is essentially horizontal. In principle, groundwater flow and contaminant transport in a porous medium domain are three dimensional. However, when considering regional problems, one should note that because of the ratio of aquifer thickness to horizontal length, the flow in the aquifer is practically horizontal. This leads to an aquifer flow model written in horizontal two dimensions only.

3. ESTIMATION OF MODEL COEFFICIENTS

In the first phase of the study, a deterministic two dimensional (2-D) groundwater model is applied in which 2-D advection-dispersion transport equation is numerically solved [Fetter, 1999] for generating the concentration of the contaminant in the groundwater at different spatial distance and at various times.

Two-dimensional deterministic advection–dispersion equation with the flow parallel to the negative direction of x axis can be written as [Fetter, 1999]

$$\frac{\partial C}{\partial t} = D_x\left(\frac{\partial^2 C}{\partial x^2}\right) + D_y\left(\frac{\partial^2 C}{\partial y^2}\right) - v_x\left(\frac{\partial C}{\partial x}\right) \quad (1)$$

Where C = concentration of solute in M/L^3, D_x = longitudinal dispersion coefficient (L^2/T),
D_y = transverse dispersion coefficient (L^2/T), t = time of flow (T),
$v_x = K\, h_x/ n_e$ = steady state average velocity in homogeneous media (L/T),
K = hydraulic conductivity (L/T),
h_x = hydraulic gradient, and
n_x = effective porosity.

The coefficients (i.e., parameters) mentioned in equation (1) other than K and D_x were kept fixed for the sake of convenience of computation. Initial condition for solute concentration is taken as C_0 (t = 0) = 0. The same source is kept at the boundary for a 10 day time period. With this strategy and specified initial condition numerical estimate of solute concentrations for 10 different spatial locations at 5 different time intervals are computed. Time intervals used in the computation are: t = 1, t = 3, t = 5, t = 7, t = 10 day. The size of the aquifer selected for computation of solute concentration is 10 m x 5 m (2-D aquifer). Values of hydraulic conductivity used in the computation are ranged from 30-230 m/day.

Estimation of two coefficients

Simulation of the aquifer mentioned is continued further to estimate two parameters such as hydraulic conductivity, K (m/day) and longitudinal dispersion coefficient, D_x(m^2/day). In this case, 30 concentration values at different locations and at different time intervals for two actual outputs of K and D_x are used to train the network.

4. ANN ARICHITECTURE

Numerical outputs generated are used as the input required for training of the proposed ANN model. Training of these data was established using ANN package BIKAS. Supervised training has been carried out to establish the functional approximation of complex excitation and response of generated data. Training cycle is concluded on the basis of minimization of the error between the actual and network response. Estimation of the associated parameters was carried out using trained pattern of the concentration data that represents real world aquifer systems. By using the training pattern hydraulic conductivity of the artificial simulated aquifer was also tested which was required for assessing the performance of the ANN model.

A simple 3 layer MLP network is utilized to train the network to find the complex relationship of output **K,** and the concentration values. Scale Conjugate [Reifman, 1994] training algorithm with unipolar sigmoid transfer function was used in this study. The ANN had 2 neurons in the input layer, 5 in the hidden layers and one in the output layer as shown in figure 1. This is also referred as 2-5-1 architecture. The input dataset was divided into two categories – the training and the test set. 50% of the data was taken for the training and the remaining 50% was put in the test set. The data was put uniformly in the training and the test set. The ANN was trained for 500000 iterations.

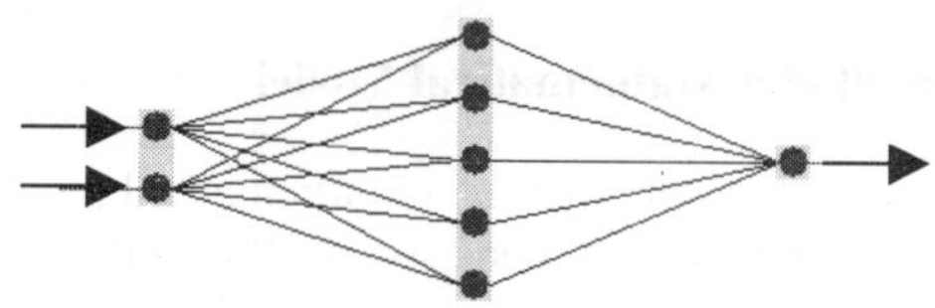

Fig. 1: A 2-5-1 ANN used in this study

The other ANN used in this study was similar to the one shown in figure 1 but it had a 5-1-1 architecture. The training algorithm used to train this ANN was resilient back propagation [Riedmiller, 1994] with tan hyperbolic transfer function. The ANN was trained for 100000 iterations using the default values offered by the software for all other parameters. Training algorithm, transfer function, number of neurons in the hidden layer and the number of iterations were determined empirically in both these ANNs.

5. RESULTS AND DISCUSSION

After training and testing the ANN, the distribution of percentage errors (deviation from expected values) was prepared. One such distribution of error is shown in figure 2 (2 input one output case). It shows that in most of the cases the error is less than 3%. The highest error observed was 5.9%. The error frequency distribution curve for 5 input 1 output data looks similar to figure 2 except that in that case the maximum error observed was 0.6%.

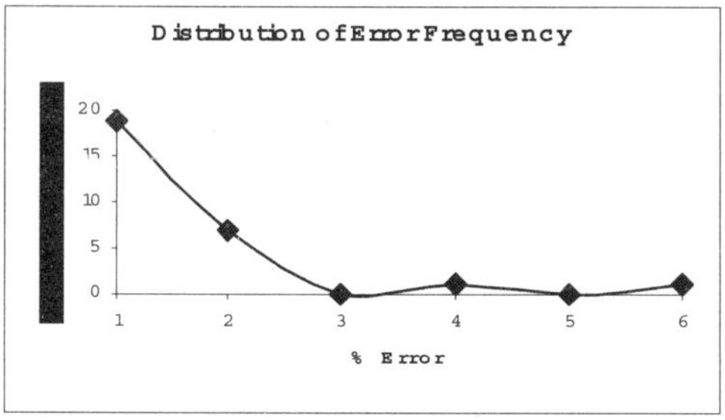

Fig. 2: Error frequency distribution

Statistics of the mean error for different ranges of K are shown in Table -1.

Table 1 Statistics of estimated error for ifferent ranges of K

Error	Range of K (m/day)				
	30-230	30-80	80-130	130-180	180-230
Max	5.65	1.88	2.23	3.0	2.98
Mean	2.05	0.23	0.38	0.32	0.39
StDev	0.47	0.28	0.41	0.43	0.45

Statistics of two parameter estimation are shown in Table -2.

Table 2 Statistics of two parameter estimations

Parameter	Range (K & D_x)	± % noise	Estimated Error		
			Max	Mean	MSE
K	190-240	5 20	2.48 6.78	0.99 2.35	2.65 3.18
D_x	0.03-0.08	5 20	0.0034 0.0087	0.0009 0.0025	0.0014 0.0029

In this case, maximum error of 180-230 range has been reduced by about 90%. Therefore, it cane be reasonably assumed that more the flexibility in numerical computation to generate the bulky data, better will be the training pattern and consequently error also will be possible to minimize. Two different levels of noise are allowed; one up to ± 5% and other up to ± 20%. The parameter ranges of K and D_x considered in two parameter estimation are 180-230 m/day and 0.03 – 0.08 m^3 /day respectively.

6. CONCLUSION

The paper presents an attempt of ANN in groundwater modelling. The prediction error of the estimate of K from the MLP network increased with the actual K value in the range from 30-230 m/day. Furthermore it was observed that neural network model algorithm embedded in package BIKAS provide reliable estimates of parameters even under uncertain and noisy conditions. Prediction of two parameter estimation also produced reliable estimates with lesser percentage errors for a moderately noisy system.

7. REFERENCES

1. Coulibaly, P., et al., 2001, Artificial neural network modeling of water table depth fluctuations, Water Resources Research, 37(4), pp. 885-896.
2. Fetter, C.W., 1999, Contaminant Hydrogeology, Prentice-Hall Inc., New Jersey, 1999.
3. Kitandis, P and Vomvoris, E.G., 1983, A Geostatistical approach to the problem of groundwater modelling (steady state) and one-dimensional simulation, Water Resources Research, 19(3), pp. 677-690.
4. Maier, H.R. and Dandy, G.C., 2000, Neural networks for the prediction and forecasting of water resources variables; a review of modelling issues and application, Environmental modelling & Software, 15, pp. 101-124.
5. Martin Riedmiller, 1994, Advanced Supervised Learning in Multi-Layer Perceptrons - From Backpropagation to Adaptive Learning Algorithms, Computer Standards and Interfaces, 16, pp 265 – 278.
6. Ranjithan, S., Eheart, J. W., and Garrett, J.H., 1993, Neural network-based screening for groundwater reclamation under uncertainty, Water Resources Research, 29(3), pp. 563-574.
7. Reifman, J., and Vitela, J. E.,1994, Accelerating Learning of Neural Networks

with Conjugate Gradients for Nuclear Power Plant Applications, Nuclear Technology, 106, pp. 225-241.

8. Rogers, L.L. and Dowla, F.U., 1994, Optimization of groundwater remediation using artificial neural networks with parallel solute transport modelling, Water Resources research, 30(2), pp. 457-481.

9. Yeh, W., W-G, 1986, Review of parameter identification procedures in groundwater hydrology; The inverse problem, Water Resources Research, 22(2), pp. 95-108.

Reliability, Safety and Hazard
Editors: P.V. Varde et al.

DEVELOPMENT OF AN ANN BASED ALGORITHM FOR PERSONNEL MONITORING TLD BADGE SYSTEM FOR X AND GAMMA RADIATION

K. Srivastava, V.K. Dhar[#], A.K. Bakshi, G. Varadharajan and R.K. Kher

Radiological Physics and Advisory Division
[#]Astrophysical Sciences Division
Bhabha Atomic Research Centre, Mumbai 400 085

ABSTRACT

To develop an algorithm that gives direct response of the TLD badge in terms of the ICRU operational quantities and to improve the overall accuracy in the personal dose estimation, a new approach based on Artificial Neural Network (ANN) was considered. Study was conducted to determine the response of the TLD badge system based on $CaSO_4$:Dy Teflon TLD discs for different doses and dose combinations delivered by various pure and mixed fields of X and gamma radiations. Response obtained using the prevalent algorithm and neural network predicted algorithm were compared and it was found that ANN predicted algorithm results in better accuracy especially in the mixed field of two radiations compared to the prevalent algorithm.

Keywords: Personnel Monitoring, TLD Badge, $CaSO_4$:Dy Teflon TLD Disc, Artificial Neural Network

1. INTRODUCTION

Personnel Monitoring is an integral part of any radiation protection programme and accurate measurement of personal dose is required to ensure the compliance of dose limits for radiation workers. In India, a three element TLD Badge system (Vohra et el., 1979) based on $CaSO_4$:Dy Teflon TLD discs is being used for Individual Monitoring of about 62,000 radiation workers. The indigenously developed TLD system has fully replaced the film badge system, which was in use since 1952, in a phased manner as it offers better dosimetric characteristic and meets the requirement on accuracy, reliability and ruggedness of a personal dosimeter.

The methodology of estimation of personal dose equivalent has therefore remained a subject of constant review along with the changing philosophy of radiation protection. International Commission on Radiation Units and Measurements (ICRU), through their publications ICRU-39, 43, 47 & 57 (ICRU reports 1985, 1988, 1992 and 1998) has provided an unified method of estimating effective and equivalent dose by measurement of personal dose equivalents Hp (10) and Hp(0.07), respectively. At present the calibration of personal dosimeters and the recording and reporting of dose are done in terms of old quantities -whole body dose and skin dose. It is now being actively considered to implement the ICRU operational quantities for the recording and reporting of occupational dose.

Presently the personal doses are estimated using an algorithm (Dere et al., 1993) was tested partially by us (Pradhan et al., 1999) during an earlier study. However, in the present study it has been tested in mixed fields of X and gamma rays covering the wide energy range from 28 keV to 1.25 MeV encountered in the occupational monitoring. The prevalent algorithm provides three different formulas in different energy ranges of photons depending upon the ratios of TL response of disc under different filter regions of the TLD badge, which is very important to ascertain the energy and type of radiation. It has been observed that the uncertainties in the dose estimation sometimes even exceed the acceptable limit for few photon fields, especially in the mixed field of low energy X and Gamma radiation using the prevalent algorithm.

Therefore to improve the overall accuracy in the dosimetry for broad energy range of photons and to report the doses in terms of ICRU operational quantity - Personal Dose Equivalent Hp(10), a new approach based on Artificial Neural Network (ANN) was considered. This is a feasibility study to explore the capability of ANN for the development of algorithm for the TLD Badge. For this, BIKAS - ANN simulator package was used and Resilient Backpropagation method was found to be most suitable. In this study, the response of the TLD badge system in terms of Hp(10) was compared using the prevalent algorithm and ANN predicted algorithm for various combinations of pure and mixed fields of X and gamma radiation. Software package MATLAB was also used to ensure the reliability of results obtained using BIKAS for similar network configuration and results of two software were found to be in agreement.

2. MATERIALS AND METHODS

TLD Badge System

TLD Badge comprises of a TLD card and a plastic cassette as shown in Fig 1a & 1b respectively (Pradhan et. el., 2002). TLD card has three $CaSO_4$:Dy Teflon discs (13.3 mm diameter and 0.8 mm thick) that are mechanically clipped onto circular holes (12 mm dia) punched in an aluminium plate of size 52.5 x 30 x 1 mm^3. An asymmetric 'V' cut on the card permits its loading into the cassette in only one orientation and ensures the proper positioning of three discs as- i) Disc D1 under combined metal filter of 1 mm thick Cu and 1 mm thick Al; ii) Disc D2 under 1.6 mm thick plastic and iii) Disc D3 under circular open window, on either side of cassette. In case of combined metal filters, copper filter is rectangular in shape of size 30 mm x16 mm .. mm on the front side of the badge and of circula in shape of diameter 15.6 mm on the rear side which is in contact with body. Aluminium filter on both sides is circular in shape and of diameter 12.6 mm and is embedded on Cu filter on the outer side. TLD cassette has a S.S Crocodile clip fixed with silver plated eyelets to the slider part for affixing to the user's clothing.

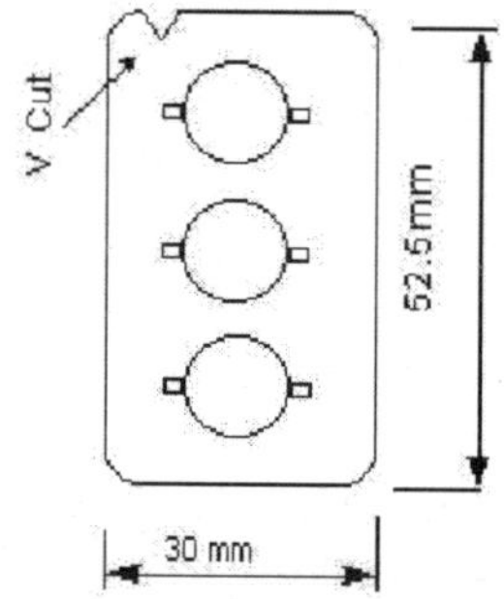

Fig. 1a: TLD Card

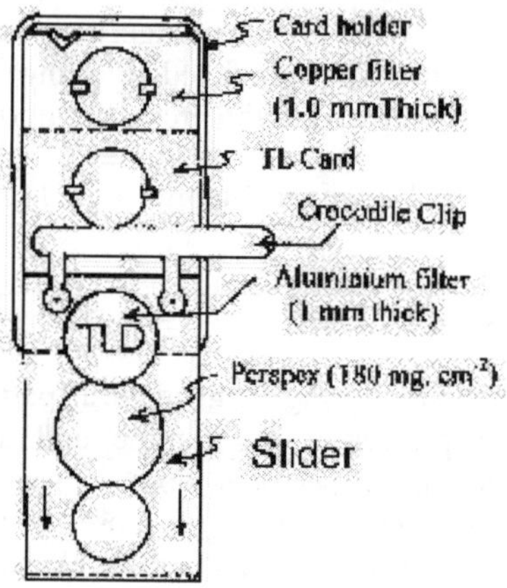

Fig. 1b: TLD Cassette

The main objective of TLD cassette or holder is – (1) to make the photon response energy independent and (2) to provide the ratios of responses that can be used for energy discrimination. Since $CaSO_4$: Dy is non tissue equivalent material having higher value of Z_{eff} = 15.3 (Z_{eff} for tissue = 7.4) its response is highly energy dependent, especially in the low photon energy range (Laxmanan et al., 1989). Fig 2 gives the relative response of Disc1(under metal filter) and Disc D3 (no filter) with photon energy. The TL response shown at Y-axis is normalized with respect to response of ^{137}Cs (660 keV photon). One can see that at 28.4 keV the relative response of Disc 3 is about 11.2 times whereas for Disc D1 the relative response varies only between 0.37 (for 28.4 keV X-ray) and 2.5 (for 60 keV gamma photon of ^{241}Am) in the energy range from 28.4 keV to 1.25 MeV. However the incorporation of the combined metal filters over Disc D1 along with selection of appropriate dose evaluation algorithm makes the energy response within ± 30% in the photon energy range of 32 keV to 1.25 MeV.

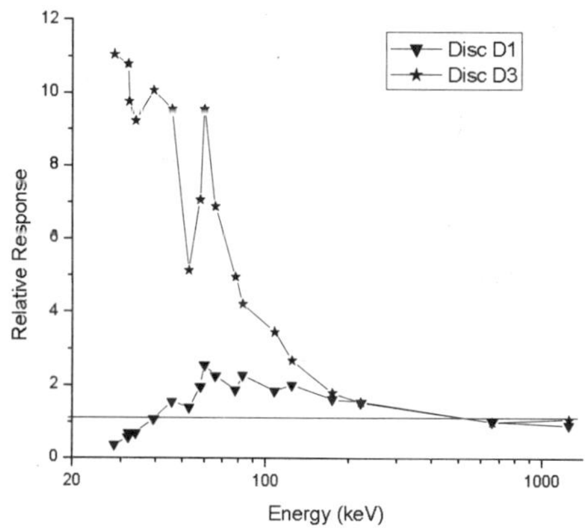

Fig. 2: Energy Response of TLD Badge

Radiation Fields and Exposure Geometry:

X-ray irradiations were carried out using a Deep Therapy X ray machine (Philips RT-250) for photons in the energy range 32 keV to 225 keV. Low energy X-rays in the energy range from 28 keV to 53 keV were delivered from newly installed Industrial X-ray Unit YXLON MG325. Gamma exposures were given from ^{241}Am, ^{137}Cs and ^{60}Co discrete gamma ray sources for 60 keV, 662 keV and 1.25 MeV monoenergetic photon beams. Output in terms of air kerma (mGy) free-in-air were measured using a 30 cc PTW ionization chamber and UNIDOS Dosimeter traceable to PTB standard with an accuracy within ± 3% for all the above photon beams. TLD badges were exposed to mixed fields of X & gamma rays in the ratios 1:3, 1:1, 3:1 apart from being exposed to pure X-ray and gamma rays. All exposures were given to the dosemeters attached on the front side of an ISO water phantom (Dimension 30 cm x 30 cm x 15 cm, with 1 cm wall thickness all around except the front wall which is 0.25 cm in thick), at a distance of 4 m for X-rays and 1 m for gamma rays. In order to consider their angular response, TLD Badges were exposed at normal incidence as well as at angles of 20°, 40° and 60° for few photon fields in the energy range 32 keV to 108 keV using wide spectra and ^{241}Am and ^{137}Cs sources.

Readout of the TLD cards were taken on a N_2 gas based semiautomatic TLD reader after 24 h of exposure. Individual calibration factor was applied for each TLD disc of the TLD card for better precision of the result. Pre-exposure annealing of TLD cards was carried out at 230 °C for 4 hours. Values of Hp(10) were obtained by using conversion coefficient provided by by ISO (ISO, 1999) for narrow spectrum, wide spectrum and for mono-energetic beams. For each experimental point, average readout of at least 5 TLD cards was taken and the variation in the readouts were generally within ± 5%(1σ).

The experimental data of the last few IAEA-RCA External Dosimetry Intercomparison exercises conducted by IAEA (from 1991-2002) were also included for training and testing of the ANN based algorithm. For this data the delivered Hp(10) values were provided by IAEA within ± 2.5 % accuracy limits (Massand et al., 1998 & 2000).

3. NEURAL NETWORK APPROACH

Artificial Neural Networks (ANN) were originally introduced as very simplified models of brain function, so that it is instructive to consider the broad analogies between ANN's and biological ones. Of course, the latter are considerably more complex, and many ANN models today bear very little resemblance to biological ones (Hertz, et al., 1991). In recent years there has been great deal of interest in ANN applications. The ability of ANN to handle non-linear data interactions, and their robustness in face of high noise levels have encouraged their successful use in diverse areas as physics, biology, medicine, agriculture, computer and nuclear research.

ANN is a relatively loose term referring to mathematical models which consist of processing nodes (or neurons) with multiple inter-connections. Each neuron has an output and several inputs. The input of a neuron is assigned an adjustable weight. The neuron calculates the weighted sum of its inputs and passes the weighted sum through a transfer function to generate an output. The most widely used transfer function is the sigmoid transfer function (though any non-linear function can be used) and is defined as :

$$F(x) = 1 / (1 + e^{-x}) \qquad \text{..........(1)}$$

We have employed the BIKAS ANN Simulator developed by BARC-IIT (Kanpur) collaboration under BRNS project (2000/36/5-BRNS). The network used by us is resilient backpropagation with 3:25:1 configuration, i.e. 3 inputs corresponding to the TL readout of three Discs D1, D2 and D3, with one hidden layer with 25 neurons and one output corresponding to value of delivered dose to the irradiated dosemeters. Also it has been seen that using radial basis (RBF) as the aggregation function instead of summation function gives better convergence. A typical RBF feature, i, has a gaussian (bell-shaped) response, $\phi_s(i)$, and dependents only on the distance between the state, S, and the feature's prototypical or center state, C_i, and relative to the feature's width, $\sigma_{i,}$ and is defined as:

$$\phi_s(i) = \exp(-(S-C_i)^2) / 2\sigma_i^2 \qquad (2)$$

Various ANN algorithm were tried for training the network but Resilient backpropagation method as discussed below gives better error convergence.

Resilient Backpropogation Algorithm

This is an advancement of the standard backpropagation algorithm and follows the above methodology in principle except that the weight change happens as follows:

$$\nabla W_{ij} = - \nabla_{ij}(t) \text{ if } (\partial E(t)/\partial W_{ij}) > 0 \qquad (3)$$

$$\nabla W_{ij} = - \nabla_{ij}(t) \text{ if } (\partial E(t)/\partial W_{ij}) < 0 \qquad (4)$$

$$\nabla W_{ij} = 0 \text{ if } (\partial E(t)/\partial W_{ij}) = 0 \qquad (5)$$

where,

$$\nabla_{ij}(t) = \eta^+ \nabla_{ij}(t-1) \text{ if } (\partial E(t-1)/\partial W_{ij})^* (\partial E(t)/\partial W_{ij}) > 0 \qquad (6)$$

$$\nabla_{ij}(t) = \eta^- \nabla_{ij}(t-1) \text{ if } (\partial E(t-1)/\partial W_{ij})^* (\partial E(t)/\partial W_{ij}) < 0 \qquad (7)$$

$$\nabla_{ij}(t) = \nabla_{ij}(t-1) \text{ if } (\partial E(t-1)/\partial W_{ij})^* (\partial E(t)/\partial W_{ij}) = 0 \qquad (8)$$

where ∇_{ij} are the update values. The weights are adjusted by calculating ∇W_{ij} using equations (3) to (5). At the beginning, all update values are set to an initial value ∇_0 and weights are randomized. Values of η^+ and η^- are fixed to be 1.2 and 0.5 respectively for the optimum performance. Though Reidmiller (Reidmiller et al, 1994) suggested that the values of $\nabla_0 = 0.1$ and $\nabla_{max} = 50$, he has compared several networks and found resilient backpropagation algorithm to be insensitive to the initial selection of the parameters ∇_0 and ∇_{max} or on the network architecture whereas the backpropagation algorithm is quite sensitive to the selection of the initial values of parameters α and β and the network architecture. Because of these advantages we have preferred resilient backpropagation for the present study. Fig 3 shows the neural network configuration used by us. We have also used the similar configuration in the ANN module of the MATLAB package. The results generated using MATLAB are in agreement with the BIKAS simulator thereby verifies the reliability of the BARC-IIT (Kanpur) package.

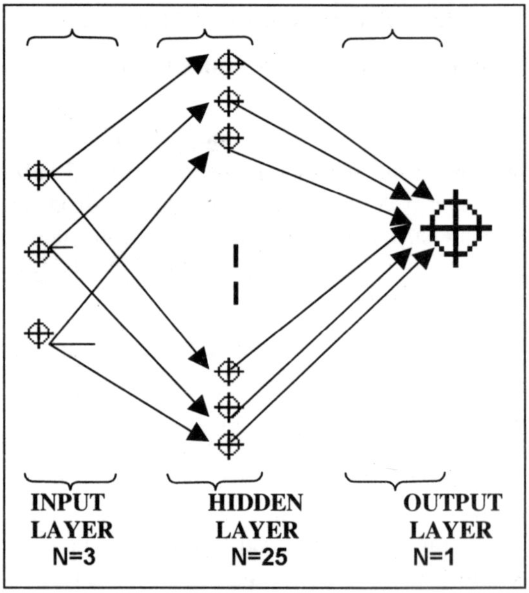

Fig. 3: Neural Network Configuration used for the development of the Algorithm

4. RESULTS AND ANALYSIS

Prevalent Algorithm for Dose Estimation

The prevalent algorithm for the evaluation of whole body dose developed by Dere et al. was based on the response of loose CaSO4:Dy Teflon disc with different filter combination (simulating the TLD badge) measurement on a solid Perspex phantom(size 30 x 30 x15 cm^3) with in-air calibration of the TLD reader. This algorithm uses three different formulas for dose evaluation of pure / mixed X-Gamma fields; i) response of disc under metal filter region of TLD badge for photon energy range above 200 keV, ii) A 4th order polynomial for photon energy range 40-200 keV and iii) Average of the response of D2 and D3 for photon energy below 40 keV.

Fig.4 shows the graph of response of TLD badge with photon energies using prevalent algorithm (Algorithm-1) in the energy range 28.4 keV to 1.25 MeV for pure photon field. The response is within ± 30 % except at very low energy of 28.4 keV for which response is 1.34, slightly exceeding the limit. This shows that incorporation of metal filter along with application of dose evaluation algorithm reduces the energy dependence within ± 30 % (compared to response of Disc 1 in Fig.2 which varies from 0.37 to 2.5 for same energy range).

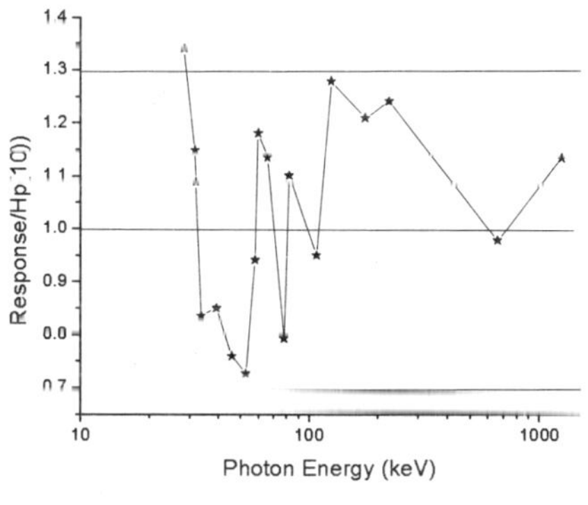

Fig. 4: Response of TLD Badge with photon energy using Algorithm-1.

For the TLD badge, the disc ratios due to response under different regions of the cassette is very important to ascertain the type and energy of incident radiation or radiations. These ratios were found to vary not only with energy and radiation type but also with proportions of two radiations in case of mixed fields and angle of incidence.

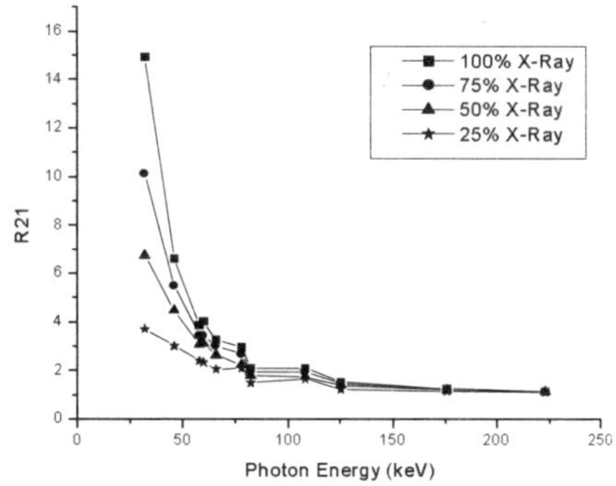

Fig. 5: Variation of R21 with different proportions of X ray (from 32 to 226 keV) and ^{137}Cs Gamma ray for mixed field.

Fig 5 shows the variation of R21 (ratio of Disc 2 to Disc 1 response) with energy for mixed fields of X and Gamma in proportion of 3:1, 1:1, 1:3 along with that for pure X-rays. The maximum variation is 4.2 times (R21 reduces from 14.9 for 100 % X-ray to 6.73 for 25% x-ray for same dose) at low energy of 32 keV. Figure 6 shows the variation of R21 with angle of incidence at 20°, 40° and 60° compared to normal incidence for X-rays in the energy range 32 keV to 108 keV. The angular dependence of R21 is less, however the R21 values at 60^0 angle were always found to be higher than R21 value at normal incidence and the maximum variation is 1.36 for 60 keV photon of ^{241}Am.

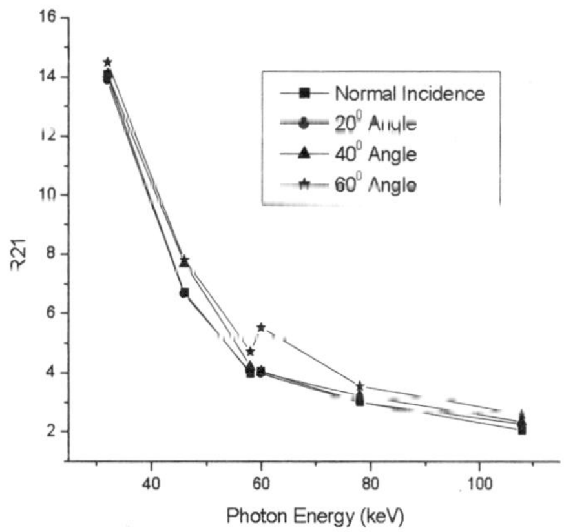

Fig. 6. Variation of R21 with angle of incidence in the photon energy range from 32 to 108 keV

A new algorithm was developed based on ANN approach using Resillient backpropagation method as described earlier in the text. This algorithm was tested extensively for different dose ranges of pure and mixed field radiations of X and gamma and results were found to be within acceptable limits. Fig.7 shows the response obtained using the prevalent algorithm (Algorithm –1) and ANN based Algorithm (Algorithm-2) for the experimental data used in IAEA-RCA intercomparison exercise in a Trumpet Curve. Trumpet Curve (Bohm et al., 1994) is the graphical representation of the permissible accuracy interval around the conventional true dose (delivered dose) as a function of personal dose equivalent Hp(10)). This is an internationally accepted method for indication of performance in the field of individual monitoring and the performance of a dosimetry system is considered to be satisfactory if ratios of estimated dose to delivered dose lies in the trumpet curve for 95% of dosimeters tested (IAEA -TECHDOC -1126).

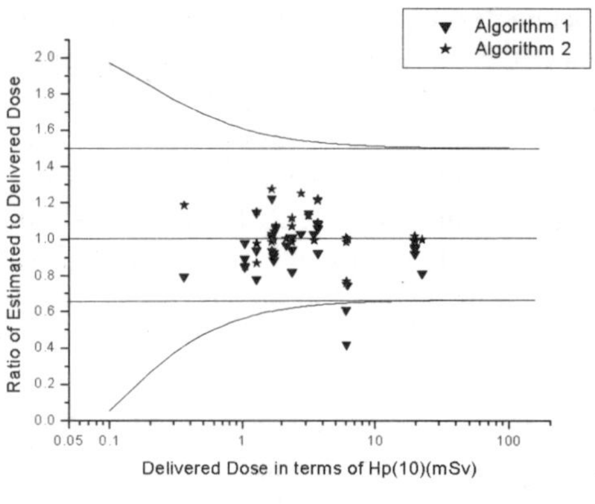

Fig. 7: Response of TLD Badge shown in trumpet curve using the Algorithm 1 and Algorithm-2

It was observed that at least four points are outside the trumpet curve for Algorithm –1, which belongs to the category of mixed field (45 keV+134 keV and 45 keV + ^{60}Co source) whereas all points are within the trumpet curve for Algorithm-2. However in the Fig 7 only two points are seen outside the trumpet curve due to overlap of data for similar delivered dose on X-axis. For this data, as per the ANSI criteria (ANSI 2001), the average performance quotient $\overline{P}$ =– 0.109 (standard deviation σ_p =0.20) for Algorithm -1and $\overline{P}$ = 0.011 (σ_p = 0.134) for Algorithm-2, which is much less than the limiting value of 0.4 for $\overline{P}$ for both cases. Therefore it was found that algorithm based on ANN gives better accuracy (on an average the % error was found to be of the order of 20%) for the dose response in the mixed field of X and gamma radiation as compared to the prevalent algorithm. The same data was tested using MATLAB also for similar network configuration to ensure the reliability of result obtained using BIKAS package. Results obtained using two software were found to be in agreement.

5. CONCLUSIONS

The feasibility of using Artificial Neural Network for the development of algorithm for the TLD badge system was explored in this study. Personal dose evaluation in terms of Hp (10) is possible within the prescribed accuracy limit using algorithm-2, for both pure and mixed photon fields. Algorithm-1 has the limitation in terms of accuracy (beyond ±30%) for the evaluation of dose in mixed fields at some energy points. Use of artificial neural network improves the accuracy in the dose evaluation as compared to the prevalent algorithm in the photon energy range from 28keV to 660 keV. However the accuracy and precision of dose evaluation using algorithm-2 may be further improved with more extensive training of ANN using large set of input data points for all possible combinations of the radiation fields encountered in personnel monitoring.

6. ACKNOWLEDGEMENTS

Authors are grateful to Shri H. S. Kushwaha, Director, HS&EG, Dr. B. S. Rao, Head, RPAD and Shri R. Koul, Head, ApSD for their encouragement during this work. Authors are thankful to Dr. B. P. Dubey, RCnD, Dr.P.V.Varde, RRSD and Dr. A.K.Tickoo, ApSD for fruitful discussions and suggestions on the use of neural network. Authors are also thankful to Shri U.R.Kini and Shri K.L.Popli, PMS, RPAD for sharing the data on IAEA intercomparison for the study. Also we acknowledge the help extended by RSS, RSSD in providing the various irradiation facilities for the experiment.

7. REFERENCES

1 ANSI (2001), American National Standard for Dosimetry-Personnel Dosimetry Radiation Performance- Criteria for Testing, American National Standards Institute (ANSI) Inc., New York, HPSN 1311 - 2001, 11/2000 Version.

2 Dere A.V, Popli K.L, Nagpal J.S., Kapoor D.K and Patel, P.H (1993); Dose Computation Algorithm for Individual Thermoluminescence Dosimetry, Bull. Radiat.Prot. 16 (1&2) 54-56.

3 ICRU (1985), International Commission on Radiation Units and Measurements, Determination of Dose Equivalents Resulting from External Radiation Sources, ICRU report 39, Bethesda, Maryland, USA.

4 ICRU (1988), International Commission on Radiation Units and Measurements, Determination of Dose Equivalents from External Radiation Sources-Part-2, ICRU report 43, Bethesda, Maryland, USA.

5 ICRU (1992), International Commission on Radiation Units and Measurements, Determination of Dose Equivalents Resulting from External Photon and Electron Radiations, ICRU report 47, Bethesda, Maryland, USA.

6 ICRU (1998), International Commission on Radiation Units and Measurements, Conversion Coefficients for use in Radiological Protection Against External Radiation, ICRU report 57, Bethesda, Maryland, USA

7 International Organization for Standardization (ISO) (1999), Calibration of Area and Personal Dosemeters and the Measurement of their Response as a Function of Energy and Angle of Incidence, Part-3, ISO-4037-3, Geneva, Switzerland).

8 Hertz j. Anders K. and Richerd G. P. (1991); Introduction to theory of neural computation, Eddison Wesley Reading,

9 Laxamanan A. R., Popli K. L. and Kher R. K. (1989); Photon Energy Dependence of $CaSO_4$:Dy TLD under during this work. Authors are thankful to Dr. B.P. Dubey, RCnD, Dr.P.V.Varde, RRSD and Dr. A.K.Tickoo, ApSD for fruitful discussions and suggestions on the sue of neural network. Authors are also thankful to Shri U.R.Kini and Shri K.L.Popli, PMS, RPAD for sharing the data on IAEA intercomparison for the study. Also we acknowledge the help extended by RSS, RSSD in providing the various irradiation facilities for the experiment.

10 Different Metal Filters in Terms of the New ICRU Quantities, Rad. Prot. Dosim, 28, 4, 273-276

11 Martin Reitmiller (1994): Computer Standards and Interfaces, 16, 265-278)

12 Massand, O.P., Kini, U.R., Popli, K.L and Gambhir, S.P. (1998); IAEA/RCA External Dosimetry Intercomparison, BARC report, Report No.BARC/1998/E/033

13 Massand, O.P., Kini, U.R., Popli, K.L. and Sneha, C. (2000); IAEA/RCA External Dosimetry Intercomparison, BARC report, Report No.BARC/2000/E/035

14 Pradhan, A. S., Adtani, M. M., Varadharajan G., Bakshi, A. K., Srivastava Kshama and Bihari, R. R (2002); Hand Book on the use of TLD Badge Based on $CaSO_4$:Dy Teflon TLD Discs for Individual Monitoring, BARC/2002/E/025.

15 Pradhan, A. S., Bihari, R.R., Srivastava K. and Bakshi A.K. (1999); Measurement of Personal Dose Equivalents Hp(10) and Hp(0.07) by TLD Badge – Response in Mixed Fields of Low Energy X-rays and High Energy Gamma Rays, Proc. 24th IARP Conference on Radiation Protection in Nuclear Fuel Cycle, KAPS, pp 223-229.

16 Vohra, K. G., Bhatt R. C., Bhuwan Chandra, Pradhan A.S., Laxamanan A.R. and Shastry S.S. (1979); A Personnel Dosemeter TLD Badge based on CaSO4:Dy Teflon TLD Discs, Health Phy. Vol. 38 pp 193-197.

Reliability, Safety and Hazard
Editors: P.V. Varde et al.

ARTIFICIAL NEURAL NETWORK FOR SOLVING SAFETY RELATED ENVIRONMENTAL PROBLEMS: RAINFALL PREDICTION

D. Datta

Health Physics Division Bhabha Atomic Research Centre, Mumbai 400 085

ABSTRACT

Prediction of weather parameters such as weather category, wind direction, rainfall, humidity, etc. with a high degree of reliability is one of the safety issues in any nuclear installation. A supervised learning algorithm specialized for such prediction in multilayer feed-forward connectionist networks has been applied. The algorithm developed for predicting the outcome of such a future event is based on a class of incremental learning procedures so called as Temporal difference. The paper presents an application of handling a nonlinear multi-step rainfall prediction problem using an admixture of backpropagation and Temporal difference.

Keywords: Multilayer Perceptron, BP algorithm, TD algorithm, Connectionist networks

1. INTRODUCTION

Predictive representations represent the state of a dynamical system as a vector of predictions about future action-observation sequences. In predictive state representations (PSRs) introduced by Littman and others [Littmann et al., 2002], each prediction is an estimate of the probability of some sequence of observation given a sequence of actions. This paper introduces a scheme for solving the problem of prediction –i.e., of using past experience with an incompletely known system to predict its future behaviour. For example, through experience, one might learn to predict for a particular weather condition whether there will be rain or not, or what should be the wind direction at the next few hours. Prediction of weather condition is an essential input to predict further the impact of effluent release into the environment..

In the literature, prediction problems have been conveniently relegated to the more broad class of functional approximation problems. The Multilayer perceptron (MLP) with its ability to perform an arbitrarily close approximation to any continuous nonlinear mapping [Cybenko, 1989], has been utilised for the purpose. Given a set of examples, the backpropagation (BP) [Rumelhart, 1986] algorithm can be called upon to learn the mapping at the example points. When these conventional methods are applied for prediction problems, they do not take into account the Temporal Sequence of sensory inputs but instead address a structural credit assignment issue. Hence they prove unsuitable for prediction problems.

Temporal difference (TD) learning is a technique developed by Sutton [Sutton, 1988] for prediction problems. TD learning is a general method for learning predictions whenever multiple predictions are made of the same event over time, value functions being just one example. The most pertinent of the more general uses of TD learning have been in learning models of an environment or task domain [Dayan, 1993; Sutton, 1995, Sutton, 1999]. In these works, TD learning is used to predict future values of many observations or state variables of a dynamical system.

The essential idea of TD learning can be described as "learning a guess from a guess". Before TD networks, the two guesses involved were predictions of the same quantity at two points in time, for example, of the discounted future reward at successive time steps. TD networks generalize TD methods by allowing the second guess to be different from the first. The current TD network learning algorithm uses 1-step backups; the target for a prediction comes from the subsequent time step. In conventional TD learning, the TD (λ) algorithm is often used to do more general, nstep backups. Rather than a single future prediction, n-step backups use a weighted average of future predictions as a target for learning [Sutton, 1988]. TD (λ) is really a spectrum of algorithms, controlled by the continuous valued parameter λ ε [0, 1]. TD (λ = 0) uses the earliest possible prediction as the target for learning while TD (λ = 1) uses the latest possible prediction as the target. For other values of λ, the targets for learning are distributed among all of the predictions along the way. To be more precise, it must require making a distinction between the extensive definition of a prediction, expressing its desired relationship to measurable data and its TD definition, expressing its desired relationship to other predictions. In all previous work, the two guesses involved were predictions of the same quantity at two points in time, for example, of the future value of any release of radioactive material into the environment (atmospheric and aquatic) at successive time steps.

TD methods offer several advantages – the training inputs can be taken directly from the temporal sequence of sensory examples, they are incremental and therefore, easier to compute. They tend to make more efficient use of their experiences they coverage faster and produce better predictions. Sutton's Random walk [Sutton, 1988] also estimates the probability of a balanced random walk sequence ending in some particular states. So far, TD methods have not been applied to mutilayer feed forward connectionist architectures. The combined effect of both backpropagation and TD methods improve prediction accurately. Backpropagation addresses a structural credit assignment issue, while TD methods address a temporal credit assignment issue. The paper explores an application of this new learning methodology in prediction problem using the combination of backpropagation and TD methods.

In conventional backpropagation, learning is driven by the error between predicted and actual outputs whereas in TD methods, it is driven by the difference between temporarily successive prediction, i.e., learning occurs whenever there is a change in prediction over time. TD learning is widely used in reinforcement learning methods to learn moment-to-moment predictions of total future reward (value functions). In this setting, TD learning is often simpler and more data-efficient than other methods. For example, one can predict at the end of each month of a certain year, the quantum of rainfall in any of the month of the next year. While backpropagation compares each prediction to the actual rainfall in the prescribed month of the next year, TD methods compare each month's prediction with that made on the following month. If a 30% value is predicted in any month and an 80% value is predicted in next month, then the TD method increases predictions for months similar to the previous month.

2. MULTI-STEP PROBLEMS

Multi-step prediction problems are those in which correctness of the prediction is not revealed until more than one step after the prediction is made, but parallel information pertaining to its correctness is revealed at each step. TD methods can be distinguished from supervised learning methods only in these problems; thus the former improve over conventional techniques only on Multi-step problems. The rainfall prediction problem mentioned above is such a problem because in conclusive evidence relevant to the correctness of prediction of rainfall of any month becomes available in the form of new observations in subsequent months. The paper focuses the multi-step problems. The problem of making a prediction for a particular fixed amount at a later time is addressed in this paper.

3. THEORY OF TD METHODS

TD Networks

TD networks learn to predict future observations that will be generated by their environment (a dynamical system, such as prediction of release of gaseous and liquid effluent at various time steps, prediction of rainfall at every month steps, etc). At each of a series of discrete time steps t, the environment generates an observations $o_t \varepsilon \varphi$, and the agent takes an action $a_t \varepsilon A$. The action and observation events occur in sequence, a_{t-1}, o_t, a_t, o_{t+1}, a_{t+1}, o_{t+2}, with each event potentially dependent only on those preceding it in the sequence. This sequence is called *experience*. The work presented into this paper describes the usage of the TD networks with backpropagation networks (BPN) to predict action-conditional functions of future experience in partially-observable environments.

A TD network is a network of nodes each of which represents a single scalar prediction. Each node has links to other nodes and/or the observation from the environment. These links represent the targets for learning - what the node should predict about the data. Each of these nodes is some question about the system, and accordingly this set of nodes and links is called the *question network*. A separate set of interconnections determines the actual computational process- the updating of the predictions of each node from their previous values and the current action and observation. A typical format of the type of question network is shown in Figure 1.0 In this example, the question being asked by the node labeled y^1 at time t is "If the next action is a_1, what is the probability that the next observation o_{t+1} will be 1?" Similarly, the question asked by the node labeled y^4 at time t is "If the next action is a_2, what is the expected value of the prediction from node y^1 at time t + 1?"

Each node in a TD network can be thought of as a function approximator. The inputs to each node are defined by a set of interconnections called the *answer network*. The prediction made by each node is a function of these input features. These predictions provide the answers to the questions asked by the question network.

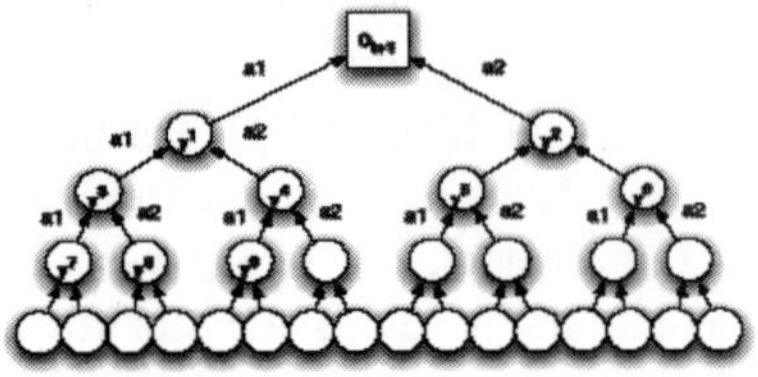

Fig. 1: Symmetric action-conditional question network.

Formally, the prediction for i^th node at time step t is denoted as $y_t^i \,\varepsilon\, [0,1]$, $i = 1,...,n$. The column vector of predictions $y_t = (y_t^1,....,y_t^n)^T$ is updated by

$$y_t = \sigma\,(\mathrm{W}_t x_t) \tag{1}$$

where $x_t \,\varepsilon\, R^m$ is a feature vector, $\mathbf{W_t}$ is a n x m matrix of modifiable weights, and σ is the S-shaped logistic function $\sigma(s) = 1/(1+\exp(-s))$. The feature vector is a function of the preceding action, observation, and node values.

$$x_t = x(a_{t-1}, o_t, y_{t-1}) \,\varepsilon\, R^m \tag{2}$$

In the present experiments, $\mathbf{x_t}$ has one component for each unique combination of the current observation and previous action and n continuous valued components equal to the predictions $\mathbf{y}_{t-1}$.The operation of a TD network can be described into the following steps.

(i) Choose an action a_{t-1} and receive an observation from the environment O_t.
(ii) Calculate the input vector x_t as a function of the previous predictions y_{t-1}, and the new observation O_t.
(iii) Create the new predictions of y_t (Equation (1)).
(iv) Compute the targets for the previous predictions y_{t-1} using y_t and the observation O_t according to the question networks links and the action conditions.
(v) Update the weights W according to the difference between targets and the predictions.

4. COMPUTATIONAL SCHEME

TD methods as such cannot be directly applied because each prediction is of a different event and thus there is no clear desired relationship between them.

More formally, let x_t be a vector of observation available at time t (month) in a sequence of observations represented by {x_1, x_2, ….., x_{12}}, and y_t denotes the prediction of the outcome of sequence. Let { P_1, P_2, ….., P_{12}} is a sequence of predictions produced for each sequence {x_1, x_2, ….., x_{12}}. Each sequence of predictions is an estimate of y. If P_t^0 is the actual rain in t^{th} month, then one can easily written as

$P_t^d = P_{t+1}^{d-1}$, for all t, $1 \leq d \leq 12$ (3)

Where P_t^d is an estimate of rain in d months later. Using these recursive equations expressing the desired relationship between predictions at different times in the sequence it is assumed that the predictions are based on a vector of modifiable weights, w^m. For each observation, an increment to w, Δw is determined. Therefore, we can write

$$w^m = w + \sum_{t=1}^{12} \Delta w_t \qquad (4)$$

The conventional TD (λ) family of learning procedures uses an exponential weighting with recency, in which alternatives to the prediction of observation vectors occurring k steps in the past are weighted according to λ^k, for $0 \leq \lambda \leq 1$,

$$\Delta w_t = \alpha\,(P_{t+1} - P_t) \sum_{k=1}^{t} \lambda^{t-k}\, \nabla_k P_k \qquad (5)$$

where, α = learning rate

$(P_{t+1} - P_t)$ = difference between temporally successive predictions

λ^{t-k} = provides the recency factor

and $\nabla_w P_k$ is the vector of partial derivatives of P_k with respect to each component of w.

For fixed step problem, the rule for updating w can be written as

$$\Delta w_t^d = \alpha\,(P_{t+1}^{d-1} - P_t^d) \sum_{k=1}^{t} \lambda^{t-k} \nabla_w P_k^d \qquad (6)$$

In a MLP, $\nabla_w P_t$ is calculated by a backpropagation process as described below:

Learning algorithm in BP

$$\nabla_w P_t = \frac{\partial P_t}{\partial w_{l,j,i}}, \qquad (7)$$

Let, $u_{0,i}$: ith component of the training vector,

$u_{l,j}$: output of the jth node in layer l

$w_{l,j,i}$: weight which connects the ith node in layer l-1 to the jth node in layer l.

N_l = number of nodes in layer l

L = no layers, and

$P_t = u_{L,1}$

Assuming a sigmoid nonlinearity for the final output layer, equation (7) can be rewritten as

$$\frac{\partial P_t}{\partial w_{L,1,i}} = \frac{\partial u_{L,1}}{\partial w_{L,1,i}} - u_{L,1}(1-u_{L,1})u_{L-1,i} \qquad (8)$$

Equation (8) is used to compute all the partial derivatives of equation (7). Assuming the form of error at time step t, e_t as expressed into equation (9),

$$e_t = \sum \lambda^{t-k} \nabla_w P_k^d \qquad (9)$$

Now, the error at next time step, i.e., e_{t+1}, can be well established by the following recurrence relation

$$e_{t+1} = \nabla_w P_{t+1}^d + \lambda e_t \qquad (10)$$

The reason behind taking the exponential form of e_t is that it can be computed incrementally. Thus using only current information, the value of e_t for t, e_{t+1} can be computed incrementally.

Like all learning methods, TD methods can be viewed as Gradient Descent in the space of the modifiable parameters (weights) [Sutton, 2005]. Formally minimization of the criterion as given in equation (11) is achieved.

$$J(w) = \sum_{p=1}^{M} \sum_{k=1}^{NP} \lambda^{NP-k} (y - P_t(x_p(k)))^2 \qquad (11)$$

where, M is the number of examples,
NP is the number of steps in the pth example = 12

y	is the actual outcome of the pth example
$P_t(x_p(k))$	is the output of the network when presented with $x_p(k)$

$\lambda \in [0,1]$ places more emphasis on predictions temporally close to the outcome.

It is important to note that, in the definition of J(w), the real aim is for each prediction to match the expected value of the subsequent outcome, not the actual outcome occurring in the training set. TD methods can perform better than supervised learning methods because the actual outcome of a sequence is often not the best estimate of its expected value.

5. RESULTS

Temporal sequences of rainfall (actual) patterns for a period of 17 years are as shown in Figure 2.0.

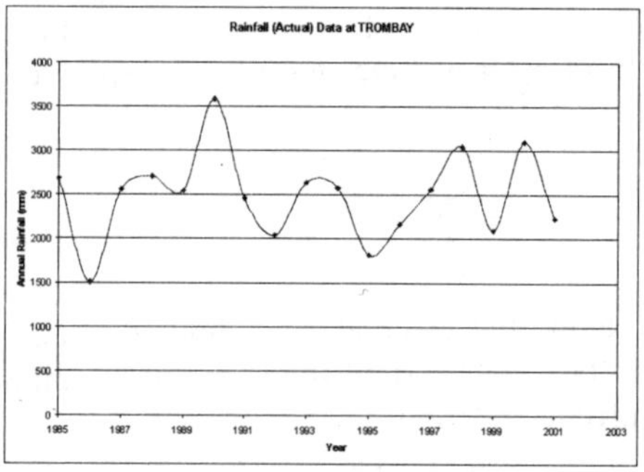

Fig. 2: Yearly Variation of Rainfall (actual) at Trombay

A 3 layer MLP was trained with this temporal sequence. Two such networks were developed – one trained with the conventional backpropagation technique and the other with the combination of methods described above. The λ value was set at 0.1 so as to place more emphasis on recent predictions temporally close to the actual outcome. A momentum term was included in both the networks to speedup the convergence process. It has been observed that, after a certain number of iterations (100000), the backpropagation model gets trapped in local maxima and leads to sub-optimal predictions (Figure 3.0), whereas the new method gets converged to near-optimal predictions after 7500 iterations (Figure 4.0). This proves the generally accepted theory that backpropagation minimizes error on the training set and it does not necessarily minimize error for future experience. It is TD (λ) that converges to what can be considered the optimal estimates for matching future experience.

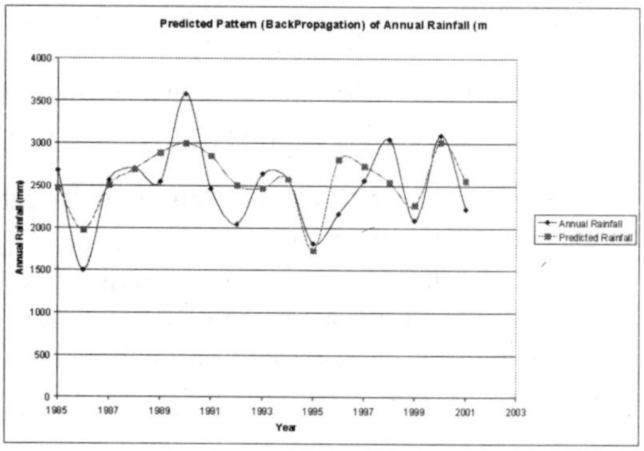

Fig. 3: Backpropagated Predicted Pattern of Annual Rainfall (mm) at Trombay

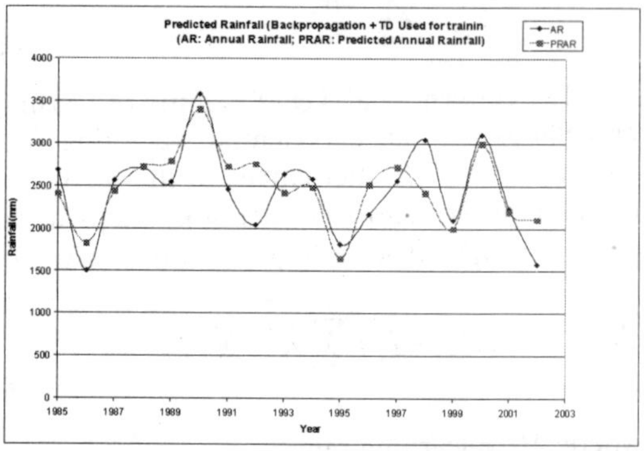

Fig. 4: Predicted Annual Rainfall (mm) with combined effect of Backpropagation and TD Learning

6. CONCLUSION

TD networks suggest a largest set of possibilities for learning to predict. Results show that even in a completely observable setting there may be significant advantages to TD methods when learning TD-defined predictions. Results show that TD coupled with backpropagation methods can learn dramatically faster than other methods. It may be fruitful explore the expensive potential of TD-defined predictions. Based on the current work, it can also be stated that the combination of TD and backpropagation learning method may be very well applicable to predict the release of gaseous and liquid effluent into the environment around any nuclear facility at any particular year. This predicted release limit then can be applied to compute the dose received by the members of the public. The methodology suggested may be a scientific basis of dose apportionment in the field of radiation protection practices.

7. REFERENCES

1. Cybenko, G., 1989, Approximation by Superposition of a Sigmoidal function, Mathematics of Control, Signals and Systems, 2(4), pp. 303-314.
2. Dayan, P., 1993, Improving generalization for temporal difference learning: The successor representation, Neural Computation, 5(4), pp. 613-624.
3. Littman, M., Sutton, R.S., and Singh, S., 2002, Predictive representations of state, Advance in Neural Information Processing Systems, 14, Cambridge, MA:MIT Press.
4. Rumelhart, D.E., and McClelland, J.L. (eds.), 1986, Parallel Distributed Processing, 1, M.I.T. Press, Cambridge, M.A.,
5. Sutton, R.S., 1988, Learning to predict by the methods of temporal differences, Machine Learning, 3, pp. 9-44.
6. Sutton, R.S., et al., 1999, Between MDPs and Semi-MDPs, A framework for temporal abstraction in reinforcement learning, Artificial Intelligence, 112, pp. 181-121.
7. Sutton, R.S., 1995, TD Models, Modelling the world at a mixture of time scales, In A. Prieditis and S. Russel (eds.), Proceedings of the 12th International Conference on Machine Learning, pp. 531-539, Morgan Kaufmann, San Francisco.
8. Sutton, R.S., and Tanner, B., 2005, Tempral difference networks, Advances in Neural Information Processing Systems, 17, Cambridge, MA: MIT Press, pp. 1377-1384.

Reliability, Safety and Hazard
Editors: P.V. Varde et al.

RELIABILITY ANALYSIS OF SECONDARY SHUTDOWN SYSTEM OF RAPS-3&4: FINDINGS AND SOLUTION

Brajesh Kumar, P. Chattopadhyay and Hemant Kumar

Technical Services Unit, Rajasthan Atomic Power Station-3&4
PO. Anushakti via Kota 323 303

ABSTRACT

The purpose of Reliability Analysis is to validate system design, demonstrate adequacy of maintenance program and testing requirements by means of comparison of reliability analysis results of various safety systems with limiting values assumed in plant safety report. Fault tree analysis is the most widely used method for developing system models. The reliability analysis of Secondary Shutdown System (SSS), done at site has revealed many important facts, which necessitated change in trip logic circuitry to increase redundancy and thus reliability of the system. Also, based on this analysis, the surveillance testing frequency has been proposed to change to reduce the chances of wear and tear of SSS components. This paper describes the methodology adopted for reliability analysis of SSS, and discusses its findings and suggested remedies.

Keywords: Shutdown System, Fault Tree, Unavailability, Local Coincidence, Logic circuit

1. INTRODUCTION

Consequences of a nuclear accident are far reaching. Therefore, for nuclear power plant safety, philosophy of 'defence in depth' is followed, which states simply that plant safety should not depend on only one physical barrier or system. The terms "prevention", "protection", "mitigation" and "accommodation" are manifestation of the 'defence-in-depth' concept. That is to say, first, try to prevent accidents (by operating the plant with in safe parameter range); second, provide protection (to stop their progress); and third, provide mitigation for (to reduce their effects if they occur), fourth, keep large areas of no man's land around the plant and have emergency preparedness plan to accommodate ill effects of accident. In case of an accident, four safety functions are required in a nuclear reactor. These are Reactor shut down, Decay heat removal, Containment of radioactivity, and Monitoring the state of the plant

Shutdown is one of the most important safety functions in a reactor because it reduces the amount of energy that has to be removed from the fuel after an accident. It is usually accomplished through rapid insertion of a neutron-absorbing material into the core. Another way is to remove moderator from the core, which is essential to the chain reaction. Pressurised Heavy Water Reactors have Positive void coefficient of reactivity. Therefore, it is absolutely necessary to have highly reliable shutdown system. Hence, in PHWR type of nuclear power plants, two independent shutdown systems are provided to increase redundancy. In 220 MWe Indian standard PHWR nuclear power plants, there are two fast acting shut down systems, namely, Primary Shutdown System

(PSS) and Secondary Shutdown System (SSS). Keeping in view, importance of these systems, their design includes various features like 'fail safe to power failure', preventing 'system failure on single component failure', avoiding 'common cause failures', providing 'redundancy and coincidence in trip logics' to make them reliable and robust.

2. SYSTEM DESCRIPTION

Primary shutdown system is the prime means for reducing reactor power from full power state to sub-critical state during normal as well as accident conditions. It consists of 14 cadmium sandwiched SS rods and winch type drive mechanism at the top of calandria. During normal operations, these rods are parked out side the reactor with the help of electromagnetic clutch for each rod. On sensing a rector trip command, the electromagnetic clutches de-energizes and the rods drop into the reactor under gravity. The rods are taken out for restarting the reactor with the help of worm and worm wheel and drive motor.

Secondary Shut down System is back up to the primary shut down system. The system consists of twelve vertical tubes located in calandria shell. Shut down is achieved by filling these tubes with solution of lithium pentaborate (a neutron absorber) in heavy water. These twelve tubes are grouped in four banks, each bank comprising of 3 tubes and provide similar worth. Three banks are sufficient to provide negative reactivity worth for bringing the reactor to shut down state. In any bank, the poison reaches up to the core height within 1.8 seconds of initiation of the system. This is done by pressuring the system with helium gas of pressure 10-14 kg/cm^2. Each bank has its independent gas tank and liquid poison tank feeding three shut off tubes through SS piping. The high pressure gas tank is connected to the poison tank through two fast acting solenoid valves (normally closed) connected in series. These valves are in closed position during normal operation and opened when secondary shutdown system is required to actuate.

The series arrangement of the valve permits on line testing of the system, while retaining the availability of the bank during testing. Pressure balance line connecting top of shut off tube and liquid poison tank also has duplicated fast acting solenoid valves connected in series. These valves remain open during normal operation and get closed on shutdown demand. Valves closing time is selected in such a way that the valves in pressure balance line close earlier than the opening of valves connecting high pressure gas tank to poison liquid tank. Shut off

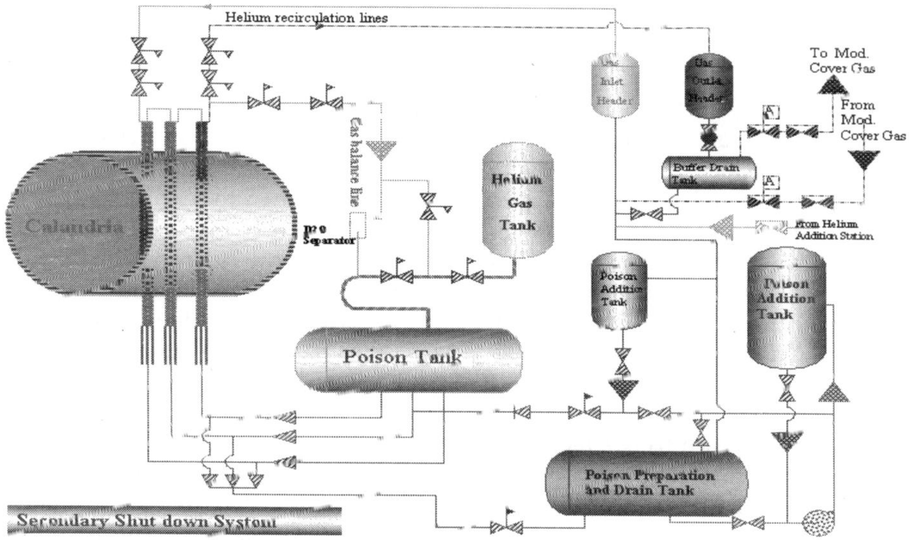

Fig. 1: Secondary Shutdown System

Tube atmosphere is connected to moderater cover gas through Helium recirculation lines. This necessitated introduction of two more pairs of fast acting solenoid valves in the cover gas lines to and from the shutoff tubes. These valves are to be closed for providing positive isolation of shutoff tubes from cover gas circuits during SSS actuation.

3. TRIP LOGIC CIRCUIT

SSS trip logic circuit is designed keeping in view following salient features :

1. Local coincidence for trip actuation.
2. System testability in poised condition.

SSS actuates on local coincidence logic, meaning SSS trip will occur only if 2 out of 3 channels actuate on sensing same trip parameter.

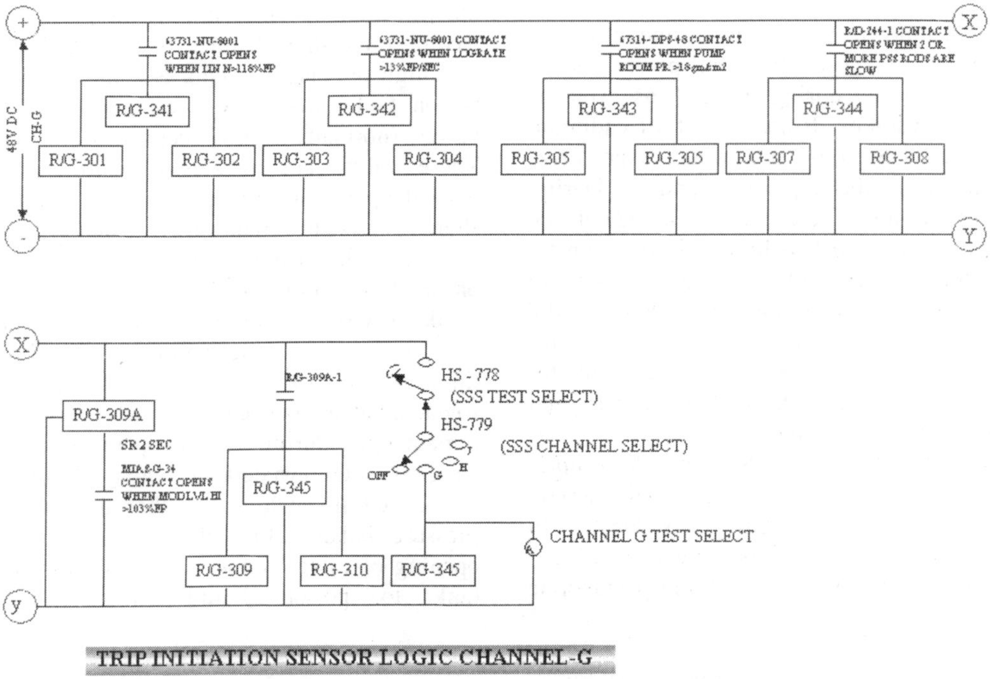

Fig. 2: Trip Initiation Sensor Logic Ch-G

For SSS actuation, there are three independent trip logic channels G, H, and J. To design logic circuit for local coincidence, for each trip parameter, there are three different sensors; each sensor is having three independent relays for sending signal to three trip channels (total 9 relays for three channels). Each trip channel receives signal from all three sensors but through three independent relays forming a 2/3- logic made up of three relays getting signal from three sensors measuring same trip parameter.

For allowing testing, each bank's four pairs of solenoid valves are divided in two groups such that two valves of a pair go to different group. Further, to avoid bank actuation while channel testing, each group of a bank is connected with different channel. Therefore, for actuation of any bank, two channels (connected with it) have to actuate. CH-G trip logic circuit operates group-1 valves of banks 1, 3 & 4. CH-H trip logic circuit operates group-2 valves of banks 1, 2 & 4. CH-J trip logic circuit operates group-1 valves of bank 2 and group 2 valves of bank 3. Therefore, all three channels should trip for actuation of three/four SSS banks. To make SSS actuate on 2/3 channels coincidence, a latching relay in each channel is provided which trips and latches the channel if other two channels are in tripped condition for more than 120-milli sec. This relay also latches own channel, if channel is in tripped condition for more than 100-milli sec.

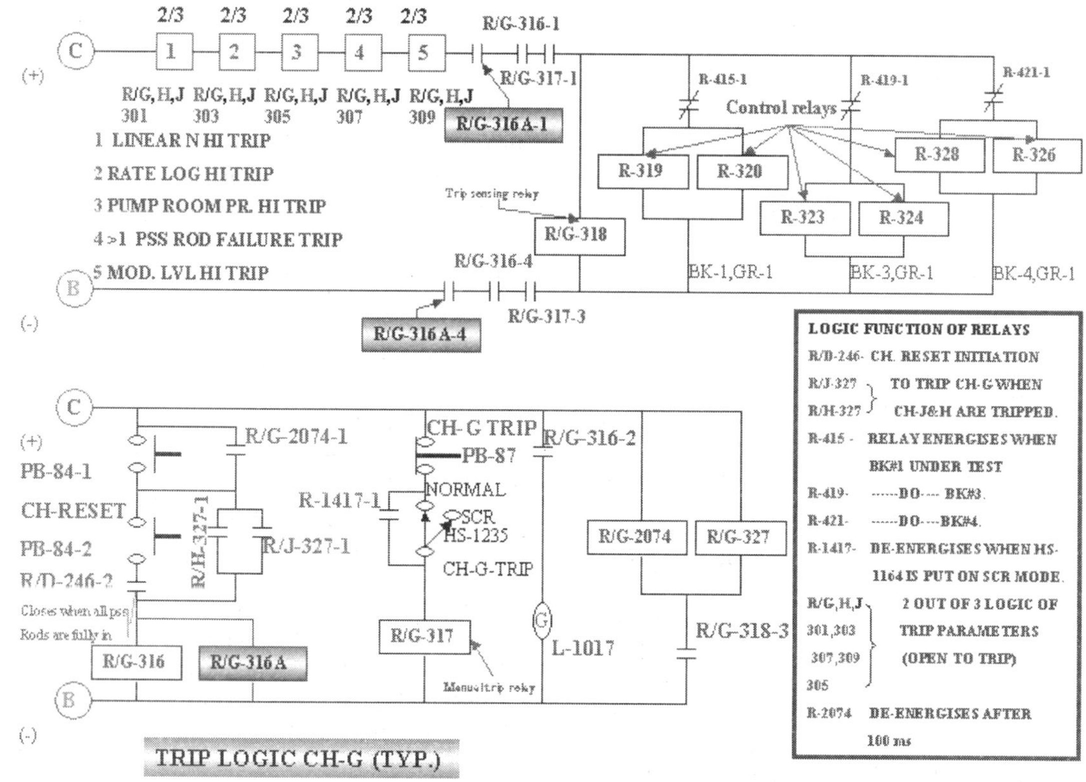

Fig. 3: Trip Logic Circuit Ch-G (typ.)

4. SYSTEM MODELING FOR RELIABILITY ANALYSIS

Reliability Analysis is assessment of reliability of a system from its basic elements. It is done for Safety and Engineered Safety Systems. Its purpose is to demonstrate adequacy of maintenance program and testing requirements by means of comparison of reliability analysis results of various safety systems with limiting values assumed in plant safety report. Reliability measure used is mean unavailability of system. Unavailability of any system is found out by analyzing all possible failure logics/ modes. System modeling is done for failure analysis. Fault tree analysis is the most widely used method for developing system models. **Fault Tree** is a reliability analysis technique, which is a deductive model (deduction: reasoning from general to specific), first postulates that the system has failed in a certain way and then attempt is made to find out what modes of system/component behaviour contribute to this failure. A fault tree starts with the system's undesired event (system failure) at the top of the graphical representation, and then proceeds downwards identifying the events/combination of events in terms of component failure, which cause the top event. Thus providing a graphical model of the various combinations of events that will result in the occurrence of the predefined undesired event. All the causes/ components failures are connected to the top events through logical OR/AND gates.

For SSS "Failure of 2/4 banks to actuate" is kept as top event. Poison fails to remain inside and poison fails to enter, both of these conditions are considered for failure of any bank.

SSS Fault tree was developed incorporating all possible process, mechanical, structural, instrumentation and logic failures, which can occur at any time in the system

Generic and Bayesian updated plant specific data was used to quantify fault tree.

Though there has been no incidence of Common Cause Failures (CCF), the CCF groups have been considered on the potential for such failures. The CCF parameters have been considered based on the generic values for the Alpha Factor Model.

The generic human error probability values given in SWAIN Handbook have been used for latent human actions. Dynamic human actions are quantified using Human Cognitive Reliability (HCR) model based on TEC-DOC-592.

5. SOFTWARE USED FOR ANALYSIS

Reliability analysis of SSS was done using "RISK SPECTRUM" PSA Professional 2.00 software of REL CON, AG of Sweden.

For Bayesian updating of plant specific data, a software called PSTERIOR was used.

6. FINDING NO. 1 AND ITS SOLUTION

SSS fault tree run was taken to generate minimal cut sets. Minimal cut set analysis showed channel-latching relays coming as single order cut set. This revealed that all three SSS channels should trip in order to actuate all four banks if any one latching relays R/-G, H, J-316 fails in stuck close position.

It was thought to incorporate an extra relay parallel to latching relay in each channel to provide redundancy to latching relay and hence, meeting single component failure criterion.

Fault tree analysis of SSS with and without thought out modification, was done to see the impact of change on system unavailability. Which came out to be as given below:

1. 3.26 E-04 yr/yr- without modification.
2. 6.86 E-05 yr/yr- with thought out modification.

For calculation, testing frequency of tested model type basic events was used as one week. Total accounting period was taken as 4 reactor years, considering 4-reactor years operation for each unit.

Results clearly, showed marked improvement in system reliability. Finding and results were discussed with designer and It was proposed to incorporate an extra relay parallel to latching relay in each channel.

7. FINDING NO. 2 AND ITS SOLUTION

As per regulatory guidelines, SSS manual trip test was being performed on daily basis. One trip channel was tested in each shift. Testing involved tripping of individual channel and recording opening or closing time of one group of Solenoid Valves of 2 or 3 banks, depending upon the channel tested. Site apprehension in daily manual trip testing was, the frequent operation of such fast acting valves may result in wear & tear of plunger, seat or stem. This could have resulted in passing of the valves and thus increasing chances of system not getting actuated as per design intent, when it is called for. This could also have increased down time and man-rem expenditure on maintenance.

A proposal was submitted by site to regulatory body regarding change in test frequency from daily to weekly. Proposal was supported by Reliability Analysis of SSS to find out the change in system reliability when testing frequency is changed from daily to weekly.

Reliability Analysis of SSS was done twice, once by keeping frequency of all basic events of tested model type 'daily' and second time by keeping the frequency 'weekly'. System unavailability figures came out as given below:

$6.845*10^{-5}$ yr/yr based on weekly basis testing.

$5.899*10^{-5}$ yr/yr based on daily basis testing.

These results indicated that by changing the test frequency from daily to weekly does not substantially affect the reliability figures.

The proposal for change in testing frequency of SSS, from daily to weekly, was accepted and approved by regulatory bodies.

8. CONCLUSION

Secondary Shutdown system has to work against gravity and poison is in the form of liquid. Therefore, in the case of Secondary Shutdown System, a deliberate force (hence trip signal) is needed constantly, for its insertion and for the poison to remain in the core as long as it is required. Hence, trip signal should exist as long as poison has to remain in the core. This is done by Trip channel latching. By providing a redundant latching relay in each channel, trip logic circuit is not only meeting single component failure criterion for 2 out of 3 Trip channels Local coincidence scheme it is also providing redundancy in channel latching.

Testing frequency of Secondary Shutdown System

Trip Channels has been changed from daily to weekly at RAPS-3&4 and KAIGA-1&2. So far no failure has been detected.

9. REFERENCES

1. PSA Level-I studies of Kakrapar Atomic Power Station.
2. IAEA TechDoc-478 on Component Reliability Data for use in Probabilistic safety Assessment.
3. Detailed Review Report on KAPS Level-1 PSA by Committee on PSA for Nuclear Facility.
4. Secondary shut down System Design Manual DM/RAPP-3&4/31340/148/R-1.
5. Operational Flow sheets and Engineering Drawings of RAPS-3&4 Secondary Shutdown System.
6. Theory manual of PSA software Package, RISKSPECTRUM PSA Professional 2.00 of REL CON AB of Sweden.

Reliability, Safety and Hazard
Editors: P.V. Varde et al.

SAFETY IMPROVEMENT OF MAPS CONTAINMENT ISOLATION SYSTEM BASED ON RELIABILITY ANALYSIS

D. Sayee Lakshman, K. Padmanabhan, V.A. Subramani and S. Krishnamurthy

Madras Atomic Power Station, Kalpakkam 603 102

ABSTRACT

The Containment isolation at MAPS is provided by two dampers in series in the supply and exhaust of RB ventilation system ducts. The pneumatically operated isolation dampers are activated by Boiler room pressure high signal or containment high activity signal through respective relay logics and solenoid valves. Scenario of one of the exhaust dampers closing spuriously during reactor operation and during the period of corrective action, the requirement of damper to be kept open was taken up for study. To review the Containment Isolating System(CIS) reliability Risk Spectrum Fault Tree(FT) methodology was used to determine the Probability of Failure on Demand(PFD) and spurious failure rates of CIS dampers. Suggested improvements in the logic design to reduce the PFD and spurious failure rate are discussed.

1. INTRODUCTION

The Madras Atomic Power Station consisting of two identical units of 220 MW(e) each is located at Kalpakkam village on the shore of Bay of Bengal, 88 kms. south of Chennai in Tamil Nadu state. Unit-1 was in operation from July-1983 and Unit-2 was from Sept-1985 and generated 19,921MU and 19,661MU respectively till August - 2005.

The containment system of a Nuclear Power Plant (NPP) is a safety system provided to limit/mitigate the consequences of postulated accidents in order to protect plant personnel, public and environment. The containment serves as the final physical barrier for the radioactivity contained in the reactor core and hence has an important role in protection of the plant personnel, public and environment during Design basis event

All piping system or other openings that penetrate the containment envelope shall have appropriate provisions to bring about containment isolation when required. In this respect, the Reactor Building Ventilation system exhaust ducts are provided with isolation dampers (the exhaust duct is connected to the stack). There are two dampers in series in the supply and exhaust duct to ensure reliable isolation. The dampers are pneumatically operated and actuated by Boiler room high-pressure signal or containment high activity signal through respective relays and solenoid valves. This system is known as containment isolation system, which includes the control and instrumentation.

Scenario of one of the exhaust damper closing during Unit operation:

During normal unit operation, the two exhaust dampers and supply dampers will be in open condition. These dampers will close automatically in 3 seconds due to Boiler room pressure high signal or containment activity signal during LOCA condition. If any one of the exhaust damper closes spuriously during unit operation (i.e. without any LOCA condition) then boiler room pressure will rise. In order to keep the unit operating it is required that the damper has to be kept open during the period of corrective action. In this scenario, the reliability analysis of the containment isolation system is done using Risk spectrum Fault tree methodology to determine the probability of failure on demand and spurious failure rates of containment isolation dampers.

Functioning of system interlocks

High activity containment logic (logic b):

Triplicate (QT-33, 34 and 35) gross beta – gamma instrumentation monitors the gross beta-gamma activity in the Reactor building exhaust. `Very high' activity, which is 150 times the normal background subject to a maximum of 250 mr/hr, actuates the logic `b'. Two out of three input signals to QT–33 to 35 will actuate the logic b. All the relays in the logic circuit are energised to `open' and de-energised to close. When logic `b' actuates Relay (R-4749) which in turn de-energises, the respective relay and Solenoid valve resulting in closure of the four isolation dampers.

High pressure containment logic (logic-c):

Pressure monitoring is done from boiler room. A rupture in the coolant circuit will manifest itself in reactor building pressure rise and reactor building temperature rise. Pressure monitoring is a triplicate system. Boiler room pressure is measured by a triplicate differential pressure switches PS-19, 20 and 21. These switches are located in the boiler room. The reference side is taken from atmosphere in service building. When two out of three input signals of boiler room pressure with reference to atmosphere in service building exceeds 0.018 kg/cm^2 and primary heat transport system temperature is more than 101°C, the logic `C' will actuate and also trips the reactor. When logic `C' actuates it de-energises Relay-4748 which in turn de-energises the respective relays and solenoid valves and then closing all the four isolation dampers.

Reliability analysis of the containment isolation system: Reliability Analysis using Risk spectrum Fault tree methodology (Fig.3) is done to evaluate the relative contribution of the relays. The containment isolation system logic diagram is shown in the Fig-1. It was observed from the Fault tree analysis the probability of failure on demand of containment isolation system is 6 X10^{-4}. If one damper is disabled, then the probability of failure on demand of other three dampers is 1.4X10^{-3}. It was observed from the minimal cut sets that the contribution of R-4748 and R-4749 is 49% each. Thus indicating the dominant contribution of these two relays in the system probability of failure on demand. If any one relay is not getting de-energised (single relay failure), then all the four dampers will not close (unsafe failure) on Reactor building high pressure or Containment activity high logic.

During unit operation any one relay spuriously operates then all the four dampers will close (safe failure) resulting in reactor building pressure rise and the operator has to shutdown the unit as per Technical specification requirement, even though there is no situation which calls for closure of the reactor building isolation dampers.

FT RESULTS OF THE EXISTING SYSTEM

No	PROB.	%	COMPONENT	
1	3.00E-04	49.01	R-4749 FLR	
2	3.00E-04	49.01	R-4748 FLR	
3	8.26E-07	0.13	TE-479 FLR	TE-484 FLR
4	8.26E-07	0.13	TE-479 FLR	TE-902 FLR
5	8.26E-07	0.13	TE-484 FLR	TE-902 FLR
6	2.73E-07	0.04	R-4611 FLR	TE-902 FLR
7	2.73E-07	0.04	R-4612 FLR	TE-479 FLR
8	2.73E-07	0.04	R-4612 FLR	TE-484 FLR
9	2.73E-07	0.04	R-4611 FLR	TE-479 FLR
10	2.73E-07	0.04	R-4610 FLR	TE-484 FLR
11	2.73E-07	0.04	R-4619 FLR	TE-902 FLR
12	2.73E-07	0.04	R-4622 FLR	TE-902 FLR

FT RESULTS OF THE PROPOSED SYSTEM

No	PROB.	%	COMPONENT	
1	8.26E-07	5.87	TE-484 FLR	TE-902 FLR
2	8.26E-07	5.87	TE-902 FLR	TE-479 FLR
3	8.26E-07	5.87	TE-484 FLR	TE-479 FLR
4	2.73E-07	1.94	R-4622 FLR	TE-902 FLR
5	2.73E-07	1.94	R-4610 FLR	TE-484 FLR
6	2.73E-07	1.94	R-4625 FLR	TE-479 FLR
7	2.73E-07	1.94	R-4612 FLR	TE-479 FLR
8	2.73E-07	1.94	R-4619 FLR	TE-902 FLR
9	2.73E-07	1.94	R-4611 FLR	TE-902 FLR
10	2.73E-07	1.94	R-4625 FLR	TE-484 FLR
11	2.73E-07	1.94	R-4610 FLR	TE-902 FLR
12	2.73E-07	1.94	R-4622 FLR	TE-479 FLR

FLR – Failure

MINIMUM CUTSETS OF THE EXISTING CONTAINMENT ISOLATION SYSTEM

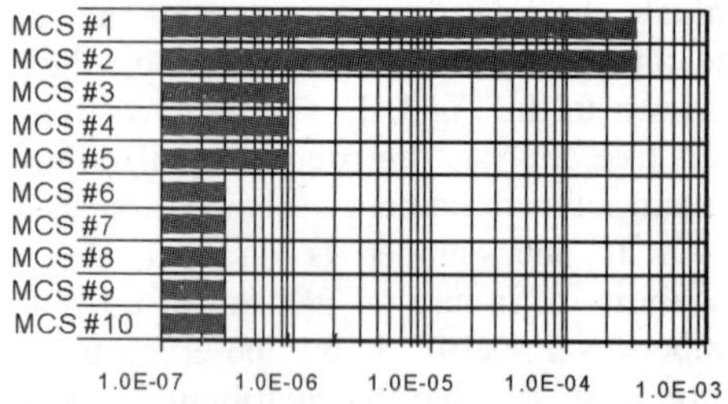

MINIMUM CUTSETS OF THE PROPOSED CONTAINMENT ISOLATION SYSTEM

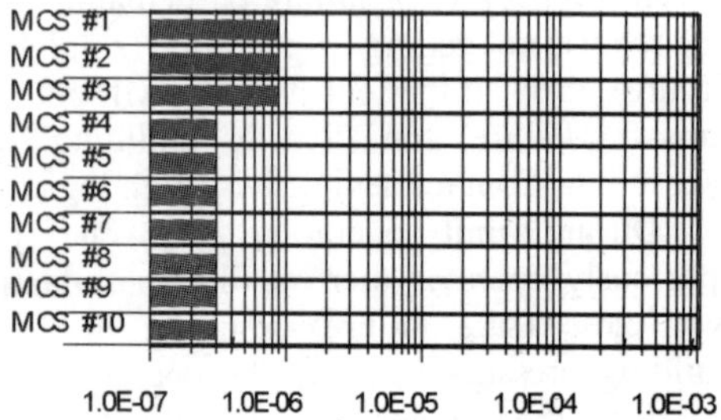

2. PROPOSED LOGIC SCHEME

The existing logic scheme contains five relays, which contributes for closure of the four isolation dampers. In the proposed logic scheme only four relays were rearranged in such a way that (Fig. 2) the reliability of the system improved to 1.4 $X10^{-5}$. The scenario of spurious operation of one damper getting closed and if the damper is kept open during corrective action, while unit is operating, the probability of failure on demand increases to 8.3 X 10^{-4}.

Comparison of the probability of failure on demand

Condition	Existing scheme PFD	Proposed scheme PFD
All 4 dampers are operable.	6 X 10^{-4}	1.4 X 10^{-5}
One damper closing spuriously and kept open during unit operation (only three dampers are operable).	1.4 X 10^{-3}	8.3 X 10^{-4}
One relay spuriously de-energised	1 - All the four dampers will close.	$1.64X10^{-3}$ - One supply and one exhaust damper will close.

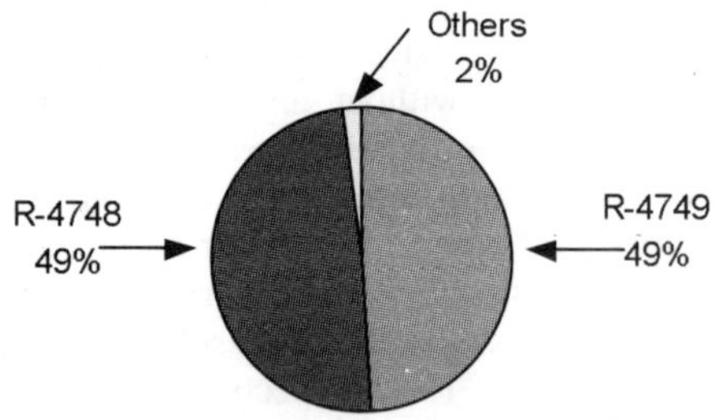

PERCENTAGE OF COMPONENTS CONTRIBUTION (EXISTING SCHEME)

From the Pie chart, it indicates the failure contribution in the existing scheme is 98% from the 2 first order cut sets of single component failure of R-4748 and R-4749. In the proposed scheme, the first order cut sets are eliminated and the failure contribution is balanced, as there is no predominant contribution from any single component.

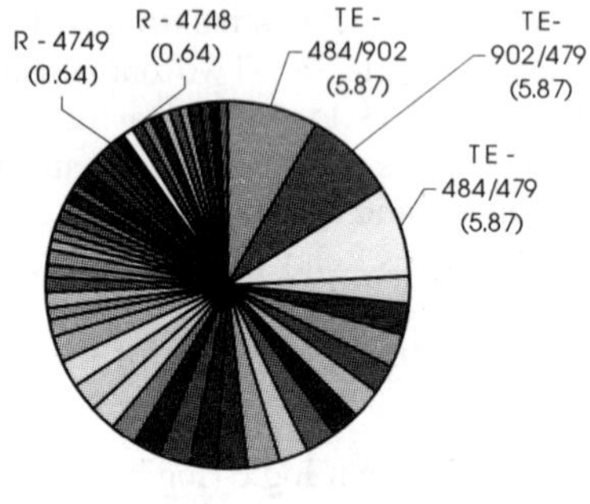

PERCENTAGE OF COMPONENTS CONTRIBUTION (PROPOSED SCHEME)

Advantages of the proposed scheme

- The reliability of the system improves from 10^{-4} to 10^{-5}
- Spurious operation of any one relay will result in closing of only one supply and one exhaust damper.
- The possibility of spurious operation of Reactor building isolation dampers is caused by four relays instead of five relays in the existing scheme.

3. CONCLUSION

The containment isolation system relay logic modification is proposed based on reliability analysis to improve the probability of failure on

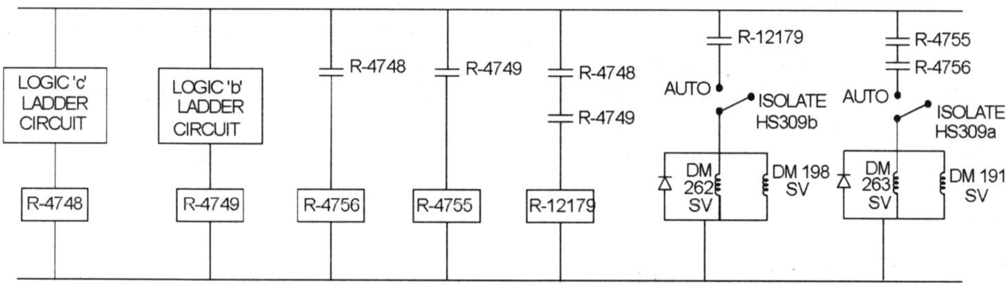

EXISTING LOGIC SCHEME OF RB ISOLATION DAMPERS (REF:- 67314-2009 ED)

Fig.1

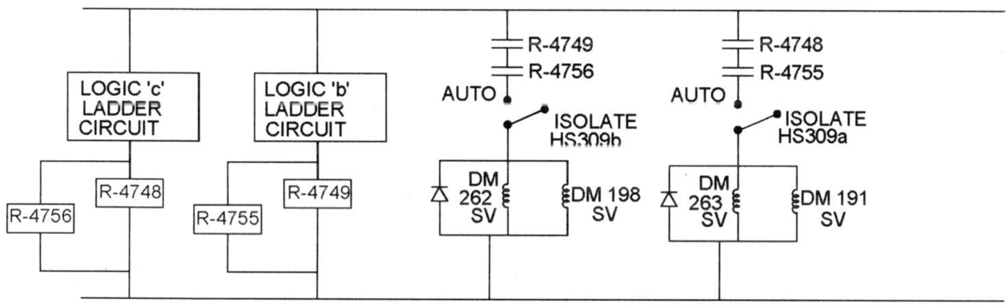

PROPOSED LOGIC SCHEME OF ISOLATION DAMPERS

Fig.2

FAULT TREE FOR EXISTING SYSTEM

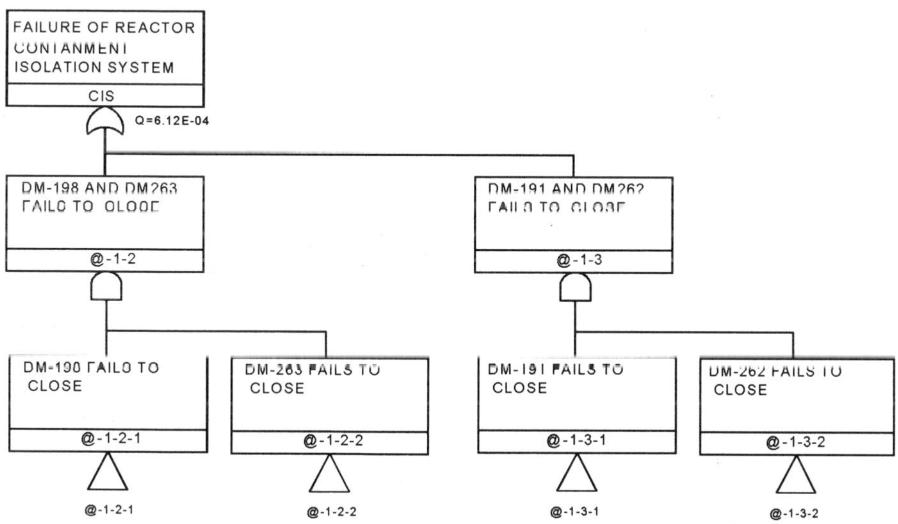

Fig.3

demand to isolate the containment as well as to decrease the spurious failure rates of the dampers. Essentially single component failure (relay) leading to system failure has been eliminated to ensure overall safety and operation reliability.

4. REFERENCES

1. 67314 – 2009 Engineering Drawing for Containment isolation dampers.
2. Plant specific failure data obtained from MAPS Monthly and Annual Performance reports.
3. Risk spectrum professional version – 1.20

Reliability, Safety and Hazard
Editors: P.V. Varde et al.

RELIABILITY STUDIES OF RAPS-1&2 SAFETY SYSTEMS

K.K. Jha and L.K. Jain

Rajasthan Atomic Power Station-1&2, Rawatbhata

ABSTRACT

This paper covers the Reliability assessment methodology for various safety systems of RAPS-1&2. The detailed Reliability assessment is done as part of 'Application for Renewal of Authorization (ARA) and Periodic safety Review (PSR). The Fault Tree (FT) analysis method is used for this evaluation. Both plant-specific and generic data are used for reliability calculations. The quantitative evaluation of Fault trees is done by the computer program 'Risk spectrum'. The calculated values of Safety system Reliabilities indicate their good operational health and adequacy of maintenance program & surveillance requirements.

1. INTRODUCTION

The Reliability analysis of RAPS-1&2 safety systems is done on a regular basis to assess their performance in a quantitative way. Detailed Reliability studies is done as part of 'Application for Renewal of Authorization (ARA) and Periodic safety Review (PSR). Various safety systems for which Reliability analysis is done are - Reactor Shut down system, Emergency core cooling system, Emergency feed water system, Fire fighting water system, Emergency power supply system (Class-I, Class-II & Class-III), Shutdown cooling system, Containment system and Dousing system.

The Reliability Analysis of a system involves the development of a logical model which represents the relationship between its various components and elements, and interaction among various components for occurrence of a specified system state. Output of the analysis provides qualitative information on various success/failure modes of the system, and quantitative measure of its performance which reflect its ability or inability to do the required function.

2. SYSTEM MODELING

There are several system modeling methods of reliability analysis like Failure mode and effect analysis (FMEA), Failure mode effect and criticality analysis (FMECA), Preliminary hazard analysis (PHA), Fault hazard analysis (FHA), Fault tree analysis, Parts count approach, Double failure matrix (DFM), Success path model etc. Out of all these available methods, The "Fault tree" method is the most widely used for system modeling for reliability analysis. This method has been used in the Reliability analysis of RAPS –1&2 safety systems. The systems were modeled systematically down to the last component on process as well as on instrumentation logic side. The logic modeling included complete control circuits, sensors (i.e. level transmitters, pressure transmitters, relays etc.).

3. THE FAULT TREE METHODOLOGY

The reliability analysis technique based on the fault trees accounts for almost all concerned important features of the systems such as operation modes, status monitoring, test and maintenance strategies.

The Fault tree analysis is a deductive failure analysis method, which can be simply described as an analytical technique whereby an undesired state of a system is specified and the system is then analysed in the context of its environment and operation to find all credible ways in which the undesired state could be brought about. The fault tree itself is a graphical model of various series parallel combinations of faults that will result in the occurrence of predefined undesired event. The faults can be events that are associated with component hardware failures, human errors or any other pertinent events which can lead to the undesired event.

RAPS-1&2 safety system Fault trees have been constructed on the basis of rules given in the Fault Tree Hand Book (NUREG-0492). [4]

4. FEATURES OF SYSTEM FAULT TREES

Following are the salient Features of System Fault trees :

a) The system failure criteria had been clearly identified as the top event of Fault tree.
b) The fault trees were constructed using the immediate cause concept, starting with the Top event and down to the basic events (primary component failure or root case event).
c) Basic Event coding was decided before developing the fault tree. It ensured the standardization in the Reliability Analysis Reports. The coding was compatible with the Computer code used and related the basic event clearly to the component failure mode, component identification and specific system in which the component was located.
d) The System boundaries were clearly identified.
e) The Tech spec requirements were taken into consideration during FT construction and RA evaluation. Due care was taken so that the minimal cut sets do not show any combinations of events that violate the technical specification requirements.
f) Following component failure modes were reflected in the Fault Trees:
   i) Contribution due to demand failure.
   ii) Contributions due to outage for testing and maintenance.
   iii) Contribution due to failure during mission time.
   iv) Contribution due to run time failures i.e. fail to run
g) A number of documents were referred for FT construction. Discussion with O&M personnel and concerned System Engineers also helped a lot in FT construction. Some of the documents referred were as follows :
   i) Operational Flowsheets
   ii) Electrical drawing
   iii) Design manuals
   iv) Test OFS
   v) Technical Specifications
   vi) Station policies

5. SUPPORT SYSTEMS CONSIDERED

Support systems are those systems, which are required to operate for proper functioning of the main systems. Following support systems were considered available for Reliability Analysis.

a) 48V DC Control power supply for the operation of logic circuitry and relays.
b) 118 V AC Control power supply for the operation of relays and instruments like transmitters, controllers, IAMs etc.
c) 250V DC Control power supply for the operation of DC motors, Circuit breakers and breaker's control circuit.
d) 415 V AC Class-III power supply for the operation of various pumps and coolers.
e) 415 V AC Class-II power supply for the operation of various pumps & motorized valves.
f) Service air supply for the operation of various valves.
g) Instrument air supply for the operation of various Instruments.
h) De-mineralised water supply required on the primary side of various Heat exchangers.
i) L.P. & H.P. process water supply required on the secondary side of various Heat exchangers.

6. SYSTEM BOUNDARIES

These are the physical boundaries and component detail upto which the system has been analysed. Modeling for different components/conditions has been carried out within these boundaries only. The System boundaries were defined separately for Process, logic and power supply parts of the system.

a) For electrical power supply, the boundary was up to the bus feeding the component.
b) For support systems, the boundary was up to the main connection of support systems.
c) For the actuation signals, the boundary was up to the sensor/detector level.

7. KEY MODELING ASSUMPTIONS

Following assumptions are made for System modeling and reliability calculations to ensure consistency.

a) Failure of pipeline, elbows and other mechanical fittings are not considered.
b) Valve mechanical failures are not further deduced i.e. valve actuator failures and other failures are included in the valve mechanical failures.
c) Similarly Pump mechanical failures are not further deduces. Pump mechanical failure is considered as a Basic event.
d) Normal 'Auto' position of Hand switches are considered for logic development. Other positions –Open / Close - are not modeled.
e) Since the failure of the relay contact is a direct result of the failure of its corresponding relay, therefore relay failure is considered as a basic event wherever its corresponding contact causes a failure.
f) Spurious signals were considered only when component is expected to receive an additional signal during the course of system operation.
g) Similarly, the failures of Instruments are not deduced further. Instrument failure is considered as a basic event.
h) 'Reset' logic and 'overload protection' of Motorised valves were not deduced in the fault trees.
i) Trips of circuit breakers of pump motors due to protective signals were not considered in the fault trees.
j) Wherever test override facility exist, Modeling of related logic is not done and unavailability during test period is not considered.
k) For Reliability Analysis of a Power supply systems, other power supplies were considered available.
l) Failure of power supply / air supply leading to valve safe operation were not considered.
m) Failures due to Human errors were not considered.
n) To limit the size of fault tree deductions were not made for support systems.

8. DATA SOURCES

Following data sources were used for component failure data collection:-

a) Plant specific Data sources :

Surveillance Test OFS, Monthly Performance Reports, Incident Reports, Maintenance History Cards, Control Engineer log book, PM schedules etc. were utilized for processing the reliability parameters such as component failure data, component repair data, component test data, component maintenance data, testing frequency etc. Interaction with maintenance personnel helped in getting information about average equipment downtime.

b) Generic Data sources :

Following documents were referred for assigning generic data, whenever inadequacy of plant specific data was observed.

(i) IEEE – 500 [2]
(ii) IAEA-TEC DOC-478 . (Component Reliability data for use in PSA). [3]

9. PARAMETERS ESTIMATION

Following methodology was adopted for estimation of reliability parameters :

a) Clubbing of Data

For the component / equipment of a system operating in the same environment; the Reliability Parameters were calculated by using the appropriate formulae for clubbing. This was done to improve the population base for the respective reliability parameter calculation. Following Data was collected for clubbing

i) Total observed failures of similar components.
ii) Total operating time of similar components.

iii) Total no. of demands i.e. test demands + actual demands of these components.
iv) Total downtime due to maintenance of similar components.

b) Bayesian updating of data

Generic and plant-specific data were combined with the Bayesian technique and the posterior estimates so obtained were used for the analysis. This was done because of the inadequacy and uncertainty in plant specific failure data.

10. THE SOFTWARE PACKAGE

'Risk spectrum (Version-1.2) computer code' was used for reliability calculations. It is one of the most widely used package for reliability calculations and other PSA applications. This software contains all advanced modules for carrying out a Level-1 PSA.

Some of the results it gives for fault trees are:-

a) Top event probability
b) List of minimal cutsets
c) Probabilities of cutsets
d) Percentage contribution of each cutest in total unreliability
e) Percentage contribution of each basic event in total unreliability.

11. RESULTS & DISCUSSION

The results of RAPS-2 safety System Unreliabilities for the period 2001-2003 are depicted in Table-1. The calculated unreliability values of most of the systems are better than their Design values.

The systems for which, Design values of Unreliabilities are not specified in the Safety Report, comparision has been made with unreliabilities of similar systems in other PHWRs. The unreliability values are found to be comparable to the unreliabilities of similar systems in other PHWRs.

12. CONCLUSION

The calculated values of Safety system Unreliabilities indicate their good operational health and adequacy of maintenance program & surveillance requirements.

Table 1 RAPS-2 safety System Unreliabilities for the period 2001-2003 [1]

Sr. No	System Name	Unreliability (Yr/Yr)	Design Unreliability (Yr/Yr)
1.	Reactor Protective System	6.22×10^{-6}	10^{-4}
2.	Emergency core cooling system	8.17×10^{-4}	5.44×10^{-4}
3.	Shut down cooling system	1.88×10^{-4}	Not Specified
4.	Dousing System	1.58×10^{-6}	2.55×10^{-4}
5.	R/B isolation system	1.8×10^{-5}	10^{-2}
6.	Class-I power supply system	2.8×10^{-14}	Not Specified
7.	Class-II power supply system	9.34×10^{-7}	3.5×10^{-5}
8.	Class-III power supply system	1.84×10^{-4}	6.23×10^{-5}
9.	Auxiliary Boiler feed water system	1.93×10^{-4}	Not Specified
10.	Fire fighting system	4.24×10^{-3}	Not Specified

13. REFERENCES

1. Application for Renewal of Authorisation for RAPS-2. (April-2004)
2. IEEE – 500
3. IAEA-TEC DOC-478. (Component Reliability data for use in PSA).
4. Fault Tree Hand Book (NUREG-0492).
5. Probabilistic Risk Assessment in Nuclear Power Industries by Fullwood & Hall.
6. Reliabilityu and Risk analysis by Norman J. Mc Cormick

Reliability, Safety and Hazard
Editors: P.V. Varde et al.

APPLICATION OF SYSTEM RELIABILITY ANALYSIS FEEDBACK TO IMPROVE SYSTEM AVAILABILITY

Prabodh Kumar, S.K. Tiwary and A.K. Gupta

Kakrapar Atomic Power Station, (Nuclear Power Corporation of India Limited), Kalpakkam

ABSTRACT

Though all the perceived operational and maintenance requirement are normally well taken care-of at design stage itself, still the scope for improvement either in system design or in operational/maintenance practices, never ends in order to further improve the system's performance/reliability.

This paper brings out a case study as to how Reliability Analysis feedback has been gainfully utilized in identification of and reduction of individual component maintenance down time, which was indirectly degrading the reliability of one of the Reactor Shutdown system (namely Secondary Shutdown System-SSS), at KAPS.

Keywords : Primary Shutdown System, Secondary Shutdown System, Liquid Shutoff tubes, Mechanical Shut Off Rods, Bank down time

1. INTRODUCTION

Reliability Analysis is a tool that helps in improving the system at all stages starting with its conceptualization till it is put to use. At an operating station, the process can be effectively utilized to understand various factors affecting the system reliability/performance from varied points of view – namely 'Design', 'Operation' and 'Maintenance' and apply the feedback to improve it further. In this context, it is imperative for the analyst to be objective in performing system reliability analysis. Greater is the depth of analysis, more precisely it will be possible to identify & rectify the system weaknesses which have direct/indirect bearing on reliable performance of the system.

Kakrapar Atomic Power Station (KAPS) is a 220 MWe (twin Unit) PHWR Station, located at Kakrapar, Dist-Surat, Gujarat, India. Each Reactor has 2 independent and diverse shutdown systems, namely Primary Shutdown System (PSS) & Secondary Shutdown System (SSS). While the PSS has 14 mechanical shut off rods which fall into the Calandria, SSS has 12 shut off tubes which are filled with liquid poison on a demand for Reactor Trip.

At KAPS, the Reliability Analysis is being routinely carried out for all Safety Related Systems and the feedback is being effectively utilized.

About the Reactor

Kakrapar Atomic Power Station (KAPS) is a standard 220 MWe twin unit Indian PHWR based on CANDU design. The station is located at Kakrapar, in District Surat, in Gujarat.

Shutdown Systems

Each reactor has two fully independent, self sufficient fast active shutdown systems. One called as Primary Shutdown System (PSS) [1] and the other one as Secondary Shutdown System (SSS) [2]. As the name suggest, the first one is the main shutdown system and the second one acts as backup to it. Principally, PSS is gravity assisted system while the SSS acts against gravity.

The PSS comprises of 14 mechanical shut off rods made up of Cadmium layer, sandwiched in SS tubes and all 14 rods are normally kept hung outside the core with the help of rope, sieve, electromagnetic clutches & gear arrangement. These rods drop under gravity due to their own weight on energisation of the electromagnetic clutches on a reactor trip demand (Auto or Manual) and the Rx trip is achieved within 2.2 seconds.

In case 2 or more PSS rods fail to cross 90% position within 2.2 seconds or on LOCA signals, the SSS gets actuated. SSS can also be tripped manually. SSS comprise of 12 liquid shutoff tubes distributed vertically in the core. These 12 tubes are clubbed in 4 groups of 3 tubes each. Each group is called a Bank. Each Bank has one liquid poison storage tank and one Helium gas tank arranged in a U loop configuration. The bank also has 8 main SVs (SVsX01 to X08) as shown in the sketch given in Annexure-1, which normally remain energise. On SSS actuation demand (Auto or Manual) all of these 8 SVs get de-energise and the liquid poison gets filled up in the shutoff tubes under Helium pressure. The reactor trip is then achieved within 1.8 seconds. SSS is expected to actuate on 'AUTO' only in rarest of rare event. SSS, therefore remains poised during unit operation and shutdown states. Both PSS & SSS are designed to achieve a Reactor trip state in shortest possible time. For long term shutdown, boron solution in D2O is injected into Moderator. The SSS injection is said to be successful if atleast 3 banks have actuated successfully, within designated time (i.e. 1.8 sec).

Operational / Maintenance Practices

While a limit of 30 minutes is defined for maintenance outage for any one bank at a time, in the station Technical Specifications [3] during unit operation, no such limit is defined for unit shutdown period. However, maintenance of one bank only is permitted at a time, with all other reactivity devices (PSS & adjuster rods) kept 'Fully IN', in the core during shutdown.

The system unavailability figures are calculated on routine basis, as part of System Reliability Analysis programme. The first study on Reliability Analysis was done in the year 1997 [4]. The target unavailability of SSS is defined as 10-3 yr/yr.

The SSS unavailability is a function of following factors –

i. Demand failure
ii. Unavailability due to planned test
iii. Unavailability due to planned maintenance
iv. Unavailability for sampling.
v. Common cause failure.

During initial days of unit operation the emphasis was on keeping at least 3 banks available all the time, during shutdowns. As re-poising the bank under maintenance at the earliest was not required by any specific guide lines /stipulations.

The specific examples of activities which were affecting the SSS Reliability are -

1. Each bank was required to be drained into poison preparation tank, re-circulated and sampled, every month (now every 6 month) [3]. After ensuring the minimum concentration of boron in the poison sample, the bank is required to be re-poised, before draining any other bank for the same purpose. Please refer Annexure-1 for schematic line diagram of one bank of SSS. The total duration for the activity depends on the priority given to each step in the activity. However, the job was not given high priority in those days, as it was not violating any set norm/ standard/stipulations. The average bank down time used to be about 16 hrs (i.e. 2 shifts) for one bank [4,5].

2. Another example is outage of the bank for calibration of RVs mounted on the Helium Tank. These RVs are required to be calibrated annually & for this purpose the Helium tanks need to be de-pressurised. Here, again, the down time of the banks used to be about 1 day on an average per RV [4,5], as the same RVs were used to be cleaned, re-calibrated and installed back. In some cases requiring overhauling of the RV, the time used to increase even further. The process was considered to be normal till the System Reliability (Unavailability) Analysis was carried out in the year 2001 at the station.

ANNEXURE#1

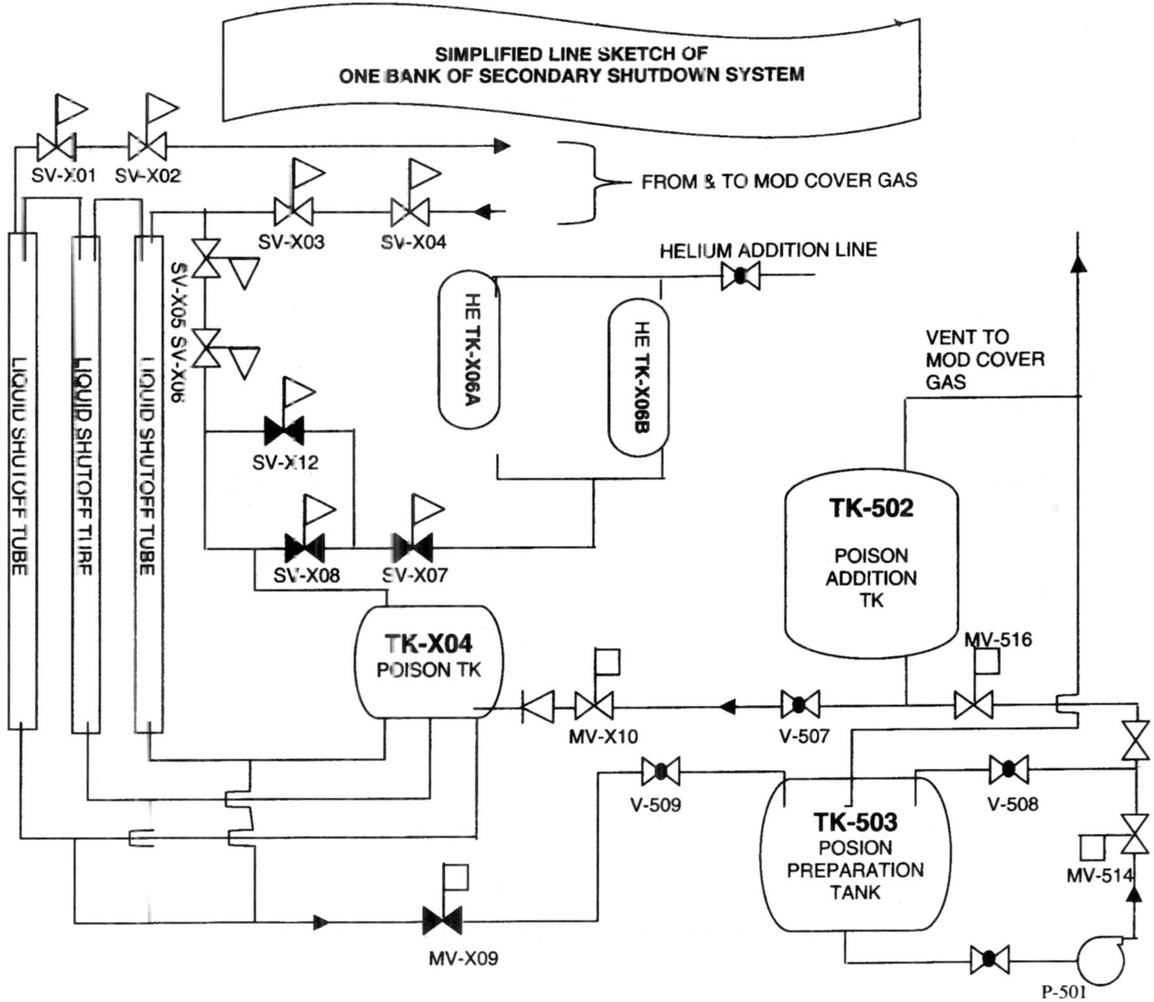

SIMPLIFIED LINE SKETCH OF
ONE BANK OF SECONDARY SHUTDOWN SYSTEM
SV-X01
SV-X02
FROM & TO MOD COVER GAS
SV-X03
SV-X04
HELIUM ADDITION LINE
SV-X05 SV-X06
HE TK-X06A
HE TK-X06B
VENT TO
MOD COVER
GAS
LIQUID SHUTOFF TUBE
LIQUID SHUTOFF TUBE
LIQUID SHUTOFF TUBE
SV-X12
SV-X08
SV-X07
TK-502
POISON
ADDITION
TK
TK-X04
POISON TK
MV-516
MV-X10
V-507
V-509
V-508
TK-503
POSION
PREPARATION
TANK
MV-514
MV-X09
P-501

ANNEXURE#2

SECONDARY SHUTDOWN SYSTEM FAILURE FAULT TREE

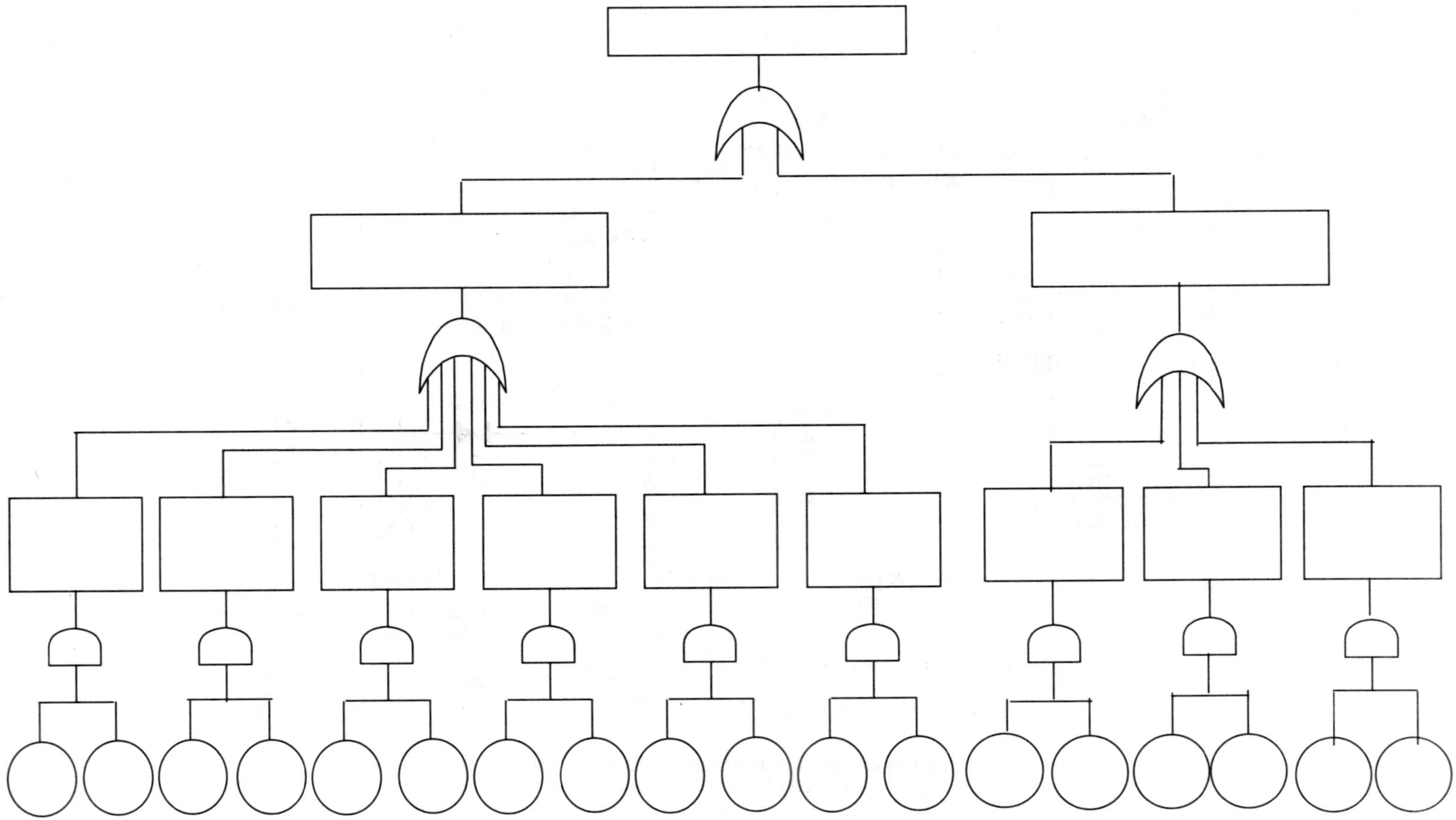

FTI- fail to insert in desired time.

2. OBSERVATIONS

During this analysis it was observed that these two were the major contributor for any single bank unavailability during shutdown period and were affecting all 4 banks uniformly. This avoidable downtime of each of the banks, individually, was silently and adversely affecting the availability of the system as a whole. Please refer System Fault Tree used for calculating system unavailability figures in Annexure-2.
It was noticed that the system unavailability could be significantly improved if only these two factors could be controlled.

3. CORRECTIVE ACTIONS

This feed back was discussed in appropriate forum within the station and the following were decided –

1. The banks will be drained for sampling at such time when all the resources were readily available and the bank could be poised back within minimum possible time, without involving any breaks e.g. Tea / Lunch etc in between various steps of the activity to the extent possible. This reduced the bank down time for boron sampling from an average 16 hrs to 4 hrs [6].

2. Spare RVs were procured which were calibrated in advance and were kept ready before de-pressurising a bank for its RV calibration. The old RVs are now simply replaced with pre-calibrated ones. This reduced the bank down time from about 24 hrs to about 2 hrs on this account [6].

4. RESULTS

This study brought-in the awareness among operation, maintenance & performance monitoring groups that the overall availability of the system is dependent on availability of the redundant equipment.

Both the above practices helped significantly in improving the system reliability / reducing the system down time.

5. CONCLUSION

Station is utilizing the System Reliability Analysis technique to identify the latent weaknesses in various important systems and to improve upon the existing design and operational / maintenance practices.

The reliability analysis results are further being utilized in Level-I Probabilistic Safety Assessment (PSA) [7], which is also being updated routinely.

6. REFERENCES

1. PSS Design Basis Report (KAPP-1-2/31350/DBR/009/R-0, April 1990).
2. SSS Design Manual (KAPS-1-2/31340/DM/039/R-0, October 1990).
3. Technical Specifications for KAPS Operation (Rev-2, Dec 1999).
4. Reliability Analysis (January 27, 1997).
5. Annexure-10 of Application for renewal for authorization (ARA) for KAPS, (R-1, June 2001).
6. Reliability Analysis Report on KAPS Safety Systems (Ref KAPS/SD/ENC(P) /2004/S/130 dtd 5/5/04).
7. Level-I PSA issued vide KAPS/01560/02/M/38 dtd 22/04/2002.

AUTHOR INDEX